Annotated Teacher's Edition

SCIENCE PLUS®

TECHNOLOGY AND SOCIETY

LEVEL GREEN

Project Directors

International: **Charles McFadden**
Professor of Science Education
The University of New Brunswick
Fredericton, New Brunswick

National: **Robert E. Yager**
Professor of Science Education
The University of Iowa
Iowa City, Iowa

Project Authors

Earl S. Morrison *(Author in Chief)*
Alan Moore *(Associate Author in Chief)*
Nan Armour
Allan Hammond
John Haysom
Elinor Nicoll
Muriel Smyth

This new United States edition has been adapted from prior work by the Atlantic Science Curriculum Project, an international project linking teaching, curriculum development, and research in science education.

HOLT, RINEHART AND WINSTON
Harcourt Brace & Company
Austin • New York • Orlando • Atlanta • San Francisco • Boston • Dallas • Toronto • London

ACKNOWLEDGMENTS

Project Advisors

Herbert Brunkhorst
Director, Institute for Science Education
California State University,
San Bernardino
San Bernardino, California

David L. Cross
Science Consultant
Lansing School District
Lansing, Michigan

Jerry Hayes
Associate Director, Science Outreach
Teacher's Academy,
Mathematics and Science
Chicago, Illinois

William C. Kyle, Jr.
*Director, School Mathematics
and Science Center*
Purdue University
West Lafayette, Indiana

Mozell Lang
Science Education Specialist
Michigan Department
of Education
Lansing, Michigan

Annotated Teacher's Edition Writers

Shirley Key, Ph.D.
Multicultural Consultant and Writer
Missouri City, Texas

Mary Beth McManus
Brockton, Massachusetts

Lynn Roudabush-Novak
Midlothian, Virginia

Pamela Russ, Ph.D.
Multicultural Consultant and Writer
Turlock, California

Feature Writers

Margy Kuntz
San Anselmo, California

Vanessa McCallum
Germantown, Tennessee

David Stienecker
Forest Hills, New York

Teacher Reviewers

Robert W. Avakian
Alamo Junior High
Midland, Texas

Barry Lynne Bishop
San Rafael Junior High
Ferron, Utah

Paul Boyle
Perry Heights Middle School
Evansville, Indiana

Renae Cartwright
Cedar Park Middle School
Cedar Park, Texas

Kenneth Creese
White Mountain Junior High
Rock Springs, Wyoming

George Henrie
Delta Middle School
Delta, Utah

Kenneth Horn
Fallston Middle School
Fallston, Maryland

Roberta Jacobowitz
C. W. Otto Middle School
Lansing, Michigan

Pamela Jones
Birmingham Covington School
Birmingham, Michigan

Janet Porter
Humboldt High School
Humboldt, Tennessee

James B. Pulley
Liberty High School
Liberty, Missouri

Steve Siegel
McCormick Junior High
Cheyenne, Wyoming

Patricia Soto
G. W. Carver Middle School
Miami, Florida

Larry Tackett
Andrew Jackson Middle School
Cross Lanes, West Virginia

Nancy Wesorick
Sunset Middle School
Longmont, Colorado

Academic Reviewers

Kenneth Brown, Ph.D.
Professor of Chemistry
Department of Chemistry
Northwestern Oklahoma State
University
Alva, Oklahoma

Patricia Buis, Ph.D.
Assistant Professor
Department of Geology and Geological
Engineering
University of Mississippi
University, Mississippi

Linda K. Butler, Ph.D.
Division of Biological Sciences
University of Texas
Austin, Texas

Mark Coyne, Ph.D.
Assistant Professor of Agronomy
University of Kentucky
Lexington, Kentucky

Andrew A. Dewees, Ph.D.
Department of Biological Sciences
Sam Houston State University
Huntsville, Texas

Albert B. Dickas, Ph.D.
Professor of Geology
University of Wisconsin
Superior, Wisconsin

Frederick R. Heck, Ph.D.
Associate Professor of Geology
Ferris State University
Big Rapids, Michigan

Smith L. Holt, Ph.D.
Dean, College of Arts and Sciences
Oklahoma State University
Stillwater, Oklahoma

James Kaler, Ph.D.
Professor of Astronomy
University of Illinois
Urbana, Illinois

Doris I. Lewis, Ph.D.
Professor of Chemistry
Suffolk University
Boston, Massachusetts

R. Thomas Myers, Ph.D.
Professor of Chemistry Emeritus
Department of Chemistry
Kent State University
Kent, Ohio

Ramiro Sanchez, Ph.D.
Department of Chemistry
University of Houston
Clear Lake, Texas

Thomas Troland, Ph.D.
Associate Professor
Department of Physics and
Astronomy
University of Kentucky
Lexington, Kentucky

Blue-Ribbon Committee

The following teachers constituted a special committee of current *SciencePlus* users who provided information and insights on how to improve the *SciencePlus* program. Their input was invaluable.

Patricia Barry
Milwaukee, Wisconsin

Carol Bornhorst
Chula Vista, California

Catherine Carlson
Plano, Texas

Harry Dierdorf
New Brighton, Pennsylvania

Jeff Felber
Spring Valley, New York

Barbara Francese
Madison, Connecticut

Kenneth Horn
Forest Hills, Maryland

Doug Leonard
Toledo, Ohio

Betsy Mabry
Enid, Oklahoma

Debbie Melphi
Syracuse, New York

Lynn Roudabush-Novak
Midlothian, Virginia

Donna Robinson
Rapid City, South Dakota

Sandy Moerke-Schaefer
Marysville, Washington

Marvin Selnes
Sioux Falls, South Dakota

Patricia Soto
Coral Gables, Florida

Margaret Steinheimer
St. Louis, Missouri

Sandy Tauer
Derby, Kansas

Joy Ward
Chicago, Illinois

Gary Weaver
Vacaville, California

Nancy Wesorick
Longmont, Colorado

Brenda West
Scott Depot, West Virginia

Project Associates

We wish to thank the thousands of science educators, teachers, and administrators from the scores of universities, high schools, junior high schools, and middle schools who have contributed to the success of *SciencePlus*.

CONTENTS ANNOTATED TEACHER'S EDITION

OWNER'S MANUAL

UNIT INTERLEAF INFORMATION

CONTENTS

iv

Unit 2 PATTERNS OF LIVING THINGS 60

v

Unit 3 · IT'S A SMALL WORLD 128

Unit 4 · INVESTIGATING MATTER 198

Unit 5 CHEMICAL CHANGES 250

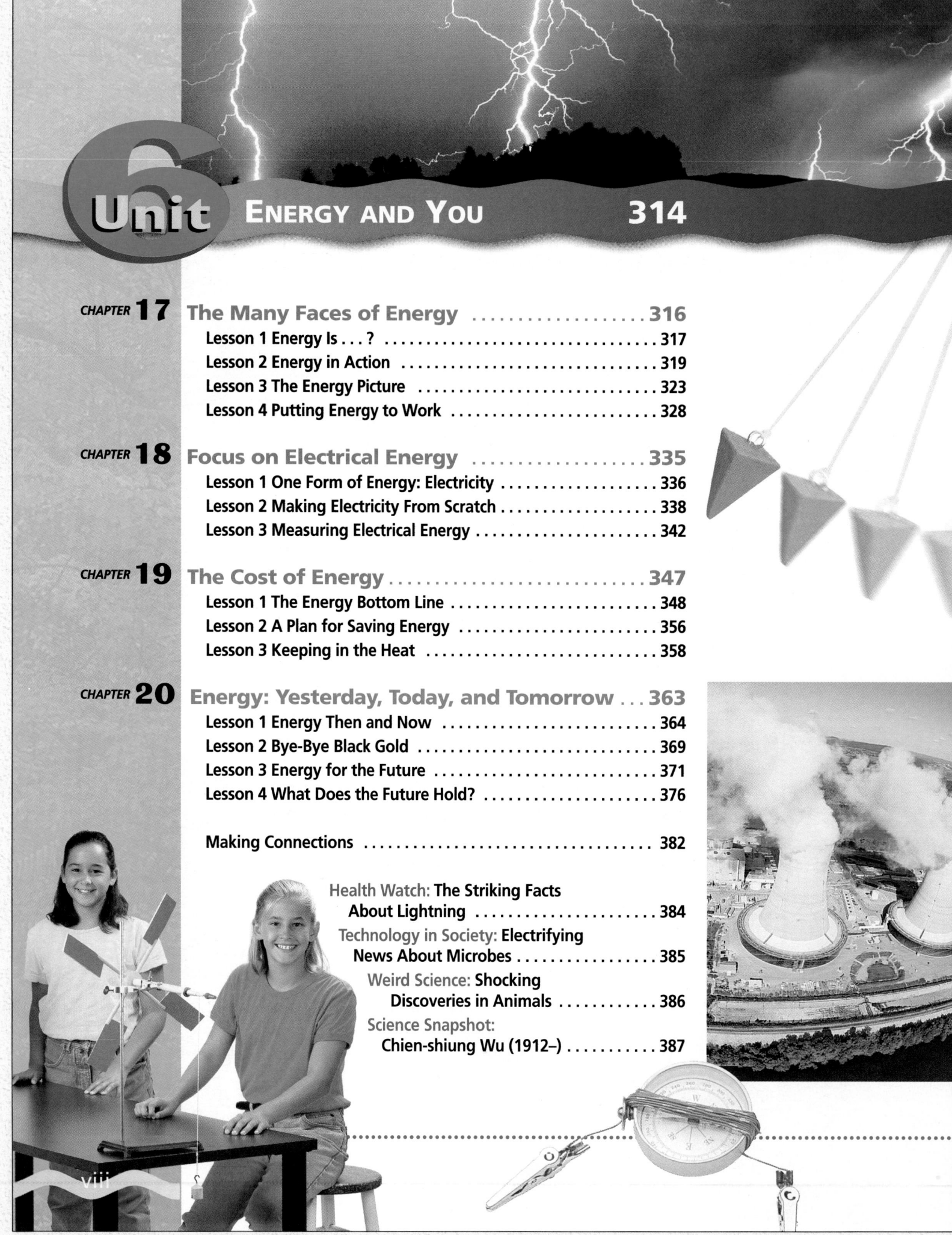

6 Unit ENERGY AND YOU 314

viii

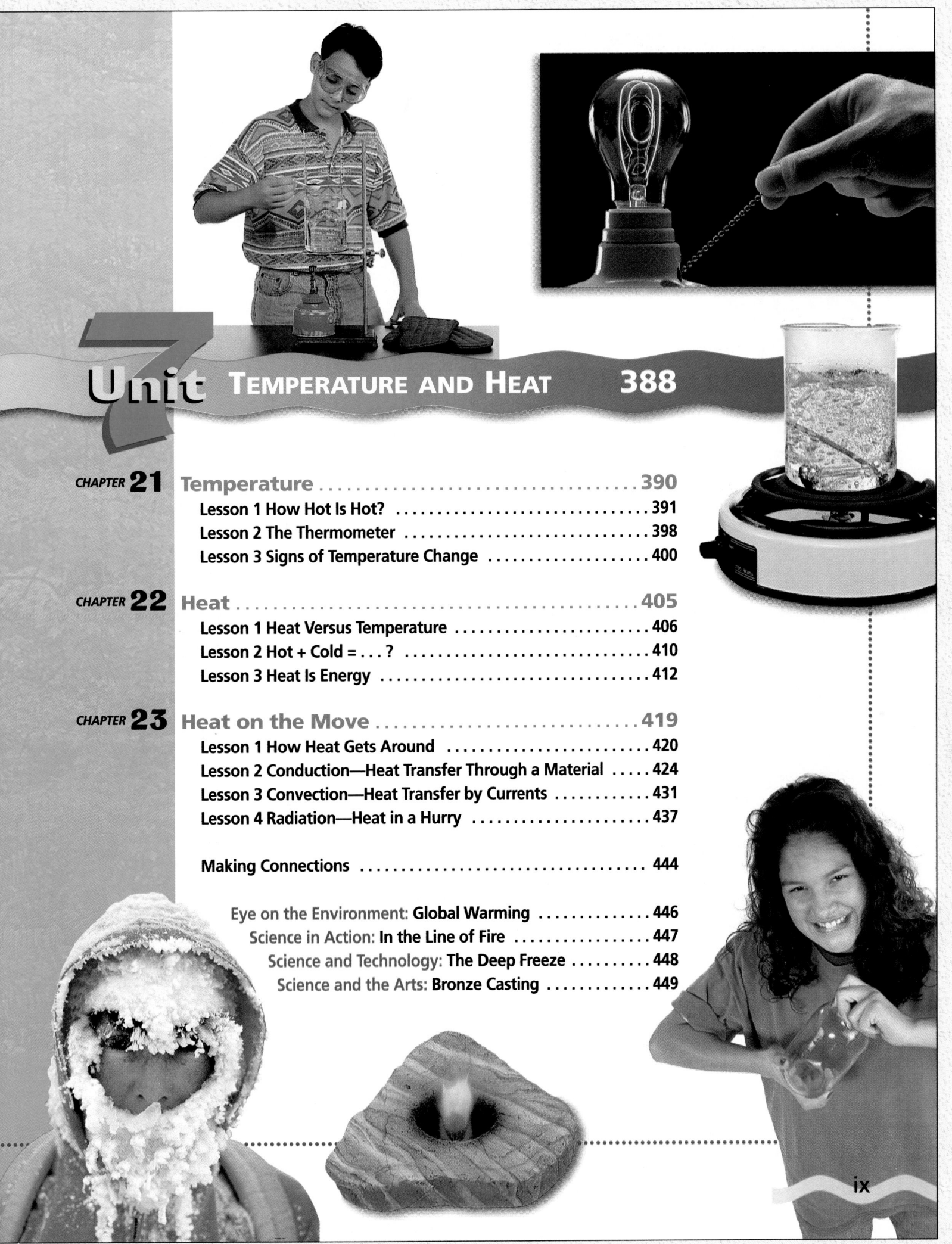

ix

8 Unit OUR CHANGING EARTH 450

SOURCEBOOK

xi

SciencePlus Owner's Manual

THE SCIENCEPLUS PHILOSOPHY

Welcome to *SciencePlus*, an innovative approach to science education. *SciencePlus* is unlike any science program you have used before. It is designed from the ground up to teach science in precisely the way that students learn best—by thinking, talking, and writing about what they do and discover. *SciencePlus* is activity- and inquiry-based. In other words, it is both hands-on and minds-on. *SciencePlus* is lively, engaging, and relevant to the students' world. *SciencePlus* is loaded with thought-provoking activities designed to challenge students' thinking skills while introducing them to realistic methods of science.

SciencePlus works for students and teachers alike. Students enjoy and benefit from its varied and active approach, while teachers find it to be teachable under real conditions. Laboratory-type activities require about 30 to 40 percent of class time. The remainder of class time is taken up by a rich variety of learning activities.

SciencePlus is tailored to meet the needs of middle school–aged students. It accommodates every type of student you are likely to encounter: the basic student, who requires substantial guidance; the average student, who is not yet fully equipped with abstract-thinking skills; and the advanced student, who needs only to be pointed in the right direction to succeed.

At every stage, the *SciencePlus* program emphasizes concept and skill development over memorization of facts. By doing science activities and then thinking about the results, students learn the *whys* and *hows,* not just the *whats* and *whens,* of science. Ultimately, students come to see science as a system for making sense of the world.

Origins

The *SciencePlus* program was originally developed by the Atlantic Science Curriculum Project (ASCP) of Canada to replace the traditional recall-based curriculum, which had proven to be ineffective. *SciencePlus* represents a ground-breaking effort, the culmination of many years of labor by dozens of talented, dedicated science educators.

The *SciencePlus* development team was guided every step of the way by the latest insights into how children actually learn. The *SciencePlus* program has been thoroughly tested on real students in realistic settings, refined, and then retested. The result is a program that works! Teachers using *SciencePlus* have reported dramatic gains in student comprehension and retention of scientific concepts. Above all, students enjoy using *SciencePlus* and develop a heightened interest in science.

A Continuing Tradition

This edition of *SciencePlus,* now in its second edition as an American publication, continues the tradition of excellence begun in Canada many years ago. The program has undergone many changes in response to recommendations of curriculum reform movements, such as Project 2061 and the NSTA Scope, Sequence, and Coordination. In addition, the revision of this program was heavily influenced by *SciencePlus* teachers, whose combined and varied experiences helped us make the program even more relevant and easier to implement.

An Interactive, Effective Program

SciencePlus employs proven teaching strategies: guided and open-ended investigations, small-group discussions, exploratory writing and reflective-reading tasks, games, picture and word puzzles, and independent long-range projects. This variety motivates and helps maintain the interest of students and teachers alike.

SciencePlus develops scientific process skills as an essential goal. As the curriculum progresses, the students will master increasingly complex tasks. For example, students will move from directed to open-ended inquiry and from reading and completing tables and graphs to constructing them from experimental data they have collected on their own.

In general, each of the units in *SciencePlus* is self-contained and may be taught as a separate instructional module. *SciencePlus* contains a balance of physical, biological, earth/space, and environmental science topics.

Guiding Principles

The guiding principles of *SciencePlus* are simple and few:

- **Anyone can learn science**
  The image of science as the private domain of the superintelligent is wrong and damaging. Science is for everyone. Children exposed to science for the first time take to it naturally. It is only later, after science explorations have been replaced by fill-in-the-blank worksheets and recall drills, that love for science is replaced by boredom and even dread. *SciencePlus* can rekindle the sense of wonder and fascination that lies dormant within your students.

- **Science is a natural endeavor**
  Whether we realize it or not, each of us applies science nearly every day. Hardly a day passes that we don't ask ourselves, "How does this work?" or "Why does that happen?" or "What happens if . . . ?" Scientists differ from other people only in that it is their profession, rather than their avocation, to figure out "how," "why," and "what if."

Unfortunately, stereotypes about science and scientists abound. Many students feel that only "nerds" or "geeks" enjoy science. This falsehood may do as much to turn people away from science as any curricular shortcoming. *SciencePlus* actively refutes these stereotypes. It portrays science as a rewarding, quintessentially human undertaking. Scientists are portrayed as normal people, not aloof geniuses who talk in equations.

- **Science is its own reward**
  There is no feeling quite like the thrill of discovery or the sense of accomplishment that comes from rising to a difficult challenge. Science can be thought of as a voyage into the unknown. This voyage can be exciting and rewarding for all.

Aims

SciencePlus is designed to help you further develop each of the following in your students:

- Understanding of the interrelationships among science, technology, and society
- Understanding of important science concepts, processes, and ideas
- Use of higher-order thinking skills
- Ability to solve problems and apply scientific principles
- Commitment to environmental protection
- Interest in independent study of scientific topics
- Social skills
- Communication skills

To accomplish these goals, a wide variety of teaching strategies are employed. The common denominator among these is their emphasis on doing. At all times, students are to be active and involved.

> *SciencePlus* can rekindle the sense of wonder and fascination that lies dormant within your students.

Science, Technology, and Society

Science and technology are flip sides of the same coin; each supports the other. Neither should be studied in isolation. *SciencePlus* explores the relationship between science and technology, and the effect of both on our society as a whole. Even people who never again set foot in a laboratory after leaving school can benefit from an understanding of science and technology and how both relate to each other and to society at large.

Our society, complex as it is, will become even more so in the years to come. Science and technology will play roles in every aspect of life. In the future, "high-tech" will be more than a catch phrase—it will permeate every aspect of life. To prepare students for the challenges of the future, they must become science literate. They must be given the tools that will enable them to become responsible and productive individuals in a highly technological world.

Science for All

SciencePlus is designed to put the "process" back into science education and, in so doing, to provide students with the mental skills they need to truly understand and apply science. Today, as never before, a thorough grounding in science is absolutely essential; without it, students—the citizens of tomorrow—cannot expect to be fully conversant in and responsive to the complex issues of the twenty-first century.

No program can teach itself. You, with your energy, enthusiasm, and ability, are the key to a successful outcome. *SciencePlus* will help you help your students develop all of the skills they need to learn independently. *SciencePlus* fosters a spirit of joint exploration. Let the journey begin.

> To prepare students for the challenges of the future, they must become science literate.

THE SCIENCEPLUS METHOD

BUILDING A BETTER UNDERSTANDING OF SCIENCE

The *SciencePlus* program is based on the Constructivist Learning Model (CLM). Constructivism is based as much on common sense as on the results of research. With the CLM, students "construct" an understanding of concepts step by step. Students begin by identifying what they already know about a topic. Any misconceptions they may have about a topic are exposed at this point. Identifying these misconceptions is a critical part of the process. Next, students do hands-on activities to experience the subject matter directly. Their experiences cause them to amend, add to, or scrap altogether the mental model they already have of the subject in question.

Constructivism is based on a few key steps:

1. Invitation
The Invitation stimulates students' curiosity and engages their interest. At this stage, students note the unexpected, pose questions, or define a problem.

2. Exploration
Explorations engage students in the search for solutions or explanations. Students look for alternative sources of information, collect and evaluate data, and clarify their findings through discussion and debate.

3. Proposing explanations and solutions
At the conclusion of the Explorations, students propose their response to the problem or question posed in the Invitation. The class is exposed to a variety of possible responses, and students have the opportunity to consider each.

4. Taking action
Students make decisions about a course of action based on the various proposals offered. If the class reaches consensus, then this stage may bring about closure of the lesson. It may happen, though, that this stage identifies new questions to explore.

Another way to think of the Constructivist Learning Model is in terms of the five *E*s: Engage, Explore, Explain, Elaborate, and Evaluate.

> *For an in-depth discussion of Constructivism, see "The Constructivist Learning Model," by Bob Yager*, The Science Teacher, *September 1991.*

With the Constructivist Learning Model, students "construct" an understanding of concepts step by step.

CONCEPTUAL FRAMEWORK

The information and chart that follow outline the conceptual framework for *SciencePlus*, Level Green. The information on this page will help you interpret the chart that follows on the next two pages.

Concept Focus

Each unit focuses on one major scientific concept. This concept is developed through a thematic approach. Although other concepts are covered in each unit, this column indicates the major focus.

Content Integration

Content Focus
The content of each unit integrates the traditional disciplines of life, earth, and physical science. This column lists the main discipline covered in each unit.

Supporting Content
The secondary content focus listed in this column supports the major content focus of each unit and helps achieve integration.

Thematic Focus

SciencePlus uses five themes as its organizational framework. These themes allow students to see content relationships and promote conceptual understanding. This column lists the main themes of each unit.

Science, Technology, and Society (STS)

This column lists three STS topics introduced in the unit that allow students to consider the development and impact of technology, to examine social issues, and to explore possible solutions.

Process Skills

All Major Skills Covered
SciencePlus places a strong emphasis on process skills. There are 11 major science process skills, and when a check mark appears in this column, all 11 skills have been used in the unit. The 11 process skills covered in *SciencePlus* can be found on page T47.

Process Skills Focus
Of the 11 skills used in *SciencePlus*, the 3 skills listed in this column are emphasized in the unit.

CONCEPTUAL FRAMEWORK

CONCEPT FOCUS	CONTENT INTEGRATION				PROCESS SKILLS	
	Content Focus	Supporting Content	Thematic Focus	STS	All Major Skills Covered	Process Skills Focus
Unit 1 **Science and Technology** The definition of science, the scientific method, and the roles of science and technology in everyday life	•Life science •Earth science •Physical science	Technology	•Changes Over Time •Structures •Systems	•Invention of the telephone and light bulb •Treating diabetes with insulin •Solving problems using scientific methods	X	•Observing •Hypothesizing •Inferring
Unit 2 **Patterns of Living Things** The classification of organisms as living, nonliving, or dead; the various adaptations of animals and plants; and the building blocks of life	Life science	•Earth science •Physical science	•Structures •Changes Over Time •Cycles •Systems	•Using a microscope for detailed observation •Monitoring the activities of honeybees •Examining patterns of human growth	X	•Communicating •Comparing •Analyzing
Unit 3 **It's a Small World** Microorganisms and their effects on our health, our environment, and our food supply	Life science	•Earth science •Physical science	•Changes Over Time •Structures •Cycles	•The role of cleanliness in promoting good health •Making bread with yeast •Analyzing fertile soil	X	•Observing •Contrasting •Analyzing
Unit 4 **Investigating Matter** Matter and its physical properties, the metric system, and the states of matter	Physical science	•Life science •Earth science	•Structures •Cycles •Energy	•Producing paper •Developing and using measuring devices •Examining useful properties of plastics	X	•Measuring •Hypothesizing •Classifying
Unit 5 **Chemical Changes** The role of chemicals in our lives, chemical and physical changes, chemical reactions, and acids and bases	Physical science	•Life science •Earth science	•Changes Over Time •Structures •Energy	•Identifying chemicals in our daily lives •Handling chemicals safely •The role of chemical changes in digestion	X	•Observing •Communicating •Classifying

CONCEPTUAL FRAMEWORK

CONCEPT FOCUS	CONTENT INTEGRATION				PROCESS SKILLS	
	Content Focus	Supporting Content	Thematic Focus	STS	All Major Skills Covered	Process Skills Focus
Unit 6 **Energy and You** The definition of energy, forms of energy, electricity, and society's use of energy	Physical science	• Life science • Earth science	• Energy • Systems • Changes Over Time • Structures	• Methods of generating electricity • Analyzing past, present, and future energy resources • Measuring rates of energy consumption	X	• Organizing • Inferring • Predicting
Unit 7 **Temperature and Heat** The difference between temperature and heat, how they are measured, and how heat moves	Physical science	• Life science • Earth science	• Changes Over Time • Structures • Energy	• Using a thermometer • Measuring the energy content of food • Testing insulating materials	X	• Measuring • Contrasting • Predicting
Unit 8 **Our Changing Earth** The evidence for geologic changes and the processes of geologic change, including weathering by water, wind, and ice	Earth science	• Life science • Physical science	• Structures • Energy • Changes Over Time	• Detecting landslides and earthquakes • Using seismic waves to "see" underground • Evaluating water resources	X	• Observing • Comparing • Organizing

COMPONENTS OF SciencePlus

SciencePlus is no ordinary textbook. *SciencePlus* is a student-friendly text: lively; abundantly illustrated with clever, colorful illustrations; and loaded with engaging activities. Every effort has been taken to make this text the sort of book that students will actually want to use.

Units, Chapters, Lessons

SciencePlus contains eight units, which are further divided into chapters and lessons on closely related subject matter. Each lesson includes a wide variety of activities and explorations designed to develop the lesson content.

Each chapter begins with a ScienceLog page that invites students to express what they already know about the subject of the chapter. Highly graphic and playful questions help expose misconceptions students may have about the topics to be covered. This important process sets the stage for firsthand exploration of the subject area. The answers to the ScienceLog questions are the equivalent of hypotheses or predictions, which are either supported or disproved as the students explore the chapter content. Either way, the students become invested in what they are about to learn.

Explorations

Scattered throughout each unit are a series of Explorations—hands-on, inquiry-based activities. These Explorations allow students to see scientific principles in action. The Explorations are essential for inducing real learning in students. As students do the Explorations, they have the opportunity to compare their mental models of scientific principles with the real things. As weaknesses are exposed, students adjust their thinking to accommodate what they have learned.

Most of the Explorations are designed so they can be done cooperatively in small groups. In this way, the Explorations model real scientific experiences, in which scientists work together to solve problems. By working in cooperative groups, students develop important skills such as communicating ideas and sharing responsibility. And the cooperative groups not only make science more interactive and more fun, but also provide valuable opportunities to develop social skills. They are also more economical to implement.

For more information on cooperative learning, see pages T41 through T46 of this Annotated Teacher's Edition.

Most of the supplies needed for the Explorations consist of very common equipment and materials. In most cases, they can be gathered from the home. For help in gathering supplies, the *Teaching Resources* package contains a comprehensive *Materials Guide,* which contains both a master materials list and individual unit materials lists. In addition, the second page of the Home Connection parent letter in each *Teaching Resources* booklet provides a checklist of materials that you can use to invite parent donations. Using this portion of the parent letter will help you keep your budget low and at the same time get parents involved in the *SciencePlus* experience.

Assessment

The *SciencePlus* Pupil's Edition contains a variety of methods for checking student learning.

- **Chapter Review—Challenge Your Thinking**

  To make sure your students comprehend the new information, each chapter concludes with Challenge Your Thinking questions. These questions challenge students to apply newly learned material in a variety of ways. Many of the questions are like brain-teasers in that they are unusual and highly creative. Because the questions require more than simple recall, students find them fun to figure out.

  The chapter review pages conclude with an invitation to students to rewrite their answers to the ScienceLog questions located at the beginning of the chapter. This exercise gives students the opportunity to confront and discard any misconceptions they may have had at the outset of the chapter.

- **Unit Review—Making Connections**

  The Making Connections pages consist of two parts: The Big Ideas and Checking Your Understanding. The Big Ideas asks students to formulate their own summary of the unit, using a list of questions as a guide. The Checking Your Understanding poses a selection of comprehensive questions designed to gauge students' understanding of the unit's subject matter.

  Also included on the Making Connections pages is a short description and table of contents of the corresponding SourceBook unit. This reference directs students to the appropriate unit of the SourceBook, where they can extend their knowledge of the concepts they just explored.

All answers to the Challenge Your Thinking and Making Connections questions are located in the Wrap-Around Margins of this Annotated Teacher's Edition.

For more information about assessment, see pages T56 through T64 of this Annotated Teacher's Edition.

ScienceLog

For more information about using the ScienceLog, see pages T37 and T38.

Special Features

A common complaint among students is that the material they learn is not relevant. The end-of-unit features in *SciencePlus* help show students how science is an integral part of their lives.

In keeping with the spirit of *SciencePlus*, all of the features are interactive, with thought-provoking questions and research ideas that encourage students to explore the topic further. Each unit concludes with four features selected from the menu below.

- **Science in Action** features illustrate working scientists today. These features focus on the excitement and variety that characterize science careers. Students will learn about real experiences and what the scientists like most and least about their work. They may even learn about the scientists' personal career aspirations as well as their greatest disappointments.

- **Science Snapshots** are glimpses into the lives of famous scientists. Such scientists are featured to give students a sense of the colorful history of science. Present and future scientists really do stand on the shoulders of giants!

- **Health Watch** features provide insights into the variety of connections between science and health.

- **Eye on the Environment** features focus on the effects that science and technology have on our environment—both good and bad.

- **Science and Technology** features explore the science on which new technologies are based. In many cases, explorations in science open the doors to new technology. Highlighting the impact of this technology on our everyday lives shows students how relevant scientific research can be.

- **Technology in Society** features take the logical step beyond the technology that has been created through the use of science. These features explore the impact of technology on our society and put forth the inevitable question: Do the benefits outweigh the potential problems?

- **Science and the Arts** features show that science and art are not necessarily polar opposites. The connection between science and society is reinforced by showing students how artists have been inspired by nature and how scientific methods or principles have enhanced or empowered their work.

- **Weird Science** features showcase some of the odd, outlandish, and even unbelievable creatures and phenomena that make up our unique world.

SOURCEBOOK

The SourceBook is an in-text science reference located at the back of the Pupil's Edition. This unique reference includes information that is organized to match the units of *SciencePlus*. Each SourceBook unit both reinforces and extends beyond the material presented in the Pupil's Edition, providing an excellent resource that students can use to add depth to their understanding of the topics presented in *SciencePlus*.

Because the SourceBook is not designed for hands-on exploration, it lends itself to reading at home or outside the lab. It's handy for homework assignments or for when you have a substitute teacher. Each unit of the SourceBook ends with a Unit CheckUp for checking students' comprehension. Worksheets are also available in the *Teaching Resources* booklets.

> **The SourceBook is referenced in many ways throughout the Pupil's Edition.**

The *SciencePlus* Annotated Teacher's Edition will help you achieve the full potential of *SciencePlus*. The Annotated Teacher's Edition consists of two major parts: the Unit Interleaf and the Wrap-Around Margins. Each Unit Interleaf consists of a six-page insert preceding each unit. The Wrap-Around Margins provide on-page annotations and teaching suggestions.

Using the Unit Interleaf

Each Unit Interleaf consists of six pages of information to help you plan your lessons. Each interleaf has the following planning and preparation aids.

- **Unit Overview** is a quick overview of the concepts and content presented in the unit.

- **Using the Themes** provides descriptions of the themes that are relevant to the unit, including short explanations of how they apply and how to integrate them into your teaching.

- **Using the SourceBook** is a concise description of the content in the accompanying SourceBook unit.

- **Bibliography** consists of three subsections categorized for teachers; for students; and for films, videotapes, software, and other media.

- **Unit Organizer** is a comprehensive chart that identifies the objectives, time requirements, and teaching resources that are available for all of the chapters and lessons in the unit.

- **Materials Organizer** is a chart that identifies the materials required to carry out each Exploration and Activity in the unit.

- **Advance Preparation** identifies what you may need to do in advance to prepare for the Explorations and Activities.

- **Unit Compression** provides recommendations on how to compress or reduce the time spent teaching the unit without sacrificing the integrity of the unit.

- **Homework Options** is a chart that lists the various homework opportunities, suggestions, and worksheets available to support each of the chapters in the unit.

- **Assessment Planning Guide** is a chart that identifies the assessment opportunities and worksheets available at the lesson, chapter, and unit levels.

- **Using the *Science Discovery* Videodiscs** provides an overview and barcodes for the Science Sleuth mystery and Image and Activity Bank selections that correspond to the unit.

Using the Wrap-Around Margins

The entire Pupil's Edition of *SciencePlus* has been reduced in size and placed on the pages of this Annotated Teacher's Edition. The margins on the reduced pages have been filled with teaching suggestions and commentary to help you teach each lesson with maximum effectiveness.

UNIT OPENER

Each unit opener has the following teacher's information to help you quickly engage your students in the subject of the unit.

- **Unit Focus** provides interactive suggestions for introducing students to the unit.

- **Connecting to Other Units** is a table that shows at a glance the integration between that unit and other units in the program.

- **Using the Photograph** provides suggestions for using the unit photograph to start students thinking and asking questions about the subject of the unit topic.

- **Answer to In-Text Question** provides the answer to any question that is posed to the students on the unit opener pages.

CHAPTER OPENER

Each chapter has the following information to help you prepare to teach the chapter.

- **Connecting to Other Chapters** is a chart that identifies the basic topics explored in each chapter of the unit. This chart allows you to see the logical progression of concepts through the unit.

- **Prior Knowledge and Misconceptions** describes how to use the ScienceLog questions and suggests other ways to access students' prior knowledge or misconceptions about the chapter topics.

LESSON

Each lesson of the unit has extensive commentary and teaching notes to help you in teaching the material. The lesson information is divided into three parts: Focus, Teaching Strategies, and Follow-Up. In addition, a Lesson Organizer is provided at the beginning of each lesson.

Lesson Organizer

The Lesson Organizer is provided at the beginning of each lesson. The Lesson Organizer provides easy access to the following information about the lesson:
- Time Required
- Process Skills
- Theme Connections
- New Terms
- Materials
- Teaching Resources

FOCUS

The Focus consists of Getting Started and the Main Ideas. Getting Started provides suggestions for an activity and discussion to begin each lesson. The Main Ideas list the main concepts of the lesson.

TEACHING STRATEGIES

Teaching Strategies provide a variety of methods for successfully guiding your students through each lesson. The following categories of strategies are provided to serve your needs.

Meeting Individual Needs

Today's classrooms are places of diversity—populated by students with different ability levels and from a variety of ethnic backgrounds and cultures. The following categories of teaching suggestions are provided to help you tailor your instruction to meet the needs of all of your students.
- Gifted Learners
- Second-Language Learners
- Learners Having Difficulty

CROSS-DISCIPLINARY FOCUS

Suggestions and activities are provided that cross the boundaries between science and other disciplines, such as art, health, mathematics, social studies, music, language arts, foreign language, and industrial arts. By doing these activities, students will experience the interrelatedness of science with other areas of study and will come to understand that science is not just for scientists.

Multicultural Extension

Under this heading you will find activities and suggestions that serve to highlight cultural diversity and show its positive influence on science as well as other disciplines. These activities build on the multicultural elements already included in the Pupil's Edition to ensure ample opportunities to show students how culture and science are integrated to the benefit of us all.

Integrating the Sciences

These strategies help you explore concepts from the standpoint of more than one scientific discipline. In this way, you can help students to see how life, earth, and physical sciences work together to help us understand the world around us.

Theme Connection

Each Theme Connection consists of a Focus question and its answer. By asking the Focus question, you can help students organize their learning within an overarching theme to help them make better sense of the ever-increasing pool of scientific knowledge.

ENVIRONMENTAL FOCUS

Many lessons lend themselves to an environmental focus. In these cases, information is provided both to inform and to invite class discussion.

Homework

Homework suggestions are provided as well as recommendations as to which activities and worksheets make good homework assignments.

Portfolio

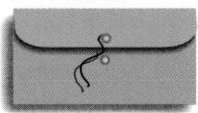

Students are encouraged to choose a portion of their work to include in a personal Portfolio. The suggestions provided in Teaching Strategies help you guide your students in developing this Portfolio. You may then choose to use the Portfolio as part of your final assessment.

Cooperative Learning

Suggestions are provided to help you organize cooperative-learning activities. Grouping strategies, individual and group responsibilities, and other important topics are covered.

Guided Practice

Questions are provided to help you lead group discussions related to the concepts presented in the lesson. The questions help invite and encourage group participation.

Independent Practice

Suggestions are provided for additional activities that students can do independently to help them solidify their understanding of difficult concepts.

Answers to Questions

All questions that have been posed in the course of the lessons, either within the running text or in the review materials, are answered in the Wrap-Around Margins for your convenience. In only a few instances, where the answer requires a graph or chart, will you have to turn pages to refer to the answer. In addition, all questions in the running text have lettered annotations on the reduced Pupil's Edition pages for quick and easy reference.

Did You Know...

Science facts are provided to add special insights and interest to your lessons.

Safety Alert/Waste Disposal

Additional information is provided when there are special safety and waste-disposal concerns.

FOLLOW-UP

The Follow-Up consists of four parts:
- Reteaching
- Assessment
- Extension
- Closure

This information provides effective methods to close out the lesson and to evaluate whether students have grasped the main concepts. If you find that your students need additional help, a reteaching strategy is provided. The extension provides an idea or activity for learning more about a concept covered in the lesson.

Review Pages

All questions appearing in each Challenge Your Thinking and Making Connections are fully answered in the Wrap-Around Margins. All answers for the SourceBook Unit CheckUps are provided in the back of this Annotated Teacher's Edition. Annotations in the Wrap-Around Margins reference the appropriate page numbers.

Features

All of the special-interest features at the end of each unit include background information, answers to questions, and a selection of teaching strategies, including ideas for extending, discussing, debating, and analyzing the feature.

TEACHING RESOURCES

All of the worksheets you will need or want are bound into eight convenient unit booklets. Now you do not have to spend precious time compiling worksheets from various sources. Each *Teaching Resources* booklet has all the worksheets you need to teach an entire unit.

Each *Teaching Resources* booklet contains the following types of worksheets:

Home Connection

The Home Connection begins with a two-page parent letter that introduces the unit of study to the parents and includes home activities to encourage parents' participation. The last page lists the supplies needed to do the Explorations and Activities in the unit. This page makes it easy for you to invite donations in order to keep your budget low.

Chapter Worksheets

- **Resource Worksheets** Worksheet versions of charts, graphs, puzzles, and activities from the Pupil's Edition
- **Exploration Worksheets** Worksheet versions of the Explorations in the Pupil's Edition
- **Discrepant Event Worksheets** Demonstrations and activities that spur students' curiosity and motivate further exploration
- **Theme Worksheets** Worksheets that draw thematic connections between concepts
- **Transparency Worksheets** Review, reinforcement, and assessment worksheets that correspond to the overhead transparencies
- **Math and Graphing Practice Worksheets** Worksheets that reinforce important math concepts and graphing skills
- **Review Worksheets** Worksheet versions of the Challenge Your Thinking review pages in the Pupil's Edition
- **Chapter Assessment** Tests consisting of a variety of questioning strategies to check students' comprehension. Challenge questions that require students to use higher-order thinking skills are included.

Unit Worksheets

- **Unit Activity Worksheets** Extension and review activities, usually in the form of a crossword puzzle or word search
- **Unit Review Worksheets** Worksheet versions of the Making Connections review pages from the Pupil's Edition
- **Self-Evaluation of Achievement** Checklist that helps students evaluate their progress throughout the unit
- **End-of-Unit Assessment** Test that consists of a variety of question types, including word usage, short response, correction/completion, short essay, numerical problems, and interpreting illustrations, graphs, and charts. At least two Challenge questions are also provided to check higher-order thinking skills.
- **Activity Assessment** Performance-based test that requires students to use materials and data to solve a problem
- **Spanish Resources** Spanish blackline masters consisting of Home Connection letters, Big Ideas, and unit glossaries

SourceBook Worksheets

- **SourceBook Activity Worksheets** Hands-on laboratory explorations and activities to illustrate and reinforce the content in the SourceBook
- **SourceBook Review Worksheets** Worksheet versions of the Unit CheckUp review pages in the SourceBook
- **SourceBook Assessment** Tests consisting of multiple-choice, true-false, and short-answer questions

The *SciencePlus* program includes the following support materials to make your teaching both effective and efficient.

Teaching Transparencies

Full-color transparencies that highlight key points from each chapter and organize important content with charts, graphs, and puzzles from the Pupil's Edition and *Teaching Resources*

Getting Started Guide
(for Levels Red and Blue)

Blackline master booklet designed to orient both teachers and students who are new to the *SciencePlus* method of teaching and learning science

Assessment Checklists and Rubrics

Blackline master booklet that provides a variety of assessment checklists and rubrics to make your assessment tasks efficient and timely. Electronic versions of these blackline masters are available on the *SnackDisc*.

Materials Guide

Complete listing of all the supplies and equipment needed to teach the entire level, as well as each individual unit, of *SciencePlus*

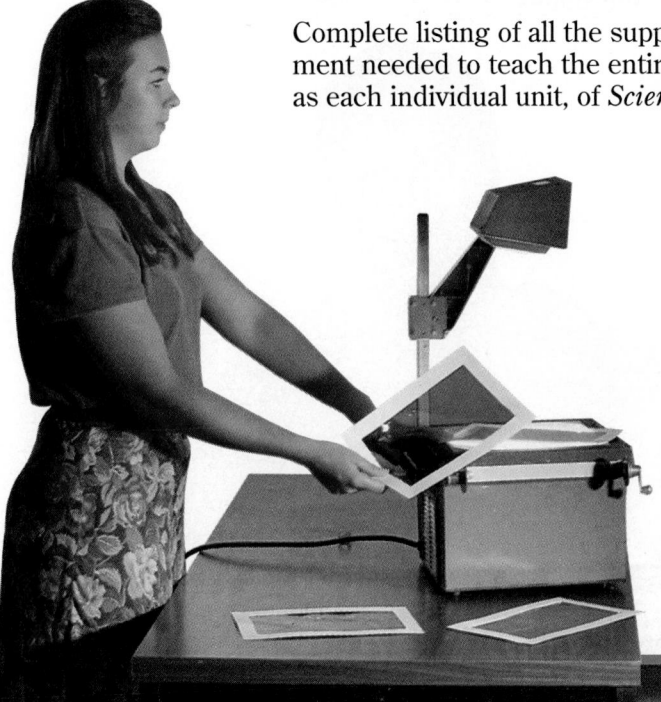

Test Generator

Test item software for Macintosh® and Windows® allows you to create your own tests from an extensive bank of tests and test items. In addition, you can edit the items and even add your own original items. A comprehensive *Test Item Listing* is provided to preview the test items in hard copy.

English/Spanish Audiocassettes

Audiocassette tapes that provide important preview information for each unit in both English and Spanish. Each unit begins with an attention-grabbing skit or story and then takes students on a visual tour of the unit.

Videodisc Resources

Instructions, barcodes, and worksheets for using the *Science Discovery* videodiscs with the *SciencePlus* program. For more information about the *Science Discovery* videodisc program, see page T31 of this Annotated Teacher's Edition.

SnackDisc

CD-ROM disc for Macintosh® and Windows® that contains a vast array of worksheets and information, most of which is customizable to fit your specific needs. The following is a partial listing of the items on the *SnackDisc*.
- Safety First! including safety review, test, and contracts
- Student Review Worksheets, providing review of math, graphing, and the metric system
- Science Sites, showing points of scientific interest on maps of the United States, Canada, and Mexico
- Home Connection letters in English and Spanish
- Assessment Checklists and Rubrics
- Lab Inventory Checklists
- District Requirements Reference Chart
- *SciencePlus* Communicator

Science Discovery is a comprehensive videodisc program that blends imagination and state-of-the-art technology into an exciting program that will add a new dimension to your science teaching. The program includes the following:

- Science Sleuths videodisc
- Science Sleuths Teacher's Guide
- Science Sleuths Resource Directory
- Image and Activity Bank videodisc
- Image and Activity Bank Directory
- Image and Activity Bank Quick Reference Card

Science Sleuths

The Science Sleuths videodisc consists of 24 open-ended science mysteries, one for each unit of the *SciencePlus* textbook series. Each mystery begins with a dramatization in which a problem is presented and the students are asked to serve as consultants (sleuths) to figure it out. Problems include such mysteries as Exploding Lawn Mowers, Dead Fish on Union Lake, and The Misplaced Fossil. The videodisc then becomes a "videophone" that connects students to the laboratory, where they can request information such as the testing of samples, experiments, interviews, tables, newspaper clippings, news broadcasts, and many other bits of information from an extensive menu.

The students work as a class or in small groups to explore the mystery, develop hypotheses, and support their position using the data from the videodisc. They try to solve the mystery using as few of the video segments as possible. In this way they either challenge themselves or other groups to see who can solve the mystery most efficiently.

In working as Science Sleuths, students improve their problem-solving and reasoning skills while applying their science knowledge from the textbook. Students work in a realistic scientific mode in which information comes from a variety of sources. Students must also judge the accuracy of each source and separate raw data from interpretation and inference.

Science Sleuths Teacher's Guide

The Science Sleuths Teacher's Guide contains the teaching plans for using the videodisc.

Science Sleuths Resource Directory

The Science Sleuths Resource Directory contains the barcodes and frame numbers for using each mystery. There are five directory sets so that you can have five working cooperative groups at a time.

Image and Activity Bank

The Image and Activity Bank videodisc consists of a still and motion image database designed to reinforce and extend concepts presented in *SciencePlus*. The still images include hundreds of photographs and computer graphics related to life, earth, and physical sciences. The motion images include demonstrations, experiments, and selected motion footage.

Image and Activity Bank Directory

The Image and Activity Directory provides descriptions, frame numbers, and barcodes for the Image and Activity Bank videodisc. A separate Reference Card provides barcodes and frame numbers for selected topics.

> *For complete directions on how to use* **Science Discovery** *with* **SciencePlus,** *see the Videodisc Resources booklet described on page T30.*

USING SciencePLUS

THEMES IN SCIENCE

The traditional division of science into three branches—life, earth, and physical—often leads students to believe that nature is similarly arranged. Too often, students view science as a system of separate, unrelated abstractions or as a compilation of facts and difficult-sounding terms. But science is simply the study of nature, and there are certain underlying principles, or themes, that unite the study of all areas of science. The themes provided here are not meant to replace the traditional teaching of scientific disciplines, but rather to create a framework for the unification of these disciplines.

The following themes are emphasized in *SciencePlus*. These themes are intended to integrate facts and ideas to provide a context for discussing the textual matter in a meaningful way. You can employ these themes as an organizational tool to reinforce understanding of the subject matter.

Energy

Energy puts matter into motion and causes it to change. Energy is what makes the universe and everything in it dynamic. The study of dynamic systems in any field of science requires an understanding of energy: its origins, how it flows through systems, how it is converted from one form to another, and how it is conserved. Energy provides the basis for all interactions, whether biological, chemical, or physical. Thematically, energy connects all disciplines.

Systems

A system is any collection of objects that influence one another. The parts of a system can be almost anything—planets, organisms, or machines, for example. And a system may be very small, such as a cell nucleus; very large, such as a galaxy; or very complex, such as the human body. All of the science disciplines involve the study of some kind of system or systems. Understanding a system involves knowing what its important parts are and how those parts work together.

Structures

Structure provides a basis for studying all matter, from the most basic forms to the most complex. Structure is closely related to function, so scientists study the structures of things to learn how they work. For example, the structure of an eye reveals much about the process of vision. And the key to a diamond's strength lies in the tight lattice structure of carbon atoms within the diamond.

Changes Over Time

A change over time is not a single alteration, but a progression of alterations that occurs over the continuum of time. Biological evolution is an example of change over time. Evolution has gradually changed the characteristics of organisms ranging from the starfish to the giraffe. Changes over time occur in physical and earth science as well. For example, the chemical change of rust forming on an iron nail can be traced through time, as can the movements of the Earth's plates. An understanding of how changes occur over time allows students to appreciate not only the present state of the world around them but also what it may have looked like in the past and what it may look like in the future.

Cycles

A cycle is a pattern of events that recurs regularly over time or a circular flow of materials in a system. Cycles occur throughout nature and appear in all of the scientific disciplines. The water cycle, the rock cycle, and the life cycle of a particular organism are examples of cycles at work. A defining feature of a cycle is that it has no beginning or end; therefore, the study of any part of a cycle is incomplete without consideration of the other elements in the cycle.

Using the Themes

A major strength of the thematic approach is that seemingly different processes, structures, or systems can be shown to have underlying similarities. Although many thematic organizations are possible, each Unit Interleaf in this Annotated Teacher's Edition suggests at least three major themes that can be discussed in relation to the unit and chapter material. Focus questions are also provided in the Wrap-Around Margins to promote discussion and an understanding of how the themes relate to specific text material.

Although major themes have been identified in each Unit Interleaf and in the Wrap-Around Margins, it is up to you to decide which themes are most appropriate. The direction that your class discussion takes will most likely guide you in your choices.

In your class discussions, use the themes to provide a framework of understanding for your students. For example, whether you are studying photosynthesis or the way in which the forces of nature have shaped the physical appearance of the Earth, the theme of **Energy** can be discussed. Similarly, one subject can be addressed from the viewpoint of many different themes. For example, in discussing an organism such as a zebra, **Energy** can be applied in a discussion of how the zebra takes in food from the environment; **Changes Over Time** can be discussed in relation to how the zebra's structures are adaptations to its environment; and **Cycles** can be introduced in a discussion of how the zebra's migration habits are based on the seasons.

Theme Connection

Changes Over Time
Focus question: When succession occurs at an abandoned farm, in what order would the following plants appear?
a. taller plants—goldenrod and grasses
b. weeds that thrive in sunny locations—wild carrots and dandelions
c. slower growing trees—oak and hickory
d. young saplings—aspens
(Answer: b, a, d, c)

> In your class discussions, use the themes to provide a framework of understanding for your students.

INTEGRATING THE SCIENCES

Throughout *SciencePlus*, students are asked to look at things from many different viewpoints. Since *SciencePlus* is not organized according to life, earth, and physical sciences, these disciplines are integrated throughout the program. Each discipline comes into play as needed in covering a main concept or theme in *SciencePlus*. In this way, science disciplines are blended together so that the emphasis is on student comprehension of the "big picture" rather than on isolated components of different areas of science.

Additional integration suggestions are provided in the Wrap-Around Margins. Look for the Integrating the Sciences within the Teaching Strategies of the lessons. These suggestions will help you build connections between the science disciplines so that students can get a fuller, more in-depth view of the concept at hand.

Integrating the Sciences

Life, Earth, and Physical Sciences
Every field of science employs its own classification system in order to make sense of diversity. Students will learn later in this unit how biologists apply a classification system to living things. You might want to explain to the class how some other disciplines classify things in their own field.

CROSS-DISCIPLINARY CONNECTIONS

SciencePlus is also integrated in terms of non-science disciplines. For example, students are asked to write a headline for a science story, to write advertising copy highlighting geological features for a travel agency, and even to make dioramas of animal habitats and ways of life. In doing these activities, students see firsthand the connections between science and a variety of other disciplines, including history, geography, social studies, and others. And the students even have fun making the connections.

Although the Pupil's Edition contains a wide variety of cross-disciplinary activities and Explorations, additional suggestions are included in the Wrap-Around Margins of this Annotated Teacher's Edition. Look for the Cross-Disciplinary Focus in the Teaching Strategies of the lessons. Each Cross-Disciplinary Focus provides a suggestion for an activity that highlights the interrelatedness of science and disciplines such as mathematics, social studies, art, music, health, language arts, foreign language, and industrial arts.

CROSS-DISCIPLINARY FOCUS

Foreign Languages
The naming system used by scientists makes use of many Latin words. Have students make a dictionary of some of the commonly used Latin words, examples of each word's use, and the common name for the animal. For example, *Homo* means "man" and is used in such species names as *Homo erectus* and *Homo sapiens*, the name given to modern humans.

SCIENCE, TECHNOLOGY, AND SOCIETY

STS is an approach to teaching science in which the impacts of scientific, technological, and social matters are explored. STS teaches science from the context of the human experience and in so doing leads students to think of science as a social endeavor. STS emphasizes personal involvement in science. Students become active participants in the scientific experience. For this reason, *SciencePlus* incorporates an STS approach.

STS contrasts with the traditional "basic science" approach to science education, in which students follow a carefully sequenced program of the basic models, concepts, laws, and theories of science. The fundamental flaw of the basic science approach is that the student is more or less a passive participant in the learning process.

There are three parts to STS:

 Science concept and skill development, and knowledge of the nature of science

This component of STS introduces science as a system for learning about the natural world and gives students the foundation they need to actually practice science in and out of the classroom. The ultimate goal of STS, and of *SciencePlus*, is to turn students into scientists, at least for the duration of their science education. To accomplish this, students first learn the methods of science and the skills that scientists draw on. Whenever possible, the major ideas of science are introduced from the standpoint of those who developed them. In this way, students come to see the reasoning that went into the development of these ideas.

 Knowledge of the relationship of science and technology, and engagement in science-based problem solving

Students who understand the real-world applications of science are better able to appreciate and enjoy it. This component of STS reinforces the practical value of science. Students see that science is a system for solving practical problems. Students themselves become practical scientists—first identifying problems and then developing solutions to them. Students learn to analyze, to plan, to organize, to design, and to refine models and designs.

 **Engagement in science-related social issues and attention to science as a social institution**

This component of STS deals with the ways in which science serves human needs. The benefits may be tangible or intangible, but either way, they are real. To emphasize the social responsibility of science, scientists are shown to be concerned about the impact of their work on society as a whole. Advances in science and technology sometimes lead to thorny ethical issues. Such issues are often used as a focus for discussion and investigation. From these investigations, students draw conclusions and form reasoned opinions.

Studies have shown that students begin the study of science full of curiosity and enthusiasm. But after a few years of a traditional curriculum, the curiosity is squelched and the enthusiasm has all but vanished. By contrast, under the STS approach, student interest builds through the years and is maintained long after formal study ends. The key to this success is the active involvement in science that STS imparts. The end products of the STS approach are students who appreciate and understand science and who are equipped to deal sensibly with the complex issues that will become commonplace in the decades ahead.

> *For a fuller discussion of STS, see the NSTA position paper, July 1990.*

STS emphasizes personal involvement in science. Students become active participants in the scientific experience.

One of the most important skills that students can acquire is the ability to communicate what they have learned, both orally and in writing. *SciencePlus* challenges students to develop and communicate their mastery of new ideas in novel ways—for example, by writing for one another or for some audience other than the teacher. Students' comprehension is enhanced when they are called upon to reformulate in their own words what they have learned.

Throughout *SciencePlus*, students are called on to communicate what they have learned in many different ways. Students may be called on to interpret a passage, illustrate a paragraph, write a headline, label a diagram, or write a caption for a photo or drawing. These strategies complement the inquiry approach followed throughout *SciencePlus*.

SciencePlus challenges students to communicate ideas in novel ways.

Reading and Writing in the Classroom

The time spent in helping students prepare to read is critical in fostering comprehension. Two strategies are particularly important in the pre-reading phase of instruction: building on prior knowledge and establishing a purpose for reading.

The Power of Prior Knowledge

The amount of prior knowledge that students have about a topic directly influences their comprehension of that topic. The more students know about something, the easier it is for them to grasp new information about it. Helping students identify the information they already know about a topic assists them in relating the new information to existing knowledge.

Research strongly suggests that students tend to retain misconceptions they may have about a topic. If the text information seems to conflict with their preconceptions, students may ignore or reject new information. It is therefore extremely important to identify these misconceptions so that they may be dispelled.

Reading to Understand

One way to establish a purpose for reading is to make a study guide with questions that students can answer as they read. You can also help students learn to make their own study guides. First teach students to preview a unit by looking at all the unit headings, illustrative material, and terms and phrases in boldface or italics. This technique helps students gain a feel for the unit and helps them build a basic structure for the new information. Then show them how to use the captions, headings, and highlighted words to devise study-guide questions to answer after reading the unit. Following this method, students will read with a purpose in mind, a purpose of their own devising. You may also wish to consider using the English/Spanish Audiocassettes, which are an excellent means for previewing the unit with students.

Writing to Understand

Studies show that writing is an effective tool for improving reading. As students write, they are creating a text for others to read. The most important advice to give students about scientific writing is to strive for clarity and accuracy. These characteristics can often be achieved with simple vocabulary and short sentences. One useful approach might be to have them imagine that they are writing for a younger audience.

Keeping a Journal

One highly successful tool for improving students' performance in science is the journal. The *SciencePlus* version of the journal is called the ScienceLog. The ScienceLog has many functions. First and foremost, it is an ongoing record of students' learning. Students begin the study of a new topic by recording prior knowledge of that topic. Any misconceptions that students may have are thus exposed. As the lesson progresses, students record any and all new findings. In many cases, students find that what they learned in the activities contradicts their preconceptions.

Much of the work that students do should be recorded in their ScienceLog. The ScienceLog is a constant reminder to students that learning is occurring. Students can look back and compare their early work with later work to see and take pride in the progress they have made. To supplement their other work, you may want to ask students to briefly summarize what they have learned each week. This makes a very handy capsule history of their work.

What makes a good ScienceLog? Insofar as is possible, the ScienceLog should be neat and easy to follow. Students may organize their ScienceLogs in any of a number of ways—chronologically, by unit, by chapter, or by lesson, to name a few. Some kind of heading should set off each major entry.

A spiral-bound notebook or hard-bound lab-type notebook makes a good ScienceLog. Or you may make copies of the sample ScienceLog pages in the *Teaching Resources* or *SnackDisc* and distribute them to your students to use if you wish. These ScienceLog pages are located at the beginning of the Unit Worksheets for Unit 1.

Portfolios: What, Why, and How

Definitions and descriptions vary as to what constitutes a portfolio. Very simply, a portfolio is a collection of work that is done by the student during the course of the year. Usually, the students themselves have a say as to what goes into the portfolio, but selections should represent the objectives outlined by the curriculum. The purpose of the portfolio is to represent the student's mastery of skills and knowledge within the subject area. This may sound simple enough, and, in fact, it can be very simple. On the other hand, portfolios can be elaborate collections of work that are limited only by the teacher's and students' imaginations.

Your initial decision to use portfolios in your assessment leads to a host of other decisions that must be made. However, before you launch into making these decisions, it is a good idea to seek input from other teachers as well as from your school administration. You may even want to discuss your ideas with parents and other members of your community. This will make it easier to get feedback later as to the impact portfolios are having on student learning and attitudes.

You will also need to plan a system for organizing and managing the portfolios. You will need to establish guidelines for what types of information will be admitted into the portfolios, how and by whom the selections will be made, and when materials can be added or revised. These decisions will be based on your individual preferences and on the level and attitudes of your students. The following guidelines may help you in developing your plans.

> The most important advice to give students about scientific writing is to strive for clarity and accuracy.

- Allow students to select for themselves the sample materials that best represent their level of understanding and mastery. Although you may require that certain projects or materials be placed in the portfolios, it is advised that the students play an active role in selecting their best work. This gives students ownership and encourages them to take increasing responsibility for the quantity and quality of their work.
- Allow students to revise the selections in their portfolios at any time. The portfolios should evolve as your students' skills improve.
- Identify well in advance when you will be assessing the portfolios. Students should be given plenty of warning before the portfolios are collected for assessment.
- Determine in advance the criteria you will use for grading the portfolios. Share this criteria with the students up front. You may want to make a criteria checklist that each student can place inside his or her portfolio. This checklist would provide both you and your students with a ready reference to the grading criteria.
- You may want to keep the portfolios in the classroom or in some other area, allowing frequent but controlled access to them. This will decrease the chances of portfolios being lost. This may require a significant amount of room, depending on the nature of the portfolios.

- Make available examples of high-quality portfolios so that students can see examples of excellent work. This, of course, may be possible only after you have used portfolio assessment for at least one course.

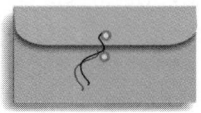 as a Portfolio

The ScienceLog provides an excellent opportunity for portfolio assessment. By directing your students to include in their ScienceLogs representative samples of their work, as well as comments and reflections about their samples, you can add depth to your evaluation method. To help you utilize portfolios effectively in your teaching, specific strategies are provided in Portfolio boxes in the Wrap-Around Margins of this Annotated Teacher's Edition.

PORTFOLIO
Students may wish to include their written composition in their Portfolio.

Too often, students are able to master the individual elements of a topic without truly grasping the "big picture." If students fail to understand how the elements fit together or relate to one another, they cannot truly comprehend the topic. Concept mapping is a very effective method of helping students see how individual ideas or elements connect to form a larger whole. Concept maps are a highly effective tool for helping students make those logical connections.

The most effective concept maps are those that students construct on their own. Used in this way, concept maps are both a self-teaching system and a diagnostic tool. To construct a proper concept map, the student must first examine closely his or her mental model of the topic at hand. Any flaws or shortcomings in that model will be reflected in the concept map.

Concept maps are flexible. They can be simple or highly detailed, linear or branched, hierarchical or cross-linked, or they can contain all of these major elements. Students can construct their own maps from scratch or can finish incomplete maps. Concept maps can take almost any form as long as they are logically arranged.

Making Concept Maps

The steps involved in making a concept map are outlined below. To provide guidance to your students in making concept maps, direct them to page xvii in the Pupil's Edition.

1. Make a list of the concepts to be mapped. Concepts are signified by a noun or short phrase equivalent to a noun.
2. Choose the most general, or the main, idea. Write it down and circle it.

3. Select the concept most directly related to the main idea. Place it underneath the main idea and circle it. If two or more concepts bear the same relationship to the main idea, they should be placed at the same level.
4. Draw a line between the related concepts, leaving a space for a short action phrase that shows how the concepts are related. These are linkages.
5. Continue in this way until every concept in the list is accounted for.

The simple concept map below shows the relationship among the following terms: plants, photosynthesis, carbon dioxide, water, and sun's energy. More detailed maps are shown on the next page.

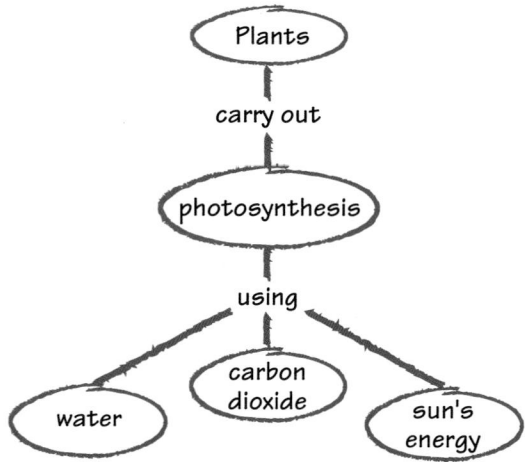

For any given topic, there is no single "correct" concept map. Not all maps are equally valid, however. Good concept maps have most or all of the following characteristics:
- start with a single, general concept—a big idea—and work down to more specific ideas
- represent each concept with a noun or short phrase, each of which appears only once
- link concepts with linkage words or short phrases
- show cross-linkages where appropriate
- consist of more than single path
- include examples where appropriate

> Concept maps are a highly effective tool for helping students make logical connections.

Using Concept Maps

Concept maps can be applied in many ways.

- to gauge prior knowledge of a topic
- as end-of-lesson, chapter, or unit evaluation
- as pre-test review
- to help summarize special presentations, such as films, videos, or guest speakers
- as an aid to note taking
- for reteaching

You may also want to use partially completed concept maps as pop quizzes or as devices for summarizing particularly difficult class sessions. Also, be sure to use the concept map in each Making Connections review.

Evaluating Concept Maps

Again, there is no single correct concept map. However, you should consider the following criteria as you evaluate your students' concept maps.

- how comprehensive the map is (Are all relationships shown?)
- how clearly the concepts are linked (proper relationship between concepts, use of linkage terms between all concepts)
- overall clarity of presentation (Could the map be simpler? Is it redundant? Is it logically arranged? Are linkage terms used properly?)

Used properly, concept maps can increase comprehension, improve retention, and sharpen study skills in your students. They are a valuable addition to any student's arsenal of learning strategies.

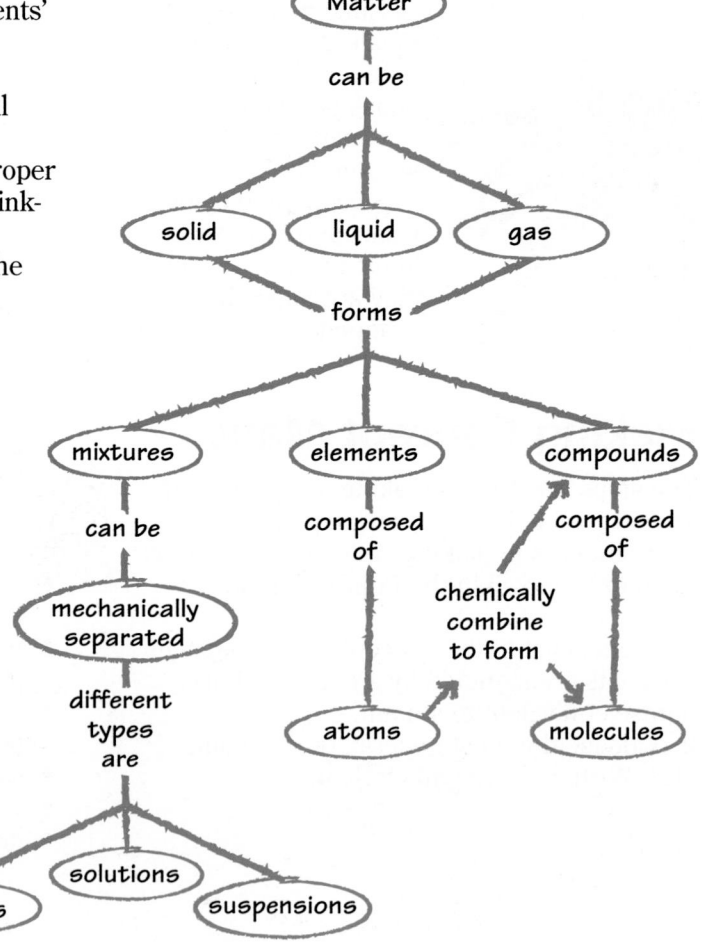

The most effective concept maps are those that students construct on their own.

Cooperative learning is a teaching technique that brings students together to learn in small, heterogeneous groups. In these groups, students work interdependently without constant and direct supervision from the teacher. Assignments are structured so that everyone contributes. Challenges as well as rewards are shared. Brainstorming, lively discussion, and collaboration are the hallmarks of the cooperative-learning classroom.

Cooperative learning is an ideal complement to the *SciencePlus* approach. The discussions, explorations, research projects, games, and puzzles of *SciencePlus* can all be used in a cooperative-learning format.

What It's Not!

- Cooperative learning is **not** the same as ability grouping, where a teacher divides up the class in order to instruct students with similar skills.
- Cooperative learning is **not** having students sit side by side at the same table to talk while they complete individual assignments.
- Cooperative learning is **not** assigning a task to a group in which one student does the work and the others get equal credit.

Benefits of Cooperative Learning

Traditionally, teachers structure lessons so that students work individually to achieve learning goals or compete against one another to be "the best." While these formats can be useful, cooperative learning provides an important alternative.

- **Cooperative learning models the scientific experience.**
  Students working in groups learn about the joys as well as the frustrations involved in scientific inquiry. Cooperative learning models real scientific experience, in which scientists work together, not in isolation, to solve difficult problems. With cooperative learning, the classroom becomes a fertile environment for ideas and novel solutions.

- **Cooperative learning empowers and involves students.**
  Cooperative learning raises students' self-esteem because they are learning something on their own through cooperation, rather than being handed prepackaged knowledge. It helps students become self-sufficient, self-directed, lifelong learners. In a cooperative-learning environment, students are less dependent on you for knowledge.

- **Cooperative learning serves the heterogeneous classroom.**
  With group work, everyone has the chance to participate, and everyone has a role to play. As students join forces to achieve a common goal, they come to recognize commonalities that cut across differences related to ethnicity, socioeconomic background, and gender. Likewise, cooperative learning provides an excellent vehicle for students of differing ability levels to work together in a positive way. Basic students can interact successfully with average and advanced students, and in so doing can learn that they, too, have something to offer.

- **Cooperative learning strengthens interpersonal skills.**
  Group tasks are structured so that students must cooperate to succeed. Students quickly understand that they will "sink or swim" together by how constructively they interact. Consequently, students develop important interpersonal and social skills that help them function in a group setting and that will ultimately benefit them socially, at work, and in other situations.

- **Cooperative learning develops appropriate social skills.**
  When doing cooperative group work, students channel their energies into constructive tasks while satisfying their fundamental need for social interaction.

- **Cooperative learning is an effective management tool.**
  Establishing cooperative learning in the classroom requires you to relinquish some control, so the students themselves can become responsible for building their own knowledge. Working in groups to probe and investigate ideas, answer questions, and draw conclusions about observations allows students to discover and discuss concepts in their own language. When students learn through cooperation, the knowledge derived becomes their own, not just a loan of your ideas or those from the textbook.

- **Cooperative learning increases achievement.**
  Since the 1920s, there has been extensive research on cooperative-learning techniques. Results clearly indicate that cooperative learning promotes higher achievement for all grade levels in all subject areas.[1]

[1] Johnson, Johnson, Holubec, and Roy. *Circles of Learning, Cooperation in the Classroom.* Association for Supervision and Curriculum Development. ©1984

Using Cooperative Learning With *SciencePlus*

Cooperative learning is an essential component of the *SciencePlus* philosophy. To help you take full advantage of this important component, several cooperative-learning activities have been highlighted in each unit of this Annotated Teacher's Edition. These activities are broken down to provide suggestions for group size, group goal, positive interdependence, and individual accountability.

This symbol designates cooperative-learning activities in the Wrap-Around Margins of this Annotated Teacher's Edition. Cooperative Learning worksheets are are also provided in the *Teaching Resources* booklets.

Group Size Although group size will vary depending upon the activity, the optimum size for cooperative learning is between three and four students. For students unaccustomed to this learning style, keep the group size to about two or three students.

Group Goal Students need to understand what is expected of them. Identify the group goal, whether it be to master specific objectives or to create a product such as a chart, a report, or an illustration. Identify and explain the specific cooperative skills required for each activity.

Positive Interdependence A learning activity becomes cooperative only when everyone realizes that no group member can be successful unless all group members are successful. The "we're all in this together" part of group work is the positive interdependence. Encourage positive interdependence by assigning each student some meaningful role, or allow students to do this themselves. You can also encourage positive interdependence by dividing materials, resources, or information among group members.

Individual Accountability Each group member should have some specific responsibility that contributes to the learning of all group members. At the same time, each group member should reach a certain minimum level of mastery.

Meeting Individual Needs With Cooperative Learning

Cooperative learning is an effective tool for meeting the individual needs of your students. Cooperative learning builds relationships among students where relationships might not have developed before. Students are forced to interact with each other as individuals with common goals. In so doing, students learn more about each other's personal characteristics, and as a result, many stereotypes are destroyed.

There is no single set of cooperative-learning strategies that will work with all students in all situations. However, the following strategies may provide you with insight and guidance in developing your own set of strategies that will work for your students.

Balance the needs of students of all levels and learning styles

Your ultimate goal will be to ensure that all students are able to work effectively in any group. Initially, however, you may wish to develop special grouping strategies to foster the growth of learners having difficulty and second-language learners, and to assure gifted students that their grade will not be affected by slower learners. More information is provided about grouping strategies on the next page.

Have a clearly defined goal

When you tell students what is expected of them, be sensitive to their special needs. Be sure that each student understands the group goal and his or her own personal responsibility.

It is also important that assignments be specifically appropriate for groups. In other words, simply having students fill in the blanks of a worksheet or answer the end-of-unit questions is not creating an adequate cooperative-learning assignment. Students need tasks that cannot be easily completed alone. Students should see that if they work together, the end product will be better and more complete than if they had worked alone.

Answer questions only when the whole group has the same question. It's a good idea to designate one person per group as the liaison between you and the group.

Praise success

If students seem unmotivated or feel that their individual tasks are unimportant to the success of the group, you may wish to consider offering group rewards. Reward the groups as they successfully complete each activity. Reward successful project results as well as positive interaction and effective group-process skills. However, rewards should not become automatic. They should be used only for the short term. For the long term, students should take pride in their group's achievements and should benefit from the knowledge that these achievements contribute to their success as individuals.

Encourage interpersonal problem solving within groups

Pulling a disruptive student out of a group is sometimes necessary, but be sure that the isolation is only temporary. Difficult students need the support of others. Build into each group the spirit of encouraging each other. Suggest that groups evaluate their own performance after an activity is finished. Encourage students to suggest solutions to problems without criticizing individuals.

Grouping Strategies

With cooperative learning, you can either place students in particular groups or assign students to groups at random. There are advantages to both approaches. At first, however, it is recommended that you assign students to particular groups.

> *The* ScienyPlus SnackDisc *CD-ROM and the* SciencePlus *Assessment Checklists and Rubrics booklet contain several checklists to help you implement cooperative-learning groups.*

Assigned Grouping

Composing groups yourself lets you create groups that are heterogeneous in terms of academic ability, gender, ethnicity, and cultural background. Heterogeneous groups are preferred because cooperation among diverse students not only teaches the widest range of interpersonal skills but also promotes frequent exchange of explanations and greater perspective in discussions. This increases depth of understanding and retention of concepts.

To create *effective* heterogeneous groups, balance each group with students who have different strengths. First decide who your resource students are. These are students you think will facilitate group work—either because of their academic ability or because of their interpersonal skills. Assign at least one resource student to each group. Distribute students who may be disruptive and students who lack academic skills evenly throughout the groups. Avoid putting close friends together to prevent cliques from disrupting teamwork. Put students who have limited English proficiency in groups with bilingual students who can act as translators.

Random Grouping

Random grouping can be especially effective with experienced cooperative learners or if you plan to change group membership often. To create random groups, you can simply have students count off from 1 to 4. All of the 1s form a group, all of the 2s form another group, and so on.

There are many other fun ways to assign groups randomly. For example, you can hold a lottery in which students pick numbers out of a hat. Numbers 1 to 4 form one group, numbers 5 to 8 form another, and so forth. You can also use cards naming sets of a particular type of item. All students who draw items belonging to the same set form a group. For example, all students whose cards name types of flowers belong in one group, students whose cards name farm animals belong in another group, those with cards naming heavy-metal bands form a third group, etc. Students have a fun and lively time discovering who belongs in the same group.

You can also combine lesson content with assigning groups. First decide how many students you want in each group. For each group, write a different scientific term or principle on a flashcard. Then for each group's term, list on separate cards the definition of the term, a synonym, or an example of what the term means. Mix up the cards and hand them out as students come into the room or once students are seated. Students use the cards as clues to find the others in their group.

Assigning Roles

Assign roles at first; students can choose their own later.

Assigning roles is very important, especially at first. Consider the behavior patterns of students, and assign roles that will complement those patterns. Group work needs to be structured so that everyone has a part to play. In other words, there needs to be positive interdependence. Just as members of a surgical team work together, with each person contributing his or her own special skill, students work effectively in teams when everyone has a unique role that is vital to the group's success.

Some examples of useful roles are listed here. Use as many as you need, modify them, combine them, or invent roles yourself.

- **Facilitator** The facilitator is a leadership role. The facilitator keeps an activity running smoothly by presiding over the work flow. He or she manages the group so that all members have a chance to talk, questions are answered, students listen to one another's ideas, and ideas are substantiated with reasons and explanations.
- **Recorder** The recorder records data and answers questions posed to the group.
- **Reporter** The reporter explains the group's findings to the teacher or the entire class.
- **Safety officer** The safety officer makes sure safety practices are followed and notifies the teacher of any unsafe situations.
- **Checker** The checker makes sure that everyone has finished his or her worksheet or other individual assignment.
- **Materials manager** The materials manager gathers activity materials at the outset, monitors their use during the activity, and organizes the cleanup and return of materials to their proper place after an activity.

Again, assign roles carefully, especially at first, taking into account students' behavior patterns. A shy student might be most comfortable as a recorder, while a student who likes attention might make the best reporter. The facilitator is a role that some students will always want and others will avoid. Be careful not to stereotype. Sometimes the most unlikely students will make the best leaders.

Assessment

Assessment within a cooperative-learning setting is not as difficult as it may seem. Like any other assessment, you must determine in advance what you would like to assess and to what degree. You will also need to develop some slightly different monitoring skills. In addition to the following information, the *SciencePlus* program includes several checklists and rubrics to make this task simple and efficient.

A variety of checklists designed for monitoring and assessing cooperative group work are provided on the SciencePlus Snackdisc *CD-ROM and in the* SciencePlus Assessment and Rubrics *booklet. Both student and teacher checklists are provided.*

Monitor Groups

Resist the temptation to get caught up on paperwork as the groups do their work—this is the time to observe, monitor, and coach. As you monitor the groups, you can reinforce cooperative behavior with a formal observation sheet. Appropriate checklists are available on the *SnackDisc* CD-ROM and in the *Assessment Checklists and Rubrics* booklet. Record how many times you observe each student using a collaborative skill, such as contributing ideas or asking questions.

If a group seems hopelessly confused or "stuck," you can intervene to guide students to a solution. But make sure students have the opportunity to reason through problems themselves first. Consider the following differences between direct supervision and the kind of monitoring that supports cooperative learning.

DIRECT SUPERVISION	SUPPORTIVE MONITORING
Lecturing	Giving feedback
Disciplining	Encouraging problem solving
Telling students what to do	Providing resources
Leading discussions	Observing

What to Assess

What should you assess in a cooperative-learning activity? Individual success? Group success? Cooperative skills? Actually, many teachers find it useful to evaluate all three. And there are many ways to assess each of these areas.

Individual success can be evaluated by asking students to fill out answers to a worksheet as they progress through an activity; by having them record, analyze, and submit data; or by having them take a quiz. Some activities are structured so that each student turns in a product, such as a report or a poster, that can be individually graded. The individual accountability portions of the cooperative-learning features in this Annotated Teacher's Edition also provide ways to assess individual performance.

Group success is evaluated according to how well the group accomplished its assigned task. Was the task completed? Were the results accurate? If not, were errors explained and accounted for? Criteria such as these provide a framework for group evaluation.

Cooperative skills are evaluated based on your observations of students' behavior in their groups. Evaluating students' use of cooperative skills will motivate students to use them. If you intend to grade cooperative skills, it is helpful to use a formal observation checklist as you monitor students at work. Log the frequency with which group members exhibit cooperative skills or disruptive behavior.

Weigh each of the three grades as you wish to compute a single overall grade. Stress the factors that you consider most important. Use cooperative learning to meet the needs of your students. Use it and enjoy!

PROCESS SKILLS

Process skills are a means for learning and are essential to the conduct of science. For this reason, *SciencePlus* is strongly process oriented.

Perhaps the best way to teach process skills is to let students carry out scientific investigations and then point out to them the process skills they used in the course of the investigations. The Lesson Organizer at the beginning of each lesson identifies the process skills that are emphasized in the corresponding text.

SciencePlus makes regular use of many different process skills, as highlighted below. These and other process skills are called upon in *SciencePlus* in virtually every lesson.

Observing

An observation is simply a record of a sensory experience. Observations are made using all five senses. Scientists use observation skills in collecting data.

Communicating

Communicating is the process of sharing information with others. Communication can take many different forms: oral, written, nonverbal, or symbolic. Communication is essential in science, given its collaborative nature.

Measuring

Measuring is the process of making observations that can be stated in numerical terms. In *SciencePlus*, all measurements are given in SI units.

Comparing

Comparing involves assessing different objects, events, or outcomes for similarities. This skill allows students to recognize any commonality that exists between seemingly different situations. A companion skill to comparing is contrasting, in which objects, events, or outcomes are evaluated according to their differences.

Contrasting

Contrasting involves evaluating the ways in which objects, events, or outcomes are different. Contrasting is a way of finding subtle differences between otherwise similar objects, events, or outcomes.

Organizing

Organizing is the process of arranging data into a logical order so it is easier to analyze and understand. The organizing process includes sequencing, grouping, and classifying data by making tables and charts, plotting graphs, and labeling diagrams.

Classifying

Classifying involves grouping items into like categories. Items can be classified at many different levels, from the very general to the very specific.

Analyzing

The ability to analyze is critical in science. Students use analysis to determine relationships between events, to identify the separate components of a system, to diagnose causes, and to determine the reliability of data.

Inferring

Inferring is the process of drawing conclusions based on reasoning or past experience.

Hypothesizing

Hypothesizing is the process of developing testable explanations for phenomena. Testing either supports a hypothesis or refutes it.

Predicting

Predicting is the process of stating in advance the expected result of a tested hypothesis. A prediction that is accurate tends to support the hypothesis.

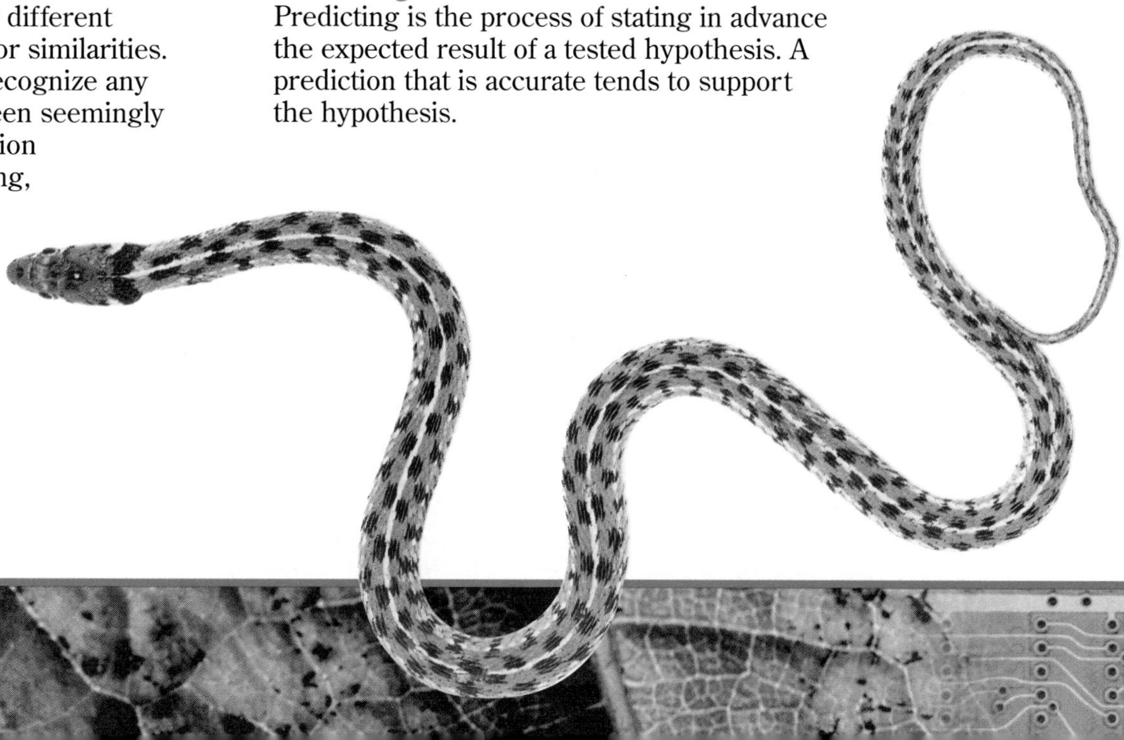

CRITICAL THINKING

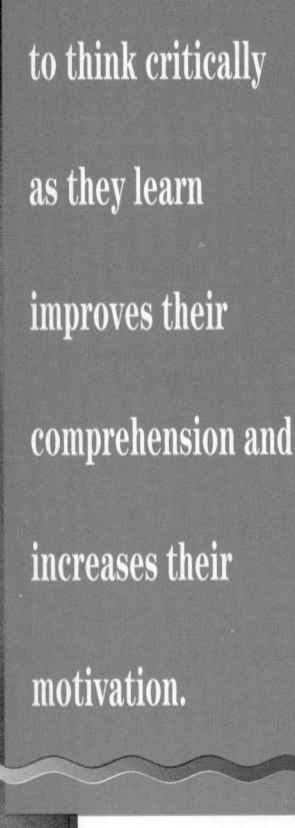

Requiring students to think critically as they learn improves their comprehension and increases their motivation.

Critical-thinking skills are essential for making sense of large amounts of information. Too often, science lessons leave students with a set of facts and little ability to integrate those facts into a comprehensible whole. Requiring students to think critically as they learn improves their comprehension and increases their motivation.

Loosely defined, critical thinking is the ability to make sense of new information based on a set of criteria. Critical-thinking skills draw on higher-order thinking processes, especially synthesis and evaluation skills. Critical thinking takes a number of different forms.

Validating Facts

This type of critical thinking involves judging the validity of information presented as fact. Too often, people will accept as valid almost any statement, no matter how outrageous, as long as it comes from a supposedly authoritative source. It is important for scientists to treat all untested data with suspicion, no matter how reasonable it may seem.

Students may validate facts in a number of ways: by observing, by testing, or by rigorously examining the logic of the so-called fact. *SciencePlus* presents students with many opportunities to critically evaluate facts and hypotheses.

Making Generalizations

A scientist must often be able to identify similarities among disparate events. Generalizations are drawn based on a limited set of observations that can be applied to an entire class of phenomena. One does not have to test every substance known to make the generalization that solid substances melt when heated. Generalizations allow scientists to make predictions. Once the rule is known, future outcomes can be forecast with a high degree of confidence.

It is important that students base their generalizations on an adequate amount of information. A generalization that is formed too quickly may be wrong or incomplete or may lead the student down a dead-end path.

Making Decisions

Many students would not regard science as a field requiring decision-making skills. But in fact, scientists must make decisions routinely in the course of their work. Any time a scientist works through a problem or develops a model, a whole series of decisions must be made. A single faulty decision can throw the entire process into disarray. Making informed decisions requires knowledge, experience, and good judgment.

Interpreting Information

Having all the information in the world is useless unless one also has the tools to interpret that information. Scientists must know how to separate the meaningful information from the "noise." Information can come in any form detectable by the five senses. It is important that scientists and students alike interpret information to determine its meaning, validity, and usefulness.

ENVIRONMENTAL AWARENESS

No species affects its surroundings as dramatically as does the human species. Thanks to recent highly publicized events—such as Chernobyl, destruction of the rain forests, the depletion of the ozone layer, and the greenhouse effect—people have come to realize the global impact that human actions can have. It is incumbent upon the educational system to promote environmental awareness among students. *SciencePlus* addresses environmental issues in a way that students can easily grasp.

Environmental issues run the gamut from global to local. While large-scale problems get headlines, they can be hard to grasp for many students, who may have never directly observed their impact. In most cases it is best to introduce your students to local issues to start building their awareness. Local issues not only are more relevant to their lives, but also are more likely to lead to direct involvement.

Environmental awareness serves two purposes: it promotes understanding of the living world and the place of humans within it, and it produces a positive change in students' behavior toward the environment. *SciencePlus* pursues both goals. You may involve students directly in environmental issues by using the suggested activities in the Pupil's Edition.

In addition, this Annotated Teacher's Edition contains teaching strategies and special Environmental Focus boxes throughout the Wrap-Around Margins. See the sample below. The Environmental Focus boxes contain relevant environmental information that you can use to add depth to the topic at hand and to encourage discussion or some other action.

> Environmental awareness produces a positive change in students' behavior toward the environment.

ENVIRONMENTAL FOCUS

Many organisms produce poisonous chemicals called *biotoxins* that are used as a defense against predators or in capturing prey. The most-studied biotoxins are snake venoms because they are so easy to obtain. Have students do some research or contact a local government agency to find out about poisonous organisms in their area. Students can then create a booklet of information about biotoxins from poisonous plants and animals found in their area.

Respect for your students' cultures will go a long way in helping each of your students achieve and maintain a positive self-concept.

The success of our nation would not be possible without the contributions of the many cultures and ethnic groups that make up our country. Multicultural instruction serves to ensure that all students have the chance to learn, to succeed, and to become whoever they would like to become, regardless of race, gender, socioeconomic background, or disability. Multicultural instruction affirms the positive nature of this country's diversity by helping students develop an open mind, a positive self-concept, and a realistic understanding of the world that surrounds them.

Meeting the needs of culturally diverse students is perhaps the most demanding challenge faced by today's teachers. We must constantly strive to arouse adolescent curiosity, minimize risks of failure, and be as relevant as possible to our individual students' needs. The more culturally relevant we make our science programs, the better we will be at serving our changing class populations. These challenges are especially difficult with middle-school students because they are also going through the physical and emotional changes associated with adolescence.

By their very nature, middle-school science programs have the potential for promoting the full development of individual learners—especially when science classrooms are perceived as places of inquiry and discovery. When such environments exist in science classrooms, students become more successful learners.

Relevance, Positive Self-Concept, and Multiculturalism

Let your students help you incorporate multicultural learning into the classroom by allowing them the freedom to express their feelings and attitudes during your classes. With a program tailored specifically to the personal experiences of your students, you will find that your students are more curious about the world around them. With an increased level of relevance, learning is more important to young thinkers.

Likewise, it is very important to be sensitive to the cultural identities of your students. You must view these cultural identities on an equal footing. Having a healthy respect for your students' cultures will go a long way in helping each of your students achieve and maintain a positive self-concept.

Using Multicultural Instruction

A strong program of multicultural instruction can begin by implementing a few of these basic strategies. While none of the strategies are exclusively multicultural, they can provide proper contexts and situations that capitalize on the cultural backgrounds of students.

- Recognize and convey to students that all languages are equally valid. The learning of English, however, increases the range of opportunities available to the students.
- Draw special attention to the diversity of role models in *SciencePlus*. At every opportunity, provide information about past and current scientists from diverse cultures.

    The power of such role models should not be underestimated. Role models may create interest and motivation, and may even influence a student's pursuit of a career.
- Use cooperative learning to diversify student groups. You will find that students develop more of an open-minded awareness as well as more positive, accepting, and supportive relationships with peers. Labels concerning ethnicity, gender, ability, social class, and handicaps cease to exist. For more detailed information about cooperative learning, see pages T41–T46.
- Peer and cross-age tutoring is an excellent strategy for fostering better understanding among individuals. Peer tutoring involves students tutoring students their own age. Cross-age tutoring involves older students tutoring younger students. These strategies are beneficial both to the tutors and to the students being tutored.

    In using either of these strategies, be very careful when pairing students. Although this is an excellent opportunity to integrate students, both the tutor and the student must be willing participants.

- Take every opportunity to relate science to personal experiences. Invite your students to discuss any of their own experiences that may apply. You may discover some very relevant connections and analogies, and the learning process will become more interactive and personalized to the class. This process might also add to the richness of the class by highlighting the cultural differences among your students.
- Allow students to select independent projects relevant to their own world. These projects should permit students to create new, positive avenues of self-expression from their own experiences. Students find opportunities to select, design, and articulate their own interest within science programs while developing their creative-thinking and problem-solving skills. In addition, these activities promote the development of positive attitudes toward general academics, social interactions, and the study of science.

Multicultural Instruction and *SciencePlus*

Science is for everyone, and *SciencePlus* is designed to serve the multiethnic and multicultural classrooms of today. Students, regardless of their ethnic backgrounds, will not have to look hard to find positive role models in the pages of *SciencePlus*. In addition, content that shows events, concepts, and issues from diverse ethnic and cultural perspectives is provided. As students work through *SciencePlus*, they will come to understand that science is a human endeavor that has been advanced by the contributions of many cultures and ethnic groups.

To add depth to your multicultural instruction, the Wrap-Around Margins of this Annotated Teacher's Edition periodically include a feature called Multicultural Extension. This information provides activities to help you focus on cultural diversity, highlighting the individuality and contributions of different ethnic groups.

Meeting the needs of culturally diverse students is perhaps the most demanding challenge faced by teachers today.

Obviously, to teach effectively you must be able to reach every individual in your class. This is seldom easy, given the diverse nature of most of today's classrooms. In addition, certain students present special challenges. Dealing adequately with these students requires special preparation and strategies. In many cases a minimal amount of preparation is sufficient to make the classroom a place where all can learn. Some of the more common situations you are likely to encounter are discussed below.

Learners Having Difficulty

Learners having difficulty are those who, for any of a number of reasons, are liable to perform poorly and who have a high probability of dropping out. *SciencePlus* is engaging and interesting throughout, appealing to all students. Throughout *SciencePlus*, clear easy-to-read prose and straightforward, attractive graphics reduce the potential for students to become bored. The style of *SciencePlus* is intentionally friendly and unintimidating. Field-testing has shown that the performance of students considered at-risk in science increased substantially when using *SciencePlus*.

Additional activities and teaching suggestions for learners having difficulty are provided under the learners Meeting Individual Needs: Learners Having Difficulty heading in the Wrap-Around Margins of this Annotated Teacher's Edition.

Meeting Individual Needs

Learners Having Difficulty

Using a hot plate, heat a few spoonfuls of sugar in a nonstick frying pan, stirring frequently. Once the sugar has caramelized, pour it onto wax paper to cool. Provide students with a cube of sugar and a piece of the caramelized sugar for comparison. Students should describe the metamorphosis that has taken place, compare the appearances of the two substances, and relate the change to the formation of metamorphic rock.

Second-Language Learners

Because *SciencePlus* places so much emphasis on doing science rather than reading about it, the program is ideal for students who are not proficient in English. Science is a universal language—the language of curiosity and logical reasoning. Many *SciencePlus* activities are easy to follow and require a minimum of reading. Lengthy explanations are seldom called for. You need only to get students started in the right direction; thereafter their intuition and common sense take over. The cooperative approach emphasized in *SciencePlus* helps to give second-language learners the extra support they need.

Additional activities and teaching suggestions for second-language learners are found in the Wrap-Around Margins of this Annotated Teacher's Edition. Look for the heading Meeting Individual Needs: Second-Language Learners.

Meeting Individual Needs

Second-Language Learners

Have students tell the story of a volcanic eruption using a comic-strip format. Students can create or gather a series of drawings or photographs to illustrate the sequence of events. Then encourage them to write a simple dialogue or commentary to accompany the pictures.

The Spanish Resources section of the *Teaching Resources* booklets also contains useful information, including blackline masters of parent letters translated into Spanish, as well as unit summaries and unit glossaries in Spanish.

Also available are *English/Spanish Audiocassettes,* which provide important preview information to assist Spanish-speaking students and students who are auditory learners.

Gifted Learners

The difficulty of teaching gifted students lies in keeping them interested, motivated, and challenged. Gifted students who are inadequately challenged may become bored, withdrawn, or even openly disruptive. *SciencePlus* includes many activities suitable for even the most advanced student. Open-ended activities, in particular, are especially suited for gifted students.

The *SciencePlus* approach emphasizes creative problem solving. In many cases there is no single right answer to a problem or question, so students' answers can reflect their individual abilities. This approach is ideal for gifted students, as they may extend the activities to fit their interests and talents.

Additional activities and teaching suggestions for gifted students are found in the Wrap-Around Margins of this Annotated Teacher's Edition. Look for the heading Meeting Individual Needs: Gifted Learners.

Meeting Individual Needs

Gifted Learners
Have students create their own guidebooks of local plants. Students could either sketch or photograph each plant. They can then list the common name of the plant along with its scientific name and some of its characteristics.

Physically Impaired Students

Make your classroom as easy to move about in as possible. Remove or bypass any obvious barriers. Encourage your students to assist physically impaired students. If the student uses a wheelchair, make the aisles wide enough to accommodate the chair. Make sure that the student can reach any equipment he or she needs. You may wish to enlist the aid of other students in the class to assist the disabled student as necessary.

As much as possible, adapt the classroom to make it possible for physically impaired students to engage in the same activities as other students. Use a mobile demonstration table so that it can be moved to different areas of the room for maximum visibility.

Visually Impaired Students

Seat students with marginal vision near the front of the room to maximize their view of both you and the chalkboard, or assign a student to make copies of what you write. You could also assign a student to explain all visual materials in detail as they are presented.

Students who are completely blind should be allowed to become familiar with the classroom layout before the first class begins. Promptly inform these students of any changes to your classroom layout. Whenever possible, provide blind students with Braille or taped versions of all printed materials. Blind students may also use hand-held devices for converting written text into speech.

Hearing-Impaired Students

If you have hearing-impaired students in your class, remember to always face the class while speaking. Minimize classroom noise, and arrange seating in a circle or semicircle so that hearing-impaired students can see others. This arrangement facilitates speech reading. Speak in simple, direct language and avoid digressions or sudden changes in topic. During class discussions, periodically summarize what students are saying and repeat students' questions before answering them. Use visual media such as filmstrips, overhead projectors, and close-captioned films when appropriate. You might arrange a buddy system in which another student provides copies of notes about activities and assignments.

A student who is completely deaf may require a sign-language interpreter. If so, let the student and the interpreter determine the most convenient seating arrangement. When asking the student a question, be sure to look at the student, not at the interpreter. If the student also has a speech impairment, group assignments for oral reports may be advisable.

> In many cases a minimal amount of preparation is sufficient to make the classroom a place where all can learn.

Speech-Impaired Students

Mainstreaming speech-impaired students is generally not very difficult. Patience is essential when dealing with speech-impaired students, however. For example, resist the temptation to finish sentences for a student who stutters. At the same time, do not show impatience. Also pay attention to nonverbal cues, such as facial expression and body language.

Be supportive and encouraging. You need not leave the speech-impaired student out of normal classroom discussions. For example, you may call on a speech-impaired student to answer a question and then allow the student to write out his or her response on the chalkboard or overhead projector. Use multisensory materials whenever possible to create a more comfortable learning environment for the speech-impaired student.

Learning-Disabled Students

Learning disabilities are any disorders that obstruct a person's listening, reasoning, communication, or mathematical abilities, and they range from mild to severe. An estimated 2 percent of all adolescents have some type of learning disability. Learning disabilities are the most common type of disability. Provide a supportive and structured environment in which rules and assignments are clearly stated. Use familiar words and short, simple sentences. Repeat or rephrase your instructions as needed.

Students may require extra time to complete exams or assignments, with the amount of extra time being dependent on the severity of their disability. Some students may need to tape-record lectures and answers to exam questions. For those who have difficulty organizing materials, you might provide chapter or lecture outlines for them to fill in. Having peer tutors work with learning-disabled students on specific assignments and review materials can be effective.

Computer-assisted instruction is an extremely useful tool for some learning-disabled students. This mode of instruction can even help these students develop good learning skills. For learning-disabled students, computers serve as a tireless instructor with unlimited patience. In addition, students receive simplified directions; proceed in small, manageable steps; and receive immediate reinforcement and feedback with computerized instruction.

Students With Behavioral Disorders

Behavioral disorders are emotional or behavioral disturbances that hinder a student's overall functioning. The behaviorally impaired may exhibit any of a variety of behaviors, ranging from extreme aggression to complete passivity.

Obviously, no single teaching strategy can accommodate all behavioral disorders. In addition, behavioral psychologists disagree on the best way to deal with students who have behavioral disorders. As a general rule, try to be fair and consistent, yet flexible, in your dealings with behaviorally disabled students. Make sure to state rules and expectations clearly. Reinforce desirable behavior or even approximations of such behavior, and ignore or mildly admonish undesirable behavior.

Because learning disabilities often accompany behavioral disorders, you might also wish to refer to the guidelines for learning disabilities.

> Computer-assisted instruction is an extremely useful tool for some learning-disabled students.

MATERIALS AND EQUIPMENT

SciencePlus is designed to be teachable even by those with a limited budget for materials. Most activities use common household items that can be brought to class by students or parents or that can otherwise be easily obtained. For a comprehensive listing of the materials and equipment you will need to carry out the activities of *SciencePlus*, see the *Materials Guide*. The *Materials Guide* includes both a master list of materials and a unit-by-unit list.

In addition, the second page of the parent letter contained in the *Teaching Resources* booklets lists the supplies needed for each unit. This page makes it easy for you to invite donations from parents to keep your budget as low as possible.

For teachers who would rather have the convenience of purchasing supplies through the mail, Science Kit is the official *SciencePlus* materials and equipment supplier. For your convenience, Science Kit offers kits that contain the materials and equipment you will need to teach each unit of *SciencePlus*. Or, if you prefer, you can order needed materials and equipment individually as necessary.

Science Kit ordering information is available from your local HRW representative.

SCIENCEPLUS TEACHER'S NETWORK

The *SciencePlus* Teacher's Network (SPTN) is part of the Atlantic Science Curriculum Project (ASCP), linking teaching, curriculum development, and research in science education. The ASCP, which produced the Canadian version of *SciencePlus*, has existed for over 15 years, beginning as a collaborative, grass-roots activity to improve science teaching in middle schools and junior high schools. The goal of the SPTN is to continue this collaborative effort, linking teachers with teachers through newsletters, electronic bulletin boards, conferences, and small-group and regional meetings.

The *SciencePlus Communicator* is the official U.S. publication of the SPTN. The communicator is written by and for *SciencePlus* teachers. It provides a forum for *SciencePlus* teachers to share their classroom experiences with other *SciencePlus* teachers. Each edition includes copy masters, activities, current research, assessment ideas, and much more that can help make your teaching more effective. The *SciencePlus Communicator* links you to a community of like-minded teachers whose breadth of experience can provide encouragement and support.

ASSESSING STUDENT PERFORMANCE

A COMPREHENSIVE APPROACH TO ASSESSMENT

Developing strategies for assessing student progress is an important step in realizing the goals of *SciencePlus*. Students pay the most attention to those aspects of a lesson on which they know they will be graded. Teachers who want their students to be successful should therefore teach with continual assessment in mind.

In *SciencePlus*, there is no distinct boundary between teaching and assessing. You will find that most of the tests and assessment activities in this program are designed to teach as well as to evaluate comprehension and performance. This emphasis can help correct the preoccupation with measuring and sorting students. The suggestions here are intended to aid you in your primary task: teaching.

ASSESSMENT AIDS IN *SCIENCEPLUS*

SciencePlus includes a wide variety of assessment aids to help you measure your students' mastery of the concepts and processes covered in this course. In addition to the assessment materials contained in the Pupil's Edition, including the Challenge Your Thinking chapter reviews, Making Connections unit reviews, and SourceBook Unit CheckUp reviews, the following materials are available for comprehensive and convenient assessment of your students.

In the *Teaching Resources* booklets
Blackline-master tests are available in the *Teaching Resources* booklets in the following categories:
• **Chapter Assessment**
• **Activity Assessment**
• **End-of-Unit Assessment**
• **SourceBook Assessment**
   In addition, a checklist is available to help you implement self-assessment into your classroom. The **Self-Evaluation of Achievement** checklist will allow you to add this facet of assessment to your other assessment methods. The checklists are tailored to each unit in the Pupil's Edition.

In the *Assessment Checklists and Rubrics* booklet
This booklet contains over 40 different checklists for student self-evaluation, peer evaluation, and teacher assessment. Checklists are also provided to aid ongoing assessment.
   This booklet contains a variety of assessment rubrics that serve as models for grading writing assignments, portfolios, reports, presentations, experiments, and technology projects.
   Several progress reports are also provided to help you keep track of your students' progress.

On the *SnackDisc* CD-ROM 
All of the checklists and rubrics contained in the *Assessment Checklists and Rubrics* booklet are also available on the *SnackDisc*. This CD-ROM for Macintosh® and Windows® makes these assessment materials extremely easy to access and fully customizable. Now you can quickly and easily create assessment checklists and rubrics that fit your own criteria and class situation. You can add to the recommended criteria or replace the criteria with your own.

In the *Test Generator* and *Test Item Listing* booklet

The *Test Generator* for Macintosh® and Windows® contains five categories of tests for each unit of the Pupil's Edition: Chapter Assessment, End-of-Unit Assessment, SourceBook Assessment, Activity Assessment, and Extra Assessment Items. The Extra Assessment Items consist of questions found only in the *Test Generator* and *Test Item Listing.* All of the other tests have blackline masters in the *Teaching Resources* booklets.

The *Test Generator* is easy to use, includes graphics, and is fully customizable. You can easily change the questions or add questions of your own.

The *Test Item Listing* booklet is a handy way to preview the tests and questions contained in the *Test Generator.*

In the *Annotated Teacher's Edition*

Each Unit Interleaf in this Annotated Teacher's Edition contains a comprehensive **Assessment Planning Guide.** This guide identifies, in chart form, all of the assessment components available in the program. A sample Assessment Planning Guide is shown below.

The Wrap-Around Margins of this Annotated Teacher's Edition also contain valuable assessment options. Each lesson contains a Follow-Up section that includes an optional assessment strategy. See page T28 for an example of these strategies.

Assessment Planning Guide

Lesson, Chapter, and Unit Assessment	SourceBook Assessment	Ongoing and Activity Assessment	Portfolio and Student-Centered Assessment
Lesson Assessment Follow-Up: see Teacher's Edition margin, pp. 303, 313, 323, 338, and 343 **Chapter Assessment** Chapter 14 Review Worksheet, p. 35 Chapter 14 Assessment Worksheet, p. 38* Chapter 15 Review Worksheet, p. 46 Chapter 15 Assessment Worksheet, p. 48* **Unit Assessment** Unit 5 Review Worksheet, p. 52 End-of-Unit Assessment Worksheet, p. 54*	SourceBook Unit Review Worksheet, p. 64 SourceBook Assessment Worksheet, p. 69* 	Activity Assessment Worksheet, p. 59* **SnackDisc** Ongoing Assessment Checklists ♦ Teacher Evaluation Checklists ♦ Progress Reports ♦	Portfolio: see Teacher's Edition margin, pp. 298, 309, 320, 323, and 334 **SnackDisc** Self-Evaluation Checklists ♦ Peer Evaluation Checklists ♦ Group Evaluation Checklists ♦ Portfolio Evaluation Checklists 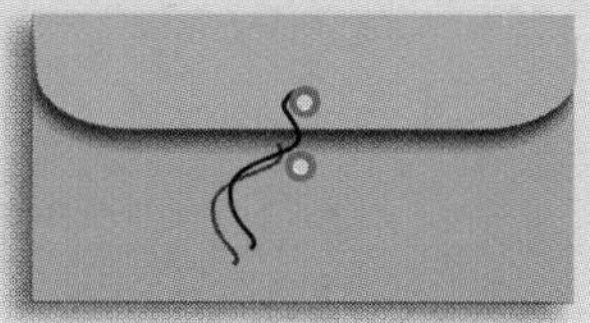 ♦

* Also available on the Test Generator software
♦ Also available in the Assessment Checklists and Rubrics booklet

Assessment should be ongoing and should measure performance on exams and quality of class work. Homework, lab work, and ScienceLog entries should all be factors in assigning grades.

The authors strongly discourage reliance on recall-based assessment strategies. Teachers who currently rely heavily on such assessment strategies may find it difficult at first to adopt new methods of assessment. However, once the transition is made, the reward—in the form of improved student performance and motivation—will more than offset the inconvenience.

ScienceLog Assessment

SciencePlus provides many opportunities for students to demonstrate their understanding of specific concepts. The first page of each chapter is devoted to getting students to think about what they already know about the concepts in the chapter. Students are asked several questions and are encouraged to write down what they already know or think they know. After students complete the chapter, they are

given the opportunity to revise their ScienceLog entries in the Challenge Your Thinking chapter review. Here students will confront any misconceptions they may have had in the beginning. In this way students actually assess their own prior knowledge and make adjustments accordingly.

Although you should not grade these ScienceLog entries beyond checking that students have done them, viewing students' initial entries can give you a good idea of their understanding of the main concepts. In this way, these entries can provide you with an excellent diagnostic tool to determine where to start your teaching and what concepts will need the most emphaisis.

By checking your students' revised ScienceLog entries, you can get a good idea of the progress that has been made as well as what topics may need to be revisited to ensure understanding.

In addition, the ScienceLog serves as a companion notebook for most of the written work that is assigned throughout the program. You may want to have students hand in their entire ScienceLog periodically to check their work and progress.

Portfolio Assessment

SciencePlus is ideally suited to the use of Portfolios as one method of assessing your students' performance and accomplishments. For more information about Portfolios and their use in *SciencePlus*, see page T38 of this Annotated Teacher's Edition.

For help in assessing student Portfolios, several checklists and rubrics have been provided in the *Assessment Checklists and Rubrics* booklet and on the *SnackDisc*.

Assessing Scientific, Psychomotor, and Communication Skills

In *SciencePlus*, knowledge and understanding are closely linked to the development of important process skills such as observing, measuring, graphing, writing, predicting, inferring, analyzing, and hypothesizing. All learning tasks are designed to help develop these skills. The teacher can assess such skill development by inspecting student work and by observing student performance.

The sample tables below are suitable models for evaluating student performance.

ASSESSING SCIENTIFIC BEHAVIOR					
Behavior	Poor	Satisfactory	Good	Very Good	Excellent
• Cooperates with others in small groups					
• Observes and records observations					
• etc.					

ASSESSING TECHNICAL SKILLS			
Task	Yes	No	Uncertain
• Is able to read thermometer correctly			
• Is able to use spring scale to measure force			
• etc.			

Assessing Scientific Attitudes

It can be useful to survey your students about the types of science-related hobbies and interests that they pursue outside of class. In a direct way, this provides feedback on the success of your school's science program. A successful science program is reflected in a student body with outside interests in science. Ask your students to keep a tally of any science-related activities they undertake outside of class. These could include reading or writing about science and technology, science-related projects, visits to museums, attending lectures on scientific and technological topics, and viewing science programs on television.

Assessing Environmental Awareness

SciencePlus was written with a commitment to environmental awareness. Many activities that promote such awareness are included. The teacher is provided with suggestions on extending this theme through creative projects, cleanup or recycling projects, and so on. Tasks such as these promote environmental consciousness. The care that students take in carrying out these activities is a measure of their awareness of environmental issues.

For developing and assessing individual students' interest in and attitudes toward science, the assignment of elective reading and independent projects is essential. In addition to the numerous project ideas and extension activities included in the units of *SciencePlus*, the Science in Action features provide a range of project ideas from which students may choose. Students may also want to read further about a topic in the SourceBook or in another science-related book or magazine.

Student work on elective projects should count as a significant part of overall assessment. This type of work provides the surest indication of a student's interest and proficiency in science, especially the student's ability to study, plan, and research independently.

> A successful science program is reflected in a student body with outside interests in science.

A Balanced Assessment

The authors of *SciencePlus* recommend a balance between the different forms of assessment. As a general rule, the following proportions are suggested:

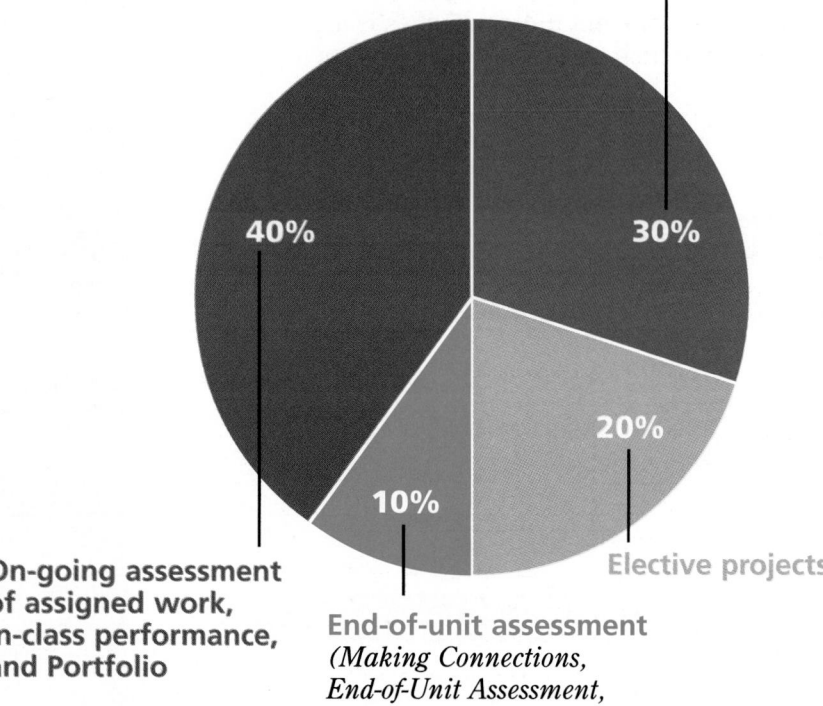

End-of-chapter assessment
(Challenge Your Thinking, Chapter Assesssment, Test Generator)

30%

20%

Elective projects

40%

10%

On-going assessment of assigned work, in-class performance, and Portfolio

End-of-unit assessment
(Making Connections, End-of-Unit Assessment, Test Generators)

Assessing Science Projects

SciencePlus offers students abundant opportunities for independent investigation. Many open-ended, curiosity-stimulating questions are posed to students. Some of these questions are natural starting points for science projects. Students using *SciencePlus* have often developed successful science-fair projects based on questions in the text. The Wrap-Around margins of this Annotated Teacher's Edition also contain additional project ideas where appropriate.

Undertaking a project provides students with a host of positive experiences. Students learn to organize, plan, and piccc togcthcr many separate ideas and pieces of information into a coherent whole. Undertaking a project also allows students to experience the sense of accomplishment that comes from tackling and completing a difficult task. It has even been argued that no science education is complete without having undertaken and completed a major project.

Many students will resist the idea of undertaking a major project because they feel that it is too much work or that they are simply not up to the task. The following suggestions may help you overcome students' reluctance:

- Allow students select their own project ideas.
- Encourage students to be creative.
- Provide a clear set of guidelines for developing and completing projects.
- Help students locate sources of information, including people in science-related fields who might advise students about their projects.
- Allow students the option of presenting their completed projects to the class.
- Emphasize the satisfaction students will derive from completing their projects.
- Inform students of the general areas on which assessment may be made, such as scientific thought, originality, and presentation.
- Do not emphasize the details of assessment. "Scoring points" should not be a major incentive.

Do not allow preconceived notions about how the project should be done to detract from the students' interest, enjoyment, and satisfaction in doing original work. Rather than forcing all students to fit their project work into a mold suited to scientific research, establish three sets of criteria, as described in the tables on pages T62 to T64.

When developing tests, you should bear in mind the sort of skill required to answer each assessment item. The manner of testing determines what is learned: tests that require tick-mark responses teach tick-marking. Tests that require verbal, graphic, illustrative, and numeric responses develop writing, speaking, graphing, drawing, and mathematical skills. A superior test draws on as many skills as possible.

The assessment-item development model below is designed to help you construct comprehensive tests that will meet the educational goals of *SciencePlus*. The model features four categories (verbal, graphic, illustrative, and numeric) and twelve types of items. The variety of tests and test items that accompany this program have been developed in accordance with this model.

The manner of testing determines what is learned: tests that require tick-mark responses teach tick-marking.

ASSESSMENT-ITEM DEVELOPMENT MODEL

VERBAL

Word Usage: The words given are to be used in a prescribed situation.

Correction/Completion: Incorrect or incomplete sentences and paragraphs are given for correction or completion.

Short Essays: Information is given or a question is posed for short-essay response.

Short Responses: Answers to these questions require a tick mark, a line, or a single word, phrase, or sentence.

ILLUSTRATIVE

Illustrations for Interpretation: Illustrations (drawings or photographs) are presented for interpretation.

Illustrations for Correction or Completion: An incorrect or incomplete illustration is given for correction or completion.

Answering by Illustration: A question requiring a drawing as the expected answer is asked.

GRAPHIC

Graphs for Interpretation: A graph of a relationship between two variables is given for interpretation.

Graphs for Correction or Completion: An incorrect or incomplete graph is given for correction or completion.

Graphing Data: Data are given to be graphed.

NUMERIC

Data for Interpretation: A data table is given for interpretation.

Numerical Problems: A problem requiring a numerical solution is given.

Rubric for Reports and Presentations

SCIENTIFIC THOUGHT (40 POINTS POSSIBLE)				
40–36	35–31	30–26	25–21	20–10
Complete understanding of topic; topic extensively researched; variety of primary and secondary sources used and cited; proper and effective use of scientific vocabulary and terminology	Good understanding of topic; well-researched; a variety of sources used and cited; good use of scientific vocabulary and terminology	Acceptable understanding of topic; adequate research evident; sources cited; adequate use of scientific terms	Poor understanding of topic; inadequate research; little use of scientific terms	Lacks an understanding of topic; very little research, if any; incorrect use of scientific terms

ORAL PRESENTATION (30 POINTS POSSIBLE)				
30–27	26–23	22–19	18–16	15–5
Clear, concise, engaging presentation; well-supported by use of multisensory aids; scientific content effectively communicated to peer group	Well-organized, interesting, confident presentation supported by multisensory aids; scientific content communicated to peer group	Presentation acceptable; only modestly effective in communicating science content to peer group	Presentation lacks clarity and organization; ineffective in communicating science content to peer group	Poor presentation; does not communicate science content to peer group

EXHIBIT OR DISPLAY (30 POINTS POSSIBLE)				
30–27	26–23	22–19	18–16	15–5
Exhibit layout self-explanatory, and successfully incorporates a multisensory approach; creative use of materials	Layout logical, concise, and can be followed easily; materials used in exhibit appropriate and effective	Acceptable layout of exhibit; materials used appropriately	Organization of layout could be improved; better materials could have been chosen	Exhibit layout lacks organization and is difficult to understand; poor and ineffective use of materials

Rubric for Experiments

SCIENTIFIC THOUGHT (40 POINTS POSSIBLE)

40–36	35–5
An attempt to design and conduct an experiment or project with all important variables controlled	An attempt to design an experiment or project, but with inadequate control of significant variables

ORIGINALITY (16 POINTS POSSIBLE)

16–14	13–11	10–8	7–5	4–2
Original, resourceful, or novel approach; creative design and use of equipment	Imaginative extension of standard approach and use of equipment	Standard approach and good treatment of current topic	Incomplete and unimaginative use of resources	Lacks creativity in both topic and resources

PRESENTATION (24 POINTS POSSIBLE)

24–21	20–17	16–13	12–9	8–5
Clear, concise, confident presentation; proper and effective use of vocabulary and terminology; complete understanding of topic; able to arrive at conclusions	Well-organized, clear presentation; good use of scientific vocabulary and terminology; good understanding of topic	Presentation acceptable; adequate use of scientific terms; acceptable understanding of topic	Presentation lacks clarity and organization; little use of scientific terms and vocabulary; poor understanding of topic	Poor presentation; cannot explain topic; scientific terminology lacking or confused; lacks understanding of topic

EXHIBIT (20 POINTS POSSIBLE)

20–19	18–16	15–13	12–11	10–6
Exhibit layout self-explanatory and successfully incorporates a multisensory approach; creative and very effective use of materials	Layout logical, concise, and can be followed easily; materials used appropriate and effective	Acceptable layout; materials used appropriately	Organization of layout could be improved; better materials could have been chosen	Layout lacks organization and is difficult to understand; poor and ineffective use of materials

Rubric for Technology Projects

SCIENTIFIC TECHNICAL THOUGHT (40 POINTS POSSIBLE)				
40–36	**35–31**	**30–26**	**25–21**	**20–10**
An attempted design solution to a technical problem; the problem is significant and stated clearly; the solution reveals creative thought and imagination; underlying technical and scientific principles are very well understood	An attempted design solution to a technical problem; the solution may be a standard one for similar problems; underlying technical and scientific principles are recognized and understood	A working model; underlying technical and scientific principles are well understood; model is built from a standard blueprint or design	Model is built from a standard blueprint or design or from a kit; underlying technical and scientific principles are recognized but not necessarily understood	Model is built from a kit; underlying technical and scientific principles are not recognized or understood

PRESENTATION (30 POINTS POSSIBLE)				
30–27	**26–23**	**22–19**	**18–16**	**15–5**
Clear, concise, confident presentation; proper and effective use of vocabulary and terminology; complete understanding of topic; able to extrapolate	Well-organized, clear presentation; good use of scientific vocabulary and terminology; good understanding of topic	Presentation acceptable; adequate use of scientific terms; acceptable understanding of topic	Presentation lacks clarity and organization; little use of scientific terms and vocabulary; poor understanding of topic	Poor presentation; cannot explain topic; scientific terminology lacking or confused; lacks understanding of topic

EXHIBIT (30 POINTS POSSIBLE)				
30–27	**26–23**	**22–19**	**18–16**	**15–5**
Exhibit layout self-explanatory, and successfully incorporates a good sensory approach; creative and very effective use of material	Layout logical, concise, and easy to follow; materials used in exhibit appropriate and effective	Acceptable layout of exhibit; materials used appropriately	Organization of layout could be improved; better materials could have been chosen	Layout lacks organization and is difficult to understand; poor and ineffective use of materials

SCIENCE PLUS

TECHNOLOGY AND SOCIETY

TO THE STUDENT

This book was written with you in mind!

There are many things to try, to create, and to investigate—both in and out of class. There are stories to read, articles to think about, puzzles to solve, and even games to play.

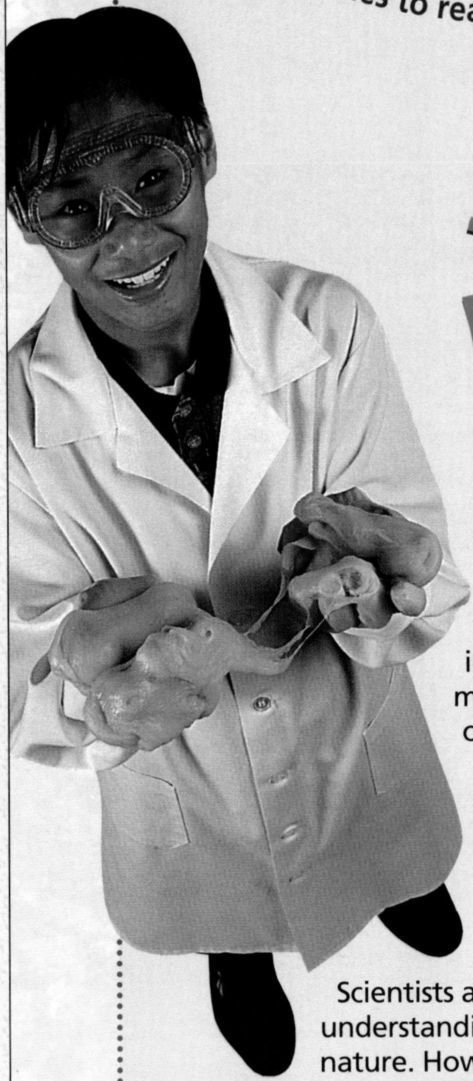

GET INVOLVED!

The best way to learn is by doing. In the words of an old Chinese proverb:

Tell me—*I will forget*

Show me—*I may remember*

Involve me—*I will understand*

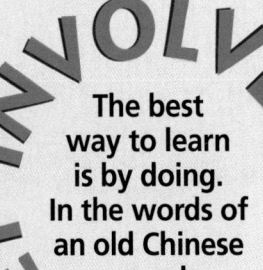

The activities in this book will allow you to make some basic and important scientific discoveries on your own. You will be acting much like the early investigators in science who, without expensive or complicated equipment, contributed so much to our knowledge.

What these early investigators had, and had in abundance, was curiosity and imagination. If you have these qualities, you are in good company! And if you develop sharp scientific skills, who knows?— you might make your own contributions to science someday.

Scientists are usually interested in understanding things that happen in nature. However, the discoveries that scientists make are often used by inventors and engineers. Using science in this way has resulted in our most sophisticated technology, including such things as computers, laser discs, nuclear reactors, and instant global communication.

SCIENCE & TECHNOLOGY

There is an interaction between science and technology. Science makes technology possible. On the other hand, the products of technology are used to make further scientific discoveries. In fact, much of the scientific work that is done today has become so technically complicated and expensive that no one person can do it entirely alone. But make no mistake, the creative ideas for even the most highly technical and expensive scientific work still come from individuals.

A built-in reference section is located at the back of this book. It's called the SourceBook. **CHECK IT OUT!**

Keep a ScienceLog

A journal is an important tool in creative work. In this book, you will be asked to keep a type of journal, called a ScienceLog, to record your thoughts, observations, experiments, and conclusions. As you develop your ScienceLog, you will see your own ideas taking shape over time. This is often the way scientists arrive at new discoveries. You too may log some discoveries as you develop your own journal.

GO FOR IT!

Science is a process of discovery, a trek into the unknown. The skills you develop as you do the activities in this book—like observing, experimenting, and explaining observations and ideas—are the skills you will need in order to be a part of science in the future. There is a universe of scientific exploration and discovery awaiting those who take up the challenge.

SAFETY FIRST!

The study of science is challenging and fun, but it can also be dangerous. Don't take any chances! Follow the guidelines listed here, as well as safety information provided in the particular Exploration you are doing. Also, follow your teacher's instructions and don't take shortcuts—even when you think there is little or no danger.

Accidents can be avoided. The major causes of laboratory accidents are carelessness, lack of attention, and inappropriate behavior. These things reflect a person's attitude. By adopting a positive attitude and by following all safety guidelines, you can greatly reduce your chances of having an accident. Even a minor accident in a science laboratory can cause major injuries, so be very careful.

SAFETY GUIDELINES

GENERAL

Always get your teacher's permission before attempting any laboratory explorations. Read the procedures carefully, paying particular attention to safety information and caution statements. If you are unsure about what a safety symbol means, look it up here or ask your teacher. You cannot be too careful when it comes to safety! If an accident does occur, inform your teacher immediately, regardless of how minor you think the accident is.

EYE SAFETY

Wear safety goggles when working around chemicals, acids, bases, or any type of flame or heating device, and any other time when there is even the slightest chance that harm could come to your eyes. If any substance gets into your eyes, notify your teacher immediately. Treat any unknown chemical as if it were a dangerous chemical. Never look directly into the sun with an optical device, and never use direct sunlight as a light source for a microscope.

SAFETY EQUIPMENT

Know the location of and how to use the nearest fire alarms and any other safety equipment, such as fire blankets and eyewash fountains, as identified by your teacher.

NEATNESS

Keep your work area free of all unnecessary books and papers. Tie back long hair and secure loose sleeves or other loose articles of clothing such as ties and bows. Remove dangling jewelry. Don't wear open-toed shoes or sandals in laboratory situations. Never eat, drink, or apply cosmetics in a laboratory setting; food, drink, and cosmetics can easily become contaminated with dangerous materials.

SHARP/POINTED OBJECTS

Use knives and other sharp instruments with extreme care. Never cut objects while holding them in your hands. Place objects on a suitable work surface for cutting.

HEAT

Wear safety goggles when using a heating device or a flame. Whenever possible, use an electric hot plate instead of a flame as a heat source. When heating materials in a test tube, always slant the test tube away from yourself and others. Wear oven mitts, when instructed to do so, to avoid burns.

ELECTRICITY

Be careful with electrical wiring. When using a microscope with a lamp, do not place the cord where it could cause someone to trip. Do not let cords hang over a table edge in a way that could cause equipment to fall if the cord is accidentally pulled. Do not use equipment with damaged cords. Be sure your hands are dry and that the electrical equipment is in the "off" position before plugging it in. Turn off equipment when you are done.

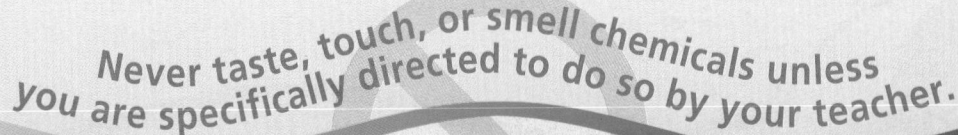

Never taste, touch, or smell chemicals unless you are specifically directed to do so by your teacher.

If you are instructed to note the odor of a substance, wave the fumes toward your nose with your hand. Never put your nose close to the source. Never mix chemicals unless you are told to do so by your teacher.

CHEMICALS

Wear safety goggles when handling any potentially dangerous chemicals, acids, or bases. If a chemical is unknown, handle it as you would a dangerous chemical. Wear an apron and latex gloves when working with acids or bases or when told to do so in the Exploration or Activity. If a spill gets on your skin or clothing, rinse it off immediately with water for at least 5 minutes while calling your teacher.

ANIMAL SAFETY

Always obtain your teacher's permission before bringing any animal into the school building. Handle animals only as your teacher directs. Always treat animals carefully and with respect. Wash your hands thoroughly after handling any animal.

PLANT SAFETY

Do not eat any part of a plant or plant seed used in the laboratory. Wash hands thoroughly after handling any part of a plant.

GLASSWARE

Examine all glassware before using. Be sure that it is clean and free of chips and cracks. Report damaged glassware to your teacher. Glass containers used for heating should be made of heat-resistant glass.

CLEANUP

Before leaving, clean up your work area. Put away all equipment and supplies. Dispose of all chemicals and other materials as directed by your teacher. Make sure water, gas, burners, and electric hot plates are turned off. Hot plates and other electrical equipment should also be unplugged. Wash hands with soap and water after working in a laboratory situation.

concept mapping

A Way to Bring Ideas Together

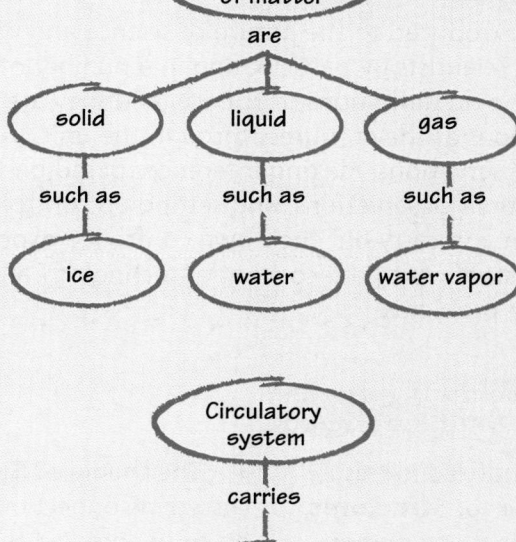

What Is a Concept Map?

Have you ever tried to tell someone about a book or a chapter you've just read, and you find that you can remember only a few isolated words and ideas? Or maybe you've memorized facts for a test, and then weeks later you're not even sure what topic those facts are related to.

In both cases, you may have understood the ideas or concepts by themselves, but not in relation to one another. If you could somehow link the ideas together, you would probably understand them better and remember them longer. This is something a concept map can help you do. A concept map is a visual way of choosing how ideas or concepts fit together. It can help you see the "big picture."

How to Make a Concept Map

1. **Make a list of the main ideas or concepts.**

   It might help to write each concept on its own slip of paper. This will make it easier to rearrange the concepts as many times as you need to before you've made sense of how the concepts are connected. After you've made a few concept maps this way, you can go directly from writing your list to actually making the map.

2. **Spread out the slips on a sheet of paper, and arrange the concepts in order from the most general to the most specific.**

   Put the most general concept at the top and circle it. Ask yourself, "How does this concept relate to the remaining concepts?" As you see the relationships, arrange the concepts in order from general to specific.

3. **Connect the related concepts with lines.**

4. **On each line, write an action word or short phrase that shows how the concepts are related.**

Look at the concept maps on this page and then see if you can make one for the following terms: **plants, water, photosynthesis, carbon dioxide, and sun's energy.**

The answer is provided below, but don't look at it until you try the concept map yourself.

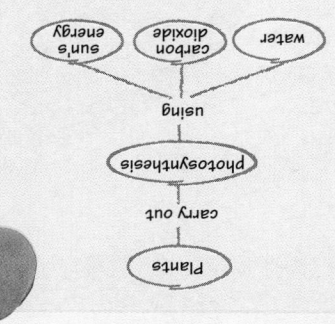

Unit 1 SCIENCE AND TECHNOLOGY

Bibliography for Teachers
Friedl, Alfred E. *Teaching Science to Children: An Integrated Approach.* 2nd ed. New York City, NY: McGraw-Hill, 1991.

Morowitz, Harold J. *The Thermodynamics of Pizza.* New Brunswick, NJ: Rutgers University Press, 1991.

Strahler, Arthur N. *Understanding Science: An Introduction to Concepts and Issues.* Buffalo, NY: Prometheus Books, 1992.

Bibliography for Students
Markle, Sandra. *Science to the Rescue.* New York City, NY: Atheneum, 1994.

Norman, Winifred L., and Lily Patterson. *Lewis Latimer.* New York City, NY: Chelsea House Publishers, 1994.

Towle, Wendy. *The Real McCoy: The Life of an African-American Inventor.* New York City, NY: Scholastic Inc., 1993.

VanCleave, Janice. *Janice VanCleave's 201 Awesome, Magical, Bizarre, and Incredible Experiments.* New York City, NY: John Wiley & Sons, Inc., 1994.

Unit Overview

In this unit, students are introduced to the nature of science and what scientists do. In Chapter 1, students act as scientists by participating in a number of hands-on experiments. They develop their own definitions of science, and they are encouraged to add to or change their definitions as they progress through the unit. In Chapter 2, students practice making careful observations, making inferences based on their observations, using a model to explain their observations, and setting up controlled experiments. In Chapter 3, students design and carry out their own controlled experiments. This chapter concludes by introducing students to the concept of technology and how it incorporates science in solving practical problems.

Using the Themes

The unifying themes emphasized in this unit are **Changes Over Time, Structures,** and **Systems.** The following information will help you incorporate these themes into your teaching plan. Focus questions that correspond to these themes appear in the margins of this Annotated Teacher's Edition on pages 10, 21, 31, 44, and 45.

The theme of **Changes Over Time** is evident in the processes of science itself and therefore can be incorporated into all three chapters in this unit. Science is an active process in which facts are continually accumulated, reevaluated, and reprocessed. In addition, scientific investigations provide the means for recognizing changes that have occurred over time in the natural world.

In Chapters 2 and 3, opportunities arise to emphasize structural similarities between classroom objects and objects in the natural world. In these cases, the theme of **Structures** can be a useful teaching tool.

The theme of **Systems** can be discussed in connection with the use of models to understand things that are not readily visible. Models are representations of the components that make up a system, and they can illustrate how these components work together.

Films, Videotapes, Software, and Other Media
Careers in Math and Science
    Film or videotape
    Coronet/MTI
    108 Wilmot Rd.
    Deerfield, IL 60015
Hello!
    Software (Macintosh, IBM, or Apple IIGS)
    NGS Kids Network
    National Geographic Society
    Educational Services
    P.O. Box 98018
    Washington, D.C. 20090-8018
Inferring in Science, Observing in Science, and ***Questioning in Science***
    Videotapes
    Agency for Instructional Technology
    1800 N. Stonelake Dr., Box A
    Bloomington, IN 47402

Using the SourceBook

Unit 1 focuses on science and its methods. Students are introduced to science as a search for answers to the many questions raised by our human curiosity. The major fields of science are highlighted, with attention to various careers in each field. Finally, scientific methods are examined and illustrated by actual investigations performed by working scientists.

Unit Organizer

Unit/Chapter	Lesson	Time*	Objectives	Teaching Resources
Unit 1 Opener, p. 2				Science Sleuths: The Traffic Accident English/Spanish Audiocassettes Home Connection, p. 1
Chapter 1, p. 4	**Lesson 1, The Nature of Science, p. 5**	3	1. Describe who scientists are and what they do. 2. Discuss the nature of science.	Image and Activity Bank 1-1 Resource Worksheet, p. 3 Theme Worksheet, p. 4 Resource Worksheet, p. 6 Resource Worksheet, p. 7 ▼
	Lesson 2, Defining Science, p. 13	1 to 2	1. Talk about the human qualities of a scientist. 2. Discuss what scientists do. 3. Formulate personal definitions of the nature of science.	
End of Chapter, p. 16				Chapter 1 Review Worksheet, p. 8 ▼ Chapter 1 Assessment Worksheet, p. 11
Chapter 2, p. 18	**Lesson 1, The Science of Observation, p. 19**	3 to 4	1. Distinguish between qualitative and quantitative observations. 2. Distinguish between an observation and a property. 3. Describe an object in terms of its properties. 4. Discuss reasons why people do not always make the same observations.	Image and Activity Bank 2-1 Exploration Worksheet, p. 14 Exploration Worksheet, p. 18 Exploration Worksheet, p. 21
	Lesson 2, From Observations to Inferences, p. 25	4	1. Distinguish observations from inferences. 2. Classify questions according to whether or not they lead to scientific discoveries. 3. Suggest investigative questions.	Image and Activity Bank 2-2 Exploration Worksheet, p. 22 ▼
	Lesson 3, Is There a Gremlin in the Drink Machine? p. 31	1	1. Discuss the meaning of a model. 2. Create a model that can explain the operation of a machine.	
	Lesson 4, Testing Ideas, p. 33	2 to 3	1. Write a hypothesis that can be investigated. 2. Identify the cause and effect in a hypothesis. 3. Identify the variables that must be controlled in order to conduct a fair test (controlled experiment).	Image and Activity Bank 4-1
End of Chapter, p. 37				Chapter 2 Review Worksheet, p. 27 Chapter 2 Assessment Worksheet, p. 30
Chapter 3, p. 39	**Lesson 1, From Hypothesis to Experiment, p. 40**	5 to 6	1. Design an experiment to test a hypothesis. 2. Perform an experiment according to a suggested format.	Image and Activity Bank 3-1 Discrepant Event Worksheet, p. 33 Exploration Worksheet, p. 34 Exploration Worksheet, p. 37 Graphing Practice Worksheet, p. 40 Transparency 6
	Lesson 2, Technology— Brainchild of Science, p. 47	2	1. Distinguish between science and technology. 2. Provide examples of technology in everyday life.	Image and Activity Bank 3-2
End of Chapter, p. 52				Chapter 3 Review Worksheet, p. 42 Chapter 3 Assessment Worksheet, p. 45
End of Unit, p. 54				Unit 1 Activity Worksheet, p. 48 ▼ Unit 1 Review Worksheet, p. 49 Unit 1 End-of-Unit Assessment, p. 52 Unit 1 Activity Assessment, p. 57 Unit 1 Self-Evaluation of Achievement, p. 60

* Estimated time is given in number of 50-minute class periods. Actual time may vary depending on period length and individual class characteristics.
▼ Transparencies are available to accompany these worksheets. Please refer to the Teaching Transparencies Cross-Reference chart in the Unit 1 Teaching Resources booklet.

Materials Organizer

Chapter	Page	Activity and Materials per Student Group
1	9	**Let's Do Some Science!:** 75 cm of adding-machine tape; small piece of transparent tape; metric ruler; scissors
	13	**Science Is . . . Fun:** newspapers; magazines; scissors; glue
2	20	**Exploration 1:** 2–3 matches; small ball of modeling clay; candle; metric ruler; jar lid or aluminum pie plate; watch or clock; safety goggles
	21	**Exploration 2:** 3 toothpicks; 3 jar lids or small bowls; 5 mL measuring spoon; 10 mL each of confectioners' sugar, baking powder, and plaster of Paris; 30 mL of water; 100 mL graduated cylinder; 3 small pieces of masking tape; pen; safety goggles; optional items: eyedropper; 10–15 mL each of baking soda, cornstarch, vinegar, and iodine starch-test reagent; an additional 5 mL of confectioners' sugar, baking powder, and plaster of Paris; 20 mL of water (additional teacher materials: three 250 mL beakers; 3 small pieces of masking tape; 25 mL of 0.1 M sodium thiosulfate; newspaper; see Advance Preparation below.)
	23	**Exploration 3:** metric ruler (additional teacher materials: 8 matches, support stand with clamp, watch or clock with second hand; safety goggles for you and any students who assist or are nearby during the demonstration described in Test 4)
	27	***Exploration 4, The Dancing Disk:** empty glass bottle; enough cold water to fill the bottle; sink; thin plastic or plastic-foam disposable plate; scissors; **The Reappearing Coin:** coin; shallow dish; 10 cm of transparent tape; about 100 mL of water; **The Curious Cup:** drinking glass; enough water to fill the glass; small piece of paper; bucket or sink; **Ice Heist:** 10–20 ice cubes; 250 mL beaker; about 10 g of salt; 30 cm of string; watch or clock with second hand; **Lights Out! #1:** aluminum pie plate or petri dish; candle; large jar; 2–3 matches; small ball of modeling clay; about 50 mL of water; safety goggles; **Lights Out! #2:** large beaker or jar; candle; 2–3 matches; small ball of modeling clay; 30 mL of baking soda; 15 mL of vinegar; 1 L container; watch or clock with second hand; safety goggles (additional teacher materials: newspaper; see Advance Preparation below.)
	30	**Classifying Statements:** washer; 30 cm of string; optional item: support stand with clamp
	37	**Challenge Your Thinking, question 2:** 1 index card per student
3	43	**Exploration 1:** paper; 1–2 paper clips; scissors; metric ruler
	45	**Exploration 2, Experiment 1:** meter stick; **Experiment 2:** meter stick; small ball of modeling clay

* You may wish to set up these activities in stations at different locations around the classroom.

Advance Preparation

Exploration 2, page 21: Prepare a classroom set of labeled samples of confectioners' sugar, baking powder, and plaster of Paris in 250 mL beakers. Students will need to use these samples after they have tested their mystery powders. Newspaper and 25 mL of 0.1 M sodium thiosulfate will be used for disposing of the iodine-stained materials; see page 21.

Exploration 4, page 27: The newspaper will be used for disposing of the solids; see page 27.

Unit Compression

Chapter 1, which encourages students to think about science and what scientists do, should be considered essential because it helps students understand the purpose and relevance of studying science. Note that the section titled Making the "Best" Airplane is intended to be an at-home activity. Also, you may wish to have students read some sections of this chapter ahead of time to speed in-class discussion.

In Chapter 2 opportunities for compression are more prevalent. To conserve time, Exploration 2 could be omitted, and Exploration 3 (except Test 4) could be assigned as homework. A number of strategies could be employed to shorten Exploration 4, such as assigning one group to each activity or performing some of the activities as demonstrations. In addition, Lesson 3 could be omitted and discussed later in conjunction with the particle model of matter covered in Unit 4.

Chapter 3 could be compressed by omitting Exploration 2 or by assigning it as a homework or enrichment activity.

Homework Options

Chapter 1	See Teacher's Edition margin, pp. 7, 11, 12, 14, and 17 Resource Worksheet, p. 6 SourceBook, pp. S2 and S5
Chapter 2	See Teacher's Edition margin, pp. 21, 23, 26, 27, 31, 34, 35, and 37 SourceBook, pp. S12 and S16
Chapter 3	See Teacher's Edition margin, pp. 45 and 48 Graphing Practice Worksheet, p. 39
Unit 1	Unit 1 Activity Worksheet, p. 48 Activity Assessment Worksheet, p. 57 SourceBook Activity Worksheet, p. 61

Assessment Planning Guide

Lesson, Chapter, and Unit Assessment	SourceBook Assessment	Ongoing and Activity Assessment	Portfolio and Student-Centered Assessment
Lesson Assessment Follow-Up: see Teacher's Edition margin, pp. 12, 15, 24, 30, 32, 36, 46, and 51 **Chapter Assessment** Chapter 1 Review Worksheet, p. 8 Chapter 1 Assessment Worksheet, p. 11* Chapter 2 Review Worksheet, p. 27 Chapter 2 Assessment Worksheet, p. 30* Chapter 3 Review Worksheet, p. 42 Chapter 3 Assessment Worksheet, p. 45* **Unit Assessment** Unit 1 Review Worksheet, p. 49 End-of-Unit Assessment Worksheet, p. 52*	SourceBook Review Worksheet, p. 63 SourceBook Assessment Worksheet, p. 68* 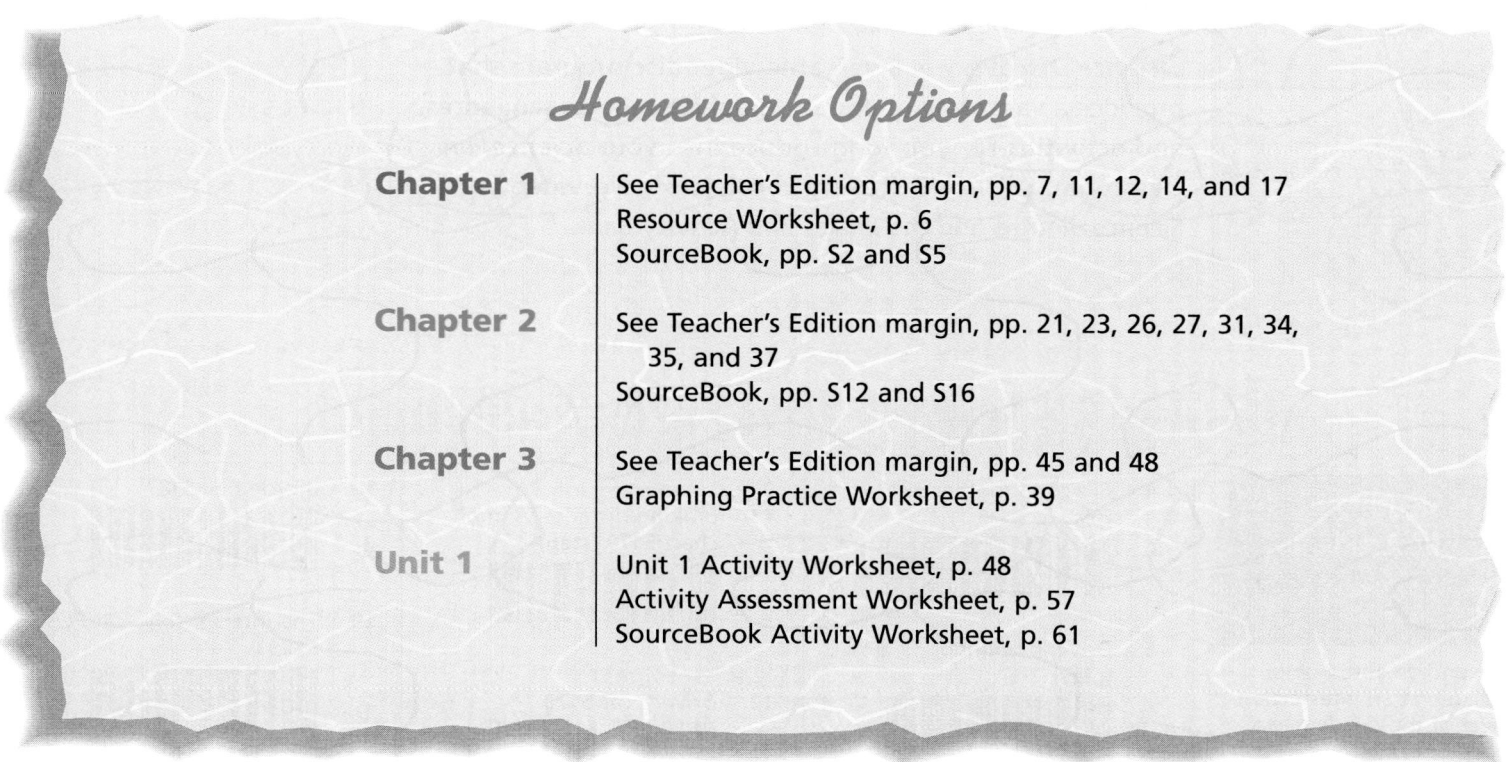	Activity Assessment Worksheet, p. 57* **SnackDisc** Ongoing Assessment Checklists ♦ Teacher Evaluation Checklists ♦ Progress Reports ♦	Portfolio: see Teacher's Edition margin, pp. 6, 11, 14, 28, 42, 44, and 49 **SnackDisc** Self-Evaluation Checklists ♦ Peer Evaluation Checklists ♦ Group Evaluation Checklists ♦ Portfolio Evaluation Checklists ♦

* Also available on the Test Generator software

♦ Also available in the Assessment Checklists and Rubrics booklet

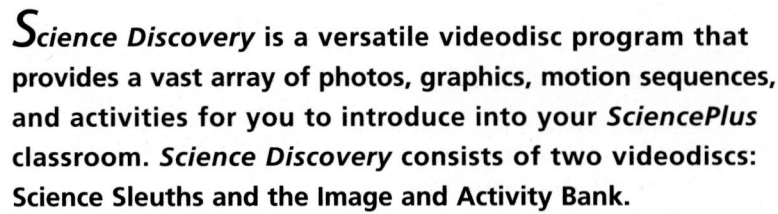

Science Discovery is a versatile videodisc program that provides a vast array of photos, graphics, motion sequences, and activities for you to introduce into your *SciencePlus* classroom. *Science Discovery* consists of two videodiscs: Science Sleuths and the Image and Activity Bank.

Using the *Science Discovery* Videodiscs

Science Sleuths: The Traffic Accident
Side A

A bicycle and an automobile have collided at an intersection. The driver of the car claims to have been obeying the speed limit when a cyclist appeared out of nowhere. The cyclist claims that he braked when he saw the car approaching, but that the driver swerved to hit him. The Science Sleuths must analyze the evidence for themselves to determine what really happened.

Interviews

1. Setting the scene: Insurance claims adjustor **218 (play ×2)**

2. Car driver **1155 (play)**

3. Bicyclist **2225 (play)**

4. Bicyclist's mother **3313 (play)**

5. Driver's neighbor **3939 (play)**

6. Witness **4653 (play)**

Documents

7. Bicycle safety manual **5157 (step ×3)**

8. Police report **5162 (step ×2)**

9. Map of accident site **5166**

10. Rainfall chart **5168**

Sleuth Information Service

11. Braking distance vs. mph chart **5170 (step)**

Still Photographs

12. Driver's car **5173**

13. Car tires **5175**

14. Accident impact mark **5177**

15. Damage to bicycle **5179 (step)**

16. Bicycle hand brakes **5182 (step)**

17. Bicycle lights **5185**

18. Bicycle bell **5187**

19. The intersection—car POV **5189**

20. The intersection—bike POV **5191**

21. The skid marks **5193**

Image and Activity Bank
Side A or B

The following is a selection of still images, short videos, and activities available for you to use as you teach this unit. For a larger selection and detailed instructions, see the Videodisc Resources booklet included with the Teaching Resources materials.

1-1 · The Nature of Science, page 5
Weather balloon launch 218
Students launch a weather balloon to take atmospheric readings.

Chemist at work 216
A chemist performs experiments in the laboratory.

Franklin, Ben 2137
Ben Franklin flies a kite to study electricity.

Geologist; lava-flow studies 1106
Photographic records are important in geological studies and reports.

2-1 The Science of Observation, page 19
Quantitative vs. qualitative 460–461
Measure the height, weight, and volume of an object to make a quantitative observation. (step) Note the properties of an object to make a qualitative observation.

◀ Step Reverse    Play ▶    Pause ‖    Step Forward ▮▶

2-2 From Observations to Inferences, page 25

Observation and inference 472
Statements of observation and inference

Air-pressure demonstration; breaking a ruler 13127–13359 (play ×2) (Side A only)
Air pressure on the paper provides enough weight to resist the applied force and to break the ruler.

Refraction 41525–41683 (play ×2) (Side A only)
Water bends light so that the penny is no longer visible.

Burning of oxygen by candle 49570–49932 (play ×2) (Side A only)
The candle goes out when it burns up all of the oxygen in the glass.

Candle extinguished by carbon dioxide 49933–50332 (play ×2) (Side A only)
Vinegar and baking soda produce carbon dioxide. Because it is heavier than air, the carbon dioxide remains in the bottle. It can extinguish a flame by excluding oxygen.

2-4 Testing Ideas, page 33

Problem solving, technical 478–479 (step)
Model for technical problem solving

Light bulb, incandescent 2112
This light bulb contains a carbon filament that is different from the tungsten filament found in regular light bulbs. Carbon filaments demonstrate a different light spectrum from that of tungsten filaments and thus produce a different color of light.

3-1 From Hypothesis to Experiment, page 40

Graphing data 4919–5828 (play ×2) (Side A only)
Place the independent variable (time) on the x-axis and the dependent variable (weight) on the y-axis. Be as accurate as possible in plotting the data points. Use a ruler to draw a best-fit line to represent all data points.

Scientific method 473–476 (step ×3)
State a hypothesis. (step) Devise an experiment. (step) Collect observations and data. (step) Make a conclusion.

3-2 Technology—Brainchild of Science, page 47

Machine invention time line 1837
Machines that have been invented since 3500 B.C.

Clock, pendulum 1843–1844 (step)
A pendulum is hung so that, under the influence of gravity, it moves back and forth in a regular arc.

Windmill; water pump 1787
Wind helps transport water. As the wind rotates the windmill's blades, the windmill generates suction that can pull water through pipes.

Steam engine 27504–27802 (play ×2) (Side A only)
This coal-fired steam engine generates steam to drive a piston.

Space Needle, Seattle 1997
Built in 1962 for Century 21, a world's fair, the Space Needle is an engineering marvel.

Hologram 44449–44965 (play ×2) (Side A only)
A hologram of an automobile is illuminated.

Satellite; Landsat 1638

Space-shuttle takeoff 20048–20256 (play ×2) (Side A only)
The fisheye lens shows the space shuttle taking off.

Fiber optics; telephone communication 42272–42512 (play ×2) (Side A only)
The fiber-optic process conveys light by reflecting it within tiny glass tubes.

Remote sensing 24951–26905 (play) (Side B only)
Infrared and computer-enhanced photos are used to study land use. Fields that have stagnant water are breeding grounds for mosquitoes.

Unit 1 SCIENCE AND TECHNOLOGY

UNIT FOCUS

With an air popper, pop some popcorn. Ask students to observe the shape and size of the corn kernels as they are heated. Then ask: What causes the kernels to change? *(Explain that the inside of the kernel is filled with starch and a small amount of water. As the kernel is heated, the water changes to a gas and expands. This causes so much pressure inside the kernel that it bursts and the starch tissue is blown outward.)*

Discuss with students how this activity demonstrates science and the role of scientists, especially in terms of making observations and inferences.

A good motivating activity is to let students listen to the English/Spanish Audiocassettes as an introduction to the unit. Also, begin the unit by giving Spanish-speaking students a copy of the Spanish Glossary from the Unit 1 Teaching Resources booklet.

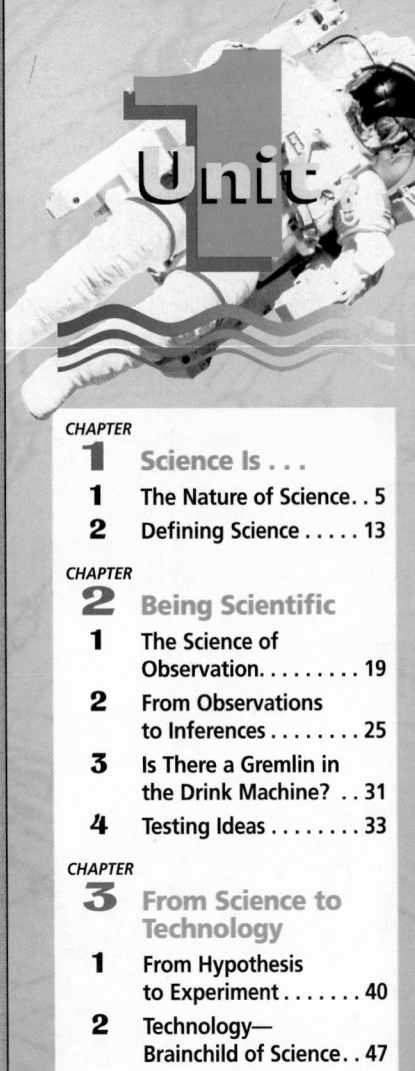

The waters off the Florida coast serve as a laboratory for scientist Barbara Bernier.

Connecting to Other Units

This table will help you integrate topics covered in this unit with topics covered in other units.

Unit 2 Patterns of Living Things	Understanding causation is useful in learning how organisms respond to their environments.
Unit 3 It's a Small World	Problem-solving techniques are used to determine the effects of microbes on humans.
Unit 4 Investigating Matter	Knowledge of scientific models is useful when exploring the particle model of matter.
Unit 5 Chemical Changes	Strong observational skills are necessary when distinguishing physical and chemical properties.

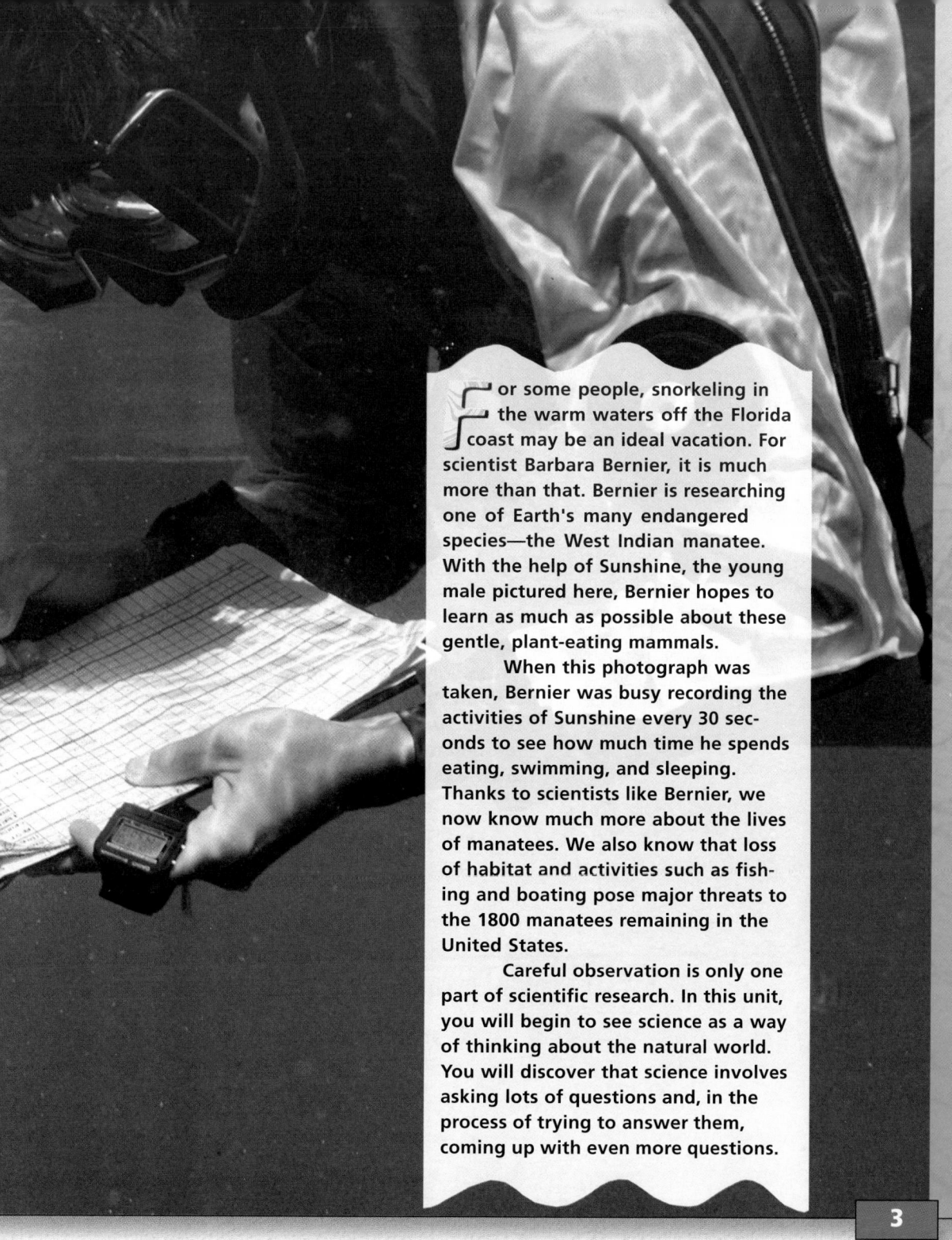

For some people, snorkeling in the warm waters off the Florida coast may be an ideal vacation. For scientist Barbara Bernier, it is much more than that. Bernier is researching one of Earth's many endangered species—the West Indian manatee. With the help of Sunshine, the young male pictured here, Bernier hopes to learn as much as possible about these gentle, plant-eating mammals.

When this photograph was taken, Bernier was busy recording the activities of Sunshine every 30 seconds to see how much time he spends eating, swimming, and sleeping. Thanks to scientists like Bernier, we now know much more about the lives of manatees. We also know that loss of habitat and activities such as fishing and boating pose major threats to the 1800 manatees remaining in the United States.

Careful observation is only one part of scientific research. In this unit, you will begin to see science as a way of thinking about the natural world. You will discover that science involves asking lots of questions and, in the process of trying to answer them, coming up with even more questions.

3

Connecting to Other Units, continued

Unit 6 Energy and You	The process of invention is crucial to the development of alternative energy sources.
Unit 7 Temperature and Heat	Observation and inference are involved in studying heat and the greenhouse effect.
Unit 8 Our Changing Earth	Our knowledge of the Earth's history has developed through scientific reasoning and careful inference.

CHAPTER 1
Science Is . . .

Connecting to Other Chapters

Chapter 1
explores the nature of science
and what it means
to be a scientist.

Chapter 2
invites students to practice scientific
thinking, including observing,
inferring, and hypothesizing.

Chapter 3
guides students through the process
of experimentation and discusses the
nature of technology.

Prior Knowledge and Misconceptions

Your students' responses to the ScienceLog questions on this page will reveal the kind of information—and misinformation—they bring to this chapter. Use what you find out about your students' knowledge to choose which chapter concepts and activities to emphasize in your teaching.

In addition to having students answer the questions on this page, you may wish to have them complete the following activity: Have them write a short skit featuring a middle-school student and a scientist. In the skit, the middle-school student interviews the scientist to find out what he or she does and what he or she likes about the profession. You may wish to have volunteers act out a few of the skits. Emphasize that there are no right or wrong answers in this exercise. Collect the papers, but do not grade them. Instead, read them to find out what students know about science and scientists, what misconceptions students may have, and what is interesting to them about science.

CHAPTER 1
Science Is . . .

1. **What do you think of when you think of science?**

2. **Have you ever done science? What did you do?**

3. **Which of the people shown on this page are scientists? How can you tell?**

4. **Draw a scientist.**

ScienceLog

Think about these questions for a moment, and answer them in your ScienceLog. When you've finished this chapter, you'll have the opportunity to revise your answers based on what you've learned.

4

LESSON 1
The Nature of Science

Science Is . . .

"I am an explorer. Some paths that I travel have been traveled before. Other paths are uncharted—leading to new insights and to new discoveries."

Who was the author of this statement? Was it an early explorer? an astronaut? Actually, the statement was made by a scientist. It describes what you will be doing in this unit and throughout this book. This unit is about science: what it is, what scientists do, and how they do it. More importantly, this unit is about how you become a scientist each and every day.

This park ranger uses special equipment to measure air quality in Bryce Canyon National Park, Utah.

These scientists are studying bioluminescence (light produced by living things) in piddock mollusks at the seashore in Plymouth, England.

This scientist is collecting a greater false vampire bat in the Tun Khee Noke Cave in Southeast Asia.

This botanist is examining a fungus in the Amazon rain forest.

Investigators at the Institute of Rice Research develop and grow new varieties of rice.

5

LESSON 1 ORGANIZER

Time Required
3 class periods

Process Skills
observing, analyzing, predicting

Theme Connection
Changes Over Time

New Term
none

Materials (per student group)
Let's Do Some Science!: 75 cm of adding-machine tape; small piece of transparent tape; metric ruler; scissors

Teaching Resources
Resource Worksheets, pp. 3, 6, and 7
Theme Worksheet, p. 4
Transparency 1
SourceBook, p. S2

LESSON 1
The Nature of Science

Getting Started

Display several open baby-food jars filled to the rim with uncooked rice. Ask students to try to find a way to lift a jar using only one sharpened pencil and without spilling any of the rice. After students have had a chance to experiment, you can demonstrate the solution. Insert the pencil about halfway into the jar with the pointed end down. Press down on the rice with your fingers. Insert the pencil a little farther down, and then press on the rice again. Lift the jar by holding only the pencil. Friction between the grains of rice and the pencil makes this work. (It may take a few attempts!) Explain to students that in trying to solve this problem they have been acting as scientists.

Main Ideas

1. Viewpoints about science vary from person to person.
2. Science encompasses our knowledge of the physical universe and the ways our knowledge is acquired.

TEACHING STRATEGIES

Science Is . . .

Ask a volunteer to read aloud the quotation on page 5. Ask: Why is a scientist an explorer? Why does a scientist sometimes travel paths that have already been traveled? Why does a scientist sometimes travel uncharted paths? *(Accept all reasonable responses.)* Tell students that in exploring science they will take some paths that have already been traveled and they may even blaze trails of their own and discover something new!

Yes, No, or Maybe?

Have students read the list of statements about science on this page. Students may work in pairs or in small groups to explain in their ScienceLog why they agree or disagree with each statement.

When they have finished, discuss their responses as a class. Some of the main points that you could emphasize include the following:

- Science is a job for anyone who has the interest and determination.
- Scientific ideas are constantly evolving and changing.
- Science is more than just doing experiments.
- Science is not just a collection of facts.

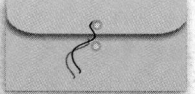

PORTFOLIO
Suggest that students record their best statement about science in their Portfolio. They may want to amend or add to their statement as they complete the unit.

SAFETY TIP Begin the year with a safety contract. The contract should include the following basic rules: always follow the instructions in the book and those given by the teacher, always wear safety goggles in the lab, never remove chemicals from the lab, and avoid rowdiness in the lab. A customizable safety contract and safety license are available on the *SciencePlus SnackDisc.* Add any other rules that you think are important. After each student has signed the contract, send it home for a parent or guardian to sign as well. Then issue a "safety license" to each student.

Did You Know...

Although the word *science* has been used in English since the 1300s, the term *scientist* wasn't coined until about 1840. Scientific investigators such as Benjamin Franklin and Chevalier de Lamarck, therefore, were not known as scientists until long after their death.

What is science? Who does science? How do they do it? Is science important? In the pages that follow are a number of activities that will help you answer these questions. You will find that doing these tasks with a partner or in a small group is more enjoyable than doing them by yourself.

Yes, No, or Maybe?

A science class made a list of statements about what they believed to be true about science. Which statement do you think is the best? Which one do you agree with the least? In your ScienceLog, list the numbers of the statements that you agree with, disagree with, and are uncertain about.

This scientist is examining the DNA taken from a nematode worm.

This paleontologist is at work cleaning a fish fossil.

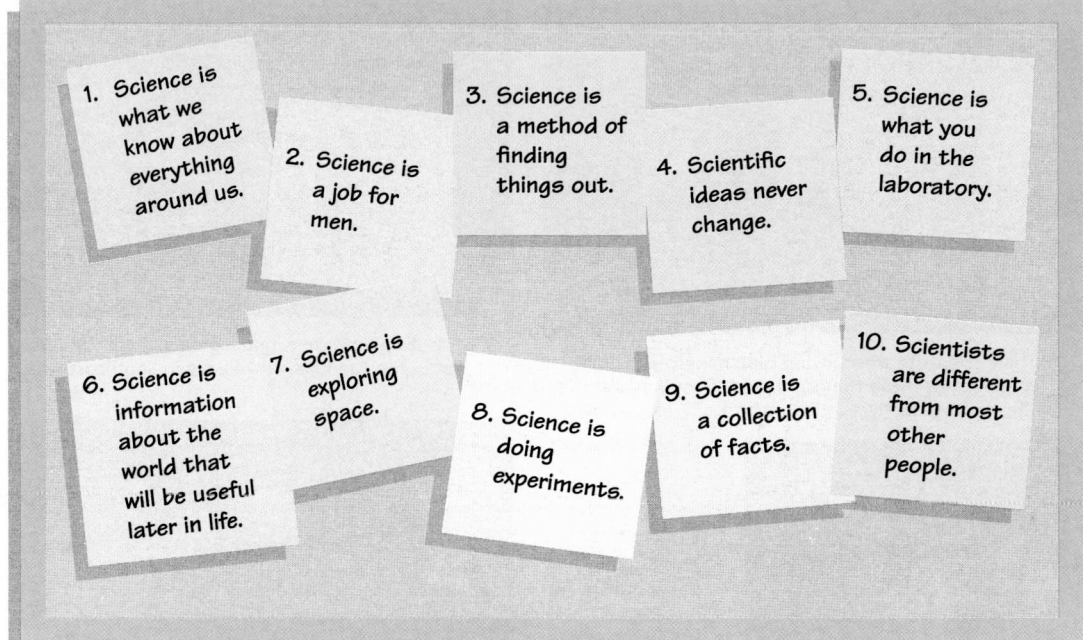

1. Science is what we know about everything around us.

2. Science is a job for men.

3. Science is a method of finding things out.

4. Scientific ideas never change.

5. Science is what you do in the laboratory.

6. Science is information about the world that will be useful later in life.

7. Science is exploring space.

8. Science is doing experiments.

9. Science is a collection of facts.

10. Scientists are different from most other people.

It's your turn! Create your own statement about science, and then share it with your classmates. Do they agree with your statement? Do you agree with theirs? After completing the next four activities, you may wish to change your statement about science in some way.

6

Science Case Histories

While reading about the activities of the scientists described below, make a list of the things these scientists did. Afterward, share your list with your classmates.

Death of a Star!

An event that took place 160,000 years ago was first seen by an astronomer working in South America. At 11 P.M. on February 23, 1987, Ian Shelton was doing some routine work at a small observatory on a mountaintop in Chile. Unexpectedly, he noticed a surprisingly bright star in a photograph he had just taken. After a closer look, Ian concluded that the photograph showed a supernova—an exploding star. The star had exploded 160,000 years ago but was so far away that its light was only now reaching the Earth. Ian had discovered the largest recorded supernova in almost 400 years. Because Ian was the first to make the observation and to recognize its significance, the supernova was named after him.

Ian Shelton with his telescopic camera

 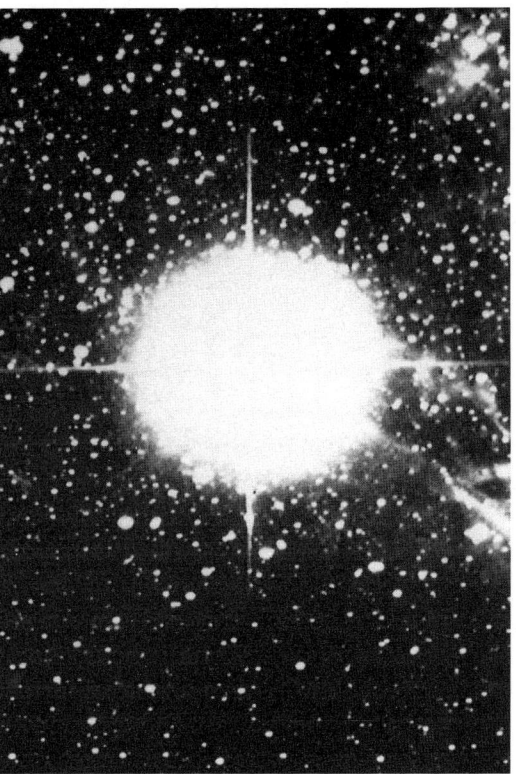

The Shelton supernova 3 years before discovery (left) and shortly after discovery (right)

7

Science Case Histories

After students have read pages 7 and 8, ask them to discuss, either in small groups or as a class, the things that the scientists did that led to their discoveries.

Answer to
Death of a Star!

Sample list: Shelton quite unexpectedly observed a bright star; he studied a photograph of the star more closely; and finally, he concluded that it was a supernova.

Homework

You may wish to have students read the material on pages 7 and 8 as homework. This will speed in-class discussion.

CROSS-DISCIPLINARY FOCUS

Mathematics

Students may not understand why the light from a star that exploded 160,000 years ago reached the Earth in 1987. Explain that light travels at a speed of about 300,000 km/s. Ask students to calculate how far away the star was from the Earth when it exploded. Suggest that students begin by calculating how fast the light traveled per hour, per day, per month, and per year. Then have students multiply the number of kilometers per year by 160,000 years. *(9,460,800,000,000 km/year × 160,000 years = 1,513,728,000,000,000,000 km. This number is read as 1 quintillion 513 quadrillion 728 trillion kilometers.)*

Multicultural Extension

The Mysteries of Comets

Ancient people often had only their powers of observation to help them explain natural events. For example, the ancient Babylonians thought that comets were heavenly beards, the Greeks thought that comets were flowing hair, and the Arabs thought that they were flaming swords. Have interested students make drawings of how some of these early peoples viewed comets or of how they themselves view comets. Students can then do research to find out what comets actually are and why they appear as they do.

Meeting Individual Needs

Gifted Learners

Create two teams to debate whether large sums of money should be spent on the research and development of systems to prevent comets and meteors from striking the Earth. Review with students some basic rules of debate. Tell students that to prepare for the debate, they should try to anticipate what their opponents' arguments will be. Then they should think of comments to discredit or weaken those points.

▲ Banting and Best with a dog successfully treated with insulin

A Lifesaving Discovery

The year 1923 must have been a highlight in the life of Frederick Banting because that was the year he shared the Nobel Prize in medicine with J.J.R. Macleod for discovering insulin. Why was the discovery deemed important enough to win a Nobel Prize? Insulin is one of the many chemicals in the human body. It is necessary for the body's use of sugar. People whose bodies cannot produce it soon die. Lack of insulin causes the disease known as diabetes. Before Banting's discovery, there was no effective treatment for diabetes.

Banting's interest in this area of research began in 1920, while he was a doctor in private practice. He read about the possible connection between diabetes and a small organ called the pancreas. An idea for an experiment to test the connection came to him. To perform the experiment, Banting "borrowed" the laboratory of J.J.R. Macleod.

Under the direction of Macleod, Banting and his assistant, Charles Best, tested Banting's idea, but the experiment did not support the original hypothesis. However, new observations suggested a new experiment. Finally, early in 1922, Banting announced the discovery of insulin, a hormone secreted by the pancreas. Today, diabetics the world over lead normal, healthy lives because of this vital discovery.

This girl is learning how to use an ▶ insulin pump that will be strapped to her abdomen. The pump will give her a continuous flow of insulin through a needle inserted just under the skin.

Let's Do Some Science!

Making a Möbius Strip

 Make a Möbius strip by following the directions illustrated here.

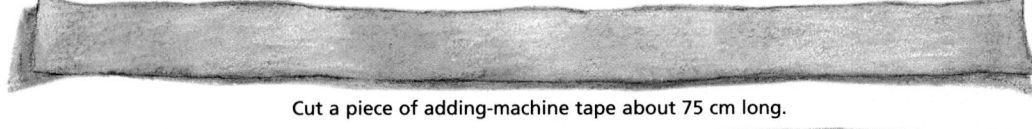

Cut a piece of adding-machine tape about 75 cm long.

Twist once.

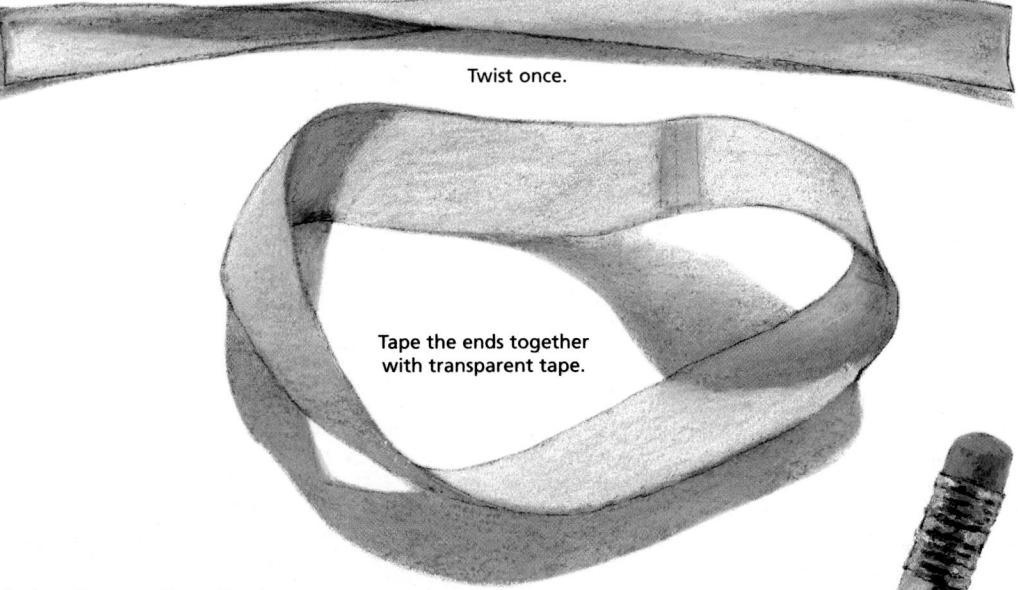

Tape the ends together with transparent tape.

Make Some Predictions

• How many sides does a Möbius strip have? With a pencil, draw a line down the center of the strip. Do not stop where the strip has been taped together.

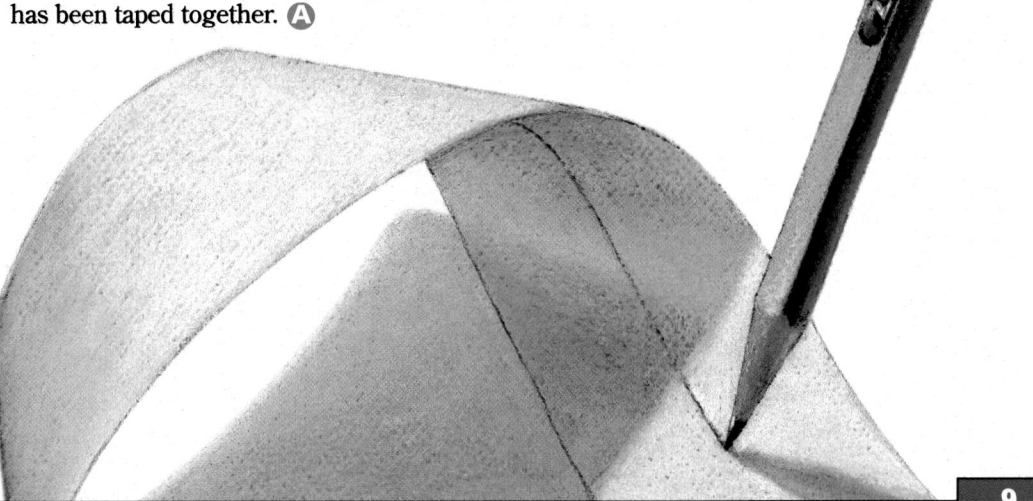

Let's Do Some Science!

In these activities, students discover the peculiar nature of Möbius strips. Have students work in pairs to make and investigate their Möbius strips. Students may be interested to know that the Möbius strip was named after August Möbius, a German astronomer and mathematician who lived from 1790 to 1868. Möbius was the first to describe the mathematical properties of one-sided surfaces, such as the well-known Möbius strip.

★ **A Resource Worksheet is available to accompany Let's Do Some Science! (Teaching Resources, page 3).**

Make Some Predictions

Encourage students to make a prediction in response to each of the questions asked and then to test their predictions using their Möbius strip.

Answer to
In-Text Question

Ⓐ Students will discover that a Möbius strip has only one side. A pencil line drawn down the center will eventually meet its starting point.

Ⓐ If students cut along this line, they should end up with one large loop of paper with an added twist.

Ⓑ By cutting again, students will obtain two interlocking loops of paper. Students may suggest that they are doing what scientists do— making predictions, testing predictions, forming ideas, and discovering new facts.

INDEPENDENT PRACTICE

Challenge students to experiment further with their Möbius strip. First, ask students to speculate about what would happen if the two loops were cut one more time. They should predict what would happen and then test their predictions. After the students have designed and carried out their experiments, have them share their results with one another.

Theme Connection

Changes Over Time

Use the theme of Changes Over Time to show students how scientific understanding evolves over time as scientists accumulate information by investigating, questioning, and evaluating ideas. **Focus question:** How did your ideas about the Möbius strip change as you progressed through the exercise? *(Most students will probably say that before they performed their investigations and made observations, they viewed the strip as having two sides.)*

⭐ **A Theme Worksheet is available as a teacher demonstration to accompany this Theme Connection (Teaching Resources, page 4).**

• How many pieces of paper do you think you will have if you cut along this line with a pair of scissors? Try it. Ⓐ

• How many pieces of paper do you think you will have if you cut once more along the center of the paper strip? Again, try it! What does this activity tell you about science and about what scientists do? Ⓑ

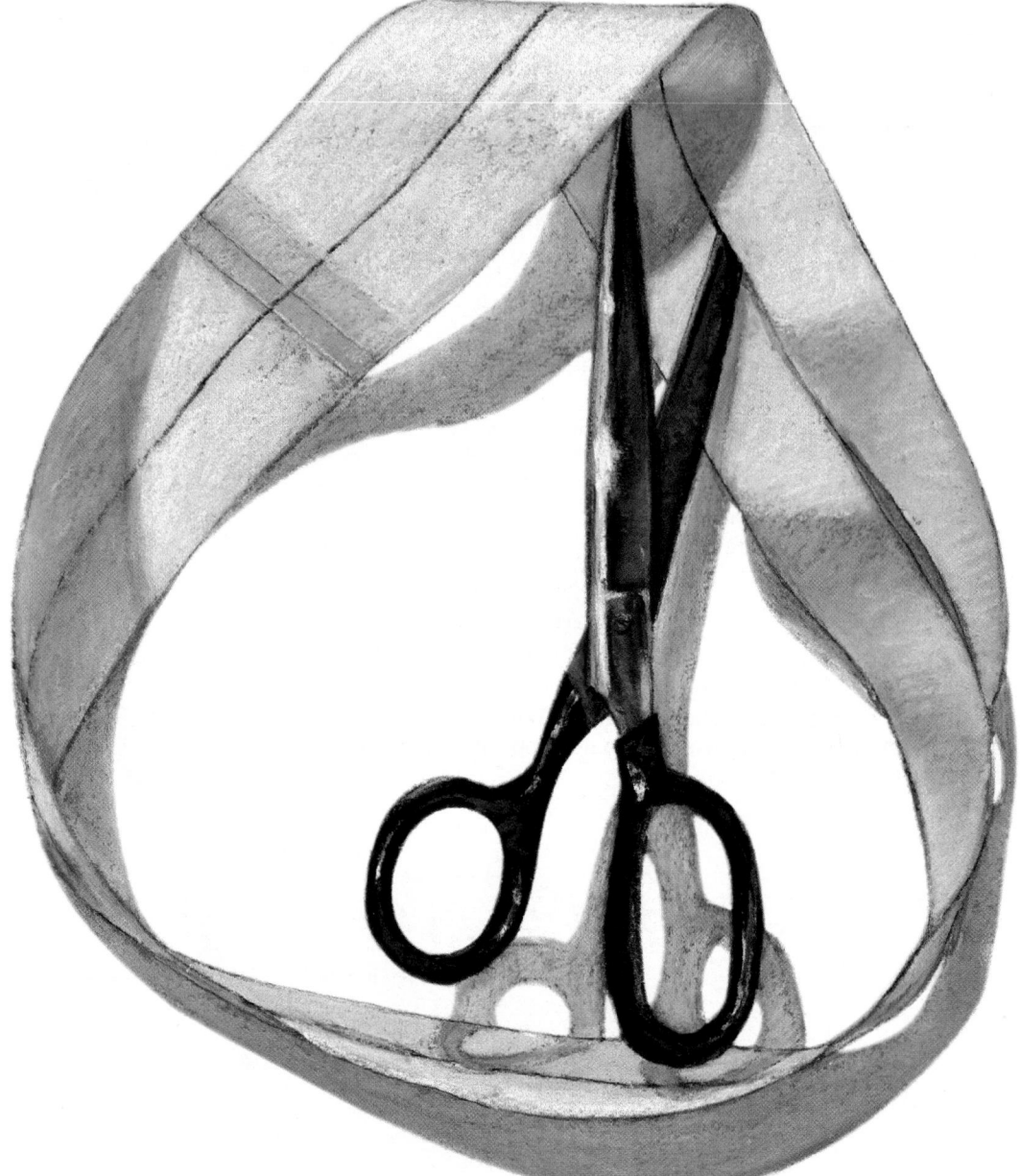

10

Making the "Best" Airplane
An At-Home Activity

At home, construct the model airplane shown in the drawing.

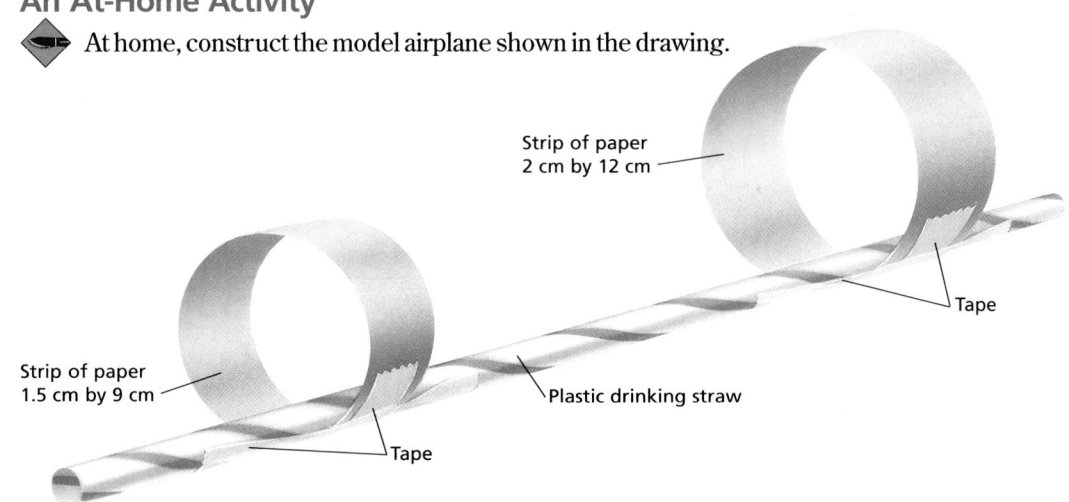

Strip of paper
2 cm by 12 cm

Tape

Strip of paper
1.5 cm by 9 cm

Plastic drinking straw

Tape

Then discover the best way to make your plane fly. For example, try placing the loops of paper at different positions along the straw. What else can you try to make the plane fly better? Make a report to the class about your findings. Include a diagram of your most successful design, and include data on the distance your plane traveled. **C**

What does this activity tell you about the nature of science? **D**

11

This airplane works very well and is easy to make. Encourage students to do this activity at home. Ask them to write a report that includes diagrams of their designs and their data. Allow time for students to share their results.

GUIDED PRACTICE Ask students if they had difficulty controlling variables that they were not testing. For example, it may have been difficult to throw the model airplane at the same angle and with the same force each time. Ask students if they tested each change in variable a number of times and if they found that the model airplane went the same distance each time. If not, they may have been unable to control the variables that were not being tested.

Answers to
In-Text Questions

C Students might try varying the width of the loops, the force and angle at which the plane is launched, and the type of straw used.

D Sample answer: Science is a process, not just a collection of facts. Scientists experiment and test their ideas to obtain reliable results. Scientists may have to repeat the same experiment many times. Scientists control (keep constant) variables other than the one being tested. Scientists also keep records and communicate their ideas to others.

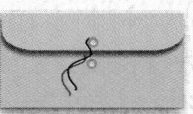

PORTFOLIO
Have students evaluate how their actions in designing their model airplanes were like those of a scientist. Suggest that students make a note of three or four ways in which they could improve their investigative skills in future experiments and investigations.

Homework

You may wish to provide students with the Resource Worksheet that is available to accompany the at-home activity Making the "Best" Airplane (Teaching Resources, page 6).

A The cryptogram reads, "Science is built up with facts, as a house is with stones. But a collection of facts is no more a science than a heap of stones is a house. Poincaré, 1885."

B Accept all reasonable answers. Students may suggest that the system of finding, organizing, and linking together scientific facts distinguishes science itself from just a collection of scientific facts. Similarly, the design and construction of a building distinguish it from a simple collection of stones.

⭐ **A Resource Worksheet (Teaching Resources, page 7) and Transparency 1 are available to accompany Cracking the Code.**

CROSS-DISCIPLINARY FOCUS

Industrial Arts

Shanequa wishes to improve the design of her model airplane. Which of these variables would NOT affect the success of her new design?

a. changing the color of the paper loops *(Correct)*

b. moving the paper loops along the straw

c. replacing the straw with a pipe cleaner

d. adding a third paper loop

Primary Source

Description of change: excerpted from *Science and Method*
Rationale: to focus on the nature of science

Homework

You may wish to assign the activity Cracking the Code as homework.

FOLLOW-UP

Reteaching

Have students list the steps that they took in their investigations with the Möbius strip or the "best" airplane. Ask students: How did you act like a scientist in this investigation?

Cracking the Code

How is solving a cryptogram similar to what a scientist may do? Try cracking this code, which conceals a definition of science written by the French scientist and mathematician Jules-Henri Poincaré. **A**

HXRVMXV RH YFROG FK DRGS UZXGH, ZH Z SLFHV RH DRGS HGLMVH. YFG Z XLOOVXGRLM LU UZXGH RH ML NLIV Z HXRVMXV GSZM Z SVZK LU HGLMVH RH Z SLFHV.
KLRMXZIV, 1885

To help you, here are two decoded words.
YV HXRVMGRURX
BE SCIENTIFIC

First, copy the cryptogram into your ScienceLog. Then use the clue above to solve it. After you decode this statement about science, discuss what you think it means with a classmate. **B**

Assessment

Challenge students to design an experiment to determine the best length for the tail of a kite. Have students list the ways that they would be acting like a scientist. *(They could try several tail lengths, test each length several times, and keep records of how the kite behaves with each different length. Students would be acting as scientists by conducting experiments, repeating their experiments several times, and keeping records of their results.)*

Extension

Challenge students to make a paper strip similar to a Möbius strip except that it forms two interlocking loops when cut. *(The strip must be twisted 360° instead of 180°.)*

Closure

Invite students to create a cryptogram for their own definition of science. Collect these puzzles, photocopy them, and share them with the class.

LESSON 2 · Defining Science

After finishing the previous activities, you should now be ready to share your discoveries about science with your class.

First, construct your own definition of science. You can start with the phrase "Science is . . ." At home, compare your definition with one from a dictionary. At school, compare your definition with those of your classmates. Are all of the definitions the same? Why do you think there are so many views of science?

Throughout this unit and this book, you will be examining many aspects of science. After you finish the unit, try writing another definition of science based on your new knowledge and experience. You will find it interesting to compare this definition with the one you just wrote. To help with this task, record the phrase "Science is . . ." at the top of a page in your ScienceLog. As you proceed through the unit, look for words, phrases, and sentences that convey something about what science is, and add them to your "Science is . . ." page. The information on this page can include anything you discover about the nature of science or the tasks that scientists do. Here is a statement to start with; let's hope that you agree with it.

Science Is . . . Fun

Create another "Science is . . ." page in your ScienceLog. From newspapers and magazines, collect pictures that convey something about science, and make a collage. This collage will be your definition of science for the world to see.

Another way to define science is through verse. You might, for example, write a poem entitled "Science Is." Here is one written by Tammy Doane, a middle-school student.

Science Is

Things to learn
Leaves to burn
Trees, flowers and all
The first snowflakes to fall
An opening flower, a croaking frog
The moss growing on a log
The sun sets at what time?
How much metal in a dime?

LESSON 2 ORGANIZER

Time Required
1 to 2 class periods

Process Skills
analyzing, communicating

New Terms
none

Materials (per student group)
Science Is . . . Fun: newspapers; magazines; scissors; glue

Teaching Resource
SourceBook, p. S5

LESSON 2 · Defining Science

Getting Started
Before students begin this lesson, ask them to name three people they think of as scientists. Then have students write a short paragraph about what scientific things these people do that qualify them as scientists. Student answers may reveal some common misconceptions about who scientists are. At the end of the lesson, you may wish to have them review their descriptions and discuss how their ideas about scientists have changed.

Main Ideas
1. Science can be described in many different ways.
2. Scientists are ordinary people.

TEACHING STRATEGIES

After students have read the introductory paragraphs for this lesson, have them write their own definition of science in their ScienceLog. Ask volunteers to share some of their definitions with the rest of the class. Then have a student read a dictionary definition of science, and discuss with the class how the dictionary definition compares with their own definitions.

Science Is . . . Fun
Explain to students that a collage is an art form in which bits of objects, such as newspaper, photos, cloth, and pressed flowers, are combined to create a symbolic effect. You may wish to have students display their collages on the wall.

Ask a volunteer to read the poem written by Tammy Doane. Point out to students that the poem illustrates some important characteristics of science, such as things to learn, things to do, and things to investigate.

Science Is . . . People

INDEPENDENT PRACTICE As an exercise in critical thinking, have students consider the following questions while reading the interview. Write the questions on the board, and have students respond to them in their ScienceLog.

- How are scientists similar to and different from you or me? *(Scientists have made a career of doing science. Otherwise, they are ordinary people.)*
- If you were Gabriela, what questions would you have asked Irene Long? *(Accept all reasonable responses.)*
- What is one thing about scientists that you learned from this interview? *(Students may mention that many careers are involved in the space program, for instance.)*
- Do you agree with everything Irene Long said? Explain your answer. *(Answers will vary.)*

GUIDED PRACTICE Stimulate class discussion by asking students the following questions:

- What characteristics should a person have in order to become a research scientist? *(Sample answer: an inquiring mind, the ability to look at things critically, a willingness to dedicate several years of study and sometimes long hours to experimental work, and the ability to be motivated by discovering something new)*
- Why is the study of science important for a person who does not intend to become a scientist? *(By understanding more about science, students will be better prepared for the technology of the future. Science also teaches students how to think critically, a skill that can be useful in other disciplines and in everyday life.)*

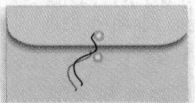

PORTFOLIO

Suggest that students include the collage they made for Science Is . . . Fun in their Portfolio.

Science Is . . . People

Here are some questions that middle-school student Gabriela Sanchez asked Dr. Irene Long. Dr. Long is the director of NASA's Biomedical Operations and Research Office at the Kennedy Space Center in Florida. She is one of a team of doctors that provides medical attention for the astronauts in the event of an emergency or when a mission is cut short. Her team also collects data on the medical condition of returning astronauts.

Gabriela: How did you become interested in space travel?

Dr. Long: When I was about nine years old, I was watching a television program—*Man and the Challenge*—about Lt. Col. John Paul Stapp, who was studying the effects of acceleration on the human body. I thought that was fascinating. I thought, "Gosh! He's a physician and he's involved in space. I'd like to do that!"

Gabriela: What kind of research are you conducting now?

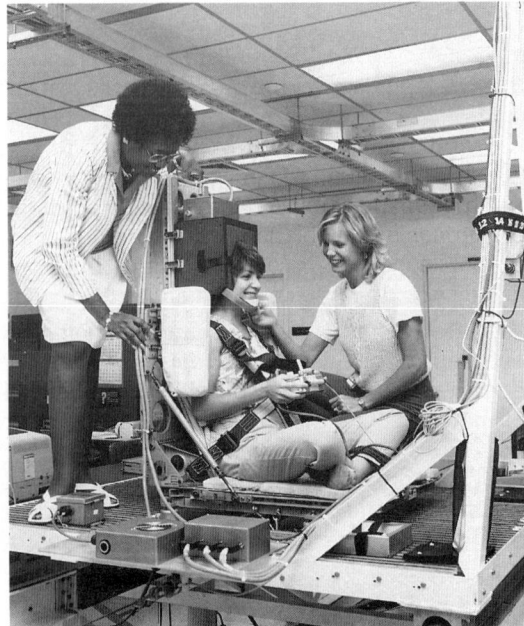

Dr. Long is shown here setting up a space science summer program for college students.

Dr. Long: The majority of my work is geared toward occupational medicine and environmental health. We do research on the effect of protective equipment. We make sure that the astronauts' equipment is actually working. Most of our work is geared toward launching and landing because we are located at the space port.

Gabriela: Do you think that space travel to places such as the planet Mars will be achieved in the near future?

Dr. Long: I hope so. I think that it is technically possible. I just can't say exactly when it will happen. I feel that we do have pressing problems here on Earth, but with appropriate management we can succeed in both areas and both areas can benefit. I don't feel that one area excludes another. It's a management challenge, but I think that human beings can handle more than one challenge at once. I do it all the time. I think one of the greatest things learned from extensive space travel is international cooperation. When you have a common goal, the goal becomes more important than any differences you might have.

Dr. Long has been a pioneer in research for occupational medicine and environmental health for humans in space.

14

Integrating the Sciences

Earth, Life, and Physical Sciences
As students are forming definitions for science in general, ask them to define Earth science, life science, and physical science. Encourage students to investigate the different branches of each science (such as geology, oceanography, meteorology, and astronomy within Earth science) and to find out more about any areas in which they are interested.

Homework

From a recent newspaper or magazine, have students select an article that describes a scientist and his or her work. Have the students bring the article to class along with answers to the following questions: What field of science is being described? What problem is the scientist attempting to solve? What kind of training do you think the scientist needed? *(Answers will vary.)*

Gabriela: Have you realized your dream of flying aboard the space shuttle?

Dr. Long: No. Very early on, many of us focused on flying in space. You realize, however, that the space program involves thousands of people. Not everyone has to be an astronaut to feel the satisfaction of contributing to the flight. When astronauts fly, they are representing all of us who helped them get there, and you can really feel good about that.

Gabriela: To what one thing would you credit your great professional success?

Dr. Long: Support from my parents was the major thing. In addition to that, I defined my goals early on. Then I had to find out what I needed to do to reach those goals. But realizing my dreams took a lot of hard work. I made it happen. Every time someone told me I couldn't do something, I went out and did it.

Gabriela: What advice would you give to a young person today who has dreams of working in the space program?

Dr. Long: I would encourage students to do some research on all of the different careers in the space program. You might have the impression that the only people who are in the space program are the people who fly, but hundreds of careers are involved in supporting roles. For instance, I knew that I wanted to be a doctor, and I knew that I wanted to be involved in space, so I looked for a career that had both. It is possible to have multiple goals because they can often fit together. We have lawyers, engineers, secretaries, administrators—there are so many!

Gabriela: What do you think will be the future of the space program?

Dr. Long: I think that both manned and unmanned flights will complement each other. It will depend on the destination. But the space program is important because it stimulates technology. When you add humans to the picture, you begin to understand how people can survive in other environments. This then stimulates medical technology also.

If you'd like to learn more about the different disciplines of science, turn to page S5 of the SourceBook.

15

Reteaching

Invite a scientist to your class to talk to students about what he or she does. Have students write questions for the scientist on slips of paper in advance.

Assessment

Ask students to come up with a title for each of Dr. Long's responses which summarizes the main ideas of each response. (Sample answers: Inspiration Is Born; Protecting the Health of Astronauts; Mars—Working Toward a Common Goal; Soaring High With Teamwork; Having a Goal; So Many Careers in Space; Our Future in Space)

Extension

Ask students to research and report on the life and work of a scientist who interests them.

Closure

Have students conduct their own interview of a scientist. Students should make a list of questions to ask the scientist. A local college, university, or science-based company would be a good source for scientists. Students could make an audio or video recording of the interview if equipment is available and the subject consents.

Meeting Individual Needs

Learners Having Difficulty

Assemble a group of objects for students to investigate, observe, and take apart if they wish. These objects might include a flashlight, a stapler, a lawn sprinkler, a manual typewriter, and a zipper. After students examine each object, ask them to speculate about how each works. To get them started, suggest that they make a list or draw a diagram of the different parts in each object. Then they can proceed by writing down the possible function of each part. Ask students to summarize the ways in which they behaved like scientists during this activity.

ENVIRONMENTAL FOCUS

Dr. Long believes that space travel to other planets will be possible. Discuss with students environmental problems that human beings would encounter in establishing a permanent colony on Mars. (Students may note that Mars has a very different atmosphere from that of Earth and that Mars's temperature is very different from Earth's temperature.)

Answers to
Challenge Your Thinking, pages 16–17

1. Answers will vary, but encourage students to compare their actions with the actions of the scientists discussed in the textbook.

2. The solution key is as follows:

A = N	H = U
B = O	I = V
C = P	J = W
D = Q	K = X
E = R	L = Y
F = S	M = Z
G = T	

The solution to the cryptogram is the quotation that appears in Lesson 1 on page 5.

3. One possible method for solving the puzzle would be to use the clue that the word *paths* appears twice. The only word with five letters that appears twice is *CNGUF;* thus, those letters must spell *paths.* By applying these known letters to replace other letters in the puzzle and making some logical associations to fill in the remaining letters, the puzzle can be solved.

Some scientific skills involved might be logic, imagination, recalling from memory, testing, and hypothesizing.

4. Reports should show how a news article relates to science. In presenting an article to the class, students should demonstrate an understanding of the scientific research or application being discussed.

5. The cover page is intended to allow students to express their own understanding of science. Students should include items that show science as a way of learning about themselves and the world around them.

6. Students are expected to dispel misconceptions about science and scientists. This includes the stereotype of scientists as social misfits or mad geniuses. Ideal answers will acknowledge that scientists often do experiments in laboratories, keep records of data, take measurements, and label specimens, but that science involves more than just experiments and is not limited to laboratories.

CHALLENGE YOUR THINKING

1. It's Not a Problem
When was the last time you used scientific thinking (outside of your science classroom)? Describe the circumstances and the type of problem you were trying to solve.

2. Cryptic Quotes
Solve the cryptogram below to reveal a quote. Here's a clue to help you get started: the word *paths* appears twice. Solving this quote may give you a case of *déjà vu.* Show your solution key.

V NZ NA RKCYBERE. FBZR CNGUF GUNG V GENIRY UNIR ORRA GENIRYRQ ORSBER. BGURE CNGUF NER HAPUNEGRQ—YRNQVAT GB ARJ VAFVTUGF NAQ GB ARJ QVFPBIREVRF.

3. A Clever Solution
Describe the process you used to solve the cryptogram in the previous question. In what ways did your solution method involve scientific thinking?

 You may wish to provide students with the Chapter 1 Review Worksheet that accompanies this Challenge Your Thinking (Teaching Resources, page 8). If you choose to use this worksheet in class, Transparency 2 is available to accompany it.

4. Alert the Media!

Check the daily newspapers for news about science. Write a brief report on an article that interests you, and share what you find out with your class.

5. Picture This

In your ScienceLog, create a cover page for this chapter by drawing a diagram or creating a collage of pictures that illustrates what you learned in this chapter.

Walk This Way— Mastodon Tracks Discovered

University scientists recently discovered the perfectly preserved, 11,000-year-old tracks of a male mastodon. By examining the tracks, scientists estimate that this huge, hairy relative of today's elephant stood 3.5 m tall and weighed 6 tons!

6. Crazy About Science?

Above is one person's drawing of a scientist. What ideas about science and scientists are suggested in the drawing? What ideas presented do you think are true? What ideas do you think are untrue?

ScienceLog

Review your responses to the ScienceLog questions on page 4. Then revise your original ideas so that they reflect what you've learned.

ScienceLog

The following are sample revised answers:

1. Answers will vary. Students may state that science is a process involving observation and experimentation, for instance. Ask students to explain how their answers have changed since they began this chapter.
2. Answers will vary. For instance, students may note times when they wondered why something happened, made observations about the world, or made measurements of something.
3. These people are all making observations and collecting data, either on their own or with the aid of equipment. Therefore, all of them could be described as scientists.
4. Accept all reasonable responses. Encourage students to be creative.

Homework

Have students interview five students and five adults, asking each person for a definition of science. Then have the students rank the definitions from best to worst. Students should explain why they ranked the definitions in the order that they did.

17

CHAPTER 2

Being Scientific

Connecting to Other Chapters

Chapter 1
explores the nature of science and what it means to be a scientist.

Chapter 2
invites students to practice scientific thinking, including observing, inferring, and hypothesizing.

Chapter 3
guides students through the process of experimentation and discusses the nature of technology.

Prior Knowledge and Misconceptions

Your students' responses to the ScienceLog questions on this page will reveal the kind of information—and misinformation—they bring to this chapter. Use what you find out about your students' knowledge to choose which chapter concepts and activities to emphasize in your teaching.

In addition to having students answer the questions on this page, you may wish to do the following: Ahead of time, press some clay against one side of the interior of an empty soup can, and then cover the top with aluminum foil and tape. Have students watch as you roll the can across a desk or table (it will have a jerky motion). On the board, make a list of students' statements about the demonstration. Then have students discuss the list and divide it into groups of similar statements. *(Although students may not know the terms* observation *and* inference, *they should recognize that some statements are factual, while others involve some interpretation of the event.)* Listen carefully to the discussion to find out what students know about scientific thinking, what misconceptions they may have, and what is interesting to them about scientific thinking.

CHAPTER 2

Being Scientific

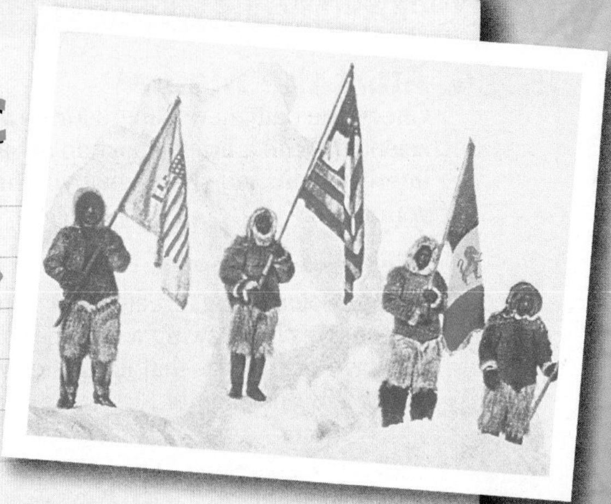

1 Observe this photograph. What do you think is happening? What do you think will happen next?

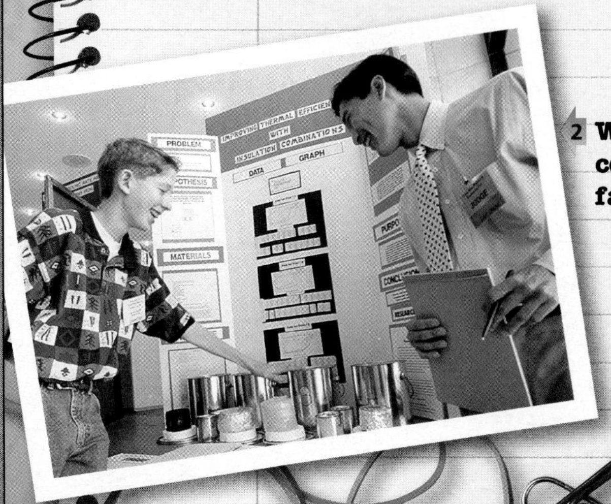

2 When you perform a controlled experiment for a science-fair project, what does that mean?

3 Look at the objects below for 15 seconds, and then close your book. Describe the picture with as much detail as you can remember. Did you write down the same answer as the person sitting next to you?

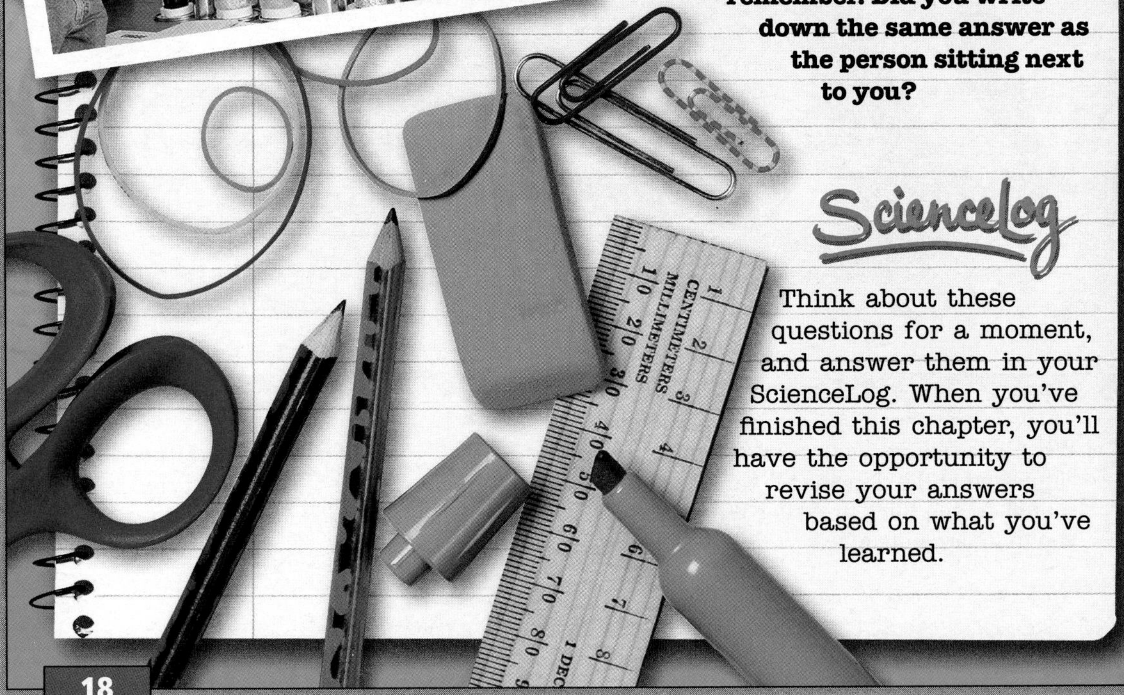

ScienceLog

Think about these questions for a moment, and answer them in your ScienceLog. When you've finished this chapter, you'll have the opportunity to revise your answers based on what you've learned.

LESSON 1 · The Science of Observation

Like science itself, many science-fiction stories contain scientific facts. In the story that follows, you will read about Kriavalinan Z (known to his friends as Zed), who has just arrived on Earth from the planet Nebulos.

A Stranger Has Landed

After the complicated maneuvers that were necessary to avoid detection, Zed finally landed on Earth. He stretched himself to his full height of 198 cm and yawned, relaxing after the difficult journey. Next came a quick meal. He eagerly began preparing to leave the spacecraft that had been his home for the last three and a half years.

As Zed collected his belongings, he felt a thrill of excitement. Would Earth be similar to Nebulos, which was trillions of kilometers away? Would it be a refuge or a place of danger? Now Zed was ready to leave his spacecraft. He proceeded slowly toward the door, his face a picture of concentration, wonderment, and anticipation. Zed pressed the door activator with the six fingers that made up his left hand. The door slid open, and Zed saw Earth up close for the first time.

The first thing Zed noticed was all of the color—the vivid reds and yellows of flowers; the waving greenery of the trees; and the pale, delicate blue of the sky, which was very much like the color of his skin. "Before venturing farther," Zed thought, "I must observe and try to make sense of this world. Later—perhaps much later—I will make contact with the beings who populate this lovely planet."

In these introductory paragraphs of a science-fiction story, you read about someone who explored an unfamiliar world in a scientific way. Making sense of something begins with making observations—seeing, feeling, hearing, tasting, and smelling.

In Exploration 1, imagine that you are Zed and that you are observing a strange object for the first time. What can you discover about a candle by making observations?

19

LESSON 1 ORGANIZER

Time Required
3 to 4 class periods

Process Skills
observing, classifying, inferring, analyzing

Theme Connection
Structures

New Terms
Observation—making sense of something by seeing, feeling, hearing, tasting, and smelling

Properties—characteristics that distinguish one material from another
Qualitative observation—an observation that does not involve measurements or numbers
Quantitative observation—an observation that involves measurements or numbers

Materials (per student group)
Exploration 1: 2–3 matches; small ball of modeling clay; candle; metric ruler; jar lid or aluminum pie plate; watch or clock; safety goggles

continued ▶

FOCUS

Getting Started

Stand a small cylindrical object (such as a paper towel tube) on top of a plate. The diameter of the plate should be slightly greater than the height of the cylinder. Ask students which distance is greater, the height of the cylinder or the diameter of the plate. Have a ruler handy so that the distances can be measured. Use this demonstration to explain to students that they cannot always believe what they see. Stress that accurate measurements are important when making scientific observations.

Main Ideas

1. We learn about the world around us through observations.
2. Observations involving measurements are called quantitative observations.
3. Observations that do not involve measurements or numbers are called qualitative observations.
4. Observations can involve all of the senses—sight, touch, taste, smell, and hearing.

TEACHING STRATEGIES

A Stranger Has Landed

Call on a volunteer to read aloud the story on this page. Ask students to describe any science-fiction stories that they have read and enjoyed.

Use the science-fiction story *A Stranger Has Landed* to set the stage for Exploration 1. Discuss with the class how Zed is being scientific. *(By observing)*

CROSS-DISCIPLINARY FOCUS

Language Arts
Ask: In the opening paragraphs on page 19, Zed could be described as
a. delirious.
b. agitated.
c. astounded.
d. inquisitive. *(Correct)*

Although Exploration 1 is designed to stimulate discussion about the difference between qualitative and quantitative observations, have each student write the answer to question 5 individually before the discussion.

★ **An Exploration Worksheet is available to accompany Exploration 1 (Teaching Resources, page 14).**

Answers to
For Discussion

1. Observations should be classified according to the sense used.

2. Quantitative observations may include length of the flame, length of the candle, length of the wick, and rate of burning. Qualitative observations may include the color of the candle, the smell of burning wax, and the way the flame flickered.

3. Quantitative observations include 198 cm, 3.5 years (or light-years), and 6 fingers. Qualitative observations include red and yellow flowers, green trees, blue sky, and blue skin.

4. **a.** Properties that help distinguish ice from wax: Wax can be scratched easily with your fingernail, but ice is harder to scratch; wax does not melt when placed on your desk, but ice does; and wax is not cold to the touch, but ice is.

   b. Properties that ice and wax have in common: Both melt when heated; both float on water; both break easily; and both stay solid below 0°C.

   c. Quantitative properties that help distinguish wax from ice include the melting point of each and the mass of a certain volume of each.

   d. Yes, there is a difference. An observation is something you perceive about an object, but a property is something you use to distinguish it from other objects.

5. Answers will vary depending on the object chosen. Have students write their descriptions as riddles, keeping the names of their objects a secret. Have the class participate in a game to guess each object and to indicate whether the observations used are qualitative or quantitative.

Test Your Powers of Observation

You Will Need

- matches
- modeling clay
- a candle
- a ruler
- a jar lid or aluminum pie plate
- a watch or clock

What to Do

1. Before lighting the candle, make as many observations about it as possible. Can you list 5? 10? 20? Your time limit is 7 minutes. Write your observations in your ScienceLog.

2. Now place the candle on the lid. Modeling clay can be used to hold the candle in place. Light the candle. In the time it takes the candle to burn down (or 10 minutes, whichever comes first), make as many more observations as possible. Use the ruler to help make some observations. Do not, however, burn the ruler or anything other than the candle!

For Discussion

1. Share your observations with your friends. Did they make observations that you didn't? Classify each observation according to whether it was made by sight, touch, hearing, taste, or smell.

2. Did you make any observations using the ruler? Observations of this type are called **quantitative observations**. Quantitative observations involve measurements and numbers. Perhaps you measured how far the candle burned down in 10 minutes or timed how long it took for the candle to burn down. These are examples of quantitative observations. On the other hand, if you observed the color of the candle, the way the flame flickered, or noted the smell of burning wax, you were making **qualitative observations**. Qualitative observations *do not* involve measurements or numbers.

   Decide which of the observations made by your class were quantitative and which were qualitative. Perhaps you can suggest some more quantitative observations that you could have made.

3. Reread "A Stranger Has Landed" on the previous page, and find statements that express quantitative observations and those that express qualitative observations.

4. Many of your observations of the candle may have described the wax that makes up the candle. Characteristics that help distinguish wax from other materials are called **properties**. For example, wax can be distinguished from ice by its ability to burn: wax will burn; ice will not burn.

   a. What other properties of wax would help distinguish a piece of wax from a piece of ice?

   b. Actually, ice and wax have many properties in common. For example, they are both solids. Can you think of other properties that they share?

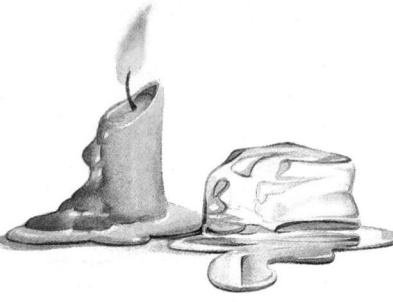

   c. You can make quantitative observations about the properties of wax that would help distinguish it from ice. What observations would you suggest?

   d. Is there a difference between an observation and a property? Explain your reasoning.

5. Write a description of an object as seen through Zed's eyes. Include both quantitative and qualitative observations. Have your classmates guess what you are describing.

ORGANIZER, *continued*

Exploration 2: 3 toothpicks; 3 jar lids or small bowls; 5 mL measuring spoon; 10 mL each of confectioners' sugar, baking powder, and plaster of Paris; 30 mL of water; 100 mL graduated cylinder; 3 small pieces of masking tape; pen; safety goggles; optional items: eyedropper; 10–15 mL each of baking soda, cornstarch, vinegar, and iodine starch-test reagent; an additional 5 mL of confectioners' sugar, baking powder, and plaster of Paris; 20 mL of water (additional teacher materials: three 250 mL beakers; 3 small pieces of masking tape; 25 mL of 0.1 M sodium thiosulfate; newspaper; see page 21 and Advance Preparation on page 1C.)

Exploration 3: metric ruler (additional teacher materials: 8 matches, support stand with clamp, watch or clock with second hand; safety goggles for you and any students who assist or are nearby during the demonstration described in Test 4)

Teaching Resources

Exploration Worksheets, pp. 14, 18, and 21
Transparencies 3 and 4
SourceBook, p. S12

Using Observations to Identify Mystery Powders

Three beakers labeled *A*, *B*, and *C* each contain a white powder. One powder is confectioner's sugar, one is baking powder, and one is plaster of Paris.

The following activity will help you determine which is which. But do not try to distinguish them by taste! Scientists never taste unknown substances because doing so could be dangerous or even deadly.

You Will Need

- 3 toothpicks
- 3 jar lids or small bowls
- a small measuring spoon with a capacity of about 5 mL
- samples of mystery powders *A*, *B*, and *C*
- water
- masking tape
- a pen

What to Do

1. Using the masking tape and the pen, label your lids or bowls *A*, *B*, and *C*.
2. Put about 5 mL of powder *A* into lid *A*, 5 mL of powder *B* into lid *B*, and 5 mL of powder *C* into lid *C*.
3. Add about 5 mL of water to each lid.
4. Stir each mixture with a toothpick.
5. Observe carefully, and write down your observations. Can you tell from these observations which powder is which?

If you can, you are using some information that you already have about the powders and how they should react. If you can't, ask your teacher for labeled samples of each substance. Repeat the activity with those substances. Then compare these results to the results you recorded earlier. Now you should be able to identify the mystery powders.

6. When you have finished, wash your hands, and clean up your area.

What is the most important property that enabled you to distinguish the powders from one another?

Extension

You can make this task more challenging by adding more powders to those mentioned above. Two possibilities are baking soda and cornstarch. In addition to testing each powder with water, try using vinegar and an iodine solution. Wear goggles, latex gloves, and an apron for this activity.

Now test a classmate's powers of observation. Prepare a mixture of any two powders. Can he or she name the powders in your mystery mixture?

How does this Exploration tell you what science is? **B**

Homework

Have students choose three objects in their room at home and write three quantitative and three qualitative observations about each object. In class, have volunteers read their observations aloud so that other students can guess what each object is.

Theme Connection

Structures

After Exploration 2, ask the following question. **Focus question:** What natural processes are similar to the hardening of the plaster of Paris? *(Sample answer: the drying and hardening of mud and the formation of sedimentary rocks as a result of evaporation)* Point out to students that plaster which is hardened one layer at a time would show a layered structure similar to many types of sedimentary rocks.

Milliliters are used to measure both liquids and powders. Exact quantities are not essential. The mystery powders should be placed in 250 mL beakers labeled *A*, *B*, and *C*. A teaspoon is equivalent to 5 mL, or 2 dropperfuls.

After students see the reaction of each powder with water, point out the following properties that can be used to distinguish the powders: Confectioners' sugar dissolves in water, baking powder fizzes in water, and plaster of Paris may feel warmer after adding water and will eventually harden.

Extension

Students should observe that vinegar causes both the baking soda and the baking powder to bubble and that iodine turns both cornstarch and baking powder black (indicating that both contain starch). Materials required for creating and testing mixtures will vary depending on the substances chosen by students.

WASTE DISPOSAL ALERT   Treat the iodine-stained materials with enough 0.1 molar sodium thiosulfate to decolorize them, wrap them in old newspaper, and put them in the trash. Wrap all other materials from this Exploration in old newspaper and put them in the trash as well.

⭐ **An Exploration Worksheet is available to accompany Exploration 2 (Teaching Resources, page 18).**

Answers to *In-Text Questions*

A The way the powders react with water is the most important property for distinguishing them.

B Students might respond that Exploration 2 helps to explain what science is because it encourages paying attention to procedures, observing carefully, recording observations, drawing conclusions, repeating experiments, and drawing more conclusions.

Do You See What I See?

GUIDED PRACTICE   Draw a box on the chalkboard. Inside the box, write this phrase:

> Paris
> in the
> the spring

Quickly show this to students, and then erase it. Ask them to write what they have read. Many people read, "Paris in the spring," rather than "Paris in the the spring." Explain to students that if they want to become scientists, they must practice to develop a keen sense of observation.

Meeting Individual Needs

Learners Having Difficulty

Scientists studying the Hawaiian Islands made several observations that are listed below. Ask students to identify which of the observations are qualitative and which are quantitative.

a. The lava flowing out of the volcano is red. *(Qualitative)*

b. The lava has a temperature of 1000°C. *(Quantitative)*

c. The oldest volcano is 2.4 km from the newest volcano. *(Quantitative)*

d. The ash cloud billowed above the volcano and then moved north. *(Qualitative)*

Did You Know...

Important observations do not have to be made in a laboratory. They can be made at any time under any circumstances. While he was taking a bath, Archimedes, a famous mathematician of ancient Greece, discovered that an object's volume could be measured by measuring the amount of water that the object displaces.

Do You See What I See?

To learn more about thinking like a scientist, turn to page S2 of the SourceBook.

How can several people observe the same event and not agree on what they saw? Witnesses to crimes or traffic accidents frequently disagree on the details they give to the police. Why is this so? One explanation is that witnesses see the event from different angles or perspectives, depending on their position when the event occurred.

Another important reason is simply that most people have not developed the skill of accurate observation. For example, can you make an accurate sketch of the steering wheel of the family car? Do you know the color of the car's upholstery? Just as a baseball player has to develop ball-handling skills and a pianist must work on instrumental and performing skills, you must take time to develop good observational skills.

Sometimes there are other factors that prevent you from making good observations. Can you always rely on your eyes? Think about this question as you do the first three tests in Exploration 3. The final test will give you an opportunity to compare your observational skills with those of your classmates.

22

Eyeball Benders

TEST 1

Which line is longer—*A* or *B*? After recording your answers for Figures 1–6 in your ScienceLog, check them by using a ruler to measure lines *A* and *B* in each figure.

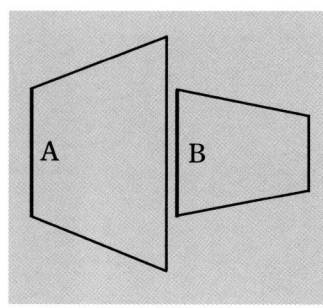

1

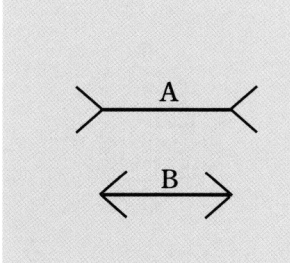

2

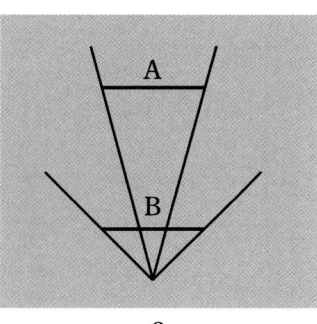

3

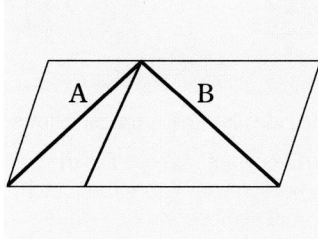

4

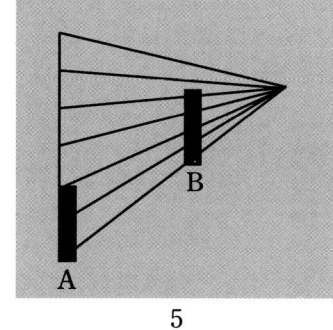

5

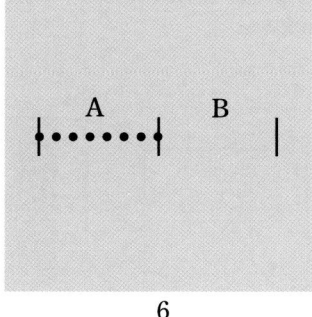

6

TEST 2

1. Are these lines parallel?

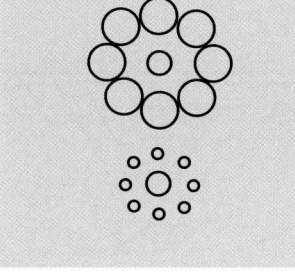

2. Are the circles in the centers of the figures the same size?

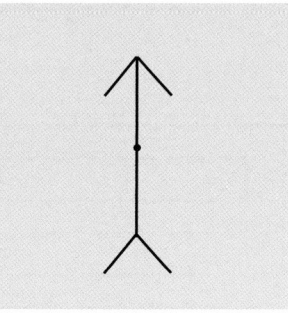

3. Are the line segments above and below the dot equal?

Exploration 3 continued ▶

23

The purpose of this Exploration is to investigate situations in which our observational skills may not be reliable. The need for making quantitative observations is demonstrated as well. Have students work in pairs to do Tests 1, 2, and 3. It is suggested that Test 4 be done as a demonstration.

 Transparencies 3 and 4 are available to accompany Exploration 3.

Answer to
Test 1

In all cases, *A* and *B* are equal in length.

Answers to
Test 2

1. The lines are parallel.

2. The circles in the center are the same size.

3. The line segments are the same length.

Homework

You may wish to assign Exploration 3, with the exception of Test 4, as a homework assignment.

TEST 4

This test may be done as a teacher demonstration. Any students that are near you during the demonstration should wear safety goggles. Compare observations from those at the front of the room with those from farther away.

 An Exploration Worksheet is available to accompany Test 4 (Teaching Resources, page 21).

Answers to
In-Text Questions

Ⓐ Artists often use perspective to create the illusion of depth on a flat piece of paper. A small object, for example, appears to be far away.

Ⓑ No; lines are connected in a way that creates illusions of shapes which could not exist in three dimensions.

Ⓒ Answers should describe the burning of the matches in as much detail as possible, using sight, hearing, and smell. Students closer to the demonstration should notice more details than those farther away.

TEST 3

(Just for Fun!)

You have probably seen these optical illusions before. How are they used in drawings? Examine other pictures to discover how artists Ⓐ have used different methods to create the illusions of depth, height, and distance.

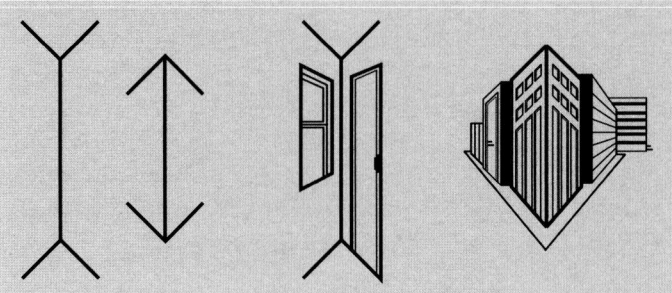

The figures below "reverse." Gaze steadily at each one to see what "happens."

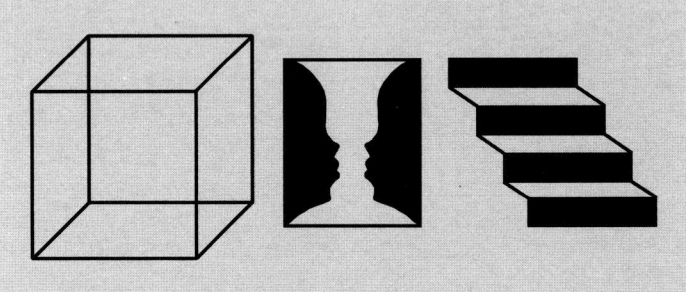

Are the figures below possible? If not, explain what is wrong with Ⓑ each one.

TEST 4

A match is held upright in a test-tube clamp as shown. Another match is lit and brought near the first one. As the first match bursts into flames, begin your observations.

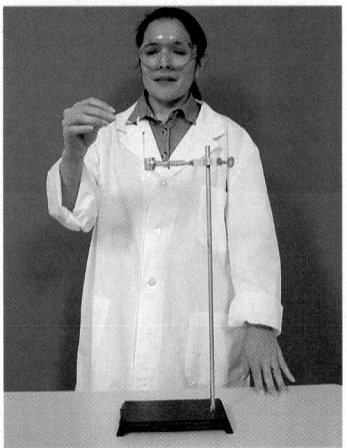

The rules for Test 4 are as follows:

1. Everyone, except for the person who lights the match, makes observations while seated.

2. Begin making observations at the instant the first match bursts into flame, and end 1 minute after the match goes out.

3. The test may be repeated three times.

Make brief notes about your observations. What did you see? hear? smell? How many observations did you make? How many different observations were made by others? Why didn't everyone make the same observations? What observations did the people sitting near the front make that those farther back missed? Ⓒ

24

FOLLOW-UP

Reteaching

Have students review the procedure and their results from Exploration 2. Have students evaluate their problem-solving skills, their use of scientific equipment, and the quality of their observations by listing three qualities they could improve.

Assessment

Read the following sentences to students: Snow is cold, white, fluffy, and slippery. If you examine it closely, you may be able to see single snowflakes, each with six points. A flake may be

only 4 mm across, but billions of them can form a snowdrift. If the temperature goes above 0°C, the flakes melt.

Then have students write three qualitative observations and three quantitative observations about snow. *(Cold, white, fluffy, slippery; snowflakes have six points, snow melts at 0°C, and each flake is about 4 mm across.)* Then have students suggest three properties of snowflakes. *(Snowflakes have six points, come in different shapes and sizes, are usually about 4 mm across, melt at 0°C, and are white.)*

Extension

Show students a video or film of a magic show. Discuss how magicians can confuse our powers of observation.

Closure

Challenge students to write a story about what a creature from outer space would observe on Earth. Stories should include both qualitative and quantitative details of the things that the creature sees, feels, hears, tastes, and smells.

LESSON 2
From Observations to Inferences

Zed approached the object with caution. What on Earth could it be? Whatever it was, there it sat by the edge of the road—taller than it was wide, and round, with bumps on it. By standing next to it, Zed estimated the object to be 5 zks high (60 cm, in metric). "Nice colors," Zed thought, "red and silver. Perhaps it's a piece of art. Or maybe it's for sitting on—although it doesn't look very comfortable!"

Can you guess what Zed saw? Here's another clue: You wouldn't want to park a car in front of one. **D**

The description you just read includes seven observations. Try to identify them. Which one is quantitative? There are also two statements that are not observations. They are known as **inferences**. Try to identify them. **E**

How do inferences differ from observations? Use the following examples to help you decide:

Observation	Inference
Marie is away today.	Perhaps Marie has the flu.
Erik didn't do as well on this test as he usually does.	Erik probably didn't study for the test.
My flowers grew better at this end of the garden.	The soil must be richer at this end of the garden.

Is each of the inferences the only possible explanation for the observation? Suggest at least three more observations and some possible inferences for each one. **F**

Inferences are not observations; rather, they are statements that attempt to explain or make sense of observations. Now you know that making observations is not the same as *interpreting* them and arriving at conclusions.

LESSON 2 ORGANIZER

Time Required
4 class periods

Process Skills
observing, inferring, analyzing

New Term
Inference—a statement that attempts to explain or make sense of an observation

Materials (per student group)
Exploration 4, The Dancing Disk: empty glass bottle; enough cold water to fill the bottle; sink; thin plastic or plastic-foam disposable plate; scissors;

The Reappearing Coin: coin; shallow dish; 10 cm of transparent tape; about 100 mL of water; **The Curious Cup:** drinking glass; enough water to fill the glass; small piece of paper; bucket or sink; **Ice Heist:** 10–20 ice cubes; 250 mL beaker; about 10 g of salt; 30 cm of string; watch or clock with second hand; **Lights Out! #1:** aluminum pie plate or petri dish; candle; large jar; 2–3 matches; small ball of modeling clay; about 50 mL of water; safety goggles;

continued ➤

FOCUS

Getting Started
To demonstrate that observations can often be biased by what we expect to see, try the following activity: Before class, select a group of 4–5 students, and instruct them to enter the room as soon as class starts. Tell them to throw a magazine on the floor, drop a plastic cup, or perform some other minor disturbance and then exit the room. Then ask other students to list observations of what happened. *(Most students will be able to list the obvious occurrences, but the minute details of appearance will probably be omitted.)*

Main Ideas
1. Making inferences is a way of making sense of what we observe.
2. Investigative questions often lead to solutions.

Teaching Strategies for Lesson 2 begin on the next page. ➤

Answers to *In-Text Questions*

D A fire hydrant

E Zed *observes* that the object is (1) by the edge of the road, (2) taller than it is wide, (3) round, (4) bumpy, (5) 60 cm high (quantitative), (6) red and silver, and (7) uncomfortable looking. Zed *infers* that it may be (1) a piece of art and (2) for sitting on.

F No; more than one inference is usually possible. Accept all reasonable responses. Sample answer: From observing that someone had a second helping of pears, you might infer that the person likes pears or is very hungry.

Point out to students that there are subtle differences between the meanings of the terms *infer* and *conclude*. To infer usually implies theorizing or predicting from what may be incomplete evidence. To conclude implies identifying a logical result through the process of reasoning, after the facts are observed.

Confusing Observations With Inferences

Encourage students to write down some of their own incorrect inferences. Sayings, superstitions, and prejudiced beliefs are good examples. Then have students discuss and compare their lists with each other.

Answers to
In-Text Questions

Ⓐ Discuss with students the three inferences on this page and the reasons why they are probably incorrect. Accepting incorrect inferences might prevent accurate observation of future events. The following are sample reasons:
- The coolness of Jorge's shoes does not increase his athletic ability. Athletic ability relates to body type and conditioning.
- The dog could have been mean because it was hungry or scared. The fact that the dog was found near the highway does not mean that it was stray or that all stray dogs are mean.
- The sister may get what she wants for her birthday because she asked for it. The age of a sister or brother probably has nothing to do with receiving birthday presents.

Ⓑ Sample inferences include the following:
- Lightning never strikes in the same place twice.
- It seems to rain more on weekends than on weekdays.
- Whenever something can go wrong, it usually does.
  Discuss with students why their sample inferences may be based on inaccurate observations. (The inference that lightning never strikes the same place twice may seem true because lightning strikes are so rare that the chances of lightning striking twice in the same place seem slim.)

Confusing Observations With Inferences

What we see is often determined by what we expect to see. For example, suppose you hear behind you the roar of an engine and the squeal of tires. What do you expect to see when you turn around? A hot rod? A police car? An accident? Unless you turn around and look carefully, you may never know whether your expectations were correct.

Scientists too have expectations of what will happen in their experiments, but they must be careful not to allow their expectations to influence their observations. Here is a good set of rules to follow when making observations.

1. Make a list of what you plan to observe.
2. Record the observations for each experiment at the time they are made.
3. Repeat experiments to check the observations made in each one.
4. Have others repeat the same experiment.

Here are three more inferences.

- Jorge has those cool basketball shoes. That's what makes him such a great player.
- My neighbor found a mean dog near the highway. I think all stray dogs are mean.
- My sister always gets what she wants for her birthday because she is the youngest.

Do you think these inferences are based on careful observations? Try to suggest a few reasons why the observations that led to these inferences may have been inaccurate. Do you think that accepting these inferences would affect the accuracy of future observations? Ⓐ

Now try to list some more inferences similar to the ones above; that is, list statements that are likely to be based on inaccurate observations. Compare your list with those of others. Ⓑ

26

ORGANIZER, *continued*

Lights Out! #2: large beaker or jar; candle; 2–3 matches; small ball of modeling clay; 30 mL of baking soda; 15 mL of vinegar; 1 L container; watch or clock with second hand; safety goggles (additional teacher materials: newspaper; see page 27 and Advance Preparation on page 1C.)
Classifying Statements: washer; 30 cm of string; optional item: support stand with clamp

Teaching Resources
Exploration Worksheet, p. 22
Transparency 5

It's Your Turn to Be Scientific

By now you should be an expert at making observations and inferences. In the following activities you will observe some unusual things. For each activity, record two or three observations and at least one inference. Record your responses in a table like the one at right. Do not write in this book.

Activity	Observations	Inferences
The Dancing Disk		

The Dancing Disk

Fill an empty glass bottle with cold water. Cut a quarter-sized disk from a plastic or plastic-foam disposable plate. Pour about half of the water out of the bottle. Place the disk on top of the bottle. Grasp the bottle tightly in both hands. Observe what happens to the disk. Make an inference to explain your observations. **C**

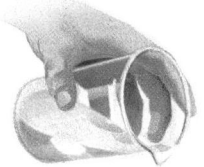

The Reappearing Coin

Using transparent tape, tape a coin to the bottom of a shallow dish, such as a margarine container. Place the container on the table. Now move backward just far enough so that you cannot see the coin over the rim of the dish. Have a friend pour water into the dish. What do you observe? What inferences can you make from your observations? **D**

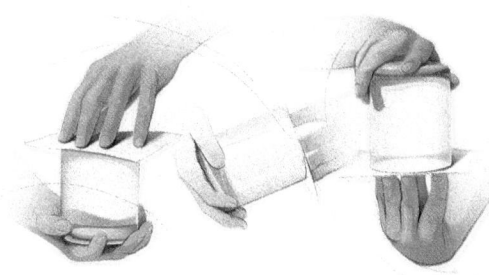

The Curious Cup

Completely fill a glass with water. Place a piece of paper that is slightly larger than the mouth of the glass over the top. Holding the paper in place, turn the glass upside down over a container or sink, and then remove your hand. What do you observe? What can you infer? **E**

Ice Heist

Drop the end of a string into a beaker. Fill the beaker halfway with ice cubes. Sprinkle some salt onto the ice cubes, and wait 30 seconds. Pull on the string. What happens? Make an inference to explain your observations. **F**

Exploration 4 continued ▶

27

Answers to *In-Text Questions*

C Observation: The disk bounces on top of the bottle. **Inference:** The hands warm the bottle and the air inside it. The air expands and, as it escapes, causes the disk to move.

D Observation: As water is poured into the dish, the coin becomes visible. **Inference:** Light entering and leaving the water is bent (refracted), allowing the coin to be seen again.

E Observation: When the glass is turned upside down, the water stays in the glass. **Inference:** Air pressure on the outside of the paper holds it in place and keeps the water in the glass.

F Observation: When the string is pulled up, the ice cubes are attached to it. **Inference:** The salt causes the ice to melt. The salt water refreezes, thus freezing the string to the ice.

Cooperative Learning
EXPLORATION **4**

Group size: 3 to 4 students

Group goal: to perform an experiment and develop an inference to explain it

Positive interdependence: Assign each student a role such as reporter (to record the group's inferences), timer (to time activities and keep the group informed of the time remaining at each station), materials manager (to handle all materials and conduct experiments), or director (to lead group discussions and communicate with the teacher).

Individual accountability: Have each student recount one activity in his or her ScienceLog. Students should explain how the experiment was conducted, what inferences they made, what the correct inference was, and what they learned from performing the experiment.

These activities can be set up at six different stations around the room. The purpose of this Exploration is not to arrive at a correct answer, but to have students make inferences. Accept all thoughtful answers.

SAFETY TIP Issuing "safety tickets" is a great way to enforce safe lab habits. If students break the safety rules, you can revoke their "safety license." Hold their license until the students review all of the safety rules and pass a safety test. If they receive more tickets, conduct a conference with the students' parents or guardians.

WASTE DISPOSAL ALERT The liquids from this Exploration can be poured down the drain, and the solids can be wrapped in old newspaper and put in the trash.

★ **An Exploration Worksheet (Teaching Resources, page 22) and Transparency 5 are available to accompany Exploration 4.**

Homework

Have students complete The Curious Cup and The Reappearing Coin as homework.

Answers to
In-Text Questions

Ⓐ Observation: Bubbles may be seen escaping from the jar. When the flame is extinguished, the water level rises in the jar. **Inference:** The heat from the candle expands the air, and some air escapes, causing bubbles. When the flame is extinguished, the air in the jar cools. The cooling air contracts, allowing water to enter the jar.

Ⓑ Observation: When the gas in the 1 L container is "poured" over the candle, the flame is extinguished. **Inference:** The gas produced in the container (carbon dioxide) is heavier (denser) than air. It flows into the jar, displaces the air, and extinguishes the candle's flame.

Answers to
Follow-Up

1. Student responses will vary; selection of the best inference will depend on class opinion.

2. New questions could include the following: Will the experiment work if the glass is half full? How long will the paper remain on the glass? How large a glass can I use? What would happen if cardboard were used instead of paper? Students should suggest appropriate experiments to find answers to their new questions.

3. Students should agree with the statement. Most students probably made some inferences for the activities in this Exploration that they later learned were untrue.

4. Sample answer: Making inferences helps us to make sense of what we observe. New questions and experiments are often suggested by an inference.

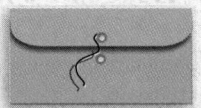

PORTFOLIO
Suggest that students include their completed table from Exploration 4 in their Portfolio. Suggest that they evaluate the quality and accuracy of their observations.

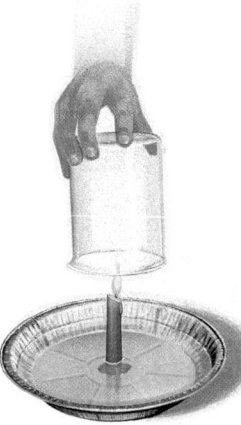

Lights Out! #1

Stand a candle upright in an aluminum pie plate, holding it in place with modeling clay. Fill the plate halfway with water. Light the candle, and then quickly place an empty jar over the flame. What do you observe? What can you infer? **Ⓐ**

Lights Out! #2

This time, place a candle in a large beaker or jar. Hold the candle in place with a piece of modeling clay. Light the candle. Add 30 mL of baking soda to a 1 L container. Pour in 15 mL of vinegar. Wait 1 minute. Hold the container almost horizontally over the candle but do not pour the liquid. What happens? What can you infer? **Ⓑ**

Follow-Up

1. For the six activities you just did, combine your observations and inferences with those of your classmates to make a class chart like the one below. Place a check mark (√) beside the best inference. Do not write in this book.

2. Inferences often raise new questions and suggest more experiments. Consider the Curious Cup activity. What new questions might be raised by it? What experiments might you try in order to answer these questions?

3. Consider this statement: Inferences are not necessarily true. Do you agree or disagree? Use the activities in this Exploration to support your answer.

4. What do you consider to be the value of making inferences?

Activity	Observations made by the class	Inferences made by the class
The Dancing Disk		
The Reappearing Coin		
The Curious Cup		
Ice Heist		
Lights Out! #1		
Lights Out! #2		

Meeting Individual Needs

Learners Having Difficulty

By changing the size of the jar, Lights Out! #1 can be modified to give students further experience in controlling variables. Ask: How can changing a different variable affect the water level in the jar and the amount of time the candle stays lit? *(The larger the jar, the more oxygen there is available, and the longer the candle will stay lit. As oxygen is used up, the water level increases to take up the leftover space.)*

Asking Investigative Questions

Think back to the discussions that you and your classmates had while doing the activities in Exploration 4. Were any of the statements you made or the questions you asked similar to the following?

Set 1

- Why don't we try . . . ?
- What happens if you . . . ?
- Why doesn't it . . . ?
- Perhaps the reason it does that is . . .

Questions and statements like these lead to scientific investigations and discoveries. However, not all questions and statements are investigative. Perhaps you have heard questions or made statements such as the following:

Set 2

- Who cares if . . . ?
- If you say so, I guess it's true.
- Nobody will ever understand that!
- I heard it on TV, so it must be right.
- You know, it just happens that way.

Does this television viewer have a scientific attitude? Explain.

Nine out of ten people surveyed drink new Bop Pop!

Hey, Nina! We'd better buy some Bop Pop!

29

Asking Investigative Questions

GUIDED PRACTICE Choose one of the activities from Exploration 4, and have each student write a question about it. Record some of their responses on the chalkboard. Then have students read the section titled Asking Investigative Questions. When they are finished reading, have students write new questions or statements that are similar to those in Set 2. Record some of these as well, and encourage students to compare the two lists.

Integrating the Sciences

Earth and Physical Sciences

Over 4000 years ago the Babylonians observed that the sun moved from east to west during the day, but the shadows it cast moved in the opposite direction. They used this information to tell time by making the first shadow clock, or sundial. Have interested students make their own sundial and use it to tell time.

CROSS-DISCIPLINARY FOCUS

Mathematics

Ask: In a survey, 148 out of 200 people chose Bop Pop over the leading brand of soft drink. About what percentage of the people prefer Bop Pop?

- **a.** 65%
- **b.** 70%
- **c.** 75% (Correct)
- **d.** 80%

Answer to Caption

The viewer does not have a scientific attitude because he is simply accepting information given to him. If he had a scientific attitude, he would make his own observations and inferences.

Questions and statements like those in Set 1 help when you are searching for answers. They suggest activities such as the following:

- learning more about something
- finding a reason or an explanation
- repeating an experiment
- collecting further observations
- trying something else

Asking questions is part of being scientific. Why . . . ? How . . . ? What if . . . ? When . . . ? Will it ever happen again? Such questions guide scientists as they seek answers to scientific problems. Do you think that the questions and statements in Set 2 would help in your search for answers? Why or why not? **Ⓐ**

Classifying Statements

1. State whether or not each of the following statements or questions indicates that the speaker is being scientific. (Remember that investigative questions or scientific statements ask about things, try to explain things, or suggest other experiments to try.)

  a. That's impossible!

  b. Would you show me that again?

  c. Do you expect me to believe that?

  d. How should I know?

  e. Could we try it a different way?

  f. What would happen if we dropped it from a greater height?

  g. What was supposed to happen?

  h. I don't think it would work with cold water.

2. Write three questions or statements that show an unscientific attitude. Refer to one or more of the activities in Exploration 4 for help with this task.

3. Now write three questions or statements that show a scientific attitude, again referring to Exploration 4.

4. If you attach a weight such as a washer to a string, the washer will swing back and forth on the string like the pendulum of a clock.

You can time how many swings it makes in 10 seconds. What would happen to the number of swings if you added more washers to the string? Try it! Then write an investigative question about this experiment.

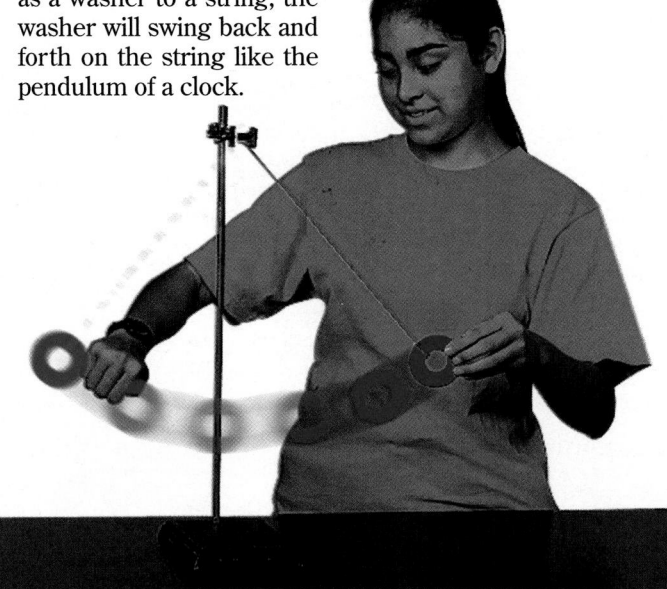

which makes them float until the bubbles escape. Scientific statements might begin with the following phrases: What would happen if . . . or Let's try . . .)

Extension

Have students go outside for 30 minutes and make as many observations about the natural world as they can. From these observations, have them write at least three investigative questions. Ask students to make inferences or to think of possible explanations to answer their investigative questions.

Closure

Select several full-color advertisements from magazines. Display them at the front of the classroom. Then divide the class into two teams. Have each team take turns analyzing the advertisements and writing examples of unscientific questions, attitudes, or inferences based on the advertisements. The team that comes up with the most examples wins the game.

Is There a Gremlin in the Drink Machine?

What would you think if you were asked this question? Ms. Garcia's class was equally puzzled. Let's listen in on their interpretation of how a drink machine works. Turn the page to see the class's explanation in action.

> This is what I think happens when you drop your money into the machine. A gremlin grabs the money, counts it, and then drops it into a tin box. If there is not enough money, the gremlin keeps it. If there is too much, the gremlin makes change and then pushes the change through a slot in the wall of the machine. There are a number of paths the gremlin can follow, each one blocked by a gate. When you make your selection, a gate is lifted, and the gremlin scurries along the path, grabs a can, and throws it down the chute. And what happens when the gremlin falls asleep on the job? A sign automatically appears on the outside of the machine saying "OUT OF ORDER," and no more money will enter the gremlin's tin box.

The Gremlin Model

The explanation above takes into account many observations and inferences about the operation of a drink machine. In science, such an explanation is called a model. A **model** is a picture or representation of the real thing, and it is supported by observations and inferences.

1. Make a list of the observations and inferences that support the gremlin model, using a table with the headings "Observation" and "Inference (suggested by observation)." Then suggest an observation that does not support the gremlin model.

2. Draw or write a description of your model of a drink machine. Be sure that *your* model is supported by observations.

31

LESSON 3 ORGANIZER

Time Required
1 class period

Process Skills
observing, inferring

Theme Connection
Systems

New Term
Model—a picture or representation of a real object, supported by observations and inferences; often used to help visualize something that cannot be directly observed

Materials (per student group)
none

Teaching Resources

FOCUS

Getting Started

Set up a display of models, such as models of the solar system, the Earth, and organs of the body. Ask students why models are useful. (*Models allow us to visualize something that we cannot see, and they help us to understand how things are structured and how their different parts work together.*)

Main Idea

Models are useful because they help us to understand and investigate real objects.

TEACHING STRATEGIES

GUIDED PRACTICE   Read Ms. Garcia's model to the class. Ask: Do you think this is a good explanation for what occurs inside a drink machine? Have students list the inferences made and the observations that must have led to these inferences.

Theme Connection

Systems
Focus question: What similarities exist between the model of the soda machine and a model of the solar system? (*Both models give information about systems that are not readily visible. For example, a model of the solar system might show the number and relative size of the planets, the relative distance between planets, or how the planets revolve around the sun.*)

Answers to The Gremlin Model are on the next page. ▶

Homework

The Gremlin Model makes an excellent homework activity.

Reteaching

Have students construct a model of the sun and the Earth that illustrates why there is night and day. Have balls and flashlights available for students to use.

Assessment

Have students scientifically describe to Zed how a common object, such as a stapler or pencil sharpener, works. Divide the class into groups, and allow each group to devise and present its own explanation. Then hold a class vote to determine the best explanation.

Extension

Have students speculate about how a photocopier works and write a description of a possible model. Then have students research how an actual photocopier works and compare their model with the actual operation of the machine.

Closure

Ask students to explain why models are changed over time. (Models are changed to reflect the acquisition of new information and new understanding.)

Answers to
The Gremlin Model, pages 31–32

1. For a sample table, see below. Student observations that do not support the gremlin model include the following:
   - No sound of scurrying feet can be heard.
   - No one is ever observed feeding the gremlin.
   - When door of machine is open, no gremlin is seen.

2. Accept all reasonable responses.

3. a. Those who believed in a flat Earth probably based their belief on their observation that the Earth appears flat when they stand on its surface. They might have reasoned that if the Earth were not flat, you would be upside down if you walked far enough.
   b. Student answers will vary. Students might point out how a ship slowly disappears over the horizon, or they might display a photograph of the Earth taken from space.

3. a. Models help us make sense of what we observe. But sometimes a model may prove to be incorrect as more observations are made. For example, at the time of Columbus, some people thought that the Earth was flat. This was their model of the Earth. What observations do you think supported this model?

   b. There are people today who still claim that the Earth is flat. They belong to the Flat Earth Society. Would you join this organization? What would you say to convince a member of the Flat Earth Society that the Earth is really spherical?

32

Observations	Inference (suggested by observation)
Money is put into the machine.	The gremlin takes the money.
Change is returned.	The gremlin made change.
The sound of coins on metal is heard.	The gremlin dropped the money in a tin box.
You make a selection.	The gate is lifted.
A few seconds elapse.	The gremlin is scurrying up the path.
The sound of a rolling can is heard.	The gremlin threw the drink down the chute.
You receive your drink, or an out-of-order sign appears.	The gremlin is ready for more money or is asleep.

A Tale of Two Kittens

Jane and Larry Quinn loved cats. Jane and Larry even talked to them as though they were people. Jane and Larry also experimented with their cats—not in a way that would harm them, of course. They wanted them to be as healthy as possible. They reasoned that if people often need extra vitamins, then cats might also. Therefore, they predicted that added vitamins would improve their cats' health. To test this **prediction**, they created their own mixture of cat food with a special vitamin supplement.

1
People need extra vitamins sometimes, and I bet cats sometimes need extra vitamins, just like people. I predict our cats will be healthier if we give them extra vitamins.

To test their prediction, the Quinns obtained a special vitamin supplement that they mixed with the ordinary cat food their cats ate.

Let's mix vitamins with the cat food. They'll love it!

2
Look at how big and healthy our cats are, Wu Feng. This vitamin supplement really works.

But how do you know? Your cats might have grown just as well with ordinary cat food.

Wait! Your results prove nothing! Even if you hadn't used your supplement, the Persian would have been larger; Siamese cats don't grow as large as other cats.

Oh no! He's right! We can't tell how the cats might have grown without the supplement.

3
The Quinns had just received two new kittens. They fed the special food to the Persian, and they fed the food with no extra vitamins to the Siamese. When they showed the cats to Wu Feng 3 months later, he seemed convinced, but only for a moment.

LESSON 4 ORGANIZER

Time Required
2 to 3 class periods

Process Skills
hypothesizing, observing, predicting

New Terms
Controlled experiment—an experiment in which all variables but one are kept constant; also called a fair test
Hypothesis—a prediction that links cause and effect and that can be tested by an experiment

Prediction—the expected outcome of a future event. Predictions can be based on observations, experiences, or scientific reasoning.
Variable—a factor that can affect the outcome of an experiment

Materials (per student group)
none

Teaching Resource
SourceBook, p. S16

FOCUS

Getting Started
Tell students you are going to perform a scientific test. Select two different brands of paper towel. Use a whole sheet of the first brand to absorb 100 mL of water from a bowl. Use half of a sheet of the second brand to absorb 100 mL of corn oil or syrup from another bowl. Show students the results. Tell them that you conclude that the first brand of paper towel is the most absorbent. Ask students if they agree with you. Guide the class in a discussion of why the test was unfair and how they might improve it.

Main Ideas
1. A fair test is a controlled experiment in which all variables except the one being investigated are controlled (or kept constant).
2. Variables are factors in an experiment that can influence the outcome of the experiment.
3. A hypothesis is a possible explanation for an observable event. It can be tested by simple observation or by a controlled experiment.

TEACHING STRATEGIES

A Tale of Two Kittens
This story is about two students who are being scientific by designing an experiment to prove a point. It provides an excellent opportunity to look again at the nature of science and to introduce the concepts of controlled experiments, hypotheses, and variables.

SAFETY TIP Take this opportunity to remind students that A Tale of Two Kittens is for discussion purposes only and that students should not experiment with live animals without your approval and supervision.

1. The Quinns share the following traits with scientists:
   - They are observant.
   - They are curious.
   - They experiment.
   - They collect and record data.
   - They try not to jump to conclusions.
   - They base their conclusions on a controlled experiment.
   - They have a scientific attitude.

2. Initially, they jumped to conclusions. They expected the vitamin supplement to improve their cats' health. Since their cats were healthy, they inferred that the supplement was responsible. In their first experiment, they did not control all of the variables. They failed to compare cats who received no vitamins with those who did. In their second attempt, they improved by comparing one cat who received the supplement with one cat who did not receive the supplement. However, they still had more than one variable that changed because they compared two different kinds of cats. Also, they did not monitor the amount of food that the two cats received.

3. Wu Feng was not convinced that the vitamin supplement worked. There is no way to measure the effects of the supplement on the two cats because Persian cats naturally grow larger than Siamese cats. The Quinns had still not controlled all of the variables.

4. The Quinns did control the variables of age and amount of food and water. The variables that were not controlled include the type of cat and the amount of exercise.

5. In the third experiment, the controlled variables were the kind of cat, the amount of food and water, the living conditions, the amount of exercise, the fur color, and the age of the cats. The variable that changed was the vitamin supplement, which was given to only one of the cats. The Quinns' second experiment was not a controlled one because they allowed more than one variable to affect the outcome of the experiment.

Larry and Jane stuck to their procedure and recorded the mass of each kitten every week. Jane even took pictures to record the experiment.

The Moral of the Tale

1. How did the Quinns demonstrate the traits of a good scientist?

2. What mistake did they make at first (frames 1 and 2)? How did they improve on their experiment (frame 3)?

3. In frame 3, why was Wu Feng not convinced that the vitamin supplement worked?

4. In the Quinns' second experiment (frame 3), a number of factors were kept the same, or controlled, in order to keep the experiment fair and reliable. Such factors are called **variables**. Which variables did the Quinns control? What are two variables that were not controlled?

5. In frames 5 and 6, the Quinns described an experiment in which only one variable was changed. They constructed a **controlled experiment**. What are the variables that were controlled? Which variable was changed? Why wasn't their second experiment (frame 3) a controlled experiment?

6. One part of a controlled experiment is called a *control*. Setting up a control makes it possible to check the results of the experiment against a standard. What part of the experiment in frames 5 and 6 was their control?

6. The control was the cat that did not receive the vitamin supplement in its food. By starting the experiment with kittens of the same mass and weighing both kittens every week, the weight of the kitten being fed the supplement could be compared with the weight of the kitten not being fed the supplement.

Homework

You may wish to have students complete questions 1–4 of The Moral of the Tale as homework.

Hypothesis—A Special Kind of Prediction

On page 33, you read how the Quinns predicted that vitamins would improve their cats' health. They tested this prediction by devising an experiment. A prediction that can be tested by an experiment is called a **hypothesis**. In most hypotheses, you can identify the following:

- the variable being changed in the experiment. This is what the hypothesis predicts as the *cause* for any differences observed during the experiment.
- the expected outcome of the experiment. This is what the hypothesis predicts will be the *effect* of changing a variable.

What were the cause and the effect in the Quinns' experiment? Ⓐ

A Link Between Cause and Effect

One day, Mr. Kumar gave his class the following task:

1. Write an investigative question.
2. Suggest a hypothesis for the question.
3. List the variables that should be controlled in order to test the hypothesis.

Elizabeth's group decided to use the following as their investigative question: What affects plant growth?

Now listen in as they decide on a hypothesis.

Rubin: Water, sunlight, and fertilizer are needed by plants.

Susan: Well, my brother's plants get plenty of all of those things, but they still don't do very well. I think the plants don't like that awful music he plays.

Duncan: Maybe plants grow better when listening to rock music. That's what I like listening to.

Elizabeth: So our hypothesis could be that plants exposed to rock music grow better than those not exposed to rock music.

Identify the cause and effect in the hypothesis suggested by Elizabeth. Ⓑ

35

Hypothesis—A Special Kind of Prediction

After discussing the task that Mr. Kumar gave to his class, students should be able to identify the cause and effect in Elizabeth's hypothesis. It may be helpful to point out that a hypothesis is not the same thing as a fact; a hypothesis is a statement that must be tested.

Answers to *In-Text Questions*

Ⓐ **Cause:** One kitten receives a vitamin supplement. **Effect:** The kitten is healthy.

Ⓑ **Cause:** The plants are exposed to rock music. **Effect:** The plants grow better.

Homework

Propose the following scenario to students: You hear a friend say that aspirin will make plants grow better. You wish to do a controlled experiment to find out if this is true. Describe your experiment, being sure to state your hypothesis, and list the variables you would control.

ENVIRONMENTAL FOCUS

Tell students to imagine that they have been hired by their community to investigate a puzzling problem. The fish have been dying in the town's lake. Residents suspect that dangerous pesticides from a nearby farm have seeped into the lake, causing the fish to die. Challenge students to come up with a fair test to determine if the residents are correct.

Answer to
In-Text Question

Ⓐ Other variables that should be controlled include the type of plant and the amount of water each plant receives. The control in their experiment will be the plant grown with no exposure to rock music.

Answers to
Now Answer These

1. Sample hypothesis: If I increase the height from which the washer is dropped, it will not increase the number of swings the pendulum will make in 10 seconds.

2. All factors that can affect the outcome of the experiment need to be controlled. For the hypothesis above, the number of washers and the time the pendulum is allowed to swing should be controlled.

3. a. Yes. Cause: Different brands of soap are used. Effect: The amount of the stains removed differs.

b. Possible responses: Do all soap powders remove stains equally well? What affects the ability to remove stains?

c. The variables to control include the temperature of water, the kind of stain, the size of the stain, the amount of water, how much the cloth is shaken while in the water, and the amount of soap powder used.

4. Student answers will vary; make sure that all variables except for the one being tested are controlled.

After some discussion, the group came up with the following ways to control the variables:

- Start with the same size of plant.
- Use the same type of soil.
- Let each plant receive the same amount of sunlight.

Are there other variables that should be controlled? What are they? What will the control be in the experiment? Ⓐ

Now Answer These

1. On page 30, you wrote an investigative question about pendulums. Now write a hypothesis for this question.

2. For your hypothesis, what variable(s) should be controlled in order to make this a controlled experiment?

3. Jenna's group, below, suggested this hypothesis for another investigative question: Brand X soap powder removes stains better than Brand Y does.

a. Does the hypothesis contain both a *cause* and an *effect*?

b. What could have been their investigative question?

c. What variable(s) would you control in order to test their hypothesis?

4. Now it's your turn. In small groups, suggest the following:

a. an investigative question

b. a possible hypothesis

c. variables that should be controlled

36

FOLLOW-UP

Reteaching

Using poster board and red, blue, and yellow watercolor paints, have students create a hypothesis that answers the question, How can I make the colors green, purple, and black? Have students identify the cause and effect in their hypothesis. Then allow them to mix the paints and see if their hypothesis is correct.

Assessment

Share the following with students: Michael thought that the fizziness of carbonated lemonade depended on the

amount of sugar. Identify the cause and effect in his hypothesis. *(Cause: The amount of sugar varies. Effect: More sugar provides more fizziness.)* In testing his hypothesis, what should he change? *(The amount of sugar)* What should he keep the same? *(The amount and temperature of the water, the number of lemons, and the amount of carbonated water)*

Extension

Have students think of a practical problem they would like to solve, such as finding out what dishwashing soap

cuts grease the best. Have them conduct a controlled experiment to solve the problem. Check students' proposed procedures before allowing them to proceed.

Closure

Bring to class a small bottle of glycerin, a bottle of dishwashing soap, several plastic bowls, and bubble wands. Have students discover the best recipe for a bubble solution. Then have students list the investigative questions and hypotheses they used in discovering the best bubble solution.

CHALLENGE YOUR THINKING

1. Can I Have the Recipe?

Read the two recipes below for making a blueberry pie, and then answer the following questions:

a. If you follow Grandma's recipe, will your pie taste the same as hers? Explain.

b. Are Grandma's recipe and the cookbook recipe qualitative or quantitative? Explain.

Grandma's Recipe
Simply mix three handfuls of berries with two handfuls of sugar, a handful of flour, and a pinch of salt.

Cookbook Recipe
Line a 23 cm pie plate with pie crust.
Combine in a mixing bowl:
200 mL sugar
60 mL flour
6 mL tapioca
25 mL lemon juice
Stir mixture gently into 950 mL fresh blueberries.
Pour mixture into pie crust, and dot with 15 mL butter.

2. Who Am I?

On an index card, make a table with the following headings: "Quantitative traits" and "Qualitative traits." Use this table to describe yourself. Choose one person to read the cards (but not the names) to the class so that other students can try to guess who is being described. The best description wins!

3. Thinking Like a Scientist

On page 5, the following statement was made: "This unit is about how you become a scientist each and every day." Do you think this is true? Support your position with examples.

37

Meeting Individual Needs

Gifted Learners

Have students research the models of the solar system that were accepted before and after the time of Copernicus. Students could make models that compare the ideas of Ptolemy and Copernicus. These models can then be displayed for the rest of the class to enjoy.

Meeting Individual Needs

Second-Language Learners

To help students understand experimental setups, have them review the experimental setups in this unit that they were asked to design themselves and then sketch all of the variables in each. Students should draw and label the controlled variables as well as those that are changed. Have students label their drawings with words that you have written on the chalkboard for them.

Answers to *Challenge Your Thinking*

1. a. It will probably not taste the same because her measurements are not precise and are therefore difficult to duplicate exactly. For example, her "handful" and someone else's would probably be different.

b. Both are quantitative, but Grandma's measurements are based on one of her handfuls, rather than on a known unit of measure. To make her measurements more accurate, you could measure the amount her hand can hold in metric units. Then it is more likely that her results could be duplicated.

2. Answers will vary. If students have difficulty deciding who is being described, discuss with the class what is lacking in the traits given. For example, qualitative traits often cannot be measured and therefore may vary according to who is making the observation.

3. Answers will vary. Most students can probably give examples of how they act like scientists in their daily lives by observing, inferring, measuring, devising experiments, and gaining new knowledge.

Answers to Challenge Your Thinking continued ►

Homework

Have students visit a bedroom, bathroom, and kitchen in their home. In each room, have them make an observation and then propose an inference that explains the observation.

Answers to
Challenge Your Thinking,
continued

4. Articles based on scientific surveys or statistical studies of large populations are the best articles for this exercise. Students should be able to put into their own words the inferences contained in the article and to explain why the statement is an inference as opposed to an observation or a conclusion.

5. a. Students should observe that before the detergent is added, the color stays in little globs. As the detergent is added, the globs of color shoot out toward the edge of the pie plate. The colors swirl up and mix with the milk, especially where the detergent was added.

b. Sample inference: The dishwashing detergent somehow causes the food coloring to move and mix with the milk.

c. Sample question: Does the same thing happen when water is used instead of milk?

d. Sample hypothesis: If water is used instead of milk, the same thing will happen because the food coloring and dishwashing detergent will produce the same results.

6. A prediction states the expected outcome of a future event, while a hypothesis provides an explanation for the outcome by linking a cause and an effect in a statement that can be tested. Sample prediction: This plant will grow more quickly than that one. Sample hypothesis: Plants given fertilizer grow more quickly than those not given fertilizer.

 You may wish to provide students with the Chapter 2 Review Worksheet that accompanies this Challenge Your Thinking (Teaching Resources, page 27).

---

4. Current Events

Choose a newspaper article that has a scientific focus. Underline statements in the article that you think are inferences. Find out if a classmate agrees with you.

5. On Your Own

Try this experiment at home:

- Fill a pie plate halfway with milk.
- Add several drops of different colors of food coloring around the edge of the pie plate.
- Add several drops of dishwashing detergent to the center of the pie plate.
- Observe for 3 minutes.

a. Record as many observations as you can.

b. Suggest several inferences about what you observed.

c. Suggest one investigative question that would help you to discover more about the phenomenon.

d. Suggest a hypothesis based on your question.

6. How Predictable!

In your own words, explain the difference between a prediction and a hypothesis. Use examples if necessary.

ScienceLog

Review your responses to the ScienceLog questions on page 18. Then revise your original ideas so that they reflect what you've learned.

The following are sample revised answers:

1. Answers will vary. Students should be able to support their inferences and predictions based solely on the details shown in the photograph. Observations should not be confused with inferences or conclusions. Encourage students to ask investigative questions based on their observations.

2. A controlled experiment is one in which only one variable—the one being tested—changes. A controlled experiment is a way to test a hypothesis. There are usually many variables, or factors, that can affect the outcome of an experiment. By keeping constant all variables except the one being tested, scientists can more easily judge the validity of a hypothesis.

3. The details included in each student's answer should be both quantitative and qualitative. Ideal answers would record the relationships between the objects in the picture so that someone who has never seen the picture could recreate it.

CHAPTER 3

From Science to Technology

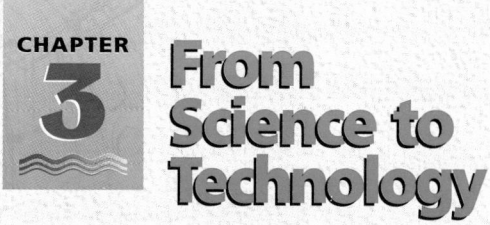

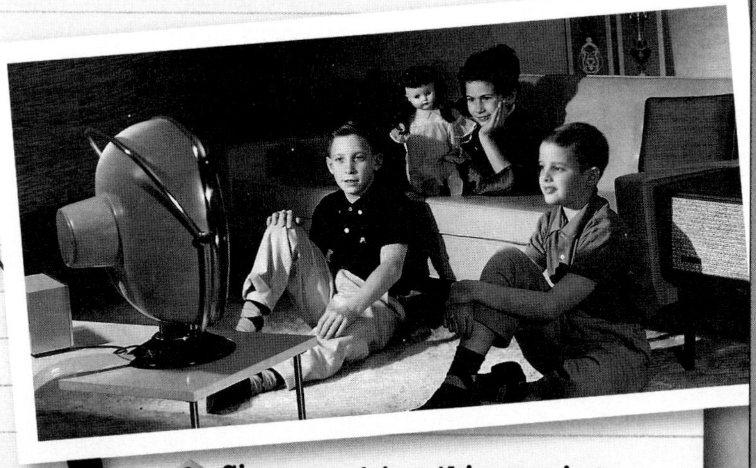

1 **What is technology? How does it differ from science?**

2 **Since you got up this morning, how has technology made your life different from that of your parents or your grandparents when they were your age?**

3 **Do you think science contributed to the development of these products? Explain.**

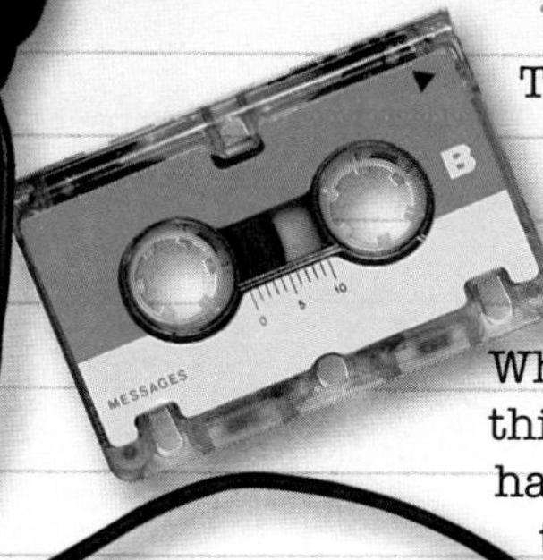

ScienceLog

Think about these questions for a moment, and answer them in your ScienceLog. When you've finished this chapter, you'll have the opportunity to revise your answers based on what you've learned.

39

★ **A Discrepant Event Worksheet is available as a teacher demonstration to accompany the material on this page (Teaching Resources, page 33).**

CHAPTER 3

From Science to Technology

Connecting to Other Chapters

Chapter 1 ***explores the nature of science and what it means to be a scientist.***

Chapter 2 ***invites students to practice scientific thinking, including observing, inferring, and hypothesizing.***

Chapter 3 ***guides students through the process of experimentation and discusses the nature of technology.***

Prior Knowledge and Misconceptions

Your students' responses to the ScienceLog questions on this page will reveal the kind of information—and misinformation—they bring to this chapter. Use what you find out about your students' knowledge to choose which chapter concepts and activities to emphasize in your teaching.

In addition to having students answer the questions on this page, you may wish to do the following: Show students a hand-held video game or tape player. Tell them to imagine that they are the inventor of the device. They should write their story, including how they got the idea for the device, what they had to learn in order to design it, and the difficulties they had when creating the final product. Emphasize that there are no right or wrong answers in this exercise. Collect the papers, but do not grade them. Instead, read them to find out what students know about technology, what misconceptions they may have, and what is interesting to them about technology.

LESSON 1

From Hypothesis to Experiment

FOCUS

Getting Started

Present students with three beakers—one filled with 50 mL of white vinegar, one with 50 mL of water, and one with 50 mL of salt water. Tell students that the beakers contain three different substances. Have students discuss different ways they might test the substances to find out what each one is. Remind students that touching and tasting unknown substances are not safe laboratory procedures. Record their comments on the chalkboard, and help students form hypotheses from their suggestions. Explain that in this lesson, they will have a chance to test different hypotheses through experiments.

Main Ideas

1. Being scientific involves performing experiments in order to answer questions and test hypotheses.
2. The relevant variables within an experiment (except the one being investigated) must be controlled in order for the experiment to be considered a fair test.

TEACHING STRATEGIES

Ask students to read the opening paragraphs of this lesson. After they have finished reading, ask: What possible hypothesis might the students have come up with that they could test? *(Answers will vary. Sample hypothesis: The surface area of the paper affects how fast the paper falls through the air.)*

LESSON 1

From Hypothesis to Experiment

Being scientific involves observing, measuring, inferring, hypothesizing, classifying, and more! Being scientific involves doing experiments in order to answer questions and test hypotheses. Throughout this book, you will be asked to be scientific by designing experiments. Here's the place to start. Join Alexander's group as they discuss the following problem: What determines how long it takes for paper to fall through air?

The group decides to try Ralph's hypothesis and test the effect of surface area on how fast paper falls through the air.

In groups of three, examine the plan on the next page that was made by this group of students. Decide whether it is a good plan. Is the test a fair one—in other words, is it a controlled experiment? What variables are being controlled? Are the results reliable? Can they be repeated?

LESSON 1 ORGANIZER

Time Required
5 to 6 class periods

Process Skills
observing, hypothesizing, measuring, analyzing

Theme Connections
Structures, Changes Over Time

New Terms
none

Materials (per student group)
Exploration 1: paper; 1–2 paper clips; scissors; metric ruler
Exploration 2, Experiment 1: meter stick; **Experiment 2:** meter stick; small ball of modeling clay

Teaching Resources
Exploration Worksheets, pp. 34 and 37
Graphing Practice Worksheet, p. 40
Transparency 6

Investigating Paper Falling Through Air

Hypothesis

Paper with a large surface area falls more slowly through air than paper with a small surface area.

Our Plan

For the test to be fair, we made sure that all pieces of paper used had the same mass. To do this, we folded one sheet of paper in half and secured it with a touch of glue. A second, identical piece of paper was folded twice. This paper had one-quarter of the surface area of an unfolded piece of paper. A third piece of paper was not folded at all. Altogether, we used three pieces of paper that had the same mass but different surface areas. We measured the time it took for each paper to fall a distance of 1.5 m.

1. Which variable mentioned in the hypothesis is changed in the experiment?

2. What variables are controlled? **B**

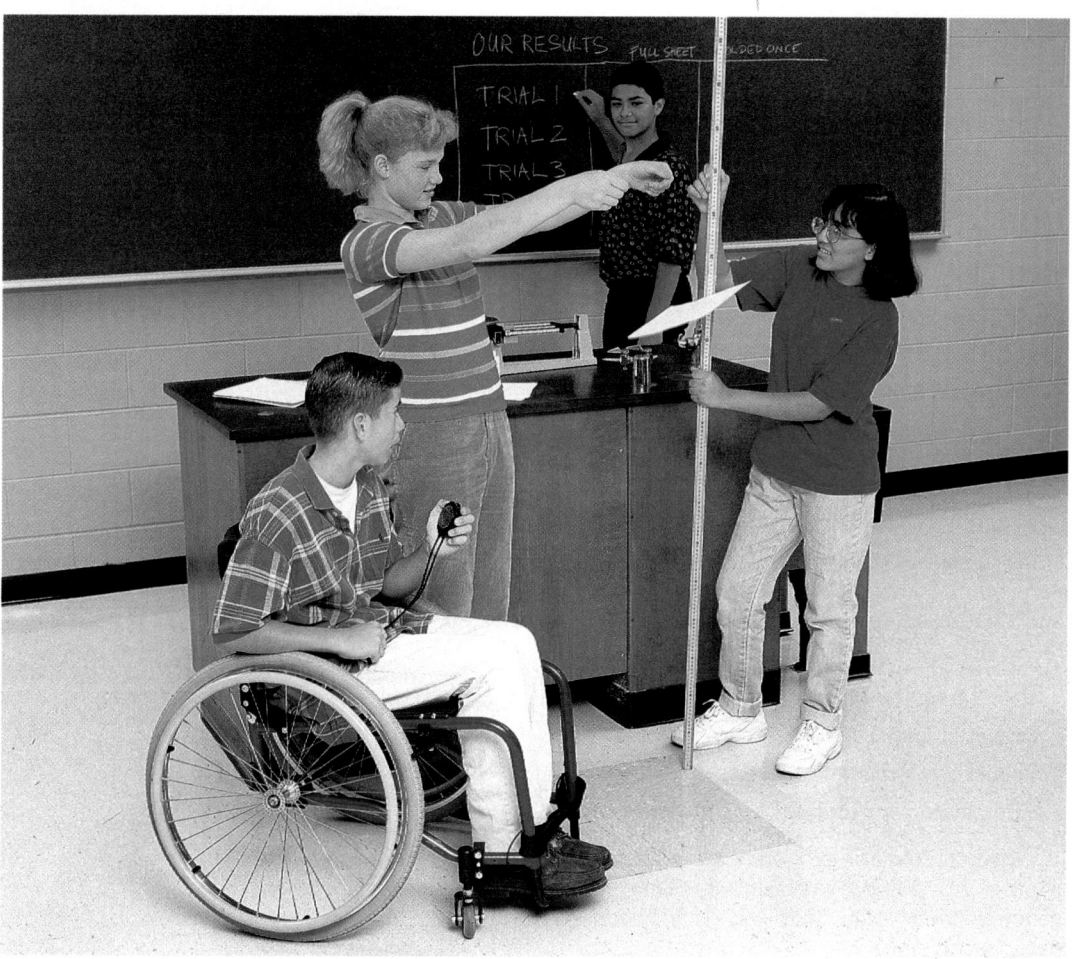

This reading presents an actual experimental design so that students can observe how a hypothesis, a plan, a data table, a graph, and conclusions are used. Students should use this as a model to follow when designing their own experiments. This lesson is set up as a discussion task. Ask students to discuss the hypothesis, the experimental plan (including variables tested, variables controlled, and measurements made), the results, and the conclusions.

GUIDED PRACTICE   Ask students: What are the measurements that the students must make? (*Mass of the paper, surface area of the paper, time it takes the paper to fall, and the distance of 1.5 m*)

INDEPENDENT PRACTICE   Allow time for students to share the results from their own experiments. Encourage discussion of the two questions found at the bottom of page 42. (*Answers will vary.*)

⭐ **Transparency 6 is available to accompany Investigating Paper Falling Through Air.**

Answers to
In-Text Questions

A The surface area of the paper is the variable mentioned in the hypothesis that is changed.

B The controlled variables are the mass of the paper and the distance that the paper falls.

Homework

You may wish to assign Investigating Paper Falling Through Air as homework.

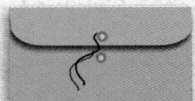

Ⓐ Doing more than one trial is good scientific technique. It verifies results, shows that results can be repeated, and allows results to be averaged to obtain better answers. The fact that the results can be repeated is illustrated by the four trials.

Ⓑ A graph is a "picture" of the results. It is easier to see relationships in a graph than in a data table, and sometimes a graph can be used to make predictions for surface areas not measured in the experiment.

Ⓒ Yes. The data support the conclusion. The greater the surface area of the paper, the slower the paper falls through the air. The inference is that the paper with more surface area encountered more air resistance.

PORTFOLIO

Prepare a class data chart using all of the students' results from the experiments that they designed. Suggest that students record both the class data and their own results in their Portfolio. Students can then evaluate how their own results compare with the class results.

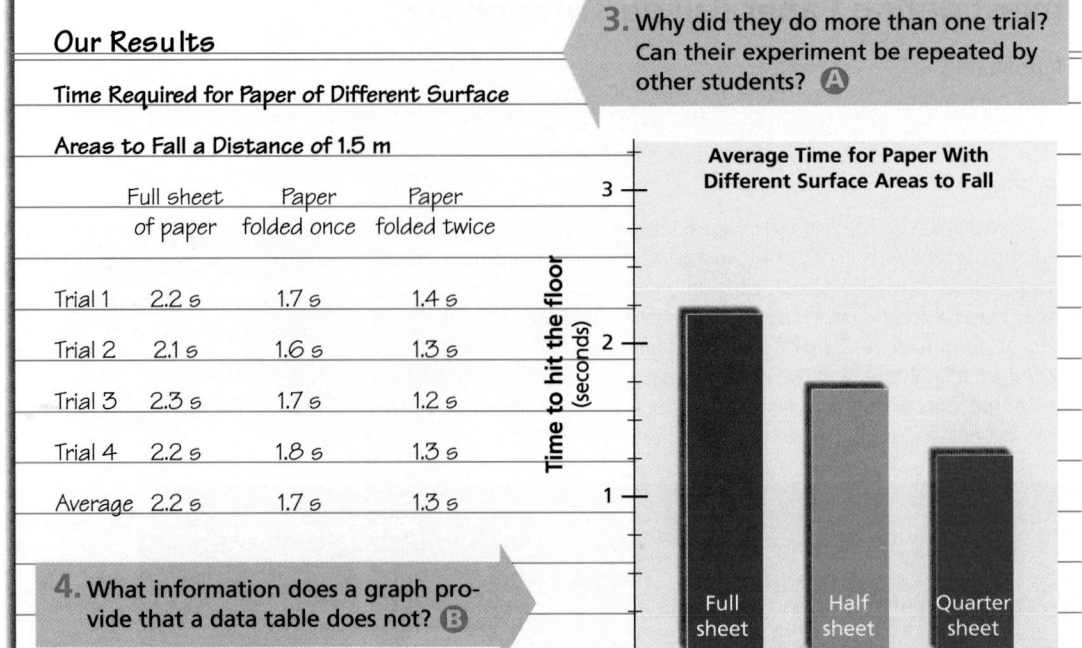

3. Why did they do more than one trial? Can their experiment be repeated by other students? **Ⓐ**

Our Results

Time Required for Paper of Different Surface Areas to Fall a Distance of 1.5 m

	Full sheet of paper	Paper folded once	Paper folded twice
Trial 1	2.2 s	1.7 s	1.4 s
Trial 2	2.1 s	1.6 s	1.3 s
Trial 3	2.3 s	1.7 s	1.2 s
Trial 4	2.2 s	1.8 s	1.3 s
Average	2.2 s	1.7 s	1.3 s

4. What information does a graph provide that a data table does not? **Ⓑ**

Average Time for Paper With Different Surface Areas to Fall

(Bar graph: y-axis "Time to hit the floor (seconds)" from 1 to 3; x-axis "Surface area of paper" with bars for Full sheet, Half sheet, Quarter sheet)

Conclusion

Our data show that our hypothesis is correct. Perhaps the object with greater surface area has more air resistance.

5. Do their data support their conclusion? What inference was made? **Ⓒ**

Now it's your turn to design an experiment to investigate this problem. Think back to the discussion among the members of Alexander's group. They mentioned several variables that could affect the speed of paper falling through air. Surface area was the variable that Alexander's group investigated. Divide into groups, and choose another variable to investigate. In doing so, follow these steps:

- State a hypothesis.
- Design an experiment.
- Collect observations and data.
- Summarize your findings by drawing a conclusion.

After the experiment, compare your results with those of another group that investigated the same variable. Did each group conduct a fair test? Did each group arrive at the same conclusion?

If you're still not sure about the steps that scientists take in studying a problem, turn to page S12 in the SourceBook to read more about it.

42

Investigating Objects Falling Through Air

Examine the illustration below. If the hammer and the feather were dropped at the same time from the same height, which would hit the floor first? Write a hypothesis to explain what is being tested. What variables are being controlled? If this experiment were performed in a vacuum, there would be a surprising result. Do you know what that result would be? **D**

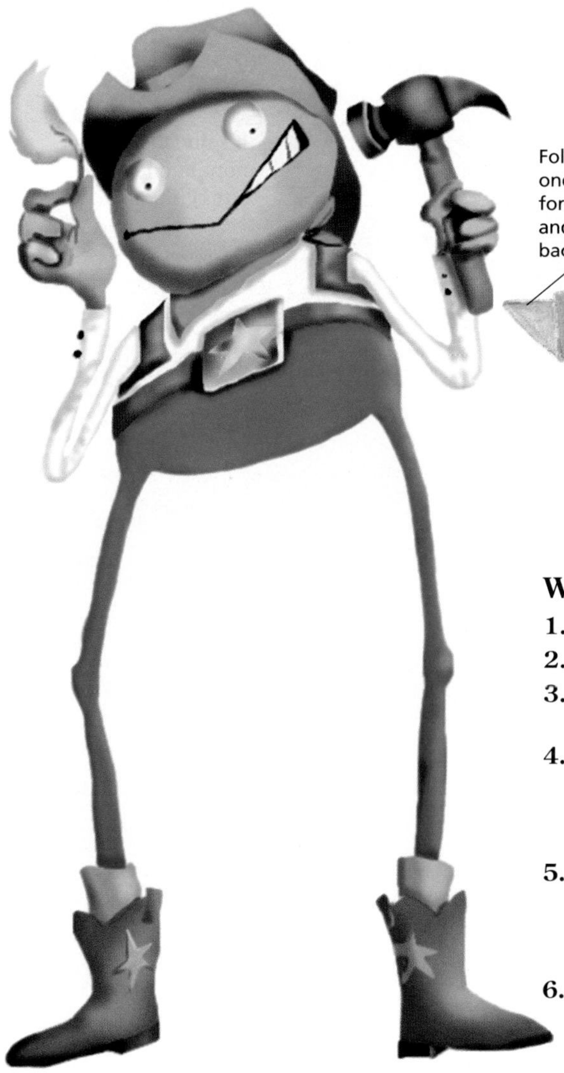

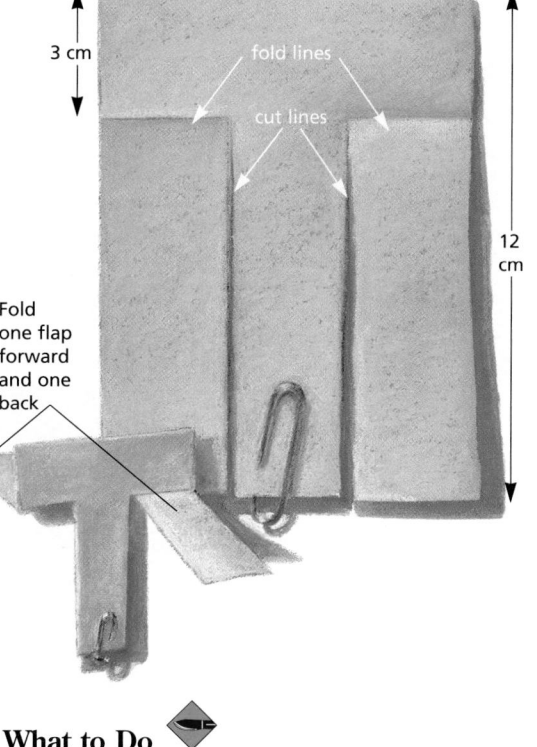

Air and the "Paper Thing"

A "paper thing" (P.T.) is an interesting tool for investigating how certain types of objects fall through air.

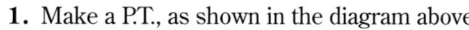

— 9 cm —

3 cm

fold lines

cut lines

12 cm

Fold one flap forward and one back

What to Do

1. Make a P.T., as shown in the diagram above.
2. Drop your P.T., and observe its motion.
3. List all of the variables you can think of that could affect its motion. **E**
4. On a piece of paper, write one hypothesis about how you could change the way in which the P.T. falls. Exchange your hypothesis with another group. **F**
5. Design and then try an experiment to gather evidence that either supports or contradicts the hypothesis that you received from the other group. **G**
6. Share your plan and the results with the group that gave you the hypothesis.

Exploration 1 continued ▶

43

If time permits, you may wish to have each group test an original hypothesis in addition to the hypothesis received from another group. This would enable students to test a hypothesis of particular interest to them, such as one that they cannot predict to be true or false.

Cooperative Learning
EXPLORATION 1

Group size: 4 to 5 students
Group goal: to make observations in order to support or to disprove a hypothesis
Positive interdependence: Assign each student a role such as materials coordinator (to gather necessary supplies and equipment), craftsperson (to prepare materials), timer, recorder, or focuser (to run the experiment and keep the group focused and on track).
Individual accountability: Each student in the group should be able to state whether he or she agrees with the hypothesis and why. Students could express their statement in their ScienceLog or Portfolio.

★ An Exploration Worksheet is available to accompany Exploration 1 (Teaching Resources, page 34). This worksheet provides a cutout of the P.T.

Answers to
In-Text Questions

D If this experiment were performed in a vacuum, both objects would reach the bottom at the same time because there is no air resistance in a vacuum.

E Some variables that could affect its motion are the height from which it is dropped, room temperature, size of paper clip, and type of paper.

F Sample hypothesis: A larger paper clip would make the P.T. fall faster.

G Experimental designs should test only the variable mentioned in the hypothesis.

Exploration 1 continued ▶

Answers to
Investigating Objects Falling Through Air

The hammer will probably win the race because it will overcome air resistance better than the feather. Hypothesis being tested: Heavier objects fall through the air faster than lighter objects. The variables being controlled are the angle and the height from which the objects are dropped.

Answers to
Making a Better P.T.

To make a better P.T., students might try changing the type of paper used to make the P.T. or the size or shape of the P.T., or students may create an entirely new design. A smaller P.T. will fall faster and spin faster. Putting a larger paper clip on the P.T. will also make it fall and spin faster. A larger P.T. may simply drop to the floor without spinning.

Making a Better P.T.

What makes a better P.T.?

In making a better P.T., would you make the same one for each of these definitions of "better"?

What is your definition of a better P.T.? Make your better P.T., and test it against those of your classmates. Who has the "best" P.T.?

Theme Connection

Structures
Show students the spinning fruits of a maple tree. **Focus question:** What structural similarities do you see between the maple fruits and the P.T.? *(Sample answer: Both are winged and spin around a central weight.)* Encourage students to build a model of a maple fruit.

PORTFOLIO
Students may wish to record their P.T. design or the designs of others in their Portfolio. Encourage them to produce accurate diagrams to compare and contrast the P.T. designs. Students can then evaluate how changes in the design affected the performance of each P.T.

Meter Stick Experiments

EXPERIMENT 1

In the first part of this experiment, you are going to compare the reaction times of the people in your group. What factors might have an effect on reaction time? List them in your ScienceLog.

What to Do

Person *A* holds the meter stick between the fingers of person *B*. Person *A* drops the meter stick without warning, and person *B* catches it.

The distance that the meter stick falls before person *B* catches it will be a measure of that person's reaction time.

In groups of four or five, devise rules that should be followed in order to make this a fair test. These rules should include your controlled variables. Test the reaction times of the members in your group using your set of rules. **B**

Exchange your rules with another group. Try their rules. Do you like their rules better? Answer this question by writing down how you would improve their set of rules.

Use your new, revised set of rules to test one or more of the following hypotheses:

- Reaction time is slower when the person is seated.
- Girls have faster reaction times than do boys.
- Right-handed people have faster reaction times than do left-handed people.

Be sure to record data to support the conclusions you will be making.

Exploration 2 continued ▶

45

INDEPENDENT PRACTICE Students may wish to test some other hypotheses, such as the following:

- Reaction time is better in the morning than in the late afternoon.
- Reaction time is better after vigorous exercise.
- Reaction time varies with age.
- Reaction time varies between the right hand and the left hand.

★ An Exploration Worksheet is available to accompany Exploration 2 (Teaching Resources, page 37).

Cooperative Learning
EXPERIMENT 1

Group size: 4 to 5 students
Group goal: to develop a hypothesis to explain reaction times
Positive interdependence: Assign each member of the group a different role such as recorder, timer, investigator, editor, and orator. Students should present their findings to the rest of the class.
Individual accountability: Each student should prepare a data sheet that shows the group's results. Students should also list any conclusions that were drawn from the results of the investigation. Have groups record their data in the form of a short report.

Answers to In-Text Questions

A Sample factors: length of finger, age of student

B Some possible rules for each group to follow include the following:
- Hold your fingers at the same centimeter mark each time.
- Keep your fingers a certain distance from the meter stick.
- Do not make any distracting noises during the trials.
- Give each person the same number of trials.
- Either average the results or use the best time.

Exploration 2 continued ▶

Homework

You may wish to have students perform the first part of Experiment 2 as homework. If they do not have a meter stick at home, a ruler, a broomstick, or some other similarly shaped object may be used instead.

Theme Connection

Changes Over Time
Focus question: If you were to repeat the dropping and catching of the meter stick 25 times, how would your results change? *(Students may respond that they would improve with each successive trial, that they may achieve fairly constant results after several trials, or that their reaction time may slow down as they become tired. Many variables are involved in predicting the outcome of the experiment.)*

Answers to
In-Text Questions

 In Experiment 2 students should find that the two fingers always end up at the center of gravity of the meter stick. Both fingers must move toward the center of the meter stick to keep it balanced. If only one finger moves, the meter stick will eventually fall.

B Adding clay to one end of the meter stick changes the center of gravity and the finger positions. Sample hypotheses: (a) If just the right amount of clay is placed at the 0 cm mark, then your fingers will end up at the 20 cm mark. (b) If one-third as much clay is placed at the 0 cm mark, then your fingers will end up at the 40 cm mark. Students should conclude that their observations do or do not support their hypotheses.

⭐ **A Graphing Practice Worksheet is available to accompany Lesson 1 (Teaching Resources, page 40).**

FOLLOW-UP

Reteaching
Have students set up experiments to test other variables that affect paper falling through air, such as the type of paper or the temperature of the air.

Assessment
Share the following story with your students: A boy picks up a rock and sees several sow bugs scurry away under some dead leaves and other stones. He wonders if the sow bugs sense the light and always go toward darkness or if they sense the dryness of the air and always go toward moisture. Ask: How could he test these hypotheses? *(He could capture several sow bugs, put them in a box that has a hole to let in light on one side, and then observe whether the sow bugs gather on the light or the dark side of the box. To test whether they go toward moisture, he could cover the hole in the box so that both sides of the box are dark. Then he*

EXPERIMENT 2

What to Do
Hold a meter stick with two fingers, as shown.

Now move your fingers toward the center of the meter stick. Can you move just one finger? Try this again, and carefully observe what happens. Where do your two fingers always end up? **A**

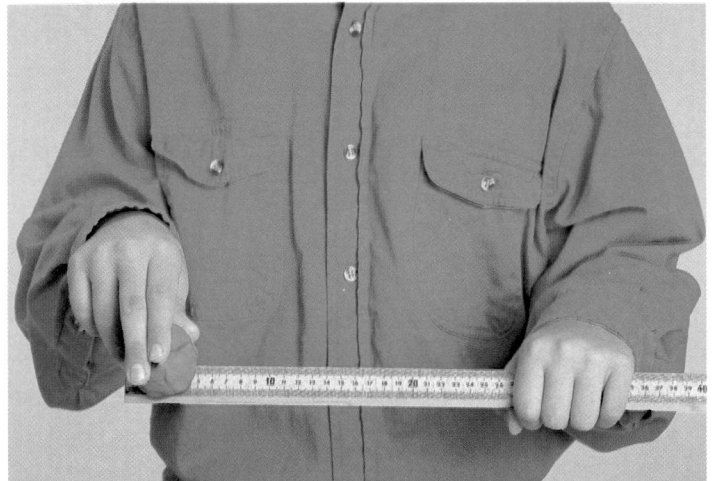

A Problem to Solve
By adding modeling clay to the meter stick, think of a way for your fingers to end up (a) at the 20 cm mark on the meter stick and (b) at the 40 cm mark on the meter stick. Devise a hypothesis for this experiment. Do your results support your hypothesis? What would be an appropriate conclusion? **B**

could wet one side of the box and observe whether the sow bugs gather on the moist or the dry side of the box.)

Extension
This would be an opportune time to discuss possible projects for a science fair. Encourage a project that includes a controlled experiment and follows the format used in this lesson for setting up an experiment. Students should not just write a report or make a collection.

Closure
Tell the class that a scientist is investigating whether human beings can fly by themselves. Have the class list the variables the scientist will have to control. Then have students suggest possible experiments that this scientist might conduct.

LESSON 2
Technology—Brainchild of Science

If you had to pick the single most important mechanical invention of all time, what would it be? Thousands of years ago, an unknown person made an important observation. Objects without sharp corners (what we call *round*) can be moved more easily by rolling than can objects with corners. This was a very valuable scientific discovery. Wouldn't that early scientist be surprised to see what uses are now being made of that observation! There are wheels on almost everything that moves. Rollers and bearings, which are offshoots of the wheel, also make motion easier. The invention of wheels and rollers was a major technological advance that led to even more inventions.

In 1902, Marie Curie made a scientific discovery—a new element, which she called radium. Little did she know that her discovery would someday be used in the treatment of cancer!

In 1987, Ian Shelton noticed a supernova, as you read earlier. His scientific discovery may not lead to any new invention or solve any practical problem faced by humankind, but studying the supernova adds to our knowledge of the universe.

Science is a human activity that involves the pursuit of knowledge about the natural world. The application of knowledge to solve practical problems and to make new inventions is called **technology**.

This collage shows how a basic idea—the wheel—has been refined and adapted for different uses over the centuries. How might the wheel on the right have been used? How might wheels be improved in the future? **C**

LESSON 2 ORGANIZER

Time Required
2 class periods

Process Skills
observing, analyzing

New Terms
Science—a way of asking and answering questions in order to learn about the natural world
Technology—the use of scientific knowledge to solve practical problems and to make new inventions

Materials (per student group)
none

Teaching Resources
none

LESSON 2
Technology—Brainchild of Science

FOCUS

Getting Started
Bring a piece of equipment such as an answering machine or a bicycle wheel into the classroom. Ask: Why was this developed? What purpose does it serve? Lead a discussion about how technology has been used to solve practical problems. Ask: Is technology a good thing? Have students record their thoughts in their ScienceLog.

Main Ideas
1. Technology is the use of scientific knowledge and innovation to solve practical problems.
2. Technology plays a vital role in society.

TEACHING STRATEGIES

GUIDED PRACTICE Have students read page 47 to discover the meaning of the word *technology*. Discuss with students the importance of making scientific discoveries. (*Scientific discoveries bring greater understanding and knowledge of the natural world and assist in the development of technology to solve human problems.*)

Answer to *Caption*

C The wheel on the right was designed for use on a lunar rover. Future improvements might involve lighter and stronger materials and changes in the shape and size of wheels.

Integrating the Sciences

Life and Physical Sciences

One day a man named George de Mestral took his dog for a walk in the mountains near his home in Switzerland. He noticed the dog's fur was covered with burs. As he struggled to pull off these seeds, he wondered what made the burs cling so strongly. He looked at the seeds under a microscope, and what he discovered gave him an idea for an invention—Velcro!

Have students look at burs and Velcro under a compound microscope to see how they are similar. (Both have tiny hooks that cling to fur or cloth.) Nature often provides inventive minds with models that can be used in nature to ignite an idea. Challenge students to find some things in nature that could serve as models for inventions.

Homework

You may wish to have students complete questions 1–4 on this page as homework.

Discuss with a classmate your responses to the following questions. They all involve the idea of technology.

1. What are two technological advances that you have recently heard about? On what scientific discoveries are they based?

2. Does all new knowledge lead to the solving of practical problems—that is, to new technology? Give examples to support your answer.

3. You might get rich if you were to invent a better mousetrap, but could you build one without using science? What would you need to know about the habits and behavior of mice in order to design your new mousetrap?

4. Have you solved a practical problem today? Perhaps you opened a can of juice without an opener. If so, you might have used your knowledge about levers and about the strength of metals. Perhaps you helped your parents build a garden shed. A great deal of knowledge and skill is needed to solve this practical problem. Are there any practical problems that you are trying to solve right now? Name them.

5. Do research on an inventor. Did he or she follow the steps shown below? Explain.

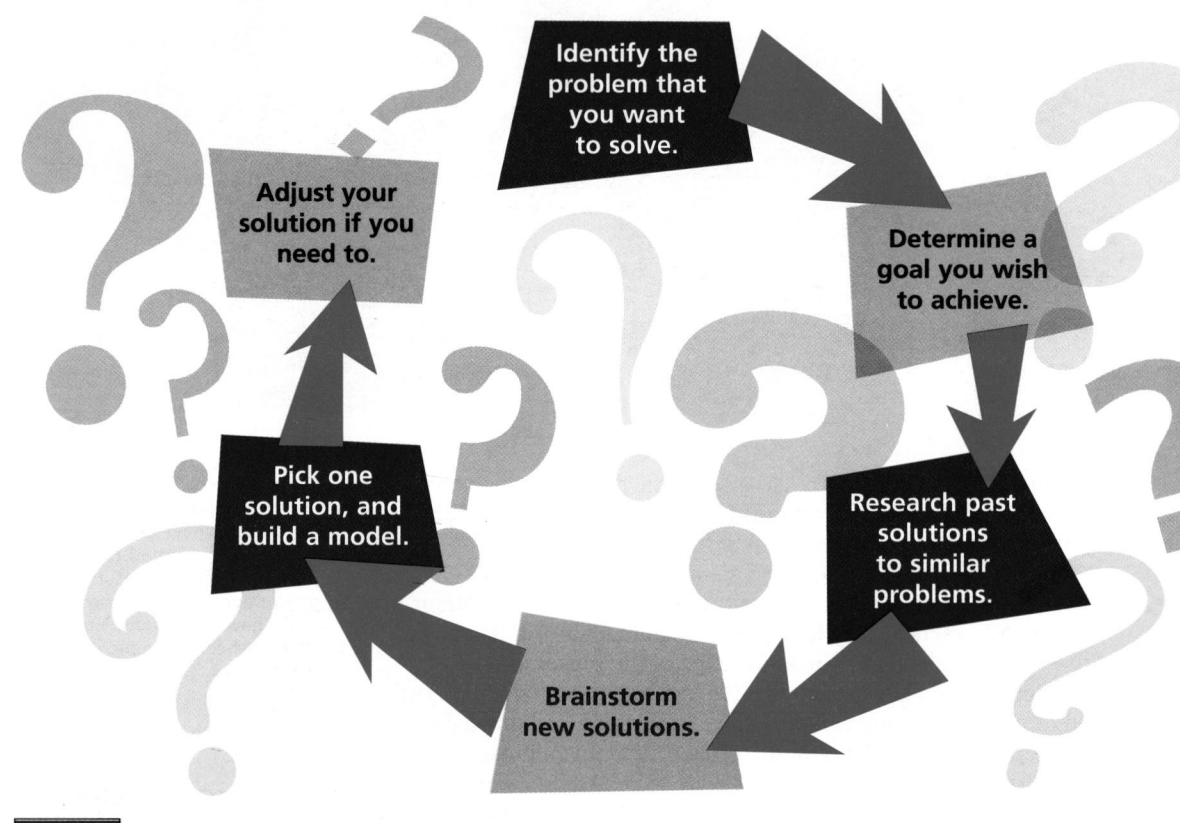

48

Multicultural Extension

Technology in Saudi Arabia

Much of Saudi Arabia is dry, sandy desert. Although this land may be unsuitable for farming, it contains huge, underground reserves of oil. Have interested students research and report on how the discovery of oil in Saudi Arabia allowed the development of technology and industry in that country.

Did You Know...

Before Johannes Guttenberg invented the printing press in 1453, books had to be copied by hand; they were both rare and very expensive. Consequently, the majority of people in Europe at that time never learned to read. The printing press made books more widely available and inexpensive, and literacy rates in Europe soared in the decades that followed.

Technology That Keeps Us Talking

How often do you use the telephone? If you're like most people, you use it nearly every day. You probably know that Alexander Graham Bell invented the telephone, but what else do you know about him? As you read this brief biography, make a list of the inventions that are described. Beside each invention, suggest the practical problem that it was intended to solve. **Ⓐ**

Alexander Graham Bell— Scientist and Inventor

When Alexander Graham Bell was a boy in Scotland in the 1850s, he liked to experiment with sound. For instance, he built a model head with a movable tongue that could say "Mama." He could manipulate the throat of his dog, Skye, so that Skye sounded as if he were speaking. This interest in sound was passed on to Bell by his father and grandfather, both of whom were also named Alexander.

By the time Bell was 23, his two brothers had died of tuberculosis, which was caused partly by the cold, damp Scottish climate. Bell was also showing signs of the disease, so in 1870 the family moved to Canada for the healthier climate. In 1872, Bell moved to Boston and opened a school to train teachers of deaf people. Bell also taught deaf students himself.

Alexander Graham Bell was interested in many things having to do with sound. For instance, after seeing lines of people waiting to use the telegraph, he invented a method of sending as many as eight signals over the same wire at the same time.

In 1881, when President James Garfield lay mortally wounded, Bell invented a metal detector in the hope of locating the bullet that was lodged in the president's body. However, the metal bedsprings interfered with the readings, and Garfield died before the bullet could be found.

Another invention was intended to add to Bell's personal comfort. When his room became too hot in the summer, Bell built an air-cooling system—the first air conditioner!

However, Bell's major interests continued to be sound and its transmission over long distances. It is fitting that he became best known for inventing the telephone. In doing so, Bell drew from previous inventions and his own knowledge of sound, electricity, and the human ear. His reasoning might have been along these lines:

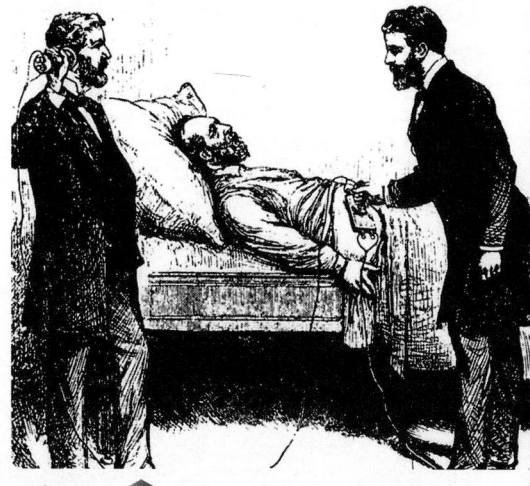

Bell examining President Garfield with his homemade metal detector

"The electric current can be turned back into speech with the new receiver I have invented."

"Sound waves can be turned into an electric current and carried along a wire. The telegraph shows me this."

"I can use a device similar to an eardrum to pick up sound vibrations. In fact, a recent invention called a phonautograph does precisely that."

`49`

The Phone Call Heard Around the World

GUIDED PRACTICE   After students have read pages 49 and 50, ask: How might an inventor like Bell think? What kinds of qualities do inventors need? *(They must have active imaginations. They must be able to look at everyday objects in new ways. They must be curious about the world around them and about how and why things work. They must be able to identify problems and want to solve them. They must be able to picture things in their mind's eye and to visualize the inner workings of gadgets.)*

CROSS-DISCIPLINARY FOCUS

Social Studies

Ask students to find magazine and newspaper articles related to science and technology. Have students share their articles with the rest of the class and discuss them in terms of how science and technology affect society. For example, what are the implications of new medical technologies? Should research, such as that done in genetic engineering, be monitored or regulated? How can new technologies help solve environmental problems? After the discussion, students could create a bulletin-board display with their articles.

The Phone Call Heard Around the World

Bell's reasoning worked. In a dramatic moment in 1876, he spoke his first words into his new invention, and the words were heard by his assistant in another part of the house.

Bell became famous and wealthy from the telephone and his other inventions. With this money, he continued to sponsor schools for the deaf and laboratories for hearing research. He died in 1922.

Below are two of Bell's many inventions. The photophone, left, uses light to create a visual record of sound. The phonautograph, right, is a device for recording sound.

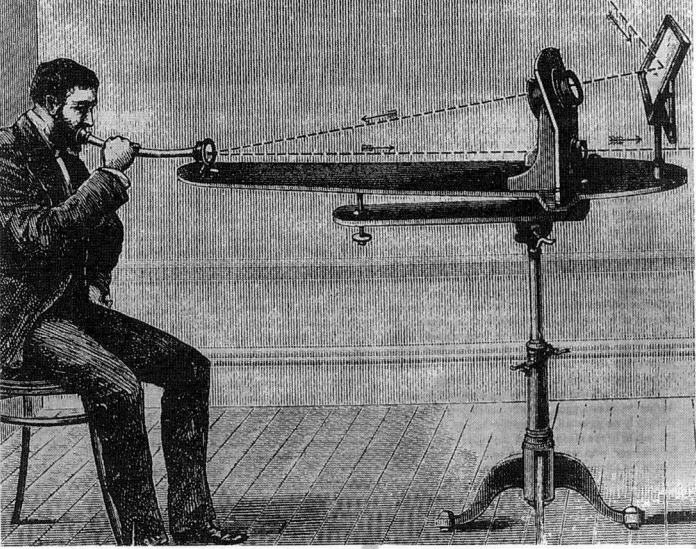

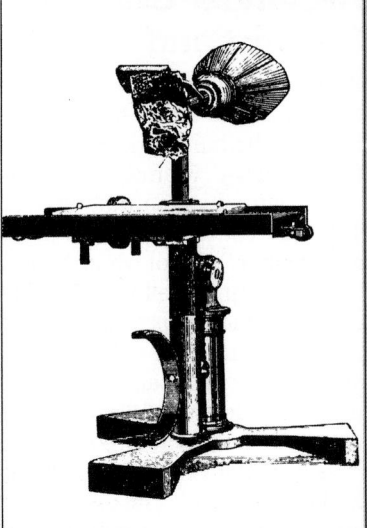

Bell, after making the first telephone call in 1876

Multicultural Extension

Technology Around the World

For technology to be useful, it should be harmonious with the needs and lifestyles of the people in a community and with the natural world in which the people live. For example, in this country most people can get a drink of water by going to the sink and turning on a faucet to fill a glass. This may seem simple, but the faucets are connected by kilometers of underground pipes to pumping stations, water filters, and reservoirs. All of these together make up the technology that is used to supply water. In countries where the technology is different, people may use containers to carry the water to their homes and to store it until needed. Have students research and report on a technology used in another country. If possible, arrange for a former Peace Corps volunteer or a member of a foreign consulate to speak with the class about the technologies of a different country or culture.

Be an Inventor!

Julia invented a way to wash those hard-to-reach spots on her bicycle. She used a plastic bottle filled with soapy water. When the bottle was squeezed, the warm, soapy water squirted out and loosened the dirt. This invention solved Julia's problem.

You, too, can be an inventor and dream up an invention for a problem of your own. When you do, share it with your friends. Make a display that includes a written description of your invention. Tell how you constructed it, what it does, what problem it solves, and so on.

What Makes It Tick?

How does an aerosol can work? How does a light bulb work? Who invented these things? What practical problems did these inventions solve? Do research on an invention or a medical discovery, and report your findings to your classmates.

A Final Word

In a way, we are all like Zed, the Quinns, or Alexander Graham Bell—we observe, infer, test ideas, and try to make sense of the world.

Whether we realize it or not, almost everybody does science. Throughout this book, you will do science many times. Within these pages you may even find an idea for an award-winning science-fair project.

- Gardeners experiment with plants that best suit the local soil and climate.
- Musicians experiment with different musical instruments and techniques.
- Pet owners experiment with techniques for keeping their pets in the best possible health.
- Cooks experiment with different techniques and ingredients to prepare the perfect dish.
- Teachers experiment with new ways of teaching.
- Store owners experiment with ways of arranging merchandise to attract more customers.
- Joggers experiment with different running techniques and ways of getting into shape.

- You, throughout this book, will conduct many experiments to satisfy your curiosity, discover new ideas, verify old ideas, and arrive at conclusions.

Be an Inventor!

This activity could be designed as a class competition. Exhibited inventions could be examined for scientific knowledge and evaluated in terms of how well they perform the task for which they were designed.

Here are some guidelines to help students with their inventions:

1. **Identify** Every successful invention solves a problem or fulfills a need. Think of something you need, or talk to friends and family members about what would make their jobs or their lives easier. What things do you use that you wish worked better?

2. **Brainstorm** Think of as many ideas as possible for solving the problem or fulfilling the need. Let your imagination run wild.

3. **Research** Learn everything you can about the subject related to the problem or need. Read books and talk to people who know about the subject. For example, if you want to invent a new rattrap, you should find out all you can about rats and how and where they cause problems.

4. **Select** Choose one idea for your invention. Ask yourself if your idea is really original and practical. If possible, use available tools and materials to make the invention. Check students' proposed procedures for safety before allowing them to proceed.

5. **Evaluate** Test your invention to see if it solves the problem. Invite potential users to try your invention. Ask for their opinions.

51

Reteaching

Write *science* on one side of the chalkboard and *technology* on the other. Then have students call out words or phrases that relate to each term. Place words that relate to both terms in the middle of the board.

Assessment

Indicate whether each of the following statements refers primarily to science or to technology:

a. With the microscope, I can see smaller things than I can with the naked eye. *(Technology)*

b. Astronomers speculate that there are objects called black holes in space. *(Science)*

c. I use my bicycle to deliver papers. *(Technology)*

d. Scientists are learning more about the AIDS virus every day. *(Science)*

Extension

Invite students to write a science-fiction story about how science and technology will affect them in the future. They can imagine what kinds of technologies will be available and how new products will affect their lives.

Closure

Invite a guest speaker to your classroom. The speaker should be an inventor, a research scientist, or someone who uses state-of-the-art technology in his or her career. Encourage students to ask questions about the role of science and technology in the speaker's work.

Answers to
Challenge Your Thinking, pages 52–53

1. Accept all reasonable responses. (These expressions are intended to promote discussion.) Although a saying may, at first glance, seem completely untrue, many of these expressions are based in fact. Remind students that appearances are often misleading. For example, when waiting for a pot of water to boil, staring at the pot makes the time required for the water to boil seem longer. Point out to students that scientific tests are helpful because they test our observations.

a. Some investigative questions are as follows: Does the age of a dog affect its ability to learn a new task? Are flies attracted by the smell of different substances? Is the amount of light at its lowest level just before sunrise?

b. Encourage students to use their imagination to answer this question. Remind them to control their variables. For example, several dogs of the same breed and the same age could be taught a skill. Their results could be compared to the success of younger dogs of the same breed.

2. Answers will vary, but a few examples of new inventions since 1900 are computers, televisions, radios, and sophisticated communication and transportation devices.

3. Student answers will vary, but the obvious candidates for change are in the areas of computers, communication, and transportation. Accept all reasonable ideas.

4. Student answers will vary but could include ideas about new inventions. Answers should also show how the knowledge gained through scientific inquiry makes advances in technology possible.

5. The connections that students draw should be reasonable and thoughtful. For example, between *you* and *problems,* students may write "use science to solve."

CHALLENGE YOUR THINKING

1. Words of Wisdom
Most of our everyday expressions are based on observations. Not all of these observations are accurate. Which of the following expressions do you think are true or partly true? Which could be scientifically tested?

* You can catch a bird by putting salt on its tail.
* You can't teach an old dog new tricks.
* You can catch more flies with honey than with vinegar.
* It's always darkest before the dawn.
* A watched pot never boils.
* There is always a calm before a storm.
* A stitch in time saves nine.
* A rolling stone gathers no moss.
* You reap only what you sow.

a. Write an investigative statement for four of the expressions above.

b. Design an experiment to test one or more of the expressions above.

2. Time Travel
Imagine being transported back in time to 1900, when the first automobiles were traveling the roads. What are some other ways in which the technology of 1900 differs from that of today?

52

 You may wish to provide students with the Chapter 3 Review Worksheet that accompanies this Challenge Your Thinking (Teaching Resources, page 42).

3. Life Imitates Art

Jules Verne, a famous science-fiction writer, wrote about technological advances many years before they were invented. His stories included the submarine, the helicopter, and even the fax machine. Imagine that you are a science-fiction writer and that you are writing about events in the year 2075. What new advances in technology would you include in your story?

4. Speak for Yourself

The title of Lesson 2 is Technology—Brainchild of Science. In your own words, describe what these words mean to you, and give examples.

5. Draw Me a Map

Copy the words below into your ScienceLog. Draw lines between the words that you think are connected in some way. Along each line that you have drawn, write a few words to explain the connection. Do not write in this book!

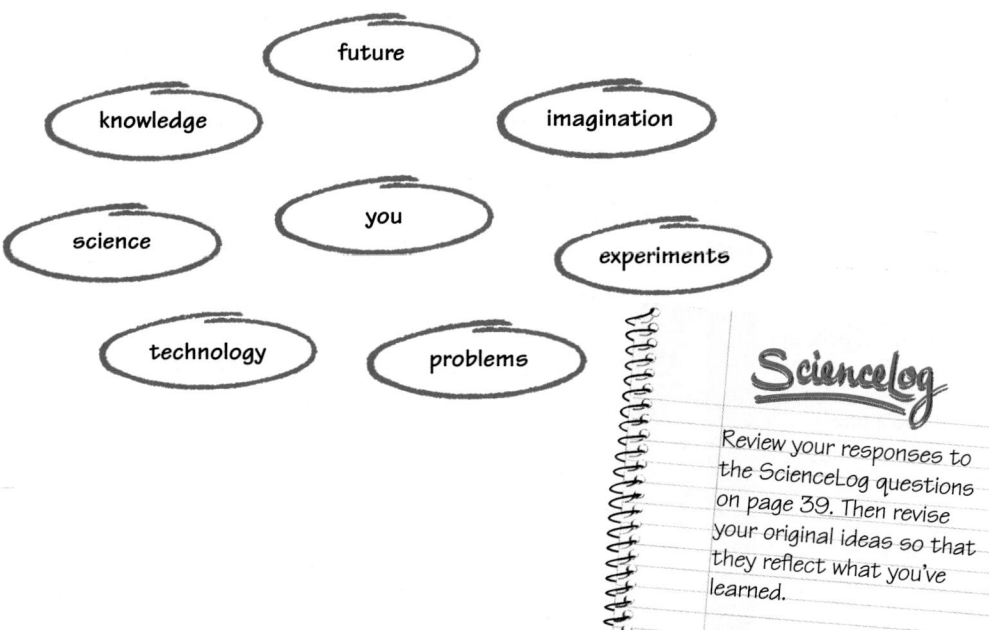

future

knowledge

imagination

science

you

experiments

technology

problems

ScienceLog

Review your responses to the ScienceLog questions on page 39. Then revise your original ideas so that they reflect what you've learned.

ScienceLog

The following are sample revised answers:

1. Science is a way of understanding the natural world. Technology is the application of scientific knowledge to solve practical problems.
2. Encourage discussion about how technological advances affect society as a whole. Allow students to voice their opinions about the advantages and disadvantages of new technology.
3. The development of technology relies on and contributes to scientific knowledge. For example, understanding how to transmit and amplify sound allowed scientists to create technologies such as tape recorders, speakers, and headphones. In turn, these technologies help scientists to conduct research in other areas of science.

The Big Ideas

***The following is a sample
unit summary:***

Being scientific is a way of looking at the world. It includes observing, questioning, inferring, hypothesizing, testing, and drawing conclusions. It requires imagination, persistence, and curiosity. (1)

An observation is something that is noted about the physical world. An inference is an idea, suggested by observations, that explains an occurrence in nature. A conclusion is an inference that seems to be true based on available evidence. (2) These steps help people to act scientifically because they are part of a process of understanding the physical world. (3)

An investigative question is one that can be studied systematically to find out its answer. (4)

In order to test a *hypothesis,* a *controlled experiment* is set up by allowing only one *variable* to change; the other variables remain constant. (5)

Experiments are important because they allow us to test the accuracy of our observations and the validity of relationships. (6)

The steps involved in designing an experiment include the following:
• Ask an investigative question.
• Choose one variable to change.
• Make sure that all other variables remain constant.
• Set up several trials.
• Report the results. (7)

Technology is the application of knowledge to solve practical problems. Science may provide this knowledge. (8) Science and technology help each other by sharing knowledge learned in the pursuit of each. (9)

Unit 1

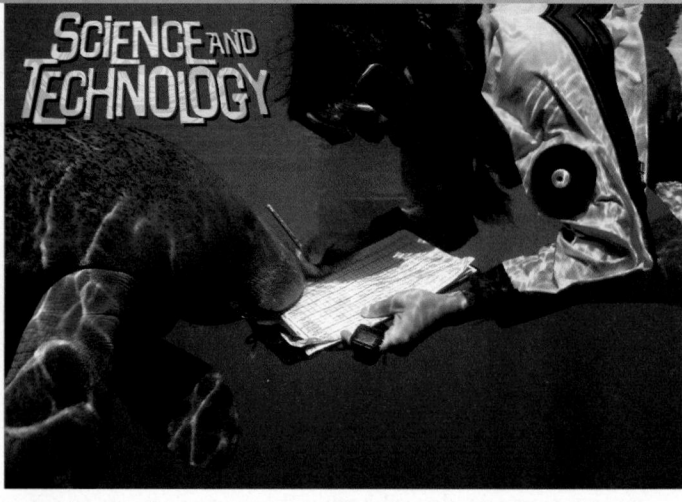

SCIENCE AND TECHNOLOGY

SOURCEBOOK

Turn to the SourceBook to find out more about who scientists are and what they actually do. You can learn about the different branches of science and read an actual case study of two scientists who used the scientific method to study the behavior of coyotes.

**Here's what you'll find
in the SourceBook:**

54

The Big Ideas

In your ScienceLog, write a summary of this unit, using the following questions as a guide:

1. What does it mean to be scientific?
2. What is an observation? an inference? a conclusion?
3. How does each of these help people do science?
4. How does an investigative question differ from a question that is not investigative?
5. How are the terms *hypothesis, variable,* and *controlled experiment* related?
6. What is the purpose of experiments?
7. What are some of the steps in designing experiments?
8. How are science and technology related?
9. How do science and technology help each other?

Checking Your Understanding

1. Yvonne's group made the statements below as they did the candle activity at the beginning of this unit. Which statements are inferences? Which are observations?
   a. The candle is blue.
   b. The candle is 5 cm tall.
   c. A pool of liquid forms on top of the candle as it burns.
   d. This liquid is made of the same substance as the candle.

e. The candle flickers as it burns.

f. Blowing hard on the candle causes it to go out.

g. Blowing hard on a candle causes it to go out because you blow all of the air away from it.

h. Candles need air to burn.

2. Below is another cryptogram. Decipher the cryptogram to discover a quotation. Hint: One of the words in the quotation is technology.

VN VW NSZ UHVAVUA UK WUYZ NSMN OVKZ UA
ZMFNS JUEOB CZ CZNNZF VK AUN KUF WPVZAPZ
MAB NZPSAUOUDL.

After you crack the code, write a paragraph or two stating whether you agree or disagree with the quotation and why.

3. Consider the following statement: Scientists discover; inventors invent. Explain in writing what you think this statement means. Indicate whether you agree or disagree, and why.

4. Once upon a time, there was a young boy who lived in the country. This boy noticed that every morning, just before dawn, the roosters began to crow. He hypothesized that the roosters' crowing caused the sun to rise. Design an experiment to test his hypothesis.

5. If an experiment repeatedly disproves a hypothesis, which of the following actions would be a correct response for a scientist? Justify your response in writing.

a. Ignore the results of the experiment.

b. Keep trying new experiments until the hypothesis is supported.

c. Reject the old hypothesis and form a new one.

d. Conclude that the experiment had some sort of flaw.

6. Copy the concept map at right into your ScienceLog. Do not write in this book! Then complete the map using the following words: *the world, observing, scientists, inferring, experimenting, variable, hypothesis,* and *models.*

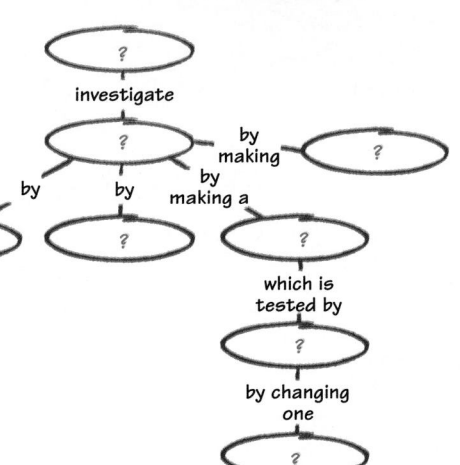

1. a. Observation
   b. Observation
   c. Observation
   d. Inference
   e. Observation
   f. Observation
   g. Inference
   h. Inference

2. The cryptogram reads, "It is the opinion of some that life on Earth would be better if not for science and technology."
   Answers will vary. Get students started by having them make a list of pros and cons for science and technology. For example: Technology can be used to stop or lessen pollution (pro). Technology can cause pollution (con).

3. Sample answer: Scientists and inventors share many skills, such as performing careful analysis and problem solving. While scientists usually seek new explanations for things they observe, inventors usually seek new solutions to existing problems. Student opinions will vary but should be clear and complete.

4. Answers will vary, but students should indicate that the variable to change is the time at which the rooster crows.

5. A scientist should make sure that his or her experiment is a good one. New experiments to test the hypothesis can be tried, but eventually new hypotheses may need to be formed. Therefore, choices (c) or (d) would be correct.

6. For a completed concept map see page S234.

★ You may wish to provide students with the Unit 1 Review Worksheet on page 49 of the Unit 1 Teaching Resources booklet.

Homework

The Unit 1 Activity Worksheet makes an excellent homework activity (Teaching Resources, page 48). If you choose to use this worksheet in class as a group activity, Transparency 7 is available to accompany it.

Background

Advertisers, movie directors, and music-video makers morph images together to create a seamless transformation from one image to the next. In the case of complex morphs, such as the ones used in some films, computer-graphics specialists and artists work together to create the image.

Software now exists that enables people to create morphs on their home computers. Some students may have one of these graphics programs and may be able to demonstrate how it works for the class.

Make a Morph!

Students may wish to work in small groups to create their morphs. Encourage students to display their images for the class to observe. As a fun extension to this activity, have students create small flip books with their morphs. Tell students to cut 15 to 20 pieces of graph paper, 5 cm × 10 cm each. Tell them to stack these pieces and staple them together at one end of the stack so that if the flip book were viewed vertically, the staple would be at the top, and the bottom of each page would be unstapled. Students should then copy the first image from the previous exercise onto the last page of the flip book at the bottom center of this page. The last image should then be drawn at the bottom center of the first page. The student should then create enough transitional images to fill the other pages. Once the flip book is finished, students should be able to flip (back to front) through the pages of their flip books to animate their morphs.

Extension

Invite a computer-graphics specialist to speak to your class. You may wish to contact a local university or community college, an advertising firm, or a computer company to locate a suitable speaker.

Morphing Magic

A blob of molten metal transforms into a human body. A speeding car warps into the shape of a running tiger. Grinning human faces turn into growling panthers. How do they do that?

Computer Wizardry

Morphing, which is short for *metamorphosing*, is a special video effect in which one image shifts and changes into another. In order to get that clean shift, animators

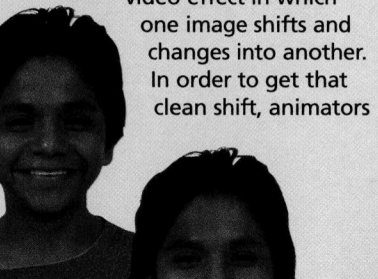

manipulate images on digital videotape. For a really complicated morph, a staff of animators may use several computers and computer programs to create the images required.

Morphs start with a beginning image and a final image. Then animators create a series of middle images to sandwich in between. In creating these images they must decide which parts of the first image will

transform into which parts of the final image. Using computer models as well as physical models, they carefully adjust each image to show a slight change from the previous image. When these separate images are run together in a video sequence, the end result is a seamless-looking morph.

Miraculous Medical Morphs

Reconstructive surgery performed on a person's face dramatically changes the way he or she looks. Morphing technology can also help plastic surgeons show patients how surgery will change their appearance. Changes are more obvious to patients when they can see the image

shift instead of seeing only before-and-after pictures.

A Parade of Faces

One of the first morphs of moving images was in the Michael Jackson video "Black or White," released in 1991. Animators morphed faces of people from different ethnic backgrounds. First, they gathered images from models who were singing and dancing to the music. On the computer, they lined up the faces so that the eyes, nose, and mouth of one image could morph into the same features on another face. Then they decided which feature should shift first. The end result was a long sequence that merged faces from many different races.

Make a Morph!

Draw a simple image on a piece of graph paper. Then draw the final image on a second piece of graph paper. Next, decide what the transition images will look like. You should create two or three transition images on separate sheets of graph paper. Use the lines of the graph paper to line up the images as they change. Display your morph on a poster for your class to see.

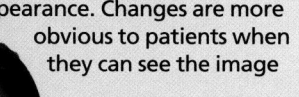 **This morph started with only two photographs, the one on the upper left and the one on the lower right.**

Meeting Individual Needs

Gifted Learners

Have students create a film script that incorporates special effects. The film should be at least 5 minutes long. Students should create the script, provide direction, and design special effects. If possible, have students make and show the film to the class and explain the techniques they used to create the special effects. Students should submit a plan to you for a safety review before proceeding.

Science Snapshot

Lewis Latimer (1848–1928)

Few inventions are perfect when they are first made. For example, Thomas Edison's light bulbs did not last very long because their carbon filaments were quite fragile. Lewis Latimer, a draftsman and engineer, made major improvements to the carbon filaments, making light bulbs more practical and better suited to everyday use.

Fragile Fibers

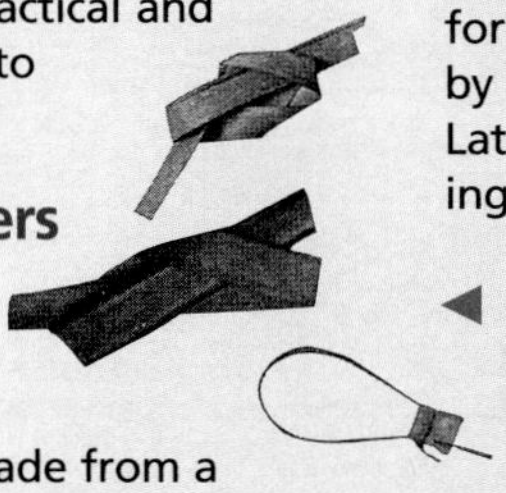

The thin filaments used in early light bulbs were made from a variety of natural fibers, including bamboo and cotton. However, before these fibers could be used in light bulbs they had to be carbonized, which is a process of heating the fibers to reduce them to strands of charcoal. Early processes of carbonizing filaments were unreliable and produced filaments that did not "burn" very long.

Lewis Latimer tackled this problem and soon invented a superior process of carbonizing natural filaments. Latimer's system produced uniform filaments that lasted much longer. He also invented an ingenious system of attaching the filaments to the connecting wires inside light bulbs. Both his inventions found widespread use and helped make electric lighting successful.

Bright Ideas

Lewis Latimer became an inventor in a roundabout way. He began by working as a draftsman in a patent office. He drew diagrams of inventions that people wanted to patent. He even helped prepare Alexander Graham Bell's patent application for the telephone! Surrounded by inventions and inventors, Latimer began to do some inventing and engineering of his own.

◀ **All kinds of fibers were tested as possible filaments for early light bulbs. Shown here are just a few.**

Lewis Latimer was the son of George Latimer, a slave who escaped from Virginia and fled to Boston in the 1830s. With the help of powerful abolitionist friends, George bought his freedom for $400 in 1842. Six years later, Lewis was born. No one expected that the son of a former slave would one day light up the nation.

City Lights

Latimer played an extremely important role in the development of the electric industry. Working for the United States Electric Lighting Company, he supervised the first electric lighting systems in New York, Philadelphia, and London.

In 1884, Latimer joined Edison's Electric Light Company (now known as the General

▶ **Replica of Edison's first successful incandescent light bulb, invented in 1879. This light bulb had a filament of carbonized sewing thread, which "burned" for 40 hours.**

Electric Company). He first worked in the engineering department but later transferred to the legal departments where he provided expert testimony to defend Edison's patents. He even wrote the first textbook about electric lighting systems. In fact, he published many writings, from patent applications to poetry. In his long career as an inventor and engineer, Latimer left behind a shining legacy of innovative and creative ideas.

Inventions and Innovations

One of Latimer's innovations was the use of *parallel circuits* instead of *series circuits* in outdoor lighting. Find out about the difference between these two types of circuits, and illustrate each one. What do you think is the primary advantage of parallel circuits over series circuits?

Background

Lewis Latimer was born in Chelsea, Massachusetts, on September 4, 1848. When he was 10 years old, he went to work to help support his mother and four younger siblings. At age 16, he joined the United States Navy and served on the side of the Union in the Civil War.

In 1865, after the war ended, Latimer looked for work in Boston. He was hired by an office of patent solicitors. Latimer was a hard worker and was talented at his job. He went on to become the chief draftsman in the firm.

Latimer's inventions included a water closet for railroad cars, the "Latimer lamp," a process of manufacturing electric-light carbons, and a globe supporter for electric lamps.

Cooperative Learning
LEWIS LATIMER (1848–1928)

Group size: 3 to 4 students

Group goal: to read about Lewis Latimer and research his life and accomplishments

Positive interdependence: Assign each student a role such as principal investigator, recorder, orator, or checker. The principal investigator can be in charge of obtaining the sources for research and leading discussions. All of the group members will then decide which facts about Lewis Latimer's life are the most important to relay. The recorder can take notes or develop an outline of facts and opinions. The checker can be responsible for reviewing the notes and making recommendations for further research or more discussion, if necessary. The orator can then compile the notes into a report and present the report to the class.

Individual accountability: Have students review their own work in their ScienceLog. They should comment on their contributions to the group's progress and final results. You may find it useful to provide students with a Self-Evaluation Checklist, which is available on the SnackDisc.

Discussion Questions

1. What do you think are some attributes of great inventors? *(Work hard, ask questions, pay attention to detail, like to solve problems, use their imaginations)*
2. What questions might have led Lewis Latimer to formulate hypotheses regarding the invention of the Latimer lamp? *(How can I make the bulb burn longer? What makes filaments weak or strong?)*

Answer to
Inventions and Innovations

In a series circuit, there is one path for the current to follow. The current passes directly through each bulb in turn. If one bulb burns out, the circuit is broken and the flow of electricity stops. A parallel circuit has two or more separate paths for the current to follow. If one bulb in a parallel circuit burns out, the other bulbs will continue operating. Student illustrations should clearly show these differences.

Background

Inventors have approached the problem of designing a "roadable" plane in different ways. The most common designs include removable airplane components that can be attached to the body of a car. These designs require that wings, propellers, and sometimes even engines be stored or towed. A second design uses the body of an airplane and detachable airplane parts. Finally, another design allows the vehicle to be driven or flown without any alterations to the body of the vehicle.

The Aircar, made by helicopter designer Kenneth Wernicke, is one of the few successful flying cars ever designed. About the size of a motor home, the Aircar can fit in a parking space. Its wings are small and stumpy, but provide a surprising amount of lift. The Aircar has successfully passed tests in wind tunnels. However, manufacturers of the Aircar still have at least one obstacle to overcome before mass production is possible—extreme production costs. Currently, the Aircar carries a price tag of at least $100,000.

Discussion

Ask students to imagine what their city or town would be like if flying cars were commonplace. Have them consider the following related issues: public safety, pollution, traffic control, and economics. Encourage students to recognize that widespread changes in transportation can have many different effects on society.

Extension

Challenge students to design model airplanes. Provide students with materials such as paper, paper clips, rubber bands, balloons, scissors, yarn, hole punchers, craft sticks, and tissue paper. Before they begin, ask students to identify some characteristics of a successful model airplane. Students may focus on the speed of the plane, the weight of the plane, or the distance the plane can fly. You may wish to have the students test their models to find the fastest, lightest, or farthest-flying model. Involve students in a discussion about any difficulties they encountered while creating their designs.

Highway to Skyway

Imagine being able to soar over traffic jams at 480 km/h per hour, never having to stop for a traffic light. If the weather turns bad, you can land on the nearest highway and drive the rest of the way. Sound exciting? For a long time, the idea of a winged automobile has captured inventors' imaginations.

Cars Make Poor Airplanes

Since 1930, 70 patents have been issued for flying cars, and at least 17 different models have been flown. People seem to like the idea of driving short distances and then soaring through the air for long-distance trips or pleasure rides. So why haven't manufacturers rushed to produce flying cars? Quite simply, flying cars are bulky and hard to use. Some involve complete makeovers to turn cars into flying machines—swapping engines and attaching wings and propellers, for instance. Others are more lightweight and flight-ready, but they are too fragile to be safe in traffic.

Cars are designed to be roomy and sturdy enough for rough roads and occasional dents. Airplanes, on the other hand, have to be lightweight and aerodynamic. No design has ever been successful at switching effortlessly between the function of a car and the

▲ **This experimental flying car was first flown in 1947. It had a detachable wing section that could be left at the airport or other location while the car was in use. Many other designs have been developed since this early model.**

function of a plane, but inventors keep searching.

Flying cars face obstacles beyond design problems, however. These vehicles would be very expensive, and thousands of landing strips would have to be built. Also, can you imagine the sky filled with flying cars? A simple fender-bender would have tragic results.

A High-Tech Image

In some ways, however, the marriage between cars and planes has already occurred. Many high-tech car features, such as antilock brakes, aerodynamic shapes, and rear spoilers, have been borrowed from airplane designs.

Technology for Technology's Sake

Whether flying cars would benefit society as a whole is an interesting question. Think of examples of technology that strike you as a waste of resources or perhaps as an unnecessary luxury. Do you think that developing new technology, such as flying cars, would be a waste of time?

Technology for Technology's Sake

Write "Pro" and "Con" on the board. Students should write statements in favor of flying cars under the first column and statements against flying cars under the second column. Encourage students to clearly explain their ideas. Allow students to debate the idea that some technologies provide unnecessary luxuries.

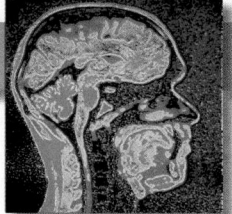

Searching for Cures

Starve a fever, feed a cold. Have some hot chicken soup and you'll feel better. A glass of warm ginger ale will settle your stomach. Have you heard advice like this before? While these remedies might help treat the symptoms of an illness, they can't treat the cause of the disease—whether it's a head cold or something more serious. In most cases, it takes extensive work by medical researchers and other scientists to identify and provide cures for illnesses.

Taking Steps Toward Treatment

Medical researchers investigate the symptoms, causes, and possible cures of different diseases. Treatments must be thoroughly tested before doctors can prescribe them for their patients. This orderly search for knowledge is known as the scientific method. Medical investigators use this method to study everything from the common cold to cancer.

▼ Researcher does work in a high-security biological laboratory at the CDC

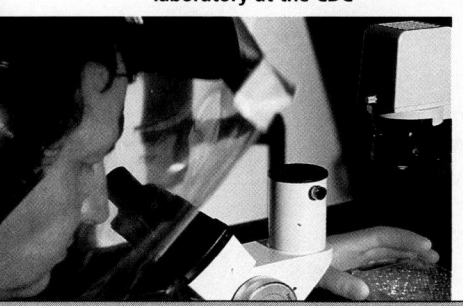

Legionnaires' Disease: A Case Study

At the 1976 Pennsylvania State Convention of the American Legion, 182 Legionnaires were struck with a mysterious illness, and 29 people died. The symptoms included headache, chills, high fever, severe cough, chest and stomach pain, and difficulty breathing. Experts from the Centers for Disease Control (CDC) in Atlanta were called in to help find out what had caused the disease, how it could be treated, and how to prevent an outbreak from happening again.

A network of doctors, nurses, and researchers joined forces. In an effort to find out what was causing the outbreak, some of them collected and tested fluid and tissue samples from sick patients and people who had died from the disease. Others sent out questionnaires and interviewed American Legion members who had been to the convention.

A Breakthrough

Finally, through hotel guest lists and questions posed to the Legionnaires, scientists identified the hotel that was the headquarters for the convention as the most likely source of the infection. Then, months later, a researcher at the CDC discovered a previously unknown bacteria in tissues taken from people who had died from the illness. The

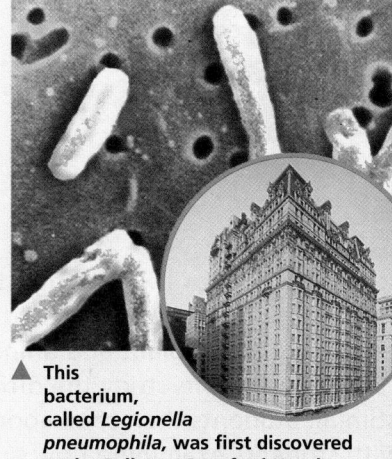

▲ This bacterium, called *Legionella pneumophila,* was first discovered at the Bellevue-Stratford Hotel in Philadelphia. It caused the epidemic of Legionnaires' disease.

bacteria seemed to have spread from the cooling water used in the hotel's air-conditioning system. Thanks to the work of those scientists and others who continue to research the disease, doctors can now test for Legionnaires' disease. If it is caught in time, they can treat the disease with an antibiotic.

Some Research of Your Own

Epidemics happen all over the world. In 1995, for example, parts of Africa suffered from an epidemic of the Ebola virus. Find out more about medical epidemics and the steps that health workers must take to fight widespread disease.

59

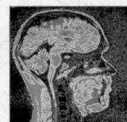
Background

Epidemiology is the study of epidemics. Epidemiologists study how diseases spread, how diseases can be treated, and ways to limit the spread of diseases. At the United States Centers for Disease Control and Prevention, epidemiologists conduct research and provide assistance to areas suffering from epidemics in the United States and in other countries.

Throughout history, more people have been killed by epidemics than by wars. Diseases that have caused epidemics include the bubonic plague, smallpox, cholera, malaria, influenza, AIDS, Ebola, typhus, and Legionnaires' disease.

Legionnaires' disease is a form of pneumonia caused by a rod-shaped bacterium called *Legionella pneumophila.* Researchers also determined that this bacterium was responsible for similar outbreaks in Washington, D.C.; Pontiac, Michigan; Benidorm, Spain; and Nottingham, England. The bacteria exist in soil and drinking water. Scientists are still uncertain about exactly how the disease is transmitted to people, although they suspect it can be transmitted through air-conditioning vents.

A large number of people still contract Legionnaires' disease each year. Physicians can diagnose it by testing for antibodies in the blood or by locating bacteria in mucus from the lungs. Doctors usually treat the disease with the antibiotic erythromycin. To date, there is no vaccine to prevent the disease.

Discussion Questions

1. What is an epidemic? *(a disease that spreads rapidly among a large number of people)*

2. How did researchers use the scientific method to find out what caused the epidemic of Legionnaires' disease? *(They made observations about the symptoms of the disease. They analyzed data about the victims, including what hotels they stayed in. They tested blood and tissue samples from patients. They identified a previously unknown bacterium and hypothesized that it caused the disease. They inferred that the disease spread through contaminated water in leaking air-conditioning vents.)*

Multicultural Extension

Medicine Practices

Most students will be familiar with the way diseases are treated by Western doctors. However, different treatments are practiced in different cultures. Have students investigate how other cultures use different methods to treat diseases. Areas of research could include acupuncture and herbal remedies.

Answers to *Some Research of Your Own*

Students should note that epidemics are often controlled and prevented by increased sanitation, quarantine of victims, and vaccination against disease.

Unit 2 PATTERNS OF LIVING THINGS

Unit Overview

In this unit, students explore some of the characteristics of living things. In Chapter 4, students classify objects into three groups—those that are living, those that are dead, and those that have never been alive. Then they examine plant and animal motion as a characteristic of living things. In Chapters 5 and 6, students study the different ways that animals and plants grow and respond to stimuli. Students then read about actual studies of animal behavior. Chapter 6 concludes with students examining how animals have adapted to their environments both physically and behaviorally. In Chapter 7, students have the opportunity to use microscopes to observe some differences between plant and animal cells.

Using the Themes

The unifying themes emphasized in this unit are **Structures, Changes Over Time, Cycles,** and **Systems.** The following information will help you incorporate these themes into your teaching plan. Focus questions that correspond to these themes appear in the margins of this Annotated Teacher's Edition on pages 67, 84, 98, 102, 107, 108, and 120.

In Chapter 4, the theme of **Structures** may be incorporated by having students compare the structural and functional components of animals and human inventions. This introduces the idea of structural adaptations, which are discussed further in Chapter 6, Lesson 4.

Changes Over Time is a recurrent theme in this unit. In Chapter 5, students may be encouraged to compare different patterns of growth, such as those of plants with those of mammals. In addition, long-term climatic

changes can elicit responses over the course of generations. A focus question addressing this topic can be found in Lesson 4 of Chapter 6.

Since many of these changes are cyclic in nature, the theme of **Cycles** may be useful as well. For example, in Chapter 6, daily activity cycles are examined in relation to environmental changes such as the availability of food. Later in Chapter 6, it may be helpful to discuss the response patterns that occur in many animals as a result of changing seasons.

Finally, as your students progress through Chapter 7, use the focus question in that chapter to help students visualize cells as components of organismal **Systems.** This discussion will allow students to consider the similarities and differences among cells and will give them an appreciation of the role that cells play in living organisms.

Using the SourceBook

Unit 2 focuses on the classification of living organisms, the characteristics and functions of cells, and the organization of cells in living things. Students discuss evolution and genetics as these topics relate to modern classification schemes. The unit concludes with a study of human organ systems and genetic instructions.

Bibliography for Teachers

Cooke, John. *The Restless Kingdom: An Exploration of Animal Movement.* New York City, NY: Facts On File, Inc., 1991.

Wheelwright, Jeff. *Degrees of Disaster: Prince William Sound: How Nature Reels and Rebounds.* New York City, NY: Simon & Schuster, 1994.

Young, John K. *Cells: Amazing Forms and Functions.* New York City, NY: Franklin Watts, 1990.

Bibliography for Students

Bryan, Jenny. *Movement: The Muscular and Skeletal System.* New York City, NY: Dillon Press, 1992.

Harlow, Rosie, and Gareth Morgan. *Energy and Growth.* New York City, NY: Warwick Press, 1991.

Parker, Steve. *Animal Babies: A Habitat-by-Habitat Guide to How Wild Animals Grow.* Emmaus, PA: Rodale Press, 1994.

Pope, Joyce. *Two Lives.* Austin, TX: Steck-Vaughn Co., 1992.

Films, Videotapes, Software, and Other Media

Describing How Living Things Are Alike
Software (Apple II, Macintosh, MS-DOS)
Queue, Inc.
338 Commerce Dr.
Fairfield, CT 06430

Creepy Crawlers: Legless Locomotion
Film or videotape
Coronet/MTI
108 Wilmot Rd.
Deerfield, IL 60015

Life: How Do We Define It?
Videotape
Britannica
310 South Michigan Ave.
Chicago, IL 60604-9839

Discovering the Cell
Videotape
National Geographic Society
Educational Services
P.O. Box 98019
Washington, D.C. 20090-8019

Unit Organizer

Unit/Chapter	Lesson	Time*	Objectives	Teaching Resources
Unit Opener, p. 60				Science Sleuths: The Plainview Vandals English/Spanish Audiocassettes Home Connection, p. 1
Chapter 4, p. 62	**Lesson 1, Signs of Life, p. 63**	2	1. Classify objects as living, nonliving, or dead. 2. Compare the characteristics of living and nonliving things. 3. Distinguish between plants and animals.	none
	Lesson 2, Moving Around, p. 65	1 to 2	1. Compare and contrast the movements of plants and animals. 2. Observe and describe the motion of several animals. 3. Identify several forms of animal locomotion that have been copied by human technology.	Image and Activity Bank 4-2 Exploration Worksheet, p. 3 ▼ Transparency Worksheet, p. 6 ▼
	Lesson 3, Locomotion on Land, p. 68	3	1. Observe different forms of animal locomotion. 2. Describe and compare the locomotion of a variety of animals.	Image and Activity Bank 4-3 Exploration Worksheet, p. 8 Exploration Worksheet, p. 9 Exploration Worksheet, p. 11
End of Chapter, p. 73				Chapter 4 Review Worksheet, p. 13 Chapter 4 Assessment Worksheet, p. 16
Chapter 5, p. 75	**Lesson 1, Human Growth, p. 76**	2	1. Compare growth rates in human beings. 2. Identify the parts of the human body that grow least or most during various growth stages.	Image and Activity Bank 5-1 Transparency 11
	Lesson 2, Other Growth Patterns, p. 80	3	1. Design an experiment to test the effect of different conditions on seed growth. 2. Explain why large size may pose problems to living things. 3. Identify different types of growth patterns, including continued growth, renewal, reproduction, regeneration, and harmful growth.	Image and Activity Bank 5-2 Activity Worksheet, p. 19 Exploration Worksheet, p. 20 Graphing Practice Worksheet, p. 21 Activity Worksheet, p. 23 ▼
End of Chapter, p. 89				Chapter 5 Review Worksheet, p. 25 ▼ Chapter 5 Assessment Worksheet, p. 29
Chapter 6, p. 91	**Lesson 1, Stimulus and Response, p. 92**	2	1. Recognize the relationship between a stimulus and a response. 2. Observe an earthworm's response to a stimulus and classify this response as positive or negative.	Image and Activity Bank 6-1 Exploration Worksheet, p. 32
	Lesson 2, You Be the Scientist, p. 96	2	1. Infer the correct relationship between a stimulus and a response in animal behavior. 2. Comprehend how scientists may work cooperatively over long periods of time to solve scientific problems. 3. Explain the role of biological clocks in the behavior of living things.	none
	Lesson 3, Seasonal Behavior of Animals, p. 100	4	1. Describe examples of migration and the factors affecting it. 2. Identify how different animals regulate body heat. 3. Compare and contrast warmblooded and coldblooded animals and their responses to temperature changes. 4. Set up an experiment to determine the effect of body size on rate of heat loss.	Image and Activity Bank 6-3 Discrepant Event Worksheet, p. 34 Exploration Worksheet, p. 35 ▼ Activity Worksheet, p. 36
	Lesson 4, Adaptation of Structure, p. 105	2	1. Recognize the relationship between an animal's adaptation and the animal's ability to survive. 2. Compare the time of adaptation with the life span of an individual organism.	Image and Activity Bank 6-4 Transparency 15 Theme Worksheet, p. 37
End of Chapter, p. 110				Chapter 6 Review Worksheet, p. 39 Chapter 6 Assessment Worksheet, p. 42
Chapter 7, p. 112	**Lesson 1, Taking a Closer Look, p. 113**	1	1. Differentiate among the parts of a microscope. 2. Use a microscope properly to observe tiny objects.	Image and Activity Bank 7-1 Transparency 18 Transparency 19 Resource Worksheet, p. 44 ▼ Transparency Worksheet, p. 45 ▼ Exploration Worksheet, p. 47
	Lesson 2, Skilled Observations of Cells, p. 116	3	1. Prepare wet mounts of several kinds of cells. 2. Identify the main parts of plant and animal cells. 3. Compare and contrast typical plant and animal cells.	Image and Activity Bank 7-2 Transparency Worksheet, p. 48 ▼ Exploration Worksheet, p. 50 Exploration Worksheet, p. 52 Transparency 21
End of Chapter, p. 120				Chapter 7 Review Worksheet, p. 54 Chapter 7 Assessment Worksheet, p. 57
End of Unit, p. 122				Unit 2 Activity Worksheet, p. 59 Unit 2 Review Worksheet, p. 60 Unit 2 End-of-Unit Assessment, p. 64 Unit 2 Activity Assessment, p. 69 Unit 2 Self-Evaluation of Achievement, p. 72

* Estimated time is given in number of 50-minute class periods. Actual time may vary depending on period length and individual class characteristics.

▼ Transparencies are available to accompany these worksheets. Please refer to the Teaching Transparencies Cross-Reference chart in the Unit 2 Teaching Resources booklet.

Materials Organizer

Chapter	Page	Activity and Materials per Student Group
4	66	**Exploration 1**: 3 or 4 of the following: spider, ant, sow bug, grasshopper, earthworm, caterpillar, millipede, and lizard; shoe box; sheet of plastic wrap and piece of cloth or cardboard to cover shoe box; large rubber band; straight pin; metric ruler; watch or clock (See Advance Preparation below.)
	68	**Limbs and Feet:** live snail; glass jar with lid; live spider
	69	**Exploration 2:** live, crawling insect; piece of paper
	72	**Exploration 4:** earthworm; sheet of stiff paper or cardboard; about 10 mL of water (See Advance Preparation below.)
5	76	**Measuring Human Growth:** 2 meter sticks
	80	**In-Text Activity:** metric measuring tape; leaf from seedling and leaf from full-grown tree (same species); graph paper
	81	**Exploration 1:** 10–20 quick-growing seeds such as mung beans, alfalfa, carrot, or radish seeds; a variety of seed-growing materials, such as several paper towels, 2–3 small plastic cups, 2–3 small jars, and a few cupfuls of soil, sand, sawdust, or cotton; 1–2 L of water; alcohol thermometer
6	92	**Exploration 1:** earthworm; shallow pan; about 100 mL of water; flashlight; about 10 mL of vinegar; several paper towels; safety goggles
	103	**Exploration 3:** 2 glass bottles (with lids), one twice as large as the other; 2 alcohol thermometers; pitcher of hot water; watch or clock; safety goggles; oven mitts
7	114	**How to Use a Microscope:** microscope; prepared microscope slide
	115	**Exploration 1:** microscope; microscope slide with coverslip; eyedropper; letter e from a newspaper; forceps; a few drops of water; tissue
	116	**Exploration 2:** microscope; 2 microscope slides with coverslips; about 10 mL of water; eyedropper; one-quarter of an onion; forceps; drop of iodine starch-test reagent; prepared microscope slide of human cheek cells; safety goggles; lab aprons (See Advance Preparation below.)
	118	**Exploration 3:** microscope; 8–10 microscope slides with coverslips; knife; 2 straight pins; 4–5 drops of iodine starch-test reagent; about 25 mL of water; eyedropper; thin sections of potato with skin, banana peel, and orange peel; tiny piece of beef liver; safety goggles; lab aprons; optional materials: small pieces of lettuce, green pepper, carrot, tea, nutmeg, pepper, mustard, coffee, and ginger

Advance Preparation

Lesson 2, page 65: 14 days before the lesson, start growing pea or bean seedlings near a window.

Exploration 1, page 66: A day or two before the activity, ask students to consider the environments of the animals to be observed and to duplicate that environment as closely as possible in a container.

Exploration 4, page 72: Earthworms should be obtained just before this exercise. Live earthworms can be caught in moist soil or purchased at a bait store. They can be kept alive for a few days in a refrigerator.

Exploration 2, page 116: Cut onions into quarters before class.

Unit Compression

Perhaps the most important goals of this unit are (1) to teach students a respect for living things and (2) to give students confidence in their ability to observe their surroundings. For that reason, all Explorations and other activities in this unit that involve direct observations of organisms should be considered essential. Examples include Exploration 2 of Chapter 4 (page 69) and Exploration 1 of Chapter 6 (page 92). These Explorations also provide an opportunity for you to observe students as they progress toward achieving these two goals.

If time limitations become a problem, Chapter 6 can be shortened without severely disrupting the conceptual flow of the unit. For instance, Lesson 2, You Be the Scientist, could be omitted or assigned as homework. Similarly, you may wish to skip Lesson 4, Adaptation of Structure, since it deals with responses that occur over the course of many generations.

Chapter 7 should not be omitted since it introduces students to the proper use of a microscope. However, to save time, you may wish to provide students with slides of the letter e for Exploration 1 on page 115. (Instead of making wet mounts of the letter e, simply use transparent tape to attach copies of the letter to microscope slides.) You can spend time teaching students how to make wet mounts for Exploration 2 on page 116. For more details, see the margins of this Annotated Teacher's Edition on those pages.

Homework Options

Chapter 4	See Teacher's Edition margin, pp. 64, 69, and 72 Exploration Worksheet, p. 9 SourceBook, pp. S24 and S34
Chapter 5	See Teacher's Edition margin, pp. 77, 78, 80, 82, 86, 87, and 89 Activity Worksheet, p. 19 Graphing Practice Worksheet, p. 21 Activity Worksheet, p. 23 SourceBook, p. S59
Chapter 6	See Teacher's Edition margin, pp. 94, 97, 101, 103, 104, 107, and 110 Activity Worksheet, p. 36 SourceBook, p. S30
Chapter 7	See Teacher's Edition margin, pp. 113, 114, 119, and 121 Resource Worksheet, p. 44 SourceBook, pp. S36 and S37
Unit 2	Unit 2 Activity Worksheet, p. 59 SourceBook Activity Worksheet, p. 73

Assessment Planning Guide

Lesson, Chapter, and Unit Assessment	SourceBook Assessment	Ongoing and Activity Assessment	Portfolio and Student-Centered Assessment
Lesson Assessment Follow-Up: see Teacher's Edition margin, pp. 64, 67, 72, 79, 89, 95, 99, 104, 109, 115, and 119 **Chapter Assessment** Chapter 4 Review Worksheet, p. 13 Chapter 4 Assessment Worksheet, p. 16* Chapter 5 Review Worksheet, p. 25 Chapter 5 Assessment Worksheet, p. 29* Chapter 6 Review Worksheet, p. 39 Chapter 6 Assessment Worksheet, p. 42* Chapter 7 Review Worksheet, p. 54 Chapter 7 Assessment Worksheet, p. 57* **Unit Assessment** Unit 4 Review Worksheet, p. 60 End-of-Unit Assessment Worksheet, p. 64*	SourceBook Review Worksheet, p. 75 SourceBook Assessment Worksheet, p. 79* 	Activity Assessment Worksheet, p. 69* **SnackDisc** Ongoing Assessment Checklists ♦ Teacher Evaluation Checklists ♦ Progress Reports ♦	Portfolio: see Teacher's Edition margin, pp. 66, 77, 97, 108, and 117 **SnackDisc** Self-Evaluation Checklists ♦ Peer Evaluation Checklists ♦ Group Evaluation Checklists ♦ Portfolio Evaluation Checklists 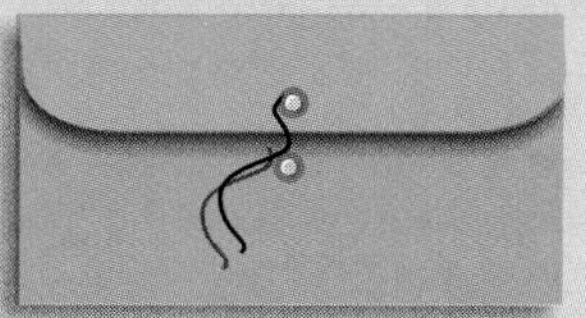 ♦

* Also available on the Test Generator software
♦ Also available in the Assessment Checklists and Rubrics booklet

Science Discovery is a versatile videodisc program that provides a vast array of photos, graphics, motion sequences, and activities for you to introduce into your *SciencePlus* classroom. *Science Discovery* consists of two videodiscs: Science Sleuths and the Image and Activity Bank.

Using the *Science Discovery* Videodiscs

Science Sleuths: The Plainview Park Vandals
Side A

Teenagers are accused of vandalizing an urban park. A park ranger gathers a variety of evidence that leads her to believe that the teenagers are the culprits. The Science Sleuths must analyze the evidence for themselves to determine the true cause of the vandalism.

Interviews
1. Setting the scene: Teenagers accused of vandalism **5200 (play ×2)**

2. Park administrator **5763 (play)**

3. Jogger in the park **6524 (play)**

4. Alien theorist **6900 (play)**

5. School guidance counselor **7922 (play)**

6. News broadcast on alien sightings **8379 (play)**

Documents
7. Park brochure **9369 (step ×2)**

Literature Search
8. Search on the words: PLAINVIEW, VANDALISM, WILDLIFE, LAKES, PARKS, DOGS **9373 (step)**

9. Article #1 ("Wildlife Problems Causing City Financial Crunch") **9376 (step ×2)**

10. Article #2 ("More Than a Squirrel") **9380 (step ×2)**

11. Article #3 ("Police Stymied by Vandals") **9384 (step)**

12. Article #4 ("Youth Gangs: New Menace?") **9387**

Sleuth Information Service
13. Animal track chart **9389 (step ×2)**

14. Map of park **9393**

15. Map of county **9395**

16. Storks **9397 (step ×2)**

Still Photographs
17. Mudslide **9401**

18. Cut and broken trees **9403 (step)**

19. Pile of sticks **9407**

20. Footprints found in park **9409**

21. Outflow drain clogged by sticks and mud **9411**

22. The lake **9413**

23. UFO photos shown on news broadcast **9415 (step ×2)**

Image and Activity Bank
Side A or B

A selection of still images, short videos, and activities is available for you to use as you teach this unit. For a larger selection and detailed instructions, see the Videodisc Resources booklet included with the Teaching Resources materials.

4-2 Moving Around, page 65
Cheetah 49001–49629 (play ×2) (Side B only)
Cheetahs are well known for their stealthy stalking and for their ability to run up to 117 km/h.

Ducks swimming underwater 53147 (play ×2) (Side B only)

Gibbon swinging 46945–47336 (play ×2) (Side B only)
To brachiate means to move by swinging from branch to branch with the arms. Gibbons and some other arboreal monkeys are the only animals that do this.

Kestrel hovering 44528–45001 (play ×2) (Side B only)
The round shape of the upper surface of a kestrel's wings creates turbulence that reduces the air pressure above the wings. The pressure below the wings lifts the bird.

Honeybee; dance 51738–52188 (play ×2) (Side B only)
When a bee finds a new source of nectar, it returns to the hive and performs a coded dance that conveys the direction and distance of the source of nectar to the other bees.

◀ Step Reverse    Play ▶    Pause ❚❚    Step Forward ❚▶

Geese flying 45002–45513 (play ×2) (Side B only)
Geese flock except when nesting. This behavior protects them from predators and teaches migration routes to young geese. Some species fly in formation, a pattern that makes flight easier for all but the leader, so leadership rotates.

Venus' flytrap 39856–40125 (play ×2) (Side B only)
A Venus' flytrap uses changes in pressure to open and close its traps.

4-3 Locomotion on Land, page 68
Gecko; foot 3693
Geckos can climb on vertical walls because of the plates on their feet. The plates are covered with microscopic hairs that use friction to stick to surfaces.

Grunt, blue-striped; scales 3542
Some scales are colored to camouflage the fish. Many scales have growth rings that can be used to estimate the age of certain fishes.

Ciliates 43820–44122 (play ×2) (Side B only)
Ciliates are among the most successful creatures in the microscopic world. They are driven through the water by the movement of a coat of beating hairs called cilia. Cilia also create currents that waft tiny food particles into the organisms' gullets.

5-1 Human Growth, page 76
Embryo, human; 40 days; 4220–4222 (step ×2)
The organ in the middle of this embryo is its heart. Note the tail (which will disappear later), the limb buds, and the eye. (step) This fetus is about 11 weeks old. Bone formation begins as cartilage structures start to become bone. Arms, legs, and digits are formed.

Note the jawbone and the lack of skull bones. (step) This fetus is between its third and fourth months. Note the more developed skull. The bridge of the nose has formed, appendages are fully formed, and joints are beginning to develop.

5-2 Other Growth Patterns, page 80
Plants growing 36764–37213 (play ×2) (Side B only)
Leaves and stems of most plants are above ground, where light, oxygen, and carbon dioxide are available. Roots are usually in the soil, which supplies water and minerals. When conditions are adequate, plants will grow.

Budding; sea-star arms 4150–4152 (step ×2)
This sea star had nine arms originally and is growing nine additional new ones. (step) This sea star is growing an arm to replace one that has been lost. (step) The object on the lower right side of this frame is a severed sea star arm that will regenerate into a complete new sea star. These one-armed sea stars are called comets.

Planaria 3126
Flatworms are simple animals with eyespots. They have bilateral symmetry, which means the left half of their body is the same as the right half.

6-1 Stimulus and Response, page 92
Stimulus; on a worm 37215–37898 (play ×2) (Side B only)
An earthworm reacts to tactile stimulation.

6-3 Seasonal Behavior of Animals, page 100
Ptarmigan, willow 3792–3793 (step)
This camouflaged willow ptarmigan is sitting on her ground nest. She has brown summer plumage. (step) This willow ptarmigan has winter plumage.

Grebes; courting 50536–51140 (play ×2) (Side B only)
Grebes perform elaborate dances during the mating season. The dances stimulate them to release reproductive hormones. Bird courtship often involves ritualized movements related to nestbuilding and feeding the young.

Gazelles, Thompson's; head-butting 51141–51592 (play ×2) (Side B only)
Competition for females becomes fierce during mating season when males of many horned-animal species butt each other.

6-4 Adaptation of Structure, page 105
Bird Identification 3876–3881 (step ×5)
General body shapes; (step) size; (step) bill shape; (step) legs and feet; (step) special markings; (step) tail shape and markings

Skimmer feeding 47833–48053 (play ×2) (Side B only)
A skimmer drags the lower half of its bill through the water to feel for prey. When a prey animal is struck, the bill snaps shut.

Spoonbill nesting 51593–51737 (play ×2) (Side B only)
Spoonbills build large platform nests on island trees. Birds nest in trees to escape predators.

Lizard, frilled; display 50083–50535 (play ×2) (Side B only)
Here the frilled lizard is not displaying its ruff. (play) Here the frilled lizard raises its ruff to intimidate predators. (play) The ruff makes the lizard appear much larger and more formidable than it really is.

Platypus, duckbill 3914
With webbed feet, a broad flat tail, no external ears, and a sensitive skin-covered beak, the platypus is adapted to aquatic life. It

burrows into river banks and lays and broods its eggs in a grass-lined den. Males have a poisonous spur on their foot.

7-1 Taking a Closer Look, page 113
How to use a microscope 26906–29346 (play ×2) (Side B only)
This sequence identifies all the parts of a microscope and demonstrates how to change lenses and adjust the diaphragm. It also shows how to focus, how to avoid air bubbles, and how and why to focus while scanning.

How to make a slide 29347–30380 (play ×2) (Side B only)
To make a wet-mount slide, place a small piece of the specimen on the slide using tweezers. Add enough water to cover it, and then lay the coverslip on the droplet. Be careful not to trap any air bubbles.

7-2 Skilled Observations of Cells, page 116
Cell under the microscope 2670–2672 (step ×2)
Three possible things that might occur while looking through a microscope, and what to do about them

Cells; cheek 2669
Cheek cells as seen under a powerful, modern microscope

Cells, animal and plant 2692–2705
Structures of an animal cell (step ×5) Structures of a plant cell (step ×8)

Unit 2 PATTERNS OF LIVING THINGS

UNIT FOCUS

Ask students to define the word *biology.* *(Accept all reasonable answers.)* Explain to students that biology is the study of living things. Tell students that in this unit they will learn about some of the characteristics that all living things have in common.

A good motivating activity is to let students listen to the English/Spanish Audiocassettes as an introduction to the unit. Also, begin the unit by giving Spanish-speaking students a copy of the Spanish Glossary from the Unit 2 Teaching Resources booklet.

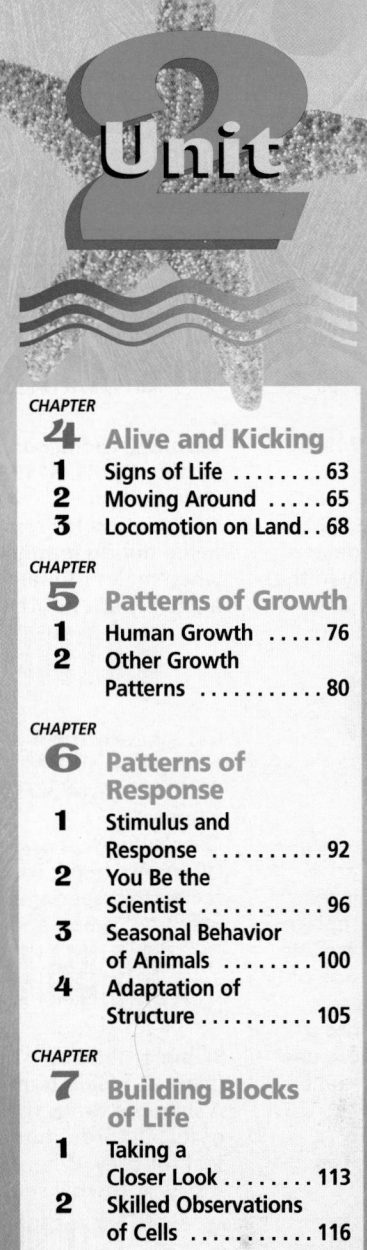

Unit 2

PATTERNS OF LIVING THINGS

The basilisk lizard, a native of Central and South America, is able to run across the surface of water.

Connecting to Other Units

This table will help you integrate topics covered in this unit with topics covered in other units.

Unit 1 Science and Technology	The ability to correctly use a microscope is essential to many scientific investigations.
Unit 3 It's a Small World	Students will need to understand the structure and function of cells in order to study microorganisms.
Unit 6 Energy and You	Endothermic organisms utilize energy from food in order to maintain a constant body temperature.
Unit 7 Temperature and Heat	The rate at which an animal loses heat depends on both its size and type of body covering.
Unit 8 Our Changing Earth	Biological adaptations are responses to the type of environment in which organisms live.

According to European legends, basilisks were creatures born from eggs that were laid by cocks and hatched by snakes. These fierce creatures were said to possess magical powers that allowed them to destroy plants and animals with just a look. Named for its resemblance to the legendary creature, the basilisk lizard has a long snakelike tail and a flap similar to a cock's comb on its head.

Despite its appearance, the basilisk is a harmless lizard. It lives near tropical rivers and feeds on insects and plants. Although it usually walks on all fours, the basilisk rears up on its hind legs when frightened and races toward the water. The basilisk can sprint across the water's surface for several meters before sinking and swimming away. Its tail helps it balance on two legs, and its widespread toes are fringed with long scales that help it skim the surface of the water. The basilisk's method of escaping danger is an example of how living things respond and adapt to their environment.

In this unit, you will examine the behavior and structures of various living things to see how they respond to their environment. Compared with some of the things that you will encounter in this unit, the fact that a lizard can run on water may not seem strange to you at all.

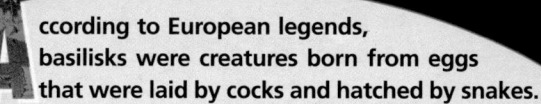

Using the Photograph

Ask: Why would the ability to move across the surface of the water be beneficial to this organism? *(Accept all reasonable responses. Students may suggest that the lizard is able to escape predators, for instance.)*

ENVIRONMENTAL FOCUS

Ask: Why is it important to save animals or plants that are in danger of extinction? *(All organisms depend on other organisms for their survival—often in very subtle ways. As a result, when one organism becomes extinct, the whole environment can be affected.)* Have students think of an animal or plant and describe why that organism is important to the environment in which it is found.

Connecting to Other Chapters

Chapter 4
examines the characteristics of living things, their motion, and the differences between plants and animals.

Chapter 5
investigates a variety of patterns of growth, in both humans and other organisms.

Chapter 6
explores the responses that organisms can have to stimuli and to long-term changes in their environment.

Chapter 7
introduces students to both the structure of cells and the proper use of a microscope.

Prior Knowledge and Misconceptions

Your students' responses to the ScienceLog questions on this page will reveal the kind of information—and misinformation—they bring to this chapter. Use what you find out about your students' knowledge to choose which chapter concepts and activities to emphasize in your teaching.

In addition to having students answer the ScienceLog questions on this page, you may wish to have them write about the following hypothetical situation: Imagine that you are an interplanetary explorer who has been sent from a civilization at the far end of the galaxy to make a chart of all the life-forms on Earth. How would you distinguish between living and nonliving objects? Write down all of the criteria that an object would have to meet before you would classify it as living. Remind students that there are no right or wrong answers to this exercise. Collect the papers, but do not grade them. Instead, read them to find out what students

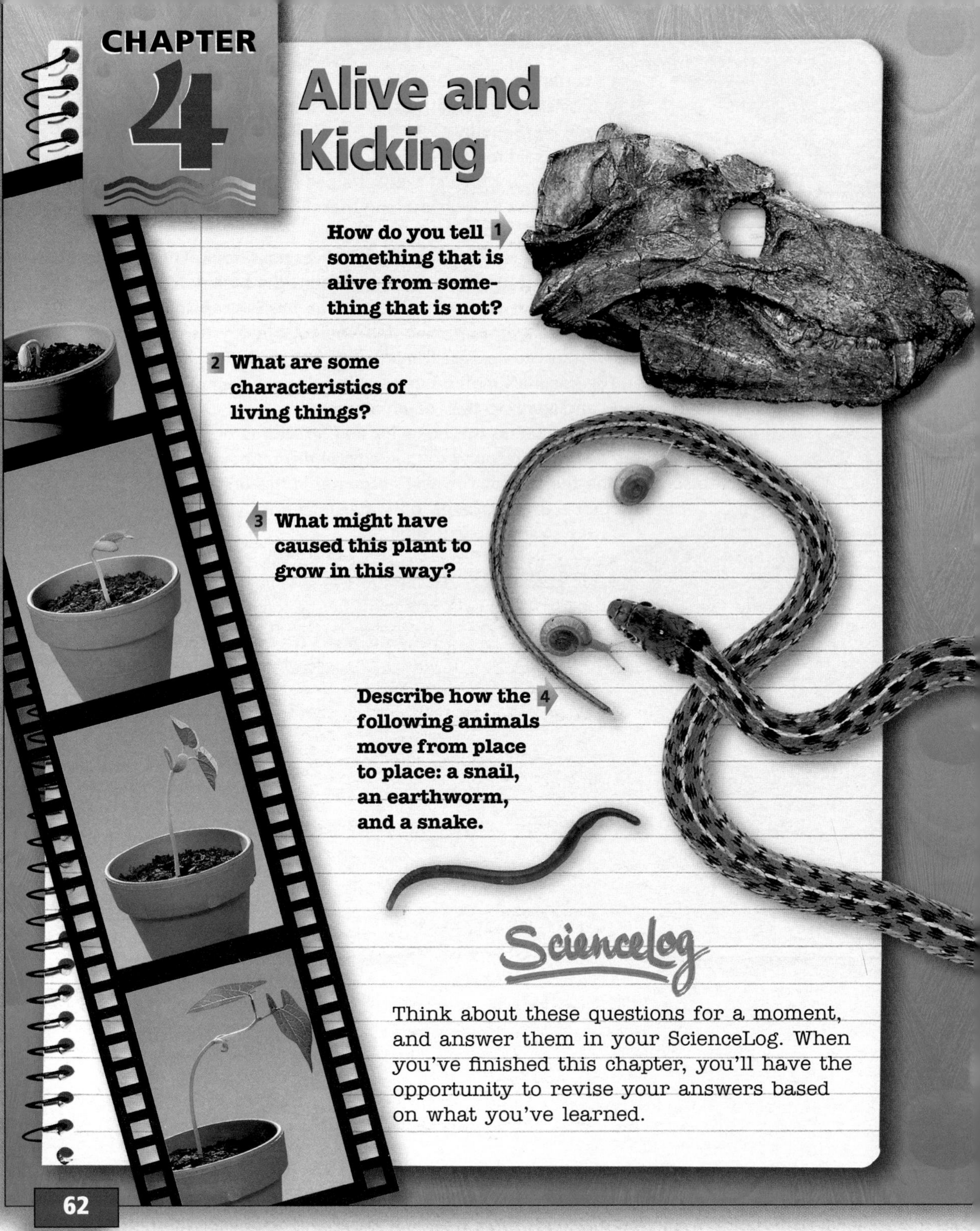

CHAPTER

4

Alive and
Kicking

How do you tell 1 something that is alive from something that is not?

2 What are some characteristics of living things?

3 What might have caused this plant to grow in this way?

Describe how the 4 following animals move from place to place: a snail, an earthworm, and a snake.

ScienceLog

Think about these questions for a moment, and answer them in your ScienceLog. When you've finished this chapter, you'll have the opportunity to revise your answers based on what you've learned.

62

know about the characteristics of living things, what misconceptions they may have, and what they find interesting about this topic.

Did you know that there are over 2 million different kinds of animals and plants in the world? That's a lot of living things! But how can you tell whether everyday things are living or nonliving? What are the signs of life? Nonliving things, such as water or a rock, have never been alive. Something that is dead was once alive but no longer has any of the signs of life. A fish being cooked in a frying pan is dead, but the metallic pan in which it is cooked is nonliving. How would you classify a leather wallet—as living, dead, or nonliving?

Look at the object or objects pictured in each of the postage stamps on this page. Sort the objects on the stamps into three lists: living, dead, and nonliving. Write down your reason(s) for classifying each object as you did. **B**

63

LESSON

1 Signs of Life

FOCUS

Getting Started

Write the headings Living, Dead, and Nonliving on the board. Ask students to name items in the classroom that would fall under each heading. Then ask students how they know whether an object is living, was once living, or was never alive. Encourage discussion about what makes something alive and allow students to express their ideas freely.

Main Ideas

1. All living things share certain characteristics.
2. Plants and animals are living things.

TEACHING STRATEGIES

GUIDED PRACTICE After discussion of the classification activity, discuss the following situation with students: Santos knows that the tree in his backyard is a living object. Describe at least three signs of life that he might observe. *(Sample answer: It reproduces by means of seeds; it grows; and it responds to stimuli by losing leaves in the fall.)*

Answers to
In-Text Questions

A Since leather was once part of a living organism, it can be considered dead.

B Answers will vary depending on the reasons students give, but lists may include the following: living: first man on the moon, hippopotamus, flowering plant, penguins, antelope, turtle, tree frog, tree; dead: Copernicus, *Homo habilis*, pioneers; nonliving: amethyst, rock formation, minerals.

LESSON 1 ORGANIZER

Time Required
2 class periods

Process Skills
observing, classifying

New Terms
Dead—no longer having the signs of life
Living—having the signs of life
Nonliving—having never been alive

Materials (per student group)
none

Teaching Resources
SourceBook, p. S24

You Be the Judge

The following statements were made by some students who were discussing the differences between living and nonliving things. Do you agree or disagree with their ideas? Explain in writing how you would respond to each student.

Velma: Smoke is alive because it can hurt you, and volcanoes are living because they rumble.

Olga: Clouds and wind are living because they have energy, and energy makes things live.

Blair: A river is living because it moves; soil is nonliving because it does not move.

Paul: A candle flame is living because it gives off heat and light. It wiggles around and can make new flames.

It is not a simple matter to explain how living, dead, and nonliving things differ. Living things share, not just one, but a number of common characteristics, or signs of life. Write down the signs of life that you think are common to all living things. Ⓐ

Roberto: Crystals are alive. I have seen them grow larger.

Lifestyles of the Green and Leafy?

Look again at the list you made of the stamps that showed living things. Are all of the objects alive in the same way? Divide these living things into two further groups to show two different ways in which things may be alive.

You may have chosen to divide the living things into plants and animals. With the help of your classmates, prepare a brief statement that explains how a plant differs from an animal.

> If you'd like to learn more about how living things are classified, turn to page S24 in the SourceBook.

Answers to *You Be the Judge*

Although all of the characteristics mentioned might be qualities of some living things, students should see that none of these examples is living. A living thing possesses a combination of important traits like those shown here.

Answer to *In-Text Question*

Ⓐ Students may suggest that all living things eat, move, grow, and reproduce.

Answers to *Lifestyles of the Green and Leafy?*

Students may respond that all objects are alive in the same way because they share the characteristics of life, such as growth, reproduction, and movement. Or some students may suggest that the organisms pictured are alive in different ways because they live in different environments, eat different foods, and reproduce differently.

Many students will divide the organisms into plants and animals, although other responses are acceptable.

Statements about how a plant differs from an animal may include the following: Most plants are rooted in soil and cannot move quickly; most animals can move quickly. Most plants have green leaves and make their own food; animals get their food from other organisms. Animals have nervous systems, digestive systems, and sensory organs; plants do not.

Homework

The Closure activity on this page makes an excellent homework assignment.

FOLLOW-UP

Reteaching

Choose one photograph of a plant and one photograph of an animal. Ask students to describe the signs of life associated with each living thing.

Assessment

Distribute a set of magazine pictures of a variety of plants, animals, and inanimate objects (dead and nonliving) to small groups of students. Have them classify the pictures as living, nonliving, or dead and explain their reasoning.

Extension

Have interested students choose an organism and research the following: how the organism moves, what and how it eats, how it reproduces, and how it grows.

Closure

Have students make a collage of pictures showing the signs of life. Students may bring in photographs or drawings that illustrate growth, movement, reproduction, and energy use, for instance.

Moving Around

After looking at the stamps on page 63, you may have decided that one of the characteristics of living things is the ability to move. Do plants have this characteristic? Take a look at the photographs at right of a mimosa plant. The leaves of the mimosa sometimes close up very suddenly. Can you give a reason for this kind of motion? (Hint: The mimosa is also known as the "sensitive plant.") **B**

Plant motion is the result of a plant's reaction to changes in its environment. Because plants usually move rather slowly, we do not easily notice it. Can you think of any other instances in which a plant moves? Read the following to find out how one student investigated other types of plant movement.

Mimosa plant—before (top) and after (bottom) a stimulus

Plants on the Move

Henry decided to conduct some experiments in order to learn more about plant movement and growth. Using four identical seedlings, he performed the experiments shown below. Examine the illustrations, and then answer the following questions:

1. For each experiment, make a prediction about what you think will happen. In what direction do you think the plants will grow? Why do you think so?

2. In each experiment, Henry set up a *control*. Which are the controls? Why is this step important?

3. If you carried out Experiment 2, what characteristics of living things do you think your plants would demonstrate?

4. Try these experiments for yourself at home! Were your predictions correct?

Experiment 1

Experiment 2

65

LESSON 2 ORGANIZER

Time Required
1 to 2 class periods

Process Skills
observing, measuring, contrasting

Theme Connection
Structures

New Term
Locomotion—the ability of an animal to move on its own from place to place

Materials (per student group)
Exploration 1: 3 or 4 of the following: spider, ant, sow bug, grasshopper, earthworm, caterpillar, millipede, and lizard; shoe box; sheet of plastic wrap and piece of cloth or cardboard to cover shoe box; large rubber band; straight pin; metric ruler; watch or clock (See Advance Preparation on page 59C.)

Teaching Resources
Exploration Worksheet, p. 3
Transparency Worksheet, p. 6
Transparencies 8 and 9

FOCUS

Getting Started
Obtain a *Mimosa pudica* (a sensitive plant) from a nursery. Students can observe how the leaves close when touched. The mimosa's leaves also close in response to irritation, smoke, or darkness.

Main Ideas
1. Plants move in response to environmental changes.
2. Animals move on their own from place to place.
3. Animals move in a variety of ways.
4. Many forms of animal locomotion have been used as models for human inventions.

TEACHING STRATEGIES

Answer to
In-Text Question

B The mimosa plant responds by closing its leaves when touched, perhaps to avoid damage.

Plants on the Move
If possible, start growing pea or bean seedlings near a window 14 days before beginning this lesson so that students can observe how the seedlings move toward light. Let students turn the seedlings several times during this period to see how the plants respond to the change.

Answers to
Plants on the Move

In both experiments, the plants should grow toward a light source. Plant B is the control for Experiment 1, and plant A is the control for Experiment 2. Controls are used to gauge the effects of changing an experimental variable. The plants in both experiments show several characteristics of living things, including growth and motion.

Animals on the Go

GUIDED PRACTICE Ask: How is plant movement different from animal movement? (*Plant movement is usually much slower than animal movement. Also, only one part of a plant usually moves, and the entire plant does not move from one place to another under its own power. Animal locomotion involves the coordinated movement of many or all parts of the body, and the entire animal moves from one place to another.*)

EXPLORATION 1

Ask students to consider the environment in which the animal lives so that they can duplicate it in their containers. If possible, return the animal to its natural environment at the end of the Exploration.

Have each group carefully observe three or four animals. Student tables should be completed with objective and thorough observations. You may wish to discourage students from assigning human characteristics such as "scared" or "happy" to the animals.

Cooperative Learning
EXPLORATION 1

Group size: 3 to 4 students
Group goal: to analyze the locomotion of several animals
Positive interdependence: Assign each student a role such as primary investigator, recorder, pacer, or materials coordinator.
Individual accountability: Each group member should be able to summarize the group's findings and rate his or her own performance.

 An Exploration Worksheet (Teaching Resources, page 3) and Transparency 8 are available to accompany Exploration 1.

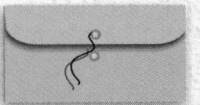 **PORTFOLIO**
Create a class data chart of student observations from Exploration 1. Suggest that students record the data chart in their Portfolio. Then they can compare their own observations with those of the entire class.

Animals on the Go

One of the first things you may notice about an animal is that it is able to move about on its own. This is called **locomotion** (*loco* is from the Latin word meaning "place"). Most animals have skeletons and muscles that work together to provide freedom of movement.

EXPLORATION 1

Observing Locomotion

You Will Need

- a shoe box
- a large rubber band
- a cloth or piece of cardboard
- several small animals
- plastic wrap
- a pin
- a watch or clock

What to Do

For this experiment, you can use any of the following small animals: a spider, an ant, a sow bug, a grasshopper, an earthworm, a caterpillar, a millipede, or a lizard. Be very gentle so that you do not harm the animal. Record your observations in a table like the one shown below.

Put one of the animals in a shoe box or other small container with a cover of plastic wrap. Poke several small holes in the wrap with a pin. Cover the container with something that will keep out light. (A cloth or piece of cardboard will do.)

Remove the covering carefully, and watch the animal closely. How does the animal move when the cover is first removed? Record your observations. Continue to observe the animal for 2 minutes. Clearly and in as much detail as possible, describe the way the animal moves. Estimate how far the animal moves during the 2 minutes.

Repeat the experiment using other animals, and compare the results. Using your data, fill out a lab report that includes the information in the table.

Wash your hands after working with any animal. When you have finished the Exploration, be sure to return the animal to the place you found it.

Questions

1. What do you think made the animals move when the cover was first removed?
2. Why do you think the animals continued to move?
3. Which body parts did each animal use for locomotion?
4. If all of the animals you observed were in a 2-minute race, which one would win?
5. What characteristic of this animal would make it the winner of such a race?
6. According to your data, which animal would come in last? Can you suggest reasons why?

Piece of cloth

Plastic wrap   Shoe box   Rubber band

Name of animal	How the animal moved when the cover was first removed	Description of animal movement	How far the animal moved in 2 minutes

Answers to *Questions*

1. The animals probably moved in response to the sudden increase in light.

2. Answers will depend on the animals used. The animals probably continued to move in order to get out of or into the light.

3. Answers will depend on the animals used. (For clarity, have students draw and label the different body parts used by the animal.)

4. Answers will vary depending on the animals used.

5. Students should point out locomotive characteristics that enhance speed. In general, animals that can fly or leap move the fastest.

6. Earthworms, snails, and other animals without legs probably move the slowest.

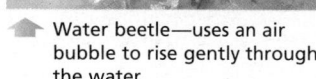

Water beetle—uses an air bubble to rise gently through the water

Bat—extends flaps of skin to catch the air

Animal Innovators

Can you imagine yourself flying through the air or swimming like a fish through the water? The animals shown here fly or swim in unique ways that make it look easy. In fact, the different ways their bodies work remind us of some of our own inventions. On this page are pictures of animals that have different ways of moving in air or water. For each animal on this page, suggest human inventions that seem to imitate the animal's movement.

Hummingbird—beats its wings so fast that they are only a blur; can hover and even fly backward

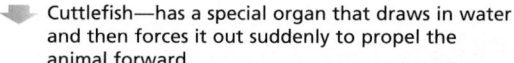

Cuttlefish—has a special organ that draws in water and then forces it out suddenly to propel the animal forward

Beaver—webs of skin between its toes help to push the water

67

Theme Connection

Structures
Focus question: How do the structures of modern aircraft resemble the flying structures of birds? *(Both birds and aircraft must be lightweight yet durable. They must have a long wingspan to catch the wind, and their weight must be evenly distributed. Both birds and aircraft must have landing gear appropriate for the type of landing surface they use.)*

Integrating the Sciences

Earth and Life Sciences
The wide variety of animal innovations (some of which are pictured on this page) are largely adaptations to the particular environment in which that animal lives. Have students make a list of at least five animal innovations other than the ones pictured on this page. In addition, students should describe the environment in which each animal lives and explain why the innovation makes the animal well suited to its environment.

Some possible student answers for human inventions could include the following: bat—hang glider or airplane; water beetle—diving bell or hot-air balloon; hummingbird—helicopter or hovercraft; cuttlefish—boat engine or jet aircraft; and beaver—diving fins or flippers.

FOLLOW-UP

Reteaching

Instruct students to choose an animal that they did not investigate in Exploration 1 and to follow the steps outlined in that Exploration to study the method of locomotion of this animal. Students can use drawings to describe how the animal moves.

Assessment

Have students write their own definition of the word *movement* as it applies to living things, and have them explain why movement is important to organisms. *(Movement is a change of body position or location. It allows organisms to respond to their surroundings.)*

Extension

Some of humanity's early attempts to fly involved attaching wings to a person's body. Have students do research to find out about one of these early attempts to fly, and have them write a report that explains how humans imitated the locomotion of animals in these attempts.

Closure

As a class, combine all of the data collected in Exploration 1, and classify the animals studied according to how they moved.

★ **A Transparency Worksheet (Teaching Resources, page 6) and Transparency 9 are available to accompany the material on this page.**

FOCUS

Getting Started

Tell students to sit on the floor, take off their shoes (leaving on their socks), and try to make their feet move forward by wiggling their toes. Explain to students that this is a rough approximation of how snakes move.

Main Ideas

1. Many animals use legs, arms, or other specialized limbs for locomotion.
2. Locomotion can be accomplished with one limb, many limbs, or no limbs at all.

TEACHING STRATEGIES

Limbs and Feet

You may wish to provide students with snails and spiders that are already in jars with perforated lids. Spiders can be refrigerated for 30 to 60 minutes to slow down their leg movements.

 Spiders can bite, and the mucous trail of snails is mildly toxic. Wash snail trails off surfaces with soap and water.

Answers to
In-Text Questions and Caption

(A) Students may note a rippling or shifting motion of the snail's foot.

(B) Apes can appear taller or carry objects for short distances by walking on two legs.

(C) Flightless birds include the ostrich, cassowary, emu, rhea, kiwi, and penguin.

(D) A snail's foot is located underneath the shell; it extends from the shell and is visible when the snail moves.

LESSON 3
Locomotion on Land

▲ Two-footed animals

Limbs and Feet

Many animals rely on the use of limbs for locomotion. A limb can be a leg, an arm, or a wing, all of which are extensions from the body used for locomotion. Some animals move about on

Where is the snail's foot? (D)

many limbs. Other animals have only one limb; slugs and snails belong to this group. Their one limb is known as a foot. Put a snail in a glass jar, and observe how it uses its foot to move itself. Describe what you see. (A)

Most animals that walk on land have four or more limbs. Some of these animals (for example, bears, rabbits, and squirrels) can stand for a time on their hind legs to feed or to look around for danger. Gorillas, chimpanzees, and other great apes can walk on either two or four limbs. Walking on two limbs is a little awkward for the great apes. Why would they choose to do this? (B)

Birds have four limbs; however, the two front limbs have developed into wings. Birds stand and walk on their two hind limbs. Did you know that not all birds can fly? Try to name three flightless birds. (C)

The kangaroo stands on its two hind limbs, using its heavy tail to keep its balance. People stand and walk on two limbs. This is called an *erect posture*. Human babies take time to develop this skill. See if you can find out from your parents how old you were when you were able to balance on two legs for the first time.

Insects have six legs. Imagine how much more complicated locomotion is with six legs than with two legs. There are hundreds of thousands of different kinds of insects, and there are many variations in the way they move around on their legs.

Spiders are fascinating eight-legged animals. A spider can walk on the thin threads of the web it spins. Some spiders, such as the tarantula above, are excellent jumpers. Watch a spider and see if you can figure out the order in which it moves its legs.

68

LESSON 3 ORGANIZER

Time Required
3 class periods

Process Skills
observing, comparing, predicting

New Terms
none

Materials (per student group)
Limbs and Feet: live snail; glass jar with lid; live spider
Exploration 2: live, crawling insect; piece of paper
Exploration 4: earthworm; sheet of stiff paper or cardboard; about 10 mL

of water (See Advance Preparation on page 59C.)

Teaching Resources
Exploration Worksheets, pp. 8, 9, and 11
Transparency 10
SourceBook, p. S34

Solve the Insect Movement Mystery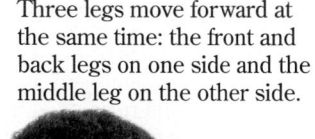

Put a crawling insect on a piece of paper. Watch carefully to see how the insect moves. Decide which of the following things happens:

a. Both of the front legs move forward at the same time.

b. Both of the middle legs move forward at the same time.

c. Both of the back legs move forward at the same time.

d. All three legs on one side move forward at the same time.

e. On one side, the front and middle legs move forward at the same time.

f. On one side, the front and back legs move forward at the same time.

g. All of the legs move at different times.

h. Three legs move forward at the same time: the front and middle legs on one side and the back leg on the other side.

i. Three legs move forward at the same time: the front and back legs on one side and the middle leg on the other side.

EXPLORATION 3

How's Your Horse Sense?

Below is a sequence of pictures showing a horse walking. The pictures are arranged to show the order in which steps are taken. Study the set of pictures carefully. Then answer the following questions.

Questions

1. As the horse walks, how many hooves touch the ground at a time? Are all of the horse's hooves off the ground at any one time?

2. Describe the way the horse moves its legs throughout the walking sequence—for example, "right foreleg first, left hind leg second," and so on.

3. Have you ever seen a baby crawl? How does the horse's walking gait compare to the movements of a crawling baby?

Exploration 3 continued ▶

69

CROSS-DISCIPLINARY FOCUS

Mathematics

Provide students with the *Guinness Book of World Records* and have them look up the speeds of the fastest animals, including the fastest land animal, the fastest bird, and the fastest aquatic animal. Have them make a bar graph of these speeds. Then have students calculate their own running speed and add it to the graph.

Homework

The Exploration Worksheet that accompanies Exploration 3 may be assigned as homework (Teaching Resources, page 9). Transparency 10 is also available if you use this worksheet in class.

EXPLORATION 2

Students should discover that while an insect is walking, it always has at least three legs supporting its body. The supporting legs are positioned like a triangle. For example, during a step, an insect may move the middle leg on one side with the front and rear legs on the other side. During the next step, the other three legs are moved together.

★ **An Exploration Worksheet accompanies Exploration 2 (Teaching Resources, page 8).**

EXPLORATION 3

It is very difficult to observe the sequence of a horse's leg movements while the horse is galloping. When a horse gallops, at one point in the stride, all four legs are in the air simultaneously. Students might be interested to know that this was not proven until 1877. Photographer Eadweard Muybridge arranged 24 cameras with very rapid shutter speeds to photograph a moving horse. He attached strings to the camera shutters. These strings were tripped by the horse as it passed through them. The images on this page were adapted from photographs like those of Muybridge.

Answers to *Questions*

1. At least two hooves usually touch the ground during a walk. At no time are all four hooves off the ground while the horse is walking.

2. The walking sequence is right hind leg, right foreleg, left hind leg, left foreleg.

3. They are similar in that both alternate limbs from front to back and from side to side.

Exploration 3 continued ▶

Answers to *Comparing Structures*

1. For a labeled diagram of the horse's leg, see page S234.

2. The structure of the human wrist and hand is very complex. Four fingers and a thumb are used to grasp objects, while the wrist joint allows for a great deal of flexibility. The thick bones of the horse's upper leg support a lot of weight, while the hoof is flat and provides traction.

Meeting Individual Needs

Second-Language Learners

Have students trace the human and horse bones shown in Exploration 3 onto a separate piece of paper and color homologous bones the same color. Have them label the bones in the horse's leg with the names of the corresponding human bones.

Legless Wonders

Have students study the pictures on pages 70–71 and read the text aloud. Explain to students that snakes typically move in an S-shaped wave by moving their bodies from side to side and pushing off against small objects such as pebbles and plants. Some snakes move in a straight line, particularly when creeping up on prey.

Answers to *In-Text Questions and Caption*

Ⓐ Just as a human has difficulty getting traction on ice, a snake has difficulty getting traction on a very smooth surface such as glass.

Ⓑ The snake would have difficulty moving along a glass tabletop because the friction is reduced. Its scales would not be able to gain traction on the surface.

Comparing Structures

1. The illustrations below show a horse's forelimb and its human counterpart. Find the equivalents of the fingernails, fingers, hand, wrist, forearm, elbow, and upper arm in the horse's leg.

2. How is each structure in each forelimb well suited to its function?

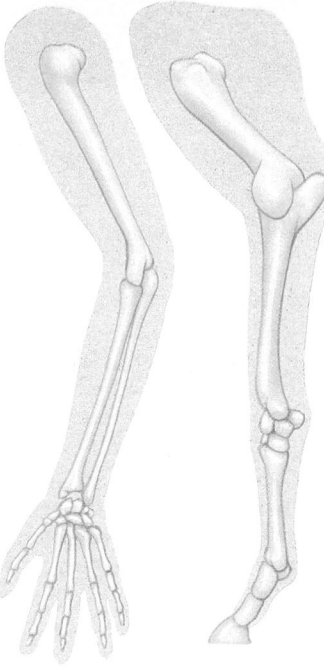

Legless Wonders

You have seen examples of animals that swim, fly, and walk. Animals that fly have wings, animals that swim have fins or flippers, and animals that walk have legs. Of the animals that walk, some have only two legs, while others have hundreds! Some animals even have no legs at all.

Snakes are closely related to lizards and alligators, yet they differ in one major way: snakes have no legs. Even so, they have no trouble moving about. Snakes are able to move easily through tunnels, crevices, or thick vegetation. Many snakes can even climb trees. How do they do it?

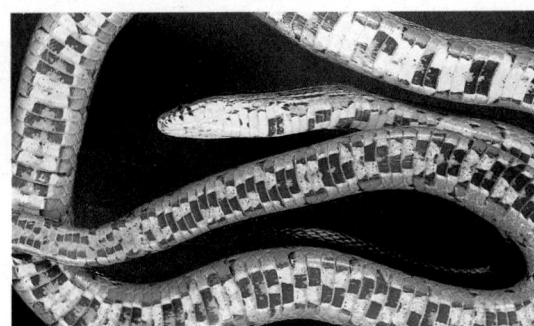

A snake's belly is covered with tough flaps of skin called scales.

Think for a moment about how you move as you walk. Your feet push against the ground, your leg muscles pull you forward, and then you push off.

A snake's scales work somewhat like your feet do. The scales catch on rough surfaces just as your feet do. Once the scales have something to push against, the snake moves forward. However, on very smooth surfaces such as glass, snakes have difficulty moving about. Why do you think this is so? Hint: What happens when you try to walk on ice? **Ⓐ**

Snakes often move by first bending their body into a series of S-shaped loops. The loops push off from projections, such as pebbles and small plants, to move the snake forward. As the snake moves, it resembles an S-shaped wave in motion.

Would this snake have any trouble moving along a glass tabletop? **Ⓑ**

Did You Know...

Stories about the speed at which some snakes can travel are greatly exaggerated. There are legends of African black mambas overtaking galloping horses. Even though such stories are not true, some snakes can reach speeds of 10 km/h (6 mph).

Meeting Individual Needs

Gifted Learners

Not all snakes move in the same way. In fact, snakes employ a variety of methods of locomotion depending on factors such as the environment, the rate at which they wish to travel, or whether they need to be stealthy. Have interested students research various methods of snake locomotion and their advantages. You may wish to have some students research the locomotion of arboreal or aquatic snakes. Students should present their findings to the class in the form of an oral report.

Snakes have several other methods of locomotion. Some snakes, such as the python, bend their body in a series of tight coils. Then they use these coils as an anchor so that the front of the body can extend forward. Pythons can also move in a straight path, as shown below. Can you think of any reasons why it is advantageous for pythons to move this way and not in S-shaped curves? ⓒ

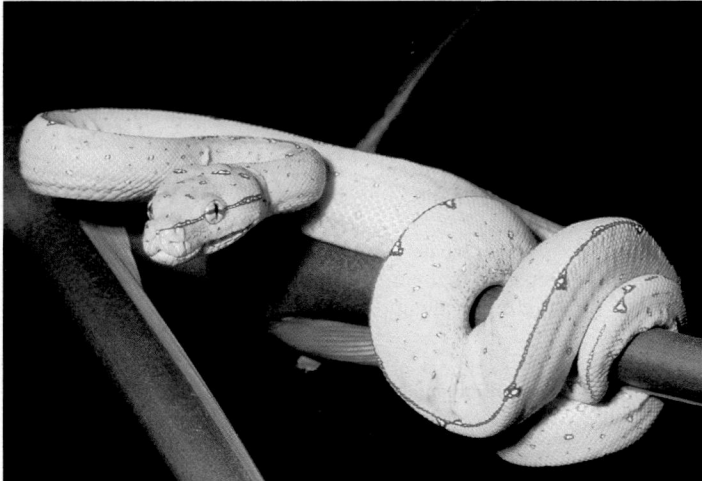

◀ Green tree python

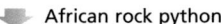

◀ African rock python

The scales on the python's belly are attached to muscles that move the scales forward and backward. In straight-line motion, one segment of the snake's skin is pulled forward, and the scales of that segment are held in place by friction with the ground. The snake's body then moves forward until it "catches up with" its skin. The snake thus moves forward even though it shows little sign of internal movement.

71

Answer to
In-Text Question

ⓒ Moving in a straight line might be less conspicuous, allowing the python to sneak up on prey.

ENVIRONMENTAL FOCUS

An ecosystem consists of all of the living organisms in a certain area as well as their physical environment. For a long time, it was thought that snakes had no ecological purpose—that is, they had no significant role in the balance of an ecosystem. We now know that this is untrue. Have interested students find out about a local species of snake, including what it eats, what eats it, and why it is important to the local ecosystem. Have them give an oral report or display their findings on a poster.

Meeting Individual Needs

Learners Having Difficulty
You may wish to have a nonpoisonous snake available in the classroom for students to feel. They may be surprised to learn that snakes are not wet and slimy, but that they have tough, scaly skin that resists drying out. Have students notice how the snake moves and how the snake's skin feels against their own skin as it uses its muscles and scales to push itself along.

EXPLORATION 4

Live earthworms can be found in moist soil or purchased at a bait store. The earthworms can be kept alive for a few days in a refrigerator, but take care not to dehydrate them.

The sound made by an earthworm moving on a stiff piece of paper may be loud enough to pick up with a microphone.

You may wish to introduce the terms *cross section* and *lengthwise (lateral) section* so that students can scientifically describe the muscle action.

 An Exploration Worksheet accompanies Exploration 4 (Teaching Resources, page 11).

Answers to
In-Text Questions

A Students may note that the bristles are used to gain traction on the surface below the earthworm.

B The earthworm uses its muscles together with its bristles to move the segments of its body.

C Students should note that when one set of muscles contracts and the other set relaxes, the earthworm moves forward. Each segment lengthens when the ring muscles contract and shortens when the lengthwise muscles contract. (Refer students to the description at the bottom of the page.)

Homework

The Reteaching activity described below makes an excellent homework assignment.

FOLLOW-UP

Reteaching

Have students make a "flip book" to illustrate the motion of a snake or another creature. Students should draw a similar-sized diagram on each page of a notebook or pad of paper, changing it slightly from one page to the next. When flipped steadily with the fingers, the sequence of diagrams should illustrate the animal's movement.

Listening to an Earthworm!

You Will Need
- an earthworm
- a piece of stiff paper
- 10 mL of water

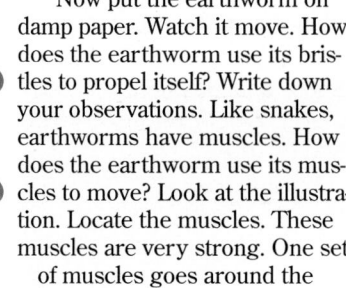

What to Do

Put an earthworm on a piece of stiff paper. Can you hear it make any noise as it moves?

Hold the paper up level with your eyes, and try to look between the animal and the paper. Do you see the bristles? Rub your finger back and forth along the lower side of the earthworm. Do you feel the bristles? When the stiff bristles are extended, they hold the animal in place on the ground. When the bristles are pulled in, the animal can slide along. When the earthworm moves along the paper, its bristles sometimes make a scraping noise.

Notice the rings on the body of the worm. The inside of an earthworm's body is divided into sections called *segments,* which show up on the outside as rings. There are four pairs of bristles for every segment. How many bristles does your earthworm have?

Now put the earthworm on damp paper. Watch it move. How does the earthworm use its bristles to propel itself? Write down your observations. Like snakes, earthworms have muscles. How does the earthworm use its muscles to move? Look at the illustration. Locate the muscles. These muscles are very strong. One set of muscles goes around the worm in rings.

Now locate the muscles that run lengthwise along the worm's body. The two sets of muscles work against each other. What happens when one set of these muscles contracts and the other set relaxes? When the ring muscles contract, what happens to the length of the segment? When the lengthwise muscles contract, what happens to the length of the segment? See if you can piece together this information about the muscles and bristles to write a description of how earthworms move. Compare your description with the one provided below. **C**

▶ **An inside and lengthwise look at an earthworm**

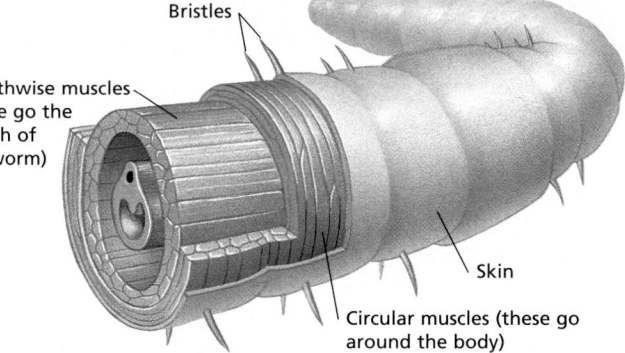

Bristles

Lengthwise muscles (these go the length of the worm)

Skin

Circular muscles (these go around the body)

Dan's Description

1. When the earthworm moves, the ring muscles in the segments at the front contract, and the lengthwise muscles relax.
2. At the same time, the bristles at the front end are pulled in, and the bristles at the back end hold the animal in place. In this way, the earthworm is "squeezed" forward.
3. The bristles at the front end are now extended to keep the front end in place, and the bristles at the back end are pulled in.
4. The ring muscles in the segments at the back relax, and the lengthwise muscles contract, causing the rear part of the body to slide forward.

Assessment

Have students explain how a snake's movement is similar to an earthworm's movement. *(A snake uses scales to gain traction while its muscles move the scales, resulting in locomotion. In this sense, an earthworm's use of bristles is similar to a snake's use of scales.)*

Extension

Search for detailed illustrations or photographs of a centipede and a millipede. Ask students to answer the following: (a) Count or estimate the number of legs on each animal. (b) How many legs are found on each body segment of each animal? (c) Do more legs help an animal move faster? Why or why not? (d) What is the advantage of having more legs?

Closure

Invite a veterinarian to your class to talk with students about animal locomotion. Ask your guest to explain some of the methods of treatment used when motion is impaired (in the case of a broken leg, for instance). Encourage students to ask questions.

1. Wanted: Nonliving or Living

What characteristics do all living things have in common? Sometimes a nonliving thing has a characteristic of a living thing. How many examples of this can you list? Make a list. Can you top 20 examples? If a nonliving thing has a characteristic of something that is alive, why is it classified as nonliving?

2. Classify It

A biologist made the following classification of all objects:

a. living—having all of the signs of life

b. dead—having once had all of the signs of life

c. nonliving—having never had all of the signs of life

Where does a wooden chair fit in? How about other things—water, mushrooms, a pie? Classify each of the following into one of the three groups: oyster, moss, yeast cake, salt, sugar, bones, hibernating bear, pencil, wool sweater, seaweed, sponge, pearl, volcano, kernel of corn, clams, barnacle on a rock, bean seed, baked beans, freshly picked strawberries, pine cone, lichens on a rock, electric fan, cactus, paper.

3. Curious Questions

Your class has just been asked to write the section called "Curious Questions" in the book *Mr. Know-It-All's Science Facts*. Choose one question from the list below. The question should be answered clearly in a paragraph of 150 words or less and should be understandable to an 11- or 12-year-old. If an illustration would help, draw one!

a. How do you tell a plant from an animal?

b. In how many different ways can animals move?

c. How does an insect use its six legs to move?

d. How does a horse use its four legs to move?

73

⭐ **You may wish to provide students with the Chapter 4 Review Worksheet that accompanies this Challenge Your Thinking (Teaching Resources, page 13).**

Answers to *Challenge Your Thinking*

1. All living things share characteristics such as growth, reproduction, food consumption, movement, response to stimuli, ingestion, respiration, and excretion.

Examples of nonliving things that exhibit certain signs of life include machines that move and mechanical devices that respond to stimuli, such as smoke detectors or burglar alarms.

For something to be alive, it must have *all* of the characteristics of living things.

2. a. Living: mushrooms, oyster, moss, yeast cake, hibernating bear, seaweed, sponge (if natural and still living in water), kernel of corn (contains an embryo), clams, barnacle, bean seed, seeds in the strawberries, seeds in the pine cone, lichens, cactus

b. Dead: pie, wooden chair, bones, the wood in the pencil, wool sweater, baked beans, paper

c. Nonliving: water, salt, sugar, the lead in the pencil, pearl, volcano, electric fan

3. a. Unlike animals, most plants make their own food, and most contain pigments for making food. Also, plants are typically unable to move from place to place; they can move only parts of their body.

b. Swimming, burrowing, walking, running, hopping, crawling, soaring, and flying are some examples of animal locomotion.

c. Insects usually walk by moving the middle leg on one side at the same time that they move the front and hind legs on the other.

d. A horse walks by alternating its legs from front to back and from side to side. One possible sequence could be the following: left rear, left front, right rear, right front.

Answers to Challenge Your Thinking continued ▶

Answers to
Challenge Your Thinking,
continued

4. Students should look for a combination of the following characteristics: growth, reproduction, food production or consumption, movement, response to stimuli, having a life cycle, ingestion, digestion, respiration, and excretion. The steps that the students suggest should allow for careful observation of the subject and should include exposing the subject to a variety of stimuli. The photograph shows a colony of cellular slime mold and is described in more detail below in Did You Know.

5. Accept all reasonable responses. For instance, between the words *animal* and *living things*, students might put "is an example of." Between *limbs* and *human,* students might put "are used for locomotion by a."

The following are sample revised answers:

1. Students might respond that living things move, reproduce, use energy, die, respond to stimuli, and grow. Nonliving things may show some, but not all, of these characteristics.

2. Characteristics of living things include growth, reproduction, energy usage, response to stimuli, and movement.

3. The plant is growing toward the light.

4. A snail uses a flexible foot that moves in a wavelike motion. An earthworm has tiny bristles and sets of opposing muscle groups that allow it to move in a push-pull motion. A snake moves by using scales to grip the surface and by pushing against objects.

4. That's Life?
You are a scientific explorer, and you have just come across the mysterious blob shown at right. Do you think it is alive? What characteristics would you look for? Write down a series of steps you would take to answer these questions.

5. It's All Connected
Copy the following terms into your ScienceLog. Do not write in this book! Draw lines between the words that you think are connected in some way. Then explain why you chose to connect the words that you did. Write your explanations along the lines that you drew.

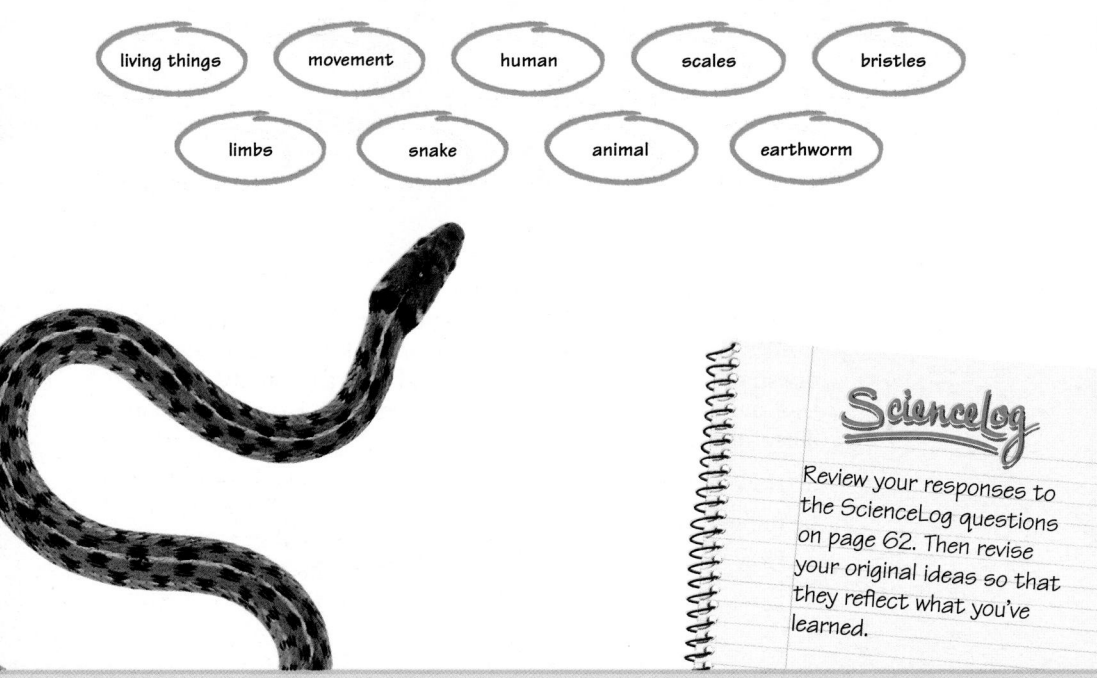

living things · movement · human · scales · bristles · limbs · snake · animal · earthworm

Review your responses to the ScienceLog questions on page 62. Then revise your original ideas so that they reflect what you've learned.

74

Did You Know...
The organisms in the photograph at the top of this page are a colony of cellular slime mold. Individual slime mold cells resemble amoebas but can form mobile colonies when local food supplies are exhausted. This colony has found a suitable area in which to relocate and is starting to differentiate and produce spores. There are about 70 known species of cellular slime molds. They usually live in soil, where they ingest bacteria.

CHAPTER 5 Patterns of Growth

1 How tall will you be as an adult? Why do you think so?

2 Do living things continue to grow as long as they are alive? Are there different kinds of growth? Name some.

3 What are some things that plants need for healthy growth?

ScienceLog

Think about these questions for a moment, and answer them in your ScienceLog. When you've finished this chapter, you'll have the opportunity to revise your answers based on what you've learned.

75

CHAPTER 5 Patterns of Growth

Connecting to Other Chapters

> **Chapter 4**
> examines the characteristics of living things, their motion, and the differences between plants and animals.

> **Chapter 5**
> investigates a variety of patterns of growth, in both humans and other organisms.

> **Chapter 6**
> explores the responses that organisms can have to stimuli and to long-term changes in their environment.

> **Chapter 7**
> introduces students to both the structure of cells and the proper use of a microscope.

Prior Knowledge and Misconceptions

Your students' responses to the ScienceLog questions on this page will reveal the kind of information— and misinformation—they bring to this chapter. Use what you find out about your students' knowledge to choose which chapter concepts and activities to emphasize in your teaching.

In addition to the ScienceLog questions on this page, you may wish to pose questions such as the following for discussion:

1. How can we predict human growth?
2. Which parts of the body grow first?
3. Is size an advantage in nature?

Emphasize that you are not looking for right or wrong answers to these questions. It may be helpful to record student responses on the board. The responses may help you gauge what students already know about growth and size as an adaptation to the environment, what misconceptions students may have, and what is interesting to them about this topic.

FOCUS

Getting Started

Tell students to imagine that they are still the same height as when they were born. Ask: How would life be different? After a moment, ask students to imagine that people never stop growing. Ask: How tall do you think you would be if you lived to be 100 years old?

Main Ideas

1. Growth is a fundamental characteristic of all living things.
2. Not all people grow at the same rate or reach the same size.
3. The proportion of different body parts to the total size of a person changes as the person develops from an embryo to an adult.

TEACHING STRATEGIES

Measuring Human Growth

Have students work in pairs to determine their measurements and to fill in a data chart like the one on page 77. Ask each student to predict how tall he or she will be as an adult.

You may wish to demonstrate how to make accurate body measurements. To measure height, tape a large sheet of paper to the wall. Have a student stand up straight against the paper. Place a meter stick on the student's head so that the meter stick is perpendicular to the wall. Mark the paper where the meter stick touches it. Measure leg length from the floor to the side hip joint, where the leg bends. Measure head height from the bottom of the chin to the crown.

Locomotion is vital to most animals, but growth is a fundamental characteristic of all living things—plants and animals. One thing you know for certain is that you are growing. But how much do you grow? Do all parts of your body grow at the same time or at the same rate? Do they grow in the same proportions?

Measuring Human Growth

With the help of a friend, find the following: What proportion of your total height is taken up by your legs? What proportion of your total height is taken up by your head? Record your results in a table like the one on the next page. How do you think your results would differ if you were measuring a 6-month-old baby? Compare your measurements with those of an adult.

How tall will you be as an adult? You can find out if you have a record at home of your height when you were very young. If you are a girl, double your length at 18 months. If you are a boy, double your length at 2 years. How old do you think you were when you were one-half of the height you are now? You might be surprised to discover how long ago that was!

A Class Profile

A. A few people

LESSON 1 ORGANIZER

Time Required
2 class periods

Process Skills
observing, measuring, predicting, analyzing

New Terms
none

Materials (per student group)
Measuring Human Growth:
2 meter sticks

Teaching Resources
Transparency 11

	Height/length (cm)	Proportion (%)
Total		
Legs		
Head		

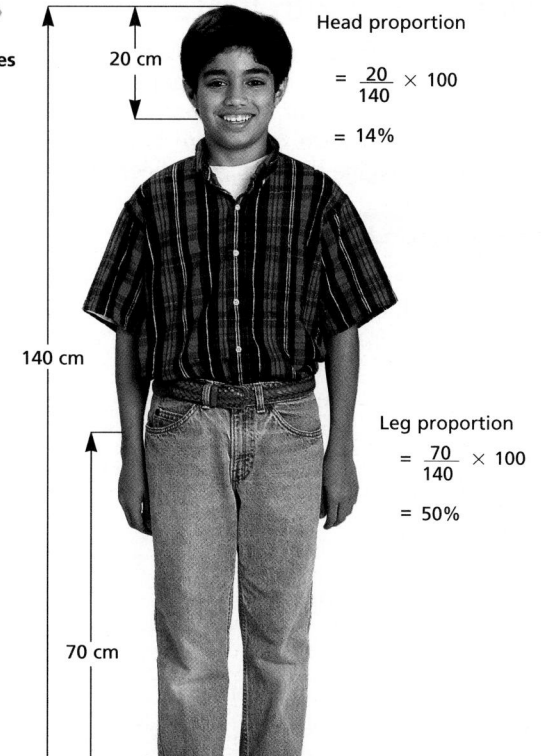

How to calculate percentages

Head proportion

$$= \frac{20}{140} \times 100$$

$$= 14\%$$

Leg proportion

$$= \frac{70}{140} \times 100$$

$$= 50\%$$

20 cm

140 cm

70 cm

Your Class Profile

Picture all of the members of your class standing in line from the shortest to the tallest person, as in the illustration below. Compare the heights of all members of your class. Is your class "profile" similar to the one in the illustration below? How many people are in group *A?* in group *B?* in group *C?*

Not all people grow at the same rate, nor do all people reach the same height. There is nothing unusual about this. For any characteristic you could name—height or hair color, for example—each person differs at least slightly from every other person. In fact, such variation can be found in any group of living things. The amount of variation depends on the characteristic measured. If you made a class profile of shoe size, would you expect the shoe-size profile to be similar to the height profile? Try it.

B. Most people

C. A few people

77

It is important that students realize that a range of measurements is quite normal in any group of living things. Since many of the girls may have already had their growth spurt, they may be taller than the boys. Additionally, students vary considerably in the age at which they begin their growth spurt. Tell students that they will learn more about growth spurts in Activity 3 on page 79.

If you feel that some of your students may be embarrassed by this activity, an alternative would be to have students collect similar data from their parents.

GUIDED PRACTICE Direct students' attention to the figure that shows how to calculate percentages. Make sure students understand the meaning of the term *proportion* as well as how it is calculated. Ask: Why is a proportion expressed as a percentage? How else can a proportion be expressed? *(A proportion is a part in relation to its whole. Thus, it can be expressed as a fraction, a decimal between 0 and 1, a ratio, or a percentage.)*

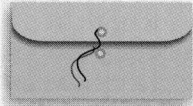

PORTFOLIO

Suggest that students record their results from Measuring Human Growth and Your Class Profile in their Portfolio. Encourage students to record their data using graphs, charts, or diagrams. Students may then assess the scientific skills they used in collecting, organizing, and presenting their data.

Homework

Have students gather the following information from three individuals: height, head size, and leg length. One individual should be a child between four and eight years old, another should be a classmate, and the third should be an adult. Have students prepare a chart similar to the Human Growth Stages diagram on page 78.

Human Growth Patterns

Human Growth Patterns

As you know, people grow to different heights and weights. But did you know that as we grow, the proportions of our head, body, and limbs change? This pattern of change is similar for everyone. Imagine that you have been selected by a team of scientists to write a report on their findings about human growth patterns. The results of the scientists' study are shown on these two pages. The directions in the following Activities will help you write your report.

ACTIVITY 1

The characters in the figure at right show the human body in seven stages, from before birth to adulthood. Each stage is drawn to the same height to allow you to compare the growth of the different parts of the body. Answer the following in your report:

1. How do the head and leg proportions change through the different stages?

2. When do the legs undergo the greatest change?

3. What part of the body seems to develop the earliest?

4. Describe the changes in the proportions of a developing person from stage 1 to stage 7.

Human Growth Stages

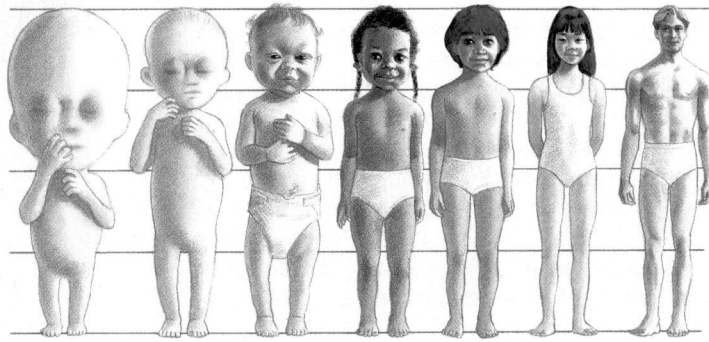

−7 months −4 months Birth 2 years 5 years 12 years Adult

Growth Rates

0–2 years   3–4 years   5–7 years

ACTIVITY 2

The table at right compares average height, head size, brain weight, and total weight at three different ages. The comparisons are shown as a percentage of the adult values. Answer the following in your report:

1. Which feature is the most developed at birth?
2. Which feature is the least developed at age 5?
3. An 18-year-old is 160 cm tall and has a body weight of 70 kg. Using the chart, figure the height and body weight of this person at birth and at age 5.

Changes in Size and Height

Feature	At birth	At 5 years	At 18 years
head size	60%	90%	100%
brain weight	25%	90%	100%
height	30%	65%	100%
total body weight	5%	30%	100%

ACTIVITY 3

The figure below compares the growth rate of girls and boys. Answer the following in your report:

1. When do girls get a large growth spurt?
2. When do boys get a large growth spurt?
3. Who, on average, is taller at adulthood?

200 cm

175 cm

150 cm

8–10 years   11–15 years   16–20 years

79

FOLLOW-UP

Reteaching

Have students select magazine pictures of people from birth to age 20, or have them collect photos of themselves at various stages of growing up. Have students carefully examine the people in the pictures, noting characteristics such as head size, leg length, body weight, and head-to-body proportions. Then have students label each picture with the approximate age of each person shown.

Assessment

Tell students to pretend that they have just received a letter from a friend across the country. That friend is curious about why height is changing a lot for some students in her seventh-grade classroom and not changing at all for others. She mentioned two friends who were once the same height as she was— one is now shorter than she is and the other is now taller than she is. Have students write a letter to this friend explaining growth rates, growth spurts, and proportions.

Extension

Ask students to compare human life spans to those of some animals. For example, the average human life span in the United States is about 75 years. The average life span of a lion is 23 years; a bear's is about 20 years; and that of a cat, dog, or monkey is about 12 years. Have students make a bar graph showing the life spans of these animals and others that they wish to find out about.

Closure

Ask students to complete a class profile entitled "Our Class in 10 Years." (Refer to pages 76 and 77 for instructions.) Students should use a bar graph to show what the expected heights of class members will be at that time. Ask students to compare that bar graph to the current class profile to answer the following questions: How do the profiles compare? What changes do you notice?

Answers to
Activity 2

1. Head size
2. Total body weight
3. For the 18-year-old who is 160 cm tall and weighs 70 kg, the height at birth was 0.30 × 160 cm = 48 cm; the body weight at birth was 0.05 × 70 kg = 3.5 kg; the height at age 5 was 0.65 × 160 cm = 104 cm; and the body weight at age 5 was 0.30 × 70 kg = 21 kg.

Answers to
Activity 3

1. Between 8 and 15 years old
2. Between 11 and 20 years old
3. On the average, men are taller than women at adulthood.

LESSON

2

Other Growth Patterns

FOCUS

Getting Started

Ask students how fast living organisms can grow. Tell them that giant kelp (*Macrocystis pyrifera*) can grow 36–41 cm per day and reach a length of 61 m!

Main Ideas

1. As in humans, the proportion of different body parts changes as plants and animals grow to adulthood.
2. Many factors, such as moisture and temperature, affect the health and growth of organisms.
3. Forms of growth include tissue renewal and repair, regeneration, reproduction, cancer, and galls.

TEACHING STRATEGIES

If possible, take students outdoors for the tree-comparison activity. You may wish to provide students with the circumference measurements of a seedling and of a full-grown tree. To find the area of the leaves, students may outline the leaves on graph paper and then count the squares within the outlines.

Answers to
In-Text Questions and Caption

Ⓐ Students may measure the length, width, or area of the leaves, for instance.

Ⓑ The stem circumference changes more than the leaf size. At an early age, leaves develop faster than stems to provide the plant with sites of photosynthetic activity where food is made for growth.

Ⓒ No. Every tree grows differently.

Homework

The Activity Worksheet that accompanies the beginning of Lesson 2 may be assigned as homework (Teaching Resources, page 19).

Humans arc not the only things that grow. All plants and animals grow. Measuring growth is not always as easy as you might think, however.

With a partner, compare the leaves of a seedling with those of a full-grown tree of the same type. What different methods can you think of to measure the leaves? Using the method that you think will work best, measure the change in leaf size. **Ⓐ**

Now compare the sizes of the stems of the seedling and the tree by measuring the distance around the stems (the circumference). As the plant grows to maturity, which changes most—the size of the leaves or the size of the stem? Which part of the plant do you think develops the most at an early age? Why? **Ⓑ**

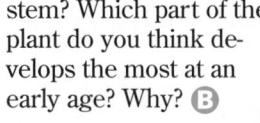
Compare the maple seedling with the full-grown maple tree. You can look at the leaves and stems to describe the maple's growth pattern. Do you think all trees have the same growth pattern? **Ⓒ**

LESSON 2 ORGANIZER

Time Required
3 class periods

Process Skills
observing, measuring, analyzing

Theme Connection
Changes Over Time

New Terms
Cancer—an uncontrolled growth of cells within an organism
Gall—a harmful growth found on the leaves, stems, or roots of a plant
Germinate—to sprout or grow from a seed or bud

Ovary—the female reproductive organ that produces eggs
Regeneration—the growth of new body parts to replace lost or damaged ones
Reproduction—the process by which organisms produce offspring of their own kind
Species—a group of closely related organisms that can reproduce successfully

continued

Conditions for Growth

If you were to fall into icy water, you would probably die within 10 minutes. An orchid dies when the temperature falls below 28°C. Tropical fish suffer when the water temperature drops suddenly. A cactus plant thrives in the parched desert. A rosebush blooms when the air is moist.

Many conditions affect the lives of plants and animals.

Moisture and temperature are two such conditions, as the examples above indicate. What other conditions do you think influence the lives of plants and animals?

EXPLORATION 1

Changing Conditions

Imagine that you are a botanist working for a seed company. You are responsible for testing different conditions for growing seeds and for developing directions to go on the label. Discover what conditions are needed to produce the best results.

Make a report for the company files, and write a draft of the instructions for the seed package. In your report, include the following information: the sets of conditions you tested, what happened to each group of seeds, and which conditions were the best. One word you might find useful in writing the report is *germinate*. This word means "to begin to grow from a seed."

In performing your tests, be sure to change only one variable. Remember that a variable is a condition that can or does change. Variables that you could change include temperature, moisture, and amount of light. All other variables should remain the same. For example, when testing the effect of changing the amount of light available, all other variables, such as moisture, temperature, and the method of seed germination, should remain the same. Some methods for germinating seeds are shown at right. Use the

same type of seed for each set of conditions. Try to include bean, carrot, or radish seeds.

One group of students counted the number of seeds that germinated under the following conditions:

- Experiment 1: no light, low light, and bright light
- Experiment 2: 15°C, 20°C, and 25°C
- Experiment 3: one watering at planting, one watering at planting and then every day after that, and one watering at

planting and then every other day after that

They used several seeds for each condition. Why is this important? Before you begin, make a prediction about what you think the result of each experiment will be.

Ways of Growing Seeds

81

Answer to
In-Text Question

Ⓓ Plants and animals may also be affected by light, food supply, weather, pollution, population density, type and number of predators, and shelter.

EXPLORATION 1

Provide the students with untreated mung beans or alfalfa, carrot, or radish seeds. Allow at least 10 days to permit complete germination. Students should make careful observations and record the number of days it takes for the seeds to germinate, as well as the germination percentage for each group of seeds. Variables to be tested should include the amount of water, heat, and light. Students may use the sample experiments described in the text, or they may design their own experiments. Be sure to check student designs for safety before allowing them to proceed. Students should discover that seeds germinate best when they are kept warm, moist, and in the dark.

Answers to
Exploration 1

It is important to use several seeds for each condition because an individual seed may not germinate due to damage or a genetic flaw. Accept all reasonable predictions.

🖐 Cooperative Learning
EXPLORATION 1

Group size: 3 to 4 students
Group goal: to test different conditions for growing seeds and to develop an understanding of the term *variable*
Positive interdependence: Assign students roles such as materials coordinator, recorder, leader, and investigator.
Individual accountability: Have each student define the term *variable* and explain why it is important to identify variables when conducting an experiment.

⭐ An Exploration Worksheet is available to accompany Exploration 1 (Teaching Resources, page 20).

Large vs. Extra Large

Discuss with students the disadvantages of extreme size in nature. For example, the food requirements of an animal depend on the volume of the animal's body. On the average, an African elephant is 5 m tall and an Asian elephant is 3.5 m tall. Proportionally, the African elephant is 1.4 times taller than the other, but it is also longer and wider. Its volume and weight are more than 1.4 times the volume and weight of the Asian elephant.

Answers to
In-Text Question and Caption

Ⓐ Although the African elephant is not twice as tall as the Asian elephant, it spends almost twice as much time eating food because it is longer and wider than the Asian elephant. The average weight of a male African elephant is about 5400 kg, while the average weight of a male Asian elephant is about 3600 kg.

Ⓑ Students should observe that the African elephant has much larger ears and tusks than the Asian elephant and that the African elephant's back dips in the middle.

Large vs. Extra Large

Is bigger always better? Large size can have its problems. An African elephant grows much larger than an Asian elephant does. The African elephant also eats much more than the Asian elephant does. This means that the African elephant must spend most of its waking time finding food and eating, just to stay alive. The Asian elephant, on the other hand, spends only about half of its waking time eating. Why does the African elephant need to eat so much more food? Is the African elephant twice as big as the Asian elephant? Ⓐ

Can you tell the Asian elephant from the African elephant? (Hint: Size is not the only clue.) Ⓑ

Did You Know. . .

The largest animal that has ever lived on the Earth is the blue whale. Blue whales can reach a length of 30 m (100 ft.) and weigh up to 200 metric tons.

Homework

Have students research a large animal. In a written report, students should describe the advantages and disadvantages of the animal's immense size.

How Old?

Some living things show their age in obvious ways. You probably know that tree rings can be counted to determine the age of a tree. What part of each living thing shown would you examine to find out how old it is?

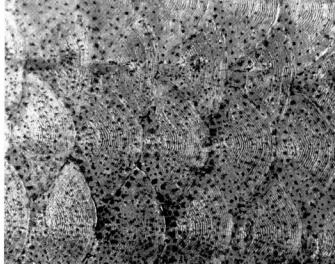
Whitefish scales

Clam

Waterbuck

Horned puffin

The Tale of a Tree

Tree rings are often used to tell the age of a tree. In the spring, rapid growth produces large, thin-walled cells that form a wide, light-colored band of wood. In the summer, late in the growing season, newly formed cells are smaller, more dense, and have thicker walls. These smaller cells form a dark-colored band of wood. In a typical year, an annual ring consists of a dark-colored band of wood surrounding a light-colored band. Look at the tree section shown.

1. How old was this tree when it was cut in 1995?
2. Identify the following:
   a. the scar left by a fire
   b. growth that occurred during wet years
   c. growth that occurred during dry years
3. In the tropics, why is counting tree rings an unreliable way to tell the age of a tree?

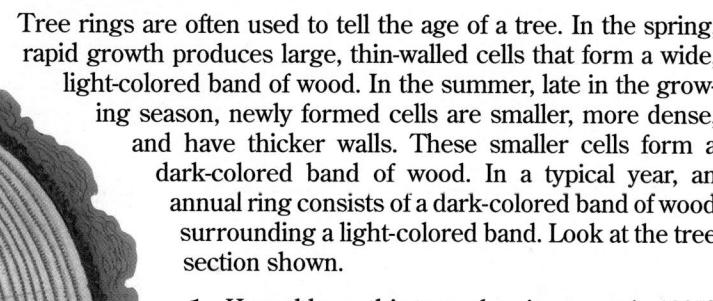

83

CROSS-DISCIPLINARY FOCUS

Language Arts

Many trees live to be hundreds of years old. Have students complete a creative-writing assignment in which they conduct an imaginary interview with a tree whose life has spanned several hundred years. Which events does the tree find particularly memorable? What are its fondest memories? What are its most unpleasant memories? Encourage students to share their interview with the class.

How Old?

GUIDED PRACTICE If possible, bring some clam shells or a cross section of a log to class. Explain that both the rings of a tree and the bands on a clam shell are caused by seasonal variations in growth.

Answers to
How Old?

To determine the ages of the organisms shown, count the growth rings on the fish scale, the ridges on the clam's shell, and the rings on the antelope's horn. Changes in the markings on the puffin's bill can be used to distinguish a juvenile from an adult puffin.

Integrating the Sciences

Earth and Life Sciences

Layers of rock, like rings in a tree, are similar to pages in a history book. They reveal a sequence of events that took place in the Earth's past. In a tree, you can not only count the rings to determine its age, but also note any changes in the size or color of the rings. Have students compare layers of the Earth with the rings in a tree. Encourage them to research how environmental changes might be recorded in the rock layers of the Earth's crust. *(For example, a period of volcanic activity might be recorded as a layer of ash.)*

Answers to
The Tale of a Tree

1. 35 years old

2. **a.** The dark, irregular patch is a fire scar.
   b. Wide rings indicate wet years.
   c. Narrow rings indicate dry years.

3. In the tropics, seasonal variation, especially in temperature, is minimal. Therefore, trees will not have the distinct rings that trees in more temperate regions have. (You may wish to point out to students that not all types of trees exhibit growth rings. The age of tropical palm trees, for example, cannot be determined by counting growth rings.)

Ⓐ The thumbnail grows at a faster rate. Your hair is another part of you that continues to grow throughout your life.

Renewal

Ask students what happens when they have a cut, scratch, or burn that has removed part of the skin. Discuss how the skin is replaced as the wound heals. Tell students that all living things, both plants and animals, are able to regenerate some body parts. But some animals, unlike humans, are able to regenerate major parts of their bodies. Have students read aloud pages 85–86 about some animals that have amazing regenerative abilities.

GUIDED PRACTICE   Ask students if they have ever tried catching lizards or salamanders. They may have discovered that the tail falls off when grasped. They also might have noticed that the tail is more brightly colored than the rest of the body. Ask students how the ability of some lizards and salamanders to lose their tail helps to protect them. *(The bright colors of the tail draw a predator's attention away from the vital parts of the animal. The tail then breaks off, allowing the animal to escape.)*

Theme Connection

Changes Over Time
Focus question: How does the growth of a tree compare with the growth of a kitten? *(As long as environmental conditions meet the needs of the tree, it will continue to grow. By contrast, the kitten has a preset size that it can reach, given suitable environmental conditions.)*

Continued Growth

The upper front teeth of mice and related animals, such as beavers, rabbits, and squirrels, continue to grow throughout the animals' lives. These animals keep their teeth sharp and at an appropriate length by gnawing. Parts of your body grow continuously too. It takes about one year for a fingernail to grow from the base to the edge at the tip of your finger. If it takes a year for each fingernail to grow, which must grow at a faster rate, the nail on your thumb or on your little finger? What other part of you continues to grow throughout your life? Ⓐ

Renewal

Another kind of growth is the renewal that follows an injury. When you scrape your elbow, some of the skin is destroyed. Your body immediately begins a healing process to repair the damage. But there is a limit to your body's ability to heal itself. After severe burns, for example, renewal may not occur naturally; skin grafting may be necessary. Some animals have a remarkable capacity for renewal. For example, salamanders can replace a lost tail with a new one. This replacement process is called **regeneration**.

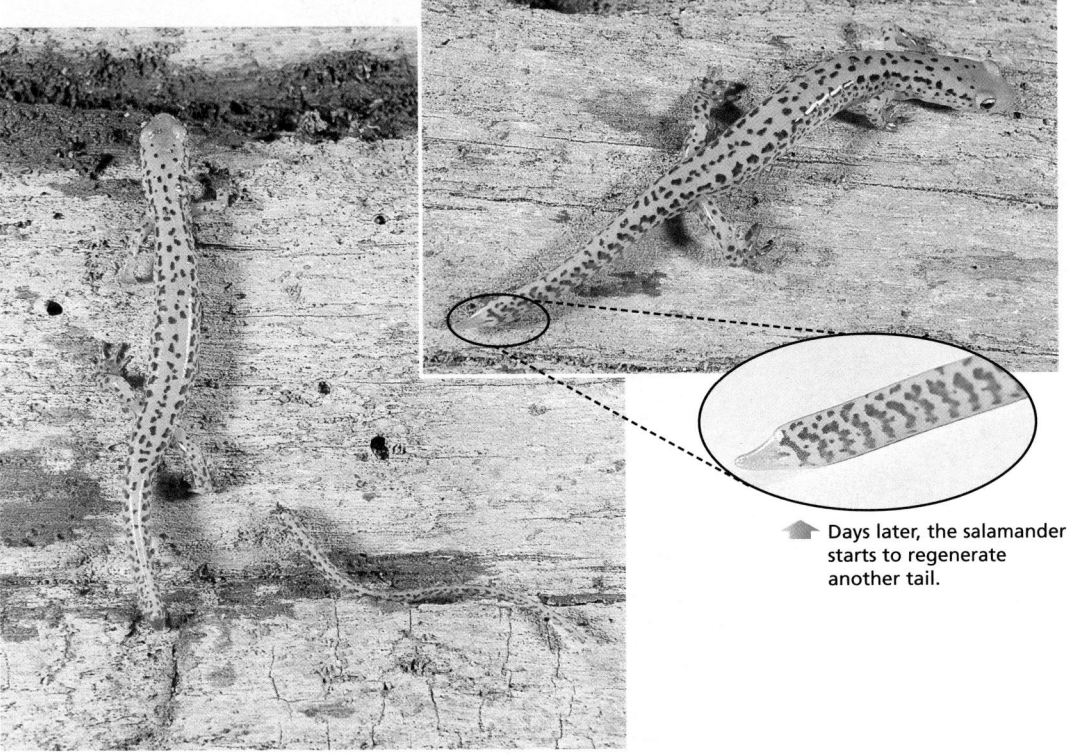

⬆ Days later, the salamander starts to regenerate another tail.

⬆ This long-tailed salamander just lost its tail.

The Regeneration Professionals

Lobster ▶

A lobster gets its claw caught in a rock crevice. As the lobster struggles to get free, its claw breaks off at a joint. Several months later, a new, full-sized claw has replaced the one that was lost.

Tadpole ▶

Tadpoles that lose a leg can grow a new leg in its place! Adult frogs cannot regenerate lost legs.

Lobster

Ask students if they have ever seen lobsters or crabs with one claw that is smaller than the other. Explain to students that these animals have lost their original claw and are in the process of growing a new one. Crustaceans, such as lobsters and crabs, have the ability to self-amputate their legs. If caught or injured, a leg can be broken off at a specific point at the base of the leg. A special muscle in the leg contracts excessively to cause a rupture at the breaking point. Wounds heal more quickly at this point, preventing excessive blood loss.

Tadpole

INDEPENDENT PRACTICE The changes that a tadpole undergoes to become an adult frog are referred to as metamorphosis. Have students think of other organisms that undergo metamorphosis. *(For example, many caterpillars become butterflies or moths.)* Then ask the students to write a narrative from the perspective of an organism experiencing the changes involved in its own metamorphosis.

Planaria

Planaria study kits can be ordered from a biological supply company. These kits allow students to study not only regeneration, but also the reponses of planaria to stimuli such as light and electricity.

Starfish

GUIDED PRACTICE Ask: Why is it important for fishermen to know as much as possible about local aquatic organisms? *(They will be less apt to make mistakes that can upset the balance of the ecosystem, such as cutting up starfish.)* Why is it important for them to know about these animals' prey and predators? *(They will be able to harvest the most fish possible, without upsetting the balance of the ecosystem.)* Tell students that an adult starfish can eat 8–10 clams or oysters a day. Thus, an overpopulation of starfish can quickly decimate a clam or oyster bed.

Homework

Overpopulation of starfish has threatened Australia's Great Barrier Reef. Have students conduct research on what measures have been taken by Australian officials to protect the reef. Students should report their findings in a brief written summary.

Planaria ➡

Planaria are tiny flatworms found under logs and stones in fresh water. Some are about the size of a large ant. If they are accidentally cut in half, two planaria will result.

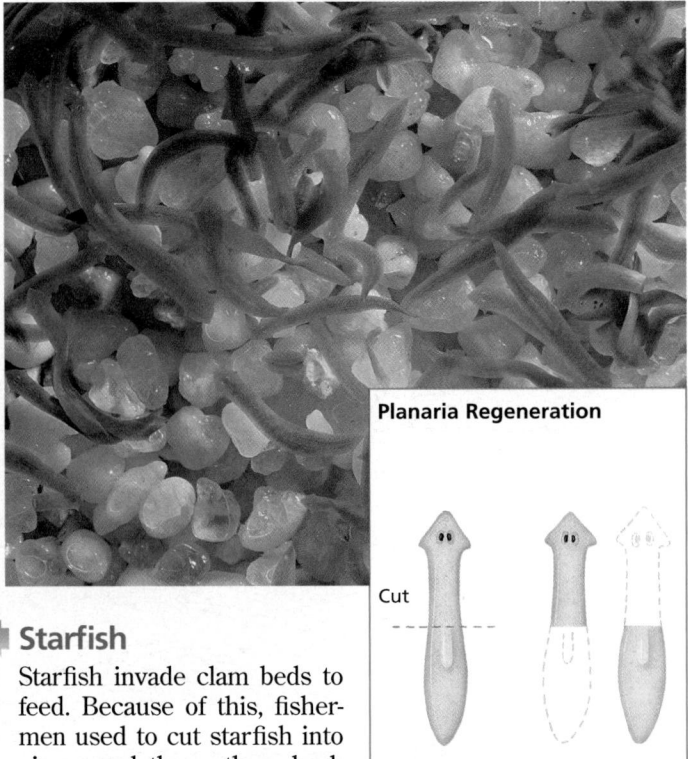

◀ Starfish

Starfish invade clam beds to feed. Because of this, fishermen used to cut starfish into pieces and throw them back in the water to destroy them. Unfortunately, this caused a population explosion because most of the pieces grew into complete starfish.

Planaria Regeneration

Cut

These are examples of how animals regenerate lost body parts. Can you think of ways in which plants regenerate? Ⓐ

Sea Cucumber ➡

If attacked, the sea cucumber may expel some of its internal organs to distract the attacker. It later regenerates these lost parts.

86

Harmful Growth

Can growth ever be out of control and be harmful to living things? Under normal circumstances, cell division is carefully controlled. (All living things are made up of tiny units known as *cells*.) Sometimes, however, cells lose these controls and continue to divide. These dividing cells can damage or destroy surrounding tissues in a condition that is called **cancer**. Not all continually dividing cells cause cancer, though. A wart is an example of a noncancerous growth caused by continually dividing cells.

Two forms of harmful growth: left, a gall on a plant stem; right, human skin cancer

Another kind of harmful growth, called a **gall**, is found on the stems and leaves of many plants. These "sores" are caused by insects that have laid their eggs on a plant and irritated the plant's cells, making the cells grow and form a gall.

Reproduction

All plants and animals grow old and eventually die. A new generation must take the place of the old generation. Each animal and plant reaches a stage of maturity when it is able to produce a new generation of its **species**. A species is a group of closely related organisms that can reproduce successfully. This kind of replacement is called **reproduction**. Often, reproduction is accomplished by special structures of the animal or plant.

Some small living things reproduce by dividing to form two new individuals. However, in many plants and most animals, reproduction requires something contributed by two different individuals, one male and one female. Fertilization occurs when sperm from a male successfully merge with eggs from a female. A new generation develops from the fertilized eggs.

A female frog lays her eggs in the water before they are fertilized. The male then releases sperm that fertilize the eggs. Why is this called

B external fertilization? Look at the pictures of the stages of development of a frog's fertilized eggs. Write a caption for each stage of development shown. Make the captions as descriptive as possible. Do not write in this book! **C**

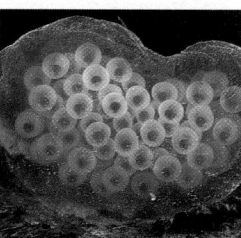

1. Write your own caption.

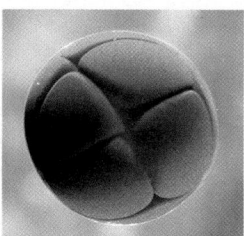

2. Write your own caption.

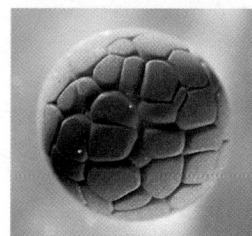

3. Write your own caption.

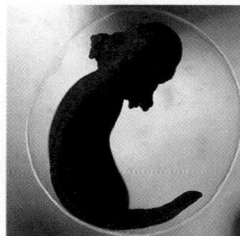

4. Write your own caption.

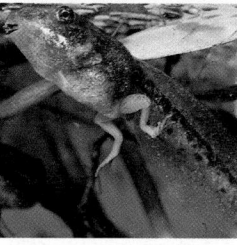

5. Write your own caption.

6. Write your own caption.

87

Homework

You may wish to assign the Graphing Practice Worksheet and Activity Worksheet that accompany the end of Lesson 2 as homework (Teaching Resources, pages 21 and 23). If you choose to use the Activity Worksheet in class, Transparency 12 is available to accompany it.

Harmful Growth

Ask students to look for and bring in samples of different kinds of galls on leaves or stems. These can be found during any season. (Oak trees often have leaf galls.) Galls can be caused by a variety of organisms, such as insects, fungi, and bacteria. After the infecting organism enters the plant, the plant responds by growing a gall. Normal tissue is transformed into cancerous tissue, which undergoes rapid growth. The galls provide the invading organisms with shelter and food. Open up a gall caused by an insect and look for the developing insect inside or for the hole that the insect used to leave the gall.

Reproduction

Explain to students that plants and animals reproduce in a variety of ways. One way is sexual reproduction (usually requiring two parents), which results when two specialized cells from two distinct sexes are joined. Examples of this are fertilized eggs in animals and seeds in flowering plants. Another way is asexual reproduction, which results when cells in a single parent divide. Examples include runners in plants and spores and budding in yeasts.

Answers to
In-Text Questions

B The fertilization takes place outside of the frog's body, or externally.

C Answers will vary but may be similar to the following: (1) mass of frog eggs, (2) a fertilized egg beginning to divide, (3) the fertilized egg at a later stage of division, (4) a tadpole forming within the egg, (5) the tadpole developing hind legs, and (6) the tadpole with all four legs developed.

Answers to
In-Text Questions

Ⓐ The fertilized eggs of both chickens and frogs contain yolks that are surrounded by jellylike protective layers. However, chicken eggs are fertilized internally and contain a larger yolk than frog eggs. Chicken eggs are laid on land and have a hard, waterproof shell. Frog eggs are laid in water and have a jellylike coating. In addition, a frog emerges from its egg as a tadpole whose physical features differ significantly from those of an adult frog.

Answers to
Find Out

a. The platypus is an aquatic, egg-laying mammal. Its eggs are fertilized inside the female. After 2–3 weeks, the female deposits 1–3 leathery eggs in a nest that she builds within a burrow. The eggs are 16–18 mm long, and they hatch in about 12 days. Like all mammals, the young are nursed by the milk-bearing mother. By the time they are 17 weeks old and about 35 cm long, the young have a coat of hair, and they emerge from the burrow.

b. The eggs of an ostrich are fertilized and form hard shells inside the female. Ostrich eggs are 15 cm long and weigh up to 1.35 kg. Three to five females deposit 15–60 eggs in a group nest, which consists of little more than a hole in the ground. The eggs are incubated for 40 days. During this time the father of all the eggs sits over the eggs by night, and the various mothers take turns sitting over the eggs by day. The young ostriches are able to travel with the adults about a month after they hatch.

c. Crocodiles eggs are fertilized and form brittle shells inside the female. The female crocodile digs a nest near the water's edge for the incubating eggs. The number of eggs that the female produces depends on her size and age, but there may be more than 100 in a single mating cycle. The female guards the nest during the eggs' incubation period of 2–3 months. Just before the young crocodiles are ready to hatch, they emit squeaking sounds that signal the mother to remove the cover of dirt and leaves from the nest. The young are 20–25 cm long when they hatch. They are self-sufficient when they hatch, but

Other Ways to Grow

In many animals, eggs are fertilized inside the female's body. An egg forms in a part of the chicken called the **ovary**. It is fertilized internally by the rooster. Then the new chicken begins to develop. The developing chick and the yolk become surrounded by the egg white, a membrane, and the shell. Once the egg is laid, the rest of the development of the chick takes place externally, outside the female's body.

Examine the illustrations at right. Compare the chick's development with that of the frog shown on page 87. What are some of the similarities that you can see? What are some of the differences? Ⓐ

Find Out

Do birds usually care for their eggs? What sort of care do the eggs need? If you take care of a hen's egg until it is ready to hatch, what will you have to do? How long will you have to look after the egg? You may need to do a little research to find out. You may even decide to do a project to find out more about eggs and the development of young animals.

You and several students may work together to prepare a report. Get together with your group, and divide up the tasks that need to be done. Who will do the research? Who will write the report? Who will present the report to the class?

Choose one of the following animals, and investigate the development of the animal's young:

a. platypus

b. ostrich

c. crocodile

d. sea turtle

Consider the following as you investigate:

- what the eggs look like
- whether fertilization is internal or external
- how many eggs are laid at one time
- where and how the young develop
- what care is provided for the young at each stage of development
- how long it takes for the young to develop

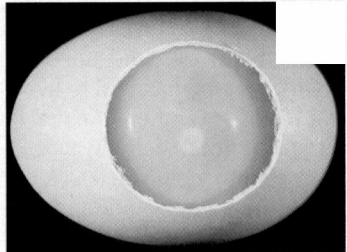
Inside view of a newly laid chicken egg

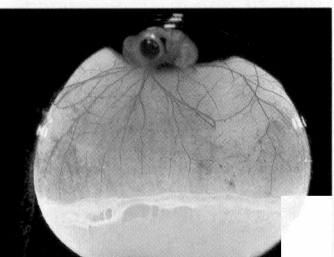
Chicken embryo after 5 days of incubation

Hatching chicken after 21 days of incubation

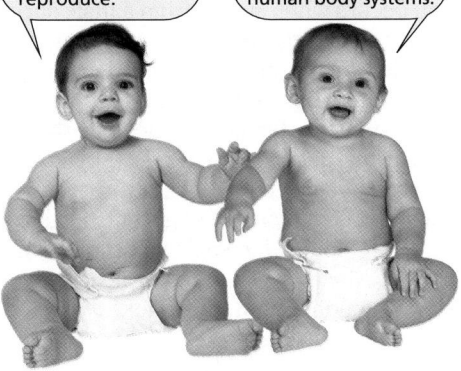

The reproductive system contains special organs that allow humans to reproduce.

Check out page S47 of the SourceBook for more information on this and other human body systems.

they remain near their mother for protection for a few months.

d. Sea turtles can lay fertile eggs for up to 4 years after mating. The soft-shelled eggs of sea turtles are fertilized inside the female and are about 5 cm in diameter. A sea turtle can lay up to 150 eggs in a single clutch and several clutches in one season. Using her hind flippers, the female digs a pit for her eggs on a sandy beach and then covers her eggs with sand. The incubation period varies from 2–3 months, depending on the temperature. When the young hatch, they are about 6 cm long and have scales on their shell and skin that disappear within the first 2 months. After they hatch, the young dig to the surface and rush to the safety of the sea.

Follow-Up for Lesson 2 is on the next page. ▶

CHALLENGE YOUR THINKING

1. An Eggsercise

Take a look at the photograph of the chicken egg at right. Use the photograph and what you have learned in this chapter to answer the following questions:

a. What is the purpose of the eggshell?

b. What is the purpose of the yolk? How long does it have to last?

c. Where in the egg would the young chick have developed if the egg had been fertilized?

d. How does the structure of an egg protect a developing chick?

2. Check It Out

Look at the bar graph illustrating the height of students in Ada's class, and then answer the following questions. Ada is 130 cm tall.

a. How many students are in Ada's class?

b. How many boys are shorter than Ada?

c. How many girls are as tall as Ada or taller?

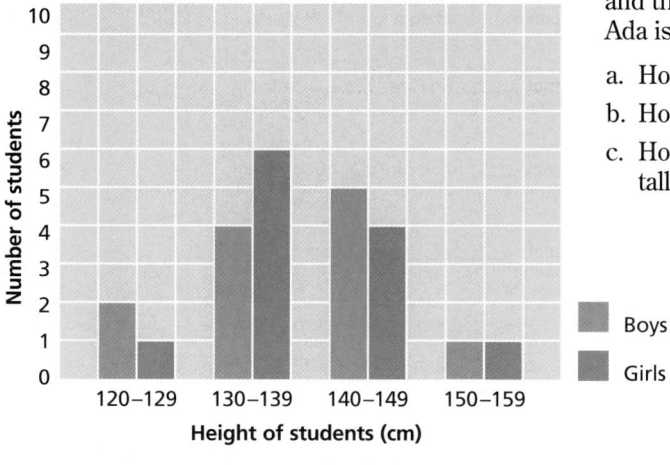

Boys
Girls

3. Inquiring Minds

Add the answers to the following "Curious Questions" that you worked on earlier in the unit.

a. Why does a baby have such a big head?

b. Is bigger always better?

c. How can you tell the ages of plants and animals?

d. Can an animal lose a part of its body and grow a new one to replace it?

e. Do all parts of humans grow at the same rate?

89

---

1. a. The eggshell protects the developing embryo. (It is also porous to allow for gas exchange.)

b. The yolk is the food for the developing embryo. It must last until the chicken hatches.

c. The young chick would have developed from the yolk.

d. The shell provides protection against external dangers. The embryo is surrounded by the egg white, which cushions the embryo during development and protects the embryo against disease-causing organisms that penetrate the shell.

2. a. 24
b. 2
c. 10 (11 including Ada)

3. a. The brain develops rapidly because of its immediate importance in carrying out bodily functions. A baby, therefore, has a large head with respect to total body size.

b. In general, when an organism is bigger, it usually requires a greater amount of energy to sustain itself.

c. In animals, age can be determined by the size, shape, and proportions of body parts; in plants, age can be determined by the total size of the plant, the leaf size, the circumference of the stems and branches, and the number of rings in the trunk.

d. Some animals, including planaria, lizards, salamanders, crabs, lobsters, starfish, and tadpoles, can regenerate body parts.

e. Different parts of the body grow at different rates, depending on the stage of growth.

Answers to Challenge Your Thinking continued ▶

FOLLOW-UP

Reteaching

Have students design an experiment to test the effects of two variables not tested in Exploration 1 on page 81. For safety, have students submit their designs to you before proceeding. Remind students to set up the experiment so that only one variable at a time is changed.

Assessment

Have individuals or student groups prepare reports on different forms of growth, such as tissue renewal and repair, regeneration of lost body parts, cancer, or galls.

Extension

Have students try growing new plants from edible vegetables. An excellent resource for this activity is *Gardens From Garbage: How to Grow Plants From Recycled Kitchen Scraps* by Judith F. Handelsman (Brookfield, CT: The Millbrook Press, 1993).

Closure

Have students write a short story describing how large size creates special problems for an organism.

Homework

The Closure activity described at left makes an excellent homework assignment.

Answers to
Challenge Your Thinking,
continued

4. a. Answers will vary. Sample answers: (A) Where does fertilization take place? (B) What type of eggs are produced? (C) For eggs with shells, are the shells brittle or soft? (D) Does the fertilized egg develop inside or outside the female? (E) Are the eggs cared for? (F) For how long and in what way are the young cared for?

b. The human, dog, and eagle probably have the best chances of survival. The salmon, frog, and turtle are probably least likely to survive. In general, animals with internal fertilization and parental care of eggs and young have the best chance to survive and to grow up.

c. If a frog lays many eggs, there is a better chance of many young surviving to adulthood. A bird, however, is limited by the number of eggs that it can care for.

d. Fish produce more eggs than turtles because fish eggs are fertilized externally and are not surrounded by a shell, which makes them more vulnerable to predators. Producing a greater number of eggs ensures that some will survive.

The following are sample revised answers:

1. Female students can estimate their adult height by doubling their length at 18 months. Male students should double their length at 2 years.

2. Student answers could include the following: proper temperature, light, fertile soil, water, and nutrients.

3. Some living things continue to grow as long as they are alive. Some may not grow larger, but instead replace normal tissue that has been damaged due to aging or injury. Cancerous growth, regeneration, and reproduction are types of growth.

---

4. Parenting Skills

Animals with backbones are called *vertebrates*. They develop from fertilized eggs. Their ways of reproducing, however, show many differences. In fact, you could sort them into groups by answering a few questions about their fertilized eggs and what happens to them.

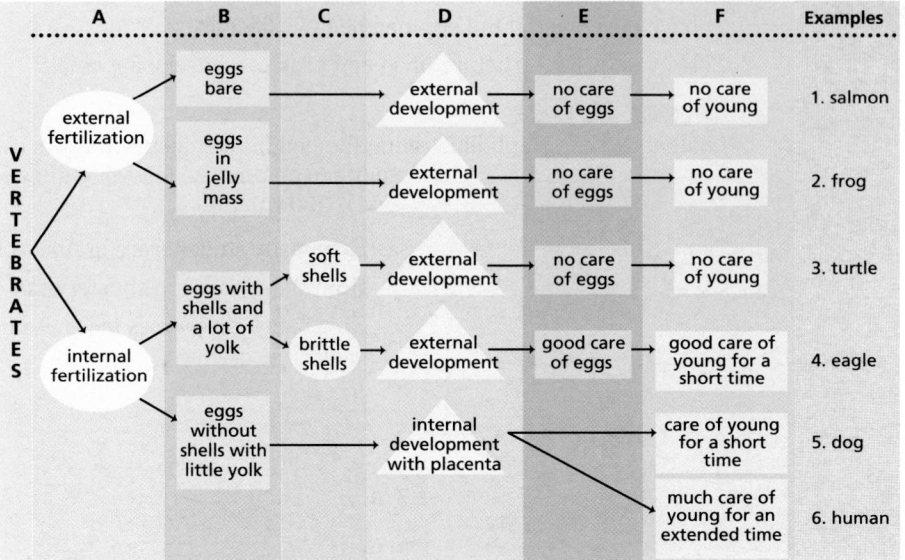

a. The letters *A* through *F* in the diagram represent headings that could be in the form of questions. For each letter, supply the question that is answered by that section of the chart. For example, the question for *A* could be, Where does fertilization take place?

b. Based on the information in the chart, which of the animals named above do you think would have the best chance to survive and to grow up? Which would be the least likely to survive? Give reasons for your choices.

c. Why do you think a frog produces many more eggs at one time than a bird does?

d. Which do you think produces more eggs at one time, a fish or a turtle? Why?

ScienceLog

Review your responses to the ScienceLog questions on page 75. Then revise your original ideas so that they reflect what you've learned.

 You may wish to provide students with the Chapter 5 Review Worksheet that accompanies this Challenge Your Thinking (Teaching Resources, page 25). Transparency 13 is also available for your use.

Can plants tell time? What about bees? **1**

2 Why is it important for a living thing to be able to respond to its environment?

4 What features of this chameleon make it well suited to its environment?

ScienceLog

Think about these questions for a moment, and answer them in your ScienceLog. When you've finished this chapter, you'll have the opportunity to revise your answers based on what you've learned.

3 Why do some animals migrate?

CHAPTER

6

Patterns of Response

Connecting to Other Chapters

Chapter 4
examines the characteristics of living things, their motion, and the differences between plants and animals.

Chapter 5
investigates a variety of patterns of growth, in both humans and other organisms.

Chapter 6
explores the responses that organisms can have to stimuli and to long-term changes in their environment.

Chapter 7
introduces students to both the structure of cells and the proper use of a microscope.

91

Prior Knowledge and Misconceptions

Your students' responses to the ScienceLog questions on this page will reveal the kind of information—and misinformation—they bring to this chapter. Use what you find out about your students' knowledge to choose which chapter concepts and activities to emphasize in your teaching.

In addition to having students answer the ScienceLog questions on this page, you may wish to have them respond in writing to the following information: Show students pictures of giraffes. Point out that giraffes are the tallest of all mammals, growing to a height of 5.5 m (18 ft.). Much of a giraffe's great height is due to its tremendously long neck. Tell students that giraffes have not always had long necks; it is a trait that the species has acquired over time. Ask students to form a hypothesis that explains how and why long necks became a characteristic of modern giraffes. Remind students that there are no right or wrong answers to this exercise. Collect the papers, but do not grade them. Instead, read them to find out how much students know about adaptation of structure, what misconceptions they may have, and what they find interesting about this topic.

LESSON 1 Stimulus and Response

LESSON 1 Stimulus and Response

FOCUS

Getting Started

Ask students to think about what happens to their mouth when they smell their favorite food. *(Their mouth waters.)* Explain that smell can be a stimulus to the salivary glands. The production of saliva is the body's response to this stimulus. Ask students why the body might produce saliva in response to the smell of food. *(Saliva production is the first step in the digestion of food.)*

Main Ideas

1. A stimulus can cause a response in a living organism.
2. Responses to stimuli may be positive or negative.
3. Most plants respond slowly to stimuli, but some can respond very quickly.

TEACHING STRATEGIES

Have a student read this page aloud. Then ask students to name examples of animals reacting to stimuli.

Answers to
In-Text Questions and Caption

Ⓐ The common meaning of irritability is that a person is quick to anger; the scientific meaning is similar, but it applies to any reaction, not just anger. The stimuli depicted on page 91 are the pencil tip, the fragrance of the flower, the weather changing, and the appearance of the insect.

Ⓑ Cats might respond negatively to a dog or a loud noise, for instance.

Teaching Strategies for Exploration 1 are on the next page. ▶

Anything that causes living things to react is called a **stimulus**. The reaction to a stimulus is called a **response**. All living things respond to stimuli. The scientific term for the ability to respond to stimuli is *irritability*. How does the scientific meaning of this word compare to the common meaning? What were the stimuli that caused the responses depicted in the photos on page 91? Ⓐ

Animals respond to a number of stimuli, such as odor, temperature, light, taste, touch, gravity, and electric shock. In order to test the responses of living things, you should experiment with one stimulus at a time. That stimulus will be the variable in the experiment. There are two ways of responding to stimuli: positively and negatively. A living thing responds *positively* by moving toward a stimulus. It responds *negatively* by moving away from a stimulus.

The mouse responds positively to the cheese and negatively to the cat. The cat responds positively to the mouse. Suggest or draw a scene in which the cat is responding negatively to something. Ⓑ

92

EXPLORATION 1

Experimenting With Stimulus and Response: An Earthworm Responds

It is easy to test an earthworm's response to a number of stimuli. As you carry out these experiments, record your findings on a data sheet similar to the one on page 93.

The earthworm has no eyes, ears, or nose. Can it still sense light, sound, and odor, in addition to other stimuli?

You Will Need

- an earthworm
- a shallow pan
- paper towels
- water
- a pencil
- a flashlight
- vinegar

What to Do

Observe and record all responses in your ScienceLog using a data sheet like the one shown on the next page. Use the questions provided to guide you.

You are going to observe and investigate the earthworm's response to touch (pencil test), light (flashlight test), and smell (vinegar test). Clean up your area, and wash your hands with soap and water when you are finished.

LESSON 1 ORGANIZER

Time Required
2 class periods

Process Skills
observing, communicating, analyzing

New Terms
Irritability—the ability of a living thing to respond to stimuli
Response—the reaction to a stimulus
Stimulus—anything that causes a living thing to react

Materials (per student group)
Exploration 1: earthworm; shallow pan; about 100 mL of water; flashlight; about 10 mL of vinegar; several paper towels; safety goggles

Teaching Resources
Exploration Worksheet, p. 32

Data Sheet: An Earthworm Responds

Purpose:

Write a statement outlining what you hope to learn by performing these tests.

Stimulus: Touch

1. Describe how the earthworm responds to touch.

2. Is the response positive or negative?

3. What happens when you touch the front end, the back end, and the sides?

4. Did you make any other observations?

Stimulus: Light

5. Describe how the earthworm responds to light.

6. Is the response positive or negative?

7. What happens when you shine a light on the front end, on the back end, and on the sides?

8. Did you make any other observations?

Stimulus: Vinegar

9. Describe how the earthworm responds to having the towel soaked in vinegar brought near it.

10. Is the response positive or negative?

1. Place the worm on damp paper towels in the shallow pan. (The worm will die if it is not kept moist.)

2. Gently touch (do not poke!) the side of the worm with the tip of a pencil. Then touch the worm on its back end.

Exploration 1 continued ▶

93

Tell students to let the earthworms rest between stimulations because they may not react if they receive too much stimulation in a short time span. Be sure to practice this Exploration before the lesson begins to ensure enough time between stimulations. The earthworms should be returned to their natural environment when the activity is completed.

Cooperative Learning
EXPLORATION 1

Group size: 3 to 4 students
Group goal: to test the responses of earthworms to stimuli
Positive interdependence: Assign each student a role such as chief investigator, reporter, materials coordinator, or timekeeper.
Individual accountability: Each group member should be able to explain how an earthworm senses stimuli such as touch, light, and odor.

⭐ An Exploration Worksheet is available to accompany Exploration 1 (Teaching Resources, page 32).

Answers to
Data Sheet:
An Earthworm Responds

The earthworm responds negatively to all three stimuli.
Touch: Students should observe that when the sides and back end are touched, the worm recoils and squirms more vigorously than when the front end is touched.
Light: When the light shines on its head, the worm slowly moves away. It does not respond when the light shines on its back end and sides.
Vinegar: The worm slowly moves away when the vinegar is brought near it.

Meeting Individual Needs

Second-Language Learners

To encourage students' interest and to develop their observational skills, display living organisms in the classroom. The beetle *Tenebrio* is an excellent organism to study in the classroom because it is very easy to maintain. The larvae of this beetle are called meal worms and can be purchased at pet stores. To start a culture, obtain about 10 meal worms and place them in a large jar with a screen cover. Feed them moist oatmeal or bran and small pieces of raw carrot. Challenge students to find out as much as possible about the meal worms, such as how they grow and change, how they move, and how they respond to stimuli. Have students design their own experiments to find out how *Tenebrio* larvae and adults respond to different stimuli.

Answers to Questions

1. **a.** No, some parts are more sensitive to touch than others.
   b. The sides and back end of an earthworm seem to be more sensitive to touch than the front end.

2. **a.** The front part of the worm seems most sensitive to light.
   b. The worm responded negatively to the light.

3. Earthworms are not usually found on the surface of the ground because their food source is found in the soil. Also, earthworms are more vulnerable to predators on the surface.

4. Worms are likely to be found on the surface at night because there is no sunlight. Worms also take in oxygen through their moist skin; after a hard rain, worms will come to the surface because there is not enough oxygen available in the water-soaked soil.

5. The earthworm can sense certain chemicals in the air, a process that is similar to our ability to smell.

6. The earthworm's negative response to touch might help it escape from predators. Also, because earthworms respond negatively to light, they remain underground during the day. This keeps them from drying out in the sun and keeps predators from seeing them.

Homework

Have students write a brief paragraph about why it is important that only one stimulus is tested at a time. *(Students should understand that only one stimulus is tested at a time so that results can be clearly linked to a specific experimental variable.)*

3. Gently touch the worm on its front end several times.

5. Soak a piece of paper towel in vinegar. Uncover the worm, and bring the towel near it. **Do not touch the worm with the towel.**

4. Let the earthworm rest for a while under a damp paper towel. Remove the towel, and shine a beam of light on the front end of the worm. Give the worm another rest, and shine the light on other parts.

Questions

1. **a.** Are all parts of an earthworm equally sensitive to touch?
   b. If there is a difference, which parts respond more to touch than others?

2. **a.** Which part of the worm seems most sensitive to light?
   b. Did the worm move toward the light (positive response) or away from the light (negative response)?

3. Give two reasons why earthworms are not usually found on the surface of the ground.

4. When are you likely to find earthworms on the surface? Why are they there at that time?

5. How could the earthworm sense the vinegar without touching it?

6. How are the earthworm's responses useful for its way of life?

94

Plant Responses

Plants also respond to certain stimuli. You saw earlier in this unit that plants grow toward the light when they are placed near a sunny window. The movement of plants is, in most cases, very gradual.

However, some plants show surprisingly quick responses. You may have watched the leaves of the mimosa plant close in response to touch. The leaves of the Venus' flytrap quickly snap shut when insects (or anything else) touch sensitive hairs on the leaves' surface. Wood sorrel, a common plant, folds its leaves in the evening. The prayer plant is a houseplant that also folds its leaves in the evening. The wood sorrel and the prayer plant are responding to the same type of stimulus. Do you know what it is? Ⓐ

Stimulating Things to Investigate

Now try to answer one or two of the following questions about stimulus and response. You may have to go to a nursery or to the library for the answers. Or you could conduct your own research.

1. What are the parts of the mimosa that are sensitive to touch?

2. If you watch a mimosa over a period of time, you will see the leaflets open in the morning and close at night. If a mimosa were placed in a dark closet, do you think

Why does this Venus' flytrap respond to touch? (Hint: The soil of its native habitat is poor in certain minerals.) Ⓑ

Can you think of other plants that respond to the same stimulus as the wood sorrel shown here? Ⓒ

the leaflets would still open and close?

3. Is wood sorrel sensitive to touch?

4. Is there a definite temperature that will cause a crocus flower to close or open?

5. How does an earthworm respond to food? to temperature changes?

6. Does a sow bug seek light or moisture?

7. How do fish respond to the stimuli of sound and light?

8. As seedlings grow, do their roots always grow downward and their stems upward?

9. How would a prayer plant respond if you kept it in the dark all day and in the light all night?

Plant Responses

Ideally, one or more of the plants mentioned in the text should be available for the class to see. The leaves of the mimosa close up quickly when they are touched; this is caused by certain cells losing water and collapsing. The Venus' flytrap responds to touch when the leaf hairs are stimulated. The soil in which the Venus' flytrap grows is very poor in nitrogen and phosphates.

Answers to
In-Text Question and Captions

Ⓐ The prayer plant and wood sorrel respond to light levels.

Ⓑ It obtains minerals (nitrogen and phosphates) from the insects that are caught in its leaves.

Ⓒ Morning glories close their blossoms at night.

Answers to
Stimulating Things to Investigate

1. The leaves

2. No

3. No

4. Crocus flowers open as temperatures warm in the spring.

5. An earthworm moves toward food and away from very hot or very cold temperatures.

6. Moisture

7. Negatively

8. Yes

9. It would eventually switch to unfolding its leaves in the evening.

FOLLOW-UP

Reteaching

Ask students to set up the following experiment to find out which habitat earthworms prefer: Fill one jar halfway with moist soil and the other halfway with moist sand. Place three earthworms in each jar. Gently cover the earthworms in the soil-filled jar with moist sand and those in the sand-filled jar with moist soil. Wait a day, and then carefully dig with spoons to find out in which layer the earthworms have settled. *(In both jars, the earthworms*

should be in the soil rather than in the sand.) Ask: What might account for this behavior? *(The earthworms gain more nutrients from the soil.)*

Assessment

Have students write and perform a short skit that describes an animal reacting to a variety of stimuli. The skit should explain both the reactions themselves as well as the benefits that the animal receives by reacting in the indicated ways.

Extension

Have interested students do research to find out more about other insectivorous plants such as the sundew and the pitcher plant. Ask: How are these plants similar to the Venus' flytrap? How are they different?

Closure

Have students use illustrations to describe the difference between the positive and negative responses of organisms to a stimulus.

LESSON 2 You Be the Scientist

FOCUS

Getting Started

Ask students if they know of any animals that dance. Explain that scientists have discovered that honeybees do a dance called the "waggle dance" that tells the other bees the location and distance of food sources. This behavior was first studied by Karl von Frisch, whose work students will study in this lesson.

Main Ideas

1. Scientists ask questions, test ideas, and communicate their results.
2. Scientific knowledge usually develops over many years as each scientist builds on the work of others.
3. Some plants and animals have an internal sense of time.

TEACHING STRATEGIES

Have students work in groups to answer the Thinking as a Scientist . . . questions on pages 96–99. Tell students that in each scenario, scientists noted certain animal responses, and they tried to figure out what the stimuli were. Have students explain (a) the different responses observed in the animals, (b) how to find out what stimuli provoked the responses, and (c) what conclusions can be drawn from the test results.

Answers to
Thinking as a Scientist . . . (1906)

Answers will vary, but the arrival of the bees could be explained by the odor or color of the jam, the odor of any of the other foods, or the time of day. (Forel also noticed that the bees came at breakfast time even when no food was present. It seemed to Forel that the bees had some type of internal clock.)

LESSON 2 You Be the Scientist

It is not always easy to explain how and why an animal behaves a certain way. Many stimuli are at work influencing an animal's behavior. Finding out which specific stimulus is causing the animal to behave in a particular way can take hours, days, or sometimes even years!

Here are five short accounts of scientists who solved some riddles of animal behavior by looking at bees and flying squirrels. Imagine that you are the scientist. First read about the observation or experiment, and then explain what was observed. Use the questions provided to help you think about these observations.

Scientists Observe and Experiment . . . With Bees

1906

In the summer of 1906, Auguste Forel, a Swiss scientist, noticed honeybees coming to share his jam when he ate breakfast outdoors.

Thinking as a Scientist . . .
How would you explain the arrival of the bees? Is your explanation the only one possible?

96

LESSON 2 ORGANIZER

Time Required
2 class periods

Process Skills
hypothesizing, analyzing

Theme Connection
Cycles

New Terms
none

Materials (per student group)
none

Teaching Resources
none

1909

A few years after Forel's observation, a German scientist, von Buttel-Reepen, began studying the activity of bees. He noticed that bees always came for the nectar of buckwheat flowers in the morning and never in the afternoon. The bees somehow knew when the flow of nectar started and stopped. Perhaps it was the fragrance of the flowers that attracted the bees.

However, von Buttel-Reepen found the flowers just as fragrant in the afternoon as in the morning. He might have suggested any one of the following statements to explain why the bees arrived only in the morning:

a. The bees kept some scouts to watch for and report on the nectar flow.

b. The bees responded to the sun's position.

c. The bees had their own sense of time.

Thinking as a Scientist . . .

How would you test whether or not it was the fragrance of the flowers that attracted the bees?

How would you explain why the bees arrived only in the morning? How would you test your explanation?

Answers to
Thinking as a Scientist . . . (1909)

To test whether it was the fragrance of the flowers that attracted the bees, a scientist could try to detect any difference in the flower's scent at different times of the day or use an extract of the flowers to see if the scent of the flowers alone would attract the bees.

Students might suggest that the bees responded to the warming of the air in the morning. They could test this hypothesis by observing the bees on a morning when the temperature does not change significantly.

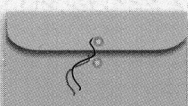 ## PORTFOLIO

Encourage students to record their responses to Thinking as a Scientist . . . in their Portfolio. They may review their responses and evaluate them for critical-thinking and problem-solving skills. You may wish to provide students with a Self-Evaluation Checklist (available on the SnackDisc).

Homework

The Thinking as a Scientist . . . questions on this page make an excellent homework activity.

1927

Students might conclude that bees can tell time. Things that might have helped the bees tell time include light levels, temperature changes, the sun's position, the level of humidity in the air, cosmic rays, and electrical charges in the air.

Answers to
Thinking as a Scientist . . . (1955)

Renner and von Frisch's experiment showed that the bees were not using the sun's position to tell time, nor were they confused by other external conditions. The two scientists inferred that the bees must indeed have an internal clock.

Answers to
Thinking as a Scientist . . . (1955–1960), page 99

The squirrels began running regularly, but earlier, each day because they gradually lost their sense of time without the light. This indicated that the squirrels' internal clock was based on sunlight.

Theme Connection

Cycles
Focus question: Does sunlight have an effect on our behavior? *(The development of artificial light sources has allowed humans to become less dependent on the availability of sunlight when setting their daily routines. Yet research reveals evidence that human behavior is still closely tied to solar rhythms.)* Have students find out how the cycle of light and darkness caused by the Earth's rotation affects the human body's biological rhythms.

CROSS-DISCIPLINARY FOCUS

Social Studies
Have students use a large piece of poster board or paper to translate the information on pages 96–99 into a time line. Have them label each date with a concise summary of the appropriate observations or experiments.

1927

Ingeborg Beling, working in Germany in 1927, studied the problems that had interested Forel and von Buttel-Reepen. By putting food out at the same time every day, she trained a group of bees to go to a feeding station at a certain time. She marked the bees with colored dots so that she could tell them apart.

One day, Beling removed the food and watched. The bees still came to the empty feeding station at their usual time. This led her to test all of the possible stimuli that might help the bees tell time.

Thinking as a Scientist . . .

What conclusion(s) do you draw from this experiment? Name things that you think might have helped the bees tell time. (Beling listed six different things.)

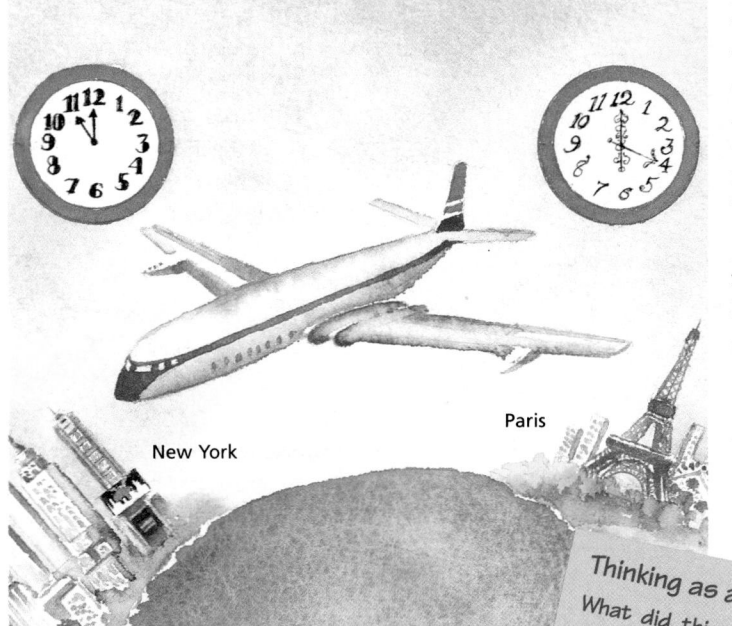

New York

Paris

1955

In 1955, Max Renner and Karl von Frisch conducted an unusual experiment to test the effects of other stimuli on bees. A hive of bees was trained to eat once every 24 hours in a special room in Paris. The hive of bees was then flown overnight to New York and placed in a room just like the one in Paris. The bees came out of the hive and began to eat 24 hours after their last feeding, as if they were still in Paris. They did not seem to react to the difference between the sun's position in New York and the sun's position in Paris. It made no difference to the bees that New York time is 5 hours behind Paris time.

Thinking as a Scientist . . .

What did this experiment show about the bees' sense of time?

Multicultural Extension

Time
Some animals seem to have internal clocks and calendars, but human beings have to use mechanical devices and charts to determine what time or day it is. Many different groups throughout human history have developed calendars. Suggest that students research one of the following calendars: Babylonian, Egyptian, Roman, Mayan, Hebrew, Chinese, Aztec, or Islamic. When was the calendar developed? On what was it based? Is the calendar still in use? How did the use of a calendar affect the daily lives of the people who used it?

Observations and Experiments . . . With Flying Squirrels

1955–1960

From 1955 to 1960, Patricia DeCoursey studied the behavior of flying squirrels at the University of Wisconsin. She put the squirrels in cages outdoors and observed their daily activity. The exercise wheel was the center of most of the squirrels' activity. DeCoursey found that the squirrels began their daily run on the exercise wheel around sunset.

In another experiment, flying squirrels were kept in a dark room below ground level. They received no light from the sun. Their activity over a period of 25 days is shown in the figure to the right. The dark line indicates the time of day when the squirrels ran on the exercise wheel.

DeCoursey concluded that flying squirrels have their own sense of time that is set by a particular stimulus. In fact, it may be that all living things have such "built-in clocks."

Flying-Squirrel Activity

(Graph: y-axis "Days of experiment" from 5 to 25; x-axis "Time of day" with markings 12 Noon, 4, 8, 12 Midnight, 4, 8, 12 Noon)

Thinking as a Scientist . . .
Did the squirrels begin running
a. at the same time each day?
b. regularly, but earlier each day?
c. regularly, but later each day?
d. at no particular time each day?
To what stimulus were the squirrels responding?

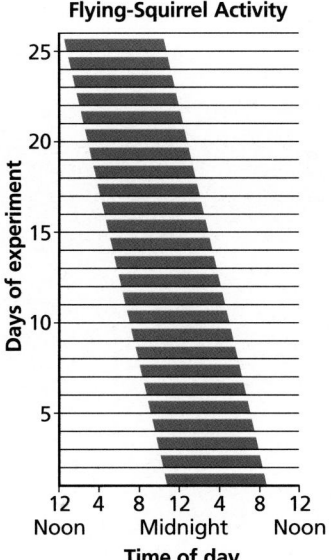

EXPLORATION 2

Animal Timekeeping

You have been asked to make a contribution to the school's monthly newspaper. Your topic is *biorhythms*. The scientific research you have just been reading about is a good starting point for your report. Here are some things to think about as you write your article:

- Do humans have built-in clocks?
- What animals have built-in clocks?
- What stimuli do you think set an animal's clock?

EXPLORATION 2

Suggest that students do library research on biorhythms to help them write their article. Also, they may wish to conduct some of their own experiments to learn about biorhythms. For example, they could measure their own temperature and pulse rate at different times during the day and night to find out when they have their lowest and highest temperature. *(Students should discover that there are daily patterns, not only in activities such as sleep, but also in things such as body temperature and pulse rate.)*

Cooperative Learning
EXPLORATION 2

Group size: 2 to 3 students
Group goal: to investigate biorhythms and write a newspaper article summarizing the findings
Positive interdependence: Assign each student a role such as chief reporter (to do primary research), synthesizer (to make sure the research is concise and thorough and possibly do further research), or editor (to correct grammar, check that the writing is clear and understandable, and make sure the report is legible).
Individual accountability: Have students perform a self-assessment. They can record their thoughts about their personal contributions to the results of the project, and they can explain what they gained by doing the project.

99

FOLLOW-UP

Reteaching

Many insects show daily variations in their level of activity. Ask students to perform the following investigation: Collect a cricket and leave it in a terrarium or in a small container for a few days. Make the container as similar to the cricket's natural environment as possible. Note any evidence of daily rhythms or sense of time, such as feeding, moving, and chirping. Release the animal when the experiment is completed.

Assessment

Have students describe a series of experiments designed to test why a certain bird sings only in the morning. *(The described experiments should test different variables one at a time.)*

Extension

Have students do research on sense of direction in migratory birds. Students might want to compare the sense of direction in birds with that in other migratory animals such as butterflies.

Closure

Have students pretend that they are world-famous scientists living by the sea. They have recently noticed that pairs of dolphins come to shallow areas near the coast. Those dolphins are sometimes joined by a third dolphin. Have students write three investigative questions about this behavior.

Seasonal Behavior of Animals

FOCUS

Getting Started

Ask: What time of year have you seen migrating birds? *(Migrations often occur when the temperature becomes noticeably cooler or warmer.)* Ask: Where were the birds going? *(Birds may migrate in search of warmer temperatures or more food, for instance.)*

 The Discrepant Event Worksheet on page 34 of the Unit 2 Teaching Resources booklet describes a teacher demonstration that makes an excellent introduction to this lesson.

Main Ideas

1. Migration in animals is a response to seasonal changes.
2. Endothermic animals maintain nearly constant body temperatures even when the surrounding temperature changes.
3. The body temperatures of ectothermic animals change according to the temperature of their surroundings.

TEACHING STRATEGIES

Answer to
In-Text Question

Ⓐ Examples of animals that migrate include some insects, such as monarch butterflies; many birds and fish, including waterfowl, robins, swallows, herons, hawks, salmon, and eels; and some mammals, such as wildebeests and mule deer.

INDEPENDENT PRACTICE Have students do some research to answer the questions.

Answers to Questions are on the next page. ▶

Song of the Humpback

Down deep
they say the humpbacks
sing to each other
eerie notes
as mysterious
as the blue depths themselves,
the green shadows
that haunt that other world.
But the song I know
is forty tons of leviathan
breaking the gray pacific surface,
forty tons of grace,
a massive living wave
that rolls slowly,
the great water wing
towering into the air
before the crescendo
of descent
when it slaps the ocean,
leaves the air
full of silver spray
and a great swirl of brine
like a coda,
the signature
of a maestro.

—James Michael Robbins

Humpback whales are well known for their eerie, mournful-sounding "songs." These songs are thought to be some form of communication, but their exact function is not known. Throughout the year, humpbacks sing the same song again and again. The following year they will sing an entirely different song.

These graceful, gentle giants spend the summer in the cold waters of the northern Pacific Ocean and in the Arctic Ocean. In the fall they travel south to the warmer waters off the coasts of California and Mexico.

Like the humpback whale, many other animals move from place to place, often over long distances. They, or their descendants, return again and again to the same places. This type of behavior is called **migration**. For most migrating animals, one of these places is ideal for the young to hatch or be born, while another place is best for growing, maturing, or waiting out a cold winter. What other animals can you think of that migrate? Ⓐ

Questions

1. Why do you think the humpback whales travel from north to south every year?
2. What might be the purpose of the humpbacks' songs?
3. Why might humpbacks change their song every year?
4. Why do you think animals migrate?
5. Do people ever migrate? Give examples.
6. Do you think that migration always takes place over large distances? Why or why not?

100

LESSON 3 ORGANIZER

Time Required
4 class periods

Process Skills
inferring, hypothesizing, predicting, analyzing

Theme Connection
Changes Over Time

New Terms
Dormancy—a sleeplike state during which bodily functions are suspended or greatly reduced
Ectothermic, or coldblooded—having a body temperature that matches that of the surroundings

Endothermic, or warmblooded—maintaining a nearly constant body temperature
Migration—the seasonal movement of animals to another place in order to grow, mate, or nurture their young
Perspire—to secrete water from glands in the skin (sweat)

continued ▶

Keeping Warm

The animals in these photographs all have a common need to keep warm in winter. Their body temperature must remain steady no matter what the air temperature is. Even a small change in their body temperature has serious consequences. Animals that keep their body temperatures at nearly constant levels (usually above the temperature of their surroundings) are called **endothermic** animals. These animals are commonly known as *warmblooded*. Try to name at least six other endothermic animals. **B**

In cold weather, endothermic animals have difficulty maintaining their body temperature. Before reading any further, look at the photographs below, and write down all of the ways that you think endothermic animals survive winter. **C**

⬆ Walrus

⬆ Grosbeak

⬅ Grizzly bear

⬆ Woodmouse

101

Answers to
Questions, page 100

1. Humpback whales travel south every year to breed and give birth to their young in warm waters. They then travel back north to feed in the food-rich waters of the Arctic Ocean.

2. The songs may serve as locating signals or indicators of territorial claims to feeding grounds, or they may have some reproductive function.

3. The songs may change every year to signal changes in relationships or in territorial claims.

4. Animals migrate for a number of reasons—to find food, to escape from bad weather, or to raise young, for instance.

5. Some students may have relatives who spend the winter in certain areas of the country. In addition, some nomadic groups, such as the Bedouins of northern Africa, migrate with the change of seasons.

6. Some animals migrate only a short distance—by climbing up a mountain, for example. There they may find food or shelter that is unavailable in a low-lying area.

Answers to
In-Text Questions

B All birds and mammals are endothermic. Some examples include polar bears, seals, penguins, beavers, wolves, whales, bats, kangaroos, hawks, and humans.

C The thick hide and layer of blubber under the walrus's skin helps protect it from the cold. Birds are covered with interlocking feathers, which cover a layer of soft down. They fluff up their feathers to trap a layer of warm air close to their bodies. The bear seeks a shelter where it can sleep for the winter. Mice prepare for winter by storing extra food during the fall.

Homework

Have students research the migration of an animal and prepare a brief oral report on it.

Ⓐ The snowshoe hare has shorter ears than the other two hares. Students may infer that since the ears are thin and well supplied with blood, a lot of body heat can be lost there. The jack rabbit that lives in hot, southern areas has large ears that give off heat and help keep it cool. The snowshoe hare has short ears that minimize heat loss.

Ⓑ As you wave your wet hand through the air, it feels cooler because the water evaporates and lowers the temperature of your hand.

Ⓒ Many mammals, including domestic cats, pant. Birds also open their mouths to rid their body of excess heat. (Suggest that students observe a dog's tongue when it pants—the tongue increases in width, thus increasing the surface area to help rid the body of excess heat.)

Keeping Cool
Ask students to distinguish between the terms *warmblooded* and *coldblooded*. Point out that coldblooded does not mean that the animal has cold blood; instead, the temperature of coldblooded animals varies with the temperature of the environment. Body temperature is not constant in coldblooded animals. By contrast, it is held constant by internal controls in warmblooded animals. You may wish to point out that some warm-blooded (endothermic) animals, such as chipmunks, enter a dormant state in winter. During that time their body temperature does not remain constant, but actually falls below normal.

GUIDED PRACTICE Ask: Why does warm spring weather act as a stimulus for many dormant ectothermic animals? *(The warm spring weather raises their body temperatures, stimulating them to increase their activity. Frogs, toads, and other amphibians require a moist skin at all times, so warm, wet, spring weather will increase their activity.)*

In winter, endothermic animals must eat more than usual to renew their energy supply and maintain their body temperature. Moving around is another way to produce body heat. Some animals grow thick coats in the winter. Endothermic animals keep warm by other means as well. The blood vessels near the surface of their skin contract, or become smaller. As a result, more blood is kept in the warmer, inner parts of the body, reducing heat loss. You can imagine how useful this is for whales, dolphins, and other endothermic animals that live in water that is often quite cold. Look at the pictures below of North American hares. Compare the ear length of the hare that lives in the north with the ear length of its southern cousins. Use what you know about how animals stay warm to explain the differences between them. (Hint: The ears are thin and well supplied with blood.) Ⓐ

🔺 Snowshoe hare   🔺 Black-tailed jack rabbit   🔺 White-tailed jack rabbit

Keeping Cool
What are the different things that endothermic animals do to keep cool in hot weather? First, think of yourself. When you get hot, you **perspire**—your skin secretes water from many tiny glands. As the water evaporates, it takes heat away from your body, helping it to stay cool. This process is also called sweating. Try this experiment: wave your hand through the air; then wet your hand and wave it through the air. Which method felt cooler? Ⓑ

Another method that some endothermic animals use to keep cool is panting. Think of a dog on a hot day. Every time a dog exhales, heat is removed from its body. Breathing rapidly removes heat more quickly. At the same time, evaporation from the tongue also helps to cool the dog.

🔺 This dog is trying to be "cool." What other endothermic animals pant? (Surprise! The cheetah, one of the big cats, pants.) Ⓒ

102

CROSS-DISCIPLINARY FOCUS

Language Arts
In "Song of the Humpback" on page 100, what is the "song" according to the author?
a. the coda
b. a forty-ton freighter
c. the movement of the humpback whale *(Correct)*
d. mysterious, eerie vocalizations of the humpback whales

Theme Connection

Changes Over Time
Focus question: How does the Earth's motion around the sun affect the migration of animals? *(Because the Earth is tilted on its axis, the North Pole may point toward or away from the sun. When the North Pole is pointed away from the sun, the Northern Hemisphere has shorter, colder days. The position of the Earth in relation to the sun thus brings about changes in temperature and food supplies, which may cause the migration of animals.)*

Which Loses Heat Faster—A Mouse or a Mountain Lion?

What do you think? Make a prediction first, and then test your idea.

You Will Need

- 2 glass bottles (with lids), one of which can hold twice as much as the other
- 2 alcohol thermometers
- a pitcher
- hot water from the tap
- oven mitts

What to Do

1. Fill each bottle with hot water at the same time.

2. Record the temperature of the water in each bottle. Put on the lids.

3. Measure the temperature of the water in each bottle every 5 minutes for 30 minutes.

Caution: Be careful not to burn yourself with the hot water. Use oven mitts as necessary to handle the bottles of hot water.

Was there any difference in the rate of cooling of the small and large containers? Try to explain the difference. Does the result support your prediction about the mouse and the mountain lion? **D**

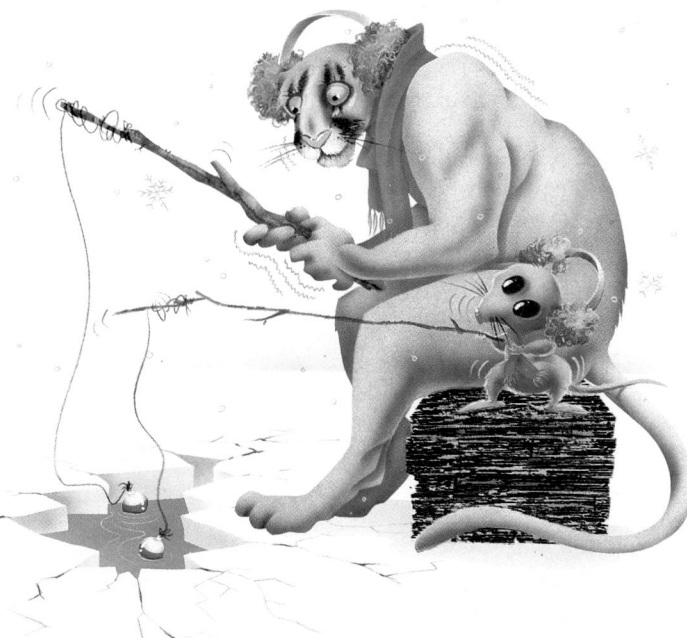

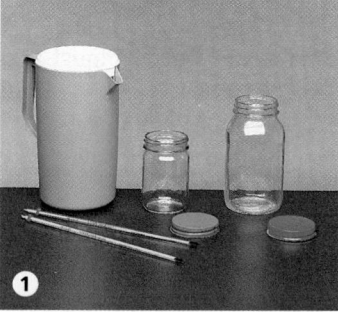

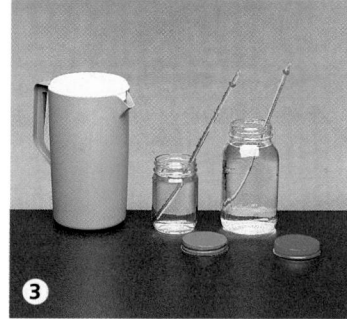
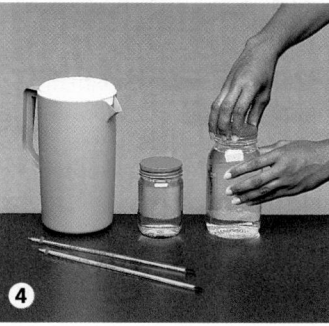

103

Within 5 to 10 minutes, the heat loss in the jars should become noticeable. Students should observe that the temperature decreases faster in the smaller jar. Have students graph both rates of temperature loss.

SAFETY ALERT Caution students to be careful when working with hot water. Use a hot plate or hot water from the faucet for the source of hot water. Use water that is below 55°C to avoid scalding. Do not allow students to distribute the hot water.

Cooperative Learning
EXPLORATION 3

Group size: 3 to 4 students
Group goal: to decide if the size of an organism affects the rate of heat loss
Positive interdependence: Assign each student a role such as task leader, materials manager, recorder, or timekeeper.
Individual accountability: Have each student answer the following question in his or her own words: Which loses heat faster—a mouse or a mountain lion?

★ An Exploration Worksheet (Teaching Resources, page 35) and Transparency 14 are available to accompany Exploration 3.

Answer to
In-Text Question

D The larger volume of water contains twice as much heat energy as the smaller volume, but the surface area of the larger bottle is not twice that of the smaller bottle. Therefore, the rate of cooling in the larger "animal" is slower than that in the smaller one. Heat content is a function of volume, but heat loss is a function of surface area.

Homework

Have students research an endothermic animal, focusing on what measures that animal employs to keep warm in winter. Students should present their findings in the form of a poster.

Multicultural Extension

Nomadic Tribes

A *nomad* is a person who moves from one place to another as a way of life. Most nomadic groups move through a certain area based on a cycle of activities or seasons. There are many different kinds of nomads, including those who move in search of game, edible vegetation, and water and those who move to find water and pasture for their livestock. Ask students to do research on some of these groups, such as the people of Lapland, the African Pygmies, the Australian Aborigines, or the Bedouins of Northern Africa and Arabia. Student research should include the reasons for the groups' migrations, as well as any contributions that the groups have made to the environment and to the culture of their homeland.

A The human's body temperature stays constant. The snake's body temperature varies greatly due to the change in the temperature of the environment.

B A snake's body temperature would be higher if the temperature of the environment were above 37°C.

C A frog might spend the winter buried below the frost line, where there is no danger of freezing.

Answers to
How I Survive—A Personal Account

Answers will vary. Students may create a short story, cartoon, or poem about their animal. The following is a sample poem about being ectothermic by a student named Michelle Newman:

The Limitations of Being Ectothermic
My body temperature changes—
I freeze in the winter and fall.
In the spring and the summer it ranges,
I'm never happy at all.
One minute I'm cold, and
The next I'm as warm as can be.
I'm ever so glad when the time comes,
When I sleep in the state of dormancy.
It's ever so peaceful and quiet;
I relax for the rest of the year.
When I wake, I go search for my water and food,
But I really wish I could stay here.

Homework

An Activity Worksheet may be assigned as homework after students have finished Lesson 3 (Teaching Resources, page 36).

Summer and Winter Retreats

Humans and many other animals are endothermic. However, there is another group of animals known as **ectothermic** animals. Also known as *coldblooded*, these animals react to the temperature of their surroundings very differently than do endothermic animals. For example, use the graph below to compare the body temperature of a snake with that of a human. What differences exist between a snake's and a human's body temperatures over a year? Why do you think the snake's temperature changes so much? **A**

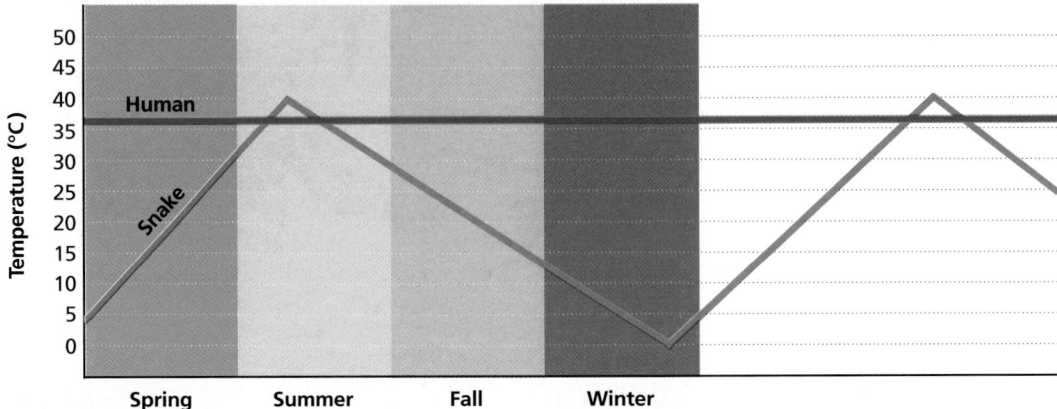

Body Temperature of Humans and Snakes Compared

The snake is an example of an ectothermic animal. Other examples are frogs, toads, turtles, and fish. In what circumstances would a snake's body temperature be higher than that of a human? **B**

The body temperature of an ectothermic animal closely matches the temperature of its surroundings. As their body temperature drops, animals such as frogs and salamanders begin to move more slowly. Their heart beats fewer times per minute, and their breathing rate slows down. As winter temperatures approach, they enter a sleeplike state called *dormancy*. During the winter, ectothermic animals must find a place to rest that is away from the extreme cold that could kill them. Where do you suppose a frog spends the winter? **C**

How I Survive—A Personal Account

Choose an animal that you would like to represent. The animal may be either endothermic or ectothermic. Write a story as if you are that animal, and explain how you react to changes in temperature in order to survive throughout the year. You might want to include a poem or a series of cartoons with your account. You may also need to do some research to find out more about the animal. Here is a title used by one student: "Why I Like Being Ectothermic."

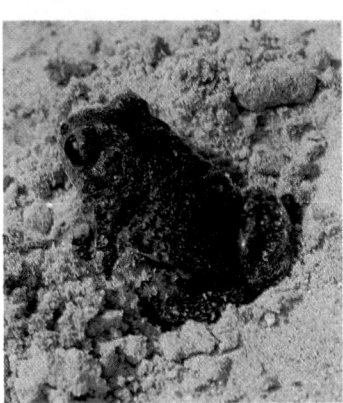

⬆ The spadefoot toad survives dry periods by digging a burrow and entering "summer sleep," or estivation.

FOLLOW-UP

Reteaching

Have an interested student research and give a report on the migration of the monarch butterfly. Other students should then write a short summary of the report, including where the butterfly goes and what causes it to migrate.

Assessment

Have students compare the body temperatures of a cheetah and a turtle. Suggest that they make their comparison in the form of a graph or a short essay.

Extension

Ask students to investigate how an unusual animal, such as a bee, bat, or whale, survives the winter. *(Many bees, for instance, huddle together in the hive, emerging only rarely, if at all.)*

Closure

Have students draw illustrations that show an ectothermic animal and an endothermic animal responding to the specific conditions of their environments.

LESSON 4
Adaptation of Structure

How well suited are you to your surroundings? What if you moved farther south or farther north? How would you deal with tropical heat or polar cold? Humans can add clothing or turn on the air conditioner, but plants and animals do not have these options. In the previous lesson, you read about how other animals are able to adapt to seasonal changes in their environment. But how do plants and animals adapt to changes that occur slowly, over a very long time?

When long-term changes occur in the environment, plants and animals must adapt or risk dying out. For most kinds of animals and plants, this kind of adapting takes a long time and continues as conditions change. Can you think of slow changes that might force a species of animal or plant to adapt? Can you think of any kinds of animals that might have died out because they couldn't adapt? **D**

Where would you find a cactus growing naturally? Where would you look for a frog? Frogs and cactuses need very different conditions to survive. List the ways in which each is adapted to its environmental conditions. What adaptations would each need to have if their environments were reversed? **E**

Anza-Borrego Desert, California

Bull frog

LESSON 4 ORGANIZER

Time Required
2 class periods

Process Skills
observing, analyzing

Theme Connections
Structures, Changes Over Time

New Term
Adaptation—an inherited trait that makes an organism better suited to its environment

Materials (per student group)
none

Teaching Resources
Theme Worksheet, p. 37
Transparency 15
SourceBook, p. S30

LESSON 4
Adaptation of Structure

FOCUS

Getting Started
If possible, show students photographs of closely related species that live in different habitats, such as sea turtles and land tortoises. Ask students to examine the photographs and point out how the physical structures of the animals differ. Then ask students to speculate about how the different structures might help the animals survive in their particular environments.

Main Ideas
1. Living things adapt gradually to their environment over a long period of time.
2. The structure of living things reflects the special conditions of their environment.

TEACHING STRATEGIES

In the previous lesson, students learned how animals respond to seasonal changes. In this lesson, they will learn that environmental changes can also trigger adaptations. Students may be confused by the difference between adaptations that an individual can make during its lifetime and adaptations that occur in a species over the course of a generation or more. For example, when winter arrives, individuals can migrate to warmer climates. However, if there is a long-term change in climate, such as a permanent decrease in temperature, individuals with thicker fur or a greater capacity to store fat would be more likely to reproduce successfully, and succeeding generations would demonstrate those traits.

Answers to In-Text Questions are on the next page. ▶

Ⓓ Slow climatic changes might force animals to adapt. Dinosaurs and mastodons are examples of animals that may have become extinct because they couldn't adapt.

Ⓔ Cactuses grow in dry areas, and frogs are usually found in ponds or lakes. If their environments were reversed, cactuses would need adaptations to cope with excessive moisture, and frogs would need adaptations to cope with dry conditions.

Bird Adaptations

Divide the class into small groups for this matching exercise. Encourage a friendly competition to determine which group can find all of the correct answers in the shortest amount of time.

Explain that the task is to identify which bird beak, foot, or wing best fits the lifestyle of the bird that is described.

Answers to
Bird Adaptations

Starting in the upper-left corner and moving clockwise, the correct answers are 6, 16, 9, 1, 11, 15, 10, 7, 12, 13, 5, 8, 2, 14, 3, and 4.

 Transparency 15 is available to accompany the material on this page.

Meeting Individual Needs

Second-Language Learners

Take students on a field trip to a local zoo. Give them the opportunity to practice their English skills by asking them to discuss how the various animals they see are adapted to their native enviroment. Ask zoo employees for information on how the zoo provides appropriate food and a suitable environment for each of the animals.

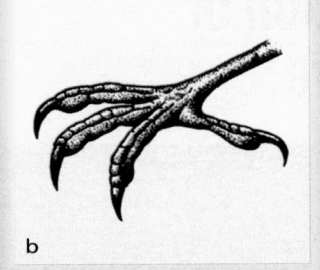

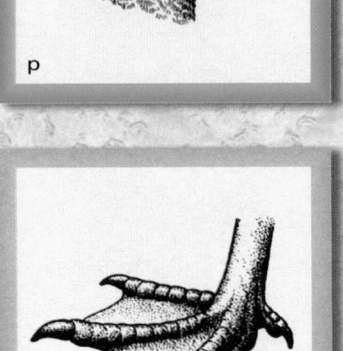

Bird Adaptations

Over time, a species may change so that its members are adapted to special conditions and needs. For example, a seemingly endless variety of adaptations make birds well suited to their environments. Match the descriptions below to the illustrations.

1. The **owl** has a short, sharp, curved beak that it uses to pierce the skull or slash the throat of its prey.
2. The feet of the **curlew** have three spread-out toes and a small hind toe. This design allows the bird to land and walk on very soft ground.
3. A **duck**'s foot, with a tough membrane between the toes, is a natural paddle, making swimming a breeze.
4. The aptly named **crossbill** uses its beak to separate the scales of pine cones and get at the seeds underneath.
5. The **ostrich** cannot fly, but its two-toed feet help it to run very swiftly.
6. The **golden eagle** can soar for hours as it searches for prey because its wings are adapted to catch the slightest updraft.
7. The **osprey**'s feet are well adapted for grasping and carrying off prey.

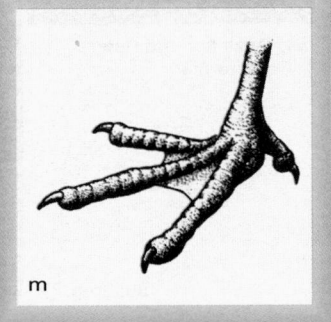

Integrating the Sciences

Earth and Life Sciences

Adaptations occur slowly over time. Sometimes, however, cataclysmic events change the environment so quickly that only those animals already possessing traits that allow them to live in the altered environment will survive. Many scientists believe that such an event occurred about 65 million years ago when a large meteorite struck the Earth, forcing a large percentage of the Earth's species (including the dinosaurs) into extinction.

Did You Know...

The ostrich is the fastest bird on land; it can reach speeds of up to 64 km/h (40 mph).

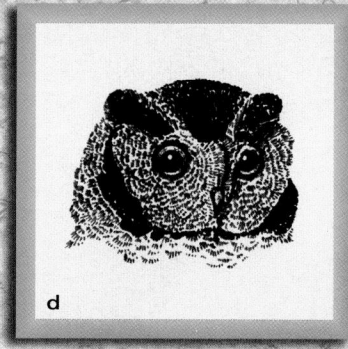

d

e

f

8. The **sandpiper**'s long and slender beak is well suited for probing sand and mud for small animals.

9. The **peregrine falcon**'s pointed wings and bullet-shaped body enable it to swoop down on its prey at speeds of over 300 km per hour.

10. Although flightless, the **penguin**'s streamlined body and short but powerful wings enable it to "fly" underwater.

11. The **scissortail flycatcher** is an aerial acrobat. It has a broad, flattened bill with bristles at the base that help it catch insects in flight.

12. The **ptarmigan** is found in the Arctic and other places that are covered with snow for most of the year.

13. The flightless **kiwi** feeds by probing the leaf litter of the forest floor for insects and other tiny animals.

14. The **woodpecker** uses its sharp bill to bore holes in trees in search of insects.

15. The **nighthawk** feeds by flying through the air with its mouth open, catching insects in midflight.

16. The **evening grosbeak** can curl its toes tightly around a branch, allowing the bird to perch securely in trees.

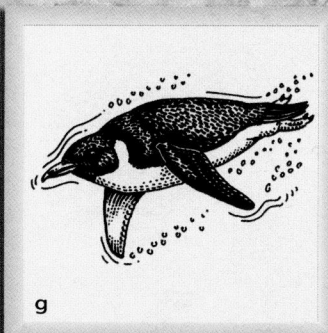

g

h

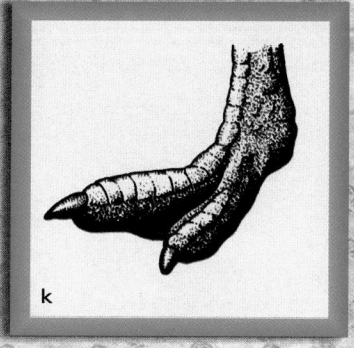

k

j

i

107

Homework

Humans are changing their own environment at a rapid rate. Have students write an essay that includes three predictions about environmental changes that may occur within the next 200 years and predictions of what human adaptations these changes might cause.

Meeting Individual Needs

Gifted Learners

Have students keep detailed records of birds seen in the vicinity of the school during each month of the year. Many different field guides for birdwatching are available at bookstores and libraries. Have students keep a tally on a large wall chart, listing the number of sightings of each bird species and the date of the sighting. Encourage them to illustrate their charts with drawings of the birds. From their chart, students should try to determine which birds migrate, which ones are the first to return in the spring, and which ones stay all winter.

Theme Connection

Structures

Focus question: How are animals physically adapted to their habitat? Interested students can construct dioramas to illustrate how different animals are structurally adapted to their habitat and way of life. If possible, have them visit a museum of natural history first to observe dioramas of animals in their natural settings. Then they could construct dioramas in boxes by using clay to model the animals and construction paper, paint, twigs, stones, and other materials to model the plants and the landscape. They could construct a diorama of a real or imaginary habitat (such as one that might be found on another planet). Emphasize that the real or imaginary animals should be suited to the habitat in which they are placed.

Meeting Individual Needs

Learners Having Difficulty

In order to highlight the connection between adaptation of structure and environmental influences, you might wish to have students make a chart of all the birds listed on pages 106 and 107. The chart should include the area of the world in which each bird lives, the specific adaptation of interest, and a brief description of why that adaptation is an advantage for the bird's environment.

Insect Adaptations

Ask students to carefully observe the pictures of the insects on page 108. Have them list the adaptations that each insect has and explain how these structures help the animal to survive.

Answers to
Insect Adaptations

The honeybee has a stinger for discouraging its enemies; a long, coiled tube with which to suck nectar; and baskets on its hind legs for carrying pollen.

The leaf katydid has wings that look like leaves. When the katydid is sitting on a branch, it is well hidden from its enemies.

The stink bug uses camouflage for protection against predators and also gives off a foul odor when disturbed.

The harlequin beetle has hardened front wings to protect the more delicate hind wings and has well-developed antennae for sensing food.

PORTFOLIO

Suggest that students include in their Portfolio their descriptions and sketches of imaginary plants and animals from Plants and Animals Made to Order on page 109.

Theme Connection

Changes Over Time

Focus question: If species from an equatorial region were to move to one of the coldest regions of the Earth, what new adaptations would help it to survive in this new habitat? *(Sample answer: It would need protection from the cold, such as thick fur. It would also need to be able to cope with different food sources and amounts of daylight.)* The Theme Worksheet on page 37 of the Unit 2 Teaching Resources booklet accompanies this Theme Connection.

Insect Adaptations

Insects have also adapted to their surroundings over time. Many insects have special body parts for gathering food and for defense. What special body parts do the insects below have? How do these parts help the animals?

⬆ Honeybee

⬆ Leaf katydid

Turn to page S30 of the SourceBook to learn more about how environmental forces shape the development of a species over time.

⬇ Harlequin beetle

⬆ Stink bug

Plants and Animals Made to Order

Below is a list of hypothetical plants and animals faced with special conditions that require the plant or animal to have certain adaptations in order to survive. Imagine that you are the plant or animal that must live in the conditions described. What might you look like? Describe and sketch your special adaptation(s). Try to think of adaptations that real plants and animals have for conditions similar to these. There may be many possibilities.

1. an animal that lives in an area with very heavy snowfall but that doesn't migrate or become dormant
2. a plant that lives in a region with little rainfall
3. an animal that spends most of its life in trees
4. an animal (other than a bird) that flies
5. a tree that must withstand an arctic climate
6. an animal that needs to see above and below water equally well and at the same time
7. a plant that must be able to spread its seeds around
8. a small, slow animal with weak teeth and small claws that needs a good defense system
9. a plant that lives in trees
10. a seashore animal that must stay moist during low tide

Head like a lizard

Ears like a cat

Tail like a monkey

Legs like a kangaroo

Feet like a bird

Which of these creatures is real? Ⓐ

Lays eggs like a snake

Hair like an otter

Tail like a beaver

Webbed toes like a frog

Bill like a duck

109

FOLLOW-UP

Reteaching

Divide the class into groups, and have each group choose three birds from the list on pages 106–107. Then have students compare the adaptations of the three birds chosen by their group.

Assessment

Challenge students to design a new animal. Students should decide what kind of body the animal has, what kind of feet it has, how it will defend itself, and so on.

Extension

Insects exhibit a remarkable variety of adaptations to their environment. Have students choose one kind of insect and describe its specialized parts and behavior. Encourage them to include illustrations with their reports.

Closure

Have students write an essay, poem, short story, or song describing how a specific plant or animal is suited to its environment.

Answer to *Caption*

Ⓐ The platypus (bottom) is the real animal.

Answers to *Plants and Animals Made to Order*

The following are sample answers:

1. Adaptations: thick, white fur or feathers, fat deposits; real examples: polar bear, arctic hare, ptarmigan
2. Adaptations: extensive root systems, massive waxy trunks that store water, thick waxy leaves; real examples: succulent, cactus
3. Adaptations: hands or claws for grasping bark, the ability to maneuver through trees either by flying with wings or by climbing; real examples: monkey, squirrel, bird
4. Adaptations: wings, muscles for flapping wings, lightweight body; real examples: bat, many insects
5. Adaptations: needle-like leaves that can withstand freezing, leaves that drop off during extreme cold and then grow back; real examples: arctic willow, Douglas fir
6. Adaptations: large, protruding eyes that can focus upward and downward; real examples: frog, water beetle, mudskipper
7. Adaptations: seeds dispersed by wind or by sticking to animal fur; real examples: maple tree, burdock
8. Adaptations: ability to escape from predators by burrowing or spraying a noxious substance, a shell or spines; real examples: mole, skunk, turtle, porcupine
9. Adaptations: ability to cling to trees and extract minerals and water without having roots that grow in the ground, ability to catch and hold rainwater in leaves; real examples: orchid, ball moss, some lichens
10. Adaptations: hard covering for protection against drying out and predators, ability to cling to rocks or burrow into the sand to prevent being washed out to sea with the tides; real examples: periwinkle, barnacle, shellfish

Answers to
Challenge Your Thinking, pages 110–111

1. Answers will vary depending on the animal chosen. Encourage students to choose animals from various environments around the world. Students may make drawings, puppets, or even costumes to help illustrate the different structures and behaviors of each animal.

2. Answers will vary. Look for logical reasoning in student responses. Students might choose different types or colors of candy and place the candy on an ant mound, for instance. Student observations should describe the activity of the ants and the variables they tested. Remind students to limit their testing to one variable at a time.

3.

Stimulus	Response	+	–
candy in store	child reaches	X	
child reaches	woman pulls child away		X
cat	dog chases cat	X	
dog	cat flees from dog		X
dog and cat in front of car	driver screeches to a halt		X
sunshine	man puts on sunglasses		X
truck backing up	boy on bicycle waits		X
heat and sunshine	man on bench wipes brow	X	
red light	car stops		X

4. Students' explanations should describe the different behaviors as responses to one or more stimuli. The students' explanations should include two separate variables that the students can test. Manipulating one variable at a time will allow students to look for changes in the animals' behavior that might support or refute their explanations.

Homework

The Closure activity on page 109 makes an excellent homework assignment.

CHALLENGE YOUR THINKING

1. An Animal's Tale (Tail?)

Each member of the class should choose a specific animal and tell a story, from that animal's point of view, about its preparations for winter. If you choose to be an animal that migrates, tell the class how, when, and where you migrate and the reasons for your migration. If you choose to be an animal that stays put, describe your preparations.

You will have to do research. Look in books on the subject, and ask informed people. Try to make your story about 150 words long.

2. I Can't Say No to Sweets

Describe how you would test the responses of ants to the stimuli of sugar, moisture, and heat. Use labeled diagrams. Make a prediction about what the ants' responses will be.

3. Good Reflexes

Look at the picture below. Pick out all of the examples of stimuli and responses to these stimuli that you can find. Tell if each response is positive or negative.

4. Strange Behavior

Read the conversation below. Write at least two possible explanations for each animal behavior mentioned. Then choose one of the behaviors, and explain how you would test one of your explanations.

Pamela: I see ants going up and down the stems of our rosebush. Why do they do that?

Mike: I have a mystery too. A lot of crows fly into the trees behind our apartment building late in the afternoon each day.

Ani: Some kind of animal dug up big patches of our lawn last night.

Raoul: There's a bird that pecks at one of our windows early every morning. Why does it do that?

Review your responses to the ScienceLog questions on page 91. Then revise your original ideas so that they reflect what you've learned.

111

Sciencelog

The following are sample revised answers:

1. Some plants may appear to have a sense of time because they respond to the presence of light during the day and to the absence of light at night. Bees also appear to have a sense of time because they can be trained to feed on a periodic schedule that is independent of variables such as the position of the sun.
2. Survival depends on the ability to respond to the environment. For example, organisms must be able to respond positively to food and negatively to predators.
3. Many animals migrate in response to seasonal changes in their environment.
4. A chameleon's coloring helps it to blend in with its surroundings. To catch insects, a chameleon has a long, sticky tongue, grasping tail and feet, quick reflexes, and large eyes that can move independently of each other. (A popular misconception about chameleons is that they have the ability to change color to blend in with their environment. Although they do change color due to light and temperature variations and as a fear response, they cannot deliberately blend in with their surroundings. Their camouflage is like that of the polar bear—an acquired trait that has taken generations to develop.)

⭐ You may wish to have students complete the Chapter 6 Review Worksheet that accompanies this Challenge Your Thinking (Teaching Resources, page 39).

Multicultural Extension

Migration Patterns

Ellis Island, located in Upper New York Bay, was once the largest immigration station in the United States. Between 1894 and 1943, over 12 million European immigrants passed through its doors. Discuss with students migration patterns in the United States. What are some of the key factors that influence migration patterns? *(Weather, political and economic issues, religion, quality of life, family, etc.)* You might wish to have students interview an older member of the community to find out why people settled in their geographic region. Encourage students to share their findings with the class.

Connecting to Other Chapters

Chapter 4
examines the characteristics of living things, their motion, and the differences between plants and animals.

Chapter 5
investigates a variety of patterns of growth, in both humans and other organisms.

Chapter 6
explores the responses that organisms can have to stimuli and to long-term changes in their environment.

Chapter 7
introduces students to both the structure of cells and the proper use of a microscope.

Prior Knowledge and Misconceptions

Your students' responses to the ScienceLog questions on this page will reveal the kind of information—and misinformation—they bring to this chapter. Use what you find out about your students' knowledge to choose which chapter concepts and activities to emphasize in your teaching.

In addition to having students answer the ScienceLog questions on this page, you may wish to pose questions such as the following:

1. What is a cell?
2. What are some examples of objects that are made up of cells?
3. How can we study cells?
Emphasize that you are not looking for right or wrong answers to these questions. If you wish, use these questions to prompt a class discussion and to encourage students to brainstorm answers. It may be helpful to record student responses on the board. The responses may help you gauge what students already know about cells and

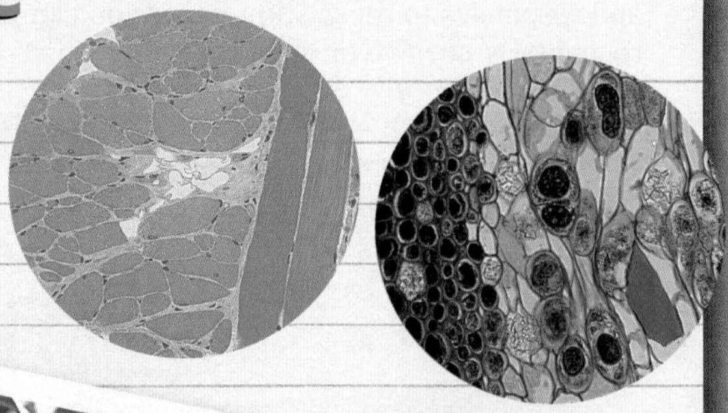

1 **Do all living things have cells?**

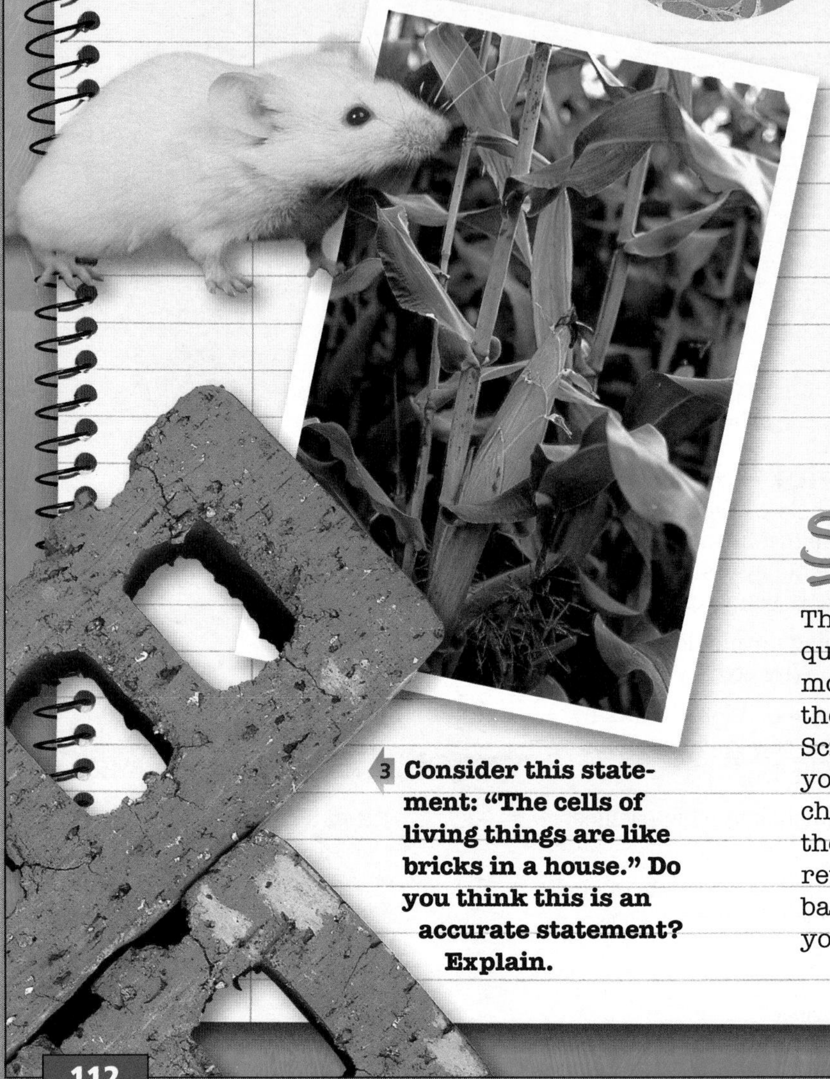

2 **Which of these are animal cells? Which are plant cells? How can you tell?**

3 **Consider this statement: "The cells of living things are like bricks in a house." Do you think this is an accurate statement? Explain.**

ScienceLog

Think about these questions for a moment, and answer them in your ScienceLog. When you've finished this chapter, you'll have the opportunity to revise your answers based on what you've learned.

112

cellular structure, what misconceptions students may have, and what is interesting to them about this topic.

LESSON 1 · Taking a Closer Look

When you approach a building from a distance, you first notice the shape of the whole building. As you draw nearer to it, you become aware of its different parts: walls, windows, doors, staircases, chimneys, and so forth. When you are only a few feet away, you realize that the walls are made up of bricks. Each brick forms part of the structure of the entire building.

1×

3×

60×

Likewise, animals and plants have an overall shape or structure, as well as different parts. When you examine the structure of animals and plants more closely, you see that they are made up of smaller units. These tiny "bricks" that make up the structure of living things are called **cells**. Most cells are too small to be seen with the naked eye, but they are clearly visible under a microscope.

1×

3×

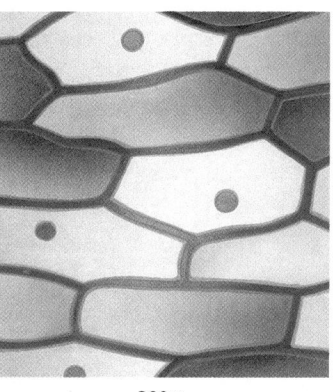
300×

▲ Taking a closer look at an onion

113

LESSON 1 ORGANIZER

Time Required
1 class period

Process Skills
observing, analyzing, comparing

New Terms
Cells—the smallest units that make up the structure of living things
Wet mount—a thin sample of cells placed in water between a microscope slide and a coverslip

Materials (per student group)
How to Use a Microscope: microscope; prepared microscope slide

Exploration 1: microscope; microscope slide with coverslip; eyedropper; letter e from a newspaper; forceps; a few drops of water; tissue

Teaching Resources
Resource Worksheet, p. 44
Transparency Worksheet, p. 45
Exploration Worksheet, p. 47
Transparencies 16–19
SourceBook, p. S36

FOCUS

Getting Started
Place a piece of waxed paper on top of a piece of newspaper. Using an eyedropper, place one drop of water on the waxed paper. Have the students look at the newsprint through the drop of water to see how the appearance of the newsprint changes. Explain that the drop of water acts like a lens in a microscope. Ask students to compare how a large drop and a small drop change the appearance of the print.

Main Ideas
1. Cells are the fundamental units of living things.
2. Microscopes magnify tiny objects and reveal their structural details.

TEACHING STRATEGIES

GUIDED PRACTICE Ask students to read the introductory paragraphs on page 113 and examine the illustrations. Encourage students to sketch other objects at several degrees of magnification, like the illustrations on this page. Examples of appropriate objects include a newspaper, a forest, or a television screen.

Homework

The Activity Worksheet on page 47 of the Unit 2 Teaching Resources booklet makes an excellent homework assignment. If you choose to use this worksheet in class, Transparency 16 is available to accompany it.

 Transparencies 18–19 are available to accompany the material on this page.

How to Use a Microscope

Each group of students should have access to a monocular microscope. Tell them to carefully follow the directions outlined on page 114. Have them locate the important parts of the microscope and practice focusing and adjusting the light source. Students should pay close attention to steps 6 and 7. Emphasize that they should always start with the lowest power objective lens, turning the coarse-adjustment knob so that the lens moves away from the stage. Only then should they turn the fine-adjustment knob to focus the lens.

Answer to
In-Text Question

 a. eyepiece
b. coarse-adjustment knob
c. fine-adjustment knob
d. base
e. tube
f. nosepiece
g. objective lens
h. stage
i. diaphragm
j. light source

Homework

You may wish to assign the Resource Worksheet that accompanies How to Use a Microscope as homework (Teaching Resources, page 44). Transparency 16 is also available if you use this worksheet in class.

A Transparency Worksheet (Teaching Resources, page 45) and Transparency 17 are available to accompany the material on this page.

Meeting Individual Needs

Second-Language Learners

There are many opportunities for students to draw rather than write about the concepts in this unit. For example, students could draw various types of movements in plants and animals, human growth patterns, conditions necessary for seed growth, different plant and animal adaptations, and plant and animal cells. For vocabulary practice, have students label their illustrations.

How to Use a Microscope

The instrument that allows us to see cells and examine small objects is the microscope. As you read the instructions for its use, try to match the names of the parts of the microscope with the labels on the diagram. **A** The diagram below shows only one example of a microscope. The microscope that you use in your own classroom may be slightly different. For example, it may have a mirror instead of its own light source.

1. Be sure that the microscope is steady, with its base sitting on a flat, stable surface. Look into the eyepiece at the top. This eyepiece fits into the tube or barrel of the microscope.

2. Place the microscope slide on the stage, with the material to be viewed over the hole in the stage. Use the stage clips to hold the slide in place.

3. Turn on the light source, and adjust the diaphragm as necessary to control the amount of light that passes through the object.

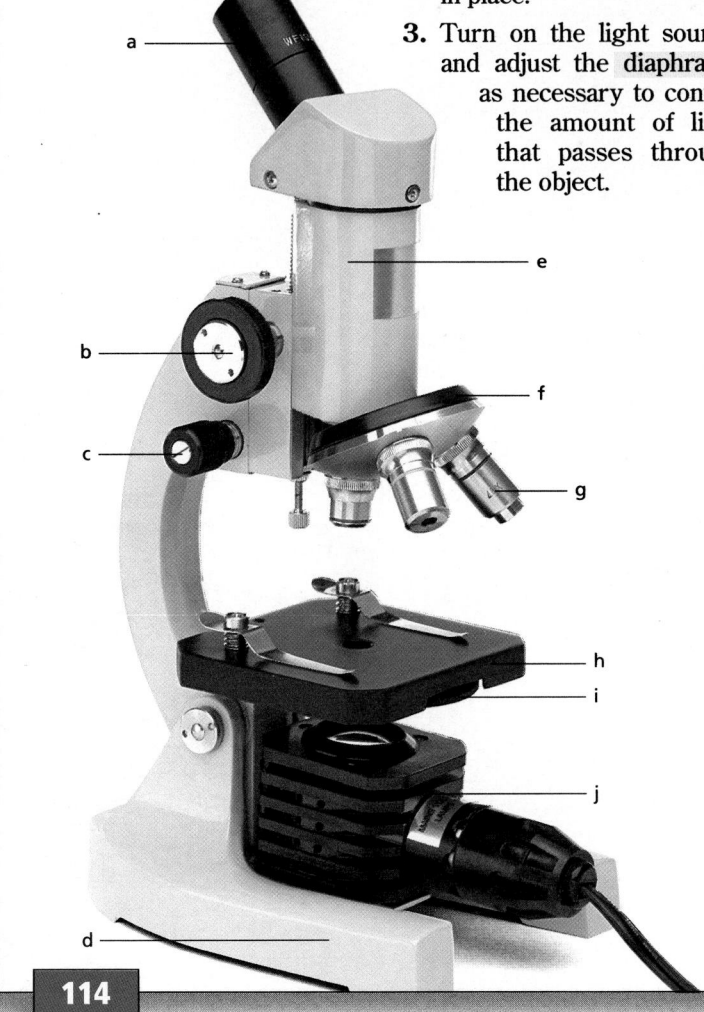

4. There are two or three objective lenses attached to the nosepiece at the base of the tube. One is low power ($5\times$ or $10\times$), one is medium power ($20\times$), and the others are medium or high power ($20\times$ to $50\times$). Turn the shortest objective lens under the nosepiece until the lens clicks into place. This is low power. Always use low power first. The longer lenses will provide higher power.

5. While watching closely from the side, turn the coarse-adjustment knob to move the low-power lens down close to the stage.

6. Look through the eyepiece, and turn the coarse-adjustment knob away from the stage until your specimen is in focus.

7. Turn the fine-adjustment knob to make the focus clearer, if necessary.

8. Without adjusting the focus, change the objective lens from low power to medium or high power. Be sure that the objective lenses do not hit the stage when you rotate them into place. This could scratch the fragile lenses or break the slide.

CROSS-DISCIPLINARY FOCUS

Health

Share the following with students: The microscope was invented in the 1600s, but it wasn't until the 1800s that improvements in the production of glass made the microscope exceptionally valuable to doctors. Pathogens such as bacteria could then be studied.

Homework

Have students research the history of the microscope and make a list of at least five different fields of science in which the microscope is used.

Looking at Cells

When you examine the cells of a plant or animal through a microscope, the sample you use must be very thin so that light can shine through it. Also, the sample must be kept moist. The illustrations on the right show how to make a **wet mount**.

You Will Need

- a microscope
- microscope slide(s)
- coverslip(s)
- an eyedropper
- a small piece of newspaper with a letter *e*
- forceps
- water
- tissue

Watch Out for These!

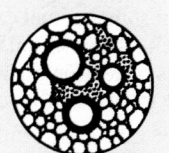

Dark, round circles are air bubbles.

Dark, jagged lines are coverslip edges.

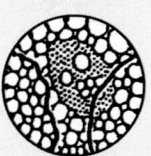

Thin, irregular lines are caused by water drying up.

What to Do

In this activity, you will see that things look different, not just larger, under a microscope.

Making a Wet Mount

1. Put a drop of water on a slide.

2. Place the piece of newspaper with the letter *e* on the water.

3. Lean a coverslip against a pencil. Slowly lower the pencil, and finally remove it. Avoid getting air bubbles under the coverslip. Absorb excess moisture around the coverslip with tissue.

4. Now view the slide through the microscope, first with the low-power lens and then with the high-power lens. Draw what you see.

Analyze Your Work

1. If you move the slide to the left, in which direction do the objects in the field of view move?

2. When you change from the low-power lens to the high-power lens, how does your field of view change?

> To calculate total magnification, multiply the magnification of the eyepiece lens by the magnification of the objective lens you are using. A 10× eyepiece lens and a 10× objective lens will give a total magnification of 100×.

Tell students that they will make wet mounts of the letter *e* cut from a newspaper. Students should examine the slide under a microscope and make a sketch of the microscopic view. The sketch should be labeled with the power of magnification.

Students will observe that the microscopic image is both backward and upside down compared with the real object. Students will notice this because each time they move the slide downward on the stage, the image will appear to move upward, and each time the slide is moved to the right, the image will move to the left.

Answers to *Analyze Your Work*

1. To the right

2. The field of view becomes smaller due to the higher magnification.

★ **An Exploration Worksheet is available to accompany Exploration 1 (Teaching Resources, page 47).**

Multicultural Extension

Uses of Cell Walls

Tell students that every culture in the world relies in some way on the cell walls of dead organisms. Materials such as thatch, reeds, and wood are composed of cell walls that remain after an organism has died. Challenge students to pick a present or past culture and to identify its uses of these materials or other materials made of durable cell walls.

FOLLOW-UP

Reteaching

Make copies of a small diagram of a microscope. Have students make flashcards with the copies, labeling one part of the microscope on each copy and writing the name of the labeled part on the back of the copy.

Assessment

Have students imagine that a student who missed class needs instruction on how to use a microscope. In their own words, have them explain the correct procedure for making a wet mount and observing it with a microscope.

Extension

Provide students with a stereomicroscope and instructions on how to use it. Have them compare the stereomicroscope with the one used in Exploration 1.

Closure

Have students make sketches of a wet mount of a piece of human hair at various magnifications.

FOCUS

Getting Started
The videotape *Discovering the Cell* from the National Geographic Society makes an excellent introduction to this lesson. (For a complete reference, see page 59A.)

Main Ideas
1. Cells have internal structures that can be examined with a microscope.
2. The structures of plant cells and animal cells exhibit similarities and differences.

TEACHING STRATEGIES

EXPLORATION 2

Cut onions into quarters ahead of time. Demonstrate again how to make the wet-mount slides. You may wish to refer students to the diagrams of typical plant and animal cells on page 117 so that they can look for the various cellular structures in the samples while observing their own wet mounts with the microscope.

The nuclei and other structures of the cheek cells may appear deep blue in color due to hematoxylin or methylene blue stain. The cytoplasm may have a reddish color due to eosin stain.

 SAFETY ALERT Iodine is an eye irritant and is somewhat corrosive to the skin. Have students wear safety goggles to prevent contact with eyes and wash the affected area with water if iodine touches the skin.

WASTE DISPOSAL ALERT The procedure for disposing of the iodine-stained materials is given on page 21.

 An Exploration Worksheet is available to accompany Exploration 2 (Teaching Resources, page 50).

Answers to What's the Difference? are on the next page. ▶

LESSON

2

Skilled Observations of Cells

Now that you know how to use a microscope and prepare microscope slides, it's time to put these skills to work in the following Exploration.

EXPLORATION 2

What Are Plants and Animals Made Of?

You Will Need
- a microscope
- microscope slides
- coverslips
- water
- a knife
- eyedroppers
- one-quarter of an onion
- forceps
- iodine solution
- a prepared slide of human cheek cells

What to Do

1. Make a wet mount of a piece of onion skin.
   a. Remove the outside layer of one-quarter of an onion.
   b. Remove the thin skin from the inside of the layer.
   c. Cut off a small piece for your slide.
   d. Place the onion skin on a slide, add a drop of water, and place a coverslip on top.
2. Adjust the diaphragm and the light source to get the best lighting. Observe the onion skin under low power and then high power.
3. Prepare a second wet mount, but this time add a drop of iodine solution to the onion skin. How does this change what you see? Ⓐ
4. As you look at the onion skin, do you observe a pattern of very small parts that all look very much alike? These are the cells. Draw two or three onion-skin cells under low power and then under high power. Make note of the magnification used each time.
5. Draw all of the features that you see in one cell. Label the cell and the **cell wall** (the thick casing of the cell).
6. Clean and dry the slides and coverslips.
7. Look at the prepared slide of material taken from the inside of a person's cheek. Notice the color due to the stain used. Instead of iodine, other stains are often used.
8. What do you observe through the microscope? Make drawings under low and high power as you did for the onion material. Don't forget to note the magnification used. Show all detailed features as accurately as possible.

What's the Difference?
You have seen some plant cells—in the skin of an onion. You have also seen some animal cells—human cheek cells. Consult your drawings of cells, and use your memory to answer the following questions:

1. What characteristics do plant and animal cells share?
2. Would you be able to distinguish onion-skin cells from human cheek cells? If so, how would you do it? Think about this as you carry out the next Exploration.

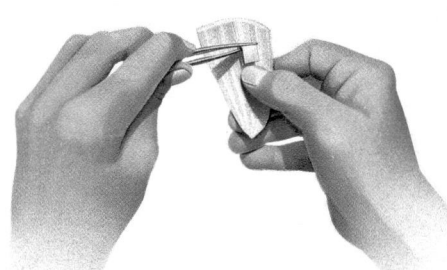

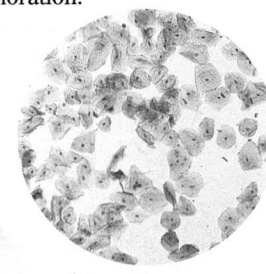

▲ Human cheek cells

LESSON 2 ORGANIZER

Time Required
3 class periods

Process Skills
observing, comparing, contrasting

Theme Connection
Systems

New Terms
Cell membrane—the thin outer edge of a cell
Cell wall—a thick casing surrounding a plant cell
Chloroplast—the part of a plant cell where photosynthesis takes place
Cytoplasm—the material in a cell that lies outside the nucleus
Mitochondria—the parts of a cell that provide energy to the cell
Nucleus—the part of a cell that controls its life activities
Vacuole—storage area in a cell

Materials (per student group)
Exploration 2: microscope; 2 microscope slides with coverslips; about 10 mL of water; eyedropper; one-quarter of an onion; forceps; drop of iodine starch-test reagent; prepared microscope slide of human cheek cells; safety goggles; lab aprons (See Advance Preparation on page 59C.)

continued

Main Cell Features— Consider Your Observations

What holds a cell together and separates it from other cells? Did you notice an outer edge that gives support and shape to the cell? In plants, that outer edge tends to be thick and noticeable. In Exploration 2, you learned that it is called the cell wall. How would you describe the outer edge of the animal cells you have seen? In animals, that outer edge is called the **cell membrane**. How does it differ from a cell wall? Do your drawings show this difference? (Plants also have a cell membrane, located just inside the cell wall.) **B**

Have you noticed an area inside the cell that is darker and more noticeable than the rest of the cell, especially when stained? It should be shown in your drawings. This dark spot inside a cell is called the **nucleus**. It has been compared to a "head office" because it controls the life activities of the cell. Did you see a nucleus in both animal and plant cells? The material in a cell outside the nucleus is called **cytoplasm**.

The **chloroplasts** in the cells of green plants use sunlight to make sugars. The **mitochondria** are the cell's powerhouse, releasing a constant supply of energy to both plant and animal cells. **Vacuoles** are large, centrally located storage areas in plant cells and in some animal cells. Look for the nucleus (plural, *nuclei*) and other features as you continue to observe cells.

Observe the diagrams at right of a plant cell and an animal cell. Do you know which is the animal cell and which is the plant cell? Use what you have learned so far to label the different parts of each cell. Do not write in this book. **C**

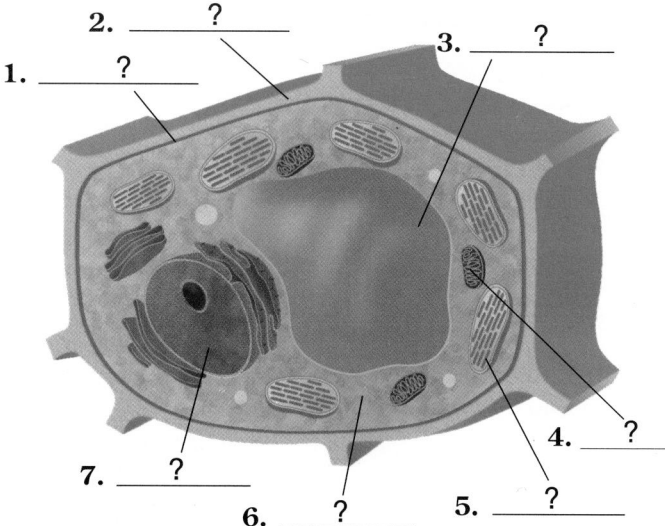

1. _____ ?
2. _____ ?
3. _____ ?
4. _____ ?
5. _____ ?
6. _____ ?
7. _____ ?

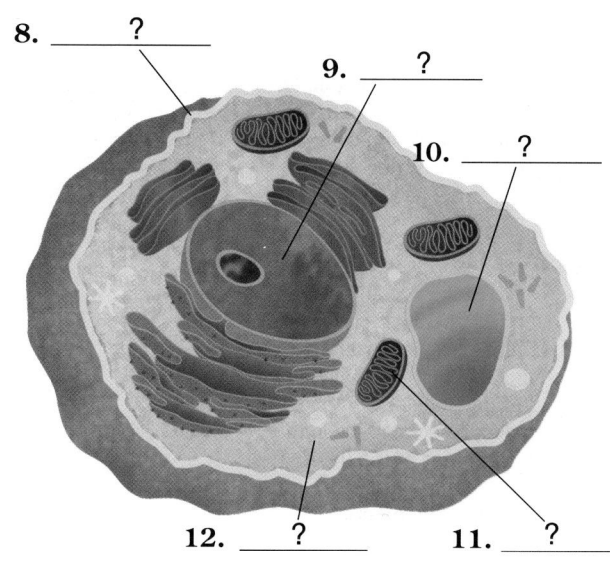

8. _____ ?
9. _____ ?
10. _____ ?
11. _____ ?
12. _____ ?

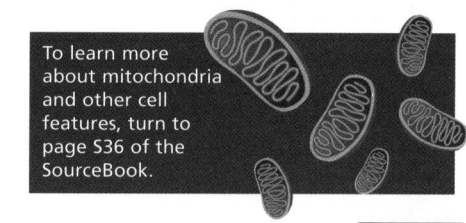

To learn more about mitochondria and other cell features, turn to page S36 of the SourceBook.

117

ORGANIZER, *continued*

Exploration 3: microscope; 8–10 microscope slides with coverslips; knife; 2 straight pins; 4–5 drops of iodine starch-test reagent; about 25 mL of water; eyedropper; thin sections of potato with skin, banana peel, and orange peel; tiny piece of beef liver; safety goggles; lab aprons; optional materials: small pieces of lettuce, green pepper, carrot, tea, nutmeg, pepper, mustard, coffee, and ginger

Teaching Resources
Exploration Worksheets, pp. 50 and 52
Transparency 21
SourceBook, p. S37

Encourage proper technique when making wet mounts of the sample cells. You may wish to prepare sample slides ahead of time. Make sure students draw and label their cells.

Emphasize the importance of making the liver very thin. After students have observed the liver cells, the slides and coverslips can be cleaned with dishwashing detergent and water.

SAFETY ALERT Be sure that students are aware of the potential hazards and proper handling of slides and coverslips.

WASTE DISPOSAL ALERT The procedure for disposing of the iodine-stained materials is given on page 21.

Answer to
In-Text Question

Ⓐ Sample answer: eggs, yogurt, any fruits or vegetables, chicken

Answers to
Questions (Part 1)

Sample answers:

1. Nuclei, vacuoles, chloroplasts, and other round structures may be observed. These structures serve to direct the cell's activities and store or make food.

2. The potatoes, for instance, showed a blue to purple-black color.

3. The potato should test positive for starch. Other sources of starch include pasta, rice, and oatmeal.

Answers to
Questions (Part 2)

1. Sample answer: Cells are like tiny compartments. Both plant and animal cells have nuclei and are enclosed by a cell membrane.

2. Plant cells have a rigid cell wall. The shape of plant cells is roughly polygonal, but animal cells have various shapes. Plant cells are often green.

 An Exploration Worksheet is available to accompany Exploration 3 (Teaching Resources, page 52).

Edible Cells

There are cells of many living things in your kitchen. How many cells can you think of? Ⓐ Make a list. You may use some of them in this Exploration.

PART 1

Plant Cells

You Will Need

- a microscope
- microscope slides
- coverslips
- a knife
- 2 straight pins
- iodine solution
- water
- eyedroppers
- pieces of plant material: potato, lettuce, green pepper, banana peel, orange peel, and carrot (optional: tea, nutmeg, pepper, mustard, coffee, and ginger)

What to Do

1. Cut or scrape off sections as thin as possible from the potato and from the banana and orange peels. Try to include a piece of skin along with scrapings of the flesh of the potato. Use the straight pins to pull the material apart. If your specimens are too thick, you may find that high power does not show details well.

2. Mount each type of plant on two slides. Use only water to mount one slide, and use a drop of iodine as stain on the other.

3. Draw the cells you observe, and label each drawing.

4. Clean up your lab area and wash your hands after you are finished.

Questions

1. Did you observe any round objects (other than air bubbles)? What do you think these might be? Of what use might they be to the plants?

2. When the iodine was used, did you notice a color on some slides that didn't appear on the other slides? What was the color?

3. Iodine is used as a test for the presence of starch, one of the substances in foods. Adding iodine to starch results in a blue to purple-black color. Some of the substances you examined do contain starch particles. Which plants observed gave a positive test for starch? Can you think of other sources of starch in your diet?

PART 2

Animal Cells

You Will Need

- a tiny piece of beef liver
- the same equipment used in Part 1

What to Do

1. Put a tiny piece of liver on a slide.

2. Pull apart the piece of liver using two straight pins so that the material is as thin as possible.

3. Make a wet mount. Carefully examine the sample under low and high power.

4. Add a drop of iodine to a second piece of liver as a stain, and observe this slide under high and low power as well.

5. Make two drawings of a liver cell, one with iodine and one without. Show as much detail as you can.

Questions

1. Now that you have observed cells closely, write a two-sentence answer to the question, "What are cells like?"

2. In Exploration 2, you compared plant and animal cells. Write any further observations you have that would help you distinguish between plant and animal cells.

A human liver cell at 160×

Did You Know...

Some organisms are made up of a single cell, but the human body is composed of more than 10 trillion cells.

Integrating the Sciences

Life and Physical Sciences

Ask students what they think the word *respiration* means. Students may equate respiration with breathing. Point out that respiration is also the chemical process that takes place in the mitochondria of cells to release energy from food. Breathing is the physical process of taking in oxygen and expelling carbon dioxide. The processes are related in that breathing provides the oxygen required by cellular respiration and removes the carbon dioxide produced by cellular respiration.

Plant or Animal?

Sort these photos into plant or plant-like cells and animal or animal-like cells. State your reasons for your decisions. (Hint: Pay attention to the outer boundary of each cell. Does it appear rigid, thick, and definite in shape or delicate and flexible?)

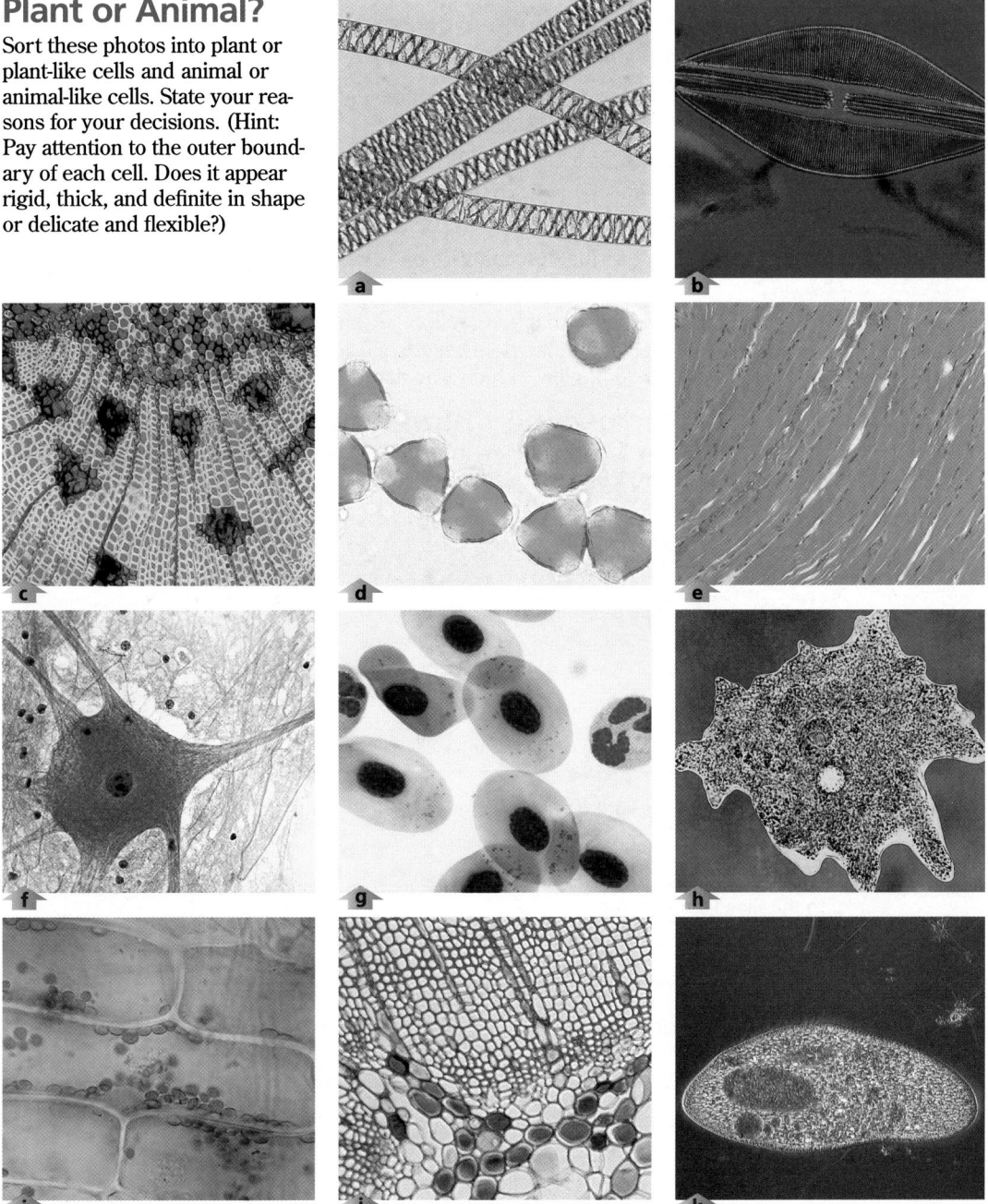

Reteaching

Prepare slides of elodea cells or other leaf surfaces. Ask students to look at the slides and draw what they see. Make sure that students carry the microscope properly, start with the lowest power objective lens, adjust the diaphragm or mirror, and correctly focus the coarse and fine adjustments.

Assessment

Obtain a water sample from the bottom of a pond or marsh. (Do not allow unsupervised students to do so. Anyone handling the pond water should wear latex gloves and wash his or her hands with soap and water when finished.) Many microscopic plants and animals may be visible. Several types of filamentous algae may appear, along with various protozoans and tiny roundworms. Ask students to identify, with the aid of the microscope, examples of plant cells and animal cells in the water sample. *(The one-celled organisms may show locomotion, may appear to be hunting for food, and may respond to stimuli. Students may see an organism excreting waste from its body. Students may also observe reproduction by division.)*

Extension

When a great number of similar cells work together to serve a particular function in organisms, those collections of cells are called tissues. Ask students to research and sketch the cells making up muscle, bone, and nerve tissue.

Closure

Have students record thoughts about microscopes in their ScienceLog. They might comment on how a microscope is operated, what it is used for, how long it has been used, and what influence it has had on science.

Homework

The Extension activity above makes an excellent homework assignment.

 Transparency 21 accompanies the material on this page.

Answers to *Plant or Animal?*

Some of the cells are difficult to classify from the photographs. Accept all reasonable answers that refer to visible characteristics of the cells.

- **a.** plantlike (spirogyra)
- **b.** either plantlike or animal-like (diatom)
- **c.** plant (vascular tissue)
- **d.** plant (pollen grains)
- **e.** animal (muscle)
- **f.** animal (nerve cell)
- **g.** animal (red blood cells)
- **h.** animal-like (amoeba)
- **i.** plant (elodea)
- **j.** plant (root-tip cells)
- **k.** animal-like (paramecium)

Answers to *Challenge Your Thinking,* pages 120–121

1. a. Cells are like bricks in that they form a structure. Many bricks can be joined together to form a wall; likewise, many cells can be joined together to form a tissue. Cell walls of plants are rigid and give support to the organism, just as bricks give support to a building.

b. Cells are unlike bricks in that each cell is made up of even smaller structures that carry out functions for the cell. These cell functions are the activities of life.

c. Microscopes are important because they allow us to examine the tiny structures that make up an organism.

d. The objects are also inverted.

e. Down and to the left

2. Answers will vary. (If students rate themselves as "Fair" or "Could be better," have them write a few sentences describing how they could improve their laboratory activities.)

3. The cell shown is a plant cell. The rigid, thick boundary is a cell wall, which is only found in plants. See page S41 of the SourceBook for a labeled drawing of the photograph.

4. The answers are cytoplasm, vacuole, chloroplast, mitochondria, cell wall, cell membrane, nucleus, and cells. The mystery word is *patterns*.

ENVIRONMENTAL FOCUS

Tell students the following: Every minute billions of cells within your body die, only to be replaced by billions of new cells. In a healthy body, cells reproduce at exactly the same rate at which they die. However, some agents make cells reproduce uncontrollably, causing a disease known as cancer. One of these cancerous agents is ultraviolet radiation, which is emitted by the sun and ultraviolet lamps. People who spend excessive amounts of time in the sun run the risk of developing skin cancer.

CHALLENGE YOUR THINKING

1. A Cellular Questionnaire
Quiz yourself with the following questions:

a. How are the cells that make up living things like bricks?

b. How are cells not like bricks?

c. How do microscopes aid in the study of living things?

d. Describe how things look different under a microscope (other than just looking larger).

e. Suppose that you are looking through a microscope at a tiny animal. It swims up and to the right and moves out of your field of view. Which way do you move the slide to follow it?

2. Microscope Champions
Copy the table below into your ScienceLog. Then rate yourself as a microscope user. Do not write in this book.

	Activity	Good	Fair	Could be better
a.	I carry the microscope carefully to my work area.			
b.	I begin observing with the lowest power objective lens.			
c.	I carefully adjust the light level.			
d.	I avoid touching the objective lens to the slide I am viewing.			
e.	I focus the microscope by first using the coarse-adjustment knob and then using the fine-adjustment knob.			
f.	I properly prepare the microscope for storage when my work is finished.			

Theme Connection

Systems
Cells of larger organisms perform specific functions, depending on where the cells are located. These cells are known as specialized cells. Specialized cells make up tissues, which in turn make up the larger organs of the body. **Focus question:** What similarities do all cells, whether individual organisms or specialized cells of larger organisms, share? *(Even though cells may be shaped differently, they all take in food, produce waste, reproduce, and die.)*

3. Look and See

Examine the photograph of the cell at right. Is it a plant cell or an animal cell? How can you tell? Copy the photograph into your ScienceLog, and use the following words to label your drawing: *cell wall, vacuole, chloroplast,* and *nucleus.* Where would the cell membrane be located if it were visible? Label this also. Do not write in this book.

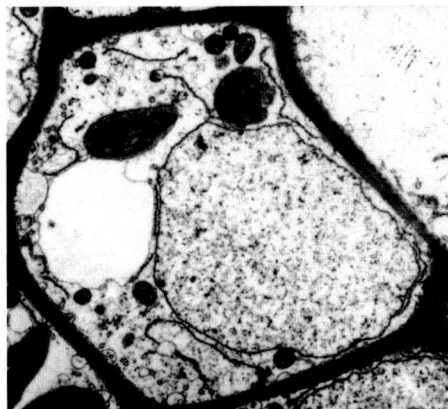

4. Mystery Word

Fill in the correct "cellular" words below to find out what this unit is all about.

living matter found outside the nucleus of the cell

space used for storage (larger and more common in plant than in animal cells)

green, often oval, structure in plant cells; used in food production

oval bodies containing folded membrane; sites of cell activity where energy is released

thick outer covering of the cell; provides strength and protection

outer edge of an animal's cell

dense, round or oval body; controls the life activities of the cell

building blocks of all living things

ScienceLog

Review your responses to the ScienceLog questions on page 112. Then revise your original ideas so that they reflect what you've learned.

121

ScienceLog

The following are sample revised answers:

1. Yes, all living things are composed of cells. Cells are too small to be seen without a microscope.

2. The photograph on the left shows mouse cells; the photograph on the right shows corn cells.

3. Cells are like bricks in that they form a structure. Many bricks can be joined together to form a wall; likewise, many cells can be joined together to form a tissue. Cell walls of plants are rigid and give support to the organism, just as bricks give support to a building.

   Cells are unlike bricks in that each cell is made up of even smaller structures that carry out functions for the cell. These cell functions are the activities of life.

⭐ **You may wish to have students complete the Chapter 7 Review Worksheet that accompanies this Challenge Your Thinking (Teaching Resources, page 54).**

Homework

Have students research the cells of the human body. Students should list at least five types of specialized cells and include a brief description of what each type of specialized cell does.

The Big Ideas

The following is a sample unit summary:

Signs of life include movement, energy consumption, growth, reproduction, response to stimuli, and adaptation to the environment. (1)

Plant motion is usually very slow; animal motion can be slow or fast. Animals can move about on their own from place to place, whereas plants cannot. (2)

The growth patterns of plants and people are similar in that the proportions of different body parts change as they grow larger and become more complex. (3) Their growth patterns differ in that plants can continue to grow larger throughout their lives, given suitable environmental conditions. People stop growing larger when they reach adulthood. (4)

Growth can be developmental, continuous, regenerative, or cancerous. (5) Developmental growth gives rise to an adult organism. Reproduction is a type of developmental growth that gives rise to entirely new organisms. Regenerative growth restores lost or damaged structures. Continuous growth replaces structures that wear away, such as hair and nails. Cancerous growth is uncontrolled cell growth and is harmful. (6)

"Response to stimuli" refers to an organism's ability to react positively or negatively to its surroundings. (7) All living things respond to changes in the environment or within the organism itself in order to survive. (8)

Recurring periods of activity, such as sleeping and eating, suggest that animals have a sense of time. Research indicates that an organism's internal clock is set by external stimuli. For example, migration may be triggered by seasonal weather changes. (9)

Adaptation in species occurs over long periods of time and relies on individuals with certain traits being able to reproduce more successfully than other individuals of the same species. In this way, advantageous traits are passed on to the next generation. For example, a species of fish might evolve camouflage, which increases its chances of catching prey, or a species of plant might evolve long slender leaves, which increase its exposure to sunlight. (10 and 11)

Unit 2

SOURCEBOOK

In this unit of the SourceBook, you will learn more about how cells work, how characteristics are passed on from one generation to another, and how living things change over time.

Here's what you'll find in the SourceBook:

122

PATTERNS OF LIVING THINGS

The Big Ideas

In your ScienceLog, write a summary of this unit, using the following questions as a guide:

1. What are some signs of life?
2. How do plant and animal movement differ?
3. In what ways are the growth patterns of plants and people similar?
4. In what ways are they different?
5. What are some of the different forms of growth?
6. What is the importance of each?
7. What does "response to stimuli" mean?
8. Why is this response valuable?
9. What evidence is there that animals have a sense of time?
10. What are some adaptations of plants and animals over long periods of time?
11. How do plant and animal species adapt over long periods of time?
12. How are plant and animal cells similar? How do they differ?
13. Describe the procedure for making a wet mount.

Checking Your Understanding

1. Would a seed be an example of a living thing? Why or why not?
2. What are some of the reasons that animals migrate? Name at least three animals that migrate.

Plant and animal cells are similar in that they both contain cell parts such as a nucleus and mitochondria. Although both types of cells are enclosed by a cell membrane, plant cells also have a cell wall that provides a rigid structure for the plant. (12)

To make a wet mount, put a drop of water on a slide and add the sample to the water. Lower the coverslip onto the slide with a pencil, and remove the excess water with tissue. (13)

Answers to
Checking Your Understanding,
pages 122–123

1. A seed is a living plant embryo that is waiting for the right conditions to begin development.

2. Reasons for migrating include avoiding unfavorable weather and the need to find food or resources. Three types of animals that migrate are whales, bats, and birds.

3. The life cycle of a butterfly consists of four distinct stages—egg, larva, pupa, and adult. These stages differ in appearance, movement, and

3. Look at the illustration to the right showing the life cycle of a butterfly. How does the butterfly's growth pattern differ from that of a human?

4. 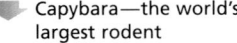 Construct a concept map using the following terms: *ectothermic, bear, endothermic, animals,* and *frog.*

5. What are some of the different ways in which animals move?

6. Are all forms of growth useful? Explain.

7. Is bigger always better? Explain.

8. How do snakes demonstrate adaptation to their environment?

9. In your own words, compare an earthworm's locomotion to that of a snake.

10. How might a sense of time be helpful to animals? to plants?

11. Add the following to your list of "Curious Questions." Suggest answers to each.

   a. Where do snakes, frogs, and salamanders go in the winter?

   b. How do endothermic animals keep their body temperature constant?

   c. What advantages do endothermic animals have over ectothermic animals?

   d. What causes animals to start migrating? to stop migrating?

   e. What cues do animals use to guide them as they migrate?

Capybara—the world's largest rodent

propels itself forward by alternately contracting and extending its muscles. The muscles move the earthworm's bristles, which push the earthworm along.

10. A sense of time in animals and plants helps them survive the difficulties of their environments. For example, animals might sleep during the day to avoid predators or migrate in the winter to find food. Some plants close up their leaves at night to conserve heat and water or drop their leaves in the fall to reduce the chance of freezing in the winter

11. a. Some snakes, frogs, and salamanders become dormant in the winter. They may spend the winter underground.

   b. Endothermic animals use the energy in food to regulate their temperature. Mammals have a layer of fat under their skin that serves as insulation. Animals with hair or feathers can fluff up this outer covering to trap air and insulate themselves.

   c. Endothermic animals can typically function normally even when their environment is cold. They can also live in areas that experience extreme temperatures, whereas ectothermic animals may not be able to live there.

   d. Environmental cues such as day length may cause animals to start migrating. A combination of learned and innate behavior may signal the end of migration.

   e. Cues from the environment include landmarks, wind direction, and the position of the sun.

 You may wish to have students complete the Unit 2 Review Worksheet that accompanies this Making Connections (Teaching Resources, page 60).

feeding habits. The growth pattern of a human, from egg to adult, is a more gradual process and does not include distinct stages like those of the butterfly.

4.

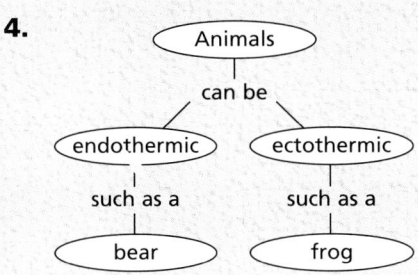

5. Crawling, walking, swimming, and flying

6. Uncontrolled growth is not useful. Cancerous growth, for instance, can be harmful or even fatal.

7. Large size has its disadvantages. Larger organisms, such as the African elephant, must spend more time finding and eating food than do their smaller counterparts, for instance.

8. The scales of a snake help the animal grip the ground so that it can push itself forward.

9. A snake moves either by flexing its body in an S-shaped curve or by using its scales to move forward in a straight line. An earthworm

Homework

The Unit 2 Activity Worksheet on page 59 of the Unit 2 Teaching Resources booklet makes an excellent homework assignment.

Background

The body temperature of an *ectothermic* animal, commonly called "cold-blooded," is about the same temperature as its surroundings. Amphibians, reptiles, and most fish are ectothermic. When the sun warms the body of an ectothermic animal, its body temperature can rise above that of an *endothermic,* or "warmblooded," animal.

Endothermic animals include all birds and mammals. The body of an endothermic animal produces most of its heat by metabolizing food. In many endothermic animals, a layer of fat beneath the skin and a covering of feathers, fur, or hair also help the animal maintain a constant body temperature. The principal means of reducing body heat are panting and sweating.

Many ectothermic animals partially control their body temperature through their behavior. For example, ectothermic land animals may bask in the sunlight to become warmer or move to the shade to cool down. Fish may swim closer to the surface of the water to warm themselves. If they become too warm, they may swim to deeper, cooler water.

Extension

Obtain goldfish or tropical fish for students to observe. If you have a tropical aquarium, point out the heater that keeps the water at a constant temperature. Invite students to use what they know about ectothermic animals to explain why this heater is necessary. *(The heater keeps the temperature of the water warm throughout the year, simulating the natural habitat of tropical fish.)*

Warm Brains in Cold Water

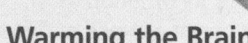

Brain
Heater muscles
Eye

Of the world's 30,000 kinds of fish, only a few carry around their own brain heaters. *Brain heaters?* Why would a fish need a special heater just for its brain? Before you can answer that question, you have to think about how fish keep warm in the cold water of the ocean.

A Question of Temperature

Most fish and marine organisms are *ectotherms.* The body temperature of an ectotherm closely matches the temperature of its surroundings. An ectotherm may use the heat from the sun or from sun-warmed waters to keep warm. But when the surrounding temperature falls, the body of an ectotherm cools and, as a result, slows down. That makes it difficult to look for food or escape from enemies.

Endotherms, on the other hand, are animals that maintain a steady body temperature

▼ Specialized muscles help this swordfish hunt for prey in cold waters.

▶ **Why do you think it's important to protect the brain and eyes from extreme cold?**

regardless of the temperature of their surroundings. People are endotherms. Other mammals—such as dogs, elephants, and whales—as well as birds are also endotherms. But only a few kinds of fish, such as tuna, are endotherms. As such, they can hunt for prey in extremely chilly water. Yet these fish pay a high price for their ability to inhabit very cold areas—it requires a lot of energy.

Being endothermic requires far more energy than being ectothermic does. Some fish, such as swordfish, marlin, and sailfish, have adaptations that let them heat only part of their bodies. Instead of using large amounts of energy to warm the entire body, they warm only their eyes and brains. That's right—they have special brain heaters!

Warming the Brain

In a "brain-warming" fish, a small mass of muscle attached to each eye acts as a thermostat. It adjusts the temperature of the brain and eyes as the fish swims through different temperature zones. These "heater muscles" help maintain delicate nerve functions that are important to finding prey.

Heater muscles allow the swordfish, for example, to swim both near the warm surface of the ocean and in depths of 485 m, where the temperature drops to near freezing. This adaptation has an obvious advantage: it gives the fish a large range of places to look for food.

Ectotherms in Action

Contact a local pet store that sells various kinds of fish. Find out what water temperature is best for different fish from different regions of the Earth. Some fish to find out about include goldfish, discus fish, and angelfish. Why do you think fish-tank temperatures must be carefully controlled?

Answers to *Ectotherms in Action*

Student answers will vary depending on the fish they select. For instance, a Japanese goldfish is best suited for cool temperatures near 18°C (64°F), a South American discus requires warm temperatures of about 28°C (82°F), and a South American angelfish fairs best in a temperature range of 24–26°C (75–79°F). The temperature in fish tanks must be controlled to ensure the fish's survival.

Meeting Individual Needs

Learners Having Difficulty

Have students create posters that illustrate the advantages and disadvantages of being endothermic and of being ectothermic. Encourage students to include photographs or drawings of various organisms on their posters and to classify the organisms according to how they regulate body temperature.

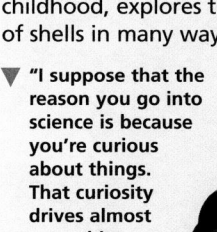
Hands-On Paleontology

Have you ever run your fingers over the slippery surface of a shell or admired its strange markings or unusual shape? Dr. Geerat Vermeij (ver MAY), a paleontologist and professor at the University of California at Davis, certainly has! In fact, Dr. Vermeij, who has been blind from glaucoma since childhood, explores the secrets of shells in many ways.

▼ "I suppose that the reason you go into science is because you're curious about things. That curiosity drives almost everything else you do."

Understanding Shell Evolution

Dr. Vermeij investigates *evolution,* the process by which living organisms change over time. "Things do not remain the same; things really do change," he explains. "It's a very important idea." By studying shells and fossils, Dr. Vermeij works to find out how the animals that built the shells have evolved over millions of years.

"Many shells are built like tanks—incredibly thick—and their openings are small . . . They're built to withstand the very substantial crushing efforts of enemies, like crabs or fish." In other words, the shell helps the animal defend itself.

The Investigation Begins

"My interest in science started out as an interest in shells. When I was in fourth grade—I had just come from the Netherlands—a teacher in the United States had brought some shells from Florida. I found those shells so extraordinary that I was completely taken in by them."

Although Dr. Vermeij can't see, he still makes many observations. In fact, he says, "Observation is the central activity. And, obviously, I use my fingers an awful lot, but I use every other sense I can too."

"I think for me the truly key idea is that you keep all of your senses open to as much of the world as you can and think about everything—think about things you hear and feel and see."

A Science Surprise!

Dr. Vermeij really enjoys the element of surprise that comes with science. For example, while he and his wife were in Guam, they discovered a very unusual crab. It looked like a solid piece of coral. They put it into a bucket that held some shells. The next morning all of the shells were broken. "It turned out that this crab was, in fact, a massive shell-crusher; it was so utterly unexpected!"

As Dr. Vermeij says, "It isn't just facts, it's how you get the facts and ask questions" that makes science so thrilling.

Your Own Observations

What shells are found in your area? Begin your own shell collection with shells from your neighborhood. Then use field guides to help identify the shells that you find. Be sure not to disturb living organisms or the environment when you add to your collection.

125

Background

Paleontologists study organisms from prehistoric times. By studying fossils and comparing them with living organisms, paleontologists can learn about the evolution of different species.

There are three main branches of paleontology—invertebrate, vertebrate, and paleobotany. Invertebrate paleontologists study fossils of animals without backbones, such as insects, while vertebrate paleontologists examine fossils of animals with backbones. Paleobotanists study fossils of plants.

Students may be interested in knowing more about Dr. Vermeij's blindness. Glaucoma, a disease of the eye, is the leading cause of blindness worldwide. The main symptom of the disease is increased pressure of the fluid in the eye. Normally, this fluid provides nourishment to the eye's tissues and exits the eye through channels leading to blood vessels. If a blockage occurs, the fluid cannot exit, and pressure in the eye increases. When the pressure increases too much, it destroys the optic nerve, and blindness results.

Discussion Questions

1. What does a paleontologist study? *(By investigating organisms from prehistoric times, a paleontologist studies how organisms have changed over time.)*

2. How might Dr. Vermeij's disability affect his work? *(It could help him detect shapes, textures, and minute details of shells that other scientists might overlook.)*

Extension

Provide students with an opportunity to role-play by asking for student volunteers to be blindfolded. Once they are blindfolded, ask them to reach into a discovery box of unknown items and to make observations about an item they find. Some sample items to include are different types of clean shells, sanitized animal bones and feathers, and dried, nontoxic flowers. Tell each student to make a list of the observations they made for each object. Ask: Besides touching the surface of each object, what ways did you use to investigate the objects? *(Possible answers include shaking, tapping, and holding the objects to estimate weight.)*

Your Own Observations

Encourage students to bring in any shells or fossils they have found or bought. (Only sanitized shells should be brought to class.) Let students work in groups to identify their shells or fossils by using field guides and then to share their discoveries with the class.

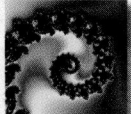

Background

John James Audubon's work captured the essence of the wildlife in North America. He was one of the first artists to paint birds in their natural surroundings. In Audubon's paintings, birds are often posed dramatically on branches or among plants.

The accounts of North American bird species that Audubon created in conjunction with other naturalists are still some of the most complete and accurate ever published. Largely as a result of Audubon's influence, the birds of North America have been portrayed and described in far more detail than those of most other parts of the world.

The name *Audubon* has been adopted by the National Audubon Society, an organization dedicated to preserving the wildlife that Audubon spent his life studying and painting.

Discussion Questions

1. Audubon's paintings are very realistic. Why is this important? *(Because the details of features such as beaks, feet, feathers, and body size are scientifically accurate, the paintings can be used by scientists and other naturalists as references.)*

2. Audubon worked as a scientist and an artist. What are some occupations that require a person to work in both capacities? *(Architect, medical illustrator, computer-animation artist)*

3. What might be the value of a drawing or a painting rather than an actual photograph of a particular bird in a bird guide? *(Distinguishing features can be emphasized.)*

Extension

Take students on a walk to identify as many different species of birds as they can by using field guides. Check the library for field guides specific to your geographic region.

SCIENCE and the *Arts*

A Naturalist Finds His Niche

John James Audubon was an art school dropout. He was also a bankrupt businessman. By relying on his own special talents, however, Audubon became an internationally recognized artist and naturalist.

Audubon's Early Life

While growing up in France in the late 1700s, Audubon loved hunting and the outdoors much more than school. Since he showed an interest in drawing, his father enrolled him in art school. Much to his family's disappointment, Audubon failed because he was impatient with the instruction and was interested in painting only birds.

Business Failures Bring Creative Freedom

After moving to the United States in 1803, Audubon tried unsuccessfully to start a number of businesses. Audubon eventually landed in jail because he was unable to pay his debts. He was left with only his clothes, his gun, and his drawings, which at the time were considered worthless. Audubon's many failures in business finally convinced him to pursue his true passion, the study and painting of birds. To make ends meet, he painted portraits and taught drawing and music.

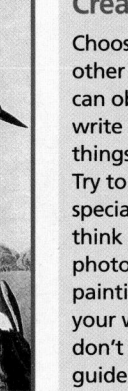

▶ Watercolor of kingfishers painted by Audubon in 1832

Success and Recognition

Audubon had always devoted a lot of time and energy to exploring the American countryside. In this new phase of his life, Audubon embarked on several flatboat expeditions on the Mississippi River that were especially productive. His goal was to publish paintings of all the bird species in North America. Because no American publisher would accept the project, Audubon went to England in 1826. There he published a huge book of paintings called *The Birds of America*, which included 435 species of birds. His book was a great success. When he returned to the United States in 1831, Audubon was hailed as a great artist and naturalist.

▶ Audubon's 1825 painting of wood ducks vividly shows the differences between the males (left) and the females (right) of this species.

Create a Field Guide

Choose plants, animals, or other living things that you can observe firsthand. Then write descriptions of the living things that you have chosen. Try to explain any behavior or special adaptation that you think is peculiar. Include a photograph, drawing, or painting to go along with your written descriptions, and don't forget to give your field guide a catchy title.

Create a Field Guide

Encourage creativity, accuracy, detail, and consistency in student illustrations and descriptions. Students should emphasize markings that can be used for field identification. Students should make their field guide on unlined paper and place it in a notebook.

EYE ON THE ENVIRONMENT

Hot-Water Dumps

A power plant uses water from a lake to cool its reactors. This water is then released back into the lake. Soon the lake begins to change—for the worse. The number of plant species drops by 75 percent. The number of fish species drops by a third. What could cause this devastation?

▼ **Power plants can heat natural waters, causing thermal pollution.**

► **Cooling ponds such as this one can cool the water before it is released into the river.**

Cooling With Water

Power plants, steel mills, oil refineries, and other industries use large amounts of water to cool their machines. The electric-power industry alone uses about 730 billion liters of water per day. During the cooling process, all of this water absorbs heat.

Some factories take the water they need for cooling from natural bodies of water, such as lakes and rivers. Then the heated water is returned to its source. The dumping of heated water into freshwater or saltwater environments is called *thermal pollution*. The temperature of the water does not have to increase by much for thermal pollution to occur. An increase of only 3°C can eliminate species from the area and devastate the ecosystem.

Upsetting the Environment

When heated water gets dumped into a pond, lake, river, or ocean, it prevents some animals from eating, reproducing, or even swimming. Heated water also holds less oxygen. Lack of oxygen may kill off certain plants and animals that live in the water, altering the existing food web.

In addition, thermal pollution may encourage the growth of other organisms. Because these organisms can tolerate the heat, they may replace the original organisms. As a result of these changes, the food web is disrupted. The original ecosystem may disappear.

Finding Solutions

Good news! There are solutions to thermal pollution. In some cases, factories construct large, shallow ponds to hold the heated water. After the water cools, it can be safely released back to its original, natural source. Some factories treat heated water in cooling towers. The heated water is sprayed into the air and cooled by evaporation. Other factories construct artificial lakes so that they don't have to rely on natural sources of water at all.

Warming Water

Fill a 1 L container halfway with room-temperature water and place a thermometer in the water. Slowly add 10 mL of boiling water to the container. Record the change in temperature after you add the water. Continue to add boiling water, 10 mL at a time, until the temperature rises by 5°C. How much boiling water would be needed to raise the temperature of a small pond (about 30,000,000 L) by 5°C?

127

Answers to
Warming Water

- The temperature should rise 1.4°C for every 10 mL of boiling water added. To increase the temperature 5°C, the total amount added should be 35.7 mL. Because the water will cool off slightly during the activity, students should observe a 5°C rise after adding 40 mL of water. (Remember: Students are adding water in 10 mL increments, so their answer will be a multiple of 10.)

- To solve the pond problem, students should set up the following ratio and solve for *x*.
  x/30,000,000 L = 40 mL/500 mL
  x = (40 mL/500 mL) $\times$ 30,000,000 L
  x = 2,400,000 L
  Therefore, 2,400,000 L of boiling water will raise the temperature of a 30,000,000 L pond 5°C.

EYE ON THE ENVIRONMENT

Background

In most power plants, steam turns turbines in generators to produce electricity. The steam used for this purpose is created in boilers that are heated by burning fossil fuels or by a nuclear reactor. After the steam has driven the turbines, cool water is used to condense the steam and turn it back into liquid water. Then it can be returned to the boiler for reheating. The coolant water is usually taken from a nearby natural body of water. If the coolant water is returned to its source while still warm, it could cause thermal pollution.

If power plants do not plan carefully when they are going to begin operation or when they are going to temporarily shut down for repairs, the temperature of the natural body of water may rise sharply. This creates a severe form of thermal pollution known as thermal shock. Thermal shock causes the sudden death of large numbers of fish and other organisms.

Meeting Individual Needs

Second-Language Learners

Have students illustrate the process of thermal pollution. They may make a drawing, collage, or model using a variety of materials. Encourage students to use labels in both English and their native language to identify areas of their illustration.

Integrating the Sciences

Earth and Life Sciences

Heating and cooling affect the kinds of organisms that can survive in an environment. Have students analyze a small site, such as a city block or an area on the school grounds, to discover what factors affect temperature changes in the area. Students should consider climate, season, elevation, vegetation, proximity and availability of water, and traffic in the area. Then have students document the living organisms that are visible in the ecosystem they have been studying. Ask them to speculate why these organisms are able to survive there.

Unit 3 — IT'S A SMALL WORLD

Unit Overview

In this unit, students draw on their personal experiences to learn about the dynamic world of microorganisms. In Chapter 8, students learn about the diversity and classification of microorganisms by observing them in their natural environment and by growing cultures in the classroom. In Chapter 9, students examine both the beneficial and harmful relationships that exist between people and microorganisms. By studying the work and methods of Louis Pasteur, students examine how scientists approach problem solving. In Chapter 10, students learn about mechanisms for controlling the growth of harmful microorganisms. By studying Alexander Fleming's work, students see how these mechanisms have protected people against the spread of disease. Then students learn how microorganisms cause food to spoil and how this can be prevented from happening.

Using the Themes

The unifying themes emphasized in this unit are **Changes Over Time, Structures,** and **Cycles.** The following information will help you incorporate these themes into your teaching plan. Focus questions that correspond to these themes appear in the margins of this Annotated Teacher's Edition on pages 129, 153, 169, and 191.

Changes Over Time can be discussed in a variety of contexts in this unit. Students should keep in mind that one of the most important functions that microorganisms perform is the breaking down of organic material both inside and outside the human body. Point out to students that this process is essential to life on Earth.

When learning about fungi in Chapter 9, a discussion of **Structures** will help students understand how fungi spread through an environment. Emphasize that the small size of fungi's spores contributes to their pervasiveness.

Finally, the spread of microorganisms from sources to carriers may be discussed as a series of **Cycles.** In your discussions of this idea, invite students to consider the positive and negative implications of microorganisms moving freely throughout the environment.

Using the SourceBook

Unit 3 focuses on microorganisms and the various diseases they can cause both in animals and in plants. Students learn how the structure and function of bacteria and viruses affect their interactions with other organisms. Finally, the unit details our natural defenses against diseases and major immune system disorders, including AIDS.

Bibliography for Teachers

Dixon, Bernard. *Power Unseen: How Microbes Rule the World.* New York City, NY: W.H. Freeman & Company Limited, 1994.

Bodanis, David. *The Secret Garden: Dawn to Dusk in the Astonishing Hidden World of the Garden.* New York City, NY: Simon and Schuster, 1992.

Bibliography for Students

Lovett, Sarah. *Extremely Weird Micro Monsters.* Santa Fe, NM: John Muir Publications, 1993.

Sagan, Dorion, and Lynn Margulis. *Garden of Microbial Delights: A Practical Guide to the Subvisible World.* Boston, MA: Harcourt Brace Jovanovich, Publishers, 1988.

Films, Videotapes, Software, and Other Media

The Microscope
Videotape
Britannica
310 S. Michigan Ave.
Chicago, IL 60604

The Microscope and Its Incredible World
Videodisc
Barr Media
12801 Schabarum Ave.
Irwindale, CA 91706

Organizing Protists and Fungi
Software (Apple II, MS-DOS, or Macintosh)
Queue, Inc.
338 Commerce Dr.
Fairfield, CT 06430

Protists: Form, Function, and Ecology (2nd ed.)
Videotape
Britannica
310 S. Michigan Ave.
Chicago, IL 60604

Simple Organisms: Bacteria (rev.)
Film and videotape
Coronet/MTI
108 Wilmot Rd.
Deerfield, IL 60015

Unit Organizer

Unit/Chapter	Lesson	Time*	Objectives	Teaching Resources
Unit Opener, p. 128				Science Sleuths: The Biogene Company Picnic English/Spanish Audiocassettes Home Connection, p. 1
Chapter 8, p. 130	Lesson 1, "But I Don't See . . . ," p. 131	2 to 3	1. Identify some of the precautions that should be taken to avoid the contamination of food by microorganisms. 2. Define *microorganism*. 3. Briefly describe the development of the microscope. 4. Estimate the size of microorganisms by using the diameter of the field of view of a microscope.	Image and Activity Bank 8-1 Transparency Worksheet, p. 3 ▼ Exploration Worksheet, p. 5 Math Practice Worksheet, p. 6
	Lesson 2, Our Tiny Neighbors, p. 139	4 to 5	1. Identify a variety of microorganisms that live in water. 2. Describe some of the environmental factors that could affect populations of microorganisms. 3. Describe some of the characteristics that are used to classify microorganisms.	Image and Activity Bank 8-2 Exploration Worksheet, p. 7 Exploration Worksheet, p. 10
End of Chapter, p. 146				Chapter 8 Review Worksheet, p. 13 ▼ Chapter 8 Assessment Worksheet, p. 17
Chapter 9, p. 148	Lesson 1, Tall (but True) Tales, p. 149	3	1. Explain the relationships that exist between microorganisms and humans. 2. Describe what some soil microorganisms look like. 3. Identify some of the harmful, as well as beneficial, effects of microorganisms that live in the soil. 4. Explain the relationship between microorganisms and the food sources of marine organisms.	Image and Activity Bank 9-1 Exploration Worksheet, p. 20 Theme Worksheet, p. 21 Exploration Worksheet, p. 22 Exploration Worksheet, p. 25
	Lesson 2, A Balanced View, p. 158	3 to 4	1. Outline the problem-solving method used by Pasteur to discover the relationship between bacteria and food spoilage. 2. Explain how society depends on the technology of pasteurization. 3. Describe how knowledge of microorganisms can lead to their control and use in foods.	Image and Activity Bank 9-2 Transparency 24 Resource Worksheet, p. 28
	Lesson 3, The Staff of Life, p. 164	3	1. Describe the conditions necessary for the active growth of yeast cells. 2. Explain how molds grow and reproduce. 3. Identify some harmful effects of molds.	Image and Activity Bank 9-3 Exploration Worksheet, p. 30 ▼ Exploration Worksheet, p. 33 Activity Worksheet, p. 34 Transparency Worksheet, p. 35 ▼
End of Chapter, p. 168				Chapter 9 Review Worksheet, p. 37 Chapter 9 Assessment Worksheet, p. 40
Chapter 10, p. 170	Lesson 1, Germ Warfare, p. 171	3	1. Describe how substances produced by some microorganisms can be used to control harmful microorganisms. 2. Explain the purpose and importance of antibiotics.	Exploration Worksheet, p. 43
	Lesson 2, Microorganisms and Food, p. 178	3	1. Explain how microorganisms are able to multiply so rapidly. 2. Describe how temperature affects the reproduction rate of microorganisms. 3. Analyze methods of food preservation.	Image and Activity Bank 10-2 Transparency 27 Activity Worksheet, p. 46 Activity Worksheet, p. 47
	Lesson 3, Surrounded by Microorganisms, p. 183	1 to 2	1. Describe how microorganisms travel through the air. 2. Explain the importance of proper hand-washing techniques. 3. Develop a set of rules for reducing the risk of infection by microorganisms.	Activity Worksheet, p. 48 Discrepant Event Worksheet, p. 49
	Lesson 4, Let's Eat Out Tonight, p. 185	1 to 2	1. Explain what to do if unsanitary conditions are encountered in a restaurant. 2. Describe how to register a complaint about unsanitary conditions in a restaurant.	Graphing Practice Worksheet, p. 50 Activity Worksheet, p. 52 ▼
End of Chapter, p. 190				Chapter 10 Review Worksheet, p. 53 ▼ Chapter 10 Assessment Worksheet, p. 56
End of Unit, p. 192				Unit 3 Activity Worksheet, p. 59 ▼ Unit 3 Review Worksheet, p. 61 Unit 3 End-of-Unit Assessment, p. 64 Unit 3 Activity Assessment, p. 72 Unit 3 Self-Evaluation of Achievement, p. 75

* Estimated time is given in number of 50-minute class periods. Actual time may vary depending on period length and individual class characteristics.

▼ Transparencies are available to accompany these worksheets. Please refer to the Teaching Transparencies Cross-Reference chart in the Unit 3 Teaching Resources booklet.

Materials Organizer

Chapter	Page	Activity and Materials per Student Group
8	138	**Exploration 1:** microscope; transparent plastic ruler (metric)
	139	**Exploration 2:** several small containers with lids; several samples of water from various outdoor sources; microscope; several microscope slides with coverslips; one eyedropper for each sample; spoonful of boiled rice grains; wax pencil; safety goggles (See Advance Preparation below.)
	141	**Exploration 3:** water samples from Exploration 2; microscope; several microscope slides with coverslips; one eyedropper for each sample; 1 L of distilled water or boiled tap water; several shallow glass or plastic dishes with covers; few fibers of lens paper or cotton batting; 4–6 g of dry grass, cut into 2 cm segments; 1000 mL beaker; hot plate; metric balance; wax pencil or several small pieces of masking tape; safety goggles (See Advance Preparation below.)
9	151	**Exploration 1:** magnifying glass (10×)
	151	**Exploration 2:** 50 mL of soil from a garden or flowerpot containing living plants; 500–600 mL glass container with cover; 300 mL of distilled water; microscope; several microscope slides with coverslips; eyedropper (See Advance Preparation below.)
	154	**Exploration 3:** enough leaves, grass clippings, and sawdust to make a 15–20 cm layer in a bucket; small bag of soil; small bag of pasteurized or composted manure; several handfuls of kitchen waste, such as potato peelings, apple cores, banana peels, lettuce leaves, coffee grounds, and tea bags; a few small, household-waste items made from plastic, metal, wood, or cloth; large plastic bucket with air holes; metric ruler; large plastic garbage bag; alcohol thermometer; several small flags (labels attached to pieces of wire); microscope; several microscope slides with coverslips; at least 10 L of water; safety goggles; lab aprons; latex gloves (additional teacher materials: electric drill or hammer and nail; see Advance Preparation below.)
	156	**Exploration 4:** index cards; markers
	164	**Exploration 5:** 5–6 mL of dry yeast; 100 mL graduated cylinder; 6 test tubes or similar containers; 6 mL of white sugar; set of metric measuring spoons; ice cubes; 3 alcohol thermometers; 3 thermometer holders; three 250 mL beakers; about 1 L of water; hot plate; marker or several small pieces of masking tape; microscope; several microscope slides with coverslips; eyedropper; watch or clock; safety goggles; oven mitts (See Advance Perparation below.)
	167	**Exploration 6:** a prepared wet-mount slide of mold; microscope (See Advance Preparation below.)
10		none

Unit Compression

As your students progress through this unit, there are numerous places where material could be either read ahead of time or assigned as homework in order to speed in-class discussion. Use what you learn about your students' prior knowledge and interests to help you select material for this purpose.

If a lack of time becomes a problem, several Explorations could be omitted or performed as demonstrations. One example is Exploration 3 on page 141. Also, you may wish to omit all or part of Lesson 3 in Chapter 9. Keep in mind, however, that this deprives students of several opportunities to make detailed observations of some common microorganisms.

Advance Preparation

Exploration 2, page 139: To increase their likelihood of observing the microorganisms pictured in the text for this Exploration, you may wish to accompany students as they collect samples from several of these areas: a pool formed from melting ice or snow, a pond, mud from a garden, or a neglected vase of flowers. In addition, this activity calls for boiled rice, which will need to be prepared in advance.

Exploration 3, page 141: If distilled water is unavailable, 1 L of tap water needs to be boiled ahead of time for each student group. You may also wish to cut the grass and prepare the infusion in advance.

Exploration 2, page 151: If distilled water is unavailable, 300 mL of tap water needs to be boiled ahead of time for each student group.

Exploration 3, page 154: Using an electric drill or a hammer and nail, make plenty of air holes in each bucket.

Exploration 5, page 164: You may wish to have students gather and set up the materials (except for the water and ice) the day before they actually perform the Exploration.

Exploration 6, page 167: Prepare wet mounts of bread mold in advance. To help prevent allergic reactions, you may wish to prepare these slides somewhere other than in your classroom. If you prefer, prepared slides of various types of molds can be purchased from most biological supply companies.

Homework Options

Chapter 8   See Teacher's Edition margin, pp. 135, 136, 137, 143, and 146
Math Practice Worksheet, p. 6
SourceBook, pp. S66 and S70

Chapter 9   See Teacher's Edition margin, pp. 150, 153, 156, 161, 163, 166, and 167
Theme Worksheet, p. 21
Activity Worksheet, p. 34
SourceBook, pp. S71 and S74

Chapter 10   See Teacher's Edition margin, pp. 173, 176, 179, 180, 183, 186, 187, 189, and 190
Activity Worksheets, pp. 46, 47, 48, and 52
Graphing Practice Worksheet, p. 50
SourceBook, p. S78

Unit 3   Unit 3 Activity Worksheet, p. 59
SourceBook Activity Worksheet, p. 76

Assessment Planning Guide

Lesson, Chapter, and Unit Assessment	SourceBook Assessment	Ongoing and Activity Assessment	Portfolio and Student-Centered Assessment
Lesson Assessment Follow-Up: see Teacher's Edition margin, pp. 138, 145, 157, 163, 167, 177, 182, 184, and 189 **Chapter Assessment** Chapter 8 Review Worksheet, p. 13 Chapter 8 Assessment Worksheet, p. 17* Chapter 9 Review Worksheet, p. 37 Chapter 9 Assessment Worksheet, p. 40* Chapter 10 Review Worksheet, p. 53 Chapter 10 Assessment Worksheet, p. 56* **Unit Assessment** Unit 3 Review Worksheet, p. 61 End-of-Unit Assessment Worksheet, p. 64*	SourceBook Review Worksheet, p. 78 SourceBook Assessment Worksheet, p. 82* 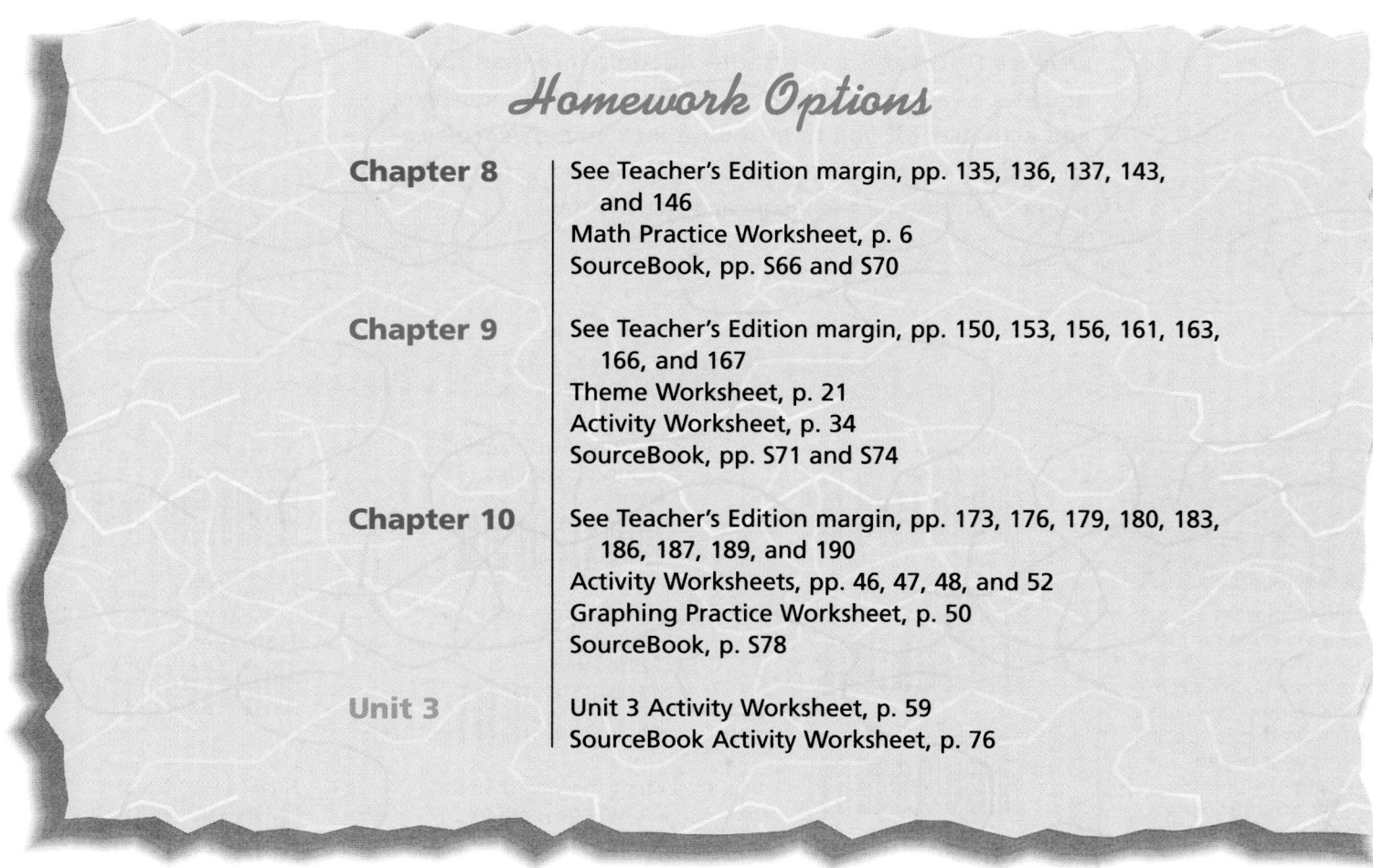	Activity Assessment Worksheet, p. 72* **SnackDisc** Ongoing Assessment Checklists ♦ Teacher Evaluation Checklists ♦ Progress Reports ♦	Portfolio: see Teacher's Edition margin, pp. 134, 140, 151, 162, 168, 182, 184, and 185 **SnackDisc** Self-Evaluation Checklists ♦ Peer Evaluation Checklists ♦ Group Evaluation Checklists ♦

* Also available on the Test Generator software
♦ Also available in the Assessment Checklists and Rubrics booklet

Science Discovery is a versatile videodisc program that provides a vast array of photos, graphics, motion sequences, and activities for you to introduce into your SciencePlus classroom. Science Discovery consists of two videodiscs: Science Sleuths and the Image and Activity Bank.

Using the *Science Discovery* Videodiscs

Science Sleuths: The Biogene Picnic
Side A

Many of the employees attending the Biogene company picnic have suddenly become sick. The emergency-room doctor needs to know the cause of the illness in order to treat the sick people and to alert public health officials in case of a possible epidemic. The Science Sleuths must analyze the evidence for themselves to determine the true cause of the illness.

Interviews
1. Setting the scene: Emergency-room physician **9425 (play ×2)**

2. Biogene salesperson (a very sick picnicker) **10360 (play)**

3. Picnic supervisor **10898 (play)**

4. Biogene company chef **11701 (play)**

5. Toxicologist **12788 (play)**

6. Gastroenterologist **13507 (play)**

7. Bacteriologist **14031 (play)**

Documents
8. Picnic supervisor's chart **14499 (step ×2)**

9. Hospital admissions list **14503 (step)**

10. Health department's lab results on patients **14506 (step ×2)**

11. Health department's lab results on picnic food **14510 (step)**

12. Chart showing chemical-poisoning symptoms **14513 (step ×2)**

Literature Search
13. Search on the words: FLU, FOOD POISONING, POISON, SOLVENTS, SPILLS, TOXIC WASTE **14517**

14. Article #1 ("Protecting Your Family From Food Poisoning") **14519 (step ×2)**

15. Article #2 ("It's Flu Time!") **14523 (step)**

16. Article #3 ("Action on Gasoline Spill") **14526 (step)**

Sleuth Information Service
17. *E. coli* **14529 (step)**

18. *Salmonella* **14532 (step)**

19. Plague **14535 (step)**

20. Toxigenic blue-green algae **14538 (step)**

Still Photographs
21. Snapshots taken at picnic **14543 (step ×11)**

Image and Activity Bank
Side A or B

A selection of still images, short videos, and activities is available for you to use as you teach this unit. For a larger selection and detailed instructions, see the Videodisc Resources booklet included with the Teaching Resources materials.

8-1 "But I Don't See . . . ," page 131
Bacteria, sources of 4149
Food should be cooked and stored properly to prevent growth of harmful bacteria.

8-2 Our Tiny Neighbors, page 139
Volvox 44123–44316 (play ×2) (Side B only)
This volvox is almost as big as the head of a pin. It's composed of hundreds of cells with tiny tails that all beat as one.

Spirogyra 2731
Spirogyra is a filamentous alga. It is commonly found in freshwater ponds, where its filaments can form a dense mat. Individual filaments look like fine threads and contain many identical cells. Any cell may produce spores.

Algae, marine 2730
Many algae are one-celled and microscopic in size, but some of the marine species (such as this one) form large, multicellular seaweeds.

◀‖ Step Reverse    Play ▶    Pause ‖    Step Forward ‖▶

Kelp 2741

Kelp grows worldwide in sea water 10.5 to 15 m deep. Some species grow as long as 22.5 m. All species are attached to the bottom by a large mass of fibers.

Kelp 2755

Kelp has air bladders that float. The bladders ensure that the plant's blades are close enough to the surface to receive light for photosynthesis.

Bacteria; classification 2808

Three types of bacteria

Cyanobacterium; nostoc 2728

Cyanobacteria used to be considered algae. They have chlorophyll and photosynthesize, but their physical structures are different from those of algae. Cyanobacteria are abundant in all aquatic ecosystems.

Cyanobacteria in geysers 2729

Cyanobacteria are among the most hardy organisms. Some grow in hot springs (shown here). Others grow under the ice in antarctic lakes; in lakes choked with salt; or on moist soil, tree bark, and snow banks.

9-1 Tall (but True) Tales, page 149

Mold, slime 2762–2673 (step)

Slime molds are individual cells that act like a single organism. As decomposers, they ooze along the forest floor over rotting logs and other decaying matter, engulfing particles. (step) Slime

molds can construct reproductive bodies out of individual cells and release spores. The spores become cells that are attracted to each other by a secretion, forming a new slime mold.

Mold, slime 2765–2766 (step)

Slime molds consume much of the organic matter in their paths as they ooze along. Fortunately, they move very slowly. (step) This is a plasmodium, a glistening sheet of slime mold with no definite form. A plasmodium can grow as large as a meter across. It will continue to grow as long as food and water are available.

Classification: the five kingdoms 2807

The five kingdoms of living organisms

Puffball 2772

Puffballs have white interiors. These interiors change to yellow or black as spores are produced. When mature, a hole forms on the top, and spores "puff" out of the puffball when it's struck by a raindrop.

Fungus, coral 2775

There are many types of coral fungus; this species is the most colorful. Coral fungi grow on forest floors where the soil is rich with humus. The spores are produced on the branchlike structures.

Mushroom, gilled 2776

The radiating blades (gills) on this mushroom bear the mushroom's spores. The most common type of mushroom has a cap, gills, and a stalk.

Mushroom, amanita 2779

There are many species of amanita mushrooms. Some are edible, but the species shown here is highly poisonous.

Fungus, rust 2788

This fungus grows on wet leaves. It can be very specialized; many types grow on only one species of plant. The powdery rust is the fungus's spores. Rust fungus is destructive to plants and is considered a pest by gardeners.

Spore dispersal; fungus 43337–43816 (play ×2) (Side B only)

This fungus is releasing spores in the wind. Each spore could start a new fungus.

9-2 A Balanced View, page 158

Pasteur's experiment; pasteurization 2809–2810 (step)

This experiment indicates that bacteria cause food to spoil. (step) This experiment demonstrates that pasteurization kills unwanted bacteria.

Cheese making 2813

Four stages of cheese making

9-3 The Staff of Life, page 164

Mold 2767–2770 (step ×3)

This mold is a decomposer grown from a spore on a pumpkin pie. The spore matures and releases more spores, all of which can create more mold. (step) The mold growing on this orange is a type of penicillium. Penicillium produces the drug penicillin. (step) Before preservatives were used, mold would appear on bread after several days. (step) Mold needs moisture and food to grow. This mold is growing under a rotting pine log, which provides both of these conditions.

Paramecium; movement 2814–2821 (step ×7)

This protozoa is navigating around an object by trial and error.

10-2 Microorganisms and Food, page 178

Microorganism; growth as temperature rises 2823

Bacteria grow rapidly up to a certain temperature. Above that temperature, bacteria are hindered or killed by heat. Each microorganism has a preferred range of temperatures.

Food preservation; history 2824–2827 (step ×3)

Time line of developments in the food-preservation industry

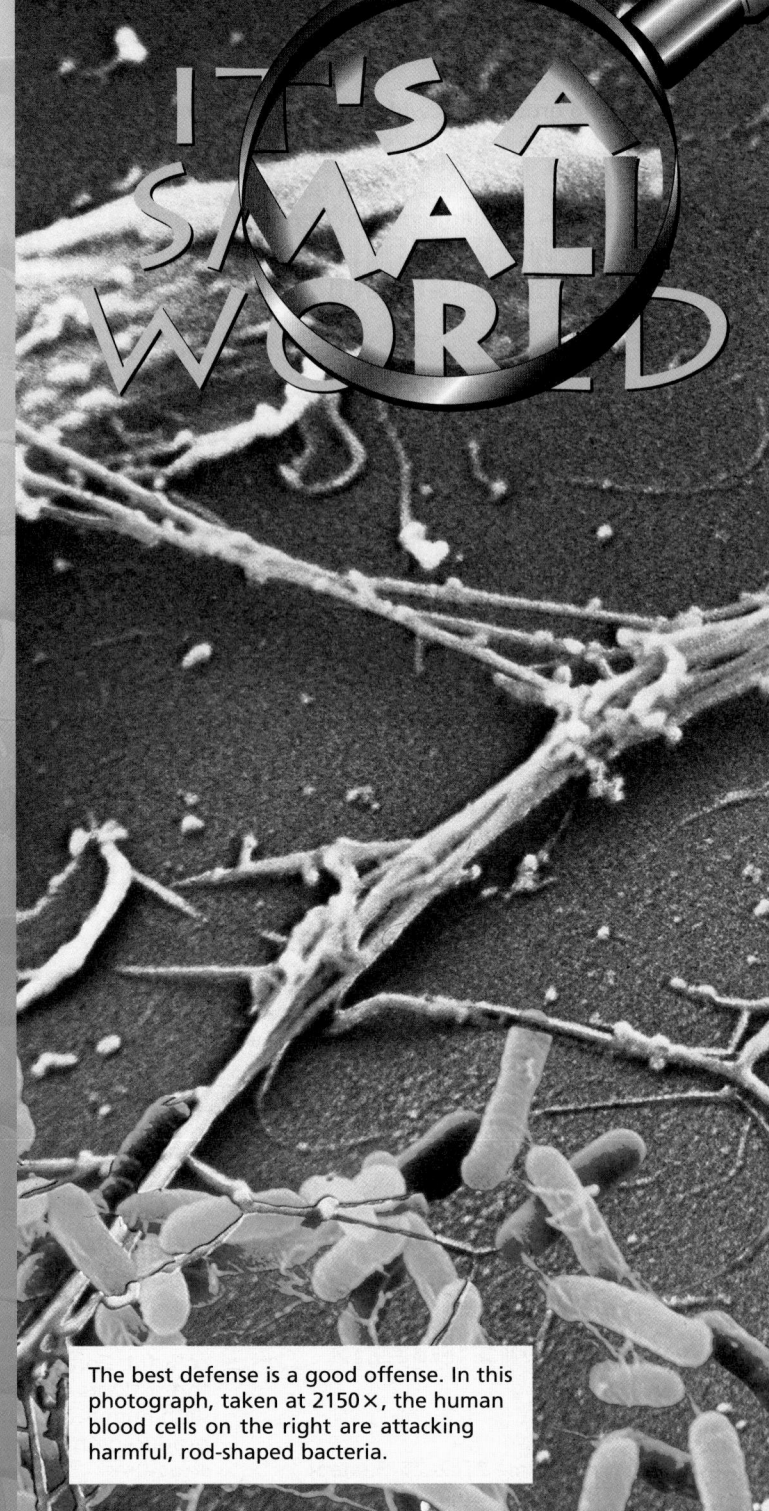

UNIT FOCUS

Ask students to describe the symptoms of an illness. Then involve the class in a brief discussion of what causes people to become ill. Help students arrive at the primary cause of disease, accepting such terms as *bugs, viruses,* and *germs.* Then ask students if they have ever seen a germ that causes an illness. Have them explain why they cannot see these organisms with the unaided eye. *(They are too small.)* Explain to students that this unit will help them to understand and redefine such terms as *bugs, viruses,* and *germs.*

A good motivating activity is to let students listen to the English/Spanish Audiocassettes as an introduction to the unit. Also, begin the unit by giving Spanish-speaking students a copy of the Spanish Glossary from the Unit 3 Teaching Resources booklet.

The best defense is a good offense. In this photograph, taken at 2150×, the human blood cells on the right are attacking harmful, rod-shaped bacteria.

Connecting to Other Units

This table will help you integrate topics covered in this unit with topics covered in other units.

Unit 1 Science and Technology	Careful scientific observations and reasoning have led to the discovery and understanding of microorganisms.
Unit 2 Patterns of Living Things	Most microorganisms show all of the signs of living things, but viruses do not.
Unit 4 Investigating Matter	An understanding of metric units helps students conceptualize the size of microorganisms.
Unit 5 Chemical Changes	Chemicals, such as salt, are used to preserve food against contamination by microorganisms.
Unit 7 Temperature and Heat	The activity and growth of microorganisms are very sensitive to changes in temperature.

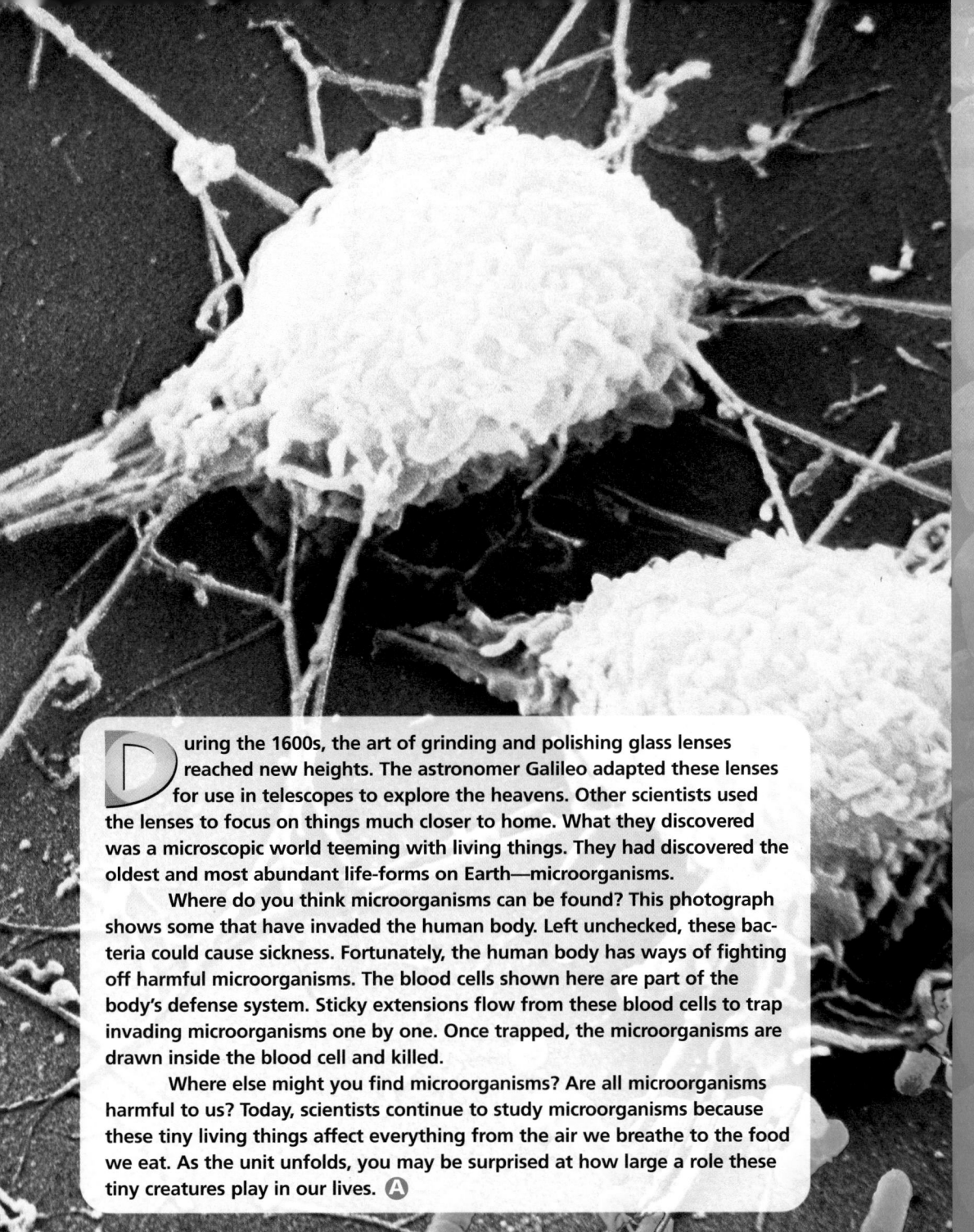

During the 1600s, the art of grinding and polishing glass lenses reached new heights. The astronomer Galileo adapted these lenses for use in telescopes to explore the heavens. Other scientists used the lenses to focus on things much closer to home. What they discovered was a microscopic world teeming with living things. They had discovered the oldest and most abundant life-forms on Earth—microorganisms.

Where do you think microorganisms can be found? This photograph shows some that have invaded the human body. Left unchecked, these bacteria could cause sickness. Fortunately, the human body has ways of fighting off harmful microorganisms. The blood cells shown here are part of the body's defense system. Sticky extensions flow from these blood cells to trap invading microorganisms one by one. Once trapped, the microorganisms are drawn inside the blood cell and killed.

Where else might you find microorganisms? Are all microorganisms harmful to us? Today, scientists continue to study microorganisms because these tiny living things affect everything from the air we breathe to the food we eat. As the unit unfolds, you may be surprised at how large a role these tiny creatures play in our lives. **A**

129

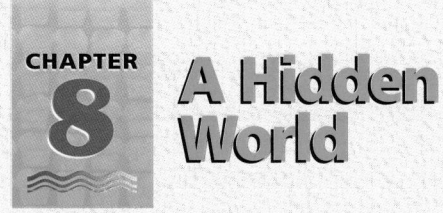
Connecting to Other Chapters

Chapter 8
explores the diversity of microorganisms and the conditions they need to thrive and reproduce.

Chapter 9
examines both the harmful and beneficial effects that microorganisms have on humans.

Chapter 10
takes an in-depth look at antibiotics and the regulation of microorganisms in food preparation and storage.

Prior Knowledge and Misconceptions

Your students' responses to the ScienceLog questions on this page will reveal the kind of information—and misinformation—they bring to this chapter. Use what you find out about your students' knowledge to choose which chapter concepts and activities to emphasize in your teaching.

In addition to having students answer the questions on this page, you may wish to have them answer one or more of the following questions:

1. What is a microorganism? Give three examples.
2. Are microorganisms more like tiny plants or tiny animals? Provide as much evidence as you can to support your answer.
3. What do microorganisms look like? Sketch a few examples.
4. Where are microorganisms found? Where are they not found?

Emphasize that there are no right or wrong answers in this exercise. Collect the papers, but do not grade them. Instead, read them to find out what students know about microorganisms, what misconceptions they may have, and what about microorganisms is interesting to students.

CHAPTER 8 A Hidden World

1 How small are the smallest living things?

2 What do you think of when you think of "germs"?

3 What kinds of living things are visible only through a microscope?

ScienceLog

Think about these questions for a moment, and answer them in your ScienceLog. When you've finished this chapter, you'll have the opportunity to revise your answers based on what you've learned.

Jason was late getting home for dinner. His grandmother, who was in the kitchen, said, "We ate, and your granddad and Melanie went to a baseball game. Your plate's in the fridge. Why are you so late, anyway?" Then, seeing his expression, she added, "What's wrong?"

Jason's grandmother always knew when something was wrong, especially when Jason really didn't feel like talking about something. Jason and his sister, Melanie, lived with their grandparents.

"You might as well tell me," she said.

"Bad day," Jason said. "Oh, not at school, at work." Work was helping old Pete run the local lunch counter called Pete's Place. Pete had been catering to the school kids of the community for years, and Jason was happily employed there three days a week after school and on Saturdays. He was saving his money for a new electric guitar.

"The health inspector came in this afternoon," Jason said. "He looked at everything and wrote down a lot of stuff. He even watched how I washed my hands before I made sandwiches." Jason hesitated. The next part was not easy to admit, especially to his grandmother.

LESSON 1 ORGANIZER

Time Required
2 to 3 class periods

Process Skills
observing, analyzing

New Terms
Microbiology—the study of microorganisms
Micrometer—one-thousandth of a millimeter
Microorganism—a tiny living thing, usually single-celled, that can be seen only with the aid of a microscope

Materials (per student group)
Exploration 1: microscope; transparent plastic ruler (metric)

Teaching Resources
Transparency Worksheet, p. 3
Exploration Worksheet, p. 5
Math Practice Worksheet, p. 6
Transparency 22
SourceBook, p. S66

FOCUS

Getting Started

Provide each student with a metric ruler. Ask: What is the smallest unit of measurement on the ruler? *(Millimeter)* Explain to students that the smallest bacteria are 1000 times smaller than a millimeter and that 10,000 small viruses could be placed inside one of those small bacteria.

Main Ideas

1. Improper preparation and storage of food can lead to contamination by microorganisms.
2. Microorganisms are tiny living things that can be seen only with the aid of a microscope.
3. Simple microscopes have been used for hundreds of years to observe microorganisms.
4. The size of a microorganism can be estimated by comparing it with the diameter of the field of view of a microscope.

TEACHING STRATEGIES

Before students begin reading the lesson, ask them whether they have ever thrown away food because it was spoiled. Involve students in a discussion of why they think food spoils and what can be done to prevent it.

Have students pause after reading page 132. Call on a volunteer to read Guidelines for Food Service Personnel aloud while another student demonstrates the instructions.

Ask students to speculate about why the health inspector was concerned about the way Jason washed his hands. *(Sample answer: Jason's hands needed to be very clean so that he wouldn't spread germs on the food he was preparing.)*

Encourage students to express whether they agree with what the inspector did.

After students have finished reading this page, assess their understanding by asking questions such as the following:

1. What did the inspector mean by "the public isn't safe in Pete's Place"? *(Pete's Place is so unsanitary that people eating there might get sick.)*

2. Why might wiping the cutting board with a sponge be a bad idea? *(Microorganisms can thrive in a moist environment such as a sponge.)*

3. Why did the inspector threaten to close Pete's Place if the place isn't fixed up quickly? *(Because people's health is jeopardized by the conditions at Pete's Place)*

4. Have students respond to Jason's final question and suggest where he might go to find out about microorganisms. *(Accept all reasonable responses. Students may suggest that Jason do library research, call a local university, or check on-line computer databases.)*

Invite a volunteer to read the list on the back of the sheet that Jason gave his grandmother. Challenge students to explain what the inspector meant by each of the items. *(Accept all reasonable responses.)*

"He said I didn't know how to wash my hands properly and gave me this." Jason pulled a crumpled paper from his pocket and handed it to his grandmother. She looked at Jason with concern, smoothed the paper, and read:

GUIDELINES FOR FOOD SERVICE PERSONNEL
Instructions for Washing Hands

- Wet your hands with warm water.
- Cover your hands with soap.
- Rub your hands together in a circular motion for 30 seconds.
- Remember to rub the areas between your fingers and to clean under your fingernails.
- Hold your hands with the fingers pointing down as you rinse them.
- Dry your hands well with fresh paper towels.

After reading the guidelines, Jason's grandmother asked, "What else did the inspector have to say?"

Jason replied, "He told Pete a lot, I guess, and left a copy of his report. Pete's really upset because he thinks somebody complained and that's why the inspector came today. Mr. Grover, the inspector, says that's usually what happens when he's told to go out late in the day, like today. He generally drops by every few months and takes a quick look around, but this time he was really picky.

"But that's not the worst part. Mr. Grover said that the public isn't safe in Pete's Place and that he'll give Pete only so long to fix up the place—six weeks probably—or it will have to be closed for good. I'll lose my job. How will I get my guitar, then?"

His grandmother said quietly, "You will have to help Pete make things right."

Jason said, "There's a long list of things from what I saw Mr. Grover writing. He was muttering to himself, something about 'germs having a field day' there. He went over the list with Pete. I know the fridge doesn't work well, and the sliding door on the freezer comes off sometimes, but what's wrong with the way we wash dishes? And why shouldn't I make sandwiches on the cutting board? I always use the sponge to clean the board before and after. He said a lot of other things too that Pete and I don't understand. I wrote down some of them and told Pete I'd try to find out about them. They are on the back of the sheet I gave you."

His grandmother looked at the words:

microorganisms
optimum growth temperature
Salmonella
safety standards
organisms per milliliter

As Jason got his plate of dinner from the fridge and put it in the microwave oven to heat, he thought, "But I don't see the need for all this fuss. I wonder where I can get information on microorganisms . . ."

Microorganisms— Our Invisible Companions

How well did you understand the criticisms made by the health inspector in the story on the previous page? Examine the picture of Pete's Place, below, and see if you can find some additional problems not mentioned by Jason. How can Pete and Jason gain an understanding of the health inspector's criticisms in order to run the lunch counter safely? Ⓐ

Pete and Jason were unaware of their problem because they couldn't see it. The inspector mentioned **microorganisms**. They are the subject of this unit. *Micro* means "very small," too small to be seen with the naked eye. Organisms are living things. Microorganisms are tiny living things that can be seen only if you use a microscope.

133

 A Transparency Worksheet (Teaching Resources, page 3) and Transparency 22 are available to accompany the material on this page.

Microorganisms—Our Invisible Companions

Have students pause after reading the first paragraph, and allow them time to examine the picture.

INDEPENDENT PRACTICE Ask students what they see in the picture that might contribute to health problems. Have students make a list of their ideas in their ScienceLog.

Answer to
In-Text Question

Ⓐ Additional problems might include the following: dirty dishes are stacked up; dirty washrags and towels are scattered around; cupboards containing cleaners and other supplies are standing open; chemicals are stored next to towels and cooking pots; milk and other food products that should be refrigerated are sitting out where they can spoil; food is sitting in uncovered pots and dishes; the cutting board is not covered; the chef is using his hands to put food on the grill; the garbage can is not covered; a mop is standing where food is being prepared; and both the waitress and the chef have uncovered hair.

To gain a better understanding of the inspector's criticisms, Pete and Jason can examine the list that the health inspector made and research each item on the list.

Microorganisms—Our Invisible Companions continued ▶

Microorganisms—Our Invisible Companions, continued

Have students finish reading the rest of the page. You may wish to call on a volunteer to read the poem aloud and then have students discuss its meaning.

Please note that questions 1–4 on page 134 are meant to evoke prior knowledge about microorganisms. Point out to students that these questions do not necessarily require correct answers. You may wish to have students work in small groups to develop and discuss their answers.

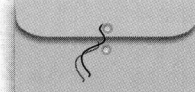 PORTFOLIO

Have students include their responses to the questions on page 134 in their Portfolio. At the end of the unit, have students update their original answers to reflect what they have learned.

Primary Source

"The Microbe," reprinted without alteration from *More Beasts for Worse Children* by Hilaire Belloc

CROSS-DISCIPLINARY FOCUS

Language Arts

The word *sanguine* in the poem "The Microbe" means
a. confident and optimistic. *(Correct)*
b. bitterly unhappy.
c. sad.
d. intelligent.

Answers to
In-Text Questions

Ⓐ The inspector meant that there was a good possibility that germs could thrive and multiply in Pete's Place.

Ⓑ Other terms that are commonly used when referring to microorganisms include *microbes, germs, viruses, bacteria, microscopic organisms,* and *cooties.*

Ⓐ What did the inspector mean when he muttered "germs having a field day"? People commonly refer to some microorganisms as germs. What are some other terms that are used to refer to Ⓑ microorganisms? You will find one in the following poem by Hilaire Belloc.

The Microbe

The Microbe is so very small
You cannot make him out at all.
But many sanguine people hope
To see him through a microscope.
His jointed tongue that lies beneath
A hundred curious rows of teeth;
His seven tufted tails with lots
Of lovely pink and purple spots,
On each of which a pattern stands,
Composed of forty separate bands;
His eyebrows of a tender green;
All of these have never yet been seen—
But Scientists, who ought to know,
Assure us that they must be so . . .
Oh! let us never, never doubt
What nobody is sure about!

1. Do you suppose microorganisms really look like this?

2. What do you know about any of the different types of microorganisms?

3. Use drawings, a poem, or a paragraph to show what you know about the following:
   a. what microorganisms look like
   b. how small they are
   c. where you could find them
   d. how they act

4. Are microorganisms important to you? Explain.

▶ If you could become small enough to explore the inside of the human body, what microorganisms would you see? Would they all be harmful?

Answer to
Caption

Responses will vary. An explorer in a human host might see beneficial bacteria in the digestive tract or harmful microorganisms in the blood if the host is ill.

Back to the Microscope

Jason wanted to learn about microorganisms, so he attended a microscope workshop at the Discovery Center nearby. At the Discovery Center, Jason saw a collection of old microscopes and read the following information, which was part of the display:

Meet the Pioneers

Imagine being able to go back in time almost 350 years. If you traveled to Italy, you might meet a group of scientists, including one named Galileo, who formed a club called the Academy of the Lynxes. The members of this club used glass lenses to study things either very small or very far away, and they made detailed drawings of what they saw. Galileo and his friends conducted their studies with lenses made in the Netherlands, where the glass lens industry first developed. It was there that Anton van Leeuwenhoek (LAY ven hook) became the first person to use lenses to study microorganisms that live in water.

▲ Anton van Leeuwenhoek

Anton van Leeuwenhoek

Anton van Leeuwenhoek (1632–1723), a linen merchant in Delft, Holland, taught himself to grind lenses. He was so skillful at this craft that his instruments had a magnification of up to 275×. This is remarkable since his instruments, which were simple microscopes, had only one lens. Leeuwenhoek came to be known as the founder of **microbiology**. If you take the word *microbiology* apart, you can find its meaning. **C**

Over a period of almost 50 years, Leeuwenhoek wrote hundreds of letters to the Royal Society in London describing his observations.

He observed and drew what he called an "abundance of very little and odd animalcules." In this unit you are going to investigate some of Leeuwenhoek's "animalcules."

Like many scientific pioneers, Leeuwenhoek was an amateur scientist, not a professional. This does not in any way diminish his accomplishments. *Amateur* means "one who loves." Amateurs do what they do simply for the joy of it, not because they are paid to do so. Many important contributions to science have been made by amateurs all over the world.

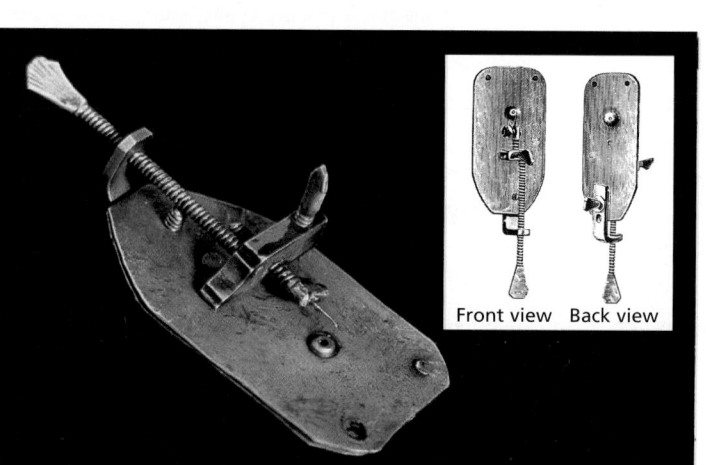

Front view   Back view

These drawings show one of the microscopes that Leeuwenhoek made. Look for the lens. Try to imagine where the object was placed, how the image was brought into focus, and how the instrument was held. **D**

Back to the Microscope

INDEPENDENT PRACTICE   Encourage students to review the information on microscopes in Chapter 7.

SAFETY TIP   If using a microscope with a mirror, remind students to avoid reflecting sunlight directly into the eyepiece.

Meet the Pioneers

Point out to students that the passion these early scientific pioneers had for their subjects helped them to make significant contributions to existing scientific knowledge. Anyone with a keen interest in science can contribute valuable observations and inferences.

Anton van Leeuwenhoek

Ask a student volunteer to read this section aloud. Have students discuss this section by comparing Leeuwenhoek's microscope with modern microscopes.

Answers to
In-Text Question and Caption

C *Micro* means very small, and *biology* is the study of living things; thus, *microbiology* is the study of very small living things.

D The object was placed on the pointed tip in front of the lens; the image was brought into focus by turning the screws; and the instrument was held in front of one eye with the tip pointing up.

Homework

Have students compare the function of a microscope with that of a telescope. Ask: What do their names mean? *(Microscopes and telescopes both contain lenses. They are both used to magnify the images of very small objects. However, a microscope (literally, "to see small things") is used to view objects that are physically very small, such as a microorganism. A telescope (literally, "to see from far away") is used to view objects that seem small because they are very far away, such as a star.)*

How Small Is Small?

Some students may have difficulty with this section because they have not yet had extensive exposure to the metric system. Therefore, it may be helpful to provide a brief overview of the metric system. This material is covered in Chapter 12, Lesson 1, SI: The Metric System of Measurement, which begins on page 211.

Answer to
In-Text Question

Ⓐ From top to bottom, the answers are blue whale, hummingbird, ant, most bacteria, and influenza virus.

Answer to
Try This

It would take 500 cells that are 2 μm wide to cover a distance of 1 mm.

Homework

The table on page 136 lists a few of the metric prefixes that are useful in studying microorganisms. Have students research other prefixes that they might encounter in scientific research. You may wish to have students research units that are smaller than a nanometer or larger than a meter.

How Small Is Small?

What unit would be most convenient for measuring the size of something viewed under a microscope? When you want to describe the size of something, you usually measure it and use some convenient unit (such as meters, centimeters, or millimeters) to express the size. Normally you have no difficulty seeing something the size of 0.1 mm (about the thickness of a hair) without a microscope. If you had a microscope that magnified 100×, what would you be able to see with it? Did you calculate 0.001 mm, or one-thousandth of a millimeter? This unit is called a **micrometer** (μm). The Greek letter *μ* is the abbreviation for the metric prefix *micro*.

Think of it this way: Divide 1 m (about the height of a doorknob above the floor) into 1000 parts. Each part is 1 mm (about the thickness of a dime). Divide 1 mm into 1000 parts and each part is 1 μm. Under 100× power, an object as wide as a micrometer looks about as large as the thickness of a hair. A micrometer is a convenient unit for expressing the size of small things like microorganisms. A nanometer is used for measuring even smaller things—a nanometer (nm) is one-billionth of a meter. In Unit 4, you will learn more about the metric system.

Examine the following table. Which units do you think would best measure different things? Fill in the table using the following list: hummingbird, influenza virus, blue whale, ant, most bacteria. Do not write in this book. Ⓐ

Try This

One type of cell is 2 μm wide. How many of these cells would it take to cover a distance of 1 millimeter?

A Staggering Size Fact

More than 250,000 bacteria could fit on the period at the end of this sentence.

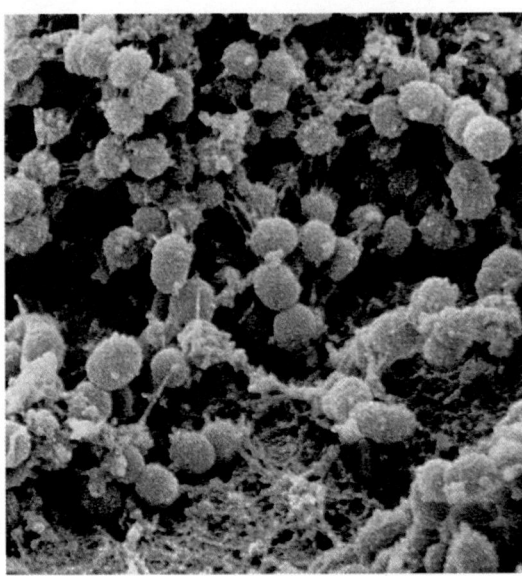

⬆ Billions of bacteria live in our mouths. Some bacteria, like those shown here at 31,000×, grow on teeth and can cause cavities and gum disease.

Metric unit of length	What things are measured with this unit?
meter (m)	
centimeter (cm) = one-hundredth of a meter	
millimeter (mm) = one-thousandth of a meter	
micrometer (μm) = one-millionth of a meter	
nanometer (nm) = one-billionth of a meter	

Measuring Cells

The following exercise will help you find a way to estimate the sizes of cells. Below is a diagram of two people floating in a round swimming pool. A rope 5 m long is stretched across it. This rope measures the diameter of the pool. Can you tell the approximate heights of the people marked *A* and *B*?

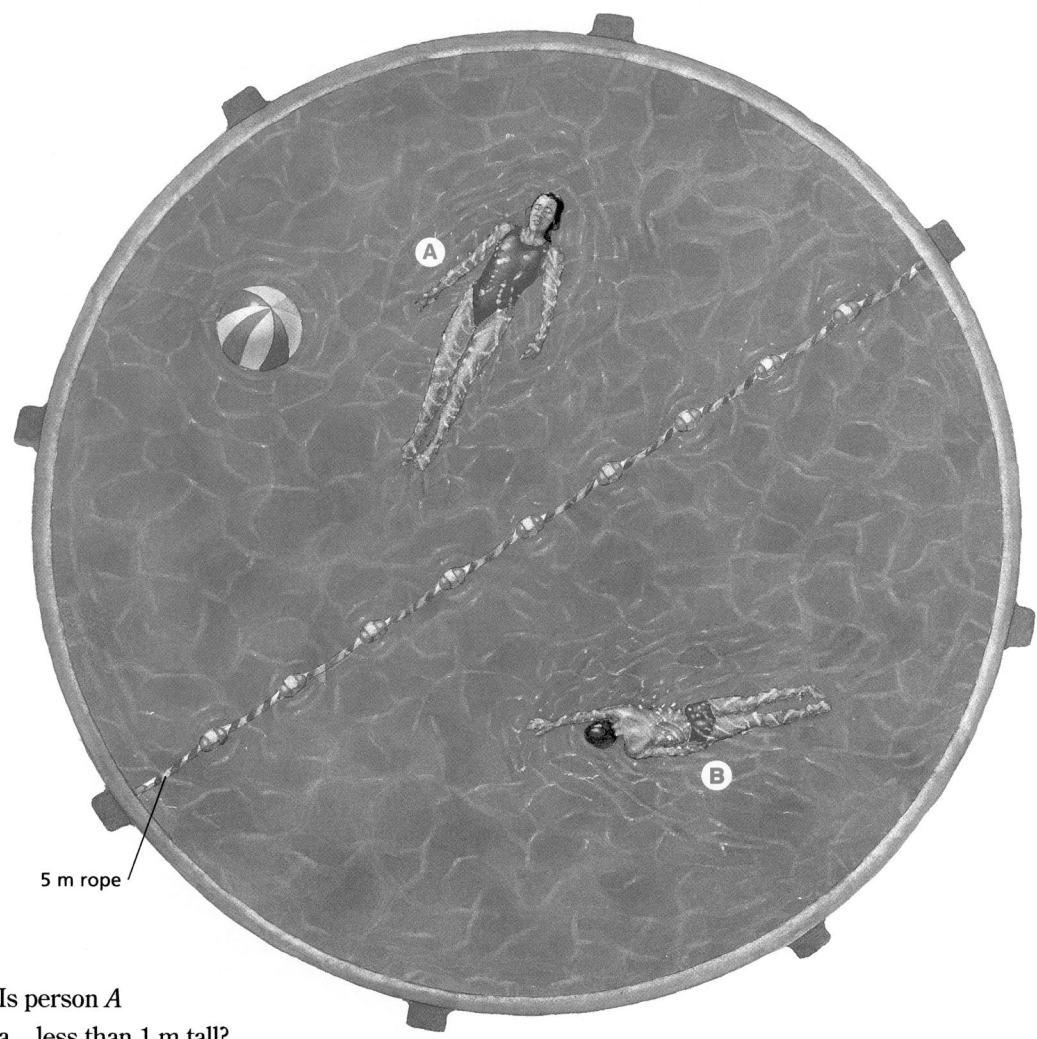

5 m rope

Is person *A*

a. less than 1 m tall?

b. 1 m tall?

c. 1.5 m tall?

d. 2 m tall?

Answer the same question for person *B*.
How did you arrive at your height estimates?

Answers to
Measuring Cells

Person *A* is approximately 1.5 m tall. Person *B* is between 1 m and 1.5 m tall. The knots in the rope are 0.5 m apart and provide one way to estimate the height of each person. For example, person *A* is roughly 3 knots tall, or 1.5 m.

CROSS-DISCIPLINARY FOCUS

Mathematics

If the diameter of a field of view is not known, another way of determining its size is by timing how quickly an object crosses from one side of the field to the other. Tell students that they are studying a microorganism that moves 1 mm every minute. Ask: If it takes 5 minutes for the microorganism to move across your field of view, what is the diameter of the field? *(Assuming that the microorganism crosses the diameter of the field, the field must be 5 mm wide.)*

★ **An Exploration Worksheet (Teaching Resources, page 5) is available to accompany Exploration 1 on page 138.**

Homework

The Math Practice Worksheet on page 6 of the Unit 3 Teaching Resources booklet makes an excellent homework activity.

EXPLORATION 1

Ask students if they have ever seen height markers on the doors of convenience stores or other shops. Inform students that, just as a store owner can estimate the height of a customer entering or exiting the shop by comparing the customer with the marker, a scientist can estimate the size of cells viewed under the microscope by comparing their size with marks on a transparent metric ruler.

Cooperative Learning
EXPLORATION 1

Group size: 2 to 3 students
Group goal: to determine the diameter of a microscope's field of vision
Positive interdependence: Students may be assigned roles such as primary investigator (to make observations using the microscope), recorder (to keep track of the group's activities and observations), and mathematician (to calculate the answers to the questions at the end of the Exploration).
Individual accountability: Randomly select one student from each group to answer the following question: How would you estimate the size of a cell seen under a microscope? *(The size of the cell can be compared to the millimeter marks on a transparent metric ruler seen under a microscope.)*

Answers to
In-Text Questions

Ⓐ Sample answer: More detail, but less of the total area, can be seen.

Ⓑ If the diameter of the field of view at low power (100×) were 2.8 mm, the diameter of the field of view at high power (400×) would be 2.8 mm ÷ 4 = 0.7 mm. The microscope pictured on this page is set on medium power (200×) and has a field of view of 1.4 mm.

FOLLOW-UP

Reteaching

Have students make the following mathematical conversions:
a. 0.2 μm = __ mm *(0.0002 mm)*
b. 10 μm = __ mm *(0.01 mm)*
c. __ μm = 0.005 mm *(5 μm)*

The Diameter of the Field of View

You can use the same approach that you used in the previous exercise to estimate the size of cells seen under a microscope. Just as you needed to know the distance across the swimming pool, you must first know the distance across the circular area seen under the microscope. Once you have measured the diameter of this circular area (the field of view), you can determine the sizes of cells by comparing various distances.

You Will Need

• a microscope
• a transparent plastic ruler (metric)

What to Do

1. Place the ruler on the microscope stage.
2. Using low power (100×), focus on the millimeter lines along the ruler's edge.
3. Count and record the greatest number of millimeters on the ruler that can be seen across the field of view. This is the diameter of the field of view.

Think About It

When you switch to high power, the magnification increases. What effect does this have on the area that you are able to see? Ⓐ Suppose the magnification changes from 100× (using low power) to 400× (using high power). The magnification *increases* by a factor of $\frac{400}{100} = 4$. The size of the field of view *decreases* by a factor of four. This means the field of view will be $\frac{1}{4}$ of the size it was under low power. This mathematical relationship is called an *inverse proportion*.

If the diameter of the field of view at low power (100×) was 2.8 mm, what would it be at high power (400×)? Ⓑ

Calculate the diameter of the field of view at high power for the microscope you are using. To do this, you need to refer to your observation in step 3 (the diameter of the field of view at low power).

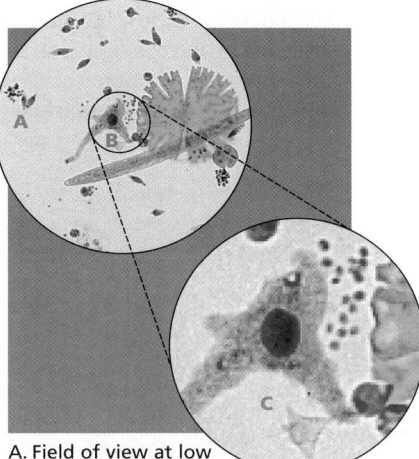

A. Field of view at low power (100×)
B. The part of A that will be seen at high power (400×)
C. Magnification at high power enlarges the image of the small area, B; C is the new field of view.

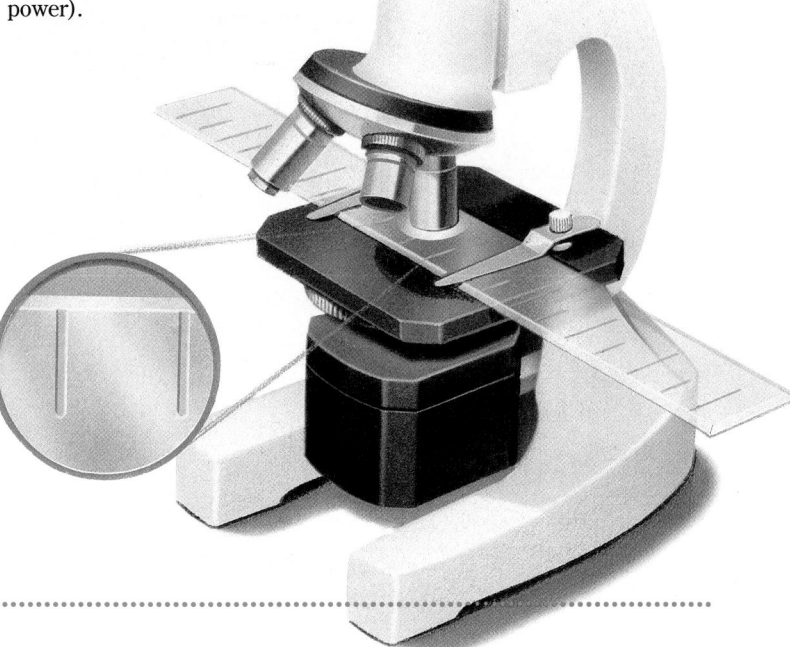

Assessment

Have students write poems or short stories about microorganisms. Call on volunteers to share their poem or story with the class. Then suggest that students organize all the poems and stories in a book, make a cover, and decide on a title. The book may be displayed in the classroom or library.

Extension

Challenge interested students to do further research on the development of microscopes. Students might focus on the contributions of amateur scientists other than Anton van Leeuwenhoek or on how different lenses are used together to produce a clear, magnified image. Have students share what they learn with the class.

Closure

Have students describe problems they have encountered that they think may have been caused by microorganisms. *(Students' responses might include food spoilage, childhood diseases, and other illnesses.)*

As you know, plants and animals—indeed all living things—are composed of cells. Some living things are composed of only a few cells, while others are composed of billions or even trillions of cells. Any living thing that is composed of more than a single cell is said to be *multicelled*—you are multicelled, for example. There are some living things, though, that consist of only a single cell. As you might expect, many microorganisms fall into this category.

In the pages that follow, you will have the opportunity to view many different kinds of microorganisms.

EXPLORATION 2

Water Neighbors

You Will Need

- water samples
- a microscope
- microscope slides and coverslips
- eyedroppers
- rice grains (boiled)
- several small containers with lids
- a wax pencil

What to Do

1. A day before using the microscope, collect samples of water from several of these sources: a pool formed from melting ice or snow, a pond, mud from a garden, or a neglected vase of flowers.

   Caution: Do not go near a pond or body of water unless accompanied by an adult. Wash your hands with soap and water after making your collections.

2. Label the containers to identify the sources of the samples. Include some mud and plant parts in the bottom of each sample when possible.

- Lid on loosely
- Marked water level
- Water sample
- Plant parts
- Bottom mud

3. Add 2 or 3 grains of boiled rice to each sample. Let the samples stand loosely covered in their containers in soft light (not in direct sunlight).

4. After one day, follow the technique illustrated at right, which shows how to gather material from the sediment on the bottom of each container and how to make a wet mount. Examine the material in the wet mount for the presence of microorganisms. Also examine the upper and middle layers of the water in each container for microorganisms.

5. Examine your samples at low power first and then at high power. Notice that the thin film of liquid is like a deep pool for tiny organisms. If a microorganism you are watching

a. Squeeze the bulb of the eyedropper before putting it into the container.
b. Insert the tip of the dropper near the material to be sampled.
c. Release the bulb slightly, and material will enter the dropper.

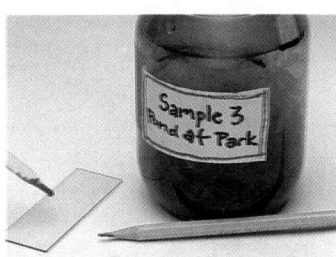

d. Add a drop of the sample to the center of the slide. Place the coverslip over the sample.

Exploration 2 continued ▶

LESSON 2 ORGANIZER

Time Required
4 to 5 class periods

Process Skills
observing, comparing, analyzing

New Terms
Algae (singular, *alga*)—plantlike and usually single-celled organisms
Culture—to grow microorganisms by providing optimum living conditions
Kingdom—the most general category used to classify organisms
Monera—a kingdom of microorganisms, mainly bacteria

Protista—a kingdom consisting mainly of single-celled microorganisms
Protists—certain single-celled organisms, such as euglena and paramecium, with plantlike or animal-like characteristics or both
Virus—a microorganism that does not behave like a living thing unless it comes into contact with living material; must be viewed with an electron microscope

continued ▶

FOCUS

Getting Started

Students may be interested to know that fossils of microorganisms, such as bacteria and algae, have been found in rocks estimated to be 3.5 billion years old.

Main Ideas

1. Water provides an ideal habitat for a variety of microorganisms.
2. The ability of microorganisms to survive and reproduce depends on conditions in their environment.
3. Difficulty in classifying microorganisms prompted scientists to expand the number of kingdoms to include protists and bacteria.

TEACHING STRATEGIES

EXPLORATION 2

The following is a list of likely habitats for specific microorganisms:

- euglena: duck ponds
- paramecium: garden pools with decaying matter; vases of wilting flowers
- amoeba: clear ponds; on or near sphagnum moss or water plants
- vorticella: clear ponds; sticks of dead wood (in white fluffy patches); underside of duckweed; on submerged stones
- stentor: clear ponds (on aquatic plants in the spring); around sphagnum moss

 Cooperative Learning
EXPLORATION 2

Group size: 3 to 4 students
Group goal: to observe microorganisms from different habitats
Positive interdependence: Assign roles such as explorer (to collect water samples and other materials), reporter (to record answers and notes), illustrator (to draw pictures of microorganisms), and director (to keep group focused).
Individual accountability: Ask each student to answer questions 1–4 on page 140 in his or her own words.

Exploration 2 continued ▶

Students should not go near a pond or body of water to collect specimens unless accompanied by their teacher. Also, for additional safety, you may wish to collect the water samples yourself. Remind students to wash their hands with soap and water after working with their samples.

Because the results of this Exploration can be unpredictable, it's a good idea to have students obtain a variety of samples from different locations. Suggest that students include a portion of sediment as well as some surface water in their samples. Point out that different kinds of microorganisms are likely to live in different parts of the sample materials.

As students begin examining their samples, have them watch for any movement. This is the best indication of the presence of microorganisms.

★ **An Exploration Worksheet (Teaching Resources, page 7) is available to accompany Exploration 2 on page 139.**

Answers to *Questions to Ponder*

1. Living; evidence that they are living may include movement under their own power, reactions to other things in their environment, and evidence of feeding on other organisms. Also, they are composed of the basic unit of life—the cell.

2. A variety of organisms will probably be observed.

3. Answers will vary. Sample answer: Some microorganisms live in water, some live in sediments, and others live on plants and decaying matter. Students may describe where their samples were collected.

4. Answers will vary. There may be organisms not pictured here that are in students' samples, and students may not encounter all of the organisms pictured. Students should identify specific aspects of the microorganisms that are similar to or different from those pictured.

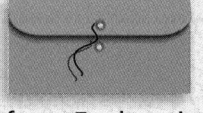 **PORTFOLIO** Suggest that students include their drawings from Exploration 2 in their Portfolio.

"takes a dive" downward in the pool, you will have to make an adjustment in focus in order to see it clearly again. The depth of the field of view at high power is very limited.

6. Draw each kind of microorganism you discover in your samples. Pictures of some living things commonly found in such surroundings are shown below. Be sure to label the drawing of each microorganism and to indicate the sample and the magnification used. Keep the samples for the next Exploration.

7. Using what you learned in Exploration 1, calculate the size of one or two of the microorganisms that you see. Compare your observations with those of your classmates.

8. Clean up your work area. Wash your hands with soap and water before leaving.

Questions to Ponder

1. Are the objects you drew living or nonliving? Explain your answer. Give evidence to support your answer.

2. Do all of your samples contain the same kinds of microorganisms?

3. In what different living conditions do you find microorganisms?

4. Are any of the organisms that you drew similar to those pictured here? How are they similar? How are they different?

Some common water neighbors

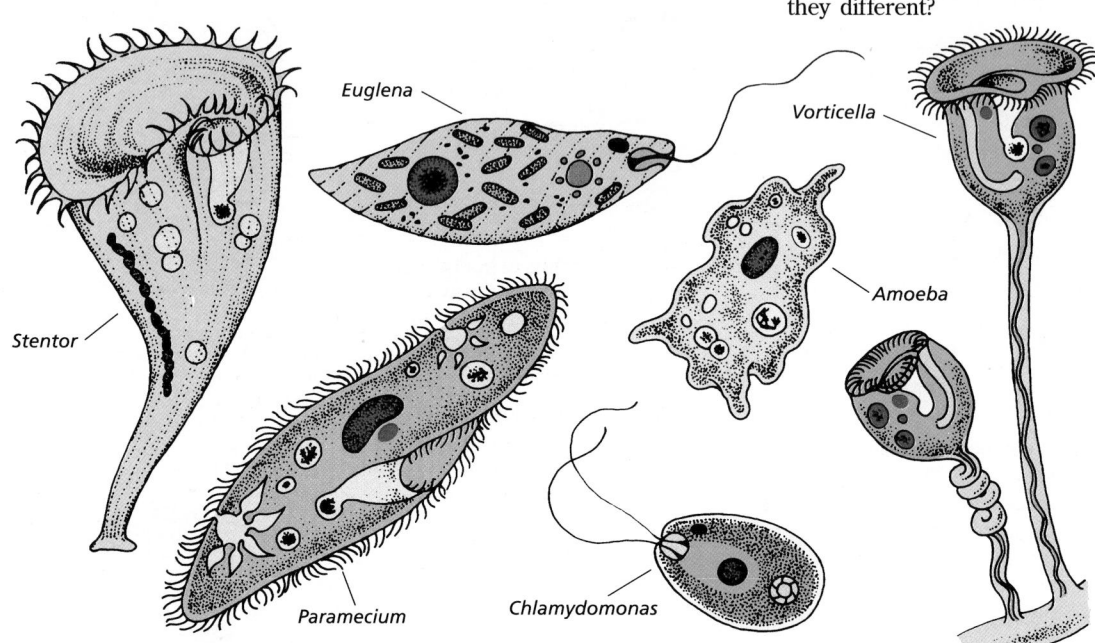

Stentor
Euglena
Vorticella
Amoeba
Paramecium
Chlamydomonas

Time for a Tea Break!

How do you make tea? If you know how, then you know what an *infusion* is, even though you may not call it by that name. Boiling water is poured over tea leaves, and the mixture is left for a few minutes to "infuse." Material from the tea leaves enters the water and gives it color and taste, resulting in a beverage that "hits the spot." The first step of the next Exploration is to prepare an infusion for microorganisms.

ORGANIZER, *continued*

Materials (per student group)
Exploration 2: several small containers with lids; several samples of water from various outdoor sources; microscope; several microscope slides with coverslips; one eyedropper for each sample; spoonful of boiled rice grains; wax pencil; safety goggles (See Advance Preparation on page 127C.)
Exploration 3: water samples from Exploration 2; microscope; several microscope slides with coverslips; one eyedropper for each sample; 1 L of distilled water or boiled tap water; several shallow glass or plastic dishes with covers; few fibers of lens paper or cotton batting; 4–6 g of dry grass, cut into 2 cm segments; 1000 mL beaker; hot plate; metric balance; wax pencil or several small pieces of masking tape; safety goggles (See Advance Preparation on page 127C.)

Teaching Resources
Exploration Worksheets, pp. 7 and 10
SourceBook, p. S70

Life in a Hay Infusion

The microorganisms pictured on the previous page thrive under certain conditions. In this Exploration, you will learn how to **culture** microorganisms, which means to give them the living conditions that will allow them to grow best.

You Will Need

- a microscope
- 1 L of distilled water or boiled tap water
- microscope slides and coverslips
- eyedroppers
- shallow glass dishes with covers
- a few fibers of lens paper or cotton batting
- dry grass
- water samples from Exploration 2
- a 1000 mL beaker
- a hot plate
- a balance
- a wax pencil or tape

What to Do

1. Boil 4–6 g of the grass (spikes and stems, cut into 2 cm segments) in 1 L of water. If possible, use distilled water. If you must use tap water, boil it before adding the grass.

2. The water should become brown. Let the infusion stand for 2 days, and then pour it into several shallow dishes.

3. Add material from several samples used in Exploration 2. Label the dishes carefully.

4. Leave the dishes at room temperature in a fairly well-lighted place, but not in direct sunlight. Cover them loosely.

5. Every day, if possible, remove drops from the dishes and examine them for microorganisms. Look at drops taken from the top and the bottom of the liquid or from near the pieces of grass. Make note of the sample name, what you observe, and the area of the liquid from which the sample was taken. Adjust the light level of the microscope to get the best image.

6. Draw any microorganisms you have not previously seen.

7. Put a few fibers of lens paper (or cotton batting) in a drop of sample before adding the coverslip. Observe and record any effects.

8. Describe the movements and activities of any organisms that you see. Be especially careful to record how they move and how they get their food.

9. Clean up your area and wash your hands when you have finished.

Thinking About Your Observations

1. Did the living conditions favor some living things more than others? Give evidence to support your answer.

2. You tried to create good living conditions for the organisms. Which steps did you take that encouraged the living things to thrive? Explain the purpose of each step.

3. You have been growing cultures of microorganisms. What does *culture* mean? How is it related to the word *cultivate*? What do people cultivate or culture in their home environments?

4. Consider how you could use the cultures you started to answer the following questions:

   a. What effect would freezing have on the microorganisms?

   b. How would a salt solution or vinegar affect these living things?

   c. What would happen if small pieces of lettuce or spinach were added to a culture?

   d. What would happen if . . . ? (Add your own questions.)

   Elise was asked question (a) above, and she said, "Freezing will kill all of the microorganisms." Her answer is called a *prediction*. Write your own predictions to answer the other questions in item 4. You might decide to design an experiment that would test one of your predictions. Obtain permission from your teacher before carrying out any experiment. **Ⓐ**

Encourage students to report any changes that they observe, and challenge them to explain why the changes occurred.

Caution students to be sure that they check the water level in the dishes every day and replace any water lost through evaporation. Replacement water should be taken from the infusion, if possible. If no infusion water is left, distilled water, spring water, or boiled tap water may be used.

In step 5, students may discover that the amount of light needs to be decreased with the diaphragm when observing these microorganisms.

The fibers from the lens paper (or cotton batting) used in step 7 slow down the movements of the microorganisms. This will make it easier for students to observe the microorganisms in order to make their drawings.

Answers to *In-Text Questions*

Ⓐ Sample answers:

   a. Freezing would slow the growth of microorganisms.
   b. A salt solution or vinegar would kill the microorganisms.
   c. The microorganisms would begin to consume the vegetables.
   d. Accept all reasonable responses.

Before students perform their own experiment, have them present a written description of the steps that they propose. Check for safety before approving.

Answers to *Thinking About Your Observations*

1. Answers will vary. Amoebas, if present, will appear in the greatest numbers after about 2 weeks. Changes in the number and kind of microorganisms in the infusion will depend on several factors, including temperature, light, and the decaying plant material that is present.

2. Sample answer: I used hay, which provided food for the bacteria. The bacteria then became food for other microorganisms. I boiled tap water to get rid of the chlorine, which is poisonous to microorganisms. I let the infusion stand for 2 days to let the bacteria grow so that there would be a plentiful supply of food. Then I left the dishes at room temperature in a well-lighted place out of direct sunlight, which created conditions for them to thrive. Finally, I left the dishes covered only loosely, which prevented some evaporation and provided air.

3. As a noun, a *culture* is a population of living things grown in a laboratory setting. As a verb, to *culture* means to provide an ideal environment in which to grow living things. *Culture* as a verb and *cultivate* have the same meaning, such as to *culture* plants in an outdoor garden. Sample answers: Houseplants, gardens, flowers

4. Students should design experiments in which only one variable is tested for each sample. In addition, each design should include a control for comparison. Sample answers are given in Answers to *In-Text Questions* above.

Plant, Animal, or . . . ?

After students have read page 142, you may wish to refer back to page 140. Point out that the organisms shown in the illustration on page 140 are all protists. Ask students if they observed any of these microorganisms while doing Explorations 2 and 3. Have students who respond positively identify the protists that they observed. Then ask students which of the organisms on page 140 looks like the one the two scientists are arguing about on this page. *(Euglena)* You may wish to point out that the euglena has a "tail" that it uses to propel itself through the water, enabling it to move around like an animal. But it also contains chloroplasts, which enable it to carry out photosynthesis like a plant. Although euglenas sometimes consume food for energy and sometimes use photosynthesis for energy, not all protists can use both methods.

At this point you may wish to discuss the five-kingdom classification system. These five major groups are monerans (bacteria), protists (algae and protozoans), fungi, plants, and animals. Make a chart on the chalkboard and ask students to name as many members of each kingdom as they can.

Plant, Animal, or . . . ?

You have seen that one-celled microorganisms move, take in food, and respond to stimuli such as temperature and light. In these and other ways, they fit our description of living things. When you were studying Unit 2, you discussed members of the plant and animal **kingdoms**. Using the microscope, you may have seen some things that you thought had characteristics of both plants *and* animals. In fact, deciding what to call them has sometimes caused problems.

Doctor Pro and Doctor Con are having an argument. Can you help them reach an agreement?

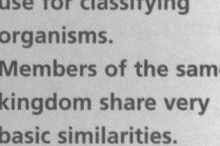

KINGDOM—
The most general category scientists use for classifying organisms. Members of the same kingdom share very basic similarities.

It's a plant. It's green and uses sunlight to make its own food.

No, it's an animal! It moves around and sometimes uses food not made by itself.

DR PRO

DR CON

There *is* a way to settle the argument and, at the same time, respect the opinions of both scientists. How would you settle this dispute? At one time scientists actually debated this issue. Eventually, they agreed to put certain one-celled organisms into their own kingdom, separate from the plant and animal kingdoms. This kingdom is called **Protista**. *Euglena* and *Paramecium* are **protists**.

142

Meeting Individual Needs

Gifted Learners

Explain to students that microorganisms have organelles (stuctures within cells) that carry out such life processes as digestion, locomotion, and energy production. Suggest that they choose a member of the kingdom Protista, such as a paramecium, and research the major organelles that make up its internal structure. Have students make a poster-sized diagram of the protist and label the organelles by name and function.

An Exploration Worksheet (Teaching Resources, page 10) is available to accompany Exploration 3 on page 141.

Algae

Was the water sample you used earlier from a pond? Did you observe what appeared to be microscopic plants? Did they look like any of the following pictures? You might obtain some green scum from the surface of a pond and make a wet mount of it so that you can examine it with a microscope. The tiny plantlike organisms like those found in pond water provide food for a variety of animals and are the first link of the food chain.

Be Careful: Be sure to have an adult accompany you to a pond or any body of water.

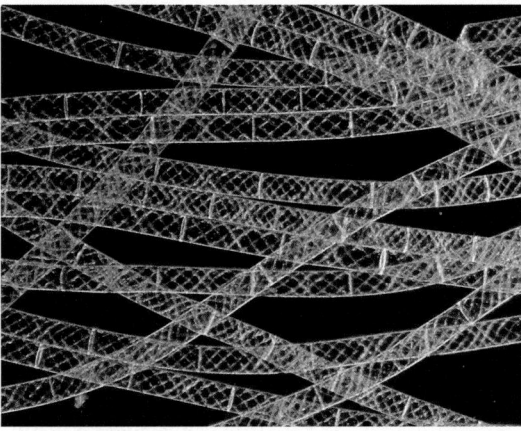
▲ *Spirogyra*

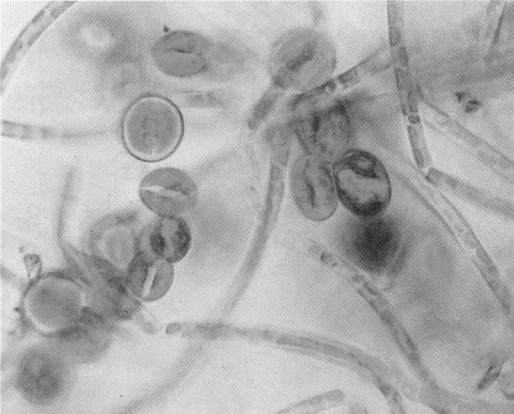
▲ *Chlorella* (the round cells)

▲ *Protococcus* (growing over a tree)

▲ *Prasiola* and *Ulothrix* (covering rocks)

These plantlike organisms are called algae (singular, *alga*). Algae can be found in places other than ponds. Look at the pictures above to find out where. The photos show a few of the many forms that algae can take. One type of algae, kelp, even grows to a length of more than 50 m!

Algae

Some students probably observed algae in Exploration 2. If so, ask them to describe what the algae looked like, and have them draw pictures on the chalkboard of what they saw. You may wish to point out to students that most algae are green. Challenge them to offer an explanation for why this is so. (*Algae contain chlorophyll and carry out photosynthesis.*)

Algae are usually classified as members of the kingdom Protista. They differ from other protists in that they contain chlorophyll and carry out photosynthesis. There are about 20,000 species of algae. Most algae are aquatic, but some grow in soil or on tree bark. Point out that a *food chain* describes how certain living things depend on one another for food energy.

Integrating the Sciences

Earth and Life Sciences

The two most abundant substances in most ocean-floor sediments are silica and calcium carbonate. The silica comes primarily from microorganisms such as diatoms and radiolarians. The buildup of these organisms is so great that *diatomaceous earth* is mined for commercial use. Have students do research to find out what diatomaceous earth is used for and how the physical properties of diatoms make diatomaceous earth useful.

Homework

Certain protists are necessary for the survival of some organisms. A good example of this is the termite. Have students research the role that protists play in a termite's survival. (*Certain protozoans, a type of protist, live in the digestive track of the termite's body. Termites eat cellulose, but it is the protozoa that are responsible for converting the cellulose into an edible form for the termites. Otherwise, the termites would die.*)

Did You Know...

Some marine algae, such as kelp, are usually included in the kingdom Protista even though they are not unicellular. This is because the physical structure of these multicellular algae is very similar to the structure of single-celled algae.

Bacteria—Also Our Neighbors

You may wish to share the following information with students: Bacteria live just about everywhere—in the air, water, soil, food, and even in and on the bodies of most living things. They have been found in the frigid polar regions and in the near-boiling waters of hot springs. A single drop of pond water may contain over 50 million bacteria. A spoonful of soil may contain many billions of bacteria. There are even billions of bacteria living on and in the human body. About 5000 different species of bacteria have been identified by scientists. Many more have yet to be classified.

Bacteria are essential in the production of some foods, including many dairy products such as butter, buttermilk, yogurt, and cottage cheese. Vinegar is also a product of bacterial action. Bacteria are responsible for most food spoilage.

Some kinds of bacteria cause diseases in people by destroying healthy cells. These diseases include cholera, leprosy, gonorrhea, pneumonia, syphilis, tuberculosis, typhoid fever, whooping cough, and *Salmonella* poisoning. Certain bacteria produce toxins that cause diseases such as diphtheria, scarlet fever, tetanus, and botulism.

Bacteria are one of the two major groups of decomposers (the other group is fungi). Without these organisms, dead animals, plants, and other organisms would accumulate and tie up valuable inorganic and organic resources.

Bacteria—Also Our Neighbors

Bacteria are another type of microorganism. What do you know about bacteria? How large are they? How numerous are they? What do they look like? The following photos show the main types of bacteria, described by shape:

- sphere-shaped bacteria
- rod-shaped bacteria
- spiral-shaped bacteria

▽ Sphere-shaped bacteria

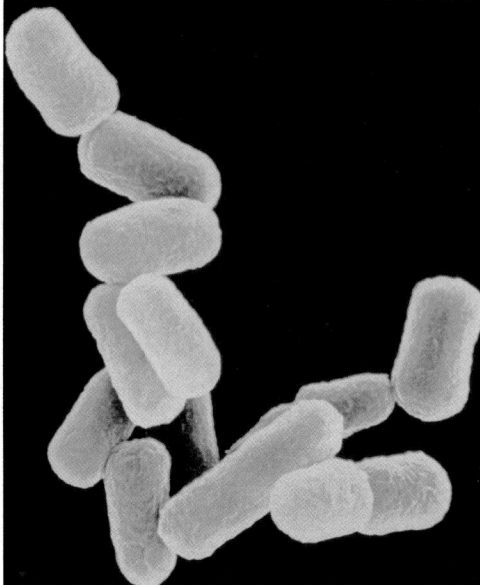

▽ Rod-shaped bacteria

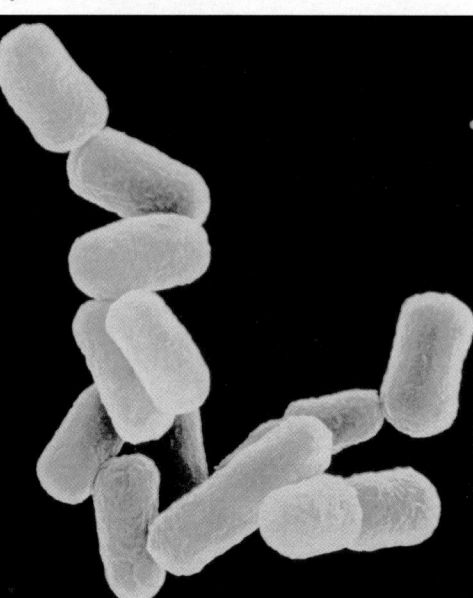

▽ Spiral-shaped bacteria

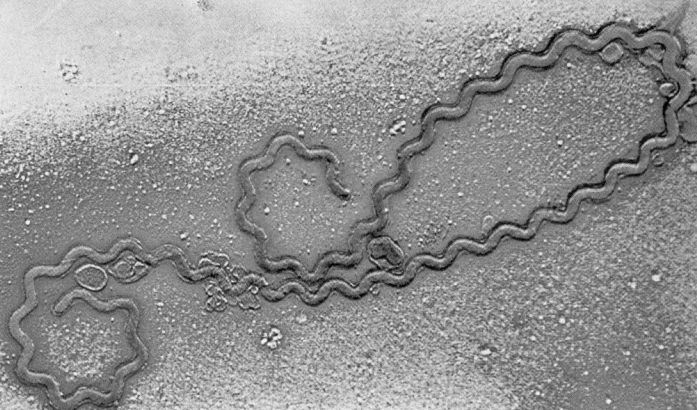

Looking at these pictures may make you suspect that you saw bacteria while you were looking at the wet-mount slides you prepared earlier. Bacteria are very small, even when magnified at 400×, and appear as tiny quivering shapes in the field of view.

Bacteria are not classified as animals, plants, or protists. They are in a separate kingdom called **Monera**. They can be cultured easily, but since some cause disease, you will not be culturing them. They can be easily viewed in prepared slides.

To discover more about how disease-causing microorganisms are spread, turn to page S66 of the SourceBook.

144

Not So Simple

What types of microorganisms have you met so far? What are the names of those found in the water samples? of those found in the green scum on the surface of ponds? of those that are just barely visible even when magnified 400 times? Ⓐ

There is another type of microorganism that is much smaller than any you have seen so far. If an average bacterium (singular of *bacteria*) were the size of a watermelon, then one of these microorganisms would be about the size of an aspirin tablet. A special type of microscope—an *electron microscope*—is needed to see them. Have you guessed what microorganisms are being described? If you said **viruses**, you are correct.

The word *virus* means "poison" in Latin. Viruses cause many diseases: the common cold, influenza (flu), measles, chickenpox, and AIDS, to name a few. Viruses differ from bacteria and other microorganisms in that they do not clearly behave like living things. Most viruses resemble crystals. Outside of living material, they are totally inactive: they do not move, eat, breathe, or show any other signs of life. When viruses come into contact with living material, however, they can multiply, causing symptoms of disease.

So far, your work has involved microorganisms in water. They have been held in tiny ponds between glass pieces for convenience and may have died in the process. The conditions were artificial for them. Can you imagine how microorganisms lead their normal lives? The next chapter will help you do this.

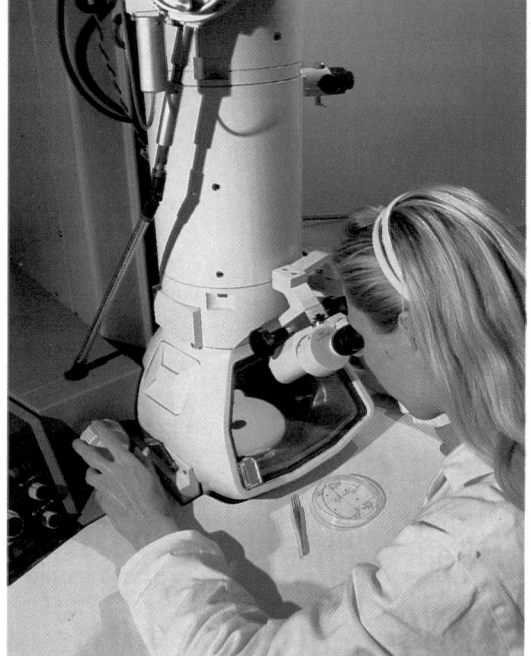

An electron microscope. A beam of tiny particles known as *electrons* is projected down the microscope column and through the sample of material. The particles hit a fluorescent screen, which produces the image.

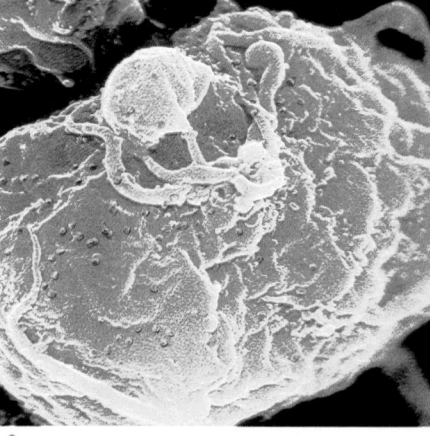

HIV (the virus that causes AIDS). The virus (green) can be seen budding from the membrane of an infected cell.

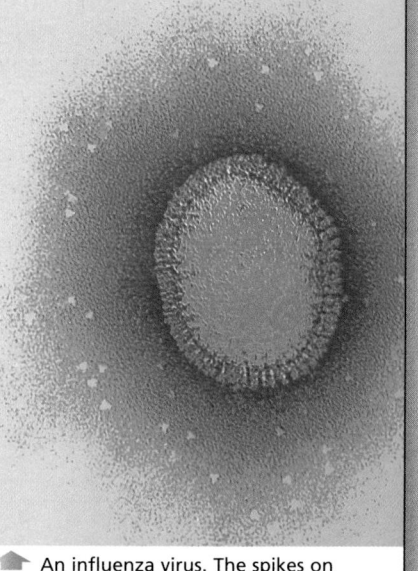

An influenza virus. The spikes on the outside of the virus attach to a host cell.

145

Not So Simple

Have students read the page silently. Then direct them to the photographs of viruses. Point out that these were taken with an electron microscope because viruses are too small to be seen under a light microscope. Students may find it interesting to know that viruses are measured in nanometers (nm). Remind students that one nanometer is equal to one-millionth of a millimeter. Spherical viruses, for example, range in size from 15 nm to about 200 nm in diameter.

Answers to
In-Text Questions

Ⓐ Students should identify the microorganisms they observed in the water samples as protists, including euglena, paramecium, and amoeba; the organisms in green pond scum as algae; and the organisms that are just barely visible even when magnified 400 times as bacteria, or monerans.

Did You Know...

In 1849 a cholera epidemic swept through some regions of the Southwest, killing 600 people in 6 weeks. Cholera is caused by a bacterium found in water or food that has been contaminated with the feces of people who have the disease.

FOLLOW-UP

Reteaching

Have students construct a concept map using the following words: *microorganisms, kingdoms, bacteria, algae, protists, paramecium, Monera, virus, euglena, Protista,* and *microscope.*

Assessment

Have students make posters to show at least one representative microorganism from each of the following groups discussed in the lesson—protists, bacteria, and viruses.

Extension

Have students find out what a diatom is and identify some of the food chains that contain diatoms.

Closure

Take a class field trip outdoors to look for a common alga called *Protococcus.* It frequently grows on tree trunks or concrete foundations. It may look like a green stain or moss. Instruct students to collect a sample of the alga by scraping a little of the material onto a damp paper towel. Have them make a wet mount with their sample and look at the cells with a microscope. Add a drop of iodine starch-test reagent to stain the cells. (For waste disposal procedures, see Waste Disposal Alert on page 21.) Ask students to draw a cell in as much detail as possible and to label their drawings.

Answers to *Challenge Your Thinking*

1. The techniques used so far have involved culturing microorganisms and viewing them with the aid of a microscope at different powers of magnification. Some organisms, such as bacteria, are smaller than others and must be viewed with the high-power objective lens. Viruses are too small to be seen using a light microscope. Other organisms, such as algae and fungi, can form colonies that are visible to the naked eye.

2. Answers will vary, but microorganisms can be found almost everywhere—in air, soil, water, food, plants, animals, and even on the head of a pin.

3. a. Conditions that are good for microorganisms include fresh, unchlorinated water, food, oxygen, sunlight, and moderate temperatures. Conditions that are bad for microorganisms include a lack of fresh water, food, oxygen, and sunlight. Extreme changes in temperature and water chemistry also harm microorganisms.

  b. Knowledge of the conditions that help or harm microorganisms is useful because microorganisms can spoil food and cause illness. However, microorganisms also play a key role in many food chains.

4. a. See graph on page S235.

  b. Euglenas do not grow old in the way that humans do. Instead, they divide and continue as two new organisms. However, if conditions are poor (such as when there is a lack of food), the euglenas may fail to divide or may die from lack of nourishment.

  c. Other possible explanations: The conditions (such as light or temperature) might have favored the paramecia and not the euglenas; the paramecia might have eaten the euglenas; the paramecia might be faster and more likely to get food than the euglenas; the paramecia might release poisons that kill the euglenas; and the light may have been too low to allow the euglenas to photosynthesize.

  d. Answers will vary, but one possible approach would be to separate the euglenas from the paramecia and then test the effect of different variables (such as food and light) on each.

CHALLENGE YOUR THINKING

1. Look Closely
What techniques have you used so far to view microorganisms? How did the type of microorganisms you saw vary for each technique? Why?

2. Where Are They?
Where might you look to find microorganisms? Make a list of at least 10 places.

3. A Little Culture
a. List all of the conditions that were good for the microorganisms that you collected. What conditions seemed to harm the microorganisms?

b. Why would knowledge of the conditions that helped or harmed microorganisms be useful?

4. Microbe Farmers
A group of students watched the microorganisms in a hay infusion for 10 days. Each day they recorded the average numbers of paramecia and euglenas they saw. Their data is shown in the table at left.

Day	Number of paramecia	Number of euglenas
1	15	3
2	15	6
3	18	10
4	23	15
5	28	7
6	25	1
7	15	0
8	40	1
9	too many to count	0
10	too many to count	1

a. Graph their data.

b. To explain the changes in the number of the different protists, the students wrote the following explanation: "We think that maybe the euglenas were really old and at the end of their lives, or maybe they were sick or not given the right foods." Do you think that their explanation could be right? Explain.

c. What other possible explanations could there be to account for the increase in the number of paramecia and the decrease in the number of euglenas?

d. Design an experiment to test the hypothesis that you think is the best explanation of the data.

Homework

Ask: What is the typical size of the smallest feature that can be seen with a light microscope? with an electron microscope? You may wish to have students research why an electron microscope can resolve smaller features than a light microscope can. *(In order to see an object clearly, the wavelength of the light used must be much shorter than the size of the object. Because the shortest visible wavelength is about 0.4 µm, light microscopes can resolve objects only about as small as bacteria. The wavelength of "electron light," however, is much shorter than that of visible light. Electron microscopes can resolve features smaller than 0.0002 µm, or 0.2 nm, the size of a typical virus.)*

5. Space Exploration

You are a space scientist exploring the planet PRMP-5. From the material collected by your assistant, you must decide whether there are any life-forms on PRMP-5 that are similar to those found on Earth. After reading your assistant's notes, which are shown at right, classify each sample according to what you learned in this chapter.

Sample A:
single-celled, rod-shaped; smaller than a protococcus, but larger than a virus

Sample B:
had to use the electron microscope to view these organisms; might not be alive at all though—don't seem to show any of the normal signs of life; resemble crystals

Sample C:
microscopic organism that makes its own food using sunlight; single-celled; moves about by using a whiplike propeller

6. Under the Weather

A friend of yours had the flu. He said anything that could make him feel that bad must have been alive. What would you say?

Review your responses to the ScienceLog questions on page 130. Then revise your original ideas so that they reflect what you've learned.

147

5. Sample A: moneran
   Sample B: virus
   Sample C: protist

6. Answers will vary, but students should understand that both viruses and bacteria can cause illnesses, although only viruses can cause a cold or flu. While bacteria are classified as living things, there is some debate over the nature of viruses. Most viruses resemble crystals and show no signs of life outside a host organism.

★ **You may wish to provide students with the Chapter 8 Review Worksheet that is available to accompany this Challenge Your Thinking (Teaching Resources, page 13). Transparency 23 is also available.**

ScienceLog

The following are sample revised answers:

1. The smallest living things are so small that they can be seen only with a microscope. Viruses, which may or may not be living things, can be seen only with an electron microscope.

2. The word *germs* refers to microscopic organisms (especially bacteria) that cause disease. Microorganisms are tiny living things that are found almost everywhere. Microorganisms live in water, in soil, and even in our bodies.

3. Microscopic organisms include bacteria, protists, fungi, and tiny animals.

ENVIRONMENTAL FOCUS

Ask: What factors in the environment of the hay infusion on page 141 might have contributed to either the growth or the decline of a certain type of microorganism? *(Environmental factors include temperature, food supply, sunlight, amount of chlorine, and presence of predators.)*

Connecting to Other Chapters

> **Chapter 8**
> explores the diversity of microorganisms and the conditions they need to thrive and reproduce.

> **Chapter 9**
> examines both the harmful and beneficial effects that microorganisms have on humans.

> **Chapter 10**
> takes an in-depth look at antibiotics and the regulation of microorganisms in food preparation and storage.

Prior Knowledge and Misconceptions

Your students' responses to the ScienceLog questions on this page will reveal the kind of information—and misinformation—they bring to this chapter. Use what you find out about your students' knowledge to choose which chapter concepts and activities to emphasize in your teaching.

In addition to having students answer the questions on this page, you may wish to have them write a short narrative about the life of a microorganism of their choice. Make sure that they mention the environment of the microorganism, as well as what it eats, what it looks like, and what effects it has on its surroundings. They should not mention what microorganism is being described. You may wish to have a few volunteers read their narrative to the class and to see if the class can guess what type of microorganism is being described. Emphasize that there are no right or wrong answers in this exercise. Collect the stories, but do not grade them. Instead, read them to find out what students know and what misconceptions they may have about the lives and effects of microorganisms. Point out that in the next lesson, students will have the chance to read a few more of these narratives.

CHAPTER
9
Friend or Foe?

1 Are microorganisms an important ingredient in soil? Explain your answer.

2 What conditions would encourage microorganisms to grow on your skin?

3 "News flash! All microorganisms have been eliminated from the Earth." Would this be good news or bad news? Explain your answer by using examples.

The Daily Planet News
All the News That Hasn't Happened Yet

All Microorganisms Have Been Eliminated From the Earth!

ScienceLog

Think about these questions for a moment, and answer them in your ScienceLog. When you've finished this chapter, you'll have the opportunity to revise your answers based on what you've learned.

148

Tall (but True) Tales

[Alice] said: "one can't believe impossible things."

"I dare say you haven't had much practice," said the Queen. ". . .Why, sometimes I've believed as many as six impossible things before breakfast."

—Lewis Carroll, Through the Looking-Glass

Read the passages below. Do they seem almost impossible? What do you think these living things are? Where might they live? You might try to illustrate each situation with sketches.

A The inhabitants of this world number perhaps 72 million. They avoid oxygen and live sealed under a rubbery webbing, where they thrive in near-tropical heat and humidity. The living is easy because their food comes from below, welling up through cracks. Sometimes, however, pieces of the surface break loose, carrying thousands of the dwellers, as if on rafts, to faraway places and almost certain death.

B The creature lay, mouth up, in a forest of tall, red stalks. It lay waiting, with its large front claws ready to receive the small flakes of crunchy food that were drifting slowly downward toward its open mouth. Suddenly the stalks began to tremble, and the surface began to vibrate. A mighty force pulled the creature upward. It lost its grip on the nearest stalk and was taken up into a whirling mass of particles and gas. When the motion stopped, the creature found itself in an even better place than before—a place with great stores of food. The creature settled down and started eating again peacefully.

C The signal had been received. With food supplies depleted, it was time to move. Billions of baglike creatures began to crawl, changing shape like blobs of partially set jelly, all moving as one. They kept out of dark shadows when possible but otherwise did not allow anything to affect their progress. Many died along the way, but their passing was unheeded by the others. Upon arrival, they climbed on one another to form a pyramid and rested awhile. Then they began to form a taller structure, a tower.

To hold the tower firmly together, many individuals produced a kind of fast-acting glue. When the glue hardened, it cemented their dying bodies together like bricks. From within the tower, some of the survivors crawled up over the others and reached the top. Once there, they sealed themselves inside capsules with enough water and food for a long journey and were heaved from the tower into the air. Some of them would find a place to establish another colony.

149

LESSON

1

Tall (but True) Tales

FOCUS

Getting Started

Display several decaying leaves and several fresh leaves for students to examine. Ask for volunteers to explain why leaves decay. *(Due to soil bacteria and fungi)*

Main Ideas

1. The lives of microorganisms differ from those of other living things.
2. Microorganisms are intimately involved in the lives of people.
3. Microorganisms that live in soil can be beneficial, harmful, or both.
4. Fungi are microorganisms that are abundant in soil and that help break down organic material.
5. Some common microorganisms in sea water are called *plankton* and are the first link in many food chains.

TEACHING STRATEGIES

Have students read each description on this page, pausing after each one to speculate about its subject. You may wish to list student ideas on the chalkboard for reference when students read the explanations on page 150.

Primary Source

Description of change: excerpted from *Through the Looking-Glass,* by Lewis Carroll

Rationale: excerpted to point out the incredible nature of many scientific facts

LESSON 1 ORGANIZER

Time Required
3 class periods

Process Skills
observing, analyzing, hypothesizing

Theme Connection
Changes Over Time

New Terms
Compost—a natural fertilizer made from decomposing matter
Fungi (singular, *fungus*)—the fifth kingdom; includes mushrooms, molds, and puffballs; composed of many tiny organisms growing together

Plankton—a microscopic aquatic plant or animal; a food source for nearly all aquatic animals

Materials (per student group)
Exploration 1: magnifying glass (10×)
Exploration 2: 50 mL of soil from a garden or flowerpot containing living plants; 500–600 mL glass container with cover; 300 mL of distilled water; microscope; several microscope slides with coverslips; eyedropper (See Advance Preparation on page 127C.)

continued

INDEPENDENT PRACTICE Before
students read page 150, have them
make sketches of what they think each
description is about. Encourage students
to share and discuss their sketches
before they read the explanations on
this page.

Meeting Individual Needs

Second-Language Learners
Have students distinguish among the
terms *epidemic, pandemic,* and *endemic.*
Explain to the class that although the
meaning of these terms may vary
slightly, an *epidemic* is generally an outbreak of a disease that affects many
people at once. If the disease is a permanent part of a region, then the disease is *endemic* to that region. If the
disease affects the entire world, it is
pandemic.

Homework

Have students give a presentation on
another type of microscope of their
choosing. They should draw a diagram and give a brief report on its
strengths and weaknesses. Other
types of microscopes include scanning
probe microscopes, scanning tunneling microscopes, atomic force microscopes, and ion microscopes.

CROSS-DISCIPLINARY FOCUS

Social Studies
Share the following information with
students: During the Middle Ages and
the Renaissance, epidemics of the
plague had a profound effect on the
world view of Europeans. At its height
in the 1300s, the plague destroyed
nearly two-thirds of the population in
some areas. William Shakespeare's
Hamlet, which was first performed in
1601, reflects the European preoccupation with the plague in that much of its
imagery refers to disease.

Those left behind dried up and died—all of
their energy had been used to make certain that a few would have the chance to begin
life anew in another place. Most of the population was sacrificed to permit a few to escape.

These may sound like creatures from a
science-fiction story, but they are real and are
very close to us. Description (A) is about the
microorganisms found on a person's forehead.
The tiny rafts are skin flakes, which are constantly falling off as new skin cells form below
them. The food is perspiration, oils, and other
substances that the skin normally produces.

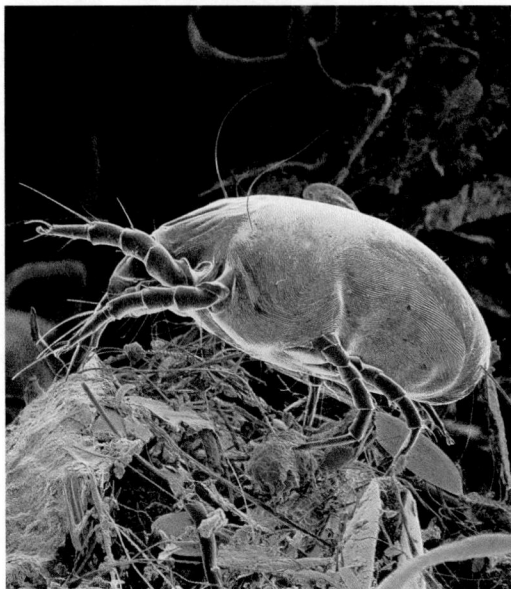

The creatures of description (B) are called
dust mites and are commonly found in carpets
and mattresses—even in clean homes. They eat
the skin flakes that we shed. The dust mites in
the paragraph are being drawn into a vacuum
cleaner from a red carpet.

A colony of slime mold is described in (C).
These protists are quite common. You might
even find them on your lawn. When a colony's
food source is used up or destroyed, the whole
group migrates to find new food supplies.
Sometimes many members of the colony are
sacrificed to allow a few individuals to escape
and start new colonies.

Bacteria that might be found on your skin

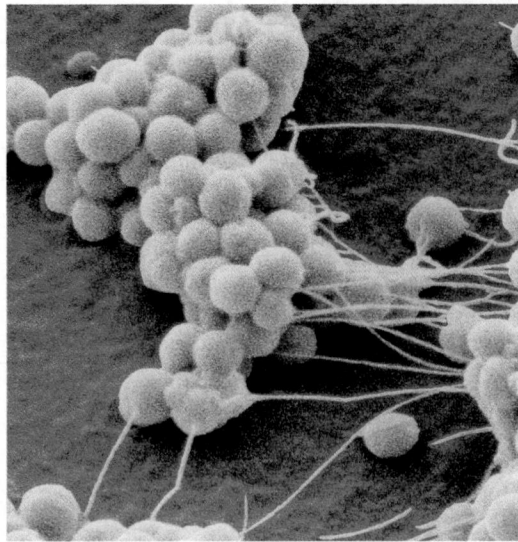

One of the many microscopic dust mites, like those in
your home, is shown here. This mite has been magnified about 500×. They use their serrated (saw-toothed)
front claws to collect flakes of human skin. Dust mites
are the cause of most dust allergies.

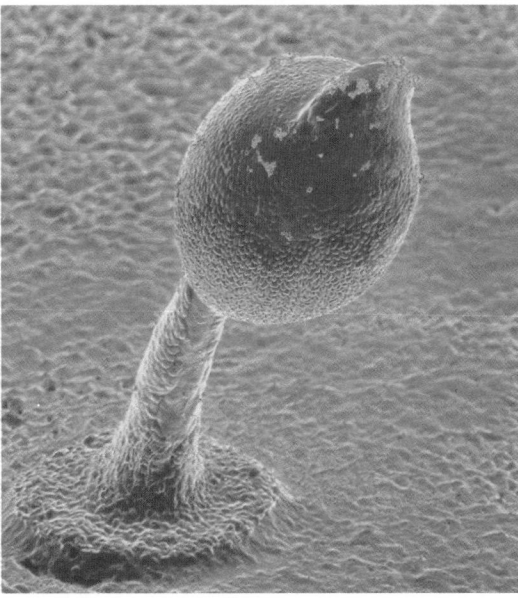

A colony of slime mold

ORGANIZER, continued

Exploration 3: enough leaves, grass
clippings, and sawdust to make a
15–20 cm layer in a bucket; small bag
of soil; small bag of pasteurized or
composted manure; several handfuls of
kitchen waste, such as potato peelings,
apple cores, banana peels, lettuce
leaves, coffee grounds, and tea bags; a
few small, household-waste items
made from plastic, metal, wood, or
cloth; large plastic bucket with air
holes; metric ruler; large plastic
garbage bag; alcohol thermometer;
several small flags (labels attached to
pieces of wire); microscope; several

microscope slides with coverslips; at
least 10 L of water; safety goggles; lab
aprons; latex gloves (additional teacher
materials: electirc drill or hammer and
nail; see Advance Preparation on
page 127C.)
Exploration 4: index cards; markers

Teaching Resources
Exploration Worksheets, pp. 20, 22,
  and 25
Theme Worksheet, p. 21
SourceBook, p. S71

EXPLORATION 1

An Up-Close View

You Will Need

• a magnifying glass (10×)

What to Do

1. Examine your hands with a magnifying glass. Note the different areas: the back, the palm, between the fingers, around and under the nails, and the knuckles.

2. Describe your hand as if you were a microorganism living on it. Include the following:

   a. favorable areas where living would be easy

   b. areas where you would have difficulties

   c. how easy it would be to get from one area to another

   d. probable dangers

   e. what might happen to you if the hand touched someone's face or a dusty table

   f. your food supply

How can microorganisms be so near and so numerous without our being aware of them? The fact that they go unnoticed suggests that not all microorganisms pose a threat to us. Many microorganisms are actually helpful to us.

You have learned that microorganisms are easy to find in water. Are they also common in soil? If so, what types are you likely to find? The next Exploration will help you answer these questions.

EXPLORATION 2

Microorganisms in Soil

You Will Need

• a sample of soil from a garden or from a flowerpot that contains living plants
• a 500–600 mL glass container with cover
• distilled water
• a microscope
• microscope slides and coverslips
• an eyedropper

What to Do

1. Mix about 50 mL of soil with about 300 mL of distilled water in the glass container.

2. Leave the container loosely covered at room temperature for 5 or 6 days in a spot where there is dim light.

3. Use the microscope to examine the mixture for microorganisms. Carefully adjust the light level.

4. Look for moving organisms. You may also see objects that do not move but have a definite shape or form a pattern in the field of view.

5. Draw examples of what you see. Remember to include the magnification used. Estimate the sizes of the organisms.

6. Write a description of each organism that you saw.

7. Clean up your area and wash your hands with soap and water when you are finished.

151

EXPLORATION 1

Challenge students to examine their hands from the point of view of a microorganism that might live there. Encourage them to be creative as they write their descriptions. Suggest that they include drawings or poems. Some students may want to write in a serious tone, while others may prefer a more humorous style.

If students are having difficulty getting started, suggest that they reread the descriptions on the two previous pages.

ENVIRONMENTAL FOCUS

Ask: What would Earth be like without microorganisms? *(Microorganisms are at the bottom of many food chains, so many food chains would break down. Microorganisms are also responsible for the decomposition of dead organisms, and plankton performs much of the photosynthesis that takes place on Earth. Students should understand that balanced ecosystems include producers, consumers, and decomposers, and that decomposers are essential for the continuation of life on Earth.)*

 An Exploration Worksheet (Teaching Resources, page 20) is available to accompany Exploration 2.

EXPLORATION 2

Distilled water should be used to prevent the introduction of additional microorganisms into the soil. If distilled water is unavailable, tap water that has been boiled in advance may also be used.

SAFETY TIP For additional safety, you may wish to collect the soil samples yourself.

GUIDED PRACTICE You may want to prepare students for some of the evidence of microorganisms that they may observe in the soil samples. For example,

they may see strands of hyphae from fungi, the cell walls of diatoms, and spores of various kinds. Point out that they may observe things like pollen grains and insect parts that are not microorganisms but rather microscopic parts of larger organisms.

Since it is very likely that students will observe parts of fungi, be prepared to talk about fungi. Draw on students' previous knowledge of molds, toadstools, and mushrooms. Point out to students that they will learn about the fungi kingdom later in the lesson.

As students examine the samples of

soil, you may wish to monitor them and have individuals describe their observations. Encourage students to look through each other's microscopes, especially if someone observes something particularly interesting or unusual. When students finish the Exploration, encourage them to share and discuss their descriptions and observations.

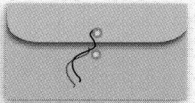 **PORTFOLIO**
Suggest that students include their descriptions and drawings from Explorations 1 and 2 in their Portfolio.

Good Guys or Bad Guys?

Have students read this paragraph silently or call on a volunteer to read it aloud. Encourage students to brainstorm answers to the questions. Accept all reasonable responses. You may wish to write some of the ideas on the chalkboard for reference as students proceed through the lesson.

A Recipe for Soil

Call on a volunteer to read each of the letters aloud. Be sure that students understand the questions before they begin writing letters of reply for Rhea's column. After they finish, have several students read their letters to the class. Encourage discussion of the responses, and help resolve any differences.

Answers to
A Recipe for Soil

Although students' letters will differ in style and organization, the following examples may serve as models:

1. Dear Inquirer,
   Good garden soil is not just small pieces of rock. It also contains the decaying remains of animals and plants, which help to keep the soil moist and full of nutrients. This decaying matter also keeps the soil particles loose by creating air spaces.
   Good soil also contains many living things. Some of them, such as worms and insects, are visible to the naked eye. Others, called microorganisms, are visible only with the aid of a microscope. These microorganisms "swim" in the thin film of water that surrounds the soil particles.

2. Dear Anxious,
   Some microorganisms that live in the soil can be harmful to people. The kind of bacteria that causes tetanus, or lockjaw, is an example. Its spores can live in the soil for years. When the conditions are just right, such as when the bacteria enter a person's body, they come alive. Rusty nails are one place where the tetanus bacteria may live. Gardeners and others who work outdoors should be inoculated to protect themselves from the disease. To avoid ingesting soil particles and the harmful microorganisms they may contain, it is important to wash all vegetables well before eating them.

Good Guys or Bad Guys?

People sometimes buy sterilized potting soil for their house-plants. In this soil, all living things have been killed. Does this mean that all soil microorganisms are undesirable? How do farmers deal with this problem? Do we benefit from microorganisms, or should we fear them all? You will find answers to these questions as you proceed.

A Recipe for Soil

Rhea writes a column for a gardening magazine and needs your help. She has had time only to make rough notes for her column, "Answers to Readers' Questions." She asks you to write her column from her rough notes. Below you will find the questions from readers and Rhea's notes.

> **1.** **Dear Rhea,**
>   **I enjoy reading your column and have used a number of your ideas. In one of your replies, you advised using good garden soil. When you say "good" garden soil, what do you mean?**
>   **An Inquirer**
>
> Notes: Soil—not just small rocks—includes lots of living things: worms and insects (can see), microorganisms (need microscope to see), plant and animal remains. Moisture very important—soil organisms need water (especially micros, they "swim" in it); also need air spaces for roots, bugs, worms.

> **2.** **Dear Rhea,**
>   **You recently described how to pot plants in sterilized potting soil. I know that sterilization kills microorganisms. Are there soil microorganisms that can harm me?**
>   **Anxious**
>
> Notes: Some can!—wash vegetables well—DON'T eat soil particles—can have harmful amoebas (dysentery) and viruses (hepatitis) in the soil. Also can have tetanus bacteria (lockjaw)—see photo—notice round ends and spores—these can live years—become active in right conditions—rusty nails can carry tetanus too—gardeners (also others) need tetanus shots to provide protection (immunity) from disease.

In 1 mL (a pinch) of soil there may be as many as 2 billion microorganisms.

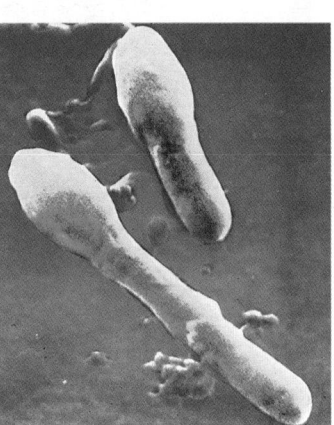
Tetanus is caused by these rod-shaped bacteria.

Multicultural Extension

Soil Usage and Treatment

In most cases, agriculture involves some treatment of the soil. This treatment may be as simple as plowing or as complicated as the manufacture and application of chemical fertilizers. Have students research a method of agriculture, such as swidden (slash-and-burn) or organic agriculture, and investigate the changes that the soil goes through and how this type of agriculture affects the lifestyle of the people who use it. Have students share their results with the class.

Another Microorganism

Rhea has the following suggestions for keeping garden flowers free of molds and mildew. She asks you to be her editor and to cut what she has written to two or three sentences. Pick out the main ideas, and rewrite the information in your ScienceLog. Use your own words, and make sure you include what is important.

Good gardening methods help to prevent problems with molds and other fungi on the leaves of flowering plants (such as phlox). These microorganisms have less of a chance to grow if the plants are kept neat and trimmed. Cut off leaves if the plants are very thick or show signs of mildew (gray, powdery spots). Let the air move freely around the plants, and see that the area is not too damp. Be sure to pick up any dead leaves so that they do not become a place where fungi can grow.

Microorganisms to the Rescue

Rhea is writing an article titled "Microorganisms—Make Use of Them." She asks you to provide suitable short titles for the following paragraphs. Note what she says about the benefits of microorganisms. Do not write in this book.

_____
(paragraph title)

Take advantage of the help that soil microorganisms can give. They help break down dead plant and animal matter and return substances to the soil for plants to reuse. Also, important gases (like carbon dioxide, which is needed by plants) are released into the air. Dead leaves and other plant materials can be used to produce a natural kind of fertilizer with the help of microorganisms.

_____
(paragraph title)

Many gardeners put unwanted plant materials (such as vegetable peelings) in a special area to be broken down over a period of time by the action of microorganisms. They may add packages of bacterial culture to help the decaying action. This pile of decomposing matter is called *compost* and is a natural kind of fertilizer. Heat is produced as the material decays. The temperature inside a compost heap may reach 75°C.

Try making your own compost! Use the Exploration that follows as a guide.

FUNGI
THE FIFTH KINGDOM

Fungi (singular, *fungus*) include mushrooms, molds, and puffballs. Fungi are different enough from other living things to be classified in a separate kingdom. They play an important role as decomposers of dead material. Fungi are especially abundant in soil but are found almost everywhere else as well—and it's a good thing too. Without fungi, the world might soon become crowded with dead matter.

Decision Please!

Prepare the pro and con arguments for this debate: **Resolved: Microorganisms are bad guys.** Add to these arguments as you progress through the unit.

Theme Connection

Changes Over Time
Focus question: How do microorganisms contribute to the formation of fossil fuels such as coal and petroleum? *(Microorganisms partially decompose plant and animal remains. Then, over time, sediments build up on top of the remains, and chemical changes occur. Those chemical changes may result in the formation of coal and petroleum.)* A Theme Worksheet (Teaching Resources, page 21) is available to accompany this Theme Connection and makes an excellent homework activity.

Homework

Have students choose a member of the kingdom Fungi to investigate. Encourage students to concentrate on the life cycle of the fungus that they chose. Sketches of the various stages in the life cycle of a fungus may help students describe the growth and reproduction of these organisms.

Another Microorganism

In this activity, students are asked to read a passage and identify two or three main ideas. When students finish writing, have them share their ideas with each other. Be prepared to help resolve any differences. Encourage students to write their responses in their own words.

Answers to *Another Microorganism*

Sample answer: To help prevent molds and fungi from growing on garden flowers, cut some leaves from very thick plants and any leaves that show signs of mildew (gray, powdery spots). Allow air to circulate freely around the plants, and keep the area from getting too moist. Pick up any dead leaves so that they won't provide a place for fungi to grow.

Microorganisms to the Rescue

By writing short titles for each of the paragraphs, students are given the opportunity to identify the main idea in each one. When students complete the activity, call on volunteers to share their titles with the class.

You may wish to direct students' attention to the information about fungi in the right-hand column of page 153. Point out that the molds and mildews discussed on this page, as well as many of the organisms that help plants to decay and to eventually become compost, belong to this kingdom.

Answers to *Microorganisms to the Rescue*

Sample answers:
First paragraph: Microorganisms Enrich the Soil; Microorganisms, Fertilizer, and the Air; and Take Advantage of Microorganisms
Second Paragraph: From Plant to Compost; Making Compost; and Microorganisms Turn Plants Into Fertilizer

Note that many packaged soils are steril-ized and so will not contain live bacteria to aid in decomposition. Soil from a gar-den or schoolyard is likely to work bet-ter for this Exploration. A large stick or paint paddle may be used to stir the compost mixture.

 Make sure that any manure used for this Exploration is com-posted. Do not allow students to add pet manure, cooked foods, or meat of any kind to the compost bucket.

An Exploration Worksheet (Teaching Resources, page 22) is available to accompany Exploration 3.

Making Compost

You Will Need

- several bags of leaves, grass clippings, and sawdust
- 1 bag of soil
- 1 small bag of composted manure
- kitchen waste (for example, potato peelings, apple cores, banana peels, lettuce leaves, coffee grounds, tea bags)
- other types of household waste, including items made from plas-tic, metal, wood, or cloth
- a large plastic bucket with holes
- a metric ruler
- a plastic garbage bag
- an alcohol thermometer
- flags (labels attached to pieces of wire)
- a microscope, slides, and coverslips
- water
- latex gloves

What to Do

Caution: Do not add any hazardous materials such as sharp objects or harsh chemicals to your compost.

1. Your teacher will supply you with a bucket that has plenty of air holes cut into it at vari-ous heights. Mix enough leaves, grass, and sawdust to make a 15–20 cm layer at the bottom of the bucket. The amount of ingredients that you will need depends on the size of the bucket.

2. Add a 5–10 cm layer of composted manure. It may be purchased from a gar-dening store. If composted manure cannot be obtained,

special starter pills for com-posting can be purchased from gardening stores.

3. Add a 5–10 cm layer of soil.

4. Repeat the layers.

5. Water thoroughly, and let stand for 1 week.

6. After 1 week, stir your com-post mixture well, and add kitchen waste. Do not use pet manure, cooked foods, or meat of any kind. Bury the kitchen waste in the top layer of the bucket. As these food scraps decompose, stir the mixture, and add more.

To keep your mixture moist at all times, cover the bucket with a plastic bag.

7. In addition to kitchen waste, add a few small items made of plastic, metal, wood, or cloth. (Normally, you would not add these to compost.) Mark these manufactured ingredients with small flags. Over the next four to six weeks, record any changes that you observe in these items. Examine these mate-rials carefully. Record how long it takes for different materials to change.

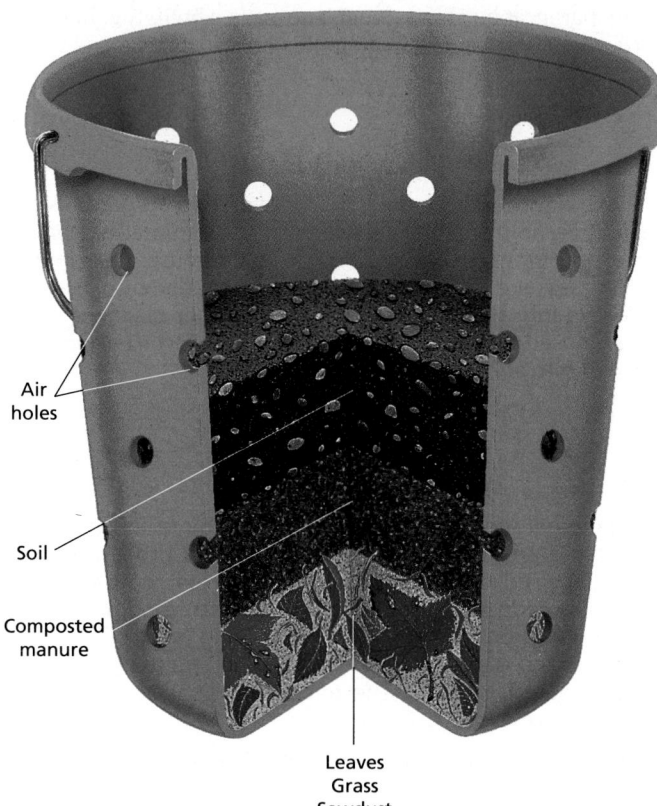

Air holes

Soil

Composted manure

Leaves Grass Sawdust

8. Keep a running list of observations in your ScienceLog as you experiment with your compost over the next several weeks. Twice a week, carefully insert a thermometer through the air holes, and record the temperature at different depths.

9. You may wish to make and observe wet mounts of materials from different levels of the compost. Record the type and number of different organisms that you see. Do you see a relationship between the number of organisms present and the temperature? Explain your observations. **(A)**

10. Wash your hands with soap and water before leaving the lab.

Questions

1. Why is it necessary to cut air holes in the bucket? What would happen if this step were omitted?

2. Why is it important to keep the compost mixture moist?

3. Why is it a good idea to add soil to a new compost heap?

4. How do you explain the temperatures that you recorded for various depths in the bucket?

5. Explain why certain materials decomposed more rapidly than others. Did the rates of decomposition match your expectations?

Saltwater Neighbors

If you were drawing a picture of the ocean, what color would you make the water? Would you make it brick red? If you were sailing at night, would you expect the sea around you to glow? What might cause these things to happen? You may be surprised to learn that microorganisms can cause both of these unusual effects.

One cupful of sea water may contain millions of microscopic plants, animals, and protists. These organisms drift with the water currents and are called **plankton**, from a Greek word meaning "wanderer." Life throughout the oceans is linked to the plankton near the surface. It takes trillions of tiny plankton to provide energy for the oceans' organisms—from the smallest seahorse to the largest blue whale. The next Exploration will let you be a part of the action as you discover the links between marine organisms.

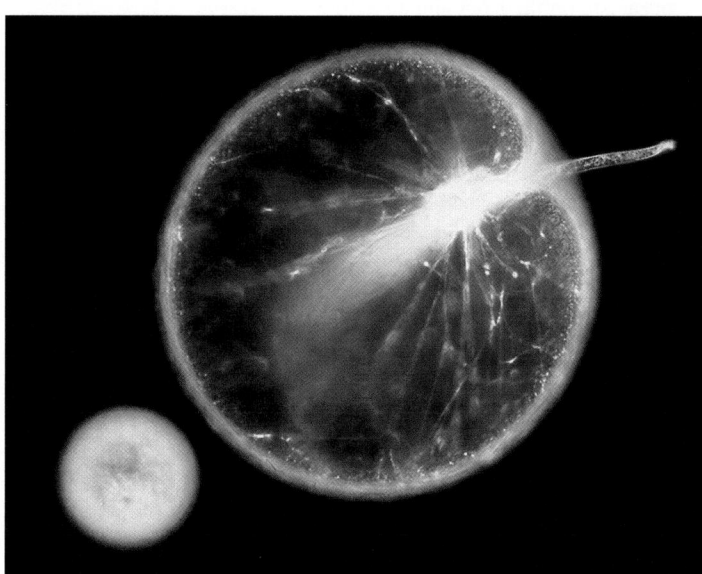

When disturbed, the bioluminescent microorganism *Nocticula* (at right) emits a tiny flash of light. Look up the word *bioluminescence* to find its meaning. **(B)**

Saltwater Neighbors

Students may find it surprising that blue whales subsist primarily on food that is so much smaller than they are. Point out that large, toothless whales have plates of *baleen* attached to their palates that serve to filter out the plankton from the water. This allows them to ingest the huge quantities of plankton necessary to sustain themselves.

Answers to
In-Text Question and Caption

(A) Students will probably notice different types of organisms living in regions of the compost with different temperatures. Students should observe that the warmer the compost is, the more organisms are present.

(B) *Bio* means "life" and *lumin* means "light," so *bioluminescence* is the production of light by living organisms. Fireflies and some deep-sea animals are examples of bioluminescent organisms.

Answers to
Questions

1. Air holes in the bucket allow the entry of oxygen, which is required by the microorganisms, fungi, and small animals that live in the soil. If this step is omitted, the plant material and kitchen waste in the bucket will decay much more slowly.

2. The organisms that break down waste require moisture in addition to oxygen to thrive.

3. Adding soil to a new compost heap ensures that a variety of microorganisms are present to begin the process of decay.

4. Heat is given off during the chemical reactions that take place in a compost heap. The highest temperatures occur in the center of the bucket because the surrounding material acts as insulation and keeps the heat from escaping. As the internal temperature of the compost heap increases, the microorganisms break down waste and multiply at a greater rate.

5. Answers will vary depending on the materials used. In general, organic materials will decompose more quickly than inorganic materials.

Make sure students understand that to make their cards, they will have to read the right side of the table first to find out what kinds of things their marine organism likes to eat. Remind students that only one person per group should act as "diatoms and other plant plankton."

 ## Cooperative Learning
EXPLORATION 4

Group size: 5 to 8 students

Group goal: to study the interactions within oceanic food chains

Positive interdependence: Each student group will represent a group of organisms that depends on another group of organisms for food.

Individual accountability: Randomly select one of the questions in Thinking It Over on page 157 to ask each student.

 An Exploration Worksheet (Teaching Resources, page 25) is available to accompany Exploration 4.

Homework

Challenge students to design their own game to illustrate the role of microorganisms in terrestrial food chains.

Ocean Wanderers

You Will Need

- index cards
- markers

What to Do

1. The table at right includes the names of some of the living things found in the oceans. There are also the names of some other organisms that depend on them for food. Use the information in the table to prepare a set of cards. Each card should have the name of an organism on the front and all of the foods that the organism eats on the back. There may be more than one card for each type of organism so that there are enough cards to go around. In addition to the organisms in the table, one person should represent diatoms and other plant plankton. Diatoms and other plant plankton make their own food from sunlight. The front and back of a sample card are shown below.

2. Select a card. You now represent that organism. Check the back of your card for a list of foods that you like to eat. After you have read your card, hold it so that the name of the organism you represent is clearly visible to others.

Marine organism	Eaten by
Diatoms and other plant plankton	A. Plant-eating plankton, such as immature crabs, barnacles, sea worms, and small fish B. Baleen whales, such as blue whales, right whales, and humpback whales C. Mature fish such as herrings, sardines, and mackerel
Plant-eating plankton, such as immature crabs, barnacles, sea worms, and small fish	A. Animal-eating plankton, such as arrow worms and comb jellies B. Baleen whales, such as blue whales, right whales, and humpback whales C. Mature fish, such as herrings, sardines, and mackerel
Animal-eating plankton, such as arrow worms and comb jellies	A. Baleen whales, such as blue whales, right whales, and humpback whales B. Mature fish, such as herrings, sardines, and mackerel
Mature fish, such as herrings, sardines, and mackerel	A. Tuna B. Sharks C. Squid D. People

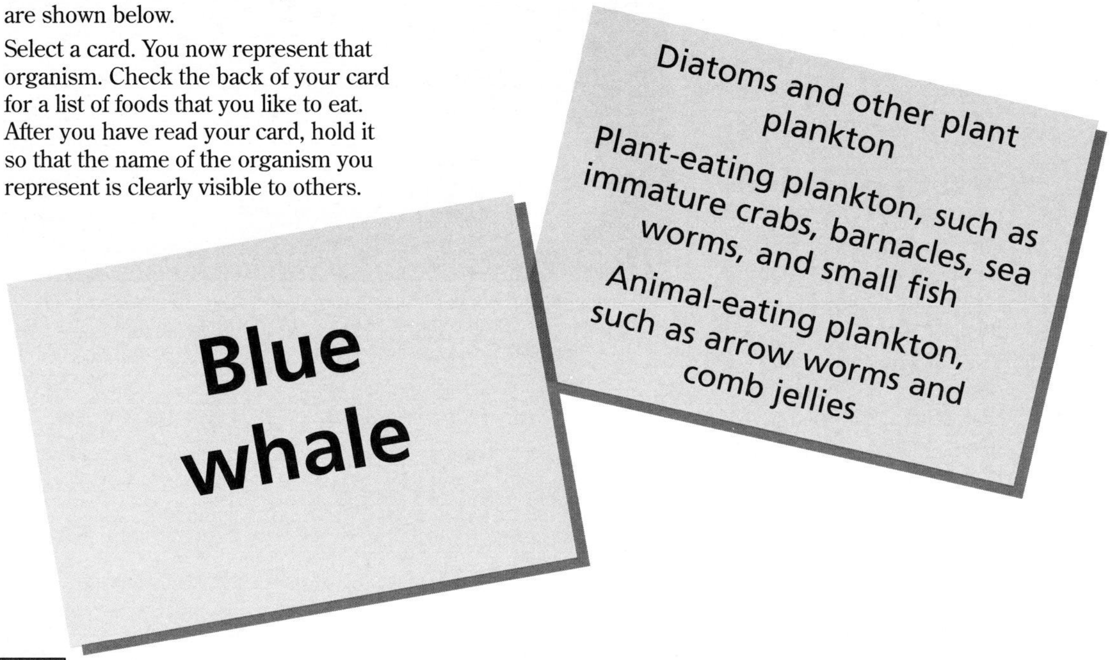

Diatoms and other plant plankton

Plant-eating plankton, such as immature crabs, barnacles, sea worms, and small fish

Animal-eating plankton, such as arrow worms and comb jellies

Blue whale

3. To catch your prey, find a person who represents one of the food items on your list. Hold your card in your right hand, and stretch out your left arm to touch the shoulder of that person.

4. By the time everyone has "eaten," the whole group should be connected.

5. Have two people volunteer to draw a diagram showing the names and positions of each marine organism. The volunteers do not have to take cards, but they should make sure that each organism has eaten an appropriate meal.

6. Copy the diagram into your ScienceLog as a record of the Exploration.

Thinking It Over

Use the diagram in your ScienceLog to answer the following questions:

1. What kind of living thing was first in the diagram?

2. Which organism(s) used the first organism in the diagram as food?

3. Which organism(s) used the second organism in the diagram as food?

4. If the first organism in the diagram were removed from the ocean, what effect would

this have on the other organisms in the diagram? Make a prediction.

5. Write a sentence to show your understanding of the role that plankton play in the ocean. Be sure to include the following words in your sentence: plankton, animal, plant, link, food, depend, and ocean.

Further Thoughts

1. How do the plant plankton get their food? (Hint: it's the same source of food that plants on land use.)

2. What could possibly happen to threaten the animal and plant plankton's existence?

Answers to *Thinking It Over*

1. Diatoms and other plant plankton

2. Plant-eating plankton, baleen whales, and mature fish

3. Animal-eating plankton, baleen whales, and mature fish

4. None of the other organisms in the diagram would be able to exist without plankton.

5. Accept all reasonable answers. Sample answer: Tiny *plants* and *animals* that live in the *ocean* are called *plankton* and are *linked* to many other organisms that *depend* on them for *food*.

Answers to *Further Thoughts*

1. Plant plankton get their food through photosynthesis.

2. Sample answer: The temperature or salinity of the water could change so much that the plankton are unable to survive. Also, a new predator could appear in the ecosystem.

FOLLOW-UP

Reteaching

Have students make posters to show how microorganisms can be helpful and harmful. Display the posters around the classroom or in the school library.

Assessment

Have students write an imaginary interview with microorganisms. What kind of environment would they like? Encourage students to be creative. When they finish, have volunteers share their interview with the class.

Extension

Have students design an experiment to determine whether microorganisms will decompose paper. Make the following suggestion to get students started: Place small pieces of paper in a dilute solution of plant fertilizer dissolved in water. (The fertilizer will provide nitrogen for the bacteria.) Then add a little unsterilized soil. Be sure to have students set up a control. Have them observe and record what happens over a period of 1–2 weeks and then report their findings. Remember to check each student's proposed procedure for safety.

Closure

As an extension of Exploration 2, students can take a supervised field trip outdoors to examine soil from areas such as the following: under a tree or shrub where many leaves have fallen and rotted; in an area where the soil is hard and dry; along a roadside; in a field of grain; and in a damp area such as near a pond or drainage ditch. Remind students not to go near a pond without adult supervision and to wash their hands with soap and water when they return inside. Have students record the differences that they find, especially in the types and abundance of each microorganism. Challenge students to make inferences about the differences they observe.

LESSON

2
A Balanced View

Louis Pasteur

By now you've probably learned a great deal about microorganisms. You know something about their beneficial and harmful effects in the soil. You have learned that some microorganisms are needed. You have also discovered ways to avoid the possible harmful effects they may have. Methods of dealing with harmful microorganisms and of using helpful microorganisms became possible through scientific discoveries. French scientist Louis Pasteur (1822–1895) was one pioneer whose discoveries advanced our knowledge of microorganisms. His work laid the foundation for dealing wisely with microorganisms, especially those found in food.

FOCUS

Getting Started

Ask students to think of milk, cheese, and yogurt. Challenge students to explain how these foods are related. *(Yogurt and cheese are made from milk by the action of microorganisms.)*

Main Ideas

1. Scientists approach problem solving in an organized and systematic manner.
2. Our understanding of microorganisms can be used to control them for our use.
3. Harmful microorganisms can be controlled without harming beneficial microorganisms.

A System for Problem Solving

Louis Pasteur was a chemist and a problem solver. People in the silk and wine industries often called on him for help in solving problems. Pasteur himself was interested in the problem of food spoilage. He came to believe that food spoilage might be the work of microorganisms.

Louis Pasteur was a good problem solver. Study the diagram below, which suggests how he sometimes approached problems. Have you ever tackled problems in a similar way?

TEACHING STRATEGIES

Ask students if they have ever heard of Louis Pasteur. Call on students to share their knowledge with the class. Some students may know that Pasteur is responsible for the process of pasteurization. Some may also be aware of his work in developing a rabies vaccine.

A System for Problem Solving

Have students read the paragraph silently. When they finish, direct their attention to the diagram. Call on volunteers to read each of the steps aloud. Pause after each step has been read to provide students with an opportunity to discuss it. Be sure that they understand each step before going on to the next. When the diagram has been reviewed to your satisfaction, encourage students to discuss the way they solve problems. Help them to decide if their methods are similar to Pasteur's.

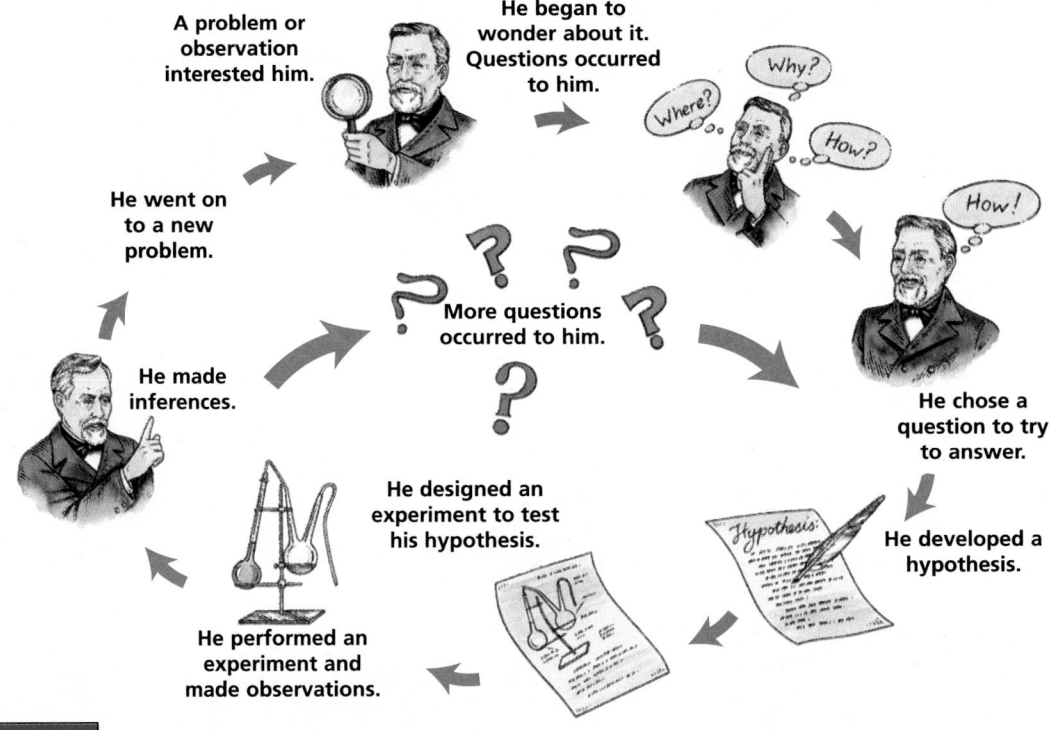

A problem or observation interested him.

He began to wonder about it. Questions occurred to him.

Why? Where? How?

How!

He went on to a new problem.

More questions occurred to him.

He made inferences.

He chose a question to try to answer.

He designed an experiment to test his hypothesis.

Hypothesis

He developed a hypothesis.

He performed an experiment and made observations.

 Transparency 24 is available to accompany Lesson 2.

LESSON 2 ORGANIZER

Time Required
3 to 4 class periods

Process Skills
analyzing, inferring, comparing

New Term
Pasteurization—a method of gently heating a substance to kill unwanted microorganisms while sparing the beneficial ones

Materials (per student group)
none

Teaching Resources
Resource Worksheet, p. 28
Transparency 24

Investigations

Four of Louis Pasteur's investigations are outlined below. Read them, and use the diagram on the previous page to identify which part of the problem-solving process he used at each step.

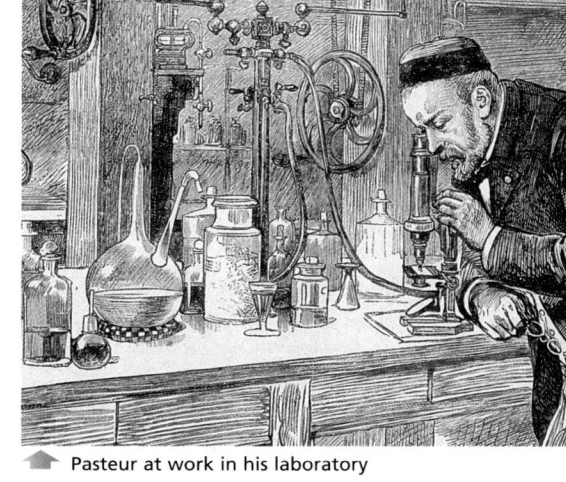
▲ Pasteur at work in his laboratory

1. a. Pasteur found bacteria in the slime at the bottom of a jug of sour milk.
   b. He put a drop of the slime into a sugar solution.
   c. The solution turned sour, and bacteria grew in it.

2. a. Pasteur boiled sugar solutions and then tightly covered them.
   b. These solutions did not spoil.

3. a. Pasteur took covered flasks of boiled broth and opened them in four places: Paris, the open country, a tall hill, and a glacier.
   b. He obtained the following results:

Number of flasks of broth	Place	Number of flasks in which the broth spoiled
20	Paris (dusty, crowded yard)	20
20	the open country	8
20	high on a hillside	5
20	a mountain glacier	1

4. a. Pasteur found large numbers of microorganisms in wine that had developed an unpleasant flavor.
   b. He gently heated additional wine to between 55°C and 60°C.
   c. The microorganisms were not present in the wine that had been heated, and the wine did not taste unpleasant.

Conclusions

Four of Pasteur's discoveries are listed below. Read them. Look again at the investigations outlined above. Choose the investigation that most likely helped him make each discovery.

1. Bacteria can enter food from the air.
2. Gentle heating may kill unwanted bacteria in food but may not change the flavor of food.
3. Bacteria in food come from somewhere else and can be killed by heating.
4. Bacteria can cause food to spoil.

159

Integrating the Sciences

Life, Earth, and Physical Sciences
Remind students that microorganisms are responsible for the decay of animal and plant matter. Ask: What do you think life would be like if nothing ever decayed? You may wish to divide the class into small groups and have each group develop a story to illustrate their ideas. Then have a volunteer from each group share the group's story with the rest of the class. *(Without biological decay, plant and animal remains would continue to accumulate as living organisms died. Eventually the entire planet would be covered with a layer of plant and animal remains, which would interfere with the natural life processes of the entire planet.)*

Pasteurization

GUIDED PRACTICE Have students read page 160. Then ask: Why do you think the method of gently heating a substance to kill unwanted microorganisms is called pasteurization? *(It is named after Louis Pasteur, the scientist who discovered the process.)* How does pasteurization differ from sterilization? *(Sterilization kills all living things in a substance. Pasteurization kills only harmful organisms while allowing beneficial organisms to survive.)*

Answer to
In-Text Question

Ⓐ Refrigerators and freezers help to keep dairy products safe until we consume them. At Pete's Place, food that should have been refrigerated was left sitting out, the refrigerator did not work properly, and the freezer door was broken.

Answer to
Now You Know

A simple answer might be the following: Germs get into the milk and cause it to turn sour. A more scientific answer might be the following: Microorganisms called bacteria get into the milk from the air or through contamination of the milk container. When the conditions are right, they begin to digest the milk and multiply, causing the milk to go sour.

Pasteurization

Pasteur's method of gently heating a substance to kill unwanted microorganisms while sparing beneficial ones has been given the name **pasteurization**. Its use in treating milk has protected generations of people from serious diseases. The treatment kills most of the harmful bacteria that may be in milk but does not damage those that give milk its desired flavor.

We depend on technology to pasteurize dairy products. We also depend on technology to keep the pasteurized dairy products safe until we consume them. You should be able to name some inventions used for that purpose. At Pete's Place, how had this technology broken down? Ⓐ

Now You Know

If you were asked why milk goes sour, how would you answer the question in a simple way? in a more scientific way?

Problems to Ponder

Get together with a friend to answer these questions. You may have to do some library research to find the answers.

1. Fresh milk from healthy cows is free of infectious microorganisms. How can it become contaminated with microorganisms that cause spoilage or disease?

2. After pasteurization, how must milk be stored in order to keep it safe to drink?

3. In the last lesson, you learned about sterilization of soil. Soil can also be pasteurized. How might you pasteurize soil?

4. Why might someone choose to use pasteurized soil rather than untreated soil or sterilized soil?

5. Ultraviolet radiation is used to pasteurize milk. What is ultraviolet radiation? How does it do the job?

Lait de la Vie Eternelle*

The nomadic Mongols who moved across Asia and Europe in the thirteenth century ate it. Galen, a doctor who lived during the second century A.D., recommended it highly to his patients. In India, where it was introduced about 2500 years ago, it was considered food fit for the gods. In 1902, Elie Metchnikoff, a microbiologist who later won a Nobel Prize, declared that people could hope for longer lives if they ate it. In France, it has been called *lait de la vie eternelle*.

What is this wonder food? The answer is *yogurt*. Even today, it is considered healthful and is very popular. Have you ever eaten frozen yogurt? Some people prefer it to ice cream. Yogurt is digested in about 1 hour. Regular cow's milk takes about 3 hours to digest. Yogurt is recommended for people who cannot digest milk.

Yogurt is made from milk by the action of certain microorganisms. This process is known as *fermentation*. Yogurt actually contains helpful bacteria. These bacteria are helpful because they either destroy or weaken almost all of the harmful bacteria that are likely to come into contact with them.

When you make yogurt, you put microorganisms to work for your benefit. The next activity will show you how.

These Mongolian nomads use fermented horse's milk as a staple in their diet.

*Milk of Eternal Life

161

Homework

The questions in Problems to Ponder make an excellent homework activity.

Answers to Problems to Ponder

1. As soon as it leaves the cow, milk may be contaminated by microorganisms from many sources, including the cow itself, the milking equipment, the air, the people handling the milk, and the containers in which the milk is stored.

2. Milk needs to be covered and stored at a cool temperature.

3. Soil can be pasteurized in two ways. A compost pile with good bacterial action will produce enough heat to kill most harmful organisms, but not enough heat to kill the beneficial ones. Also, soil may be heated to 75°C in an oven.

4. Pasteurized soil wouldn't contain the harmful organisms of untreated soil, but it would contain the helpful bacteria that sterilization would destroy.

5. Ultraviolet radiation is an invisible form of light energy. The sun is a major source of ultraviolet radiation. The energy of ultraviolet radiation can kill living cells, so it can be used to kill microorganisms.

Lait de la Vie Eternelle

Have students read this introduction to the activity that follows. When they finish, ask them to respond by a show of hands if they have ever eaten yogurt. Call on several volunteers to describe how yogurt tastes and whether they like it.

Multicultural Extension

Use of Microorganisms

Encourage students from other countries and cultures to share with the class how their culture uses microorganisms to make certain kinds of foods. For example: What kinds of bakery products that require yeast are common in their culture? What cheeses are made in their native country? How is yogurt used in cooking? Encourage students to share appropriate recipes with the class or to organize them into a booklet for class members to read at their leisure.

Making Yogurt at Home

 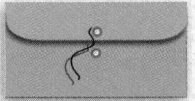
INDEPENDENT PRACTICE As students follow and analyze the directions, challenge them to observe and make notes of any evidence of the changes made by bacteria.

PORTFOLIO

Suggest that students include their comments about Making Yogurt at Home in their Portfolio.

Answers to *Points to Ponder*

1. Washing your hands, washing and scalding utensils, and heating the milk to 82°C kill harmful bacteria.

   Cooling the milk to 48°C, adding the yogurt culture, and incubating the mixture at 40–46°C help the yogurt-making bacteria thrive.

   Removing the yogurt from the thermos and storing it in the refrigerator preserve its taste and halt bacterial growth.

   Making another batch of yogurt from the first batch demonstrates that the bacteria only need the right conditions in order to begin reproducing again.

2. The bacteria in the yogurt will remain alive indefinitely if they are not killed. They can, therefore, be used to start new batches.

3. Answers will vary depending on individual results, but homemade yogurt can be just as creamy and thick as commercial yogurt, and it costs much less, especially if the first batch is used to make others.

4. **a.** Additives may include corn syrup, fruit pectin, sorbic acid, potassium sorbate, flavoring, coloring, starch, and citric acid.
   b. Many commercial yogurts are pasteurized. Because some bacteria may be destroyed, commercial yogurts may not be as good to use as "starters."

Making Yogurt at Home

You Will Need

- 1 L of milk (whole, 2%, or skim)
- 1 envelope of yogurt culture
- a heat source
- a pan
- a container with cover
- an alcohol thermometer
- a thermos (1 L size, with a wide mouth if possible)
- a metric measuring cup

Caution: Do not use any laboratory equipment to perform this activity. Obtain permission from a parent or guardian before doing this activity.

What to Do

1. Wash your hands well. Wash all utensils, and rinse them with boiling water.
2. In a pan, heat the milk to 82°C.
3. Let the milk cool to 48°C.
4. Add the yogurt culture, and mix well.
5. Pour the mixture into the thermos. The mixture should be between 40°C and 46°C for the incubation period. (If you are not familiar with the word *incubation,* look it up in a dictionary.)
6. Keep the mixture in the thermos with the lid on for 4 hours. If the yogurt still seems thin, let it stand awhile longer in the thermos, and then check it again.
7. Remove the yogurt from the thermos, and store it in a clean, covered container in the refrigerator.
8. To make the next batch of yogurt, simply add 45 mL of the yogurt you just made to lukewarm milk, and incubate. You do not need to use another envelope of yogurt culture.

Note: To make firmer yogurt, mix 45 mL of skim milk powder and 5 mL of unflavored gelatin with a little water, and add this to the milk before heating.

Points to Ponder

Along with others who made yogurt, discuss and record answers for some of the following:

 A Resource Worksheet (Teaching Resources, page 28) is available to accompany the material on this page.

1. Suppose you are using this method to make yogurt at home. A nine-year-old neighbor is visiting and wants to know what you are doing. Explain why each step is necessary in terms of what is happening to microorganisms.

2. Why do you need to buy only one envelope of yogurt culture, even if you want to continue making batches of yogurt?

3. a. How much did it cost to make the yogurt?
   b. How much would it cost to buy the same amount of yogurt in the supermarket?
   c. How do the smoothness and taste compare with those of store-bought yogurt?

4. Look at the labels on yogurt containers from the store.
   a. Pure yogurt contains only milk and bacteria. List other ingredients (additives) in commercial yogurts.
   b. Have commercial yogurts been pasteurized? If so, what effect do you think this has on the healthful quality of the yogurt? Would it be possible to make a good-quality yogurt by using some of the pasteurized yogurt as a "starter"? Why or why not?

Making homemade yogurt is easy.

Say Cheese!

What kind of cheese do you like best? Name as many different kinds of cheese as you can. As you read the following informative article about cheese, look for answers to these questions:

1. How can cheese be made from pasteurized milk?
2. What health precautions must be taken when making cheese?
3. What causes some cheeses to be harder than others?
4. What kinds of microorganisms help to make cheese?
5. How do cheese makers keep out unwanted microorganisms?

A Really Cheesy Story

Natural bacteria present in milk can be used to make cheese, but today the milk is normally pasteurized first for safety. Then specially cultured bacteria are added to the milk. If nonpasteurized milk is used, some countries require the cheese to remain in storage for at least 2 months before it can be sold to the public. This protects the public from disease-causing organisms.

The hardness of cheeses varies. The longer the bacteria work, the harder the cheese is. Cheese is said to "ripen" as bacteria continue to work, and the flavor and the odor change. You may have heard of "processed" cheeses. Processed cheeses are made by blending together various natural cheeses to achieve a desired taste. The resulting blend is then pressed into shape.

Molds are also important in the manufacture of some cheeses, such as Roquefort, blue cheese, and Gorgonzola. Brie cheese is mold-ripened from the outside and must be eaten within a few days of ripening.

Dairies and cheese makers must keep out unwanted microorganisms while keeping the needed microorganisms healthy. For example, certain molds that are necessary when making a few special cheeses would spoil other cheeses. Control of these molds and their spores is very important. Extreme care and cleanliness are necessary. Dairies do not welcome visitors in their plants because the cultures are likely to become contaminated with undesirable microorganisms. The successful operation of dairies and yogurt- and cheese-making plants depends on maintaining a delicate balance between helpful and harmful microorganisms.

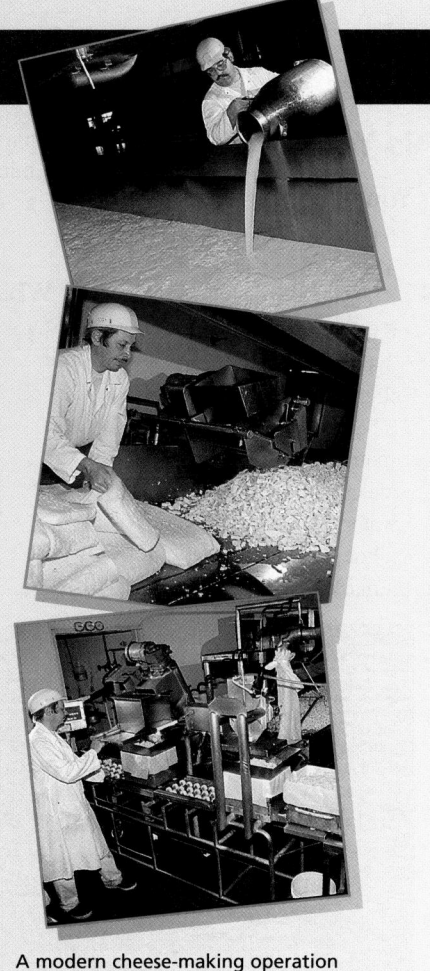

A modern cheese-making operation

163

Say Cheese!

Keep track of the kinds of cheeses that students are familiar with by listing the responses on the chalkboard. Then have students read the article. Encourage them to write the answers to the questions as they read. When they finish reading, check their understanding of the article by involving them in a discussion of their answers.

Answers to Say Cheese!

1. Cheese can be made from pasteurized milk by adding specifically cultured bacteria to the milk.
2. Pasteurize the milk first, or if unpasteurized milk is used, store the cheese for at least 2 months. Extreme cleanliness is also necessary.
3. The longer the bacteria work, the harder the cheese becomes.
4. Bacteria and molds help to make cheese.
5. Extreme care and cleanliness are necessary. Dairies do not allow visitors to wander through their plants.

Homework

The Extension activity on this page makes an excellent homework activity.

FOLLOW-UP

Reteaching

Have students create a problem-solving diagram for Making Yogurt at Home. Each diagram should show the steps required to complete the activity.

Assessment

Have students write a summary of the lesson. Remind them that a summary is a shortened version of what they have read. It should include the main ideas and the most important facts, presented in their own words. It should be short and to the point.

Extension

Have students do some research to discover how people kept food from spoiling before modern packaging and refrigeration techniques were developed. Have them use their information to make a mural or bulletin-board display.

Closure

Suggest that students visit a local grocery store and make a list of all the dairy products that are made with the help of microorganisms, including different kinds of yogurt, sour cream, and cheese. Have students share what they discover with the class. Also have students see how many products they can find that are pasteurized.

FOCUS

Getting Started

Do the following activity to demonstrate how yeast cells produce gas. Place one package of yeast in a bottle with some warm water and a little sugar. Place a balloon over the top of the bottle and tie it securely with a piece of string. Place the bottle in a warm location, and allow the class to observe what happens. Students should notice the balloon gradually filling with gas. Ask them to try to explain what is happening. *(The yeast uses the sugar for food. The yeast causes the sugar to change into alcohol, carbon dioxide gas, and energy. The gas in the balloon is carbon dioxide.)*

Main Ideas

1. Yeast is an important ingredient in some food products.
2. Yeasts and molds require specific growing conditions.
3. The growth of yeasts and molds can be controlled by controlling their environment.

TEACHING STRATEGIES

GUIDED PRACTICE   Call on a volunteer to read the bread recipe to the class. Involve students in a discussion of which ingredient might be alive. Encourage students to speculate about why yeast is used in bread.

Answer to
In-Text Question

Ⓐ We would not expect there to be any living microorganisms in the bread because they would have been killed during baking. Among the ingredients, however, yeast consists of living organisms.

Teaching Strategies for Exploration 5 are on the next page. ➤

LESSON

3

The Staff of Life

What could be a more basic meal than milk, cheese, and bread? We have learned about the connections between microorganisms and milk, cheese, and yogurt, but what microorganisms would you find in bread? Ⓐ

Actually, in baked bread we would not expect to find living microorganisms. Why not? Look at the recipe for Best White Bread. Which ingredients of this bread might be alive?

Best White Bread
1 package of dry yeast
60 mL of warm water
500 mL of milk, scalded
30 mL of sugar
10 mL of salt
15 mL of shortening
1.5 L of sifted all-purpose flour

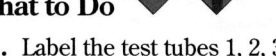

Exploration 5

Is Yeast Alive?

You Will Need
- 5 to 6 mL of dry yeast
- a graduated cylinder
- 6 test tubes or similar containers
- 6 mL of white sugar
- 3 alcohol thermometers
- thermometer holders
- 3 beakers
- metric measuring spoons
- ice cubes
- water
- a hot plate
- a marker or masking tape
- a microscope
- microscope slides and coverslips
- an eyedropper

What to Do
1. Label the test tubes 1, 2, 3, 4, 5, and 6.
2. Prepare three beakers as shown below. Add ice to beaker A to bring it to the correct temperature. Use the hot plate to bring beakers B and C to the correct temperature.
3. Divide the yeast into six parts (about 1 mL each). Put one part into each test tube.
4. To tube 1, add 5 mL of water. Put tube 1 into beaker A.
5. To tube 2, add 5 mL of water. Put tube 2 into beaker B.
6. To tube 3, add 5 mL of water. Put tube 3 into beaker C.
7. To tube 4, add 5 mL of water and 2 mL of sugar. Put tube 4 into beaker A.
8. To tube 5, add 5 mL of water and 2 mL of sugar. Put tube 5 into beaker B.
9. To tube 6, add 5 mL of water and 2 mL of sugar. Put tube 6 into beaker C.
10. The tubes must stay upright so that their contents won't spill into the baths. Let the tubes stand (do not shake them) in the water baths for at least 10 minutes. Observe.
11. Remove tubes from the baths, and observe them closely. Record all observations.

Caution: Use oven mitts when handling heated glassware. Turn off the hot plate when you are finished.

| Temperature: 0°C | Temperature: 30°C to 35°C | Temperature: 100°C |

LESSON 3 ORGANIZER

Time Required   3 class periods

Process Skills   observing, measuring, inferring, analyzing

New Terms
Hyphae (singular, *hypha*)—threadlike or stemlike parts of a mold
Spore cases—swellings or spheres at the ends of the hyphae

Materials (per student group)
Exploration 5: 5–6 mL of dry yeast; 100 mL graduated cylinder; 6 test tubes or similar containers; 6 mL of white sugar; set of metric measuring spoons; ice cubes; 3 alcohol thermometers; 3 thermometer holders; three 250 mL beakers; about 1 L of water; hot plate; marker or several small pieces of masking tape; microscope; several microscope slides with coverslips; eyedropper; watch or clock; safety goggles; oven mitts (See Advance Preparation on page 127C.)
Exploration 6: prepared wet-mount slide of mold; microscope (See Advance Preparation on page 127C.)

Teaching Resources
Exploration Worksheets, pp. 30 and 33
Activity Worksheet, p. 34
Transparency Worksheet, p. 35
Transparencies 25 and 26

12. Examine a drop from each test tube under the microscope. It may be necessary to dilute the drop with a little water of the same temperature to view the yeast best. Do you see what appear to be yeast cells?

13. Draw several yeast cells.

14. In a table like the one below, record any observations about the differences in the yeast cells from each of the test tubes. Do not write in this book.

Test tube	Substances in tube	Temperature of water bath	Time of testing	Observations
1	1 mL of yeast 5 mL of water			
2				
etc.				

Thinking About Your Experiment

Using the information you have recorded in the table, consider the following questions:

1. What were the variables that you changed in this Exploration?

2. In what ways did you try to make the testing "fair"; that is, how did you control all possible variables?

3. Did you find any evidence that yeast is a living microorganism? If so, what conditions seem to be good for its survival and growth?

4. Did you observe any special characteristics of yeast when it was active? Was any gas produced? How might this gas be valuable?

5. Describe yeast that you saw under the microscope. Then compare what you discovered with the following description of yeast:

> Yeasts are single-celled fungi. They cannot manufacture their own food. They live on sugar. As they consume the sugar, they produce carbon dioxide gas as a waste product in such amounts that it can cause bread dough to swell up, or rise. Yeasts grow best at 29–32°C. The highest temperature that yeast cells can stand is between 50°C and 60°C. Since yeast spores are likely to be in any kitchen, yeasts can contaminate and spoil some foods if care is not taken.

New Yeast Cells Form by a Process Called Budding

A single yeast cell

A bud begins to form.

The bud grows.

The new bud breaks away from the original cell.

EXPLORATION 5

You may wish to have students gather and set up the materials for this Exploration (except for the water and ice) the day before they actually perform it. Dry yeast can easily be obtained from a supermarket. The "quick-acting" variety will produce more dramatic results. Note: For best results, the thermometer should not come in contact with the bottom of the beaker next to the heat source.

Students may observe some yeast cells in the budding stage. If so, ask students what they think is happening. Help them to reach the conclusion that this is the way yeast reproduces.

 ## Cooperative Learning
EXPLORATION 5

Group size: 3 to 4 students
Group goal: to examine yeast cells and determine the conditions that are most suitable for their growth
Positive interdependence: Assign each student a role such as materials organizer, group leader, recorder, or safety manager.
Individual accountability: Call on students randomly to answer one of the questions in Thinking About Your Experiment.

An Exploration Worksheet (Teaching Resources, page 30) and Transparency 25 are available to accompany Exploration 5 on page 164.

Answers to *Thinking About Your Experiment*

1. Manipulated variables include the amount of sugar and the temperature of the water baths.

2. The same amount of water and yeast were put into each of the test tubes; the testing time was the same for all the tubes; and the containers that were used were the same type and size.

3. The bubbles indicate that the yeast is giving off a gas (carbon dioxide); the yeast seems to be growing by producing new cells.
   The best conditions for the growth and survival of yeast include a moist environment with a temperature of 30–35°C and a food supply such as sugar.

4. Students may have observed budding and the grouping of yeast cells into colonies. Gas bubbles should be visible. The bubbles are trapped by bread dough, making the dough rise.

5. Details of the yeast cells will not be visible, but their general shape and their apparent size in the field of view may be visible. The description provided in the text should help to explain some of the students' observations. For instance, the production of carbon dioxide gas explains the presence of bubbles.

Answer to
Caption

Explain that the circular patterns on the surface of some of the yeast cells are places where buds have broken off or will eventually form. Help students to understand that most types of yeast grow new cells by forming buds that increase in size and then break away from the parent. You may wish to review the illustrations of budding on page 165.

Answers to
Do Some Research

1. The date by which yeast should be used is on the package. The viability of yeast can be checked by placing it in some warm water with a little sugar. If it is usable, it will begin to grow and produce gas bubbles within a few minutes. Old yeast will not do this, and therefore bread made with it will not rise properly.

2. Foods that are made with sugar are most likely to suffer yeast spoilage.

3. There may be a peculiar "yeasty" odor, bubbling, or discoloration of the food.

4. Use clean utensils and keep foods in airtight containers. Store food in cool, dry areas.

Molds

INDEPENDENT PRACTICE Ask students to conduct an investigation to find where mold grows. Have students make note of the environmental conditions in which the mold is found.

SAFETY ALERT Caution students to avoid any contact with molds and to avoid inhaling their spores. Molds can cause allergic reactions and other health problems.

Answer to
In-Text Question

Ⓐ Sample answer: under rocks, in-between tiles on a bathroom wall, and on spoiled food (especially dairy products)

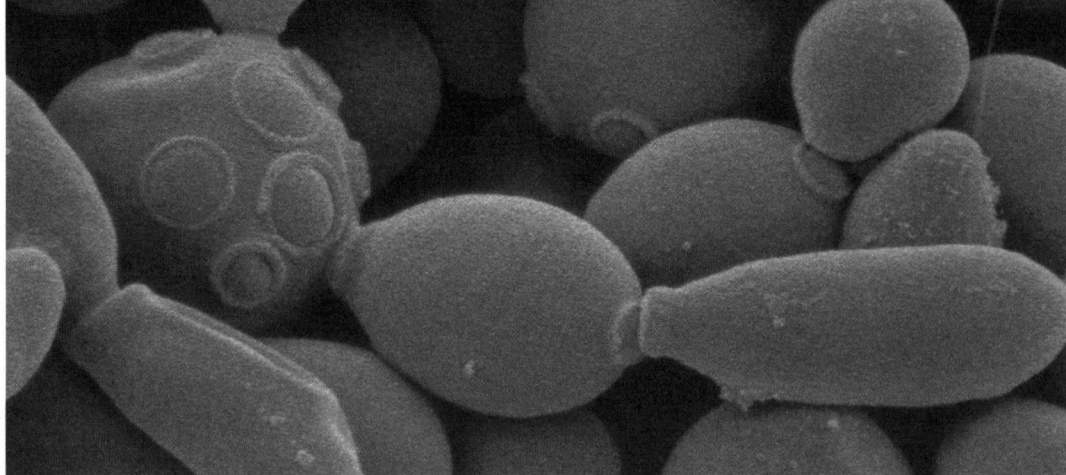

⬇ Can you spot the new buds on these yeast cells?

Sourdoughs

Gold prospectors in the West were once called sourdoughs. This is because they carried a *sourdough* with them wherever they went. A sourdough is a mixture of flour, water, and yeast that has been soaked in warm water. This is "fed" with a little sugar and milk and then kept cool for later use. The same culture can be kept for years if it is fed now and then. Prospectors could use their sourdough cultures to make bread—a useful skill at a time when bakeries were few and far between.

Do Some Research

1. The length of time that a product can last before becoming unusable is called its shelf life. How long is the shelf life of dry yeast? Check at the grocery store. Find out how you can tell if the yeast is too old to be usable. What might result if you used yeast that was too old?

2. What types of food do you think would be most likely to spoil because of yeasts?

3. What signs would make you suspect that a food product had been spoiled by yeast growth?

4. Suggest how to prevent spoilage of food by yeasts.

166

Molds

You read about molds earlier in connection with gardening. However, you are probably more familiar with molds around the home. In what different places have you seen them? Mold is a Ⓐ microorganism that grows very quickly. You have probably seen this happen with food. What does a mold look like up close? The next Exploration will show you.

What other types of food are made with the help of microorganisms? Turn to page S74 of the SourceBook.

Did You Know...

There are about 350 different species of yeasts. Some cause infections in animals and people, while others are used in medicine, in baking, and in the production of wine, beer, and cider. The most-used species of yeast is probably *Saccharomyces cerevisiae*. A strain of this species is used in making bread.

Homework

Have students help prepare a meal at home or observe someone else preparing a meal. Have them develop a list of practices that were used to avoid health problems.

Bread Mold

You Will Need

- a prepared wet-mount slide of bread mold
- a microscope

What to Do

1. The slide will be prepared for you. Observe the mold under low power and high power.

2. Draw and label the parts of the mold. Label the following parts, if you see them:

   - threadlike or rootlike parts, which are known as **hyphae** (singular, *hypha*)
   - swellings or spheres at the ends of the hyphae, which are known as **spore cases**

3. Can you suggest the functions of the hyphae and of the spores inside the spore cases?

4. Compare your discoveries with the photos and information that follow.

Caution: Some molds are harmful. Molds must be grown in sealed containers and disposed of safely. Molds should never be grown without adult supervision.

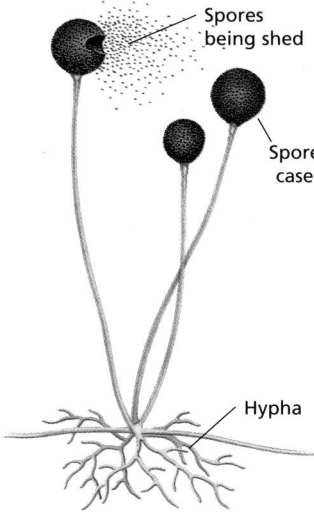

Spores being shed

Spore case

Hypha

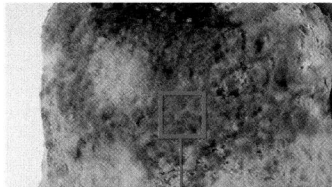

Bread mold

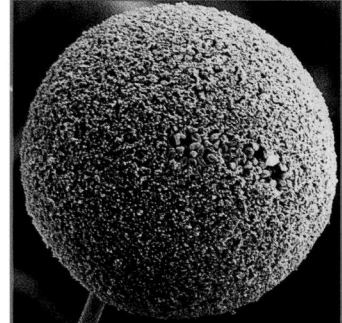

A scanning electron micrograph of a spore case from common bread mold

The illustration above shows the rootlike hyphae. Like yeasts, molds cannot make their own food. This is one way that fungi differ from members of the plant kingdom, which are able to produce food. The hyphae absorb the food they need from whatever the mold is growing on. The spores found in the spore cases are the reason for the mold's fast growth. Each spore can start a new mold if the conditions are right. There are millions of spores in a bit of mold that you see with the naked eye. Molds can be many colors. What are the colors of the molds you have seen on food? **B**

Why You Need to Be Careful

Allergic reactions caused by molds are quite common. As molds grow, they produce toxins, or poisons. Some infections and a few cases of blindness have been connected to contact with molds. It is wise to avoid any contact with molds and to avoid inhaling their spores.

167

FOLLOW-UP

Reteaching

Have students draw and label pictures to show what yeasts and molds look like. Then have them write a short article explaining how yeasts and molds are alike and how they are different.

Assessment

Ask students the following questions:

1. What is the relationship between yeast cells and bread? *(Yeast causes bread to rise.)*

2. How do yeast and bread molds grow and reproduce? *(Yeast grows and reproduces by budding, and bread molds grow and reproduce by forming hyphae.)*

Extension

Suggest to students that they make some bread at home and bring a loaf to class. There are easy-to-follow directions on most bags of flour. Encourage students to describe how they made the bread and the role of yeast in the process. Remind students to perform this activity only with adult supervision.

Closure

Have students design a hypothetical experiment in which they culture bread mold. Ask them to think of all the variables involved and indicate those variables that will be changed.

Homework

The Activity Worksheet on page 34 of the Unit 3 Teaching Resources booklet makes an excellent homework activity. Please note that this is a research activity and thus cannot be completed in a single evening.

EXPLORATION 6

SAFETY ALERT To safeguard against any allergic reactions to the molds, slides should be already prepared, and the molds should be covered. You may wish to have students read Why You Need to Be Careful at the bottom of the page before they begin the Exploration. When all students have had a chance to participate in the Exploration, dispose of the molds properly. Immerse the slides in a bleach solution consisting of one part laundry bleach to three parts water. Leave the slides immersed for at least an hour. Be sure no molds are left in the classroom. To dispel any possible confusion, you may wish to point out to the class that bread mold is not the same organism as the yeast that causes bread to rise.

 An Exploration Worksheet (Teaching Resources, page 33) is available to accompany Exploration 6. Additionally, a Transparency Worksheet (Teaching Resources, page 35) and Transparency 26 are available to accompany the material on this page.

Answer to *In-Text Question*

B Student answers may include white, blue, green, and black.

Answers to Challenge Your Thinking, pages 168–169

1. a. The paramecium is maneuvering around or avoiding an obstacle in its way.
   b. Movement, response to stimulus
   c. The paramecium is responding negatively because it is attempting to avoid an obstacle.

2. a. The wine was damaged.
   b. The wine was damaged by microorganisms found in it.
   c. People and other animals could be infected by microorganisms in a similar manner.

3. Bacteria-killing chemicals are used by the molds as a defense against bacterial attack. These antibacterial chemicals could be used to control bacterial diseases in humans.

4. a. If you heated the soil to between 77°C and 81°C in an oven, most of the harmful organisms would be killed, but the helpful bacteria would survive. To sterilize soil, the temperature would have to be high enough to kill all of the organisms.
   b. The seedlings that grow from seeds will not become infected with any plant diseases already in the soil. Competing weed seeds in the soil would also be killed.
   c. Once the tiny plants form, it is necessary to add something to put helpful microorganisms back into the soil. One solution is to add compost. In good compost, harmful microorganisms have been killed by the heat generated by the compost pile; only helpful microorganisms remain.
   d. Soil can be pasteurized by heating it to between 77°C and 81°C. A compost pile with good bacterial action will produce enough heat to kill most harmful organisms, but not enough heat to kill beneficial ones. Also, soil may be heated in an oven. Pasteurized soil is better than completely sterilized soil because pasteurization kills the bad bacteria while it leaves the good bacteria in the soil; sterilization kills both good and bad bacteria.

5. Accept all reasonable responses. Students should refer to the role that microorganisms play in food chains, in soil, and in food preparation.

1. The Paramecium Shuffle

Look at the picture below, which shows the activity of a paramecium over a period of time.

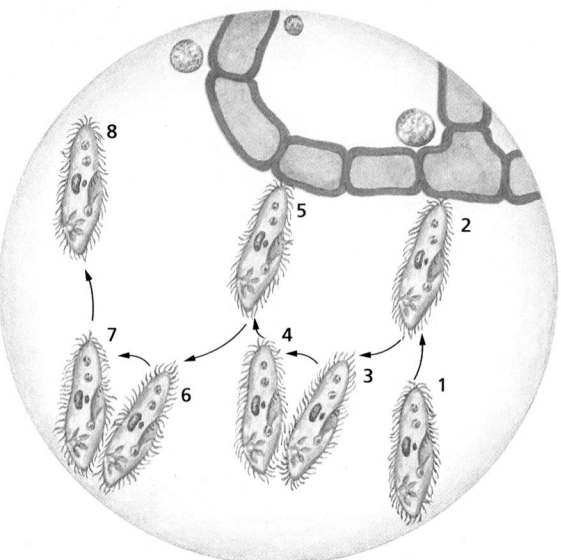

a. What is the paramecium doing?
b. What characteristic or characteristics of living things is the paramecium demonstrating?
c. Is the paramecium responding positively or negatively? Explain.

2. A Pasteurized Solution

Louis Pasteur showed that wine was damaged by microorganisms found in it. He described the wine as "diseased" and suggested that people and other animals could be infected by microorganisms in a similar way.

From the paragraph above, pick out and write down the following:

a. an observation made by Pasteur
b. an inference made by Pasteur
c. a conclusion made by Pasteur

3. Mold Can Be Beautiful

Some molds produce chemicals that kill bacteria. How might this be useful to the molds? How might this be useful to humans?

 You may wish to provide students with the Chapter 9 Review Worksheet that is available to accompany this Challenge Your Thinking (Teaching Resources, page 37).

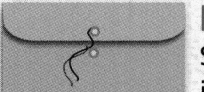 **PORTFOLIO** Suggest that students include their drawings of mold from Exploration 6. Students may wish to comment on how their observations differed from the photographs and illustrations presented in the text.

4. If You Can't Stand the Heat . . .

People often germinate seeds in sterilized soil. This is soil that has been heated to kill all of the living things in it. Look at the chart to the right to see what temperatures are needed to kill various kinds of organisms.

a. If you were preparing your own soil, what procedure for heating the soil might you follow?

b. Why is it an advantage for seeds to germinate in sterilized soil?

c. Once the tiny plants have emerged, what should the gardener add? How could this be done?

d. Soil can also be pasteurized. How might you pasteurize soil? Why might pasteurized soil be preferable to untreated or sterilized soil?

Organisms	Temperature required to kill (°C)
nematode worms	50
dangerous bacteria, fungi	65
soil insects, most viruses	70
most weed seeds	77
helpful bacteria, some viruses	82

5. Good Things in Small Packages

A noted scientist once said something like this: The role played by very small things in nature is very great.

Do you agree? Support this statement with as many instances as you can, drawing from the examples you have examined recently.

ScienceLog

Review your responses to the ScienceLog questions on page 148. Then revise your original ideas so that they reflect what you've learned.

169

ScienceLog

The following are sample revised answers:

1. Microorganisms are an important ingredient in soil because they provide plants with nutrients by breaking down waste.

2. It is normal and healthy for certain microorganisms to live on our skin and in our bodies. If you didn't clean your skin, however, harmful microorganisms might grow there because skin is naturally warm and moist.

3. Students should understand that although microorganisms cause many illnesses, life on Earth would not be possible without the benefits derived from microorganisms.

Theme Connection

Structures

When spore cases open, spores are released. Spores are very tiny structures. **Focus question:** How might the small size of spores contribute to the wide distribution of fungi? *(Because the spores are so small, they are able to stay suspended in the air for a long period of time, they can be blown long distances by wind, and they can easily pass through small cracks and filters.)*

Connecting to Other Chapters

> **Chapter 8**
> explores the diversity of
> microorganisms and the conditions
> they need to thrive and reproduce.

> **Chapter 9**
> examines both the harmful and
> beneficial effects that microorganisms
> have on humans.

> **Chapter 10**
> takes an in-depth look at antibiotics
> and the regulation of microorganisms
> in food preparation and storage.

Prior Knowledge and Misconceptions

Your students' responses to the ScienceLog questions on this page will reveal the kind of information—and misinformation—they bring to this chapter. Use what you find out about your students' knowledge to choose which chapter concepts and activities to emphasize in your teaching.

In addition to having students answer the questions on this page, you may wish to have them complete the following activity: Tell students to think about any experiences that they have had when visiting a hospital or seeing a doctor or nurse. Ask: Did you notice any differences in the medical environment compared with your everyday environment? *(Students might have noticed that hospital workers wore rubber gloves and uniforms and used only sterilized equipment. The hospital environment is likely to have been very clean and to have smelled strongly of disinfectants. Some students might have noticed the use of paper or plastic coverings on drinking glasses and examination tables or the absence of carpet except in the lobby.)* You may wish to have a volunteer list the differences on the board. Then ask the class to suggest some rea-

sons for them. *(All of these differences are the result of the need to prevent the growth of microorganisms.)* Emphasize that there are no right or wrong answers in this exercise. Listen to the class responses to find out what students know about proper sanitation, what misconceptions students may have, and what is interesting to them about germ warfare.

CHAPTER

10

Keeping Germs in Their Place

1 Is it true that taking antibiotics will help you get over a cold or the flu? Explain.

2 Of what importance are 5°C and 65°C when you prepare and store food?

3 How can harmful microorganisms be controlled?

4 How can you avoid food poisoning when you eat out?

ScienceLog

Think about these questions for a moment, and answer them in your ScienceLog. When you've finished this chapter, you'll have the opportunity to revise your answers based on what you've learned.

170

LESSON 1 — Germ Warfare

We think of wars as human struggles, but battles of terrible violence take place every day right beneath our eyes, unseen by us. The following is a dramatization of one such microscopic battle:

> The invading S troops were met by the defenders, and a battle raged over the limited food supply. The blind defenders, many of which were half-starved, moved as fast as they could over the rough terrain. They went to do battle with the newcomers, which were shaped like tiny submarines and fitted with 15,000 or more wriggling hairy extensions. Once within reach, the defenders took the S troops by surprise and sprayed them with murderous streams of poison. The S troops suffered many casualties but returned fire with their own brand of poison. When not busy with the actual battle, members of both sides ate any food they came across to gain an advantage, or they divided in two to replace slaughtered members of their group.

HOW SMALL IS IT?
One micrometer (1 µm) is one-millionth of a meter. One millimeter is equal to 1000 µm!

The battle in this story took place on a human hand. It involved the bacteria normally found on a person's hand and *Salmonella* (the S troops), a type of bacteria often found in kitchens. *Salmonella* bacteria are especially common on eggs and poultry and can cause a serious type of food poisoning. They are about 2 µm long—about one-tenth of the size of a tiny dust particle that you might barely be able to see in a shaft of sunlight. *Salmonella* bacteria need moisture to thrive and will die after a week without it.

Do you think the health inspector approved of Jason's use of a sponge in Pete's Place? You may have to look again at the story before you decide. **A**

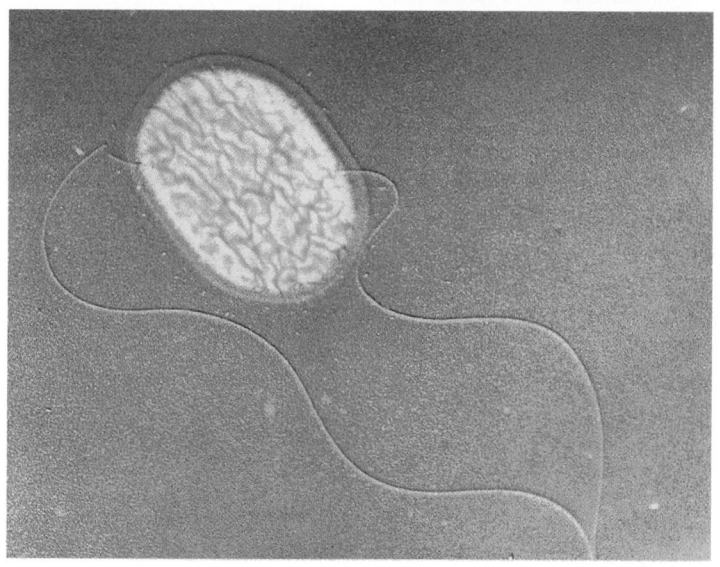

A *Salmonella* bacterium, commonly found in kitchens and on food, has extended strands that spin like a twirling lasso to propel the organism.

LESSON 1 ORGANIZER

Time Required
3 class periods

Process Skills
predicting, inferring, classifying

New Term
Antibiotic—a substance produced by one kind of microorganism that has a harmful effect on or kills other kinds of microorganisms; literally means "against life"

Materials (per student group)
none

Teaching Resources
Exploration Worksheet, p. 43
SourceBook, p. S78

LESSON 1 — Germ Warfare

FOCUS

Getting Started
Before students begin reading, you may wish to read the story on page 171 aloud and encourage them to picture the scenario as they read along. When you have finished, ask students to guess what the story was about. After they have expressed their ideas, tell students that the story was about a battle between two types of microorganisms on a human hand.

Main Ideas
1. There are naturally occurring mechanisms that control the growth of harmful microorganisms.
2. Antibiotics are important substances produced by microorganisms.

TEACHING STRATEGIES

Ask students if they know of any special precautions that should be taken to guard against harmful bacteria when preparing poultry dishes at home. *(Frozen poultry should be thawed in a refrigerator rather than at room temperature. Anything that has come in contact with the uncooked meat should be washed with hot, soapy water.)*

Answer to
In-Text Question

A Most students should agree that the inspector would disapprove of the use of a sponge because it provides a warm, moist environment for bacteria.

Divide the class into groups of three to four students to discuss and answer the questions. When they are finished, reassemble the class to share and discuss the students' ideas and responses.

Answers to
Exploration 1

1. a. All of the objects are likely to be moist.

b. The dishwasher is least likely to be contaminated because of the repeated use of hot water and detergent.

c. The cloth or sponge is most likely to be contaminated in every home because it is likely to remain damp from one meal to the next, to be stored in a warm place, and to not be disinfected after it is used.

2. a. *Salmonella* are transferred to Jack's forehead. Jack may clean his hands, but should he touch his forehead again, his hand will be recontaminated and *Salmonella* may be transferred to anything he touches.

b. *Salmonella* are transferred to the carrot. If the stew boils long enough, the heat will probably kill the *Salmonella*.

c. The *Salmonella* will move from Cindy's fingers to the potatoes, where they will multiply as the food cools.

d. Some *Salmonella* are transferred to the countertop.

e. *Salmonella* are transferred to the door handle.

3. Food would be contaminated the most in situation (c). Situation (b) or (e) is least likely to contaminate food.

4. The poisons, or toxins, that are produced by *Salmonella* make people ill.

5. Possible responses: *Salmonella* from uncooked poultry may have gotten on the lettuce. The lettuce may have been placed on a contaminated cutting board or washed in contaminated water. The person handling the lettuce may have had *Salmonella* on his or her hands.

6. Student responses should demonstrate an understanding of the conditions favorable to the growth of harmful bacteria—moisture, warmth, and food supply.

Harmful Invaders

What to Do

1. The chart below shows the results of a study of several hundred "clean" homes. The chart shows the percentage of homes found to have dangerous bacteria on certain objects in the kitchen. *Salmonella* would probably be the most dangerous, but there are likely to be a dozen other types present as well.

Assemble in small groups to discuss the following questions. Record the results of your discussion.

a. What condition is probably common to all of these objects?

b. According to this study, which object is least likely to be contaminated? Why do you think that might be?

c. Which object was contaminated in every home? Why do you think this was so?

Bacteria in a Kitchen

Object	Percentage of homes
dish towel	97.5
hand towel	98.6
sink faucet	94.0
sink surface	97.2
draining board	99.4
dishwasher	89.6
refrigerator	90.4
cloth or sponge for wiping counters	100.0

2. Assume that *Salmonella* bacteria are on people's fingers while they prepare food. Predict what might happen when

a. Jack rubs his forehead while he is checking to see if the roast is done.

b. Anne uses her fingers to push a piece of carrot back into the stew.

c. Cindy's fingers touch a mound of mashed potatoes, fluffy with milk and butter, as she spoons it into a bowl. Then she leaves the bowl to cool on the kitchen counter.

d. Max drums his fingers on the countertop while he waits for the microwave signal.

e. Carlos opens the fridge door.

3. In which of the situations (a–e) of question 2 would food probably be contaminated the most? the least?

4. When people suffer from *Salmonella* food poisoning, what do you think it is that makes them sick?

5. In 1987, between 400 and 1000 runners in the Honolulu Marathon got sick after eating lettuce contaminated with *Salmonella*. Almost 300 had to drop out of the race. How could the lettuce have become contaminated?

6. Look at the pictures to the right. Are they surprising? You may be relieved to know that more bacteria are helpful than harmful. Nevertheless, you should try to minimize the dangers from the harmful ones. What suggestions can you make to do this?

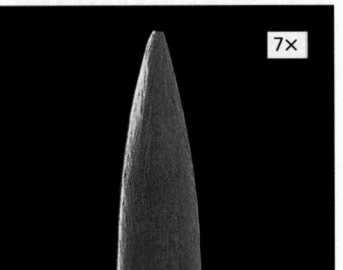

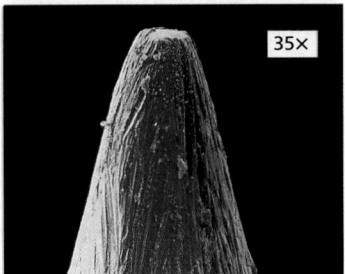

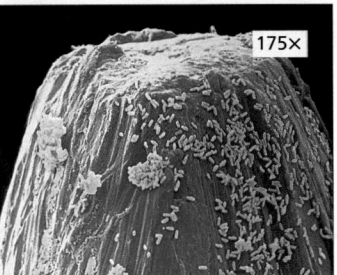

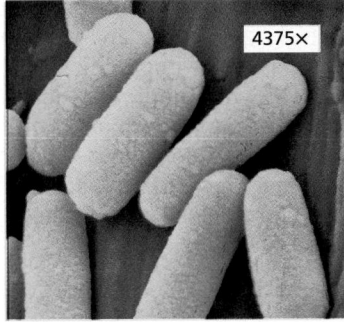

These are close-ups of a "clean" pin. As the magnification increases, clusters of household bacteria are revealed.

 An Exploration Worksheet (Teaching Resources, page 43) is available to accompany Exploration 1.

Battling Germs With Antibiotics

Microscopic battles occur almost everywhere. As you know, many microorganisms live in the soil. They eat one another in a kind of food chain—the smallest bacteria become food for slightly larger protists, and so on. They also squirt fluids to kill others that come too close to them. These fluids include the **antibiotics** (*antibiotic* means "against life") that we have come to depend on to fight bacterial infections. Although antibiotics kill bacteria, they do not work against viruses.

The potential of soil microorganisms as a source of antibiotics is great. There are many thousands, if not millions, of different types of soil microorganisms. Only a few have been tested for their antibiotic potential. Often, a certain kind of microorganism is found in only a small area. Scientists collect and investigate microorganisms from the far corners of the Earth when the opportunity arises. This is one reason why scientists are concerned about the destruction of the vast rain forests of South America.

Antibiotics— An Amazing Discovery

How long have antibiotics existed? The first widespread use of these "wonder drugs" occurred during World War II. However, antibiotics were discovered years earlier, in 1928. The discoverer was Alexander Fleming. The main points of Fleming's story are on the next page. Would you be as quick to draw conclusions as he was?

Alexander Fleming at work in his lab

Multicultural Extension

Bacteria-Fighting Foods

Share the following information with students: Although Alexander Fleming discovered antibiotics in 1928, many cultures have been using the antibiotic properties of food and other naturally occurring materials for hundreds of years. Many cultures use yogurt, mosses, and plants to fight systemic and local bacterial infections. American Indian, Chinese, Armenian, and African cultures have all used these materials.

Homework

Tell students that they are directors of a food bank that is sponsoring a food drive. Ask: What kinds of foods are the safest to receive? *(The safest foods would be canned goods and packaged dried goods because it is difficult for many microorganisms to grow in the absence of air or water.)*

Battling Germs With Antibiotics

Before students begin reading this section, ask them to respond with a show of hands if they have ever heard of antibiotics. Most students should respond affirmatively. Call on several of them to share their knowledge with the class. Many may know that antibiotics are used as medicines to treat certain diseases and infections.

INDEPENDENT PRACTICE   Have students read the first section on page 173 silently. When they finish, divide the class into small groups and encourage them to discuss what they have read and to ask questions.

Be sure students understand that antibiotics are produced by many different kinds of microorganisms, not just bacteria. You may wish to point out that the most commonly used antibiotics in medicine come from bacteria and molds. Some students may have heard of one of the most prominent bacterial antibiotics, *bacitracin*. Bacitracin was isolated in 1945 from the bacteria *Bacillus subtilis* in a contaminated wound. Bacteria known as actinomycetes have produced many of the most important antibiotics, such as tetracycline, streptomycin, and erythromycin. Among antibiotics produced from molds, perhaps the most important and well known is penicillin, which is produced by several species of *Penicillium,* a mold.

Antibiotics—An Amazing Discovery

Have students read the story of Alexander Fleming silently. When they finish, involve them in a discussion of the main points of the story, as listed on page 174. You may wish to have the points read one at a time, with students addressing each of the questions as they occur. Encourage students to honestly express what they think they would have done if they had been in Fleming's position.

Fleming's Discovery

Direct students' attention to the photo-micrograph of *Penicillium notatum.* Be sure students understand that this is the same kind of mold that Fleming observed in his experiment. Point out that this mold is related to the common, greenish molds that grow on bread, cheese, and citrus fruits.

Students may find it interesting to know that Fleming had actually thrown away his contaminated culture dishes. When a lab assistant passed by, Fleming retrieved the dishes to show the assistant what had happened. This second look sparked Fleming's curiosity, and then he realized that in the areas invaded by the mold, the bacteria no longer grew. Fleming began testing molds of all kinds and eventually discovered the one that contaminated his original cultures. However, the technology of the day prohibited him from isolating the substance he called penicillin. As a result, little attention was paid to his discovery for nearly a decade.

The beginning of World War II prompted a renewed interest in finding natural substances that attacked bacteria. At Oxford University, a team of scientists headed by Howard Florey and Ernst Chain began experimenting with some samples of Fleming's original cultures. By then, the technology was available to isolate penicillin from the mold. Because of the war raging in Europe, large-scale production of the antibiotic was carried out in the United States. By 1943, penicillin was being used to help save the lives of men wounded in battle. In 1945, Fleming, Florey, and Chain were awarded the Nobel Prize in physiology and medicine.

Fleming's Discovery

- Alexander Fleming was growing cultures of bacteria in a hospital lab.

- One of his cultures was very poor—the bacterial colonies appeared in only one part of the dish. What would you have done with the culture in the picture at right?

- Would you have thrown it out? Many scientists would have. Would you have saved it? Why?

- Fleming looked at the culture closely and wondered about it. Examine it closely, as he did. What do you notice? Do you see a bit of mold growing at the top with no colonies of bacteria growing near it?

- Do you suppose he solved the problem by concluding that somebody left the cover off the dish and ruined the culture and that it should therefore be disposed of?

- Maybe the culture was ruined, but it also posed an interesting problem. Fleming probably asked himself, Did the mold have something to do with killing the bacteria?

- Fleming identified the fungus as *Penicillium notatum,* similar to a type of mold that grows on cheese. What would you have tried next if you had been Fleming?

- Fleming grew the mold in other bacterial cultures and found that it was able to kill other types of bacteria also.

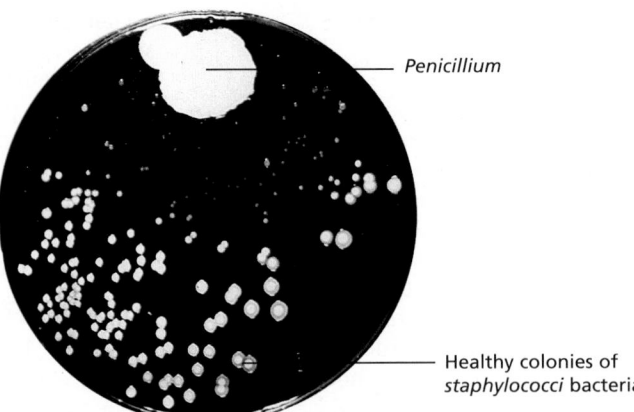

Penicillium

Healthy colonies of *staphylococci* bacteria

Here is Fleming's original petri dish, in which *Penicillium* have killed bacteria.

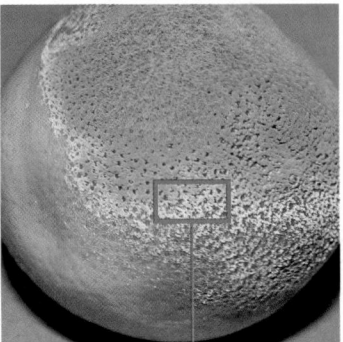

Sometime later, the chemical penicillin was discovered. It is made by the mold *Penicillium notatum* and is a valuable antibiotic that is effective against many disease-causing bacteria. How would you say Alexander Fleming and Louis Pasteur were similar to each other in the way they approached problems? Ⓐ

The green and white fuzz growing on this orange is the fungus *Penicillium.*

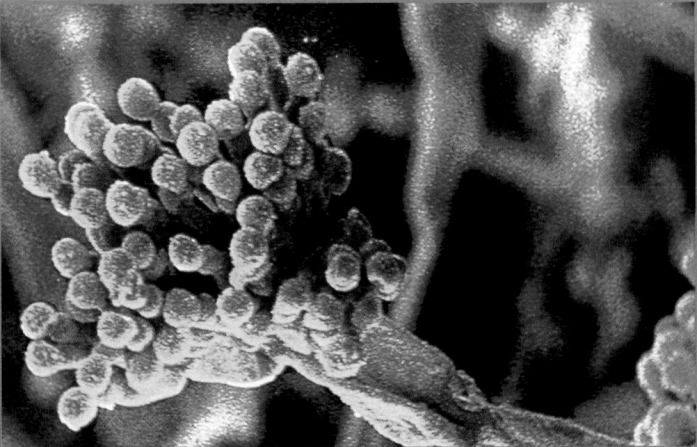

Do you recognize the parts of this fungus? The antibiotic called penicillin is produced in the round spores.

Answer to
In-Text Question

Ⓐ Both carefully followed a problem-solving process. They asked questions and drew conclusions and inferences from experimental data. Both exhibited a scientific attitude. (Remind students of the discussion of scientific attitudes in Unit 1, Science and Technology.)

Natural Bacterial Control

Even after pasteurization, a liter of milk may contain several million bacteria, but these bacteria are controlled by certain properties of the milk. For example, milk contains a natural antibiotic. In addition, the tiny particles that make up milk are very difficult for the bacteria to feed on. After a week or so, however, the growth of bacteria begins to get out of control. The bacteria cause substances in the milk to form little lumps (curds), and waste material from the bacteria changes the milk's taste.

Points to Ponder

1. Suggest words that describe the taste of milk after it has been changed by bacteria.

2. You go to the store for a carton of milk. How can you tell if the milk on the shelf should still be expected to have its bacteria under control?

3. What can you do to keep the growth of bacteria under control once you get the milk home?

Antibiotics and Disease

What antibiotics can you use for a common cold? None. Colds are caused by viruses. Viruses are not killed by antibiotics. For this reason, if you have a flu (influenza) virus and no bacterial complications, your doctor will not prescribe an antibiotic. It would do you no good because the flu is caused by a virus. Look at the following lists of infections. For which might a doctor prescribe an antibiotic?

• AIDS	• boils
• poliomyelitis	• strep throat
• common cold	• blood poisoning
• measles	• tetanus
• smallpox	• diphtheria
• chickenpox	• tuberculosis
• mumps	• typhoid fever
• meningitis	• meningitis

All of the diseases on the right are caused by bacteria and all on the left by viruses. There are many diseases that cannot be treated by antibiotics but can be treated some other way. Why is meningitis in both columns? It is a dangerous condition that may result from either a bacterial or a viral infection. There are also viral as well as bacterial causes of pneumonia.

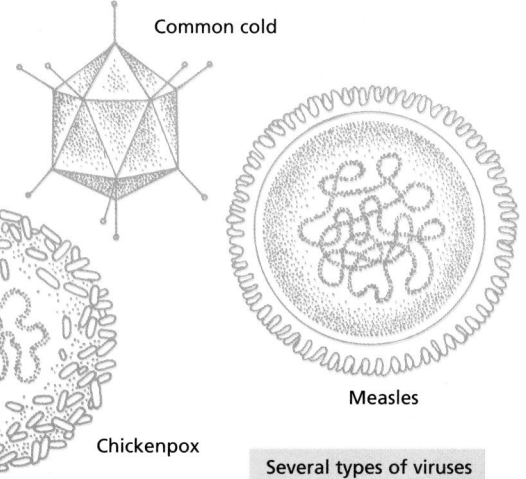

Common cold

Chickenpox

Measles

Several types of viruses

Natural Bacterial Control
GUIDED PRACTICE After students have read the paragraph, involve them in a discussion of how they can "help" food take advantage of its natural defenses against invading microorganisms. *("Help" can be achieved by providing the right storage conditions, including cleanliness, proper storage containers, and proper temperature.)*

Answers to
Points to Ponder

1. Answers may vary, but possible responses may include sour, sharp, pungent, acidic, spoiled, bad, rancid, fermented, tainted, and off.

2. Look at the freshness (sell-by) date on the milk container. Some students may correctly point out that you should check to be sure the milk is refrigerated and that the container is cool to the touch.

3. At home, milk should be stored in a clean, closed container in the refrigerator. It should be kept in the refrigerator at all times.

Wise Use of Antibiotics

Answers to
Wise Use of Antibiotics

Case A: Just because the patient is feeling better does not mean that all of the bacteria have been killed; the remaining bacteria can multiply quickly. Taking the antibiotic again may not be effective because the bacteria may have become resistant to the antibiotic. In order to stop the infection completely, antibiotics may need to be taken even after the patient begins feeling better. Also, the antibiotic may have killed helpful bacteria as well.

Case B: This patient should not have taken her mother's old antibiotics. The antibiotics may have been specific to the bacteria that caused the tooth infection. Also, the infection may be caused by a virus.

Case C: This patient may not need antibiotics because some bacteria in our throat and intestines help to prevent the growth of disease-causing microorganisms. Also, the infection may be caused by a virus.

Case D: Meningitis can be caused by either bacteria or a virus. The little boy probably has viral meningitis.

Case E: Sometimes a new strain of a common bacteria develops, and it may not be sensitive to the usual antibiotics. When this happens, a doctor must prescribe a different type of antibiotic.

ENVIRONMENTAL FOCUS

Point out to students that the cholera epidemic mentioned on page 145 was just one of many epidemics that have devastated populations. Have students research one of these epidemics. Make sure that they consider what organism was responsible for the epidemic, as well as the role of human behavior. You may wish to have students present their findings to the class. *(Some historical events to consider include the influenza epidemic of 1918, the bubonic plagues of the fourteenth and nineteenth centuries, and epidemic outbreaks of cholera, dysentery, malaria, and tuberculosis.)*

Wise Use of Antibiotics

Dr. Rogers was preparing a talk for a local group who had asked her about using drugs wisely. She included the following examples of what doctors often experience when they speak with patients.

Case A
Mrs. C.: Doctor, I took the antibiotics you gave me for six days and then stopped taking them because I felt just great. That was last week. Why am I sick again? I feel worse than before I took the medicine!

Case B
Paul N.: My mother had some antibiotics left over from last year when her tooth was infected. She gave them to me because I had a fever and coughed a lot last night, but I still feel rotten.

Case C
Mr. W.: Can you give me an antibiotic? My throat is a bit sore, and I don't want to be sick for the bowling tournament.

The doctor asked the group if they knew what was unwise about the way these people used antibiotics. Do you understand the mistakes these people made?

Dr. Rogers gave further examples of many people's mistaken belief that antibiotics can cure any illness.

Case D
Julia W.: The little boy next door is sick with meningitis, but the doctor says antibiotics may not help him. Why not? My cousin had meningitis when she was 12, and antibiotics saved her life.

Can you suggest an explanation for this?

Case E
Nu: My sister has an infection in her lungs, and the antibiotic given to her didn't help. Her doctor is going to try another antibiotic, but it may not work either, even though the infection is caused by bacteria.

What could be the problem?

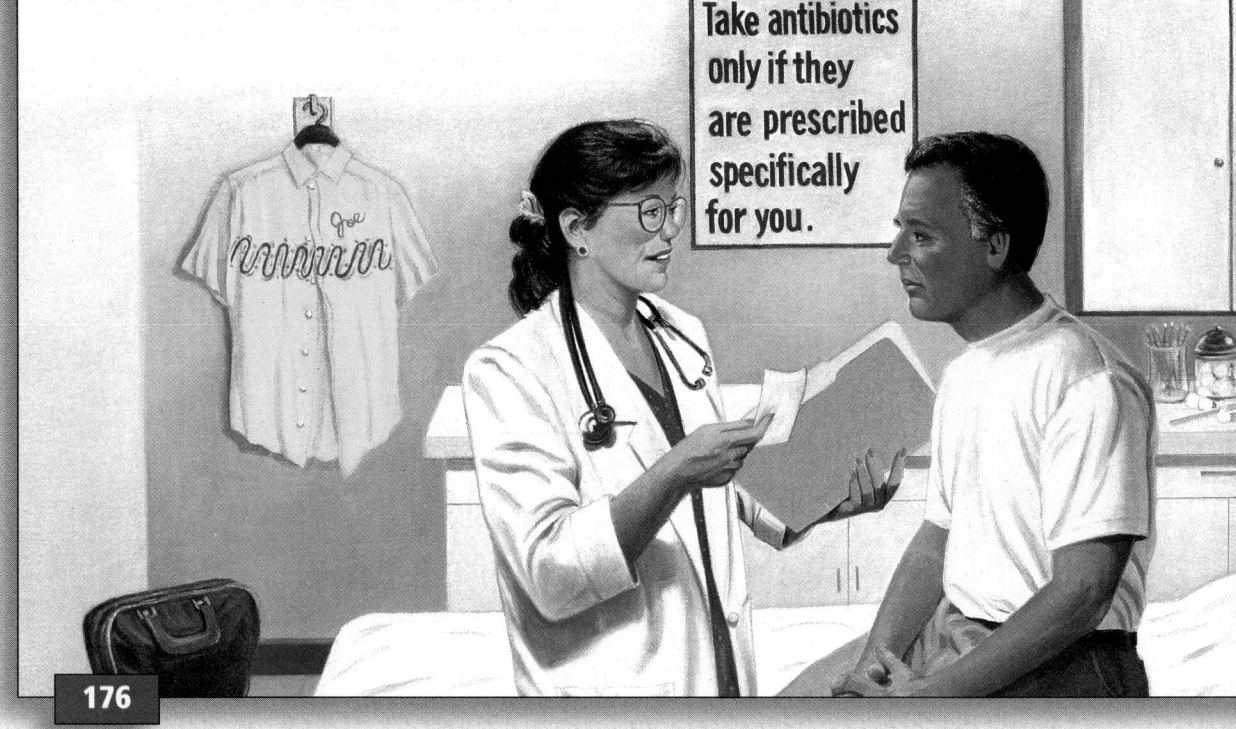

Take antibiotics only if they are prescribed specifically for you.

176

Homework

The questions in Wise Use of Antibiotics make an excellent homework activity.

Integrating the Sciences

Earth and Life Sciences

The presence of fossilized algae in rocks that are about 3.5 billion years old is an important piece of evidence in reconstructing the history of the Earth's climate. Ask: What clues about the Earth's climate do you think these fossils provide? *(Algae can survive only in a narrow range of temperatures, so the temperature of the Earth must have been relatively cool and constant 3.5 billion years ago.)*

The following information may help you answer the questions:

- An antibiotic may kill not only the disease-causing bacteria but also beneficial ones.
- Some bacteria are resistant to a particular antibiotic.
- Some bacteria in our bodies manufacture vitamins that are important in preventing nutritional deficiency diseases.
- Some bacteria in our throat and intestines help by attacking and preventing the growth of disease-causing microorganisms.
- Some illnesses, such as meningitis, may be the result of either a bacterial or a viral infection.
- Sometimes a new strain of a common bacteria develops, and it may not be sensitive to the usual antibiotics. New antibiotics are needed.
- Bacteria normally present in the body tend to compete with invaders for the food supply and thus help to control the invaders' growth.
- Doctors are careful to prescribe antibiotics only when they are really needed.

An International Concern

Suppose that a large international event is going to be held in your area. You are a member of the organizing committee. You have been asked to prepare information (in the form of a chart) for participants. The chart should show details about the main types of microorganisms that might cause an illness to spread among a large group of people. Include organisms that cause food poisoning as well as those that cause bacterial and viral diseases like the flu, hepatitis, tuberculosis, and AIDS. Include the following information in your chart:

a. the types of microorganisms involved

b. how the illness is contracted

c. the results of having the illness

d. how to guard against the microorganisms named

Form a group with three other students to obtain information from the library, consumer groups, doctors and nurses, departments of health, and the food industry.

Add the information you learn from your investigation to your list of pro and con arguments for the debate:

"Resolved: Microorganisms are bad guys."

You don't always need a prescription for antibiotics. Your own body produces them. Get the details on page S79 of the SourceBook.

177

Reteaching

Have students make a dictionary of terms that describe microorganisms. Suggest that they include drawings to help explain some of the terms. Display the dictionaries in the classroom or library.

Assessment

Set up microscope stations with prepared slides, and have students identify the following:

a. parts of molds

b. bacteria

c. yeast cells

d. euglena

e. algae

As an alternative, use drawings or photos instead of microscope slides. Have students identify the kingdom to which each organism belongs.

Extension

Bacteria are sometimes attacked by a kind of virus called a bacteriophage. Challenge interested students to do research on these organisms. Have them present their information to the class in the form of a "chalk talk," a poster, or a written report. The following questions might help students get started: What do bacteriophages look like? When were bacteriophages discovered? How do bacteriophages attack bacteria?

Closure

Have students draw a cartoon strip to illustrate the following paragraph: An egg has many tiny holes in its shell through which the egg "breathes." Bacteria can enter the shell through these holes. Once inside, however, the bacteria encounter an elaborate defense system. A thick membrane keeps the bacteria out until they dissolve a portion of the membrane to make a tiny opening in it. The egg white then attacks the bacteria, killing many of them with liquids that tear the bacteria apart. The egg white also has a way of enclosing food substances that would otherwise have nourished the invading bacteria. Therefore, the bacteria that survive the poison attack starve to death and never reach the nutrient-rich egg yolk.

An International Concern

Encourage students to share and discuss their information before they compile their charts. Be prepared to help resolve any differences that arise. Note: *Salmonella, Staphylococcus aureus,* and *Clostridium botulinum* together cause 80 to 90 percent of the food-poisoning cases in North America.

Cooperative Learning
AN INTERNATIONAL CONCERN

Group size: 4 students

Group goal: to prepare a chart showing the main sources of food poisoning and illness due to microorganisms in North America

Positive interdependence: Assign students roles such as coordinator, reporter, recorder, and editor.

Individual accountability: Randomly select one aspect of the project for each student to discuss, such as the ways that a certain illness can be contracted.

Microorganisms and Food

LESSON

2
Microorganisms and Food

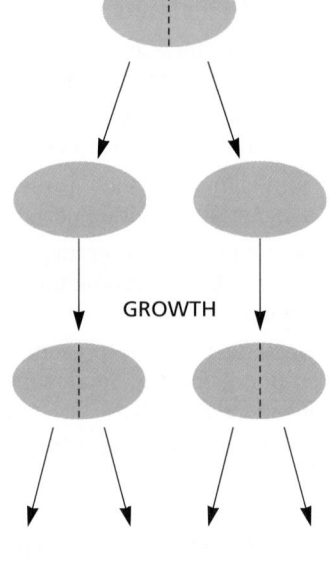

FOCUS

Getting Started

Bring in an empty egg carton and a bag of rice. Have a student place one grain of rice in the first cup in the carton. For each subsequent cup, he or she is to double the number of grains of rice found in the previous cup. (The student will quickly discover that too many grains of rice are needed.) Explain that this illustrates how bacteria multiply so quickly.

Main Ideas

1. Microorganisms multiply rapidly in the proper growing conditions.
2. Temperature and time are important variables in controlling the growth of microorganisms.
3. Many methods of food preservation are the result of new technology.

TEACHING STRATEGIES

Divide to Multiply

GUIDED PRACTICE   Before students read, remind them of how quickly the yeast cells multiplied in Exploration 5 of Chapter 9.

Because of the size of the numbers involved in these calculations, suggest that students work with calculators.

Answer to
In-Text Question

Ⓐ On the 30th day, there would be 2^{29} = 536,870,912 pennies ($5,368,709.12).

Answers to
Divide to Multiply

1. Bacteria in 1 hr. = 16
   in 3 hr. = 4096
   in 6 hr. = 16,777,216

2. 16,777,216 × 100,000 = 1,677,721,600,000

Divide to Multiply

When microorganisms have the right conditions, they multiply rapidly. Many have the ability to multiply by division, which sounds impossible but simply means that one cell can divide to form two new cells (sometimes called *daughter* cells). What do you think happens to the parent? For a glimpse of what this means in terms of quantity, try the following exercise.

Suppose you were offered work for 30 days. You would receive 1 cent for your first day's work. For the second day, your pay would be 2 cents, which would be doubled for the third day, and so on. Each day you would get paid double what you received the previous day. Would you take the job? How much would you Ⓐ be paid on the 30th day? This is the kind of math that bacteria are good at. Try these:

1. If a cell of bacterium X divides in two every 15 minutes, how many cells would there be in 1 hour? 3 hours? 6 hours?

2. If there were 100,000 bacterium X individuals at the beginning, how many would there be in 6 hours?

Why is it important to understand the idea of multiplication by division? The situation below will show you what can happen if this idea is not understood.

Suppose a stew is served at Pete's Place for dinner. The stew is cooked at 100°C for a few minutes—long enough to kill most (but probably not all) of the microorganisms in it. Some stew is left over. It has cooled a bit, and a few airborne microorganisms have fallen into it. They find perfect growing conditions. These microorganisms and the ones that were not killed during cooking start to multiply.

GROWTH

LESSON 2 ORGANIZER

Time Required
3 class periods

Process Skills
observing, analyzing, comparing

New Terms
Maximum growth temperature—the temperature above which microorganisms stop growing and die
Minimum growth temperature—the temperature below which microorganisms stop growing

Optimum growth temperature—the ideal temperature for the growth of microorganisms

Materials (per student group)
none

Teaching Resources
Activity Worksheets, pp. 46 and 47
Transparency 27

Imagine, for example, that 10 bacteria get into the stew at 7 P.M. from Jason's thumb as the stew is lifted from the stove, put on the counter, and covered. Suppose these cells divide every half-hour (even though this could actually happen every 15 minutes in a really warm room). By 8 P.M. there will be 40 bacterial cells, by 9 P.M. 160, by 11 P.M. 2560, and so on. By the next day at 11 A.M., there will be over 40,000,000,000 bacterial cells in the stew! This many would probably not change the appearance of the food. If the stew is warmed up for lunch, it might taste a bit "off" to some people, and it might cause an upset stomach in others. But because it was heated before eating, most of the bacteria will have been killed, and any problems would be caused mainly by the toxins (poisons) made by the bacteria while they were living.

1. What did you learn from the situation just described?
2. How might you have altered this situation so that it posed less of a health hazard?
3. Describe a similar scenario, based on either your own experience or your imagination.

Temperature and Food Spoilage

Temperature is an extremely important factor in food preservation. There are three sets of data supplied below in table and graph form. What information can you discover from each?

Data Set 1

Two of the most important temperatures related to food spoilage are the temperature above which organisms cease to grow and the temperature that best suits the organisms for growth. Match the following terms, which scientists use, with the temperature descriptions above: *optimum* (best) *growth temperature* and *maximum growth temperature*. Now can you explain what *minimum growth temperature* is?

The graph on the right shows what happens to the growth rate of microorganisms (how fast they multiply) as the temperature rises. Refer to points *A–F* as you answer the following questions:

1. Can you identify the three growth temperatures named above and show their location on the graph?
2. During what temperature interval is the increase in growth rate the greatest (*A–B, B–C*, etc.)?
3. During what temperature interval is there a dramatic decrease in the growth rate?

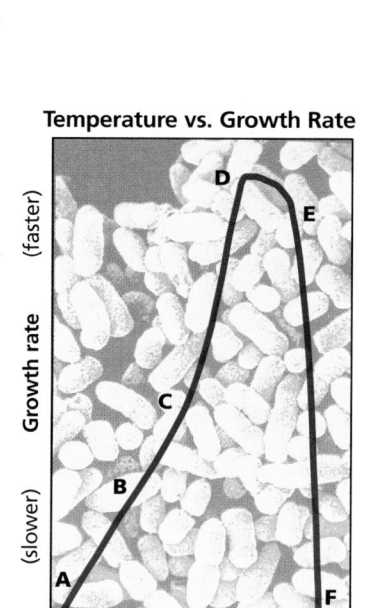

Temperature vs. Growth Rate

Growth rate (faster) / (slower)

Temperature (lower) / (higher)

179

★ Transparency 27 accompanies the material on this page.

CROSS-DISCIPLINARY FOCUS

Mathematics

The rule of 69 is a convenient way to estimate how long it takes a population of microorganisms to double, assuming exponential growth. First determine what the percentage growth is in some amount of time. Then divide 69 by that number. Ask: If a bacteria population experiences 5 percent growth each minute, how many minutes will it take for the population to double? *(69 divided by 5 is 13.8, so it would double in a little less than 14 minutes!)*

Answers to In-Text Questions

1. Students should mention that it is important to store food properly in order to keep it from spoiling.

2. The leftover stew should have been placed in the refrigerator. Jason should have either washed his hands or been wearing rubber gloves, and he should have been very careful not to get his thumb in the stew.

3. Student responses should demonstrate an understanding of how food can spoil and what can be done to prevent it.

Answers to Data Set 1

The maximum growth temperature is the temperature above which microorganisms stop growing and die.

The optimum growth temperature is the ideal temperature for the growth of microorganisms.

The minimum growth temperature is the temperature below which microorganisms stop growing, but do not necessarily die. For example, food that has been chilled might still contain microorganisms that could become active when the food is warmed. Also, any toxins present in the food before it was chilled would still be harmful.

1. On the graph, point *A* is the minimum growth temperature. Point *D* is the optimum growth temperature. Point *F* is the maximum growth temperature.

2. Increase in the growth rate is greatest during the interval *C–D*.

3. A dramatic decrease takes place during the interval *E–F*.

Homework

Ask: Why do you think a cure for the common cold has not been found? Students may need to consult a reference before answering. *(Students may suggest a number of reasons. A virus can often mutate, so medicines developed to fight the virus may not have any effect on new strains. Additionally, numerous viral infections can cause a host to exhibit symptoms similar to the common cold. What is referred to as the common cold is probably caused by any one of hundreds of different viruses.)*

Data Set 2

The data in the table are based on results from experimental work with *E. coli,* a bacterium that is common in human intestines. Its presence is also used as an indicator of the amount of sewage present in wells and other water supplies.

Answers to
Data Set 2

According to the table, the optimum growth temperature is 37°C. That is the temperature at which the microorganisms multiply in the least amount of time. Maximum growth temperature is between 45°C and 50°C. The microorganisms stop growing at some point between these two temperatures.

Minimum growth temperature is at some point below 10°C. Room temperature is usually considered to be 20°C. According to the table, the microorganism does not multiply very rapidly at this temperature. Your body temperature, however, is 37°C—the optimum growth temperature for the microorganism.

Answers to
Data Set 3

1. **a.** The diagram shows the dangerous and safe temperature ranges for preparing and storing food.
    b. Remembering this information could help you prevent serious illness from food poisoning. Accept all reasonable title suggestions.

2. Most refrigerator temperatures fall below 5°C, within the safe range on the diagram.

3. You should cover the food and place it in the refrigerator until the person arrives. Then take it out of the refrigerator and reheat it. Otherwise, the warm food will provide an ideal place for bacteria to grow. It could become contaminated and cause food poisoning.

4. Assuming that the temperature remains at 37°C, the bacteria will double 14 times over the course of 4 hours. In that time, 10 bacteria will multiply into 163,840 bacteria. The rapid rate of bacterial reproduction shows the need to be careful when preparing and storing food.

Data Set 2

The table below refers to the growth of a type of microorganism that is common in your intestines. The second column in the table shows how long it takes for one microorganism to divide and become two microorganisms.

Temperature (°C)	Time to double (minutes)
10	130
20	60
25	40
30	29
37	17
40	19
45	32
50	no growth

For this microorganism, what are the optimum growth temperature and the maximum growth temperature? The information for the minimum growth temperature is not provided, but approximately what would this temperature be? How would room temperature and our body temperature affect the growth of this microorganism?

The maximum, optimum, and minimum growth temperatures are not the same for all microorganisms. It is important to know, however, that the ones that spoil food or cause food poisoning generally grow over a wide range of temperatures, including any temperatures that people are likely to experience.

Data Set 3

The diagram below gives information on storing and preparing food.

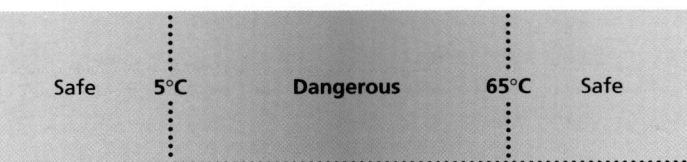

1. a. What does this diagram tell you?
    b. How is this information useful? What might be a good title for the diagram?

2. What is the temperature inside your refrigerator? Relate this to the information in the diagram.

3. If you are holding dinner for someone who is going to be at least 2 hours late, what should you do with the food while you wait? Why?

4. Use the information from Data Set 2 to calculate the possible increase in numbers of bacteria over a 4-hour period, starting with 10 bacteria at 37°C. How does this show the need to take precautions when preparing and storing food?

Food-Preservation Miniprojects

Food has to be stored and preserved, often for months, because we like to eat a variety of foods throughout the year. Select a miniproject from the ones below to investigate some ways of preserving food.

1. Ask older members of your family, or others, about the ways that foods were preserved in the past. You could make a time line to show when inventions such as the refrigerator were introduced.

2. Look at foods in your home. What preservation method is used for each? Include packaged foods, unpackaged foods, and foods kept cool. For a single food, can there be more than one preservation method? Give examples.

Multicultural Extension

Food Preservation

Many cultures preserve certain foods to store them for long periods of time. For example, many Asian cultures dry foods such as fish, fruit, and even poultry; some eastern European cultures pickle foods; and some Scandinavian cultures salt foods. Invite students from other parts of the world to describe how their native foods are preserved in ways not commonly used in this country. If possible, have those students bring examples of such foods to class.

Homework

The Activity Worksheets on pages 46 and 47 of the Unit 3 Teaching Resources booklet make excellent homework activities.

Meeting Individual Needs

Gifted Learners

Have students debate the following proposition: Students should not be allowed to keep food or drinks in their locker.

3. Do the same as for project 2, but this time look at foods in the supermarket. Also, find out how it is possible to sell fresh produce from Mexico, South America, Spain, etc.

4. What foods are used on spaceflights? Find out what you can about the quantities of food and drink that are needed and how they are preserved.

5. As a class project, make a display such as a picture collage about advances in food processing and preservation. Research and include the following: UHT processing, aseptic packaging of liquids, freeze-drying, and vacuum packaging.

6. Research arguments against food preservatives being added to foods. Sometimes preservatives are branded as "poisons in our food."

7. Shown below is the list of ingredients from a box of cereal. Can you imagine the purpose of each ingredient? Collect ingredient labels from other packaged foods. Materials used as preservatives or for color or artificial flavor are called *additives*. From your collection, pick out the names of some additives. Try to find out more about these substances, especially the ones that are used as preservatives.

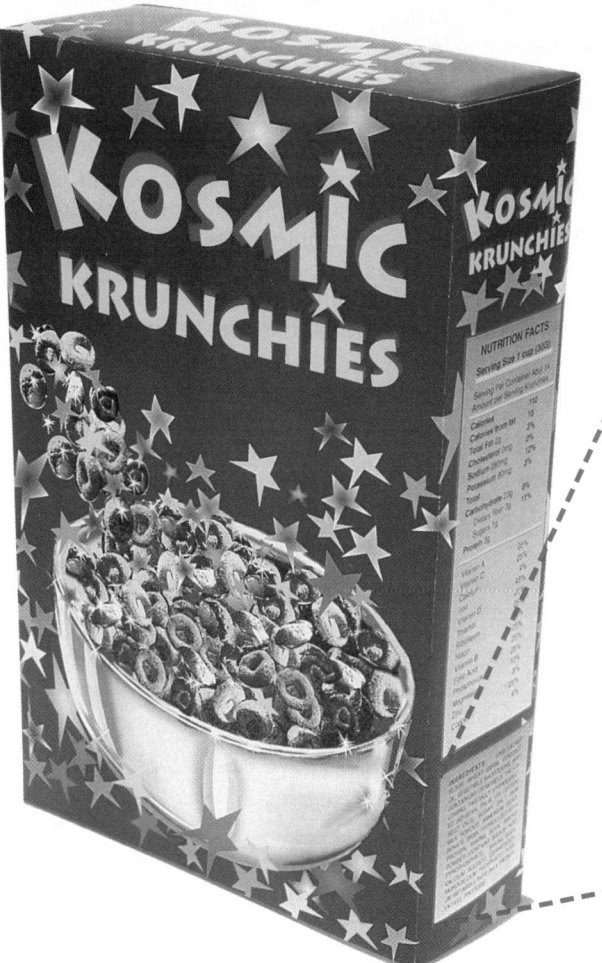

INGREDIENTS: UNBLEACHED FLOUR, WHEAT GERM, COCONUT OIL, VEGETABLE SHORTENING, (MAY CONTAIN ONE OR MORE OF THE FOLLOWING: PARTIALLY HYDROGENATED SOYBEAN, PALM, OR COTTONSEED OILS), SUGAR, SALT, SKIM MILK POWDER, AMMONIUM BICARBONATE, BRAN, HYDROLYZED SOYA PROTEIN, BAKING SODA, BAKING POWDER (CONTAINS SODIUM ACID PYROPHOSPHATE, BAKING SODA, CALCIUM SULFATE, CORNSTARCH, MONOCALCIUM PHOSPHATE), SODIUM METABISULPHITE (AS A PRESERVATIVE), PROTEASE

Answers to *Food-Preservation Miniprojects,* *pages 180–181*

1. Early methods of food preservation included drying, fermentation, and pickling. Examples of more recent developments in food preservation are listed below.

   1861: Ice and salt are used to freeze fish.
   1864: Pasteur uses heat to kill bacteria.
   1874: The refrigerator is invented.
   1906: Freeze-drying becomes widely used.

2. Students will find a variety of preservation methods used in the foods they have at home. Encourage discussion about the different preservation methods encountered.

3. Students may be able to find out from a manager of the supermarket or the produce department how produce from other countries must be picked, shipped, and stored to be kept fresh for markets here in the United States.

Answers to *Food-Preservation Miniprojects,* *continued*

4. Foods used on space flights must meet daily nutritional requirements and be easy to prepare and eat in a weightless environment. About 1.5 kg of food is allowed per person per day, and this includes almost 0.5 kg of packaging. Most foods are canned, dried, or exposed to radiation to preserve them.

5. UHT, or Ultra High Temperature, processing involves sterilizing foods for canning at higher temperatures and in less time than usual. This process reduces the loss of vitamins and flavor caused by long exposure to heat.

   Aseptic packaging involves heat-sterilizing foods and then storing them in sterilized containers, which allows them to be stored without refrigeration.

   Freeze-drying involves removing water from foods while they are frozen. The foods are heated just enough to cause the water to vaporize without thawing the food.

   Vacuum packaging involves removing air and other gases while sealing the food package. The removal of gases, especially oxygen, retards the growth of harmful bacteria.

6. Opponents of preservatives argue that the need for them could be eliminated by proper sanitation and storage methods. The use of nitrates and nitrites in curing meat provides an excellent case study for students.

7. There are many types of food additives other than preservatives. Some are designed to add nutrients, flavor, or color to a product. Others are designed to protect the appearance and texture of a product. Additives often serve more than one function in a product. (Suggest that students concentrate on one type of additive.)

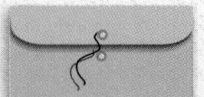
Answers to
Map It Out

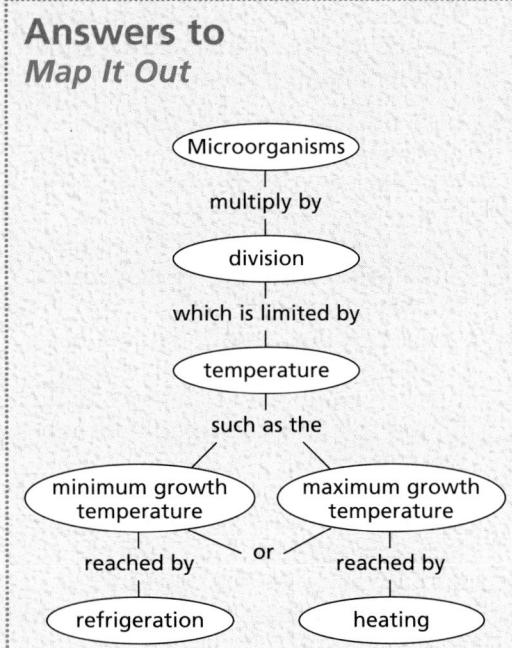

FOLLOW-UP

Reteaching

Challenge students to use the data table in Data Set 2 on page 180 to make a graph. (You may wish to suggest that the horizontal coordinates show temperature and that the vertical coordinates show the time in minutes.)

Assessment

Suggest that students use the data in Data Set 2 on page 180 to make a table that shows how many microorganisms exist after 5 hours at each temperature. Headings may include Number at start, Temperature, and Number after 5 hours.

Extension

Some students may enjoy thinking of a way to show how rapidly microorganisms can multiply. Possibilities include making a poster diagram, creating a mural or bulletin-board display, or making a model using clay figures or other materials.

A Science Pioneer
Lloyd A. Hall (1894–1971)

Most food is perishable. This means that sooner or later it will be contaminated by microorganisms and become inedible. Of course the longer that food stays safe to eat, the better. In 1925, a young man named Lloyd A. Hall developed a process of chemical sterilization that helped keep food safe to eat.

Before Hall's research on chemical sterilization, meat-packers added spices to their meat products and, in the process, contaminated their meats with molds, yeast, and bacteria. In Hall's sterilization process, these spices were exposed to a gas that was poisonous to microorganisms. The treated spices could then be added to the meats without fear of contamination. Hall's process was also used to sterilize medicines, hospital supplies, bandages, cosmetics, and many other products.

Hall's chemical sterilization method is still widely used today.

Hall's interest in chemistry began at East Side High School in Aurora, Illinois, where he was an outstanding student. He graduated among the top 10 students in his class. He attended Northwestern University on a scholarship and graduated in 1916. He went on to do graduate work in chemistry at the University of Chicago.

In 1925, Hall started work with Griffith Laboratories as chief chemist and director of research; he held these positions until he retired in 1959. While at Griffith Laboratories, he specialized in food-preservation techniques and developed processes for curing meats, seasonings, and baked goods.

Hall was granted over 100 patents on products and methods that he invented. Hall also authored dozens of scientific papers.

Research Topic

Find out more about Lloyd A. Hall, and write a paper on one of the food-preservation processes he developed.

Map It Out

Check your understanding of the ideas in this chapter by using the following words to complete the concept map at right: *microorganisms, maximum growth temperature, refrigeration, heating, division, temperature,* and *minimum growth temperature.*

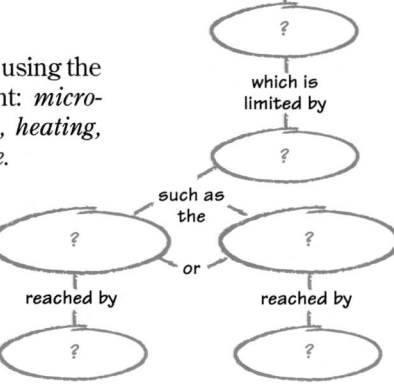

Closure

Ask students to imagine that they have made a large pot of chili for a backyard party to be held the next day. Suggest that they form small groups and write some instructions on what to do with the chili to protect it from contamination by microorganisms and to ensure that it is safe to eat.

Surrounded by Microorganisms

You know that microorganisms are found in water and soil. You also know from your reading that microorganisms are sometimes found in the air. In fact, microorganisms are all around us. Without precautions, food can easily become contaminated.

Microorganisms in the Air

"Empty" air can be teeming with microorganisms, as the three accounts below demonstrate.

1. Fungus spores are always around us. They can be carried by atmospheric currents from state to state, or even across entire continents, in only a few days. They enter buildings easily. The spores stick to damp areas, where they germinate, grow, and feed. Some fungi feed on substances in concrete; others feed on paint or the glue on wallpaper. One type of fungus even feeds on the poisonous preservatives used on wood.

   As fungi feed on house materials, they may absorb substances they cannot use. Some of these will be released into the air as gases. Have you ever heard someone describe a building as musty smelling? This 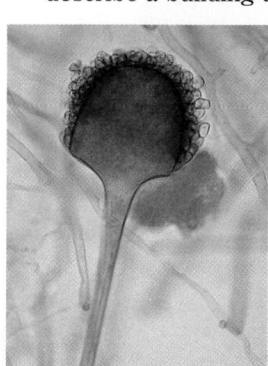 is a typical result of fungal growth—the smell comes from the gases released. You might notice this smell at a damp time of the year or in a building that has not been aired out for a while.

This photo shows a fungus that has emerged from a spore. The rounded part is the main body, and the long tube is the hypha arm that will dig into materials like the plaster or brick of a house wall in search of food.

With an abundance of food nearby, this dust mite grazes on the contents of a vacuum-cleaner bag.

2. When a vacuum cleaner is used, dust particles, dust mites, and many other microorganisms are picked up, but the tiniest particles are blasted out the back of the machine at high speed. The person pushing the vacuum cleaner is hit with these tiny particles, which travel like shots from a cannon. All of the particles small enough to come through the filter are put back into the air. One scientist who studies microorganisms in the air considers the home vacuum cleaner to be just about the best dust-spreader around.

3. A sneeze can travel 60 km per hour (about the same rate as a gale-force wind, which can break branches off trees). A tissue held over the nose fails to contain the viruses that shoot out, partly because the tissue is full of tiny holes (its softness depends on the fact that the fibers are so far apart).

183

FOCUS

Getting Started

Gently blow a small amount of wheat flour into the beam of an overhead projector or flashlight. Ask students to observe how the particles appear in the light. Explain that microorganisms travel through the air in a similar manner.

Main Ideas

1. Microorganisms travel through the air.
2. Adjusting personal habits can help people protect themselves against the harmful effects of microorganisms.

TEACHING STRATEGIES

Before students begin reading, have them discuss what they know about where microorganisms live.

Microorganisms in the Air

GUIDED PRACTICE   Have students read the three scenarios, pausing after each one to discuss its implications in their daily lives. Ask students questions similar to the following:

- If fungal spores are all around, how can you keep them inactive? *(By making their environment unfavorable to their survival, such as by keeping surfaces dry)*
- When should you try to avoid using a vacuum cleaner? *(Before dusting; when people who are allergic to dust are around)*

Homework

The Activity Worksheet on page 48 of the Unit 3 Teaching Resources booklet makes an excellent homework activity.

 The Discrepant Event Worksheet on page 49 of the Unit 3 Teaching Resources booklet describes a teacher demonstration that works well as an accompaniment to Lesson 3.

LESSON 3 ORGANIZER

Time Required
1 to 2 class periods

Process Skills
analyzing, classifying, hypothesizing

New Terms
none

Materials (per student group)
none

Teaching Resources
Activity Worksheet, p. 48
Discrepant Event Worksheet, p. 49

Washing Your Hands

INDEPENDENT PRACTICE Before students read this section, ask them to take out their ScienceLog and write down as much as they can remember of the hand-washing technique presented in Lesson 1 of Chapter 8.

Making Rules for Life

If students are having difficulty organizing their rules, you may wish to suggest categories similar to the following: how to avoid spreading illnesses; how to prevent contamination of one food by another; how to keep microorganisms in food from multiplying; how to kill microorganisms already present in food; how to avoid contaminating food during handling; and how to avoid contamination of food by pets.

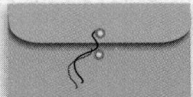 **PORTFOLIO**

Encourage students to include their list from Making Rules for Life in their Portfolio. Students may wish to illustrate their list with drawings.

FOLLOW-UP

Reteaching

Suggest the following scenario to students, and have them respond to the questions that follow it: Paula scraped mold from the surface of a piece of mozzarella cheese and at once ate some of the cheese. Then she grated the rest to use on a pizza.

Explain why eating the cheese might not have been a good idea. *(Microorganisms or their toxins might remain in the cheese.)* Would the cheese cooked on the pizza be safe to eat? *(Most likely, although some microorganisms might still be alive.)*

Assessment

Ask students to name some of the things people can do to reduce the chances of contamination and infection. *(Wash their hands, use a tissue and turn away from others when sneezing, avoid using a vacuum cleaner without a mask if they are sensitive to dust, keep surfaces dry and clean to discourage growth, and cover and refrigerate foods.)*

One second after the blast, the viruses, in the globules of spray, are about 30–40 cm away from the nose. The liquid evaporates quickly, but the viruses continue to drift slowly through the air. If someone is standing in their path, the viruses may land on him or her. A sneeze propels viruses great distances and in all directions. As a result, it is very difficult to know where you picked up a cold virus or other airborne viruses. Fortunately, we are often able to resist viral attacks because of immunity, which is usually a result of previous exposure.

So you see that microorganisms cannot be avoided. This doesn't mean that we should just give up trying to control them. There are ways to reduce the chance of contamination and infection, and most of them are quite simple.

Washing Your Hands

You probably never realized before how easily foods are contaminated. Keeping your hands as clean as possible probably makes a lot of sense to you now. You should have no difficulty in agreeing with the rules given to Jason at the beginning of this unit.

Everyone realizes the importance of soap in the cleaning process. But just what does it do? Dirt, bacteria, and other microorganisms do not dissolve in water, and water alone does not easily penetrate the many tiny folds in our skin. This is where soap comes in. Soap makes water "wetter," that is, more able to penetrate into tiny spaces to pick up dirt and microorganisms. Soap also dissolves oil and grease so that water can carry them away.

Making Rules for Life

In your ScienceLog, compile a list of rules you think should be followed by yourself and others to avoid problems caused by microorganisms. Take into account all that you have learned in this unit or know from your own experience. Talk to others to get ideas. Organize your rules into groups according to the situations to which they may apply.

184

Extension

Suggest that students determine which diseases are spread primarily through coughing and sneezing. *(Colds, influenza, measles, mumps, pneumonia, tuberculosis, and whooping cough)* Encourage them to construct a chart listing the diseases and their symptoms, the amount of time each disease is contagious, and any precautions that a person could follow to avoid getting the diseases.

Closure

Suggest that students try the following activity: Smear a thin layer of petroleum jelly on a microscope slide. Place the slide on a shelf or windowsill for a day or two. Then examine the slide with a microscope. Ask: What do you observe? What conclusions can you draw about the way microorganisms travel through the air? *(Students may observe microorganisms and conclude that they travel through the air regularly.)*

Let's Eat Out Tonight

Whether you go to a fast-food place or to an elegant restaurant, aren't you concerned about the cleanliness of the place and the care the employees take to control microorganisms? Sometimes eating out can be an unpleasant experience. Have you had any such experiences?

For each of the following dining situations, choose what your response would be. If none of the responses apply, add your own. Explain your responses in your ScienceLog.

What Would You Do If . . .

1. . . . you saw a piece of cream pie sitting uncovered on a counter, and it was served 20 minutes later as part of your order?
   a. Take a chance and eat it.
   b. Refuse it and leave.
   c. Ask for another piece from the enclosed, refrigerated display case.
   d. Ask for the manager.
   e. Other (describe)

2. . . . you saw the serving spoon resting (handle and all) on the potato salad at a salad bar?
   a. Ask for a clean spoon.
   b. Go home and make your own salad.
   c. Avoid the potato salad,

but take other foods from the salad bar.
   d. Speak to someone in charge.
   e. Other (describe)

3. . . . you went to an ice-cream stand on a warm summer day and saw the scoop dipped in the same container of water, time after time, between uses?

a. Go somewhere else.
b. Ask if there is a clean scoop.
c. Eat the ice cream and hope for the best.
d. Buy a packaged ice-cream treat.
e. Other (describe)

Just bring one dessert. We can share.

185

LESSON 4 ORGANIZER

Time Required
1 to 2 class periods

Process Skills
analyzing, classifying, observing

New Terms
none

Materials (per student group)
none

Teaching Resources
Graphing Practice Worksheet, p. 50
Activity Worksheet, p. 52
Transparency 28

FOCUS

Getting Started
Ask the class to imagine that they have gone to a salad bar, a cafeteria, or an elegant restaurant for a meal. Have students give examples of unsanitary conditions that they might encounter at each place.

Main Ideas
1. People can protest the unsafe preparation and handling of food in public eating establishments.
2. There are government agencies that are responsible for overseeing the safety of food preparation in public eating establishments.

TEACHING STRATEGIES

Have students read the introductory paragraphs silently. Then ask them if they can recall a time when they ate out and the food was either prepared or served in an unsanitary way. Call on students to share their experiences.

What Would You Do If . . .
INDEPENDENT PRACTICE Have students work in groups to answer the questions. Encourage each group member to express his or her own opinion. Students may also wish to think of an original response of their own. When students have finished, discuss the questions as a class.

What Would You Do If . . . continued ▶

 **PORTFOLIO**
Suggest that students include their responses to the questions on pages 185–186 in their Portfolio. Encourage students to give reasons for their responses and to note any differences that they had with others in their group.

What Would You Do If . . . ,

continued

If students are having trouble deciding on a course of action, you may wish to offer one of the following hints:

1. Because cream pie is made from dairy products, it can easily become contaminated with microorganisms if left unrefrigerated.

2. Because a great number of people have used the serving spoon, its handle is contaminated with microorganisms. Since it is resting on the potato salad, it is likely that the salad has been contaminated too.

3. Each time the scoop is dipped into the water, it leaves behind a residue of melted ice cream that will provide food for microorganisms, especially in the heat of the day.

4. A sneeze will spread microorganisms some distance from their source.

5. The cloth is likely to contain microorganisms that will contaminate the table and any cutlery placed on it.

6. Flies spread many kinds of harmful microorganisms and can indicate a problem, as they are often attracted to decaying food.

Answers to
What Would You Do If . . . ,
pages 185–186

In each scenario, the proper response would be to avoid the suspect food entirely. Some students might feel strongly enough about the situation to bring the problem to the attention of an employee or manager. Still others might feel that the chance of other foods being contaminated is too great, and they might opt for eating at another restaurant, eating at home, or buying packaged food instead. Point out that an informed response depends on a number of factors, such as the reputation of the dining establishment, the pervasiveness of the problem, and the willingness of the student to eat food that may be contaminated.

4. . . . you saw a waiter or waitress trying to cover a sneeze when passing the warm foods at the buffet table?

   a. Hurry to get some food before the microorganisms have time to multiply.

   b. Speak to the manager.

   c. Warn everyone in your group to avoid the foods on that table.

   d. Eat and enjoy.

   e. Other (describe)

5. . . . your waiter or waitress wiped the table with a damp cloth and then put the cutlery on the table before going for some clean place mats?

   a. Ask for clean cutlery.

   b. Wipe the cutlery with a clean napkin.

   c. Leave.

   d. Forget it—a few germs won't hurt you.

   e. Other (describe)

6. . . . you see several flies in the restaurant?

   a. Wave them away when they approach your plate.

   b. Ask if someone can get rid of them.

   c. Look for another restaurant.

   d. Look for other signs of carelessness.

   e. Other (describe)

These are all imaginary events, but they could happen. What responses did you choose? In such situations, people have choices to make. They can accept the situation as it is and ignore the possible results. They can leave and go somewhere else to eat. They can attempt to point out the problem to someone in charge, with the hope that a change for the better will occur. If the welfare of others concerns you, then you might want to speak to those in charge to encourage them to make changes.

186

Homework

Food producers in the United States cannot sell any food product that does not meet the minimum health standards set by the government. Have students do research to find out what some of these standards are. Ask students if they think these guidelines are too strict or too lax.

Did You Know. . .

Bathrooms were not built into homes until the nineteenth century. Before then, bathrooms were kept separate from the main house because biological waste is a dangerous source of disease-causing organisms.

Another Choice

There is another choice for a customer who is concerned about food safety in a restaurant. The person could send a written complaint to a government department that inspects eating establishments. This is what happened at Pete's Place (more than once, perhaps).

Get Involved!

1. Find out the names and addresses of government departments (local, state, and federal) that regulate food safety.

2. Find out what the function of each department is or if they all have the same purpose.

3. Imagine that you are a dissatisfied customer, and write a letter of complaint about one of the situations discussed above or about one of your own experiences.

It's Not All Bad!

Even though there are many unpleasant or unhealthful things that can happen when you eat out, it would be a mistake to dwell on them. After all, bad dining experiences are actually few and far between. Most people in the restaurant business are trained professionals who are as worried about cleanliness as you are. Their livelihood depends on keeping their customers happy and healthy. Think about the last time you went to a restaurant. Did you get the impression that care was taken to keep conditions clean and healthful? What, in particular, gave you this impression?

187

You may wish to have students work in small groups to complete this Exploration. Have students refer to the descriptions and illustrations on pages 131–133 for more information.

Answers to
You Be the Expert

A wide range of answers is possible for this exercise. Ideal answers will address each of the points outlined in the inspector's report, assess the validity of each, and suggest the ease with which each can be addressed. Answers should specify that Pete and Jason should follow all of the safety guidelines regarding hand-washing, dishwashing, and keeping the kitchen as clean as possible. Also, all perishables should be refrigerated, and all dirty dishes should be kept separate from clean ones. Finally, people who work in the kitchen should keep their hair covered so that it will not fall into the food. Students' letters should be written in a respectful and polite tone.

Homework

Influential books can sometimes spur legislation on the standards of quality and cleanliness in food production. Two such books are *Silent Spring* by Rachel Carson and *The Jungle* by Upton Sinclair. Have students find out what these books are about and what actions were set into motion because of them.

Meeting Individual Needs

Second-Language Learners
Provide students with pictures of several different kinds of food, such as a carton of milk, a container of yogurt, a package of meat or poultry, and eggs. Then have students write a simple sentence or phrase that states how the item should be stored to keep it safe from contamination by microorganisms. Allow them to use a bilingual dictionary if necessary. When students finish, review their work for science content, with only minimal emphasis on language proficiency.

You Be the Expert
Think about the following situation:

Pete would like your help. He has just received the inspector's report shown on the next page. The letter that came with it said that unless some dramatic changes were made immediately, Pete's Place would probably be closed permanently by the Department of Health.

Pete wants you to go over the report with him and discuss the reason for each of the criticisms. He wants to make whatever changes are necessary and to make them quickly so that the Department of Health will allow Pete's Place to stay open. He also wants to convince the department officials that his place provides a valuable service to the community.

What to Do

1. Go over the inspector's report.
2. Explain why each criticism is on the inspector's list. Decide whether the criticisms are valid.
3. Decide how to respond to each item on the list of violations. Which violations can be fixed quickly and easily? Which may take more time or effort?
4. Compose a letter to the Department of Health to indicate the immediate or long-range actions that will be taken. Give any other relevant information that will show the cooperative action being taken.

Homework

The Graphing Practice Worksheet on page 50 of the Unit 3 Teaching Resources booklet makes an excellent homework activity.

Eating Establishment Inspection Report

RECORD TYPE	FACILITY NUMBER	INSPECTION DATE	RE-INSPECTION TIME	INSPECTION TIME	REGION	EMPLOYEE CODE
1 3	4 9	10 DD MM YY 15	16 MM YY 19	20 HRS MIN 23	24 25 26	30
0 0 1						

TYPE OF FACILITY	31 CODE 33	MUNICIPALITY 34 NO 35	QUALITY RATING 36

RECORD TYPE	FACILITY NAME
1 3	10
0 0 2	*PETE'S PLACE* 34

FACILITY ADDRESS STREET 35 59	60 FACILITY ADDRESS TOWN OR CITY 77

Sir/Madam: The following are findings from a recent inspection of your premises by Department of Health staff. Areas where deficiencies exist are indicated and explained below. We would appreciate your cooperation in correcting any deficiencies noted.

	SATISFACTORY	UNSATISFACTORY
1. **Licensing**	☑	☐
2. **Construction:**		
(a) Operation	☑	☐
(b) Floors, Walls, Ceilings	☐	☑
(c) Lighting	☑	☐
(d) Ventilation	☑	☐
3. **Equipment:**		
(a) Design	☑	☐
(b) Construction	☑	☐
(c) Maintenance	☐	☑
4. **Cleaning and Sanitizing of Equipment and Utensils:**		
(a) Mechanical	☐	☐
(b) Manual	☐	☑
5. **Water Supply**	☑	☐
6. **Sewage Supply**	☑	☐
7. **Food Protection:**		
(a) Storage	☐	☑
(b) Covered	☐	☑
(c) Refrigeration	☐	☑
(d) Handling	☐	☑

	SATISFACTORY	UNSATISFACTORY
8. **Personnel:**		
(a) Hygiene	☐	☑
(b) Dress	☑	☐
(c) Habits	☐	☑
9. **Washroom Facilities**	☑	☐
10. **Handwashing Facilities**	☐	☑
11. **Garbage Disposal**	☐	☑
12. **Rodent and Insect Control**	☑	☐
13. **Dry Goods Supplies**	☑	☐
14. **Miscellaneous (identify)**		
(a) _____	☐	☐
(b) _____	☐	☐
(c) _____	☐	☐

EXPLANATION OF DEFICIENCIES AND RECOMMENDATIONS *#2: new window screens needed; and another light, over sink; dusty floor (corners).*
#3: no thermometer in fridge (temperature 6°C—too high).
#4: dirty dishes on counter; need 3 sinks.
#7: open milk carton on counter; sandwich on wooden block; damp dish towels hanging under counter; chipped dishes; food spilled on refrigerator shelf; open mayonnaise bottle on counter; cracked wooden meat block; drops of liquid on counter; must not dry-sweep the floor; keep disinfectants out of food storage area; also keep mop out of food area;
#8: personnel must wash hands properly; hang outer clothes away from food area.
#7: cover all baked goods on display.

OTHER COMMENTS *It is most important to eliminate any use of damp sponges for wiping up; and use of dish towels (dishes will air-dry quickly when washed properly).*

2/5/97
Date of Inspection

G. Grover
Inspector

OWNER OPERATOR COPY

189

FOLLOW-UP

Reteaching

Ask students to make notes about both the safe and unsafe practices they notice in a restaurant the next time they eat out. Have them make a written report of their findings.

Assessment

Ask students to imagine that they have noticed unsanitary conditions in a restaurant. Ask: What choices do you have in terms of responding to the situation? If you choose to bring the situation to the attention of others, who are some people you could contact? *(You could speak to the manager of the restaurant, or you could write a letter to public-health officials.)*

Extension

Remind students of the hand-washing technique described on page 132. Invite a nurse or doctor to your classroom to describe either the 3-minute hand-washing technique or the 7-minute scrub that is required of personnel in the operating rooms of hospitals. The guest could also explain other sterilization procedures used in the hospital environment.

Closure

Tell students to imagine that they are in charge of a food sale to raise money for a class trip. Ask them to draw up a list of reminders about food safety for the students who will be helping with the sale.

Homework

The Activity Worksheet on page 52 of the Unit 3 Teaching Resources booklet makes an excellent homework activity. If you choose to use this worksheet in class, Transparency 28 is also available.

Did You Know...

Of the 13 known vitamins, five are produced inside the human body. Three of those five are made by bacteria: vitamin K, biotin, and pantothenic acid.

Answers to *Challenge Your Thinking,* pages 190–191

1. a. An inspector might consider sources such as the hygiene of restaurant workers, the condition of food entering the restaurant, and the conditions in the restaurant.

b. From the figure, a human carrier who does not not wash his or her hands, for example, might infect food in the factory. If farm animals ate *Salmonella*-infected food from the factory, the animals would also become infected.

c. *Salmonella* can be avoided with proper hygiene and can be killed by cooking food properly.

d. A cycle is a set of events that recur in the same sequence. By starting at any point in the figure, a path of *Salmonella* contamination can be traced back to its source. The figure actually consists of two smaller cycles that are connected by the food-processing factory.

2. a. Take a sandwich without any substances (such as eggs, dairy products, and most dressings) that would encourage the growth of microorganisms. Use vinegar, lemon juice, pickles, and salt because salt and acid discourage the growth of microorganisms. Peanut butter, tomato, or cucumber sandwiches would be safe because they contain no meat, dairy, or egg products.

b. Sample answer: meat or fish sandwiches and cream pie or other desserts. Meat or fish sandwiches could spoil quickly. Cream pie or other desserts with milk or eggs offer excellent conditions for microorganisms to multiply, especially at warm temperatures.

3. 1. b; molds and bacteria
2. a; yeasts
3. b and c; mainly bacteria
4. a; molds and bacteria
5. b; viruses
6. b; bacteria
7. b and d; molds and bacteria
8. c; protists, bacteria, and molds
9. b; protists, bacteria, and molds
10. c and d; molds and bacteria
11. b; bacteria

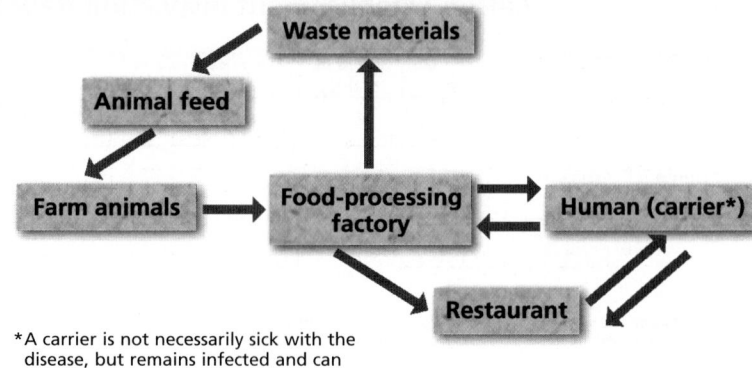

*A carrier is not necessarily sick with the disease, but remains infected and can pass on the infection.

1. The *Salmonella* Cycle

The figure above shows how *Salmonella* bacteria can be passed on to infect many people.

a. If you were a health inspector, where would you look for a possible source of *Salmonella* bacteria in a restaurant?

b. How might a human carrier cause farm animals to become contaminated with *Salmonella* bacteria?

c. *Salmonella* bacteria are very common, but we can avoid becoming ill from them. How?

d. Why is this figure called a cycle?

2. Pack a Lunch

a. Describe what kind of sandwich you would take with you for an outing on a summer day when you had no way to keep the food cold. Tell why you would choose this kind of sandwich.

b. Name two foods that you would not recommend taking because of the risk of food poisoning. Explain why you named these foods.

 You may wish to provide students with the **Chapter 10 Review Worksheet** that is available to accompany this Challenge Your Thinking (Teaching Resources, page 53). Transparency 29 is also available.

Homework

Have students consider the figure at the top of page 190. Ask: In which key area would improved sanitation be most effective in reducing *Salmonella* contamination? *(The food-processing factory is the best answer because it is the link between the two smaller cycles in the diagram.)*

3. Mysterious Microbes

Here are some principles you have learned recently:

a. Some microorganisms can be used to help us.

b. We need to control some microorganisms to prevent them from causing harm.

c. The system of microorganisms in nature seems to keep itself in balance.

d. Human actions can cause unwelcome changes in the natural balance of microorganisms.

For each of the specific examples below, choose the principle above that best applies.

1. Antibiotics can kill some bacteria.

2. A little bit of sourdough can make good bread.

3. Microorganisms are found in large numbers on human hands.

4. Soil microorganisms are the source of many antibiotics.

5. Measles is caused by a type of microorganism that cannot be killed by an antibiotic.

6. *Salmonella* bacteria on eggs and poultry often cause food poisoning.

7. It is wise to prescribe antibiotics only when absolutely necessary.

8. Soil microorganisms are a part of food chains.

9. Dishcloths and sponges in most homes contain dangerous microorganisms.

10. Bacteria may become resistant to an antibiotic and thus make the antibiotic ineffective.

11. Refrigeration is necessary to keep dairy products from spoiling.

Identify which microorganisms (protists, bacteria, viruses, yeasts, or molds) are being referred to in each of the preceding statements.

ScienceLog

Review your responses to the ScienceLog questions on page 170. Then revise your original ideas so that they reflect what you've learned.

191

The following are sample revised answers:

1. Colds and the flu are caused by viruses. Antibiotics will not cure a cold or flu because antibiotics are effective against only bacterial infections.

2. To reduce the risk of food spoilage, food should be prepared and stored above 65°C or below 5°C. The microorganisms that cause food to spoil multiply most quickly between these two temperatures.

3. Some of the ways that harmful microorganisms can be controlled include good personal hygiene, careful preparation and storage of food, and proper use of antibiotics.

4. People should be aware of unsafe practices in the preparation and storage of food whether they are at home or in a restaurant. When eating out, avoid food that is not stored properly or that does not look, smell, or taste fresh.

Meeting Individual Needs

Learners Having Difficulty

To help students understand the effect of preservatives on microorganisms, have them perform the following experiment: Dissolve a bouillon cube in 250 mL of hot water. Divide the water equally among three glasses. Add a spoonful of sugar to the first glass and label it *Sugar*. Add a spoonful of vinegar to the second glass and label it *Vinegar*. Label the third glass *Control*. Place the glasses in a warm place for two days and then examine them. Which is the cloudiest? Why? *(The control should be cloudiest because it did not contain sugar. The bouillon in the glass labeled* Vinegar *should be clearest because vinegar slows down bacterial growth the best.)*

Theme Connection

Cycles

Have students refer to the *Salmonella* diagram on page 190. **Focus question:** How many cycles can you identify in the diagram? *(At least two)* From what sources can a human be infected by *Salmonella? (From a food-processing factory or a restaurant)* Encourage students to draw a diagram of their own in which a human transmits *Salmonella* to another human by some means not indicated in the diagram on page 190.

The Big Ideas

The following is a sample unit summary:

Microorganisms are tiny living things that can be seen only by using a microscope. There are many different kinds of microorganisms, including bacteria, algae, protists, and viruses (although viruses are not always classified as living things). (1)

Microorganisms can be found everywhere—in soil, air, and water; on nonliving and living things; and even inside living things. (2)

Microorganisms are helpful in many ways. They function as decomposers; they provide food for other organisms and are thus a vital link in the food chain; their waste products create foods such as cheese, bread, vinegar, and yogurt; and their natural defense systems provide substances that humans can use to combat diseases.

Microorganisms can be harmful because they can spoil food and can cause diseases in plants and animals. (3)

Louis Pasteur's experiments showed that bacteria can cause food to spoil and that although bacteria live everywhere, their growth can be controlled by altering the conditions of their environment. For example, Pasteur demonstrated that bacteria can be killed by heating. (4)

An antibiotic may kill not only the disease-causing bacteria but also beneficial ones. Also, new strains of bacteria that are resistant to known antibiotics may develop. (5)

Bacteria thrive in a range of temperatures known as the optimum growth range. Above and below this range, bacteria either die or fail to grow and reproduce. Knowing this, we can keep the temperature of our foods out of this range. (6)

Methods for controlling microorganisms include sterilization, pasteurization, improving personal hygiene, and controlling the temperature of foods. (7)

All food-preservation methods involve the elimination of microorganisms or the control of their growth. (8)

Although there are many harmful microorganisms, most are beneficial to us. Therefore, control of all microorganisms is not necessary. (9)

Unit 3

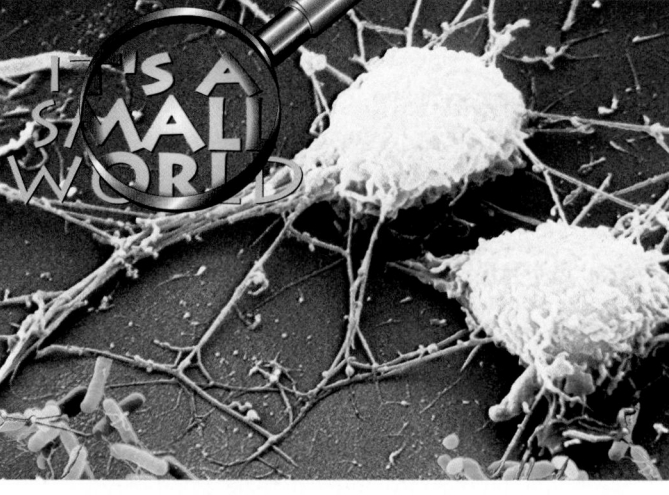

SOURCEBOOK

Turn to the SourceBook to learn more about how microorganisms cause disease, how they live and multiply, and how they are classified. You will also learn about the human body's defenses against microorganisms.

Here's what you'll find in the SourceBook:

UNIT 3
Microorganisms and Disease
  S66
More About Bacteria
  and Viruses S72
The Body's Defenses S78

The Big Ideas

In your ScienceLog, write a summary of this unit, using the following questions as a guide:

1. What are microorganisms?
2. Where are microorganisms found?
3. How are microorganisms helpful? harmful? Give examples.
4. How did Louis Pasteur contribute to our understanding of the causes of disease?
5. Why is it important to understand how to use an antibiotic wisely?
6. How are bacteria affected by temperature? How can people use this to their advantage?
7. What are some methods of controlling microorganisms?
8. What do all food-preservation methods have in common?
9. Is it desirable to control all microorganisms?

Checking Your Understanding

1. Refrigeration does not kill most microorganisms, so how is it possible to preserve foods simply by keeping them cold?

192

 You may wish to provide students with the Unit 3 Review Worksheet that is available to accompany this Making Connections (Teaching Resources, page 61).

Homework

The Unit Activity Worksheet on page 59 of the Unit 3 Teaching Resources booklet makes an excellent homework activity. If you choose to use this worksheet in class, Transparency 30 is also available.

2. Imagine that you have made a large pot of stew. Naturally, you would like to prevent it from spoiling. Would it be better if you

   a. kept the stew in one large pot or several smaller ones?

   b. placed the pot in the refrigerator immediately after taking it off the stove or waited until the stew had cooled?

   c. stored the stew covered or uncovered?

   d. froze the stew or refrigerated it?

3. The illustrations below show three bacterial cultures that were once healthy but that have become contaminated by mold. Each culture has been contaminated by a different type of mold. Which mold might be worth investigating for its special properties? Explain.

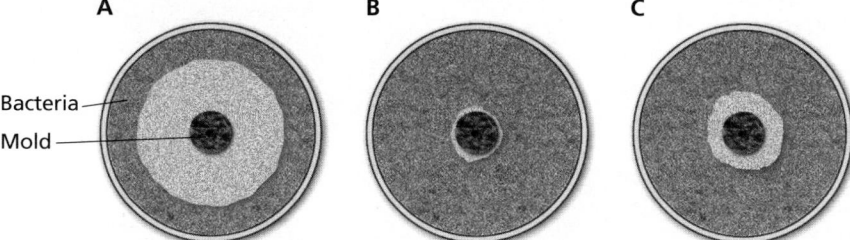

A   B   C

Bacteria

Mold

4. *(concept map)* Make a concept map using the following terms: *viruses, bacteria, molds,* and *antibiotics.*

5. Most scientists do not consider viruses to be living things. Decide for yourself whether viruses are living or nonliving, and then write down an argument to support your position.

6. Below are a few statements about microorganisms. Indicate whether you agree with, disagree with, or are uncertain about each statement. Then briefly give a reason for your choices. Do not write in this book.

Statement	Agree	Disagree	Uncertain	Reason
a. Microorganisms cause more harm than good.				
b. If microorganisms were to vanish, you probably wouldn't notice.				
c. Microorganisms are essential to us.				
d. Without microorganisms, you couldn't make a pizza.				
e. Gardeners benefit from microorganisms.				

193

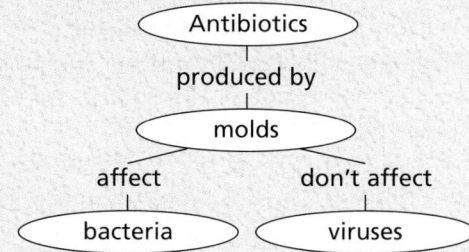

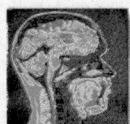
Background

Antibiotics are drugs used to treat diseases caused by bacteria. Antibiotics interfere with the disease-causing (*pathogenic*) bacteria by either damaging the cell membrane or disrupting chemical processes in the cell.

Antibiotics must be used properly to be effective. Some possible side effects from antibiotics include allergic reactions, damage to organs and tissues, and destruction of helpful microorganisms.

Modern scientific study of antibiotics began in the 1800s. Louis Pasteur discovered that bacteria spread infectious diseases, and Robert Koch developed methods for isolating and growing different kinds of bacteria. A breakthrough in treating bacterial diseases came in the early 1900s when Alexander Fleming discovered penicillin, an antibiotic formed from mold. Streptomycin, a fungal antibiotic, was discovered by Selman A. Waksman in 1943. Doctors now use antibiotics to treat diseases such as strep throat, bacterial meningitis, and tuberculosis.

Extension

Have students call or visit a pharmacy and investigate the following questions:

- Why is it necessary to finish all of the medication when taking antibiotics? *(If medication is stopped early, some bacteria that are resistant to the medication can survive and reproduce. These resistant bacteria are likely to produce other resistant bacteria. The illness caused by the new population of resistant bacteria will be harder to treat with the same medication.)*

- The instructions on some types of antibiotics state, "Take on an empty stomach," while others state, "Take with food or milk." Why? *(Some antibiotics should be taken on an empty stomach because they can be better absorbed without food. Also, some antibiotics are broken down and rendered less effective by stomach acids that are produced during digestion. Some antibiotics should be taken with food because they are more easily absorbed with food. Food can also act as a buffer for antibiotics that might otherwise cause stomach upset.)*

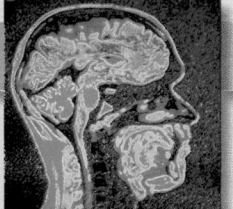
Frogs in the Medicine Cabinet?

Frog skin, mouse intestines, cow lungs, and shark stomachs—sounds like the ingredients for a nasty witch's brew, doesn't it? Actually, these animal parts are being tested in an effort to create more effective medicines to combat harmful bacteria.

Leapin' Lilypads—It's Infection Protection

In 1896 a biologist named Michael Zasloff was studying African clawed frogs. He noticed that cuts in the frog's skin always healed quickly and never became infected. Zasloff decided to investigate further. He found that when the frog was cut, its skin released a milky liquid that killed invading bacteria. Within the milky liquid was an infection fighter called an *antibiotic*.

Scientists have found other animals whose bodies contain similar infection fighters. For example, the stomach and

tissues of sand sharks (also called dogfish) contain chemicals that kill bacteria and other microorganisms. Scientists have also found useful antibiotics in moths, pigs, mice, cows, and even in the small intestines of humans!

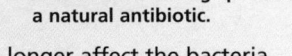

▲ African clawed frogs produce a natural antibiotic.

POW! Punching Holes in Bacteria

Many antibiotics, such as penicillin, kill bacteria by keeping the microorganisms from growing or reproducing. Unfortunately, bacteria can become resistant to these drugs, which means that the drugs no longer affect the bacteria. Scientists then face the challenge of developing new antibiotics that can overcome the resistant strains of bacteria.

Antibiotics from animals pack a different punch. These substances bore holes through the membranes that surround bacterial cells. Then the cells disintegrate and die. Bacteria have a hard time altering their cell membranes, so they are less likely to become resistant to the animal antibiotics.

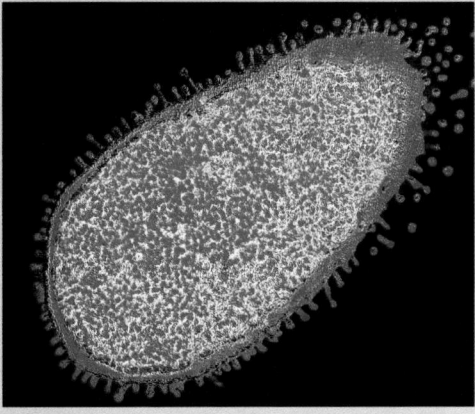

◄ This electron micrograph shows the effect of an antibiotic on a *Salmonella* bacterium. The antibiotic is causing the bacterium's outer membrane to distort and eventually burst, killing the bacterium.

What's in Dog Spit?

A healthy dog licks cuts, scrapes, and minor wounds to clean them. A mother cat licks her kittens clean. Ask your local veterinarian or do some research to find out how the saliva of dogs, cats, and other animals fights bacteria.

194

Answers to
What's in Dog Spit?

Dogs, cats, humans, and some other animals have an antibacterial enzyme in their saliva. When animals lick a wound, the enzymes break down the bacteria and help the wound heal.

The Biggest Living Thing on Earth

The largest organism in the world lives underground. It ruthlessly attacks its prey, leaving death and decay in its path. It can withstand fire and torrential rain. And it's getting bigger every day!

The Humongous Fungus

Relax! You have nothing to fear from this terrifying organism. It's actually a *fungus,* a parasite that feeds on trees. Scientists call it *Armillaria,* and it's been found in enormous proportions in the forests of Washington and Michigan.

Just how big *is* the world's biggest living thing? The first humongous fungus that was found covers more than 150,000 m²—that's about 33 football fields! It was discovered near Crystal Falls, Michigan, in 1992. Soon after, an even bigger discovery was made by forest researchers near the slopes of Mount Adams in Washington State. They found a fungus almost 30 times the size of the one in Michigan. This mammoth *Armillaria* covers more than 4 km²!

The Life of a Fungus

A fungus begins as one of millions of microscopic spores produced by a single mushroom. Wind carries the spore away from the mushroom. When the spore lands in the proper environment, it begins to grow. Underground, the spore forms a network of threadlike branches called *rhizomorphs.* The tip of a rhizomorph has special substances that break down wood. When the rhizomorph finds a root, the fungus immediately begins to digest it. Once the roots of a tree are digested by the fungus, the whole tree dies.

After a rain, the fungus produces honey-colored mushrooms above the ground and on the trunks and roots of trees. Each mushroom has spores underneath its cap. When those spores are released, the reproductive cycle of the fungus begins again.

▼ These mushrooms are visible signs of the vast organism beneath the soil.

◄ Notice how *Armillaria* invades the tree from the roots.

A Fungus Among Us!

Explore a fungus on your own. Use an edible mushroom from the grocery store. Do not use wild mushrooms—some of these can be poisonous. Observe the way a mushroom looks. Then tap the mushroom on a piece of paper to release the spores. Carefully slice open the mushroom to see what it looks like inside. You can illustrate your findings in your ScienceLog.

195

Background

The branch of science that deals with fungi is *mycology.* Mycologists study many different kinds of fungi. Yeasts, mushrooms, molds, ringworm, and mildew are some examples.

Fungi cannot make their own food because they lack chlorophyll. They secrete enzymes to digest their food outside their body and then absorb the nutrients.

Examples of organisms in the same phylum as *Armillaria* are edible mushrooms, toadstools, puffballs, rusts, and smuts. *Armillaria* is a parasite that attacks the roots of plants and can spread in three ways: reproducing by spores released from the honey-colored mushrooms, spreading by root-to-root contact, and extending rhizomorphs from infected tree stumps.

Substances called fungicides are designed to kill fungi that are harmful to plants. Fungicides can be sprayed or dusted on plants to kill the parasitic fungi.

A Fungus Among Us!

The students should be able to see spores on the paper. They look like specs of black pepper. The day before the activity, you may want to stand a mushroom on a piece of paper to catch spores. After the lesson, you can show your mushroom "print" to students. You may want to introduce the following terms to students as they do the activity: cap (top part of the mushroom), stalk (stem-like base of the mushroom), and gills (tissues underneath the cap which contain the spores.)

Background

Many applications of useful bacteria are in use or under development. For instance, some bacteria are currently being used to help break down toxic substances such as cyanide.

Certain bacteria store energy in the form of a biodegradable plastic called PHB. Originally, researchers found that the plastic was stiff and brittle. It also had a high melting point, which made it difficult to work with. However, scientists found that by feeding the bacteria a mixture of glucose and organic acids, the bacteria could produce a stronger, more flexible plastic with a lower melting point.

Genetically engineered bacteria have been developed to help create more successful fruit and vegetable crops. This technology could provide fresher and more nutritious food for consumption around the world. Opponents of genetically engineered food express safety concerns involving the introduction of new genetic material into the environment.

Extension

Have students work in groups to role-play the following scenario: A major oil company is holding a meeting to discuss the advantages and disadvantages of supporting bacterial research. *(Students need to select a name for their company, explain the jobs they perform, and present the minutes of their meeting to the class. It would be to the company's advantage to support research because it could protect the environment by providing new and better ways to clean up oil spills. One disadvantage of the research could be the possibility of harming an ecosystem by introducing new organisms that could disrupt the food web.)*

Bacteria at Your Service

Wanted: *Hard worker to clean up trash. Must have experience curing diseases. Ability to make plastics a plus. Only microorganisms need apply.* What could possibly be flexible enough to perform so many different and useful activities? Bacteria!

Cleaning Up Our Act

Without bacteria, the Earth would be littered with the dead remains of plants and animals. That's because bacteria decompose once-living matter. Some bacteria break down other substances as well. These microorganisms offer solutions to some of our toughest pollution problems.

When an oil spill occurs, it often causes severe damage to the environment. However, scientists have now produced

▼ **These bacteria store energy as plastic granules, which are then used to create biodegradable plastic products.**

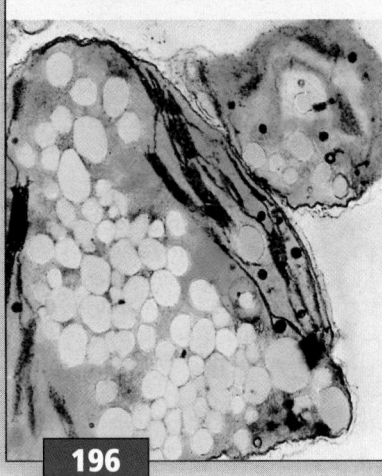

▶ **These bacteria are used to clean up oil spills.**

bacteria that actually feed on oil! As the bacteria eat the oil, they break it down into harmless substances. Scientists hope to use the bacteria to clean up large oil spills, but they must first be certain that introducing a large number of these bacteria into the ocean will not harm the environment.

Other kinds of bacteria decompose toxic wastes that cause diseases or birth defects in humans. Scientists have even cultured bacteria that break down explosive wastes.

Producing Plastics and Pills

Did you know that some bacteria act like factories? For example, one kind of bacteria manufactures biodegradable plastic! These amazing bacteria store energy as plastic granules, just as animals store energy as fat and plants store energy as starch. When certain substances are added to the bacteria's diet, the plastic they produce is flexible enough to be made into consumer products.

Scientists have found other bacteria that can produce important drugs used to fight disease. In fact, some bacteria even produce antibiotics that can fight infections caused by other bacteria! For more information about this, read page 194.

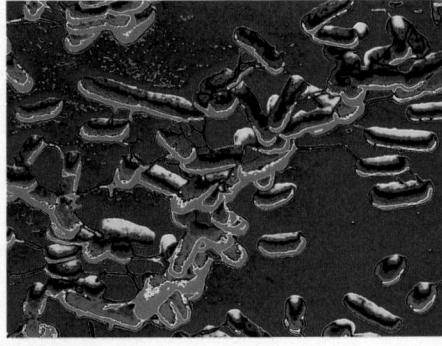

Helping Plants Grow

Some bacteria can change the way plants grow, flower, and reproduce. Genetic engineers have designed bacteria that can help make plants pest-resistant and cold-resistant. Some bacteria even keep foods fresh longer on the grocery store shelves. From the garbage dump to the grocery store, bacteria are at our service!

Hard-Working Bacteria

Observe bacteria returning nutrients to soil. Place 2–3 cm of soil at the bottom of a glass jar. Place scraps, such as potato peel or bread crusts, on top of the soil. Put the scraps close to the sides of the jar so that you can see the changes as they occur. Place a second soil layer on top of the scraps. Screw the lid on the jar and observe the scraps over a few weeks. Describe what happens to the scraps over time.

Hard-Working Bacteria

Students can use any size jar with a lid. They can recycle scraps from their lunches and add items such as paper and gelatin. After layering the soil, make sure that students screw the jar lids on tightly. Place the jars in a dark place to simulate being underground. Observe the jars on a daily or weekly basis. Students should observe the rate at which bacteria decompose the scraps. You may wish to explain that some bacteria require oxygen while others do not.

Meeting Individual Needs

Second-Language Learners
Encourage students to observe pictures or microscope slides of the different shapes of bacteria. Have them draw and label the pictures.

Science Snapshot

Dr. Jonas Salk (1914–1995)

Some of the victims would never walk again. Others could no longer breathe without the aid of a tanklike iron lung. Many lost their lives. The cause of these tragedies was the polio virus. Fortunately, Dr. Jonas Salk produced a successful vaccine against this deadly disease, earning international recognition for his research and the gratitude of a nation.

Preparing for Battle

Born in New York City in 1914, Jonas Salk came from a working-class family. He entered medical school in New York in 1939. Eight years later, Dr. Salk was in charge of viral research at the University of Michigan, where he developed the first *vaccine* against polio.

▼ **The iron lung helped polio victims breathe. Some people had to spend most of their lives in an iron lung.**

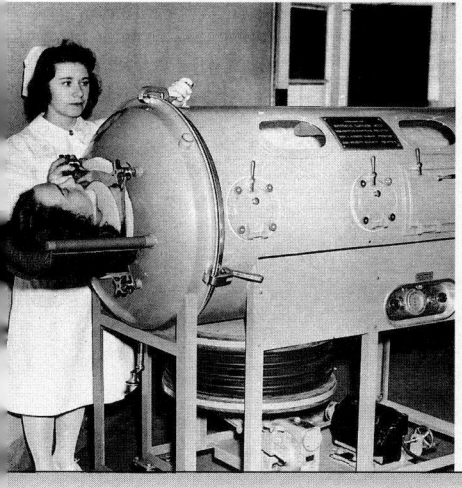

Dr. Salk took the live virus and killed it with a chemical called *formaldehyde*. The dead virus was then injected into the body as a vaccine. Vaccines work by tricking the body into producing defenses against a virus before the infection occurs. Then later, if the infection does occur, the body's defenses are already in place and active. The virus is defeated before it gets a chance to attack!

Heroic Efforts

In 1952, the worst year of the polio epidemic, nearly 58,000 cases were reported, and 3000 deaths occurred. That year, Dr. Salk began conducting field tests of his vaccine; he included his whole family in these early tests. By 1954, he and his research team had vaccinated 1,830,000 school children in a huge research effort. By 1962, fewer than 1000 new cases of polio were reported. Dr. Salk had succeeded!

In the 1960s, an oral vaccine that used an altered living virus was developed by the scientist Albert Sabin. This vaccine eventually replaced the Salk vaccine.

Continuing the War Against Viral Disease

During the last years of his life, Dr. Salk challenged yet another microscopic opponent—HIV, the virus that causes AIDS. He and his research team searched for

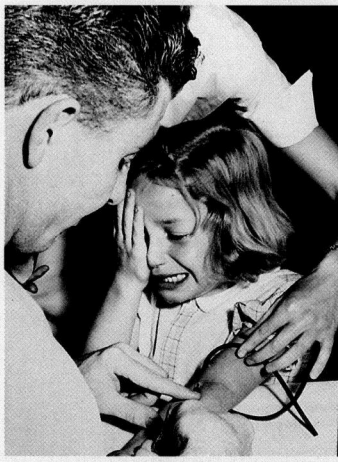

▲ **Dr. Jonas Salk giving a young girl the polio vaccine**

ways to help AIDS victims fight the existing infection. Dr. Salk died in 1995, having spent his final days continuing the fight against infection and disease.

A Part of History

Talk with relatives or friends who were alive during the polio epidemic. They probably remember the terrible fear caused by polio. Find out if they ever had the disease or if they knew people who suffered from it. Ask them what their reaction was when the vaccine was found to be effective. Be sure to save their responses on a cassette tape or in your ScienceLog—it's a firsthand account of medical history!

197

Science Snapshot

Background

Polio is caused by a virus that attacks nerve cells of the spinal cord and can cause paralysis. The virus enters the body through the nose and mouth and travels to the intestines. Then the blood or nerve fibers carry the virus to the central nervous system. Once the virus invades a cell, it multiplies rapidly, killing the cell. Paralysis results when many nerve cells are destroyed.

In a mass trial in 1954, Salk's vaccine was given to nearly 2 million children. The project was funded by the March of Dimes Birth Defects Foundation (formerly called the National Foundation for Infantile Paralysis). The vaccine was administered by injection and was declared safe and effective in April 1955. Salk received a congressional gold medal for "great achievement in the field of medicine."

In 1961 physicians started using the Sabin vaccine. Unlike the Salk vaccine, which uses whole, dead viruses, the Sabine vaccine consists of live, weakened forms of the polio virus. Advantages of the Sabin vaccine are that it lasts longer and can be administered orally instead of by injection.

In the United States, children should receive the polio vaccine at 4 months of age and a polio booster at 4–6 years of age. In 1952, more than 21,000 cases of polio were reported. By the 1980s, the number of new cases had dropped to fewer than 10 per year.

Extension

Have students find out about other vaccines that have been developed for diseases like influenza and diphtheria. Have them present their results with a poster. Students may be able to examine their own records of inoculation and compare them with those of other students in the class.

CROSS-DISCIPLINARY FOCUS

Social Studies

Have students create a calendar in honor of famous scientists. Provide students with a list of scientists. Then have students document the scientist's date of birth and most important accomplishments. Each student should record his or her findings on the calendar on the day the scientist was born. For example, Jonas Salk was born on October 28, 1914. The calendar entry for October 28 should be: "Jonas Salk, 1914, polio vaccine."

A Part of History

Have students share the information they learned from talking to polio survivors. Then lead class discussion by talking about the experiences of polio victims, their treatment, and their reactions to the creation of the vaccine.

Unit 4 — INVESTIGATING MATTER

Unit Overview

In this unit, a discussion of matter and its properties is presented in ways that allow students to draw from their own experience and knowledge. In Chapter 11, students formulate a working definition of matter and use creative-thinking skills to observe, analyze, and describe the properties of matter. In Chapter 12, students review the metric system and are asked to apply what they have learned to the quantitative analysis of mass and volume. In Chapter 13, students investigate the different states of matter. The unit concludes with a discussion of the particle model of matter and challenges students to develop models of their own.

Using the Themes

The unifying themes emphasized in this unit are **Structures, Cycles,** and **Energy.** The following information will help you incorporate these themes into your teaching plan. Focus questions that correspond to these themes appear in the margins of this Annotated Teacher's Edition on pages 204, 217, 233, 234, 239, and 248.

Structures may be discussed from two points of view. In Chapter 13, call students' attention to the connection between matter's physical properties (for example, its state) and its underlying structure. In connection with Lesson 3 of Chapter 13, discuss some of the apparent oddities in the way that matter behaves (such as the ability to dissolve a substance in a liquid without noticeably changing the volume of the liquid). These oddities can be explained by the structure of matter at the particle level.

As students explore changes of state, this information can be integrated into their previous knowledge of natural **Cycles** such as the water cycle.

A discussion of snow melting in the spring, for instance, can facilitate their comprehension of both changes of state and the cycles that occur in nature.

Energy can be explored by observing how a substance changes its energy level as it undergoes physical changes. Although not discussed explicitly in the text, this notion is an important part of understanding the differences among the various states of matter.

Using the SourceBook

Unit 4 includes discussions of molecules; the molecular basis of the states of matter; mixtures, compounds, and solutions; elements and atoms; and subatomic components. Throughout the unit, the molecular basis of the properties of matter is emphasized.

Bibliography for Teachers

Barber, Jacqueline. S*olids, Liquids, and Gases: A School Assembly Program Presenter's Guide.* GEMS Series, ed. by Lincoln Bergman and Kay Fairwell. Berkeley, CA: Lawrence Hall of Science, 1986.

Davies, Paul, and John Gribbin. *The Matter Myth.* New York City, NY: Simon & Schuster/Touchstone, 1992.

Trinklein, Frederick E. *Modern Physics.* Austin, TX: Holt, Rinehart and Winston, Inc., 1992.

Bibliography for Students

Berger, Melvin. *Solids, Liquids and Gases: From Superconductors to the Ozone Layer.* New York City, NY: G. P. Putnam's Sons, 1989.

Cooper, Christopher. *Matter.* Eyewitness Science series. New York City, NY: Dorling Kindersley, Inc., 1992.

Darling, David. *From Glasses to Gases: The Science of Matter.* New York City, NY: Dillon Press, 1992.

Films, Videotapes, Software, and Other Media

Measurements: Length, Mass, and Volume
Software (Apple II)
Focus Media, Inc.
485 South Broadway, Suite 12
Hicksville, NY 11801

The Molecular Theory of Matter (2nd ed.)
Videotape
Britannica
310 S. Michigan Ave.
Chicago, IL 60604-9839

Particles in Motion: States of Matter
Film
National Geographic Society
Educational Services
P.O. Box 98019
Washington, DC 20090-8019

The World of Molecules (rev. 2nd ed.)
Film or videotape
Churchill Media
12210 Nebraska Ave., Dept. 200
Los Angeles, CA 90025-9816

Unit Organizer

Unit/Chapter	Lesson	Time*	Objectives	Teaching Resources
Unit Opener, p. 198				Science Sleuths: The Blob! English/Spanish Audiocassettes Home Connection, p. 1
Chapter 11, p. 200	Lesson 1, About Matter, p. 201	2 to 3	1. Specify whether a particular property of matter is biological, physical, or chemical. 2. Explain the differences between chemical and physical properties.	Exploration Worksheet, p. 3 ▼
	Lesson 2, Matter's Useful Properties, p. 205	1	1. Identify properties of particular materials that make the materials useful. 2. Classify these properties as physical, chemical, or biological. 3. Investigate the properties and uses of paper through research and experimentation, and prepare a report.	Image and Activity Bank 11-2 Resource Worksheet, p. 5 ▼
End of Chapter, p. 208				Chapter 11 Review Worksheet, p. 6 Chapter 11 Assessment Worksheet, p. 9
Chapter 12, p. 210	Lesson 1, SI: The Metric System of Measurement, p. 211	1 to 2	1. Review which units of measurement are appropriate to measure length, volume, and mass. 2. Sequence the metric prefixes in order of magnitude. 3. Convert from one metric unit to another.	Image and Activity Bank 12-1 Math Practice Worksheet, p. 12 ▼ Resource Worksheet, p. 13 ▼ Resource Worksheet, p. 14 ▼
	Lesson 2, Volume, p. 214	3 to 4	1. Define volume. 2. Use a graduated cylinder to measure volume. 3. Measure the volume of solids and gases by using the displacement method.	Image and Activity Bank 12-2 Exploration Worksheet, p. 15 Exploration Worksheet, p. 17
	Lesson 3, Mass, p. 222	3	1. Define mass. 2. Use a balance to measure mass. 3. Compare the relative size of a gram to that of a kilogram. 4. Explain the relationship between the mass and the volume of a substance. 5. Make two different types of balances.	Image and Activity Bank 12-3 Exploration Worksheet, p. 21 ▼ Exploration Worksheet, p. 24 ▼ Resource Worksheet, p. 26 ▼ Exploration Worksheet, p. 27
End of Chapter, p. 226				Chapter 12 Review Worksheet, p. 28 Chapter 12 Assessment Worksheet, p. 31
Chapter 13, p. 228	Lesson 1, States of Matter, p. 229	1 to 2	1. Classify matter as either solid, liquid, or gas. 2. Discuss the properties that characterize each state of matter.	Image and Activity Bank 13-1 Transparency Worksheet, p. 33 ▼ Theme Worksheet, p. 35 Resource Worksheet, p. 37
	Lesson 2, Changes of State, p. 232	3	1. List the terms relevant to changes of state and correctly use them in sentences. 2. Provide examples of each change of state. 3. Define melting point, freezing point, and boiling point as properties of matter. 4. Explain how the dew point of a vapor (gas) can be found.	Image and Activity Bank 13-2 Activity Worksheet, p. 38 Resource Worksheet, p. 39 ▼ Exploration Worksheet, p. 41
	Lesson 3, A Model of Matter, p. 237	2	1. Explain why models are useful. 2. Use the particle model of matter to explain some properties of solids, liquids, and gases.	none
End of Chapter, p. 242				Chapter 13 Review Worksheet, p. 45 ▼ Chapter 13 Assessment Worksheet, p. 48
End of Unit, p. 244				Unit 4 Activity Worksheet, p. 52 ▼ Unit 4 Review Worksheet, p. 53 Unit 4 End-of-Unit Assessment, p. 56 Unit 4 Activity Assessment, p. 61 Unit 4 Self-Evaluation of Achievement, p. 64

* Estimated time is given in number of 50-minute class periods. Actual time may vary depending on period length and individual class characteristics.

▼ Transparencies are available to accompany these worksheets. Please refer to the Teaching Transparencies Cross-Reference chart in the Unit 4 Teaching Resources booklet.

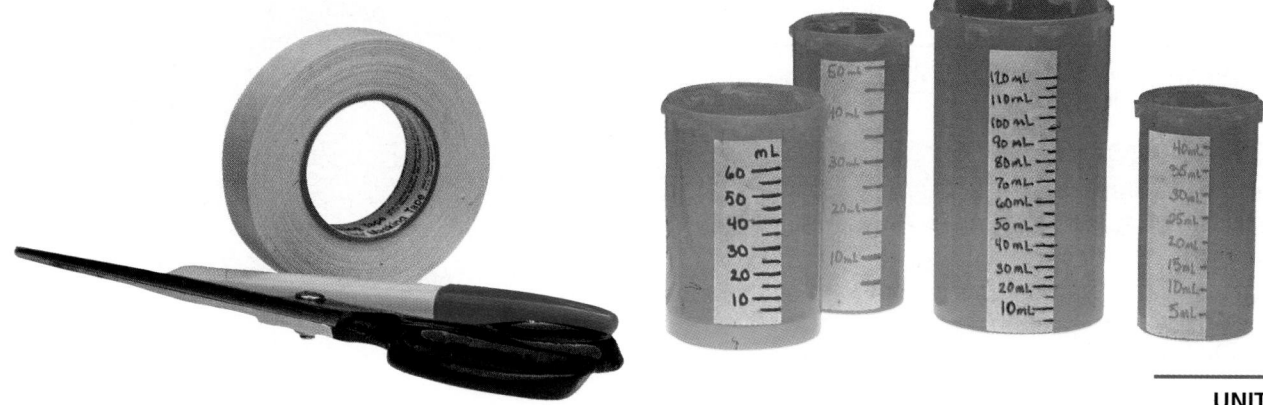

Materials Organizer

Chapter	Page	Activity and Materials per Student Group
11	203	***Exploration 1:** 500 mL of water; 500 mL of vinegar; 4 beakers (any size); ice cube; 2 eggs; balloon filled with air; 2 sugar cubes; candle; jar lid; small ball of modeling clay; matches; earthworm; geranium; air freshener; safety goggles; lab aprons
12	212	**Metric Concentration:** 21 index cards
	212	**Metric Spin:** 21 index cards; sheet of stiff paper; scissors; paper fasteners
	215	**Exploration 1:** 100 mL graduated cylinder; 1000 mL of water; white paper; empty 1.89 L ($\frac{1}{2}$ gal.) plastic milk jug; scissors; masking tape; empty transparent pill bottle; marble or pebble; metric ruler; marker
	218	**Exploration 2, Activity 1:** 2 L plastic container; crayon; **Activity 2:** 1000 mL beaker; graduated cylinder or metric measuring cup; bowl or aluminum pie plate; 1 L of water; funnel; crayon; **Activity 3:** 3.785 L (1 gal.) plastic container with screw-on top; large pan; 1 m of rubber tubing; 5 L of water; marker; 1000 mL beaker or graduated cylinder; 1 drinking straw per student
	222	**Exploration 3:** metric balance; stick of modeling clay; several small classroom objects
	223	**Exploration 4:** metric balance; 1 L container; 250 mL each of sand, soil, cereal, grain, sawdust, powdered detergent, liquid detergent, vinegar, and water; safety goggles; lab aprons
	225	**Exploration 5:** index card; pin; small box of paper clips; 20 cm of thread; lightweight classroom object; metric ruler; metric balance
13	230	**Solid or Liquid?:** 3 test tubes; nail; 20 mL of water; a few drops of food coloring; 20 mL of salt; **Liquid or Gas?:** 2 plastic syringes without needles; 5 mL of water; sponge; **Solid, Liquid, or Gas?:** 4 test tubes; 20 mL of salt; 50 mL of water; a few drops of perfume; a drop of food coloring
	234	**Exploration 1, Activity 1:** 100 mL of water; 250 mL beaker; hot plate; oven mitts; safety goggles; lab aprons; trivet; several 5 g pieces of paraffin wax; alcohol thermometer; **Activity 2:** can with a shiny surface; stirring rod or spoon; 2 or 3 ice cubes; 250 mL of water; alcohol thermometer
	238	**Pause for Thought (1):** 5 mL of salt; test tube with stopper; 20 mL of water
	243	**Challenge Your Thinking, question 4:** 2 sticks each of 2 different colors of modeling clay; 250 mL beaker or jar; metric ruler

* You may wish to set up this activity in stations at different locations around the classroom.

Advance Preparation

Metric Concentration, page 212: You may wish to make the cards for this activity ahead of time, or you may have students make their own.

Metric Spin, page 212: You may wish to make the cards and spinners for this activity ahead of time, or you may have students make their own.

Unit Compression

The conceptual progression within this unit has been designed to be as concise as possible, but if necessary, there are several Explorations and other activities that could be omitted without significantly disrupting this progression. Keep in mind, however, that these Explorations and activities should be skipped only if absolutely necessary.

- Exploration 2 from Chapter 11 (page 206)
- Exploration 2 from Chapter 12 (page 218)
- Activity 4 of Exploration 3 from Chapter 12 (page 223)
- Exploration 5 from Chapter 12 (page 225)
- Still More Experiments! from Chapter 13 (page 236)
- Writing About Matter from Chapter 13 (page 241)

In some cases, severe time constraints may make it necessary to omit an entire lesson from this unit. If this occurs, you may wish to skip Lesson 2 from Chapter 11, which provides students with a more detailed look at matter's properties, or Lesson 3 from Chapter 13, which covers the particle model of matter. Also, if your students already possess a working knowledge of the metric system, Lesson 1 from Chapter 12 could be omitted with minimal consequence.

Homework Options

Chapter 11	See Teacher's Edition margin, pp. 202, 204, and 205 SourceBook, pp. S86 and S87
Chapter 12	See Teacher's Edition margin, pp. 211, 212, 216, 220, 224, 225, and 227 Math Practice Worksheet, p. 12 SourceBook, pp. S90 and S92
Chapter 13	See Teacher's Edition margin, pp. 230, 236, 240, and 245 Theme Worksheet, p. 35 Activity Worksheet, p. 38 SourceBook, pp. S88 and S99
Unit 4	Activity Worksheet, p. 52 Activity Assessment Worksheet, p. 61 SourceBook Activity Worksheet, p. 65

Assessment Planning Guide

Lesson, Chapter, and Unit Assessment	SourceBook Assessment	Ongoing and Activity Assessment	Portfolio and Student-Centered Assessment
Lesson Assessment Follow-Up: see Teacher's Edition margin, pp. 204, 207, 213, 221, 225, 231, 236, and 241 **Chapter Assessment** Chapter 11 Review Worksheet, p. 6 Chapter 11 Assessment Worksheet, p. 9* Chapter 12 Review Worksheet, p. 28 Chapter 12 Assessment Worksheet, p. 31* Chapter 13 Review Worksheet, p. 45 Chapter 13 Assessment Worksheet, p. 48* **Unit Assessment** Unit 4 Review Worksheet, p. 53 End-of-Unit Assessment Worksheet, p. 56*	SourceBook Unit Review Worksheet, p. 66 SourceBook Assessment Worksheet, p. 70* 	Activity Assessment Worksheet, p. 61* **SnackDisc** Ongoing Assessment Checklists ♦ Teacher Evaluation Checklists ♦ Progress Reports ♦	Portfolio: see Teacher's Edition margin, pp. 203, 207, 219, 224, and 240 **SnackDisc** Self-Evaluation Checklists ♦ Peer Evaluation Checklists ♦ Group Evaluation Checklists ♦ Portfolio Evaluation Checklists 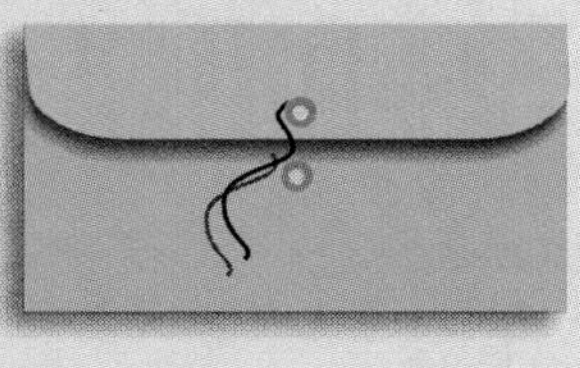 ♦

* Also available on the Test Generator software
♦ Also available in the Assessment Checklists and Rubrics booklet

Science Discovery is a versatile videodisc program that provides a vast array of photos, graphics, motion sequences, and activities for you to introduce into your *SciencePlus* classroom. *Science Discovery* consists of two videodiscs: Science Sleuths and the Image and Activity Bank.

Using the *Science Discovery* Videodiscs

Science Sleuths: The Blob Side A

A mysterious blob has washed up on a public beach. Because the authorities cannot identify it, and because they fear it is a dangerous substance, they decide to close the beach. The Science Sleuths must analyze the evidence for themselves to determine the true nature of the mysterious substance.

Interviews

1. Setting the scene: State beach authority **14556 (play ×2)**
2. Beachcomber **15371 (play)**
3. Surfers **16210 (play)**
4. Chemist **16735 (play)**
5. Oil-spill expert **17449 (play)**

Documents

6. Fax from marine biologist **18308**
7. Advertisement for gas chromatograph **18310 (step)**

Literature Search

8. Search on the words: CHESTER BAY, GARBAGE, OIL, SOLVENTS, SPILL, TIDES **18313 (step)**
9. Article #1 ("More Whales in Bay") **18316**
10. Article #2 ("Shipper and Local Authorities to Pay for Cleanup") **18318**
11. Article #3 ("States Considering Stricter Regulations on Pleasure Boats") **18320 (step)**
12. Article #4 ("Treasure Seekers Crowd the Bay") **18323 (step)**
13. Article #5 ("Oil Spills: A Modern Dilemma") **18326 (step ×2)**

Sleuth Information Service

14. Map of Chester Bay **18330**
15. Weather chart for past week **18332**
16. Chester Bay tide chart **18334**
17. Density chart of 10 common substances **18336**
18. Melting and boiling points of 10 common substances **18338**
19. Chart showing evidence of marine life at Chester Beach **18340 (step ×3)**
20. Jellyfish **18347 (step ×2)**
21. Whales **18351 (step ×3)**
22. Ambrein **18356 (step)**
23. Fatty oils **18359 (step)**
24. Benzoic acid **18362 (step)**

Sleuth Lab Tests

25. Blob volume **18365 (play)**
26. Blob weight **18499**
27. Cutting blob open **18501 (play)**
28. Blob scintillation count **18750**
29. Melting point **18752 (step ×3)**
30. Combustion point **18761 (play)**
31. Gas chromatography **18759**
32. Bacterial culture of blob and four control objects **18345**

Still Photographs

33. The blob **18904**
34. Blob on beach **18906**
35. Microscopic smear from blob **18908**

Step Reverse ◀| **Play** ▶ **Pause** || **Step Forward** |▶

A selection of still images, short videos, and activities is available for you to use as you teach this unit. For a larger selection and detailed instructions, see the Videodisc Resources booklet included with the Teaching Resources materials.

11-2 Matter's Useful Properties, page 205

Materials, common 1935
Common materials include plastic in pipes, metal in fences, wood in lumber, and hair in rope. Each material has different properties and functions and is used for specific applications.

Stone carving 1932
Stone is hard and brittle. To carve stone, a harder material must be used.

Tin cans 2321
Tin cans are actually tin-plated iron cans. Iron rust is prevented as long as the tin layer remains intact.

Copper 2323
Coil of copper tubing

Rubber 2324
Natural rubber is obtained from the juice of certain tropical plants. Synthetic rubber is made by a chemical process. These tires are made from synthetic rubber.

Iron rods 2325–2326 (step)
Iron rods are used to reinforce concrete in buildings.

Paper 2327
Paper can be made from many kinds of plants. The cellulose fibers that the plants use for support are separated, cleaned, and formed into sheets during the paper-making process.

Cement 2334
Cement is mostly made up of rock, clay, and water.

12-1 SI: The Metric System of Measurement, page 211

Metric measurements 450
Metric prefixes and symbols and their values

Metric conversions 447–449 (step ×2)
Units of length, volume, mass, and weight

12-2 Volume, page 214

Buret 213
A color-marked piece of paper makes it easier to see the meniscus. Read the bottom of the meniscus at eye level.

12-3 Mass, page 222

How to use a triple-beam balance 8588–10323 (play ×2) (Side A only)

Balance, analytical 214
This analytical balance is measuring the weight of a volumetric flask and its contents. This balance has a digital readout and is extremely accurate.

Balance, analytical 215
This balance is measuring a substance on weighing paper. The glass doors are closed when the reading is taken so that other particles or air currents won't affect the accuracy.

Balance, simple 456–459 (step ×3)
How to make and use a feather-weight balance

Volume 455
Compare the volume of 1 g of water and 1 g of water vapor in similar containers.

13-1 States of Matter, page 229

Mixing drink; solution 2512–2516 (step ×4)
Drink powder is added to water and stirred.

Liquefaction 11579–11973 (play ×2) (Side B only)
Sand that appears solid turns to a liquid when vibrated. This kind of saturated soil is very hazardous during earthquakes.

Smokestacks 2610
Smokestacks release materials into the atmosphere.

13-2 Changes of State, page 232

Moth balls 2320
Moth balls sublimate from a solid state directly into a vapor state.

Dry ice 2340
Dry ice is carbon dioxide in its solid state.

Condensation 1450
Condensation of water as seen on a window

Rime 1487
Rime consists of ice crystals formed from fog as it comes into contact with a subfreezing surface.

Ice-crystal formation 14416–14545 (play ×2) (Side A only)
Ice-crystal growth on nuclei as seen under a microscope

Boiling water 48655–48867 (play ×2) (Side A only)
As water boils, it rapidly changes from its liquid state to its gaseous state.

Red-hot molten iron 41307–41524 (play ×2) (Side A only)
Molten iron emits infrared radiation (heat) and visible light.

States of matter 2355–2357 (step ×2)
Heat changes water from its solid state to its liquid state to its gaseous state.

State, changes of 2354
Descriptive terms

Evaporation and boiling 2358

Melting and boiling points 2362
Boiling and melting points of some common substances

Science and Technology, page 249

Recycling 2654–2658 (step ×4)
Examples of recyclable cardboard, newspapers, and plastics

Recycling 20401–22072 (play ×2) (Side B only)
Recycling preserves our resources and lessens the burden on our landfills.

Unit 4 INVESTIGATING MATTER

Write the following question on the chalkboard: What can be heavy or light, soft or hard, visible or invisible, colored or clear, smooth or rough, and living or nonliving?

After students have had a chance to respond, point out that matter is the only thing that could have all of these qualities. Have students suggest an example of matter to illustrate each of the qualities in the question. Keep a list of their ideas on the board.

A good motivating activity is to let students listen to the English/Spanish Audiocassettes as an introduction to the unit. Also, begin the unit by giving Spanish-speaking students a copy of the Spanish Glossary from the Unit 4 Teaching Resources booklet.

Unit 4 INVESTIGATING MATTER

Ben Livingston, an artist in Austin, Texas, uses his knowledge of the changing nature of matter to produce works of art.

198

Connecting to Other Units

This table will help you integrate topics covered in this unit with topics covered in other units.

Unit 1 Science and Technology	The measurement of matter is one of the key skills necessary for all accurate scientific investigations.
Unit 2 Patterns of Living Things	When organisms exchange materials with their environments (such as when water transpires from plant leaves), changes of state are often involved.
Unit 5 Chemical Changes	The chemical changes that matter undergoes depend on the matter's physical properties.
Unit 6 Energy and You	Changes of state are the result of transfers of energy.
Unit 7 Temperature and Heat	Patterns of convection and conduction depend on the properties and state of matter.

The hot studio hums with the sound of high-voltage light fixtures. Filled with neon, argon, and krypton gas, the lights cast fluorescent colors all around the studio. Meanwhile, the artist shapes the glass tubes for his next masterpiece. Through experience he knows how to heat the glass until it begins to bend. If he continues to heat the glass, he can coil it at will or stretch it into a thin thread. If he heats a tube for too long, he might scorch his feet with drops of fiery glass.

Creating art from colored lights takes imagination, but the artist must also be familiar with the characteristics, or properties, of the material that he uses. How glass behaves at different temperatures is a property of glass. In this unit, you will investigate various properties of matter, and you will learn to use the metric system to measure some of these properties. Learning how matter changes from one state to another will be enlightening!

199

Connecting to Other Chapters

> **Chapter 11**
> *explores the nature of matter, including its physical, chemical, and biological properties.*

> **Chapter 12**
> *introduces the metric system and ways to measure volume and mass.*

> **Chapter 13**
> *examines the three principal states of matter—solid, liquid, and gas.*

Prior Knowledge and Misconceptions

Your students' responses to the ScienceLog questions on this page will reveal the kind of information—and misinformation—they bring to this chapter. Use what you find out about your students' knowledge to choose which chapter concepts and activities to emphasize in your teaching.

In addition to having students answer the questions on this page, you may wish to assign a "free write" to assess prior knowledge. To do this, instruct students to write for 3 to 5 minutes on the subject of matter. (If students are unfamiliar with the scientific use of this word, it may be helpful to have them describe other ways that they have heard the word used.) Tell them to keep their pens moving at all times, writing in a stream-of-consciousness fashion. Emphasize that there are no right or wrong answers in this exercise. It may be best to ask students not to put their names on their papers. Collect the papers, but do not grade them. Instead, read them to find out what students know about matter, what misconceptions they may have, and what about matter is interesting to them.

CHAPTER
11 Meet Matter

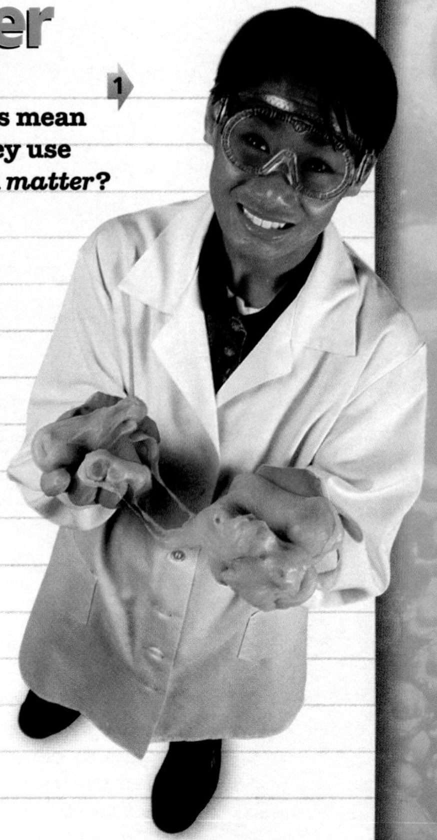

1 What do scientists mean when they use the word *matter*?

2 What do water, ice, and steam have in common? How do they differ? Is all of the matter in this picture visible? Explain.

ScienceLog

3 What happens to sugar when you add it to hot tea? Is it destroyed? Does it change into something else?

Think about these questions for a moment, and answer them in your ScienceLog. When you've finished this chapter, you'll have the opportunity to revise your answers based on what you've learned.

200

I say Watson old boy, what's the **matter** with you?

Well, Holmes, it's rather a silly **matter**.

Go on, old boy.

You would, would you?

Well, I suppose I'd like to solve a few of these crimes by myself every once in a while.

Yes, I would. However, I fear that I haven't got enough gray **matter**.

That's ridiculous, old fellow! Solving mysteries is quite simple really!

How so, Holmes?

Mind over **matter,** Watson, mind over **matter**!

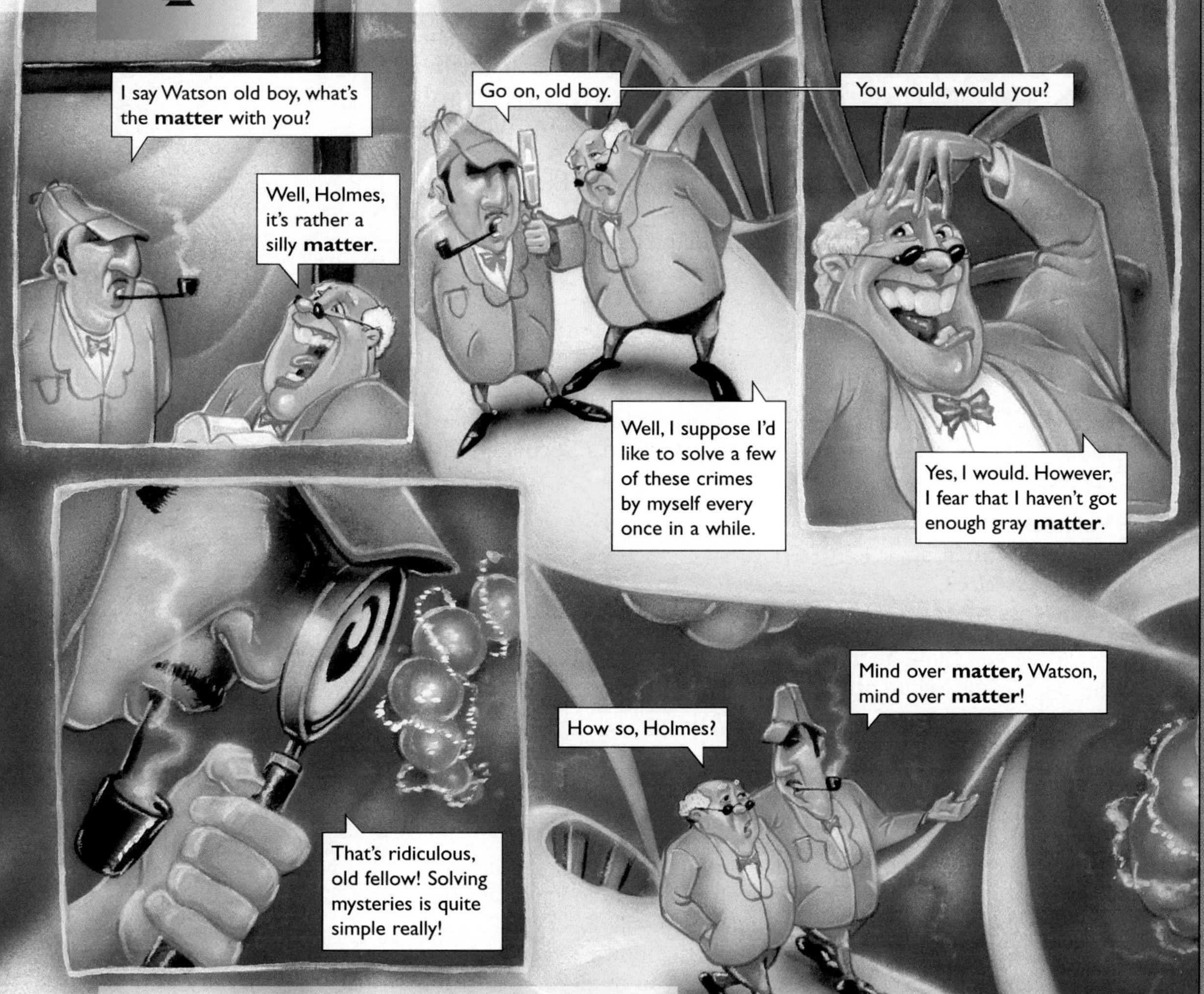

Matter. The word is used in so many different ways. How many different meanings of the word *matter*

Ⓐ can you find in the conversation above? In this unit, you are going to use the word *matter* in yet another way. You will find that . . .

• it's a very important matter . . .
• that you use your gray matter . . .
• to learn many facts about matter . . .
• in order to discover what is matter . . .
. . . for this unit is about matter.

201

LESSON
1
About Matter

FOCUS

Getting Started
Set up a display of several different objects. If possible, include solids, liquids, and gases in your display. Ask: What do all of these objects have in common? (*Accept all reasonable responses.*) After students have had a chance to answer, explain that all of the objects are types of matter.

Main Ideas
1. Matter is all around us.
2. Matter can be identified and characterized by its properties.
3. The properties of matter can be chemical, physical, or biological.
4. The chemical properties of matter are determined by its components. Chemical changes transform matter into a different substance.
5. The physical properties of matter (color, shape, boiling point, etc.) may be observed or measured without changing the identity of a substance. Change of state is one example of physical change.

TEACHING STRATEGIES

GUIDED PRACTICE   Ask students if the word *matter* is used in the scientific sense in the cartoon. Then encourage students to write sentences in their ScienceLog using the word *matter* in at least three different ways.

Answer to
In-Text Question

Ⓐ Matter is used in reference to a problem, a subject of discussion, a specific material, and a difficult situation.

LESSON 1 ORGANIZER

Time Required 2 to 3 class periods

Process Skills
observing, inferring, classifying

Theme Connection Structures

New Terms
Biological properties—characteristics of matter that are unique to living things
Chemical properties—characteristics of matter that describe how a substance behaves in chemical changes
Matter—anything that has mass and occupies space
Physical properties—characteristics of matter that do not involve chemical

changes; examples include size, shape, color, and texture

Materials (per student group)
Exploration 1: 500 mL of water; 500 mL of vinegar; 4 beakers (any size); ice cube; 2 eggs; balloon filled with air; 2 sugar cubes; candle; jar lid; small ball of modeling clay; matches; earthworm; geranium; air freshener; safety goggles; lab aprons

Teaching Resources
Exploration Worksheet, p. 3
Transparency 31
SourceBook, p. S86

The Riddle of Matter

These riddles offer a unique way to focus the students' interest on the properties, or characteristics, of matter. It also sets the stage for Exploration 1, in which students are asked to make observations and inferences about different types of matter.

Primary Source

Description of change: excerpted
Rationale: excerpted to focus on properties of matter mentioned in *The Hobbit*

Meeting Individual Needs

Learners Having Difficulty

Challenge students to create a "matter gallery." Students can photograph or sketch different examples of matter in their homes and neighborhoods. Some students may want to write poems or riddles about different forms of matter. The gallery can be labeled and displayed for the class to enjoy.

The Riddle of Matter

In the book *The Hobbit,* by J.R.R. Tolkien, the hero is a hobbit named Mr. Bilbo Baggins. Hobbits are little people who populate Middle Earth, the magical world of *The Hobbit.* At one point in the story, Bilbo is lost in a dark cave. He encounters a creature named Gollum who lives in the cave. Gollum agrees to show Bilbo the way out of the cave if Bilbo can guess all of Gollum's riddles and ask him one that Gollum cannot solve. However, if Bilbo fails, he may end up as Gollum's next meal.

Bilbo asks Gollum to go first. So Gollum asks:

What has roots as nobody sees,
Is taller than trees,
Up, up it goes,
And yet never grows?

What has these characteristics or properties? Bilbo guessed right—mountains. Bilbo also answers Gollum's second riddle correctly:

Voiceless it cries,
Wingless flutters,
Toothless bites,
Mouthless mutters.

While you think about this riddle, here is another one. (This one isn't from Gollum or Bilbo.)

Can be tasted or not tasted,
Can be seen or not seen,
Can be changed, can be wasted,
Of every color—from black to green.

The answer to the last riddle is all around you: the air you breathe, the pencil you hold, and even you. Now do you know the answer? It is **matter**. In Exploration 1, you will observe 12 more examples of matter. For each one, write down several observations and inferences about matter that you think are true. You may discover an answer to this final riddle: *What is matter?*

202

Homework

You may wish to assign question 1 of A Matter for Discussion on page 203 as homework.

What Is Matter?

In your ScienceLog, make a data table like the one at right. Observe the examples of matter listed, and record your observations or inferences in your table. Indicate with a letter *O* or *I* whether you are making an observation or an inference. See if you can record at least five observations about each example of matter. What inferences about matter do these observations suggest? Do this Exploration in groups of three. Do not taste any material unless specifically instructed to do so by your teacher.

Afterward, share your observations and inferences with your classmates. Have you solved the riddle, *What is matter?*

Example of matter	Observations/ inferences
a beaker of water	
a beaker of vinegar	
an ice cube	
an egg	
a balloon filled with air	
a sugar cube	
a sugar cube in a beaker of water	
a burning candle	
an earthworm	
a geranium	
an egg in a beaker of vinegar	
an air freshener	

A Matter for Discussion

1. After much discussion, Ms. Chester's class came up with the following list of inferences about matter. Write an observation to support each of these inferences.

   a. Matter may be either living or nonliving.

   b. Matter may be either solid, liquid, or gas.

   c. Matter may have an odor or no odor at all.

   d. Matter can undergo many kinds of changes.

   e. Often, the original matter changes so much that it appears to be destroyed.

   f. Matter can be detected using your senses—sight, hearing, taste, touch, and smell.

   g. In some types of changes, although the matter may look different, it is really still the same.

   h. Each kind of matter has its own distinctive characteristics or properties that identify it.

2. Living things are examples of matter. In Unit 2, you studied the properties, or characteristics, of living things. Review these properties, and suggest four that distinguish living things from nonliving things. These properties are sometimes called **biological properties**.

What properties are shared by the wheat and the deer, but not by the quartz? Ⓐ

203

Answer to
Caption

Ⓐ The properties shared by the wheat and the deer, but not by the quartz, include growth, movement, reproduction, and the use of energy from the environment. Students may also point out physical properties: the wheat and deer are not translucent like the quartz.

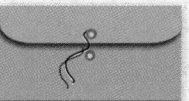 **PORTFOLIO**
Have students include their observations and inferences from Exploration 1 in their Portfolio. Students may also wish to include observations and inferences of objects of their own choosing.

 An Exploration Worksheet (Teaching Resources, page 3) and Transparency 31 are available to accompany Exploration 1.

Distribute the examples of matter among several activity stations. Have students work in groups, spending 3 to 5 minutes at each station.

SAFETY ALERT Review with students all safety guidelines about touching, tasting, and smelling matter.

Answers to
A Matter for Discussion

1. **a.** The earthworm and the geranium are living. The egg is matter that was once living. All other examples are nonliving.

   b. Solids: candle, ice cube, eggshell, balloon, sugar cube, air freshener
   Liquids: water, vinegar, melted wax
   Gases: air inside the balloon, air inside the egg, bubbles on the egg submerged in vinegar
   Note: Students may also point out that living things are made up of solids, liquids, and gases.

   c. The air freshener and the vinegar have an odor.

   d. Kinds of changes: candle burning, ice cube melting, sugar cube and eggshell dissolving

   e. Matter appears to be destroyed when the candle burns and when the eggshell and sugar cube dissolve.

   f. Answers may vary, but possible responses include the following:
   • sight: all examples except the air in the balloon
   • hearing: none of the examples
   • taste: vinegar, water, sugar cube
   • touch: Students may note temperature, texture, and wetness.
   • smell: vinegar, air freshener

   g. Whether an ice cube is frozen or melted, it is still water; when a sugar cube dissolves, it is still sugar; when solid air freshener disappears into the air (sublimates), it is still the same kind of matter.

   h. The classification of matter is based on its properties. Any two examples of matter that have the exact same properties are the same kind of matter.

2. The biological properties of living things include the ability to grow, move, reproduce, and respond.

Answers to A Matter for Discussion continued ▶

3. When you examined the sugar cube in Exploration 1, you made two kinds of observations. First, you observed the shape and color of the sugar cube. You may also have made quantitative observations about it by measuring its length and its mass. These observations describe **physical properties** of a sugar cube. Second, you observed a change—something happened to the sugar cube when it was placed in water. This observation also describes a physical property of sugar: sugar is not destroyed by water; it can still be tasted in the water. You can thus make two kinds of observations about physical properties: observations describing the matter itself and observations describing how the matter changes.

   a. Describe the physical properties of an ice cube. In your description, include examples of both types of observations about physical properties.

   b. Describe the physical properties of water, including examples of both types of observations about physical properties.

4. Another property of matter that you can observe is called a **chemical property**. This property usually describes a situation in which the identity of the matter is changed. For example, tearing a piece of paper does not change or destroy the identity of the paper, but burning it does. After the paper has burned, it has undergone a chemical change in which new substances have been formed. The fact that paper burns is a chemical property of paper.

   a. In Exploration 1, what two examples of chemical properties did you observe?

   b. What chemical property of iron causes millions of dollars worth of damage every year to objects made of iron?

   c. List three physical properties of wax and one chemical property of wax.

5. List as many properties of an apple as you can. Include biological, physical, and chemical properties.

6. Create your own definition of matter. The next section may suggest a way of improving your definition.

Be a Riddlemaster

You've had a while to guess the answer to Gollum's second riddle. Did you guess "wind"? Do you think wind is an example of matter? (By the way, Bilbo does find his way out of the cave, and he goes on to further adventures.) Ⓐ

Now it's your turn. Make a riddle about one of the examples of matter in Exploration 1. As clues to your riddle, use the properties of that example of matter—but don't make it too easy! Below is a riddle that almost stumped Bilbo. Hint: The answer is one of the examples of matter in Exploration 1. Ⓑ

> A box without hinges, key, or lid,
> Yet golden treasure inside is hid

204

Matter's Useful Properties

Polly Vinyl woke up early one morning because of the rain. Her sponge house soaked up a lot of water when it rained, and water had begun to drip on her head. It was just as well—she sometimes found it difficult to sleep on her cast-iron mattress. She put on her wooden dress and her concrete shoes and prepared to go to work. Her glass bicycle needed to be repaired, so she decided to take the paper car instead.

Does this sound a little wacky to you? How would you help make Polly's life a little easier? You might suggest that she examine the properties of the materials she uses. For example, concrete may not make the most comfortable shoes, but it does have properties that make it useful for other things.

- You probably wouldn't want to live in a sponge house, but sponge is useful for cleaning your sink. What properties of sponge make it useful for cleaning?

- Although not much good for making a car, paper is useful for making a wide range of products, from bathroom tissue to cardboard boxes.

- Glass is found everywhere. Its properties make it useful in windows, bottles, insulation, light bulbs—the list goes on and on.

The properties of materials—physical, chemical, and biological—determine how the materials are used. At right is a short list of materials. With a classmate, suggest a use for each, and then name a property that is important to this use. Is it a chemical or physical property? Record your ideas in a data table like this one. Do not write in this book.

Material	Use	Important property	Type of property
corn			
gold			
oil			
aluminum			
plastic			
coal			
concrete			
wool			
(your choice)			

205

LESSON 2 ORGANIZER

Time Required
1 class period

Process Skills
observing, classifying, analyzing

New Terms
none

Materials (per student group)
none

Teaching Resources
Resource Worksheet, p. 5
Transparency 32
SourceBook, p. S87

FOCUS

Getting Started
Collect a few consumer products, such as a rubber band, safety match, plant, and paper towel, for students to examine. Ask: What properties make each product useful? *(Students should list physical, biological, and chemical properties. For instance, a rubber band is elastic and a match burns.)*

Main Idea
The properties of a material determine its use.

TEACHING STRATEGIES

Have students work individually or in groups to complete the table on this page. For possible responses, see the table on page S235.

★ **A Resource Worksheet (Teaching Resources, page 5) and Transparency 32 accompany the material on this page.**

Meeting Individual Needs

Second-Language Learners
Provide students with photographs of several objects. Explain that each object represents a kind of matter. Then have students write down the name of each object and one or two simple sentences that describe some of the properties of each object. Allow them to use a bilingual dictionary if necessary. *(Sample answer: Ice is cold, hard, and clear. When it melts, it forms water.)*

Homework

Have students copy the table on this page into their ScienceLog. Instead of the materials listed, however, have them substitute glass, wood, rubber, paper, steel, and cotton. Instruct students to complete this table using examples from around their home. *(Accept all reasonable responses.)*

EXPLORATION 2

PART 1

GUIDED PRACTICE Before students begin reading Part 1, have them suggest ways in which paper and paper products are used. Write the list on the chalkboard to demonstrate how important paper is in our everyday lives. Then ask students to look at the list and identify at least one property of paper that makes each item useful.

Cooperative Learning
PART 1

Group size: 3 to 4 students
Group goal: to prepare a report on the properties and uses of paper
Positive interdependence: Assign students roles such as group leader, recorder, reporter, or checker.
Individual accountability: Have each student write a short description of his or her efforts to complete this project. Students should include their procedure, describe any difficulties they had while researching, list those aspects of the project that they enjoyed, assess what they accomplished, and describe what they would do differently if they did the project again.

Answers to
Part 1

1. Paper is made by pressing a pulp made from wood or other fibrous materials into flat sheets. Paper mills are usually located near rivers and lakes because making paper uses a lot of water. There are thousands of uses for paper. Because paper is lightweight and can be created in a variety of shapes, sizes, strengths, and colors, it is suitable for a wide range of uses.

2. Forests can be destroyed, but reforestation efforts and thoughtful management techniques can help conserve this resource.

3. Almost all fields use paper for communication purposes, but publishing, packaging, and paper manufacturing are especially dependent on paper. Students should identify any classmates who deliver newspapers.

EXPLORATION 2

A Class Project: Properties and Uses of Paper

For this project, you will prepare a report and present it to your class. Decide which tasks each person in the group will do. Perhaps some tasks will be done by the whole group. Later, when you prepare your report, decide who will present the different parts of the report. Will you use a poster or an overhead projector to help you? Perhaps you can make a display.

PART 1

This part involves doing library research, talking to people, visiting offices and factories, or simply thinking. Consider the following questions:

1. How is paper made? Where is it made? How many uses of paper are there? What properties of paper make it suitable for each use?

2. Is there any danger of running out of trees? What can be done to conserve this resource?

3. What jobs depend on the manufacture and use of paper? Are there jobs in your community that depend on paper? Do any members of your class deliver newspapers?

Peter, Anne, and Bettie divided up the jobs that needed to be done to prepare their report. Why is this step important? Ⓐ

At the public library, Anne spoke with the reference librarian about her project. What kinds of questions would you ask your librarian? Ⓑ

Peter and Bettie obtained permission to tour their town's newspaper printing press. They brought a list of questions they planned to ask. What questions would you ask? Ⓒ

206

Answers to
Captions, pages 206–207

Ⓐ Dividing up the jobs helps ensure that group members have tasks that complement each other and do not overlap.

Ⓑ Students might ask a librarian how to find reference material in the library on their own.

Ⓒ A student could ask how this press compares to other presses and what steps are involved in the printing process.

Ⓓ The poster is clearly labeled, neat, and visually interesting. However, the poster could have provided more information about each step.

Ⓔ Ability to absorb water

CROSS-DISCIPLINARY FOCUS

Language Arts
Challenge students to write a short story about someone who wakes up one morning to find that all synthetic materials have vanished. You may wish to work with students to create a list of common synthetic materials. Display the stories where others may read them.

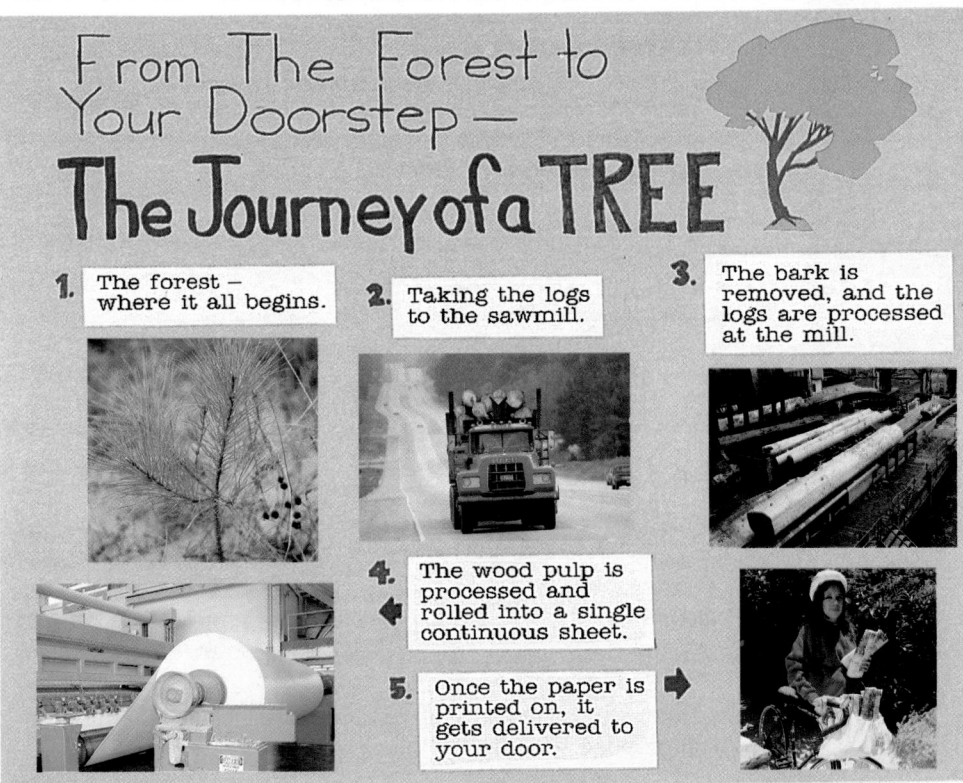

From The Forest to Your Doorstep —
The Journey of a TREE

1. The forest — where it all begins.

2. Taking the logs to the sawmill.

3. The bark is removed, and the logs are processed at the mill.

4. The wood pulp is processed and rolled into a single continuous sheet.

5. Once the paper is printed on, it gets delivered to your door.

Here is one team's poster. What did they do well? How would you improve their poster? **D**

This part can be done by devising experiments to investigate a property of paper. Here are four suggestions:

1. *Strength.* Compare the strength of a variety of brands of paper towels. What would be your test for strength? Is it a fair test? When is strength important?

2. *Ability to repel water.* Develop a test to compare how well different types of paper repel water. What is done to make paper more water-repellent?

3. *Ability to absorb water.* Test at least three brands of paper towels to determine which absorbs water the best.

4. *Ability to disintegrate in water.* Bathroom tissue should have this property so that it will not clog drains and sewer pipes. Are all bathroom tissues equally good at disintegrating? How could you find out? Did you develop a fair test?

Good luck on your project!

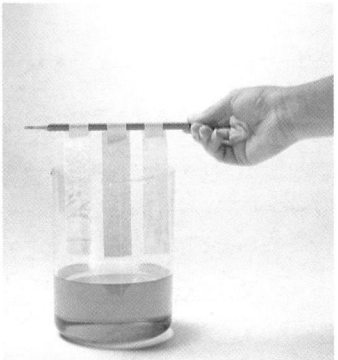

This is one group's experiment. What property of paper are they testing? **E**

207

Cooperative Learning
PART 2

Group size: 4 to 5 students

Group goal: to devise an experiment to investigate a property of paper

Positive interdependence: Assign students roles such as chief investigator, materials coordinator, timekeeper, recorder, or checker.

Individual accountability: Randomly call on one member of each group to explain what property the group investigated, how the experiment was designed, and the group's results.

PORTFOLIO

Encourage students to include their results from Exploration 2 in their Portfolio. Their report should include their experimental design, data collected, results, and conclusions.

FOLLOW-UP

Reteaching

Have students continue the riddle theme from Lesson 1 by writing riddles that identify only the useful properties of an object. Challenge the class to solve the riddles.

Assessment

Have students select an object in the classroom and make a list of its useful properties. Then have them identify the properties as either physical, chemical, or biological.

Extension

As an extension of Exploration 2, students may wish to examine the properties of a nonpaper product, such as thread, liquid detergent, or powdered drink mix. Have them compare different brands, and check all proposed procedures for safety.

Closure

Have students write a paragraph in their ScienceLog evaluating what they learned in this lesson. Their paragraphs should include concepts that were easy to understand, concepts that were difficult, and how they plan to overcome any difficulties that they experienced.

Answers to
Part 2

1. Sample answer: The strength of paper towels could be tested by finding out how many marbles one sheet will hold before it tears. Since strength is important whether a paper towel is dry or wet, this test could be performed with dry and wet paper towels.

2. Sample answer: Water repellence could be tested by pouring a known volume of water over a piece of paper and then measuring how much water is recovered. Paper can be coated with plastic or wax to make it more water-repellent.

3. Sample answer: Absorbancy could be tested by placing paper towels over similarly sized spots of water to see which one leaves the least water when it is picked up.

4. Sample answer: A few sheets of various bathroom tissues could be placed in separate bowls of water. The contents of the bowls could then be stirred to see which tissue disintegrates the quickest.

Answers to *Challenge Your Thinking,* pages 208–209

1. a. Physical properties of the pencil include the following: It is painted red, 10.5 cm long, and soft; it will float and is made of wood, metal (which can be bent easily), rubber, and graphite (which is soft and leaves marks on paper); and it has a pink eraser that is 0.25 cm long and that wears away easily. A chemical property of the pencil is that it will burn.

b. The two quantitative observations made about Mario's pencil are that the pencil is 10.5 cm long and that the eraser is 0.25 cm long. Explain to students that quantitative observations relate to the size, quantity, or amount of an object.

2. a. Salt
b. Water
c. Air

3. Spaghetti is solid and stiff, which makes it a good building material, but it also brittle, which makes it a bad building material.

4. Sample answer: I would blow up a balloon to show him that air can take up space. The fact that air is held near the surface of the Earth (instead of floating away into space) shows that air has mass.

5. Sample answer: The handle should be made of a material that is sturdy but also comfortable to hold, such as stiff rubber or plastic. The scratching part should be very stiff and fairly sharp, so a piece of plastic or metal might work well. (Student drawings should reflect their choice of construction materials.)

CHALLENGE YOUR THINKING

1. On the Lookout

a. Here is a description of Mario's pencil. List as many properties of the pencil as you can, and group them according to whether each is a physical property or a chemical property.

> My pencil is painted red and is 10.5 centimeters long. It is made of four materials. One is wood. By itself, wood floats. A pencil does not float as well. When I chew on my pencil, it leaves marks in the wood. Wood is soft enough to grind away in the pencil sharpener. You can recognize my pencil by a burn mark near one end.
>
> Enclosed in the wood is a rod made of graphite. Graphite is soft. It can easily mark a piece of paper.
>
> The eraser is held in place with a small piece of metal. Metal is used because it can be easily bent. My eraser is pink and is now only 0.25 centimeters long. When rubbed on paper, it leaves part of itself behind. But it does remove the pencil mark. That is my pencil, and if anyone sees one like it, please return it. Pencils have the property of being able to disappear.

...Well, you see it looked like...with big blue...with eighteen... and very, very...wide.

b. How many quantitative observations are included in Mario's description? List them.

2. At a Loss for Words?

Your friend Marsha has the hardest time remembering the names of things, even common household items. Fortunately, she can usually describe them very well. What common examples of matter is she describing?

a. "Pass me that stuff that forms tiny colorless cubes and flows pretty well unless it's wet, in which case it sticks together. Oh yes—it also has a strong taste. The oceans are loaded with the stuff."

b. "Oh no—I've spilled some of that stuff that's wet and clear; it hardly has any taste, and when it's really cold, it becomes solid; sometimes it falls from the sky."

c. "Look, up in the—oh, what's that stuff called? It's transparent, and you hardly know it's there. It doesn't have any taste or smell by itself. When you run or ride your bike, you can feel it, and you can actually float through it if you have the right equipment."

208

Multicultural Extension

Different Kinds of Ice

Inform students that the Inuit or Inupiat are people who live in northern Canada and Alaska. In their language, the word *ice* is not very useful. Instead, many different words are used to describe frozen water, depending on whether it is floating, salty, hard-packed, etc. Ask: Why would having specific words for different types of ice be useful? (*Different words would ease communication in a place where ice is used for many different purposes.*)

Cross-Disciplinary Focus

Art

Have students imagine that they have been asked to design a house. Ask: What properties would you look for in the materials you choose? Some of the variables to consider include strength, durability, appearance, cost, and weight. Suggest that students list the materials they would use and describe how each material would be used and why it was chosen. Suggest that students include a drawing of the house they would build.

Did You Know...

The United States has over 1000 types of soil. Each soil type has a different set of properties. This range of properties helps make it possible to grow a wide variety of crops in the United States.

3. The Spaghetti Special

Using a glue gun, John is going to build a bridge from dry spaghetti for a class project. What are some properties of this building material that may make it useful in building a bridge? What are some properties that may make building a good bridge with it difficult?

Still not sure about the difference between physical and chemical properties? Turn to page S87 of the SourceBook to find out more.

4. Let Me Explain

Your little brother, who is in the third grade, insists that air is not matter. How would you convince him that he is wrong?

5. The Science of Scratching

One step in the invention process is searching for materials with the best properties for the task the invention is to do. Be an inventor and invent a back-scratching device.

- What will you use to make the handle? Why?

- What properties do you want your back scratcher to have? What materials have these properties?

- What will the end product look like? Draw a diagram. With your teacher's permission, make the back scratcher, and compare it with others made by your classmates.

ScienceLog

Review your responses to the ScienceLog questions on page 200. Then revise your original ideas so that they reflect what you've learned.

209

You may wish to provide students with the Chapter 11 Review Worksheet that accompanies this Challenge Your Thinking (Teaching Resources, page 6).

Meeting Individual Needs

Learners Having Difficulty

Working in groups, have students play a variation of "twenty questions." One student should think of a common object, and the other students should attempt to guess what it is by asking questions about the object's properties; e.g., Does it burn? or Is it solid? The students should identify the properties as either chemical, physical, or biological.

Meeting Individual Needs

Second-Language Learners

Students may enjoy playing the "list the properties" game. Divide students into two or more teams. Then write the name or show a picture of an example of matter, such as an apple. Instruct each team to write as many words or terms as they can to describe the properties of the object. Allow them to use a bilingual dictionary if necessary. Help resolve any conflicts about the words or terms chosen. The team with the most words wins.

ScienceLog

The following are sample revised answers:

1. Matter is any substance that takes up space and has mass. Matter can be described in terms of its physical, chemical, and biological properties.

2. Water, ice, and steam are all composed of water molecules. All three substances still have all of the chemical properties of water, but they are different states of water. Air is a gas that takes up space and has mass; it is present but not visible in this picture.

3. Sugar dissolves in water or tea, but it is not destroyed by this process. It is still sugar, as you can tell by tasting the mixture.

Multicultural Extension

The History of Paper

The word *paper* comes from the word *papyrus*. Papyrus is the name of a reed that was once used by the ancient Egyptians to make a writing material. Paper as we know it today was invented in China in A.D. 105 by Ts'ai Lun, an official of the Imperial court. Ts'ai Lun made his paper from the fibers of the mulberry tree. Later, he also used fish nets, old rags, and hemp for his raw material.

Integrating the Sciences

Earth, Life, and Physical Sciences

The properties of both living things such as animals and nonliving things such as rocks are important in creating classification systems. For instance, rocks can be classified based on whether they exhibit layering in their structure. Have students make a list of 10 animals and a list of 5 types of rocks. Then have students classify the items on one of their lists based on the properties of the various items.

Connecting to Other Chapters

Chapter 11
explores the nature of matter, including its physical, chemical, and biological properties.

Chapter 12
introduces the metric system and ways to measure volume and mass.

Chapter 13
examines the three principal states of matter—solid, liquid, and gas.

Prior Knowledge and Misconceptions

Your students' responses to the ScienceLog questions on this page will reveal the kind of information—and misinformation—they bring to this chapter. Use what you find out about your students' knowledge to choose which chapter concepts and activities to emphasize in your teaching.

In addition to having your students answer the questions on this page, you may wish to use the questions below to initiate a class discussion. Allow students to express their ideas about these questions, whether right or wrong. During the course of the discussion, take note of what the students already know about measurement, mass, and volume, as well as what things intrigue them. By assessing their prior knowledge and misconceptions of these subjects, you can adapt your teaching to their particular interests and needs.

1. How do you use the word *volume*?
2. What does the word *mass* mean?
3. Why are some things measured in liters (like soda), while other things are measured in gallons (like milk)?

CHAPTER
12
Measuring Matter

1 **Do gases have mass? Do they have volume?**

There is an old riddle that asks, Which weighs more, a kilogram of lead or a kilogram of feathers? What is the answer? 2

3 **Which is the most appropriate unit for measuring your height, millimeters or centimeters? Why?**

4 **Which of these objects would you be most likely to measure in liters?**

ScienceLog

Think about these questions for a moment, and answer them in your ScienceLog. When you've finished this chapter, you'll have the opportunity to revise your answers based on what you've learned.

Imagine what the world would be like if every country used its own system of measurement! To make trade and communication easier, most countries use the Système International d'Unités (the international system of metric units), or SI.

Three of the most familiar units in the metric system are given in the table at right. You will use them many times throughout this book.

Obviously, these three units are not suitable for all measuring needs. The metric system expands the use of these units by using prefixes.

Quantity	Unit	Symbol
length	meter	m
volume	liter	L
mass	gram	g

Metric Prefixes

Prefix	Powers of 10	Symbol	Example
kilo	1000	k	kilogram (kg)
hecto	100	h	hectoliter (hL)
deca	10	da	decameter (dam)
—	1	—	meter (m), gram (g), liter (L)
deci	0.1	d	decigram (dg)
centi	0.01	c	centimeter (cm)
milli	0.001	m	milliliter (mL)

Not all of these prefixes are used equally often, but you should still know about them in case you encounter them. Notice that it is easy to convert from one prefix to another; metric prefixes are used in powers of 10. The games on the next two pages will help you become more familiar with the metric system.

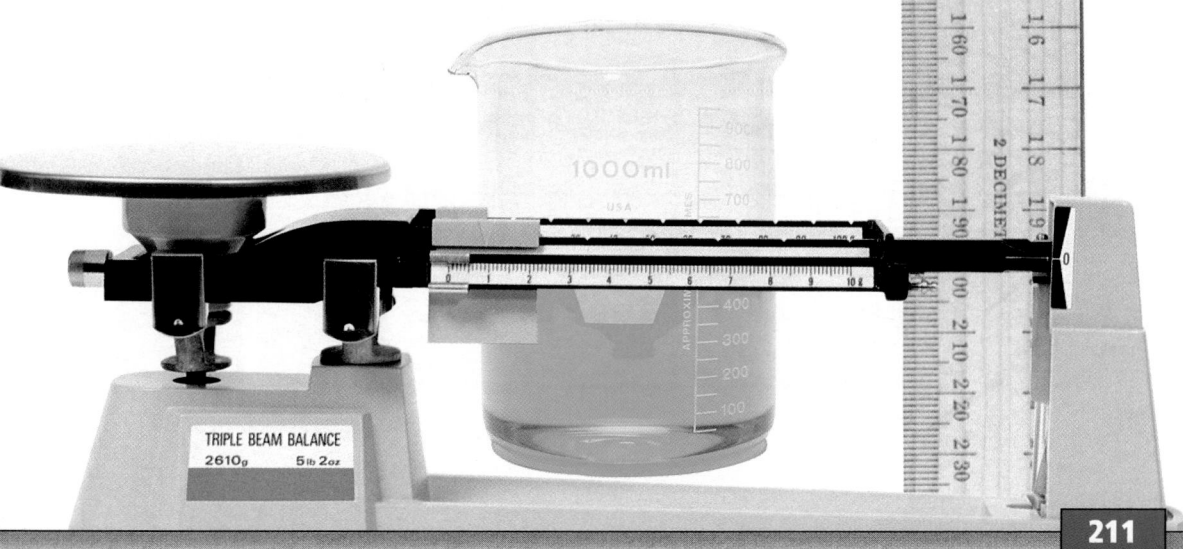

211

LESSON
1

SI: The Metric System of Measurement

FOCUS

Getting Started

Show students a product that lists a metric unit on its label (such as a 2 L soda bottle). Then ask them to indicate with a show of hands if they have used metric units. Point out that metric units are almost always used in science.

Main Ideas

1. The metric system is based on multiples of 10.
2. The magnitudes of metric units are identified by prefixes.
3. Converting from one order of magnitude to another is relatively easy.

TEACHING STRATEGIES

Before students begin the lesson, ask them to name any metric units that are familiar to them. Keep track of their responses on the chalkboard. Then ask students where they have seen metric units used. (*Product labels, speedometers, road signs, etc.*)

GUIDED PRACTICE After students have read the lesson's introduction, direct their attention to the table of metric prefixes. Allow students time to review the multiple that each prefix represents. Then tell students to close their books. Name a metric prefix, and call on a volunteer to identify the multiple that it represents. Continue until each student has had a chance to respond.

Homework

You may wish to have students complete the Math Practice Worksheet on page 12 of the Unit 4 Teaching Resources booklet as homework. If you use the worksheet in class, Transparency 33 is available to accompany it.

LESSON 1 ORGANIZER

Time Required
1 to 2 class periods

Process Skills
analyzing, comparing

New Term
Système International d'Unités—
the international system of metric units, or SI

Materials (per student group)
Metric Concentration: 21 index cards
Metric Spin: 21 index cards; sheet of stiff paper; scissors; paper fasteners

Teaching Resources
Math Practice Worksheet, p. 12
Resource Worksheets, pp. 13 and 14
Transparency 33
SourceBook, p. S90

Metric Concentration

This game will help students become familiar with metric prefixes and the multiples they represent. The game is well suited for two to four players.

You may wish to make the cards ahead of time or have students make their own. Suggest to students that they review the rules carefully to be sure everyone is in agreement before beginning.

INDEPENDENT PRACTICE Some students may wish to refine the rules or add rules of their own to come up with different versions of the game. Students who are having trouble remembering what the metric prefixes represent may wish to use the deck to practice sequencing the cards on their own.

Metric Spin

The purpose of this game is to allow students to practice converting from one metric unit to another. The game works well with two to four players.

You may wish to have students make the cards and spinner ahead of time, or you can make them yourself. Note that the cards shown on this page are examples. Any metric quantities may be used as long as they represent a variety of metric units. Suggest to students that they review the rules carefully to be sure everyone is in agreement before beginning. Be available to help resolve any disputes.

Homework

Have students prepare a table titled Metric Measurements and Conversions. Divide the table into three columns with the following headings: Object, Metric measurement, and Conversion. Students should find five objects at home that have metric measurements on them, such as food items or tools. Have them record the name of the object, the metric measurement written on the object, and one conversion of the measurement into another metric unit. *(Sample answer: can of soup, 305 g, 3050 dg)*

Metric Concentration

Use index cards to make 21 cards of identical size. Set aside three of them to be key cards. These three cards should be marked with the key units: meter, liter, and gram. Then make six cards based on each of the three key units. These cards will show prefixes with the units and appropriate symbols. A sample deck has been started below.

Once you have prepared all 21 cards, shuffle them well, and spread them out facedown on a table. One after the other, players will turn over a card for 4 or 5 seconds. If the card turned over is one of the key cards, it is set to one side to start a series. If it is not the key card, after 5 seconds it is turned back over, and the play goes on to the next player.

Once a key card has been located, players have the opportunity to start building the series for that unit. As they are picked out, the cards are placed in order on either side of the appropriate key card until all of the cards have been located and placed in their proper position. The winner is the person who places the most cards in the proper place. Can you determine how the eight blank cards should be marked?

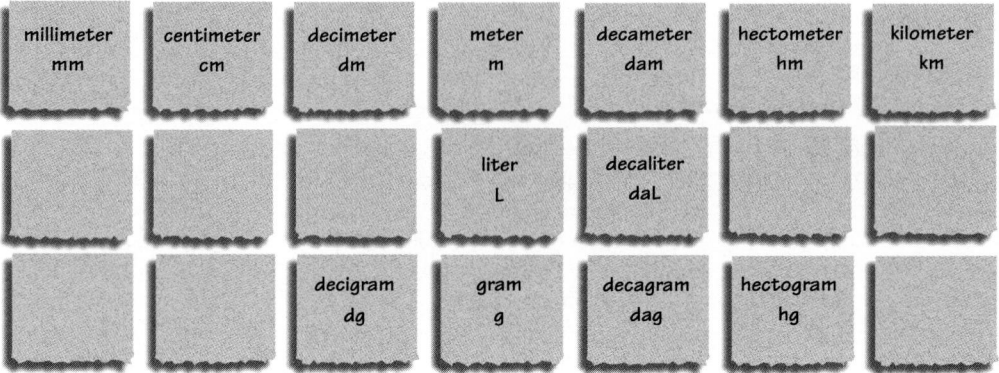

Metric Spin

Again, use index cards to make a deck of 21 cards. This time include a number with each unit. Use a variety of numbers. The cards shown below are examples of cards you might use.

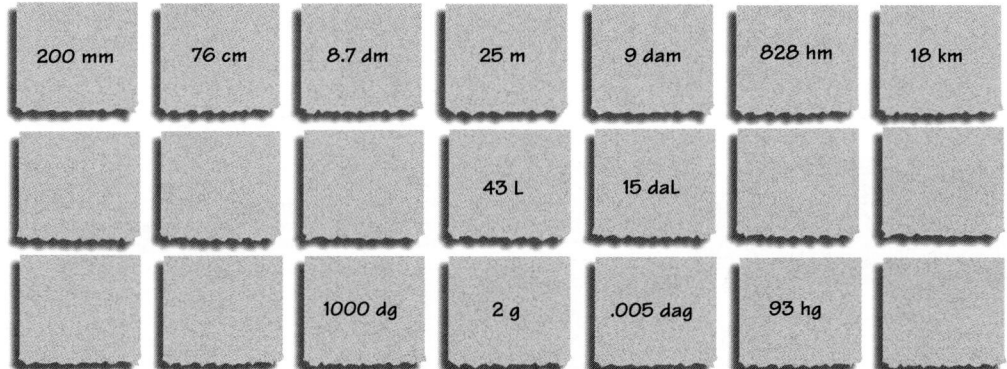

★ **Two Resource Worksheets (Teaching Resources, pages 13 and 14) and Transparency 33 are available to accompany the material on this page.**

Multicultural Extension

Other Measures of Matter

Throughout history, many different units have been developed to measure matter (e.g., the cubit, used by Egyptians, and the shibiri, used by East Africans). Have students prepare a presentation on these and other measurement units used by different cultures. Ask them to find out how the units fit the technology of the time. Their presentations could include diagrams, posters, demonstrations, or cartoons.

Next, construct a circle and spinner (an arrow) from stiff paper, and fasten the pieces together with a paper fastener. Divide the circle into a series of wedge-shaped sections. In each, write a metric prefix or an instruction, as shown in the margin. Now you are ready to play.

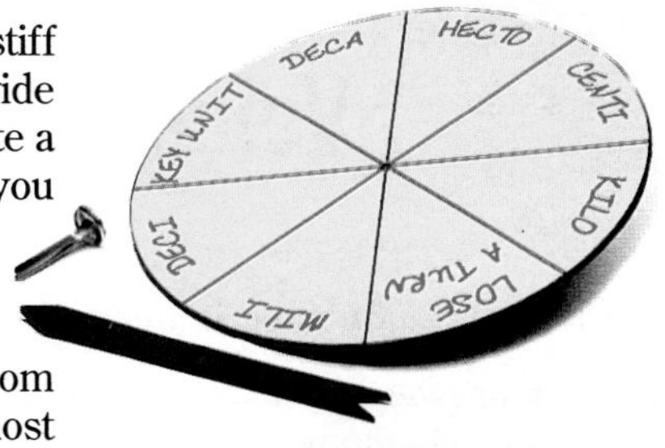

Rules of the Game

1. The purpose of the game is to change a measurement from one unit to another. The winner is the player with the most correct answers at the end of the game.

2. Before you start the game, decide if you want to play individually or in teams of two. Then decide how many times you will go through the pile of cards.

3. To take a turn, pick a card from the top of the pile and spin the arrow. Change the measurement on the card to the unit indicated by the arrow. Use the Metric Conversion Helper below to help you make the conversions.

4. Write your answer on a piece of paper. Check your answer by discussing it with the other players.

5. Players or teams continue to take turns until everyone has gone through the pile the agreed number of times.

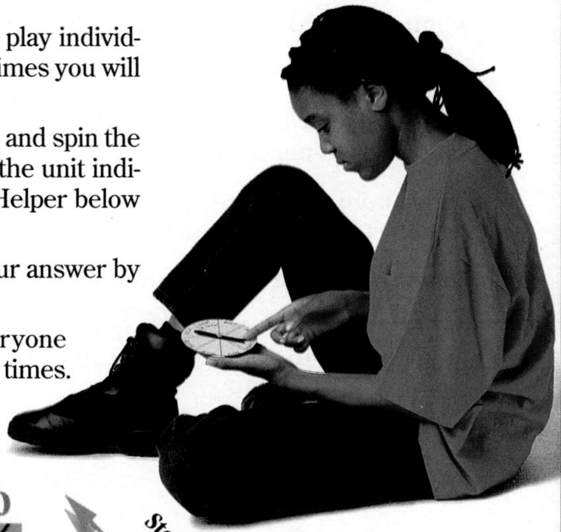

Metric Conversion Helper

The metric prefixes are shown as steps that differ by a factor of 10. Notice that going up two steps decreases a number by a factor of 10 × 10, or 100. Suppose you draw a card that has the measurement 76 m and your spin lands on *kilo*. You need to change the units of the measurement from meters to kilometers, which is three steps up, so you would divide by 10 × 10 × 10, or 1000. Divide 76 by 1000 to get the answer: 0.076 km.

Here are two more examples:

a. To change 15 kg to grams, step down three steps—multiply 15 by 10 × 10 × 10, or 1000. The answer is 15,000 g.

b. To change 5 cm to meters, step up two steps— divide 5 by 10 × 10, or 100. The answer is 0.05 m.

As you can see, you can use different units for the same measurement. For example, 100 cm is the same as 10 dm or 1 m. Why would you want to use one unit instead of another? If you use a larger unit, what happens to the size of the number? **B**

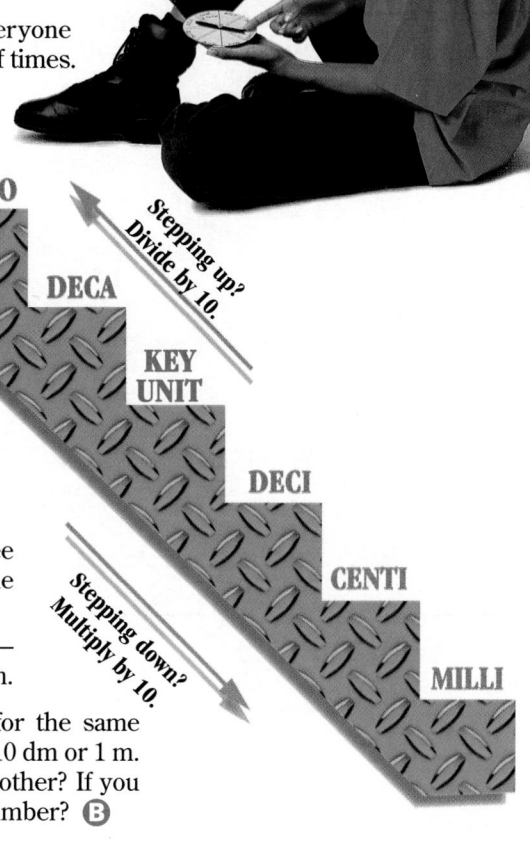

KILO
HECTO
DECA
KEY UNIT
DECI
CENTI
MILLI

Stepping up? Divide by 10.

Stepping down? Multiply by 10.

Metric Conversion Helper

GUIDED PRACTICE Ask metric conversion questions, such as How many kilometers are 10 m? *(0.01 km)* Provide time for students to figure out the answer. Then call on volunteers to share their results with the class.

If your students are accustomed to using decimals, you may wish to point out that each step in the diagram represents a change in the decimal point of the original number. As you move down the diagram, you move the decimal point one place to the right for each step. As you move up the diagram, you move the decimal point one place to the left for each step.

Answers to
In-Text Questions, pages 212–213

A The middle row should read milliliter, centiliter, deciliter, hectoliter, and kiloliter. The bottom row should read milligram, centigram, and kilogram.

B Using a different unit can help you avoid the need for decimals or unreasonably large numbers. When using larger units, the size of the number gets smaller.

Reteaching

Divide the class into two teams. Have a member from Team A call out a metric conversion question, such as How many milligrams are in 10 cg? *(100 mg)* Any member from Team B may respond. If the response is correct, Team B gets a point. Then Team B asks a question and Team A responds. Continue the game until each member has had a chance to ask and respond to a question.

Assessment

Display a variety of packaged consumer items that list metric units on their labels. (Note: Use only empty or unopened containers.) Beside each item, place a metric conversion question. For example, beside a 1.89 L milk carton, ask: How many milliliters is this? *(1890 mL)* Have students record their answers to the questions.

Extension

Have students do research to find out interesting metric distances, volumes, and masses. For example, how tall is Mount Kilimanjaro? *(5895 m)*

Closure

Have students measure familiar objects with a meter stick (e.g., their height, the width and length of a door, the height of a desk). Then ask them to convert each of the measurements to other metric units.

LESSON 2 Volume

FOCUS

Getting Started
Display objects such as a glass of water, a book, and an inflated balloon. Ask students: Which of these objects have volume? After students have had a chance to respond, explain to them that all of the objects have volume because volume is one of the properties of all matter.

Main Ideas
1. Volume has two meanings: the amount of space occupied by matter and the amount of space inside a hollow object.
2. Volume is a property of all matter.
3. A graduated cylinder is used to measure the volume of liquids and powders.
4. The liter is the base unit for measuring volume.

TEACHING STRATEGIES

Before students begin reading, encourage them to think about a definition of volume. Then have a volunteer read page 214 and compare the definition of volume in the text with the students' definitions. Finally, hold up an object such as a book or pencil and ask: Does this have volume? *(Yes)* How do you know? *(It occupies space.)* Does all matter have volume? *(Yes)*

LESSON 2 Volume

How large a furnace is needed to heat a home?

The answer depends, in part, on how much space, or volume, is to be heated in the home.

How much weight can a hot-air balloon support?

Again, the answer depends on the volume of the balloon.

How much electrical energy can be produced by a dam's hydroelectric generator?

The amount depends, in part, on the volume of water that flows through the turbines of the dam.

How many goldfish can Lynne put in her aquarium?

One of the factors determining the answer is the volume of the aquarium; another is the volume of the goldfish.

I think we need a bigger balloon.

As you can see, each of these situations involves a common factor—volume. What is volume? **Volume** is the amount of space occupied by something. When Lynne puts her goldfish in the aquarium, the fish pushes a certain amount of water out of the way. The volume of water displaced by the fish is equal to the volume of the fish.

At times, though, people use the word *volume* to mean the space inside a hollow object, such as a bottle. Volume, in this sense, is a measure of how much a container will hold. For example, the volume of Lynne's aquarium is the amount of space inside her aquarium.

Volume is one *property* of all matter—gases, liquids, and solids. The next Exploration will examine this property of matter.

214

LESSON 2 ORGANIZER

Time Required 3 to 4 class periods

Process Skills
measuring, observing, inferring

Theme Connection
Cycles

New Terms
Displacement method—a method that measures the volume of an object by measuring the amount of liquid that it displaces in a container
Lung capacity—the amount of air that the lungs can hold
Meniscus—the curved surface of a liquid

Volume—the amount of space occupied by matter or the amount of space inside a hollow object

Materials (per student group)
Exploration 1: 100 mL graduated cylinder; 1000 mL of water; white paper; empty 1.89 L ($\frac{1}{2}$ gal.) plastic milk jug; scissors; masking tape; empty transparent pill bottle; marble or pebble; metric ruler; marker
Exploration 2, Activity 1: 2 L plastic container; crayon; **Activity 2:** 1000 mL beaker; graduated cylinder or metric measuring cup; bowl or aluminum pie

continued ▶

Volume Adventures

You Will Need

- a graduated cylinder
- a 1.89 L ($\frac{1}{2}$ gal.) milk carton
- scissors
- an empty pill bottle
- masking tape
- a marble or pebble
- a marker
- a metric ruler
- a piece of white paper

What to Do

Using a Graduated Cylinder

A graduated cylinder is used to measure the volume of a liquid. In this part of the Exploration, you will practice using a graduated cylinder to measure the volume of a liquid.

First, add water to your graduated cylinder up to the 25 mL mark. Are you certain you have exactly 25 mL? Here are a few rules to follow:

a. Hold a piece of white paper behind the graduated cylinder to make the liquid level easier to see.

b. Always read the volume by examining the cylinder at eye level.

c. If the surface of a liquid is curved, use the bottom of the curve for your measurement. This curve is called the **meniscus** of the liquid.

Have you changed your mind about whether you have exactly 25 mL of water in your graduated cylinder?

Exploration 1 continued ▶

215

ORGANIZER, continued

plate; 1 L of water; funnel; crayon;
Activity 3: 3.785 L (1 gal.) plastic container with screw-on top; large pan; 1 m of rubber tubing; 5 L of water; marker; 1000 mL beaker or graduated cylinder; 1 drinking straw per student

Teaching Resources
Exploration Worksheets, pp. 15 and 17
SourceBook, p. S92

The five activities in this Exploration are ideal for small groups. The first three activities have been designed to help students develop their measuring skills by using and making a graduated cylinder. The last two activities are designed to help students prove to themselves that solids and gases occupy space.

⚡ Cooperative Learning
EXPLORATION 1

Group size: 3 to 4 students
Group goal: to make a graduated cylinder and practice several methods of measuring volume
Positive interdependence: Assign students roles such as materials coordinator, record keeper, timekeeper, or primary investigator.
Individual accountability: Have each student demonstrate one of the following tasks: how to use a graduated cylinder, one method of making a graduated cylinder, or how solids and gases occupy space.

⭐ An Exploration Worksheet (Teaching Resources, page 15) is available to accompany Exploration 1.

Using a Graduated Cylinder

Be sure that students carefully read the rules for accurately determining the volume of the water in their graduated cylinder. You may wish to have students exchange cylinders and determine the volume of water in the cylinder they receive. Have students discuss and resolve any discrepancies.

When students finish the activity, make sure they recognize that the volume of the liquid in their graduated cylinder is measured in milliliters. Ask students what other metric unit might be used to measure the volume of a liquid or a powdered solid. *(Liter)*

Answer to
In-Text Question

Ⓐ Students may wish to revise their answers based on this new information.

Making a Graduated Cylinder . . .

For best results, have students use 1.89 L (½ gal.) plastic milk cartons. A 2 L plastic soda bottle with the top cut off and the label removed will also work well for this activity. In some cases, students may need to use cardboard milk cartons for this exercise. When students make their 100 mL marks on a cardboard carton, suggest that they press just hard enough to make a dent. That way, they will be able to see where the marks are by looking for the dents on the inside of the carton.

The pill bottles that students use may hold different volumes of water. Encourage students to compare the volumes of their bottles.

Answers to
In-Text Questions

Ⓐ 1 L

Ⓑ Help students recognize that their bottle is accurate to the nearest 5 mL because this is the amount of water that they used to calibrate the bottle's scale. To make their graduated cylinder more accurate, students would have to calibrate their scale with smaller amounts.

CROSS-DISCIPLINARY FOCUS

Mathematics

Inform students that large-scale production of oil began in Texas with the opening of the Spindletop oil field near Beaumont, Texas, in 1901. When Spindletop was discovered, it gushed more than 126,792,452 L of oil into the air before it was brought under control. Ask: How many kiloliters is that? *(126,792.452 kL)* How many milliliters is that? *(126,792,452,000 mL)* If a lawnmower requires 4 L of crude oil to operate for a given length of time, how many lawnmowers could have been operated with that lost oil? *(31,698,113 lawnmowers)*

Making a Graduated Cylinder . . .

. . . From a Milk Carton

Cut off the top of an empty milk carton. Use your graduated cylinder to add 1000 mL of water to the carton. How many liters is this? Mark this level Ⓐ on the carton. To complete your homemade graduated cylinder, divide the distance from this level to the bottom of the carton into 10 equal divisions. Pour out the water, and then ask someone to add more water to your homemade graduated cylinder. Measure the water level to the nearest division marking. Have someone check your results.

. . . From a Pill Bottle

Run a strip of masking tape along the length of a pill bottle (or other small bottle). Add 5 mL of water to the pill bottle, and mark the water level. Repeat this until the bottle is full.

What volume of water will your new graduated cylinder hold? How accurate is your pill-bottle graduated cylinder? Is it accurate to the nearest 5 mL, 2 mL, or 1 mL? How could you make it more accurate? Ⓑ

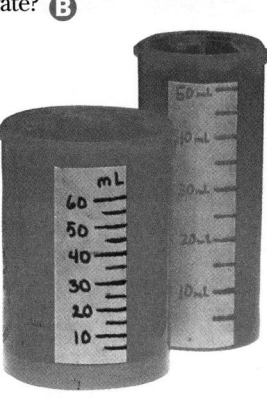
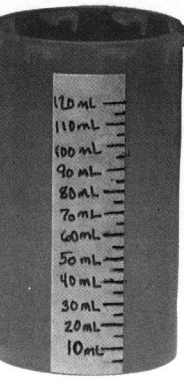
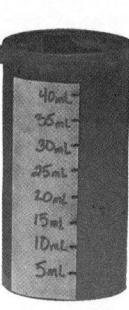

216

Homework

Have students take home their graduated cylinder and use water to measure the volume of the following objects: a bottle cap, their favorite drinking glass, a tablespoon, a flower vase, and the water in one ice cube. They should create a table to record their data. Note: If students have used their graduated cylinder for anything other than water, this activity may be unsafe due to leftover residue in the graduated cylinder.

Proving That a Solid Occupies Space

Fill a graduated cylinder half-full with water. Read the volume of water in the cylinder.

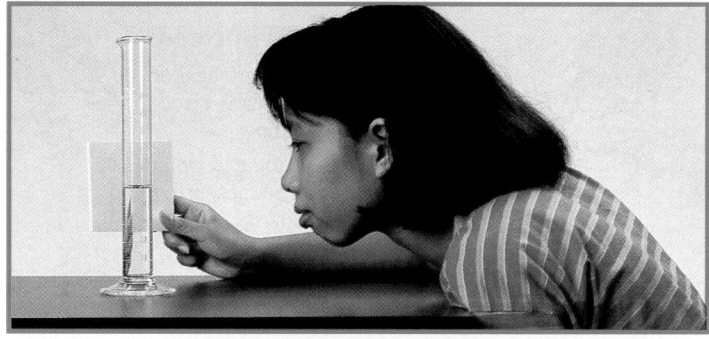

Place a marble, a pebble, or some other solid object in the cylinder. Slide the object in, using the technique illustrated here.

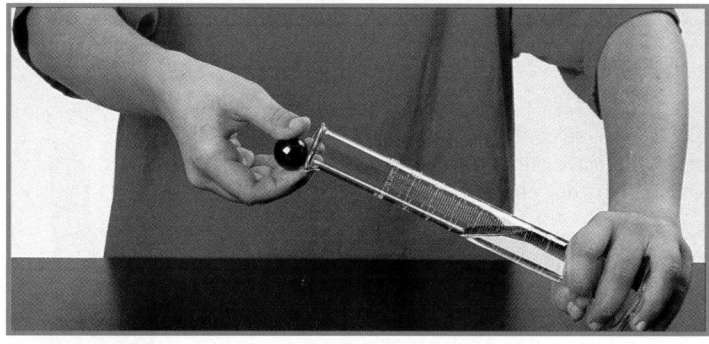

Now read the volume of water again. What is the volume of the solid object? What kind of solid could not have its volume determined by this method? **C**

Proving That a Gas Occupies Space

Push a piece of paper to the bottom of your pill-bottle graduated cylinder. Then immerse the cylinder in a beaker of water, open end first, as shown below. Why didn't the paper get wet? What must you do in order to get water to enter the cylinder? **D**

Proving That a Solid Occupies Space

Encourage students to compare their results to discover if the solid objects they used have the same or different volumes. To extend the activity, have students make a table to show the volume of each object they used.

GUIDED PRACTICE When students have completed the activity, have them discuss what happened when they placed an object in the graduated cylinder. *(The level of water in the cylinder rose.)* Why did this happen? *(The object occupied some of the space previously taken up by the water.)*

Proving That a Gas Occupies Space

Caution students not to tip the bottle as they immerse it in the water. A small piece of tape may help keep the paper in place during this activity.

Answers to
In-Text Questions

C Students should subtract the first volume reading from the second to find the volume of the solid. The volume of a solid that floats on water or that dissolves in water could not be measured by this method. Additionally, a solid too large to fit in the cylinder could not be measured.

D The air in the bottle occupies space, preventing the water from rushing in and getting the paper wet. By tipping the bottle and releasing the air, an empty space is created that is then filled by the water.

Theme Connection

Cycles
Focus question: The Great Salt Lake is a saltwater lake despite the fact that it is fed by freshwater streams. Why is this true? *(All freshwater streams contain a little bit of dissolved salt. When these streams carry their salt into a lake and the water evaporates, the salt is left behind. Most lakes empty their water into rivers that carry the salt to the ocean, but the Great Salt Lake has no outlet. Therefore, over time, the water in this lake becomes increasingly salty.)*

These two Activities further develop student awareness and understanding of volume and how it can be measured. To encourage analysis and math skills, have students work collectively to present their data in a table and to evaluate the results.

★ **An Exploration Worksheet (Teaching Resources, page 17) is available to accompany Exploration 2.**

ACTIVITY 1

SAFETY ALERT To prevent the possibility of choking on the water, caution students not to submerge their head while performing this Activity.

Since many students may be sensitive about the size of their body, a certain degree of maturity is required of the entire class. If you feel your class cannot conduct this Activity in a scientific fashion, you might wish to focus on Activity 2 instead.

Students should be careful not to fill the tub so high that it overflows, and they should remove the crayon mark when they are finished. Tell students to brush excess water off their body before leaving the tub. Remind students to record their results to bring to class.

GUIDED PRACTICE   To extend the Activity, have students discuss the variables that affect the accuracy of their results. For example, the volume of their head will not be included, and the water on their body and any water splashed out of the tub will be added to their total volume. Ask students how they could make the Activity more accurate. *(For example, they could estimate the volume of their head, be careful not to splash any water out of the tub, and let as much water as possible run off their body.)*

ACTIVITY 2

Suggest to students that they try both displacement methods and compare the results. If the results are slightly different, they may wish to calculate the average and use that as the final figure.

Body Volumes

The famous football player William "The Refrigerator" Perry, who weighed about 7.5 kg when he was born, said, "I was big when I was small!" Some people are large; some people are small. Some have large hands; others have large feet. In this Exploration, you will measure your total body volume and the volume of your hands and lungs.

ACTIVITY 1

What Is Your Body Volume?

You Will Need

- a 2 L plastic container
- a crayon

What to Do

Here is a way to find out. Take an empty 2 L plastic container and a crayon with you the next time you take a bath. Have enough water in the tub so that you can submerge yourself up to your chin.

Caution: Do not submerge your head.

The water level will rise according to your volume. Use the crayon to mark the level of the water when you are completely submerged. Now get out of the tub. Of course, the level will go down.

While you are out of the tub, bring the water level back up to the crayon mark. You can do this by filling the container with water and pouring it into the tub. Count how many times you empty the container into the tub, and then multiply that number by two. (Why?) Ⓐ This is your volume in liters.

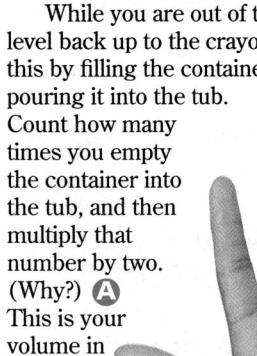

218

ACTIVITY 2

How Big Is Your Hand?

You Will Need

- a 1000 mL beaker or similarly sized container
- a graduated cylinder or metric measuring cup
- a bowl or pie pan
- a funnel
- a crayon

What to Do

You may have heard that most people have one foot that is larger than the other. Is this true of hands as well? Here is a way to find out.

You can use the same method as before. Instead of a tub, however, use a container just large enough to submerge one hand.

INDEPENDENT PRACTICE   Have students form small groups and discuss the possible reasons for obtaining different results from the two methods. Ask students to determine how experimental error could have occurred. *(For example, in the second method, some water could cling to the sides of the container and thus not be accounted for.)*

SAFETY TIP   If plastic 1000 mL beakers are available, use them instead of glass beakers to avoid the risk of breakage.

Answers to
In-Text Questions, pages 218–219

Ⓐ Students must multiply the number of times they fill the container by 2 because the container holds 2 L, not 1 L.

Ⓑ $1000 \text{ cm}^3 = 1 \text{ L}$

Submerge one hand in the beaker to the first wrinkle on the inside of your wrist. Mark the level reached by the water, and then remove your hand. Because the amount of water involved is small, use a graduated cylinder or metric measuring cup to bring the water level up to the mark. If you use a measuring cup, it may be marked in either cubic centimeters or millilitres. (Note that 1 mL = 1 cm^3. How many cubic centimeters are there in a liter?) 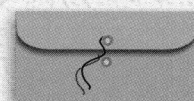 Do the same thing to find the volume of your other hand.

Another way to make the same measurement is to fill the beaker to the top so that it spills over if anything is added. Have a bowl or pie pan underneath it to catch water that spills over as you dip in each hand. Pour this water into a measuring cup or graduated cylinder to determine the volume of each hand. If necessary, use a funnel to transfer the water.

Both methods enable you to measure the volume of your hands by finding out the volume of water that each hand *displaces*, or takes the place of. For this reason, each is a type of **displacement method** for finding volume. Try both displacement methods to measure the volume of other objects as well.

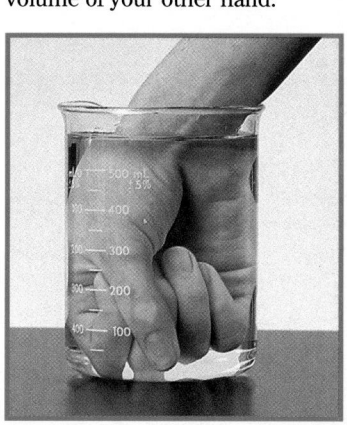

Examining Your Data

If everyone in your class did the last two activities, you now have a large amount of data about hand and body sizes. Using this data, calculate the class statistics suggested below.

a. average volume of female bodies; of male bodies
b. total volume of students in the class
c. amount of classroom air displaced by the class
d. number of people with a bigger right hand than left hand
e. average volume of the right hand; of the left hand
f. average volume of a male hand; of a female hand

Exploration 2 continued ▶

219

Examining Your Data

Remind students to make accurate records of their results. As a class, calculate the statistics asked for in this section. If the class did not do Activity 1, have students compare the data from Activity 2 instead. Students should follow along by recording these statistics in their ScienceLog.

PORTFOLIO

Have students include the class statistics in a table in their Portfolio. Students should make notes on what the table indicates and include a discussion of any problems they might have had in making accurate measurements.

Meeting Individual Needs

Gifted Learners

Pose the following questions for students to think about:
- What would society be like if there were no standard weights and measures?
- How would scientific research be affected?
- How would commerce be affected?
- Why do you think societies first developed systems of standard weights and measures?

Have students write responses to these questions. *(Accept all reasonable responses.)* Students may present their ideas as oral reports to the class.

Meeting Individual Needs

Learners Having Difficulty

Present the following situation to students: Suppose that an aquarium at a zoo has a volume of 570,000,000 mL. What metric unit listed below would you use to make this number a little easier to work with? What would the appropriate measurement be with this unit?

a. gallons; 570,000 gal.
b. liters; 5,700 L
c. liters; 570,000 L *(Correct)*
d. gallons; 5,700 gal.

ACTIVITY 3

Supply each student with a plastic straw that can be fitted onto the end of the rubber tubing as a separate mouthpiece. You may wish to have some tape handy to ensure a tight fit. Remind students not to share drinking straws.

Involve students in a discussion of what happens when they blow into the tube. *(The air from their lungs displaces the water in the jar. The water is forced out, and the air takes its place.)*

INDEPENDENT PRACTICE When students have completed the Activity, suggest that they organize their results in a table and then use the table to determine some statistics of the lung capacities of the students in their class.

Answers to
In-Text Questions

Ⓐ A container that is larger than 3.785 L could be used, for instance.

Ⓑ Lung capacity can be determined by several factors, including body size, lung condition, and how much a person exercises.

Ⓒ A long-term experiment that tests lung capacity at regular intervals could be set up. The person being tested should be starting a regular exercise program.

Homework

Have students describe the procedure for an experiment in which the following question is addressed: Does increasing your amount of daily exercise increase your lung capacity? Each student's description should outline the steps of the experiment in detail as well as list the materials and safety equipment that are necessary to complete the experiment.

ACTIVITY 3

Who Has the Largest Lung Capacity?

You Will Need

- a 3.785 L (1 gal.) plastic container with a screw-on top
- water
- a piece of rubber tubing
- a large pan
- a marker
- a 1000 mL beaker or graduated cylinder
- drinking straws

What to Do

Measuring volume can be difficult if you can't use any of the methods just discussed. One such situation is determining **lung capacity**—the amount of air your lungs can hold. How much air do you think your lungs can hold? Less than 3.785 L of air? More than 3.785 L of air?

You can find out if you have a piece of rubber tubing and a 3.785 L plastic container with a screw-on top. Completely fill the container with water, and screw on the top. Turn the container upside down, and place the neck of the container in a large pan that is half-filled with water. A partner can support the container as shown in the illustration. Unscrew the top, and insert the rubber tubing. Insert your straw into the rubber tubing.

Caution: Dispose of your straw when you have finished. Do not use anyone else's straw.

Using only one breath, blow as much air into the tube as you can. Using a marker, draw a line where the new water level is on the container. Find out how much water remains by using a beaker or graduated cylinder. Were you able to fill the container with air? How could you change

this procedure to measure your lung capacity if your lungs hold more than 3.785 L of air? Ⓐ

Compare your lung capacity with that of your classmates. What do you think are some factors that determine differences in lung capacity? Do you think exercise can increase lung capacity? Explain. What kind Ⓑ of experiment could you set up to test your hypotheses? Ⓒ

Integrating the Sciences

Life and Physical Sciences

Suggest that interested students do some research on asthma, a lung disease that affects about 10 million Americans. The following questions may help students get started:

- What causes asthma?
- How is asthma treated?
- How does asthma affect lung capacity?

Have students present their findings to the class.

Are You a Volume Whiz?

How well do you understand the concept of volume? Discuss with two classmates whether the following statements are correct, incorrect, or a bit of both.

Keep down the volume of your discussion!

a. No two objects can occupy the same space at the same time.

b. A piece of modeling clay has a larger volume when it is rolled flat than when it is rolled into a ball.

c. Saying that a marble has a volume of 8 mL is an inference.

d. The volume of sugar can be determined by the displacement method.

e. Your thumb has a volume of less than 10 mL.

f. The volume of a gas can change.

g. Fifteen milliliters of milk on your cornflakes is plenty.

h. Gases do not occupy space.

i. A liter is exactly 500 mL.

j. An appropriate unit for measuring shampoo is the liter.

k. In recipes, baking soda is measured in milliliters.

l. A solid object with a volume of 1 L will displace 1000 mL of water.

m. It is more appropriate to buy gasoline by the milliliter than by the liter.

FOLLOW-UP

FOLLOW-UP

Reteaching

Set up a display of consumer items. Using a game-show format, have student teams take turns estimating the volume of each item. The team with answers closest to the actual volumes wins.

Assessment

Set up stations around the room where various activities can be performed to assess students' ability to measure volume. Activities could include determining the volume of water in a graduated cylinder, the volume of air in a small bottle, and the volume of a solid object.

Extension

Challenge students to solve the following problem: In a container of marbles, how much space is occupied by the air between the marbles? *(Students may discover various solutions to this problem. The following is one possibility: Fill a graduated cylinder with water to its highest volume mark. Pour water from the graduated cylinder into a container of marbles until the marbles are just covered with water. Record the new volume of the water in the graduated cylinder and subtract this value from the volume of the cylinder.)*

Closure

Display several classroom items on a table (such as an eraser, a book, a pencil, and a jar). Ask students to determine a method for finding the volume of each one. *(Answers will vary. Students should display an understanding of the methods discussed so far in the unit.)*

Answers to
Are You a Volume Whiz?

a. Correct.

b. Incorrect. The volume is constant no matter what the physical shape is.

c. Incorrect. This is an inference only if the volume has not been measured.

d. Incorrect. The sugar will dissolve in the water. (There will be some increase in volume, but it will be very slight.)

e. Answers will vary depending on the size of the thumb being measured.

f. Correct. This can be demonstrated by compressing the air in a sealed syringe.

g. Incorrect. Most students will agree that this is not enough milk for a bowl of cereal.

h. Incorrect. Gases are matter and therefore, by definition, occupy space.

i. Incorrect. A liter is exactly 1000 mL.

j. Both. Some students will argue that shampoo is measured in milliliters.

k. Correct. Milliliters can be used for measuring powders.

l. Correct. 1000 mL and 1 L are the same volume.

m. Incorrect. For the amount of gasoline needed to operate a car, a milliliter is an impractical unit.

LESSON 3 Mass

FOCUS

Getting Started

Have students name items that they would purchase by volume and items that they would purchase by mass. Keep a list of their suggestions on the chalkboard under the appropriate headings.

Main Ideas

1. Mass is measured with a balance.
2. The gram is the base unit for measuring mass.
3. The standard for comparing masses is water: 1 mL of water has a mass of 1 g.
4. For a given substance, volume and mass are directly proportional.

TEACHING STRATEGIES

EXPLORATION 3

ACTIVITY 1

Display a balance in a convenient part of the classroom. Allow three or four students at a time to experiment with the balance.

Answers to
In-Text Questions and Caption

Ⓐ A balance measures the mass of an object.

Ⓑ A balance compares unknown masses with known masses.

Ⓒ The lever must be "balanced" to find the correct mass.

Ⓓ The mass of the object is less than half of a kilogram.

ACTIVITY 2

Allow two or three students to work together. Encourage them to confer and agree on the mass of the clay before they place it on the balance.

LESSON 3 Mass

We buy milk and gasoline by volume. The liter is one unit for measuring volume. As you have seen, all matter has volume.

We purchase meat, sugar, and cheese by the gram (g) or kilogram (kg). These are units of mass. All matter has mass. **Mass** is the measure of the amount of matter in an object. In Exploration 3, you will do four Activities that will help you understand more about mass.

EXPLORATION 3

Becoming a Mass Expert

You Will Need

- a triple-beam balance
- modeling clay

What to Do

ACTIVITY 1

Mass is measured with a balance. Examine a classroom balance.

- What does it do? Ⓐ
- How does it work? Ⓑ
- Why is it called a "balance"? Ⓒ

> Is the mass of this object more or less than half of a kilogram? Ⓓ

ACTIVITY 2

How large is a gram (g) of modeling clay? a milligram (mg)? a kilogram (kg)?

From a piece of modeling clay, make a ball that you think has a mass of 1 g. Check your estimate with the balance.

Now form a ball that you think has a mass of 10 g, or 1 dag (1 decagram). How close was your estimate?

Next, form a ball that has a mass of 100 g, or 1 hg (1 hectogram). Again, check your estimate.

Finally, place together 10 of the 1 hg masses made by your classmates. You now have 1000 g, or 1 kg (1 kilogram).

If you could divide your 1 g sample of modeling clay into 1000 equal parts, each part would have a mass of 1 mg (1 milligram). Now that's small!

Choose several objects from around your classroom. Estimate the mass of each one, and then check your estimate with the balance. Record your data in a table like the one below.

Object	Estimated mass	Actual mass

222

LESSON 3 ORGANIZER

Time Required
3 class periods

Process Skills
measuring, observing, predicting

New Term
Mass—the measure of the amount of matter in an object or substance

Materials (per student group)
Exploration 3: metric balance; stick of modeling clay; several small classroom objects
Exploration 4: metric balance; 1 L container; 250 mL each of sand, soil,

cereal, grain, sawdust, powdered detergent, liquid detergent, vinegar, and water; safety goggles; lab aprons
Exploration 5: index card; pin; small box of paper clips; 20 cm of thread; lightweight classroom object; metric ruler; metric balance

Teaching Resources
Exploration Worksheets, pp. 21, 24, and 27
Resource Worksheet, p. 26
Transparencies 34, 35, and 36
SourceBook, p. S90

ACTIVITY 3

Here's a massive problem. Suppose that you wish to compare your mass with that of your classmates. You have a long board. How could you use it to compare your masses? Explain how your method would work.

ACTIVITY 4

This is an activity to do at home. Examine the picture of the homemade balance below. Your challenge is to make your own homemade balance. How does it differ from the classroom balance on the previous page? **E** Instead of paper cups, you can use pails or other containers. Or perhaps you can think of another type of homemade balance to make. When you are finished, have a balance display in your classroom.

EXPLORATION 4

Comparing Masses

You Will Need

- a balance
- a 1 L container
- 250 mL of each of the following dry materials: sand, soil, cereal, grain, sawdust, powdered detergent
- 250 mL of each of the following liquid materials: liquid detergent, water, and vinegar

What to Do

ACTIVITY 1

Comparing Volumes and Masses

If you have equal volumes of two substances (for example, 250 mL of each), do you have equal masses as well? Test your prediction by placing equal volumes of the dry materials listed above in a 1 L container, one at a time. Find the mass of each with a balance. If the masses of the equal volumes are different, arrange the masses from smallest to largest.

ACTIVITY 2

Comparing Masses

If you had to choose a standard for comparing masses—that is, a substance whose mass you could use to compare all other masses—what would it be? The metric unit of mass is defined in terms of water. In other words, water is the standard for comparing masses. One milliliter (1 mL) of water has a mass of 1 g. By this definition, what must be the mass of 1 L of water? **F**

Now compare the mass of 1 L of water to the masses of 1 L of liquid detergent and vinegar. First, find the mass of your empty container. Then pour 250 mL of water into the container, and find the mass of the water and the container. (What must you do with the mass of the container in order to find the mass of just the water?) Pour out the water, **G** and, one at a time, add 250 mL of the other liquid materials to the container. (First, be sure to wipe the container dry.) Determine the mass of each, and record your findings in a data table similar to this one.

Material	Volume of material (in mL)	Mass of material (in g)	Mass of 1 L of material (in g)
water	250	250	1000
vinegar	250	?	?
liquid detergent	250	?	?

ACTIVITY 3

Students should recognize that the only comparison of masses they could make (without using a known mass) is who has a greater mass and who has a lesser mass. A student who knows his or her mass can become the standard against which other students can measure themselves. A log could be used as a fulcrum for the board to make a simple balance like a seesaw.

ACTIVITY 4

Encourage students to be creative in their designs. Set up a display of completed balances in the classroom, and encourage students to discuss the relative effectiveness of each balance and why some designs might work better than others.

★ An Exploration Worksheet (Teaching Resources, page 21) and Transparency 34 are available to accompany Exploration 3. Another Exploration Worksheet (Teaching Resources, page 24) and Transparency 35 are available to accompany Exploration 4.

EXPLORATION 4

ACTIVITY 1

By comparing the masses of equal volumes of different materials, students should discover that the masses of the materials are different. Students should conclude that even if different kinds of matter have equal volumes, they may have different masses.

ACTIVITY 2

GUIDED PRACTICE After students have completed the data table on this page, have them discuss the values they obtained. Any differences that arise will offer an excellent opportunity to discuss errors in measuring.

Answers to
In-Text Questions

E The balance on this page compares the unknown masses of two materials, but the balance on the previous page compares the unknown mass of one material to a known mass.

F 1 L of water has a mass of 1000 g.

G First, the container's mass alone must be determined, and then the mass of the container must be subtracted from the total mass. (In this case, students should be able to predict that 250 mL of water has a mass of 250 g.)

Answers to
Some Problems to Solve

1. **a.** 1 L of sawdust has a mass of 600 g because 600 mL of water has a mass of 600 g.

b. 2 L of sawdust has a mass of 1200 g; 500 mL of sawdust has a mass of 300 g; 100 mL of sawdust has a mass of 60 g.

c. The wood would have a mass of more than 600 g because 1000 mL of wood has a greater mass than 1000 mL of sawdust. A given mass of sawdust occupies a greater volume than the same mass of wood.

d. Mike's and Kathy's sawdust may have come from two different types of wood, such as a hardwood and a softwood, that had different masses for the same volume of wood. Mike's sawdust may have been more compact than Kathy's, thereby having more mass for the same volume.

e. 1 L of sand has a greater mass and therefore weighs more than 1 L of sawdust. The difference in weight depends on the types of sand and sawdust used.

2. First multiply the mass of liquids A, B, and C (from left to right) by 5 to find the mass of 1 L of each. Then compare these values to those in the table. Danny figured out that liquid A was alcohol, liquid B was glycerine, and liquid C was vinegar.

3. See the table and graph on page S236. The lines on the graph are similar in that they all show an increase in volume. They differ in the amount of the increase that they show for each substance.

Homework

You may wish to assign Some Problems to Solve as homework.

SAFETY ALERT
If you choose to use question 2 as an in-class activity, remind students that they should never taste an unknown substance.

⭐ **A Resource Worksheet (Teaching Resources, page 26) and Transparency 36 are available to accompany Some Problems to Solve.**

Some Problems to Solve

1. When Kathy did Exploration 4 using her homemade balance, she found that 1 L of sawdust was balanced by 600 mL of water.

    a. What is the mass of 1 L of sawdust (using Kathy's data)?

    b. What would be the mass of 2 L of sawdust? 500 mL of sawdust? 100 mL of sawdust?

    c. Suppose that Kathy could find the mass of exactly 1000 mL (1 L) of the wood from which the sawdust came. Would it weigh less than 600 g or more than 600 g? Why?

    d. When Mike did the activity, he found that 1 L of sawdust was balanced by 828 mL of water. Did Mike make a mistake? He examined the sawdust a little more closely and found that it was somewhat different in color from the sawdust used by Kathy. How would you explain the different results?

    e. Using your data from the activity, which has the greater mass: 1 L of sawdust or 1 L of sand? How much greater?

2. Danny had three colorless liquids to identify. He used a balance to find the mass of 200 mL of each of the liquids. By referring to the table in the margin, he was able to identify the three liquids correctly. What did Danny find?

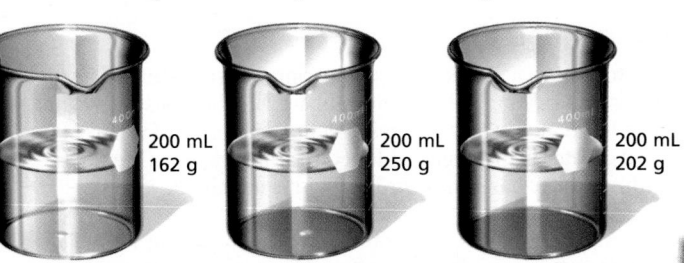

Substance	Mass of 1 L
alcohol	810 g
ethylene glycol (antifreeze)	1100 g
glycerine	1250 g
sea water	1040 g
vegetable oil	900 g
vinegar	1010 g
water	1000 g

200 mL 162 g   200 mL 250 g   200 mL 202 g

3. Complete the table below in your ScienceLog. Then plot the measurement for each substance on a copy of the graph at the right. How are the patterns of points for different substances similar? How are they different?

Volume	Mass of sand	Mass of water	Mass of sawdust
1000 mL	2800 g	1000 g	700 g
500 mL	1400 g	500 g	
200 mL	560 g		
100 mL			

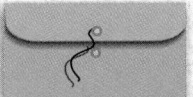

X Sand   ● Water   O Dust

Mass (g): 2800, 2400, 2000, 1600, 1200, 800, 400

Volume (mL): 0, 200, 400, 600, 800, 1000

PORTFOLIO
Have students include their answers to Some Problems to Solve in their Portfolio. Students could indicate which questions were difficult for them and why.

CROSS-DISCIPLINARY FOCUS

Language Arts
Ask students to look up the word *balance* in the dictionary and find its origin. *(It has its roots in several western European languages, including Middle English, Old French, and Latin.)* How many different definitions are offered? *(Answers may be as high as 30.)* Have students write a few sentences, a poem, or a short story that incorporates different meanings of the word.

Making a Featherweight Balance

You Will Need

- an index card
- a pin
- paper clips
- thread
- a ruler
- a very light object

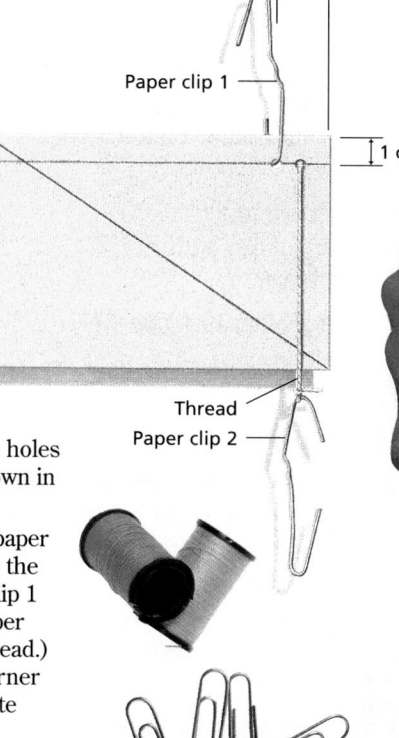

3 cm

Paper clip 1

1 cm

Thread
Paper clip 2

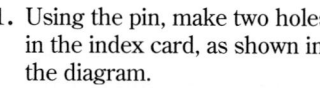

What to Do

1. Using the pin, make two holes in the index card, as shown in the diagram.

2. Make hooks out of two paper clips, and attach them to the card as shown. (Paper clip 1 anchors the balance; paper clip 2 dangles from a thread.) Draw a line from one corner of the card to the opposite corner.

3. To calibrate your balance, follow these steps.

   a. Hold on to paper clip 1.

   b. Make a mark where the thread crosses the diagonal line. Label this mark with a zero (0).

   c. Add paper clips, one at a time, to paper clip 2. For each added paper clip, mark where the thread crosses the line. You now have a calibrated balance. Each division indicates the mass of one paper clip.

4. Measure the mass of a very light object by hooking it onto paper clip 2. Record your results.

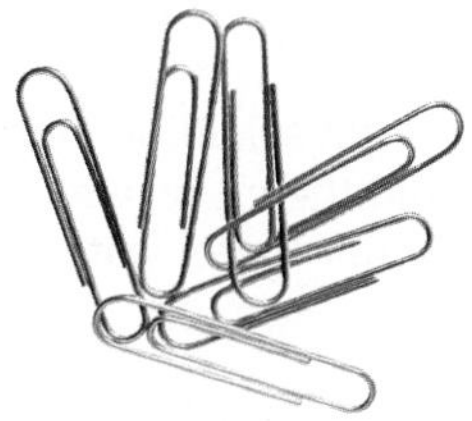

5. Check the mass of one paper clip on a classroom balance. How can you use this value to mark your calibrated balance? How much does your light object really weigh?

Questions

1. What is meant by the word *calibrate*? What did you use to calibrate your featherweight balance?

2. How is your featherweight balance similar to your classroom balance?

225

This interesting activity can be done at home or in the classroom. The balance could be calibrated with very small washers instead of paper clips.

To calibrate the balance in metric units, mark the balance with multiples of the mass of one paper clip.

★ **An Exploration Worksheet (Teaching Resources, page 27) is available to accompany Exploration 5.**

Answers to *Questions*

1. To calibrate means to mark an instrument with reference points that correspond to a known quantity. The featherweight balance was calibrated using paper clips.

2. All balances are similar in that one part or side balances another part or side. With the featherweight balance, the card balances the mass of the paper clips. With a classroom balance, an object is balanced against known masses that are placed on the opposite end of the beam or against masses that are built into the beam itself.

Homework

You may wish to assign Exploration 5 as homework.

FOLLOW-UP

Reteaching

Provide students with a list of everyday objects and ask them to estimate the mass of each object in metric units. Check to make sure that students use an appropriate metric prefix when necessary.

Assessment

Set up several stations where students can demonstrate their ability to use and read a balance. The stations could be used to do the following:

- Determine the mass of an object using a balance.
- Determine the mass of a specified volume of water or other liquid using a balance.
- Compare the masses of several objects or liquids.

Extension

Have students do research to find out the difference between mass and weight and between a balance and a scale. *(Mass is the measure of the amount of matter an object has. Weight is the pull of gravity on the mass of an object. A balance determines an object's mass by comparing it to a known mass. A scale measures weight, or the pull of gravity on the mass of an object.)*

Closure

Ask students to explain the importance of making accurate measurements. Students can do this in the form of a short essay, a short story, or a cartoon.

Answers to
Challenge Your
Thinking, pages 226–227

1. See the completed graph on page S237.

2. First, convert all of the measurements to grams.
   18 g = 18 g
   2 dag = 20 g
   0.5 hg = 50 g
   1000 mg = 1 g
   Then it is easy to arrange them from the largest mass to the smallest mass.
   0.5 hg > 2 dag > 18 g > 1000 mg

3. The scientific ideas used by the crow are that all matter has volume and that no two objects can occupy the same space at the same time.

4. Answers will vary, but sample responses include the following:
   a. *Water* is the *standard* for comparing *masses*.
   b. *1 mL* of *water* has a *mass* of *1 g*.

5. Since 1 mL of water has a mass of 1 g, sample 1 could be water. Sample 2 is too light, however, so it must be some other liquid. (If sample 2 were water, it would have a mass of 35 g.)

6. a. Meters
   b. Kilometers
   c. Meters or kilometers
   d. Milliliters
   e. Kilograms
   f. Grams or kilograms
   g. Milliliters
   h. Kiloliters
   i. Milliliters
   j. Grams or milligrams
   k. Centimeters or meters
   l. Kilograms
   m. Liters

 You may wish to provide students with the Chapter 12 Review Worksheet that is available to accompany this Challenge Your Thinking (Teaching Resources, page 28).

Primary Source
Description of change: excerpted from *Aesop's Fables*
Rationale: to focus on the nature of matter

CHALLENGE YOUR THINKING

1. What Are You Plotting?
The table on the right shows the masses of five different volumes of glycerine. Plot these values on a graph, with mass on the *y*-axis and volume on the *x*-axis. Be sure to label with units.

Mass	Volume
62 g	50 mL
125 g	100 mL
250 g	200 mL
500 g	400 mL
1000 g	800 mL

2. Metric Puzzler
Eva made balls of clay with the following masses:

   18 g, 2 dag, 0.5 hg, 1000 mg

Arrange these measurements from the largest mass to the smallest mass.

3. A Fable for You
A thirsty crow needed water. The only source available was a pitcher, but the crow could not reach the water. After much thought it started to drop pebbles, one after the other, into the pitcher. Soon the water level was high enough for the crow to take a drink. The moral of this story is, "Where force fails, patience will often succeed."

What scientific ideas did the crow use in order to get water?

What is density and what does it have to do with volume? Turn to page S92 of the SourceBook to find out.

4. Say It With Words
Use the following words to create sentences that explain what you have learned in this chapter:

a. standard, water, masses

b. 1 g, water, 1 mL, mass

226

5. Water, Water, Everywhere?

Marven had two liquids that he thought were both water. From the data in the table, would you agree or not? Explain your reasoning.

Sample	Mass (g)	Volume (mL)
1	25	25
2	25	35

6. Measuring Made Easy

Could you measure the mass of an elephant in milligrams? Would you want to? Suggest the appropriate metric unit for each of the following measurements. The first one has been done for you.

Measurement	Unit
a. height of the World Trade Center	meters
b. distance to China	
c. width of the Mississippi River	
d. volume of a penny	
e. mass of an elephant	
f. mass of a bag of potatoes	
g. volume of an ice cube	
h. volume of Lake Michigan	
i. volume of a perfume bottle	
j. mass of a fly	
k. your height	
l. your mass	
m. your volume	

Sciencelog

Review your responses to the ScienceLog questions on page 210. Then revise your original ideas so that they reflect what you've learned.

227

The following are sample revised answers:

1. Gases have both mass and volume. (Therefore, gases are a form of matter.)

2. Both a kilogram of lead and a kilogram of feathers weigh a kilogram. They have the same mass, and because they are both subject to the same force of gravity, they have the same weight.

3. The centimeter is the more appropriate unit for measuring a person's height because a height measured in millimeters would yield an unwieldy number (e.g., 1500 mm versus 150 cm).

4. A liter is a measure of volume, so the milk would most likely be measured in liters. Both the cow and the cheese have volume, but people usually are interested in the weight, not the volume, of these items.

Meeting Individual Needs

Gifted Learners

Propose the following scenario for students to consider: Suppose that you have invented a method to communicate with dolphins. You discover that they are highly intelligent beings much like ourselves. As a step toward establishing an organized society, the dolphins want to formulate a system of measurement. How would you suggest that they proceed? What system would you suggest that they use?

Suggest to students that they write an interview with one of the dolphins in which they discuss and resolve the problem. *(Accept all reasonable responses.)*

Homework

Have students participate in a metric scavenger hunt. They should collect the following items at home: an object that is 12 cm long, a container that holds 200 mL, an object with a mass of 100 g, a piece of string that is 45 cm long, something with a thickness of 1 mm, and 20 g of a dry cereal. The next day, students can use classroom tools to test the accuracy of their estimates.

Connecting to Other Chapters

Chapter 11
explores the nature of matter, including its physical, chemical, and biological properties.

Chapter 12
introduces the metric system and ways to measure volume and mass.

Chapter 13
examines the three principal states of matter—solid, liquid, and gas.

Prior Knowledge and Misconceptions

Your students' responses to the ScienceLog questions on this page will reveal the kind of information—and misinformation—they bring to this chapter. Use what you find out about your students' knowledge to choose which chapter concepts and activities to emphasize in your teaching.

In addition to having your students answer the questions on this page, you may wish to use the following demonstration to assess students' prior knowledge and misconceptions of the states of matter. Show students a pail or tray of dry ice that is sublimating.

SAFETY ALERT Do not allow students to touch the dry ice.

Have them describe and explain what is happening to the dry ice. Engage the class in a discussion of these observations. (*Accept all reasonable responses. You may wish to inform students that as dry ice sublimates, the solid carbon dioxide becomes a gas. While this is happening, energy is being transferred to the carbon dioxide from the water vapor in the air, which condenses to form a mist.*) During the

discussion, take note of what the students already know about changes of state, as well as what aspects of the topic intrigue them.

CHAPTER

13

More About Matter

1 Imagine that you could divide a liter of water in half, time and time again. What do you think would eventually happen?

2 Is the chalk in the air a solid, liquid, or gas?

ScienceLog

Think about these questions for a moment, and answer them in your ScienceLog. When you've finished this chapter, you'll have the opportunity to revise your answers based on what you've learned.

3 Where did the water on this girl's glasses come from? (No—she didn't wear them in the shower!)

228

Solid, Liquid, or Gas?

Water is a liquid, ice is a solid, and the air you breathe is a gas. You can point to many examples of solids, liquids, and gases around you, but how would you explain solids, liquids, and gases to someone who knows nothing about them?

Zed, whom you met in Unit 1, has been on Earth for several days. He is just getting to know something about the different kinds of materials here and wants someone to explain the difference between solids, liquids, and gases.

On the next page, you are the teacher. You do your best to explain solids, liquids, and gases to Zed in just three simple lessons. Zed is a good student, and after thinking over your lessons carefully, he poses some difficult questions. Read each lesson, and do the activity it describes. Then discuss with a classmate how you would answer each of Zed's questions.

229

FOCUS

Getting Started
While students are reading the lesson introduction, write the headings Solid, Liquid, and Gas on the board. When students finish reading, ask them to observe their surroundings and to suggest items for each of the categories. Write their responses under the appropriate headings. Then involve students in a discussion of the differences between solids, liquids, and gases.

Main Ideas
1. Matter can be classified into three states: solid, liquid, or gas.
2. Solids are rigid, cannot be noticeably compressed, and have distinct boundaries.
3. Liquids flow, cannot be noticeably compressed, and may have a boundary with air.
4. Gases flow, can be compressed, and have no boundary with air.

TEACHING STRATEGIES

Solid, Liquid, or Gas?
Have students complete the three mini-lessons individually or in pairs. As an alternative, have students role-play the mini-lessons for the class, one student taking the role of Zed and the other performing the activity. Involve the entire class in answering the questions.

 A Transparency Worksheet (Teaching Resources, page 33) and Transparency 37 are available to accompany this lesson.

LESSON 1 ORGANIZER

Time Required
1 to 2 class periods

Process Skills
observing, classifying, analyzing, communicating

New Term
Interface—the boundary between two different substances

Materials (per student group)
Solid or Liquid?: 3 test tubes; nail; 20 mL of water; a few drops of food coloring; 20 mL of salt; **Liquid or**

Gas?: 2 plastic syringes without needles; 5 mL of water; sponge; **Solid, Liquid, or Gas?:** 4 test tubes; 20 mL of salt; 50 mL of water; a few drops of perfume; a drop of food coloring

Teaching Resources
Transparency Worksheet, p. 33
Theme Worksheet, p. 35
Resource Worksheet, p. 37
Transparency 37
SourceBook, p. S88

Answer to
Solid or Liquid?

Help students recognize that while salt is similar to liquids in some ways, it is actually made up of many solid grains or particles. Individual grains neither flow nor take the shape of a container. To further reinforce this idea, you might point out that salt, unlike water, can be piled up. You may wish to provide students with a magnifying glass to examine an individual grain of salt.

Answer to
Liquid or Gas?

Students should conclude that a sponge is composed of an elastic solid with many holes in it. As the sponge is squeezed, the air is forced out of the holes, giving the appearance of compression. Under ordinary conditions, however, the sponge maintains its shape and does not flow like a liquid or disperse like a gas. When the sponge is squeezed, the gas inside the sponge is dispersed, not the sponge itself.

Answer to
Solid, Liquid, or Gas?

Students should recognize that although the food coloring dissolves in the water, it is still a liquid. It cannot spread out beyond the liquid in which it is placed—it forms an interface with the particles of the liquid. A gas can dissolve in a liquid and can spread out beyond the boundary of the liquid.

Meeting Individual Needs

Learners Having Difficulty

Gently shake a can of soda and open it over a sink. The soda should foam out of the top of the can. Ask students: What two states of matter exist in the substance inside the can? *(Gas—carbon dioxide; liquid—the soda)* Then pour the soda into a glass and add a few ice cubes to it. Do not drink the soda. Ask: What states of matter are found in the glass? *(Gas—carbon dioxide; liquid—soda and water; solid—ice cubes)* What state of matter will be in the glass in an hour? *(The ice will have melted and the gas will have been released from the soda. Therefore, only liquid will remain.)*

Mini-Lesson 1

Solid or Liquid?

You put a nail in one test tube and water with food coloring in another. You explain to Zed that the nail is solid and the water is liquid. (How would you explain the difference?) Zed listens intently. "Ah," he says, "so liquids flow and take the shape of their containers, while solids don't." Then he takes a box of salt, pours it into another test tube, and says, "Salt is a liquid because it flows and takes the shape of its container, right?" Now you have a problem. What do you say to Zed to convince him that salt is not a liquid, but a solid?

Mini-Lesson 2

Liquid or Gas?

You have two plastic syringes, one filled with water and the other with air. You show Zed how to close the bottom of each syringe. Then you try to push down on the plunger. The gas is easily compressed, or squeezed into a smaller space, but the water cannot be compressed easily. You explain that this is an important difference between liquids and gases. Gases can be compressed noticeably, but liquids cannot.

Zed has been listening very closely. He picks up a sponge and squeezes it. Then he says, "This sponge is a gas because it is easily compressed." Is he right? What do you say to Zed now?

Mini-Lesson 3

Solid, Liquid, or Gas?

You have three test tubes. You put some salt in the first one, some water in the second, and a few drops of perfume in the third. Once again, you explain to Zed the differences between a solid, a liquid, and a gas. What do you tell him now?

This time, you also explain to Zed that at the sides of a solid and at the top of a liquid there is a boundary with air. This boundary is called the **interface**. For solids and liquids the interface is easy to see. It is sharp and clear. But the perfume just spreads out in the air, forming no boundary. Soon you can smell it throughout the room.

Zed thinks about what has been said. Then he takes a test tube of water and puts a drop of red food coloring in it. He watches the food coloring spread out. There is no sharp boundary between the food coloring and the water. He exclaims, "Food coloring is a gas because it spreads out, forming no boundary!" What do you say to him?

Before reading any further, make sure you have clear answers to each of Zed's three perplexing questions. Then compare your answers with Zed's conclusions in the rest of the story.

230

Homework

You may wish to assign the Theme Worksheet on page 35 of the Unit 4 Teaching Resources booklet as homework.

A Resource Worksheet (Teaching Resources, page 37) is available to accompany the material on page 231.

Zed goes away after thanking you for the explanations. He knows that solids, liquids, and gases are the three states in which matter exists. He also knows that if a solid is ground into a fine powder, you may have to examine one particle to discover its rigidity.

Zed also thanks you for explaining that solids and liquids are not easily compressible and that the sponge was not solid at all, but was full of spaces between the elastic fibers. He is intrigued by the fact that solids and liquids have an interface with air, while gases do not. At the same time, while both gases and liquids flow, solids do not.

Zed now understands much more about the three forms (or states) of matter—solid, liquid, and gas. Do you?

What Have You Learned?

1. Copy the table at right into your ScienceLog. Write the words *liquid, gas,* and *solid* in the appropriate boxes of the table. One box will be blank.

2. After Mini-Lesson 1, Zed might have said, "You can see through liquids but not through solids." Give him an example or two to show that this is not necessarily the case with solids or with liquids.

3. Smoke is made of millions of tiny pieces of carbon, but as it comes out of a chimney, smoke seems to spread out and disappear. Do you think smoke is most like a solid, a liquid, or a gas? Give reasons for your answer.

4. Water can be changed into a solid or a gas.
   a. List some ways that ice and water differ.
   b. List some ways that ice and water are similar.
   c. List some ways that steam and water differ.
   d. List some ways that steam and water are similar.

Characteristics of States

	Has a boundary with air and cannot be noticeably compressed	Has no boundary with air and can be noticeably compressed
Apparently rigid		
Flows		

At-Home Investigation

Try this activity at home, but ask for permission first.

Mix 80 mL of cornstarch with 50 mL of water. Stir the mixture well. When fully mixed, it should be difficult to stir.

Now try the following investigations:

1. Form a small ball of the mixture by rolling it around in your hand. What happens when you stop?

2. Pour some water on a tabletop. Try to cause a splash by hitting the water sharply with a ruler. Pour some of the cornstarch mixture onto the table and try to make a splash.

3. Would you describe this material to Zed as a solid or a liquid? Support your decision with observations you have made.

231

FOCUS

Getting Started

Display several examples of matter changing state (an ice cube melting, water boiling, water condensing on a cold soda can, dry ice sublimating, etc.). Ask students to describe what changes are taking place in each example.

Main Ideas

1. A change of state is a physical change.
2. There are three states of matter and six possible changes of state.
3. The melting point of a substance is the temperature at which a solid changes to a liquid. The melting point is also the freezing point.
4. The boiling point of a substance is the temperature at which a liquid rapidly changes to a gas.

TEACHING STRATEGIES

Answer to
In-Text Question

 Students should note the melting of the snow and rocks, the evaporation and condensation of water vapor in both photos, and the solidification of the molten rock.

The Language of Changes of State

Use this activity to help students review and consolidate what they already know. Point out that several terms may be used to describe the same process.

★ A Resource Worksheet (Teaching Resources, page 39) and Transparencies 38 and 39 are available to accompany The Language of Changes of State. In addition, an Activity Worksheet is available to accompany this lesson (Teaching Resources, page 38).

In the fairy tale "Cinderella," a pumpkin and some mice were changed into a carriage and horses by means of a magic wand. In the classroom, you can bring about great changes without magic. For example, paraffin wax, a rigid, white solid, can easily be changed into a clear, runny liquid. This transformation is called a change of state. Later you will devise experiments to find out more about changes of state, but first it will be useful to investigate the language you need to describe such changes. Which changes of state are shown in the photos below and at right? **Ⓐ**

The Language of Changes of State

1. How many words are used to describe changes of state? As you can see from column A in the table at right, there are a surprising number! For each change of state in column A, try to find a matching description in column B. You may need to look up some of the terms. Record the results in your Science-Log. Some of the terms have similar meanings. Choose the best answer.

Column A	Column B
a. melting	(1) Over time, mothballs disappear into the air as gas.
b. condensation	(2) After a summer rain, puddles gradually disappear.
c. freezing	(3) If air is cooled to a low enough temperature, the oxygen in the air will become a liquid.
d. evaporation	(4) A meteorite hitting the ocean could produce enough heat to rapidly turn large amounts of water into water vapor.
e. solidification	(5) Solder is a useful alloy because it changes into a liquid at a lower temperature than do most metals.
f. vaporization	(6) When making homemade ice cream, coarse rock salt and ice are mixed to create temperatures low enough to harden the cream.
g. liquefaction	(7) Last night, water vapor in the air changed into dew on the grass.
h. sublimation	(8) As lava cools, it hardens into rock.
i. boiling	(9) As the candy mixture was heated, it bubbled over.

LESSON 2 ORGANIZER

Time Required 3 class periods

Process Skills
observing, analyzing, organizing

Theme Connections
Cycles, Energy

New Terms
Boiling point—the temperature at which a liquid rapidly becomes a vapor
Dew point—the temperature at which water vapor condenses into the liquid state to form dew
Freezing point—the temperature at which a liquid changes to a solid
Melting point—the temperature at which a solid changes to a liquid

Materials (per student group)
Exploration 1, Activity 1: 100 mL of water; 250 mL beaker; hot plate; oven mitts; safety goggles; lab aprons; trivet; several 5 g pieces of paraffin wax; alcohol thermometer; **Activity 2:** can with a shiny surface; stirring rod or spoon; 2 or 3 ice cubes; 250 mL of water; alcohol thermometer

Teaching Resources
Activity Worksheet p. 38
Resource Worksheet, p. 39
Exploration Worksheet, p. 41
Transparencies 38 and 39
SourceBook, p. S88

2. Many of the terms in column A that describe changes of state are opposites. Pair up as many opposites as you can.

3. Can a gas change directly into a solid? The answer is yes. The next time you see frost forming on a window, you are observing this process. What would be an appropriate term to describe this process?

4. Complete the table below. Note that some of the terms have similar or identical meanings. Use the descriptions in column B, from question 1, to determine the type of temperature change (up or down) that causes each change of state. Assume that the substance is water. Again, do not write in this book. Record your answers in your ScienceLog.

Change of state	Terms	Change in temperature
solid to liquid		
liquid to solid		
liquid to gas		
gas to liquid		
solid to gas		
gas to solid		

5. Vaporization occurs when matter changes directly into a gas from another state. For example, the change from liquid to gas that occurs when a puddle disappears is called *evaporation*. Evaporation is one type of vaporization. What are other examples of vaporization?

6. Teachers are constantly making up quizzes and worksheets for their students. With a partner, make up a quiz or a worksheet of your own that involves the vocabulary of changes of state. You could design one of the following:

- a fill-in-the-blank test
- a word search
- a crossword puzzle
- a game
- some other quiz or worksheet of your choice

Answers to
The Language of Changes of State, pages 232–233

1. **a.** 5   **f.** 1, 2, 4
   b. 3, 7   **g.** 3, 7
   c. 6, 8   **h.** 1
   d. 2   **i.** 9
   e. 8

2. Melting and liquefaction are the opposite of freezing and solidification. Evaporation, vaporization, and boiling are the opposite of condensation and liquefaction. Sublimation and vaporization are the opposite of solidification. (Keep in mind that the terms *solidification, liquefaction,* and *vaporization* may refer to more than one type of change.)

3. Accept all reasonable responses. (You may wish to introduce the term *crystallization.* If you do, be prepared for some confusion with the use of the term *crystallization* as it applies to growing a crystal from a solution. The poem titled "Crystallization" at the end of this unit could help clarify the use of this term.)

4. See the table on page S238.

5. Boiling is another type of vaporization.

6. After students finish creating their quiz or worksheet, have them exchange papers and complete the one they receive.

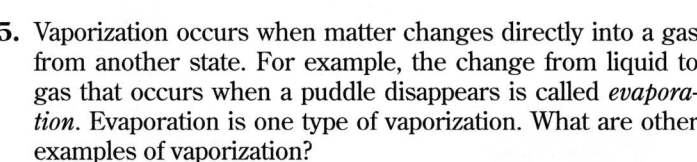

Theme Connection

Cycles
Have students diagram the following water cycle: Water leaves the surface of a tree's leaf and floats up into the air. It becomes part of a cloud and eventually falls as snow. The snow warms up, soaks into the ground, and is absorbed by the roots of a tree. **Focus question:** What changes of state take place during this cycle? *(The water evaporates from the leaf and then condenses to become part of the cloud. It freezes when it becomes snow and then melts when it warms up, becoming a liquid until it again evaporates from the surface of a leaf.)*

ENVIRONMENTAL FOCUS
The boiling point of a liquid depends on the density and pressure of the environment. The boiling point is lower when air density and pressure are lower, and it is higher when air density and pressure are higher. Ask: Why do many products provide special cooking instructions for areas at high altitudes? *(At high altitudes, air pressure is lower, and thus food products boil at lower temperatures. This can change how long they must be boiled to cook properly.)*

EXPLORATION 1

Cooperative Learning
EXPLORATION 1

Group size: 3 to 4 students
Group goal: to observe changes of state
Positive interdependence: Assign students roles such as materials coordinator, safety inspector, timer, or recorder.
Individual accountability: Ask each student to write a short description of the different states of paraffin or to define and describe dew point.

ACTIVITY 1

SAFETY ALERT Remind students not to use their thermometer for stirring.

Supply each group with only a small quantity of paraffin. This will reduce the mess that might result if any wax is dropped. Show students an example of a 5 g piece of paraffin so that they can get an idea of its size. To speed up the Activity, use water no hotter than 65°C to melt the paraffin wax. Paraffin solidifies at approximately 54°C; this may vary by several degrees, however, because paraffin is composed of several different materials, each with distinct properties.

WASTE DISPOSAL ALERT A good way to dispose of paraffin wax is to wrap it in newspaper and place it in the trash.

⭐ **An Exploration Worksheet (Teaching Resources, page 41) is available to accompany Exploration 1.**

Theme Connection

Energy
Focus question: When the solid paraffin changes to a liquid, does it gain energy or lose energy? *(From their observations in Exploration 1, students should conclude that when the paraffin is heated, it absorbs energy, which causes its temperature to rise. The energy absorbed causes the solid to change state.)*

EXPLORATION 1

Temperature Wizardry
You will now observe many of the changes of state that you have just studied.

Changing Paraffin

In this Activity, you will melt paraffin wax and then allow it to solidify. Work in small groups.

You Will Need

- water
- a 250 mL beaker
- several 5 g pieces of paraffin wax
- an alcohol thermometer
- an oven mitt
- a trivet
- a hot plate

Caution: Do not use an open flame here because of the danger of fire. Paraffin burns easily.

What to Do

1. In your ScienceLog, describe the appearance, feel, and smell of the solid paraffin.
2. Add 100 mL of water to the beaker, and heat it on the hot plate until it boils. Using your oven mitt, carefully move the beaker to a trivet.
3. Put the thermometer into the beaker, and make note of the temperature. A thermometer can be easily broken, so handle it carefully.
4. Put several small pieces of paraffin (about 5 g each) into the beaker of water. Observe and describe the paraffin as it melts.

Caution: Do not touch the melted paraffin.

5. As the temperature begins to drop, take the temperature of the water, and record it once every minute. Continue to record temperature readings until you are sure the paraffin is a solid. Compare the temperature at which your sample solidified with the temperatures obtained by other groups.

If the paraffin samples for different groups solidified at the same temperature, your class has discovered the freezing point of paraffin. The **freezing point** is the temperature at which a liquid changes to a solid. The temperature at which a solid changes to a liquid is its **melting point**.

Questions to Ponder

1. The melting and freezing points of a substance are the same temperature. How would you prove this for paraffin?
2. The freezing point of water is 0°C. What is its melting point? How can you prove it?

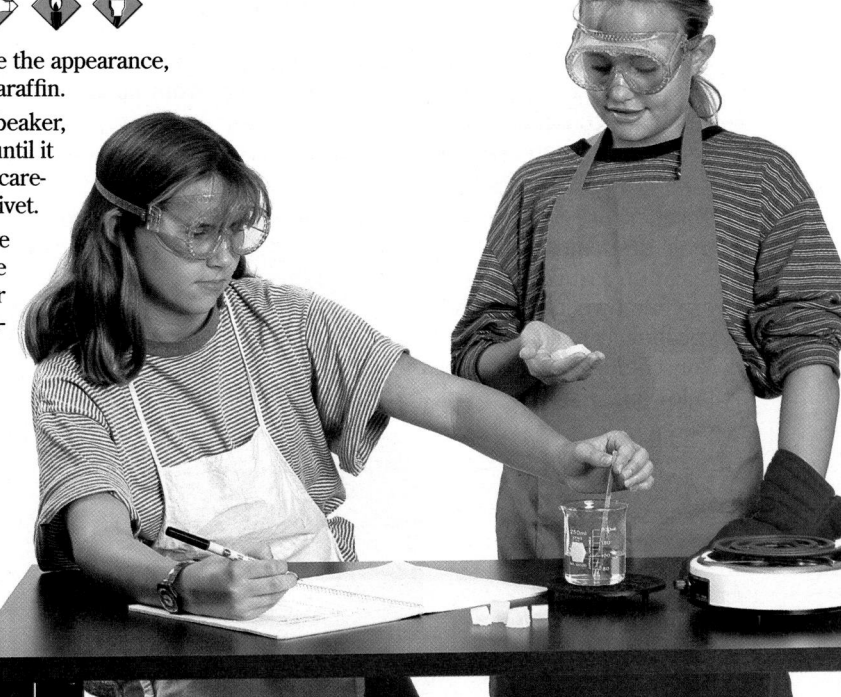

234

Answers to
Questions to Ponder,
pages 234–235

1. You could heat the water gradually and note the temperature at which the paraffin melts. Compare this temperature to the freezing point.

2. The following are possible responses: If you place enough ice cubes into water and the temperature of the water is measured repeatedly, the lowest temperature measured would be 0°C. Water cannot be cooled below this temperature using only ice cubes. Another test would be to place ice cubes into a beaker and allow them to partially melt. The lowest temperature measured for the newly melted water would be 0°C.

3. The melting points of the metals used to make cooking pots must be high enough to prevent the pots from melting when they are used.

4. Antifreeze is used as a coolant in car radiators. It remains liquid at temperatures well above those at which water would boil away.

3. Why would knowledge of melting points be important to have when manufacturing cookware?

4. Another important physical property of a substance is its **boiling point**. This is the temperature at which a substance changes rapidly from its liquid state into a gas. Rapid formation of bubbles is evidence that the liquid is at its boiling point. You may have observed water, for example, at its boiling point. Antifreeze, used in automotive cooling systems, has a higher boiling point. Why is this an important property of antifreeze?

ACTIVITY 2

Making Dew

You know that liquids can turn into gases, but can gases turn into liquids? For example, can you get a liquid out of the air that you breathe? If air is cooled to a low enough temperature or subjected to a high enough pressure, the gases in it will *condense* into liquids. (*Condensation* is the word used to describe what happens when a gas becomes a liquid.) Here are a few gases that are present in air and the temperatures at which they condense.

oxygen	−183°C
nitrogen	−196°C
helium	−269°C

There is yet another liquid that you can get out of air. This one can be condensed in your classroom. Here is how. Do this Activity with a partner because you will need more than one pair of eyes.

You Will Need

- a can with a shiny surface
- ice cubes
- a thermometer
- a stirring rod or spoon

What to Do

1. Fill the can half-full with water that is at room temperature. Note the water temperature, and record it in your ScienceLog. Remove the thermometer from the can.

2. Add two or three ice cubes to the can, and stir continuously with a stirring rod or spoon (*not* with your thermometer).

3. Keep stirring while your partner watches closely for moisture to form on the outside of the can. As soon as this happens, take the temperature of the water. At what temperature did you first observe moisture forming on the can?

Think About It

1. The temperature at which moisture first forms on the can is called the *dew point*. Why do you think it has this name?

2. If water formed on the can before you added the ice, what would this tell you about the dew point?

3. List at least three places around your home where you have noticed water condensing. For each one, explain why the condensation occurred.

4. You have seen water come out of the air, but how does it get into the air to begin with? What is this process called?

235

Integrating the Sciences

Earth and Physical Sciences

Have students keep track of the dew point and the relative humidity for a week. This information can be found in newspapers or by calling a local weather service. Then have students graph both sets of data and use the graphs to show the class how relative humidity and dew point are related.

ACTIVITY 2

This Activity explores the process of condensation by having students observe moisture condensing out of the air. Students may work individually or in pairs. The temperature at which the water will condense will depend on the relative humidity on the day the Activity is performed. The more humid it is, the higher the dew point will be.

Answers to *Think About It*

1. Dew condenses out of the air in the same way that moisture condenses on the can. The temperature at which condensation occurs is known as the dew point. At the dew point, the air is saturated and can hold no more moisture.

2. The dew point of the air in the room is the same as or higher than the temperature of the water.

3. Answers will vary, but the following are a few examples. In each case, the object on which the dew forms should be cold enough to cool the surrounding air to the dew point, causing condensation to form.
   - on a mirror after a shower
   - on cold-water pipes
   - on a bottle just removed from the refrigerator
   - on eyeglasses after coming indoors on a cold winter day

4. Water can get into the air through sublimation, evaporation, or boiling. Sublimation occurs when gaseous particles escape from the surface of a solid. Evaporation is a form of vaporization in which gaseous particles escape from the surface of a nonboiling liquid. Water evaporates from bodies of water such as oceans, seas, rivers, and lakes. Trees and other plants add a large quantity of water to the air through the evaporation of water from their leaves. Animals give off water vapor when they breathe and when perspiration absorbs body heat and evaporates. Also, when a liquid boils, it becomes a gas and enters the surrounding air.

Still More Experiments!

Have students work with one or two partners to investigate one of the hypotheses listed. Have the student teams submit plans for your approval prior to carrying out their experiments. Work through any problems with them to ensure the safe and successful completion of their investigations.

Answers to
Fifteen Hypotheses

The following hypotheses were not correct in the original list but have been rewritten here so that they are now correct.

2. Hot water takes longer to freeze than cold water because more heat must be removed.

4. As water freezes, it expands about 10 percent, so it becomes less dense than liquid water. This is why ice floats in water.

5. The atmospheric pressure changes from day to day and from place to place. Water boils at different temperatures, depending on this pressure.

9. The evaporation rate depends on surface area, temperature, and relative humidity. Water evaporates faster at higher temperatures, from larger surface areas, and at lower relative humidities.

10. Unequal masses of water boil at the same temperature as long as the conditions are identical.

Homework

You may wish to assign the activity Still More Experiments! as homework.

Still More Experiments!

One class made a list of hypotheses about changes of state. Each student then chose one hypothesis to investigate at home. Each student made a poster of his or her findings. The teacher encouraged everyone

- to make the poster neat and interesting
- to include his or her
  —hypothesis
  —plan
  —results
  —conclusions
- to make both qualitative and quantitative observations
- to use graphs and data tables when possible

Test one of the following hypotheses yourself, and report to the class in a similar way. Remember, you may find that a hypothesis is not supported by your findings. Sometimes an experiment will prove that a hypothesis is incorrect.

Fifteen Hypotheses

1. Hot water cools at a faster rate than cool water.
2. Hot water freezes in a shorter time than cold water.
3. Different amounts of water freeze at the same temperature.
4. The mass of an ice cube is the same as the mass of an equal volume of water.
5. Water boils at the same temperature every day.
6. The more salt that is added to water, the cooler the salt water can be made with ice cubes.
7. The lowest temperature that liquid water can be is 0°C.
8. Water with sugar dissolved in it evaporates more slowly than plain water.
9. Water always evaporates at the same rate.
10. The larger the quantity of water, the higher the temperature at which the water will boil.
11. The dew point changes from day to day.
12. The longer the air conditioner is on, the lower the dew point will be in your home.
13. The rate of evaporation of water depends on the amount of surface exposed to the air.
14. The melting point of ice is the same as the freezing point of water.
15. You can make matter change state by adding heat to it.

FOLLOW-UP

Reteaching

Have students look for examples of the changing states of matter in their everyday lives. Suggest that they write a brief description of each example they find.

Assessment

Divide the class into two teams. Team A should describe an event in which matter changes from one state to another.

(Water begins to bubble and becomes a vapor.) Team B then names the term that is used to describe the event. *(Boiling, vaporization)* If Team B's response is correct, then Team B gets a point. Team B then describes an event for Team A to identify. Play continues until a predetermined number of events have been described. The team with the most points wins.

Extension

Have groups of students make a mural to illustrate the terms listed on page 232. Encourage students to be creative.

Closure

Have students write a story in which they use all of the new words listed on page 232 to describe changes of state.

LESSON 3 A Model of Matter

Have you ever compared the clouds overhead to cotton balls? If so, you were using a model in which the cotton balls represented the clouds. When you look at a globe, you are using another model: the globe represents the Earth.

Some models are not objects, but ideas. Scientific models are examples of this type of model. Below, you can read about Jason's observation and how it leads to a scientific model.

Jason's Model

Jason made an important observation just after doing the Explorations on volume. If he had not done these Activities, it is doubtful that he would have made the observation at all.

In school, he measured the volume of his hand by placing his hand in a beaker of water. The water rose by an amount equal to the volume of his hand. At home, he helped his mother make pickles. First, they had to soak the cucumbers in salted water overnight. When they added the 500 mL of salt to the water and stirred, Jason observed that the volume of the liquid barely increased. If matter has volume, what happened to the volume occupied by the salt?

237

LESSON 3 ORGANIZER

Time Required
2 class periods

Process Skills
inferring, communicating, contrasting, comparing

Theme Connection
Structures

New Term
Particle model of matter—a model of matter in which all matter is made up of particles

Materials (per student group)
Pause for Thought (1): 5 mL of salt; test tube with stopper; 20 mL of water

Teaching Resources
SourceBook, p. S99

LESSON 3 A Model of Matter

FOCUS

Getting Started
Perform the following demonstration: Display a large jar or beaker full of water. Add 5 drops of food coloring to the jar and stir. Ask students if this demonstration does or does not support the idea that matter is made up of particles. Allow students to debate freely. Then point out that in this lesson they will learn more about the particle model of matter.

Main Ideas
1. A model is a representation of an object or phenomenon in the real world.
2. All matter is made up of particles.
3. The physical properties of solids, liquids, and gases can be explained by the particle model of matter.

TEACHING STRATEGIES

Jason's Model
GUIDED PRACTICE Ask a volunteer to read aloud the section titled Jason's Model. Have students speculate about why the volume of water did not increase much when the salt was added. *(Accept all reasonable responses.)* Then display a clear container filled with marbles or pebbles. Ask students whether the container is filled. Then pour water into the container. Ask students if they think this is a good model for what happens when salt is dissolved in water. *(Yes, because salt particles can occupy the spaces between the water particles)*

CHAPTER 13 • MORE ABOUT MATTER 237

238

Answers to
Pause for Thought (1)

The sum of the separate volumes of water and salt before shaking is greater than the volume after shaking. Students may infer that the particles of salt have filled in the spaces between water particles.

Answers to
Pause for Thought (2)

- You can add water to sand without increasing the total volume.
- Sand can be poured like water (although the individual particles of sand do not flow).
- Sand takes the shape of its container (although the individual particles of sand do not).
- Sand has an interface with air.
- Sand can be separated by hand into individual particles, while water cannot be; sand does not feel wet to the touch; sand does not freeze, boil, or evaporate.
- You can make a very small "water castle" by adding several drops of water together, but you cannot pile up water to the degree that you can pile up sand.
- Water does not feel gritty. One might infer that the particles are so small that you are unable to feel their edges.

Answers to
In-Text Questions

a. In solids and liquids, the particles are very close together. In gases, the particles are farther apart and can be pushed, or squeezed, closer together.

b. In solids, the particles are held in position and are not free to move. In liquids, the particles are free to slip and slide over each other.

c. In solids and liquids, the particles stick together. In gases, the particles are free to spread out. Therefore, no interface forms between a gas and the air.

Pause for Thought

You can verify Jason's observation in class. Place salt in a test tube to a depth of 1 cm. Now add water until the test tube is three-quarters full. Mark this level carefully. Stopper the test tube and shake it. What is the total volume in the test tube now? Can the volume of the salt disappear? Do you have an explanation?

Scientists have a model of matter that explains this observation. If you could magnify a single drop of water until it was as large as the Empire State Building, what would you observe? Scientists think you would observe the tiny particles that make up the water and the empty spaces in between the particles. Are you beginning to see why adding salt to water does not increase the total volume as much as expected?

In this model, all matter is made up of particles. It is known as the **particle model of matter**. The particle model is useful because it can be used to explain many of the things you have observed about solids, liquids, and gases.

Pause for Thought

You cannot see the particles that make up water, but you can use sand to represent, or be a model of, water. In what ways is sand a good model of water?

- Can you add something to sand without increasing its total volume?
- Can sand be poured like water?
- Does sand take the shape of its container?
- Does sand have an interface with air?
- How is sand unlike water?
- Can you make a "water castle" in the same way that you can make a sand castle?
- Does water feel gritty?

As you can see, a model is seldom perfect, but it does help us to understand an idea by providing a "picture" for us to see. How can you use the particle model to picture these observations?

a. Solids and liquids cannot be compressed noticeably, but gases can be.

b. Liquids can flow, but solids are rigid.

c. Gases do not have an interface with air; solids and liquids do.

Turn to page S100 of the SourceBook to learn more about me, John Dalton, and about the tiny particles that make up matter.

ENVIRONMENTAL FOCUS

The composition of a substance may not indicate whether it is a solid, liquid, or gas at room temperature. For instance, by weight, water (a liquid) is 89 percent oxygen, the ground (a solid) is approximately 49 percent oxygen, and the atmosphere (a gas) is about 25 percent oxygen. Ask: Why do these substances exhibit such different physical properties? *(Oxygen combines with different substances in very different ways. The substances produced exhibit properties different from those of oxygen alone.)*

Today we compared solids, liquids, and gases to things we are familiar with.

Tom compared a solid to the audience at a stadium, where people don't get up and roam around, but sit in one spot.

I compared a liquid to basketball players on the court. They move around, but stay within the boundary of the court.

From Sandra's ScienceLog

This section challenges students to use their imaginations to compare the particle nature of solids, liquids, and gases to familiar items and situations and to suggest additional models based on these comparisons.

GUIDED PRACTICE Provide time for students to study the comparisons made in the photographs about the particle nature of solids, liquids, and gases. After students have looked at the pictures and read the captions, ask them if they think the pictures represent good comparisons. *(Accept all reasonable responses.)*

Theme Connection

Structures

Focus question: How does the amount of space and movement among molecules of matter influence its state? *(Answers will vary, but in general, gases are composed of loosely structured particles separated by large amounts of space; liquids consist of closely spaced particles that are able to move about freely with respect to each other; and solids consist of closely spaced particles in a fixed position with respect to one another.)*

239

Answer to

Answer to
From Sandra's ScienceLog

Answers may vary, but the following is a possible response: The bees represent the particles making up a gas. These particles, like bees, can spread out and move independently of each other.

INDEPENDENT PRACTICE Challenge students to suggest comparisons of their own. Encourage them to support each comparison with an explanation. Some suggestions are as follows:
• A gas is like sand blowing in a sandstorm.
• A liquid is like runners in a marathon.
• A solid is like books arranged on a shelf.

Homework

Have students use their comparisons to make posters. Display the posters around the classroom or use them to make a mural.

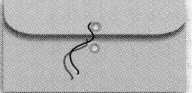

PORTFOLIO

Have students include their comparisons in their Portfolio. In addition to writing comparisons of states of matter, encourage students to write comparisons of changes of state.

Marsha compared a gas to a swarm of bees. She said . . .

What do you think Marsha's comparison was? Think of other good models for a solid, a liquid, and a gas, and write them in your ScienceLog. You might want to illustrate your ideas with drawings or pictures from magazines.

Writing About Matter

In their works, writers often include observations that are very much a part of everyday experience. In the poem "January," what common observation about matter and its changes does poet and novelist John Updike make? **Ⓐ**

January

The days are short,
The sun a spark
Hung thin between
The dark and dark.

Fat snowy footsteps
Track the floor.
Milk bottles burst
Outside the door.

The river is
A frozen place
Held still beneath
The trees of lace.

The sky is low.
The wind is gray.
The radiator
Purrs all day.

In the poem "Crystallization" below, what is being described? **Ⓑ** Which changes of state are observed?

Crystallization

It's early—and through the window I gaze
At castles and clouds and far away days
At enchanting places of shadow and light
This magic, crafted by the cold of the night

From water to ice, wispy tentacles spring
From eyes to mind, imagination takes wing
Making reality of these shimmering things

The sun now rises and with its cold rays
My fantasies melt for another day.

Be a Writer

Choose one of the changes of state listed on page 232, and include it in a piece of writing of your own. Or write about yourself as a particle of matter. What would you be doing and feeling as you went through various changes of state?

241

Writing About Matter

GUIDED PRACTICE   Lead students in a discussion of what Updike is describing in each stanza of the poem "January." Then have students evaluate the poem in terms of how it makes them feel and of the images it creates.

Answers to
In-Text Questions

Ⓐ Observations about the changes of matter alluded to in the poem include snow melting, milk freezing, water freezing, and snow forming crystals of ice on trees.

Ⓑ The poem "Crystallization" is describing a snowy winter landscape as seen early in the morning. The two changes of state alluded to in the poem are crystallization (solidification) and melting.

Primary Source

"January" is reprinted without alteration from *A Child's Calendar* by John Updike.

Be a Writer

INDEPENDENT PRACTICE   Point out to students that they may choose any form of writing—essay, poetry, short story, letter, cartoon, etc. Encourage them to be creative. Suggest that they include drawings to illustrate their writing. When students finish, call on volunteers to share their work.

FOLLOW-UP

Reteaching

Read one or more of the papers composed by students for the Be a Writer activity. Have students identify the various changes of state referred to in the papers.

Assessment

Set up a demonstration showing the following:
• an ice cube melting in water
• a sugar cube dissolving in water
    Have students use the particle model of matter to discuss and explain the dif-

ferences between the two processes. *(Students might infer that when the sugar dissolves, particles of solid sugar mix with water particles. When the ice melts, particles of solid ice become particles of liquid water.)*

Extension

Have students present a talk to the class in which they explain the particle model of matter. Their talk should describe the particle nature of solids, liquids, and gases and include everyday comparisons

to help explain the behavior of each state of matter.

Closure

Have students discuss the differences between the language of science and the language of poetry. Ask: Which language is better at expressing the feeling and mood of natural phenomena? Which language is more precise? *(Accept all reasonable responses.)*

Answers to
Challenge Your Thinking, pages 242–243

1. See A Major Meltdown Chart below.

2. Possible responses include steam forming from the hot water (evaporation); water condensing on the mirror, tap fixtures, etc. (condensation); and the mirror clearing up as the room cools and the humidity drops (evaporation).

3.

Word	Scientific meaning	Common meaning
a. melted	changed from a solid to a liquid	disappeared
b. boiling	rapidly changing from a liquid to a gas	bubbling over, out of control
c. freeze	change from a liquid to a solid	don't move, stand still
d. evaporated	changed from a liquid to gas	disappeared suddenly
e. crystallized	changed to a solid from a gas or liquid	became clear
f. solidify	change to a solid	make sure of, establish
g. condensed	changed from a gas to a liquid	made more compact

4. Using the clay balls as particles, the students should demonstrate that the "salt" balls can fit in between the "water" balls so that when salt dissolves in water, the same amount of mass is present, but the volume of the solution is less than the sum of the separate volumes. Particles of a liquid can flow past one another because the bonds between them are not as strong as those between particles in a solid. All particles remain present during melting, which is why the mass remains the same even though the volume may change. Students may be able to demonstrate that gases are less dense than liquids, for instance.

 You may wish to provide students with the Chapter 13 Review Worksheet that is available to accompany this Challenge Your Thinking (Teaching Resources, page 45). Transparency 40 is also available.

CHALLENGE YOUR THINKING

1. A Major Meltdown
Here are the melting points of a few substances. State the temperatures at which each substance would be solid, liquid, and gas.

Substance	Melting point	Boiling point
hydrogen	–259°C	–253°C
ethyl alcohol	–117°C	78°C
oxygen	–218°C	–183°C
mercury	–39°C	357°C
aluminum	660°C	2467°C
water	0°C	100°C

2. All Steamed Up
After Dominic took a shower, his bathroom was in quite a state! In fact, he noticed many changes of state. What are two changes he may have noticed?

3. What Do You Mean?
Many of the words that describe changes of state are used every day. Compare the "science meaning" with the common meaning of the italicized words in the following statements:

a. He *melted* into the crowd.

b. He was *boiling* mad.

c. *"Freeze!"* shouted the sheriff in the movie.

d. "I was following him and he *evaporated* into thin air."

e. It took a lot of thought, but suddenly the ideas *crystallized*.

f. "You should *solidify* your position by deciding now."

g. "Here is a *condensed* version of the report."

A Major Meltdown Chart

Substance	Solid temperature	Liquid temperature	Gas temperature
hydrogen	below –259°C	–259°C to –253°C	above –253°C
ethyl alcohol	below –117°C	–117°C to 78°C	above 78°C
oxygen	below –218°C	–218°C to –183°C	above –183°C
mercury	below –39°C	–39°C to 357°C	above 357°C
aluminum	below 660°C	660°C to 2467°C	above 2467°C
water	below 0°C	0°C to 100°C	above 100°C

4. Show What You Know

Here is a chance for you to make another model of matter. You will need two different colors of modeling clay and a beaker or jar. Form clay balls 1.5 cm in diameter from one of the colors of clay. Make enough balls to fill the beaker halfway. With the other color of clay, make 12 balls 0.5 cm in diameter. Now you are ready to use your models to demonstrate the following situations:

- When salt dissolves in water, the volume of the salt water is less than the volumes of the salt and water measured separately.

- The masses of the water and the salt before dissolving and after dissolving remain the same.

- A liquid like water can flow, but a solid like salt cannot.

- When ice melts, it loses its shape but not its mass.

Going Further: What other observations about the behavior of matter can you demonstrate with your model?

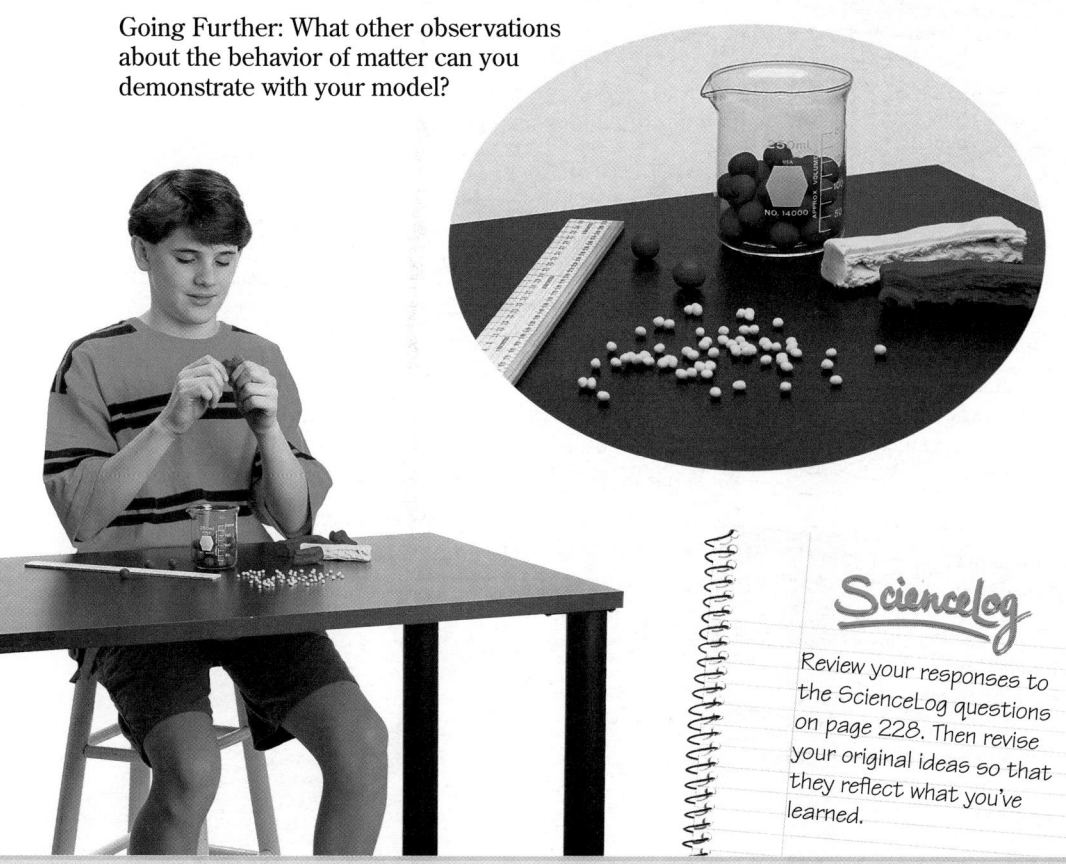

Review your responses to the ScienceLog questions on page 228. Then revise your original ideas so that they reflect what you've learned.

The following are sample revised answers:

1. Eventually, you would get to the point where all you have left are individual water particles (molecules). These cannot be divided any further while still having the properties of water.
2. The chalk in the air is a cloud of tiny, solid chalk pieces floating on air currents.
3. The water was previously dispersed in the air as water vapor, a gas. The temperature of the glass was below the dew point, so the water condensed into a liquid on the surface of the glass.

Multicultural Extension

F. J. Ferrell

F. J. Ferrell, an African American inventor, designed an instrument for melting snow, boiling the water, and harnessing the energy from the steam. Have students work in groups to discuss the following: Describe some situations in which this invention would be particularly useful. What are some situations in which it would be of little use? (Accept all reasonable responses.)

Did You Know...

When students think about a living thing, they might suspect that it is made up of solids, liquids, and gases. However, they may be surprised to learn the amount of liquid in an organism's body: the total body weight of many living things is more than 95 percent water! (Human bodies are about 65 percent water.)

The Big Ideas

The following is a sample unit summary:

Matter is anything that takes up space and has mass. Everything in the world is made of matter. (1) Matter can have many different physical properties: texture, taste, odor, hardness, and so on. Matter is categorized by the types of properties that it has. (2) Each type of matter has a unique set of properties. We make use of the properties of matter when building things, studying how things work, and even deciding what foods to eat. (3)

Matter can undergo different types of changes. A physical change is one in which physical properties—such as color, size, or state—are altered, but the actual substance remains unchanged. For example, when sugar is placed in water and the mixture is stirred, the sugar seems to disappear, but in fact it has only broken up into tiny particles too small to be seen. A chemical change, on the other hand, actually alters the structure of the matter. Burning and rusting are examples of chemical changes. (4)

Mass is the amount of matter in an object. Volume is the amount of space that an object takes up. It is also the amount of space within a hollow object. (5) Mass is measured in units called grams using a balance. Volume can be measured in liters using a graduated cylinder. (6)

The metric system uses three base units—gram, meter, and liter. These base units are paired with prefixes that are based on powers of 10. The prefixes make it possible to make very large or very small measurements in units that make sense. For example, large objects or distances may be measured using units with a prefix corresponding to 1000 or more. Small objects or distances may be measured using a prefix corresponding to 1/1000 or less.

Objects or distances with measurements that are close to the base unit need no prefix. (7)

The three states of matter—solid, liquid, and gas—are physical properties. (Some students may also suggest plasma as the fourth state of matter.) Solids are rigid, do not flow, and cannot be com-

Making Connections

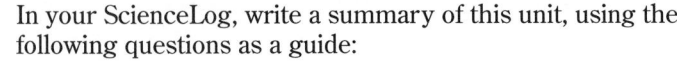

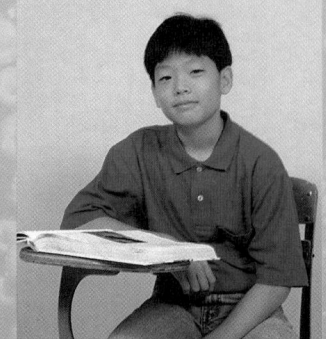

SOURCEBOOK

Turn to the SourceBook for more information on what happens to matter during physical and chemical changes, what solutions and compounds are, and how particles of matter combine to form elements, compounds, and solutions.

Here's what you'll find in the SourceBook:

UNIT 4
Matter and Molecules S86
Measuring Matter S90
Mixtures and Solutions S94
Compounds and Elements S97
Nature's Fundamental
    Particles S99

The Big Ideas

In your ScienceLog, write a summary of this unit, using the following questions as a guide:

1. What is matter?
2. What are some properties of matter?
3. How do we make use of the properties of matter?
4. What types of changes can matter undergo? How is matter affected by these changes?
5. What is mass? What is volume?
6. How can mass and volume be measured?
7. Why does the metric system use different prefixes with the base units?
8. What is meant by "state of matter"?
9. What causes matter to change states?
10. How do the terms *rigidity, interface,* and *compressibility* explain solids, liquids, and gases?

pressed. Liquids cannot be compressed either, but they do flow and take the shape of their containers. Gases expand to fill their containers and can be compressed easily. (8)

Adding heat to matter or taking heat away causes matter to change states. (9)

Properties such as rigidity, having a boundary with air, and compressibility are explained by the particle nature of solids, liquids, and gases. Particles of solids are so close together that they are rigid; the particles in solids and in liquids

are close together and form an interface with air, a gas; and particles of gas are far apart and can be compressed. (10)

 You may wish to provide students with the Unit 4 Review Worksheet that is available to accompany this Making Connections (Teaching Resources, page 53).

Checking Your Understanding

1. Jack collected the data below during an experiment to test the following hypothesis: The time it takes 100 mL of water to freeze decreases as the temperature of the freezer decreases.

Temperature of freezer (°C)	Time to freeze (min.)
−27	14
−20	20
−13	30
−8	43
0	65

a. Graph his data.

b. Explain what Jack would have to do to make this a fair test.

c. Explain what would be an appropriate conclusion to draw from this experiment.

2. Write a paragraph to answer each of the following:

a. On a cold morning you see "steam" rising from the surface of a pond. Does this mean that the pond is at the boiling point? What do you think is happening to cause this?

b. Matter is often defined as anything that has mass and takes up space. Is this a good definition? Why or why not?

3. **concept map** Copy the concept map below into your ScienceLog. Then use the following words to complete the map: *matter, physical, growing, states, solid, properties, liquid, gas, chemical, burning,* and *biological.* Do not write in this book!

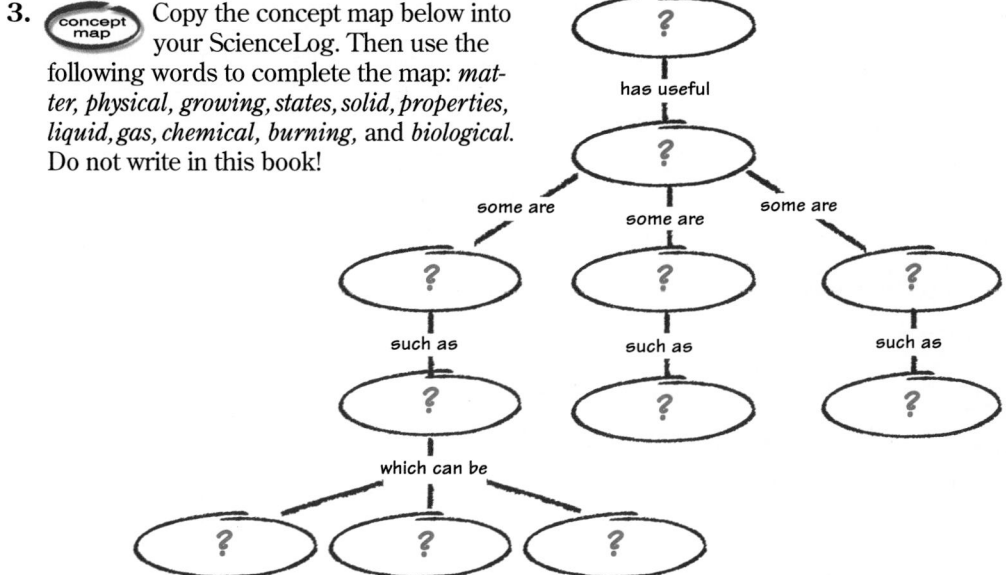

245

Answers to *Checking Your Understanding*

1. a.

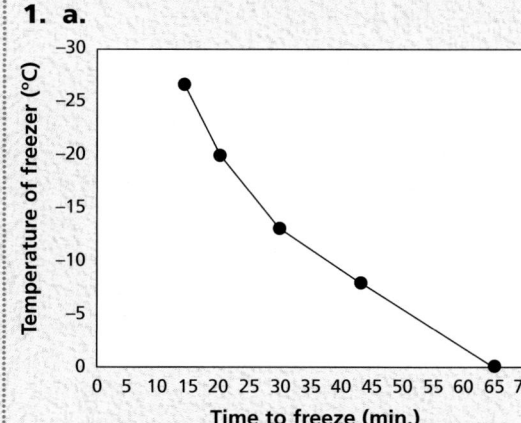

b. All variables other than the one being tested (temperature) would have to be the same for each trial. For example, the starting temperature of the water, the container holding the water, and the concentration of solutes in the water would have to be the same for each trial.

c. The time required for water to freeze varies with the temperature; the lower the temperature, the shorter the time required to freeze.

2. a. The pond is not at the boiling point; evaporation is occurring. The water releases water vapor. When this water vapor comes in contact with the cold air, it condenses and forms a cloud of tiny droplets.

b. Answers will vary, but students should note that all other common properties of matter (such as color, density, hardness, brittleness, and flexibility) vary enormously for each type of matter and may not even apply to certain substances. However, all matter has mass and takes up space.

3. See the concept map on page S238.

Homework

You may wish to assign the Activity Worksheet: A Hidden Message as homework (Teaching Resources, page 52). If you use the worksheet in class, Transparency 41 is available to accompany it.

Background

Physics is the study of energy, its changes, and its relationship to matter. Physicists study matter to discover what it is and to explain its behavior. They study energy to learn how it is released, converted, and controlled. Knowledge gained from the study of physics can be applied to other sciences, including chemistry, biology, medicine, astronomy, and geology.

Melissa Franklin is an experimental physicist, as opposed to a theoretical physicist. Theoretical physicists generally work with formulas, calculations, and predictions. Experimental physicists like Melissa actually test the theories developed by other scientists. Often, experimental physicists must build machines and develop new technologies to test these theories.

A person who wants to pursue a career in physics will find it helpful to study science and mathematics in high school. In college, physics students study chemistry and higher mathematics in addition to physics. Most people who have careers in physics have advanced degrees.

Discussion

GUIDED PRACTICE   Before students read this page, have them speculate about what a physicist does. Also, ask them whether they would like to be a physicist or physics professor. Then have students read the interview.

After students have finished reading this page, ask them to reevaluate whether they would like to be a physicist or physics professor. Encourage them to support their opinions with reasons.

Don't Be Shy!

Instead of having students call local physicists, you may wish to invite a physicist to your classroom. To facilitate discussion, have the students develop questions to ask her or him in advance.

From Big Machines to Tiny Things

In the course of one day, you might find Melissa Franklin operating a huge drill, giving a tour of her lab to a 10-year-old, putting together a gigantic piece of electronics, or even telling a joke to her students. Then you'll see her really get down to business—studying the smallest particles of matter in the universe.

Some Giant Discoveries

Melissa Franklin is an experimental physicist. "I am trying to understand the forces that describe how everything in the world moves—especially the smallest things," she explains. "I want to find the things that make up all matter in the universe and then try to understand the forces between them."

Other scientists rely on her to test some of the most important hypotheses in physics. For instance, she and her team recently contributed to the discovery of a particle called the *top quark,* which is one of six kinds of quarks. Franklin explains that a *quark* is the smallest piece of matter that makes up the very center of an *atom.* Looking at quarks, she says, would be like looking at a swarm of bees. "Each bee is pretty small, but the swarm is pretty big."

Physicists had proposed that the top quark might exist, but it had never been proven. Franklin and over 450 other scientists worked together to locate the top quark. Finding it required the use of a massive machine called a particle accelerator. Basically, a particle accelerator smashes particles together, and then scientists look for the remains of the smash. The physicists had to build some very complicated machines to detect the top quark, but the discovery was worth the effort. Franklin and the other researchers have earned the praise of scientists all over the world.

Getting Her Start

"I didn't always want to be a scientist, but what happens is that when you get hooked, you get really hooked. The next thing you know, you're driving forklifts and using overhead cranes and at the same time working on

◀ Melissa Franklin at work on one of her machines

▲ **This particle accelerator was used in the discovery of the top quark.**

really tiny, incredibly complicated electronics. What I do is a combination of all exciting things. It's better than watching TV."

It isn't just the best students that grow up to be scientists. Franklin says, "You can understand the ideas without having to be a math genius." Anyone can have good ideas, she says, absolutely anyone.

Don't Be Shy!

Melissa also has some good advice for young people interested in physics: "Go and bug people at the local university. Just call up a physics person and say, 'Can I come visit you for a couple hours?' Kids do that with me and it's really fun." Why don't you give it a try?

Meeting Individual Needs

Gifted Learners

Have students identify an aspect of physics that they have learned about, such as gravity or objects in motion. Ask students to pretend they are physics teachers and to create an engaging lesson plan for their chosen topic. The lesson plan may include an activity, quiz, or worksheet. If time permits, allow students to use their lesson plan with the class. Any activities should be checked for safety before students participate in them.

A Bat With Dimples

Wouldn't it be nice to hit a home run every time? Jeff DiTullio, a teacher at MIT, has found a way to help you get more bang from your bat.

Blowin' in the Wind

Jeff DiTullio is an aeronautical engineer—someone who studies both the way air moves and the way things move through air. For instance, airplane wings are designed by aeronautical engineers to lift a plane as it flies, and parachutes are designed to gently lower their cargo to the ground.

Like other aeronautical engineers, Jeff often works in a wind tunnel. These tunnel-shaped rooms have large fans at one end that blow air past objects in the room. Using trails of colored smoke, Jeff can observe how air passes over and around various objects.

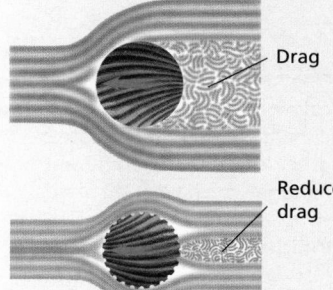

▼ **By reducing the amount of drag behind the bat, dimples help the bat move faster through the air.**

Drag

Reduced drag

An Idea for a Better Bat

If you look closely at the surface of a golf ball, you'll notice dozens of tiny craterlike dimples. When air flows past these dimples, it gets stirred up. By keeping air moving when it gets near the ball, these dimples help the golf ball to move through the air faster and farther.

Jeff decided to try this same idea on a baseball bat. His hypothesis was that dimples would let a bat move more easily through the air. To test this hypothesis, he pressed hundreds of little dimples, about 1 mm deep and 2 mm across, into the surface of a bat.

When Jeff tested his dimpled bat in the wind tunnel, he found that it could be swung 3 to 5 percent faster. That may not sound like much, but it could add about 5 m to a fly ball!

But Is It Legal?

As you might imagine, many baseball players would love to have a bat that could turn a long fly ball into a home run, but are dimpled baseball bats legal?

Every piece of equipment in the major leagues has regulations about its size and shape. A baseball bat, for instance, must be no more than 107 cm long and 7 cm across at its widest point. As Jeff points out, there

▲ **Jeff DiTullio with his dimpled baseball bat**

are no rules stating that the surface of a bat must be smooth. The rules may someday be changed to outlaw dimpled baseball bats. Until they are, though, fans of the dimpled bat will all cry the same cheer: Play ball!

Dimple Madness

Now that you know how dimples can improve baseball bats, think of other uses for dimples. How might dimples improve the ability of other items to move through the air? Draw a sketch of a dimpled object, and describe how the dimples improve its design.

Dimple Madness

Provide students with colored markers and paper to complete the sketch described in the section Dimple Madness. Once the students have completed their drawings, divide the class into groups. Have each group choose one of their dimpled products and create a poster-sized magazine advertisement for it. Ask students to present their magazine ads to the class. *(Accept all reasonable responses.)*

Background

The effect that dimples will have on an object in motion depends on several things, including the size of the object, the size of the dimples, and the medium through which the object is moving. A golf ball moving through air, for instance, will behave differently than a golf ball moving through water.

Dimples were first used on golf balls when golfers discovered that old balls with nicks and cuts sometimes flew farther than smooth, new balls. The same principle is applied in many different ways. For example, the cylindrical struts of an airplane's landing gear often have a rough surface in order to minimize the amount of turbulence generated.

Did You Know...

Many historians believe that golf began over 500 years ago when Scottish shepherds created a game using a leather ball stuffed with feathers. For clubs, they used the sticks that they carried for herding sheep!

Integrating the Sciences

Earth and Physical Sciences

Explain to students that the effect of dimples on an object in motion depends in part on the medium that the object is moving through. Then ask: Do you think that the craters on the moon allow it to move through space more quickly than it would if it didn't have craters? *(Since the moon is moving through space, there is no medium to be affected by the presence of the craters. Therefore, the presence or absence of craters has no effect on the speed of the moon.)* Explain to students that space is a vacuum and that objects moving through space do not encounter resistance.

Background

Icebergs form in both the Northern and Southern Hemispheres, but they are typically very different in size and appearance. Icebergs in the Northern Hemisphere, such as those that break off from the ice sheet covering Greenland, tend to be spiky and irregularly shaped. Icebergs in the Southern Hemisphere are much larger and usually have a fairly flat top. (The one shown on this page is an exception.)

People first considered the possibility of using icebergs as a water source many years ago. Just after 1900, for instance, Antarctic icebergs were towed by steamships to Callao, Peru. In addition to using icebergs for fresh water, people have proposed using them for air conditioning, as a source of gourmet ice cubes, and even as giant aircraft carriers!

ENVIRONMENTAL FOCUS

One of the concerns about towing icebergs to arid areas involves the infusion of large amounts of cold water into warm, equatorial ecosystems. Ask students to think about the effects that this might have. Encourage discussion. *(Aquatic plants and animals in the equatorial ecosystems might not be able to tolerate significant drops in temperature. Even if some of the plants and animals could survive the temperature change, a decline in any plant or animal population would disrupt the surrounding ecosystem.)* Encourage students to think of ways that this problem could be solved. *(Accept all reasonable responses.)*

Putting Freshwater Problems on Ice

Imagine how different your life would be if you couldn't get fresh drinking water. What would you drink? How would you clean things? Water shortages affect millions of people everyday, yet the Earth has enough fresh water to supply 100 billion liters to each person. So what's the problem?

The Ice Water Planet

Three-quarters of the Earth's fresh water is frozen in the polar ice caps. Plenty of water is there, but people can't use water that is frozen and thousands of miles away.

The ice sheet that covers Antarctica is thousands of meters thick and is almost twice the size of the United States. Hundreds of huge chunks break off its edges every year. These icebergs, which are made up entirely of frozen fresh water, float away into the sea and eventually melt. The water from one year's worth of these icebergs would be enough to supply all of Southern California for more than a century. So why not use it?

Obvious, but Not Easy

Transporting icebergs to areas that need fresh water is harder than it sounds. For one thing, many of the icebergs are huge. The largest ever recorded, for instance, was about the size of Connecticut. Even small icebergs may be 2 km long and 1 km wide.

▲ **Icebergs in Antarctica**

Researchers have thought of many methods for transporting icebergs. Most of the ideas involve towing the icebergs through the water. A few ideas involve attaching engines and propellers directly to the icebergs themselves.

However, because icebergs are so large, it takes a long time to move them. And when they finally get somewhere, they have usually melted considerably. To prevent melting, insulating materials could be wrapped around the icebergs.

A Worthy Investment

Lakes and ground water still provide the cheapest fresh water in most areas. However, when there is no lake, river, or well water available, then using icebergs becomes a possibility. Even though transporting icebergs is difficult, it still may be worthwhile to try. Irrigating a hundred square miles of desert with water from icebergs might cost as much as a million dollars, but purifying the same amount of sea water could cost over a billion dollars.

People from arid regions have spent considerable time on iceberg research. So far, no one has set up a program for harvesting icebergs. Someday, however, liquid icebergs may flow from our household faucets.

An Icy Investigation

Float an ice cube in a bowl of cold water, and record how fast it melts. Then try to insulate other ice cubes with different materials, such as cloth, plastic wrap, and aluminum foil. Which material works best? How could this material be used on real icebergs?

Theme Connection

Cycles

Focus question: How does the formation, movement, and melting of icebergs form part of a larger water cycle? *(Students should recognize that water molecules move and undergo changes of state during the formation, movement, and melting of icebergs. In addition, water that melts into the ocean can evaporate into the atmosphere, eventually falling as snow onto a glacier. When part of this glacier breaks off as an iceberg, the cycle begins again.)* You may wish to have students diagram a water cycle that includes an iceberg.

Answers to *An Icy Investigation*

Answers will vary. Students may find that some materials prevent the ice cubes from melting as quickly as they would if they were unprotected. Students should propose realistic ways of using their materials on icebergs. For a more complete discussion of this concept, see Unit 7, page 430.

The Miracle Material

▲ The exterior of this concept house being built in Pittsfield, Massachusetts consists almost entirely of plastic materials.

Imagine a material that is light enough for a baby to lift, strong enough to stop a bullet, shatterproof, rustproof, and able to withstand the heat of a blast furnace. Would you call it a miracle material? You could, but there's another name for it: plastic.

The Early Days of Plastics

We're all familiar with plastic products, from soda bottles to eyeglass frames and from CD cases to football helmets. But plastics haven't been around for all that long. The first plastic product—a billiard ball—was made by an American inventor, John Wesley Hyatt, in 1869.

Hyatt combined two types of plant extracts to produce a thick substance, or resin, which he called celluloid. His celluloid billiard ball was a good copy of the traditional ones made from ivory. Manufacturers were soon using celluloid in everything from knife handles to photographic film.

▶ The billiard ball was the first practical use for celluloid plastic.

One Material Fits All

Plastics have come a long way since the celluloid billiard ball. Today, almost all plastics are made from synthetic resins, but that doesn't mean they are all created equal. For instance, certain types of plastic are strong enough to be used in bullet-proof vests. Softer and more flexible plastics are used to make nylon stockings and pillow stuffing. Another type is so heat-resistant that it is used on the surface of spacecraft to help them withstand the extreme temperatures of reentry into the atmosphere.

Recently, engineers even built a plastic house! They replaced the wood, concrete, metal, and glass with plastic sheets, foams, and molded pieces. Since plastic houses are easily built and energy-efficient, they may be the wave of the future.

Environmental Concerns

One of the problems with plastics is that most are not biodegradable, and plastic recycling programs have not kept up with the production of plastic wastes. As a result, our landfills are filling up with discarded plastic products.

Fortunately, however, recycling programs in many areas are growing constantly. In the future, our dependence on plastics will probably grow, but so will our knowledge of how to use them in an environmentally friendly way.

Put Plastic to the Test

Obtain two coffee mugs—one that is ceramic or glass and one that is plastic. Test them to find out which one insulates the best. Fill each with hot water to see which one keeps the water hot longer. Can you think of other ways to compare the two mugs?

249

Answers to
Put Plastic to the Test

Results for this activity will depend on the size and composition of the mugs that are used. As an additional comparison, students may note that ceramic mugs break more easily than plastic mugs do.

Background

During the 1920s and 1930s, advances in chemistry led to the commercial production of three new kinds of plastic—polystyrene, acrylic, and cellulose acetate. During World War II, further advances in plastics research resulted from the shortage of many kinds of raw materials. Some of the new plastics to emerge during this period included polyethylene, silicone, and epoxy.

The rapid pace of space and nuclear research in the 1950s and 1960s led to the development of new plastics and new uses for existing ones. During this period, plastics became increasingly important in industry, medicine, and architecture.

Discussion

Ask students to make a list of all the plastic objects in the classroom. Call on a volunteer to read his or her list aloud, and invite other class members to add to it. Encourage students to identify some of the properties of the objects. *(For example, the objects could be solid, colorful, hard, or heat resistant.)* Help students recognize that the many properties of plastics make them well suited for many different uses.

Extension

Plastics can be classified as either thermosetting or thermoplastic. Suggest that students find out what these two terms mean. *(Thermosetting refers to plastic that becomes permanently hard once it is heated, and thermoplastic refers to plastic that softens when it is heated.)* Then have them make a chart to show some possible uses and important properties of each kind of plastic.

Unit 5 CHEMICAL CHANGES

Unit Overview

In this unit, students are introduced to the chemical nature of the substances they encounter in everyday life. In Chapter 14, students analyze their prior knowledge of chemicals and review chemical safety. In Chapter 15, students observe changes in substances and relate those changes to the chemical and physical properties of matter. In Chapter 15, students practice distinguishing physical changes from chemical changes and use word equations to identify the products and reactants in chemical reactions. In Chapter 16, students explore elements, compounds, the periodic table, acids, bases, combustion, and corrosion. Students also compare the mass of the products with the mass of the reactants in chemical reactions to test the law of conservation of mass. Finally, students investigate some of the chemical changes that occur in their own body by outlining the process of digestion.

Using the Themes

The unifying themes emphasized in this unit are **Changes Over Time, Structures,** and **Energy.** The following information will help you incorporate these themes into your teaching plan. Focus questions that correspond to these themes appear in the margins of this Annotated Teacher's Edition on pages 259, 279, 280, 290, 304, and 311.

Changes Over Time can be discussed in relation to environmental changes caused by chemical reactions. Examples from this unit include corrosion, cave formation, and the effects of acid rain. Understanding the processes involved in such long-term changes may help students recognize more chemical changes that take place every day.

Structures can be discussed in conjunction with Chapter 16 by focusing on the relationship between the properties of matter and its structure. By observ-

ing matter from a chemical viewpoint, the physical characteristics as well as the way a substance will behave in a given situation can be understood.

Energy can be discussed to help students recognize when a chemical change is occurring. The transfer of energy during a chemical reaction provides clues that a chemical change is occurring. Also, by considering how we get energy from chemical reactions, such as by burning fossil fuels or by digesting food, students will begin to see the transfer of energy as an integral part of chemical reactions.

Using the SourceBook

Unit 5 describes the development of atomic models. Students learn how the structure of atoms affects chemical activity, and they see how atomic structure is related to the periodic properties of the elements. Using atomic models, chemical bonding is introduced, and chemical reactions are discussed.

Bibliography for Teachers

Barber, Jacqueline. *Of Cabbages and Chemistry.* Great Explorations in Math and Science (GEMS) Series. Edited by Lincoln Bergman and Kay Fairwell. Berkeley, CA: Lawrence Hall of Science, 1990.

Bibliography for Students

Cobb, Vicki. *Chemically Active: Experiments You Can Do at Home.* Philadelphia, PA: Lippincott, 1990.

Loeschnig, Louis V. *Simple Chemistry Experiments With Everyday Materials.* New York City, NY: Sterling Publishing Co., Inc., 1991.

Woodburn, John H. *Chemistry. Opportunities in . . . Series.* Lincolnwood, IL: VGM Career Horizons, 1987.

Films, Videotapes, Software, and Other Media

Acids and Bases
Videotape
Britannica
310 S. Michigan Ave.
Chicago, IL 60604-9839

Chemistry: Elements, Compounds, and Mixtures
Film, videotape, and videodisc
Coronet/MTI Film & Video
108 Wilmot Rd.
Deerfield, IL 670015

The Chemistry Help Series
Software (Apple II family)
Focus Media, Inc.
839 Stewart Ave.
Garden City, NY 11530

Chemistry: The Periodic Table and Periodicity
Film, videotape, and videodisc
Coronet/MTI Film & Video
108 Wilmot Rd.
Deerfield, IL 670015

Periodic Table Videodisc: Reactions of the Elements
Videodisc
JCE Software
University of Wisconsin
Department of Chemistry
1101 University Ave.
Madison, WI 53706

Unit Organizer

Unit/Chapter	Lesson	Time*	Objectives	Teaching Resources
Unit Opener, p. 250				Science Sleuths: Exploding Lawn Mowers English/Spanish Audiocassettes Home Connection, p. 1
Chapter 14, p. 252	**Lesson1, Chemicals in Our Lives, p. 253**	3	1. Identify six chemicals by their specific properties. 2. State the chemical and common names for a number of substances.	Image and Activity Bank 14-1 Resource Worksheet, p. 3 Exploration Worksheet, p. 4
	Lesson 2, Safety First! p. 260	1	1. Discuss safe laboratory procedures. 2. Review the meaning of some safety icons used in this textbook. 3. Explain the warning symbols used on household chemicals.	Image and Activity Bank 14-2 Resource Worksheet, p. 10
End of Chapter, p. 262				Chapter 14 Review Worksheet, p. 12 Chapter 14 Assessment Worksheet, p. 14
Chapter 15, p. 264	**Lesson 1, Changes All Around, p. 265**	2	1. Think of questions that help to distinguish between chemical and physical changes. 2. Create a list of words associated with physical changes and a list of words associated with chemical changes.	Image and Activity Bank 15-1 Discrepant Event Worksheet, p. 17 Resource Worksheet, p. 18 Transparency 42 Transparency 43
	Lesson 2, Signs of Chemical Change, p. 270	3	1. State the type of change (physical or chemical) when given an example, and explain the reason for their choice. 2. Describe how chemical changes can be used to observe the presence of copper, carbon dioxide, and starch.	Image and Activity Bank 15-2 Exploration Worksheet, p. 19 ▼ Exploration Worksheet, p. 23 Transparency 45
	Lesson 3, Classifying Changes, p. 275	1 to 2	1. Give examples of chemical and physical changes encountered in everyday life. 2. Explain how various careers require knowledge of chemicals and chemical changes.	Resource Worksheet, p. 25 ▼
	Lesson 4, Chemical and Physical Properties, p. 278	2 to 3	1. Distinguish between physical and chemical properties. 2. Explain the relationship between the properties and uses of a substance.	Image and Activity Bank 15-4 Theme Worksheet, p. 26 ▼ Exploration Worksheet, p. 28
	Lesson 5, Products and Reactants, p. 282	1 to 2	1. Define reactants and products, and explain the relationship between the two. 2. Write word equations to summarize chemical changes.	Transparency 47 Math Practice Worksheet, p. 30
End of Chapter, p. 285				Chapter 15 Review Worksheet, p. 31 Chapter 15 Assessment Worksheet, p. 35
Chapter 16, p. 287	**Lesson 1, The Basics, p. 288**	2	1. Distinguish between elements and compounds. 2. Identify the chemical symbols for common elements, and explain the organization of the periodic table.	Image and Activity Bank 16-1 Resource Worksheet, p. 37 ▼ Resource Worksheet, p. 39 ▼
	Lesson 2, Acids and Bases, p. 293	2	1. State two properties of acids and bases. 2. Safely perform acid-base indicator tests on a variety of household substances. 3. Make an indicator solution using red cabbage.	Image and Activity Bank 16-2 Transparency Worksheet, p. 40 ▼ Exploration Worksheet, p. 42
	Lesson 3, The Nature of Burning—A Chemical Change, p. 296	2 to 3	1. Compare the theories of burning developed by Empedocles, Stahl, and Lavoisier. 2. Use experiments to test theories about burning. 3. Compare combustion and corrosion.	Image and Activity Bank 16-3 Exploration Worksheet, p. 45 Exploration Worksheet, p. 46 Exploration Worksheet, p. 47 Activity Worksheet, p. 48 ▼ Activity Worksheet, p. 49 ▼
	Lesson 4, Conservation of Mass, p. 300	2	1. Show that the mass of the reactants in a chemical reaction equals the mass of the products.	none
	Lesson 5, Chemical Changes and You, p. 302	1 to 2	1. Investigate the chemical changes involved in digestion. 2. Simulate the digestion of starch and protein.	Exploration Worksheet, p. 50 Exploration Worksheet, p. 51
End of Chapter, p. 306				Chapter 16 Review Worksheet, p. 52 Chapter 16 Assessment Worksheet, p. 56
End of Unit, p. 308				Unit 5 Activity Worksheet, p. 58 ▼ Unit 5 Review Worksheet, p. 59 Unit 5 End-of-Unit Assessment, p. 64 Unit 5 Activity Assessment, p. 70 Unit 5 Self-Evaluation of Achievement, p. 73

* Estimated time is given in number of 50-minute class periods. Actual time may vary depending on period length and individual class characteristics.

▼ Transparencies are available to accompany these worksheets. Please refer to the Teaching Transparencies Cross-Reference chart in the Unit 5 Teaching Resources booklet.

Materials Organizer

Chapter	Page	Activity and Materials per Student Group
14	254	**Exploration 1, Activity 1:** about 2 spoonfuls each of salt, baking soda, and sugar; 3 pieces of chalk; 3 or 4 pieces (about 2 cm × 2 cm each) of different types of paper, such as tissue paper, newspaper, and paper napkin; about 30 mL of distilled water; several small containers; magnifying glass; safety goggles; ***Activity 2:** all materials from Activity 1; watch glass or evaporating dish; 2 candles; aluminum pie plate; small ball of modeling clay; pair of tongs; bucket of water; several jar lids; small beaker or jar; several stirring rods; 100 mL graduated cylinder; 25 mL of vinegar; box of matches; eyedropper; about 25 mL of distilled water; safety goggles; lab aprons; latex gloves
	261	**Safety Tips:** poster board; markers
15	270	***Exploration 1, Test 1:** 1 g of copper sulfate; 10 mL of water; 10 mL graduated cylinder; iron nail; beaker; stirring rod; safety goggles; lab aprons; latex gloves; **Test 2:** 100 mL graduated cylinder; eyedropper; 10 mL of vinegar; 25 mL of milk; beaker; safety goggles; lab aprons; **Test 3:** eyedropper; eggshell; 10 mL of vinegar; bowl; safety goggles; lab aprons; **Test 4:** candle; a few matches; jar lid; aluminum pie plate; pair of tongs; small ball of modeling clay; sample of candle wax; safety goggles; lab aprons; **Test 5:** drinking straw; small beaker; 10 mL of limewater; safety goggles; lab aprons (See Advance Preparation below.); **Test 6:** 20 mL of lemon juice; several toothpicks; 2 sheets of paper; candle; a few matches; small ball of modeling clay; jar lid; bucket of water; safety goggles; lab aprons; **Test 7:** several drops of iodine starch-test reagent; eyedropper; starch (sample of potato, cereal, or bread); bowl; safety goggles; lab aprons; latex gloves (additional teacher materials: 25 mL of 0.1 M sodium thiosulfate, a few sheets of newspaper; see Advance Preparation below.); **Test 8:** 5 g of baking powder; small bowl or beaker; eyedropper; 5 mL of water; safety goggles; lab aprons
	273	***Exploration 2:** beaker; 1 L of water; 100 mL graduated cylinder; effervescent tablet; carbonated drink (bottled); few drops of vinegar; 15 mL of baking soda; 10 mL of molasses; packet of yeast; hot plate; eyedropper; stirring rod; safety goggles; lab aprons; oven mitts
	280	**Exploration 3:** 5 g of copper chloride crystals; square of heavy aluminum foil, 10 cm × 10 cm; stirring rod; 100 mL of water; small beaker; evaporating dish; safety goggles; lab aprons; latex gloves
	281	**Guess That Substance!:** 1 index card per student
16	289	**Task 3:** small mirror; piece of steel wool; 25 mL of water; zinc or magnesium ribbon; about 20 mL of 0.1 M hydrochloric acid; safety goggles; lab aprons
	292	**Elemental Investigations:** 111 index cards
	294	**Exploration 1, Activity 1:** about 10 mL each of lemon juice, apple juice, coffee, milk, water, milk of magnesia, and dilute household ammonia; tomato; 5 g of baking soda; effervescent tablet; aspirin tablet; 5 g of soap; 3 strips of litmus paper; scissors; 250 mL of distilled water; stirring rod; 250 mL beaker; eyedropper; several drops of bromothymol blue solution; plastic drinking straw; 20 drops of 0.1 M hydrochloric acid; evaporating dish or other glass container; one drop of phenolphthalein solution; several drops of 0.1 M sodium hydroxide; microscope; safety goggles; lab aprons (See Advance Preparation below.); **Activity 2:** optional materials: leaf of raw red cabbage; 250 mL of hot water; dropperful (about 2 mL) of dilute household ammonia; eyedropper; 4 to 5 drinking glasses; violet blossom; 25 mL of vinegar; watch or clock; safety goggles; lab aprons
	297	**Exploration 2:** candle; small ball of modeling clay; a few matches; aluminum pie plate; metric balance; wide-mouth jar; safety goggles; lab aprons; oven mitts
	298	**Exploration 3:** steel wool; metric balance; pair of tongs; several matches; aluminum pie plate; safety goggles
	299	**Exploration 4:** candle; jar lid; small ball of modeling clay; a few matches; jar with lid; 25 mL of limewater; 100 mL graduated cylinder; safety goggles; lab aprons (See Advance Preparation below.)
	304	**Exploration 5:** 10 mL graduated cylinder; 1 g of amylase; 600 mL of water; 3 test tubes; test-tube clamp; test-tube rack; wax pencil; several crushed, unsalted crackers; eyedropper; hot plate; 600 mL beaker; 15 mL of Benedict's solution; watch or clock; safety goggles; lab aprons; latex gloves
	305	**Exploration 6:** hard-boiled egg; knife; 2 test tubes; a test-tube rack; small beaker; stirring rod; 15 mL of water; 1 g of pepsin; 2 mL of 0.1 M hydrochloric acid; 10 mL graduated cylinder; eyedropper; wax pencil; safety goggles; lab aprons; latex gloves (additional teacher materials: 500 mL labeled container, about 300 mL of 0.1 M sodium hydroxide solution per class, a few sheets of newspaper; see Advance Preparation below.)
	306	**Challenge Your Thinking, question 3:** support stand with clamp; 100 mL graduated cylinder; 2 large beakers or medium-sized bowls; 100 mL of vinegar; 2 g of steel wool; 30 cm of thin rubber tubing; about 1 L of water

* You may wish to set up these activities in stations at different locations around the room.

Advance Preparation

Exploration 1, page 270: To prepare limewater, add 20 g of calcium hydroxide to 1 L of water, shake, and let settle. When needed, pour off some of the clear solution. For Test 7, 25 mL of 0.1 M sodium thiosulfate and newspaper will be used for disposing of the iodine-stained materials; for detailed instructions, see the Waste Disposal Alert on page 21.

Exploration 1, Activity 1, page 294: You may wish to mix the following sample materials with about 10 mL of distilled water ahead of time: tomato, baking soda, antacid tablet, aspirin tablet, and soap.

Exploration 1, Activity 2, page 294: The materials for this part of the Exploration are listed as optional because it is intended as an at-home activity. If you have time, you may wish to do this activity in class.

Exploration 4, page 299: Prepare limewater as above.

Exploration 6, page 305: All of the additional teacher materials will be used for disposal of the mixture of the egg white and gastric juice; see page 305.

Unit Compression

Chapter 14 should be considered core because it develops student understanding both of the nature of chemicals and of how chemicals can be handled safely in a classroom setting. Except for Exploration 1, many of the activities in this chapter involve reading and are conceptual in nature and may therefore be assigned as homework. This will reduce the amount of class time necessary to complete this chapter.

In Chapter 15, much of Lessons 1, 3, and 5 are also conceptual and make appropriate homework assignments. In addition, performing parts of Explorations 1–3 as teacher demonstrations will help save time during this chapter.

Chapter 16 also contains many sections that can be read ahead of time or assigned as homework in order to speed in-class discussion. If more serious time constraints arise, Lesson 3 can be omitted or skipped until a later time.

Homework Options

Chapter 14
See Teacher's Edition margin, pp. 253, 255, 261, and 262
Resource Worksheet, p. 3
Resource Worksheet, p. 10

Chapter 15
See Teacher's Edition margin, pp. 267, 269, 274, 276, 279, and 283
Resource Worksheet, p. 25
Theme Worksheet, p. 26
Math Practice Worksheet, p. 30
SourceBook, p. S130

Chapter 16
See Teacher's Edition margin, pp. 289, 290, 292, 298, 299, 301, and 305
Resource Worksheet, p. 37
Resource Worksheet, p. 39
Activity Worksheet, p. 48
Activity Worksheet, p. 49
SourceBook, pp. S108, S117, and S127

Unit 5
Unit 5 Activity Worksheet, p. 58
SourceBook Activity Worksheet, p. 74
SourceBook Review Worksheet, p. 76

Assessment Planning Guide

Lesson, Chapter, and Unit Assessment	SourceBook Assessment	Ongoing and Activity Assessment	Portfolio and Student-Centered Assessment
Lesson Assessment Follow-Up: see Teacher's Edition margin, pp. 259, 261, 269, 274, 277, 281, 284, 292, 295, 299, 301, and 305 **Chapter Assessment** Chapter 14 Review Worksheet, p. 12 Chapter 14 Assessment Worksheet, p. 14* Chapter 15 Review Worksheet, p. 31 Chapter 15 Assessment Worksheet, p. 35* Chapter 16 Review Worksheet, p. 52 Chapter 16 Assessment Worksheet, p. 56* **Unit Assessment** Unit 5 Review Worksheet, p. 59 End-of-Unit Assessment Worksheet, p. 64*	SourceBook Review Worksheet, p. 76 SourceBook Assessment Worksheet, p. 80*	Activity Assessment Worksheet, p. 70* **SnackDisc** Ongoing Assessment Checklists ♦ Teacher Evaluation Checklists ♦ Progress Reports ♦	Portfolio: see Teacher's Edition margin, pp. 253, 267, 271, 276, 281, 289, and 298 **SnackDisc** Self-Evaluation Checklists ♦ Peer Evaluation Checklists ♦ Group Evaluation Checklists ♦ Portfolio Evaluation Checklists ♦

* Also available on the Test Generator software
♦ Also available in the Assessment Checklists and Rubrics booklet

Science Discovery is a versatile videodisc program that provides a vast array of photos, graphics, motion sequences, and activities for you to introduce into your *SciencePlus* classroom. *Science Discovery* consists of two videodiscs: Science Sleuths and the Image and Activity Bank.

Using the *Science Discovery* Videodiscs

Science Sleuths: Exploding Lawn Mowers, Side A

Homeowners at Printer's Green, a new housing development, experience small explosions when they mow their lawns and a slightly larger explosion when they try to smoke out a gopher. The Science Sleuths must analyze the evidence for themselves to determine the true cause of the explosions.

Interviews

1. Setting the scene: Homeowner **18916 (play ×2)**

2. Another homeowner **19490 (play)**

3. Neighborhood student **19916 (play)**

4. Mower repairman **20479 (play)**

5. Sprinkler installer **20988 (play)**

6. Hydrologist **21434 (play)**

Documents

7. Letter from developer **22273 (step)**

8. City building records **22276 (step ×2)**

9. Brochure on Printer's Green **22280 (step ×2)**

Literature Search

10. Search on the words: INDIAN HEAD HIGHWAY, GOPHERS, METHANE, NATURAL GAS, PRINTING **22284 (step)**

11. Article #1 ("Printing Plant Buried") **22287 (step)**

12. Article #2 ("Typesetting") **22290 (step)**

13. Article #3 ("Pests in the Garden: Tunneling Mammals") **22293 (step)**

14. Article #4 ("Natural Gas Explosion Injures None") **22296**

15. Article #5 ("Natural Gas Nightmare Needs to End") **22298 (step)**

Sleuth Information Service

16. Flammability of 10 common gases **22301**

17. Annual rainfall chart **22303**

18. Geologic core sample **22305**

19. Map of development **22307**

20. Map showing explosions **22309**

21. Older map of area **22311**

22. Hydrochloric acid **22313 (step)**

23. Zinc **22316 (step)**

Sleuth Lab Tests

24. Ground-water pH in development **22319**

25. Precipitate analysis **22321**

26. Seep-water pH around precipitate **22323**

27. Sprinkler-water analysis **22325**

28. Flammable liquid analysis of ground water, seep water, and sprinkler water **22327**

29. Methane test **22329 (step)**

Still Photographs

30. Lawn mower and vicinity **22333 (step)**

31. Gopher hole **22337**

32. Precipitate **22339 (step)**

33. Other powders for comparison **22342**

◀| Step Reverse ‖‖‖‖ Play ▶ ‖‖‖‖ Pause ‖‖ ‖‖‖‖ Step Forward |▶ ‖‖‖‖

A selection of still images, short videos, and activities is available for you to use as you teach this unit. For a larger selection and detailed instructions, see the Videodisc Resources booklet included with the Teaching Resources materials.

14-1 Chemicals in Our Lives, page 253

Milk 2344
Milk is a good source of calcium. It also contains the sugar *lactose*, which is a carbohydrate.

Toothpaste, fluoride 2346
Toothpaste commonly contains fluoride compounds to help fight tooth decay.

Magnesium tablets 2347
Oxides of magnesium are often used as antacids.

Spinach 2348
Spinach is a good source of iron.

Pencil 2349
Pencil leads are made of graphite.

Iodine 2350
An alcohol solution of iodine is commonly used as an antiseptic.

Sugar 2351
Common table sugar is a carbohydrate (sucrose, a combination of fructose and glucose).

Pasta 2352
Pasta is a good source of starch, a polysaccharide carbohydrate.

Salt, iodized 2345
Table salt is a source of sodium. Iodine is often added as a dietary supplement.

14-2 Safety First! page 260

Chemical labeling 208–212 (step ×4)
This standard label shows health hazard in blue, fire in red, and instability in yellow. (step) Blank label (step) On this labeled bottle of sulfuric acid, the number 3 indicates that it is toxic or corrosive, and the number 2 in the instability square indicates an acid that may react if mixed with water. (step ×2) Chemicals should always be marked and stored in their proper place.

Emergency eyewash 4578–4918 (play ×3) (Side A only)
Rinse eyes thoroughly with the eyewash for a long time. Use goggles to avoid many problems.

15-1 Changes All Around, page 265

Karst limestone terrain 595
Limestone dissolves from weak acids in running water, creating sinkholes in the terrain.

Salt solution; evaporated 2524
Salt and other materials are left on the glass after the water from the solution has evaporated.

Heat-sensitive color paints 40760–41176 (play ×2) (Side A only)
Toys painted with heat-sensitive paints are dipped in cold and hot water.

Rubber ball in liquid nitrogen 44966–45956 (play ×2) (Side A only)
Though a rubber ball is elastic at room temperature, cooling it with liquid nitrogen slows the speed of its molecules, and it

---

becomes more rigid. Tapping the cooled ball with a hammer shatters it.

15-2 Signs of Chemical Change, page 270

Chemical reaction; hydrochloric acid and aluminum foil 51121–52308 (play ×3) (Side A only)
Concentrated hydrochloric acid drips on aluminum foil. Acids tend to dissolve metals.

Patina, copper 2527
Under normal atmospheric exposure, copper forms a layer of copper carbonate, a green substance also called *patina*, which protects the metal underneath from further corrosion.

15-4 Chemical and Physical Properties, page 278

Water; properties 2363
The physical properties of water

Reactants and products; chemical 2541
Hydrogen and oxygen react to produce water.

Car, hydrogen-powered 2596
Hydrogen tank in trunk of hydrogen-powered car

Limestone cavern 618
The large pillar is dissolved limestone re-forming on the floor and growing toward the ceiling.

Limestone cavern 620
This cave formed from a combination of limestone dissolving and an underground river wearing the limestone away.

16-1 The Basics, page 288

Compounds; summary 2367–2369 (step ×2)
Properties of compounds (step)
Common compounds include

---

water, sugar, table salt (step), chalk, nylon, and baking soda.

Matter; organizational chart 2370
Categorization of matter

Elements and compounds 2371–2372 (step)
Representation of separate elements (step) Elements combine to form compounds.

16-2 Acids and Bases, page 293

Indicator solution 50333–51120 (play ×2) (Side A only)
Boil red cabbage leaves in water to extract the natural pigments for use as an indicator. When baking soda (base) is added, the solution turns bluish-purple. When vinegar (acid) is added, it turns red.

16-3 The Nature of Burning— A Chemical Change, page 296

Burning of oxygen by candle 49570–49932 (play ×2) (Side A only)
The candle goes out when all of the oxygen in the glass is used up.

Cooling of combustion gases 49191–49569 (play ×2) (Side A only)
When combustion gases are cooled rapidly before completely burning, the carbon condenses as black soot.

Candle gases; relighting candle 49003–49190 (play ×2) (Side A only)
The gases given off by a candle are flammable. As a flame comes close to an extinguished candle, the gases near the candle ignite, and the flame traces its way back to the candle.

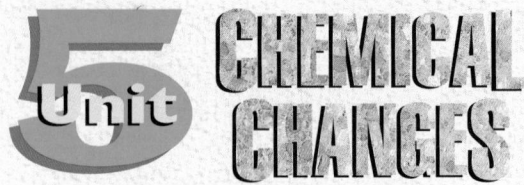

UNIT 5 CHEMICAL CHANGES

UNIT FOCUS

Begin the unit by asking students the following question: What do you think of when you hear the word *chemical?* *(Accept all reasonable responses.)* List the responses on the chalkboard. Using a common classroom item such as a piece of chalk (composed of calcium carbonate) as an example, explain that all matter is made up of different chemical compounds. Other examples include cellulose and lignin, which are the chemical compounds that form wood and paper. Ink is composed of various colored chemical pigments that are dispersed in a chemical solvent. You may wish to include other examples. Explain that students will learn more about chemicals and chemical changes as they complete the activities in this unit.

A good motivating activity is to let students listen to the English/Spanish Audiocassettes as an introduction to the unit. Also, begin the unit by giving Spanish-speaking students a copy of the Spanish Glossary from the Unit 5 Teaching Resources booklet.

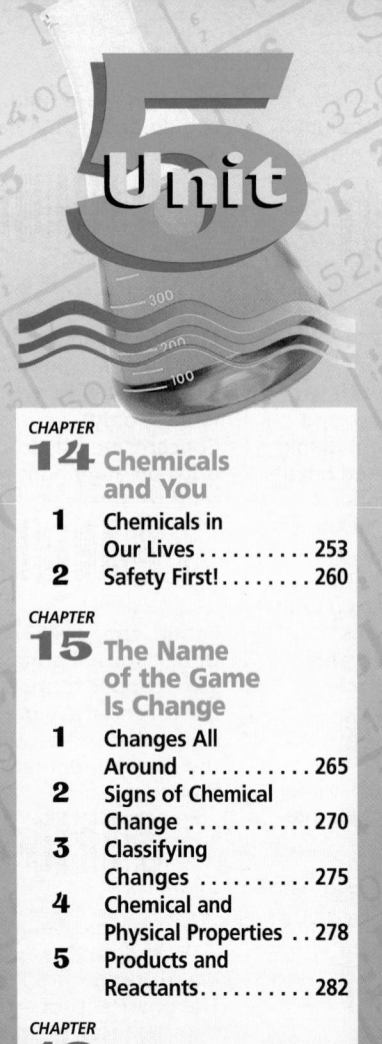

The brilliant color patterns of these tropical frogs warn predators to seek a meal elsewhere.

CHEMICAL CHANGES

Connecting to Other Units

This table will help you integrate topics covered in this unit with topics covered in other units.

Unit 2 Patterns of Living Things	The building blocks of all living organisms consist of chemical elements and compounds.
Unit 4 Investigating Matter	Many properties of a substance reflect the chemical or physical properties of its constituent molecules.
Unit 6 Energy and You	Both mass and energy are conserved in all chemical and physical changes.
Unit 7 Temperature and Heat	Heat can cause chemical changes to occur, and heat can be given off during chemical changes.
Unit 8 Our Changing Earth	Chemical reactions can cause geological change, such as when rainwater dissolves limestone to form caves.

In tropical rain forests, potential predators lurk around every corner, ready to make a meal out of an unsuspecting frog. Why, then, do the frogs shown here advertise their location with brilliant colors? These frogs boldly announce their presence to others as a warning to stay away! When threatened, they secrete chemicals that are among the most toxic substances known. A close relative of the frogs shown here can secrete enough poison to kill 20,000 laboratory mice or several adult humans.

Known commonly as poison-dart frogs, they get their name from an ancient hunting technique still practiced in Colombia by the Chocó people. The chemicals are collected from the frogs' skin, and then the frogs are set free. These powerful chemicals are used on blow-gun darts in order to kill game.

Studying the chemicals produced naturally by these frogs may one day enable scientists to produce more effective medicines artificially. Two of the most promising areas of research involve pain relievers and heart stimulants. How is it possible to produce a natural chemical artificially? The answer lies in understanding the nature of chemicals. What is a chemical? As with all other areas of science, learning about chemicals involves thinking about the world around us in a scientific way.

251

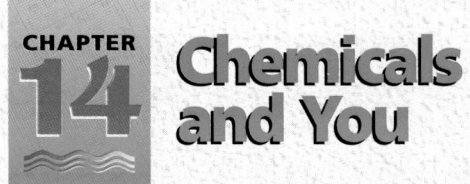

Connecting to Other Chapters

Chapter 14
introduces chemicals used in our daily lives and explains how chemicals should be safely handled.

Chapter 15
explores the differences between chemical and physical changes and properties.

Chapter 16
defines acids and bases and further explores what happens during chemical changes.

Prior Knowledge and Misconceptions

Your students' responses to the ScienceLog questions on this page will reveal the kind of information—and misinformation—they bring to this chapter. Use what you find out about your students' knowledge to choose which chapter concepts and activities to emphasize in your teaching.

In addition to having students answer the ScienceLog questions on this page, ask them to imagine the following hypothetical situation and respond in writing: As a very annoying practical joke, an old adversary of yours has taken all of the cooking ingredients in your home and placed them in unmarked containers. You cannot identify the ingredients by sight alone, and you know that you should never taste, touch, or smell any unknown substance. What can you do to distinguish among these ingredients? Assure the students that there are no right or wrong answers to this question. Collect the papers, but do not grade them. Instead, read them to find out what students know about chemical properties, what misconceptions students may have, and what aspects of chemistry are interesting to them.

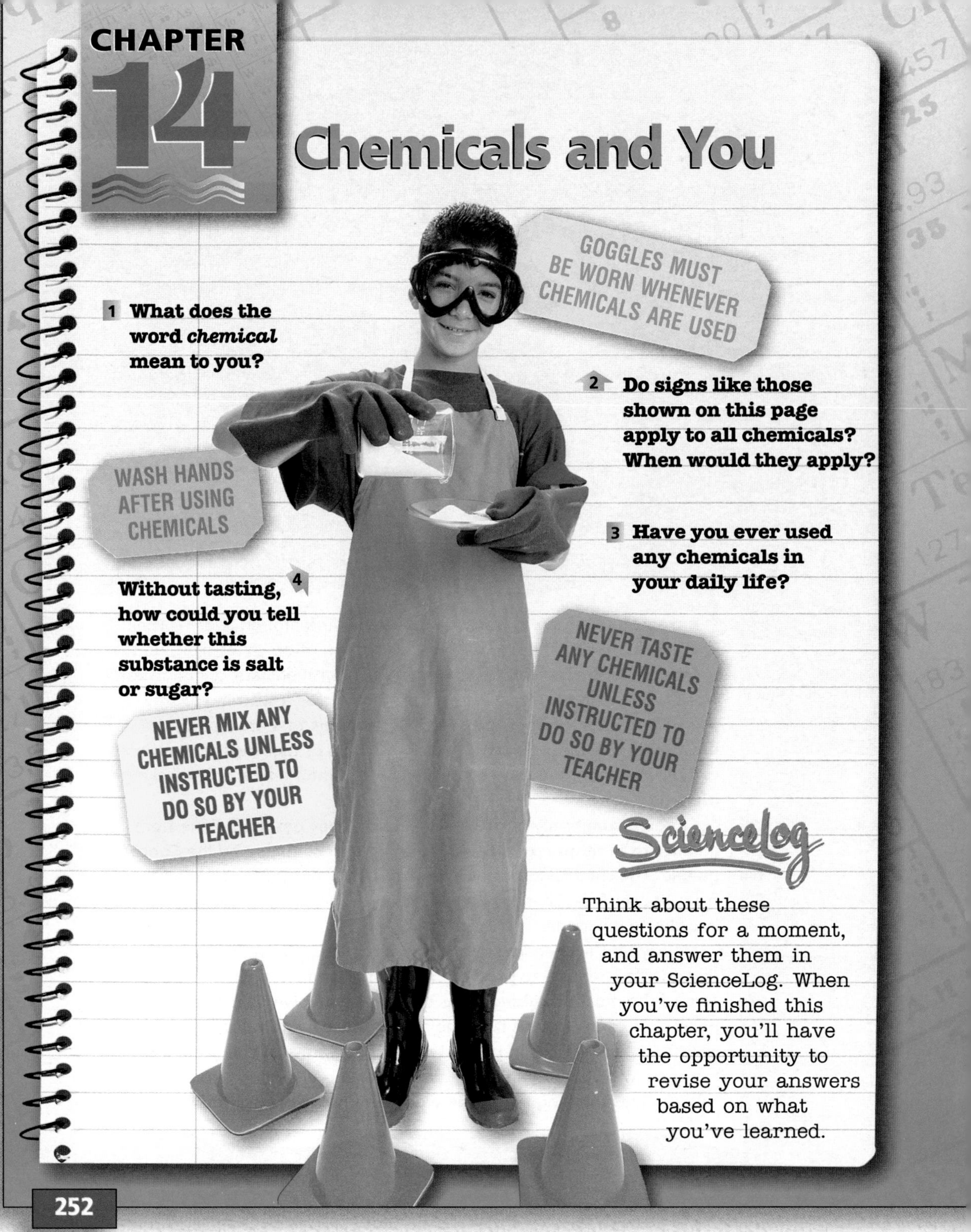

CHAPTER

14

Chemicals and You

1 What does the word *chemical* mean to you?

GOGGLES MUST BE WORN WHENEVER CHEMICALS ARE USED

2 Do signs like those shown on this page apply to all chemicals? When would they apply?

WASH HANDS AFTER USING CHEMICALS

3 Have you ever used any chemicals in your daily life?

4 Without tasting, how could you tell whether this substance is salt or sugar?

NEVER MIX ANY CHEMICALS UNLESS INSTRUCTED TO DO SO BY YOUR TEACHER

NEVER TASTE ANY CHEMICALS UNLESS INSTRUCTED TO DO SO BY YOUR TEACHER

ScienceLog

Think about these questions for a moment, and answer them in your ScienceLog. When you've finished this chapter, you'll have the opportunity to revise your answers based on what you've learned.

252

Answers to
In-Text Questions, page 253

A Accept all reasonable responses. Make sure that student answers are supported by critical thought.

B A chemical is a substance, such as water, salt, or sugar, which has distinct properties.

Chemicals in Our Lives

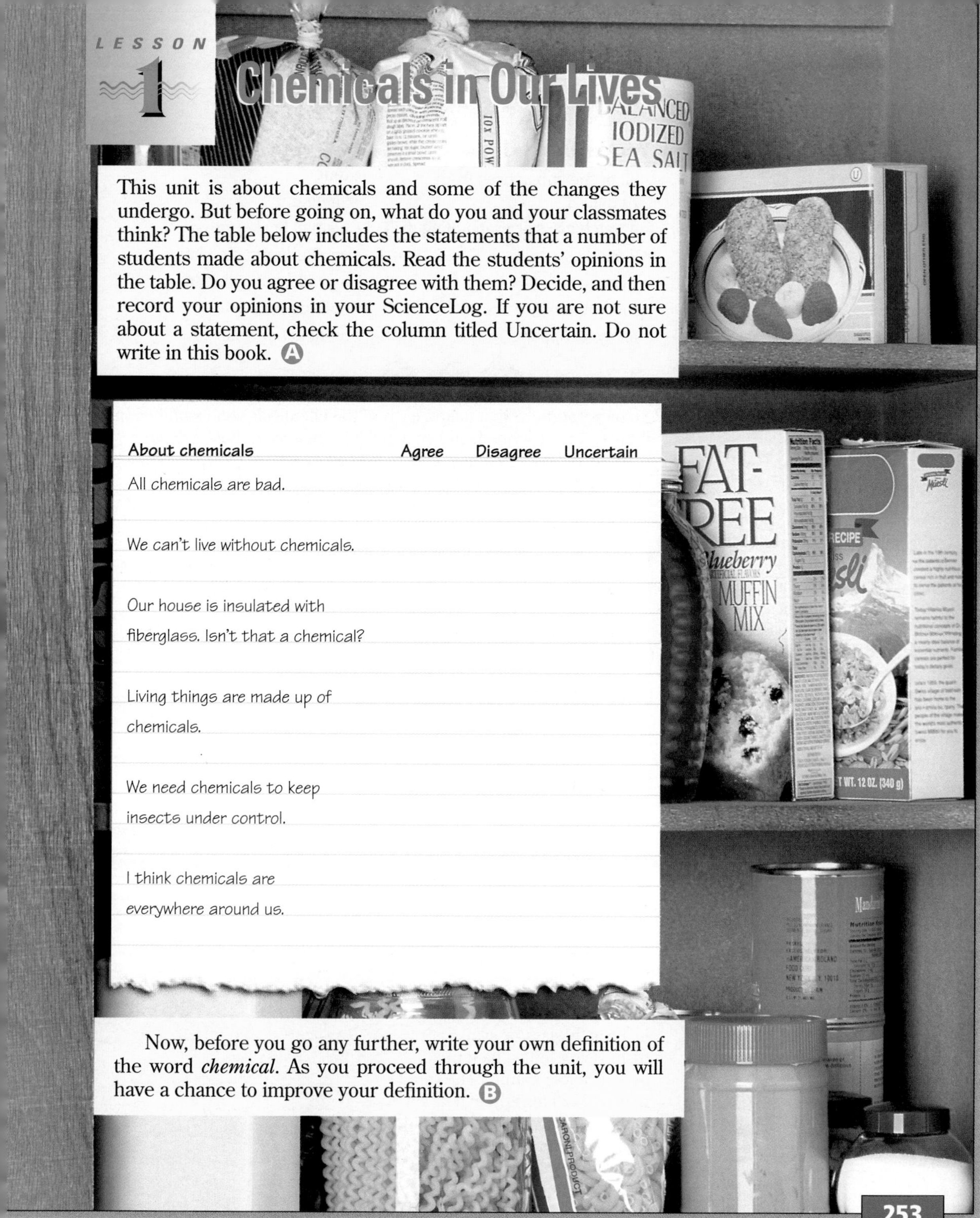

This unit is about chemicals and some of the changes they undergo. But before going on, what do you and your classmates think? The table below includes the statements that a number of students made about chemicals. Read the students' opinions in the table. Do you agree or disagree with them? Decide, and then record your opinions in your ScienceLog. If you are not sure about a statement, check the column titled Uncertain. Do not write in this book. **Ⓐ**

About chemicals	Agree	Disagree	Uncertain
All chemicals are bad.			
We can't live without chemicals.			
Our house is insulated with fiberglass. Isn't that a chemical?			
Living things are made up of chemicals.			
We need chemicals to keep insects under control.			
I think chemicals are everywhere around us.			

Now, before you go any further, write your own definition of the word *chemical*. As you proceed through the unit, you will have a chance to improve your definition. **Ⓑ**

253

Chemicals in Our Lives

FOCUS

Getting Started
Ask students to look around the classroom and to point out any chemicals that they can see. List their responses on the board. When they have finished making suggestions, write "everything in this room" as the last item on the list. Explain to students that all matter is made of chemicals.

Main Ideas
1. Chemicals play an important role in all daily activities.
2. Chemicals are uniform substances with predictable properties.
3. Each chemical has certain chemical and physical properties that distinguish it from other chemicals.

TEACHING STRATEGIES

INDEPENDENT PRACTICE The opinion table offers an opportunity for students to communicate their feelings and possible misconceptions about chemicals. The table should be completed individually. Ask students to give reasons for their opinions. Then divide the class into small groups to discuss their opinions.

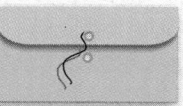

PORTFOLIO
Suggest that students include their opinion table in their Portfolio. During the course of the unit, students can make notes in the margins of the opinion table to indicate how their opinions have changed or have been reinforced.

Homework

The Resource Worksheet that accompanies the material on this page makes an excellent homework activity (Teaching Resources, page 3).

LESSON 1 ORGANIZER

Time Required
3 class periods

Process Skills
observing, analyzing, inferring, hypothesizing

Theme Connection
Changes Over Time

New Term
Chemical—a single substance with distinctive properties. All matter is composed of chemicals, and a chemical always reacts in the same way under the same conditions.

Materials (per student group)
Exploration 1, Activity 1: about 2 spoonfuls each of salt, baking soda, and sugar; 3 pieces of chalk; 3 or 4 pieces (about 2 cm × 2 cm each) of different types of paper, such as tissue paper, newspaper, and paper napkin; about 30 mL of distilled water; several small containers; magnifying glass; safety goggles;

continued ▶

EXPLORATION 1

Cooperative Learning
EXPLORATION 1

Group size: 3 to 4 students
Group goal: to explore the properties of various everyday chemicals
Positive interdependence: Assign students roles such as coordinator, materials manager, timekeeper, and reporter.
Individual accountability: Each student should be able to pick at least three chemicals from this Exploration and list several properties of each.

ACTIVITY 1

GUIDED PRACTICE Provide each group with samples of the six chemicals. Ask students if they can identify the chemicals from observation alone. *(Probably not)* Tell them that the materials are all chemicals, and pose the question: What does examining these chemicals tell you about what the term *chemical* means? *(Each chemical has certain properties that distinguish it from other chemicals.)* Ask students to devise methods for testing the properties of each chemical before proceeding to Activity 2. *(Accept all reasonable responses.)*

ACTIVITY 2

You may wish to set up six stations around the room, with a setup for one of the six tests at each station. Allow 10 minutes per station. A completed sample table is shown on page S239.

Answer to
In-Text Question

 Nothing remains in the dish after the water evaporates.

⭐ An Exploration Worksheet is available to accompany Exploration 1 (Teaching Resources, page 4).

What Are Chemicals?

ACTIVITY 1

You Will Need

- distilled water
- table salt
- sugar
- chalkboard chalk
- baking soda
- samples of different types of paper
- a magnifying glass

What to Do

Water, salt, sugar—you are probably familiar with each of these common substances. But did you know that they are all **chemicals**? Look at each chemical carefully with a magnifying glass. Are you able to identify these chemicals from your observations? Without tasting, touching, or smelling these substances, can you tell them apart? Make a list of several methods you could use. In Activity 2 you will investigate these chemicals further.

Caution: Do not touch, taste, or smell any laboratory materials unless specifically instructed to do so by your teacher.

ACTIVITY 2

You Will Need

- all of the substances from Activity 1
- a watch glass or evaporating dish
- 2 candles
- matches
- modeling clay
- an eyedropper
- several jar lids
- tongs
- vinegar
- a small beaker or jar
- an aluminum pie plate
- a graduated cylinder
- stirring rods
- a bucket of water

What to Do

In the paragraphs that follow, each of the chemicals you examined in Activity 1 is described, and an activity is suggested. Based on your reading and observations, give the result of each activity, and suggest one characteristic that may be true of each chemical. In your ScienceLog, enter your responses in a table like the one below. Clean up your area and wash your hands when you are finished.

Chemical	Result of activity	Scientific idea about this chemical
distilled water		
table salt		
sugar		
chalkboard chalk		
baking soda		
cellulose		

Water

This must be the most common chemical of all. You drink it for survival. You swim in it, sail on it, and skate on it. It falls on the Earth as rain, snow, and hail. It is one of our most important natural resources.

Place a drop of distilled water in a watch glass or evaporating dish. Using tongs, heat the dish over a candle flame. (The candle may be held upright in a jar lid with some modeling clay.) Does anything remain in the dish after the water evaporates?

ORGANIZER, continued

Exploration 1, Activity 2: all materials from Activity 1; watch glass or evaporating dish; 2 candles; aluminum pie plate; small ball of modeling clay; pair of tongs; bucket of water; several jar lids; small beaker or jar; several stirring rods; 100 mL graduated cylinder; 25 mL of vinegar; box of matches; eyedropper; about 25 mL of distilled water; safety goggles; lab aprons; latex gloves

Teaching Resources
Resource Worksheet, p. 3
Exploration Worksheet, p. 4

Salt

This chemical has been used by people for many thousands of years. At one time it was so valuable in some areas that it was used for money! Throughout the world, salt is processed in many different ways. In some places, it is mined. In others, salt is obtained through the evaporation of sea water. In still other parts of the world, such as northern Zambia, salt is produced from plants.

Do you think the salt you use at home is likely to be the same as the salt found in the home of a person from Zambia? **B**

Place a few crystals of salt in 25 mL of water, and stir. What happens? Can the salt be recovered? How? **C**

Sugar

Did you know that an average American consumes 29 kg of sugar a year? Sugar is found in all types of foods: in sweets, of course, but also in bread, fruits, and vegetables—even potatoes! Most of the sugar we use comes from sugar cane, sugar beets, and corn. It is purified by a chemical process called *refining*.

Do you think the sugar obtained from sugar beets differs in any way from the sugar obtained from sugar cane?

Without tasting, how could you distinguish salt from sugar?

Place a bit of sugar on a metal jar lid, and use tongs to hold the lid over a lit candle. What happens?

Chalk

"A classroom without chalk is like . . ." Finish this sentence yourself. Chalk occurs naturally as a soft type of limestone. The main chemical in both chalk and limestone is calcium carbonate. Chalkboard chalk is made from chemically produced calcium carbonate.

Break a piece of chalk into two pieces. Examine the powder on the broken end of each piece. Is there any difference between one piece of chalk and the other? How could you prove that chalk is chalk?

In a jar lid or small beaker, add a dropperful of vinegar to a small piece of chalk. What happens?

Exploration 1 continued ▶

Did You Know . . .

Many chemicals are very simple. It takes only two hydrogen atoms and one oxygen atom to make water. However, some of the most complex chemicals known are in our cells. For instance, nucleic acids, which are present in every living cell, consist of thousands of atoms.

ENVIRONMENTAL FOCUS

Ask: Is bottled water safer than tap water? Many people assume that bottled water comes straight from pure mountain streams. In fact, most bottled water is actually tap water that has been filtered and treated by various chemical methods to remove sodium, lead, and other metals. Even though bottled-water plants are regulated by the government, bottled water is not tested for pollutants as much as the public water supply is.

Answers to
In-Text Questions

B Pure table salt, or sodium chloride, is the same no matter where it is found or how it is produced. However, impurities may vary from place to place.

C The crystals of salt disappear as they dissolve in the water. The salt can be recovered by allowing the water to evaporate. This will leave a salt residue on the bottom of the container.

Answers to
Sugar

Beet sugar and cane sugar are chemically identical, although they may differ in their impurities.

You could distinguish salt from sugar by the shape of the granules. (Salt is cubic, and sugar is elongated.) In addition, the sugar burns and turns black when it is heated.

Answers to
Chalk

The two pieces of chalk are chemically identical. This could be proved by showing that both pieces give the same result for this activity.

The addition of a dropperful of vinegar to a small piece of chalk results in a bubbling reaction as carbon dioxide gas is released.

Homework

Have students make a list of 10 different chemicals that are needed by their body. *(Their list might include the following: elements such as calcium, iodine, and phosphorus; categories of compounds such as enzymes, carbohydrates, fats, and acids; or specific compounds such as water.)*

Answers to
Baking Soda

Adding a drop of vinegar to a small sample of baking soda causes a bubbling reaction as carbon dioxide gas is released. Each student will get a similar reaction because each student will use the same chemicals. Vinegar and baking soda, when mixed, should always react to form carbon dioxide gas.

Answers to
Cellulose

All paper crinkles up and burns; stiffer paper usually takes a few seconds longer to burn down to ashes.

Answers to
Analysis Please!

1. Students should compare their group's answers with those of other groups and resolve any differences. Make sure that they support their opinions with critical thought.

2. Accept all reasonable responses for each observation. Sample answers:
   a. Salt water is a mixture of two chemicals, salt and water.
   b. This is not always true of a mixture, but it is true of a solution. For example, an unopened bottle of soda water contains a solution of carbon dioxide and water. When the bottle is opened, bubbles form in the water. It is still a mixture, but since the bubbles are separate from the water, it is not a true solution.
   c. "Similar conditions" refers to the fact that variables such as temperature and moisture must be known in order to predict how a chemical will look and behave. For example, all paper burns, but only if it is dry.

3. The granules of salt and sugar are larger than those of chalk dust and baking soda. Salt does not burn, but sugar does. Chalk dust and baking soda both bubble when vinegar is added, but only baking soda dissolves in water.

4. Since the solid reacts with vinegar, it could be white chalk or baking soda. Both bubble when vinegar is added. However, students may discover that baking soda dissolves in water; chalk does not. Therefore, the substance is chalk.

Baking Soda

Without baking soda, cakes would be flat, refrigerators would be less sweet-smelling, and more stomachs would be upset. Baking soda is a very practical chemical!

In a jar lid or small beaker, add a drop of vinegar to a pinch of baking soda. Compare your results with those of your classmates. Did everyone get the same reaction? What does this tell you about these two chemicals?

Cellulose

You may not have heard of cellulose, but you probably use it every day. What is it, and how do you use it? Cellulose is the main chemical in the cell walls of all plants. It is also a key ingredient in paper products. As you know, there are many different types of paper. Some types are very light and flexible. Others are stiff and heavy.

Take three or four small (2 cm × 2 cm) pieces of different types of paper—tissue paper, newspaper, and a paper napkin, for example. One by one, place each piece in an aluminum pie plate, and light it with a match.

How did each of the different types of paper behave? Did they behave differently or more or less alike?

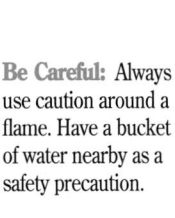

Be Careful: Always use caution around a flame. Have a bucket of water nearby as a safety precaution.

Analysis Please!

By now you have made many observations and inferences about chemicals.

1. Compare your list of discoveries from Exploration 1 with that of another group of students. Do you agree with their statements?

2. Here are some scientific ideas about chemicals. What evidence gathered from the preceding activities supports each idea below?
   a. A chemical is a single substance. This differs from a mixture, in which different substances can be mixed together in any amount. Can you find an example of a mixture that you made in this Exploration?
   b. Every part of a chemical is like every other part. Is this true of a mixture as well?
   c. A chemical always looks and behaves the same way under similar conditions. What do you think is meant by "similar conditions"?

3. The way a chemical looks and how it behaves are its *properties*. Properties are used to identify a chemical. With a classmate, discuss the ways in which you could distinguish the following chemicals from each other: salt, sugar, chalk dust, and baking soda.

4. Below is Jerome's description of one of the chemicals you examined. Which chemical is it? Has he named enough properties to positively identify the chemical?

> "This chemical is a solid. It's white. If I add vinegar to it, it fizzes and bubbles. It does not dissolve in water."

256

Nothing to Sniff At!

Here are some of the chemicals that contribute to the aroma of coffee. Which ones have you heard of before?

acetaldehyde	formic acid	methyl ethyl acetaldehyde
acetic acid	furan	methyl ethyl acetic acid
acetone	furfural	methyl mercaptan
acetyl methyl carbinol	furfuryl acetate	n-heptacosane
ammonia	furfuryl alcohol	N-methylpyrrole
cresols	furfuryl mercaptan	n-valeric acid
diacetyl	guaiacol	p-vinylguaiacol
diethyl ketone	hexanoic acid	phenol
diethyl sulfide	hydrogen sulfide	pyrazine
2, 3-dioxyacetophenone	hydroquinone	pyridine
esters	isovaleric acid	pyrrole
ethyl alcohol	methyl alcohol	resorcinol
eugenol	methylamine	trimethylamine

Turn to page S108 of the SourceBook to learn how models help scientists understand chemical properties.

There are actually many more chemicals in coffee. Imagine how many there must be in living things—in you! Just think about how many chemicals you encounter each and every day!

257

Nothing to Sniff At!

This list suggests to students the vast range of chemicals in a common beverage such as coffee. Students will probably be amazed at the strange-sounding names and the quantity of chemicals. The exercise on the following page develops this topic further.

Integrating the Sciences

Earth and Life Sciences

Many lakes become overgrown with large mats of algae called *algae blooms* when organic wastes and phosphates are dumped into the lakes by city sewage systems. Phosphates, which are used in many household detergents, accelerate the growth of algae. The algae can prevent the growth of other plant and animal life in the water. Furthermore, as the algae die and decompose, oxygen is removed from the water, suffocating the fish in the lake. Have students find out more about this problem and what can be done about it.

Chemical Close Encounters

Have students read the stories of Sasha and Harold and, through discussion, identify some "everyday" chemicals.

Multicultural Extension

Chemicals in Other Countries

If some students have lived or spent time in other countries, you may wish to ask these students to report on some of the chemical products that are used there, such as food seasonings, cosmetics, and household cleansers. Different countries have different laws regulating chemicals and pharmaceuticals, so these students might be familiar with products that are not available in the United States.

Meeting Individual Needs

Gifted Learners

Ask students to imagine living without modern household chemicals such as cleansers, pesticides, and food additives. Propose that students develop a project in which they trace how different household chemicals have been used throughout the years. Older adults in the community are excellent sources of information. Have students explain what improvements have been made in household cleaners. Students can then assemble the information in a booklet or video.

Chemical Close Encounters

Like most people, Sasha and Harold are not aware of the different chemicals they encounter each day. Even in the first 2 hours of their day, they use many chemicals. Here is a description of their morning activities. Make a list of the chemicals or items containing chemicals that they encounter.

Sasha

Taking a deep breath, Sasha gets up, goes into the bathroom, brushes her teeth with toothpaste containing fluoride, and washes her face with cleanser. Then she applies deodorant, eye shadow, and lipstick. She decides to wear nylon tights to work today, along with a rayon-cotton blouse and wool skirt. Finally, she has a quick breakfast of orange juice, cereal, and buttered toast. Then it's off to work.

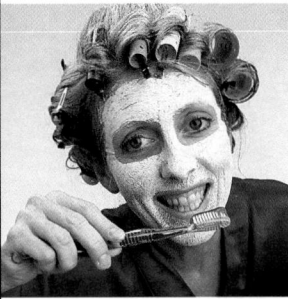   

Harold

Harold throws back the sheet and struggles out of bed. After washing his face and hands, he takes his contact lenses out of their cleaning solution. Then he rinses them in a salt solution and puts them in his eyes. Now he can see! Next, Harold shaves, using mentholated shaving cream, and then he splashes himself with aftershave lotion. He puts on a cotton shirt and slacks. For breakfast Harold has a blueberry muffin and a glass of orange juice. Finally, he heads off to work.

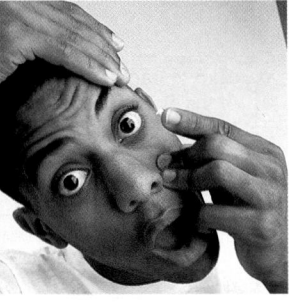   

Just like Sasha and Harold, we all eat, drink, breathe, and use chemicals all the time. Many chemicals occur naturally in the world—water and the gases in the air, for example. Other chemicals, such as nylon, plastics, and the dyes in lipstick, are made by industrial methods.

Tracking Down Chemicals

What kind of chemicals do you encounter in your day? One way to find out is to read the labels on the items you use. When you do this, you may be surprised to find more than corn in your cornflakes. Keep track of the chemicals you encounter in one day. Record them in your ScienceLog. Afterward, make a collage of labels to display some of these chemicals. Now write a brief story about your day. Include as many chemicals as you can. Try to think of chemicals not mentioned in the accounts of Sasha's and Harold's morning.

A Project

Many chemicals have both a chemical name and a common name. Can you replace the common names of the chemicals used in Exploration 1 with the chemical names? As you proceed through the unit, you will encounter more common and chemical names for substances. Keep a list in your ScienceLog, such as the one started at right, of both names for all of the substances you encounter. Also, list a use for each chemical. You can start with water.

Common name	Chemical name	Use
water	dihydrogen oxide	Plants and animals need water.
table salt		

Tracking Down Chemicals

INDEPENDENT PRACTICE Have students make a list of the chemicals that they encounter in their day, as suggested on this page. Ask each student to bring in a label from an item he or she uses, and create a class collage of everyday chemicals. Each student should write a brief story of his or her day. You may wish to have students read their story to the class. Make a composite list on the chalkboard of all the chemicals that are mentioned.

Answers to
A Project

The chemical name for table salt is *sodium chloride*. Table sugar is *sucrose*. Chalkboard chalk is *calcium carbonate*. Baking soda is *sodium hydrogen carbonate*, although it is often referred to as *sodium bicarbonate*. Water, salt, and sugar are all foods. Chalk is used as a writing instrument. Baking soda is used in cooking and for cleaning.

CROSS-DISCIPLINARY FOCUS

Art

Have students create visual accompaniments for the story they have been assigned to write on this page. The images should emphasize the vast number of chemicals they encounter during the day. Students should create illustrations of their own or use materials such as pictures from magazines to produce their visual representations.

Theme Connection

Changes Over Time

Focus question: What are some of the chemical changes that have contributed to geological changes on Earth? *(Students may point to chemical erosion, such as that which occurs in the formation of limestone caves. Some students may also comment on the chemicals that humans add to the environment. Sometimes these chemicals react negatively with the environment, such as when they pollute water sources.)*

FOLLOW-UP

Reteaching

Have students go back to the table on page 253 and review how they responded to the statements about chemicals. They should use what they have learned in this lesson to decide if they should change any of their initial responses.

Assessment

Tell students the following: Cassie and Jeff were given four white substances to identify. They examined them carefully and performed three tests on them. Describe the tests they might have performed. Include the materials required, the necessary safety considerations, and the possible results of each test.

Extension

Have students further explore the activity Tracking Down Chemicals by writing a mystery about tracking down a villainous chemical. The story should contain clues about the properties of the chemical. Have students share their mystery with the class so that everyone can take part in solving the mystery.

Closure

Invite a guest speaker who has a chemistry-related career to discuss his or her job responsibilities with the class.

LESSON 2 — Safety First!

FOCUS

Getting Started

Pretend that you are performing a lab experiment in which you violate several safety procedures. Make the demonstration humorous and exaggerated, and ask students to call out all of the safety errors you make. Have a student volunteer write your mistakes on the board. Afterward, discuss with students why each of your errors is potentially dangerous.

Main Idea

When dealing with chemicals in the laboratory, home, and classroom, safety is essential.

TEACHING STRATEGIES

Answers to
In-Text Questions and Caption

Ⓐ Calcium carbonate is chalkboard chalk. Cellulose is paper. Sucrose is table sugar. Sodium bicarbonate is baking soda.

Ⓑ The safety precautions illustrated in the top two pictures on page 260 include the following:
- Never inhale chemicals directly.
- Wear a lab apron.
- Pull back long hair.
- Avoid baggy clothing.
- Wear safety goggles.
- Use an eyewash properly.

Ⓒ Unsafe practices shown in this photograph include the following:
- Wearing baggy clothes
- Not pulling back long hair
- Reaching over a flame
- Placing nonessential objects on the laboratory table
- Sitting on the laboratory table
- Wearing sandals (which provide inadequate foot protection)
- Not wearing safety goggles
- Chewing gum

Safety Tips, *page 261*

Display students' posters around the classroom, and have students add to them as more safety techniques are discussed.

LESSON 2 — Safety First!

There is no need to put on goggles when you add sodium chloride to popcorn or when your teacher writes on the board with a piece of calcium carbonate. You don't need to wear gloves when you read your cellulose newspaper or magazine. Washing your hands is not necessary after adding sucrose to your cereal or after taking some sodium bicarbonate for an upset stomach. By now you can probably replace these chemical names with common names. For instance, sodium chloride is salt. Ⓐ

Although special precautions are not necessary in the situations described above, safety *is* important when you are handling many substances that can be found in the home, in the classroom, in the lab, and at work. The photographs at right and below illustrate one or more safety rules. Can you spot what each person is doing correctly? Ⓑ

What's Wrong Here?

The students at left are demonstrating how **not** to behave in the laboratory. How many unsafe laboratory practices can you spot? Ⓒ

260

LESSON 2 ORGANIZER

Time Required
1 class period

Process Skills
observing, classifying, analyzing

New Terms
none

Materials (per student group)
Safety Tips: poster board; markers

Teaching Resources
Resource Worksheet, p. 10

Safety Tips

In groups of two or three, discuss the safety rules concerning chemicals and other materials in your classroom. Design a chemical-safety poster. Here are some suggestions for the poster:

- Draw cartoons.
- Make diagrams.
- Take photographs.
- Write poems.
- Describe humorous situations.
- List humorous statements.

Here is a limerick that was part of one group's poster:

> There once was a girl named Di
> Who, in spite of the warnings, drank lye.
> Her teacher said NEVER
> To taste—not EVER!
> She's now in that Great Lab in the sky.

Warnings!

Detergents are poisonous. Oven cleaners are corrosive, and they can severely burn your skin. Drain cleaners contain lye, which is also corrosive. Bleaches give off a poisonous, foul-smelling gas—chlorine. Where are these chemicals stored in your home? What kind of recommendations can you make about storing and using them? What other materials around your home are potentially dangerous?

Below are some of the warning signs used throughout this book. Do you know what they mean? What warning signs are on the products in your home?

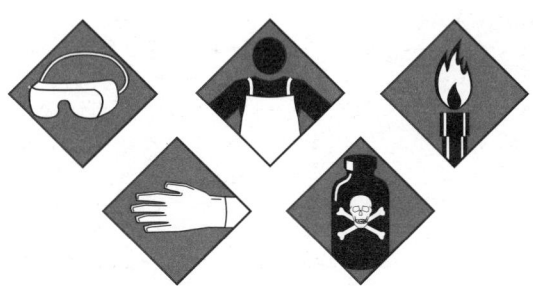

Safety Scenarios

For each of the following scenarios, suggest possible consequences of the actions taken.

1. Jon saw a liquid on a lab tabletop. "Must be water," he thought and promptly wiped it up with the sleeve of his coat.
2. The class was instructed to smell the liquid carefully. Renee held the beaker directly under her nose and inhaled.
3. Leigh noticed broken glass on the floor and began to put the glass in the trash can using her hands.
4. Ramon's class was using candles in an experiment. His desk was cluttered with books and papers and materials from another experiment.
5. Claudia wouldn't wear her safety glasses because they felt uncomfortable on her face.
6. Mike didn't feel like reading the directions for the experiment and began to mix the chemicals to see what would happen.

261

FOLLOW-UP

Reteaching

Bring in household items such as detergents, oven cleaners, and bleaches. Read the labels, and discuss each item and its potential dangers. Then have students design their own warning labels.

Assessment

Present students with the following situation: You have been appointed school safety officer. Outline six safety rules for the science laboratory. Explain why you included each rule.

Extension

Ask students to make a list of 10 everyday items. Then have them research the chemical names for these items.

Closure

Invite a nurse or an emergency medical technician to your class to discuss first-aid procedures.

Answers to
Challenge Your Thinking, pages 262–263

1. Sucrose—sugar; acetyl-salicylic acid—aspirin; ammonium hydroxide—household ammonia; polyhexanedioic acid diaminohexamide—nylon; poly-tetrafluoroethylene—Teflon

2. Carbon dioxide in the air mixed with water to form a weak acid that dissolved the limestone. In Exploration 1, the vinegar, a dilute solution of a weak acid, dissolved the chalk and released carbon dioxide.

3. Beginning in the top left corner and going clockwise, possible answers include the following:
- Agree. Pure water is a single substance, not a mixture. When water is allowed to evaporate from a container, there is no residue left in the container.
- Agree. A certain chemical will always behave in the same way under the same conditions. Heated sugar always turns black.
- Agree. Table salt is the same chemical no matter where it is found.
- Agree. Chemicals always behave in the same way under the same conditions.
- Agree. Each substance that was used was a single chemical. A mixture contains a number of different substances.
- Agree. Chemicals always behave in the same way under the same conditions. Chalk and baking soda always fizz when mixed with vinegar.

4. This activity is designed to generate thought and discussion about the role of chemicals in our society by using current events or discoveries that involve chemicals or chemical changes.

5. Students' cover pages could highlight the role of chemicals in our lives, safe and unsafe procedures for handling chemicals, or chemical reactions.

 You may wish to provide students with the Chapter 14 Review Worksheet that is available to accompany this Challenge Your Thinking (Teaching Resources, page 12).

CHALLENGE YOUR THINKING

1. What's in a Name?
As you know, many chemicals have a common name as well as a scientific name. Study the following table. Use the description and the scientific name to guess the common name of the chemical being described. Remember: don't write in this book.

Description	Scientific name	Common name
Colorless solid; used as a flavoring for food; rich in calories; extracted from plants	sucrose	
White solid; easily powdered; common remedy for headaches	acetylsalicylic acid	
Used as a cleaning agent; produces a strong smelling gas; dissolves fats and oils	ammonium hydroxide	
Light, flexible material; can be drawn into fibers; takes dyes well	polyhexanedioic acid diaminohexamide	
Dark-colored material; forms a smooth surface; few materials stick to it	polytetrafluoro-ethylene	

2. Cave Clues
Limestone is found in many parts of the United States. Where you find limestone, you also generally find caves. The limestone hills of Tennessee and Missouri are riddled with caves and sinkholes. This type of terrain is called a *karst landscape.* How do you think it was formed? (Hint: Review the experiment you performed with chalk in Exploration 1, Activity 2.)

Multicultural Extension

Ancient Chemistry
Divide the class into groups to research the practical chemistry used by different cultures of the ancient world. Each group could focus on a particular geographic region and a specific topic such as metallurgy, paints and pigments, pottery techniques, or medicine. Have each group present its findings to the class. Some possibilities include iron smelting and production of synthetic pigments by ancient Africans or the use of dyes in ancient Phoenicia.

Homework

Instruct students to sit in any room in their home and to make a list of five chemicals found there and their uses. Depending on the room the students choose, the list may include chemicals in foods, cleaning agents, fabrics, clothing, furniture, and entertainment devices. You may also wish to have students list in their ScienceLog any materials in their home that are potentially dangerous. They should identify the warning signs found on these products.

3. Oh Say What You See

Here are what some students said about the chemicals they investigated in Exploration 1. Explain why you agree or disagree with each of the following statements:

There is nothing else in pure water except more water.

Everyone's table sugar turned black when heated.

Table salt is table salt, no matter what its source is.

Both chalk and baking soda fizzed with vinegar.

Paper always burns the same way, no matter what color it is.

None of the chemicals we used are mixtures.

4. Chemical Alert!

Follow news broadcasts and newspapers for stories involving chemicals and chemical changes. Share your news stories with your class.

5. Cover Up

This unit is called Chemical Changes. Create a cover page for this unit in your ScienceLog. It can be a collage of pictures from magazines or ones that you drew yourself, but it should reflect your understanding of what this unit is about.

Sciencelog

Review your responses to the ScienceLog questions on page 252. Then revise your original ideas so that they reflect what you've learned.

263

The following are sample revised answers:

1. The word *chemical* can refer to any substance because all matter is made up of chemicals. A chemical is named according to the element or elements that make up that chemical. Many substances have both a common name and a chemical name. For example, the chemical name for table salt is sodium chloride.

2. Whenever working in a scientific setting or when dealing with unknown chemicals, safety precautions should be taken at all times. The types of safety equipment that are necessary when dealing with chemicals depend on the chemicals involved.

3. Accept all reasonable answers. Students may identify chemicals that are present in clothes, food, or household items.

4. Testing the physical and chemical properties of the substance can help to identify the substance as either salt or sugar. The shape of the granules is an example of a physical property that could distinguish salt from sugar. How the substance reacts to heat is an example of a chemical property that could tested. Sugar burns and turns black when it is heated; salt does not.

Meeting Individual Needs

Learners Having Difficulty

Read the following scenario to students: Tasha's doctor recommended that she cut back on the amount of salt she consumes. Now she reads food labels very carefully. Which of the following ingredients in a fettucine sauce would prevent her from including the sauce in her diet?

a. calcium carbonate
b. cellulose
c. sucrose
d. sodium chloride *(Correct)*

The Name of the Game Is Change

Connecting to Other Chapters

Chapter 14
introduces chemicals used in our daily lives and explains how chemicals should be safely handled.

Chapter 15
explores the differences between chemical and physical changes and properties.

Chapter 16
defines acids and bases and further explores what happens during chemical changes.

Prior Knowledge and Misconceptions

Your students' responses to the ScienceLog questions on this page will reveal the kind of information—and misinformation—they bring to this chapter. Use what you find out about your students' knowledge to choose which chapter concepts and activities to emphasize in your teaching.

In addition to having students answer the questions on this page, you may wish to have them complete the following activity: Have students write a short skit featuring a middle-school student and a third-grader. In the skit, the middle-school student explains to the third-grader the difference between a chemical change and a physical change. Encourage students to be creative in their explanations even if they are not sure of the correct answers. Remind students that there are no right or wrong answers to this activity. Have volunteers present their skit to the class, but do not grade them. Instead, use this activity to find out what students know about chemical and physical changes, what misconceptions they may have, and what about this topic is interesting to them.

The Name of the Game Is Change

2 Is ice melting a physical change or a chemical change? Explain your answer.

1 Are physical changes or chemical changes responsible for the condition of this statue? Explain.

3 Cooking is an activity that involves physical and chemical changes. List some other activities and the changes involved.

ScienceLog

Think about these questions for a moment, and answer them in your ScienceLog. When you've finished this chapter, you'll have the opportunity to revise your answers based on what you've learned.

Changes All Around

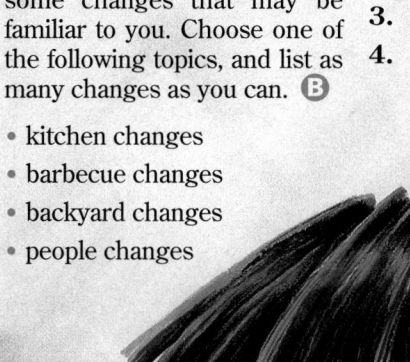

Everything Changes

There is an old saying, "The more things change, the more things stay the same." Does this make sense to you? Does everything change, or are there things that do not change? **A**

In small groups, examine some changes that may be familiar to you. Choose one of the following topics, and list as many changes as you can. **B**

- kitchen changes
- barbecue changes
- backyard changes
- people changes

What changes do you see taking place in the illustration below? Can you add any other kitchen changes? A list of changes has already been started for you: **C**

1. water boiling
2. plate breaking
3. ....
4. ....

Now that you've finished, do you have things to add to your thoughts about the saying that began this lesson? **D**

265

FOCUS

Getting Started

The teacher demonstration described in the Discrepant Event Worksheet (Teaching Resources, page 17) makes an excellent introduction to this lesson.

Main Ideas

1. Changes may be chemical or physical.
2. Bubbles, heat, light, and color changes may indicate a chemical change.
3. Chemical changes are not easily reversed.
4. Chemical changes form new substances, but physical changes do not.

TEACHING STRATEGIES

Answers to
In-Text Questions

A Sample answer: All things change, but some changes, such as changes of state, do not alter the chemical makeup of a substance.

B Sample answer: Backyard changes include the grass growing and being mowed, rain falling and soaking into the ground, and leaves changing color and falling.

C Additional changes include the following: eggs breaking and cooking, people moving, and toast burning.

D Answers will vary. Ask students to use the illustration on this page to give examples of changes in which the basic properties of a substance remain the same.

Teaching Strategies for Everything Changes are on the next page. ►

LESSON 1 ORGANIZER

Time Required
2 class periods

Process Skills
observing, comparing, classifying

New Terms
Chemical change—a change in which a new chemical is formed
Physical change—a change, such as breaking, dissolving, boiling, evaporating, or freezing, in which no new chemical is formed

Materials (per student group)
none

Teaching Resources
Discrepant Event Worksheet, p. 17
Resource Worksheet, p. 18
Transparencies 42 and 43
SourceBook, p. S130

Everything Changes, *page 265*

Other possible topics for students to consider include weather changes and changes in the classroom. Challenge students to explain why each of their examples is considered a change.

 **Cooperative Learning**
EVERYTHING CHANGES

Group size: 2 to 4 students
Group goal: to recognize changes
Positive interdependence: Each student should be responsible for one of the four categories listed.
Individual accountability: Give each student a specific example, and have him or her explain why it represents a change.

Physical and Chemical Changes Revisited

Through discussion and analysis, this exercise helps students to classify changes as either physical or chemical. Have students read the conversation between Mr. Calderon and his students and decide whether they agree with the decisions made by that class.

INDEPENDENT PRACTICE Divide the class into small groups. Have students in each group think of questions and answers to complete the table they began in their ScienceLog. The groups should then share their questions and answers with the rest of the class.

Answers to
In-Text Questions

Ⓐ For an overview of the results of Exploration 1, see pages 254–256 of this Annotated Teacher's Edition. Students should record their results in the form of a table if they have not already done so.

Ⓑ Color change, release of gases or bubbling, formation of new substances, and release of energy are all indications that a new chemical was formed.

Physical and Chemical Changes Revisited

In Exploration 1 on page 254, you looked at six substances. You also observed a change involving each one. Review what you did in the Exploration, and record the change you saw for each substance in a data table in your ScienceLog. Ⓐ

In the changes that you listed in your table, were new chemicals formed? How do you know whether a new chemical was formed during a change? These questions puzzled Mr. Calderon's class. Ⓑ

Mr. Calderon has just given his class the table shown below, which lists the changes they noticed in Exploration 1. Copy the table into your ScienceLog, and follow along with the class.

The Class's Table of Changes

Water evaporating · Salt dissolving · Sugar turning black when heated · Baking soda fizzing when vinegar added · Chalk breaking · Chalk bubbling when vinegar added · Paper burning when heated

1. Has a new chemical been formed?
2.
3.
4.
5.
6.
7.

Mr. Calderon: In row 1, I have written the question, Has a new chemical been formed? Let's put a √ under each change when you believe that a new chemical is formed as a result of the change. An X will mean that no new substance or chemical is formed. Jorge?

Jorge: I saw black smoke when we burned the paper. I think a new chemical must have been formed.

Sue: We saw bubbles when the vinegar was added to baking soda. Does that mean we should put a check under baking soda?

Hal: The burnt sugar smelled bad. Something new must have been formed in that change.

Meeting Individual Needs

Second-Language Learners

Have students make posters or scrapbooks with examples of chemical and physical changes. They can use their own artwork or select pictures from magazines. Have them write short captions for each example to describe what is taking place.

 A Resource Worksheet is available to accompany the material on this page (Teaching Resources, page 18).

Later, after much discussion about whether a new substance is formed when water evaporates, they agreed to finish the table as shown below. Do you agree with their new table?

The Class's New Table of Changes

	Water evaporating	Salt dissolving	Sugar turning black when heated	Baking soda fizzing when vinegar added	Chalk breaking	Chalk bubbling when vinegar added	Paper burning when heated
1. Has a new chemical been formed?	X	X	✓	✓	X	✓	✓
2.							
3.							
4.							
5.							
6.							
7.							

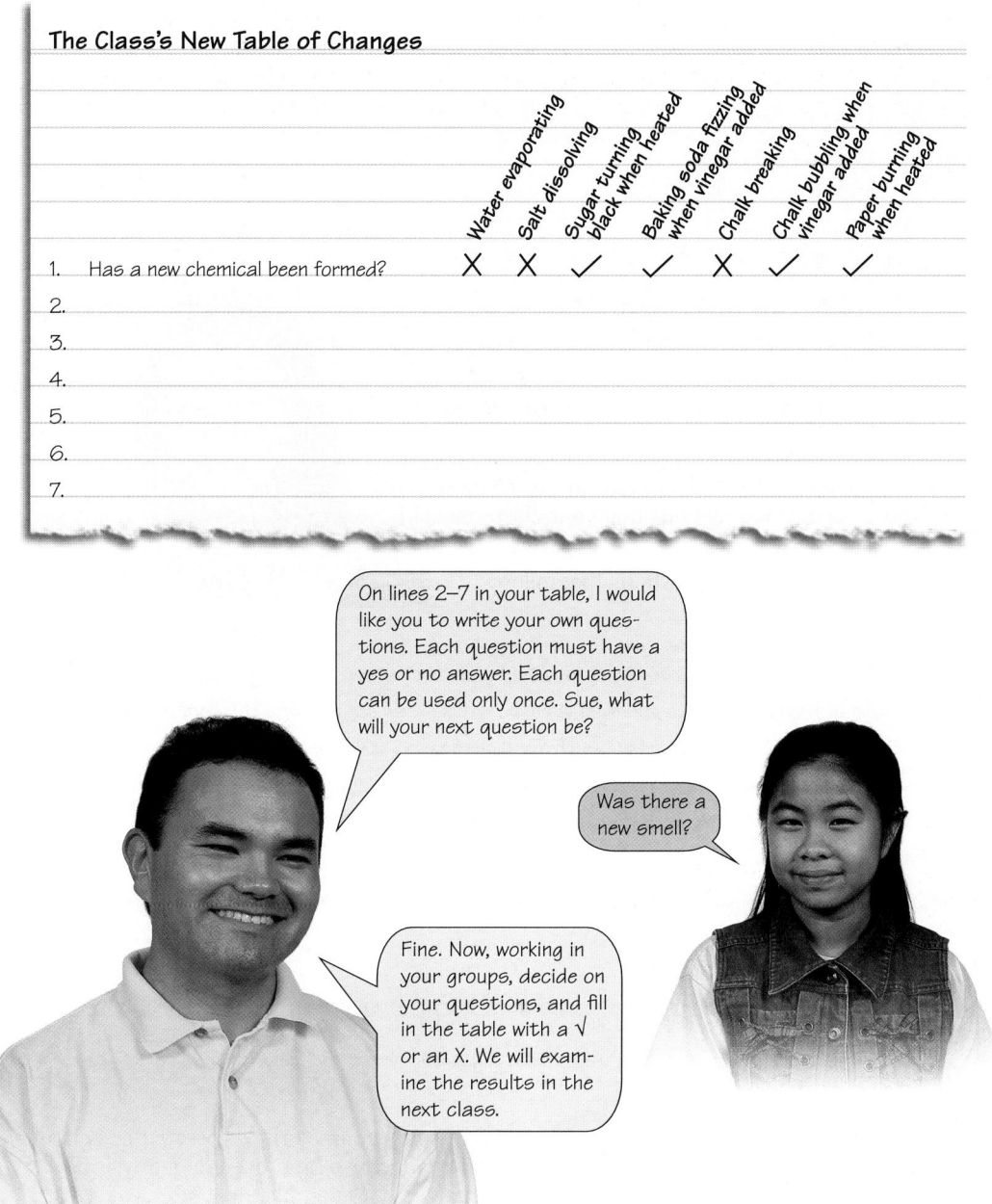

On lines 2–7 in your table, I would like you to write your own questions. Each question must have a yes or no answer. Each question can be used only once. Sue, what will your next question be?

Was there a new smell?

Fine. Now, working in your groups, decide on your questions, and fill in the table with a √ or an X. We will examine the results in the next class.

Now finish the table you copied into your ScienceLog earlier.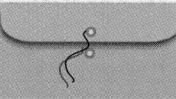

267

Integrating the Sciences

Life and Physical Sciences

Asbestos is a general name for six types of minerals that separate easily into long, flexible fibers. These fibers make useful building materials. One important physical property of asbestos is that in its fibrous form, it can be spun into yarn, cloth, or tape. In addition, it's fireproof, it's a good thermal insulator, and when wound like a rope, it's as strong as some types of steel. However, many health problems have been linked to the use of asbestos. Have students research asbestos and the health hazards related to it. In addition, have students find information about materials now used in place of asbestos.

The Class's New Table of Changes

The new table should correctly identify chemical changes with checks and physical changes with crosses. Remind students to copy tables into their ScienceLog and not to write in their book. Some sample questions are given in More Thought Required on the following page.

★ **Transparency 42 is available to accompany The Class's New Table of Changes.**

Answers to *In-Text Questions*

C Students should agree. Evaporation, dissolution, and breaking are all physical changes. A change in color, burning, or the presence of bubbles indicates a chemical change.

D Sample questions include the following: Was heat released? Was there a change in color? Were any gases released? Did any matter seem to disappear? Did any solid turn into a liquid?

PORTFOLIO

Students may wish to include in their Portfolio their table or a summary of what they learned. Suggest that students copy the table from The Class's New Table of Changes into their Portfolio to review as they complete the unit.

Homework

Tell students to make a list of five physical changes and five chemical changes that they can identify at home. Their list may include physical changes such as matter changing state, food being cut, and objects tearing or breaking. Chemical changes may include fuels burning, food cooking, digestion, and photosynthesis.

Answers to
More Thought Required,
pages 268–269

1. Students can compare their list with the list in the text.

2. From Joy's list, the three most valuable questions are the following:
   • Was there a color change?
   • Were any bubbles formed?
   • Was the change easily reversible?

3. Answers may vary, but possible responses include the following:
   • The smell of a new gas was detected. If a chemical change did not occur, you can still detect the original substances. New and different properties were observed.
   • Joy was right. The same clues cannot be used each time to identify chemical changes because the changes that are observed depend on the properties of the chemicals involved.
   • Yes, heat can cause a physical change. When an ice cube is heated, it melts. This is a physical change.
   • Melting, freezing, and dissolving are examples of physical changes in which a substance seems to disappear or a new substance seems to appear.
   • No, not all physical changes are easily reversible. For example, sawdust cannot easily be changed back into lumber.
   • Answers will vary.

More Thought Required

1. Mr. Calderon's class made a list of questions. Here are some of the questions that Joy copied into her ScienceLog. Add them to your table. Did you have different questions?

   • Was there a color change?
   • Was there any smell?
   • Was the change easily reversible?
   • Was heat given off?
   • Was light given off?
   • Was heat required for the change to happen?
   • Can the original chemical still be used?
   • Were any bubbles formed?

2. The answers to some of your questions may provide valuable clues about whether a new chemical was formed. What are the most important questions that your class can ask?

3. Read the following excerpt from Joy's ScienceLog notes, and then answer the questions that follow.

   Today we studied changes. We divided them into two types. In one type, a new chemical was formed. This type of change is called a **chemical change**. Burning is an example of a chemical change. If no new chemical is formed, then the change is a **physical change**. Breaking, dissolving, boiling, evaporating, and freezing are all physical changes.

   Our class made a list of the clues that showed that a chemical change occurred:
   • A new color appears.
   • Heat or light is given off.
   • Bubbles form.
   • The change is not easily reversible.

   The only problem is that you can't use the same clues for each chemical change. Life sure is complicated!

■ Have you and your classmates identified other ways of detecting chemical changes? What are they?

■ Was Joy right? Why can't you use the same clues to identify every chemical change?

■ Can some characteristics present in a chemical change also be present in a physical change? For example, heat often causes a chemical change. Can it also cause a physical change? If it can, give an example.

268

CROSS-DISCIPLINARY FOCUS

Language Arts

Read the following passage to students:

The *Hindenburg,* an early airship, exploded over Lakehurst, New Jersey, on May 6, 1937. It was filled with the chemical hydrogen, which ignited. Thirty-six people were killed in the accident. Modern airships and blimps use helium, a nonreactive chemical, instead of hydrogen so that tragedies like that of the *Hindenberg* can be avoided.

Which of the following supports the main idea of this passage?

a. Helium is a safe chemical to use in airships because it will not ignite. *(Correct)*

b. Airships were a popular means of travel in the 1930s.

c. The *Hindenburg* was 245 m long and 41 m wide.

d. Hydrogen and helium are two chemicals that will float in the air.

- In a chemical change, a chemical may seem to disappear, or a new one may form. Can you think of a physical change in which a chemical seems to disappear or a new substance seems to form?

- Butter melts but can become solid again. Water evaporates but also condenses. These are both physical changes. Are all physical changes easily reversible?

- At the start of this lesson, you listed a number of changes. Now underline all of the changes that you think are chemical changes.

Physical and Chemical Words

Kelly, who was also in Mr. Calderon's class, had a very good idea. She decided to make two vocabulary lists, one for each of the two types of changes, physical and chemical. Here are Kelly's lists. Copy them into your ScienceLog, and give an example of a change associated with each word. But think hard—be sure you have the right type of change. **Ⓐ**

Can you add any words and examples to the two lists? **Ⓑ**

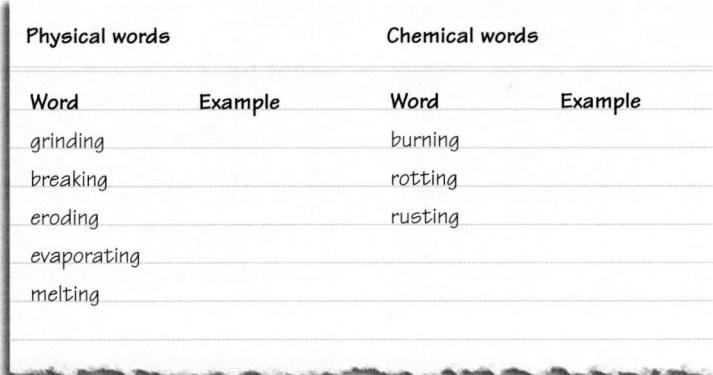

Physical words		Chemical words	
Word	Example	Word	Example
grinding		burning	
breaking		rotting	
eroding		rusting	
evaporating			
melting			

Now look up the word *state* in the dictionary. It certainly has many meanings! The following sentence uses the word *state* in a scientific way: Water exists in three states—in the solid state, as ice; in the liquid state, as water; and in the gaseous state, as water vapor or steam. **Ⓒ**

What words in the "Physical words" list above do you associate with a change of state? **Ⓓ**

▲ What physical and chemical changes are occurring in this photograph? **Ⓔ**

269

FOLLOW-UP

Reteaching

Have students write short descriptive paragraphs about some materials of their choice using the physical and chemical word lists they developed in the lesson. Ask them to be sure to describe what the materials were like before and after the changes.

Assessment

Below is a list of physical and chemical changes. Ask students to copy the list and to write the letter *C* beside each chemical change and the letter *P* beside each physical change. Students should then explain their answers.

a. A solid stretches when pulled gently. *(P)*

b. A powder dissolves in water. *(P)*

c. A liquid changes color when exposed to air for a long time. *(C)*

d. A solid shatters into small pieces when hit with a hammer. *(P)*

e. A gas pops when a burning match is brought near it. *(C)*

f. A liquid evaporates quickly in a warm location. *(P)*

g. A liquid freezes at −20°C. *(P)*

Extension

Have students research and compare chemical and physical changes that might be used by fashion designers in developing a new article of clothing. Examples include bleaching, dyeing, cutting, weaving, stretching, shrinking, and painting.

Closure

Have students form small groups to design a word search or a crossword puzzle to review the terms that describe chemical and physical changes.

Homework

The Reteaching activity on this page makes an excellent homework assignment.

 Transparency 43 is available to accompany Physical and Chemical Words.

Answers to
In-Text Questions and Caption

Ⓐ An example of a completed table is given on page S239.

Ⓑ In addition to the new entries in the table on page S239, other words and examples for physical changes might include sublimating dry ice, tearing paper, or dissolving sugar in water. Additional chemical changes might include curdling milk.

Ⓒ Some possible meanings of the word *state* include "a political division," "an emotional condition," or "to express in words."

Ⓓ Words associated with a change of state include *evaporating, melting, condensing, drying,* and *freezing.*

Ⓔ The chemical changes occurring in the photograph include wood rotting, and a pail rusting. The physical changes include wood being cut and water evaporating.

FOCUS

Getting Started

Wearing latex gloves, perform the following demonstration for students: Prepare solutions of copper sulfate (1 g of $CuSO_4$ in 15 mL of water) and sodium hydroxide (0.50 g of NaOH in 125 mL of water). In a test tube, add 40 drops of NaOH and then 10 drops of $CuSO_4$. The precipitate that forms is copper hydroxide. Write the word *precipitation* on the board. Ask students if they have ever heard this word before. *(Many may have heard it as another word for rain.)* Explain that in chemistry, precipitation has two meanings. One meaning refers to physical change from vapor to liquid, as when clouds produce rain. The other meaning refers to a chemical change—the formation of a solid in a chemical reaction. The solid formed is called a precipitate. Precipitation of this kind is a sign that a chemical change has occurred.

WASTE DISPOSAL ALERT   Combine the remaining copper sulfate solution with the contents of the test tube and the remaining sodium hydroxide solution. Mix well by stirring. Then filter the mixture. Wrap the filter paper and precipatate in old newspaper and place in the trash. Using 0.1 M hydrochloric acid, adjust the pH of the filtrate to between 6 and 8, and pour down the drain.

Main Ideas

1. Color changes, release of gases, formation of new substances, and release of energy are all signs that a chemical change may have occurred.
2. Identifying the presence of a new substance is the only true test for a chemical change.

Teaching Strategies for Exploration 1 are on the next page. ▶

We know that there are two basic types of changes, chemical and physical, but how do you tell them apart? What are the signs of a chemical change? a physical change? The following Explorations will help you learn to recognize the signs of chemical changes.

EXPLORATION 1

More Evidence of Chemical Changes

Let's examine more changes and look again for ways of recognizing whether a chemical change has occurred. You will also observe some new evidence of chemical changes here. But be careful—there are physical changes hidden among the rest. Construct a data table like the one below to record your findings. Decide whether each change is physical or chemical, and state your reasons.

Change	Physical	Chemical	Reasons or evidence
Limewater turns milky.		✓	1. Color change occurs. 2. A new substance is formed. 3. Change is difficult to reverse.

After you have finished the tests, clean up your area and wash your hands with soap and water.

TEST 1

You Will Need

- 1 g copper sulfate
- water
- a graduated cylinder
- an iron nail
- a beaker
- a stirring rod
- latex gloves

What to Do

Wear latex gloves in addition to safety goggles and an apron when handling copper sulfate. Add 1 g of copper sulfate to 10 mL of water in a beaker, and stir. Place an iron nail in the copper solution.

TEST 2

You Will Need

- an eyedropper
- vinegar
- a graduated cylinder
- milk
- a beaker

What to Do

Add 3 dropperfuls of vinegar to 25 mL of milk in a beaker.

TEST 3

You Will Need

- an eyedropper
- vinegar
- a piece of eggshell
- a bowl

What to Do

Add a dropperful of vinegar to a piece of eggshell in a bowl.

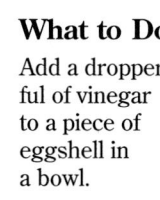

LESSON 2 ORGANIZER

Time Required
3 class periods

Process Skills
observing, comparing, analyzing, inferring

New Term
Precipitate—a solid formed as the result of a chemical change

Materials (per student group)
Exploration 1, Test 1: 1 g of copper sulfate; 10 mL of water; 10 mL graduated cylinder; iron nail; beaker; stirring rod; safety goggles; lab aprons; latex gloves; **Test 2:** 100 mL graduated cylin-

der; eyedropper; 10 mL of vinegar; 25 mL of milk; beaker; safety goggles; lab aprons; **Test 3:** eyedropper; eggshell; 10 mL of vinegar; bowl; safety goggles; lab aprons; **Test 4:** candle; a few matches; jar lid; aluminum pie plate; pair of tongs; small ball of modeling clay; sample of candle wax; safety goggles; lab aprons; **Test 5:** drinking straw; small beaker; 10 mL of limewater; safety goggles; lab aprons (See Advance Preparation on page 249C.); **Test 6:** 20 mL of lemon juice; several toothpicks; 2 sheets of paper; candle;

continued ▶

TEST 4

You Will Need

- a candle
- matches
- an aluminum pie plate
- tongs
- a jar lid
- modeling clay
- a sample of candle wax

What to Do

Place the candle in the jar lid, and hold it in place with modeling clay. Put a sample of candle wax on a pie plate, and heat it over the lit candle. (Use tongs to hold the pie plate.)

TEST 5

You Will Need

- a drinking straw
- a small beaker
- limewater

What to Do

Using a drinking straw, blow slowly into a small beaker one-quarter full of limewater.

Caution: Blow slowly so that you do not splash limewater on yourself or others. Do not drink the limewater.

TEST 6

You Will Need

- lemon juice
- a toothpick
- paper
- a candle
- matches
- modeling clay
- a jar lid
- a bucket of water

What to Do

With a toothpick, write a word on a piece of paper using lemon juice. Place the candle in the jar lid, and hold it in place with modeling clay. Heat the paper gently over the candle flame.

Caution: Do not burn the paper. Have a bucket of water nearby for safety.

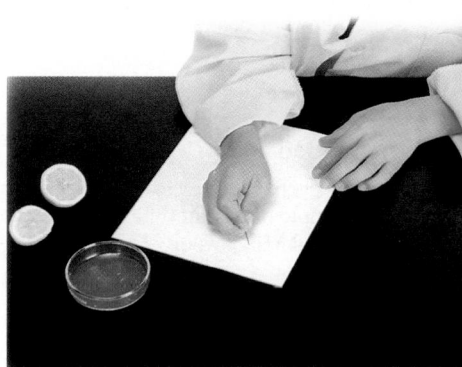

Exploration 1 continued ▶

271

ORGANIZER, *continued*

a few matches; small ball of modeling clay; jar lid; bucket of water; safety goggles; lab aprons; **Test 7:** several drops of iodine starch-test reagent; eyedropper; starch (sample of potato, cereal, or bread); bowl; safety goggles; lab aprons; latex gloves (additional teacher materials: 25 mL of 0.1 M sodium thiosulfate, a few sheets of newspaper; see the Waste Disposal Alert on page 21.); **Test 8:** 5 g of baking powder; small bowl or beaker; eyedropper; 5 mL of water; safety goggles; lab aprons

Exploration 2: beaker; 1 L of water; 100 mL graduated cylinder; effervescent tablet; carbonated drink (bottled); few drops of vinegar; 15 mL of baking soda; 10 mL of molasses; packet of yeast; hot plate; eyedropper; stirring rod; safety goggles; lab aprons; oven mitts

Teaching Resources
Exploration Worksheets, pp. 19 and 23
Transparencies 44 and 45

TEACHING STRATEGIES

EXPLORATION 1

For this Exploration, the class can be divided into groups of three or four students and assigned a specific change to investigate and report on. Or the changes can be set up at different stations around the room, and students can move from station to station to complete each activity.

Of the eight activities, only one is a physical change. You may wish to add others, such as diffusion, separation of iron from sulfur with a magnet, evaporation of alcohol, and dissolving of sugar. The answers are given in a table on page S240.

WASTE DISPOSAL ALERT The procedure for disposing of the iodine-stained materials from Test 7 is given in the margin on page 21.

PORTFOLIO
Suggest that students use their Portfolio to record their thoughts about this Exploration. They may wish to express any difficulties they had performing the activities, as well as any surprises they may have experienced.

⭐ **An Exploration Worksheet is available to accompany Exploration 1 (Teaching Resources, page 19). Transparency 44 is also available.**

Answers to *Chemical Tests*

- Copper will form a reddish brown precipitate.
- Limewater will turn milky.
- The iodine will turn blue-black.

Multicultural Extension

Chemists From Around the World
Chemistry, like all modern sciences, is an international affair. People in many countries around the world have made important contributions to modern chemistry. Have students research the life and achievements of a prominent chemist. Encourage them to choose people from all parts of the world. To help them get started, you can refer them to local colleges and universities for a list of chemists. Information gathered by students can be used to assemble a booklet or bulletin-board display of important chemists from around the world. If possible, invite a chemist to your classroom, or take students to visit a chemist's laboratory.

Meeting Individual Needs

Learners Having Difficulty
Many students are fascinated by dramatic chemical reactions such as explosions. Perform the following demonstration to capture student interest: Put 25 mL of baking soda into the corner of a zippered sandwich bag. Tie off that corner with a twist tie or string. Pour 60 mL of vinegar into the other corner of the bag, and then seal the top so that the bag is airtight. Grasp the bag at the top and untie the corner that has the baking soda in it. Tip the bag back and forth to thoroughly mix the chemicals. Tell students that a chemical change involving the vinegar and baking soda is releasing carbon dioxide. The bag will begin to inflate. Warn students that the bag might even burst. Have students research the causes of some similar reactions and present what they learn to the class. If students wish to perform an experiment to demonstrate what they have discovered, be sure that you review all experimental procedures for safety beforehand.

TEST 7

You Will Need

- iodine solution
- an eyedropper
- starch (sample of potato, cereal, or bread)
- a bowl
- latex gloves

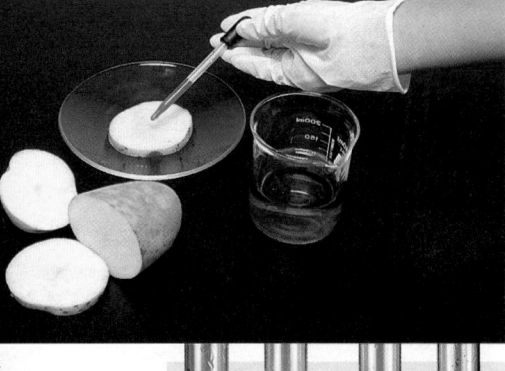

What to Do

Wear latex gloves in addition to safety goggles and an apron when handling iodine. Add a drop or two of iodine solution to a small sample of starch. (A slice of potato, some cereal, or a piece of bread may be used.)

TEST 8

You Will Need

- baking powder
- a small bowl or beaker
- an eyedropper
- water

What to Do

Add a few drops of water to a small sample of baking powder in a small bowl or beaker.

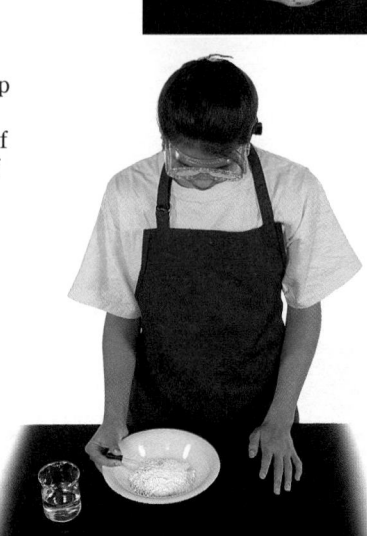

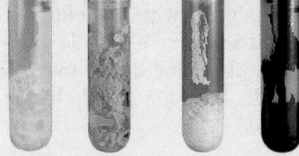

A New Word

Did you notice new evidence that a chemical change had occurred? In Tests 2 and 5, you observed a *precipitate*—a solid formed as the result of a chemical change. This is another clue that you can add to your list of signs that a chemical change has occurred. Some other precipitates are shown above.

Chemical Tests

Chemical changes can be used to help identify substances. Here is a table of some common tests for chemical changes. Copy it into your ScienceLog, and complete it by recording the result expected for each test. You have already seen these chemical changes in this unit.

Test for	Procedure	Result
copper in any substance	Place an iron nail in a solution of the substance.	
carbon dioxide	Pass the gas through limewater.	
starch	Add iodine solution.	

Did You Know. . .

Most chemical compounds occupy more space in their liquid form than in their solid form. Water, however, expands when it freezes. Therefore, ice floats in water. This is why icebergs float and lakes freeze over in winter.

Bubble Watch!

Formation of bubbles is often, but not always, evidence of a chemical change. Try these changes, and decide which of the bubbles you see resulted from a chemical change. Clean up your area, and wash your hands with soap and water when you have finished.

You Will Need

- a beaker
- water
- a graduated cylinder
- an effervescent tablet
- a carbonated drink
- vinegar
- baking soda
- 10 mL of molasses
- a packet of yeast
- a hot plate
- oven mitts
- an eyedropper
- a stirring rod

What to Do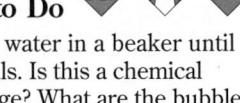

1. Heat water in a beaker until it boils. Is this a chemical change? What are the bubbles made of? How could you prove your answer?

   Caution: Wear oven mitts when handling hot objects.

2. Examine a recently opened carbonated drink. Are you observing a chemical change?

3. Drop one-quarter of an effervescent tablet into water. Is this a chemical change? What are the bubbles made of? Can you prove it?

Exploration 2 continued ▶

273

Multicultural Extension

George Washington Carver

People from many different cultures have made important contributions to modern chemistry. George Washington Carver was a particularly outstanding African American scientist in the early twentieth century. His work with agricultural products revolutionized the South's economy. Have students research Carver and describe his scientific contributions.

★ An Exploration Worksheet is available to accompany Exploration 2 (Teaching Resources, page 23).

This Exploration is best done at several activity stations, although some activities could be performed as demonstrations. If you do some activities as demonstrations, it is best to do step 5 (fermentation) first. Bubbles should be seen at the end of 40 minutes. You will obtain the best results if the water is at 37°C. Students can then perform the limewater test to prove that the gas given off by the yeast is carbon dioxide.

Explain to students that fermentation involves the use of yeast to produce alcohol from sugar. The gas given off is carbon dioxide. Although fermentation involves the use of living organisms, it still meets the criteria for chemical change.

SAFETY TIP Always perform demonstrations and activities before your students do. This ensures the safety and success of the activity.

Answers to
Exploration 2, pages 273–274

1. The boiling water is not undergoing a chemical change. The bubbles are water vapor. The vapor could be condensed on a glass beaker, and the drops that are formed could be collected. If 1 mL of the liquid had a mass of 1 g, this would provide evidence that the substance is water.

2. The bubbles are not evidence of a chemical change. They form from carbon dioxide that is dissolved under pressure in the liquid.

3. The bubbles are evidence of a chemical change. They are carbon dioxide. Students should identify a way to capture the gas (such as by collecting it in a balloon) and state that a limewater test will prove the identity of this gas.

4. The bubbles are not evidence of a chemical change; they are previously dissolved air bubbles that are released from the water as it warms.

5. Yes, this is a chemical change. A gas is given off by the yeast, and the volume of the mixture increases.

6. Yes, the bubbles are carbon dioxide. However, step 2 involves a physical change, while step 6 involves a chemical change.

Answers to
Spotting a Chemical Change

Possible answers to the table include the following: color change; bubbles (gas formation); heat, light, or both are given off; formation of a precipitate; odor changes; formation of a new substance; and change is not easily reversible. All of these clues are observed sometimes. Evidence that a new substance has formed is the best indication of a chemical change, but even this is not always directly observable.

ENVIRONMENTAL FOCUS

Many of the chemicals used in the manufacturing process are not part of the final product. Ask: What should be done with these chemicals? Is it ever possible to really throw something away? Simply burying them doesn't always work. Encourage discussion about what can be done to prevent such problems in the future.

 Transparency 45 is available to accompany the material on this page.

Homework

The Reteaching activity on this page makes an excellent homework assignment.

4. Let a cold glass of water reach room temperature. Do the bubbles indicate a chemical change?

5. In a beaker, add 10 mL of molasses to 50 mL of water. The exact amounts you use are not critical. Make a yeast mixture by stirring half of a packet of yeast into 25 mL of warm water. Add this to your molasses solution. Place the solution in a warm spot, and record all changes over the next few days. Does a chemical change occur? What evidence can you give?

6. Add a few drops of vinegar to a small sample of baking soda. Are these bubbles made of the same substance as the ones you encountered in step 2?

Spotting a Chemical Change

By now, you should be a champion "chemical-change detective." Using a table similar to the one below, make a list of all the clues that you would use to recognize a chemical change. Can evidence of a chemical change be observed in every chemical change or only in some? Do not write in this book.

What to Look for in Chemical Changes

Evidence of a chemical change	Sometimes observed	Always observed

FOLLOW-UP

Reteaching

Have students write a paragraph in which they explain how to tell when a chemical change has taken place.

Assessment

Ask students to explain how they would test for the presence of
a. carbon dioxide. (*Pass the gas through limewater.*)
b. water. (*Find the mass of 1 mL of the liquid to see if it measures 1 g.*)
c. starch. (*Add iodine starch-test reagent to the substance.*)

Extension

Have students demonstrate how starch tests, fermentation, or antacids work. Approve all student designs for safety before they proceed.

Closure

Show students two iron nails—one new and one rusty. Have students discuss whether any chemical or physical changes have occurred in either of the nails. If they decide that a change has occurred, ask them to explain which type of change occurred and how they recognized that change.

Classifying Changes

On the Menu: Changes

There are at least 10 physical and 7 chemical changes concealed in this story. Find them and list them in a table.

Ramón and Marta decided to make supper for their parents and grandmother. While Marta cut up the vegetables and cheese for the salad and boiled an egg, Ramón placed ice cubes into a glass of lemonade made from a powdered concentrate.

They had also decided to make and cook some hamburger patties and to broil some frozen french fries. Earlier in the day, Marta had made gelatin by mixing the powder in hot water and then allowing it to cool in the refrigerator until it set.

When the meat was no longer red and appeared to be completely cooked and the french fries were golden brown, they called their parents and grandmother. After the meal, it was time for dessert and coffee. Marta and Ramón boiled some water and poured it into mugs. Then they added instant coffee powder and sugar. Almost everything was ready. But where was the milk? They searched for it until they remembered that they had left it out of the refrigerator earlier in the day. It smelled sour and had already started to curdle. So they had to serve the coffee black!

Later, the scraps were placed in the garbage. Some things were added to the compost heap in the backyard. Papa washed the dishes in steaming hot water, and Ramón dried them. Marta finished an oil painting, while Mama read a book. The entire family agreed that they had enjoyed their supper thoroughly, though Abuelita did have some trouble with her stomach. However, an antacid tablet soon helped soothe the problem. All in all, it was a successful venture for Marta and Ramón.

Write your own story containing examples of hidden physical and chemical changes. Then challenge a friend to find the changes. **Ⓐ**

LESSON 3 ORGANIZER

Time Required
1 to 2 class periods

Process Skills
observing, communicating, organizing, analyzing

New Terms
none

Materials (per student group)
none

Teaching Resources
Resource Worksheet, p. 25
Transparency 46

FOCUS

Getting Started

Insert a small candle into a wide-mouthed jar so that the top of the flame is below the top of the jar. Light the candle, and then sprinkle baking soda and approximately 2 tablespoons of vinegar into the jar. Ask students to explain what happens. *(Vinegar reacts with baking soda to produce carbon dioxide. The gas pushes the oxygen out of the jar, causing the flame to go out.)* Ask: In what line of work would knowledge of this chemical change be helpful? *(Firefighting)*

Main Idea

Knowledge of chemical and physical changes applies to everyday life.

TEACHING STRATEGIES

On the Menu: Changes

Point out that some changes such as solidifying gelatin, are partly physical and partly chemical.
Sample answers are listed below.

Answers to
On the Menu: Changes

Physical changes: cutting vegetables, cutting cheese, melting ice cubes, making lemonade, making hamburger patties, dissolving gelatin, boiling water, dissolving coffee, dissolving sugar, cleaning dishes

Chemical changes: boiling egg, cooking hamburger patties, broiling french fries, milk souring and curdling, food digesting, using an antacid tablet, composting garbage

---

Answer to
In-Text Question

Ⓐ Answers will vary. Encourage creativity by asking students not to repeat any examples that have already been mentioned. You may wish to have a third student verify all of the physical and chemical changes listed by students.

Answers to
More Changes to Classify

Change/Type/Reason

- Steel wool/physical/The steel wool removed the black material but did not change it into anything else.
- Lightning/physical/Static buildup of electricity within the clouds caused the lightning to flash.
- Limestone/chemical/Bubbles show that a chemical change occurred.
- Back steps/chemical/The wood is reacting with substances in the environment to create new substances.
- Lemon/chemical/This is evidence of a precipitate and therefore is evidence of a chemical change.
- Gasoline/physical/The gasoline evaporated. The gasoline molecules are still in the air.
- Gravy/physical/The gravy solidified in the refrigerator. The change is easily reversible.
- Red meat/chemical/The color change is an indication of a chemical change.
- Cavity/chemical/Acids, other chemicals, and bacteria caused the tooth to decay. The change is not reversible.
- Paint/chemical/The paint on the car reacted with something in the environment. The color changed, and the change is not easily reversible.

Who Uses Chemical Changes?

INDEPENDENT PRACTICE This activity builds on students' growing awareness of the impact of chemical changes in their lives. When the activity is completed, have students make presentations to the class.

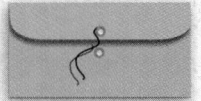

PORTFOLIO
Suggest that students include notes and comments pertaining to their presentation in their Portfolio.

Homework

The Resource Worksheet that accompanies the material on this page makes an excellent homework activity (Teaching Resources, page 25). If you choose to use this worksheet in class, Transparency 46 is available to accompany it.

More Changes to Classify

Now your knowledge of the two types of changes will be put to the test! Classify the changes listed below as either physical or chemical, and then give a reason for your decision. The first one has been done for you. Write in your ScienceLog, not in your book.

Change	Type	Reason
A newspaper yellowed after a few weeks.	chemical	Color change indicates a chemical change.
The steel wool turned the black pot a shiny silver color.		
Lightning flashed across the sky.		
Acid caused limestone to fizz.		
The back steps are rotting.		
The piece of lemon turned the tea cloudy.		
The spilled gasoline dried, but it left a bad odor in the room.		
The gravy in the refrigerator jelled.		
The red meat turned brown as it cooked.		
You got a cavity in a tooth.		
The paint on the car turned dull.		

Who Uses Chemical Changes?

Who uses chemicals and chemical changes? Almost everyone! Cooks, nurses, parents, druggists, chemists, farmers, swimming pool managers, tropical fish owners, garage owners, joggers, and on and on. Interview someone from your neighborhood. What chemicals does he or she use? What changes are involved in his or her line of work? Before your interview, consider the following:

- Whom should I interview?
- What should I ask? Make a list of questions ahead of time!
- How should I record my information?
- What form will my report to the class take?

What chemical changes might be involved in getting a new hairstyle? Ⓐ

Answer to Caption

Ⓐ Examples of products used in hairstyling that produce chemical changes include permanent waves, hair straighteners, hair conditioners, bleaches, and dyes.

Meeting Individual Needs

Gifted Learners
Have interested students form a photojournalism group to document the chemical changes that take place in the classroom. Or have student photographers form teams to record categories of change such as changes in the home, changes at school, or changes in nature. Students can compile their photos in an album and include descriptions of each change.

Can You Be Fooled by the Clues?

In earlier lessons, you learned some words associated with physical and chemical changes. You also discovered several clues by which physical and chemical changes can be recognized. Here are the clues that show that a chemical change has occurred:

- There is a color change.
- A gas is given off.
- A solid is formed.
- Heat or light, or both, is given off.
- A new substance with new identifying properties is formed.

But be careful! Can the clues fool you, as they fooled the students who made the statements below? You be the teacher, and explain (in writing) why each of the following conclusions is wrong:

Ice is often a different color than liquid water. This is an example of a chemical change.

When I open a carbonated drink, it fizzes. This shows that a chemical change has occurred.

Salt is white. When it dissolves in water, it becomes colorless and invisible. This color change is evidence that a chemical change has occurred.

Heat and light are given off by a light bulb. Therefore, a chemical change is taking place in the light bulb.

The sawdust formed when wood is cut looks quite different from the original tree. A chemical change occurs during the cutting.

277

FOLLOW-UP

Reteaching

Have students write or make posters illustrating the "life story" of an object such as a candle, a nail, a piece of wood, a leaf, a rock, or an ice cube. Ask them to identify the chemical and physical changes involved at each stage.

Assessment

Have students use a table to classify the following changes and explain the reasons for their choice:

a. When the gas emissions from some industries mix with the water in the atmosphere, acids are formed. *(Chemical; a new substance is formed)*.

b. When Rick mixed the two clear liquids, a yellow precipitate formed. *(Chemical; a precipitate indicates a chemical change.)*

c. The repeated freezing and thawing of water in the cracks in the rocks caused the rocks to break up. *(Physical; the rocks were broken apart, not changed into another substance.)*

Extension

Have students do research on the different methods used by firefighters to put out fires. Students should compile their findings in a report, with special emphasis on the firefighting methods that involve chemical changes.

Closure

Have students write a report about a career that requires knowledge of chemicals and chemical changes. They should include the educational background required, job responsibilities, and two specific chemicals that a person with this job would encounter. Some possible choices include chemical engineer, chemist, pharmacist, environmental ecologist, physician, and food scientist.

Answers to
Can You Be Fooled by the Clues?

- Ice and water look different; however, they are just different states of the same substance.
- Not all bubbles are evidence of chemical change. These bubbles were dissolved in the soft drink. This is an example of a physical change.
- This time, a color change does not indicate a chemical change. Dissolving is a physical change.
- Although heat and light are given off by the light bulb, it is not a chemical change. Electricity that is produced elsewhere passes through the filament in the bulb, causing the bulb to give off light and heat. There is no other change taking place within the light bulb.
- The sawdust cannot be changed back into wood, but this is a physical change. Both the sawdust and the wood are the same material.

LESSON

4

Chemical and Physical Properties

Getting Started

Show students a piece of limestone rock. Ask them how they
could identify this substance. *(They
could examine its physical and chemical
properties.)* Put a dropperful of vinegar
on the rock. Ask students to describe
what is happening. *(Bubbles form.)* Add
vinegar to a piece of chalkboard chalk.
(Bubbles form.) Explain to students that
these similar reactions are a clue that
the rock and the chalk may be made of
the same substance. In fact, both are
made of calcium carbonate.

Main Ideas

1. Properties can be used to identify a
   substance.
2. Properties can be classified as either
   chemical or physical.
3. Properties determine the uses of
   substances.

Mysterious Drops

Have students observe the changes
occurring on the side of a cold, empty
glass before evaluating the six plans.

SAFETY ALERT If you decide to perform any of
these tests, remind students that
they should never taste or ingest unknown chemicals.

Answers to
Mysterious Drops

Plans 2 and 4 would provide good
evidence because they rely on measurable properties of water. Plans 1 and
3 should be avoided for safety reasons.
None of these plans by themselves
would prove that the substance is
water.

Mysterious Drops

Annette's class had a problem. That is, Annette's teacher gave
them a problem to solve. A cold, empty glass was taken from a
refrigerator. A few minutes later, droplets began to form on the
outside of the glass.

"Water drops!" said one of Annette's classmates.

"Prove it," said the teacher.

How could Annette and her classmates prove that the drops
were water? What would you do?

After some discussion, the class came up with the suggestions
listed below. Working with another student, consider each suggestion, and decide whether it is a good plan of action.

Plan 1 Since it looks, feels, and smells like water, decide that it
must be water.

Plan 2 Collect enough of the drops to find their mass. If 1 mL of
the sample has a mass of 1 g, then it must be water.

Plan 3 Collect a few drops and taste them. If they taste like
water, then they are water.

Plan 4 Collect enough drops, and find the boiling point of the
sample. If it boils at 100°C, then it's water.

Plan 5 Add a few drops of water to baking soda. If no bubbles
form, then it must be water.

Plan 6 Try other chemical tests. For example, see if you can
burn a sample or try to make it react with iron, aluminum, or substances such as paper. If the drops do not
burn or react with these substances, then they are probably water.

Which plan or plans seem most likely to work? Why? Are
there some plans you should avoid? Using these tests, could you
prove that the droplets were water?

Substances are identified by their properties. In order to identify a substance, you should examine its **properties**. You should
look for the properties of salt to help you identify salt, and so on.
How would you define the term *property*? **A**

In Unit 4, you learned that matter has two basic kinds of
properties: chemical and physical. These are described for one
chemical—water—on the right.

Physical properties of
water describe how water
looks, feels, and behaves
and what physical changes
water can undergo.

Chemical properties of
water describe what
chemical changes water
can and cannot undergo.

LESSON 4 ORGANIZER

Time Required
2 to 3 class periods

Process Skills
observing, classifying, analyzing

Theme Connections
Structures, Energy

New Terms
Alloy—a mixture of two or more metals
Properties—characteristics that dis-
tinguish one substance from another

Materials (per student group)
Exploration 3: 5 g of copper chloride
crystals; square of heavy aluminum foil,
10 cm × 10 cm; stirring rod; 100 mL of
water; small beaker; evaporating dish;
safety goggles; lab aprons; latex gloves
Guess That Substance!: 1 index card
per student

Teaching Resources
Theme Worksheet, p. 26
Exploration Worksheet, p. 28

Properties of One Chemical: Aluminum

How many uses of aluminum are you aware of? As with all substances, the uses of aluminum are determined by its properties. As you read the following description of this metal, list as many of its properties as you can. Even after reading about it, you may be able to think of other properties not mentioned here. Afterward, decide which of the properties are important for the purpose of each aluminum object shown here.

Although aluminum is one of the most common chemicals, it was not refined into metallic form until 1827, long after most other metals. This is because large amounts of electricity are required to extract aluminum from its ore. Until electricity became commonly available, aluminum cost as much as $500 a pound!

Aluminum is very versatile. Because it is much less dense than most metals, aluminum is used to form lightweight alloys. An alloy is a mixture of two or more metals. Aluminum and its alloys are used in canoes, boats, cars, bicycles, and space vehicles.

Aluminum is ductile—that is, it can be drawn into a wire. It is also malleable. This means it can easily be shaped into many useful products, from aluminum siding to aluminum cans. Because it does not corrode and is not poisonous, it can be used to wrap food.

Aluminum is used to make cooking utensils because it is a good conductor of heat. It is bonded to polyester fiber and used to make sleeping bags because it reflects body heat inward to keep a person warm. Aluminum certainly is a "hot property." Can you suggest some new uses that aluminum might have in the future? **B**

Properties of One Chemical: Aluminum

GUIDED PRACTICE This reading examines the properties of aluminum and the uses that are determined by these properties. Through reading and discussion, students should be able to identify several properties of aluminum. Write their suggestions on the board. The properties named in the text are ductility, malleability, and formation of lightweight alloys. Also, aluminum is a good conductor of heat; it does not corrode; and it is not poisonous.

Answer to
Properties of One Chemical: Aluminum

Properties listed that could apply to the photographs include low density, malleability, and resistance to corrosion. Aluminum is a good material to recycle because it can be melted and remolded without altering its chemical composition.

Answers to
In-Text Questions, pages 278–279

A A *property* is a characteristic that distinguishes one substance from another.

B Aluminum could have many uses in the future, such as in building lighter and faster cars. Because a heavier car uses more fuel, a lighter car might be very important in the future.

Meeting Individual Needs

Learners Having Difficulty
Have students identify each of the properties described below as either chemical or physical.
a. A solid cracks when you tap it. *(Physical)*
b. When a gas is passed through limewater, the limewater becomes cloudy. *(Chemical)*
c. A liquid evaporates quickly when heated. *(Physical)*
d. A solid burns when ignited. *(Chemical)*

Theme Connection

Structures
Before class, make a sample of the slime described in the Theme Worksheet on page 26 of the Unit 5 Teaching Resources booklet. Present your slime to students and allow them to study it individually. **Focus question:** What physical properties does the structure of this slime exhibit? *(Answers will vary.)* You may wish instead to use the Theme Worksheet as a classroom activity or to have students make their own slime at home.

Homework

Have students make a list of the chemicals they would need to take along with them on an eight-day space-shuttle journey. Next to each chemical, students should describe its use.

The chemical reaction that is taking place is as follows:

aluminum + copper chloride →
      aluminum chloride + copper

There is also a reaction between the aluminum and the solution that produces hydrogen gas. This accounts for the bubbles that are observed.

 ## Cooperative Learning
EXPLORATION 3

Group size: 3 to 4 students

Group goal: to examine a dramatic chemical change

Positive interdependence: Assign students roles such as primary chemist, research assistant, supply coordinator, and timekeeper/reporter.

Individual accountability: Have students perform a self-assessment. They should comment on their ability to work with other group members and on their ability to grasp the purpose of the activity.

Answers to *Questions*

1. Heat was given off, the color changed, and a new substance (a precipitate) was formed, all of which indicate a chemical change.

2. The copper chloride crystals dissolved in the water.

3. Aluminum reacts with a solution of copper chloride. This is a chemical property.

4. Some of the physical properties of copper chloride are the following: it is a green crystal, and it is soluble in water.

5. Copper chloride is formed from copper and chlorine. When mixed in a solution, copper chloride plus aluminum yield copper plus aluminum chloride.

6. The red material is copper.

One Chemical Property of Aluminum

In this Exploration, you will use the chemical copper chloride to examine one of the chemical properties of aluminum. As you proceed, record all of the chemical changes that you observe.

Be Careful: Make sure you do not handle the copper chloride with your bare hands. Wear safety goggles, latex gloves, and an apron. Wash your hands after the experiment.

You Will Need
- 5 g of copper chloride crystals
- a 10 cm × 10 cm square of heavy aluminum foil
- 100 mL of water in a beaker
- a stirring rod
- an evaporating dish
- latex gloves
- an apron or lab coat

What to Do

1. Add the copper chloride to the beaker of water. Make as many observations as you can, and then stir to dissolve the copper chloride.

2. Slightly crumple the aluminum foil, and place it in the beaker. You can push it under the surface of the liquid with your stirring rod.

3. Make as many observations as you can, and then answer the questions.

4. When you have finished, wash your hands with soap and water, and clean up your area.

Questions

1. What evidence do you have that a chemical change took place?

2. Name a physical change that took place.

3. What new property of aluminum have you discovered? Is it a physical or chemical property?

4. What are some physical properties of copper chloride?

5. Name one chemical property of copper chloride.

6. Examine the red material that formed. What do you think it might be?

Theme Connection

Energy

One sign that there has been a chemical change is the release of energy. **Focus question:** What are some of the ways we use the energy that is released during a chemical reaction? *(Students may point out the we use energy from the burning of coal to produce heat and electricity. Accept all reasonable responses.)*

★ **An Exploration Worksheet is available to accompany Exploration 3 (Teaching Resources, page 28).**

Results

Summarize the results of this Exploration. Here's what you started with:

• silvery gray aluminum
• greenish blue copper chloride dissolved in water

What did you observe? Did the color of the liquid change? What happened to the aluminum? A new substance formed in the liquid. Describe its appearance. Do you think it could be copper? Pour the liquid into an evaporating dish and allow it to evaporate. Describe what is left. What do you think it could be?

Guess That Substance!

Write three properties of a substance on one side of an index card or piece of paper and the name of the substance on the other side. A classmate or your teacher will read the properties from each student's card. The winner is the person who correctly guesses the most substances.

Here is Ralph's index card:

• it's a metal
• it's ductile
• it rusts

Answer: iron or steel

Answers to
Results

The liquid changed from greenish blue to clear, and the aluminum disappeared. A reddish powder (copper) formed in the liquid. A white solid (aluminum chloride) remained after the solution evaporated.

Guess That Substance!

This is a fun way of examining chemical and physical properties of a substance. You might want to assign a specific material to each student to avoid duplications. Students may need to do a little research to fill out their cards.

PORTFOLIO

Suggest that students add their cards from Guess That Substance! to their Portfolio.

CROSS-DISCIPLINARY FOCUS

Art

Visual artists use their practical knowledge of chemicals to create paintings, sculptures, and other art objects. Painters mix pigments to create a variety of colors, and they often prepare surfaces to be painted with special coatings. Sculptors use a wide variety of materials, including metal, stone, ceramics, and composite materials. All art materials have different physical and chemical properties that the artists must be familiar with. Have students research how a well-known work of art was made. Ask students to include in their report a description of the physical and chemical changes involved in the artistic creation.

FOLLOW-UP

Reteaching

Have students classify each property listed on their cards from Guess That Substance! as either chemical or physical.

Assessment

Ask students to decide if the following statements are true or false:

a. Substances with different chemical properties are formed during physical changes. *(False)*

b. The red material formed when copper chloride and aluminum are mixed is copper. *(True)*

c. Alloys are mixtures of two or more metals. *(True)*

d. Aluminum is a malleable element. *(True)*

e. Copper chloride is a green crystal. This is a chemical property. *(False; color is a physical property.)*

Extension

Students could research the physical and chemical properties of a metal. A poster depicting the metal's uses could accompany the research. Examples include silver, iron, magnesium, and zinc.

Closure

Invite a guest speaker to your class to discuss recycling. The speaker could emphasize the chemical and physical changes that recycled products undergo. Encourage students to ask questions. Students can then discuss how to organize a recycling program in their school or neighborhood if one does not already exist.

Products and Reactants

FOCUS

Getting Started

Ask the students to guess how water could be created from other substances. Encourage volunteers to give their opinion and to support their answer. *(Water is a product of a chemical reaction between hydrogen molecules and oxygen molecules.)* Tell students that they could see water being made if they combined hydrogen gas with oxygen gas and added energy. (Encourage them not to try this at home.)

Main Idea

Reactants are the starting materials in a chemical change; products are the materials formed as a result of the chemical change.

TEACHING STRATEGIES

Summarizing Chemical Changes

This page defines the terms *reactants* and *products.* It also introduces the term *word equation.* Be sure students realize that in a word equation, what is on one side of the arrow is equivalent to what is on the other side of the arrow in terms of mass and the types of atoms. That is, reactants equal products. Have students copy the table into their ScienceLog and add to it as they proceed through the unit.

Answer to
In-Text Question

Ⓐ Students should also have noticed bubbles, an indication of another chemical reaction. In this case, the aluminum and the solution produced hydrogen gas.

 Transparency 47 is available to accompany the material on this page.

Summarizing Chemical Changes

In every chemical change, new substances are formed. These new substances are called **products** of the chemical change. When you added vinegar to baking soda, carbon dioxide gas was formed. Carbon dioxide is a product of this chemical change. From observing the change, you would not realize that water and the chemical sodium acetate were also formed as products of this chemical change.

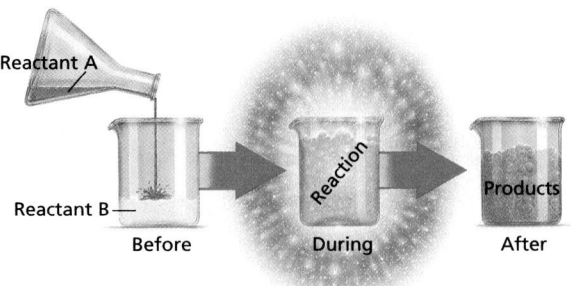

The starting substances of the chemical change are called **reactants**. Baking soda and vinegar are reactants. The drawing above represents both the reactants and the products of a chemical change.

A **word equation** shows the reactants and products of a reaction like the one above in terms of words and symbols. For example, the word equation for one of the reactions that occurs when aluminum foil is placed in a solution of copper chloride is as follows:

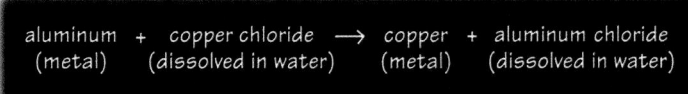

aluminum + copper chloride ⟶ copper + aluminum chloride
(metal) (dissolved in water) (metal) (dissolved in water)

What evidence did you see that another chemical reaction occurred as well? Ⓐ

Review the chemical changes you have observed so far in this unit. Try to identify some of the reactants and some of the products. (Don't worry if there are reactants and products you cannot identify. Doing so often requires special chemical tests.) In your ScienceLog, record the ones you can identify in a table like the one below. Keep adding reactants and products to this table as you proceed through the rest of this unit.

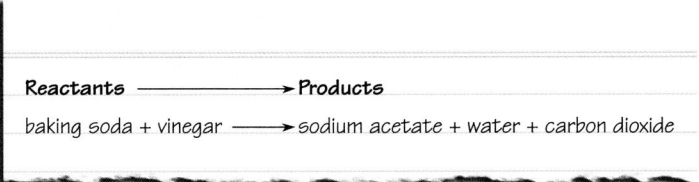

Reactants ⟶ Products
baking soda + vinegar ⟶ sodium acetate + water + carbon dioxide

LESSON 5 ORGANIZER

Time Required
1 to 2 class periods

Process Skills
observing, classifying, analyzing

New Terms
Products—new substances formed in a chemical change
Reactants—starting substances in a chemical change
Word equation—a chemical equation that identifies the reactants and products in a reaction with words and symbols

Materials (per student group)
none

Teaching Resources
Math Practice Worksheet, p. 30
Transparency 47

Special Reactants and Products

You have seen that chemical changes are part of daily life. Some interesting examples are described below. For each chemical change, try to identify the reactants and products. Then write the word equation for the reaction.

Kitchen Chemistry

Whenever you cook you are doing chemistry—mixing chemicals together to form delicious things to eat. A chemical that plays an important part in many recipes is sodium bicarbonate. It is found by itself in baking soda or mixed with cream of tartar, an acid, in baking powder. During the baking of a cake, the sodium bicarbonate in baking powder reacts with the cream of tartar to produce carbon dioxide. This reaction causes the cake to rise. Water and a chemical called sodium tartrate are also formed, but you don't notice them.

Remember—chemical changes usually result from reactions between two or more chemicals. To find out more about chemical reactions, turn to page S130 of the SourceBook.

try this at HoMe!

Make your own baking powder by adding 2.3 mL ($\frac{1}{2}$ teaspoon) of cream of tartar to about 5 mL (1 teaspoon) of baking soda. Then add some water. Watch what happens!

Table Salt From an Acid and a Base

Concentrated hydrochloric acid can cause severe burns to the skin. Sodium hydroxide, a base found in many drain cleaners, is extremely corrosive and can also burn the skin. However, when these two chemicals are combined in the right amounts, they react to form an essential part of our diet—sodium chloride, or table salt. Water is also formed in this chemical reaction. Does it seem strange that two potentially dangerous chemicals react to form chemicals that not only are harmless but also are essential for life? You will learn more about acids and bases when you try Exploration 1, on page 294, later in this unit.

Two corrosive, harmful chemicals react to form table salt, an essential part of your diet.

Caution: This reaction gives off a great deal of heat. Do not attempt to perform this reaction yourself.

283

Special Reactants and Products

The reaction between hydrochloric acid and sodium hydroxide is potentially hazardous. Explain to students that they should not attempt this reaction themselves.

Answers to
Kitchen Chemistry

Reactants: sodium bicarbonate, cream of tartar
Products: carbon dioxide, water, sodium tartrate
Equation: sodium bicarbonate + cream of tartar → carbon dioxide + water + sodium tartrate

Answers to
Table Salt From an Acid and a Base

Reactants: hydrochloric acid, sodium hydroxide
Products: sodium chloride, water
Equation: hydrochloric acid + sodium hydroxide → sodium chloride + water

Answers to
The Ingredients of Water,
page 284

Reactant: water
Products: hydrogen, oxygen
Equation: water → hydrogen + oxygen

Answers to
Chemical Caverns, page 284

Reactants: calcium carbonate, water, carbon dioxide
Products: calcium bicarbonate
Equation: calcium carbonate + water + carbon dioxide → calcium bicarbonate

Homework

The Math Practice Worksheet that accompanies the material on page 284 makes an excellent homework activity (Teaching Resources, page 30).

Answers to
Just for a Change . . . ,
page 284

- To extract aluminum, bauxite is first treated with sodium hydroxide, which dissolves the aluminum oxide found in the mineral. When this solution is treated with carbon dioxide, the aluminum oxide precipitates. Pure aluminum is separated from the aluminum oxide by electrolysis.
- Hematite and magnetite are two forms of iron oxide in iron ore. When iron ore and coke, a form of carbon, are burned in a furnace, some of the coke forms carbon dioxide: carbon (coke) + oxygen → carbon dioxide. The carbon dioxide combines with more coke to form carbon monoxide: carbon dioxide + carbon (coke) → carbon monoxide. The carbon monoxide reacts with the iron oxide to form carbon dioxide and pure iron: iron oxide + carbon monoxide → iron + carbon dioxide.

Answers to Just for a Change . . . continued ▶

The Ingredients of Water

Water, as you know, is a liquid. Heating water to its boiling point can change liquid water into steam, a gas. But it is still the same chemical—water. However, passing an electric current through water breaks it into its two components, hydrogen and oxygen, both of which are gases. This process, called *electrolysis*, is very useful because hydrogen gas is an excellent, clean source of energy. Can you guess what chemical is produced when hydro-Ⓐ gen is burned? In the future, our homes may be heated and our cars powered by hydrogen produced through electrolysis.

Chemical Caverns

Have you ever visited a cave or a cavern? Did you wonder how it was formed? Most caves are found in limestone formations and are produced by a chemical reaction. Here's how. Water from the surface seeps down into the limestone. This water contains dissolved carbon dioxide, the gas you exhale. The water and carbon dioxide mixture reacts with calcium carbonate (the main chemical in limestone) to form calcium bicarbonate. This chemical dissolves in the water and is carried away. Over time, more and more limestone is carried away. Eventually, a cave is formed.

Just for a Change . . .

Do research on a chemical change. In your research, try to determine the reactants and products of that change. In your report, include any other interesting information you find. Share your research with your classmates to see whether they can write a word equation for your chemical change. Here are some possibilities:

- Aluminum is extracted from a mineral called bauxite. How is this done?
- What is the chemical change involved in obtaining iron from iron ore? If there is a steel mill nearby, you might ask somebody who works there for help with this question.
- Plaster of Paris is made from a mineral called gypsum. What chemical change is involved in making plaster of Paris from gypsum?
- What can you learn about the reactants and products involved in making cement?
- Car batteries use a chemical change to produce electricity. What are some reactants and products in this change?
- The process of photography uses many changes. Talk to a photographer to learn about the chemical changes involved in taking and developing photographs.

According to legend, James Watt experimented with his mother's kettle before inventing the steam engine. Steam engines cause a physical change in water to produce mechanical motion. What kind of change in water does electrolysis cause? Ⓑ

This cave was formed by a chemical reaction.

CHALLENGE YOUR THINKING

1. Water World

With your help, Zed (your friend from the planet Nebulos) learned about liquids in Unit 4. Now Zed wants to know how to tell one liquid from another. On his home planet, water does not exist. Describe water to Zed so that he does not confuse it with other liquids, such as antifreeze, oil, alcohol, vinegar, or any solutions that contain water.

2. Express Yourself

Many common expressions refer to the properties of things. Read the expressions below. What properties underlie the following expressions? Does each expression make sense from a scientific standpoint?

- She's as sharp as a tack.
- This baby is as good as gold.
- That idea went over like a lead balloon.
- She's as bright as a new penny.
- That person is as hard as nails.
- He's the salt of the Earth.

What are some other expressions? Try making your own!

285

You may wish to provide students with the Chapter 15 Review Worksheet that accompanies this Challenge Your Thinking (Teaching Resources, page 31).

Answers to Challenge Your Thinking

1. At room temperature, pure water is a transparent, colorless, odorless, and tasteless liquid. Water freezes and melts at 0°C. Water has the unusual property of expanding when its temperature drops below 4°C, so the solid form is less dense than the liquid form. That is why ice floats in water. The boiling point of water is 100°C. Water can be distinguished from the other liquids by comparing physical properties such as color, boiling point, freezing point, and density. A chemical property of water that distinguishes it from other liquids is that it does not burn.

2. "Sharp as a tack"—A person who is "sharp" has a good mind. A tack is sharp because it is pointed (physical property).

"Good as gold"—Gold is the standard of many money systems. Gold is valuable because it does not easily react with other substances (chemical property) and therefore does not deteriorate easily. It is also rare.

"Like a lead balloon"—This refers to the high density of lead as compared with most elements (physical property). A lead balloon wouldn't float at all.

"Bright as a new penny"—This refers to the shiny quality (physical property) of a new copper penny. A person with this quality would be lively and enthusiastic.

"Hard as nails"—This refers to someone being tough. The metals used in nails have the property of hardness (physical property).

"Salt of the Earth"—Salt is a common substance but is very important for many living things (chemical and physical properties). A person described in this way is valuable or noble.

Other expressions could include the following:

"Oil and water don't mix."—This refers to the fact that these two liquids are not soluble in each other (physical property).

"Slower than molasses in January"—This refers to the physical property of molasses becoming thicker as the temperature decreases.

Challenge Your Thinking continued ▶

Answers to
Challenge Your Thinking,
continued

3. The rate at which an effervescent antacid tablet reacts with water could be increased by using hot water and could be decreased by using cold water or by capping the container to increase the pressure.

4. The reactants are the oxygen in the air and the starch in the potato. Sealing the potatoes in vacuum packaging would also prevent them from turning brown.

5. a. Potassium chlorate + manganese dioxide + heat → potassium chloride + oxygen + manganese dioxide

b. The everyday meaning and the scientific meaning of the word *catalyst* are similar in that both meanings refer to something that speeds up a process.

The following are sample revised answers:

1. Chemical changes similar to those that form sinkholes and limestone caves are responsible for the condition of this statue.

2. Ice melting is a physical change because the process does not form a new substance. Melting represents a change of state rather than a chemical change.

3. Student responses should distinguish between physical changes and chemical changes in whatever activity they choose to discuss.

---

3. Hurry Up and Change

There are ways that you can speed up and slow down a chemical change. When an effervescent antacid tablet is placed in water, a gas is released. Suggest three ways to change the rate at which effervescent tablets react with water. Break a tablet into three pieces and time your suggestions.

4. Spud Savers

Peeled potatoes turn brown if they are exposed to air. However, you can cover them with water to prevent this from happening. What are the reactants in this change? Can you suggest another method to prevent the potatoes from turning brown?

5. Quick-Change Artist

Certain chemicals help speed up a chemical change even though they are not used up in the change. These chemicals are called *catalysts*.

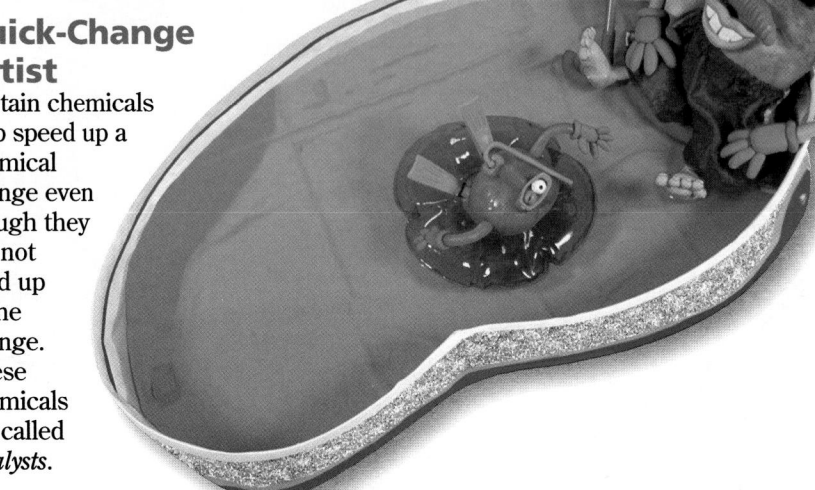

a. Oxygen gas can be prepared by heating a substance called potassium chlorate with a small amount of manganese dioxide. The manganese dioxide is a catalyst that helps the potassium chlorate to decompose into oxygen gas and potassium chloride. Write a word equation for this reaction.

b. *Catalyst* is another word with an everyday meaning and a scientific one. Check the dictionary for both meanings. How are the meanings similar?

ScienceLog

Review your responses to the ScienceLog questions on page 264. Then revise your original ideas so that they reflect what you've learned.

What is the 1
scientific
meaning of
the word
element?

2 **When something**
burns, what
happens to
the original
material?

3 **Have you encountered**
any acids or bases in
your daily life? Where?

4
Are chemical
changes involved
in digestion?
What are they?

ScienceLog

Think about these
questions for a moment,
and answer them in
your ScienceLog.
When you've finished
this chapter, you'll have
the opportunity to
revise your answers
based on what you've
learned.

287

CHAPTER

16

Studying Chemistry

Connecting to Other Chapters

Chapter 14
introduces chemicals used in our daily
lives and explains how chemicals
should be safely handled.

Chapter 15
explores the differences between
chemical and physical changes and
properties.

Chapter 16
defines acids and bases and further
explores what happens during
chemical changes.

Prior Knowledge and Misconceptions

Your students' responses to the
ScienceLog questions on this page will
reveal the kind of information—and
misinformation—they bring to this chapter. Use what you find out about your
students' knowledge to choose which
chapter concepts and activities to
emphasize in your teaching.

In addition to having students
answer the ScienceLog questions on this
page, you may wish to have them complete the following group assignment:
Divide the class into groups of three or
four students, and tell them that each
group represents a specially appointed
task force that has to develop a way to
classify all of the known elements. Make
sure that each group keeps a written
record of what procedures will be followed in order to place similar elements
together and to separate them from
other, dissimilar elements. Assure students that there are no right or wrong
answers to this exercise. Collect each
group's work, but do not grade it.
Instead, use the results to find out how
much students know about the classification process, what misconceptions
they may have, and what they find
interesting about the subject.

LESSON 1 — The Basics

FOCUS

Getting Started

Obtain the following metal chlorides: lithium chloride, potassium chloride, sodium chloride, and copper chloride. Dissolve 1 g of each metal chloride in separate 2 mL portions of water. As a demonstration, dip a clean platinum wire into each of the solutions and then place each wire into the flame of a laboratory burner. (One end of the wire can be inserted into a small cork and the other end should be bent into a small loop. The end with the loop should be dipped into the solution.) Ask students to describe what they see. *(Each compound gives a characteristic color to the flame: lithium is red; potassium is violet-blue; sodium is yellow; and copper is azure.)* Explain to students that this is known as the flame test and that it can be used to identify certain elements in a compound.

Main Ideas

1. All chemicals can be classified as either elements or compounds.
2. Ninety-one elements exist naturally on Earth.
3. The periodic table arranges elements with similar chemical properties into chemical families.

TEACHING STRATEGIES

Elements and Compounds

Explain to students that all matter on Earth is made of one or more elements. Aluminum is an example of a substance made of a single element. Table salt is a substance made of two elements, sodium and chlorine.

Answers to
Task 1

Answers to the paragraphs are as follows, listed in the order in which they should appear: compounds, compound, elements, compound, compounds, element, compounds, compounds, elements, compounds.

LESSON 1 — The Basics

Elements and Compounds

You have probably heard the legend of King Midas, who turned everything he touched into gold. For centuries, people dreamed of finding ways to make this legend a reality by turning common substances into gold. Alchemists—the earliest chemists—believed that it was possible to turn lead and other "base metals" into gold. In the course of their work, they discovered many chemical changes, but never the one that they really wanted—a way to convert other metals into gold.

We now know that the alchemists were attempting the impossible. All chemicals can be classified as **elements** or as **compounds**. Gold and lead are both elements. So are iron and aluminum. These are just a few of over 100 elements from which more complex substances are formed. Elements can combine with other elements in chemical reactions to form compounds. Compounds can be broken apart by chemical changes into the elements that compose them. However, no element can be changed into a different element by ordinary chemical changes.

By analyzing the light received from distant stars, scientists have determined that

How did this lady get her color?

the same elements found on Earth are found throughout the universe.

Here are three tasks you can try. Each one makes use of your understanding of elements and compounds. The periodic table of the elements on pages 290–291 may help you.

Task 1

In your ScienceLog, fill in each blank with the word *element* or *compound*.

Many of the __?__ of copper are blue or blue-green. Copper chloride is a green __?__. It contains two __?__, copper and chlorine. The Statue of Liberty, which is made of copper, is a green color. This green color results from the __?__ copper carbonate. Two __?__ found in the air, carbon dioxide and water, react with the __?__ copper to form this green coating.

During chemical changes, elements react with other elements to form __?__. In addition, __?__ can decompose into __?__, or compounds can react with compounds to form still other __?__. You have already seen examples of all of these types of chemical changes.

LESSON 1 ORGANIZER

Time Required 2 class periods

Theme Connection Structures

Process Skills
observing, comparing, analyzing

New Terms
Compound—formed when elements combine with other elements in chemical changes; can be broken down into elements during a chemical reaction
Element—substance that cannot be broken down during a chemical reaction
Periodic table—a chart in which all known elements are organized accord-

ing to their chemical properties

Materials (per student group)
Task 3: small mirror; piece of steel wool; 25 mL of water; zinc or magnesium ribbon; about 20 mL of 0.1 M hydrochloric acid; safety goggles; lab aprons
Elemental Investigations: 111 index cards

Teaching Resources
Resource Worksheets, pp. 37 and 39
Transparencies 48 and 49
SourceBook, pp. S108 and S117

What evidence is there that firecrackers cause a chemical change?

Task 2

All of the following chemical changes involve elements. Identify the elements, and state whether each is a reactant or a product of the chemical change.

a. A flashlight battery uses a chemical change between zinc and ammonium chloride to produce electricity.

b. Gunpowder, an early Chinese invention, contains sulfur, carbon, and other substances. Igniting gunpowder causes a flash or an explosion.

c. Passing an electric current through water produces hydrogen gas and oxygen gas.

d. Skiers sometimes use chemical hand warmers. Each hand warmer contains iron filings, carbon, water, and sodium chloride. When these substances are exposed to oxygen in the air, a chemical change occurs that produces heat.

Task 3

Here are some chemical changes you can try. In each case, both elements and compounds are involved. As you did in Task 2, identify the elements involved and state whether each is a reactant or a product.

a. Take a deep breath and exhale onto a mirror. Moisture appears on the mirror. Where does it come from? The oxygen you breathe takes part in a chemical change inside your body, forming carbon dioxide and water.

b. Wet a piece of steel wool. The rust that forms after a couple of days is a compound called iron oxide and is made up of iron and oxygen.

c. Place a piece of zinc or magnesium ribbon in dilute hydrochloric acid. The bubbles that form are hydrogen gas.

Caution: Do this activity only with your teacher present.

Chemical Science

Today's scientists know far more about the nature of chemicals and chemical changes than did early scientists. Today's scientists have even done what alchemists only dreamed of doing—making gold, though only in tiny amounts. They achieved this, not with chemical changes, but with powerful machines that alter the fundamental makeup of matter.

Ninety-one elements exist naturally on Earth. Scientists have created a few more to bring the total to over 100. But it is the 91 naturally occurring elements that make up everything in this world. That's incredible!

Susanah Hernandez, an organic chemist, is making a carbon compound.

289

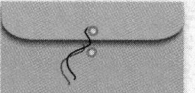
The Resource Worksheet that accompanies the material on page 288 makes an excellent homework activity (Teaching Resources, page 37). If you choose to use this worksheet in class, Transparency 48 is available to accompany it.

Homework

PORTFOLIO
Encourage students to include in their Portfolio any research or comments they made about Chemical Science.

Chemical Science

INDEPENDENT PRACTICE Have students read and discuss this material. For extra credit, have students do research on the work of alchemists and their influence on the beginnings of chemistry. Students could also research synthetic elements. These elements are produced by scientists in laboratories. Have students discover the primary uses of synthetic elements.

Homework

The Statue of Liberty is green because it is made of a metal that has oxidized. The thin green layer is called a *patina*. Have students compare and contrast patina and rust. *(Both are metals that have oxidized and changed color, but rust is destructive because it weakens iron, while a patina is constructive because it prevents further oxidation of copper or bronze and is decorative.)* Ask: Do you know of any buildings in your community that have patina roofs?

Answer to
Caption

Firecrackers give off light and make loud sounds.

Getting to Know the Elements

As you introduce the periodic table to students, emphasize that everything they see around them is composed of some combination of these elements. For example, all living things contain the following six elements: hydrogen, oxygen, carbon, nitrogen, phosphorus, and sulfur. Traces of nine other elements (sodium, magnesium, chlorine, potassium, calcium, manganese, iron, copper, and iodine) are also found in living things. Direct students' attention to Transparency 49. Ask students why hydrogen and oxygen are among the most abundant elements in our bodies. *(Hydrogen and oxygen combine to form water, and water accounts for about 65 percent of our body mass.)* Explain to students that the elements in the periodic table can be combined in different ways to form everything in the known universe.

Answer to
In-Text Question

Ⓐ Students are likely to be familiar with an element because it is common, such as nitrogen or oxygen; because it exhibits unusual properties, such as mercury or gallium; or because it has useful applications, such as uranium, platinum, or lead. Mercury has an unusually high fluidity, earning it the nickname *quicksilver,* and is often used in thermometers. Gallium is solid at room temperature but liquid at body temperature; it is used in transistors. Uranium is a source of nuclear energy. Platinum is a precious metal used in electronics. Lead is used as an X-ray absorber and in car batteries.

Getting to Know the Elements continued ►

Homework

The Resource Worksheet that accompanies the material on this page makes an excellent homework activity (Teaching Resources, page 39). If you choose to use this worksheet in class, Transparency 49 is available to accompany it.

Getting to Know the Elements

On these two pages is the **periodic table** of the elements. The large letters are the symbols for the elements. The periodic table is arranged in *families* (elements in the same column), also called *groups,* and *periods* (elements in the same row). Elements in the same family have similar chemical properties. Elements close together in the same period also have similar properties. Only 91 of these elements occur naturally. The rest are created artificially in the laboratory. Do you recognize some of these elements? What are they used for? Ⓐ

Calcium is necessary to build strong bones and teeth. Bones are made of mineral compounds that contain calcium and phosphorus. Potassium is important in nerve and muscle activity. Calcium and potassium belong to the same period.

Iron is the most abundant and least expensive metal. Because it conducts heat well, it is often used in cookware.

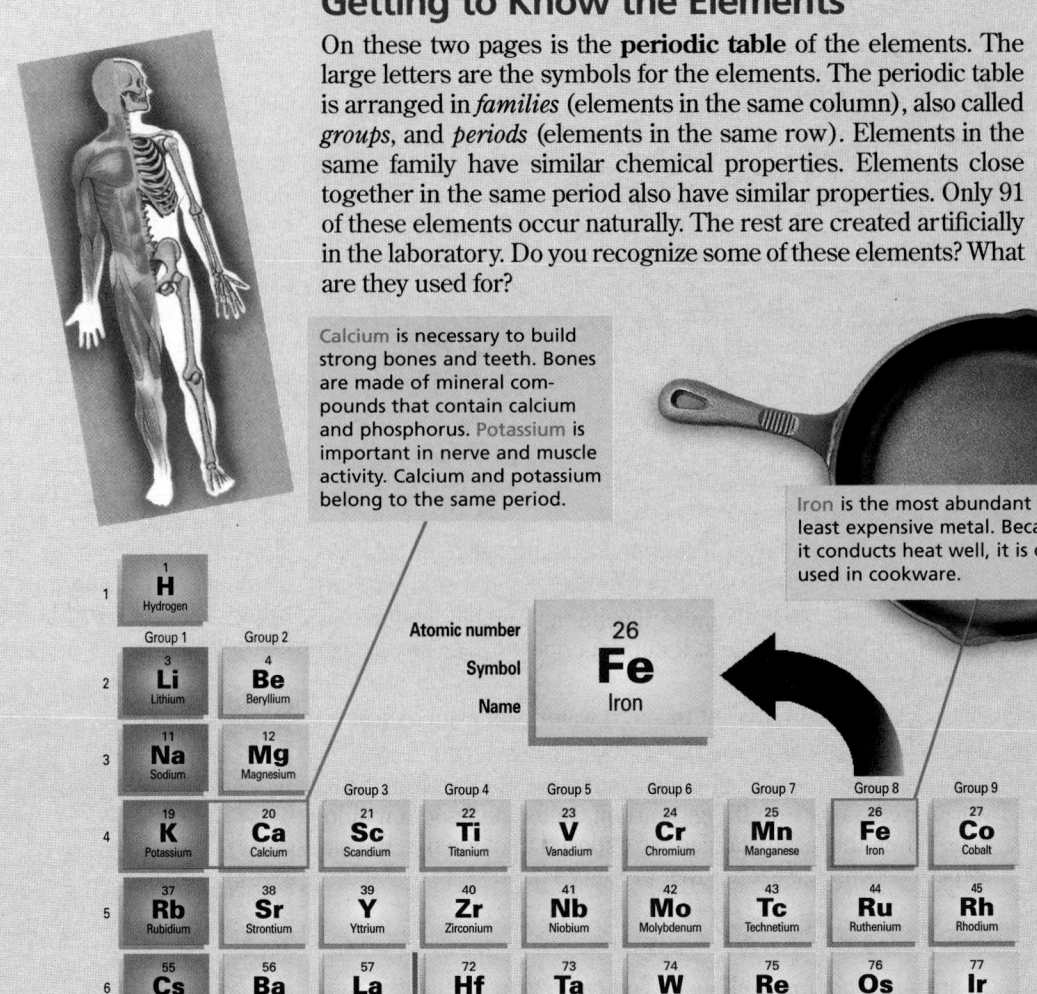

In 1903, Marie Curie and her husband received the Nobel Prize in physics for the discovery of radioactivity. Marie Curie became the first person to win two Nobel Prizes when she received a Nobel Prize in chemistry in 1911 for discovering radium and polonium.

Theme Connection

Structures

Focus question: Why are the properties of individual elements in a human body nothing like the properties of the human body as a whole? *(The elements are combined into complex compounds that form bone, skin, muscle, and so on. When elements and compounds react with one another, they form different compounds and elements that have different physical and chemical properties.)*

Did You Know...

The element gallium is often used by magicians to "bend" a spoon with their thoughts. Beforehand, a magician will break a spoon into two pieces and weld the pieces back together with gallium as a solder. During the show, the magician will pick up the spoon at the welded joint and concentrate. His or her body temperature will melt the gallium, and the spoon will bend.

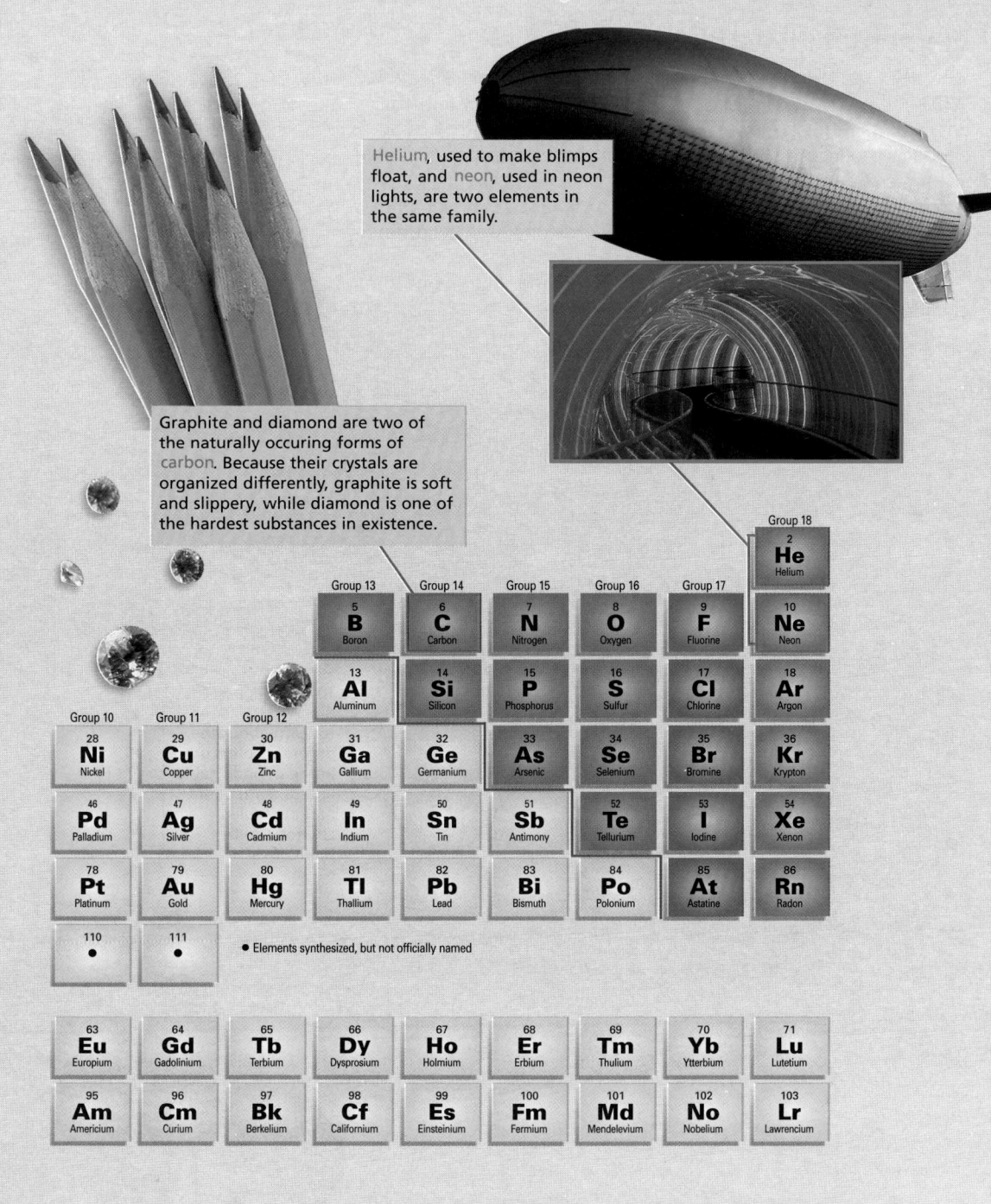

Helium, used to make blimps float, and neon, used in neon lights, are two elements in the same family.

Graphite and diamond are two of the naturally occuring forms of carbon. Because their crystals are organized differently, graphite is soft and slippery, while diamond is one of the hardest substances in existence.

			Group 13	Group 14	Group 15	Group 16	Group 17	Group 18
								2 **He** Helium
			5 **B** Boron	6 **C** Carbon	7 **N** Nitrogen	8 **O** Oxygen	9 **F** Fluorine	10 **Ne** Neon
Group 10	Group 11	Group 12	13 **Al** Aluminum	14 **Si** Silicon	15 **P** Phosphorus	16 **S** Sulfur	17 **Cl** Chlorine	18 **Ar** Argon
28 **Ni** Nickel	29 **Cu** Copper	30 **Zn** Zinc	31 **Ga** Gallium	32 **Ge** Germanium	33 **As** Arsenic	34 **Se** Selenium	35 **Br** Bromine	36 **Kr** Krypton
46 **Pd** Palladium	47 **Ag** Silver	48 **Cd** Cadmium	49 **In** Indium	50 **Sn** Tin	51 **Sb** Antimony	52 **Te** Tellurium	53 **I** Iodine	54 **Xe** Xenon
78 **Pt** Platinum	79 **Au** Gold	80 **Hg** Mercury	81 **Tl** Thallium	82 **Pb** Lead	83 **Bi** Bismuth	84 **Po** Polonium	85 **At** Astatine	86 **Rn** Radon
110 •	111 •							

• Elements synthesized, but not officially named

63 **Eu** Europium	64 **Gd** Gadolinium	65 **Tb** Terbium	66 **Dy** Dysprosium	67 **Ho** Holmium	68 **Er** Erbium	69 **Tm** Thulium	70 **Yb** Ytterbium	71 **Lu** Lutetium
95 **Am** Americium	96 **Cm** Curium	97 **Bk** Berkelium	98 **Cf** Californium	99 **Es** Einsteinium	100 **Fm** Fermium	101 **Md** Mendelevium	102 **No** Nobelium	103 **Lr** Lawrencium

291

Getting to Know the Elements, *continued*

GUIDED PRACTICE To help students start interpreting the periodic table, have each student choose a box representing a familiar element. Discuss the meaning of each item in the box. The following information is provided to help you explain the items in the table. Atomic structure is not discussed in the text, so including this material is up to you. However, atomic structure is discussed in Units 4 and 5 of the SourceBook.

- **Atomic number**—the number of protons in the atom's nucleus. Protons are positively charged particles in an atom. Have students note that the periodic table is arranged in order of increasing atomic number. Each succeeding element has one more proton in its nucleus. Note: All atoms of the same element have the same number of protons.

- **Symbol**—the internationally agreed upon symbol for a particular element. Have students discuss the symbols. Many were adapted from the Latin names for the elements, such as that of gold, silver, lead, tin, iron, tungsten, potassium, and sodium.

- **Name**—often derived from Latin or Greek words or from the name of a person or place. Some names derived from Latin or Greek words include helium, oxygen, and mercury. Elements named after people and places include mendelevium, fermium, einsteinium, californium, berkelium, curium, americium, plutonium, neptunium, nobelium, lawrencium, and francium.

Have students look for elements with similar properties and compare their locations in the table. For example, have them locate the precious metals, silver (Ag), gold (Au), and platinum (Pt). Ask them to think about what properties these metals have in common. Students may begin to see the relationship between the properties of a substance and its underlying structure.

Getting to Know the Elements continued ▶

Meeting Individual Needs

Gifted Learners
Pose one or more of the following questions for students to think about:
- What would the world be like if there were no chemical changes, only physical changes?
- How would you prepare food if the chemical changes of cooking took place in the presence of light instead of heat?

- What would society be like if all jobs involving the use of chemicals were eliminated?
- What would the Earth be like if all 91 naturally occurring elements were solids at room temperature?

Have students select one of the questions and prepare a response. Suggest that they present their ideas to the class as an oral report.

Getting to Know the Elements, continued

Cooperative Learning
GETTING TO KNOW THE ELEMENTS

Group size: 3 to 4 students
Group goal: to practice research skills and learn about the elements
Positive interdependence: Assign students roles such as primary researcher, recorder, orator, and editor (to make sure the report is coherent and error-free).
Individual accountability: Have each student briefly describe one of the elements from the family that his or her group researches.

Have groups of students investigate the elements in a particular chemical family, such as the halogens, alkali metals, or noble gases. After they have done their research, have the groups present a "family history" of their elements to the class as an oral report.

Answer to
In-Text Question

Ⓐ Answers will vary. Sample answer: iron. Iron's symbol is very different from its English name because the symbol is derived from the Latin word *ferrum*. Some of iron's physical properties include the following: it is a metal, it is solid at room temperature, and it is magnetic. Some of its chemical properties include the following: it reacts with oxygen to form rust. Iron's most interesting property may be its magnetism. It can be used to make steel and is an essential mineral for the body.

FOLLOW-UP

Reteaching

Divide the class into two teams. Have team A ask the members of team B what the symbols are for common elements chosen at random. Let each member of team A have a turn. Then reverse roles, with team B asking the questions and team A supplying the symbols. The first team to answer 20 questions correctly wins.

Assessment

Give students the following list of items. Ask them to state whether each is an element or a compound. If the item is an element, ask them to write the chemical symbol. They may use the periodic

---

Elemental Investigations

Now select an element from the periodic table and investigate it. What is the origin of its English name? Is its symbol very different from its English name? If so, why? What are its physical and chemical properties? What is its most interesting property? What are some of its uses? Ⓐ

Record your information on index cards. Then make an informative periodic table by arranging the cards on a wall according to families and periods.

Here is what Hector and Lawanda found out about the elements silver and mercury.

> ### Silver—Symbol Ag
> —name comes from Anglo-Saxon word *seolfor*
> —symbol comes from Latin *argentum*
> —known since ancient times
> —best conductor of electricity of all the elements
> —reflects 95% of the light that strikes it
> —does not react chemically with most substances
> —is not attracted to a magnet

> ### Mercury—Symbol Hg
> —symbol comes from Latin word *hydrargyrum*, meaning "silver water"
> —known in ancient times; used by alchemists in their experiments
> —is a metal, but is liquid at room temperature
> —used in thermometers, barometers, and hydrometers
> —will dissolve gold and silver to form amalgams
> —will not react with cobalt, nickel, or platinum

Turn to page S118 of the SourceBook to learn about the different colors used in the periodic table and to learn about me, the table's creator, Dmitri Mendeleev.

---

table for reference.
a. sodium chloride *(Compound)*
b. carbon dioxide *(Compound)*
c. water *(Compound)*
d. copper *(Element, Cu)*
e. potassium nitrate *(Compound)*
f. nitrous oxide *(Compound)*
g. mercury *(Element, Hg)*
h. tin *(Element, Sn)*
i. iron oxide *(Compound)*

Extension

Have students read pages S108–S113 in the SourceBook. Students can then build atomic models of various elements for a

---

class display. Some students may want to try making a model of a simple compound. Marshmallows, gumdrops, and toothpicks can be used for this task.

Closure

Have students create a concept map using the following terms: elements, periodic table, family, copper, silver, and gold.

Homework

The Closure activity on this page makes an excellent homework assignment.

B If you were asked to talk about acids, what would you say? Would you be surprised to learn the following facts?

- Vinegar is an acid.
- Milk contains an acid.
- Apples contain an acid, and so does lemon juice.
- Your stomach produces hydrochloric acid to digest food.

As you can see, there are many acids that are already familiar to you. **Acids** are a class of substances with similar properties. All acids have a sour taste. They react with metals, as well as with another class of substances called **bases**.

In your home, you can find bases in drain cleaners, in household ammonia, and in the bottle of milk of magnesia that you may have in your medicine cabinet. Bases feel slippery and have a bitter taste. They react with acids.

Acids and bases can be detected by chemical changes. There are some natural dyes that are one color in acids and another color in bases. These dyes are called **indicators**. The table above gives the expected color of three different indicators in acids and in bases.

In Activity 1 of Exploration 1,

Three acid-base indicators	Color	
	In acids	In bases
litmus	red	blue
bromothymol blue	yellow	blue
phenolphthalein	clear	pink

you will use these three indicators to answer the following questions:

a. What materials are acids or bases?

b. Can carbon dioxide dissolve in water to form an acid?

c. What happens when you mix an acid and a base?

In Activity 2, you will make your own acid-base indicator out of a red cabbage!

Caution: Some strong acids and bases can destroy skin on contact. Use acids and bases with extreme caution and only in the presence of an adult.

293

LESSON 2 ORGANIZER

Time Required 2 class periods

Process Skills
observing, comparing, classifying

New Terms
Acids—compounds that react with metals and bases and have a sour taste
Bases—compounds that react with acids, feel slippery, and have a bitter taste
Indicators—natural dyes that have one color in acids and another color in bases
Neutralization reaction—a reaction in which acid neutralizes (reacts with) a base to form a salt

Materials (per student group)
Exploration 1, Activity 1: about 10 mL each of lemon juice, apple juice, coffee, milk, water, milk of magnesia, and dilute household ammonia; tomato; 5 g of baking soda; effervescent tablet; aspirin tablet; 5 g of soap; 3 strips of litmus paper; scissors; 250 mL of distilled water; stirring rod; 250 mL beaker; eyedropper; several drops of bromothymol blue solution; plastic drinking straw; 20 drops of 0.1 M hydrochloric acid; evaporating
continued

Getting Started
Fill a large beaker halfway with lemon juice. Tell students that lemon juice is an acid. Add a few drops of universal indicator to the beaker. Then ask students what they think will happen if you add an antacid tablet to the beaker. *(Accept all reasonable responses.)* Crumble an antacid tablet and drop it into the beaker. The color will change. Explain to students that the acidic lemon juice was neutralized by the basic antacid tablet. Tell them that in this lesson they will learn how to identify acids and bases.

Main Ideas
1. Acids are a class of compounds that have a sour taste and react with bases and metals.
2. Bases are a class of compounds that have a bitter taste and react with acids.
3. Acids and bases can be detected by natural dyes called indicators.

Have students read the introduction on page 293. Start a list on the chalkboard of the acids and bases mentioned in the text, and add any others that the students may be aware of. Tell students that in the next Exploration, they will be testing some substances to see whether they are acids or bases. You may wish to demonstrate the color changes in the three indicators mentioned in the table.

★ **A Transparency Worksheet (Teaching Resources, page 40) and Transparency 50 are available to accompany the material on this page.**

Answer to
In-Text Question

B Student responses will vary, but they should realize that all acids have similar properties.

EXPLORATION 1

 SAFETY ALERT Point out the photograph on this page. Make sure students understand that the girl is blowing through the straw and *not* drinking from the beaker.

Remind students to rinse and dry their eyedropper thoroughly after each use to avoid contaminating test samples and indicator solutions.

For question 2, students can change the solution in each beaker back to blue by adding a drop of household ammonia. This shows that indicator changes are reversible.

For question 3, students may find it interesting to change the solution to pink (basic) and then to clear (acidic) by first adding the base and then adding more acid. Encourage them to finish with a solution that is just slightly pink. The crystals formed after evaporation of the liquid will be cubic. Have students examine these crystals with a microscope.

Answers to *Activity 1*

1. Answers may vary. In general, lemon juice, apple juice, a tomato, milk, and aspirin will be acidic. Baking soda, milk of magnesia, antacid tablets, ammonia, and soap will be basic. Coffee and water will be close to neutral.

2. This solution will turn yellow when students blow into it with a straw because the carbon dioxide in their breath combines with the water molecules to form a weak acid called carbonic acid.

   This reaction also takes place in the atmosphere, making normal rainwater slightly acidic. Acid rain forms when exhaust gases from cars and emissions from factories mix with water in the air in reactions similar to that which formed carbonic acid in this experiment.

3. The reaction equation is as follows: sodium hydroxide (base) + hydrochloric acid (acid) → sodium chloride (salt) + water.

⭐ **An Exploration Worksheet is available to accompany Exploration 1 (Teaching Resources, page 42).**

Investigating Acids and Bases

ACTIVITY 1

In the Classroom

You Will Need

- litmus paper
- bromothymol blue solution
- phenolphthalein solution
- substances to test, listed in step 1 below
- distilled water
- a small beaker
- a plastic straw
- dilute household ammonia
- dilute hydrochloric acid
- dilute sodium hydroxide
- an evaporating dish or other glass container
- an eyedropper
- a microscope

What to Do

1. Which materials are acidic? Which are basic? Use the litmus paper to find out. Tear each piece of litmus paper into four pieces, and use the pieces to test different materials. Try lemon juice, apple juice, a tomato, coffee, milk, water, baking soda, milk of magnesia, antacid tablets, aspirin, dilute household ammonia, and soap. Any solid material must be mixed with a bit of distilled water before being tested. Keep a record in your ScienceLog of what you find out.

2. Is carbon dioxide a part of the acid-rain problem? Fill a beaker with water, and add a few drops of bromothymol blue solution.

294

If your solution is slightly yellow, add a drop or two of dilute household ammonia until your solution is blue. Using a plastic straw, blow gently into the solution for a minute or two. What happens? What does this tell you about the acid-rain problem?

3. Do you remember how to make table salt from an acid and a base? Place about 20 drops of dilute hydrochloric acid in an evaporating dish or other glass container. Add one drop of phenolphthalein solution. Next, add dilute sodium hydroxide drop by drop, until the solution turns a very faint

pink color. If the pink color seems too intense, add a drop of dilute hydrochloric acid. This is an example of a **neutralization reaction**.

Place the dish in a quiet spot to allow the liquid to evaporate. With a microscope, look at what remains. The acid *neutralizes* the base to form salt. Water is formed as well. Write a word equation that describes how you made table salt. Label the acid, base, and salt in the equation.

4. Wash your hands with soap and water, and clean up your area when you are finished.

ORGANIZER, *continued*

dish or other glass container; one drop of phenolphthalein solution; several drops of 0.1 M sodium hydroxide; microscope; safety goggles; lab aprons (See Advance Preparation on page 249C.); **Activity 2:** optional materials: leaf of raw red cabbage; 250 mL of hot water; dropperful (about 2 mL) of dilute household ammonia; eyedropper; 4 to 5 drinking glasses; violet blossom; 25 mL of vinegar; watch or clock; safety goggles; lab aprons

Teaching Resources
Transparency Worksheet, p. 40
Exploration Worksheet, p. 42
Transparency 50
SourceBook, p. S127

ACTIVITY 2

At Home

You have already used three prepared acid-base indicators. Now you will make your own indicator using red cabbage.

You Will Need

- red cabbage leaves
- vinegar
- dilute household ammonia
- hot water
- an eyedropper
- several drinking glasses

Be Careful: Do this activity only with an adult present.

What to Do

1. Tear a red cabbage leaf into small pieces.
2. Fill a glass one-quarter full with the cabbage pieces.
3. Add enough hot water to fill the glass halfway.
4. Allow the cabbage to soak for 20 minutes.
5. Save the colored liquid and throw away the cabbage pieces. The purple cabbage juice is your indicator.
6. What color is the cabbage-juice indicator in acids? Put a dropperful of cabbage juice in a glass. Add a few drops of vinegar.
7. What color is the cabbage-juice indicator in bases? Again, put a dropperful of cabbage juice in a glass. This time, add a drop of household ammonia. What happens? Continue to add ammonia, one drop at a time. Do you get another color change?
8. Can you neutralize the basic solution that you made in step 7? Try it by adding vinegar. What happens?
9. Wash your hands and clean up any materials used when you are finished.

Violets are . . . red?

Violets are another natural indicator. Put a bluish purple violet into a glass and pour some vinegar on it. Wait 10 minutes. What happens?

ACTIVITY 3

Research

In Activity 1, you learned an important fact about acid rain. Now find out some more. What causes acid rain? Why is it harmful to the environment? How can we lessen the problem? Do research on this important problem, and use your findings to make a classroom display.

295

ACTIVITY 2

To make a purple cabbage indicator for the whole class, add a couple of torn cabbage leaves to 500–1000 mL of boiling water. When the water is a deep purple, allow it to cool and remove the leaves. Store the liquid in a refrigerator. The cabbage water will spoil in about one week, so it should be poured down the drain before then.

Explain that pH is an indication of how acidic a substance is. A range of colors is possible, depending on the pH of the substance being tested.

Answers to
Activity 2

6. In acids, the indicator is pink.

7. In bases, the indicator is blue-green. Another color change is not possible if only basic substances are added.

8. Yes, the basic solution can be neutralized by adding vinegar. The indicator will turn pink again.

Answer to
Violets are . . . red?

The violet should turn red, indicating that the solution (vinegar) is an acid.

ACTIVITY 3

Give students time to complete their research projects on acid rain. Information is available from several environmental agencies.

CROSS-DISCIPLINARY FOCUS

Language Arts

Many people around the world are becoming concerned about the effects of toxic chemicals in the environment. Toxic spills cause the death of fish and wildlife and can even leave an entire area uninhabitable. As a result of such episodes, some people have come to believe that all chemicals are bad or toxic. Have students write a letter to such a person in which the students try to convince the person that not all chemicals are bad. What examples and evidence could students use to support their position?

FOLLOW-UP

Reteaching

Ask students to write a summary of the content of this lesson using the following words: acids, bases, bitter, sour, reactants, indicators, salts, product, and neutralization.

Assessment

Ask students to explain what happens to three different indicators when an acid or a base is added.

Extension

Students could research how the pH scale relates to acid-base indicators and then compare the pH of common substances. Have them make a large poster depicting the pH scale and the pH of substances they have researched.

Closure

Stage a debate about acid rain. The presentation should include background information about the different viewpoints and the origins of the acid-rain problem.

LESSON 3
The Nature of Burning— A Chemical Change

FOCUS

Getting Started

Perform the following demonstration for students: Pour several milliliters of hydrogen peroxide (no greater than a 3% solution) into a tall jar. Add a teaspoonful of manganese dioxide and cover the jar loosely with a lid. Bubbles of oxygen will be given off for about 15 minutes. Remove the lid and insert a glowing splint into the jar. The splint will burst into flames. Ask students to explain what they think happened. *(A chemical reaction occurred between the two substances in the jar, releasing oxygen. The excess oxygen caused the splint to burst into flames.)*

WASTE DISPOSAL ALERT   Dilute the 3% hydrogen peroxide solution to 10 times its volume with water. Pour the liquid down the drain; wrap the solid in old newspaper and put in the trash.

Main Ideas

1. Burning (combustion) is a chemical change that involves a rapid reaction with oxygen.
2. Corrosion is a chemical change that involves a slow reaction with oxygen.
3. Theories can change as new observations are made.

TEACHING STRATEGIES

After students read the theories, they will notice that all of the theories seem to work. Divide the class into small groups, distribute the materials, and have them proceed through the three Explorations.

Flashback! Greece, 460 B.C.

According to Empedocles, all things were made up of one or more of the four elements: earth, air, fire, and water. The properties of a substance depended on how these elements were combined.

Flashback! Greece, 460 B.C. continued ▶

The role of elements and compounds in chemical changes was not always known. Even a very familiar change like burning mystified observers for thousands of years. Let's listen in on three scientists as they discuss what they think burning is all about.

Flashback! Greece, 460 B.C.

Empedocles is reading by the light of an oil lamp. He studies the flame and wonders, What causes burning?

Empedocles' Theory and Conclusion

"All matter is composed of four
  elements—earth, air, fire, and water.
If a piece of matter contains the
  element fire, then it will burn.
Wood contains fire—wood burns.
Oil contains fire—oil burns.
Rock does not contain fire—rock does
  not burn."

This is a good theory. It works. Or does it? How would you test his theory? According to Empedocles' theory, salt is made up of earth and water. What might have been the reasoning behind this conclusion?

Flashback! Germany, 1710

George Stahl is having dinner by candlelight. He and his wife are discussing a question that has often come up at dinnertime: What causes burning?

Stahl's Theory and Conclusion

"Phlogiston! That's right. All objects that
  burn contain phlogiston. And when the
  phlogiston is all gone or when the air
  becomes saturated with the released
  phlogiston, the burning stops.
Wood contains phlogiston—wood burns.
Coal contains phlogiston—coal burns.
Rock does not contain phlogiston—rock
  does not burn."

This is also a good theory. It explains many observations. Or does it? Can you suggest any ideas that would support or disprove this theory of burning?

I shall call this new element phlogiston!

LESSON 3 ORGANIZER

Time Required
2 to 3 class periods

Process Skills
comparing, analyzing, inferring

New Terms
Combustion—the process of chemically combining a substance with oxygen so rapidly that heat and light are produced
Corrosion—a slow, destructive chemical change, often involving metals that combine with oxygen to form a new compound

Materials (per student group)
Exploration 2: candle; small ball of modeling clay; a few matches; aluminum pie plate; metric balance; wide-mouth jar; safety goggles; lab aprons; oven mitts
Exploration 3: steel wool; metric balance; pair of tongs; several matches; aluminum pie plate; safety goggles

continued ▶

Testing Stahl's Theory

You Will Need

- a candle
- modeling clay
- matches
- an aluminum pie plate
- a balance
- a jar with a wide mouth
- oven mitts

What to Do

1. Attach the candle to an aluminum pie plate with modeling clay. Find the mass of the candle, clay, and pie plate.

2. Light the candle. Does anything seem to enter or leave the candle as it burns? After 10 minutes, blow out the candle, and find the mass of the candle setup again. Using Stahl's theory, explain why the mass is different after burning.

3. Light the candle again, and place a jar over it. What happens? How would Stahl explain this?

Caution: Wear oven mitts when holding the jar over the flame.

Flashback! France, 1775

Antoine Lavoisier is celebrating. He has devised a new theory of burning.

Oxygen causes burning! This is the best theory! It explains all observations!

297

Flashback! Greece, 460 B.C., continued

Encourage students to think of other substances that could support or disprove this theory. Have them explain their reasoning. For example, salt was probably thought to be made up of earth and water because it is a solid and when it dissolves in water, it looks just like the water it is dissolved in. You might also point out to students the similarities that this theory has with what we now know about elements and compounds: that they make up all matter and that a compound's properties are determined by the elements that are combined. Explain to students that although a theory may be incorrect, valuable information is often gained from the thought that went into the development of the theory.

Flashback! Germany, 1710, page 296

Phlogiston is pronounced *flo JIS tun*. This theory is different from Empedocles' theory in that a substance (phlogiston) is released from a burning object. However, the connection had not yet been made between oxygen in the air and burning. Again, have students think of substances that help to prove and disprove this theory.

Answers to *Exploration 2*

2. Something does seem to leave the candle because it gets smaller. According to Stahl's theory, the candle's mass is decreasing as it burns because it is losing phlogiston.

3. When a jar is placed over the candle, the candle goes out. Stahl would say that the air has become saturated with the released phlogiston, extinguishing the candle.

Flashback! France, 1775

Even though students are probably aware of Lavoisier's theory, treat it as a theory that can be either supported or disproved by observation and experimentation. Explorations 2 and 3 provide observations to support this theory of burning.

 An Exploration Worksheet is available to accompany Exploration 2 (Teaching Resources, page 45).

ORGANIZER, *continued*

Exploration 4: candle; jar lid; small ball of modeling clay; a few matches; jar with lid; 25 mL of limewater; 100 mL graduated cylinder; safety goggles; lab aprons (See Advance Preparation on page 249C.)

Teaching Resources
Exploration Worksheets, pp. 45, 46, and 47
Activity Worksheets, pp. 48 and 49
Transparencies 51 and 52

Testing Lavoisier's Theory

3. Students will observe an increase of 0.1 g or more, depending on how many times the steel wool is touched with the match. Stahl would have predicted a decrease because of the loss of phlogiston. This evidence contradicts his theory. Lavoisier would have suggested that the steel wool was combining with oxygen from the air, thus increasing its mass. This experiment supports his theory.

4. Lavoisier would have said that a candle goes out when a jar is placed over it because all of the oxygen from inside the jar is used up. When a candle burns, it loses mass. The wax combines with oxygen from the air to form a gas, which is now known as carbon dioxide.

5. The word equation for the reaction is as follows: iron (steel wool) + oxygen = iron oxide.

You Will Need

- steel wool
- a balance
- tongs
- matches
- an aluminum pie plate

What to Do

1. Place about 1 g of loosely packed steel wool in the pie plate. Find the mass of the steel wool and the pie plate.

2. Use the tongs to touch a burning match to the steel wool. The steel wool burns! Repeat this three or four times.

3. Find the mass of the pie plate and the steel wool again. Was there a change in mass? Could Stahl explain this observation with his theory? How would Lavoisier explain it?

4. According to Lavoisier, why would a candle go out after a jar is placed over it? Why does a candle have less mass after burning?

5. Complete this word equation in your ScienceLog:

Be Careful: Always use caution when handling burning materials.

iron (steel wool) + ___?___ = iron oxide

Learning About Burning

The students in the photographs below are performing simple experiments. Try the experiments yourself in the next Exploration to find out what observations the students are making.

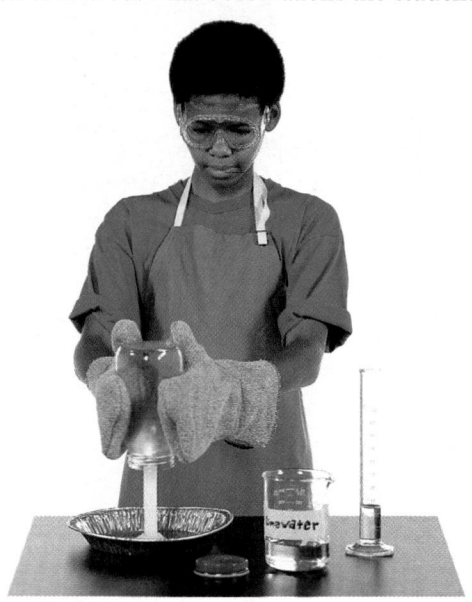

1. When a jar is held very close to a candle flame, a black deposit forms on the outside. This is a carbon compound called *soot*. The candle wax must contain carbon.

2. When a jar is placed over the candle, moisture can be seen condensing on the inside of the jar. The candle wax must be made up of a compound containing hydrogen. The hydrogen combines with the oxygen from the air during the burning process to form water.

3. If 25 mL of limewater is added to the jar from step 2, the limewater will turn milky. This provides evidence that during burning, the carbon in the candle combined with the oxygen to form carbon dioxide.

4. If 1 mL of the condensation inside the jar is collected and its mass is found to be 1 g, this provides evidence that the substance is water; the limewater test confirmed the presence of carbon dioxide. We can deduce that wax is made up of a compound containing the elements hydrogen and carbon.

5. Candle + oxygen = water + carbon dioxide

298

Attaboy Antoine! *page 299*
INDEPENDENT PRACTICE Encourage students to make a list of the major points in this lesson. They can then use this list to write a letter to Lavoisier in support of his theory.

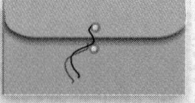

Homework

You may wish to assign the Independent Practice activity above as homework.

 **PORTFOLIO**
Encourage students to include their letter to Lavoisier in their Portfolio.

★ **Exploration Worksheets** are available to accompany Explorations 3 and 4 (Teaching Resources, pages 46 and 47, respectively).

Student Conclusions

You Will Need

- a candle
- a jar lid
- modeling clay
- matches
- a jar with a lid
- limewater
- a graduated cylinder

What to Do

1. Place a candle in the jar lid, and secure it with modeling clay. Hold the jar very close to the candle flame. What is deposited on the jar? Where did the substance come from?

2. Place the jar over the candle. What is deposited inside the jar? Where did this substance come from? How could you prove what it is?

3. After the candle has gone out, add 25 mL of limewater to the jar. Cover it quickly and shake. What happens? What does this prove?

4. What evidence did you see that a burning candle produces water? carbon dioxide?

5. Complete this word equation in your ScienceLog:

candle + oxygen = ___?___ + ___?___

6. When you have finished, wash your hands, and clean up your area.

Attaboy Antoine!

"Dear Lavoisier . . . " Write a letter to Lavoisier to congratulate him on his new theory. Tell him how your own candle experiments support his theory of burning.

Shedding Some Light on Candles

- The word *candle* comes from a Latin word meaning "to shine."
- Candles have been made from fish oil, whale oil, beeswax, tallow, tree bark, cinnamon, bayberry, and paraffin.
- Native peoples of the Pacific Northwest used the oily candlefish in an interesting way. After drying it, they would place it in the fork of a tree and light it, creating a strange-looking candle!
- North Americans use more than 90 million kilograms of paraffin each year to make candles.
- When a lighted candle casts a shadow, the darkest part of the shadow corresponds to the brightest part of the flame.

Slow, Slow Burning

A candle may burn for hours. During burning, or **combustion**, carbon and hydrogen (the two main elements in a candle) combine with oxygen to form carbon dioxide and water. Noticeable heat and light are also produced.

In Exploration 3, you saw that steel wool can burn. Iron (steel wool) combines with oxygen to produce iron oxide. This explains why the steel wool gained mass in this chemical change.

When iron is exposed to air, it reacts slowly with the oxygen in the air. Again, iron oxide is formed. This process, called rusting, is an example of **corrosion**, which can be seen all around you. How many examples of corrosion can you think of? Why do some substances corrode, while others burst into flame? How can materials be made resistant to corrosion? The process of corrosion can be used to investigate a variety of questions: Ⓐ

1. **What percentage of the air is oxygen?**

2. **Does the air you exhale contain oxygen? If so, how much?**

3. **What treatments can speed up or slow down corrosion?**

4. **Do different brands of steel wool rust at the same rate?**

Do you think you could design experiments to answer these questions? Ⓑ

FOLLOW-UP

Reteaching

Ask students to summarize what they have learned about how a theory develops and changes. Students should use examples from this unit to develop their ideas.

Assessment

Have students write an essay comparing the viewpoints of Empedocles, Stahl, and Lavoisier.

Extension

Have students investigate the phenomenon of spontaneous combustion. Ask them to explore the causes and the materials involved. Have them relate what they find out to possible fire hazards in the home and workplace.

Closure

As a class, create a time line of discoveries that led to an understanding of combustion.

Slow, Slow Burning

Assign questions 1–4 at the bottom of the page as research activities. Students can also design experiments to find the answers. Approve all student designs for safety before they are carried out.

Answers to *In-Text Questions*

Ⓐ Other examples of corrosion include the oxidation of copper and the tarnishing of silver. Some substances corrode and others burst into flames because different substances react with oxygen at very different rates. You can make materials resistant to corrosion by coating them with a nonreactive substance.

Ⓑ Sample answers:

1. One way to estimate the percentage of oxygen in the air is given in See for Yourself on page 306.

2. A similar method could be used to estimate the amount of oxygen in the air you exhale. Fill the graduated cylinder with vinegar before inverting it into the bowl of water. Using the rubber tubing, blow into the graduated cylinder to lower the level of the vinegar to the 100 mL mark. Be sure to pinch the tubing before removing it from the setup.

3. Treating a metal with salt water or an acidic solution will speed up corrosion. Corrosion can be slowed down by coating a metal with paint or another metal. Alloys can be more resistant to corrosion than a pure metal.

4. One way to test the rates of corrosion of different brands of steel wool is to graph changes in their mass over time. The brands that rust more quickly will show an increase in mass more quickly than other brands.

Homework

The Activity Worksheets that accompany the end of Lesson 3 make excellent homework activities (Teaching Resources, pages 48 and 49). If you choose to use these worksheets in class, Transparencies 51 and 52 are available to accompany them.

FOCUS

Getting Started

Have students think about the chemical changes that they have seen so far. Ask: What happens to the mass of a substance as it undergoes a chemical reaction? Is it the same as it was before the chemical reaction? Tell students that in this lesson, they will learn about how chemical reactions affect the mass of a substance.

Main Ideas

1. In a chemical change, the total mass of the reactants is the same as the total mass of the products.

2. An increase or decrease in the mass of one of the substances in a reaction is often a good indication that a chemical reaction has occurred.

TEACHING STRATEGIES

Ask students to consider why they cannot reuse candle wax once it has been burned. What do they think has happened to the wax? Tell the students that the wax cannot be reused because it has undergone a chemical reaction that has altered its chemical properties. Point out that they will learn more about chemical reactions as they read this lesson and perform the activities.

Answers to
Ravi's Group

Yes, the group's approach should have worked. After accounting for the mass of all the containers, Ravi's group should have discovered that the mass of the vinegar plus the mass of the milk equals the mass of the curdled milk.

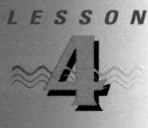

Conservation of Mass

In Exploration 2, you noticed that a candle has less mass after burning for a while. This decrease in mass is a clue that a chemical change has occurred. You also saw that after steel wool burns, it increases in mass. This gain in mass is also a clue that a chemical change has occurred.

What would you find if you were able to compare the mass of all the reactants of a reaction with the mass of all the products of the reaction? How would the masses of the reactants and products in each of the reactions below compare?

candle + oxygen → carbon dioxide + water
reactants   products

iron + oxygen → iron oxide
reactants   products

Does the mass of the reactants equal the mass of the products? How could you find out?

In groups of three, examine the ways in which some ninth-grade students answered these questions. Evaluate the approach used by each group— you may wish to try some of them yourself.

Ravi's Group

Ravi remembered that when vinegar is added to milk, a chemical change occurs—the milk curdles. His group decided to do the following:

1. Find the mass of a container of vinegar.

2. Find the mass of a container of milk.

3. Add the vinegar to the milk.

4. Find the mass of both containers and the curdled milk.

Do you think this approach worked? What did Ravi's group discover?

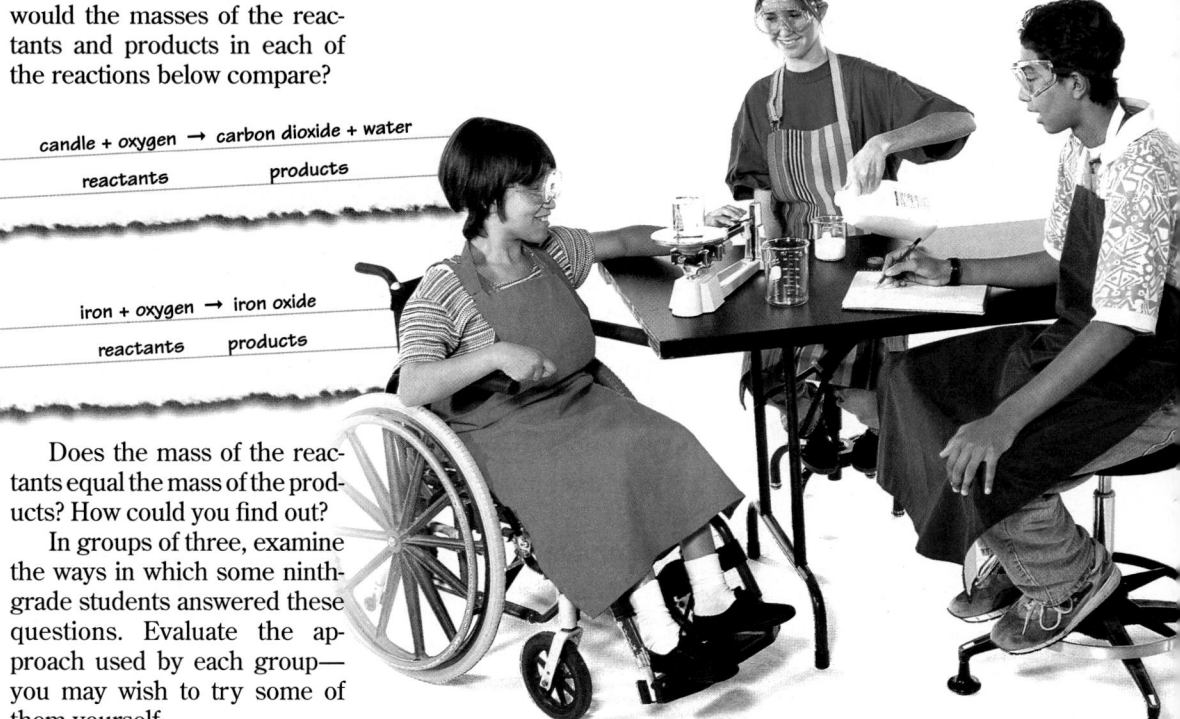

LESSON 4 ORGANIZER

Time Required
2 class periods

Process Skills
analyzing, comparing, observing

New Terms
none

Materials (per student group)
none

Teaching Resources
none

Dunja's Group

Dunja and her group followed a similar procedure to Ravi's group, except that they decided to test the reaction between water and baking powder. What did they find?

Trina's Group

Trina's group came up with a clever plan. Trina is diabetic, so she knows all about syringes. Here is their procedure:

1. Place one-eighth of an effervescent antacid tablet in a test tube. Close the tube with a one-hole stopper.

2. Place 5 mL of water in a large syringe (without a needle), and insert the end of the syringe into the stopper.

3. Place this "closed system" on a balance to find its mass.

4. Push the water into the test tube. As the reaction occurs, the plunger of the syringe will be forced outward.

5. During the reaction, leave the setup on the balance.

What conclusions did Trina's group reach? What is the main difference between Trina's plan and Dunja's plan? What do you think of Trina's approach?

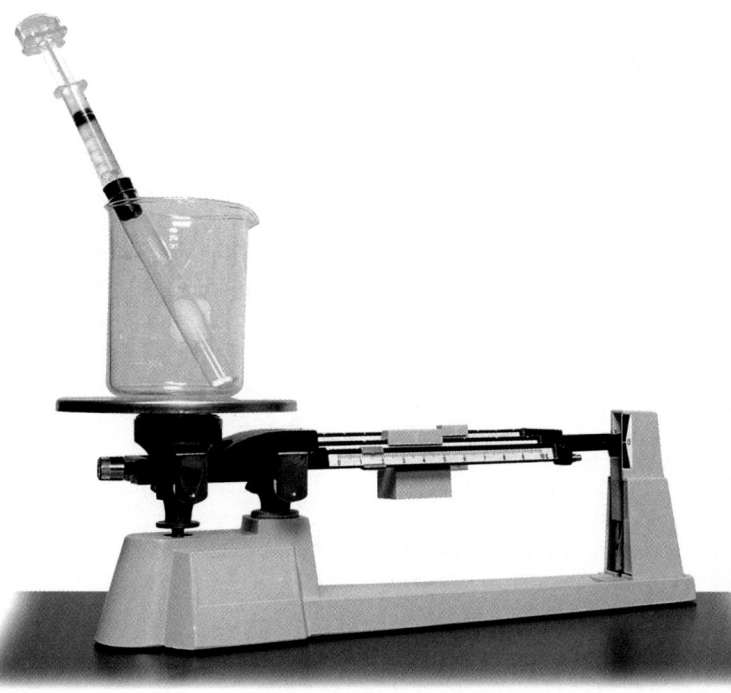

Excerpt From Alasdair's ScienceLog

Our plan worked well! We were the only group to try this idea. We placed a piece of banana in a jar and sealed the jar's lid tightly. The banana and jar had a mass of 78.4 g. Five days later, the banana had thoroughly rotted. A chemical change had occurred. The banana and jar still had a mass of 78.4 g.

Therefore, we had evidence that during a chemical change the mass of the reactants equals the mass of the products. This is called the law of conservation of mass.

Another Approach—Yours!

Can you devise another plan of action? What chemical change would you investigate? Discuss your procedure with your teacher, write it down, and try it. Then report back to your class.

301

Another Approach—Yours!

Answers will vary. One possibility is to consider a solid immersed in a liquid. This allows not only the total mass of reactants to be measured, but also the individual reactants. Some examples include iron in water or eggshells in vinegar.

FOLLOW-UP

Reteaching

Have students compare the conservation of mass in a chemical change with the conservation of mass in a physical change, such as when ice water is heated or cooled.

Assessment

Have students write a short story that includes a chemical reaction. Make sure that students address the issue of mass conservation in the story. Call on volunteers to share their story with the class.

Extension

Have students apply the law of conservation of mass to a nuclear reaction. How might this reaction differ from a chemical reaction? Is mass conserved in a nuclear reaction? Students will probably need to do some research to answer these questions.

Closure

Point out to students that, like mass, energy is conserved in chemical reactions, although it often changes form. Explain chemical energy and heat energy to the class, and then have them discuss how these terms relate to the energy in chemical reactions. They will learn more about energy and heat in the next two units.

Answers to
Dunja's Group

The baking powder will react with the water, producing carbon dioxide gas. If enough baking powder reacts, there will be a noticeable decrease in mass because the gas escapes. Thus, the group should incorrectly deduce that mass is not conserved in a chemical change.

Answers to
Trina's Group

Trina's group will notice that although gas bubbles form, indicating a chemical change, the mass remains the same. The group should deduce that mass is conserved in a chemical change. The difference between this plan and Dunja's plan is that Dunja's plan failed to account for any products that are gases, so the result of that plan is misleading.

Homework

The Assessment activity on this page makes an excellent homework assignment.

LESSON
5
Chemical Changes and You

FOCUS

Getting Started
Ask students to write a new definition of the term *chemical* based on what they have learned so far.

Main Idea
Chemical changes occur all around us and are essential to everyday life.

TEACHING STRATEGIES

The Great Breakdown— A Three-Act Play
Encourage students to participate in the play by assigning roles. The class could then act out the digestion process.

The Principal Characters

 Cooperative Learning THE PRINCIPAL CHARACTERS

Group size: 3 to 4 students

Group goal: to examine the organs of the digestive system and how they interact

Positive interdependence: Assign students roles such as coordinator, materials manager, timekeeper, and reporter.

Individual accountability: Each student should be able to write a summary of the digestion process.

GUIDED PRACTICE Set up a station for each of the digestive organs. At each station, have a picture of the organ along with a description of its function. Encourage students to discuss the organ and what it does. Ask them to discuss what unique features each organ has. At the final station, have either a picture or a model of how the organs fit together in the body. Let the students know that two of the body's vital organs, the stomach and the liver, are part of the digestive system.

The processes of life depend on many chemical changes. Living things eat, digest, respire, excrete, move, respond to stimuli, grow, and reproduce. In this lesson, you will be invited to explore at least one of these processes. In doing so, you will discover more about the chemical reactions that keep you growing and going.

After you read about the principal characters on the next page, match the numbers on the figure below to the names of each character. **A**

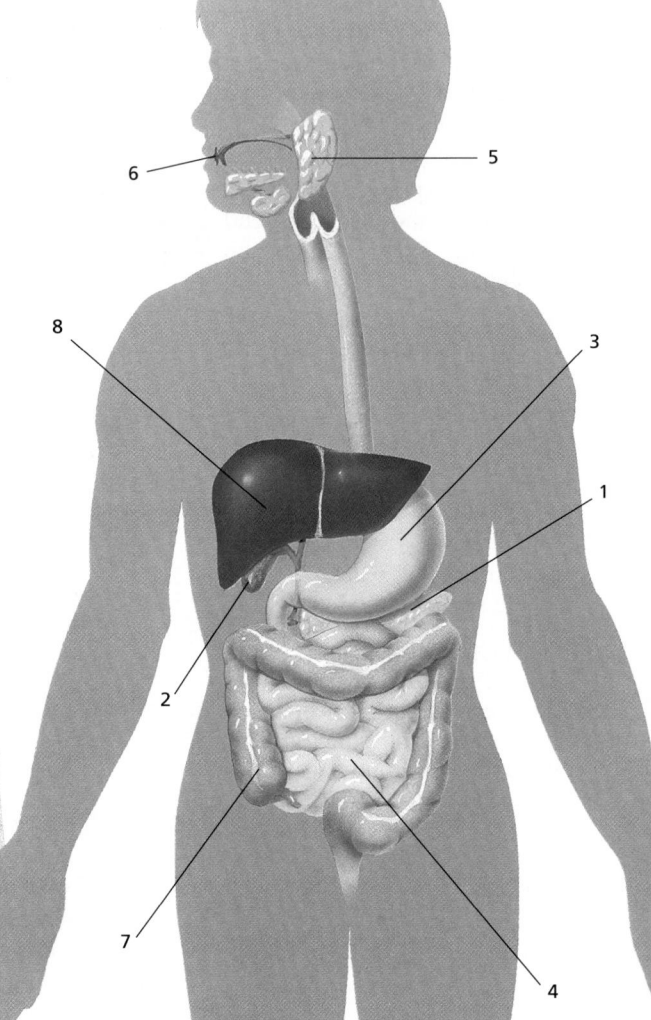

The Great Breakdown— A Three-Act Play

The Scenario
On June 6, 1822, an 18-year-old Canadian named Alexis St. Martin was accidentally shot. His doctor, William Beaumont, was able to reinsert part of a lung that protruded through St. Martin's wound, but he could not repair the wound to the stomach. A permanent hole remained. As a result, Beaumont was able to observe the temperature within St. Martin's stomach, the movement of his stomach, and the changes that the food he ate underwent as it was digested. What Beaumont observed over the next 11 years changed the way people thought about what happens to food after it is eaten. What did he discover? Read the following three-act play to find out.

The Principal Characters
What principal "characters" do you think are involved in digestion? Write down your ideas and then read the following descriptions.

302

LESSON 5 ORGANIZER

Time Required 1 class period

Process Skills
analyzing, communicating

Theme Connection Energy

New Terms none

Materials (per student group)
Exploration 5: 10 mL graduated cylinder; 1 g of amylase; 600 mL of water; 3 test tubes; test-tube clamp; test-tube rack; wax pencil; several crushed, unsalted crackers; eyedropper; hot plate; 600 mL beaker; 15 mL of Benedict's solution; watch or clock; safety goggles; lab aprons; latex gloves

Exploration 6: hard-boiled egg; knife; 2 test tubes; a test-tube rack; small beaker; stirring rod; 15 mL of water; 1 g of pepsin; 2 mL of 0.1 M hydrochloric acid; 10 mL graduated cylinder; eyedropper; wax pencil; safety goggles; lab aprons; latex gloves (additional teacher materials: 500 mL labeled container, about 300 mL of 0.1 M sodium hydroxide solution per class, a few sheets of newspaper; see page 305.)

Teaching Resources
Exploration Worksheets, pp. 50 and 51

The Mouth: The mouth is the starting place. Here the food is chopped up into smaller bits in preparation for the processes to come.

The Salivary Glands: Located at the back of the mouth, the salivary glands start producing saliva even at the thought of food. The smell and taste of food make them work even harder.

The Stomach: Things start getting interesting here. This bag-shaped region can hold 2.4 L of food. Most food passes out of the stomach within 2.5 hours, but some kinds remain much longer. The stomach acts like a churn by mixing the food with digestive juices and then forcing the mixture into the small intestine.

The Liver: This character could be referred to as the clearinghouse of the body. It removes poisons from the blood. It also produces a substance called *bile,* which is necessary for digestion. The liver's most deadly enemy is alcohol. Too much over a long period of time causes the liver to become enlarged with yellow fat—a condition called cirrhosis. However, once the intake of alcohol is stopped, the liver can regenerate itself.

The Gallbladder: The gallbladder is located just below the liver and stores the bile produced by the liver until it is needed.

The Pancreas: Located just below the stomach, it produces enzymes needed for digestion. *Enzymes* are proteins that assist in chemical reactions in the body. These enter the play during the climax.

The Small Intestine: The small intestine is not very small! It is the longest of all the characters, consisting of 600–700 cm of 4 cm wide tubing coiled below the stomach and between parts of the large intestine. Because its interior is folded into thousands of projections, it has the same surface area as a tennis court! This large area is covered with hair-like projections called villi. These absorb water-soluble nutrients and pass them into the bloodstream, which flows just inside the interior walls of the small intestine.

The Large Intestine: Our story of digestion ends with the large intestine, which is a 6.5 cm wide tube about 180–190 cm long. Food that cannot be digested is passed on from the large intestine, eventually leaving the body as a waste product.

Act 1: Digestion Begins

Even at the mere thought of food, the salivary glands start producing saliva. In one day, these glands can produce 1.5 L of saliva. Saliva contains an essential enzyme called amylase that is needed to start digesting the carbohydrates in food. Get into the act by doing the following Exploration to discover more about this stage of digestion.

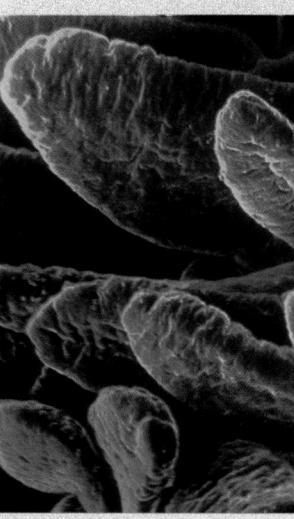

↑ The villi in the small intestine absorb the products of digestion.

CROSS-DISCIPLINARY FOCUS

Health

Have interested students find out what an appendix is and answer the following questions: What does the appendix do in humans and other mammals? Why do humans have an appendix? What is appendicitis and how is it treated? Students may report their findings on a poster or in an oral report. *(In humans, the appendix is a narrow tube attached to the large intestine. It probably has no purpose in humans, although it functions in digestion in animals such as rodents. Many scientists hypothesize that the human appendix lost its function during the course of human evolution. Appendicitis is an infection of the appendix that is usually treated by surgically removing the appendix.)*

Integrating the Sciences

Life and Physical Sciences

Cirrhosis is a disease of the liver caused by the overconsumption of alcohol. Have students research cirrhosis and the health problems related to it. In addition, have students find information about how to treat cirrhosis.

Did You Know...

The production of the gastric juices that the stomach uses in the digestive process depends to some extent on the emotional state of the individual. Both fear and sadness reduce the amount of gastric juices in the stomach, which is why we often don't feel hungry when we are afraid or unhappy.

Meeting Individual Needs

Second-Language Learners

Have students make a poster of the digestion process. Using both English and their native language, students should include labels for the digestive organs and descriptions of what is taking place. Grade the posters according to their scientific accuracy, with minimal emphasis on language proficiency.

Exploration 5 shows students a simulation of the digestion of starch. Students should learn what happens to starch when it encounters simulated saliva. To save time, you may wish to prepare the simulated saliva ahead of time.

Answer to
Exploration 5

8. The test tube containing only water and amylase solution serves as a control for the experiment.

Theme Connection

Energy
Focus question: How is the burning of fuel in a car engine similar to the way that we get energy from food? *(Both engines and humans require fuel [gasoline or food] to function. In both cases, chemical changes occur when the fuel is burned, releasing energy. Oxygen is used to burn the fuel. In addition to releasing energy, carbon dioxide and water are also produced.)* Students will learn more about the energy in food in Unit 7, Temperature and Heat.

CROSS-DISCIPLINARY FOCUS

Social Studies
The pseudoscience of alchemy was the forerunner of the modern science of chemistry. Have students research alchemy. They could focus on what an alchemist does, investigate some famous alchemists, or compare alchemy with chemistry. As a challenge, students can draw an analogy between chemistry and alchemy, and astronomy and astrology.

 An Exploration Worksheet is available to accompany Exploration 5 (Teaching Resources, page 50).

Digestion of Starch

You Will Need
- 3 test tubes
- a test-tube clamp
- a beaker
- a wax pencil
- a graduated cylinder
- water
- a hot plate
- Benedict's solution
- a starch sample (a crushed, unsalted cracker)
- an eyedropper
- a test-tube rack
- simulated saliva (1% amylase solution)
- a lab apron
- latex gloves
- a watch or clock

Caution: For this experiment, you will use simulated saliva instead of your own. To make simulated saliva, mix 1 g of amylase with 10 mL of water. Do not use your own saliva for this experiment. Wear your safety goggles, apron, and latex gloves during the entire Exploration. If you get Benedict's solution on your skin, wash it off immediately. Be careful when using the hot plate.

What to Do
1. Number the test tubes 1, 2, and 3.
2. To test tube 1, add a pinch of the crushed cracker, and cover the cracker with a dropperful (2 mL) of the simulated saliva. Shake the tube gently to mix the substances.
3. To test tube 2, add a pinch of the crushed cracker and a

dropperful (2 mL) of water. Mix the substances.
4. To test tube 3, add 2 mL of the simulated saliva and a dropperful (2 mL) of water.
5. After 15 minutes, test the substances in each test tube for sugar, as described in Step 6.
6. Using a test tube clamp, place each test tube in a bath of boiling water for 5 minutes. Remove the test tubes from the beaker, and then add 5 mL of Benedict's solution to each test tube. A yellow or reddish precipitate indicates the presence of sugar.

Test tube	Contents	Results of sugar test
1	starch sample and simulated saliva	
2	starch sample and water	
3	water and simulated saliva	

7. Record your results in a table similar to the one above.
8. In the third test tube, why did you add only water and amylase solution?
9. Wash your hands and clean up your materials before leaving the lab.

Act 2: Pepsin—More Than a Minor Player

In the stomach, starch continues to be broken down into sugar by the saliva swallowed with the food. But here the digestion of proteins begins as well. Glands in the stomach wall secrete liquids called gastric juices. These juices contain mainly hydrochloric acid and an enzyme called pepsin. During this phase of digestion, very large protein molecules are broken down into smaller units that will eventually pass into the bloodstream.

Simulating Digestion in the Stomach

In this Exploration, you will simulate the digestion of proteins by gastric juices in the stomach.

You Will Need

- a hard-boiled egg
- a knife
- 0.2 g of pepsin
- dilute hydrochloric acid
- 2 test tubes
- a test-tube rack
- a small beaker
- water
- a graduated cylinder
- an eyedropper
- a wax pencil
- a stirring rod
- latex gloves

What to Do

1. Label the test tubes 1 and 2. Cut the egg white into six small cubes that are approximately 0.5 cm × 0.5 cm. Add three of these pieces to each test tube.

2. Wear your safety goggles, apron, and latex gloves for the remainder of this Exploration. Make simulated gastric juices by mixing the following in a beaker: 10 mL of water, 0.2 g of pepsin, and 2 mL of dilute hydrochloric acid. Use a stirring rod to mix the ingredients.

3. Add 5 mL of the simulated gastric juices to one test tube and 5 mL of water to the second test tube.

4. The next day, examine the pieces of egg white in each test tube. In what ways do they appear similar? In what ways do they appear different? Explain your observations.

5. Wash your hands and clean up your materials before leaving the lab.

Act 3: Enter the Small Intestine and Liver

The play comes to its climax in the small intestine. The digestion of carbohydrates and proteins continues, and the digestion of fat begins. Here are some of the players and their roles:

- The pancreas secretes its enzymes, which act on all three food nutrients. It also produces sodium hydrogen carbonate (also known as baking soda), which neutralizes any acid. What does it mean to neutralize an acid?

- Bile, produced in the liver and stored in the gallbladder, breaks up the fat into tiny particles so that fat-digesting enzymes can work on them.

- Other enzymes in the intestinal juices complete the digestion of proteins and carbohydrates. The end products are absorbed into the bloodstream and carried to every part of the body.

- The food that cannot be digested passes through the large intestine and eventually leaves the body.

Epilogue

1. Choose one of the characters that plays a role in digestion, and find out more about it from an encyclopedia or another source. Write a description of the role it plays as though you are that character.

2. Make a flowchart of the events occurring during digestion. One has been started for you below.

305

Exploration 6 is a simulation of the digestion of proteins by gastric juices in the stomach.

WASTE DISPOSAL ALERT For disposal, place the mixture of the egg white and simulated gastric juice from each group into a labeled container. Slowly add 0.1 M sodium hydroxide solution until the pH is between 6 and 8. Slowly pour the liquid down the drain. Wrap the solids in an old newspaper, and place them in the trash.

Answer to Exploration 6

4. After one day, the pieces of the egg white in the two test tubes do not differ greatly. However, the pieces of egg white in the test tube containing the simulated gastric juices show a slight difference in texture; they appear more brittle and appear to be covered with a slimy coating.

Answer to Act 3: Enter the Small Intestine and Liver

An acid is neutralized when it reacts with a base to form a salt.

Answers to Epilogue

Accept all reasonable responses. Student descriptions and flowcharts should demonstrate an understanding of the material presented on pages 302–305.

 An Exploration Worksheet is available to accompany Exploration 6 (Teaching Resources, page 51).

Homework

The Assessment activity at left makes an excellent homework assignment.

FOLLOW-UP

Reteaching

Have students write a story or illustrate a cartoon that follows a particle of food through the digestive system.

Assessment

Have students create a concept map using the following words: digestive substances, saliva, stomach, carbohydrates, liver, gastric juices, bile, salivary glands, proteins, and fats. (Accept all reasonable responses.)

Extension

Have students do some research to find out about the importance of fiber in their diet. Why is it important? How does it aid digestion? Is it important in preventing disease?

Closure

Have students review the story of digestion and identify all of the places where chemical digestion takes place and where mechanical digestion takes place.

Answers to *Challenge Your Thinking*

1. Sample answer: Place samples of all five liquids into separate test tubes. Add phenolphthalein. The basic solution will turn pink. Now add two drops of the base to each of the other liquids. The water samples should now be basic and will turn pink with phenolphthalein. The acid solutions will still be colorless.

2. The vinegar will dissolve any lime or scale deposits that have built up due to hard water, eliminating them from the coffee maker.

3.
- Fluorine, chlorine, bromine, and iodine are in the same family.
- Helium, neon, and krypton are in the same family.
- Nitrogen and arsenic are in the same family.
- Boron, nitrogen, oxygen, fluorine, and neon are in the same period.
- Arsenic, bromine, and krypton are in the same period.

4. a. Tonya should conclude that some of the oxygen in the air inside the graduated cylinder reacted with the iron (steel wool) to form iron oxide (the reddish compound on the steel wool). As oxygen was used up in the chemical reaction, the water level in the graduated cylinder rose.

b. To simplify the calculations for part (c), have students adjust the height of the graduated cylinder so that the water level inside and outside is at the 100 mL mark.

c. Students can estimate the percentage of oxygen in the air by dividing the change in the volume of air in the graduated cylinder by the initial volume of air in the graduated cylinder. This method assumes that there was enough steel wool to exhaust the supply of oxygen in the graduated cylinder.

 You may wish to provide students with the Chapter 16 Review Worksheet that accompanies this Challenge Your Thinking (Teaching Resources, page 52).

CHALLENGE YOUR THINKING

1. Mystery Liquids

Sandra has been given five bottles labeled *P*, *Q*, *R*, *S*, and *T*. There is a colorless liquid in each. She is told that two of the liquids are dilute acids, one is a base, and the other two are water. She is also given the acid-base indicator phenolphthalein.

Write down some instructions for Sandra to help her find out whether each liquid is an acid, a base, or water.

2. A Perfect Cup

The following advice is included in the instruction manual for a coffee maker: Every month or so, run vinegar through your coffee maker. Why do you think this is suggested?

3. A Family Affair

Which of the chemicals listed below belong to the same chemical families? Which ones belong to the same periods? Refer to the periodic table on pages 290–291.

boron, oxygen, helium, fluorine, neon, arsenic, krypton, bromine, nitrogen, iodine, and chlorine

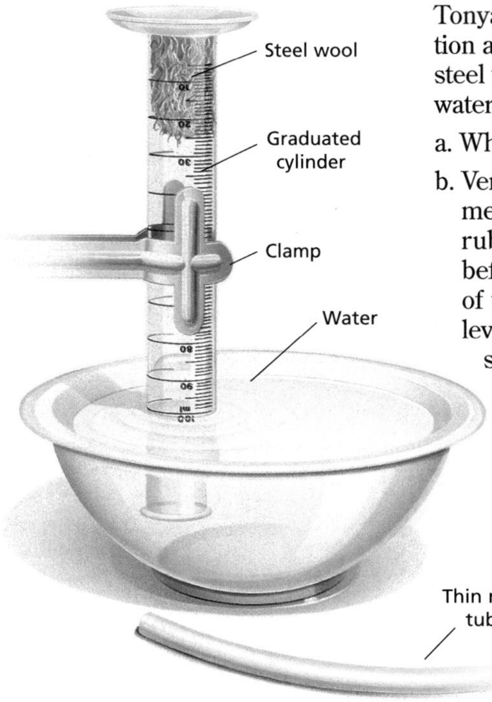

4. See For Yourself

Tonya set up the experiment shown in the illustration at left. Several days later she noticed that the steel wool had a red-yellow color and that the water in the graduated cylinder had risen 20 mL.

a. What conclusions should Tonya make?

b. Verify Tonya's results by trying the experiment yourself. Place one end of a piece of rubber tubing into the graduated cylinder before inverting the cylinder into the container of water. This will help to equalize the water level inside and outside the cylinder. You can speed up the reaction by soaking the steel wool in vinegar before placing it in the graduated cylinder. Be sure to drain the excess vinegar from the steel wool.

c. Based on this experiment, can you estimate the percentage of oxygen in the air?

Steel wool

Graduated cylinder

Clamp

Water

Thin rubber tubing

ENVIRONMENTAL FOCUS

Ask: Why is acid rain such a threat to ecosystems? *(Animals are adapted to live in an environment with a particular acidity. If that acidity changes suddenly, many of those animals can no longer survive. Problems are worse in the spring, when acidic snows melt and rush into lakes and rivers, causing an acid shock that can kill fish.)* Encourage discussion of what can be done to fight acid rain.

5. A Burning Question

In the discussion about the nature of burning, the word *theory* was used several times. What do you think it means? Check a dictionary for its meaning.

6. Basically, It's Acidic

Many colored plant parts can act as an acid-base indicator. Simply boil the plant part in water for 45 minutes to extract the color. You can use red apple peelings, beets, blueberries, red onions, and grape juice. Try at least one of these to see if it makes a good acid-base indicator. How will you test your home-made indicator? Compare your results with those of others who tried the same indicator. Can you think of other plant parts that could be tried?

7. Chemical Words

concept map — The language of acids and bases includes the following words. Make a concept map in your ScienceLog using these words. Try to add at least one other word to your map.

acid, base, bitter, sour, indicator, neutralization, compound

Sciencelog

Review your responses to the ScienceLog questions on page 287. Then revise your original ideas so that they reflect what you've learned.

307

5. A theory is a statement that attempts to explain why something happens. A theory should be supported by evidence gained from testing and observation. Because it is supported by evidence, a theory is less of a guess than is a hypothesis. However, the two words are often used interchangeably.

6. To test a homemade indicator, add a drop of an acid or base to the indicator solution. (Solid materials should be mixed with distilled water before testing.) Vinegar and lemon juice are common acids that could be tested. Bleach, ammonia, and colorless detergent are common bases. Additional plant parts that could be tried include parts of geraniums, cornflowers, raspberries, strawberries, and blackberries.

7. The following is a sample concept map to which the word *salt* has been added:

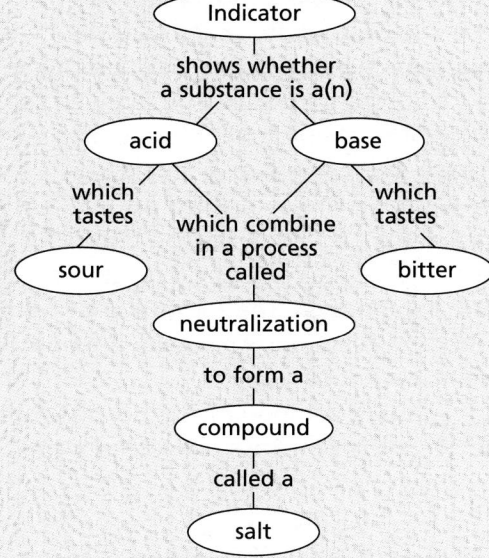

Sciencelog

The following are sample revised answers:

1. An element is a substance that cannot be broken down during chemical reactions. However, elements can combine with other elements during chemical reactions to form compounds.

2. In this chapter, students will encounter the following two reactions that involve burning: iron (steel wool) + oxygen → iron oxide; candle + oxygen → carbon dioxide + water.

3. Examples of common acids include vinegar, citrus juices, and coffee. Examples of common bases include baking soda, ammonia, and drain cleaner.

4. There are a number of chemical changes involved in digestion. Saliva contains an enzyme called amylase that begins to break down carbohydrates. Once in the stomach, gastric juices containing hydrochloric acid and pepsin break down large protein molecules. In the small intestine, enzymes from the pancreas continue to act on all food nutrients, while bile from the liver breaks up fat into tiny particles. The usable end products of digestion are carried in the bloodstream to every part of the body, while waste material passes through the large intestine and leaves the body.

The Big Ideas

***The following is a sample
unit summary:***

A chemical is a single substance with distinctive properties. All matter is made of chemicals. (1) Chemicals are used in our everyday lives in activities such as cleaning and cooking and in medicines. (2) A chemical change takes place when one or more substances are converted into different substances with different properties. (3) Some examples of chemical changes include the formation of a salt, rusting, and burning. (4) Signs that a chemical change has occurred include formation of a precipitate, production of a gas (bubbling), release of heat or light, change in color, change in odor, formation of a new substance, and difficulty in reversing the change. (5)

A physical property is one that can be observed without producing a new kind of matter. Color, size, shape, boiling point, and melting point are physical properties. Chemical properties can be observed only when a chemical change is involved. Chemical properties determine how a substance will react in a given situation. The ability to rust in the presence of oxygen is a chemical property. (6)

Elements are composed of only one type of matter. Elements combine with other elements in chemical reactions to form compounds. The formation of compounds in reactions is an example of a chemical change. (7)

Acids are compounds that react with metals and bases. Bases are compounds that react with acids. An acid and a base react to form a salt. (8) Burning (combustion) is a fast reaction in which a substance combines with oxygen to form carbon dioxide and water; heat and light are also produced. Rusting is a much slower chemical change in which a metal combines with oxygen to form iron oxide, or rust. (9) In a chemical reaction, the total mass of the products is the same as the total mass of the reactants as long as all the products and reactants are accounted for, including any gases produced during the reaction. (10)

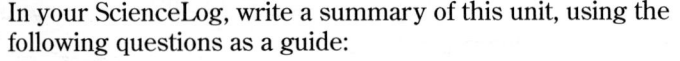

Unit 5
CHEMICAL CHANGES

SOURCEBOOK

Now that you have observed some chemical changes, read the SourceBook to learn how the structure of atoms affects chemical reactions. The SourceBook describes how atoms were discovered and how to use the periodic table to understand the chemical properties of the elements.

**Here's what you'll find
in the SourceBook:**

UNIT 5
Atomic Structure S108
Elements and Chemical
    Activity S117
Putting Atoms Together S124

308

The Big Ideas

In your ScienceLog, write a summary of this unit, using the following questions as a guide:

1. What are chemicals?
2. How do we use chemicals in our daily lives?
3. What happens during chemical changes?
4. What are some examples of chemical changes?
5. What are some signs of chemical changes?
6. How do physical and chemical properties differ?
7. How are compounds, elements, and chemical changes related?
8. What are acids and bases? How do they interact?
9. How are burning and rusting similar? different?
10. How is the mass of the reactants related to the mass of the products in a chemical reaction?
11. What are some of the chemicals involved in digestion, and where are they produced?

Checking Your Understanding

1. From the descriptions below, would you expect each change to be chemical or physical? If you can't tell, explain why.
   a. A colorless liquid disappears when heated.
   b. A colorless liquid turns yellow when heated.
   c. A flash of light occurs.
   d. When two chemicals are mixed, the temperature rises 20°C.

Some of the chemicals involved in digestion include amylase, an ingredient of saliva produced at the back of the mouth; hydrochloric acid and pepsin, produced in the stomach; bile, produced in the liver; and baking soda, produced in the pancreas. (11)

 You may wish to provide students with the Unit 5 Review Worksheet that accompanies this Making Connections (Teaching Resources, page 59).

Homework

The Unit 5 Activity Worksheet on page 58 of the Teaching Resources booklet makes an excellent homework activity. If you choose to use this worksheet in class, Transparency 53 is available to accompany it.

e. When two clear liquids are mixed together, the resulting mixture is cloudy.

f. A liquid is cooled, and a solid forms on the bottom of the beaker.

g. A bar turns black when heated.

h. A bar remains the same color when heated, but it becomes 2 mm longer.

i. A yellow solid is heated, and the room is filled with choking fumes.

j. A blue solid is dropped into water, and the water turns blue in color.

2. How could you prove or disprove the following statements?
   a. A burning match produces carbon dioxide.
   b. Water is given off when blue copper sulfate is heated.
   c. The upper surface of a plant leaf gives off water.
   d. The bubbles given off by soda water and the bubbles found in tap water are different substances.

3. Your friend Marcus has a theory: All of the caves in the world have formed in the last few hundred years. Marcus bases his theory on the following points:
   • Caves form through the action of ground water on limestone.
   • The ground water has to be acidic to dissolve the limestone.
   • Ground water begins as rainfall; therefore, rain must be acidic.
   • Acid rain is caused by people burning fossil fuels.
   • People have used fossil fuels only for the last few hundred years. Therefore, all of the caves in the world have formed in the last few hundred years!

   Is Marcus right? If not, where did he go wrong? If necessary, do a little research to answer this question.

4. In this unit you have used the following kitchen chemicals: sugar, salt, baking soda, baking powder, and starch. Develop a system for identifying these chemicals without tasting them. Base your system on your knowledge of the changes that occur when using the reactants iodine solution, water, and vinegar.

5. [concept map] Make a concept map using the following words and phrases: *chemicals, elements, compounds,* and *broken down.* You will use one word or phrase twice.

A photomicrograph of sodium hydrogen carbonate (baking soda) crystals

309

---

j. Physical—The blue solid has dissolved and can be recovered by evaporating the liquid.

2. a. Sample answer: Drop a match in a bottle and let it burn. Then add limewater and shake. If the limewater turns milky, carbon dioxide was produced.
   b. Heat enough copper sulfate in a jar to collect 1 mL of the liquid that condenses inside the jar. If 1 mL of this liquid has a mass of 1 g, then it is probably water.
   c. Collect enough of the liquid from the top of the leaf to find its boiling point and freezing point. If the liquid boils at 100°C and freezes at 0°C, then it is water.
   d. Pass bubbles from both sources through limewater, and see whether a precipitate forms. The carbon dioxide in the soda water turns the limewater cloudy.

3. Marcus was correct up until his fourth point. Rainfall becomes acidic not only from human-made pollutants but also from naturally occurring oxides in the atmosphere and in ground water. Therefore, it would not be possible to determine when the limestone was dissolved.

4. A system for identifying these chemicals could be based on the presence or absence of the chemical reactions described below.
   Iodine starch-test reagent— Although the other substances may react with iodine, only starch reacts with iodine to produce a blue-black color.
   Water—Sugar and salt are soluble in water. Baking soda is fairly soluble. Baking powder will react with water to form carbon dioxide gas.
   Vinegar—Baking soda and baking powder will react with vinegar to form carbon dioxide. Vinegar has no effect on the other chemicals.
   None of these reactions can distinguish salt from sugar. However, their physical properties, such as shape of the granules, can be used as an indication.

5. Sample concept map:

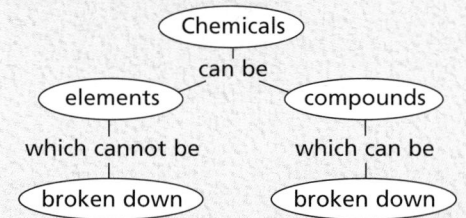

---

1. a. Physical—evaporation, change of state
   b. Chemical—color change
   c. Physical or chemical—If a flash of light occurs and no new material is produced, then the change is physical. If the flash causes something to burn or explode, then the change is chemical.
   d. Physical or chemical—A change in temperature can be caused by both physical and chemical changes.

   e. Chemical—color change; a precipitate forms.
   f. Physical—Crystallization is occurring.
   g. Chemical—A coating could have formed when oxygen combined with the metal. Or, if the bar was heated over a candle, the black color could be from soot produced by the chemical reaction of the candle's burning.
   h. Physical—The bar expanded when heated.
   i. Physical or chemical—Either the material could have vaporized, or a new gas could have been formed.

Background

The different colors in a fireworks display depend on the wavelengths of light emitted by different chemicals as they burn. Light energy of the shortest wavelength appears violet in color. Light energy of the longest wavelength appears red. When the fuse in the fireworks is lit, the gunpowder ignites and produces gases that propel the fireworks into the air. Charcoal gives the fireworks a sparkling, flaming tail. Refer to the chart below for the colors produced by various elements.

ELEMENT	COLOR
sodium	yellow
barium	green
nickel	green
copper	blue
strontium	crimson
lithium	bright red
calcium	dark red
magnesium	white

Heat is necessary to start the reaction in a fireworks display. If fireworks are not packed correctly, the thermal reaction fails, and the failed firework is called a *dud.*

Extension

To show students that each element burns a certain color, perform the following demonstration: Obtain samples of calcium chloride, strontium chloride, and sodium chloride. To prepare 0.5 M solutions, dissolve the following quantities in separate containers with enough water to make 100 mL: 5.5 g of $CaCl_2$, 13.33 g of $SrCl_2 \cdot H_2O$, and 2.9 g of NaCl. Dip a wooden splint in each solution, and use tongs to insert the splint into the flame of a portable burner to burn the chemical from the splint. Try not to ignite the splint. The splints can be dipped in the solutions again, if necessary.

The Science of Fireworks

What do the space shuttle and the Fourth of July have in common? The same scientific principles that help scientists launch a space shuttle also help pyrotechnicians create spectacular fireworks shows. The word *pyrotechnics* comes from the Greek words for "fire art." Explosive and dazzling, a fireworks display is both a science and an art.

An Ancient History

Over 1000 years ago, Chinese civilizations made black powder, the original gunpowder used in pyrotechnics. They set off firecrackers and primitive missiles with the substance. Black powder is still used today to launch fireworks into the air and to give fireworks an explosive charge. Even the ingredients—saltpeter (potassium nitrate), charcoal, and sulfur—haven't changed since ancient times.

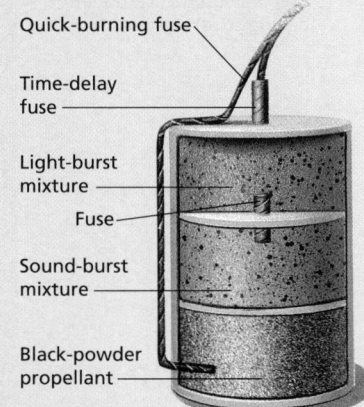

Quick-burning fuse

Time-delay fuse

Light-burst mixture

Fuse

Sound-burst mixture

Black-powder propellant

◀ **Cutaway view of a typical firework. Each shell creates a different type of display.**

Snap, Crackle, Pop!

The *shells* of fireworks contain the ingredients that create the explosions. Inside the shells, black powder and other chemicals are packed in layers. When ignited, one layer may cause a bright burst of light while a second layer produces a loud booming sound. The shell's shape affects the shape of the explosion. Cylinder-shaped shells produce a trail of lights that looks like an umbrella. Round shells produce a star-burst pattern of lights.

The color and sound of fireworks depend on the chemicals used. To create colors, chemicals like strontium (for red), magnesium (for white), or copper (for blue) can be mixed with the gunpowder.

Explosion in the Sky

Fireworks are launched from metal, plastic, or cardboard tubes. Black powder at the bottom of the shell explodes and shoots the shell into the sky. A fuse begins to burn when the

▲ **The firework on the left was packaged as a cylinder; the one on the right was packaged as a disk.**

shell is launched. Some seconds later, when the explosive chemicals are high in the air, the burning fuse lights another charge of black powder. This ignites the rest of the ingredients in the shell, causing the explosion that lights up the sky!

Bang for Your Buck

The fireworks used during New Year's Eve and Fourth of July celebrations can cost anywhere from $200 to $2000 a piece. Count the number of explosions that occur at the next fireworks show you see. If each of the fireworks cost just $200 to produce, how much would the fireworks for the entire show cost?

Answer to
Bang for Your Buck

Students should be aware that a fireworks display is costly. If a fireworks display consisted of 50 explosions at $200 per explosion, the display would cost $10,000.

Marie Curie (1867–1934)

In 1893, Marie Curie became the first woman to receive a physics degree from the Sorbonne, a prestigious university in France. In addition, she graduated at the top of her class. After earning her degree, she continued to pursue a career in science, a field that few women had previously entered. Marie Curie went on to do research that helped launch the world into the nuclear age.

A Devoted Student

At a very young age, Marie (born Manya Sklodowska in Poland) showed a remarkable talent for science and mathematics. She also became widely known for her astounding memory.

When Marie was in her teens, her family suffered financial problems, forcing her to go to work. She managed to save enough money to put herself and her sister through school. At the time, women in Poland were not allowed to attend universities, so when Marie was accepted to the Sorbonne, she moved from Poland to France. Once there, she quickly gained a reputation as a talented student and a tireless worker.

The Curie Partnership

In 1895, Marie married Pierre Curie, a chemist. They had a daughter, Irene, who later became a prize-winning scientist herself. The Curies' partnership led to a series of remarkable scientific discoveries. In 1903, Marie and Pierre shared the Nobel Prize in physics with Henri Becquerel for the discovery of *radioactivity*. Radioactive elements, such as uranium, give off high-energy particles as the elements decay.

Together, Marie and Pierre discovered the existence of the radioactive elements polonium (named after Poland) and radium, the most radioactive substance found in nature. The couple went on to learn more and more about radiation until Pierre was tragically killed by a runaway wagon in 1906.

Polonium

▲ **During radioactive decay, certain elements, such as polonium, give off high-energy particles. This process causes the element itself to change form.**

The Price of Discovery

Marie continued her work alone. In 1911, she earned the Nobel Prize in chemistry for the isolation of radium. She had spent years isolating about 1 g of the element from about a ton of pitchblende (a mineral containing uranium). Marie carried

▲ **Marie Curie with daughter, Irene, in her laboratory.**

out this difficult task with only the most primitive scientific equipment.

During her lifetime of research, Marie was often exposed to high doses of radiation, and eventually, this radiation killed her. In 1934 she died of leukemia—one of the diseases caused by exposure to dangerous amounts of radiation.

Radiation Education

Why are some elements radioactive? Why do people get sick from too much radiation? How do we use radioactive elements today? Think of some of your own questions and concerns about radiation. Then do some research to find out the answers. Share your answers with your classmates.

311

Background

Marie Curie discovered that radioactivity was linked to something inside the atoms of an element, not just to the quantity of the element. To perform her analyses, she would grind 20 kg of pitchblende at a time in an iron pot. She would dissolve, filter, crystalize, and collect it and then repeat the process. Marie and her husband isolated the element polonium in pitchblende and found that it was 300 times more radioactive than uranium. Radium, another radioactive element discovered in pitchblende, was found to be 900 times as radioactive as uranium.

Radium was used as a treatment for cancer until the mid-1950s. Doctors now use other sources of radiation because radium produces so much high-energy radiation that it can be harmful to human health. In an attempt to prove that radiation is not harmful, Pierre Curie once took a sample of radium and strapped it to his arm. Because the large burn that was produced healed, he thought he had proved radiation to be harmless. However, both of the Curies suffered from radiation sickness unknowingly. They both experienced chronic fatigue, one of the symptoms of radiation sickness.

Curium is a radioactive element that was discovered in 1944 and named after Marie Curie.

Theme Connection

Energy

Radiation is an important type of energy and has many uses. **Focus question:** Of what importance is radiation in medicine, industry, and scientific research? *(Sample answer: X rays detect broken bones and cavities, radiation kills cancer cells, nuclear power plants produce energy, and carbon dating uses radioactivity to determine the ages of objects.)*

Did You Know...

The Geiger counter (also known as the Geiger-Müller counter) is an instrument that detects radioactivity. The Geiger counter was invented in 1912 by Hans Geiger. Geiger later improved the design with the help of a fellow German physicist, Walther Müller.

Answers to
Radiation Education

All elements that have an atomic number of 83 or greater are radioactive. Because the nucleus is so large, it is unstable and emits particles of high-energy radiation. People get sick from too much radiation because the radiation interferes with the normal chemical processes in living cells.

Background

The interlocking hexagons and pentagons of Fuller's geodesic domes provide great stability because they distribute stress evenly. The buckyball (C_{60}) is one member of a large family of carbon "cages" called fullerenes. Fullerenes with fewer than 60 carbon atoms are called buckybabies. Buckytubes have more than 60 carbon atoms and are shaped like cylinders of spiraling honeycombs.

Scientists know that high temperatures are needed for buckyballs to form in nature. Thus, researchers searching for buckyballs have concentrated on intensely heated sites, such as asteroid craters and areas of lightning strikes. Buckyballs are sometimes formed in the crevices between rocks at these sites. The molecules are found in their greatest abundance in soot.

Meeting Individual Needs

Second-Language Learners

Richard Smalley, one of the discoverers of buckyballs, used a paper model to determine its structure.

Have students use pentagons and hexagons to make a model of a buckyball. For each model cut out 12 pentagons and 20 hexagons from card stock or cardboard. The edges of all the shapes should be of equal length. Using tape, assemble the models. Mark the position of carbon atoms with a permanent marker. Discuss how the buckyball is a cage that could trap another atom. (*Heating a buckyball opens the cage.*) Assess student work for scientific content, not language proficiency.

Buckyballs

Researchers are scrambling for the ball—the buckyball, that is. This special form of carbon has 60 carbon atoms linked together in a shape much like a soccer ball. Scientists are having a field day finding new uses for this unusual molecule.

The Starting Lineup

Named for architect Buckminster Fuller, buckyballs resemble the geodesic domes that are characteristic of the architect's work. Excitement over buckyballs began in 1985 when scientists used a laser to strike a piece of graphite. In the soot that remained, researchers found a completely new kind of molecule! Buckyballs may also be found in smoke from a candle flame. Some scientists claim to have detected buckyballs in outer space. In fact, one hypothesis suggests that buckyballs may be at the center of the condensing clouds of gas, dust, and debris that form galaxies!

The Game Plan

Ever since buckyballs were discovered, chemists have been busy trying to identify their properties. One interesting property is that other substances can be trapped inside a buckyball.

A buckyball can act like a cage that surrounds smaller substances such as individual atoms. Buckyballs also appear to be both slippery and strong. They can be opened to insert materials, and they can even link together in tubes.

What can buckyballs be used for? They may have a variety of uses, from carrying messages through atom-sized wires in computer chips to delivering medicines right where the body needs them. Making tough plastics and sharp cutting tools is also under investigation. With so many possibilities, scientists expect to get a kick out of buckyballs for quite some time.

Potassium atom trapped inside Buckyball

Carbon atoms

Bond

▲ **Buckyball, short for buckminsterfullerene, named after Buckminster Fuller.**

The Kickoff

A soccer ball is a great model for a buckyball. On the model, the places where three seams meet correspond to the carbon atoms on a buckyball. What represents the bonds between carbon atoms? Does your soccer-ball model have space for all 60 carbon atoms? You'll have to count and see for yourself.

Meeting Individual Needs

Gifted Learners

Challenge students to make a model of a buckytube using clay, toothpicks, papier-mâché, or other available materials that are safe to use. Have students design their buckytube from the following description: A buckytube is a cylinder formed from a sheet of carbon atoms networked together to form pentagons.

Answers to
The Kickoff

The seams represent the bonds between atoms. A standard soccer ball has 60 places where three seams meet. In other words, the soccer-ball model has places for exactly 60 carbon atoms.

Drawing With Light and Chemicals

*P*hotography was unknown at the beginning of the nineteenth century. However, some artists at that time used a *camera obscura* to create realistic pictures. The camera obscura is basically a large, dark box with a pinhole in one side. Light enters through the pinhole on one side and projects an image on the opposite wall. Artists can sit inside the box and trace the image to make a realistic drawing. Soon, however, chemical reactions replaced artists' careful drawings, and the art of photography was born.

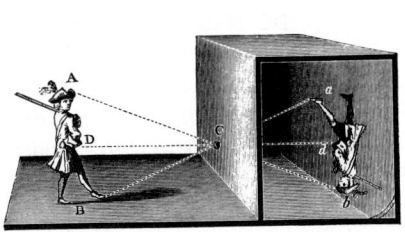

▲ **Eighteenth century engraving of a camera obscura**

Chemical Magic

In 1826, Joseph Nicéphore Niépce invented a process in

◄ **The first successful photograph, taken by Joseph Nicéphore Niépce.**

which light is used to record an image on an asphalt-coated pewter plate. The plate was placed in a small camera obscura and exposed to a city view. Where light struck the plate, the asphalt hardened. Once the exposure was complete, the plate was washed with a solvent, which dissolved any nonhardened asphalt. Although the resulting image is fuzzy, it is the first known photograph.

A Splash of Color

Modern photography uses more sophisticated chemistry. Chemicals called *silver salts* are used on color film that has three layers—each sensitive to a different color of light. When light hits the film, the places that are exposed undergo a chemical change. The film is then dipped into a chemical called *developer*,

▲ **Three layers of dyes (yellow, cyan, and magenta) are required to create a color photograph.**

which changes the silver salts into silver particles. As this happens, magenta, yellow, and cyan dyes are attached to the silver particles in the different layers. Finally, the silver is dissolved, leaving behind the colored dyes. Different combinations of these three colors produce all of the colors visible to the human eye.

Snapshot of Science

What kinds of pictures do scientific photographers take? Pretend that your job is to photograph one of the following: a beluga whale, the inside of a volcano, a blood cell, or a human fetus. Find out what special technology you would use to capture the image, and present your findings to the class.

Background

Several early photographers paved the way for modern photography. Louis Jacques Mandé Daguerre (1789–1851) created pictures called daguerreotypes using thin sheets of copper that were silver plated. Highly fragile, these plates had to be preserved behind glass. James Ambrose Cutting (1814–1867) received a patent for a photographic method that used dark red glass plates dipped in silver salts. The images appeared amber in color and were called ambrotypes. They were relatively inexpensive, and they were very popular in the United States. Hamilton L. Smith (1819–1903) made this process even more affordable by preparing images on less-fragile tin plates, hence the name tintypes.

Today, film is manufactured in many sizes to fit the type of camera being used. *Film speed* is the amount of time required for film to react to light. It is represented by a number, called an ISO number, which is printed on the film package. The speed of a film determines how much exposure to light is needed to record the image of an object. The higher the number, the faster the speed—this means that a high-speed film reacts quickly to light and needs little exposure. A photographer can take pictures in fairly dim light using film with an ISO number of 400. Film with an ISO number of 100 requires more light.

Film types include black-and-white film, color negative film, and color transparency film. Black-and-white film produces negatives and pictures in various shades of gray. Color negative film produces color negatives and prints. Color transparency film produces color slides.

Extension

1. Have students research the early days of photography from a technical and an artistic point of view. They should focus on a single photographer or photographic process.
2. Suggest that students investigate other photographic processes, such as autoradiography, medical X rays, and infrared photography.

CROSS-DISCIPLINARY FOCUS

Art

Allow students to observe 35 mm negatives showing simple black-and-white pictures. Discuss how the film captures a negative image of the subject. Ask students to copy the image onto a piece of paper. Students should shade the drawing so that they reverse the negative image into a positive image. Areas that are light on the negative should be dark on the students' drawings. Areas that are dark on the negative should be lightly shaded or white on the students' drawings.

Answers to *Snapshot of Science*

To photograph a whale, the photographer would have to use an underwater camera. Protective equipment or a long-distance lens would have to be used to photograph a volcano. A special attachment would have to be placed on a microscope if a scientist were to photograph a blood cell on a slide. If the blood cell was photographed inside the living organism, fiber optics could be used. To photograph a human fetus the photographer could use a sonogram.

Unit 6 ENERGY and YOU

Unit Overview

In this unit, students explore the concept of energy. In Chapter 17 students gain hands-on experience with energy by experimenting with energy-containing systems. They are introduced to the different forms of energy and the details of energy conversions. In Chapter 18, students explore electrical energy, including how it is generated and how it is measured. Chapter 19 takes a practical approach to saving energy; students audit their home's energy use and evaluate the costs of various energy consumption patterns. In Chapter 20, students explore the history of energy usage, and they research prospective energy resources.

Using the Themes

The unifying themes emphasized in this unit are **Energy, Systems, Changes Over Time,** and **Structures.** The following information will help you incorporate these themes into your teaching plan. Focus questions that correspond to these themes appear in the margins of this Annotated Teacher's Edition on pages 315, 321, 326, 329, 340, 351, 353, 358, and 366.

    Energy is clearly an important theme in this unit. Take every opportunity to relate the new concepts that students encounter to life and Earth sciences. Specifically, the body's intake and usage of stored energy can be compared to other living things, devices such as an automobile, the occurrence of an earthquake, or even a piece of paper burning. Also, the rate at which a body uses energy can be measured in the units that students encounter in Chapter 19.

    Systems is another important theme. In Chapter 18, students learn about the relationship between electricity and magnetism that is the basis of all electromagnetic systems.

    In Chapters 19 and 20, there are opportunities to emphasize the theme of **Changes Over Time.** This theme can be stressed in terms of the changing patterns of energy consumption both during a typical year and over the last two centuries as humans have come to rely more and more on technology.

    Finally, **Structures** is an important theme to emphasize in Chapter 19 because the principles of heat transfer can be applied both to organisms in life science and to our atmosphere in Earth science.

Using the SourceBook

Unit 6 in the SourceBook focuses on energy and the many forms in which it is manifest. Energy transformations are viewed in terms of the law of conservation of energy. Students also learn how energy flows through the environment and how this flow can be disrupted. Finally, the historical role of energy in society is discussed, and the possibilities of future sources of energy are explored.

Bibliography for Teachers

McKie, Robin. *Energy*. Science Frontiers series. New York City, NY: Hampstead Press, 1989.

National Research Council. *Fuels to Drive Our Future*. Washington, DC: National Academy Press, 1990.

Yanda, Bill. *Rads, Ergs, and Cheeseburgers: The Kid's Guide to Energy and the Environment*. Santa Fe, NM: John Muir Publications, 1991.

Bibliography for Students

Cohen, Bernard L. *The Nuclear Energy Option: An Alternative for the 90s*. New York City, NY: Plenum Press, 1990.

Gardner, Robert. *Experimenting With Energy Conservation*. New York City, NY: Franklin Watts, 1992.

Hansen, Michael C. *Coal: How It Is Found and Used*. Hillside, NJ: Enslow, 1990.

Pringle, L. *Global Warming*. New York City, NY: Arcade, 1990.

Films, Videotapes, Software, and Other Media

Electric Bill
Software (Apple II family)
Queue
338 Commerce Dr.
Fairfield, CT 06432

Energy at Work
Film or videotape
Churchill Media
12210 Nebraska Ave., Dept. 200
Los Angeles, CA 90025-9816

Energy Seekers
Film or videotape
Coronet/MTI
108 Wilmot Rd.
Deerfield, IL 60015

Unit Organizer

Unit/Chapter	Lesson	Time*	Objectives	Teaching Resources
Unit Opener, p. 314				Science Sleuths: The Energy Mystery House English/Spanish Audiocassettes Home Connection, p. 1
Chapter 17, p. 316	Lesson 1, Energy Is . . . ? p. 317	1	1. Create a working definition of energy.	Image and Activity Bank 17-1 Transparency Worksheet, p. 3 ▼
	Lesson 2, Energy in Action, p. 319	2 to 3	1. Classify energy as either stored or released. 2. Identify different forms of energy.	Image and Activity Bank 17-2 Exploration Worksheet, p. 5 ▼
	Lesson 3, The Energy Picture, p. 323	2	1. Describe how energy is converted from one form to another. 2. Explain how energy can be lost due to inefficiency. 3. Compare the efficiencies of various systems.	Image and Activity Bank 17-3 Discrepant Event Worksheet, p. 10
	Lesson 4, Putting Energy to Work, p. 328	1 to 2	1. Discuss how energy-converting devices are used to do work. 2. Explain why perpetual motion machines cannot exist. 3. Review a science-fair project intended to test the efficiency of different windmill designs.	Image and Activity Bank 17-4 Resource Worksheet, p. 11
End of Chapter, p. 333				Chapter 17 Review Worksheet, p. 12 ▼ Chapter 17 Assessment Worksheet, p. 15
Chapter 18, p. 335	Lesson 1, One Form of Energy: Electricity, p. 336	2	1. Distinguish between static and current electricity. 2. Generate and test for electricity from three different sources.	Image and Activity Bank 18-1 Exploration Worksheet , p. 18
	Lesson 2, Making Electricity From Scratch, p. 338	2 to 3	1. Describe a simple electricity generator. 2. Explain how the principles of a simple generator are adapted for large-scale use.	Image and Activity Bank 18-2 Resource Worksheet , p. 22 ▼ Transparencies 57 and 59
	Lesson 3, Measuring Electrical Energy, p. 342	1	1. Measure and compare the electrical energy consumption and production of various devices. 2. Plan an experiment to test the cost-effectiveness of different fuels.	Image and Activity Bank 18-3 Resource Worksheet, p. 23 ▼ Exploration Worksheet, p. 25
End of Chapter, p. 345				Graphing Practice Worksheet, p. 27 Chapter 18 Review Worksheet, p. 29 Chapter 18 Assessment Worksheet, p. 32
Chapter 19, p. 347	Lesson 1, The Energy Bottom Line, p. 348	2	1. Compare the amount of energy used by different appliances in one month. 2. Describe and conduct a home or school energy audit.	Image and Activity Bank 19-1 Exploration Worksheets, pp. 35 ▼ and 40 Theme Worksheet, p. 38 Transparency 63
	Lesson 2, A Plan for Saving Energy, p. 356	1	1. Suggest several ways to conserve energy. 2. Calculate the savings that result from conserving energy.	none
	Lesson 3, Keeping in the Heat, p. 358	2 to 3	1. Identify areas in a home where energy escapes. 2. Calculate the effectiveness of various insulating materials.	Image and Activity Bank 19-3 Exploration Worksheets, pp. 42 ▼, 44, 46 Activity Worksheet, p. 45
End of Chapter, p. 361				Chapter 19 Review Worksheet, p. 47 Chapter 19 Assessment Worksheet, p. 49
Chapter 20, p. 363	Lesson 1, Energy Then and Now, p. 364	1	1. Compare modern and historical energy sources. 2. Name several energy-related events in history.	Image and Activity Bank 20-1 Transparency 65
	Lesson 2, Bye-Bye Black Gold, p. 369	1	1. Classify energy sources as renewable or nonrenewable. 2. Make predictions about future sources of energy.	Image and Activity Bank 20-2 Transparency 66
	Lesson 3, Energy for the Future, p. 371	2 to 3	1. List the advantages and disadvantages of eight different energy sources.	Image and Activity Bank 20-3 Transparency Worksheets, pp. 52 ▼, 54 ▼ Exploration Worksheet, p. 56
	Lesson 4, What Does the Future Hold? p. 376	1	1. Describe how future energy supplies may affect society.	none
End of Chapter, p. 380				Chapter 20 Review Worksheet, p. 57 Chapter 20 Assessment Worksheet, p. 59
End of Unit, p. 382				Unit 6 Activity Worksheet, p. 62 ▼ Unit 6 Review Worksheet, p. 63 Unit 6 End-of-Unit Assessment, p. 66 Unit 6 Activity Assessment, p. 73 Unit 6 Self-Evaluation of Achievement, p. 76

* Estimated time is given in number of 50-minute class periods. Actual time may vary depending on period length and individual class characteristics.

▼ Transparencies are available to accompany these worksheets. Please refer to the Teaching Transparencies Cross-Reference chart in the Unit 6 Teaching Resources booklet.

Materials Organizer

Chapter	Page	Activity and Materials per Student Group
17	319	***Exploration 1, The Swinging Pendulum:** support stand with clamp; 20 cm of string; weight; **A Light Touch:** magnifying glass; small piece of newspaper, dry leaf, or small chip of wood; aluminum pie plate or lid from a jar; bucket of water; safety goggles; **The Spinning Spool:** empty spool of thread; 2 toothpicks; rubber band; washer; scissors; paper clip; **Hot Stuff:** candle; jar lid; modeling clay; matches; small aluminum pan or pie plate; a few copper sulfate crystals; small piece of plastic; drop of water; tongs or pliers; trivet or hot pad; safety goggles; **Chemical Magic!:** small alcohol thermometer; about 15 mL of vinegar; small piece of steel wool; 1 L jar with lid; small bowl or beaker; safety goggles; lab aprons; **Paper-Clip Magic:** bolt; 20 cm of insulated wire; 2 small pieces of masking tape; iron filings or a few paper clips; D-cell battery; wire strippers
18	336	***Exploration 1, Activity 1:** 1 m of fine, insulated wire with 2 alligator clips attached; wire strippers; compass; D-cell battery; **Activity 2:** 3 jars or 250 mL beakers; 3 strips each of copper and zinc metal; 25 mL of salt; 1 L of water; 2 alligator clips; low-voltage LED; 2 m of fine, insulated wire; wire strippers; cardboard bathroom-tissue tube; metric ruler; strong magnet; galvanometer from Activity 1; solar cell; light source (See Advance Preparation below.)
	344	**Exploration 2:** hot plate; portable burner (propane or propane-butane mixture); striker; two 600 mL beakers; support stand with ring clamp; wire gauze; 1 L of water; oven mitts; safety goggles; metric balance
19	358	**Exploration 3:** 2 cans, one that fits easily into the other; enough of one of the following insulating materials to line the larger can: sand, tissue paper, dried grass, cotton batting, wool cloth, plastic foam, sawdust; alcohol thermometer; small piece of plastic wrap; rubber band; metric ruler; 100 mL of 60°C water; graduated cylinder; watch or clock (See Advance Preparation below.)
	360	**Exploration 4:** small sheet of plastic wrap; clothes hanger; tape
20	371	**Exploration 2:** sheet of poster board; several colored markers

* You may wish to set up these Activities at
different locations around the classroom.

Advance Preparation

Exploration 1, Activity 2, A Liquid Cell, page 337: For best results, use steel wool to scour the copper and zinc strips before class. Also, the necessary salt solution can be made by dissolving 25 mL of salt in 1 L of water.

Exploration 2, page 355: Since students need to record data over a 10-day period for this Exploration, you may wish to have them begin collecting the data in advance.

Exploration 3, page 358: If hot tap water is not available, 100 mL of water must be heated to 60°C for each group.

Unit Compression

In Chapter 17, have students read the introductions to Lessons 1 and 3 ahead of time to speed up in-class discussion. In Lesson 3, time constraints may make it necessary for you to skip the last section, The Energy Efficiency Challenge. Similarly, the later sections of Lesson 4 (Perpetual Motion, Energy From Thin Air, and In the Hot Seat!) may be omitted without severely disrupting the flow of ideas.

Chapter 18 is not easily compressed. However, having students read Lessons 2 and 3 ahead of time will allow you to proceed more rapidly through this material in class.

The most important lessons in Chapter 19 are Lessons 1 and 2. Keeping these lessons intact is important, but you may find it necessary to omit The Cost of Comfort from Lesson 1 in order to save time. Lesson 3 may also be omitted if necessary.

Although the first two lessons of Chapter 20 should be covered if at all possible, keep in mind that most of Lesson 3 involves research that takes place outside of the classroom. If time becomes a problem, Lesson 4 of Chapter 20 may be omitted.

Homework Options

Chapter 17	See Teacher's Edition margin, pp. 318, 321, 324, 325, 329, and 332
	Resource Worksheet, p. 11
	SourceBook, pp. S136, S137, S142, and S143
Chapter 18	See Teacher's Edition margin, pp. 339, 340, 343, and 345
	Resource Worksheet, p. 22
	Resource Worksheet, p. 23
	Graphing Practice Worksheet, p. 27
	SourceBook, pp. S138, S148, and S150
Chapter 19	See Teacher's Edition margin, pp. 350, 355, and 360
	Exploration Worksheet, p. 40
	Exploration Worksheet, p. 44
	Activity Worksheet, p. 45
	Exploration Worksheet, p. 46
	SourceBook, p. S158
Chapter 20	See Teacher's Edition margin, pp. 367, 370, 376, and 380
	SourceBook, pp. S148 and S151
Unit 6	Activity Worksheet, p. 62
	Activity Assessment Worksheet, p. 73
	SourceBook Activity Worksheet, p. 77

Assessment Planning Guide

Lesson, Chapter, and Unit Assessment	SourceBook Assessment	Ongoing and Activity Assessment	Portfolio and Student-Centered Assessment
Lesson Assessment Follow-Up: see Teacher's Edition margin, pp. 318, 322, 327, 332, 337, 341, 344, 355, 357, 360, 368, 370, 375, and 379 **Chapter Assessment** Chapter 17 Review Worksheet, p. 12 Chapter 17 Assessment Worksheet, p. 15* Chapter 18 Review Worksheet, p. 29 Chapter 18 Assessment Worksheet, p. 32* Chapter 19 Review Worksheet, p. 47 Chapter 19 Assessment Worksheet, p. 49* Chapter 20 Review Worksheet, p. 57 Chapter 20 Assessment Worksheet, p. 59* **Unit Assessment** Unit 6 Review Worksheet, p. 63 End-of-Unit Assessment Worksheet, p. 66*	SourceBook Review Worksheet, p. 78 SourceBook Assessment Worksheet, p. 82* 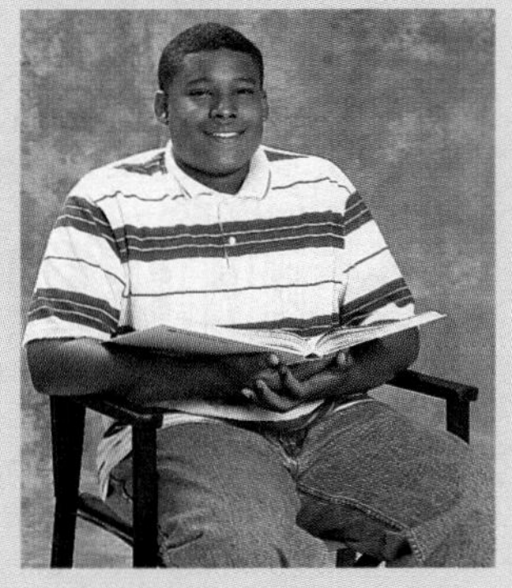	Activity Assessment Worksheet, p. 73* **SnackDisc** Ongoing Assessment Checklists ♦ Teacher Evaluation Checklists ♦ Progress Reports ♦	Portfolio: see Teacher's Edition margin, pp. 324, 341, 354, 368, 374, and 378 **SnackDisc** Self-Evaluation Checklists ♦ Peer Evaluation Checklists ♦ Group Evaluation Checklists ♦ Portfolio Evaluation Checklists 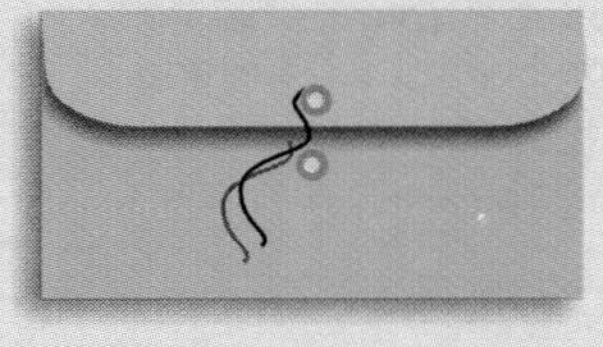 ♦

* Also available on the Test Generator software

♦ Also available in the Assessment Checklists and Rubrics booklet

Science Discovery is a versatile videodisc program that provides a vast array of photos, graphics, motion sequences, and activities for you to introduce into your *SciencePlus* classroom. *Science Discovery* consists of two videodiscs: Science Sleuths and the Image and Activity Bank.

Using the *Science Discovery* Videodiscs

Science Sleuths: The Energy Mystery House
Side A

Potential home buyers discover that the utility bills for the home of their choice are three times the normal amount for a house that size. The Science Sleuths must analyze the information for themselves to find out why the bills are so high.

Interviews
1. Setting the scene: Home buyers 22350 (play ×2)

2. Home sellers 23060 (play)

3. Real-estate agent 23737 (play)

4. Next-door neighbor 24695 (play)

5. Mailperson 25222 (play)

6. Furnace-maintenance person 25817 (play)

7. Building inspector 26402 (play)

Documents
8. Utility bills 27572

9. DOE brochure 27574 (step ×3)

10. List of appliances in house 27579

Literature Search
11. Search on the words: FUEL BILLS, FUEL EFFICIENCY, OLD HOUSES 27581

12. Article #1 ("Why Are Bills Higher?") 27583 (step)

Sleuth Information Service
13. Furnace-efficiency graph 27586

14. House heating plan 27588

15. Graph of average energy bills in neighborhood 27590 (step)

16. Energy usage of appliances chart 27593 (step ×2)

17. Annual temperature chart 27597

Still Photographs
18. Different parts of the old house 27599 (step)

19. Infrared photos of the house and other houses in neighborhood 27607 (step)

20. Attic insulation 27611

Image and Activity Bank
Side A or B

A selection of still images, short videos, and activities is available for you to use as you teach this unit. For a larger selection and detailed instructions, see the Videodisc Resources booklet included with the Teaching Resources materials.

17-1 Energy Is . . . ? page 317
Mount St. Helens eruption 1012–1021 (step ×9)
This series of clips shows how Mount St. Helens looked before, during, and after its eruption on May 18, 1980.

Tornado 14546–15296 (play ×2) (Side A only)
A tornado hit just outside of Minneapolis, Minnesota, in July 1986.

Aurora borealis 19012–19389 (play ×2)
This time-lapse photo was taken with an all-sky camera. The aurora borealis appears as luminous ribbons of light in the night sky. It is seen most vividly close to polar latitudes.

Space-shuttle takeoff 20048–20256 (play ×2)
A fisheye lens shows the space shuttle taking off.

17-2 Energy in Action, page 319
Battery, rechargeable 2123
Rechargeable batteries convert electrical energy back into chemical energy when they are recharging. Because disposable batteries are toxic, using rechargeable batteries cuts down the amount of toxic material in dumps and landfills.

Bowling-ball pendulum 29833–29974 (play ×2) (Side A only)
As long as energy is not added to the system, a pendulum does not gain energy in its oscillation. Therefore, it does not exceed its original starting point.

17-3 The Energy Picture, page 323
Energy; potential vs. kinetic 1869–1873 (step ×4)

Sailboat 2625
The kinetic energy of wind propels this sailboat.

Microwave oven 1799
Electricity creates high-energy radiation waves (microwaves) that interact with material placed in a microwave. Interaction between the microwaves and the material produces heat.

Converter efficiency 1853
Different energy converters have different efficiencies.

17-4 Putting Energy to Work, page 328
Turbine demonstration; Christmas ornament pinwheel 27261–27503 (play ×2) (Side A only)
The flames of the candles heat the air, which then rises and moves the blades. By this process, heat energy is converted to mechanical energy.

Hero's engine 25894–26316 (play ×2) (Side A only)
A Hero's engine converts steam energy into kinetic energy. The steam emerging from the bent-glass pipes creates a force in one direction; the engine reacts to the force by rotating in the other direction.

18-1 One Form of Energy: Electricity, page 336
Voltmeter 30311–30398 (play ×2) (Side A only)
Here a 1.5-volt battery is tested by a voltmeter to determine its voltage.

Magnetically induced current in wire 32745–33021 (play ×2) (Side A only)
The voltmeter shows a current generated by a constantly changing magnetic field. Note that the voltage is positive when the magnet is moved one way and negative when the magnet is moved the other way.

18-2 Making Electricity From Scratch, page 338
Dam, hydroelectric 1789–1790 (step)
The large amounts of water behind dams can be tapped to produce electricity. Dams convert the energy released by the water as it moves from a higher to a lower elevation. (step) Large turbines installed within generators capture the energy of the moving water. The rushing water pushes against the turbine blades that turn the turbine, converting mechanical energy into electricity.

18-3 Measuring Electrical Energy, page 342
Converter efficiency 1859–1862 (step ×3)
Converter efficiency of a 100 W desk lamp, an electric kettle, a curling iron, and a generator

19-1 The Energy Bottom Line, page 348
Energy use of appliances; table 1863–1864 (step)
Average energy use for several common appliances

Electric meter 1796
Electrical companies use meters to monitor the amount of electricity used in homes. Energy is measured and purchased in kilowatt-hours.

19-3 Keeping in the Heat, page 358
Insulation 2055–2056 (step)
Insulation is used in almost every home to decrease the amount of energy or heat lost to the surroundings. The insulation holds the heat within the walls of a house. (step) Fiberglass is a major component in most insulation. This man is caulking around air vents to further reduce air exchange.

20-1 Energy Then and Now, page 364
Steam engine 27504–27802 (play ×2) (Side A only)
This coal-fired steam engine generates steam to drive a piston.

Laser beam 2208
Gases such as helium and neon are used to make a laser beam more visible.

20-2 Bye-Bye Black Gold, page 369
Oil rig 2629
These rigs are pumping oil. Oil is a fossil fuel.

Pipeline, oil 2631
Oil is transported through a pipeline.

Electric car 2594–2596 (step ×2)
This is the battery compartment of an electric car. (step) The car is plugged into an electrical source to recharge.

20-3 Energy for the Future, page 371
Solar panel 2127
Entire fields of solar panels such as this generate great amounts of energy for municipal use.

Solar houses 2048–2049 (step)
South-facing glass in a passive solar home harnesses the sun's energy to create heat. These houses often use nighttime window insulation to keep the heat in.

Methanol vs. gasoline 2640
Methanol burns much cleaner than gasoline.

Wind farm 1779
As the wind turns the blades of a windmill, generators convert mechanical energy into electrical energy, which can be stored or transported for use.

Windmill field 1788
Large numbers of advanced windmills

Nuclear-fission reactor 1855
Fission is the splitting of an atomic nucleus. When uranium nuclei are split, a huge amount of energy is released.

Unit 6 ENERGY and YOU

Unit 6

Ask students to write a sentence using the word *energy*. Ask volunteers to read their sentences. Have students think about the scientific meaning of the term *energy* by asking: Where does your energy come from? *(From the food you eat)* Where does the energy in food come from? *(The sun)* What happens to the energy in food after your body burns it? *(It is changed into the energy of motion and body heat, and it is used to build and repair body cells.)*

A good motivating activity is to let students listen to the English/Spanish Audiocassettes as an introduction to the unit. Also, begin the unit by giving Spanish-speaking students a copy of the Spanish Glossary from the Unit 6 Teaching Resources booklet.

314

Riders on the roller coaster Kumba at Busch Gardens in Tampa, Florida, getting ready for a pair of corkscrew turns

 **Connecting to Other Units**

This table will help you integrate topics covered in this unit with topics covered in other units.

Unit 2 Patterns of Living Things	The growth and activity of living organisms depend on the storage and release of energy in various forms.
Unit 4 Investigating Matter	Changes of state are best understood from the perspective of matter that is gaining or losing energy.
Unit 5 Chemical Changes	All physical and chemical changes require the input or release of energy in some form.
Unit 7 Temperature and Heat	Heat is a type of energy; therefore, all temperature changes are essentially transfers of energy.
Unit 8 Our Changing Earth	Most geologic events, such as plate movements and erosion, involve energy conversions.

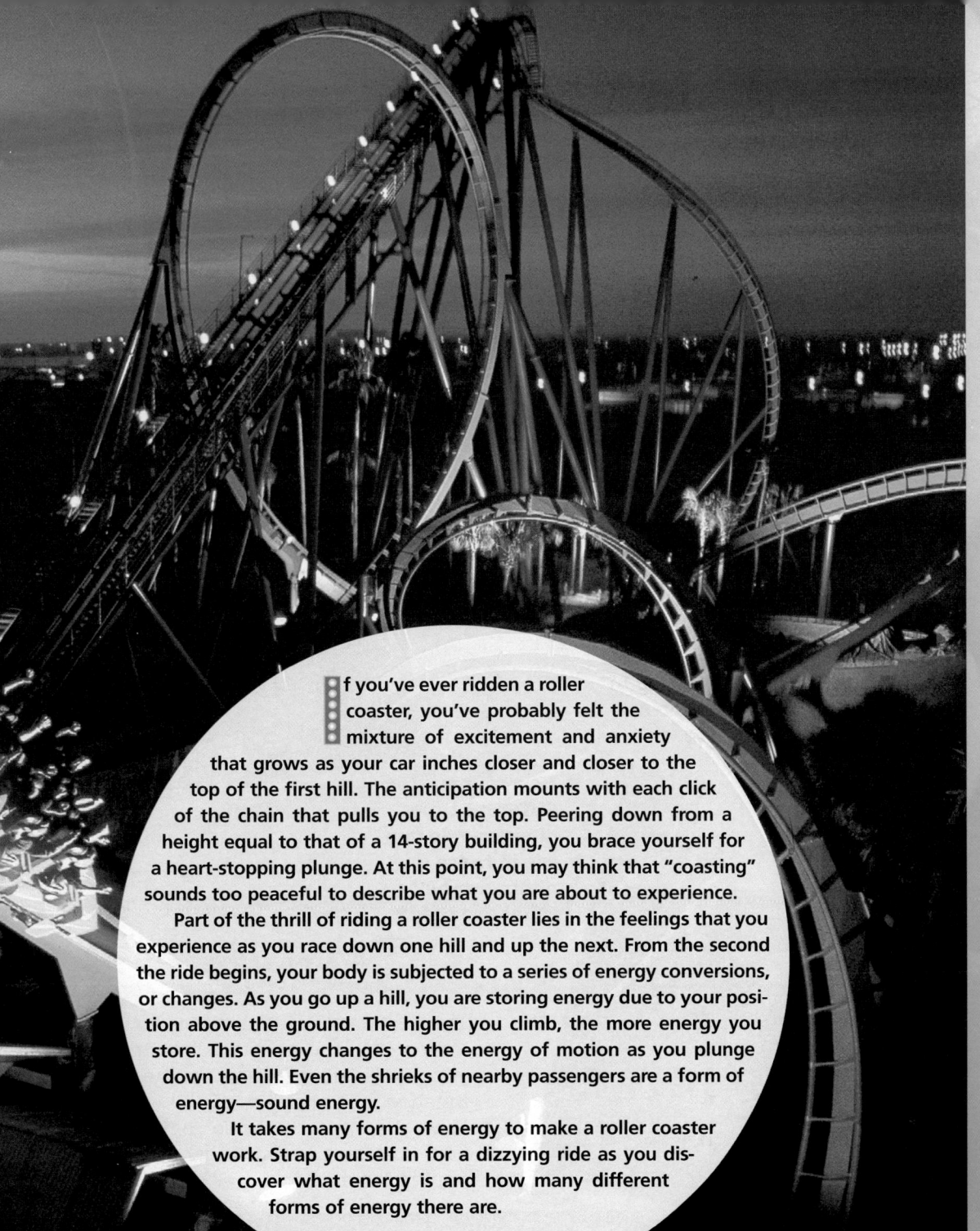

If you've ever ridden a roller coaster, you've probably felt the mixture of excitement and anxiety that grows as your car inches closer and closer to the top of the first hill. The anticipation mounts with each click of the chain that pulls you to the top. Peering down from a height equal to that of a 14-story building, you brace yourself for a heart-stopping plunge. At this point, you may think that "coasting" sounds too peaceful to describe what you are about to experience.

Part of the thrill of riding a roller coaster lies in the feelings that you experience as you race down one hill and up the next. From the second the ride begins, your body is subjected to a series of energy conversions, or changes. As you go up a hill, you are storing energy due to your position above the ground. The higher you climb, the more energy you store. This energy changes to the energy of motion as you plunge down the hill. Even the shrieks of nearby passengers are a form of energy—sound energy.

It takes many forms of energy to make a roller coaster work. Strap yourself in for a dizzying ride as you discover what energy is and how many different forms of energy there are.

315

Have students use the word *energy* to describe something about the photograph. *(Sample answer: Roller coasters need energy to carry people along the track.)* Discuss with students some of the different ways that the word *energy* can be used, such as when objects, people, or devices are described as containing a lot of energy.

Theme Connection

Energy

Focus question: How is the energy of snow that is about to slide down a mountain in an avalanche like the energy of a roller coaster perched at the top of its tallest peak? *(Accept all reasonable responses.)* Explain to students that in both cases, the movement that is about to occur represents a release of energy.

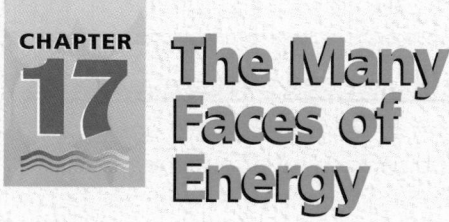

Connecting to Other Chapters

> **Chapter 17**
> gives students hands-on experience with various forms of energy and discusses energy conversions.

> **Chapter 18**
> explores electrical energy, including how it is generated, used, and measured.

> **Chapter 19**
> offers a chance to audit home energy use and evaluate the costs of energy consumption patterns.

> **Chapter 20**
> outlines the history of energy usage and explores possible future energy resources.

Prior Knowledge and Misconceptions

Your students' responses to the ScienceLog questions on this page will reveal the kind of information—and misinformation—they bring to this chapter. Use what you find out about your students' knowledge to choose which chapter concepts and activities to emphasize in your teaching.

In addition to having students answer the questions on this page, you might want to have students generate their own questions about energy. Their questions should address the types and uses of energy that they are already familiar with or that they want to learn more about. Have students choose one of their questions for the class to answer. Emphasize to the class that there are no right or wrong answers in this exercise. Listen to the responses that the class gives to the questions in order to find out what students know about energy, what misconceptions they may have, and what about energy interests them.

CHAPTER
17
The Many Faces of Energy

1 Do you see energy here? Explain.

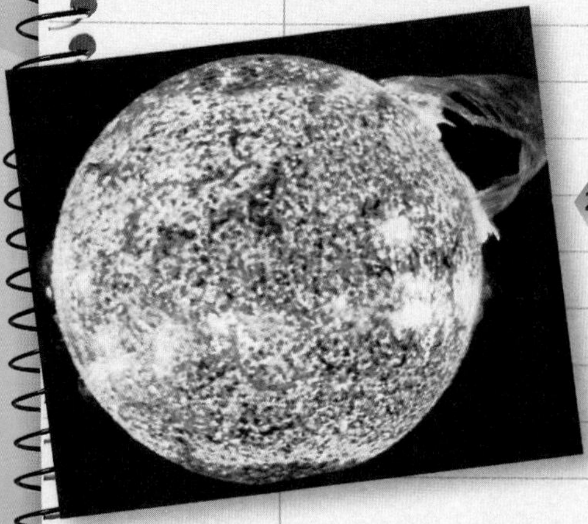

2 Name five ways that the sun provides you with energy.

3 Use the word *energy* to describe what you see happening here.

4 How many different types of energy do you see in these photographs? What are they?

ScienceLog

Think about these questions for a moment, and answer them in your ScienceLog. When you've finished this chapter, you'll have the opportunity to revise your answers based on what you've learned.

316

Energy Is...?

Energy! It's all around us! But how do you know it's there? Just what *is* energy? People talk about it every day. What do they say about it?

This old building just isn't very energy efficient!

I have zero energy today. No way I can take that test!

Dude, my batteries are out of juice!

Hey, can I have a bite of that? All this riding made me hungry!

You've already had five bananas! How can you eat all of that?

They give me quick energy!

317

FOCUS

Getting Started

Place a lit candle in the front of the classroom for students to observe. Ask students to call out some observations about the candle. Ask: Once the candle is lit, where does it get the energy to continue burning? *(Accept all reasonable answers.)* Explain that the burning is fueled by the combination of oxygen in the air and hydrocarbons in the candle wax.

Main Ideas

1. Energy is an abstract concept that is difficult to define.
2. The term *energy* is used in many different ways.

TEACHING STRATEGIES

Before the lesson begins, ask students to write a definition of *energy* in their ScienceLog. After students have completed writing in their ScienceLog, ask them to share their definition of *energy* with the class. Then ask a student to look up the word *energy* in the dictionary. Discuss the different ways that the word *energy* can be used. Point out to students that in this unit, they will learn how the word *energy* is used in science.

INDEPENDENT PRACTICE Suggest that students use their definition of energy to write a poem or an acrostic to include in their ScienceLog.

LESSON 1 ORGANIZER

Time Required
1 class period

Process Skills
observing, inferring, communicating

New Terms
none

Materials (per student group)
none

Teaching Resources
Transparency Worksheet, p. 3
Transparency 54
SourceBook, p. S136

Getting a Picture of Energy

Have students work individually to describe, in terms of energy, what is happening in the photographs. Tell them that their sentences should include descriptions of what energy does and what forms it takes. *(It makes things happen and takes the forms of wind, fuel, etc.)*

Divide the class into groups of two to four students to compare their descriptions. Students should read their statements and ask the other students in the group whether they agree or disagree. After completing this discussion, the students should update their definition of the word *energy.* Then they can share their definition with the rest of the class.

 A Transparency Worksheet (Teaching Resources, page 3) and Transparency 54 are available to accompany Getting a Picture of Energy.

Answers to
In-Text Questions

Ⓐ Sample sentences: The dancers are using a lot of *energy* in their dancing. The wave has a great deal of *energy,* allowing it to carry the surfer with ease. The *energy* of the tornado allows it to tear up houses and telephone lines. Both the sun and oil are sources of *energy* for people.

Ⓑ Accept all reasonable responses. For instance, between *energy* and *food,* students might write "for life comes from."

Homework

Have students write sentences using the word *energy* in five different ways.

FOLLOW-UP

Reteaching

Take students on a 15- to 20-minute walk around the school. Ask them to list all of the places where they observe energy making something happen.

Getting a Picture of Energy

Can you put your ideas about energy into words? The photos here might give you some hints about what energy is. Look at them, and then write a sentence or two in your Science-Log about each one. Use the word *energy* in your description of what is happening in each picture. Ⓐ

Now compare your descriptions with those written by a classmate. Using your combined ideas, compose a single statement starting with the words "Energy is . . ." When you come to the end of this unit, you will be able to check whether your definition can be improved.

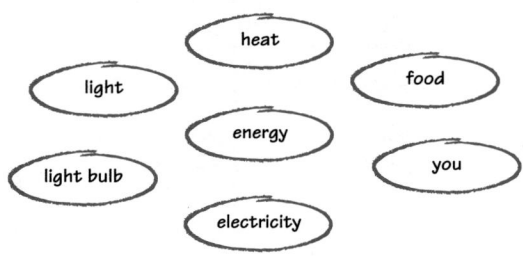

Here's an activity you can do that might help you sort out how energy relates to *you.* Copy the words below into your ScienceLog. (Please do not write in this book!) Then draw lines between the ovals that you think are connected in some way. Along each line, write a few words to explain the connection. Don't worry—you will have a chance to revise your answers at the end of this chapter. Ⓑ

```
          heat
 light            food
       energy
light bulb        you
     electricity
```

318

Assessment

Ask students to make a collage of magazine pictures that show energy at work. As a class, have students describe the part played by energy in each picture.

Extension

Ask students to keep an "energy diary" for a week. Have them record at least three ways that they use energy each day.

Closure

Invite a "mystery guest" from an energy-related industry to visit your classroom. Have students ask your guest yes or no questions about his or her occupation. Set a limit of 25–30 questions. If the class has not correctly guessed the guest's occupation by the end of the questioning, allow the guest to reveal his or her occupation and to inform the students how energy is related to his or her job.

Energy in Action

You probably realize now that energy comes in many different forms. You probably also realize that energy is necessary to us—just think of trying to get along without energy! Although energy itself has no mass, you can see what it does. Energy makes things happen.

Where is energy found? Energy is found in many places. One place is in a battery. You can't see the energy in the battery, but you can see what it does, so you know it must be there. The energy is *stored* in the chemicals inside the battery until it is needed; when it is used, it is *released*. How is the energy stored in the battery put into action? **C**

Energy is stored in many places other than batteries. Look at the picture below. Does the load of bricks contain stored energy? How could you tell? What would happen if the man let go of the rope?

Is energy about to be released? **D**

In Exploration 1, you will encounter six simple energy-containing systems. Ask yourself the following questions about each: **E**

1. Is energy causing something to happen? (This is your clue that energy is really there.)

2. Where was the energy stored before you made use of it? (You might have to make some guesses.)

EXPLORATION 1

Simple Energy Systems

1. The Swinging Pendulum

You Will Need

- a support stand with clamp
- a 20 cm length of string
- a weight

What to Do

Make your pendulum by attaching the weight to the stand with the string. Pull the pendulum to one side and hold it. Do you think it has any energy now? How could you show that it has energy? How is the energy stored? Give this type of energy a name.

Now let the pendulum go. Watch it as it swings a number of times. Do you think the pendulum has any energy when the string is vertical? How could you prove whether or not it does?

Exploration 1 continued ▶

319

FOCUS

Getting Started

Hold an inflated balloon by the end without tying it. Ask: Is this balloon storing energy? *(Yes)* Release the balloon. Ask: What did releasing the balloon do? *(Releasing the balloon released the energy stored in the balloon, causing the balloon to move around the room.)*

Main Ideas

1. Energy is stored in many forms.
2. Once energy is released, it makes many things happen.

TEACHING STRATEGIES

Answers to
In-Text Questions and Caption

C Connecting the two poles of the battery in a circuit releases electrical energy.

D Yes. The load of bricks contains energy because the bricks would fall if the man let go.

E The answers to these two questions are provided for each activity on the pages that follow.

Teaching Strategies for Exploration 1 and Answers to The Swinging Pendulum begin on the next page. ▶

LESSON 2 ORGANIZER

Time Required
2 to 3 class periods

Process Skills
classifying, hypothesizing, inferring, analyzing, observing

Theme Connection
Energy

New Terms
none

Materials (per student group)
Exploration 1, The Swinging Pendulum: support stand with clamp;

20 cm of string; weight; **A Light Touch:** magnifying glass; small piece of newspaper, dry leaf, or small chip of wood; aluminum pie plate or lid from a jar; bucket of water; safety goggles; **The Spinning Spool:** empty spool of thread; 2 toothpicks; rubber band; washer; scissors; paper clip; **Hot Stuff:** candle; jar lid; modeling clay; matches; small aluminum pan or pie plate; a few copper sulfate crystals; small piece of plastic; drop of water; tongs or pliers; trivet or hot pad; safety goggles; **Chemical Magic!:** small alcohol ther-

mometer; about 15 mL of vinegar; small piece of steel wool; 1 L jar with lid; small bowl or beaker; safety goggles; lab aprons; **Paper-Clip Magic:** bolt; 20 cm of insulated wire; 2 small pieces of masking tape; iron filings or a few paper clips; D-cell battery; wire strippers

Teaching Resources
Exploration Worksheet, p. 5
Transparency 55
SourceBook, p. S137

EXPLORATION 1

These activities introduce students to potential energy (stored energy) and kinetic energy (energy of motion). The students will conduct experiments with light, heat, chemical, and electrical energy. This Exploration works well when set up as six stations around the classroom. Two sets of materials at each station should be adequate.

 An Exploration Worksheet (Teaching Resources, page 5) and Transparency 55 are available to accompany Exploration 1.

Answers to
The Swinging Pendulum, page 319

When the pendulum is pulled to the side, it has energy stored in the position of the weight. This can be shown by releasing the weight and letting it swing. Accept all reasonable names for this type of energy—it might be called potential energy.

To prove that a pendulum has energy when it is vertical and moving, students can place an object, such as a paper tube, in its path. The object will move when it is hit by the pendulum. To prove that the pendulum has energy when it is vertical and not moving, students can cut the string. The weight will fall.

Answers to in-text question E on page 319: (1) Energy is causing the pendulum to swing; (2) the energy was stored by the position of the weight.

Answers to
A Light Touch

The newspaper, wood, or dead leaf will smolder or burn when the sun's rays are focused on it. The energy comes from the sun. Students may call this solar or radiant energy.

Answers to in-text question E on page 319: (1) Energy is causing the material to burn or smolder; (2) it was light energy before being changed to heat energy.

2. A Light Touch

You Will Need
- a magnifying glass
- a small piece of newspaper, a dry leaf, or a small chip of wood
- a lid from a jar or an aluminum pie plate
- a bucket of water

Caution: Do *not* focus the light on yourself or anyone else.

What to Do

Put a tiny piece of newspaper, dry leaf, or wood on the lid. For safety, have a bucket of water nearby. With a magnifying glass, focus the rays of the sun on the tiny piece of newspaper, dry leaf, or wood. What happens? Where does the energy come from? Can you give this energy source a name?

3. The Spinning Spool

You Will Need
- an empty spool of thread
- toothpicks
- a rubber band
- a washer
- scissors
- a paper clip

Make a spool toy like the one shown at right. (Use the paper clip to thread the rubber band through the washer and the spool.) Give the rubber band 5 turns. Does the toy move? Try 10 and then 15 turns. What happens? Where does the energy come from? How does the toy store energy?

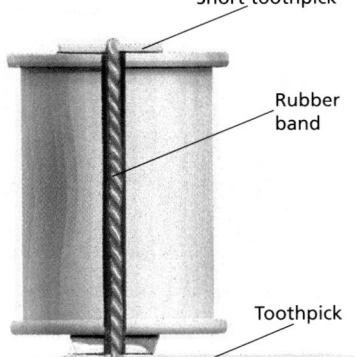

Short toothpick

Rubber band

Toothpick

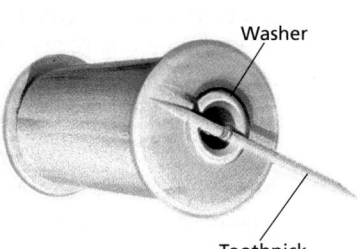

Washer

Toothpick

320

The Spinning Spool

Explain to students that the washer is needed to reduce the friction so that the toothpick can turn without getting stuck against the spool. To reduce the amount of friction further, students may rub a bar of soap over the outside of the washer. Tell students to break in half the toothpick on the end that does not have the washer. The rubber band should be about the same length as the spool. A thick rubber band works better than a thin one. To adjust the rubber band's length, cut and tie it.

Answers to
The Spinning Spool

The toy moves slowly when the rubber band is given 5 turns. When it is given 10 to 15 turns, the spool moves forward at a slow, steady pace. The energy that causes this movement is stored in the twisted rubber band. Students may call this elastic energy.

Answers to in-text question E on page 319: (1) Energy is causing the spool toy to spin; (2) the energy was stored in the twisted rubber band before it was released.

4. Hot Stuff

You Will Need

- a candle
- a small aluminum pan or pie plate
- a drop of water, a copper sulfate crystal, a small piece of plastic
- tongs or pliers
- a trivet or hot pad
- a jar lid
- modeling clay
- matches

What to Do

Grip the aluminum pan with tongs. Set the candle in the jar lid, and secure with modeling clay. Use the flame from the candle to heat the pan. Try heating the following in the pan:

a. a drop of water

b. a copper sulfate crystal

c. a small piece of plastic

Carefully record how the substances change.

Be Careful: When you have finished heating the aluminum pan, put it on top of the trivet where it can't burn anything.

How would you explain to your friends that energy is involved in these changes? Where does the energy come from?

5. Chemical Magic!

You Will Need

- a small alcohol thermometer
- vinegar
- a small piece of steel wool
- a 1 L jar with a lid
- a small bowl or beaker

What to Do

Place the thermometer inside the jar, and put the lid on. Make note of the temperature in your ScienceLog. Soak the steel wool in the vinegar. Squeeze out the excess vinegar from the piece of steel wool, and wrap it around the bulb of the thermometer. Put the thermometer and the steel wool into the jar, and put the lid on. After 5 minutes, make note of the temperature.

Does the temperature change? What types of energy are present? Does one type of energy change into another?

Exploration 1 continued ▶

Theme Connection

Energy

Focus question: How does using the sun's energy to burn paper compare with the energy changes that take place when food is processed by the body? *(The paper burns because the sun's energy starts a chemical reaction that converts stored chemical energy within the paper to heat and chemical energy. Similarly, the body uses some energy to "burn" food and release heat and chemical energy from it.)*

Homework

Challenge students to develop their own energy-containing system like those in this Exploration. They should describe the system in detail and list the types of energy it incorporates.

Hot Stuff

SAFETY ALERT Students should always use the tongs to pick up the aluminum pan and should put the pan on the trivet or hot pad when they are finished heating it. Tell students to use very small quantities of the copper sulfate and plastic.

WASTE DISPOSAL ALERT To dispose of copper sulfate, dissolve the copper sulfate in water, and treat the solution with 0.5 M NaOH solution. After the precipitate settles, the supernate should be colorless. Filter out the precipitate, and wrap the filter paper and its contents in newspaper before putting them in the trash. Using 0.5 M acid, adjust the pH of the filtrate to about 7 and pour the filtrate down the drain.

Answers to *Hot Stuff*

Students might explain to their friends that energy is involved because changes take place when each substance is heated. The energy source is the candle. Students may call this heat or thermal energy.

Answers to in-text question E on page 319: (1) Energy is making things happen—the water boils, the copper sulfate changes from blue to white, and the plastic burns or melts; (2) the energy was stored in the candle.

Chemical Magic!

You may wish to inform students that the vinegar reacts with the steel wool, causing a chemical reaction in which bubbles of hydrogen gas and heat are produced.

Answers to *Chemical Magic!*

The temperature rises. Students may refer to this as heat or thermal energy. Energy stored in the steel wool and the vinegar changes to heat when they react with each other. Students may call this chemical energy.

Answers to in-text question E on page 319: (1) Energy causes the steel wool and vinegar to heat up; (2) the energy was stored in the two substances.

Exploration 1 continued ▶

Paper-Clip Magic

This activity will work best with new D-cell batteries. It may be helpful to try this activity in advance to make sure that the desired result is distinct enough for students to observe. You may wish to inform students that the device created in this activity is called an electromagnet. It attracts substances that contain iron.

Answers to
Paper-Clip Magic

The bolt attracts the paper clips or iron filings. Students may recognize that the electricity creates a magnetic field. The energy was stored in the battery. When the cell is disconnected, the paper clips and iron filings are released. This is because the electricity stops flowing and the magnetic field is no longer being generated. Students may call this electrical energy.

Answers to in-text question E on page 319: (1) The energy causes the bolt to attract metal objects; (2) the energy was stored in the battery.

Making Sense of Your Observations

A sample completed table appears on page 93 of the Unit 6 Teaching Resources booklet.

FOLLOW-UP

Reteaching

Ask students to bring in toys that exhibit various forms of energy. Have them demonstrate the toys and explain the types of energy involved.

Assessment

Perform the following demonstrations. Ask students to name what form of energy is used, where the energy is stored, and what happens when it is released.
1. Have a student do jumping jacks. *(Chemical energy stored in the body's cells causes muscle cells to contract when this energy is released.)*
2. Light a safety match. *(Chemical energy stored in the materials of the match head causes the match to burn when this energy is released.)*

6. Paper-Clip Magic

You Will Need

- a bolt
- a 20 cm length of insulated wire
- wire strippers
- masking tape
- paper clips or iron filings
- a D-cell battery

What to Do

Wind about 20 turns of insulated wire around a bolt. Strip the insulation from the ends of the wire using the wire strippers, and connect the ends to the D-cell battery with masking tape. What happens when you move the end of the bolt near some paper clips or iron filings? Why do you think this happens? Where does the energy come from?

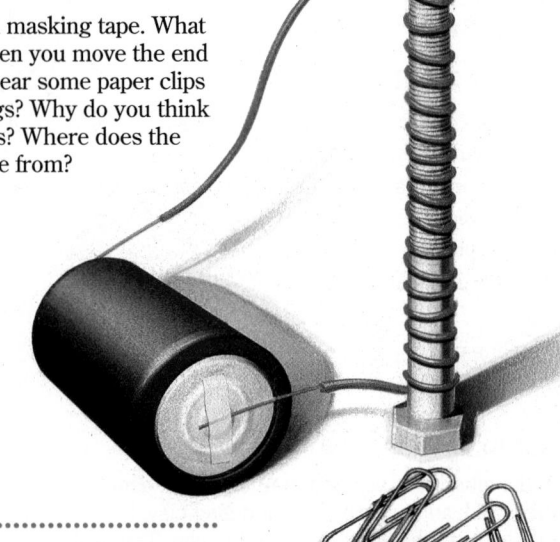

Making Sense of Your Observations

In each situation in Exploration 1, energy was involved. How do you know? Energy must have been involved because *something happened* in each case. Give a name to each type of energy you encountered.

Now make a table like the one below. The table has been completed for one situation—The Spinning Spool—as an example.

Title of situation	What happened? (What evidence is there that energy was used?)	Where did the energy come from? (Where was it stored?)	My name for this type of stored energy
The Spinning Spool	The spool moved.	in the rubber band	elastic energy
Hot Stuff			

Extension

Challenge students to make their own stored-energy toys. Be sure to check the safety of students' proposed procedures before allowing them to make their toys. Students can then demonstrate their toys to the class and indicate how stored energy is converted into energy of movement. Students should also describe any other energy conversions that take place.

Closure

Have students make a list in their ScienceLog describing the physical and chemical changes that occurred in Exploration 1. *(Physical changes include the water evaporating and the plastic melting. Chemical changes include the paper burning, the copper sulfate changing color, and the vinegar and steel wool producing bubbles.)* Then ask students to explain the changes in terms of stored and released energy.

The Energy Picture

The collage on this page illustrates many forms of energy: *light energy, heat energy, sound energy, electrical energy, nuclear energy,* and *chemical energy.* Look at the collage and find examples of each form of energy. Do any of these photographs show more than one type of energy? Is energy **Ⓐ** converted from one form to another? Explain.

As you find each form of energy, you may notice that some examples show energy in motion, while others show stored energy. Another name for energy in motion is *kinetic energy.* Stored energy is called *potential energy* because it has the potential to make something happen. What examples of kinetic and potential energy do you see here? Think back to the picture of the **Ⓑ** construction site on page 319. The bricks on top of the building have potential energy. If the worker were to drop the bricks, they would have kinetic energy. A coiled spring has potential energy. A turning wheel has kinetic energy. Now do you get the idea?

323

The Energy Picture

FOCUS

Getting Started
The Discrepant Event Worksheet (a teacher demonstration) on page 10 of the Unit 6 Teaching Resources booklet makes an excellent introductory activity for this lesson.

Main Ideas
1. Energy comes in many forms, including light, heat, sound, electrical, nuclear, and chemical energy.
2. Potential energy is stored energy; kinetic energy is energy of motion.
3. Energy can be converted from one form to another.

TEACHING STRATEGIES

Answers to
In-Text Questions

Ⓐ Examples of each form of energy include the following:
   - light energy—lightning, campfire
   - heat energy—steam from the geyser, campfire
   - sound energy—violin
   - electrical energy—lightning
   - nuclear energy—nuclear power plant
   - chemical energy—egg, marshmallows

   Most of the photographs show more than one type of energy. For instance, the picture of the man dangling from the clock includes potential energy due to the man's distance from the ground and kinetic energy in the form of the clock's moving hands and the man's moving body. Encourage students to find as many conversions as possible in the photographs. For example, the violinist converts chemical energy into kinetic energy, which is converted by the violin into sound energy.

Ⓑ Examples of kinetic energy include the moving hands of the clock and the girls jumping rope. Examples of potential energy include the chemical energy in the egg and in the grass.

LESSON 3 ORGANIZER

Time Required
2 class periods

Process Skills
observing, analyzing, classifying, inferring

Theme Connection
Energy

New Terms
Efficiency—ratio of waste energy to effective work of a system
Kinetic energy—energy of motion
Potential energy—stored energy

Materials (per student group)
none

Teaching Resources
Discrepant Event Worksheet, p. 10
SourceBook, p. S142

Energy Changes

Have students read this page silently. Discuss the terms *energy converter, input,* and *output.* Ask students to think of examples in which they can use these words to describe energy changes. It may be helpful to introduce the term *friction,* which refers to the rubbing of objects against one another, producing heat.

Meeting Individual Needs

Second-Language Learners
Have students play a game of energy bingo. Each student will need a card divided into 16 squares. Words and drawings that represent certain energy-related terms can be drawn in each square in any order. The role of the caller is to give clues that describe one of the energy terms. Ask students to help you make the cards and the corresponding clues. When students hear the clue being given, they cross off that term or picture on their card. The first student to get four squares in a row says "energy."

Homework

To test students' understanding of kinetic and potential energy, have them bring to class three pictures that illustrate potential energy and three that illustrate kinetic energy. A written explanation should accompany each picture.

Energy Changes

Think about the energy that you use. Does the energy originate in the form in which you use it, or does it start out as another form of energy? Ⓐ

Energy makes things happen only when it is converted from one form to another. For example, chemical energy in the food you eat enables you to run or walk. *You* are an energy converter. As you move, the chemical energy in the food is changed into the energy of motion. Are there any other types of energy produced? Ⓑ

A wood stove is another example of an energy converter. It works by changing the chemical energy in wood (input) into heat energy (output). Are any other forms of energy produced? Ⓒ

Energy input	Energy converter	Energy output
chemical energy (food)		kinetic energy (running, breathing, etc.) other forms?
chemical energy (wood)		heat energy other forms?

Integrating the Sciences

Life and Physical Sciences
To focus student attention on the relationship of the different forms of energy to life science, ask: What senses are involved in detecting the different forms of energy? *(Our eyes respond to light energy; our ears detect sound energy; the nerves in our skin detect heat.)*

PORTFOLIO
Suggest that students review their own class notes, laboratory notes, quizzes, tests, and other materials already collected in their Portfolio. They can identify new questions about energy, reveal new information they have learned, and correct any previous misconceptions about energy.

Problems to Ponder

Get together with a friend, and answer the following questions. Remember: two heads are often better than one when it comes to solving problems!

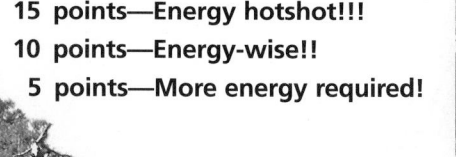

1. Look at the five energy converters shown in the photographs on this page. What is the energy input for each? How many different types of energy output can you think of? Give yourself 1 point for each energy output you identify. Now total your score. Here is the rating system:

   15 points—Energy hotshot!!!

   10 points—Energy-wise!!

   5 points—More energy required!

2. How many energy converters can you suggest for each of the following energy changes?

   a. chemical energy → mechanical energy (kinetic)

   b. electrical energy → heat energy

3. A bouncing ball comes to rest sooner or later. Where does all the energy go?

4. Do you agree with Milo and Floyd (pictured at right)? If yes, why? If no, why not?

You know, you can't create energy or destroy energy. But you can change energy from one form to another.

You don't say!

325

Homework

Have students bring in a photograph or drawing of an energy converter other than those discussed in class. They should also include a brief description of the energy input and output.

Answers to Problems to Ponder

1. **Redwood:** input: solar energy; output: potential energy in the form of chemical energy (production of food)

   Shuttle: input: chemical energy (fuel); output: potential energy (due to height), kinetic energy (due to motion), sound energy, light energy, and heat energy

   Wind farm: input: wind energy; output: kinetic energy (motion of blades), potential energy in the form of electrical energy, heat energy (friction), and sound energy

   Electric guitar: input: electrical energy, kinetic energy (movement of player's hands); output: sound energy, kinetic energy (movement of strings)

   Skiers: input: chemical energy (stored in the person's body), potential energy (due to height); output: heat energy (from the person's body and friction), sound energy, and kinetic energy

2. **a.** Cars, boats, people, airplanes, etc.

   b. Space heaters, electric blankets, toasters, stoves, light bulbs, etc.

3. The kinetic energy of the ball is converted into sound energy and heat energy due to friction.

4. Students should understand that energy does not disappear but rather is transformed.

CROSS-DISCIPLINARY FOCUS

Language Arts

Ask students to research the origin of some of the common units for measuring energy. Have them share their findings with the class. Common energy units that you may wish to suggest include the erg, joule, BTU, calorie, and dietary calorie. As a challenge, you may also wish to assign the electron-volt, kilowatt-hour, or the foot-pound.

How Efficient Are You?

Have students discuss the idea that energy cannot be created or destroyed. Most energy that seems lost actually becomes heat energy. In fact, an incandescent light would not work if most of the electrical energy were not converted into heat. When electricity flows through the tungsten filament in a light bulb, heat from friction is produced. When enough heat is produced, the tungsten filament begins to glow.

After students have drawn pictures or written passages to show why different devices are not 100 percent efficient, discuss why this is so and where the "lost" energy goes.

Answers to
In-Text Questions

Ⓐ Whenever any energy conversion occurs, some energy is changed to heat and is lost to the environment where we cannot use it.

Ⓑ Sample answer: A television wastes some energy in the form of heat. This can be verified by feeling the warmth of a television that has been on for some time.

How Efficient Are You?

Every time you change energy from one form to another, some "waste" energy is produced. In fact, no energy converter is 100 percent efficient. Why do you think that might be so? Ⓐ

Consider the following situation. If you place your hand close to a glowing light bulb, you feel heat. Thus, a light bulb produces both light *and* heat. Since the purpose of a light bulb is to produce light, any heat produced is wasted energy.

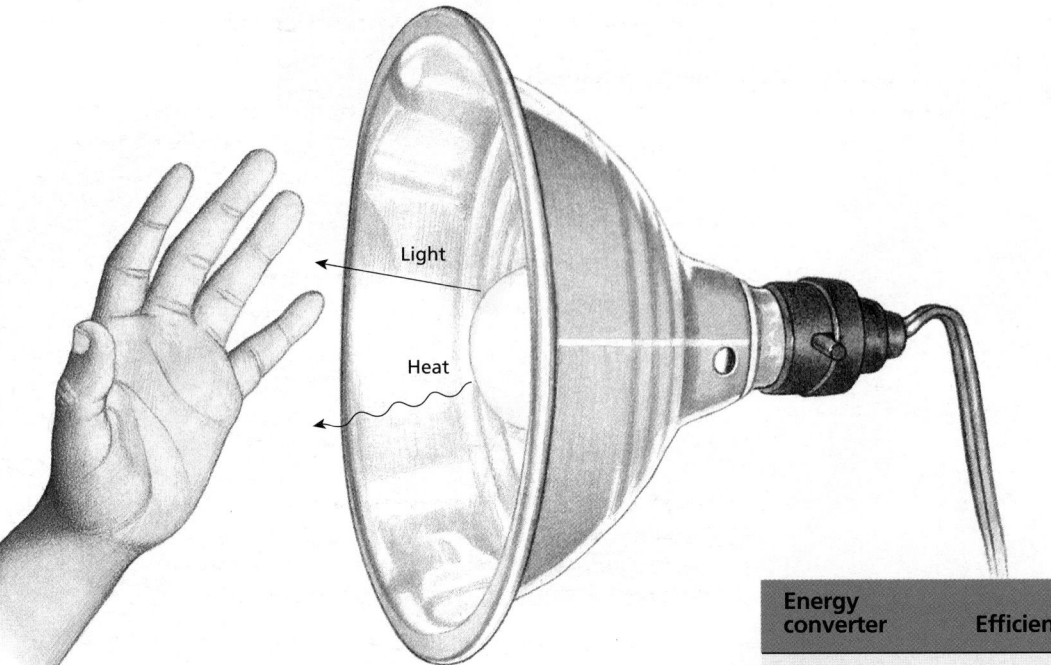

Light

Heat

Energy converter	Efficiency
windmill	35%
hydroelectric plant	95%
solar collector	62%
nuclear reactor	30%
photocell	20%
steam locomotive	9%
electric motor (large)	93%
electric motor (small)	62%
furnace (oil)	66%
furnace (gas)	85%
incandescent lamp	5%
fluorescent lamp	25%
human	10%

The **efficiency** of a device or system tells you how much useful energy is produced. You can see from the table at right that an incandescent lamp (a light bulb) is only 5 percent efficient; 95 percent of the energy is "lost" as heat.

Select another device, and draw a picture or write a passage to show why the device is not 100 percent efficient. Ⓑ

Theme Connection

Energy
Focus question: How are the energy conversions that occur in an automobile engine similar to the energy conversions that occur in your body? *(Both an engine and a human body require a source of chemical energy—the engine requires gasoline and the body requires food. Both convert this chemical energy into mechanical, sound, and thermal energy.)*

Did You Know...

One gram of either protein or carbohydrates contains about 4 dietary Calories of energy. A gram of fat contains about 9 dietary Calories—nearly twice as much as protein or carbohydrates.

The Energy Efficiency Challenge

Who or what is the most efficient energy user? Who gets the most bang for his or her energy buck? Let's hold a contest to find out. The winner of the competition is the one who goes the farthest on a certain amount of energy. To make the competition fair, all participants will be given the equivalent of 1 kilowatt-hour of energy for each kilogram of mass. One kilowatt-hour is the amount of energy used by a 100-watt light bulb in 10 hours. You'll learn more about kilowatt-hours in the next chapter.

Who do you think will go farther, the runner or the cyclist? the helicopter or the plane? the seagull or the bee? the salmon, the hummingbird, or the car? Who wastes the most energy? the least? How does each contestant waste energy? The runner and the cyclist will get hot because not all of the energy they use will be converted into motion; some of it will be converted into heat. Much of the energy given to the machines is also wasted as heat energy.

Try to predict the order in which the contestants will finish. How did you arrive at your prediction?

OFFICIAL ENERGY EFFICIENCY RACE WEIGH-IN

327

Reteaching

Divide the class into two teams. Have Team A name an energy converter and Team B name the energy input and energy output. Then have the teams reverse roles. Be sure that all members of each team have a chance to participate. The team naming the energy converter should try to stump the other team. The team that manages to stump the other team three times (and produces a correct answer themselves each time) is the winner.

Assessment

Using the pictures of energy conversions from the Homework on page 325, ask students to identify the energy converter, energy input, and energy output and to rank the converters based on their efficiency.

Extension

To reinforce what students have learned in this lesson, have them read pages S144–S145. Ask: What kinds of chemical changes are involved in cellular respiration? *(Energy is released in your body when chemical bonds in food are broken. This energy is in the form of heat and chemical energy.)*

Closure

Have a pizza party. During the party, ask students to identify the ingredients they see or taste in the pizza, and list them on the chalkboard. For each ingredient, ask students to find out the types of energy required to make the ingredient available for the pizza and the energy conversions that occurred to make that food. For example, some of the energy required could include energy from the sun for plant growth, energy to cultivate and harvest the crops, energy to preserve and transport the foods, and energy to cook the foods.

Answers to
The Energy Efficiency Challenge

The most likely order of the contestants will be, from the most efficient (wasting the least energy and traveling the farthest) to the least efficient: cyclist, salmon, runner, car, sea gull, plane, helicopter, hummingbird, and bee.

Most energy is wasted as heat that is produced either by the friction of moving parts or by an organism's cells as they process energy.

The predictions are based on the fact that some of the contestants (the cyclist, salmon, car, sea gull, and plane) are able to coast or glide, thus using less energy and creating less friction. The car and the plane have many moving parts, which create a lot of friction. Although the salmon can glide, moving through water creates more friction than moving through air.

LESSON 4
Putting Energy to Work

Getting Started

Crack open several room-temperature eggs and put them in a bowl. Have a student volunteer measure the temperature of the eggs using a thermometer, and have another student record the temperature on the chalkboard. Use an eggbeater to beat the eggs for a few minutes. Have the students measure and record the temperature of the eggs again. Ask: Why did the temperature rise? *(Some mechanical energy was converted into heat.)*

Main Ideas

1. Many different devices have been designed to convert one form of energy into another.
2. Energy converters cannot be 100 percent efficient because they lose useful energy to friction while they are operating.
3. Good experimental designs take into account all possible variables and test them one at a time.

TEACHING STRATEGIES

GUIDED PRACTICE After students have read the introductory paragraph, ask them to observe the pictures of the devices. Ask: What energy conversion is taking place in each picture? *(Sample answers include the following: the light bulb converts electrical energy into light energy and heat; the vehicles convert chemical energy to heat, sound, and mechanical energy; and the drill and mixer convert electrical energy into heat, sound, and mechanical energy.)*

LESSON 4
Putting Energy to Work

Can you imagine life without machines—without cars, appliances, or any of the other helpful devices that you use every day? Machines make our lives easier. They do work for us by converting one form of energy into another.

Just over 200 years ago, engineers designed the steam engine. This device converted the chemical energy of wood or coal into kinetic energy, the energy of motion. This invention sparked the Industrial Revolution. It changed our lives forever.

Engineers have designed many other devices to convert the energy that's available into the energy that we need. A few examples are shown at right.

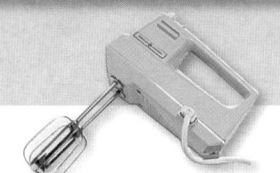

328

LESSON 4 ORGANIZER

Time Required
1 to 2 class periods

Process Skills
observing, analyzing, inferring

Theme Connection
Energy

New Terms
none

Materials (per student group)
none

Teaching Resources
Resource Worksheet, p. 11
SourceBook, p. S143

Energy Makeover

Look at the lists in the table below, and try to match each energy-converting device with the appropriate energy change.

Energy converter	Energy change
A. car engine	1. electrical energy →mechanical (kinetic) energy
B. electric motor	2. light (radiant) energy →heat energy
C. home gas or oil furnace	3. electrical energy →light (radiant) energy
D. electric generator	4. chemical energy →mechanical (kinetic) energy
E. incandescent lamp	5. light (radiant) energy →electrical energy
F. solar cell	6. chemical energy →heat energy
G. greenhouse	7. mechanical (kinetic) energy →electrical energy

Some Problems for You

1. How can you turn mechanical energy into heat energy?

If you're still not sure why energy is never lost, turn to page S143 of the SourceBook.

2. Pick any two types of energy. Is it possible to convert each type directly into each of the other types, or are several steps needed? How many ways are there of making this change? (For hints, look at the list of energy changes above.)

3. Energy is sometimes defined as "the ability to do work." Do you think this is a good definition? Why or why not? How does this definition compare with the one you wrote at the beginning of this unit?

329

Answers to *Energy Makeover*

A. 4		**E.** 3	
B. 1		**F.** 5	
C. 6		**G.** 2	
D. 7			

Answers to *Some Problems for You*

1. Mechanical energy is turned into heat when movement occurs and heat is produced due to friction.

2. You may wish to point out to students that in the examples given in Energy Makeover it is often difficult to convert each type of energy directly into the other. For example, in a car engine, chemical energy is first converted into heat and then into mechanical energy. Also, in an incandescent lamp, electrical energy is first converted into heat and then into light energy.

3. Student answers will vary but should include reasons for their answers. When energy changes from potential to kinetic energy, work is done. For example, when a ball is dropped, the work being done is the downward movement of the ball.

Homework

The Resource Worksheet that is available to accompany Energy Makeover may be assigned as homework (Teaching Resources, page 11).

Theme Connection

Energy

Focus question: What are some examples of potential energy being converted to kinetic energy in both life science and Earth science? *(A moving animal uses energy that has been stored in its body, and an earthquake is the release of energy that has been stored in layers of rock.)*

Integrating the Sciences

Earth and Physical Sciences

Share the following information with your students: Wind power has been used by humans for generations. As far back as the 1600s, wind was used to propel ships, pump water, and power small industrial shops. In early America, many windmills were used to pump ground water for farms and ranches. In the 1930s and 1940s wind turbines produced electricity for rural areas beyond the reach of power lines. Since the 1970s modern turbines have been developed and used in 95 countries to generate electricity. These turbines can produce reasonably cost-effective electricity in areas such as mountain passes and coastlines where average wind speeds are 22–38 km/h.

Perpetual Motion

Have students work in pairs or small groups to analyze the perpetual motion machine.

Answers to
In-Text Questions

 Students should understand that if a generator was linked to a motor and the motor was linked back to the generator, the motor would not run forever because some energy is lost as heat due to friction.

B The most serious flaw in the machine illustrated on this page is that it does not take into account the friction in the axle of the wheel that will slowly bring the wheel to a halt.

Energy From Thin Air

Tell students that you will play the part of the science-fair judge and that student groups will represent Melanie and Theresa. The groups should analyze the science-fair project carefully so that they are able to answer your questions. Tell students that as the judge, you will ask them questions similar to the following: What was the purpose of each experiment? Why were the experiments set up the way they were? What were the controlled and experimental variables? What were the results of the experiments? What conclusions were drawn?

Cooperative Learning
ENERGY FROM THIN AIR

Group size: 2 to 3 students
Group goal: to analyze an in-text experiment and answer a series of questions about the experiment
Positive interdependence: Assign the students in each group roles such as principal investigator, recorder, or timekeeper.
Individual accountability: Each member of the group will be quizzed individually on the questions.

Perpetual Motion

What would happen if you linked an electric generator and a motor so that the generator powered the motor and the motor drove the generator? Would the motor run forever? Why or why not? **A**

Many ingenious inventors have tried to develop *perpetual motion machines,* machines that run forever once started, without any additional input of energy. No one has ever succeeded. The reason is that no machine is 100 percent efficient—some energy always escapes. Also, energy cannot be created by a machine. Energy can only be changed from one form into another.

See if you can spot the flaws in the machine illustrated below. **B**

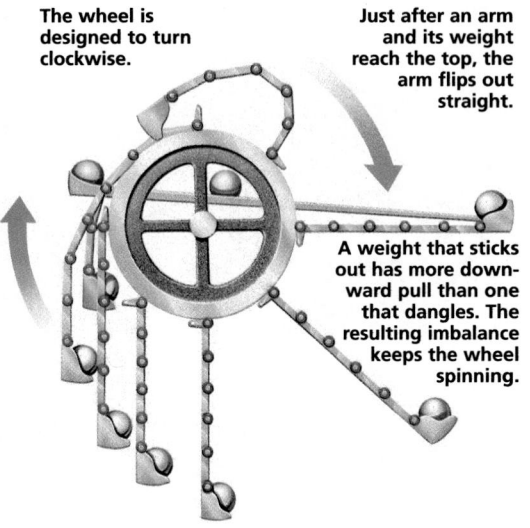

The wheel is designed to turn clockwise.

Just after an arm and its weight reach the top, the arm flips out straight.

A weight that sticks out has more downward pull than one that dangles. The resulting imbalance keeps the wheel spinning.

Energy From Thin Air

Melanie and Theresa were interested in generating electricity from wind. They decided they wanted to design the most efficient windmill possible and make it their science-fair project. After reading about their project, put yourself in Melanie's and Theresa's shoes as they face the judges at the science fair!

Energy From Wind

Introduction

- Wind is a free source of energy.
- What is the best design for a windmill?
- We decided to investigate this question systematically.
  a. Does the angle of the blades make a difference?
  b. Does the length of the blades make a difference?

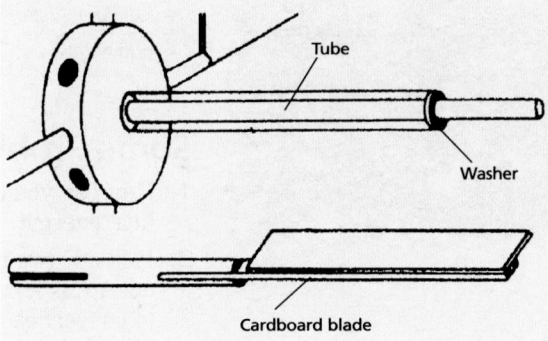

Tube

Washer

Cardboard blade

Apparatus
We found a design for a windmill in the book *Renewable Sources of Energy.* The windmill was made from Tinkertoys.

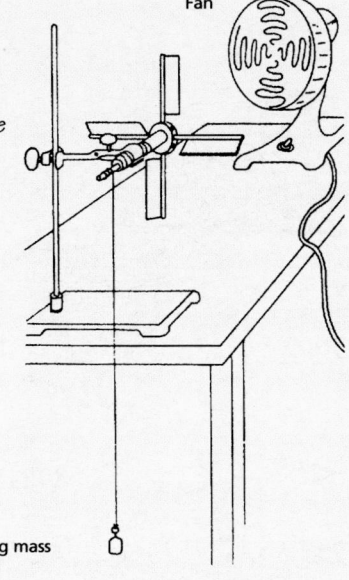

Fan

100 g mass

Meeting Individual Needs

Second-Language Learners
Have groups of two to three students work together to build the windmill described in the text. Students can first follow the design in the text and then experiment with their own designs. Challenge students to explain why some designs are more successful than others. Reward students with prizes for the most creative design, the most efficient design, the best use of materials, etc.

Multicultural Extension

Windmills in the Netherlands
Wind power is an important source of energy in the Netherlands. Discuss with students how this has affected Dutch culture and what advantages and disadvantages have resulted from this reliance on wind power. *(Windmills appear frequently in Dutch art, for instance. Windmills are clean and safe, but they are also noisy and can interfere with some television and radio transmissions.)*

Experiment One

The Effect of Blade Angle

- We counted the number of times the windmill turned per minute.
- The angle of the blade to the wind varied from 0° to 90°. Nothing else was changed.
- Experimental conditions:
  Distance of windmill from fan = 1 m
  Size of blades = 3 cm × 10 cm
  Fan setting = medium
  Counting started when windmill speed was steady.

Results

Angle in degrees	Rev. per min.
0	0
15	21
30	40
45	49
60	43
75	24
90	0

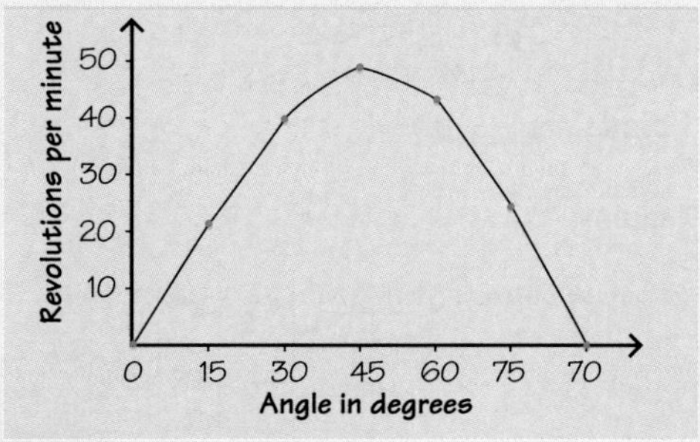

Conclusion

The best angle for windmill blades is 45° (at low speeds).

GUIDED PRACTICE The following sample questions will help you in your role as the science-fair judge. Possible answers to the questions are also given.

1. Why did you choose this design for your first experiment? *(This design allowed us to do an experiment without using expensive equipment.)*
2. What were you trying to find out in Experiment One? *(The effect of the blade angle on the speed of the windmill)*
3. In Experiment One, what were the controlled variables, and what was the experimental variable? *(The controlled variables were the distance of the windmill from the fan, the size of the blades, and the fan setting. The experimental variable was the angle of the blades to the wind.)*
4. How did you measure the speed of the blades? *(We counted the number of times the blades turned per minute.)*
5. What were you testing in Experiment Two? *(Blade length was varied to test the effect of the blade area on the average time taken for the windmill to raise a 100 g mass to a height of 1 m.)*
6. Why does the second graph have a curved line connecting the points? *(Drawing a curved line between the readings on the graph results in the best estimate of the location of points between the readings.)*

Experiment Two

The Effect of Blade Length

- We measured the average time taken for the windmill to raise a 100 g mass through 1 meter.
- We varied the blade length. Nothing else was changed.
- Experimental conditions:

  Distance of windmill from fan = 1 m

  Fan setting = maximum

  Width of blade = 3 cm

  Blade angle = 45°

Conclusions:
The longer the blade, the more powerful the windmill. But long blades bend a lot.

Results

Length of blade (cm)	5	7.5	10	12.5	15
Time 1 (sec.)	195	147	95	85	73
Time 2 (sec.)	210	151	97	83	75
Time 3 (sec.)	202	154	104	87	72
Average time (sec.)	202	152	99	85	73

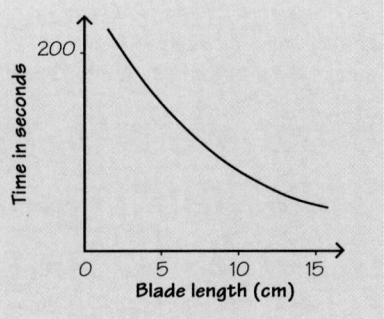

Answers to
In the Hot Seat!

a. The distance between the fan and the windmill must be held constant because distance is a variable that could influence the number of revolutions. The fan's setting must be constant for the same reason.

b. Wind speed was held constant in the science-fair project by using the same fan setting for each experiment. If students are interested in measuring the wind speed in kilometers per hour to simulate actual wind conditions, they could place an anemometer 1 m in front of the fan.

c. When the blade angle was 45°, the blades of the windmill made the greatest number of revolutions.

d. A possible response could be that using a 100 g mass caused the greatest measurable difference in performance level among the different blade lengths.

e. Because the blades are made from cardboard, the relative gain in mass is very small.

f. One reading is insufficient because if it is made incorrectly, the conclusion will most likely be incorrect also. Increasing the number of trials and taking the average of the readings would give the most precise results. Ten trials would be even more precise but might take too long.

g. Other experiments could test a change in the blade shape, wind speed, or number of blades.

Homework

You may wish to have students answer questions (a)–(g) as homework.

In the Hot Seat!

Here are some of the questions that the judges asked Melanie and Theresa at the science fair. How would you have answered them?

a. Why did you specify the distance between the windmill and the fan, as well as the fan setting?

b. How could you have measured wind speed?

c. What made you conclude that 45° is the best angle for the blades?

d. Why did you use a 100 g mass in the second experiment but not in the first?

e. As you increase the length of the blade, doesn't the mass of the blade increase, offsetting any gain in performance?

f. Why did you take the average of three readings—why not one reading, or ten?

g. What other experiments could you have done to learn more about windmill design?

My bike and I produced the least friction, and so we got the most bang for our energy buck.

FOLLOW-UP

Reteaching

Ask students to draw pictures to illustrate why a lamp, radio, television, and electric stove are not 100 percent efficient.

Assessment

Ask students to name at least five energy converters that they use every day and to describe the energy changes that occur in each.

Extension

Have students attempt to create a perpetual motion machine. They should try to make their machine as efficient as possible. They should also describe why their machine loses energy and eventually stops. Be sure to check the safety of all procedures before allowing students to build their machines.

Closure

Ask students the following questions to increase their awareness of the outcome of a scientific investigation: How does an increase in the area of the blades affect the mass of the blades in Experiment Two? *(An increase in the area of the blades also increases the mass of the blades. An increase in the mass of the blades increases the friction between the tube and the rotating dowel and decreases the efficiency of the system.)* Since the area, and not the mass, of the blades is the variable tested in this experiment, how would you take into account the increase in the mass of the blades caused by an increase in blade area? *(The effects of changing the mass of the blades could be tested separately. The results could then be used to isolate the effects of changing the blade area in subsequent experiments.)*

1. All Mixed Up

Below are three picture sequences in random order. Do the following in your ScienceLog:

a. Put the pictures in the correct order.

b. Explain what is happening in each sequence.

c. Spot where one form of energy changes into another form. Identify each time this happens.

d. Name all of the different types of energy that you see in each sequence. Choose from this list: heat, light, electrical, sound, nuclear, chemical, potential, and kinetic.

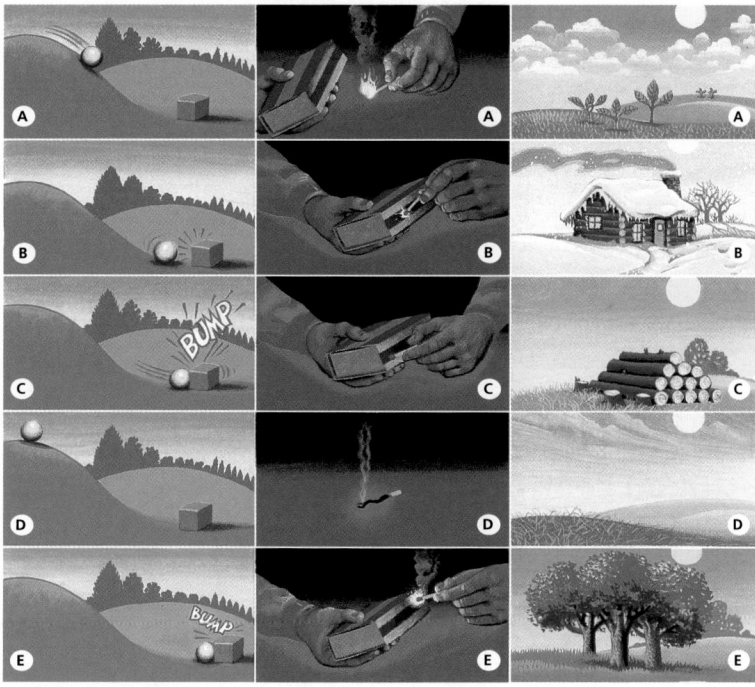

2. Survival of the Energy-Wise

Imagine that you are stranded on a deserted island. You have an ample supply of wood. To reduce your workload, you design a device to change the energy in wood into kinetic energy. What would your device look like? How would it work?

333

Answers to Challenge Your Thinking

1. **a.** Sequence one:
   D, A, C, B, E
   Sequence two:
   C, B, E, A, D
   Sequence three:
   D, A, E, C, B

b. In the first sequence, a ball rolls down a hill and bumps into a box. After this initial contact, the ball rolls back slightly. It then rolls back down against the box with a weaker impact.

In the second sequence, someone strikes a match against the outside of a match box. The match lights and then burns out.

The third sequence covers a long period of time. The first picture simply shows the sun over an empty hill. The second picture shows seedlings beginning to grow. The third picture shows a mature group of trees. The fourth picture shows stacked wood left from cutting down the trees. The final picture shows a house built from the logs in the previous picture.

Answers to Challenge Your Thinking, *continued*

c. In the first sequence, the potential energy of the ball changes to kinetic energy as the ball rolls down the slope. Some of the kinetic energy changes to sound energy and heat when the ball is in motion. When the ball hits the block at the bottom of the slope, kinetic energy is converted to sound energy and heat. Some of the kinetic energy is transferred to the block, causing it to move. After hitting the block, some of the kinetic energy of the ball reverts to potential energy as the ball rolls up the slope. This sequence of energy conversions continues until all of the initial potential energy of the ball dissipates as heat and sound.

In the second sequence, the kinetic energy of the hand is converted into heat and sound as the match head is dragged along the striking surface. The potential energy stored as chemical energy in the match changes to light, sound, and heat when the match ignites.

In the third sequence, some of the light energy from the sun changes to heat as the light reaches the Earth. The seedlings and trees convert the light energy into chemical energy. The logs and the house are the results of kinetic energy being used. The smoke from the chimney indicates that the potential energy stored as chemical energy in the logs is changing into heat, sound, and light.

d. Sequence 1: potential, kinetic, heat, and sound; sequence 2: heat, light, sound, chemical, potential, and kinetic; sequence 3: heat, light, sound, chemical, potential, and kinetic

2. Encourage creative and imaginative ideas. Students should describe in detail how their device changes the energy in wood into kinetic energy.

Answers to Challenge Your Thinking continued ▶

 The Chapter 17 Review Worksheet (Teaching Resources, page 12) and Transparency 56 are available to accompany this Challenge Your Thinking.

Answers to
Challenge Your Thinking,
continued

3. Tania made several errors in her diary. The following are sample correct answers:
    light—electrical to heat to light
    television—electrical to heat, sound, and light
    radio—electrical to sound and heat
    cooking on stove—electrical to heat
    light—electrical to heat to light
    eye catches springy toy—potential to mechanical (kinetic)
    doing homework—chemical to mechanical
    radio—electrical to sound and heat
    car—chemical to heat to mechanical and sound
    looks at clock—potential to mechanical (kinetic)
    machine oil—chemical to mechanical

4. There are many possibilities for the concept map. Look for links that are sensible. For instance, *light* and *heat* are types of *energy*. *Electricity* comes from electrical *energy* and is used to illuminate a *light bulb*. The chemical *energy* stored in *food* is a necessity to *you*.

ScienceLog

The following are sample revised answers:

1. The uppermost piece of rock has a lot of potential energy because of its position. Because the rock below it may slide down the slope, this lower rock also has potential energy.

2. There are many ways in which the sun provides us with energy. The sun provides the light energy necessary for plants to photosynthesize. This allows us to get energy by eating food. Every time we burn a piece of wood or a gallon of oil to produce heat or electricity, we are releasing energy from the sun that was stored by living organisms. The energy from the sun also drives the Earth's weather cycle and provides us with heat, wind, and rain. The energy contained in weather can be converted into kinetic, or mechanical, energy to do work or to produce electricity.

3. Accept all reasonable responses that use the word *energy* to describe what is happening in the photo-

3. Tania's Energy Diary

Tania's teacher asked the class to keep an energy diary. "Any time energy changes form, put in parentheses the kind of change that takes place," the teacher said. Check Tania's answers for her.

When I got home, I turned on the light (electrical to light) and went into the living room. The television (sound to electrical) was on, but I switched it off and turned on the radio (electrical to sound). I got up and went into the kitchen and saw my mom cooking on the stove (heat to electrical). We were going to have tuna casserole.

After this I went up to my bedroom and turned on the light (electrical to light). My eye caught Zebedee, my springy, fluffy toy (potential to movement). I sat down on the bed and did my homework (very hard work to movement). Then I remembered that I had left the radio on (sound to electrical) and went downstairs to turn it off.

I heard a car pull up (movement to chemical). It was my dad coming home from work. He was late getting home, and he looked at the clock (potential to movement). We ate our dinner in silence. After dinner I excused myself from the table and went to the store to get some machine oil (chemical to movement) for my sewing machine.

4. Linked Learning

Here's another chance for you to work on the concept map that you started at the beginning of this chapter. Again, draw lines between the ovals that you think are linked in some way. Then write a few words along each line to explain how the linked words are connected.

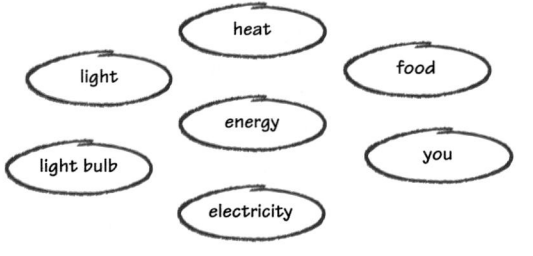

ScienceLog

Review your responses to the ScienceLog questions on page 316. Then revise your original ideas so that they reflect what you've learned.

graph. Here is a sample description: The animal gets energy from eating food, such as grass. The animal is using energy to chew the grass. The animal stores the energy from the food as chemical energy and uses some of it to heat its body.

4. The types of energy represented in the photographs include potential, kinetic, light, heat, and chemical energy.

How does this device convert your energy into light energy?

2 How often do you use electrical energy? Make a list of your electrical energy uses for one day.

3 Where does the electricity that you use come from? Imagine yourself traveling along the wire from the outlet to the source of the electricity. What would you see?

ScienceLog

Think about these questions for a moment, and answer them in your ScienceLog. When you've finished this chapter, you'll have the opportunity to revise your answers based on what you've learned.

335

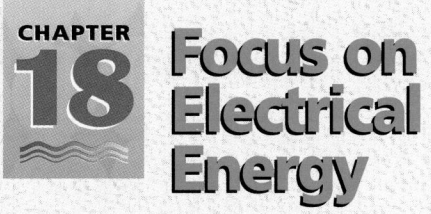

CHAPTER
18
Focus on
Electrical Energy

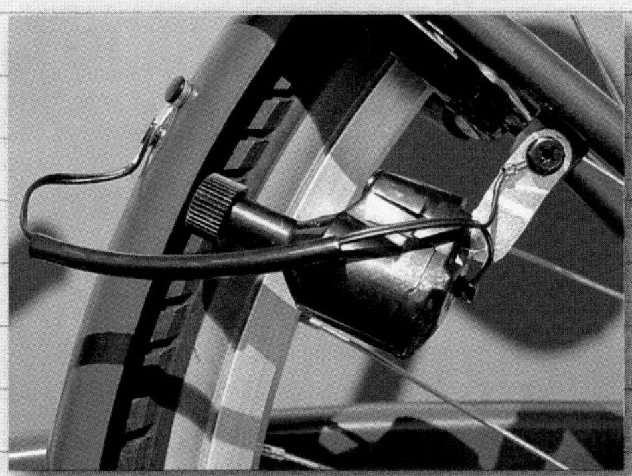

Connecting to Other Chapters

> **Chapter 17**
> *gives students hands-on experience with various forms of energy and discusses energy conversions.*

> **Chapter 18**
> *explores electrical energy, including how it is generated, used, and measured.*

> **Chapter 19**
> *offers a chance to audit home energy use and evaluate the costs of energy consumption patterns.*

> **Chapter 20**
> *outlines the history of energy usage and explores possible future energy resources.*

Prior Knowledge and Misconceptions

Your students' responses to the ScienceLog questions on this page will reveal the kind of information—and misinformation—they bring to this chapter. Use what you find out about your students' knowledge to choose which chapter concepts and activities to emphasize in your teaching.

In addition to having your students answer the questions on this page, you may wish to use the following demonstration to assess students' prior knowledge and misconceptions about

electrical energy: Before class begins, obtain two electrical wires (each about 1 m long), some electrical tape, and a large flashlight. Remove the light bulb and batteries from the flashlight. Tape one end of each of the wires to the two ends of one of the batteries. Tape the other end of one of the wires to the metal side of the bulb and the other to the bottom of the bulb. This should complete an electrical circuit. Demonstrate to the class that the bulb turns off whenever any part of the circuit is disconnected. You may wish to repeat the experiment with two batteries con-

nected end-to-end. Point out that the bulb is now brighter. Ask students to write a paragraph explaining what they think is happening in this experiment. Emphasize that there are no right or wrong answers in this exercise. It may be best to ask students not to put their names on their papers. Collect the papers but do not grade them. Instead, read them to find out what students know about electrical energy, what misconceptions they may have, and what about electricity is interesting to them.

LESSON 1

One Form of Energy: Electricity

LESSON 1

One Form of Energy: Electricity

EXPLORATION 1

Making and Testing for Electricity

A Homemade Galvanometer

In this Activity you will make a simple electricity tester, or galvanometer. After making your tester, check its operation with a D-cell battery.

You Will Need

* a 1 m piece of fine, insulated wire with 2 alligator clips attached
* wire strippers
* a compass
* a D-cell battery

What to Do

Wind about 20 turns of insulated wire over the north-south axis of a compass. Using the wire strippers, scrape the insulation off of the ends of the wire. Attach the ends of the wire to alligator clips. Line up your needle parallel with the wire coils and then touch the clips to the D-cell battery to test your galvanometer. The needle will move if a current flows—the larger the current, the larger the movement.

FOCUS

Getting Started

To demonstrate that an electric current produces a magnetic field, make a hole in the center of an index card. Put about 25 cm of copper wire through the hole and connect the ends to two D-cell batteries to form a circuit. Hold the card horizontally. Sprinkle iron filings on the card, and tap it gently with a pencil. Watch a pattern develop.

Main Ideas

1. A galvanometer or an electric bulb connected in a circuit can be used as an electricity tester.
2. Electricity can be generated in many ways.

TEACHING STRATEGIES

EXPLORATION 1

This Exploration works well with five stations set up around the classroom. Two sets of materials at each station should be adequate.

 An Exploration Worksheet is available to accompany Exploration 1 (Teaching Resources, page 18).

Explain that a compass can be used as an electricity tester because a magnetic field is created when electricity flows through a wire. The compass needle is deflected in response to this magnetic field. The amount of deflection indicates the amount of electricity present.

The Discovery of Electricity

Which form of energy do you think is the easiest to use? Which form can be most simply converted into other forms? Is electricity high on your list? You use electricity many times during the day. Can you imagine life without it? People have known about *static electricity* for thousands of years. You can make it by running a comb through your hair. Static electricity causes objects to attract one another. This type of electricity is called static because it builds up on objects and does not move. However, the kind of electricity that operates your television, radio, and blow-drier was discovered fairly recently. This type of electricity, known as *current electricity,* flows through wires. Read the following time line to learn about the discovery of current electricity.

1600 William Gilbert coins the term *electric*, referring to the force between two objects charged by friction. What do we call this type of electricity today? (Hint: It can be a hair-raising experience.)

1733 Charles Dufay, gardener to the king of France, describes the attractive and repulsive qualities of electrical forces. Benjamin Franklin labels these qualities with a plus sign (+) and a minus sign (−). Later, Franklin receives a big shock while experimenting with lightning rods.

1792 Luigi Galvani notices by chance that the legs of dead frogs twitch when the nerves are touched by metal. The discovery that electricity flows through living organisms intrigues Galvani, and he calls it "animal electricity."

1800 Allesandro Volta decides that dead frogs are unnecessary (not to mention messy) to produce an electrical current. He creates the first battery by stacking copper and zinc disks separated by moist cardboard. (Make your own battery in Activity 2A of the following Exploration.)

1819 Hans Christian Oersted notices that electrical currents flowing through a wire can deflect a compass needle. He has discovered the first current detector. (You will make your own current detector in Activity 1 of the Exploration that follows.)

1831 Taking Oersted's work even further, Michael Faraday learns that moving a magnet near a wire produces a current. This is the concept behind electrical generators. (Follow in Faraday's footsteps in Activity 2B of the following Exploration.)

336

LESSON 1 ORGANIZER

Time Required
2 class periods

Process Skills
observing, inferring, predicting, analyzing

New Terms
none

Materials (per student group)
Exploration 1, Activity 1: 1 m of fine, insulated wire with 2 alligator clips attached; wire strippers; compass;

D-cell battery; **Activity 2:** 3 jars or 250 mL beakers, 3 strips each of copper and zinc metal; 25 mL of salt; 1 L of water; 2 alligator clips; low-voltage LED; 2 m of fine, insulated wire; wire strippers; cardboard bathroom tissue tube; metric ruler; strong magnet; galvanometer from Activity 1; solar cell; light source

Teaching Resources
Exploration Worksheet, p. 18
SourceBook, p. S138

ACTIVITY 2

A. A Simple Battery

You Will Need

- 3 jars or 250 mL beakers
- 3 strips of copper metal
- 3 strips of zinc metal
- salt solution
- 2 alligator clips
- a low-voltage LED bulb

What to Do

Arrange the beakers in a triangle, as in the illustration. Put one copper and one zinc strip into each beaker. Bend the ends of each strip over the edge of the beakers so that they touch and lay on top of the metal strips in the adjoining beakers. The zinc strips should be in contact with the copper strips. Clip two sets of zinc and copper strips together with alligator clips. Leave one set of strips unattached. Fill all of the beakers from half to three-quarters full of salt solution. Touch the shorter end of the LED bulb to the unattached zinc strip and the longer end of the LED to the unattached copper strip.

What happens? Where do you think the electricity comes from?

B. A Simple Generator

A generator is a device used to make electricity. You will learn more about generators in the next lesson.

You Will Need

- a 2 m piece of fine, insulated wire
- wire strippers
- a cardboard bathroom-tissue tube
- a metric ruler
- a strong magnet
- your homemade galvanometer

What to Do

Wind about 50 turns of insulated wire around the outside of the cardboard tube. Leave an excess of about 30 cm of wire on each end of the coil. Strip the insulation from the ends of the wire, and connect them to your galvanometer. Move the magnet quickly into and out of the coil. (Keep the compass needle away from the magnet!) Did the needle move? How could you make the galvanometer needle move even farther?

Where do you think the electricity came from? What do you think you would find inside a factory-made generator?

C. A Solar Cell

Solar cells are usually made from thin layers of impure silicon. They are used to provide electricity for a variety of things, from calculators to satellites. Do you know of any other uses of solar cells?

You Will Need

- a solar cell
- a light source
- your homemade galvanometer

What to Do

Attach your galvanometer to the solar cell. Line up the needle with the coils of wire. Put the solar cell under the light. Did the needle move? Move the light closer and then farther away. Did you see anything happen? Move the light over the cell from side to side. Test your apparatus in sunlight. Where do you think the electricity came from?

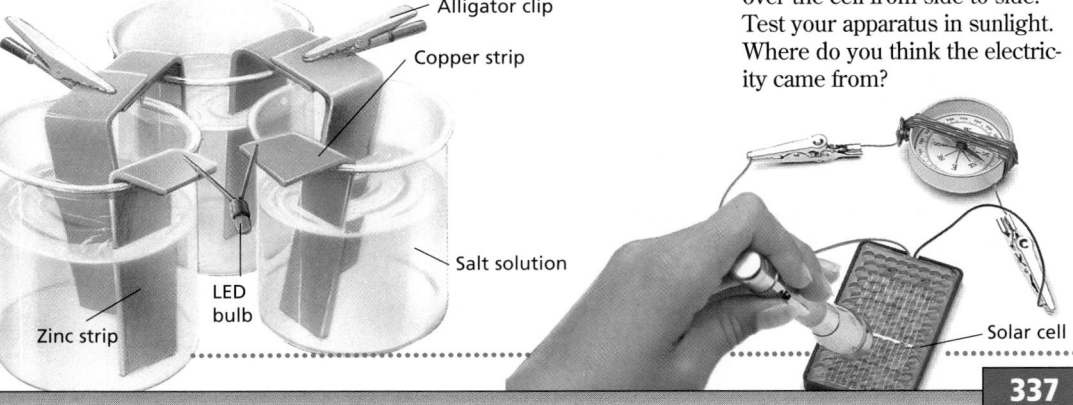

Alligator clip
Copper strip
Salt solution
Zinc strip
LED bulb
Solar cell

337

Students will use the LED to test the simple battery and will use their homemade galvanometer from Activity 1 to test the other two sets of apparatus in this Activity. Explain that there must be a complete circuit before electricity will flow. Draw a diagram of each setup, and have students trace the path of electricity that flows in the circuit. For best results, scour the copper and zinc strips with steel wool before class and between uses of the liquid cell. To make the salt solution, dissolve 25 mL of salt in 1 L of water.

Answers to
Activity 2

A. A Simple Battery
The bulb lights up. A chemical reaction is occurring and is supplying energy in the form of electricity.

B. A Simple Generator
When either pole of the magnet passed through the coil, a current was generated and the galvanometer needle was deflected. The faster the magnet moves, the more electricity is generated and the farther the needle is deflected. The electricity was generated in response to the magnetic field. A factory-made generator might have a coiled wire and a powerful magnet inside it.

C. A Solar Cell
Solar cells are used in experimental cars and in solar panels on rooftops. The movement of the needle depended on how much light energy was received by the solar cell; the energy came from light and was converted to electrical energy.

FOLLOW-UP

Reteaching

Ask students to draw pictures of three ways to generate electricity in their ScienceLog.

Assessment

Have students draw diagrams of the flow of energy through each circuit investigated in Exploration 1.

Extension

As an extension to A Solar Cell, have students make a graph showing the distance to the light source versus the needle's deflection. *(Graphs should reveal an inverse proportion: as the distance increases, the deflection decreases.)*

Closure

Ask students: How is the generator made in Activity 2 different from the electromagnet made in Chapter 17? *(In the electromagnet, the current flowing through the wire sets up a magnetic field. In the generator, the opposite happens—the movement of the magnetic field causes electricity to flow.)*

LESSON
2
Making Electricity From Scratch

FOCUS

Getting Started

Build the windmill described on page 330, and connect it to a generator (such as a simple DC motor from a toy). Connect wires from a galvanometer to the generator to monitor the electricity generated. Have students discuss their observations of this device.

Main Ideas

1. In an electrical generator, the wire coil moves within a magnetic field to produce electricity.
2. Steam and moving water turn coils in commercial generators.

TEACHING STRATEGIES

Answers to
In-Text Questions

Ⓐ Magnetic energy is changed into electrical energy in a generator. If the wire loop could be moved very quickly for a long period of time, a large amount of electricity could be produced. A stronger magnet would also increase the amount of energy produced.

Ⓑ Powering the lights in your home by turning a handle would be very impractical and inconvenient.

Ⓒ In each case, a turning object is attached to a generator: the teapot's steam turns the pinwheel; wind turns the turbine on the house's roof; the running mouse turns its wheel; and the weight of the falling water turns the wheel. Student drawings should identify and explain practical ways to generate electricity.

 Transparency 57 is available to accompany the material on this page.

Teaching Strategies for Turning Motion Into Electricity are on the next page. ▶

In Exploration 1, you made your own electricity generator. It was simple but not very practical since you had to be there to push the magnet into the coil. Think how much more practical your generator would be if you could find a way to keep the magnet moving continuously with respect to the coil. How might you do this?

The people who invented the type of generator used for producing electricity today had a really clever idea: turn the in-and-out motion into a circular motion. Can you figure out how this type of generator works? Trace the flow of the electrical current from the coil (*ABCD*) to the wires attached to the light bulb.

Do you see now that an electricity generator is really just a special type of energy converter? What change takes place? How do you think you could get a generator to produce a really large amount of electricity? Ⓐ

Turning Motion Into Electricity

The generator on this page shows one way to turn mechanical (kinetic) energy into electrical energy. This generator is turned by hand. Do you see any problems with this method of generating electricity? What if you had to power all of the lights in your home this way? Ⓑ

The devices shown on the following page show a few alternative ways to change the energy of motion into electrical energy. For example, in device *A*, as the weight falls, the shaft spins to turn the pulley and the belt. The belt is attached to an electric motor. Why a motor? Many of the little electric motors found in battery-powered devices can also be used as generators. You can either send an electrical current through the motor to cause the spindle to turn, or you can turn the spindle of the motor to generate electricity.

In your ScienceLog, explain how each of the other devices might work. If you had to pick, explain which one you would choose to generate electricity in your home. Or draw up plans for your own device to drive a generator in your home. Whatever method you choose, explain why you think your selection is best. Ⓒ

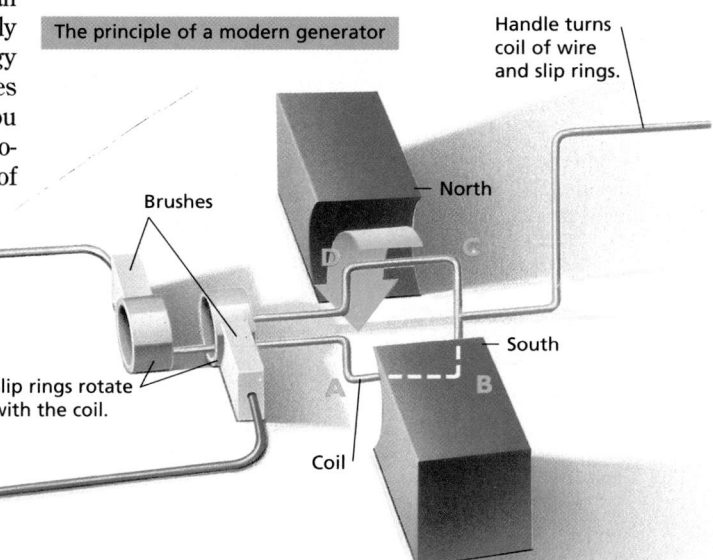

The principle of a modern generator

Handle turns coil of wire and slip rings.

North

South

Brushes

Lamp

Slip rings rotate with the coil.

Coil

338

LESSON 2 ORGANIZER

Time Required
2 to 3 class periods

Process Skills
observing, analyzing

Theme Connection
Systems

New Terms
none

Materials (per student group)
none

Teaching Resources
Resource Worksheet, p. 22
Transparencies 57, 58, and 59
SourceBook, p. S148

Motor

Belt & Pulley

B

Light Bulb

A

Mass

20 g.

Steam

C

Wind Turbine

Wood Fire

Drain pipe

E

Water Wheel

D

Generator

Mouse

Light Bulb

Turning Motion Into Electricity, *pages 338–339*

GUIDED PRACTICE   Demonstrate how a small motor can be used as either a motor or a generator. First, connect the wires from the motor to a battery to turn the spindle. Then, disconnect the battery and reconnect the wires to a flashlight bulb. Turn the spindle as quickly as possible to generate electricity, and the bulb will light up. (You may need a starter cord to turn the spindle fast enough.)

Be sure that students understand that the same device can be used as either a generator or a motor. To use it as a generator, you turn the spindle, and a current is produced in the coil. To use it as a motor, you send current through the coil, and the spindle is turned. Remind students that electricity flowing through a wire coil sets up a magnetic field around the coil. The magnetic field around the wire coil interacts with the field around the magnet and causes the spindle to turn.

Meeting Individual Needs

Learners Having Difficulty

Have students use the generator they built in Activity 2 of Exploration 1 and an additional magnet to answer the following question: In which circumstance will electricity NOT be generated?

a. if two magnets are moved simultaneously into the tube
b. if a 10 cm longer wire is used
c. if the alligator clips of the electricity tester are switched to opposite ends of the generator
d. if the magnet rests next to the tube instead of moving inside the tube *(Correct)*

Did You Know...

In typical sink or bath water, ions (charged particles) are present that allow the water to conduct electricity easily. A liquid that conducts electricity in this way is called an *electrolyte.*

Integrating the Sciences

Earth and Physical Sciences

Piezoelectricity is the generation of electricity by compressing or stretching certain types of crystals. Have students find at least one example of a piezoelectric crystal and discuss what uses such crystals might have. *(The most common example is quartz. Such materials are used in watches, speakers, and sonar systems.)*

Homework

You may wish to have students read pages 338–339 as homework in order to speed up in-class discussion.

Industrial-Strength Generators

INDEPENDENT PRACTICE Have students read the page, study the diagram, and use their ScienceLog to fill in the blanks in the story at the bottom of the page. Then call on volunteers to read their answers. Be sure that all students in the class agree.

Homework

The Resource Worksheet that accompanies Industrial-Strength Generators makes an excellent homework activity (Teaching Resources, page 22). If you choose to use this worksheet in class, Transparency 58 is available to accompany it.

Answer to
In-Text Question

Ⓐ The turbine in the generator on this page is turned by moving steam instead of by hand.

Answers to
The Story of a Steam-Powered Generator

The following answers are listed in order of their appearance in the paragraph: chemical, heat, kinetic, electrical, electrical, heat, and light. (Note that the last two answers are interchangeable.)

Theme Connection

Systems
Focus question: How does a generator demonstrate the relationship between electricity and magnetism? *(Students should understand that the movement of either the wire or the magnet generates an electric current in the wire. Thus, mechanical energy is converted into electrical energy. This electrical energy can then be converted into some other useful form of energy.)*

Industrial-Strength Generators

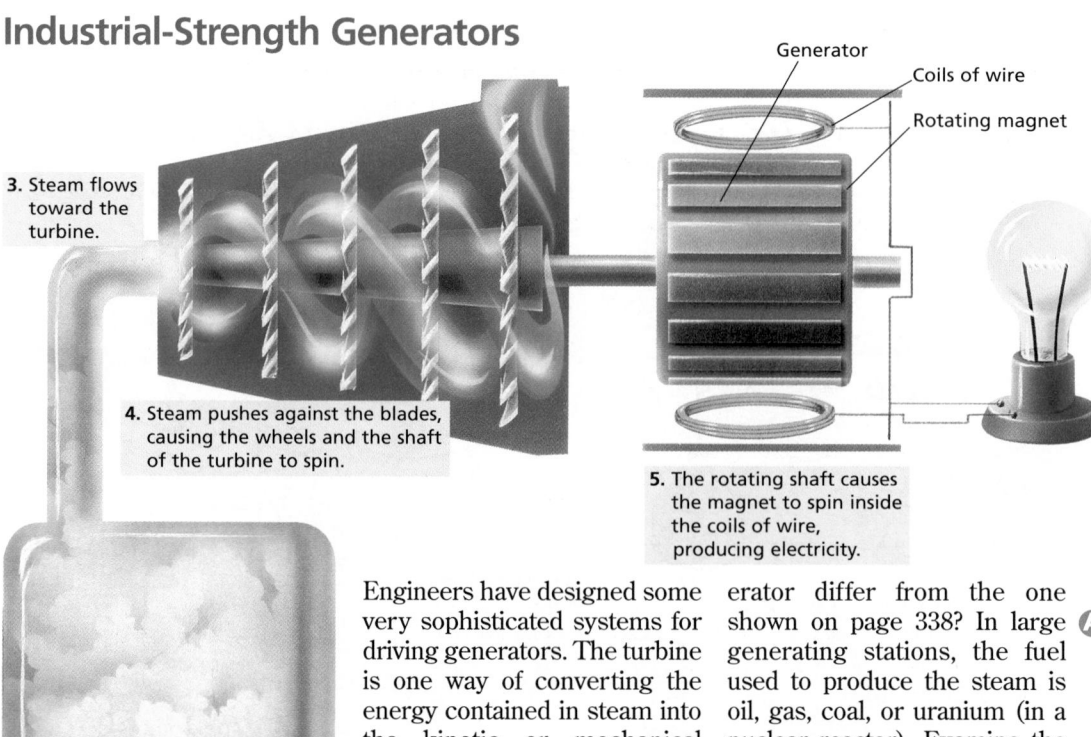

Generator
Coils of wire
Rotating magnet

3. Steam flows toward the turbine.

4. Steam pushes against the blades, causing the wheels and the shaft of the turbine to spin.

5. The rotating shaft causes the magnet to spin inside the coils of wire, producing electricity.

2. Boil water, generating steam.

1. Burn fuel.

Engineers have designed some very sophisticated systems for driving generators. The turbine is one way of converting the energy contained in steam into the kinetic or mechanical energy of a generator. The turbine fits snugly into a pipe, through which water or steam flows. The only way the water or steam can get through the pipe is to turn the turbine. The motion of the turbine drives the generators, producing electricity.

In the diagram above, you see how steam causes the turbine to turn the magnet in the generator. How does this gen- erator differ from the one shown on page 338? In large Ⓐ generating stations, the fuel used to produce the steam is oil, gas, coal, or uranium (in a nuclear reactor). Examine the diagram closely, following all of the steps.

Did you trace the steps from making the steam to lighting the lamp? If so, you will have no trouble in filling in the blanks in the following story. Write the completed story in your ScienceLog. Here, in random order, are the forms of energy that fit into the blanks: *chemical, electrical, heat, light,* and *kinetic.* Do not write in this book!

The Story of a Steam-Powered Generator

____?____ energy from gas, oil, or coal is converted into ____?____ energy, which is added to water to produce steam. A turbine changes the energy of the steam into the ____?____ energy of the rotating turbine and generator. The generator converts this energy into ____?____ energy. The lamp is an energy converter, changing ____?____ energy to ____?____ energy and ____?____ energy.

CROSS-DISCIPLINARY FOCUS

Health
Electrical devices should always be handled with care. Even relatively small electric currents can be extremely harmful to the human body. Point out to students the following possible effects of an alternating electric current on the human body. You may wish to lead a class discussion on electrical safety as well.

- 0.001 amps: tingling sensation
- 0.01 amps: pain; contraction of affected muscles (which may prevent a person from releasing a grasped object)
- 0.02 amps: contraction of the chest muscles; possible halting of breathing
- 0.1 amps: fibrillation of the heart
- a few amps: blockage of the nervous system; respiratory failure

An electric current can also kill the tissue through which it passes, sometimes resulting in severe burns.

 Transparency 59 is available to accompany The Hydroelectricity Story on page 341.

The Hydroelectricity Story

Here is a diagram of a hydroelectric power plant. Can you see how it works? What do you think the transformer does? If you don't know, do some research to find out.

With another person, write a story about a hydroelectric power plant. What kind of energy conversions take place there? Extend the story to talk about the energy conversions

that happen when the electricity leaves the plant, travels through the wires, enters your home, and finally reaches an electric stove that is being used to boil water.

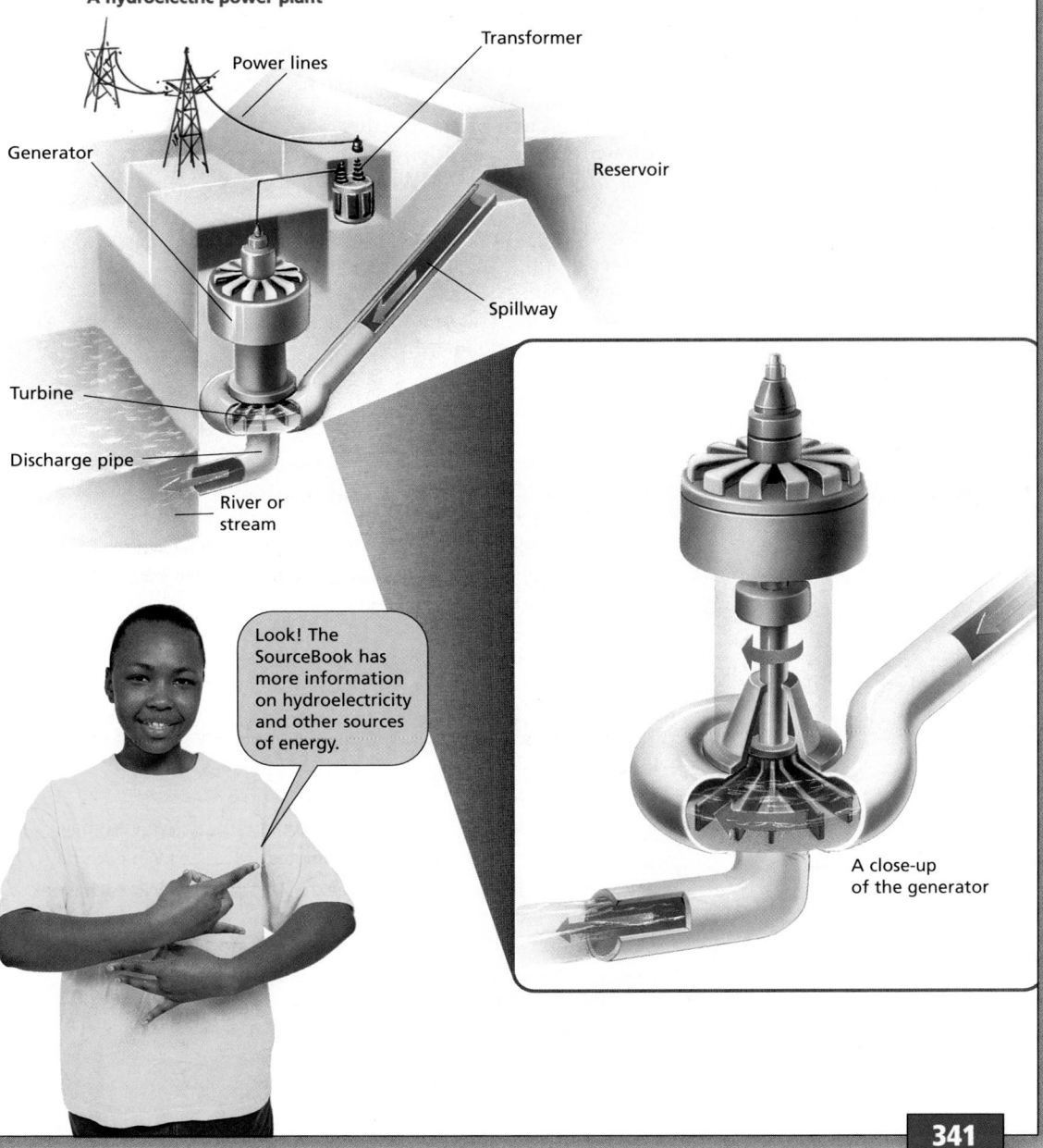

A hydroelectric power plant

Power lines

Transformer

Generator

Reservoir

Spillway

Turbine

Discharge pipe

River or stream

Look! The SourceBook has more information on hydroelectricity and other sources of energy.

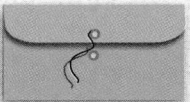

A close-up of the generator

341

The Hydroelectricity Story

Point out to students the transformer that connects the generator to the power lines in the diagram. The function of the transformer is briefly described in the sample answer that follows.

Answers to
The Hydroelectricity Story

The following is a sample answer: Hydroelectricity is produced from moving water. In the picture, water from the reservoir travels downward through the spillway to a turbine. The falling water turns the turbine, which is connected to the spindle of a generator. The spindle turns the wire coil inside the magnet to produce electricity. The transformer increases the voltage of the electricity. (Voltage can be thought of as the pressure of the electricity in the electric wires. Electricity is carried over the electric lines at a very high voltage. Less electrical energy is lost to the heat of friction when the voltage of the electricity is high during transmission. Thus, increasing the voltage is the most efficient and economical way to distribute the electricity.) Once the electricity reaches individual homes, the voltage is decreased by other transformers. Inside the home, the electricity is converted by various electrical appliances into useful forms of energy.

PORTFOLIO

Suggest that students assess their hydroelectricity story for scientific accuracy, creativity, and comprehension of the subject matter. You may wish to use the Self-Evaluation Checklists available on the SnackDisc for this purpose.

FOLLOW-UP

Reteaching

Have students illustrate the flow of energy through an industrial generator or a hydroelectric power plant. Students should describe what is happening at each step.

Assessment

Ask students to explain why an electric motor can also be a generator. Have them demonstrate how to make a generator from a small electric motor and how to test whether or not it works.

Extension

Have students write a report on the use of geothermal power to create electrical energy. In their reports, they should include information about the mechanics of converting the steam into electricity and the relative cost of the electrical energy produced.

Closure

Conduct a class contest to see who can design the most creative generator. Have students draw a design and write a short

description of how the generator works. Interested students can then build their generator to see if it will work. Be sure to check the safety of students' proposed procedures before allowing them to build their generator.

FOCUS

Getting Started

Set up a display that includes a blender, a hair dryer, a light bulb, and an iron. Point out the energy information on the appliances, and discuss what this information means.

Main Ideas

1. The energy consumed by a device is measured in kilowatt-hours.
2. Electrical appliances use different amounts of electricity.

TEACHING STRATEGIES

You may wish to clarify the distinction between power and energy usage. Power refers to how much electrical energy is required to run an appliance for one second and is measured in watts. The power rating of an appliance reflects how much energy the appliance will use in a given length of time. Energy usage is usually measured in kilowatt-hours and is found by multiplying the power rating of an appliance by the amount of time that the appliance is used.

Assist students in completing the tables on pages 342 and 343. You may wish to have students volunteer to find the data for certain appliances and then compile the information as a class. Or you may wish to supply students with the sample data in the margin on page 343.

Answers to
In-Text Questions

Ⓐ **Electric kettle:**
1500 W × 0.5 hr./day = 750 Wh/day, or 0.75 kWh/day
Curling iron:
800 W × 0.25 hr./day = 200 Wh/day, or 0.2 kWh/day
0.2 kWh/day × 6 days = 1.2 kWh/week
Generator:
1000 kW × 720 hr./month = 720,000 kWh/month, or 720 MWh/month, or 720,000,000 Wh/month

---

Have you ever looked at the metal tag on an electrical appliance such as a motor, refrigerator, electric kettle, or curling iron? On all of them—unless they are quite old—you would have seen a number followed by the letter *W*. This shows the *power* of the appliance. Power is measured in **watts** (W), a unit named after the scientist and engineer James Watt. The power *(wattage)* of an electrical appliance tells you how fast it uses electrical energy. The higher the wattage of an appliance, the faster it uses electrical energy.

If you multiply the power by the amount of time the appliance is used, you will find the total amount of energy consumed by the appliance, measured in a unit called a watt-hour (Wh). The most commonly used unit of energy, the **kilowatt-hour,** is 1000 Wh. One *megawatt-hour* (MWh) is 1,000,000 Wh.

It's time to check your understanding of measuring electrical energy. Complete the following table. The first example has been done for you. **Ⓐ**

Appliance	Power		Hours used	Electrical energy used or produced
desk lamp	100 W		Kiyoshi forgets and leaves the lamp on for 20 hours.	2000 Wh *or* 2000 ÷ 1000 = 2 kWh
electric kettle	1500 W		Ben uses the electric kettle for 10 minutes, 3 times a day.	__?__ Wh each day __?__ kWh each day
curling iron	800 W		Anna uses the curling iron for 15 minutes, 6 days a week.	__?__ Wh each day __?__ kWh each day __?__ kWh each week
generator (electricity producer)	1000 kW		John's generator runs steadily for 1 month.	__?__ Wh each month __?__ kWh each month __?__ MWh each month

LESSON 3 ORGANIZER

Time Required
1 class period

Process Skills
classifying, predicting, analyzing

New Terms
Kilowatt-hour—a measure of electrical energy (equal to 1000 watts being consumed for 1 hour)
Power—the rate at which a device uses electrical energy (expressed in watts)
Watt—a measure of the rate of energy consumption (equal to 1 joule/second)

Materials (per student group)
Exploration 2: hot plate; portable burner (propane or propane-butane mixture); striker; two 600 mL beakers; support stand with ring clamp; wire gauze; 1 L of water; oven mitts; safety goggles; metric balance

Teaching Resources
Resource Worksheet, p. 23
Exploration Worksheet, p. 25
Transparencies 60 and 61
SourceBook, p. S150

Now, which appliance do you think consumes the most electrical energy in 1 hour? Place the appliances listed below in order from the greatest amount of electricity used to the least. (B)

- an electric kettle
- a toaster
- a bright light bulb
- an iron
- a stereo
- a hair dryer
- an electric drill
- a clothes dryer
- a washing machine
- a refrigerator
- a television set
- a vacuum cleaner

Check your predictions at home by looking at the tags on the appliances. Then calculate how many kilowatt-hours each appliance would consume in an hour, and enter your findings in a table like the one at right. Do not write in this book. (C)

Appliance	Power (W)	Power (kW)	Energy used in 1 hour (kWh)

Other Sources of Energy

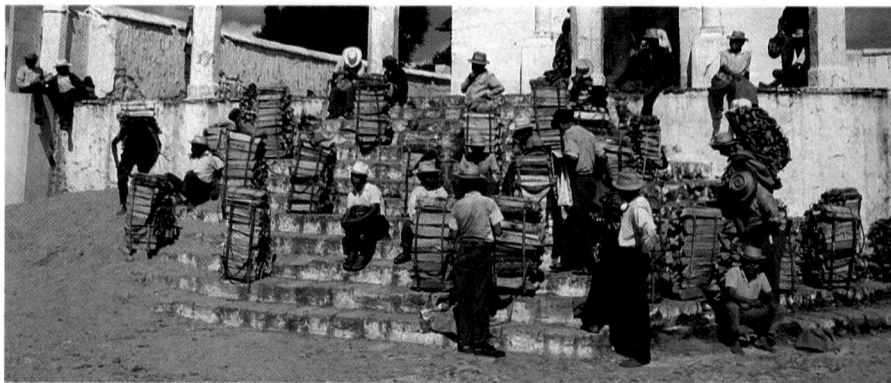

Wood sellers in Chichicastenango, Guatemala

In most countries, electricity is measured and sold by the kilowatt-hour, oil and gasoline by the liter, natural gas by the cubic meter, and propane by the kilogram. Which energy source would warm you the most: a liter of oil, a cubic meter of gas, a log of hardwood weighing a kilogram, a kilogram of propane, or a kilowatt-hour of electricity? You can check your prediction by studying the table at the end of Exploration 2, which shows the approximate amount of energy you would need to heat water for your bathtub.

In the United States, fuels such as gasoline, propane, and heating oil are usually sold by the gallon. Natural gas is sold in cubic feet (cf) or thousand cubic feet (mcf).

Answers to *In-Text Questions*

(B) The electricity consumption for the appliances will probably be within the following ranges (placed in order from greatest electricity consumption to least):
clothes dryer: 2500–3000 W
refrigerator: 1000–2000 W
electric kettle: 1200–1500 W
hair dryer: 1000–1500 W
iron: about 1100 W
toaster: 800–1200 W
vacuum cleaner: 500–900 W
electric drill: 200–300 W
bright light bulb: 150–200 W
television: 50–200 W
stereo: 30–130 W
washing machine: 70–80 W

Note: Some appliances may state their amperage and voltage, but not their wattage. In these cases, multiplying amps by volts gives watts.

(C) For a table that has been completed using sample values, see page S240.

Answers to *Other Sources of Energy*

According to the table at the end of Exploration 2 on page 344, 1 m^3 of natural gas, 1 L of oil, and 1 kg of propane all provide about the same amount of heat. One kilowatt-hour of electricity would provide one-tenth as much heat as these others, and a log weighing 1 kg would provide only about one-twentieth as much heat.

Homework

A Resource Worksheet is available as homework to supplement the material on page 342 (Teaching Resources, page 23). If you choose to use this worksheet in class, Transparencies 60 and 61 are available to accompany it.

To save time, you may wish to provide students with the typical prices listed below in the answer to question 6. Based on those prices, the students should conclude that heating the bathtub water with natural gas is the cheapest method. You may wish to have students complete Exploration 2 in groups of two to three students.

⭐ **An Exploration Worksheet is available to accompany Exploration 2 (Teaching Resources, page 25).**

Answers to *What to Do*

1. Predictions will vary. A typical result would be 5 minutes, although results may vary.

2. Sample answer: A hot plate that operates at 1100 W and takes 10 minutes to boil water would use 0.18 kWh.

3. Students should subtract the mass of the portable burner after boiling the water from the mass of the burner before boiling the water to find the mass of the propane used. A typical answer would be about 20 g.

4. Electricity typically costs about 6 cents/kWh; propane typically costs about 72 cents/kg. If a propane-butane mixture is used, use the price of propane to figure the cost. Using the above values, boiling 500 mL of water with the hot plate costs about 1 cent and boiling 500 mL of water with the portable burner costs 1–2 cents.

5. Reports should be clear and should include the comparative prices of these energy sources.

6. Oil typically costs about 12 cents/L; natural gas typically costs about 4 cents/m³. The following are sample answers for the table: natural gas: 4 cents/m³; wood: $1.90/20.5 kg; oil: 12 cents/L; electricity: 6 cents/kWh; propane: 72 cents/kg.

The Energy Economy Contest

You Will Need

- a hot plate
- a portable burner (You may use propane or a propane-butane mixture.)
- a striker
- two 600 mL beakers
- a ring stand
- wire gauze
- water
- oven mitts
- a balance

What to Do

Plan an experiment to determine which is the most cost-effective way to boil water.

1. Which do you think will cost the least to boil 500 mL of water—the hot plate or the portable burner? Predict how long it will take to boil water with the hot plate, and then try it. Turn off the hot plate when you are finished.

2. Calculate the number of kilowatt-hours used by the hot plate.

3. How much fuel will be needed to boil the water with the portable burner? Record the mass of the burner before and after it is used to determine how much fuel was consumed.

4. Find out the cost of electricity and the cost of the fuel used in your burner. Use these figures to calculate how much it costs to boil water using each method.

5. Write a report in which you pick the better energy source. Imagine that your report will air on "Market Place," a consumer-interest TV program.

6. Find out the cost of the other energy sources listed in the table below. Calculate the approximate cost of heating water for your bathtub with each of these energy sources. Write your answers in your ScienceLog, not in this book.

Approximate Amount of Energy Required to Heat Bathtub Water

Source of energy	Approximate cost
1 m³ of natural gas	
20.5 kg of wood	
1 L of oil	
10 kWh of electricity	
1 kg of propane	

FOLLOW-UP

Reteaching

Have students draw a bar graph to illustrate the amount of electrical energy used by appliances. The appliances could be arranged from the most energy consumed to the least.

Assessment

Ask students to determine how much electrical energy is needed to operate a 12 W computer and its 200 W monitor for 8 hours. *(1696 Wh or about 1.7 kWh)*

Extension

Have students list the electrical appliances used by their families. Students should then rate the appliances from most used to least. Students can extend this activity by researching the efficiency of each appliance.

Closure

Bring in some energy information tags from different appliances, and ask students to interpret the information. Students should determine how many kilowatts per hour are required to run the appliances.

CHALLENGE YOUR THINKING

1. Wavy Engineers

Imagine that you are a "wave-energy engineer." Design a device to convert the up-and-down movement of ocean waves into electrical energy.

2. True or False?

You can make electricity by sticking a strip of aluminum and a piece of copper wire into a lemon. How could you check this out? Write out a list of the steps that you would take.

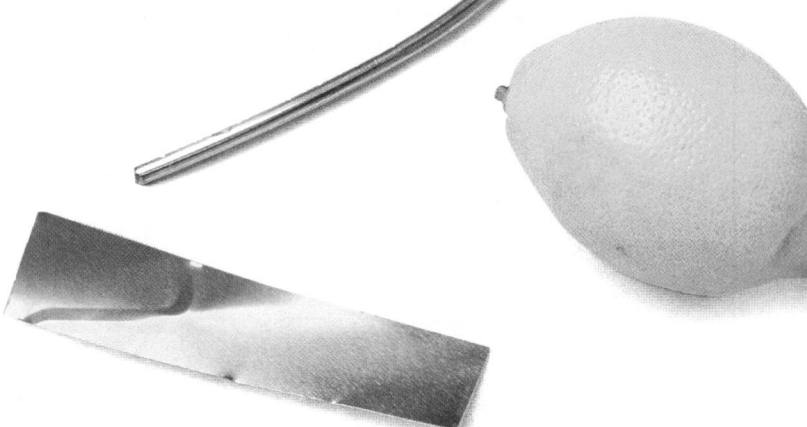

3. Electric Connections

Create a concept map using the following words: *solar cell, electricity, turbines, magnet, steam, generator, battery, wind,* and *water.* To make the map, draw lines between the words that you think are linked in some way. Then write a few words along these lines that explain how the linked words are connected.

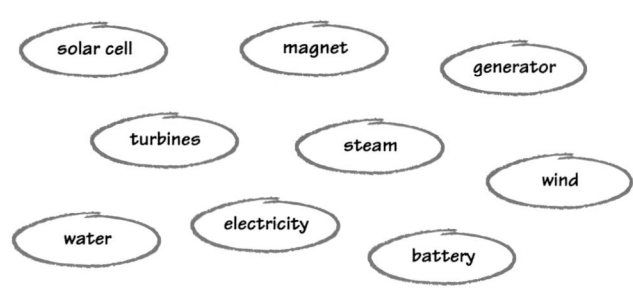

 *Homework*

A Graphing Practice Worksheet is available as homework (Teaching Resources, page 27).

 You may wish to provide students with the Chapter 18 Review Worksheet that accompanies this Challenge Your Thinking (Teaching Resources, page 29).

Answers to *Challenge Your Thinking,* pages 345–346

1. Answers will vary. Encourage students to be creative. An example of a design that has been tested is a butterfly-shaped platform that rests facedown on top of the ocean. The motion of the waves causes the "wings" of the platform to rock up and down, and this movement is converted into electricity.

2. This is a true statement. To test it, insert an aluminum strip and a copper wire into a lemon. Attach one wire from a homemade galvanometer to the aluminum strip and the other wire to the copper wire. The galvanometer's needle moves. This movement indicates an electric current. As a control, students should check for current when the aluminum strip and copper wire are not inserted into the lemon.

3. Sample concept map:

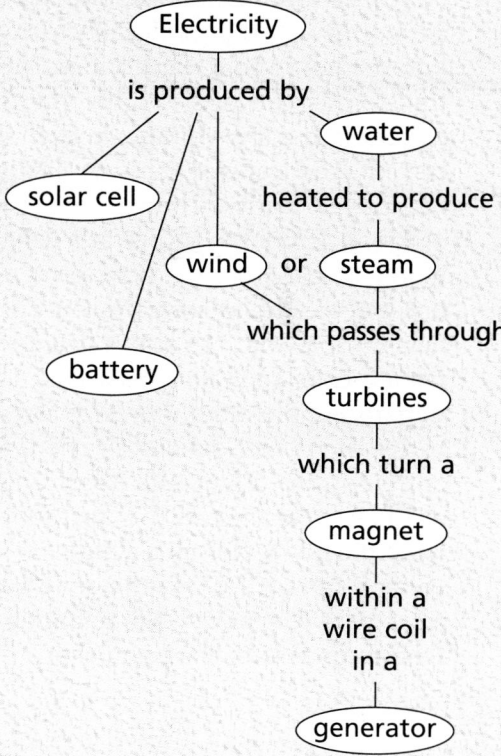

4. **D.** b.
   B. d.
   E. c.
   C. a.
   A. e.

The following are sample revised answers:

1. A person converts chemical energy from food into kinetic or mechanical energy to turn the wheel of the bicycle. As the wheel turns, it turns coils of wire in a magnetic field within the bicycle generator. The electricity produced is then transmitted to a light bulb. Heat from the friction of the electric current passing through the filament of the light bulb causes the filament to glow and to give off light energy.

2. Students should conclude that they use electrical energy very frequently. A sample list might include lights, curling iron, television, radio, CD player, toaster, and microwave.

3. Traveling along the wire from an electrical outlet to the source of electricity, one first sees a fuse box or a circuit-breaker box. From the fuse box or circuit-breaker box, one travels along the wire that brings electricity into the house to a transformer. The transformer converts a high-voltage current into a low-voltage, household current. The transformer is necessary because electricity is transmitted most efficiently from the source at a very high voltage. From the transformer, one travels along high-voltage transmission lines that are either strung from towers and tall poles or laid underground. At the source of electricity, the power generator, one finds another transformer. The transformer at the source converts the current from the generator into a high-voltage current. Within the generator, a wire coil moves in a magnetic field to generate electric current.

4. An Exciting Story

Below you will find the story of one way that sunlight can be transformed into the light from a light bulb in your home. But the story is all mixed up! Put the story in Column A in the correct order, and then match each part of the story to the appropriate energy change in Column B.

Column A	Column B
A. Electricity travels along wires to your home. Electricity passing through the filament in a light bulb lights up a room.	a. kinetic to electric
B. The coal is burned in a power plant.	b. light to chemical
C. The steam turns the blades of a turbine. This causes the generator to produce electricity.	c. heat to kinetic
D. Plants use light from the sun to make food. Over millions of years, the plants become coal.	d. chemical to heat
E. The burning coal heats water, which creates steam.	e. electric to light and heat

Sciencelog

Review your responses to the ScienceLog questions on page 335. Then revise your original ideas so that they reflect what you've learned.

CHAPTER
19
The Cost of
Energy

1 Make a list of the appliances you use the most. Which do you think uses the most energy?

2 What kind of sacrifices would you be willing to make in order to save energy?

3 People who live in colder climates are worried about keeping heat in their homes during the winter. How do you think heat escapes?

Think about these questions for a moment, and answer them in your ScienceLog. When you've finished this chapter, you'll have the opportunity to revise your answers based on what you've learned.

347

Connecting to Other Chapters

Chapter 17
gives students hands-on experience with various forms of energy and discusses energy conversions.

Chapter 18
explores electrical energy, including how it is generated, used, and measured.

Chapter 19
offers a chance to audit home energy use and evaluate the costs of energy consumption patterns.

Chapter 20
outlines the history of energy usage and explores possible future energy resources.

Prior Knowledge and Misconceptions

Your students' responses to the ScienceLog questions on this page will reveal the kind of information—and misinformation—they bring to this chapter. Use what you find out about your students' knowledge to choose which chapter concepts and activities to emphasize in your teaching.

In addition to having your students answer the questions on this page, you may wish to write the following terms on the board and use them to initiate a class discussion: kilowatt, kilowatt-hour, energy, power, and thermal resistance. Allow volunteers to suggest definitions for these terms. Emphasize to the class that there are no right or wrong answers in this exercise. During the course of the discussion, take note of what the students already know about measuring energy, what misconceptions they may have, and what about the topic interests them.

FOCUS

Getting Started

To prepare students for the activities in this lesson, do the following demonstration: Using an appliance such as a vacuum cleaner, demonstrate the function of the appliance while a student volunteer times how long the appliance is used. Using the power rating supplied on the appliance, calculate as a class how much energy was used by the appliance.

Main Ideas

1. There are many ways to save energy in the home.
2. It is possible to calculate how much energy it takes to run an appliance by multiplying the estimated number of hours it is used by its power rating.

TEACHING STRATEGIES

The Kilowatt Kids Investigate

Have students read the cartoon on pages 348–349. It sets the stage for Exploration 1 on page 350. Tell students that they will be conducting an investigation similar to the one conducted by the Kilowatt Kids.

GUIDED PRACTICE After students have read the cartoon, ask: What do you need to know to calculate the energy used by an appliance in a month? *(The power rating of the appliance, how long the appliance is used each time, and how many times it is used in a month)*

Students should begin to realize how energy can be saved (for example, by turning off lights when not in use). Students may be surprised to learn how much electricity a clothes dryer uses.

LESSON
1
The Energy
Bottom Line

Here's your chance to help your family save energy—and money. Even if your parents don't increase your allowance, they will be delighted to have you help them save on their utility bill!

The Kilowatt Kids Investigate

Here's how some other students went about it. Kelly, Kristy, Karl, and Kevin (the Kilowatt Kids) decide to work together as a team. They all have different ideas about which appliances use the most electricity in a certain time period. You may need to help them with their calculations.

348

LESSON 1 ORGANIZER

Time Required
2 class periods

Process Skills
observing, organizing, analyzing

Theme Connections
Energy, Changes Over Time

New Terms
none

Materials (per student group)
none

Teaching Resources
Exploration Worksheets, pp. 35 and 40
Theme Worksheet, p. 38
Transparencies 62 and 63

The cartoon points out some of the problems involved in trying to estimate the amount of energy used by some appliances. Tell students that when they carry out their own investigations, it may be necessary to estimate how often and for how long some appliances are used. For example, they could time how long a refrigerator runs during a half-hour period and multiply that number by 48 to estimate how long it runs during a day. However, they should take into account that it will run much more often when it is opened and closed frequently.

Answers to
In-Text Questions, pages 348–349

A By watching only 2 hours of television per day, Karl's family would use 18.90 kWh/month.
28.35 kWh/month − 18.90 kWh/month = 9.45 kWh saved per month

B Coffee maker: energy used per month = 0.04 kWh/use × 120 uses/month = 4.8 kWh/month

C Dryer: 4.80 kW × 0.62 hours = 2.976 kWh used

D Total kWh = 260 Wh + 160 Wh + 100 Wh = 520 Wh = 0.52 kWh

E The Kilowatt Kids could develop a system to estimate how long each appliance is used. This would help them be more accurate. For example, leaving a checklist of appliances in each room of their home would allow family members to record the amount of time that each appliance is used.

Appliances not listed in the text may be added to the list. For each appliance, the students should record the power rating and the estimated time that the appliance is used each day (how many times it is used each day and how long it is used each time). Then they should estimate how many times it is used each month.

Cooperative Learning
EXPLORATION 1

Group size: 4 to 5 students
Group goal: to investigate the electrical energy usage of several appliances
Positive interdependence: Assign each group member a specific role such as manager, recorder, pacer, or materials coordinator. Each group member should have at least one appliance to investigate.
Individual accountability: Each member of the group should be able to write a summary of the group's findings.

In addition to making observations, students may need to interview family members to get a good estimate of how often and how long different appliances are used each month. Allow time for team members to meet and discuss any problems they may be having. Each team should present their results to the class so that the class as a whole can complete the data chart.

⭐ **An Exploration Worksheet (Teaching Resources, page 35) and Transparency 62 are available to accompany Exploration 1.**

Homework

Note that part of Exploration 1 involves researching the energy information of appliances at home. You may wish to make sure that each student has been assigned at least one appliance to investigate.

How Much Energy Do Appliances Use?

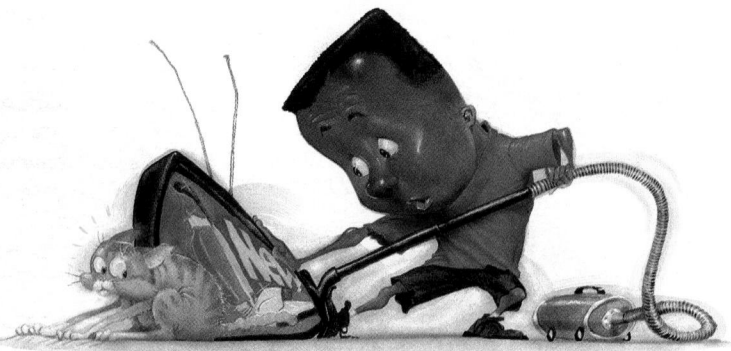

Do you think you could save your family $10 per month on the utility bill? You have seen how the 4K Consultants went about the job and how they made their calculations. Now plan your own survey of the electrical energy used in your home. Here are some guidelines.

1. Team up with three or four classmates. There are numerous appliances to investigate, enough for each team to have several.

2. Each team should select four or five appliances from the table on the right. Can you think of any others to add?

3. Make sure that the different teams have covered all of the appliances among them.

4. Each team should now be ready to plan the investigation. Choose one appliance for each team member, and prepare job sheets just as Karl, Kelly, Kevin, and Kristi did. Discuss any problems that may arise.

5. Bring your findings back to class, and fill in a table like the one at right. Work out the average energy (in kilowatt-hours) used by each appliance.

6. How much does a kilowatt-hour cost? Work out the cost per month for each appliance.

Appliance	Power (kW)	Time used (hr./month)	Energy per month (kWh)	Cost per kWh	Cost per month
air conditioner					
blender					
clock					
clothes dryer					
curling iron					
dehumidifier					
dishwasher					
freezer					
electric blanket					
electric frying pan					
electric kettle					
hair dryer					
humidifier					
iron					
lights					
microwave oven					
mixer					
radio					
range					
refrigerator					
television					
toaster					
vacuum cleaner					
washing machine					
others					

Mr. Catalano's Class Investigates

The table below shows the results that Mr. Catalano's class obtained. Are their results similar to yours?

Appliance	Average kWh used per month	Appliance	Average kWh used per month
air conditioner	200.2	iron	9.8
blender	1.0	electric kettle	23.1
clock	1.5	lights	96.4
clothes dryer	80.3	microwave oven	25.6
curling iron	0.4	mixer	0.9
dehumidifier	33.3	radio	8.2
dishwasher	30.0	range	100.7
freezer	107.5	refrigerator	117.5
electric blanket	10.4	television	39.8
electric frying pan	15.0	toaster	3.2
hair dryer	2.0	vacuum cleaner	3.8
humidifier	14.9	washing machine	5.0

For Discussion

1. Which appliances use the most electricity?
2. How could you reduce the amount of electricity your family uses?
3. Would your family support your plan? Prepare a report on all of the findings, with recommendations for your family.

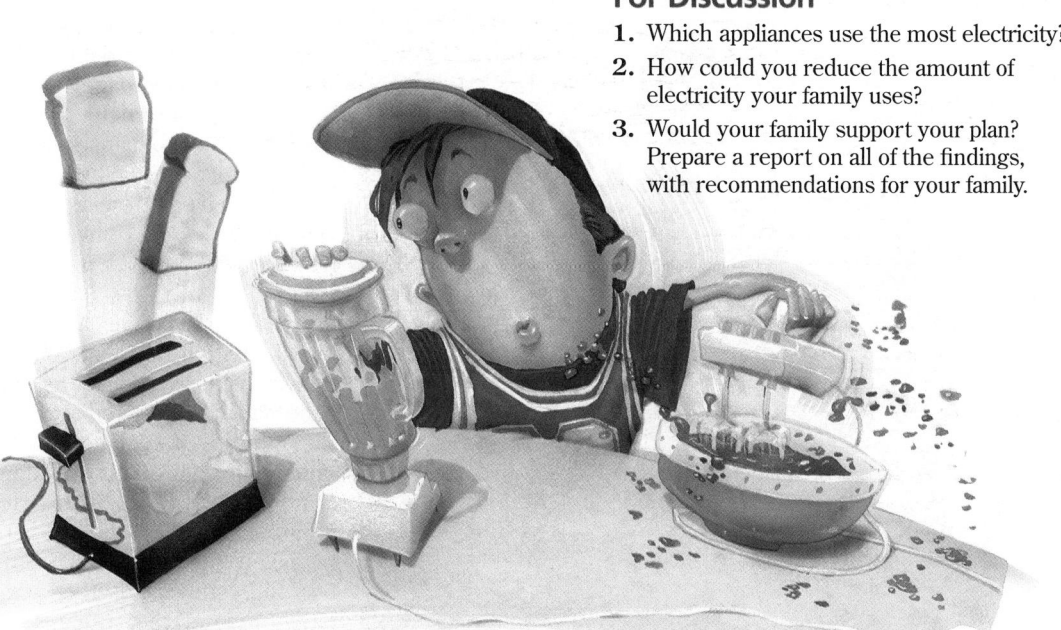

Mr. Catalano's Class Investigates

GUIDED PRACTICE Discuss with students how their results compare with those of Mr. Catalano's class. Ask: How are the sets of data the same? How are they different?

Answers to *For Discussion*

1. It is likely that air conditioners, refrigerators, freezers, stoves, clothes dryers, and lights use the most electrical energy in a home.

2. Answers may vary, but possible responses include turning up the refrigerator thermostat, keeping the refrigerator door closed as much as possible, turning off unused lights, using light bulbs with lower wattages, baking several items in the oven at the same time, raising the thermostat on the air conditioner, and lowering the thermostat on the furnace.

3. Answers may vary. Ask volunteers to share their report with the class.

Theme Connection

Energy

Point out to students that the rate at which their body uses energy can also be measured in watts. **Focus question:** How does the rate of energy use of typical appliances compare with that of the human body? *(Students may be surprised to learn that many appliances use energy at a much higher rate than the human body uses for many activities. For instance, a typical value for energy consumption while running up a flight of stairs would be 400 W. The Theme Worksheet found on page 38 of the Unit 6 Teaching Resources booklet describes an excellent activity to illustrate this comparison.)*

The Cost of Comfort

Students can read this page to find out how to calculate the amount of fuel or electricity used to heat or cool their homes. Ask for a show of hands to determine how many students use gas, oil, electricity, or a combination of these to heat their homes. Tell students who don't know what type of fuel they use to find out when they go home.

 Transparency 63 is available to accompany The Cost of Comfort.

Antonio in Action

Explain to students how to read the electricity meter shown on this page. The dials are read from left to right, and the numbers on the dials are marked in opposite directions. The first dial is read counterclockwise, the second is read clockwise, the third counterclockwise, and the fourth clockwise. Each dial represents a digit in a four-digit number. The four-digit number is then multiplied by 10 to provide the correct number of kilowatt-hours. Pointers that fall between two numbers of a dial are read as the lower of the two numbers. Therefore, Antonio's electricity meter reads 6618 kWh × 10 = 66,180 kWh.

GUIDED PRACTICE Ask: How could Antonio calculate what part of the total electricity had been used to heat his house? *(He could determine the energy used for all appliances and subtract this from the total. Or he could read the meter on a warm day, when the heater is not being used.)*

Answer to
In-Text Question

Ⓐ According to Antonio's notes, the reading was 65,830 kWh 10 days ago (November 4). 66,180 − 65,830 = 350 kWh. Therefore, 350 kWh were used in 10 days, which would be 1050 kWh (350 kWh × 3) in a month. His calculation was correct.

The Cost of Comfort

The next time an energy bill arrives in the mail, watch what happens. Is there a rush to open it as though it were a letter from a dear friend, or is it greeted with a lot of grumbling because it isn't particularly welcome?

Energy bills are not good news. They hit us where it hurts—our wallets. Energy bills account not only for the electricity that runs our appliances, but also for the energy we use to warm or cool our homes. Unless the weather is very mild, the largest part of a utility bill is usually the cost of heating or cooling. How much do you think it costs to keep warm—to heat the water and the air in your home—or to keep cool? Here's how to find out.

Some homes are heated with gas, some with oil, and some with electricity. Air conditioners use electricity. If your home is heated with gas, just read the gas meter. If your home is heated with oil, simply find out how much oil you use. If your home is heated with electricity, you can find the cost of keeping warm by subtracting the amount of electricity used to run the appliances from the total amount used. Follow the same procedure to calculate how much it costs to keep your house cool.

The "A Team" Investigates

Antonio, Akeisha, and Anila decided to investigate how much their families spend on energy.

Antonio in Action

Antonio's home uses only electricity, so all he had to do was read the meter. He read it at five o'clock on November 4 and again at five o'clock 10 days later, on November 14.

The illustration above shows how the meter looked on November 14. Copy and complete Antonio's data sheet shown on the next page in your ScienceLog. The meter wasn't easy to read because some pointers turned clockwise and some turned counterclockwise. He wished someone could have helped him.

Meters with four dials, like Antonio's, often have "Multiply by 10" written on them. When meters have five dials, it is not necessary to multiply.

Antonio calculated how much energy his family used in a month: 1050 kWh. Wow! He found that hard to believe, so he asked to see the latest electric bill, shown on the next page. Was Antonio's calculation correct? Ⓐ

352

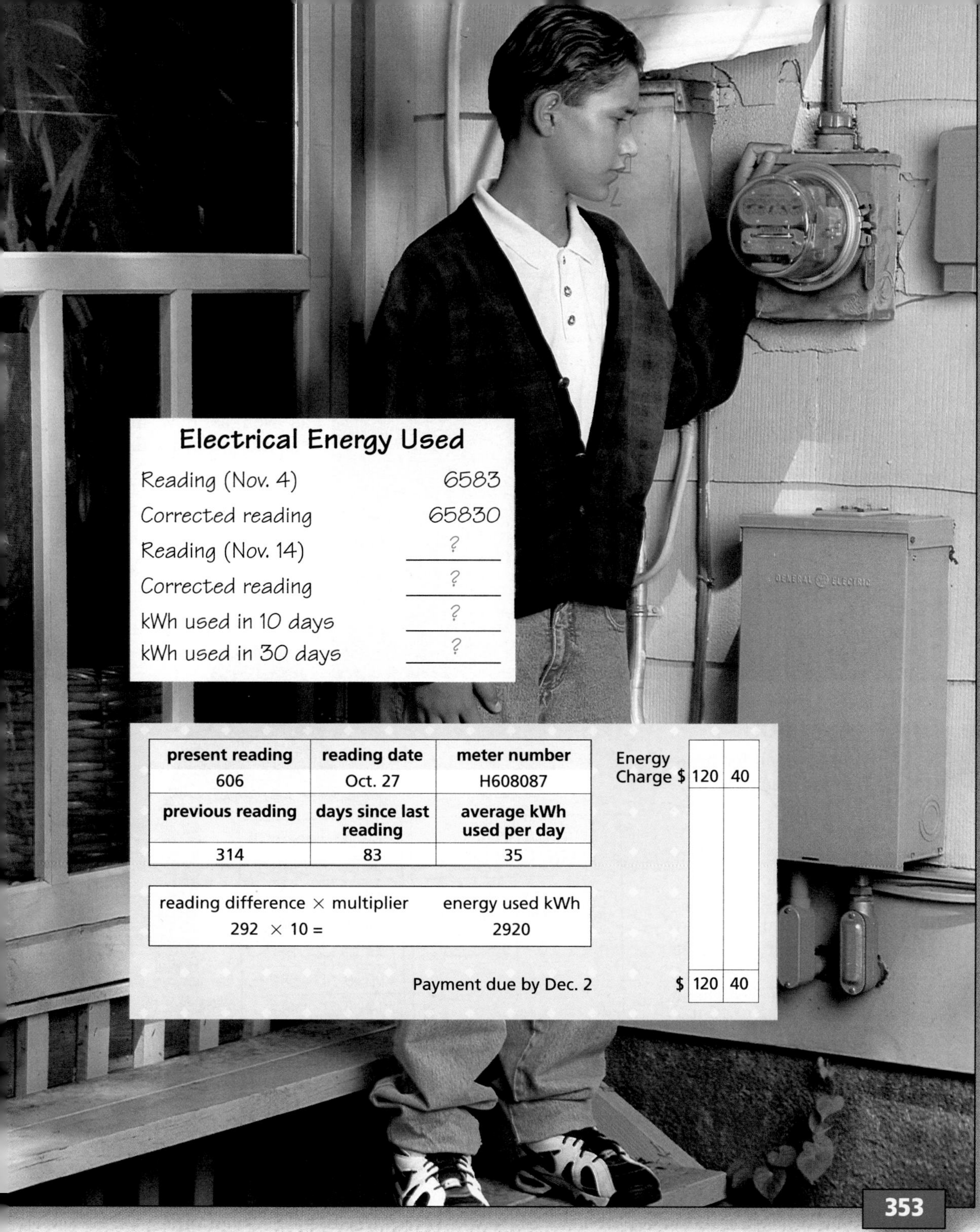

Electrical Energy Used
Reading (Nov. 4) 6583
Corrected reading 65830
Reading (Nov. 14) ?
Corrected reading ?
kWh used in 10 days ?
kWh used in 30 days ?
present reading 606
reading date Oct. 27
meter number H608087
previous reading 314
days since last reading 83
average kWh used per day 35
reading difference × multiplier 292 × 10 =
energy used kWh 2920
Energy Charge $ 120 40
Payment due by Dec. 2
$ 120 40
353

Theme Connection

Changes Over Time

Have students graph the data from the table below to show how electrical consumption for one family changed over the course of a year. **Focus question:** Why did energy consumption change over the course of a year? *(In summer, consumption was higher because the outside temperature was higher and more energy was needed to cool the home. Consumption also increased in winter because of heating needs.)*

Electrical Energy Consumption

Month	Energy used (kWh)
January	1050
February	1020
March	1070
April	1150
May	1200
June	1310
July	1300
August	1350
September	1190
October	1050
November	1040
December	1100

Sample graph:

Electrical Energy Consumption

Energy used (kWh)
1400
1200
1000
800
600
400
200
0
J F M A M J J A S O N D
Month

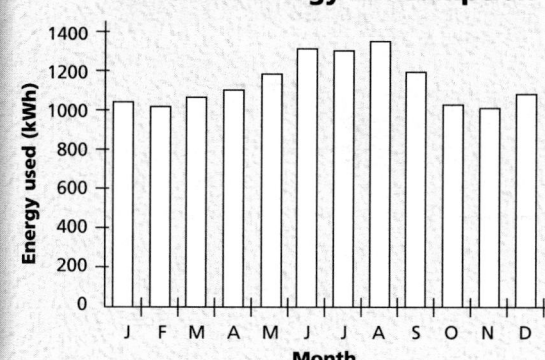

Akeisha in Action

Have students read Akeisha's plans for figuring out the amount of electricity and gas used in her home and decide whether they agree with her plans. *(To find the amount of gas used, she read the gas meter. To find the amount of electricity used, she read the electric meter.)* Remind students that 1 m³ of natural gas = 10 kWh of electricity, according to the table on page 344.

Anila in Action

Have students study the chart showing the number of liters of oil delivered on the given dates. Ask students to notice when the deliveries were most frequent. *(During the winter months)*

INDEPENDENT PRACTICE For practice, ask students to plot all of Anila's data in a bar graph in their ScienceLog. Students who use oil in their homes for heating could also gather data as Anila did and then make bar graphs based on their own data. Ask: How could Anila's graph be improved? *(The x-axis could be labeled "Weeks" and the y-axis labeled "Average Number of Liters of Oil Used." This would improve Anila's graph because the oil is not delivered at regular intervals.)* Then have students hypothesize about why the amount of oil used varied so much.

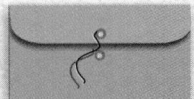

PORTFOLIO

Ask students to record the data they gathered in this lesson in their Portfolio. Suggest that they analyze their data and briefly describe an energy plan. They should indicate how well they understand different energy sources, and they should identify energy sources and conservation issues that need clarification.

Anila in Action

Anila's home uses oil for heating and hot water and electricity for everything else. She tried to find out how many liters of oil were used in 24 hours. She decided the only way to be sure was to check the oil bills. Like Antonio and Akeisha, Anila also looked at the electric meter.

Anila got some excellent data. Her dad showed her the oil bills for the whole year. She started by plotting the amount they used during the year on a bar graph and then tried to explain why it went up and down so much.

ANILA AFSHAR ENERGY CONSUMPTION

HOT AIR AND WATER AVERAGE FOR NOV.	APPLIANCES (NOV. 10TH)
14.2 LITERS OF OIL	27 kWh OF ELECTRICITY

(bar graph: y-axis 100–800; bars labeled APR. 21, JULY 4, OCT. 29)

FILL-UP DATE	LITERS OF OIL DELIVERED
APR. 21	498
JULY 4	481
OCT. 29	661
DEC. 9	584
JAN. 8	490
FEB. 6	575
FEB. 28	450
MAR. 24	462
MAY 1	526

Akeisha in Action

Akeisha's home uses both electricity and gas. The furnace and water heater use natural gas. The gas meter looks very much like an electric meter, but as you know, gas is measured in cubic meters or feet, not kilowatt-hours. Akeisha decided that she should measure how much gas her family used in a day, as well as how much electricity they used.

CROSS-DISCIPLINARY FOCUS

Mathematics

The *solar constant* is defined to be the average amount of energy from the sun that strikes a given area of the Earth. It is equal to 1.4 kW/m². Ask: How many 100 W bulbs could be lit from the energy that strikes an area of 2 m²?

a. 1.4
b. 2.8
c. 14
d. 28 *(Correct)*

Did You Know...

It takes 95 percent less energy to make an aluminum can from recycled material than from aluminum ore.

The Cost of Heating and Cooling

ACTIVITY 1

Keeping Warm

1. Find out what type of energy is used to heat your home.

2. Wait for a cold snap. Whatever form of energy is used to heat your home, find out how much of it you use over a 10-day period. Be sure to start and stop measuring at the same time of day. Multiply by 3 to estimate how much energy you would use in a month.

3. If your home is heated by gas or oil, the figure you found in step 2 is the amount of energy you use to heat your home.

4. If your home is heated electrically, you need to subtract the energy used to operate appliances from your energy total. One way to do this is by looking at an energy bill from a month in which little or no

heating or air conditioning was used. This should show you about how many kilowatt-hours are used by appliances alone. Subtract this number from the number you found in step 2 to find out how many kilowatt-hours you use for heating.

5. Find out the cost per unit of whatever form of energy is used to heat your house. Multiply the amount of energy used by the cost of the energy per unit. This gives your total heating cost. Compare your results with those of your classmates. Compile a class average.

ACTIVITY 2

Keeping Cool

1. Wait for warm weather to do this activity. Follow the procedure outlined in Activity 1. Find out how many kilowatt-hours you use in 10 days. Multiply by 3 to estimate how many kilowatt-hours you use in a month.

2. Subtract the energy used by appliances from this total by following the procedure outlined in step 4 of Activity 1. This gives you the amount of energy used by the air-conditioning system.

3. Multiply this amount by the cost per unit of electricity to figure out how much the air conditioner costs to run for a month.

Finding the Cost of Heating or Cooling

Wait for warm or cold spell.
⬇
Find the amount of energy used in 10 days.
⬇
Multiply by 3 to find energy used in 1 month.
⬇
Subtract energy used by appliances.
⬇
Multiply by cost per unit of energy.

How Much Energy Does Your School Use?

The members of "B Team" (Brent, Barb, and Beth) all live in apartments where the energy bills are included in the rent. They decided to survey their school instead. It was quite a task! How did they go about it? They began by making a list of questions. What questions would you add to theirs? Ⓐ

355

Have students complete Activity 1 or 2, depending on the season. It will take students 10 days to collect their data.

INDEPENDENT PRACTICE Have students draw class graphs of energy consumption and costs over a 1-month period. Each student could draw one bar on a graph to represent the amount of energy consumed at home and another bar to represent the cost. Then the cost of using different kinds of energy could be compared.

In warm weather, bring in some heating bills from winter months so that students can calculate heating costs per month. In cool weather, bring in electric bills from summer months so that students can calculate air-conditioning costs per month. For students who live in apartment complexes, the exercise that follows allows them to participate in calculating energy use.

Answer to
In-Text Question

Ⓐ Other questions the "B Team" could have asked include the following: Does the school use more than one type of energy? Who could we ask to find out about the school's energy bills? Who should we ask for permission to read the meters?

Homework

You may wish to assign the Exploration Worksheet that accompanies Exploration 2 as homework (Teaching Resources, page 40).

FOLLOW-UP

Reteaching

Have students make a chart in which they list all the electrical appliances used in the classroom. Ask them to estimate, on a weekly basis, the following things for each appliance: the hours used, the average kilowatt-hours consumed, and the cost of using it.

Assessment

Explain to students: You have a lamp that is on 5 hours per day. You can choose between a 60 W incandescent

bulb and a 14 W fluorescent bulb. How many kilowatt-hours of energy would you save in a month if you chose the fluorescent bulb? *(60 W × 5 hours/day × 30 days/month = 9000 Wh ÷ 1000 Wh/kWh = 9 kWh/month; 14 W × 5 hours/day × 30 days/month = 2100 Wh ÷ 1000 Wh/kWh = 2.1 kWh/month. The energy savings would be 9 kWh – 2.1 kWh, or 6.9 kWh/month.)*

Extension

Have students interview their parents to

find out which appliances they may have purchased during the last 10 years. They should also ask if the amount of electricity required to operate the appliance was considered. Students should inquire if their parents would make the same choice again after being made aware of energy-saving considerations.

Closure

Ask students to identify the five household activities that consume the most energy in their own home.

FOCUS

Getting Started

Ask students to list one appliance or use of energy that they would be willing to give up in order to save energy and money. Have students explain why they made their choice.

Main Idea

By analyzing how energy is used in the home, an energy-saving plan can be developed.

TEACHING STRATEGIES

Discuss with students why it is difficult to make sacrifices like those described in the text. *(In many cases, it means giving up comfort and convenience.)* You may wish to point out to students that much of the energy used today comes from fossil fuels, which are in limited supply. Saving energy not only saves money, but also preserves valuable natural resources. A description of fossil fuels and alternative sources of energy can be found in Chapter 20 of this unit.

Answers to *Personal Sacrifices*

The following sample answers are based on a price of 6 cents/kWh:

- If the television is on 6 hr./day, or 180 hr./month, a family could save 180 hr./month × 0.1 kWh/hr. = 18 kWh/month. This would save $1.08/month.
- If the dishes are washed once a day, then the savings would be about 30 L of water and 1.5 kWh of electricity per day, or 900 L/month and 45 kWh/month. The savings on electricity would be about $2.70 each month. Using a dishwasher only when it is full saves water and electricity. Also, some dishwashers can be programmed to dry the dishes without heat.
- At one shower per day, showering saves 0.5 kWh/day, or 15 kWh/month in addition to the water savings of 300 L/month. In one month, the money saved on electricity by showering would be $0.90.

LESSON

2

A Plan for
Saving Energy

Could you imagine yourself throwing money into the trash? Wasting energy is just like throwing away money. How difficult is it to avoid wasting energy? It depends. It takes very little effort to turn off the lights when you leave a room. It doesn't require much effort to turn off the TV when you've finished watching it. On the other hand, saving large amounts of energy requires some effort. Saving energy whenever possible makes good sense—but are you prepared to make sacrifices? Is your family?

Personal Sacrifices

- You can save about 0.1 kWh every hour by watching a black-and-white TV rather than a color one. Do you think the sacrifice is worth the savings? Find out the cost of 1 kWh of electricity. How much would the savings be in one month?

- If you wash the dishes by hand, you need about 20 L of hot water. An automatic dishwasher uses 40 L to 60 L. You could save between 1 kWh and 2 kWh on heating water every time you wash the dishes. The personal cost is time and effort! Is it worth the savings? How much would the savings be in one month if you made the effort? How could you use a dishwasher in a way that would actually save energy?

- A shower typically uses 30 L to 50 L of water, and a bath uses 40 L to 60 L. You could easily save 10 L of hot water, or half of a kilowatt-hour of electricity. But what would be the savings in money? in cleanliness? in relaxation? Is it worth it? How much money could you save in one month by showering instead of bathing?

Family Sacrifices

Here are some more situations. Ask yourself what the cost of each would be. Do your friends and family agree? Would you make the sacrifice? Would they?

- Should you leave on the house lights or a radio (0.1–0.2 kWh per hour) to discourage break-ins? If you were away on a two-week vacation, how much would this anti-theft system cost you? Would it be worth it?

LESSON 2 ORGANIZER

Time Required
1 class period

Process Skills
analyzing, comparing, contrasting

New Terms
none

Materials (per student group)
none

Teaching Resources
SourceBook, p. S158

- A frost-free refrigerator consumes about 140 kWh of energy per month, and a nonfrost-free refrigerator consumes about 90 kWh per month. Which would you recommend buying? Why do people buy frost-free refrigerators?

- Washing machines generally use about 120 L of hot water per load. Rinsing in cold water saves about 40 L of hot water (about 2 kWh). How much money could you save per month by making use of this information? Is rinsing in cold water worth it?

- Fluorescent lights are about five times as efficient as regular incandescent lights. Would you install them in your home?

- If you turned down the thermostat in your house by 1°C (from 20°C to 19°C), you would reduce your heating bill by about 5 percent. How much would this be in dollars per year? Would you do it?

- If you turned down the thermostat at night by 4°C (from 20°C to 16°C), you would reduce your heating bill by about 10 percent. It would be possible to save as much as 400 L of oil, 400 m³ of gas, or 4000 kWh of electricity per year. How much money would this be? Would you turn down the thermostat at night?

A Call to Action

Crisis! Your family is trying to reduce expenses, but the cost of energy is on the rise.

Your Response

You decide to reduce your family's total energy consumption by 25 percent. How will you achieve this goal? Prepare a list of recommendations. Then look back over this and the previous lesson to get help in estimating the savings—in both money and energy—that each recommendation will bring. Report your recommendations to your family—they may want to try them!

357

Group size: 2 to 3 students
Group goal: to evaluate whether a family should make a sacrifice in a given situation
Positive interdependence: Make sure that one person in each group is in charge of leading the discussion (group leader), one is in charge of keeping track of comments (recorder), and one is in charge of keeping the group focused (monitor).
Individual accountability: Each member of the group is responsible for keeping notes on the group's findings.

Answers to
Family Sacrifices, pages 356–357

Student opinions will vary. The following answers are based on a price of 6 cents/kWh for electricity:

- Keeping the lights or radio on for 2 weeks uses 33.6–67.2 kWh of energy. The cost would be $2–$4.
- Many people feel that the extra 50 kWh/month is worth not having to defrost the refrigerator by hand.
- If the washing machine is used four times a week, rinsing with cold water saves 32 kWh/month. The savings would be $1.92.
- Fluorescent lights also have a much longer life. Some types are irregular in shape or require special fittings, however.
- If a family spends about $800 per year on heating, the savings would be about $40.
- If a family spends about $800 per year on heating, the savings would be about $80.

Answers to
Your Response

Sample recommendations: install timers on the lights, purchase a smaller television set, use the dishwasher only when it is full, and use cold water in the washing machine.

ENVIRONMENTAL FOCUS

Point out that irresponsible energy consumption may affect more than an energy bill. Ask: What hidden, environmental costs might result from excessive energy consumption? *(Sample answer: Pollution might increase, affecting people's health and the health of natural ecosystems.)*

FOLLOW-UP

Reteaching

Ask students to make a list of ways to conserve the amount of electricity used by their television set. *(Watch less TV; watch a smaller TV; switch from a color TV to a black-and-white TV.)*

Assessment

Ask students to list ways in which electricity is wasted in their home. Then have them grade themselves on how they save or waste energy in one week.

Extension

Have students send for consumer information on energy conservation in the home from a local utility company or a county cooperative extension agency. Have them draw posters and set up an exhibit on energy conservation based on this information.

Closure

Have students list all of the electrical appliances in their homes. Ask them to estimate the savings if they no longer used the nonessential ones.

Keeping in the Heat

FOCUS

Getting Started

Bring a sweater, a scarf, a coat, a wool cap, and gloves to class. Ask a volunteer to put on these items. Ask: Why is this student getting warm? *(The student feels warm because the body produces heat that is trapped next to the body by the extra layers.)* Explain that these articles of clothing act as insulation.

Main Ideas

1. Heat energy can escape from homes in many ways.
2. Insulation reduces heat loss from homes.

TEACHING STRATEGIES

EXPLORATION 3

Call on a volunteer to define the word *insulation* so that everyone is familiar with the term. *(A material that slows down the flow of heat from one place to another)*

Ask one group to perform the experiment with no insulation (the control) while each of the other groups chooses one insulating material.

 An Exploration Worksheet (Teaching Resources, page 42) and Transparency 64 are available to accompany Exploration 3.

Theme Connection

Structures

Focus question: What helps to maintain a fairly constant temperature at the surface of the Earth? *(The layers of the atmosphere act as insulation, blocking out excessive radiation from the sun and keeping in heat at the Earth's surface.)*

Keeping in the Heat

People who live in a cold climate spend a great deal of money heating their homes. On a cold day, heat inside buildings travels through the walls and escapes to the outside. How does heat escape from your home? Can we slow down its escape? First, consider what materials your home is made of. Does the material make a difference when it comes to saving heat? Is your home well insulated? Does the type of insulation in your home matter? The following Exploration will help you answer these questions.

EXPLORATION 3

A Heat-Saving Competition

In this Exploration, you will construct an experimental "house" out of two cans. You will then test your house to see how quickly it loses heat. Compare your results with those of your classmates to find out who has the best design.

You Will Need

- 2 cans—one that fits easily into the other
- 100 mL of water
- a variety of insulating materials
- an alcohol thermometer
- plastic wrap
- a rubber band
- a metric ruler

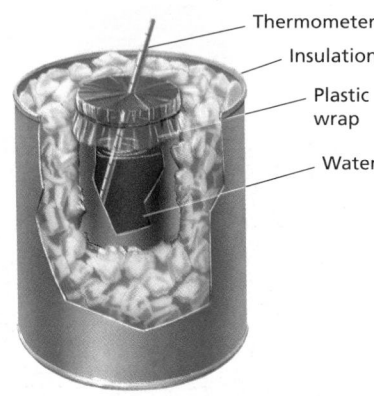

Thermometer
Insulation
Plastic wrap
Water

What to Do

To ensure fairness, all of the competitors should follow the same procedure. How can you make the competition a fair one? Look at the pictures below of two experimental setups. Would this be a fair test? What procedure must each competitor be certain to follow? **B**

Ⓐ

Caution: The edges of the cans may be sharp. Be careful not to cut yourself.

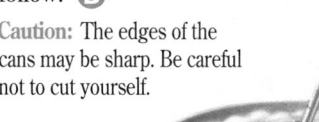

Add your insulation to the large can so that it forms a layer about 5 cm deep. Use whatever type of insulation you think will work well—sand, tissue paper, dried grass, cotton batting, wool, sawdust, or plastic foam. Put 100 mL of water at 60°C into the smaller can. Cover the smaller can with plastic wrap, and secure it with a rubber band. Poke a hole in the plastic wrap just large enough for your thermometer. Place the small can inside the large can. Fill the remaining space between the cans with more insulation. All competitors must have the same starting temperature. Why do you think this is important?

Take the temperature every minute for 10 minutes. Plot your results on a graph. Compare your results with those of a classmate by plotting your classmate's results on the same graph. The winner of the heat-saving competition is the person whose "house" has the warmest water after 10 minutes.

LESSON 3 ORGANIZER

Time Required
2 to 3 class periods

Process Skills
predicting, analyzing

Theme Connection
Structures

New Term
Thermal resistance—a measure of how well materials prevent heat from passing through them

Materials (per student group)
Exploration 3: 2 cans, one that fits easily into the other; enough of one of the following insulating materials to line the larger can: sand, tissue paper, dried grass, cotton batting, wool cloth, water, plastic foam, sawdust; alcohol thermometer; small piece of plastic wrap; rubber band; metric ruler; 100 mL of 60°C water; graduated cylinder; watch or clock

Exploration 4: small sheet of plastic wrap; clothes hanger; tape

Teaching Resources
Exploration Worksheets, pp. 42, 44, and 46
Activity Worksheet, p. 45
Transparency 64

Where Does the Heat Go?

You have seen that insulation can help keep heat from escaping from a building. Still, some heat energy will always be lost. This section will help you to find out how this happens and to prepare a home energy report for your family.

Take a look at the thermal photograph below. A thermal photograph of a house is similar to a normal photo except that it shows temperature instead of light. In fact, thermal photographs are often taken at night! See if you can interpret this one.

- Why do you think the basement is so bright?
- Why are the steps leading to the house black?
- Why are some of the windows bright?
- Why are the upstairs windows dark? Are the curtains closed in any of the rooms?
- Do you think the house has a storm door? Why or why not?
- Why is the line of the roof so bright?
- Why do you think the house appears to be striped?

How is it possible to slow down the escape of heat from a building? You found one answer in Exploration 3: use insulation. To insulate buildings, foamy or fluffy materials such as plastic foam, fiberglass batting (fluffy, thick sheets), or vermiculite (loose particles) are used. Do you know of any other types of insulation used in homes? **C**

The **thermal resistance** of insulating materials is called the *R-value*. It measures just how good materials are at preventing heat from getting through them. Most common insulating materials have an R-value of about 1.25 for every centimeter of thickness. Some plastic foams have higher R-values. The lower the thermal resistance of a material, the more easily heat travels through it. For instance, twice as much heat escapes through a wall whose R-value is 1.5 as through a wall whose R-value is 3.0. If you visit a building supplier, you will see the R-value of insulation products stamped on the packages.

- What is the R-value of a wall filled with 10 cm of fiberglass batting? **D**
- What is the R-value of a ceiling insulated with 15 cm of fiberglass? **E**

Thermal Resistance of Some Common Building Materials

Material	R-Value (per cm)
brick or concrete	0.06
gypsum board	0.24
wood	0.48
glass fiber or rockwool	1.25
cellulose fiber	1.49
expanded polystyrene ("Beadboard")	1.6
extruded polystyrene (Styrofoam)	1.9

Assign this Exploration as an at-home activity. It works best when the weather is cold and windy—drafts are less noticeable on a warm, calm day.

EXPLORATION 5

Students should calculate how much money it would cost to achieve a certain R-value rating for a given area with each type of insulation. Display the collection of insulation samples in the classroom.

Homework

The Exploration Worksheets and the Activity Worksheet that accompany Explorations 4 and 5 may be assigned as homework (Teaching Resources, pages 44–46).

FOLLOW-UP

Reteaching

Provide students with samples of different kinds of insulation materials and their R-value. Have students figure out how thick each material would need to be to insulate the walls and ceiling of their classroom. *(Answers will depend on the material used. Calculations should be based on an R-value of 18–21 for the walls and 33–38 for the ceiling.)*

Assessment

Have students pretend that they have been hired to renovate a home that was built in the 1930s. After determining that insulation levels in the home are very low, they must write a letter to the owners of the home and explain the methods used to determine the insulation levels, the results of the investigation, whether there were areas in the home that left gaps where air could escape, the importance of adequate insulation, and a recommendation for improving the situation.

How much insulation should you have in your ceilings and walls? Use too little, and you will waste heat. Use too much, and you will spend a fortune on insulation! Architects have been trying to find the best balance. Here are their recommendations.

Part of house	Recommended R-values	
	warm climates	cold climates
ceiling or roof	33	38
walls	18	21
basement	12	12

If you live in a warm or hot climate, you may be more concerned with keeping your house comfortably cool rather than warm. Insulation works just as well to keep heat *out* as to keep it *in*. A well-insulated house will have lower heating *and* cooling bills, so it makes good sense to have a well-insulated, energy-efficient house no matter where you live.

Many homes, especially older ones, are not well insulated. However, your family may be able to do something about adding insulation to your home. In Exploration 5, you can check out the best buys. Perhaps you will have some important suggestions about insulation to make to your family.

EXPLORATION 4

Find Those Gaps!

Warm air can find every gap in your home, even small cracks that you didn't know existed! About a quarter of the total heat loss in your home is from air exchange, and a quarter of that is from gaps around doors and windows.

Suppose that your family's heating bill for the winter is $400. You could save $25 just by sealing these gaps.

You Will Need

- plastic wrap
- a clothes hanger
- tape

What to Do

Make a simple draft gauge from some plastic wrap, tape, and a clothes hanger. Choose a cold day and look for those gaps!

Any gaps you find can easily be fixed with weatherstripping (between moving parts) or with caulking. Before you prepare your report, check the cost of weatherstripping and caulking.

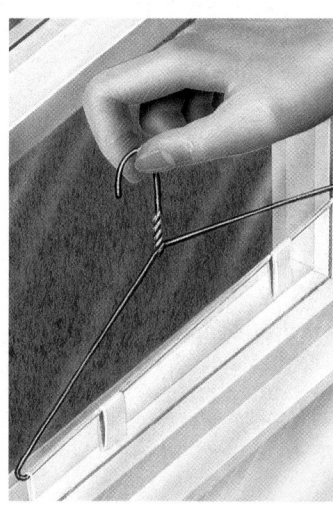

EXPLORATION 5

Best Buys in Insulation

Get together with some of your classmates, and make a display of common insulation materials. Go to a building supply store, and find out all you can about different insulation materials. Find out the R-values of each material and how much each material costs. Try to obtain small samples of different types of insulation to display. If you can't get samples, draw an illustration of each material and provide the information you found. Which material is the best buy?

Extension

As a class, work with the principal or the maintenance staff to determine the amount and type of insulation in key areas of the school. Discuss findings, and brainstorm ideas to improve existing conditions.

Closure

Have students examine photographs taken with infrared-sensitive film and identify areas of high heat loss.

1. Kilowatt-Hour Quiz

a. Which appliance uses the most energy during the same period? Put them in order from most to least energy used.

- 100 W light bulb burning for 6 hours
- television set (200 W) operating for 4 hours
- microwave oven (1450 W) operating for 10 minutes
- washing machine (512 W) operating for 1 hour
- radio (100 W) operating for 4 hours

b. Hector made a list of several appliances in his home, along with their wattages. He and his family decided to use the information to reduce their energy consumption, but they were having some problems. Use his data to help him answer the following questions.

Appliance	Power
microwave	1450 W
toaster	1100 W
oven	2600 W
hair dryer	1000 W/1500 W
light bulbs	40 W, 60 W, 150 W

(1) It takes Hector 1 hour to cook a baked potato in the oven and 8 minutes to cook it in the microwave. Which is the better energy choice?

(2) Hector's big brother Harvey likes to use the broiler setting on the oven to make toast. Hector told his brother that he would save energy if he used the toaster. Was Hector right?

(3) Helen is always running late in the morning, so she uses the high setting on her hair dryer. Instead of taking 18 minutes to dry her hair, it takes only 15 minutes. Did she choose the right setting to save energy?

(4) Helen and Harvey were having a disagreement. Harvey said that the 15 minutes a day Helen spends using the hair dryer on the high setting is a big waste of energy. Helen responded by saying that Harvey always leaves his closet light on during the day (60 W for 8 hours) and that this is a much bigger waste. Who was right?

361

Answers to Challenge Your Thinking

1. a. The following list shows the order of the appliances from most to least energy used:
- television set (800 Wh)
- light bulb (600 Wh)
- washing machine (512 Wh)
- radio (400 Wh)
- microwave oven (242 Wh)

b. Comparisons should be based on the power requirements of each appliance multiplied by the amount of time the appliance is used.

(1) Cooking a potato in a microwave uses much less energy than in an oven: microwave—1450 W × 8 min. × 1 hr./60 min. = 193 Wh oven—2600 W × 1 hr. = 2600 Wh

(2) Hector was right because an oven requires more than twice the power of a toaster.

(3) No; using the high setting on her hair dryer saves Helen 3 minutes but requires more than 1.2 times as much energy: low setting—1000 W × 18 min. × 1 hr./60 min. = 300 Wh high setting—1500 W × 15 min. × 1 hr./60 min. = 375 Wh

(4) Helen was right to point out that Harvey's closet light wastes more energy than her hair dryer: hair dryer—1500 W × 15 min. × 1 hr./60 min. = 375 Wh light bulb—60 W × 8 hr. = 480 Wh

Answers to Challenge Your Thinking continued ▶

Multicultural Extension

Packaging Practices

Explain that food to be sold is packaged in many different ways in different parts of the world. In the United States, food is wrapped in polyethylene or placed on plastic-foam trays and then wrapped. In other countries, like Indonesia, China, and Tunisia, food is wrapped infrequently. Shoppers bring bags to the market and put the food directly in these bags. Thus, much of the energy used to produce packaging containers is saved in other countries. However, this procedure increases the risk of improper sanitization. Ask students to interview their parents about their willingness to use a different kind of packaging system. Have students report their findings.

Meeting Individual Needs

Learners Having Difficulty

As added practice in designing experiments, ask students to design and perform an additional experiment to test the insulating capability of cotton, wool, polyester, paper, or mylar. Be sure to check student designs for safety considerations.

Answers to
Challenge Your Thinking, continued

2. Encourage students to be creative in their solutions to this problem. There is more than one correct answer. One possible response is to measure how long it takes to collect 1 L of water from the tap. Next, compare the temperature from the hot-water tap with that from the cold-water tap. Find out how many kilowatt-hours it takes to raise the temperature by this amount. Then multiply this figure by the number of liters lost in a day to determine how many kilowatt-hours are wasted each day.

3. Students can refer to information provided by local utility companies, government agencies, or building contractors. Students might also want to design experiments to investigate these topics.

★ **The Chapter 19 Review Worksheet (Teaching Resources, page 47) is available to accompany this Challenge Your Thinking.**

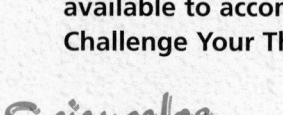

The following are sample revised answers:

1. Lists will vary. The amount of electrical energy that an appliance uses depends on the appliance's wattage. Encourage students to check for a wattage label on the appliances that they use most often. In terms of energy consumption, the wattage of an appliance may not be as important as the amount of time that the appliance is used.

2. Accept all reasonable answers. Students may mention using less electricity or lowering their thermostat in winter.

3. A lot of heat is lost when air from inside a typical house escapes through gaps around windows and doors or through cracks in ceilings and walls. The air that escapes is replaced with cold air from outside the house. Heat is also lost when the air inside the house comes in contact with cold surfaces such as the roof, the outer walls and doors, and the windowpanes. (These patterns of heat loss can also occur in an igloo.)

2. Water Torture

The hot-water tap in the bathroom drips about once every second. The family argues about how much energy it wastes. Can you design an investigation to find the answer?

3. Investigate Further!

Get together with a classmate, and pick one of the following topics to investigate. Draw up plans for your investigation, and check them out with your teacher.

- Which keeps a room the coolest: plastic blinds, wooden blinds, drapes, or shades?
- Does the color of the roof affect the temperature inside a home?
- Which commercially available type of home insulation is the best at keeping something cold?

Review your responses to the ScienceLog questions on page 347. Then revise your original ideas so that they reflect what you've learned.

362

Integrating the Sciences

Life and Physical Sciences

Share the following with your students: What's one of the greatest insulators in nature? Feathers! Feathers enable penguins to survive the fierce winters on the Antarctic icecap, the coldest place on Earth. Because the penguin is a flightless bird, its feathers are devoted entirely to keeping it insulated. The feathers do this by trapping air in a continuous layer all around the penguin's body. With this layer and a thick coat of fat just beneath the skin, the warmblooded penguins can stand around in a blizzard with temperatures as low as −40°C. The penguins can remain there for weeks without boosting their internal warmth with a meal. When humans look for the most effective way to keep themselves warm in the coldest of conditions, they often turn to eiderdown, the feathers of an arctic duck.

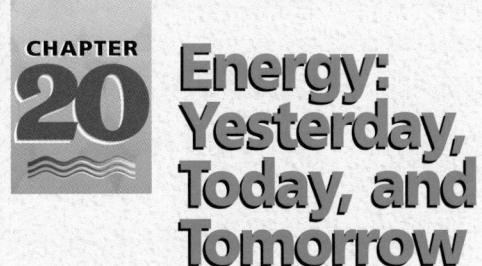

CHAPTER

20 Energy: Yesterday, Today, and Tomorrow

1 Today we use about 10 times as much energy as we did 100 years ago. How would you explain this huge increase?

Do you think we will ever run **2** out of oil? If we did, what sources of energy would you suggest for transportation? for heating? for electricity?

Can you think of a way **3** to use this trash to provide energy?

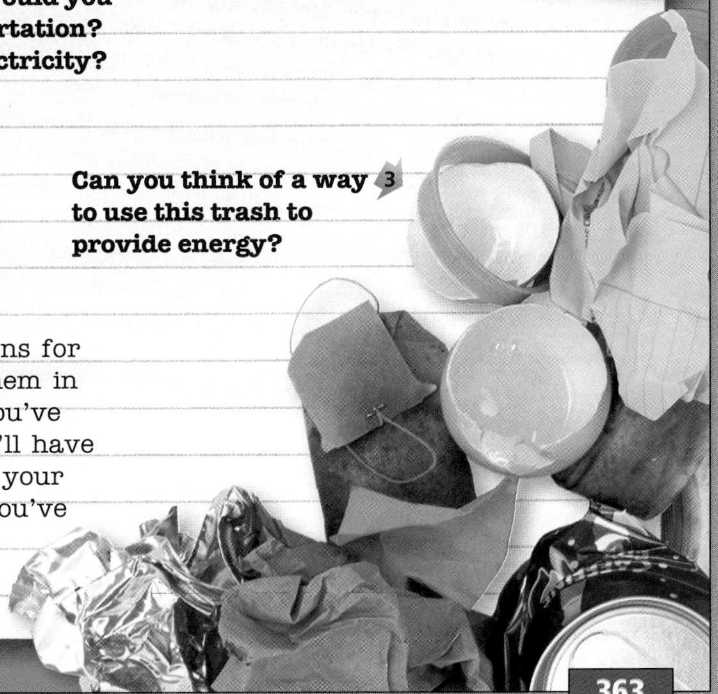

ScienceLog

Think about these questions for a moment, and answer them in your ScienceLog. When you've finished this chapter, you'll have the opportunity to revise your answers based on what you've learned.

Connecting to Other Chapters

Chapter 17
gives students hands-on experience with various forms of energy and discusses energy conversions.

Chapter 18
explores electrical energy, including how it is generated, used, and measured.

Chapter 19
offers a chance to audit home energy use and evaluate the costs of energy consumption patterns.

Chapter 20
outlines the history of energy usage and explores possible future energy resources.

363

Prior Knowledge and Misconceptions

Your students' responses to the ScienceLog questions on this page will reveal the kind of information—and misinformation—they bring to this chapter. Use what you find out about your students' knowledge to choose which chapter concepts and activities to emphasize in your teaching.

In addition to having students answer the ScienceLog questions on this page, you may wish to have them perform the following exercise: Ask the class to suggest as many sources of energy as they can, and list them on the board. Ask students to consider modern sources of energy, past sources of energy, and possible future sources of energy. Then have each student choose an energy source and write a one-page report in which they evaluate the strengths and weaknesses of the energy source. Encourage them to suggest how the energy source might benefit or harm society, and when, if ever, the energy source has been used by people. Emphasize that there are no right or wrong answers in this exercise. It may be best to ask students not to put their names on their papers. Collect the papers but do not grade them. Instead, read them to find out what students know about sources of energy, what misconceptions they may have, and what about energy sources is interesting to them.

FOCUS

Getting Started

Ask students to make a list in their ScienceLog of at least three ways that they have used energy today. Ask them to classify their answers as either human-made or natural. For example, a student could have been awakened by the sun (natural) or an alarm clock (human-made). Have students speculate on how their answers might have been different 100 years ago.

Main Ideas

1. Energy sources and usage today are very different from those in the past.
2. In the last 100 years, energy use has increased dramatically.

TEACHING STRATEGIES

Luke Booker's Diary

GUIDED PRACTICE   Have students list in their ScienceLog some of the differences between Luke's lifestyle and their own. Make a list of these differences on the chalkboard. Ask: Which of these differences have something to do with energy? Have students rate the list of differences. You could use the following scale:

0 = nothing to do with energy
1 = somewhat connected with energy
2 = strongly connected with energy

Answers to
Luke Booker's Diary

Accept all reasonable responses. Students may note differences such as a lack of electric lights or appliances in Luke's home, as well as the fact that his home is heated with a wood stove instead of a gas or electric furnace.

Students should realize that many aspects of a person's lifestyle are related to the types of energy sources that are available to him or her.

Imagine going 100 years back in time. What do you see? You walk to the nearest house. How does it differ from your home? You knock and are invited in. Is it like your home inside? What are the people doing? If it's early in the morning, what you see might be very much like what follows.

Luke Booker's Diary
January 18, 1893

"Luke! Time to get up." My mother's voice roused me from a sound sleep. Must be six o'clock. I stretched, got out of bed, and felt my way toward the window. The wooden floor was cold under my feet. I scraped the ice off the pane and peered out, but it was still dark outside. "Hurry up, Luke!" I went into the kitchen. The kerosene lamp shone warmly on the table. Already my mother had breathed some life into the dying embers in the range. The kindling crackled. "Luke, I need some wood." I wiped the sleep from my eyes and found my heavy woolen shirt and corduroy pants, which had been hung to dry. I staggered outside to the woodpile.

When I returned, the fire was well lit. It beckoned to me. I went across and ladled some warm water from the cistern that was attached to the range. It never ceased to amaze me that it was still warm. I remembered the old days. There's nothing worse than washing in cold water! "Here's your breakfast, Luke." I really quite liked hot porridge but pretended I didn't. Mother said it kept out the cold. "Time for school, Luke." I put on my coat, said goodbye, and ran outside. It was light now. A horse and buggy clattered down the road. I'd rather walk anyway!

List some of the main ways in which your life is different from Luke Booker's. Do you think lifestyle has any connection with energy?

As far back as we have evidence of their existence, human beings have harnessed the energy around them.

Exploration 1 examines the ways in which our use of energy has changed over time.

LESSON 1 ORGANIZER

Time Required
1 class period

Process Skills
analyzing, inferring

Theme Connection
Changes Over Time

New Terms
none

Materials (per student group)
none

Teaching Resources
Transparency 65

Energy Through the Ages

This Exploration shows a few of the energy-related events that have occurred throughout history. As you read the time line, think about how energy affects lifestyle. Then discuss the questions that follow.

500,000 B.C.
Early humans use fire. Using the energy released when wood burns, humans warm themselves and cook food.

8000 B.C.
The first oil lamps are used by cave dwellers in western Europe.

2400 B.C.
The sail is invented in Egypt. Sails use the wind, rather than human muscles, to move boats.

A.D. 1000
The Pueblo peoples of North America learn to use passive solar heat by building homes that take the greatest advantage of the winter sun.

1710
Thomas Newcomen invents the first practical steam engine. His design is later improved by James Watt. The steam engine revolutionizes industry by allowing the construction of machines requiring large amounts of power.

1800
Alessandro Volta invents the first chemical cell to produce electricity. It becomes known as the voltaic pile.

1831
Michael Faraday constructs the first true electrical generator. It consists of a copper disk that rotates between the poles of a permanent magnet.

Exploration 1 continued ▶

365

CROSS-DISCIPLINARY FOCUS

Language Arts
Have interested students do a report on the contributions of one of the inventors and scientists mentioned in the time line, such as Newcomen, Watt, Volta, Faraday, Edison, Daimler, Benz, Ford, or Fermi.

Have students work in groups of two or three to read the time line and then answer the questions on page 368.

You may wish to have student groups do research through readings and interviews to find out more about how food, clothing, and shelter were provided during pioneer days. Point out that the energy sources used to do most of the work at that time were human muscle, animal muscle, and wood. Groups could then summarize their findings in writing and present them to the class.

Cooperative Learning
EXPLORATION 1

Group size: 2 or 3 students
Group goal: to examine the time line and generate answers to the questions on page 368
Positive interdependence: Assign each student a role such as discussion leader, recorder, or timekeeper.
Individual accountability: Each person should be able to answer the questions on page 368 based on the group's discussions.

Exploration 1 continued ▶

★ **Transparency 65 is available to accompany Exploration 1.**

Meeting Individual Needs

Learners Having Difficulty
With two coins, you can demonstrate a simple voltaic pile like the one referred to under 1800 in the time line. Take two coins made of different metals. Clean them well with steel wool. Fold a paper towel into a pad slightly larger than the coins. Soak the pad in salt water. Place the pad between the two coins. Hold the pile between your thumb and index finger. Connect both leads of a galvanometer to the coins and watch the deflection of the needle.

GUIDED PRACTICE Ask students to think about energy events since 1950. Ask: Why do you think energy usage has increased so dramatically since 1950? *(The population has increased and consumers have purchased and used more energy-consuming goods, such as cars and electrical appliances.)* What has spurred the increased interest in energy conservation in the last 30 years? *(Rising fossil-fuel prices and the new awareness of the damage that fossil fuels can cause to the environment)* What energy trends do you foresee in the future? *(Possible responses include more interest in conservation of energy; more interest in recycling; people using fewer energy-consuming goods; and the development of alternative, nonpolluting energy sources.)*

Theme Connection

Changes Over Time

The formation of fossil fuels such as oil and the subsequent use of these fuels involve a series of energy conversions.

Focus question: List the energy conversions involved in the formation of oil and in the use of oil in a modern machine. *(Sample answer: Radiant energy from the sun is converted to chemical energy in plants. This chemical energy is stored in the remains of plants and animals. As layers of dead organisms accumulate and become buried by tons of sediment, they form oil, which is used to produce gasoline and other petroleum products. When a car burns gasoline, it converts chemical energy into heat energy, sound energy, and kinetic energy.)*

Multicultural Extension

Technology in China

China has an extremely long and colorful history of technological and cultural development. Have interested students do some research to create a time line of technological developments related to energy in China. This time line could then be shared with other students.

1859 Col. Edwin Drake drills the first oil well near Titusville, Pennsylvania. He strikes oil at a depth of 21 m.

1879 Thomas Edison invents the light bulb. The California Electric Light Company begins operating the first commercial power plant to sell electricity to private customers.

1885 Gottfried Daimler and Karl Benz, working independently, develop an improved internal combustion engine that uses gasoline as fuel.

1900 Over 3600 electric utility companies are in operation across America.

The automobile age begins when Henry Ford introduces the Ford Model T. The demand for petroleum soars as automobiles come into widespread use. More than 15,000,000 Model T automobiles will eventually be manufactured.

1941 The first commercial wind-powered electric generator begins operating in Vermont.

1942 The world's first nuclear reactor, designed by Dr. Enrico Fermi, is activated at Stagg Field on the campus of the University of Chicago.

Did You Know...

In 1865, many years before the Wright brothers made their first successful flight, an inventor named Jacob Brodbeck built and flew a flying machine powered by coiled springs. The craft soared to just above tree level before it crashed. His creation conquered many of the difficulties of human flight long before the development of the internal-combustion engine made modern aviation possible.

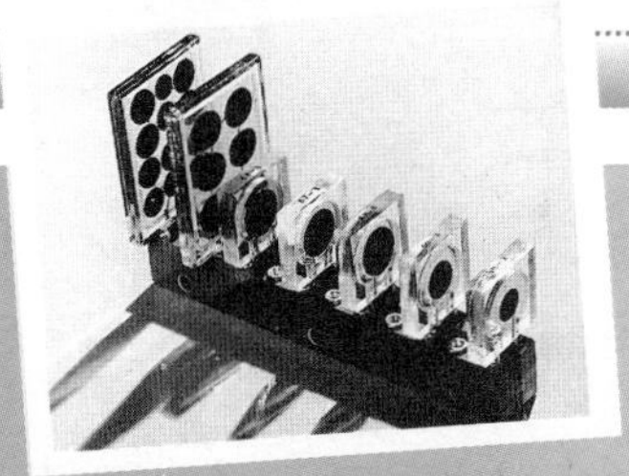

1954
Bell Laboratories develops the solar cell, which converts sunlight directly into electricity.

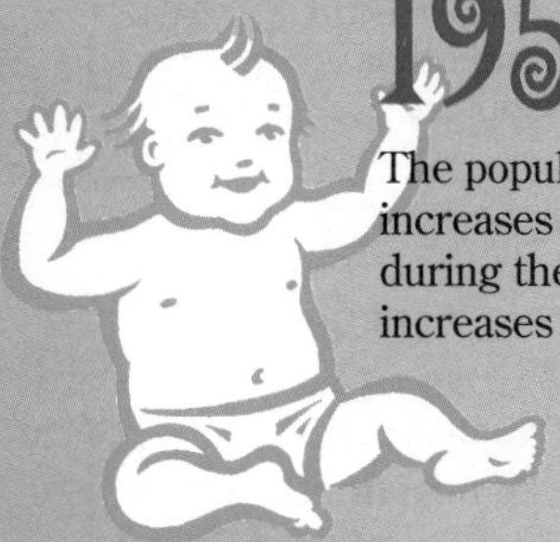

1950-1970
The population of the United States increases by about 20 percent, while during the same period, energy usage increases by 90 percent.

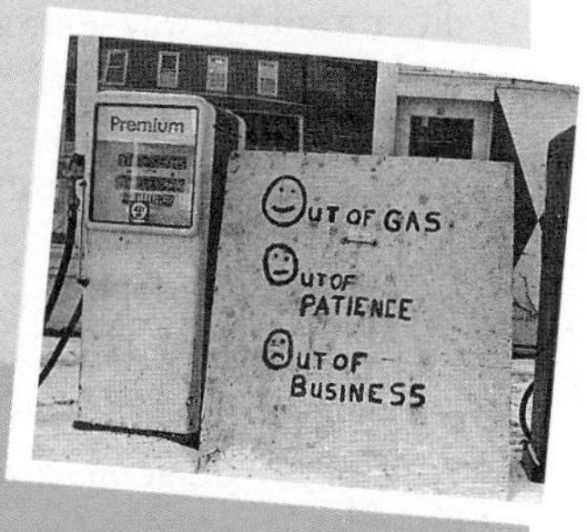

1973
The Arab oil embargo disrupts American oil supplies, causing fuel shortages all over the United States. This motivates the American government to enact the first peacetime energy-conservation measures. The embargo also signals the dawn of an era of rising fuel prices.

1979
The Trans-Alaska Pipeline is built to transport oil. A major engineering feat, the pipeline is built to lessen America's dependence on foreign oil by tapping the rich oil fields of Alaska's North Slope. The pipeline is strongly opposed by many people who fear it will damage Alaska's delicate environment.

1980s
Rising fossil-fuel prices spur increased interest in energy conservation. Fossil fuels—oil, coal, and natural gas—are shown to be a major source of greenhouse gases, adding urgency to the search for alternative sources of energy.

2000 and Beyond???

Exploration 1 continued ▶

367

GUIDED PRACTICE List these energy sources on the chalkboard: human muscle, animal muscle, electricity, and gasoline. Divide the class into groups of four or five, and ask each group to list several machines that are powered by each of the listed energy sources. Ask: What are some weaknesses of each type of machine? *(Muscle-powered machines cannot operate unless a person or animal is powering it; electric machines must usually be plugged in; and gasoline-powered machines often give off pollutants and are noisy.)*

Exploration 1 continued ▶

Meeting Individual Needs

Gifted Learners
Have interested students build a model of a telegraph machine and then demonstrate it for the rest of the class. Students should describe the energy conversions that occur when a message is sent. Students can also be prepared to answer questions from other students about the development of the telegraph and its place in history.

Multicultural Extension

Latimer's Light Bulb
Lewis Latimer, a pioneer in the development of the electric light bulb, was the only African American member of the Edison Pioneers, a group of distinguished scientists and inventors who worked with Thomas Edison. Have students trace the changes in the light bulb's design since Latimer worked on it. They should present their findings in the form of a time line.

Homework
You may wish to have students complete questions 1–4 on page 368 as homework.

ENVIRONMENTAL FOCUS

Discuss with students how our changing patterns of energy use have affected the environment. *(As the use of fossil fuels has increased, the pollution that is associated with them has increased as well. In addition, mining and drilling for oil has caused environmental damage in some cases.)* This is a good opportunity to stress the theme of Changes Over Time as it relates to environmental changes.

Answers to *For Discussion*

1. Students will probably suggest alternative energy sources, such as wind, solar, or nuclear energy.

2. Students should describe different types and levels of energy usage that coincide with those of about 50 and 75 years ago.

3. In 1945, coal was the major source of energy. In 1990, petroleum was the main source of energy. Between 1945 and 1990, use of coal decreased while use of natural gas increased. Nuclear energy was not available in 1945.

4. Energy consumption in 1990 was more than 75 percent greater than the consumption in 1960.

5. Answers will vary. Encourage students to think of the many types of energy they use so that they can justify their answers.

6. They were formed from the fossilized remains of ancient plant and animal life.

7. Possible responses include the following:
   a. Light bulbs greatly increased people's nighttime activities. Cars made transportation easier and faster, so people began traveling more often and farther from home.
   b. It is unlikely that light bulbs would have been invented because they would not have worked without electricity. Although cars are usually powered by gasoline, an inventor could have created a vehicle powered by natural gas, steam, or alcohol.
   c. Energy consumption has greatly increased since the invention of light bulbs and cars.

8. There are many connections between our lifestyle and the way we use energy. For example, many electrical appliances in homes make life easier but require electricity, so our society produces and uses a great deal of electricity.

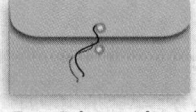 **PORTFOLIO**

Students may wish to include their answers to *For Discussion* in their Portfolio.

For Discussion

1. By the middle of the twenty-first century, much of the oil known to exist today will be gone. Where do you think our energy will come from in the future? Write your own time-line entries for the future.

2. Examine the graphs at right. How might the way you use energy differ from the way your grandparents did when they were your age? How about the way your great-grandparents did when they were your age?

3. How is the nation's energy consumption in 1990 different from that in 1945?

4. Examine the graph below. How many times more energy did we use in 1990 than in 1960?

5. Which form of energy is most important to you?

6. Why do you think oil, coal, and natural gas are called "fossil fuels"?

7. a. How have light bulbs and cars influenced our lives?
   b. Do you think light bulbs or cars would have been invented if electricity or gasoline had not already been discovered? Explain.
   c. What kind of effect do you think light bulbs and cars had on energy consumption?

8. What do you think is the connection between our lifestyle and the way we use energy?

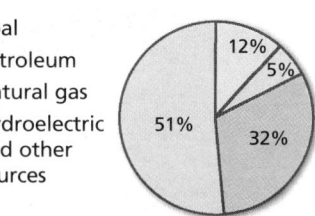

1945
32 Quads* Used

- Coal
- Petroleum
- Natural gas
- Hydroelectric and other sources

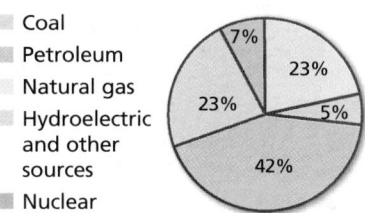

1990
78 Quads Used

- Coal
- Petroleum
- Natural gas
- Hydroelectric and other sources
- Nuclear

*One quad is equal to 293 billion kWh.

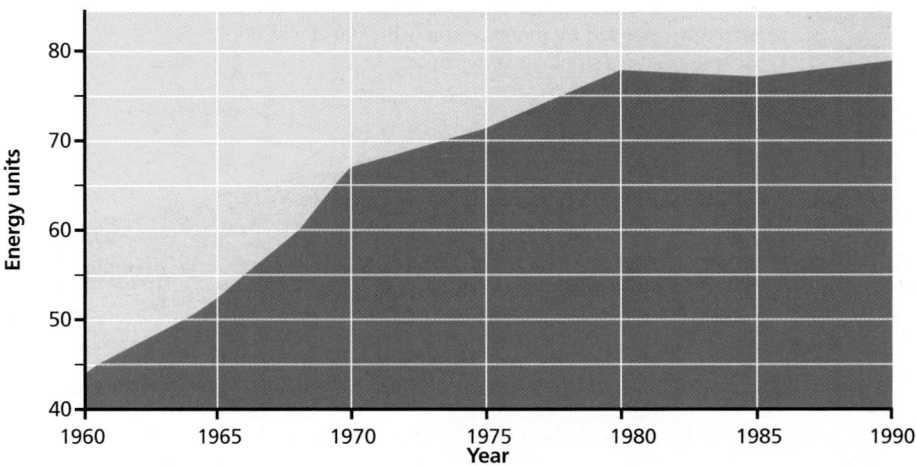

Total Energy Consumption 1960–1990

FOLLOW-UP

Reteaching

Have students pick a date along the time line and describe what forms of energy would have been used to provide the basic needs of food, clothing, and shelter at that date.

Assessment

Have students write their own story about how their life depends on energy or about how their life would be different without electricity.

Extension

Have students interview senior citizens about what life was like when they were growing up and how their lifestyles have changed as energy use has changed. Students should share their interview with the class.

Closure

Ask students to choose one invention from the time line and to describe how it has influenced their way of life.

Our demand for oil is increasing, but the supply is running out—and not just in the United States, but all around the world. Once it's gone, it's gone for good. Why do you think this is so? **A**

U.S. Oil Use — 1970 and 1990

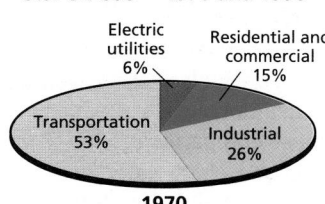

Electric utilities 6%
Residential and commercial 15%
Transportation 53%
Industrial 26%

1970

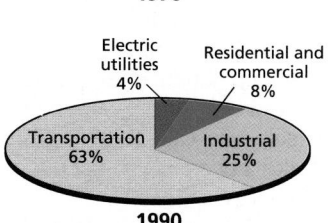

Electric utilities 4%
Residential and commercial 8%
Transportation 63%
Industrial 25%

1990

What is this stuff we call "black gold"? Where does it come from? How did it form? Imagine a time millions of years ago, long, long before humans arrived on the scene. Most of the Earth is covered by warm seas teeming with tiny plants and animals. When these tiny organisms die, their remains fall to the ocean floor, where they accumulate century after century. As time passes, sediments cover the layers of dead organisms. Over long periods of time, the sediments continue to pile up higher and higher. Eventually, they may be several thousand meters thick. Crushed by the heavy load of sediments and warmed by the heat of the Earth, the tiny plants and animals are "cooked"; the chemicals that made up their bodies are slowly changed into crude oil. Presto: black gold!

Without oil, our civilization would collapse. Oil makes our industrial world go around. It provides energy for our factories, heat for our homes, and gasoline for our cars. It is no wonder that crude oil is called black gold. Oil also supplies us with *petrochemicals,* which are used to make plastics, paints, fibers, medicines—the list is almost endless. Look around you. You are probably surrounded by things made from petrochemicals.

Renewable and Nonrenewable Resources

The graphs above compare the use of oil in the United States in 1970 and in 1990. How has our use of oil changed? Compare the figures for transportation. What would you conclude? Oil usage (on a percentage basis) has dropped in every category other than transportation. Would you conclude that less power is being generated, fewer manufacturing facilities are being operated, and fewer homes are being heated? Explain. **B**

369

FOCUS

Getting Started
Write a phrase such as "no more oil" on the board. Ask students to brainstorm what this condition might lead to. Write their ideas on the board and draw arrows to show the relationships between these ideas.

Main Ideas
1. Energy sources that can be replaced are called renewable; those that cannot are called nonrenewable.
2. The supply of oil and other non-renewable energy sources is running out.

TEACHING STRATEGIES

Ask if any students can define and give examples of petrochemicals. *(Petrochemicals are chemicals made from petroleum or natural gas; examples are plastics, synthetic fibers, and paints.)*

⭐ **Transparency 66 is available to accompany the material on this page.**

Answers to
In-Text Questions

A Nature would not be able to replace the depleted supply of oil for thousands or even millions of years.

B The graphs are based on percentages, so they could indicate either a switch from oil to other sources of energy by all areas except transportation or they could indicate an increased dependence on oil-powered transportation, or both.

CROSS-DISCIPLINARY FOCUS

Mathematics
Have students convert the pie graphs on this page into a bar graph.

LESSON 2 ORGANIZER

Time Required
1 class period

Process Skills
classifying, analyzing

New Terms
Renewable resources—energy sources that can be replaced naturally, such as wood, hydroelectricity, and solar energy

Nonrenewable resources—energy sources that cannot be replaced once they are used up, such as oil, natural gas, and coal

Materials (per student group)
none

Teaching Resources
Transparency 66
SourceBook, p. S151

Several countries in the Middle East have large, natural supplies of oil. As these resources are depleted, these countries must turn to other sources of energy. Have interested students find out what research on alternative energy sources is being done in these countries.

Cooperative Learning
RENEWABLE AND NONRENEWABLE RESOURCES

Group size: 3 to 4 students
Group goal: to discuss how the United States could deal with an oil shortage
Positive interdependence: Assign each student a role such as discussion leader, recorder, pacer, or speaker.
Individual accountability: Each member of the group should write a short summary of the group's discussion.

Answers to
In-Text Questions

Ⓐ Oil consumption in the United States is increasing faster than oil production. Ten years from now, the current trend will probably still be in effect, but fifty years from now, both oil production and consumption will probably have dropped off because the oil resources will have begun to decline.

Ⓑ Renewable resources include wind, wood, sunlight, and hydroelectricity. Nonrenewable resources include natural gas, uranium, coal, and oil.

Ⓒ Students will probably favor cutting back on energy use and developing renewable energy resources.

Homework

You may wish to have students answer the in-text questions on pages 369–370 as homework.

FOLLOW-UP

Reteaching

To emphasize the meaning of the word *renewable,* ask: Is a rosebush a renewable or nonrenewable source of

---

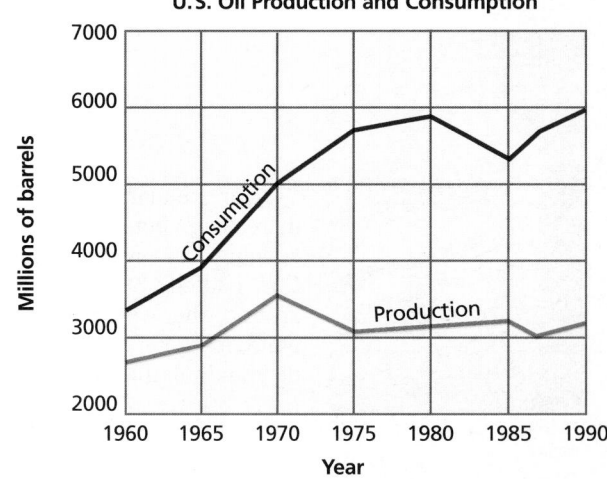

U.S. Oil Production and Consumption

Millions of barrels (y-axis: 2000, 3000, 4000, 5000, 6000, 7000)
Year (x-axis: 1960, 1965, 1970, 1975, 1980, 1985, 1990)

Consumption

Production

Now look at the graph above comparing oil production and consumption. What would you conclude from this? How might these comparisons look 10 years from now? 50 years from now? Explain. Ⓐ

Fortunately, oil is not our only source of energy. There are others, including coal, hydroelectricity, wood, and natural gas. Like oil, some of these cannot be replaced once they have been used up. Such energy sources are known as **nonrenewable resources**. Energy sources that are replaced naturally, if used wisely, are called **renewable resources**. In your ScienceLog, make a two-column list of nonrenewable and renewable sources of energy. Ⓑ

Even though the United States is one of the world's leading producers of oil, it must import much of its oil to meet its energy needs. So far, the United States has had no trouble finding enough oil to meet the demand, but oil is running out all over the world, and its price, although fluctuating, is heading generally upward. What about the future? How should we respond?

- Don't worry; something will happen.
- Cut back on energy consumption.
- Search harder for more nonrenewable resources—oil, coal, and natural gas.
- Develop more renewable resources—hydroelectricity, solar energy, tidal power, geothermal energy, and wind energy.
- Increase the number of nuclear power plants.
- Try to find new energy resources.
- Other (specify)

Take a few minutes and discuss each of these options. Based on your current knowledge, which option(s) do you favor? Why? Ⓒ

Squeezing Oil From Stone

Geologists have long known of vast deposits of oil-bearing rocks called oil shales. The United States alone has billions of barrels of oil in the form of oil shales. What's the catch? Extracting oil from oil shales is difficult and costly. The rock must be mined (usually strip mined), crushed, and heated to separate the oil from the rock. Then the waste rock must be disposed of. The cost of oil would have to be much higher to make extracting oil from oil shales an economical option.

---

roses? *(Renewable, because the roses can grow back)*

Assessment

Have students write a newspaper article to summarize the main points in this lesson.

Extension

Have interested students organize a debate to discuss the extraction of oil in environmentally sensitive regions such as wildlife refuges. Each team should do research to find out the benefits and the risks involved and then be prepared to defend one side of the issue.

Closure

Ask students to propose possible incentives for people to use less energy. Have students design an advertising campaign that focuses attention on the benefits of conserving gasoline. Encourage students to suggest simple ways in which individuals can use less gasoline in their daily life. Discuss with interested students how they might implement such an advertising campaign in their school or local community.

LESSON 3 — Energy for the Future

What is the future of our energy resources—oil, hydroelectricity, coal, natural gas, solar, geothermal, wind, and nuclear?

This is a big question because our whole way of life depends on having enough energy. It is also a big question because answering it is quite a challenge. There is a huge amount of information available on this topic, and you may find yourself having to make some difficult decisions about what to investigate.

EXPLORATION 2

A Big Question to Answer

Get together with about three friends, and select one energy source that you would like to research. Then list some of the questions about this energy source that you think will help you answer the big question: What is the future of this energy source?

In the pages that follow, you will find some information and a few important questions about each energy source. You may use these questions as starting points for your investigation.

The government publishes many papers and pamphlets on the subject of energy. Write or call the U.S. Department of Energy for more information.

Many other government agencies, such as the Nuclear Regulatory Commission, and energy-related businesses will also be pleased to provide information.

Which agencies or companies are involved with your chosen source of energy? You might want to write a letter to ask them for information.

When you have completed your project, create a display on poster board to let your classmates know what you have found out. Do they feel that the future of your energy source is a cause for optimism?

Turn to the SourceBook to get more information for your research project.

Once you have heard about the future of other energy resources from the other groups, you should be able to discuss more fully what course our country should take.

As you read about these energy sources, think about the advantages and disadvantages of each. Is any single energy source the answer to all of our energy needs?

> 345 Sixth St.
> Anytown, USA 10010
>
> Department of Energy
> 1000 Independence Dr.
> Washington, DC 20585
>
> Dear Sir or Madam:
>
> I am collecting information on energy from coal for a school project. I would really appreciate it if you would send me any pamphlets that you think might be useful.
>
> Yours Sincerely,
> Joe Cool

Exploration 2 continued ▶

371

LESSON 3 ORGANIZER

Time Required
2 to 3 class periods

Process Skills
contrasting, comparing, analyzing, predicting

New Terms
none

Materials (per student group)
Exploration 2: sheet of poster board; several colored markers

Teaching Resources
Transparency Worksheets, pp. 52 and 54
Exploration Worksheet, p. 56
Transparencies 67 and 68
SourceBook, p. S148

FOCUS

Getting Started
Encourage students to brainstorm ways that our world would be different if all of our electricity were produced from solar energy. (*In some ways, our daily lives might not be greatly affected. Overall, however, there would be less pollution produced and less damage done to the environment.*)

Main Idea
Different types of energy sources have unique advantages and disadvantages.

TEACHING STRATEGIES

EXPLORATION 2

This Exploration is a research project that requires students to use the library and other sources of information. The questions posed after each article are designed to start students thinking about the advantages and disadvantages of each type of energy. Since some of these questions require much more research than others, you may wish to allow each group to omit one question of their choice. Each group of three students should choose a different energy source to research and then display their findings.

INDEPENDENT PRACTICE Suggest that students prepare an "energy open house" so that they can share and discuss their projects. This could be scheduled during class hours, after school, or in the evening to include parents.

★ **An Exploration Worksheet (Teaching Resources, page 56) is available to accompany Exploration 2.**

Answer to
In-Text Question

D Students should realize that it is unlikely that any one source will be able to meet all of our energy needs.

Exploration 2 continued ▶

EXPLORATION **2,** *continued*

Answers to
Oil—Old Reliable in Decline

1. The United States has consumed about 80 percent of its oil reserves. This does not account for oil deposits that are expected to be discovered in the future.

2. It is expensive to get oil from under the sea, and there is a serious danger of oil leaks causing pollution and destroying marine life.

3. It takes hundreds of thousands, or even millions, of years for oil to form naturally. At present rates of consumption, the world's supply of oil will run out.

4. Examples include looking in deposits of tar sand.

Answers to
Natural Gas—A Clean but Limited Alternative

1. There are about 250 trillion cubic feet of natural gas reserves left in the United States. Deposits are common throughout the United States.

2. There are about 2000 trillion cubic feet of natural gas reserves left in the world. This does not account for natural gas deposits not yet discovered.

3. One major drawback of natural gas is that it is very difficult and expensive to transport.

4. Condensing the gas into a liquid helps relieve the transportation and distribution problems because the liquid form takes up less volume. The condensed gas is still expensive to transport, however.

Oil—Old Reliable in Decline

Even though our country is one of the world's leading oil producers, the United States must import about 55 percent of the oil it uses. Because of the great demand for petroleum products, researchers are developing oil production methods for very harsh climates such as in the Arctic and for very deep water.

Questions

1. How much oil is left in the United States?

2. What are the problems of getting it out from under the sea? Is there any risk of pollution?

3. How long does it take for oil to form naturally? Will more form that we will be able to use?

4. Where else are we looking for oil?

Oil field, Cook Inlet, Alaska

Natural Gas— A Clean but Limited Alternative

Natural gas is found on every continent. It is abundant, cheap, and produces no harmful emissions when it burns. Natural gas is tapped by drilling wells, just as oil is. Because it is such a clean fuel, natural gas is growing in importance. Like oil, though, natural gas is a fossil fuel. Current supplies will not last forever.

Questions

1. Are there untapped supplies of natural gas in the United States? If so, where?

2. How much natural gas is left to be recovered in the world?

3. What is one major drawback of natural gas?

4. It's possible to condense natural gas into liquid form. How will this help relieve natural gas transportation and distribution problems? What are some drawbacks of this approach?

372

Multicultural Extension

Lifestyles and the Price of Gasoline
Point out that gasoline is cheaper in the United States than in many other industrialized countries. Divide the class into groups of four students. Ask each group to develop a list of ways in which lifestyles in countries with high gasoline costs are likely to be different from lifestyles in the United States. *(Accept all reasonable responses.)*

Here Comes the Sun

Solar energy is the ultimate renewable resource, and it isn't difficult to make your own solar heater. In many ways it's similar to a greenhouse. Why not try to make one and test it? The illustration below can give you some ideas.

Questions

1. What is a typical design for the type of solar heater that is used to heat a home?
2. How much do solar heaters cost?
3. How much heat (in kilowatt-hours) do they produce?
4. What are the drawbacks of solar-powered systems?
5. Should every family that can afford to install a solar heater do so?

⬆ A solar-heated house

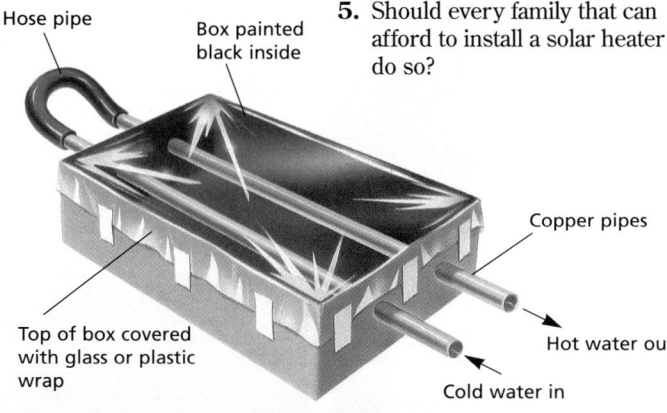
- Hose pipe
- Box painted black inside
- Copper pipes
- Top of box covered with glass or plastic wrap
- Hot water out
- Cold water in

Water Power

The source of hydroelectric power never runs out: rain falls, rivers flow, and dams fill up. Hydroelectricity seems ideal, but is it?

Questions

1. Are there many rivers that have hydroelectric possibilities?
2. Does the building of a hydroelectric dam have any negative effects on the environment?
3. How much does it cost to build a dam?
4. Are there any problems in getting electricity from the generating station to where it is needed?
5. Could hydroelectricity meet all of our energy needs? Explain.

 A hydroelectric dam, Arizona

Exploration 2 continued ▶

Answers to *Here Comes the Sun*

1. Solar collectors are insulated boxes with transparent covers. A heat-transferring fluid is heated as it circulates through pipes in the collector.

2. For an average-sized solar collector system (74 m²), which includes collectors, a storage tank, pumps, heat exchangers, plumbing, controls, insulation, and labor, the cost might be around $20,000.

3. The amount of heat produced by a collector depends on its size and efficiency. In a sunny location, a 20 m² solar collector could capture millions of kilowatts of energy in an average year.

4. They are expensive to install. Solar collectors cannot be used in all locations—they need a lot of sunlight to be effective.

5. Not necessarily; if locations do not get enough sun, they are not suitable for a solar heater.

Answers to *Water Power*

1. Not very many rivers are still available. Also, most that could be used are in remote locations.

2. Reservoirs often destroy wildlife habitats and an area's scenic beauty.

3. A large dam can cost millions of dollars to build, while a small dam built for local use may only cost a few thousand dollars. Also, storage of water may damage the environment by flooding areas and by altering the flow of rivers.

4. Many sites that are suitable for building a dam for hydroelectric power are remote, and transporting electricity is expensive.

5. Since most of the hydroelectric resources have already been tapped, this is a very limited resource. It cannot fully meet our energy needs.

Exploration 2 continued ▶

Meeting Individual Needs

Gifted Learners

Challenge students to do experiments with solar collectors. Students can build the collector pictured on page 373, or they can design their own. Check all designs for safety before students construct their collectors. Have them experiment to determine the most efficient angle for catching the sun's rays. Students should graph their results.

★ **A Transparency Worksheet** (Teaching Resources, page 52) and **Transparency 67** are available to accompany the material on this page.

Answers to
Garbage: What a Waste!

1. Answers will vary. Households in the United States throw away an average of about 24 kg of garbage every week. Many cities now have recycling programs that will accept a variety of materials, including paper, cardboard, glass, metals, and plastics. It is probable that much of the household garbage that cannot be recycled could be incinerated.

2. Incinerators use the heat from burning garbage to produce steam that turns the turbines of a generator. Students may wish to diagram how an incinerator produces steam and how toxic byproducts are handled.

3. The United States burns only a small percentage of its garbage to generate electricity. For example, only 3 percent of the garbage in the United States was burned as fuel in 1985. In contrast, 75 percent of the garbage in Switzerland, 64 percent in Japan, and 51 percent in Sweden was burned for electricity.

4. The benefits of incineration include the reduction of landfill material and the production of electricity. Although modern incinerators are equipped to minimize hazardous emissions, they do release small quantities of poisonous gases as well as particles of toxic heavy metals.

Answers to
Energy From Wind

1. Few areas have sufficient wind because an average wind speed of about 18 km per hour is required.

2. Along with the difficulties in picking a location, a storage system is needed to store the energy when the wind stops blowing. Also, windmills can be very noisy.

3. The best design for a modern windmill has lightweight blades mounted on tall towers.

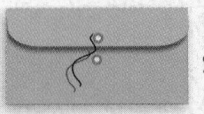 **PORTFOLIO**
Suggest that students include the letters they wrote, the resources they used, and a summary of their project from Exploration 2 in their Portfolio.

Garbage: What a Waste!

Disposal of garbage is a problem. For years people have been burying it in landfill sites. However, these fill up, and new sites are often hard to find. If only we could find a use for the garbage that we can't reuse or recycle.

The city of Detroit thinks that it has found a solution. It burns the leftover garbage and turns the heat energy produced into electricity. This sounds like a great idea, but there are some drawbacks. See if you can find out more about garbage incineration.

Questions

1. How many kilograms of garbage does your family throw away each day? How much of this can be recycled? How much is left over that could be incinerated?

2. How does an incinerator produce electricity?

3. What percentage of garbage in the United States is used to produce electricity? How does this compare with other countries?

4. Incinerators produce some chemicals that are toxic to humans and to the environment. Are the dangers from these chemicals outweighed by the benefits from incineration? What's your opinion?

Energy From Wind

Electricity was first generated using the wind in 1880. The first wind-operated power plant went on line in 1941, but even today, relatively few wind-operated power plants are in operation. This may change in the future, though, as the cost of energy continues to rise.

Questions

1. Would wind-powered generators be feasible in your part of the country?

2. What drawbacks does this energy source have?

3. What are some different designs for windmills?

Multicultural Extension

The Demand for Firewood

Share the following information: In many developing countries, firewood provides over 95 percent of the total energy needs. Adding to the existing problems in many developing nations, obtaining firewood has led to erosion and desertification in previously forested areas. Some solutions include using efficient stoves and planting fast-growing trees. A satisfactory solution to the problem has not yet been found.

Integrating the Sciences

Earth and Life Sciences

If possible, display a sample of bituminous, or soft, coal for students to examine. Have students wrap pieces of coal in a towel and use a hammer to break up the pieces on top of a newspaper. Students may be able to find fragments of fossils in the coal.

Coal—The Rock That Burns

Coal is formed from the remains of plants that died millions of years ago. Sometimes coal is found close to the Earth's surface and is strip mined; sometimes it is found far below the ground and must be mined conventionally. Strip mining often leaves ugly scars on the landscape.

The United States has huge amounts of coal. Most is used for electricity; much of the rest is exported. Unfortunately, most coal produces harmful gases when it is burned. Burning coal is a major cause of acid rain.

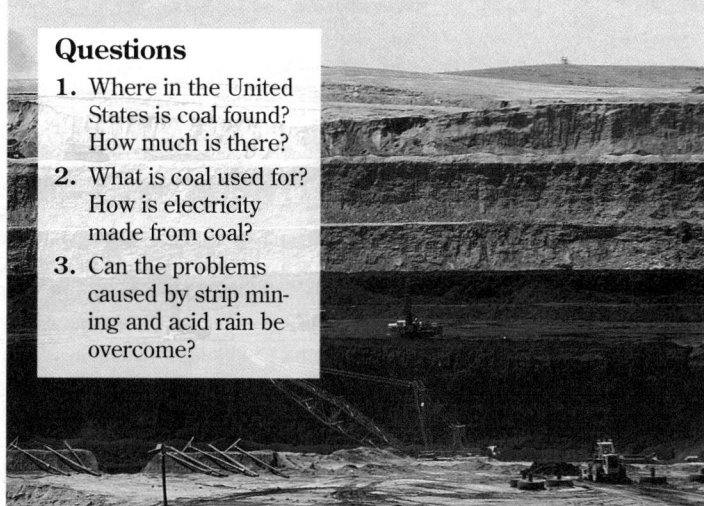

Questions

1. Where in the United States is coal found? How much is there?
2. What is coal used for? How is electricity made from coal?
3. Can the problems caused by strip mining and acid rain be overcome?

Unlocking the Power of the Atom

Radioactive substances such as uranium break down naturally all the time, one atom at a time. When they do, they give off heat as well as other forms of energy. We don't notice this process because it occurs slowly and because these substances are spread thinly throughout the Earth's crust. But when radioactive substances are concentrated, the breakdown speeds up dramatically. The breakdown of one atom can trigger the breakdown of another, and so on— much like one match setting fire to another. Nuclear reactors harness this natural breakdown by keeping the process going at a controlled rate and using the enormous amount of heat that is released to boil water. The steam generated in this way is used to drive a turbine, which in turn drives a generator.

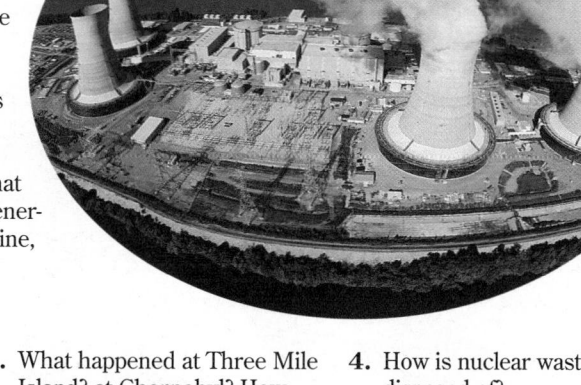

Questions

1. How do nuclear reactors work? How much power can they generate? How many are there in the United States? Are there plans for more?
2. How risky are nuclear reactors?
3. What happened at Three Mile Island? at Chernobyl? How did these accidents happen? Could what happened at Chernobyl happen here?
4. How is nuclear waste disposed of?
5. What are the pros and cons of using nuclear energy?

 A Transparency Worksheet (Teaching Resources, page 54) and Transparency 68 are available to accompany the material on this page.

FOLLOW-UP

Reteaching

Have students make a wall chart of the different energy sources presented in this lesson and the advantages and disadvantages of each.

Assessment

Have students list the characteristics of good energy resources and rate these resources with a plus or minus sign, based on these characteristics.

Extension

Have students determine which countries have the most fuel reserves and which have the greatest energy demands.

Closure

Plan a field trip to a home that is heated with solar energy so that students can see firsthand how solar heating works.

Answers to
Coal—The Rock That Burns

1. Coal deposits are found in Kentucky, Wyoming, and the eastern United States. The United States accounts for roughly 22 percent of the world's coal reserves, which is enough coal to last about 400 years at the present rate of consumption.

2. Coal is burned to produce steam to run electric generators.

3. Efforts can be made to restore a mined area to its original condition after strip mining. Stack scrubbers can be installed to clean impurities from the smoke.

Answers to
Unlocking the Power of the Atom

1. The heat from the nuclear reaction is used to create steam that turns a turbine in an electric generator. Nuclear reactors can generate millions of watts of energy. There are just over 100 nuclear reactors in the United States. Plans for new plants have declined sharply in the past few years.

2. In 1987, nuclear power plants in the United States reported 430 emergency shutdowns and 104,000 incidents of radiation exposure. Nuclear reactors have become much safer, however.

3. At Three Mile Island, a malfunctioning pump and valve, combined with a series of operator errors, caused a partial meltdown. At Chernobyl, operators attempted to correct a mistake and deliberately overrode a series of safety systems designed to prevent an accident. The chances of a similar accident occurring in the United States are slim; however, students should understand that there is always the possibility of error.

4. At present, nuclear waste is being stored until solutions are found.

5. The normal operation of nuclear power plants causes little environmental damage, and these plants produce a huge amount of electricity. However, accidents and hazardous waste can pose a great danger to the environment. Nuclear power plants are also extremely expensive to build.

LESSON 4 — What Does the Future Hold?

FOCUS

Getting Started

Ask students: By the time you are 50 years old, what do you think the Earth will be like in terms of energy production and consumption? Write student responses on the board. *(Answers will vary.)* For any problems that are mentioned, discuss possible solutions with students.

Main Ideas

1. New and improved energy sources are needed for the future.
2. Energy supplies in the future will have a dramatic impact on lifestyle.

TEACHING STRATEGIES

Have students read the introductory paragraphs and the questions that follow. Ask students to read the two articles silently. Instruct the students to read the articles twice—the first time for content and the second time to answer the questions.

Divide the class into groups of four or five to discuss the articles and questions. Reassemble the class and lead a discussion in which each group has a chance to voice its opinions.

Homework

Have students find an article that discusses some aspect of energy use, such as current trends in energy use or new technology designed for alternative sources of energy. Ask students to write a concise summary of the article's major points and to explain why they agree or disagree with the content or tone of the article.

No one knows for sure what the future holds. The two essays that follow portray very different futures. Which do you think is more realistic?

The first essay, written by scientist Glenn Seaborg, has an optimistic tone. As you read it, keep the following questions in mind:

1. Why does Seaborg think people will need even more energy in the future?

2. What is his solution to the energy problem?

3. What do you think the two most important statements in his article are?

4. Do you see any difficulties with Seaborg's ideas about energy production and use in the future?

5. Why do you think Seaborg wrote this article?

Nuclear Energy and Our Future
by Glenn T. Seaborg, Nobel Prize winner

Humans live by energy. The more energy we can put to work, the better we live. By harnessing energy, we can flick a switch to pump water, start a train, heat a home, or light a city. In the twenty-first century, most of this energy will come from the nucleus of the atom, and by using it wisely people can improve the way they produce food, use water and raw materials, handle waste, and build cities. The need for improvement is evident. In the decades ahead, the Earth's population will double and double again. But the Earth itself will not grow, and to support these extra billions we must learn to use and reuse our resources with an efficiency we rarely even approach today.

By far the most important contribution the atom will make to the world of tomorrow will be cheap electricity and huge amounts of heat for use in manufacturing processes. We should proceed with deliberate speed in the full development of breeder reactors, which create more nuclear fuel than they burn.

One must emphasize that having a large amount of cheap energy is not a panacea in itself. We must develop the technology to take advantage of this energy; it must be applied skillfully, productively and wisely by people who have the tools and training to use it.

Some scientists have considered the concept of using these huge amounts of energy in giant nuclear powered industrial complexes, which we have called "Nuplexes," the energy heart of which would be breeder reactors, with a generating capacity in the multi-million kilowatt range.

A report by the Oak Ridge National Laboratory indicates some exciting prospects for coupling large dual-purpose plants located in coastal desert

376

LESSON 4 ORGANIZER

Time Required
1 class period

Process Skills
predicting, analyzing

New Terms
none

Materials (per student group)
none

Teaching Resources

There are people who foresee energy problems in the future that can't be solved. In this second essay, Ron Scammell presents a different picture of what the world may be like for your children and grandchildren. As you read Scammel's essay, think about the following questions:

1. Which of the problems outlined in the story would not have occurred if alternative sources of energy had been found?

2. What do you think might have happened to cause the energy crisis described in this article? How might it have been avoided?

3. Do you think the scenario Scammell describes is possible? Would you change his story in any way to make it more realistic?

4. Why do you think Scammel wrote this article?

areas with highly scientific farms or "food factories."

One study in this report considers the energy heart as a huge breeder reactor which could generate 1,000,000 kW of electricity and desalt 400 million gallons (1,500,000,000 L) of water a day. At the same time, power from the plant could be used to make ammonia fertilizer and phosphorus-bearing fertilizer. It could be that many other by-products from the seawater brine could be produced and used locally or exported.

Electricity from the plant would be used for highly mechanized farming and food processing, as well as to supply light, air-conditioning, and power for transport and communications for the personnel operating the complex.

On 200,000 acres (80,000 ha) irrigated with the desalted water and fertilized with locally manufactured products, crops would be grown which were specifically bred for the area. Hardly anything would be left to chance or the whims of nature.

Such a food factory could produce half a billion kilograms of grain each year, enough to feed almost 2,500,000 people at a level of 2,400 Calories (10,040 kJ) a day. In addition, it could export enough fertilizer to other agricultural areas to cultivate another 10,000,000 acres (4,000,000 ha) of land.

Whenever and wherever people can benefit from heat and electricity, nuclear energy in its many forms can help bring about a better world.

Energy and Our Way of Life

by Ron Scammell

You are living in the year 2070. In the aftermath of a worldwide energy crisis your government has banned all energy consumption, except for the bare necessities.

Electric lighting has been limited to one 50-watt outlet per house, electrical appliances have been banned completely, and coal for heating has been carefully rationed since the exhaustion of oil and gas supplies. A typical day in such an environment begins rather uninvitingly. The house is dark and cold when you wake up in the morning. In the flickering candlelight, you search for an extra sweater to put on. The weekly ration of coal ran out and a cold winter chill snuck up on you during the night, leaving you shivering in the darkness. Performing the morning chores is no easy task. There is no toaster, microwave, stove, or running water. It takes time to run out to the community water supply, fill the bucket, stoke up the wood fire, and cook breakfast. The trip to school during the winter is a real downer; long waits on windswept corners for the rickety old community bus, and then the long walk the rest of the way. Spring and fall are much better, when you can ride your bicycle, as most people do. Private automobiles were outlawed long ago. You have heard your father talk about the days when people jumped in their own cars every morning and drove right from home to school or work. No one can forget the automobile. Its mark is everywhere; miles and miles of deserted freeways, roads, and parking lots. Some communities have tried to convert them into parks and recreational areas but it is a difficult and expensive task.

The school for your district is overcrowded to the point of bursting. But it can't be avoided. The more people who can fit into one building, the less fuel consumed for heating and the less electricity for lighting.

There's not much enjoyment to be found after school either. First the

377

Integrating the Sciences

Life and Physical Sciences
Since 1930, the yield of most major crops in the United States has increased by 200 to 400 percent. The factors causing this increase include greater use of machinery, increased use of fertilizers, better seeds, the use of herbicides and insecticides, and improved transportation. Ask students to select one of these factors and research how it is related to energy. Then engage the class in a discussion of the dependence of agriculture on energy sources. Ask: Is it likely that our farming methods will change as oil and gas supplies are depleted? How might they change?

1. If alternative sources of energy had been found, there might not have been an energy crisis. The reduction of business and industry, as well as of goods and services, would not have occurred. For example, there would not have been a lack of transportation, food, electricity, heat, and appliances such as electric lights, TVs, stereos, washers, and dryers.

2. Several things might have occurred to cause the energy crisis illustrated in Scammell's article. Perhaps people simply ran out of petroleum products and had no alternative energy sources available. Perhaps people ran out of fuel for nuclear reactors, or reactors became too dangerous or too expensive to build. Or the increased population of the Earth placed enormous demands on the energy supply, depleting it very quickly. This might have been avoided if renewable energy resources had been developed and the reliance on nonrenewable resources had been lessened.

3. Students may believe that alternative sources of energy would be developed before such a catastrophe would be allowed to occur.

4. Scammell probably wanted others to imagine what might happen in the future if alternative energy sources are not developed. Perhaps he thought this article might scare some people into taking steps to prevent his bleak vision from becoming a reality.

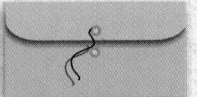

PORTFOLIO
Have students record their opinions about Seaborg's and Scammell's scenarios in their Portfolio. They might also indicate which article they found more realistic and record any predictions of their own.

lineup for two hours at the local shopping center for the day's allotment of food—that's because of the food shortage across the country caused by the breakdown in transportation. Farmers and food producers cannot guarantee delivery of their products. After that, the weekly ration of coal must be stored. Then there is the washing. It must be done by hand. Washing machines and dryers are luxuries of an earlier day. Finally, the day comes to an end, and it's time to relax. But there is not much to do. No TV. No radio. No stereo. Saturday night dances and discotheques went the way of the dinosaur long ago. Rock bands with their electric guitars, pianos, organs, synthesizers, and sound systems consume a fantastic amount of electricity. A definite impossibility in this day and age. There is a community entertainment hall a few miles away, but you don't feel like making the long trek at the end of the day.

Fuel for transportation is so scarce that only the privileged few are able to travel. Major air, rail, and sea passenger services were phased out long ago.

Because of the energy rationing, business and industry have been severely cut back. Jobs are few. The economy never really recovered after the car industry crumbled. After that it was like a house of cards collapsing, as other industries and businesses went bankrupt. There is no use dwelling on dismal thoughts of the future, or fanciful dreams of the past. You decide to stay at home at the end of the day and read a good book. But soon your eyes are tired from reading by candlelight and, anyway, the day's activities have left you exhausted.

CROSS-DISCIPLINARY FOCUS

Language Arts
In order for students to recognize points of view and statements of fact and opinion in a variety of written texts, have students reread "Nuclear Energy and Our Future" by Glenn Seaborg. Then have them answer the following reading comprehension question: Which of the following is a fact given in the passage?

a. "By harnessing energy, we can flick a switch, start a train, heat a home, or light a city." *(Correct)*

b. "By far the most important contribution the atom will make to the world of tomorrow will be cheap electricity . . . "

c. "One must emphasize that having a large amount of cheap energy is not a panacea in itself."

d. "Whenever and wherever people can benefit from heat and electricity, nuclear energy in its many forms can help bring about a better world."

Crystal Ball Gazing

No one can predict what the future holds, but much depends on what happens in the world, what the governments of the world do, and whether the people of the world make wise, well-informed decisions. The future depends on *you*.

Below are some imaginary newspaper headlines from the future. Would you be pleased to read them? Which do you connect with an improving energy picture? Which do you connect with wise decision-making? Explain your choices.

ANOTHER OIL CRISIS?
Middle East Conflict Threatens Supply

SHIP (Solar Heating Incentive Program) **Launched**

Not in Our Backyard! People Reject Plans for Nuclear Waste Disposal

Sound Off

Choose one of the headlines. Write a letter to the editor, commenting thoughtfully on the event or situation.

NUCLEAR FUSION RESEARCH
Raises Hope for Cheap Energy

Fear of Acid Rain Halts Opening of Coal Power Plant

Speed Limit of 80 km Imposed—Conservation Measure

Improved Technology Makes Nuclear Power Safer

379

Crystal Ball Gazing

Have students read the headlines. Students may disagree about the meaning of some of the headlines. Encourage an open discussion. Have them keep in mind that there are no right or wrong answers.

Sound Off

You may wish to provide the students with some newspaper clippings of real letters to the editor so that they can get a feel for the letters' tone and format. It may also be useful to have students read each other's letters and offer constructive advice.

Answer to
In-Text Question

Ⓐ Although not all of the headlines show an improving energy situation, they point out that development of renewable resources may lessen pollution and provide affordable alternatives to fossil fuels.

FOLLOW-UP

Reteaching

Ask students to recall the Getting Started activity for this lesson in which they imagined what life would be like in 50 years given the current rate of energy consumption. Have students conduct an imaginary interview with a middle school student from 50 years in the future. What does he or she need energy for in the future? How has daily life changed? What advice about energy does he or she have for the young people of today?

Assessment

Have students write a summary of each article presented in this lesson. Students should conclude each summary by stating whether they agree with the author's statements and why.

Extension

Have students write a science-fiction story or a one-act play about the future of life on Earth in relation to energy. Encourage students to read their story to the class or to act out their play.

Closure

With their parents' permission, have students interview people in the community to find out their reaction to the possibility of an energy shortage. Questions might include the following: Is there really an energy problem? If there is a problem, what can we do about it? Have students share the results of their interviews with other students.

Answers to
Challenge Your Thinking, pages 380–381

1. Student answers to the concept map will vary. A sample answer might include the following links: *Your lifestyle* depends on *energy. 100 years ago,* people used less energy. *100 years in the future, oil* may be used up and we may rely on *solar energy* to power the *television* and the *automobile.*

2. Answers will vary. Students will have to do some outside research to answer this question. The power companies in their community should be good sources of information. Student recommendations might include the development of renewable energy resources.

3. Student ideas or poems should focus on the difference between renewable and nonrenewable resources. For example, the energy we use to power most machines does not come directly from the sun.

4. Accept all reasonable responses. For instance, students might expect the use of oil-powered transportation to decline as electric cars become more widely used and as more people rely on mass-transit systems.

 You may wish to provide your students with the Chapter 20 Review Worksheet on page 57 of the Unit 6 Teaching Resources booklet.

Homework

The rising and falling of the tides occurs twice daily. Have students write a paragraph discussing how engineers could generate hydroelectric energy from this tidal motion. *(One possibility is for engineers to construct dams across the mouths of coves. Water would flow into the cove through a gate in the dam at high tide and out of the cove at low tide. In both cases, the motion of the water could turn turbines that generate electricity.)*

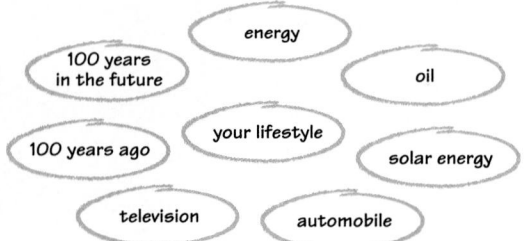

1. Energy and You

After completing this chapter, you may think that energy has more to do with your lifestyle than you once thought. Complete the following concept map. Draw lines between words that you think are connected in some way. Along each line, write a few words that explain the connection.

energy

100 years in the future

oil

100 years ago

your lifestyle

solar energy

television

automobile

2. What's the Scoop?

Do some investigative reporting. (Don't be shy!) Find out as much as you can about the electrical energy you use. Try to find the answers to questions like the following:

- What is the primary source of your electricity?
- What energy conversions take place in producing the electricity?
- Is there more than one power plant in your community?
- How much power do these plants produce?
- How many families do they serve?
- How much electricity does an average family use?
- Is the primary source of the electricity renewable or non-renewable?

380

3. A Poetic Problem!

If nature conserves the energy used, and more of it comes each day from the sun, then why all the concern that we have abused our energy supply to get our work done?

What is *your* answer to the question posed in the poem? Write down your thoughts on this matter. (Or write a poem yourself!)

4. What Kind of Pie?

Take another look at the pie charts on page 369. What do you think the pie chart for the year 2050 will look like? Draw it in your ScienceLog, and explain your thought process in making the chart.

ScienceLog

Review your responses to the ScienceLog questions on page 363. Then revise your original ideas so that they reflect what you've learned.

ENVIRONMENTAL FOCUS

Ask students the following question: How do energy changes affect our environment? Allow students to brainstorm the answer by focusing on the positive and negative effects of energy changes. *(For example, the flow of energy from the sun through living systems is what allows life to exist on our planet. On the negative side, most attempts by humans to harness energy have resulted in unwanted side effects such as excess heat loss and pollution.)*

Did You Know. . .

By the year 2000, over 40 percent of the offshore oil rigs in the United States may be unproductive. Oil companies can either remove these rigs or work with government agencies to convert the rigs into offshore reefs. If converted into a reef, the hard surface of the rig will provide a home for many reef species, from microscopic red and green algae to coral, barnacles, and fish.

ScienceLog

The following are sample revised answers:

1. Answers should mention both population increases and technological advances as causes of increased energy consumption within the last 100 years.

2. If we continue to use oil at the present rate, we will run out of oil. It takes millions of years for oil to form from fossil plants and animals. Because it takes so long for oil to form naturally, it is unlikely that enough new oil will form before we run out. Cars that run on natural gas, electricity, or solar power could replace gasoline-powered cars. Natural gas, electricity, or solar power could be used for heating as well. Electricity could be generated by a variety of different methods, including wind turbines, solar collectors, incinerators, nuclear power plants, hydroelectric dams, geothermal sources, and ocean waves.

3. Garbage could be burned to provide energy for electric generators. Although this would reduce the amount of garbage in landfills, it might also release harmful chemicals into the air.

The Big Ideas

The following is a sample unit summary:

Energy makes things happen. Energy comes in many forms. These include heat (thermal), electrical, chemical, mechanical (kinetic), and light (radiant) energy. (1)

Energy can be stored as chemical energy in the cells of living tissue and in the chemicals in a battery. Potential energy is the stored mechanical energy of a coiled spring, a taut rubber band, or an apple dangling from a tree branch. (2)

Energy makes things happen when it is converted from one form to another. (3) An energy converter changes the form of energy. Converters include steam engines, generators, automobiles, and the human body. (4)

No machine is 100 percent efficient. Some of the energy is always turned into forms that the machine cannot use. Machines cannot create energy; they can only convert energy from one form to another. (5)

A generator converts mechanical energy into electrical energy. A current flows when a magnetic field is changed, either by moving a magnet within coils of a wire or by moving coils of wire within a magnetic field. Simple generators can be powered by hand. More complicated generators can be run by steam-powered turbines that drive the generator, producing electricity. Energy from running water can also turn the turbines of a generator. (6)

Using steam-powered or water-powered generators allows electricity to be produced on a large scale. (7)

The amount of energy that a particular appliance requires in order to function is indicated by its power rating, measured in units called watts. The number of watts required multiplied by the number of hours used tells you how much energy the appliance uses over time. (8)

Heat is lost from buildings primarily through air exchange. Inside air is exchanged for outside air through cracks in the walls and gaps around doors and windows. Heat is also lost when warm air comes into contact with cold surfaces. Heat loss can be reduced by sealing cracks and by providing adequate insulation. (9)

Unit 6

SourceBook

Check out the SourceBook to learn more about how energy is converted from one form to another and about the law of conservation of energy. Learn more about the various energy sources used by humans today, as well as new sources of energy that may power the societies of tomorrow.

Here's what you'll find in the SourceBook:

ENERGY and YOU

The Big Ideas

In your ScienceLog, write a summary of this unit, using the following questions as a guide:

1. What is energy? What are some of its forms?
2. What are some ways that energy is stored?
3. How does energy make things happen?
4. What is an energy converter?
5. Why is a perpetual motion machine impossible?
6. How does a generator work? What are some ways that a generator can be powered?
7. How is electricity made on a large scale?
8. How might you determine the amount of energy that an appliance uses over a period of time?
9. Where is heat lost from a house on a cold day? How can this heat loss be reduced?
10. What is the difference between renewable and non-renewable resources?
11. Of the energy resources we use, which are renewable and which are nonrenewable?
12. What can we do to improve our energy future?

Checking Your Understanding

1. What happens when you blow up a balloon and release it? Describe what you would see in terms of energy.

Renewable energy resources can be replaced. These resources include solar energy, tidal power, wood, and wind energy.

Nonrenewable energy resources are those that cannot be replaced once they have been used up. (10) These resources include coal, oil, and natural gas. (11)

Because we depend so much on nonrenewable energy resources and these resources are running out, we need to do two things. First, we need to use less energy by conserving our current resources and by developing more efficient machines. Second, we need to develop other sources of renewable energy that are safe and inexpensive to use. (12)

2. What is the source of energy in each of the following situations?

   a. A girl pedals her bicycle down the street.
   b. A race car hurtles down the track.
   c. A candle bathes the room in flickering light.
   d. A redwood tree grows to a height of 100 m.

3. Each machine listed below wastes some energy when it operates. Complete the chart by naming the desired form of energy and the forms of waste energy that each machine produces.

Machine	Efficiency	Desired form of energy	Forms of waste energy
toaster	90%	heat	light, excess heat
steamboat	10%		
hair dryer	85%		
car engine	30%		

4. Make a concept map using the following words and phrases: *energy of motion, energy, potential energy, stored energy,* and *kinetic energy.*

5. Tony left a 100 W light bulb on in the attic for a month by mistake. How much energy did he waste?

   a. 64,000 kWh
   b. 72 kWh
   c. 36 kWh
   d. 30,000 kWh

6. A refrigerator has a power rating of 400 W. It runs about half of the time. How much electricity does it consume in a day?

7. Classify the following energy sources as renewable (R) or nonrenewable (N).

   natural gas _____   wood _____
   kerosene _____   hydroelectricity _____
   uranium _____   petroleum _____
   wind _____   sunlight _____
   coal _____   gasoline _____

8. Design a poster about the energy conservation message of your choice.

Share A Ride!

383

⭐ **You may wish to provide students with the Unit 6 Review Worksheet that is available to accompany this Making Connections (Teaching Resources, page 63).**

Homework

The Activity Worksheet on page 62 of the Unit 6 Teaching Resources booklet makes an excellent homework assignment. If you choose to use this worksheet in class, Transparency 69 is available to accompany it.

Answers to
Checking Your Understanding,
pages 382–383

1. Chemical energy from the food you eat allows you to blow air into the balloon. The air causes the balloon to expand, and mechanical energy is stored as potential energy in the stretched material of the balloon. When the balloon is released, the potential energy becomes kinetic and pushes the air out of the balloon.

2. a. Chemical energy from food is turned into mechanical energy and heat.
   b. Chemical energy from burning fuel is turned into mechanical energy and heat.
   c. Chemical energy stored in the candle is released as heat and light energy.
   d. The tree absorbs light energy, which is converted into chemical energy.

3. See completed table on page S241.

4.

Energy
comes in the form of
kinetic energy · potential energy
which is · which is
energy of motion · stored energy

5. The correct answer is (b); Tony wasted 100 W × 24 hr. × 30 days = 72,000 Wh, or 72 kWh.

6. The refrigerator consumes 400 W × 12 hr. = 4800 Wh, or 4.8 kWh.

7. Natural gas—N; kerosene—N; uranium—N; wind—R; coal—N; wood—R; hydroelectricity—R; petroleum—N; sunlight—R; gasoline—N

8. Encourage students to be creative and include illustrations and photographs in their poster. Topics could include environmental issues, energy sources, and energy conservation.

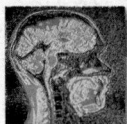

Background

The energy from a lightning bolt is transformed into heat, light, and sound. For example, in typical cloud-to-ground lightning, an initial spark shoots down from the cloud. That spark creates jagged streaks in the sky. It is quickly met by a spark from the ground. This movement of electrons heats the air dramatically, creating the shock waves that cause thunder.

Ball lightning is one form of lightning that scientists still have much to learn about. The lightning appears as a crackling, glowing sphere about the size of a grapefruit. Ball lightning has been sighted in the aisles of flying airplanes and floating down chimneys. Strangely, most reports claim that the lightning is cool instead of fiery hot. One theory to explain ball lightning suggests that the interaction between negative and positive particles in a thundercloud electrically energizes the air. The air becomes a plasma, a state of highly charged matter. The electrical field of the plasma forms a sphere, and chemical reactions inside the sphere produce the unusual characteristics of ball lightning.

Extension

Have two student volunteers perform the following demonstration: One student should rub his or her feet on a piece of carpet. (Students should leave their shoes on.) Then he or she should touch the second student on the elbow. The second student should feel a slight shock. Discuss with the class how this demonstration is similar to what occurs when lightning strikes. *(The student who rubbed his or her feet on the carpet picked up a negative charge much like that in a thundercloud. The second volunteer represents the positively charged Earth.)*

Answers to
Did You Know . . . ?

Student diagrams should indicate separate regions of positive and negative charges. A bolt of lightning should pass between these two regions.

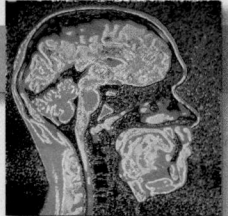

HEALTH WATCH
The Striking Facts About Lightning

If lightning strikes you, 10 to 30 million volts of electricity will pass through your body in less than a second. If this voltage doesn't kill you, it will likely cause terrible burns, brain damage, and respiratory arrest. Understanding more about lightning and about how to protect yourself from a strike can give you an edge over this fiery flash of energy.

◄ Lightning strikes when there is an imbalance of electrical charges.

A Deadly Force of Nature

Within a thundercloud, positive and negative charges build up in different parts of the cloud. Most of the bottom of the cloud becomes negatively charged. These negative charges create a "shadow" of positive charges on the Earth beneath the cloud. Opposite charges attract, so an electric current flows between the areas of positive and negative charges, creating lightning.

Lightning can travel from cloud to cloud, from cloud to ground, or even from a high structure (like a tower or a lightning rod on a building) to a cloud. The current usually flows through the air, but it can also travel through water, telephone wires, water pipes, and human bodies.

Shocking Facts

About 250 people a year are struck by lightning in the United States. Of these, about 70 percent survive the experience. Many suffer ill effects for the rest of their lives. If you are struck by lightning, your chances of survival increase substantially if you receive first aid quickly.

Reduce the Risk

How can you reduce your chances of being struck by lightning? Watch for thunderclouds, dark skies, high winds, and—of course—lightning. Get inside a building or a car, if you can. Avoid using electrical appliances or taking baths during a thunderstorm. Don't use the telephone unless you have an emergency.

If you can't go inside for safety, stay away from tall trees, telephone poles, open water, farm equipment, metal fences, and other metal objects. Drop to your knees and bend forward so that you hands rest on your knees. Don't lie flat on the ground or stand up above the surrounding landscape.

Did You Know . . . ?

Did you know that there are several different kinds of lightning strikes? Find out more about these kinds of lightning and create a labeled diagram for each type. Be sure to write a caption each diagram.

384

Integrating the Sciences

Earth and Physical Sciences

Have students investigate the formation of fulgurites. These are found in the desert where lightning has tunneled into sandy soil, superheated the sand, and caused it to form a solid, strangely branched structure. Some fulgurites reach lengths of 3 m or more. Students should bring sketches of fulgurites for display. Encourage class discussion about other effects of lightning on the Earth's surface. *(Students should recognize that lightning can start fires, kill trees, and harm animal life.)*

Electrifying News About Microbes

Your car is out of fuel, and there isn't a gas station in sight. No problem! Your car's motor runs on electricity supplied by trillions of microorganisms—and they're hungry. You pop a handful of sugar cubes into the tank along with some fresh water, and you're on your way. The microbes devour the food and release enough electricity to get you home safely.

A "Living" Battery

Sound far-fetched? Scientists from King's College in London don't think so. The electrochemists there foresee "living" batteries that will operate everything from wristwatches to entire towns. Although cars won't be using microbial batteries any time soon, the London scientists have proved that microorganisms can convert food into usable electrical energy. One test battery measuring less than 0.5 cm³ operated a digital clock for a day.

Freeing Electrons

For nearly a century, scientists have known that living things produce and use electric charges. Only within the last few decades, however, have they discovered the chemical processes involved. As living cells perform their daily activities, they break down foods

such as starches and sugars. The chemical reactions release electrons, and it is the movement of these electrons that creates an electric current. Scientists produce electricity by harvesting free electrons from single-celled organisms such as bacteria.

Feed Them Leftovers

One advantage that microbial batteries have over generators is that microbes do not require nonrenewable resources such as coal or oil. Microbes can produce electricity by consuming pollutants such as byproducts from the milk and sugar industries. For now, the London scientists are content to speculate on the microbial battery's potential. Other specialists, such as electrical engineers, are needed to make this technology more practical.

◄ Microorganisms such as *E. coli* (left) may one day be used in batteries to supply electricity for a variety of products, including automobiles.

Research Grants

Suppose that you manage a government agency and are asked to fund research on microbial batteries. Write a short report on the pros and cons of developing this energy source. Then, as a class, decide whether or not you would fund the research.

385

Research Grants

After students have written their short reports, ask several students to share their responses. Generate a master list of the pros and cons of extra funding for research into microbial batteries. Then encourage a classroom debate about the issue.

Background

The release of energy from food is called *cellular respiration*. Cellular respiration takes place in two stages, resulting in the storage of energy in *ATP (adenosine triphosphate) molecules*.

Cellular respiration involves the transfer of electrons from the glucose molecule to the energy-storing molecules of ATP. In the microbial battery, scientists harvest some of this energy and transform it into electricity that can be readily used.

ENVIRONMENTAL FOCUS

One of the benefits of the microbial battery is its ability to make use of waste products. Ask students to consider the effect this might have on the energy demands of nations that have limited access to fossil fuels. Ask: What environmental benefits are there? What problems might arise? *(Waste could be recycled, but it might be difficult to transport the energy to people in remote areas.)*

Meeting Individual Needs

Learners Having Difficulty

Have students create a flowchart that illustrates the process by which microbial batteries produce energy. If students are not familiar with flowcharts, you may wish to create a sample flowchart of a simple process, such as brushing teeth, for them to emulate.
(Sample flowchart of microbial battery:)

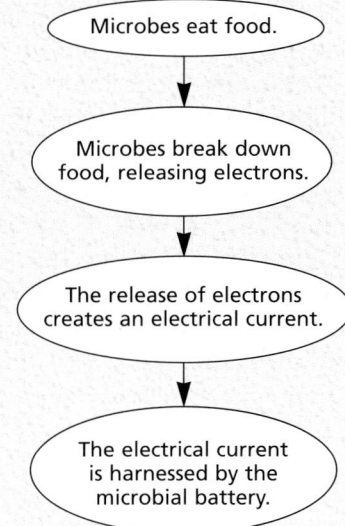

Microbes eat food.

Microbes break down food, releasing electrons.

The release of electrons creates an electrical current.

The electrical current is harnessed by the microbial battery.

Background

In addition to the electric currents that are involved in the nervous system, some body fluids such as mucus and sweat are rich in electrically charged molecules, or *ions*. These ions can also be detected by animals that sense electricity. For instance, one of Gould's experiments showed that star-nosed moles tend to attack worms in the middle of the worms' bodies—right where the worms secrete a lot of mucus.

Electric organs in many "strongly electric" animals are composed of specialized muscle-like tissue. Flattened cells called *electroplaques* are arranged like circuits to build up voltage within the organ.

Critical Thinking

Share the following information with students: When Gould performed his experiment with the batteries in the aquarium, he also included some batteries that were completely wrapped in plastic to keep them from generating a current in the water. Ask: Why was this necessary? *(The covered batteries served as a control for the experiment. Because the moles picked the uncovered batteries more than the covered ones, Gould could conclude that the electric field was what attracted the moles and not the appearance of the batteries or some other factor.)*

ENVIRONMENTAL FOCUS

Students may be interested to learn the following information: The electric elephant fish (which is named for its long snout) is another animal with the ability to detect electric fields. A scientist at London University has recently discovered that water pollution affects the electric signals that are given off by this fish. (Remember that all animals give off a weak electric field.) Thus, by monitoring the fish's electric field, the scientist can tell when the water becomes polluted!

Shocking Discoveries in Animals

*I*magine having a nose covered with dozens of wiggling tentacles. They wave around and give you a very special power. You can actually *feel* the electricity in the air. Sound strange? Not if you're a star-nosed mole!

A True Sixth Sense

A zoologist named Edwin Gould has recently discovered that star-nosed moles, like several types of aquatic animals, have a sixth sense. They are able to detect the presence of electrical fields around them. They can also detect even slight changes in those fields, and that turns out to be a very useful skill.

Gould's discovery is important because only one other mammal, the platypus, is known to have electricity-sensing abilities. Gould heard about the platypus on the radio one day. He wondered whether the star-nosed mole, which is almost blind but is still a great hunter, uses electricity too.

To test his hypothesis, he hid some batteries in an aquarium and let a few moles loose in it. Sure enough, the moles found the batteries and thought they were food! Gould conducted a variety of tests, and they all showed similar results.

Using Their Electric Abilities

All animals (including humans) use very weak electric currents in their bodies to do things like activate their muscles and send messages between their brain cells. With the tentacles on its nose, a star-nosed mole can sense those electric currents. This ability allows it to search in dark burrows and muddy waters for tasty treats such as worms.

Scientists have known about electricity in animals for hundreds of years. Some fish, such as the electric catfish and the electric eel, are considered "strongly electric" because they can generate a strong shock. Electric eels, for instance, can create a charge of over 500 volts! With it, these eels can easily stun small prey or defend themselves against predators.

"Weakly electric" fish often use their abilities for sensory purposes. Some fish recognize each other by the particular electric fields they give off. These fish usually live in dark, murky water where they cannot see very well. Other fish use electricity just as the star-nosed moles do—to hunt for prey or to avoid bumping into things.

Gould's experiments have led some scientists to wonder whether other animals might have these skills too. In the coming years, research on the electrical abilities of animals may help scientists understand how animals think, hunt, and communicate.

▶ The star-nosed mole uses its strange nose to detect electrical fields.

Find Out for Yourself

Do some research on an animal such as the electric eel, the electric catfish, the electric ray, or the stargazer fish. Find out where it lives and how it uses its electric abilities. Present your findings to the class.

▶ Like the electric eel, the electric catfish also uses its "power" to stun its prey.

Answers to
Find Out for Yourself

Sample answer: The electric rays are a group of fish that live along the bottoms of warm seas. They can generate a fairly strong shock to defend themselves or to stun their prey. The electric rays have two electric organs that are located near their head.

Science Snapshot

Chien-shiung Wu (1912–)

Her name means "coura-geous hero" in Chinese. In her career, she has conquered prejudice both against women in science and against Asians in America. Chien-shiung Wu is a devoted wife, a proud mother, and a prize-winning scientist. In many ways she has lived up to the meaning of her name.

Schooling in China and the United States

Chien-shiung Wu was born in Liuhe, about 50 km outside of Shanghai. Wu's father opened the first elementary school for girls in that area. When Wu graduated from her father's school at the age of nine, she left home to attend a presti-gious boarding school. Although she studied to become a teacher, she read chemistry, physics, and mathematics books at night.

In 1930, Wu graduated at the top of her class and was accepted to China's National Central University in Nanjing. Six years later, she came to the United States to get a graduate degree in physics.

When Wu reached the United States, she enrolled at the University of California at Berkeley. Over the next 10 years, Wu earned her Ph.D. in physics and became an expert on *fission*, the splitting of an atom's nucleus into fragments. As the United States became more and more involved in World War II,

many physicists worked on research related to nuclear power and the creation of the atomic bomb. Despite strong prejudice against Asians, Wu worked with the Division of War Research at Columbia University.

Personal Struggles

In 1937, Japan invaded China, and Wu was entirely cut off from her family until the war ended in 1945. For the next 28 years, Wu could only send letters to her parents and broth-ers a few times a year. Because of Communist China's poor relations with the United States, Wu could not return home until 1973. By that time, her parents and brothers had died. They never got to meet Wu's son, Vincent, who was born in 1947.

Dedication to Research

Despite these personal difficul-ties, Wu always worked very hard on her research. She stud-ied the release of energy and matter that occurs when atoms decay. In one of her most famous experiments, Wu over-turned a law of physics by show-ing that the tiniest particles of matter do not always have sym-metrical reactions.

◄ Dr. Wu performing a nuclear physics experiment in her lab in 1958

The "courageous hero" of nuclear physics earned the respect and praise of fellow scientists. Nobel Prize–winning physicist Emilio Segre once remarked, "Wu's willpower and devotion to work are reminis-cent of Marie Curie, but she [Wu] is more worldly, elegant, and witty."

More About Nuclear Energy

Find out more about how fission is used as an energy source. What are some of the advantages and disadvantages?

387

Science Snapshot

Background

Chien-shiung Wu's research involved the study of *parity*. In physics, parity is one property that describes a physical sys-tem. If parity is conserved, the mirror image of a process cannot be distin-guished from the process itself. In 1956, two Chinese American scientists, Tsung Dao Lee and Chen Ning Yang proposed that parity was not conserved in certain reactions in the nucleus of an atom. Wu's research provided the first proof to support the Nobel Prize–winning theory proposed by Lee and Yang.

Wu earned many awards during her career. One of these was the Woman of the Year Award presented to her by the American Association of Women in 1975. In that year, she also served as president of the American Physical Society.

Discussion Questions

1. How did Wu's research relate to the study of energy? *(Wu studied nuclear power, which involves the release of energy from the smallest particles of matter.)*
2. What personal difficulties did Wu face? *(Prejudice against women and Asians in science, and the loss of con-tact with her family due to war)*

Multicultural Extension

Science Around the World
Scientists from different nationalities and cultures often communicate with one another about their research and experi-ments. Have students create a world map of the international science community.

You may wish to tack a world map onto a bulletin board and assign each student one region or nation on the map. Students should provide a short biography of one scientist from their assigned region. On the appropriate area of the map, students should add the scientist's name, field of study, and major accomplishments.

Answers to
More About Nuclear Energy

Fission is the splitting of the nucleus of an atom. (Fusion is the combining of the nuclei of atoms.) Fission releases a great deal of energy, making it an energy source for society, but the process has drawbacks. Fission pro-duces dangerous radioactive waste, and uncontrolled fission reactions can cause meltdowns and chemical explosions.

Unit 7 TEMPERATURE AND HEAT

Bibliography for Teachers

Breckenridge, Judy. *Simple Physics Experiments With Everyday Materials.* New York City, NY: Sterling Publishing Company, Inc., 1993.

Daniels, Patricia, Allan Fallow, and Karin Kinney, eds. *Physical Forces.* Alexandria, VA: Time-Life Books, 1993.

Leggett, Jeremy, ed. *Global Warming: The Greenpeace Report.* New York City, NY: Oxford University Press, 1990.

Bibliography for Students

Johnson, Rebecca L. *The Greenhouse Effect: Life on a Warmer Planet.* Minneapolis, MN: Lerner Publications Co., 1990.

Unit Overview

In this unit, the concepts of heat and temperature are examined through numerous activities and investigations. In Chapter 21, students investigate what temperature is and how it can be measured by the physical changes that solids, liquids, and gases undergo. In Chapter 22, students explore the difference between heat and temperature in three ways. First, they observe the temperature changes caused by adding a specific amount of heat to different volumes of water. Second, they search for a pattern to predict the average temperature of a mixture of hot and cold liquids. Third, they convert mechanical energy into heat and learn how the unit of measure for energy, the joule, equates a specific amount of mechanical energy with a corresponding amount of heat. In Chapter 23, students investigate the way heat is transferred from one place to another through conduction, convection, and radiation.

Using the Themes

The unifying themes emphasized in this unit are **Changes Over Time, Energy,** and **Structures.** The following information will help you incorporate these themes into your teaching plan. Focus questions that correspond to these themes appear in the margins of this Annotated Teacher's Edition on pages 401, 411, 421, 439, 440, and 446.

Changes Over Time can be discussed in Chapter 21 in relation to human sensory perception, including how perception changes with prolonged stimulation. In Chapter 23, this theme can be related to the greenhouse effect, contributions humans may have made to the greenhouse effect, and actions humans can take to prevent global warming.

Energy is discussed in Chapters 22 and 23 in terms of what heat is and how it is transferred from one object or place to

another. Students learn that these changes result from the transfer of energy from one form into another and from one place to another.

Structures can be discussed by comparing the underlying structure of matter with the different methods of heat transfer presented in Chapter 23. In addition, students learn that the structure of a solar home is designed so that solar energy can be converted into heat.

Films, Videotapes, Software, and Other Media

Heat and Energy Transfer
  Film and videotape
  Coronet/MTI
  108 Wilmot Rd.
  Deerfield, IL 60015
Heat: Molecules in Motion
  Videodisc
  AIMS Media
  9710 DeSoto Ave.
  Chatsworth, CA 91311-4409
Heat, Temperature, and the Properties of Matter
  Film and videotape
  Coronet/MTI
  108 Wilmot Rd.
  Deerfield, IL 60015
Thermal Energy
  Videotape
  Agency for Instructional Technology
  P.O. Box A
  Bloomington, IN 47402-0120
Thermometers and How They Work
  Videotape
  Britannica
  310 S. Michigan Ave.
  Chicago, IL 60604-9839

Using the SourceBook

Unit 7 introduces the relationship between heat energy and the kinetic energy of atoms and molecules. Students learn about the different methods of heat transfer. Then the unit focuses on how heat affects the weather on Earth. Finally, the unit compares the measurement of temperature and heat and introduces the concept of specific heat.

Unit Organizer

Unit/Chapter	Lesson	Time*	Objectives	Teaching Resources
Unit Opener, p. 388				Science Sleuths: Burning Barns and Exploding Silos English/Spanish Audiocassettes Home Connection, p. 1
Chapter 21, p. 390	**Lesson 1, How Hot Is Hot? p. 391**	3	1. Explain the reason for using a thermometer to measure temperature. 2. Explain how matter behaves when it absorbs heat energy.	Image and Activity Bank 21-1 Exploration Worksheet, p. 3 Discrepant Event Worksheet, p. 4 Exploration Worksheet, p. 5 Graphing Practice Worksheet, p. 8
	Lesson 2, The Thermometer, p. 398	3	1. Explain how a thermometer works. 2. Describe how to construct and use a simple thermometer.	Image and Activity Bank 21-2 Transparency Worksheet, p. 10 ▼ Exploration Worksheet, p. 12
	Lesson 3, Signs of Temperature Change, p. 400	2	1. Describe three different kinds of thermometers.	Image and Activity Bank 21-3
End of Chapter, p. 403				Chapter 21 Review Worksheet, p. 14 ▼ Chapter 21 Assessment Worksheet, p. 16
Chapter 22, p. 405	**Lesson 1, Heat Versus Temperature, p. 406**	2	1. Describe the difference between heat and temperature. 2. Explain the relationship between the heat content of a substance and the temperature and mass of the substance.	Image and Activity Bank 22-1 Exploration Worksheet, p. 18
	Lesson 2, Hot + Cold = ? p. 410	1	1. Describe what happens to temperature when heat is transferred from one substance to another. 2. Modify an experiment in order to solve a specific problem.	Exploration Worksheet, p. 24 ▼
	Lesson 3, Heat Is Energy, p. 412	3	1. Identify heat as a form of energy. 2. Describe what a joule is. 3. Explain the role of food in meeting the energy needs of the human body. 4. Recognize that different foods contain different amounts of energy.	Image and Activity Bank 22-3 Exploration Worksheet, p. 26 Exploration Worksheet, p. 27
End of Chapter, p. 417				Activity Worksheet, p. 29 ▼ Chapter 22 Review Worksheet, p. 30 Chapter 22 Assessment Worksheet, p. 33
Chapter 23, p. 419	**Lesson 1, How Heat Gets Around, p. 420**	1	1. Identify the direction in which heat flows. 2. Explain how conduction, convention, and radiation differ.	Image and Activity Bank 23-1 Exploration Worksheet, p. 35
	Lesson 2, Conduction—Heat Transfer Through a Material, p. 424	4	1. Identify factors that determine the conduction rate of heat through a solid. 2. Identify several materials that are good conductors and several materials that are good insulators.	Image and Activity Bank 23-2 Exploration Worksheet, p. 37 Activity Worksheet, p. 39
	Lesson 3, Convection—Heat Transfer by Currents, p. 431	2 to 3	1. Explain the formation of convection currents in liquids and gases. 2. Identify examples of convection currents that occur in nature and in everyday life.	Image and Activity Bank 23-3 Exploration Worksheet, p. 40 Transparency 74
	Lesson 4, Radiation—Heat in a Hurry, p. 437	3	1. Explain how life on Earth depends directly and indirectly on the sun's radiant energy. 2. Explain the greenhouse effect and its possible consequences. 3. Describe radiation, absorption, and reflection in terms of radiant heat.	Image and Activity Bank 23-4 Resource Worksheet, p. 45 Activity Worksheet, p. 47 ▼ Transparency Worksheet, p. 48 ▼ Exploration Worksheet, p. 50 Theme Worksheet, p. 54
End of Chapter, p. 442				Chapter 23 Review Worksheet, p. 56 Chapter 23 Assessment Worksheet, p. 59
End of Unit, p. 444				Unit 7 Activity Worksheet, p. 62 ▼ Unit 7 Review Worksheet, p. 63 Unit 7 End-of-Unit Assessment, p. 67 Unit 7 Activity Assessment, p. 72 Unit 7 Self-Evaluation of Achievement, p. 75

* Estimated time is given in number of 50-minute class periods. Actual time may vary depending on period length and individual class characteristics.
▼ Transparencies are available to accompany these worksheets. Please refer to the Teaching Transparencies Cross-Reference chart in the Unit 7 Teaching Resources booklet.

Materials Organizer

Ch.	Page	Activity and Materials per Student Group
21	394	**Exploration 1:** alcohol thermometer; 3 pails or buckets; about 5 L each of cold, warm, and hot water; safety goggles
	395	***Exploration 2, Act. 1:** 500 mL Erlenmeyer flask; 500 mL of water; a few drops of red food coloring; one-hole stopper fitted with glass tubing; hot plate; metric ruler; watch or clock with second hand; safety goggles; lab aprons; oven mitts (See Advance Preparation on page 387D.); **Act. 2:** balloon; small bottle; large beaker; 500 mL of hot water; safety goggles; **Act. 3:** one-hole stopper fitted with glass tubing; 500 mL Erlenmeyer flask; support stand with clamp; 500 mL of water; large beaker; a few drops of red food coloring; safety goggles; lab aprons (See Advance Preparation on page 387D.); **Act. 4:** striker; portable burner; bimetallic strip with wooden handle; watch or clock with second hand; safety goggles; **Act. 5:** 2 large tin cans; 2 wood blocks of the same size; portable burner; 100 g mass; a variety of metal rods; straight pin; plastic straw; striker; safety goggles; oven mitts
	398	**Exploration 3:** candle; jar lid; small ball of modeling clay; matches; Pasteur pipette, sealed at one end; petri dish; 5 mL of water; a few drops of red food coloring; alcohol thermometer; 2 small beakers or jars—one filled with hot water and one filled with cold water; craft stick, painted white; metric ruler; 2 small rubber bands; safety goggles; lab aprons
22	406	***Exploration 1, Expt. 1:** alcohol thermometer; 30 mL of water; test tube; candle; jar lid; small ball of modeling clay; test-tube clamp; 100 mL graduated cylinder; matches; watch or clock with second hand; safety goggles; **Expt. 2:** alcohol thermometer; 35 mL of water; test tube; 100 mL graduated cylinder; 3 samples of 3 g of anhydrous calcium chloride; stirring rod; safety goggles; lab aprons; latex gloves (See Advance Preparation on page 387D.); **Expt. 3:** 175 mL of water; 100 mL graduated cylinder; empty metal can; alcohol thermometer; 1 m piece of cord; safety goggles; oven mitts; **Expt. 4:** 1500 mL of water; 1000 mL beaker; alcohol thermometer; hot plate; watch or clock with second hand; safety goggles; **Expt. 5:** materials from one of the previous experiments, as well as an extra candle for Expt. 1 or an additional 6 g of anhydrous calcium chloride for Expt. 2
	410	**Exploration 2:** 3 plastic-foam cups; about 200 mL each of hot and cold water; 100 mL graduated cylinder; alcohol thermometer; stirring rod; safety goggles
	412	**Exploration 3:** 8 sheets of newspaper; 75 mL of water; 100 mL graduated cylinder; small jar with lid; watch or clock; alcohol thermometer; pliers; lead sinker; hammer; safety goggles; latex gloves
	414	**Exploration 4:** small juice can; shelled peanut; straight pin; cork covered with aluminum foil; support stand with ring clamp; wire gauze; 100 mL of water; 100 mL graduated cylinder; alcohol thermometer; matches; safety goggles
23	422	***Exploration 1, Expt. 1:** 500 mL of water; 600 mL beaker; hot plate; metal spoon with a long handle; wooden spoon with a long handle; watch or clock with second hand; safety goggles; **Expt. 2:** support stand with ring clamp; wire gauze; 10 cm of masking tape; 2 alcohol thermometers; 600 mL beaker; 300 mL of water; striker; portable burner; 5 mL (1 teaspoon) of instant coffee granules; safety goggles; **Expt. 3:** lamp; 100 W light bulb; piece of glass, about 20 cm × 20 cm; metric ruler; piece of cardboard, about 20 cm × 20 cm; 2 sticks of modeling clay; safety goggles
	424	**Watching Heat Travel Through a Solid: A Demonstration:** scissors; aluminum pie plate; candle; jar lid; small ball of modeling clay; tongs; matches; safety goggles; lab aprons
	425	**Exploration 2, Ques. 1:** 4 candles of equal size; jar lid; small ball of modeling clay; matches; 4 thumbtacks; 4 rods of equal thickness, one each of copper, brass, heat-resistant glass, and iron; plastic-foam cup; watch or clock with second hand; safety goggles; **Ques. 2:** 2 candles of equal size; 2 jar lids; 2 small balls of modeling clay; matches; 2 thumbtacks; 2 iron rods of different thicknesses; 2 plastic-foam cups; watch or clock with second hand; safety goggles
	428	***Mysterious Events, The Mystery of the Scorched Paper:** copper rod and wooden rod of equal diameters; 5 cm of masking tape; candle; matches; jar lid; small ball of modeling clay; safety goggles; **Two Terrific Tricks to Try:** tongs; 10 cm × 10 cm aluminum screen; candle; wooden matches; jar lid; small ball of modeling clay; metal fork; support stand with clamp; safety goggles; **The Mystery of the Extinguished Candle:** 40 cm of electrical wire; wire strippers; candle; matches; jar lid; small ball of modeling clay; safety goggles
	429	**A Classroom Demonstration:** a few objects made of different substances, such as a glass beaker, a plastic pencil case, a tin can, and a paper notebook; alcohol thermometer and 5 cm of tape for each object
	430	**A Competition: Save the Ice Cube!:** ice cubes and a variety of materials used to keep the ice from melting
	431	***Exploration 3, Act. 1:** large (tall) jar; 1 L of water; small bottle filled with warm water; a few drops of food coloring; two-hole stopper with 2 glass tubes inserted; safety goggles; lab aprons (See Advance Preparation on page 387D.); **Act. 2:** birthday candle; 40 cm of wire; matches; 1 L glass bottle; T-shaped piece of thin cardboard covered with aluminum foil; safety goggles; **Act. 3:** aluminum pie plate; thumbtack; scissors; metric ruler; 50 cm of thread; candle; jar lid; small ball of modeling clay; matches; lamp with a light bulb; safety goggles; **Act. 4:** 1 L of water; several drops of food coloring; ice cube tray; 20 cm of wire; two 250 mL beakers; safety goggles; lab aprons; **Act. 5:** cardboard box, at least 30 cm × 30 cm × 30 cm; candle; jar lid; small ball of modeling clay; matches; scissors; 40 cm of clear plastic wrap; wooden splint; masking tape; 2 corks; metric ruler; safety goggles; **Act. 6:** 2 support stands with ring clamps; 20 cm of wire; 2 pieces of wire gauze; 2 ice cubes; 500 mL of water; two 250 mL beakers; 2 portable burners; striker; safety goggles; **Act. 7:** 2 beverage cans; support stand with ring clamp; wooden dowel; 50 cm of string; binder clip; portable burner; striker; safety goggles; **Act. 8:** candle; jar lid; small ball of modeling clay; matches; tongs; metric ruler; wooden splint; safety goggles
	439	***Exploration 4, Ques. 1:** lamp; light bulbs of different wattages; meter stick; safety goggles; **Ques. 2:** lamp; 100 W light bulb; metric ruler; alcohol thermometer; square piece of cardboard, about 20 cm × 20 cm; mirror; safety goggles; **Ques. 3:** lamp; 100 W light bulb; piece of cardboard, about 5 cm × 25 cm; large (thick) magnifying lens; alcohol thermometer; 10 cm of masking tape; safety goggles; **Ques. 4:** 2 balloons; bottle painted white; bottle painted black; lamp; 100 W light bulb; 5 ice cubes; squares of red, yellow, blue, white, and black paper, 10 cm × 10 cm each; alcohol thermometer; variety of colored cards, 5 cm × 25 cm each; 10 cm of masking tape for each colored card; safety goggles (See Advance Preparation on page 387D.); **Ques. 5:** 3 tin cans of the same size—one painted white, one painted black, and one unpainted; 3 cardboard covers to fit over the cans; 600 mL of hot water; 100 mL graduated cylinder; 3 alcohol thermometers; safety goggles (See Advance Preparation on page 387D.)

* You may wish to set up these activities in stations at different locations around the classroom.

Exploration 2, Activities 1 and 3, pages 395 and 396: Insert fire-polished glass tubing into one-hole stoppers before students enter the classroom. Make sure that the glass tubing has been fire polished to remove any sharp edges. Wearing leather gloves and using a small amount of glycerin to lubricate the glass tubing, slowly insert glass tubing into one-hole stoppers with a gentle twisting motion.

Exploration 1, Experiment 2, page 407: Using a metric balance, measure out 3 samples of 3 g of calcium chloride for each student group.

Exploration 3, Activity 1, page 431: For proper procedure for inserting fire-polished glass tubing into stoppers, see Exploration 2, Activities 1 and 3, above. Arrange the glass tubes so that one is sticking up higher than the other.

Exploration 4, Question 4, page 440: You may wish to paint the bottles with white or black paint before students enter the classroom.

Exploration 4, Question 5, page 441: You may wish to paint the cans with white or black paint and cut out the cardboard covers for the cans before students enter the classroom.

Unit Compression

In Chapter 21, Lesson 3 provides examples of ideas that are introduced earlier in the chapter, but the lesson may be omitted without disrupting the conceptual flow of the unit. If your students have some background in studying heat and the energy in food, you may wish to skip Lessons 2 and 3 of Chapter 22. Do not skip Exploration 3 on page 412, however. In the case of serious time constraints, you may wish to skip all or parts of Lessons 2–4 of Chapter 23.

Homework Options

Chapter 21
See Teacher's Edition margin, pp. 392, 396, 401, and 402
Graphing Practice Worksheet, p. 8
SourceBook, p. S173

Chapter 22
See Teacher's Edition margin, pp. 407, 409, 413, 415, and 417
Activity Worksheet, p. 29
SourceBook, pp. S162, S172, and S175

Chapter 23
See Teacher's Edition margin, pp. 423, 426, 427, 429, 433, 437, and 441
Activity Worksheet, p. 39
Resource Worksheet, p. 45
SourceBook, pp. S165, S166, S167, and S168

Unit 7
Unit 7 Activity Worksheet, p. 62
Unit 7 SourceBook Activity Worksheet, p. 76

Assessment Planning Guide

Lesson, Chapter, and Unit Assessment	SourceBook Assessment	Ongoing and Activity Assessment	Portfolio and Student-Centered Assessment
Lesson Assessment Follow-Up: see Teacher's Edition margin, pp. 397, 399, 402, 409, 411, 416, 423, 430, 436, and 441 **Chapter Assessment** Chapter 21 Review Worksheet, p. 14 Chapter 21 Assessment Worksheet, p. 16* Chapter 22 Review Worksheet, p. 30 Chapter 22 Assessment Worksheet, p. 33* Chapter 23 Review Worksheet, p. 56 Chapter 23 Assessment Worksheet, p. 59* **Unit Assessment** Unit 7 Review Worksheet, p. 63 End-of-Unit Assessment Worksheet, p. 67*	SourceBook Review Worksheet, p. 78 SourceBook Assessment Worksheet, p. 82*	Activity Assessment Worksheet, p. 72* **SnackDisc** Ongoing Assessment Checklists ♦ Teacher Evaluation Checklists ♦ Progress Reports ♦	Portfolio: see Teacher's Edition margin, pp. 396, 399, 414, 422, 429, and 433 **SnackDisc** Self-Evaluation Checklists ♦ Peer Evaluation Checklists ♦ Group Evaluation Checklists ♦ Portfolio Evaluation Checklists ♦

* Also available on the Test Generator software
♦ Also available in the Assessment Checklists and Rubrics booklet

Science Discovery is a versatile videodisc program that provides a vast array of photos, graphics, motion sequences, and activities for you to introduce into your *SciencePlus* classroom. *Science Discovery* consists of two videodiscs: Science Sleuths and the Image and Activity Bank.

Using the *Science Discovery* Videodiscs

Science Sleuths: Burning Barns and Exploding Silos, Side A

The Sutfins and the Zencks have been feuding for years. When Mr. Sutfin's silo explodes and Mrs. Zenck's barn burns down, the two farmers accuse one another of arson. The Science Sleuths must analyze the evidence for themselves to determine the true cause of the damage.

Interviews
1. Setting the scene: Deputy Sheriff 27619 (play ×2)

2. Silo owner 28247 (play)

3. Silo owner's farmhand 28944 (play)

4. Barn owner 29815 (play)

5. Neighbor of barn owner 30697 (play)

6. Grain-truck driver 31272 (play)

Documents
7. Fire department report on silo fire 31883 (step)

8. Fire department report on barn fire 31886 (step)

9. Letter from insurance company 31889 (step)

10. Police department report on fight between Mr. Sutfin and Mrs. Zenck 31892 (step ×4)

11. Memo on stockroom fire 31898

12. Letter from mayor to fire department 31900 (step)

Literature Search
13. Search on the words: ARSON, BARN, CORN, EXPLOSION, FIRES, HAY 31903

14. Article #1 ("Fire at Andy's Diner") 31905 (step)

15. Article #2 ("Storing Hay") 31908 (step)

Sleuth Information Service
16. Graph of arson fires 31911

17. Schematic diagrams of silo and barn 31913 (step)

18. Common fire hazards chart 31916

19. Weather summary chart 31918

Sleuth Lab Tests
20. Temperature in silo, barn, and compost heap 31920 (step ×2)

21. Flammability of common farm materials 31924 (play)

Still Photographs
22. Silo auger and motor 32751 (step ×2)

Image and Activity Bank
Side A or B

A selection of still images, short videos, and activities is available for you to use as you teach this unit. For a larger selection and detailed instructions, see the Videodisc Resources booklet included with the Teaching Resources materials.

21-1 How Hot Is Hot? page 391
Neptune 1622
The Great Dark Spot of Neptune is partially encircled by a thin layer of clouds made up of methane ice crystals. The spot's diameter is as large as the diameter of Earth.

Stove burner 1800
Electricity passes through metal elements to create heat. The resistance between the electricity and the metal produces heat. At high temperatures, the metal becomes red hot.

Steel, molten 2170
Molten steel gives off infrared (heat) radiation.

Solar flare 18440–18709 (play ×2) (Side A only)
Solar flares are sudden eruptions of hydrogen gas on the surface of the sun.

◀ Step Reverse    Play ▶    Pause ❚❚    Step Forward ▮▶

21-2 The Thermometer, page 398

Clinical thermometer 2067
In a mercury thermometer, the mercury expands as it is heated, causing it to rise up the tube.

Meat thermometer 2069
A meat thermometer measures the temperature of food as it cooks.

Liquid crystal foils 2068
As liquid crystal foils are exposed to heat, the foils change through a broad range of colors.

21-3 Signs of Temperature Change, page 400

Heat expansion; lab setup 239–240 (step)
Materials include a thin steel rod, votive candles, books, a drinking straw, matches, and a pencil. As heat expands the rod, it rotates the pencil, as indicated by the movement of the straw.

22-1 Heat Versus Temperature, page 406

Friction creates heat 21820–22323 (play ×2) (Side A only)
The friction between the dowel and the block of wood creates enough heat to make the block burn.

Sling psychrometer 12219–12812 (play ×2) (Side A only)
Find the wet-bulb temperature by applying water to the wet bulb and allowing it to evaporate by spinning the bulb around. The wet-bulb temperature is dependent on the relative humidity of the air.

22-3 Heat Is Energy, page 412

Friction; rubbing hands together 20380–20587 (play ×2) (Side A only)
Rubbing your hands together (kinetic energy) creates friction between the surfaces of your hands. The frictional force creates heat that warms your hands.

Calorimeter, bomb 1851
Drawing of the parts of the constant-volume bomb calorimeter

Calorimeter, bomb 2529
Bomb calorimeters measure the amount of energy, or caloric content, of coal and other fuels, food, and animal feed.

Calorie; definition 1850
The traditional measure of energy commonly used in chemistry is the calorie, with a lower-case c. 1000 calories equals one Calorie (with a capital C).

Calorie; definition 1849
Calorie, with a capital C, is used to measure the energy content of food.

23-1 How Heat Gets Around, page 420

Winter jacket 2044
Jackets provide insulation against winter cold. The jacket holds your body heat to keep you warm.

Solar corona; time-lapse sequence 18216–18439 (play ×2) (Side A only)
The corona is the hot outer atmosphere of the sun. It shows up best during an eclipse.

Solar distillation 1521–1525 (step ×4)
Materials needed to make a solar distiller for water (step) The sun shines through the glass, heats the water in the pan, and allows it to evaporate. (step) The water condenses on the lower surface of the glass. (step) The water runs down the glass into the pan. (step) The water running into the pan is free of dirt and other material.

23-2 Conduction—Heat Transfer Through a Material, page 424

Heat conductors 2072
Metals are fairly good conductors of heat, while materials such as rubber and plastic are not.

23-3 Convection—Heat Transfer by Currents, page 431

Convection currents 2073–2077 (step ×4)
Assemble the materials: a large beaker of cold water, a small bottle of hot colored water, and a two-holed stopper with two glass tubes. (step) Lower the small bottle into the cold water. (step) Observe the direction of the currents. The hot water flows up the tube into the colder water (step) The cold water tends to move down the tube into the warm water. (step) Why do these currents exist?

Candle; heat rising 2065–2066 (step)
Special photography shows the heated air above the candle as it rises. (step) An air current disturbs the heated air as it rises.

Convection; house model 2078
Warm air from the heating duct rises and pushes the cooler air toward the floor. The cold air is then forced into the cold-air duct.

Ceiling fan 2650
A ceiling fan helps push warmer air near the ceiling closer to the floor.

Air balance 1544
A convection cell is set up to maintain an air balance. This example illustrates how a sea breeze is formed.

Turbine demonstration; holiday ornament pinwheel 27261–27503 (play ×2) (Side A only)
The flames of the candles heat the air, which then rises and moves the blades. By this process, heat energy is converted to mechanical energy.

23-4 Radiation—Heat in a Hurry, page 437

Sun 1579
The sun consists mainly of hydrogen and helium gases. The sun is 150 million kilometers from Earth.

Solar heat; passive 2050
Passive solar heating converts the radiation of the sun's rays into heat. Glass-enclosed areas can accumulate a great deal of heat over the course of a sunny day.

Microwave oven 1799
Electricity creates high-energy radiation waves (microwaves), which interact with material placed in a microwave. Interaction between the microwaves and the material produces heat.

Infrared photo; space shuttle 1809
Infrared photographs display temperature differences. Red corresponds to the hottest region, while blue corresponds to the coldest.

Waves, heat; animal response 2226
Pit vipers are sensitive to changes in temperature and rely on this sense to find food.

Unit 7

UNIT FOCUS

Tell students to imagine the following scenario: They place a large and a small pot, both made of the same material, on a stove and fill both pots with water at the same temperature. Next, they turn both of the stove's burners on high. Ask: Will it take the same amount of time for the water in each pot to boil? *(Students will probably know that it will take much longer for the water in the larger pot to boil.)* Tell students that both pots receive the same amount of *heat* from the stove, but the smaller pot experiences a quicker increase in *temperature.* In this unit students will discover the reason for this difference as well as the difference between heat and temperature.

A good motivating activity is to let students listen to the English/Spanish Audiocassettes as an introduction to the unit. Also, begin the unit by giving Spanish-speaking students a copy of the Spanish glossary from the Unit 7 Teaching Resources booklet.

Can you guess what these emperor penguin chicks are doing?

Connecting to Other Units

This table will help you integrate topics covered in this unit with topics covered in other units.

Unit 1 Science and Technology	An understanding of how heat affects different substances led to the invention of the thermometer.
Unit 2 Patterns of Living Things	Living things have developed a variety of mechanisms for maintaining their body temperature in both hot and cold weather.
Unit 3 It's a Small World	Temperature is one of the many factors governing the activity of microorganisms.
Unit 4 Investigating Matter	The state of substances—solids, liquids, and gases—as well as their volume and density, depends on temperature.

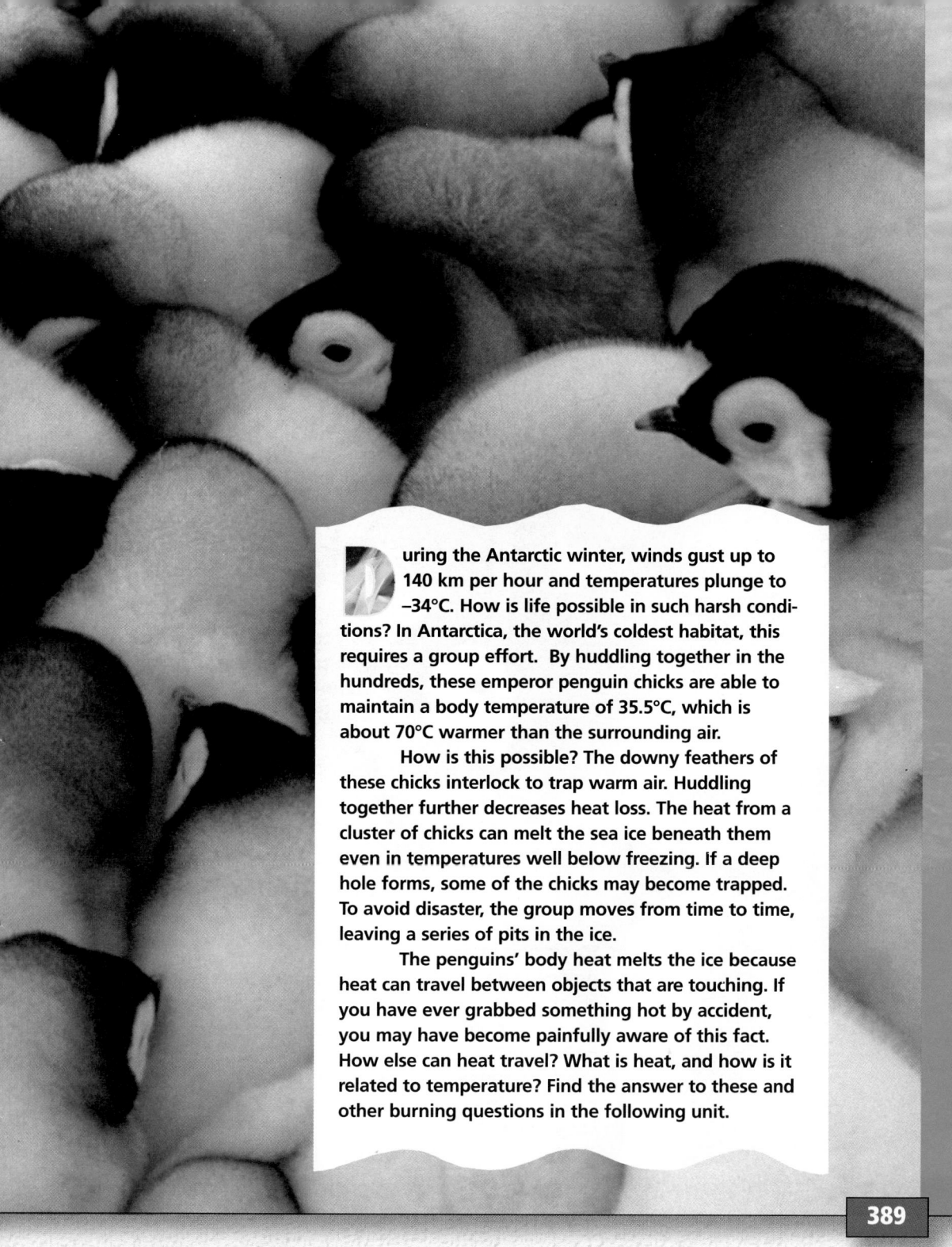

uring the Antarctic winter, winds gust up to 140 km per hour and temperatures plunge to –34°C. How is life possible in such harsh conditions? In Antarctica, the world's coldest habitat, this requires a group effort. By huddling together in the hundreds, these emperor penguin chicks are able to maintain a body temperature of 35.5°C, which is about 70°C warmer than the surrounding air.

How is this possible? The downy feathers of these chicks interlock to trap warm air. Huddling together further decreases heat loss. The heat from a cluster of chicks can melt the sea ice beneath them even in temperatures well below freezing. If a deep hole forms, some of the chicks may become trapped. To avoid disaster, the group moves from time to time, leaving a series of pits in the ice.

The penguins' body heat melts the ice because heat can travel between objects that are touching. If you have ever grabbed something hot by accident, you may have become painfully aware of this fact. How else can heat travel? What is heat, and how is it related to temperature? Find the answer to these and other burning questions in the following unit.

389

Using the Photograph

Call on a volunteer to read aloud the text on this page. Tell students that people stranded in the cold can learn from the emperor penguin chicks. Hunters, skiers, or accident victims stranded outside in the winter sometimes suffer from hypothermia. Hypothermia is a dangerous condition that occurs when body temperature drops below 32°C. One of the ways of combating hypothermia is to huddle closely with other people to receive the heat from their bodies. Ask students to give examples of heat traveling between objects that are touching. (*Answers may include a pot receiving heat from a stove, a cup receiving heat from hot coffee, a snake receiving heat from a warm rock, etc.*)

Meeting Individual Needs

Second-Language Learners
As students progress through this unit, have them make a list of the different terms that relate to heat and temperature. Suggest that they use the list to create a bilingual dictionary. Some of the terms they may wish to include are *heat, temperature, thermostat, thermometer, conductor, insulator, conduction, convection, radiation, expansion,* and *contraction.* Students may wish to add illustrations to help clarify their definitions. Students should also make a cover for their bilingual dictionary and think of a title. When students finish, review their work for science content, with only minimal emphasis on language proficiency.

Connecting to Other Units, *continued*

Unit 5 Chemical Changes	Many chemical reactions require heat to start, and many reactions give off heat once started.
Unit 6 Energy and You	Devices such as light bulbs and steam-powered generators rely on heat for their operation, as do our bodies.
Unit 8 Our Changing Earth	Many geologic processes, such as volcanic activity and tectonic movement, are driven by heat differences.

Connecting to Other Chapters

Chapter 21
allows students to estimate the temperature of objects and demonstrates how a thermometer works.

Chapter 22
introduces students to the difference between heat and temperature and discusses heat as a type of energy.

Chapter 23
introduces students to conduction, convection, and radiation as the three mechanisms for transferring heat.

Prior Knowledge and Misconceptions

Your students' responses to the ScienceLog questions on this page will reveal the kind of information—and misinformation—they bring to this chapter. Use what you find out about your students' knowledge to choose which chapter concepts and activities to emphasize in your teaching.

In addition to having students answer the questions on this page, you may wish to lead the class in the following discussion: Show the class a few weather maps. (Some sources for weather maps include atlases, almanacs, and newspapers.) Have students consider the distribution of temperatures on the maps. Ask: Where are temperatures high and where are they low? Would you like to live in such places? Why or why not? Where on the maps are temperature changes common? Where are temperatures relatively constant? Do temperatures affect where people and animals live or what types of plants will grow?

Have students write a paragraph summarizing their impressions about temperature. It might be best to ask students not to put their name on their

CHAPTER 21 Temperature

1 Would you describe the scene in the photograph as *cold, warm, hot,* or *extremely hot*? Explain.

2 Do you think you would be able to estimate the temperature of warm water within 1°C? 5°C? 10°C?

3 How does a thermometer work?

4 Have you ever noticed the seams in a concrete road or bridge? What purpose do you think they serve?

ScienceLog

Think about these questions for a moment, and answer them in your ScienceLog. When you've finished this chapter, you'll have the opportunity to revise your answers based on what you've learned.

390

paper. Emphasize that there are no right or wrong answers in this exercise. Collect the papers, but do not grade them. Instead, read them to find out what students know about temperature and its effects, as well as what misconceptions they may have. Explain to them that in this chapter they will practice estimating temperatures and study how thermometers measure temperature.

LESSON 1 — How Hot Is Hot?

LESSON 1 — How Hot Is Hot?

"That sun is blazing," said Eddie. "I bet you could fry an egg on the sidewalk."

"The radio said it was going to reach 36°C," Sonia said. "I don't think it's that hot yet. Let's each guess the temperature and see who comes the closest. There's a thermometer on the outside of the school."

"OK," agreed Eddie. "Then let's try that trivia quiz we got in class this morning. Let's see who can come closest to guessing the temperature of things."

A Burning Question

Let's join Sonia and Eddie in their guessing game. Guess the temperature of your classroom. How close was your estimate? How close were your classmates' estimates? Your next assignment is to come up with some educated guesses to the questions on the next two pages.

391

LESSON 1 ORGANIZER

Time Required
3 class periods

Process Skills
observing, analyzing, inferring, hypothesizing, predicting

New Terms
none

Materials (per student group)
Exploration 1: alcohol thermometer; 3 pails or buckets; about 5 L each of cold, warm, and hot water; safety goggles
Exploration 2, Activity 1: 500 mL Erlenmeyer flask; 500 mL of water; a few drops of red food coloring; one-hole stopper fitted with glass tubing; hot plate; metric ruler; watch or clock with second hand; safety goggles; lab aprons; oven mitts (See Advance Preparation on page 387D.); **Activity 2:** balloon; small bottle; large beaker; 500 mL of hot water; safety goggles; **Activity 3:** one-hole stopper fitted with glass tubing; 500 mL Erlenmeyer flask; support stand with clamp; 500 mL of water; large beaker; a few drops of red food coloring; safety goggles; lab aprons (See Advance

continued ▶

FOCUS

Getting Started

Have a competition to see who can come closest to guessing the temperature outside. Have students write their name and their guess on a small piece of paper. Have one student collect the papers while two more students go outside with an alcohol thermometer and measure the actual temperature. Sort through the guesses to determine the range of temperatures guessed by students. Point out that people do not necessarily have an accurate sense of temperature—thermometers are much more reliable.

Main Ideas

1. A wide range of temperatures can occur on Earth, depending on the location and the time.
2. Our sense of temperature is not always accurate.
3. Thermometers increase the accuracy with which temperature is measured or gauged.
4. As the temperature of a substance changes, its physical characteristics may change as well. These changes can be used to measure temperature.

TEACHING STRATEGIES

A Guessing Game, *page 392*
Have students read the introduction to the game silently. Then involve them in a discussion of the questions on pages 392–393.

Answers to
A Guessing Game, pages 392–393

1. About 20°C; answers will vary.

2. Highest recorded temperature:
   a. 49°C
   b. 48°C
   c. 23°C
   d. 58°C (in Al Aziziyah, Libya)

3. 117°C

4. Lowest recorded temperature:
   a. –57°C
   b. –12°C
   c. –19°C
   d. –88°C (in Vostok Station, Antarctica)

5. –163°C

6. 37°C

7. About 6000°C

8. 1530°C

9. –2°C

10. Above 55°C

11. About 400°C; this temperature causes the fluid inside an ordinary thermometer to expand so much that it will break the glass of the thermometer.

12. Dry ice has a maximum temperature of –78.5°C because it sublimates at higher temperatures.

13. About 200°C

14. About 250°C

15. About 3000°C

16. About –273°C

17. 2000 to 5000°C

18. 300,000,000 to 400,000,000°C

Answer to
In-Text Question

Ⓐ No thermometer like this one could exist because the liquid inside would have to expand to an impossible extent. Moreover, the marks on this thermometer do not represent equal temperature ranges.

Homework

You may wish to assign the questions from A Guessing Game as homework.

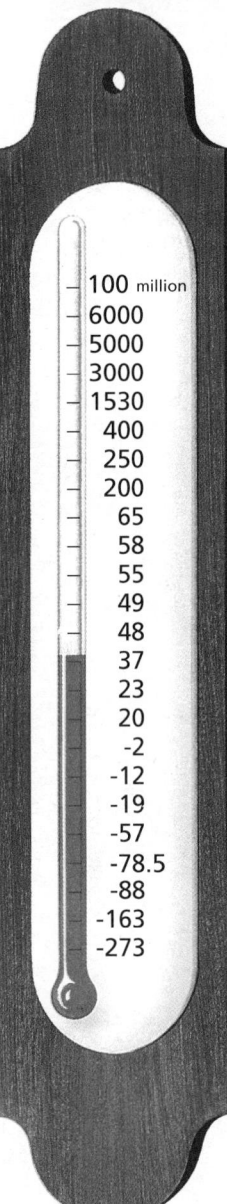

100 million
6000
5000
3000
1530
400
250
200
65
58
55
49
48
37
23
20
-2
-12
-19
-57
-78.5
-88
-163
-273

A Guessing Game

Water boils at 100°C. It freezes at 0°C. These temperatures form the basis for the Celsius temperature scale. Use them to help you make an "educated guess" for each of the questions below. All of the temperatures (exact or approximate) that you need to answer these questions are shown on the thermometer in the illustration at left. In some cases, the answer to one question will help you to answer other questions.

No single thermometer could actually span such a wide range of temperatures. Do you know why? **Ⓐ**

1. What is the temperature in your classroom when the room feels comfortable?

2. Guess the highest temperature or hottest day on record for the following locations:
   a. Gannvalley, South Dakota
   b. Phoenix, Arizona
   c. Reykjavik, Iceland
   d. the world

3. What is the highest temperature reached on the moon?

4. Guess the lowest temperature or coldest day on record for the following locations:
   a. Rogers Pass, Montana
   b. Shanghai, China
   c. the state of Florida
   d. the world

5. What is the lowest temperature reached on the moon?

6. What is the temperature of your blood?

7. How hot is the surface of the sun?

8. At what temperature does iron melt?

9. At what temperature does sea water freeze?

10. Scalding water is not as hot as boiling water, but it can cause very bad burns. How hot is "scalding hot"?

ORGANIZER, continued

Preparation on page 387D.); **Activity 4:** striker; portable burner; bimetallic strip with wooden handle; watch or clock with second hand; safety goggles; **Activity 5:** 2 large tin cans; 2 wood blocks of the same size; portable burner; 100 g mass; a variety of metal rods; straight pin; plastic straw; striker; safety goggles; oven mitts

Teaching Resources
Exploration Worksheets, pp. 3 and 5
Discrepant Event Worksheet, p. 4
Graphing Practice Worksheet, p. 8

11. What is the temperature of a burning match? You can't use an ordinary thermometer to measure the flame's temperature because the thermometer will break. Do you know why?

12. Have you heard of dry ice? Dry ice is frozen carbon dioxide. It is often used by ice-cream vendors to keep their wares cold. How cold do you think dry ice is?

15. How hot does the filament of a light bulb get?

13. When ironing a cotton shirt or blouse, you put the iron on the setting for cotton. What temperature does the iron reach at this setting?

14. A recipe calls for a very hot oven. On what temperature should the oven be set?

16. What is the lowest temperature possible, according to scientists?

17. How hot is it at the center of the Earth?

18. An atomic explosion is so hot (hotter than the center of the sun) that even the strongest materials are instantly vaporized. How hot do you think it gets?

393

Temperatures Around the World
Point out to students that the average air temperature may vary greatly in different parts of the world. Suggest they make a world map and use it to show the average summer and winter temperatures in different continents and countries. Display the maps around the classroom. Let students research different countries to determine how the temperature affects the culture of the people, the food they grow and eat, and the clothes they wear.

Cross-Disciplinary Focus

Health
Our bodies have many processes for regulating temperature, all of which involve the production or elimination of heat. Have students research one of these processes, such as fevers, goose bumps, sweating, or shivering.

Did You Know...

The temperature on the surface of Venus is approximately 500°C—over 10 times the typical temperature of the Earth's surface! This is because Venus is closer to the sun than the Earth is and because the atmosphere of Venus is much denser than the atmosphere of the Earth. Thus, Venus retains much more of the heat it absorbs from the sun.

Be sure the buckets are not so full that they overflow when a finger is submerged in the water. Suggest that students label all three buckets as shown on page 394. Students should understand that their classmates' estimates vary because people's sensitivity to heat varies.

SAFETY ALERT Hot water from the tap should be sufficient for this exercise, but make certain that the water is not above 50°C.

SAFETY TIP Make sure students wear proper shoes in the lab. No open-toed shoes or sandals should be permitted because they offer little protection from chemical spills, hot objects, and broken glass.

GUIDED PRACTICE If it is more convenient to complete this activity as a demonstration, have six or more volunteers follow the procedure outlined on this page. Students may record their estimates on the chalkboard before the actual temperature of the warm water is measured. Discuss the discrepancies between the measurement and the estimates. Elicit ideas for improving the estimates. Help students to conclude that the only sure way of obtaining an accurate temperature is by using a calibrated thermometer.

The Difficulty of Estimating Temperature

If the hot water is still hot enough and the cold water is still cold enough, you may wish to move directly into this activity from the previous one. However, it is likely that the water has approached room temperature and that you will need to change it. Keep in mind that the greater the difference in temperature between the hot and cold water, the more dramatic the results will be. Therefore, you may wish to add some ice cubes to the cold-water bucket in order to make the temperatures more extreme.

GUIDED PRACTICE As a follow-up, involve the class in a discussion of what happened, and encourage students to explain the results.

Estimating Temperature

How did you fare in the temperature quiz? In this Exploration and the next, you will refine your method of estimating temperature.

You Will Need

- an alcohol thermometer
- 3 pails or buckets: one filled with cold water, one with warm water, and one with hot water

Caution: The temperature of scalding water is 55°C. Do not use water that is above 50°C.

What to Do

1. Measure the temperature of the hot water and the temperature of the cold water with a thermometer.
2. Dip a finger into the bucket of hot water and into the bucket of cold water. Compare how each of these feels with how the warm water feels. Make an estimate of the temperature of the warm water. Write down your answer.

3. Your classmates will be doing the same thing. Compare your estimate with theirs. Why do you think estimates varied among your classmates?
4. Finally, use a thermometer to check whose answer is closest to the actual temperature of the warm water.

The Difficulty of Estimating Temperature

Here is a puzzle for you to solve. Try it on as many people as you wish, both at school and at home.

Put one hand into a bucket of hot (but not scalding) water, and the other hand into a bucket of cold water. Leave them in for about a minute.

Now put both hands into a bucket of warm water at the same time. How do they feel? Do they feel the same? If not, what is the difference?

Does everyone have the same experience? After this experience, would you trust your ability to estimate temperature?

INDEPENDENT PRACTICE Invite two students to perform the following exercise in front of the class: Have a student place his or her hand in the hot water while the other student places his or her hand in the cold water. Then have both students place those same hands in the warm water, and ask them to guess the temperature of the water. Have the class compare their responses with the actual temperature.

★ An Exploration Worksheet and a Discrepant Event Worksheet are available to accompany Exploration 1 (Teaching Resources, pages 3 and 4, respectively).

Improving Your Accuracy

A man who lives on your street tests the air temperature every morning by holding up a finger in the air. Imagine that you are selling thermometers. What would you say to convince him to buy one? Ⓐ

Estimating temperature is useful when you need only an approximation, such as when deciding what to wear outside. But if you need the exact temperature, estimates are not reliable—you must use a thermometer.

Do you know how thermometers work? Think about what happens to air when it's heated. Do you think its volume changes? What about liquids? When maple syrup is heated, it expands, taking up more space. It also becomes thinner and runnier. Finally, it starts to boil away. Changes in temperature can bring about changes in a substance. Thermometers make use of changes like these to measure temperature.

In the following Exploration, each of the Activities shows how to use changes to measure temperature. Answer the questions for each Activity in your ScienceLog.

How could the changes you will observe in Activities 1–5 be used to make a thermometer? In each case, how would you change the apparatus to make it a better thermometer? Ⓑ

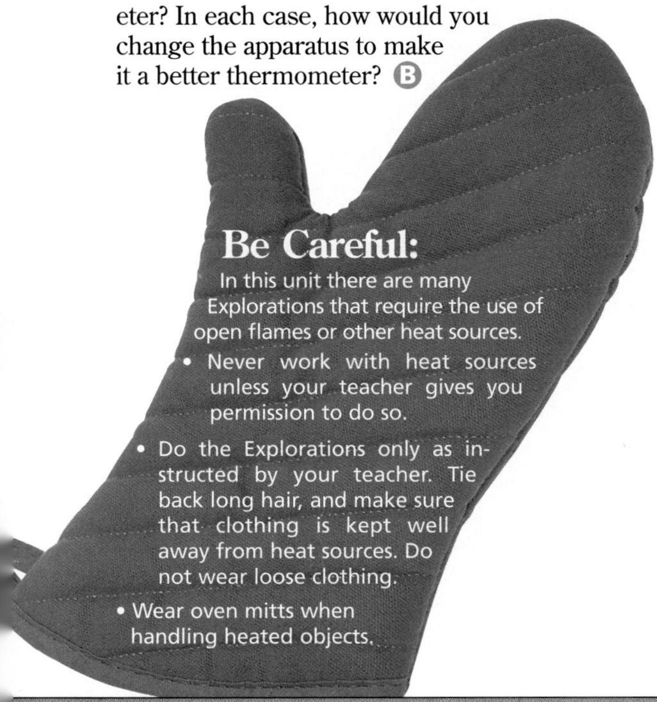

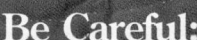

Be Careful:
In this unit there are many Explorations that require the use of open flames or other heat sources.
• Never work with heat sources unless your teacher gives you permission to do so.
• Do the Explorations only as instructed by your teacher. Tie back long hair, and make sure that clothing is kept well away from heat sources. Do not wear loose clothing.
• Wear oven mitts when handling heated objects.

EXPLORATION 2

Temperature Changes

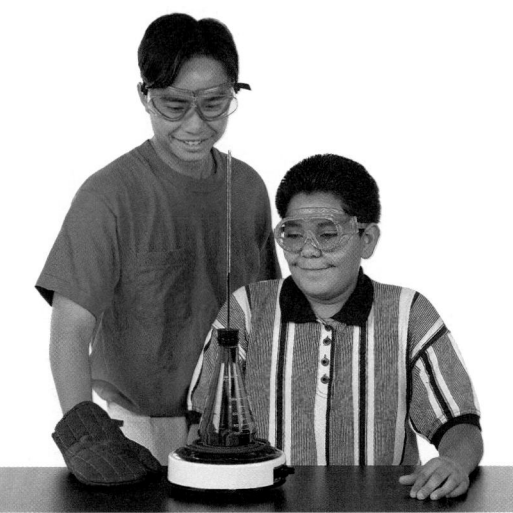

ACTIVITY 1

You Will Need

- a hot plate
- a 500 mL Erlenmeyer flask
- red food coloring
- a one-hole stopper fitted with a glass tube
- water
- oven mitts
- a metric ruler
- a clock or watch with a second hand

What to Do

Fill a 500 mL Erlenmeyer flask completely with colored water, and insert a one-hole stopper fitted with a glass tube. Heat the flask on the lowest setting of a hot plate. Wear your oven mitts to remove the flask if the water begins to overflow. Turn off the hot plate.

How long does it take the column of water to rise 1 cm? At the lowest setting of the hot plate, how far will the column of water rise? Why do you think the colored water rises?

Caution: Do not insert the glass tube into the stopper. Your teacher will do this step for you.

Exploration 2 continued ►

Improving Your Accuracy

INDEPENDENT PRACTICE Have students silently read the first paragraph, or call on a volunteer to read it aloud. Then suggest that students pair off to role-play the situation between the man and the thermometer-seller. Provide time for them to prepare a dialogue. Then invite pairs of students to present their scenarios to the class. Involve the class in a discussion of the merits of the arguments presented in each of the scenarios.

Have students read the rest of the page, pausing at the Exploration. Invite them to offer other examples of changes that take place when substances are heated.

Answers to *In-Text Questions*

Ⓐ Answers will vary, but student responses should emphasize that the man can find the exact temperature with a thermometer.

Ⓑ These questions will be answered on the pages that follow.

EXPLORATION 2

SAFETY ALERT Make sure that students avoid stirring with a thermometer or handling it roughly.

Cooperative Learning
EXPLORATION 2

Group size: 3 to 4 students
Group goal: to discuss observations, and use observations to explain how a thermometer works
Positive interdependence: Assign students roles such as group leader, materials organizer, recorder, and timer.
Individual accountability: Have students orally summarize one Activity from Exploration 2. Encourage students to explain how their group performed the experiment and what they learned from it.

Exploration 2 continued ►

Answers to *The Difficulty of Estimating Temperature, page 394*

Students should expect that the hand held in the hot water will feel cooler when placed in the warm water, while the hand held in the cold water will feel hotter. The difference can be explained by the extremes in temperature experienced by the two hands. Because everyone has the same experience, students should conclude that the only sure way of obtaining the accurate temperature is by using a thermometer.

★ **An Exploration Worksheet is available to accompany Exploration 2 (Teaching Resources, page 5).**

EXPLORATION 2, continued

The Activities described in this Exploration may be performed as demonstrations if time or materials are limited.

SAFETY ALERT Hot water from the tap should suffice for Activity 2, but make sure that the temperature of the water does not exceed 50°C.

For Activity 3, make sure that students continue their observations for a few minutes after they have let go of the flask. For more dramatic results, have students first warm the flask with their hands and then insert the tubing into the beaker. If a refrigerator is available, you may wish to cool the flasks before class.

Answers to
Activity 1 page 395

Both the amount of time it takes for the column of water to rise 1 cm and the total distance that the column of water rises will vary. The colored water rises in the glass tube when the flask is heated because water expands when it gets warmer. Students may also recognize that as the water expands, the only direction it can expand is vertically. If unconfined, the water could also expand horizontally, reducing the amount of vertical expansion. That is why thermometers tend to be very thin. In this Activity, the use of a glass tube serves a similar purpose.

In order to make a thermometer, students may suggest using a small, thin glass tube so that changes will be clearly reflected in the amount of water expansion. The tube can then be calibrated easily.

ACTIVITY 2

You Will Need
- a large beaker
- a small bottle
- a balloon
- hot water (no hotter than 50°C)

What to Do

Fasten an empty balloon over the neck of a bottle. Put the bottle in a large beaker, and fill the beaker with hot water. Explain what happens.

ACTIVITY 3

You Will Need
- a 500 mL Erlenmeyer flask
- a one-hole stopper fitted with glass tubing
- a support stand with clamp
- red food coloring
- water
- a large beaker

Caution: Do not insert the glass tubing into the stopper yourself. Your teacher will do this step for you.

What to Do

Insert the stopper fitted with glass tubing into a 500 mL Erlenmeyer flask. Using a support stand, clamp the flask upside down, and insert the glass tube into a beaker of colored water. Warm the flask by holding it in your hands for a few minutes. Explain what happens. How could you lower the level of the colored water in the tube?

Answers to
Activity 2

The hot water warms the air in the bottle. As the air is heated, it expands, causing the balloon to inflate. Gases tend to expand more rapidly than liquids when heated.

In order to make a thermometer, students may suggest putting the balloon inside a cardboard cylinder. A window could be cut into the cylinder to view the balloon's expansion, and the instrument could be calibrated accordingly.

Answers to
Activity 3

The heat from the hands warms the air in the flask, causing the air to expand. This forces air down the tube; bubbles should appear in the beaker. Once the flask begins to cool, colored water will be drawn into the tube from the beaker. To lower the level of water in the tube, students could reheat the flask with their hands.

In order to make a thermometer, students may suggest using narrow glass tubing and calibrating the instrument by marking the water level at certain known temperatures.

Homework

The Graphing Practice Worksheet on page 8 of the Unit 7 Teaching Resources booklet makes an excellent homework activity to accompany the material on this page.

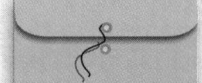 **PORTFOLIO**
Suggest that students include in their Portfolio an analysis and evaluation of one of the Activities from Exploration 2.

You Will Need

- a bimetallic strip with a wooden handle
- a butane burner
- a striker
- a watch or clock with a second hand

What to Do

Light the burner with the striker. Hold a bimetallic strip in the flame for 30 seconds. Explain what happens.

ACTIVITY 5

You Will Need

- a butane burner
- a striker
- metal rods
- a 100 g mass
- a straight pin
- a plastic straw
- 2 tin cans
- 2 wood blocks
- oven mitts

What to Do

Using tin cans and wooden blocks for support, place a metal rod over a butane burner. Rest the mass against one end of the metal rod. With a pin, attach a drinking straw to the wooden block at the other end of the metal rod. The straw should touch the metal rod. Use the striker to light the burner. Explain what happens as you heat the metal rod. Turn off the burner. Try different types of metal rods, and record your results. Wear an oven mitt to handle the heated metal rod.

Metal rod — Straw with pin
100 g mass
Wood block
Tin can
Butane burner

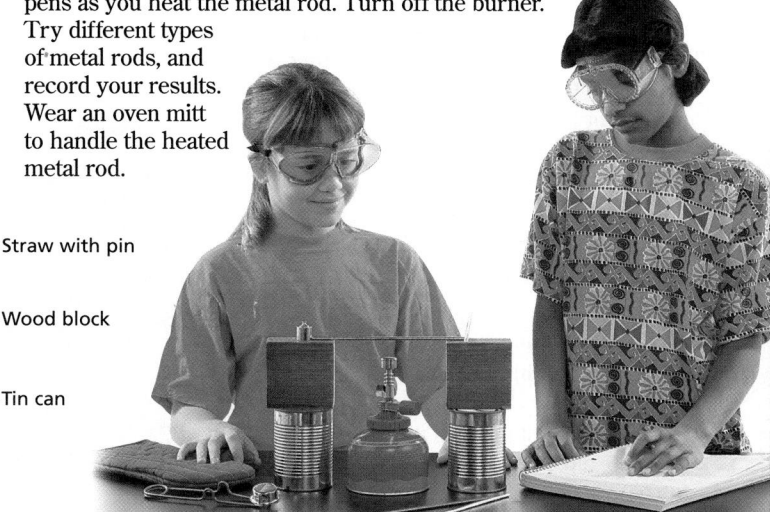

397

Answers to *Activity 4*

As the bimetallic strip is heated, it curls. When the strip is taken out of the flame, it cools and returns to its original shape. This is because one metal in the bimetallic strip expands more rapidly than the other metal when heated. Conversely, the same metal contracts more rapidly than the other metal when cooled. In order to make a thermometer, students may suggest mounting the bimetallic strip to a surface that has been calibrated. A pointer could then be attached to the bimetallic strip; the pointer would demonstrate temperature changes as the coil expands and contracts.

Answers to *Activity 5*

The metal rod will expand as it heats, partly in width but mostly in length. The 100 g mass prevents the rod from expanding in both directions. The rod pushes against the straw, and the amount the straw moves is proportional to how much the rod lengthens. The type of metal and the shape of the rod will influence how much the rod expands. In order to make a thermometer, students may suggest using a longer metal rod so that the expansion will be more visible. The rod could then be mounted on a surface and calibrated accordingly.

FOLLOW-UP

Reteaching

Have students form groups of three to role-play a scene in which they are discussing their upcoming tour of several European cities. Two students should guess the temperatures of those cities to determine how to pack. The third student could then offer suggestions for determining more accurate temperatures. *(Different regions will have different temperatures, and those temperatures will vary depending on the season. Resources for determining the probable temperatures include newspapers, travel agents, travel guides, almanacs, and encyclopedias.)*

Assessment

Tell students to imagine a hot, bright light shining on a balloon filled with air. They should write a brief paragraph in which they use the words *heat, temperature, expand,* and *contract* to describe what happens to the balloon.

Extension

Have students keep track of local high and low temperatures for a week. Then have them use the information to make line graphs showing how the temperatures varied from day to day. Display the finished graphs for others to see.

Closure

Students may enjoy making "thermometer mobiles." Each part of the mobile should resemble a small thermometer with the temperature of a particular phenomenon, such as the temperature at which water boils or normal body temperature. Hang several of the mobiles around the classroom.

FOCUS

Getting Started

Ask students to name some activities that depend on the accurate determination of the temperature. *(Answers include weather forecasting, food preparation, and the travel and transportation industries.)* Ask: How would these activities be affected if we did not have thermometers at our disposal? *(Accept all reasonable responses.)*

Main Ideas

1. Thermometers are calibrated with respect to known temperatures.
2. The construction of a useful thermometer depends on a number of design choices.

TEACHING STRATEGIES

In this lesson, students will have the opportunity to construct their own thermometer and calibrate it using the fixed-point method. Ask the class to suggest what qualities make a good thermometer. *(Answers include easy and reliable calibration, sturdiness, easy readability, good sensitivity, and a wide range of temperatures.)* Students will be able to evaluate their thermometer's performance based on some of these desired features.

The thermometers you have been using until now have been commercially manufactured. In Exploration 3 you will construct your own working thermometer!

EXPLORATION 3

The Air Thermometer

You Will Need

- a Pasteur pipette (sealed at one end)
- a craft stick, painted white if possible
- 2 small rubber bands
- red food coloring
- water
- a petri dish
- an alcohol thermometer
- a candle
- matches
- a jar lid
- modeling clay
- a metric ruler
- 2 beakers or jars

Caution: Be careful when handling the pipette. It is fragile and will break easily.

What to Do

1. Place the candle in the jar lid, and secure it with modeling clay. Heat the bulb of the sealed pipette with the flame of the candle.

2. Being careful not to touch the hot bulb, dip the open end of the pipette into a drop of water colored with red food coloring. Draw up about 5 mm of the colored water. This plug will act as a marker.

3. Test your thermometer by putting the bulb first in hot water. You can use hot water from the tap, but be careful not to scald yourself. Water that is 55°C or above will scald you. Does your thermometer seem to work? Now try putting the thermometer in cold water. How could you improve your thermometer? It might be necessary to change the position of the marker (the colored water).

4. Make a scale on the craft stick by marking it at intervals of 1 mm. Attach your thermometer to the craft stick with the rubber bands.

5. Now *calibrate* your thermometer. Put it into cold water. Mark the position of the bottom of the colored water on the craft stick. Use a commercial thermometer to find out the temperature of the same sample of water. Repeat this procedure using hot water. What change in temperature causes the colored-water marker of your thermometer to move 1 mm?

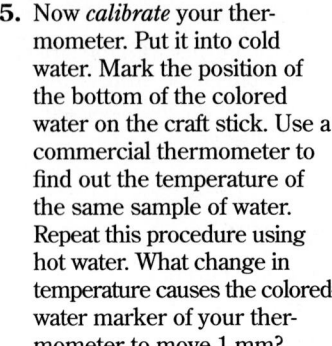

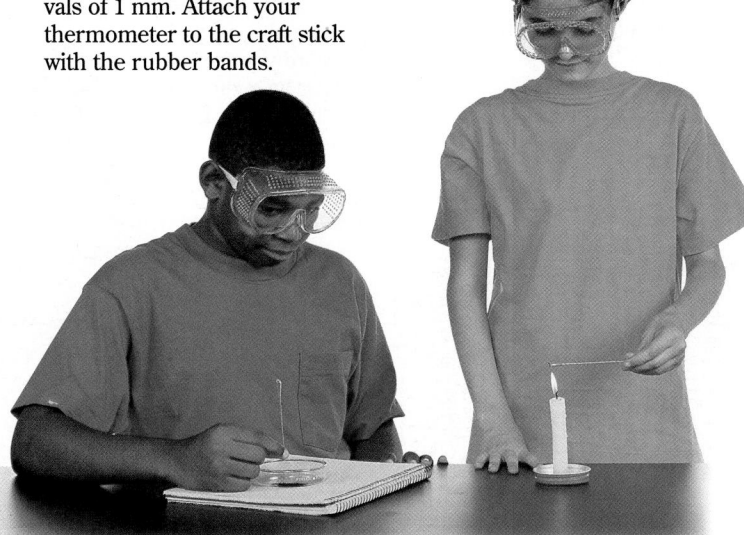

EXPLORATION 3

GUIDED PRACTICE Demonstrate how to make an air thermometer before students make their own. Invite discussion about what is happening to the air inside the pipette.

To calibrate the thermometer, divide the distance between the hot and cold temperatures into millimeters. Divide the number of millimeters into the temperature range to find out what change in temperature causes the colored marker to move 1 mm. Be sure students understand that each pipette's calibration will be slightly different because of variations in the size of the bulbs.

LESSON 2 ORGANIZER

Time Required
3 class periods

Process Skills
observing, measuring, analyzing

New Term
Calibrate—to mark an instrument with a series of units for measuring

Materials (per student group)
Exploration 3: candle; jar lid; small ball of modeling clay; matches; Pasteur pipette, sealed at one end; petri dish; 5 mL of water; a few drops of red food coloring; alcohol thermometer; 2 small beakers or jars—one filled with hot

water and one filled with cold water; two 250 mL beakers or jars; craft stick, painted white; metric ruler; 2 small rubber bands; safety goggles; lab aprons

Teaching Resources
Transparency Worksheet, p. 10
Exploration Worksheet, p. 12
Transparency 70
SourceBook, p. S173

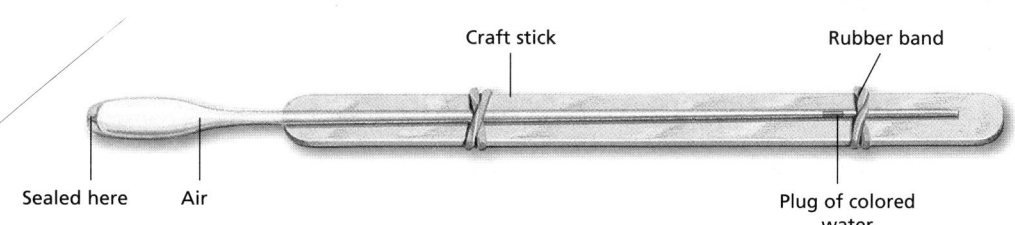

Craft stick   Rubber band

Sealed here   Air

Plug of colored water

Trying Out Your Thermometer

Using your thermometer, find out the temperatures of the following:

a. the air in the room you are in

b. an area near a radiator or hot-air vent

c. your sock

d. your hand

e. the air 5 cm from an incandescent light bulb

f. the air 5 cm from a fluorescent light

g. warm water

h. a cold drink and a hot drink

i. the air outdoors (in the sun and in the shade)

Compare your values with those obtained by your classmates.

How Good Is Your Thermometer?

Were you satisfied with how your thermometer worked? Did the marker move up and down as much as you would have liked? Would it help if you increased the amount of air in the thermometer? When you measured the temperature of the liquid, did it make a difference how deep your thermometer was in the liquid? Did it take a long time for the marker to stop moving? Would you be worried if some of the water from the marker evaporated? Do you have some suggestions for improving the design?

To help get your ideas and answers straight, write a letter to one of your fellow students. In your letter, suggest ways in which he or she might improve the thermometer.

Building a Better Thermometer

One problem with your air thermometer is that you need to calibrate it every time you use it. Why? When the pressure of the atmosphere changes, the volume of the air inside the thermometer changes. On *high-pressure* days, when the weather is clear, the air in your thermometer will occupy less space than on *low-pressure* days, when the weather is stormy. For this reason, ordinary thermometers often use liquids instead of air.

There is yet another way to make a better thermometer. It involves using a bimetallic strip. Can you devise a way to do this?

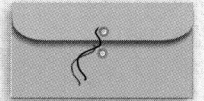

Turn to page S162 of the Source Book to find out why heat makes things expand.

399

Trying Out Your Thermometer

After students have compared their answers with those of their classmates, ask them to suggest reasons for the differences in their answers.

★ **An Exploration Worksheet accompanies Exploration 3 (Teaching Resources, page 12). In addition, a Transparency Worksheet (Teaching Resources, page 10) and Transparency 70 accompany the material on this page and on page 400.**

PORTFOLIO
Students may wish to include their thermometer from Exploration 3 in their Portfolio.

How Good Is Your Thermometer?

You might want to have students read this section prior to testing their thermometer or have them test their thermometer again with these questions in mind. Involve students in a discussion of the evaluation questions posed in the first paragraph. Students should use the points raised in the discussion when writing their letter. After students have finished writing, have them exchange their letter with a classmate. Provide time for them to read and discuss the letters.

Building a Better Thermometer

Have students read the first paragraph to see if their suggestions for improving the thermometer included either the improvement of the calibration method or the use of a liquid instead of air in the thermometer.

FOLLOW-UP

Reteaching

Have students make posters to illustrate the step-by-step procedure for making an air thermometer.

Assessment

Have a volunteer go to the board and make a list of pros and cons for his or her air thermometer. Ask: Are the pros more significant than the cons? Would you consider this a good thermometer? Explain.

Extension

Have students create a design for a better thermometer. They can then compare their idea with the designs of the thermometers they will study in the next lesson.

Closure

Ask the class to offer suggestions about ways to calibrate thermometers. One way to calibrate a thermometer is to use a previously calibrated object as a reference. Another method is to use fixed points, such as the freezing and boiling points of water, as references.

Answer to
In-Text Question

Ⓐ To make a thermometer using a bimetallic strip, attach a pointer to one end of the strip; the other end of the strip should be attached to a support. The strip should then be subjected to temperature extremes and calibrated in a manner similar to that used with the air thermometer. If students decide to make a thermometer, be sure to check their plans for safety.

Signs of Temperature Change

FOCUS

Getting Started

Bring a clinical thermometer, an outdoor thermometer, and a cooking thermometer to class. Ask students to describe these thermometers and any others they have seen and used. Ask: How do thermometers differ? Point out that in this lesson they will learn about different thermometers and how they work.

Main Ideas

1. Most thermometers measure temperature by gauging the expansion of a gas or a liquid.
2. There are many different kinds of thermometers, each designed for use in a specific setting.

TEACHING STRATEGIES

Thermometers at Work . . . In the Home

GUIDED PRACTICE   Ask students to raise their hand if they have a thermostat like the one shown on this page in their home to control their furnace. Call on a volunteer to describe how this type of thermostat is set. *(By turning it to the desired temperature)* Why is it a convenient device to have? *(It turns the furnace on and off automatically, which keeps the house at an almost constant temperature.)*

Answers to
In-Text Question and Caption

Ⓐ As the air in the room warms up, the coil unwinds, shifting the mercury to the right side of the bulb. This will break the electrical circuit, thus turning the furnace off.

Ⓑ As the temperature increases, the bimetallic strip becomes less tightly coiled (unwinds) because the copper on the inside expands more than the iron on the outside.

Signs of Temperature Change

Thermometers at Work . . . In the Home

⬆ This is what they saw. As the air in the room warms up, does the coil tighten or unwind? Ⓐ

Sarah wondered how the bimetallic strip worked. She knew that the strip was made of copper and iron joined together. At the library, she looked up the expansion of copper and found that a 1 m copper bar expands by 1.4 mm when heated from 20°C to 100°C. She also found out that a 1 m iron bar expands by 0.88 mm when heated from 20°C to 100°C. Using this information, write a note to Sarah that explains why the bimetallic strip behaves as it does when there is an increase in temperature. Ⓑ

LESSON 3 ORGANIZER

Time Required
2 class periods

Process Skills
observing, analyzing, inferring, measuring, comparing

Theme Connection
Changes Over Time

New Term
Thermostat—a thermometer that regulates temperature by controlling a heating or cooling unit

Materials (per student group)
none

Teaching Resources
none

In the Hospital

You have probably had your temperature taken when you were not feeling very well. In a hospital, taking temperatures is a regular practice. John is a student nurse who has been looking after a patient with a high fever. Imagine John trying to take the patient's temperature with an ordinary mercury thermometer. As soon as the thermometer was taken out of the patient's mouth, the mercury thread would begin to shrink. Therefore, John uses a *clinical* thermometer instead (shown below). It has a constriction, or a kink, in the fine tube just above the bulb of mercury.

When a clinical thermometer is removed from the patient's mouth, the mercury thread breaks at the constriction. Above the break, the mercury stays at the same place. After recording the patient's temperature on the chart, John shakes the thermometer to return the mercury to the bulb and places it in sterilizing fluid.

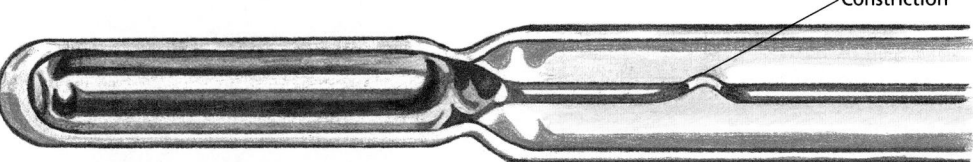

Constriction

At the School Weather Station

Chad and Elena decided to write their report for science class on weather forecasting. They went to visit their friend Sun Ling, who was studying meteorology with her high school science club. Sun Ling had set up her own weather station at her school.

"Can you explain how everything works, Sun Ling?" asked Elena.

"Sure," replied Sun Ling. "I've got to go out and take some readings right now."

Sun Ling showed Chad and Elena some of the instruments that she used in her weather station. "I have an anemometer that measures wind speed and a wind vane that shows wind direction. I also have a rain gauge and a thermometer shelter that keeps out direct sunlight."

"What's that weird U-shaped thermometer?" asked Chad.

"That's the maximum-minimum temperature thermometer. It lets me know the highest and lowest temperatures reached in a day," explained Sun Ling.

"How does it do that?" asked Elena.

"The thermometer contains both mercury and alcohol. You can see the mercury in the bottom of the U. The bulb on the left is completely filled with alcohol, but the bulb on the right is only partially filled with alcohol," said Sun Ling.

401

Answers to
Using What You Learned

1. Thermometers measure the expansion or contraction of gas or liquid caused by an increase or decrease in temperature.

2. Sample answer: A maximum-minimum thermometer is U-shaped with mercury in the middle, alcohol at both ends, and air in the bulb on the right. On each side, an index rests in the alcohol just above the mercury. As the temperature goes up, the alcohol on the left expands, moving down and past the left index. (The index stays in place.) The alcohol pushes the mercury down the left side, around the bottom, and up the right side. As the mercury moves, it pushes the right index up to record the maximum temperature. It also pushes the alcohol on the right up, compressing the air in the bulb. As the temperature goes down, the alcohol on the left shrinks, pulling the mercury with it and allowing the air in the bulb to expand. The mercury pushes the left index up to record the minimum temperature.

3. Answers will vary, but the sentence could conclude with "it regulates other temperature-related devices."

4. Sample concept map:

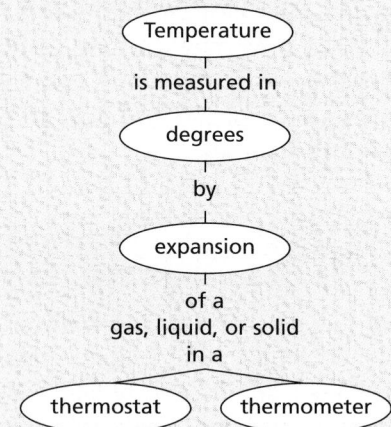

"What are those little pieces of metal?" asked Elena.

"Those are called *indexes*. When the temperature rises, the alcohol and the mercury expand. Because the alcohol in the bulb on the left has no room to expand, it pushes the mercury and the index in front of it toward the bulb on the right. When it gets cooler, the alcohol in the left bulb contracts and draws the mercury along with it. Then the mercury column on the right drops, leaving the index at the highest point reached."

"What holds the index in place?" asked Chad.

"It's held in place with a little spring."

"Oh, I think I get it. As the mercury and alcohol contract, the mercury pushes the index on the left toward the bulb on the left. The index on the left stays at the lowest temperature reached," said Elena.

"You've got it all right! To reset the thermometer for the next day's reading, I use a magnet to pull the indexes back to the mercury."

Using What You Learned

1. What changes do thermometers measure?

2. Write your own explanation of how a maximum-minimum temperature thermometer works so that someone in the third grade could understand it.

3. Using your own words, complete the following sentence: The thermostat is a special kind of thermometer because . . .

4. In this chapter you have learned about several hot topics. Copy the words below into your ScienceLog. Draw lines to connect the words that you think are related in some way. Along each line, explain why you made the connection that you did.

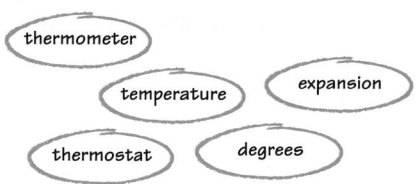

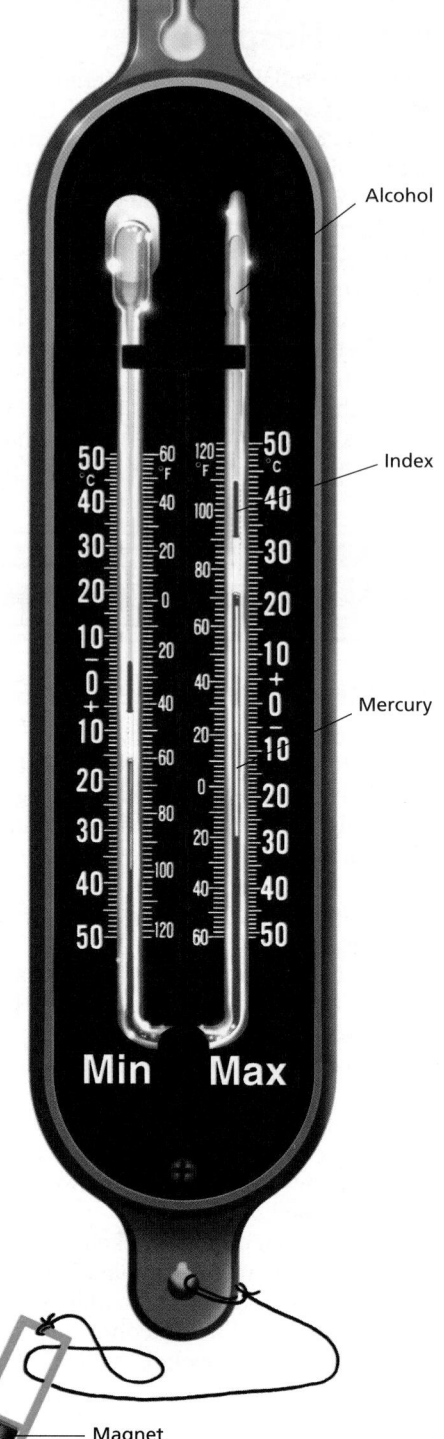

Alcohol

Index

Mercury

Min   **Max**

Magnet

FOLLOW-UP

Reteaching

Have students write a definition of *thermometer* in their own words. They should also give examples of some different types of thermometers and their uses.

Assessment

Have students illustrate the differences between a thermostat, a clinical thermometer, and a maximum-minimum temperature thermometer.

Extension

Have students research how a dry-bulb and wet-bulb thermometer works. They should discuss how this instrument measures humidity.

Closure

Have students work in groups to develop a short skit showing a situation in which the wrong type of thermometer is used to measure the temperature, such as using an air thermometer on a mountain or a bimetallic thermometer in a clinic.

CHALLENGE YOUR THINKING

1. Think Celsius

Test your knowledge of the Celsius scale. Match the descriptions below with the corresponding numbered positions on the thermometer. Do not write in this book.

a. "It's hotter than a firecracker outside today!"
b. "Come on in, the water's great."
c. Butter melts.
d. sweater weather
e. boiling water at sea level
f. boiling water on a mountaintop
g. "The water pipes froze!"
h. Ice cream stays hard.
i. "Ouch—that water's hot!"
j. dead-battery weather

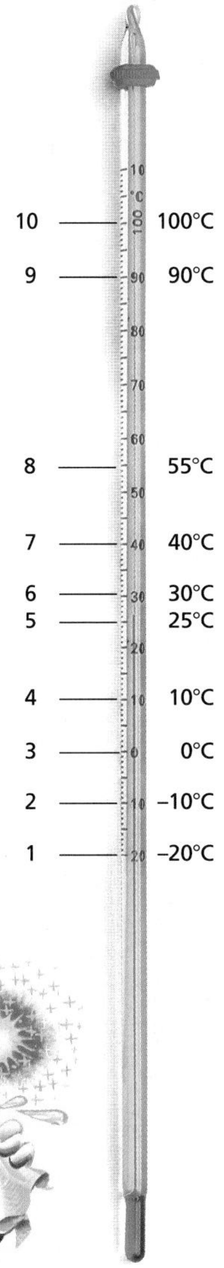

10 — 100°C
9 — 90°C
8 — 55°C
7 — 40°C
6 — 30°C
5 — 25°C
4 — 10°C
3 — 0°C
2 — –10°C
1 — –20°C

2. A Sticky Problem!

Honey becomes runnier as it warms up. Design a thermometer that uses this property.

ENVIRONMENTAL FOCUS

The seasonal variation in temperatures on the Earth is not due to the fact that the Earth is closer to the sun in the summer. In the Northern Hemisphere, the Earth is actually closer to the sun in the winter. The change in temperature is due to the fact that the axis of the Earth is tilted. Lead a class discussion to explain this phenomenon.

Answers to Challenge Your Thinking, pages 403–404

1. a. 7   f. 9
   b. 5   g. 3
   c. 6   h. 2
   d. 4   i. 8
   e. 10 j. 1

2. Answers will vary. One method would be to measure how long it takes a known volume of honey to pass through a glass-stoppered buret at various temperatures.

3. The iron would be on the outside and the copper on the inside because the copper expands and contracts more than the iron does.

4. Suggested improvements might include the following:
   • Use a smaller volume of water or liquid to lessen the time it takes for the temperature of the water or liquid to change.
   • Use tubing with a much smaller diameter to increase the distance that the liquid travels inside the tubing as its volume changes.
   • Seal the tubing to prevent atmospheric pressure from affecting the volume of the liquid.
   • Calibrate the thermometer, and use alcohol instead of water.

5. a. The friction between the tires and the road heats the tires and the air inside the tires. Because the air expands more than the tires when heated, the pressure exerted on the tires increases.
   b. Holding the cap under hot, running water loosens the cap because the metal cap expands more than the glass bottle does when heated.

You may wish to provide students with the Chapter 21 Review Worksheet that accompanies this Challenge Your Thinking (Teaching Resources, page 14). Transparency 71 is also available for your use.

The following are sample revised answers:

1. Answers will vary, but students should see signs of extreme heat in the photograph. The evidence includes the fire in the furnace, the molten metal on the pole, and the protective clothing of the two men.

2. Accept all answers. In this chapter, students were introduced to the difficulties of estimating temperature and to the use of thermometers to overcome these difficulties.

3. Thermometers rely on the expansion of solids, liquids, and gases to measure temperature. As temperature increases, so does the volume of a substance. By measuring the amount that a substance expands or contracts, thermometers indirectly measure the amount of temperature increase or decrease.

4. The seams in concrete roads and bridges allow the concrete to expand and contract as the temperature changes.

Meeting Individual Needs

Learners Having Difficulty

Students who have access to cameras might enjoy taking photos and making photo essays on heat and temperature. Allow students latitude on how they interpret the concepts. Arrange the finished essays around the classroom in a manner that suggests an art gallery.

3. The Heat Is On

Suppose that the bimetallic strip in Sarah's furnace thermostat (on page 400) is made of copper and iron cemented together. Is the copper on the inside or the outside of the coil? Why?

4. Thermometer Design

Examine the illustration at right of a thermometer designed by two students. Make some suggestions about how they might improve it. Explain why you think your changes would help.

5. There Must Be an Explanation

a. Measure the air pressure in the tires of the family car before taking a drive. After the drive is over, measure the air pressure again. You'll find that the pressure has gone up. Is there more air in the tires? What other explanations can you suggest?

b. When you are struggling to unscrew the metal cap of a bottle, it's a good idea to hold the cap under hot, running water. Why?

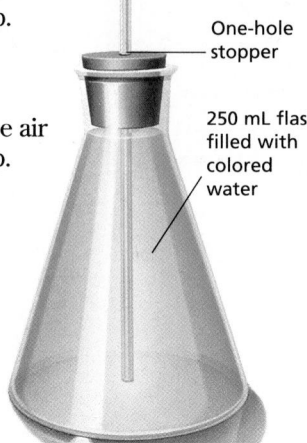

Glass tubing

One-hole stopper

250 mL flask filled with colored water

ScienceLog

Review your responses to the ScienceLog questions on page 390. Then revise your original ideas so that they reflect what you've learned.

What is the difference **1** between *heat* and *temperature*? Use both words in a sentence or two about this picture.

2 You can create heat by burning things. How many other ways of creating heat can you think of?

ScienceLog

Think about these questions for a moment, and answer them in your ScienceLog. When you've finished this chapter, you'll have the opportunity to revise your answers based on what you've learned.

3 When you exercise, you breathe more heavily and get hot. Can you explain why?

Connecting to Other Chapters

> **Chapter 21**
> allows students to estimate the temperature of objects and demonstrates how a thermometer works.

> **Chapter 22**
> introduces students to the difference between heat and temperature and discusses heat as a type of energy.

> **Chapter 23**
> introduces students to conduction, convection, and radiation as the three mechanisms for transferring heat.

Prior Knowledge and Misconceptions

Your students' responses to the ScienceLog questions on this page will reveal the kind of information—and misinformation—they bring to this chapter. Use what you find out about your students' knowledge to choose which chapter concepts and activities to emphasize in your teaching.

In addition to having students answer the questions on this page, you might want to assign a "free-write" in order to assess their prior knowledge. To do this, instruct students to write for 3 to 5 minutes on the subject of heat. Tell them to keep their pens moving at all times in a stream-of-consciousness fashion. They might also try to think of as many common phrases as they can that use the word *heat*. Examples include *heated debate* or *heat of the moment*. Emphasize that there are no right or wrong answers in this exercise. It might be best to ask students not to put their name on their paper. Collect the papers, but do not grade them. Instead, read them to find out what students know and what misconceptions they may have about heat. Tell students that in this chapter they will learn how heat and temperature are related.

Heat Versus Temperature

FOCUS

Getting Started

Show students a beaker of boiling water. Ask them to consider a large lake. Ask: Which of the two has a higher temperature? Which contains more heat? The temperature of the lake is probably around 17°C, while the temperature of the boiling water is 100°C. Explain to students that although the temperature of the boiling water is higher, the lake has more heat energy because it contains many more particles of water.

Main Ideas

1. Temperature measures the warmth of an object.
2. A temperature change depends on an object's mass and on the amount of heat applied to it.
3. Heat may be generated in a number of ways.

TEACHING STRATEGIES

Can You Tell the Difference?

When students have answered the questions, have them share and discuss their responses.

Answers to
Can You Tell the Difference?

1. heat
2. temperature
3. heat
4. temperature

Answer to
In-Text Question

Ⓐ Students should show an understanding that temperature and heat are not synonymous. In common usage, heat refers to the warmth of an object, while temperature refers to a specific measurement of that object's warmth. In the following experiments, students will learn that heat is energy and that temperature is a measure of that energy.

Heat Versus Temperature

Can You Tell the Difference?

Until now you have been studying temperature—how cold, warm, or hot something is. Is heat the same thing as temperature? What is the difference between heat and temperature? Test yourself! Using a separate piece of paper, fill in the blanks in the following questions with the right term: *heat* or *temperature*.

1. Would you get more _____?_____ in your home by burning wood or by burning oil?

2. At what _____?_____ should you set the house thermostat?

3. Is there enough _____?_____ from a burning match to boil water?

4. At what _____?_____ does water boil?

Now write what you think the difference is between heat and temperature. Compare your descriptions with those of a classmate. Ⓐ

In Exploration 1 you will conduct a number of experiments. These experiments are designed to help you clear up any uncertainties that you may have about the difference between heat and temperature.

EXPLORATION 1

A Circus of Experiments

EXPERIMENT 1

Heat From Burning

You Will Need

- a test tube
- a test-tube clamp
- a graduated cylinder
- water
- an alcohol thermometer
- a candle
- matches
- a jar lid
- modeling clay
- a watch or clock with a second hand

What to Do

Measure the temperature of 5 mL of water in a test tube. Place the candle in the jar lid, and secure it with modeling clay. With a test-tube clamp, hold the test tube over the flame of a candle for 30 seconds. Measure the increase in temperature. Repeat with 10 mL and then 15 mL of water. Record your results. Do your results make sense? How do you explain them?

LESSON 1 ORGANIZER

Time Required
2 class periods

Process Skills
observing, measuring, classifying, analyzing

New Terms
none

Materials (per student group)
Exploration 1, Experiment 1: alcohol thermometer; 30 mL of water; test tube; candle; jar lid; small ball of modeling clay; test-tube clamp; 100 mL graduated cylinder; matches; watch or clock with second hand; safety goggles; **Experiment 2:** alcohol thermometer; 35 mL of water; test tube; 100 mL graduated cylinder; 3 samples of 3 g of anhydrous calcium chloride; stirring rod; safety goggles; lab aprons; latex gloves (See Advance Preparation on page 387D.); **Experiment 3:** 175 mL of water; 100 mL graduated cylinder; empty metal can; alcohol thermometer; 1 m piece of cord; safety goggles; oven mitts; **Experiment 4:** 1500 mL of water; 1000 mL beaker; alcohol

continued

Heat From Dissolving

You Will Need

- an alcohol thermometer
- a test tube
- water
- a graduated cylinder
- anhydrous calcium chloride
- a stirring rod
- latex gloves

What to Do

Measure the temperature of 5 mL of water in a test tube. Wearing gloves and an apron, add 3 g of calcium chloride to the water. Slowly stir the mixture with a stirring rod. Now measure the water's temperature again. Repeat the procedure using 10 mL and then 20 mL of water. Record your results. Do they make sense?

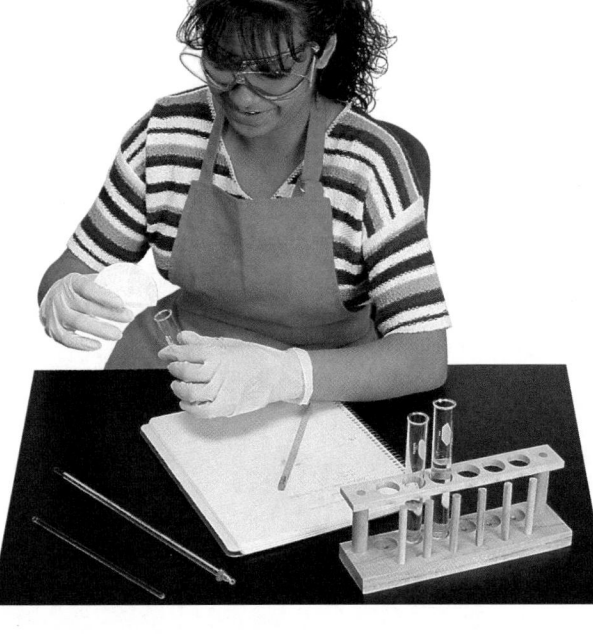

EXPERIMENT 3

Heat From Friction

You Will Need

- water
- a graduated cylinder
- an empty metal can
- an alcohol thermometer
- a piece of cord 1 m long
- an oven mitt

What to Do

Pour 25 mL of water into an empty metal can. Measure the temperature of the water. Loop a 1 m length of cord around the outside of the can. Have a classmate wear an oven mitt to hold the can in place so that his or her body heat is not transferred to the can. Pull the cord back and forth a number of times, and then measure the temperature again. Repeat the procedure using 50 mL and then 100 mL of water. Record your results. Are they what you expected? Why or why not?

Exploration 1 continued ▶

ORGANIZER, *continued*

thermometer; hot plate; watch or clock with second hand; safety goggles; **Experiment 5:** materials from one of the previous experiments, as well as an extra candle for Experiment 1 or an additional 6 g of anhydrous calcium chloride for Experiment 2

Teaching Resources
Exploration Worksheet, p. 18
SourceBook, pp. S172 and S175

Set up a workstation for each of the first four experiments. Students may complete Experiment 5 at the appropriate workstation for one of the previous experiments. Two sets of apparatus are adequate at each station. Provide each group with about 10 minutes to complete each experiment.

Cooperative Learning
EXPLORATION 1

Group size: 3 to 4 students
Group goal: to clear up any uncertainties about the difference between heat and temperature
Positive interdependence: Assign students roles such as supplies manager, timer and pacer, reporter, and director.
Individual accountability: Have each student write a short essay, draw illustrations, or give a short oral report about the difference between heat and temperature.

Exploration 1 continued ▶

⭐ An Exploration Worksheet is available to accompany Exploration 1 (Teaching Resources, page 18).

Homework

Have students answer the following: Think about the phrase "you have a temperature." Using what you have learned, explain what this phrase really means and why it is inaccurate.

CROSS-DISCIPLINARY FOCUS

Mathematics

Have students look up how to convert from one unit of energy to another. Consider the units calorie, joule, erg, and BTU. Point out that these units are also used to measure heat because heat is a form of energy.

EXPLORATION 1, continued

EXPERIMENT 1 page 406

Suggest that one student hold the test tube over the candle while another holds the thermometer in the water, keeping the thermometer from touching the test tube. Students should observe that as the amount of water is increased, the increase in temperature is less. This is because the same amount of energy is added to more water.

EXPERIMENT 2 page 407

Be sure students add the same amount of calcium chloride (3 g) to each amount of water. Students should observe that as the amount of water increases, the increase in temperature is less. The same amount of chemical energy is released, but the amount of water is larger each time.

EXPERIMENT 3 page 407

One student should hold the can while another pulls the cord. The student holding the can should wear an oven mitt to avoid transferring body heat to the can. The number of pulls should be the same each time. Again, as the amount of water increases, the increase in temperature is less.

EXPERIMENT 4

Suggest that one student hold the thermometer so that the bulb is in the water but does not touch the beaker. Again, as the amount of water increases, the increase in temperature is less.

EXPERIMENT 5

You may wish to provide students with an opportunity to complete each of the experiments again, or have students choose one experiment and report their results to the class. Students should observe that doubling the heat while keeping the volume of water constant doubles the increase in temperature.

EXPERIMENT 4

Heat From Electricity

You Will Need

- a 1000 mL beaker
- water
- an alcohol thermometer
- a hot plate
- a watch or clock with a second hand

What to Do

Pour 500 mL of water into a beaker. Measure and record the temperature of the water. Now heat the beaker of water on a hot plate for 30 seconds, and measure the increase in temperature. Repeat this procedure using 1000 mL of water. Record your results. Are they what you expected?

EXPERIMENT 5

Doubling the Heat

In each of the previous experiments, the same amount of heat was added to different amounts of water. In this experiment, the amount of heat changes, while the amount of water stays the same.

Repeat one of the previous experiments, but vary the amount of heat while keeping the amount of water the same. For example, try one of the following: burning a candle for twice as long under 10 mL of water; dissolving 6 g of calcium chloride in 20 mL of water; pulling the cord twice the number of times with 50 mL of water; heating 1000 mL of water on a hot plate for 1 minute. Keep a record of your results.

Making Sense of Your Results

First, compare your results from each of the experiments with those of other students. Then work through the questions below. Afterward, you should be able to make sense of your results.

1. In each of the first four experiments, the amount of water was changed. Was the temperature increase the same in each? Why or why not?

2. Compare Experiment 5 with the first four experiments. How does doubling the heat affect the temperature increase?

3. What would happen to the temperature increase if you doubled the amount of heat and also doubled the amount of water?

4. If you put one finger in a glass of water at 50°C and another finger in a bathtub of water at 50°C, would both fingers have the same sensation? Is the water in each container equally hot? Does the water in the glass and in the tub contain the same amount of heat? Explain your reasoning.

5. Here is a tricky question: Which do you think would require more heat—raising 1 L of water from 20°C to 30°C or raising 1 L of water from 30°C to 40°C? How could you find out?

Answers to Making Sense of Your Results

1. In each experiment, when the amount of water was increased, the amount of temperature increase was less because the same amount of heat was being added to larger amounts of water.

2. Doubling the heat should double the temperature increase.

3. There should be no change in temperature.

4. Both fingers would experience the same sensation. The water in the bathtub and the glass are at the same temperature. However, the tub contains more heat because there is more water.

5. Students may respond that raising the water from 30°C to 40°C requires more heat. However, the temperature is being raised 10°C in both cases, and the heat required is virtually the same. (In fact, slightly more heat is required to raise the water from 20°C to 30°C because the specific heat of water is higher at lower temperatures. Specific heat is discussed on page S175 of the SourceBook.) Students may suggest comparing the amount of time it takes to heat 1 L of water from 20°C to 30°C with the amount of time it takes to heat 1 L of water from 30°C to 40°C.

6. Here is another tricky question: Which do you think would melt more ice—10 mL of water at 50°C; 20 mL of water at 50°C; or 10 mL of water at 100°C?

7. Describe again in your own words the difference between heat and temperature. Compare your answer with the one you wrote at the beginning of this chapter.

8. Can you help set Dave's thinking straight about the difference between temperature and heat? In your ScienceLog, rewrite this report for him.

Dave's Report

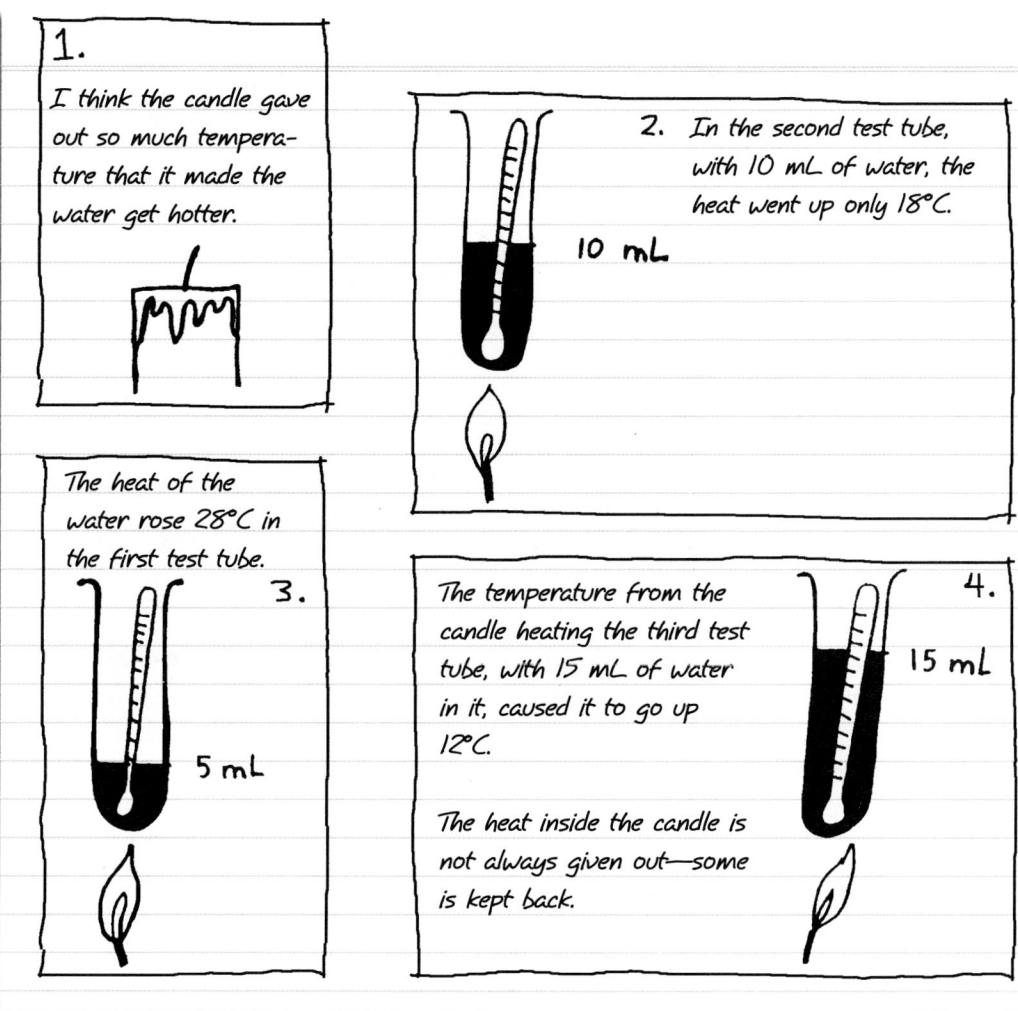

1. I think the candle gave out so much temperature that it made the water get hotter.

The heat of the water rose 28°C in the first test tube.

5 mL

3. The temperature from the candle heating the third test tube, with 15 mL of water in it, caused it to go up 12°C.

The heat inside the candle is not always given out—some is kept back.

2. In the second test tube, with 10 mL of water, the heat went up only 18°C.

10 mL

4. 15 mL

Answers to
Making Sense of Your Results,
continued

6. Accept all reasonable inferences. The total amount of ice that melts is proportional to the volume of the water multiplied by the initial temperature of the water.
   • 10 mL at 50°C melts the least amount of ice. (10 × 50 = 500)
   • 20 mL at 50°C and 10 mL at 100°C melt the same amount of ice. (20 × 50 = 10 × 100 = 1000)

7. Students' answers should reflect an understanding that temperature is a measure of how cold or hot a substance is. Heat refers to the total energy within a substance.

8. Sample answer: The candle gave out so much heat that it made the water hotter. The temperature of the water rose by 28°C in the first test tube, which contained 5 mL of water. In the second test tube, with 10 mL of water, the temperature went up by only 18°C. The heat from the candle under the third test tube, with 15 mL of water in it, caused the temperature to go up by 12°C. As the amount of water increased, the increase in temperature got smaller.

Homework
The Closure activity described below makes an excellent homework assignment.

FOLLOW-UP

Reteaching
Have students describe how the change in the temperature of a material depends on the amount of heat absorbed by the material and the mass of the material.

Assessment
Point out to students that when the heat applied to a substance is doubled, the increase in temperature is also expected to double, but that in reality this does not always happen. Challenge students to offer an explanation for why the observed increase in temperature is not always what is expected. (*Experimental errors are always present when making an observation. Also, for a given amount of heat, the change in temperature for different materials is not the same. Even as the same material heats up, its properties can change enough so that it acts like a different material at the higher temperature.*)

Extension
Point out to students that there are several different temperature scales— Fahrenheit, Celsius, and Kelvin. Suggest that they make posters to illustrate how these scales differ.

Closure
Challenge students to make bar graphs to show the results of one or more of the experiments they completed in Exploration 1. Display the completed graphs.

FOCUS

Getting Started

Start a class discussion by asking: What color would you get if you mixed equal parts of red and white paint? *(Pink)* How could you get a darker pink? *(Mix more red paint and less white paint.)* Challenge students to draw a comparison between mixing paints and mixing hot and cold water. *(When hot and cold water are mixed, the result is a temperature somewhere between the hot and the cold water, depending on how much of each is used.)*

Main Ideas

1. Hot and cold water mixed in equal volumes result in water with a temperature halfway between the two.
2. Mixing unequal volumes of hot and cold water results in water with a temperature closer to that of the greater amount of water.

TEACHING STRATEGIES

EXPLORATION 2

 SAFETY ALERT If hot tap water is used, be sure it is not above 50°C.

Suggest that students label the plastic-foam cups as shown in the photograph.

Cooperative Learning
EXPLORATION 2

Group size: 3 to 4 students
Group goal: to develop a formula for getting a certain temperature of water
Individual responsibility: Each member of the group should choose a role such as materials coordinator, chief executive, or support staff.
Individual accountability: When the Exploration is finished, encourage students to participate in a class discussion of the results and procedures.

★ An Exploration Worksheet (Teaching Resources, page 24) and Transparency 72 are available to accompany Exploration 2.

LESSON

2

Hot + Cold = ?

Have you ever wished that you could find a formula for getting the temperature of your bath or shower just right? The best way to discover the formula is to experiment. In the following Exploration, you will try mixing different amounts of hot and cold water.

EXPLORATION 2

Mixing Hot and Cold

You Will Need

- 3 plastic-foam cups: one containing hot (but not scalding) water, one containing cold water, and one empty
- a graduated cylinder
- an alcohol thermometer
- a stirring rod

Caution: Water that is above 55°C will scald you.

What to Do

1. Pour one measure (10 mL) of hot water into the empty cup.
2. Record the temperature of the hot water in the cup. Use a table like the one at the top of page 411 to record your results.

3. Record the temperature of the cold water.
4. Now add one measure (10 mL) of the cold water to the hot water. Predict the temperature of the mixture of hot and cold water. Record your prediction. Stir, and then record the actual temperature.
5. Try adding one measure (10 mL) of cold water to two measures (20 mL) of hot water. Again, make and record your prediction about the resulting temperature. Record the actual temperature. Continue to experiment by using different ratios of hot and cold water. In each case, make a prediction, and then record the actual temperature.

LESSON 2 ORGANIZER

Time Required
1 class period

Process Skills
observing, measuring, predicting

Theme Connection
Energy

New Terms
none

Materials (per student group)
Exploration 2: 3 plastic-foam cups; about 200 mL each of hot and cold water; 100 mL graduated cylinder; alcohol thermometer; stirring rod; safety goggles

Teaching Resources
Exploration Worksheet, p. 24
Transparency 72

Number of measures of hot water	Temperature of hot water	Number of measures of cold water	Temperature of cold water	Predicted temperature of mixture	Actual temperature of mixture

Searching for a Pattern

Look at the temperature of each mixture and at the proportion of hot water added. Does any pattern emerge? Now is the time to try to write your bath-water formula. What proportions of hot and cold water would be best for bath water? Decide what measurements you should make to test your formula.

Coffee Counseling

How could Juanita solve Elinor's problem? What did the previous Exploration tell you about the difference between hot and cold objects?

Dear Juanita:

I hear you have been studying what happens when you mix hot and cold liquids. Can you help me?

My problem is this: I always get up late and have to rush my morning coffee. How much milk should I add to bring it to the right temperature? I always seem to add too much or too little!

Many thanks,
Elinor

411

Answers to *Searching for a Pattern*

A pattern should emerge that depends on the ratios of hot and cold water mixed together. The temperature of a mixture is an average of the two initial temperatures. For example, two parts at 10°C and one part at 40°C would be (10 + 10 + 40) ÷ 3 = 20°C. Accept all reasonable proportions for mixing the bath water. Student formulas can be tested by measuring the temperatures and volumes of the hot water, cold water, and resulting mixture.

Answers to *Coffee Counseling*

Involve the class in a discussion of how Juanita can solve Elinor's problem, assuming that the volume of the coffee cup is 200 mL and the desired temperature of the coffee is 65°C.

FOLLOW-UP

Reteaching

Have students make a list of times during the day when they mix hot and cold substances together. Ask them to suggest the ratios that should be used to arrive at the desired temperatures.

Assessment

Have students state whether heat is moving into or out of the glass of water and what happens to the temperature of the water in each of the following situations:

a. A glass of water is placed in the refrigerator. (*Heat moves out of the water, thus lowering its temperature.*)
b. You hold a glass of water in both hands. (*Heat moves into the water, so its temperature increases.*)
c. A glass of water is left in a sunny spot. (*Heat is moving into the water, so its temperature increases.*)

Extension

Suggest that students find out the definitions of *thermal shock* and *thermal pollution*. Encourage them to interpret the definitions with illustrations.

Closure

Suggest that students make a graph to show the change in temperature as specific amounts of hot water are added to a container of cold water.

Theme Connection

Energy

Review kinetic energy with students. In addition, share the following information: The *kinetic theory* describes how heat affects different materials. In this theory, the thermal properties of a substance arise from the random motions of all the particles that make up the substance. The energy of these particles determines the substance's temperature.
Focus question: Create a diagram that illustrates the kinetic theory. (*Accept all reasonable responses.*)

ENVIRONMENTAL FOCUS

The *specific heat* of a substance measures the amount of heat needed to raise the temperature of a substance 1°C. Water, for instance, has a very high specific heat, so its temperature rises only slightly for a given amount of added heat. Air, on the other hand, has a very low specific heat. Point out to students that the temperature difference between the air in summer and the air in winter is much greater than the temperature difference between the ocean in summer and the ocean in winter.

LESSON 3 — Heat Is Energy

LESSON 3 — Heat Is Energy

FOCUS

Getting Started

Ask students to write a short phrase on a piece of paper with a pencil and then to touch the pencil eraser to their wrist. Then ask students to erase the phrase and to touch the eraser to their wrist again. Ask: Does the eraser feel any different? *(It has become warmer.)* Point out that by rubbing the eraser, they have converted mechanical energy into heat energy.

Main Ideas

1. Heat is a form of energy.
2. The *joule* is the metric unit used to measure energy. It is approximately equal to the amount of energy required to raise a 100 g mass 1 m.
3. The human body requires a certain number of Calories daily in order to function properly.

TEACHING STRATEGIES

EXPLORATION 3

Be sure that students record the temperature of the water before they begin each activity. Point out to students that some of the heat given to the sinker will be lost to the environment before it is transferred to the water, so not all of the energy added to the sinker will contribute to an increase in temperature.

WASTE DISPOSAL ALERT Lead is toxic. Do not throw sinkers in the trash. Instead, save them for later use.

Answers to *Exploration 3*

1. The temperature did rise; the heat came from the mechanical energy used to shake the jar.

2. The heat came from the mechanical energy used to hit the sinker with the hammer.

3. Amount of heat (in joules) added to the shaken water = the temperature increase (in °C) × 50 g × 4.2 J/g•°C; amount of heat added to the sinker = the temperature increase (in °C) × 25 g × 4.2 J/g•°C

LESSON 3 — Heat Is Energy

Heat is a form of energy. All objects, even cold objects, contain heat energy. Heat energy can be obtained from other forms of energy, such as the chemical energy stored in a match. Heat energy can also be converted into other forms of energy, such as the electrical energy produced from steam-powered generators.

EXPLORATION 3

Changing Your Energy Into Heat Energy

You Will Need

- a small jar with a lid
- water
- a graduated cylinder
- newspaper
- a watch or clock
- an alcohol thermometer
- a lead sinker
- a hammer
- pliers
- latex gloves

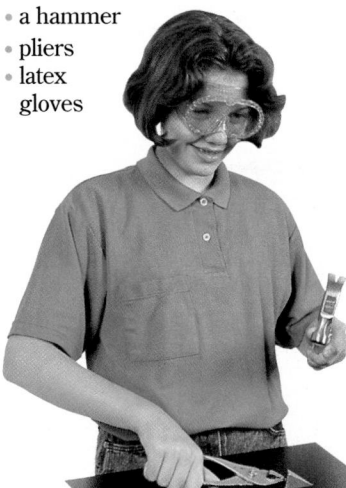

What to Do

1. Completely wrap a small jar containing 50 mL of water at room temperature in about eight thicknesses of newspaper. Shake it for about 4 minutes. Get some friends to help with the shaking. Did the temperature rise? Where do you think the heat came from?

2. Using pliers, grasp a lead sinker by the eyelet. Hit the sinker about 50 times with a hammer, and then place it in 25 mL of water at room temperature. Record the increase in temperature. Where do you think the heat came from?

3. How much heat did you add to the water and to the sinker? You will find out how to make this calculation on the next page.

Caution: Wear gloves when handling the lead sinker. When finished, dispose of your gloves and wash your hands thoroughly with soap and water. Do not hammer on a lab table or a desk. Use a hard surface such as a sidewalk or anvil.

LESSON 3 ORGANIZER

Time Required
3 class periods

Process Skills
observing, measuring, organizing

New Terms
Calorimeter—a device that measures the amount of energy in a substance
Joule—unit of energy approximately equal to the amount of energy required to raise a 100 g mass 1 m

Materials (per student group)
Exploration 3: 8 sheets of newspaper; 75 mL of water; 100 mL graduated cylinder; small jar with lid; watch or clock; alcohol thermometer; pliers; lead sinker; hammer; safety goggles; latex gloves
Exploration 4: small juice can; shelled peanut; straight pin; cork covered with aluminum foil; support stand with ring clamp; wire gauze; 100 mL of water; 100 mL graduated cylinder; alcohol thermometer; matches; safety goggles

Teaching Resources
Exploration Worksheets, pp. 26 and 27
SourceBook, p. S162

Meet Mr. Heat

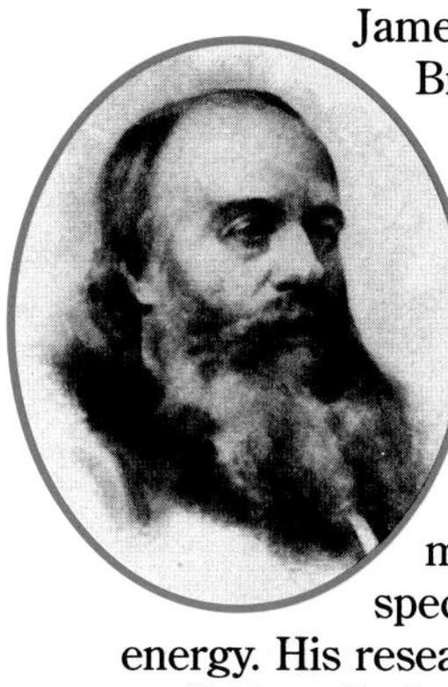

James Prescott Joule, a famous British scientist who lived during the nineteenth century, was fascinated by the idea of changing mechanical energy into heat, which you did in Exploration 3. He wasn't satisfied with merely showing that it could be done. He also wanted to know exactly how much heat is produced from a specific amount of mechanical energy. His research took 35 years!

Before Joule came along, people thought that heat was a weightless fluid called *caloric*. However, Joule was able to show that the same amount of heat was always obtained from a given amount of mechanical energy. His work and that of other scientists showed that heat is a form of energy.

One of Joule's devices, shown on the right, was used to change mechanical energy into heat energy. Can you figure out how it works by looking at the diagram?

As a tribute to his work, the unit of measure for energy is called the **joule** (J). One joule is the amount of energy required to raise a 100 g mass 1 m. Or to put it another way, a 100 g mass that has been raised to the height of 1 m has been given 1 J of energy.

In Joule's paddle-wheel apparatus, the mass is raised when the crank is turned. When the crank is released, the mass falls down, the paddle wheels spin, and the temperature of the water rises. The mechanical energy of the falling mass is transferred to the water as heat energy. It takes 4.2 J of energy to raise the temperature of 1 g of water by 1°C. A device that measures the amount of heat is called a *calorimeter*.

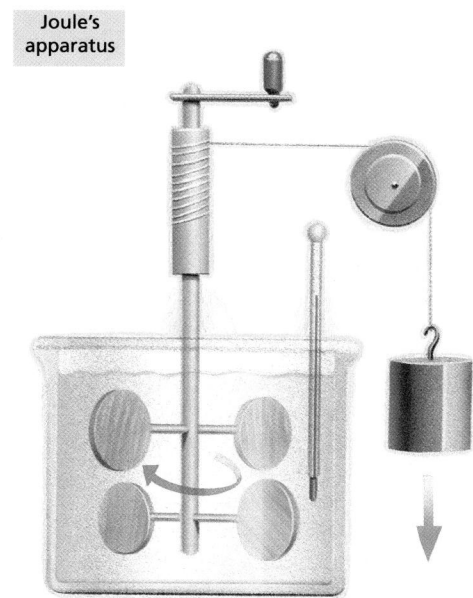

Joule's apparatus

Challenging Questions

Turn to page S172 of the SourceBook to learn more about the relationship between heat energy and temperature.

1. How many joules of energy would it take to raise a 10 kg mass 4 m? If you added this energy to 1 g of water, how much would the temperature rise?

2. How many joules are needed to raise the temperature of 200 g of water (enough to make a cup of tea) by 50°C?

3. Suppose you added a pot of boiling water (2000 g) to a bath of cool water, bringing the temperature of the bath water up to 40°C. How much heat did the hot water give to the bath water? (Hint: heat given to the bath = heat given off by the pot of boiling water.)

4. What is the rule for finding the number of joules of energy added to or given off by water?

413

An Exploration Worksheet is available to accompany Exploration 3 on page 412 (Teaching Resources, page 26).

CROSS-DISCIPLINARY FOCUS

Mathematics

How many joules are required to raise the temperature of 8 g of water by 5°C?
a. 88.2 joules   **c.** 168 joules *(Correct)*
b. 21 joules   **d.** 17.2 joules

Did You Know...

If a light bulb requires 1 joule of energy each second in order to operate, we say that it is a 1 watt light bulb. A watt (abbreviated W) is a measure of the rate of energy transfer, or power. One watt is equal to one joule/second.

Food Energy

Have students silently read the two paragraphs. You may wish to point out that the chemical process by which the body produces energy is called *metabolism.* Food provides the raw materials, or fuel, for metabolism to take place. During this process, cells break down complicated substances into simpler ones, releasing chemical and heat energy. This energy is used to keep the body functioning—to maintain breathing, heart beat, body temperature, and other basic functions. The energy released and used by the body is measured in calories or joules.

You might want to work through the answer to the question in the second paragraph along with your students. If necessary, review the definition of a joule as discussed earlier in the lesson. Point out to students that 1 kJ is the same as 1000 J.

The method for finding the amount of energy (400,000 J) required to climb a mountain that is 1000 m high if your mass is 40 kg is as follows: If it takes 1 J of energy to raise 100 g a height of 1 m, then it will take 400 J of energy to raise 40 kg (40,000 g) a height of 1 m. And it will take 400,000 J (400 g × 1000 m) to raise 40 kg to a height of 1000 m.

EXPLORATION 4

SAFETY ALERT Caution students to be particularly careful when they burn the peanut. It may be wise to have the apparatus near an open window or an exhaust hood to allow any smoke or fumes from the peanut to escape.

When students calculate the joules of heat energy produced, be sure that they use the number of degrees that the temperature rises and not the final temperature. You may find it necessary to demonstrate how to make the calculation. When students have completed the Exploration, involve them in a discussion of their results.

Instruct students to set up the apparatus as shown in the diagram. Check to be sure that they have set it up correctly and safely before they begin the activity.

Food Energy

The energy you need to walk, dance, ride a bike, or even study comes from the food you eat. Some of the energy from the food you eat is stored as carbohydrates to fuel your muscles. In many ways, your body is like a very efficient car. When you exercise hard, air goes quickly into your lungs. Oxygen from the air is carried through your bloodstream to your muscles. The carbohydrates stored there are "burned," and the energy you need is released. You get hot because some of the energy escapes as heat; your body is not 100 percent efficient!

How many peanuts would you have to eat to have enough energy to climb a mountain that is 1000 m high? If your mass is 40 kg, the amount of energy you would use climbing the mountain would be 400,000 J, or 400 kJ (kilojoules). (Remember, a joule is the amount of energy you have to use to raise a 100 g mass 1 m.) Although your body doesn't "burn" a peanut in quite the same way that a flame does, the amount of energy given off by the peanut is the same. In the following Exploration, you can find out how much energy is involved.

EXPLORATION 4

How Much Energy Is in a Peanut?

You Will Need

- an alcohol thermometer
- a graduated cylinder
- a small juice can
- water
- a shelled peanut
- a straight pin
- a cork covered with aluminum foil
- a support stand with a ring clamp
- wire gauze
- matches

What to Do

Set up the apparatus as shown in the diagram below. Pour 100 mL of water into a small can, and record the water temperature. Using a match, set fire to the peanut. How much did the water temperature rise? Note: Do not let the thermometer rest on the bottom of the can. Calculate the number of joules of heat energy produced (joules produced = 100 × temperature increase × 4.2). Remember: 1 mL of water = 1 g of water.

A Homemade Calorimeter

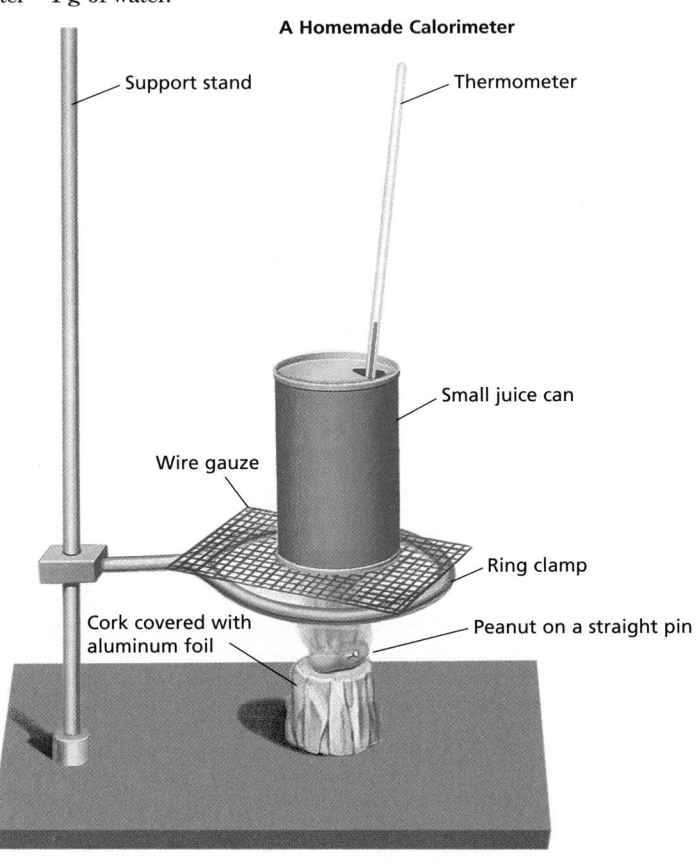

Support stand — Thermometer — Small juice can — Wire gauze — Ring clamp — Cork covered with aluminum foil — Peanut on a straight pin

 Cooperative Learning
EXPLORATION 4

Group size: 2 to 3 students
Group goal: to determine how much energy is in a peanut
Positive interdependence: Assign students roles such as chief experimenter, materials and safety manager, and recordkeeper.
Individual accountability: Have students write a self-assessment. They may want to include what they contributed to the group, how well they worked with the group, and what they learned by performing the experiment.

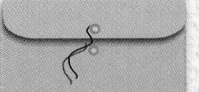 **PORTFOLIO**
Encourage students to speculate on how they might change Exploration 4 to make the results more accurate. Suggest that they include a design for an improved apparatus in their Portfolio.

 An Exploration Worksheet is available to accompany Exploration 4 (Teaching Resources, page 27).

What's Your Daily Energy Intake?

Think About It

1. A group of students did this experiment and found that the water temperature rose 25°C. How much energy did they get from the peanut?

2. In what ways is this calorie-measuring apparatus not very accurate? How could it be improved?

3. Should food be dried before measuring its energy content? Why? How would you measure the energy content of milk?

You can sometimes discover the energy content of the foods you eat by reading the package. Have a look at the side of a cereal box, for instance.

When they measure the energy content of different foods, nutritionists burn small amounts of food as you did. But they use a special device known as a *bomb calorimeter,* which gives more accurate results.

Nutritionists often use a unit known as the Calorie to measure the energy content in food. Calories are not SI units, but converting Calories to joules is easy. Simply multiply by 4200. To convert Calories to kilojoules, multiply by 4.2.

What was your energy intake yesterday? Use the Food  Energy Table on the next page to find out. Young teenagers normally require 8,000 to 10,000 kJ per day. About half of this is used to keep your body's basic systems running—your heart beating, your lungs breathing, and your brain thinking. The rest is used for movement. If you eat too little, you won't have enough energy. If you eat too much, you will put on mass—about 1 kg for every 35,000 kJ in excess of your required intake.

Five slices of cake and five soft drinks would provide enough energy to keep you going for a day, but they wouldn't be good for you. You need a *balanced* diet, one that contains the right balance of protein (found primarily in meat), carbohydrates (found in bread, potatoes, and grains), fats, and fiber. Small amounts of vitamins and minerals are also needed.

415

Homework

The solar corona is the outer part of the sun's atmosphere. Temperatures in the corona are typically millions of degrees, comparable to the center of the sun itself. Yet the corona would not feel as hot as the inside of the sun. Why not? *(The corona consists of high-energy gases that are much less dense than the sun's interior. Thus, the corona would not feel as hot as the sun's interior because fewer particles of gas are present to transfer heat energy.)*

Multicultural Extension

Hot Technologies
The invention of technologies such as ceramics and metallurgy depended on a knowledge of how heat affects different substances. The development of such technologies can often be traced to a particular region and culture. Have students research the invention of a heat-related technology to find out what knowledge of heat was required and how the technology affected the culture or cultures in the region where it first appeared.

Answers to
Think About It

1. 100 g × 25°C × 4.2 J/g•°C = 10,500 J, or 10.5 kJ.

2. Sample answer: Not all of the heat from the burning peanut reaches the can of water. Some heat is used to warm up the apparatus, and some is lost to the surrounding air. Also, the water in the can may be hotter in some places than in others. The accuracy of the calorimeter could be improved by stirring the water constantly and by covering the entire apparatus to keep in the heat from the burning peanut.

3. If food is not dried before measuring the energy it contains, then some energy will be used to heat the water in the food, and the substance may not have enough energy to combust. To measure the energy content of milk, you would first need to turn it into a dry powder.

Answer to
In-Text Question

Ⓐ Answers will vary but will probably fall within a range of about 1800 to 2500 Calories.

What's Your Daily Energy Intake?

INDEPENDENT PRACTICE Display several packages of food with nutritional labels for students to examine. Point out the type of information that is included on the labels. Have students identify the calorie content of each of the foods. The calories listed on food labels are actually dietary Calories. One dietary Calorie is equal to one kilocalorie. Point out that most labels show the number of Calories per serving. Have students identify the serving sizes on the packages and the total number of Calories in the food.

For practice, you may wish to review how to convert Calories to kilojoules (kJ) using the information from these labels. Provide students with time to compute their daily energy intake using the table on page 416. Then involve them in a discussion of their results.

Reteaching

Have students review the analogy comparing the body to a very efficient car on page 414. Then have students write their own analogies for the body and its use of food for energy.

Assessment

Suggest that students make a poster to illustrate what a joule is and how many joules are needed to raise 1 g of water 1°C. Encourage students to be creative. Their poster can be either serious or humorous. They may wish to draw cartoons. (*A joule is the amount of energy used to raise a 100 g mass 1 m. It takes 4.2 J to raise the temperature of 1 g of water 1°C.*)

Extension

For two weeks, have students keep track of the foods they eat on a daily basis. Ask them to show, in chart form, the food groups that these foods represent. Invite them to analyze their chart and to make some suggestions of foods they should add to and foods they should delete from their diet. They should also calculate their daily energy intake in joules.

Closure

Suggest that students work in small groups to prepare a TV commercial that focuses on the energy content of various foods. Calculations of the energy content of some foods should be part of the presentation. Provide time for students to present their commercial to the class.

Answer to *Caption*

Ⓐ Butter is the best energy bargain because it contains the most energy per gram; lettuce salad is the worst. (Note: Although butter is the best energy bargain, a diet high in fats increases the risks of obesity and heart disease.)

Food for Thought

1. How is the way your body gets energy similar to the way that a car gets its energy? How is it different?

2. Dieters: Look at the Food Energy Table at right. Suppose you had the following choices for dinner: a boiled potato or french fries, an apple or a piece of apple pie, green salad or carrots, an egg or meat, and a roll or a slice of bread. You could top it off with a cookie or a piece of cake. Which would you choose? Would it make much of a difference to the "balance" of your diet?

3. What are your favorite foods? Add up the kilojoules contained in a single serving of your five favorite foods. Compare your results with those obtained by other students. Do you have "high energy" or "low energy" tastes?

The Food Energy Table

Food (single serving)	Mass in grams	Energy in kilojoules
apple (1 medium)	100	380
bacon (3 crisp strips)	25	650
banana (1 medium)	125	440
beans, canned	130	670
bread (1 slice)	23	250
bread roll	38	400
butter (1 pat)	10	300
cake, iced	100	1550
candy bar	30	550
carrots, cooked	100	130
cereal, cornflakes	25	400
cereal, puffed	13	220
cheese	30	460
cookies (3 small or 1 large)	25	500
corn (1 small ear)	100	380
egg (1 large)	50	340
fish	100	800
fruit juice	180	380
hot dog	50	630
ice cream	100	890
jam	20	230
meat	100	800–1000
milk	250	670
orange (1 medium)	150	320
peanut butter	16	380
pie	160	1700
potatoes, boiled	100	270
potatoes, fried	100	1160
potato chips	20	480
salad dressing	15	260
salad, lettuce	70	40
soft drink	240	450
soup, canned	185	100–540
spaghetti	100	320
sugar	12	190
tomato (1 medium)	150	150
vegetables, green	100	100
waffle	75	860

 Which food is the best "energy bargain"? Which is the worst? Ⓐ

416

Answers to *Food for Thought*

1. Possible responses:
   A car uses gasoline as fuel and uses more fuel when it goes fast. A car emits carbon dioxide in its exhaust. The burning of fuel in a car is noisy.
   Your body uses food as fuel and uses more fuel when it goes faster and works harder. Your lungs exhale carbon dioxide. The burning of food in your body is quiet.

2. Students' answers may vary but should indicate an understanding of the importance of a balanced diet and should demonstrate an ability to calculate the energy content of food.

3. Answers will vary depending on the foods that each student chooses to evaluate. You may wish to use the questions and student responses as a springboard for a discussion of good nutritional practices.

CHALLENGE YOUR THINKING

1. Caloric Quest!

Only 200 years ago, scientists thought that heat was a weightless fluid called caloric. It seemed like a good theory because it explained so many things.

When you rub two objects together, the caloric is released, and the objects get hotter.

When you heat the end of a metal rod, the caloric flows from the hot end to the cold end.

When you burn something, it breaks up, and the caloric escapes.

Imagine that you have been transported back to those days in a time machine and that you are having an argument with these scientists about what heat is. Write a report of your experience. This is your chance to become world famous!

2. Softwood or Hardwood?

The price of softwood is $70 a cord, and the price of hardwood is $100 a cord. Three friends are trying to decide which to buy.

Daniel: I think we should buy softwood. We'll get more energy for our money.

Lucinda: No! No! That's not true! Hardwood is heavier.

Nestor: It probably makes no difference.

Does it make a difference? Design a simple home experiment to show which type of wood provides more heat per dollar. Discuss your plans with your teacher before carrying out any experiment.

417

Homework

The Activity Worksheet on page 29 of the Unit 7 Teaching Resources booklet makes an excellent homework activity. If you choose to use this worksheet in class, Transparency 73 is available to accompany it.

 You may wish to provide students with the Chapter 22 Review Worksheet that accompanies this Challenge Your Thinking (Teaching Resources, page 30).

Answers to
Challenge Your Thinking, pages 417–418

1. Accept all reasonable responses. Students could point out that all matter has mass. Therefore, no weightless fluid can exist on Earth. In response to the first scientist's statement that rubbing releases caloric, students could respond that if this were true, all of an object's caloric could eventually be used up. Students could point out that this is not the case; the heat produced by friction never

---

Answers to
Challenge Your Thinking,
continued

runs out. Responding to the second scientist, students could state that if a substance contains caloric, then it should be possible to break that substance up and release the caloric within it. However, something can be broken up into tiny pieces without ever releasing as much heat as is produced by burning. In response to the third scientist, students could say that no fluid can pass through a solid as quickly as heat can.

2. Accept all reasonable responses. One approach would be to heat two equal amounts of water by burning a sample of both types of wood. Adjust the amount of both types of wood so that the samples cost the same amount. The wood that causes a greater temperature increase in a similar amount of time provides more heat per dollar.

3. A temperature increase should occur once the hand is dipped into the cold water. When the water is colder, the temperature increase should be more. Accept all reasonable experimental designs. (Be sure to check all designs for safety before students carry out their experiments.) Factors that might affect the amount of heat gained by the water include the time that the hand spends in the water and whether any insulation such as a latex glove is present.

4. Students may use the food energy table on page 416, or they may convert from Calories to kilojoules. Remember that the Calories listed on food labels are kilocalories. To convert Calories to kilojoules, simply multiply the number of Calories by 4.2. (Point out to students that the menu on page 418 is provided as an example. Students should use this sample menu as a model to create their own menu.)

5. Although the temperature of both the bolt and the nail is 100°C, the bolt contains more heat energy because it has a greater mass.

The following are sample revised answers:

1. Students should understand that heat is a form of energy and that temperature is related to how much heat energy a substance contains. Sample sentence: The incredible heat given off by the burning trees caused the temperature of the area to rise sharply.

2. Because heat is a form of energy, all other forms of energy can be converted into heat energy either directly or indirectly.

3. Our bodies rely on chemical reactions to fuel our muscles and other organ systems. These chemical reactions give off heat. As you exercise, the frequency of these chemical reactions increases, and your body temperature rises. Some of these chemical reactions require oxygen and produce carbon dioxide. You breathe more heavily to obtain enough oxygen and to expel the carbon dioxide.

Meeting Individual Needs

Learners Having Difficulty

Some students may still have difficulty understanding the difference between temperature and heat. A good experiment to illustrate these two concepts can be carried out using a bundle of nails, two alcohol thermometers, two plastic-foam cups, a measuring cup, a pan of boiling water, a heat source, tongs, safety goggles, and oven mitts. Heat the nails in boiling water. Point out to students that the temperature of the nails in boiling water is 100°C. Add the same amount of tap water to each of the cups. Using the oven mitts and tongs, transfer one nail to one cup and the remaining nails to the other cup. Have students note the temperature change that occurs in each cup of water. Ask: Why might there be a temperature difference between the cups, even though all of the nails had the same temperature when they were taken from the boiling water? *(The larger number of nails in the second cup caused a greater amount of heat energy to be transferred to the water.)*

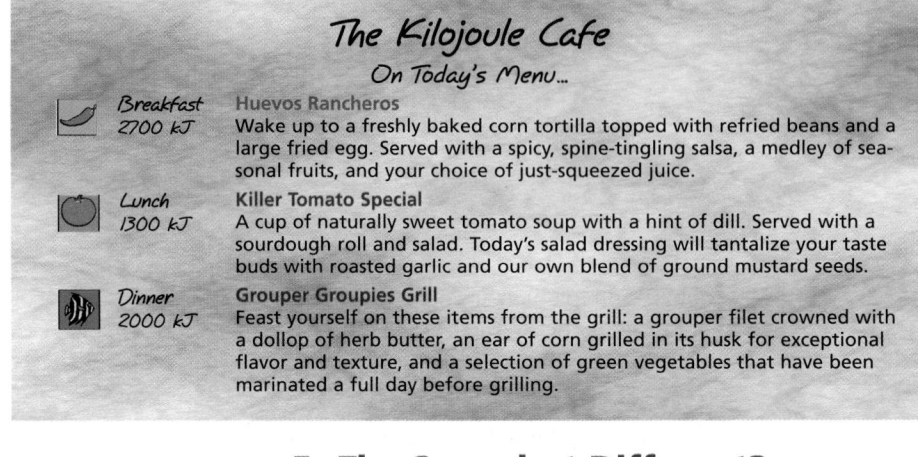

3. Shipwrecked!

How long you survive in cold water depends on how slowly or how quickly your body loses heat. Find out how much heat energy your hand loses by dipping it into a jug of cold water. Is there a temperature increase in the water? What would happen if the starting temperature of the water were higher or lower? Design some experiments to test different variables. What else might make a difference in the amount of heat gained by the water?

4. Counting Kilojoules

A member of your family has decided to go on a diet. Plan a day's menu that reduces this person's energy intake to 6000 kJ.

The Kilojoule Cafe

On Today's Menu...

Breakfast 2700 kJ — **Huevos Rancheros**
Wake up to a freshly baked corn tortilla topped with refried beans and a large fried egg. Served with a spicy, spine-tingling salsa, a medley of seasonal fruits, and your choice of just-squeezed juice.

Lunch 1300 kJ — **Killer Tomato Special**
A cup of naturally sweet tomato soup with a hint of dill. Served with a sourdough roll and salad. Today's salad dressing will tantalize your taste buds with roasted garlic and our own blend of ground mustard seeds.

Dinner 2000 kJ — **Grouper Groupies Grill**
Feast yourself on these items from the grill: a grouper filet crowned with a dollop of herb butter, an ear of corn grilled in its husk for exceptional flavor and texture, and a selection of green vegetables that have been marinated a full day before grilling.

5. The Same but Different?

This beaker of boiling water contains a 25 g bolt and a 2 g nail. Is the temperature of each object the same? Which object has more heat energy? Explain.

ScienceLog

Review your responses to the ScienceLog questions on page 405. Then revise your original ideas so that they reflect what you've learned.

418

Integrating the Sciences

Earth and Physical Sciences

Two students, A and B, are discussing the issue of whether global warming would cause a rise in the level of the world's oceans. You may wish to read the following excerpt to stimulate a class discussion on the role of heat:

A: Heat would melt the polar icecaps, creating more water.

B: The average temperature near the poles (–8°C in summer months) is well below the melting point. A rise in temperature would not melt much ice.

A: Glacial ice is constantly flowing into the water and breaking off to form icebergs. This process is often halted by the freezing of the surrounding ocean, but warmer waters would melt this ice barrier, and more glacial ice would flow into the oceans.

B: Polar icecaps receive very little precipitation. If the air were warmer, more precipitation would fall, replacing the ice lost to the oceans.

A: When water heats up, it expands, raising water levels.

B: But water contracts as it warms from 0°C to 4°C. That would lower water levels.

CHAPTER

23 Heat on the Move

Do you think heat can **1**
move through solids?
through liquids? through
gases? Use examples
from this photograph to
explain your answers.

2 Explain the following everyday facts:

- A bathroom rug feels warmer to your feet
  than a tile floor.
- If you put your hand above a light bulb, it
  feels hotter than at the side of the bulb.
- The windows of a closed car feel cool, but
  the car's interior is warm.

3 Water at the top of this test tube can
boil even though there is ice at the
bottom of the tube. How is this possible?

Boiling
water

Ice

ScienceLog

Think about these questions for a moment,
and answer them in your ScienceLog.
When you've finished this chapter, you'll
have the opportunity to revise your
answers based on what you've learned.

419

Connecting to Other Chapters

> **Chapter 21**
> allows students to estimate the
> temperature of objects and demon-
> strates how a thermometer works.

> **Chapter 22**
> introduces students to the difference
> between heat and temperature and
> discusses heat as a type of energy.

> **Chapter 23**
> introduces students to conduction,
> convection, and radiation as the three
> mechanisms for transferring heat.

Prior Knowledge and Misconceptions

Your students' responses to the
ScienceLog questions on this page will
reveal the kind of information—and
misinformation—they bring to this chapter. Use what you find out about your
students' knowledge to choose which
chapter concepts and activities to
emphasize in your teaching.

In addition to having students
answer the questions on this page, you
may wish to have them complete the
following activity: Have each student
think of a situation in which heat moves
from one place to another and create a
riddle describing the process and the
way the heat is transported. You may
wish to have volunteers read their riddle
to the class, and have the class guess
what is being described. Emphasize that
there are no right or wrong answers in
this exercise. Collect the riddles, but do
not grade them. Instead, read them to
find out what students know about heat
transfer, what misconceptions they may
have, and what about this topic is interesting to them. Tell students that in this
chapter they will focus on the three
ways that heat is transferred from place
to place.

SAFETY ALERT Caution students not to try
the experiment described in
question 3 at home.

Meeting Individual Needs

Gifted Learners

Explain to students that all objects consist of atoms or molecules that are in
constant motion. The temperature of an
object depends on how rapidly its atoms
and molecules move. For example, if the
atoms move quickly, the object feels
hot. Challenge students to do some
research on the molecular basis of heat,
also known as kinetic theory. Encourage
them to share what they learn by making diagrams or models to use in an oral
report to the class.

FOCUS

Getting Started

Show students a pot with a metal handle. Then show students a pot with a plastic handle. Point out that only the handle is plastic; the rest of the pot is metal. Ask: What is the advantage of a plastic handle? *(Most students will know that plastic handles do not get as hot as metal ones.)* Why don't plastic handles get as hot as metal handles? *(Heat cannot travel easily through plastic.)*

Main Ideas

1. Heat always flows from a hotter place to a cooler place.
2. *Conduction* occurs when heat flows through a substance or between substances that are in contact with one another.
3. *Convection* occurs when heat is transported as a result of the movement of the heated material.
4. *Radiation* occurs when heat travels through empty space or through a transparent material, without heating the space or the material between the heat source and the heated object.

TEACHING STRATEGIES

Giving Heat a Helping Hand

After students have read page 420, involve them in a discussion of the examples of heat moving from one place to another. Encourage students to offer examples of their own. Help students reach the conclusion that people use the movement of heat in many different ways. Students should understand that it is heat, not cold, that is able to move from place to place.

Giving Heat a Helping Hand

You experience heat moving from one place to another dozens of times every day. Sometimes you want to help it on its way. Sometimes you want to slow it down.

In the kitchen, for instance, you want heat to move from the stove to the food in the pot, but you do not want heat to move from the pot through the handle to your hands.

In winter, you want heat from the radiator or heater to move throughout the room. However, you do not want to lose heat to the outside through the windows and walls.

On a sunny winter day, the sun's rays heat the rooms through the windows. At night, you close the drapes to keep in heat.

When you go outside on a cold day, you may bundle up in a coat, a hat, and gloves to help retain body heat. In the summer, you wear light-colored clothes to reflect the sun's rays and to help keep yourself cool.

420

LESSON 1 ORGANIZER

Time Required
1 class period

Process Skills
observing, measuring, comparing

Theme Connection
Structures

New Terms
Conduction—heat transfer by direct contact between heated particles
Convection—the transfer of heat in a liquid or gas as groups of heated particles move from one region to another

Radiation—heat transfer through empty space or through a transparent material, without heating the material that it has passed through

Materials (per student group)
Exploration 1, Experiment 1: 500 mL of water; 600 mL beaker; hot plate; metal spoon with a long handle; wooden spoon with a long handle; watch or clock with second hand; safety goggles; **Experiment 2:** support stand with ring clamp; wire gauze; 10 cm of masking tape; 2 alcohol ther-
continued ▶

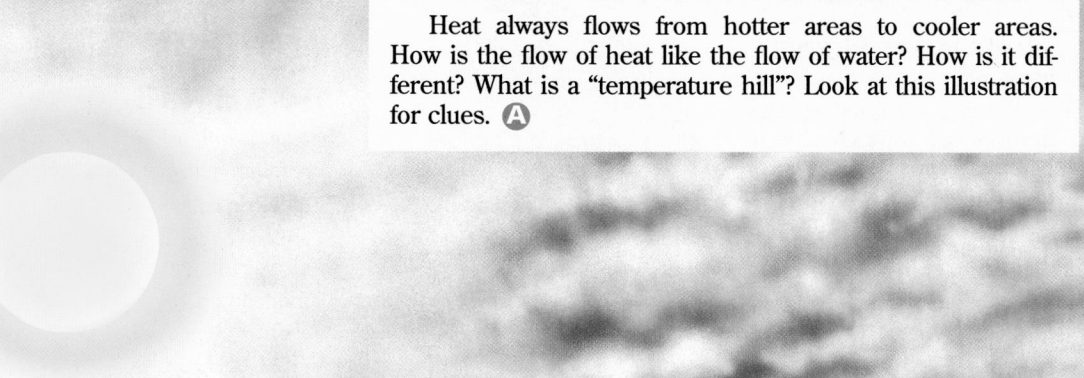

Heat always flows from hotter areas to cooler areas. How is the flow of heat like the flow of water? How is it different? What is a "temperature hill"? Look at this illustration for clues. **Ⓐ**

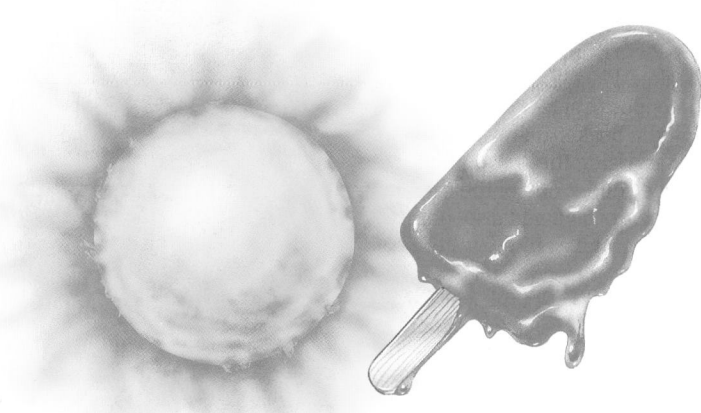

Heat can travel through solids, liquids, and gases. The way it travels through solids is generally quite different from the way it travels through liquids or gases. Heat cannot travel through empty space. However, heat sources give off *radiant energy*, such as light, that can travel through empty space. As objects absorb radiant energy, they become hotter. This is how we get heat from the sun.

⬆ Which way does the heat travel? **Ⓑ**

421

Ⓐ Both heat and water can flow from areas of high concentration to areas of low concentration. Heat, unlike water, can flow through solids.

A "temperature hill" is an area of high temperature. (An area of low temperature could be called a *temperature valley.* Heat will flow from a temperature hill into a temperature valley, similar to the way water will flow down a hill into a valley. These temperature hills and valleys are sometimes called *sources* and *sinks,* respectively.)

Ⓑ Heat travels as radiant energy from the sun to the popsicle.

Multicultural Extension

What Do You Use for Heat?
Point out that in different parts of the world, different types of energy are used to heat people's homes. Many countries in Europe rely on nuclear energy to produce electricity for heat. Iceland relies primarily on geothermal energy because Iceland is a volcanically active island. The energy needs of some mountainous countries, such as Austria and Switzerland, are provided for by hydroelectric power. Much of China relies on coal. Some countries in Africa regularly burn wood or grass for heat. Identify some heating sources that are used in the United States. Discuss with students how the types of energy used by people are closely related to the resources that are available to them.

ORGANIZER, *continued*

mometers; 600 mL beaker; 300 mL of water; striker; portable burner; 5 mL (1 teaspoon) of instant coffee granules; safety goggles; **Experiment 3:** lamp; 100 W light bulb; piece of glass, about 20 cm × 20 cm; metric ruler; piece of cardboard, about 20 cm × 20 cm; 2 sticks of modeling clay; safety goggles

Teaching Resources
Exploration Worksheet, p. 35

Theme Connection

Structures
Focus question: How does the structure of solids, liquids, and gases affect heat transfer? *(The particles that make up a solid are tightly packed together. Therefore, heat energy in a solid is easily transferred from one particle to the next as the particles collide and take on heat energy. In liquids and gases, particles are less tightly packed together and expand more easily when they are heated. Also, the heated particles become lighter and move upward, thus becoming even more sparsely distributed as they heat up.)*

You may wish to have students perform the activities individually or in small groups at workstations.

EXPERIMENT 1

SAFETY ALERT Caution students to be extremely careful around boiling water. After the water has reached boiling, students should turn off the hot plate and insert the spoons into the boiling water slowly and carefully so that they do not splash hot water on themselves or others. To prevent any accidental burns from the hot metal spoon, caution students to touch the top of the spoon lightly at first.

Students should observe that the metal spoon has become hot, while the wooden spoon has remained about the same temperature. They should conclude that some materials conduct heat well, while others do not.

EXPERIMENT 2

SAFETY ALERT Remind students not to ingest anything in a laboratory situation.

The heated water moves to the cool side of the container, setting up a current. The coffee granules are swept into the current and begin to dissolve as the cooler water warms. Eventually, the temperature readings on the two thermometers should be about the same. Students should conclude that heat spreads throughout a liquid by setting up currents in the liquid itself. These are called convection currents.

EXPERIMENT 3

The light bulb should be on for several minutes before the cardboard is removed, but it should be turned off before determining the warmth of the glass. Students should observe that when the cardboard is removed, it feels warm, while the glass has remained cool. Students should be able to feel the heat through the glass. They should conclude that heat travels through the transparent material (the glass) to the cardboard without heating the transparent material through which it passed.

How Heat Flows

These three experiments will help you investigate the different ways in which heat flows. Observe how heat travels in each of the experiments. In your ScienceLog, write down the differences you see among the three ways in which heat travels.

EXPERIMENT 1

You Will Need

- a large beaker of water
- a hot plate
- a metal spoon with a long handle
- a wooden spoon with a long handle
- a watch with a second hand

What to Do

Bring the water to a boil, and then turn off the hot plate. Place both spoons in the hot water. After 30 seconds, carefully feel the tops of the spoon handles. What do you observe? How do you explain your observations?

EXPERIMENT 2

You Will Need

- a spoonful of instant coffee granules
- 2 alcohol thermometers
- masking tape
- water
- a large beaker
- a striker
- a support stand with a ring clamp
- wire gauze
- a butane burner

What to Do

Tape two thermometers to the sides of a beaker so that the bulbs are 2 cm from the bottom of the beaker. Place the thermometers on opposite sides of the beaker. Fill the beaker halfway with water. Using the striker, light a burner under one of the thermometers, and add a spoonful of coffee granules to the water near the other thermometer.

Watch the granules and the thermometer readings. What is happening? Turn off the burner. How do you explain your observations?

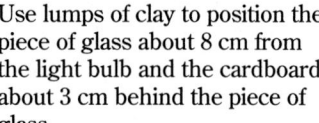

EXPERIMENT 3

You Will Need

- a lamp with a 100 W light bulb
- a square piece of glass
- a metric ruler
- a piece of cardboard the same size as the glass
- modeling clay

What to Do

Use lumps of clay to position the piece of glass about 8 cm from the light bulb and the cardboard about 3 cm behind the piece of glass.

Do you feel the heat of the light bulb through the glass and the cardboard? Remove the cardboard. Do you feel the heat through the glass? Does the glass get hot? Do you have an explanation for your observations?

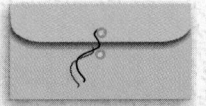 **PORTFOLIO**
Suggest that students include in their Portfolio a record of their observations from Exploration 1. Encourage students to add an analysis of what they learned about the way heat flows.

 An Exploration Worksheet is available to accompany Exploration 1 (Teaching Resources, page 35).

Multicultural Extension

Origin Myths and Folktales

In many cultures, myths and folktales have been used to explain the origin of the universe. Some of these beliefs involve intense heat or intense cold. Have students research a culture's myths and folktales about the beginning of the world and present their results to the class. Some examples include Norse mythology and the modern big-bang theory, in which the universe not only began with great heat, but will end either in a hot death or a cold death.

Which Type of Heat Transfer?

In each of the experiments on the previous page, heat traveled in a different way.

- In Experiment 1, heat traveled through a solid substance. Heat transfer between particles of a substance or substances in contact with each other is called **conduction**.

- In Experiment 2, heat traveled throughout the liquid when the liquid circulated as a result of being heated. This method of heat transfer is called **convection**. Heat can also travel through a gas by convection. Can you think of an example? Ⓐ

- In Experiment 3, heat traveled through a transparent material without heating the transparent material. This method of heat transfer is called **radiation**.

Now apply these ideas to your everyday experiences. How does food in a pot get warm? How do appliances such as an oven, a toaster, and a microwave heat food? How does your bed become warm? How is water heated for your bath or shower? Can you think of other examples of heat transfer by conduction, convection, and radiation? Ⓑ

A Problem for Mike and Sue

Here is another way to look at the three methods of heat transfer.

Mike, who is at the front of the bus, wants to pass a package to Sue, who is at the back of the bus. However, Mike is blocked by people standing in the aisle. How can Mike get the package to Sue? Ⓒ

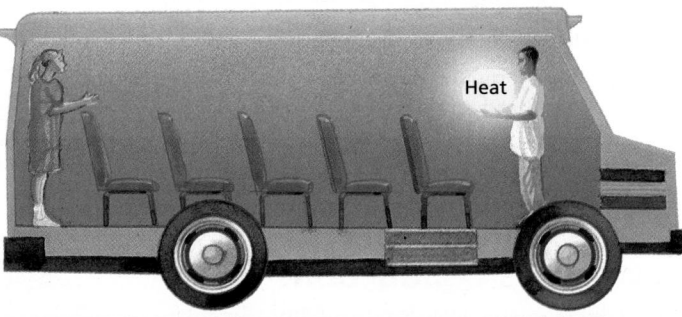

In this case, there are no passengers between Mike and Sue. How can Mike get the package to Sue? Ⓓ

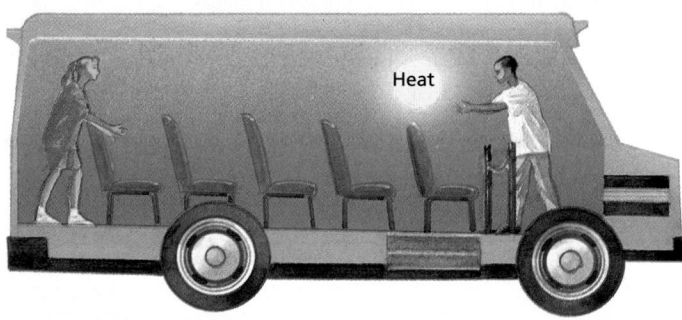

Now the aisle between Mike and Sue is roped off. How can Mike get the package to Sue? Ⓔ

Which cartoon represents heat transfer by convection? by conduction? by radiation? Ⓕ

423

LESSON 2 Conduction– Heat Transfer Through a Material

FOCUS

Getting Started

Fill a paper cup about halfway with water. Rest the cup over a portable burner until the water boils. Ask: Why is it possible to boil the water without burning the cup? *(Heat is conducted through the paper to the water so that both the cup and the water remain at about the same temperature. Since the water boils at a lower temperature than that at which the cup burns, the water boils first.)* If the cup were filled with air, it would burn. The air transports heat away from the cup, but it is not efficient enough to keep the paper from burning.

Main Ideas

1. Some materials are better conductors of heat than others.
2. The conduction rate of a material is affected by its composition and size, by the temperature of the heat source, and by the distance of the material from the heat source.
3. Increasing the thickness of an insulator increases its effectiveness.

TEACHING STRATEGIES

Watching Heat Travel Through a Solid: A Demonstration

As students observe the wax melting, encourage them to explain what they think is happening. *(Heat is being conducted along the aluminum strip from the source of heat.)*

Answers to
In-Text Questions

Ⓐ The metal spoon and small spoon get hotter faster.

Ⓑ The speed of heat transfer is affected by the type of material and its thickness.

LESSON 2 Conduction— Heat Transfer Through a Material

All matter is made of incredibly tiny particles. When conduction takes place, heat travels by transferring energy from particle to particle. Heat can also travel from one object to another by conduction. As long as the objects are touching each other, heat can travel from the particles of one object to the particles of the second object. Does this remind you of Mike and Sue's first situation? Each passenger was like a particle in a solid.

Watching Heat Travel Through a Solid: A Demonstration

1. Cut a strip of aluminum from an aluminum pie plate.
2. Pour a line of melted candle wax along the aluminum strip. Let it harden.
3. Place a candle in a jar lid, and secure it with modeling clay. Hold the aluminum strip with tongs. Heat one end of the strip with a candle. Explain what happens next.

How Fast Is Conduction?

If you dip a spoon into a cup of hot chocolate, the heat travels from the hot chocolate through the spoon to your fingers. Which gets hotter faster: a metal spoon or a plastic spoon? a large spoon or a small spoon? Ⓐ

Heat travels at different rates. What factors affect the speed of heat transfer in a solid? Ⓑ

LESSON 2 ORGANIZER

Time Required
4 class periods

Process Skills
observing, measuring, predicting

New Term
Insulator—a material that slows down the flow of heat

Materials (per student group)
Watching Heat Travel Through a Solid: A Demonstration: scissors; aluminum pie plate; candle; jar lid; small ball of modeling clay; tongs; matches; safety goggles; lab aprons

Exploration 2, Question 1: 4 candles of equal size; jar lid; small ball of modeling clay; matches; 4 thumbtacks; 4 rods of equal thickness, one each of copper, brass, heat-resistant glass, and iron; plastic-foam cup; watch or clock with second hand; safety goggles;
Question 2: 2 candles of equal size; 2 jar lids; 2 small balls of modeling clay; matches; 2 thumbtacks; 2 iron rods of different thicknesses; 2 plastic-foam cups; watch or clock with second hand; safety goggles

continued ▶

EXPLORATION 2

A Closer Look at Conduction

Investigate each of the following questions. In your ScienceLog, record your observations, and write a sentence or two describing your discovery.

QUESTION 1

Do different substances conduct heat at different rates?

You Will Need

- 4 rods of equal thicknesses: one each of copper, brass, heat-resistant glass, and iron
- 4 candles of equal size
- a jar lid
- modeling clay
- matches
- 4 thumbtacks
- a plastic-foam cup
- a watch with a second hand

What to Do

Attach a thumbtack to each rod with candle wax, and then arrange the apparatus as shown below. Make sure that each thumbtack is the same distance from the end of each rod. Place a candle in the jar lid, and secure it with modeling clay. Time how long it takes for the thumbtack to drop from each of the four rods. Use a new candle with each rod. (Why is it important to do this?) Do all substances conduct heat at the same rate?

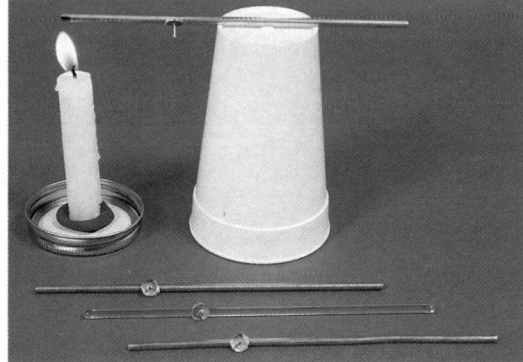

QUESTION 2

Does the thickness of an object affect the rate of conduction?

You Will Need

- 2 iron rods of different thicknesses
- matches
- 2 candles of equal size
- 2 thumbtacks
- 2 jar lids
- 2 plastic-foam cups
- modeling clay
- a watch with a second hand

What to Do

Attach one thumbtack to each rod with candle wax. Make sure the thumbtacks are the same distance from the end of each rod. Use the same setup that you used in Question 1 to test how long it takes for the thumbtacks to drop. Before you start, make a prediction. Do you think heat will travel faster along a thick rod or a thin rod? Now try it and find out.

QUESTION 3

Can you use heat conduction to make a 30-second timer?

Imagine that you are stranded on a desert island, and the battery in your watch is running out of energy. However, you have a supply of candles, thumbtacks, plastic-foam cups, and all sorts of metal rods. Design a 30-second timer. Compare your timer design with those of your classmates.

425

ORGANIZER, continued

Mysterious Events, The Mystery of the Scorched Paper: copper rod and wooden rod of equal diameters; 5 cm of masking tape; candle; matches; jar lid; small ball of modeling clay; safety goggles; **Two Terrific Tricks to Try:** tongs; 10 cm × 10 cm aluminum screen; candle; wooden matches; jar lid; small ball of modeling clay; metal fork; support stand with clamp; safety goggles; **The Mystery of the Extinguished Candle:** 40 cm of electrical wire; wire strippers; candle; matches; jar lid; small ball of modeling clay; safety goggles

A Classroom Demonstration: a few objects made of different substances, such as a glass beaker, a plastic pencil case, a tin can, and a paper notebook; alcohol thermometer and 5 cm of tape for each object
A Competition: Save the Ice Cube!: ice cubes and a variety of materials used to keep the ice from melting

Teaching Resources
Exploration Worksheet, p. 37
Activity Worksheet, p. 39
SourceBook, pp. S165 and S167

EXPLORATION 2

 Cooperative Learning
EXPLORATION 2

Group size: 3 to 4 students
Group goal: to determine what factors affect heat transfer in a solid
Positive interdependence: Assign students roles such as timer, synthesizer, director, and materials coordinator.
Individual accountability: Have each student summarize his or her group's findings in data charts.

QUESTION 1

Have students use a different candle to produce the wax for attaching the thumbtacks. Otherwise, one candle may be shorter than the others. Have students allow the wax to cool and harden before they begin. It is important to use a new candle for each rod because as a candle burns, it shortens, so the flame will be farther from the rod. Students should observe that the order of conductivity, from the fastest to the slowest, is copper, brass, iron, and glass. They should conclude that some materials conduct heat more rapidly than others.

QUESTION 2

Have students allow the wax to cool and harden before they begin. Students should observe that heat is conducted more rapidly through the thin rod than through the thick rod. They should conclude that the thickness of a material affects the rate at which the rod conducts heat.

QUESTION 3

A timer can be made by determining where to attach a thumbtack to a metal rod with wax so that the wax will melt and the thumbtack will fall in 30 seconds. Students should recognize that the thumbtack must be placed at different positions on rods made of different materials or on rods of different diameters.

 An Exploration Worksheet accompanies Exploration 2 (Teaching Resources, page 37).

1. Good conductors include copper, brass, and iron (metals). Glass (non-metal) is a poor conductor.

2. A good cooking pot should be made of a good conductor, such as copper, iron, or aluminum, so that it will quickly and easily transfer heat from the burner to the food that is being cooked. The handle of a good cooking pot, on the other hand, should be made of a poor conductor, such as plastic or wood, so that a person can use it without being burned.

3. The results of the experiment suggest that metals are good conductors of heat. Everyday products that use good conductors include cooking utensils, clothes irons, and water heaters.

Meeting Individual Needs

Learners Having Difficulty

Have students imagine the following scenario: They place a metal spoon at room temperature in one of two identical cups of hot cocoa and wait a few minutes. Which of the following best describes the result?

a. The heat from the spoon travels to the cocoa, making the cocoa with the spoon hotter.

b. The spoon does not affect the cocoa; therefore, both cups of cocoa remain the same temperature.

c. The metal spoon absorbs heat from the cocoa. As a result, the cocoa without the spoon is hotter. *(Correct)*

d. The metal spoon absorbs heat from the cocoa. As a result, the cocoa with the spoon is hotter.

Thinking About and Using Your Findings

Here are some questions to help you check your understanding. Try to answer all of them!

1. On the basis of your observations, which substances are good conductors of heat? Which substances are poor conductors of heat?

2. Apply your findings to explain how cooking pots are made. How many different types of material are used in the pots in your kitchen? What material would you use to make a pot? to make the handle?

3. Six similar rods made of different substances were placed in 70°C water. At the end of 1 minute, the top of each rod was felt. The results are shown in the table at right.

Which kind of material appears to be a good conductor of heat? There are many everyday products that use certain materials because they are good conductors of heat. Can you name five materials that you have used?

copper	hot
plastic	cold
brass	hot
rubber	cold
iron	hot
glass	cold

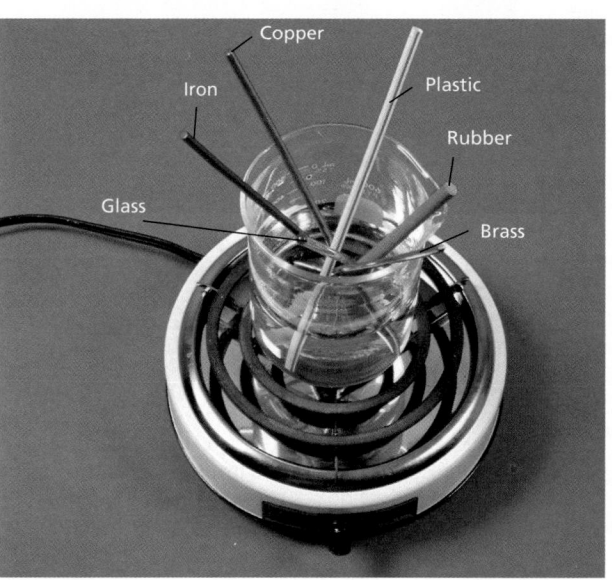

Homework

The Activity Worksheet on page 39 of the Unit 7 Teaching Resources booklet makes an excellent homework assignment once students have completed Exploration 2 on page 425.

4. A friend has just made some instant soup for you by pouring boiling water into each of the cups shown below. Which cup would you prefer to hold? Would the thickness of the cup make a difference? What other factors might make a difference? Now list the cups in order of preference.

Plastic-foam   China   Glass   Paper   Metal

5. Wool sweaters, fur, and feathers are very light yet warm. Why do you think these materials are so warm?

6. To measure conduction rates, Pat made a clever device from a large can. She placed rods of different substances, lengths, and thicknesses on a tripod made from the can so that the rods met in the center. She used some candle wax to attach a thumbtack to the outer end of each rod. Then she placed a candle under the tripod to heat the inner end of each rod equally. Pat had many difficulties, and her teacher said that the class could not rely on Pat's results. How would you improve Pat's experiment?

▲ Why do you think these lambs need plastic coats to keep warm? **Ⓐ**

Wood
Glass
Aluminum
Steel
Plastic
Copper

427

Answers to
Thinking About and Using Your Findings, continued

4. Most students will probably choose the plastic-foam cup as the one they would prefer to hold. How hot a cup is will depend partly on its thickness. Some students may recognize that the more porous the cup is, the less hot it will be. (This is because air is trapped within the pores, and air is a poor conductor of heat.) Students should realize that poor conductors make the best containers for hot liquids. The order for the cups, from coolest to hottest (least conductive to most conductive), is plastic-foam, china, glass, paper, and metal.

5. Air is a poor conductor of heat. Small air pockets in wool, fur, and feathers help to keep heat from passing through these materials. As a result, body heat becomes trapped between a person's body and a garment made from wool, fur, or feathers.

6. Students should recognize that the main problem with Pat's experiment is that more than one variable is being tested. If different substances are being tested, then the rods should all be the same thickness and length.

Answer to
Caption

Ⓐ Help students to recognize that the lambs' coats are very thin. There is a lot of space between the strands of wool, allowing the lambs' body heat to create convection currents and escape into the air. As a result, the lambs' body heat is carried away. The plastic coats help keep the heat from escaping.

Mysterious Events

The seemingly mysterious results of these activities provide students with an interesting way to apply their knowledge of conductivity in order to explain what they are observing.

Cooperative Learning
MYSTERIOUS EVENTS

Group size: 2 to 3 students
Group goal: to apply their knowledge of conductivity
Positive interdependence: Assign students roles such as leader, materials and safety manager, and recorder.
Individual accountability: Have students choose one mysterious event and explain how an understanding of conductivity helps solve the mystery.

Answers to
The Mystery of the Scorched Paper

Students should observe that the paper next to the wood scorches first. They should conclude that this happens because the copper conducts the heat away from the heat source more rapidly than the wood does.

Answers to
Two Terrific Tricks to Try

Students should observe that when the screen is placed over the flame, it pushes the flame down. The flame does not burn above the screen. They should conclude that the heat from the flame is being conducted over the surface of the screen.

It is almost impossible to get the entire match to burn because the prongs of the fork conduct heat away from the flame. However, if you heat the fork first, the match can burn completely.

Mysterious Events

The Mystery of the Scorched Paper

Join a copper rod or pipe and a wooden rod of the same diameter by wrapping one layer of masking tape around them. Place a candle in a jar lid, and secure the candle with modeling clay. Hold the ends of both rods, and heat the joint very slowly over the candle until the tape is scorched. Do not allow the tape to burn. Which side of the tape is scorched? Explain.

Two Terrific Tricks to Try

Place a candle in a jar lid, and secure the candle with modeling clay. Use tongs to lower a piece of aluminum screen slowly over the flame of the candle. What happens? Why?

Place a fork in a clamp attached to a support stand. Place a wooden match between two prongs of the fork. Light the match. Can you get the whole match to burn? Try other matches, and place them in different positions between the prongs. After observing what happens, can you explain the trick?

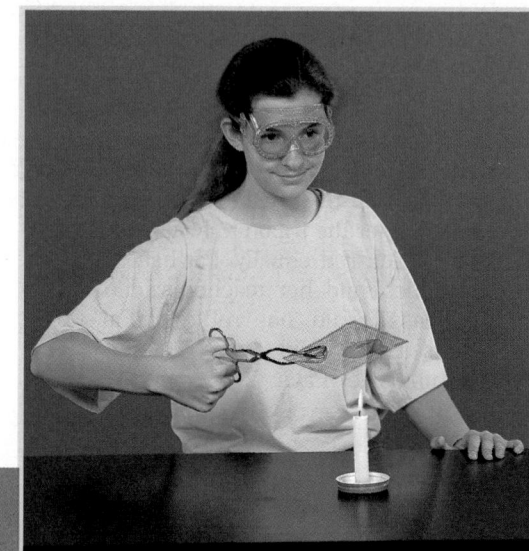

The Mystery of the Extinguished Candle

Strip about 30 cm of insulation from about 40 cm of electrical wire. Make a closely spiraling coil from the end of the wire without insulation, and hold onto the insulated part. Place a candle in a jar lid, and secure the candle with modeling clay. Place the coil over the flame of the candle. What happens? Light the candle again. This time, heat the coil above the flame until the coil glows. Now slowly lower the coil over the flame. Explain the results in your ScienceLog.

A Classroom Demonstration

Select a few objects made of different substances, and tape a thermometer to each object.

Leave the objects in the classroom overnight so that they all reach the same temperature. Here are some examples of different substances to use:

- glass (beaker)
- ordinary plastic (pencil case)
- metal (tin can)
- paper (notebook)

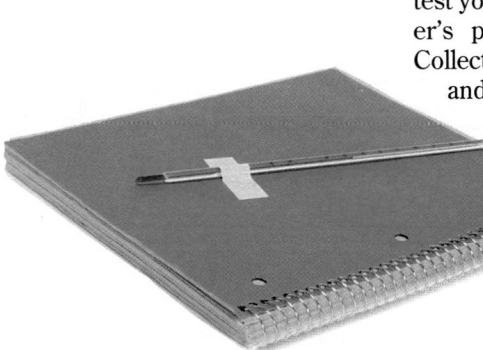

Now hold each object against your cheek. Do some feel warmer than others? Why do some of the objects feel warmer even though the thermometer shows that they are all at the same temperature?

Explanation Please!

How do you explain the results of the demonstration? What do you think causes this effect?

Design an experiment to test your ideas. Get your teacher's permission in advance. Collect the materials you need, and do the experiment. Record the results in your ScienceLog. Good luck!

Explanation Please!

The following is one possible way of testing the hypothesis: Assemble a variety of objects, such as a metal cup, a piece of wood, and a piece of glass. Leave all objects out in the room for several minutes. Check their temperatures with an alcohol thermometer until all have the same temperature. Preheat an oven to 50°C. Place the objects in the oven, and note how long it takes for each object to reach 50°C. The objects that reach 50°C in the least amount of time are the best conductors of heat. Be sure to approve all proposed experiments for safety before students begin.

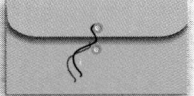

PORTFOLIO

Encourage students to record their responses to Explanation Please! in their Portfolio.

Homework

You may wish to assign the questions from Explanation Please! as a homework activity. Be sure to check all designs for safety before allowing students to carry out their experiments.

Ⓐ Student responses might include rubber, plastic, or wood. Other examples of insulators, in increasing order of effectiveness, are glass, porcelain, epoxy, and empty space.

A Competition: Save the Ice Cube!

If students perform this activity at home, have them write a report on their procedure and their results. Invite several students to share their reports with the class. Then involve all class members in a discussion of the procedures and results.

If you wish to hold the competition in the classroom, you may want to make the following materials available: ice cubes, newspaper, paper towels, cotton and wool cloth, plastic-foam, and containers made of different materials (metal, glass, plastic-foam). To make this a fair test, use ice cubes that are about the same mass. As students perform their experiments, encourage them to keep careful notes on their procedures and their results.

Answer to
Caption

Exploration 3, Activity 1, on page 431 is a good model because it shows how water circulates by convection. Water in lakes and oceans also circulates by convection. (Students may note that this depends on seasonal variations in water and air temperatures. During warm seasons, the water near the surface is warmer than the water below, and convection does not occur.)

Slowing the Flow of Heat

You do not always want to improve the flow of heat. Sometimes you want to slow it down. For example, you want to slow down the flow of heat from your house to the outside in the winter. In the summer, the reverse is true. You would also want to keep heat from flowing too quickly from a hot pan to your hand when you pick it up. Materials that slow down the flow of heat are called *insulators*. Think of as many insulators as you can. Ⓐ

A Competition: Save the Ice Cube!

Who can prevent an ice cube from melting completely for the longest time? Invent a way to keep an ordinary ice cube frozen for as long as possible (without appliances, of course). Use any materials you wish. Write up your plan. Record the size of your ice cube, the materials you used, the methods you used, and the length of time it took for the ice cube to melt completely.

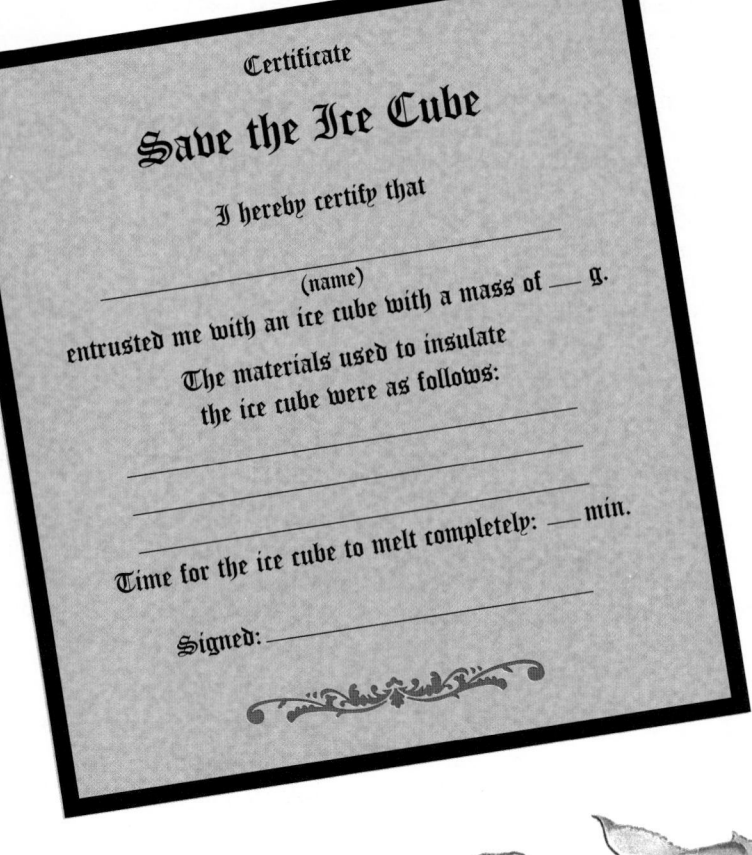

Certificate

Save the Ice Cube

I hereby certify that

_____
(name)

entrusted me with an ice cube with a mass of ___ g.

The materials used to insulate the ice cube were as follows:

_____

_____

_____

Time for the ice cube to melt completely: ___ min.

_____

Signed: _____

The next Exploration provides a good model for movement of water in lakes and oceans. Do some research to find out why it's a good model.

FOLLOW-UP

Reteaching

Have students make a survey of common materials to find out which are good conductors of heat and which are poor conductors. Have them share what they discover by organizing their information into a chart.

Assessment

Ask students to explain the following situation based on their knowledge of heat: Hot-water pipes are surrounded by fiberglass wool. Then they are wrapped with an aluminum-foil cover. The aluminum foil feels cool to the touch. *(The fiberglass wrapping keeps in the heat. The aluminum foil feels cool because it conducts heat away from your hand.)*

Extension

Have students research Benjamin Franklin's invention of the iron stove (the Franklin stove) in the 1740s. Encourage students to explain how the stove uses conduction and convection to heat a room.

Closure

Have students imagine that they will be going on a hiking trip high in the mountains. The weather will be cold. Have students make a list of things they will take to protect themselves from the weather. Next to each item, ask them to write a brief explanation of how it will slow the flow of heat away from the body, thus helping to keep the body warm.

Convection— Heat Transfer by Currents

You may recall from page 423 that heat transfer in fluids (either liquids or gases) occurs largely by convection. In the Exploration that follows, you will discover more about convection.

EXPLORATION 3

Exploring Convection

Form a small group and choose one of the following Activities. Practice it, figure out how it works, and then present it to the other groups. Let the other groups ask questions about your Activity until they understand how it works.

ACTIVITY 1

You Will Need

- a large jar of water at room temperature
- a small bottle filled with warm, colored water
- a two-hole stopper with two glass tubes inserted—one tube should almost touch the bottom of the bottle and the other should poke just inside the bottle

Caution: Your teacher should be the one who inserts the tubes into the stopper and the stopper into the small bottle. Be careful not to scald yourself. Water that is above 55°C will scald you.

What to Do

Carefully lower the small bottle into the room-temperature water while keeping your fingers over the ends of the glass tubes. Remove your fingers, and trace the direction of any currents formed. What do you think causes these currents?

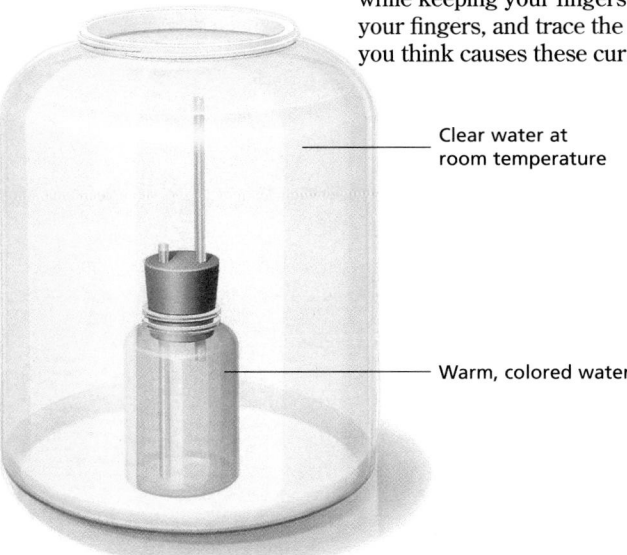

Clear water at room temperature

Warm, colored water

Exploration 3 continued ▶

431

LESSON 3 ORGANIZER

Time Required
2 to 3 class periods

Process Skills
observing, classifying, predicting

New Terms
none

Materials (per student group)
Exploration 3, Activity 1: large (tall) jar; 1 L of water; small bottle filled with warm water; a few drops of food coloring; two-hole stopper with 2 glass tubes inserted; safety goggles; lab aprons (See Advance Preparation on page 387D.); **Activity 2:** birthday

candle; 40 cm of wire; matches; 1 L glass bottle; T-shaped piece of thin cardboard covered with aluminum foil; safety goggles; **Activity 3:** aluminum pie plate; thumbtack; scissors; metric ruler; 50 cm of thread; candle; jar lid; small ball of modeling clay; matches; lamp with a light bulb; safety goggles; **Activity 4:** 1 L of water; several drops of food coloring; ice cube tray; 20 cm of wire; two 250 mL beakers; safety goggles; lab aprons; **Activity 5:** cardboard box, at least 30 cm × 30 cm × 30 cm; candle; jar lid; small ball of

continued ▶

FOCUS

Getting Started

To demonstrate how convection currents occur when liquids are heated, perform this demonstration for students. Fill a heat-resistant beaker with water and add some paper dots from a hole punch. Gently heat the beaker over a burner. Ask students to explain what they see. *(The paper dots are carried by convection currents that occur because the warm water rises and the cooler water sinks.)*

Main Ideas

1. Hot or warm air tends to rise, and cold or cool air tends to fall.
2. In a liquid or gas system, convection currents will occur if the temperature is colder at the top of the system than at the bottom of the system.

TEACHING STRATEGIES

GUIDED PRACTICE Point out that this lesson explores how heat travels by convection. Ask: Through what kinds of substances does heat travel by convection? *(Liquids and gases)* Why doesn't convection occur in solids? *(Convection depends on the motion of the material. In a solid, the material is not free to move as it is in a gas or liquid.)* Call on a volunteer to explain how convection works. *(Convection occurs when heat travels as a result of the movement of the material being heated.)*

EXPLORATION 3

Set up one workstation for each of the Activities in the Exploration. Assign one Activity to each group of students. Have each group demonstrate their Activity to the class.

Exploration 3 continued ▶

 An Exploration Worksheet accompanies Exploration 3 (Teaching Resources, page 40).

EXPLORATION 3, *continued*

Cooperative Learning
EXPLORATION 3

Group size: 3 to 4 students
Group goal: to explore convection
Positive interdependence: Assign students roles such as leader, materials and safety manager, timer and pacer, and recorder and presenter.
Individual accountability: Encourage each student to summarize how their group's experiment helped them to understand convection.

ACTIVITY 1   page 431

Students should observe that a current of colored water flows out of the taller tube as clear water enters the shorter tube.

Sample conclusion: The hot liquid in the bottle rises out of the taller tube; the cooler liquid in the jar flows into the shorter tube.

ACTIVITY 2

Students should observe that the burning candle goes out when the foil-covered cardboard is not in the neck of the bottle; it stays lit when the foil-covered cardboard is in place.

Sample conclusion: Without the foil, the heated air in the bottle rises quickly, taking the remaining air with it. With the foil in place, the hot air escapes past one side of the foil while colder fresh air flows in on the other side.

ACTIVITY 3

Students should observe that if the spiral is cut counterclockwise from the outside in, it should turn clockwise in rising air currents and counterclockwise in falling air currents. If the spiral is cut clockwise from the outside in, the results should be reversed.

Sample conclusion: The air pushing on the slanted flat surfaces of the spiral gives it a sideways push, causing it to rotate.

EXPLORATION 3, *continued*

ACTIVITY 2

You Will Need
- a 1 L bottle
- a birthday candle
- wire
- a T-shaped piece of thin cardboard
- aluminum foil
- matches

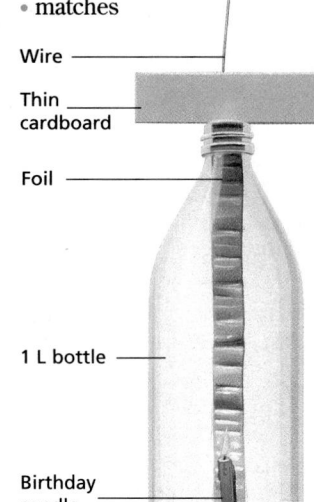

Wire —
Thin cardboard —
Foil —
1 L bottle —
Birthday candle —

What to Do

Attach the candle to the wire. Light the candle, and lower it into the bottle. How long does the candle burn? After the candle goes out, allow fresh air to enter the bottle. Try again, but this time attach the wire with the candle to the T-shaped piece of cardboard covered with foil. Light the candle, and lower it along with the foil into the bottle. Does the candle stay lit this time? Explain your results.

432

ACTIVITY 3

You Will Need
- an aluminum pie plate
- thread
- scissors
- a thumbtack
- a candle
- a jar lid
- modeling clay
- matches
- lamp with a light bulb

What to Do

With a thumbtack, make a pinhole for the thread in the center of the pie plate. Cut out a circle 15 cm in diameter from the bottom of the pie plate. Then cut the circle into a spiral, and hang it from a piece of thread.

Place the spiral in several locations: over a candle (not too close!), next to the candle, above a light bulb, over a radiator, near the bottom of a window, and in different places in the classroom or schoolyard. Before using the candle, place it in a jar lid, and secure the candle with modeling clay.

Record which direction the spiral turns (clockwise or counterclockwise) when it is put in various places. Can you explain the direction of the spiral's movement for each location?

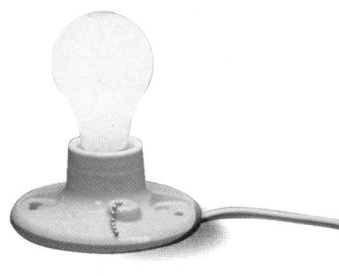

ORGANIZER, *continued*

modeling clay; matches; scissors; 40 cm of clear plastic wrap; wooden splint; masking tape; 2 corks; metric ruler; safety goggles; **Activity 6:** 2 support stands with ring clamps; 20 cm of wire; 2 pieces of wire gauze; 2 ice cubes; 500 mL of water; two 250 mL beakers; 2 portable burners; striker; safety goggles; **Activity 7:** 2 beverage cans; support stand with ring clamp; wooden dowel; 50 cm of string; binder clip;

portable burner; striker; safety goggles; **Activity 8:** candle; jar lid; small ball of modeling clay; matches; tongs; metric ruler; wooden splint; safety goggles

Teaching Resources
Exploration Worksheet, p. 40
Transparency 74
SourceBook, p. S165

You Will Need

- 2 beakers
- an ice cube tray
- food coloring
- wire
- water

What to Do

Make a tray of ice cubes using water that has been dyed very dark with food coloring. Wrap wire around one of the ice cubes so that it will sink. Place the wrapped ice cube and an unwrapped ice cube into separate beakers of clear water at the same time. As the ice cubes melt, observe how the color spreads in both of the beakers. Draw diagrams of the currents that develop.

ACTIVITY 5

You Will Need

- a cardboard box
- a candle
- scissors
- a jar lid
- modeling clay
- clear plastic wrap
- a metric ruler
- a wooden splint
- masking tape
- matches
- 2 corks

What to Do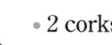

Arrange the apparatus as shown. Observe the candle flame and the direction of smoke for different pairs of open holes. Explain what is happening in each case.

Caution: The cardboard box is a fire hazard. Make sure there is at least 15 cm between the top of the candle flame and the top of the box and 10 cm between the flame and the sides of the box.

ACTIVITY 6

You Will Need

- 2 ice cubes
- 2 support stands with ring clamps
- wire
- 2 beakers of water
- 2 pieces of wire gauze
- 2 butane burners
- a striker

What to Do

Wrap some wire around one ice cube to make it sink. Place the wrapped ice cube and the unwrapped ice cube into separate beakers of water at the same time. Place the beakers on the support stands, and heat them with a burner. Predict which ice cube will melt first. How do you explain what happens?

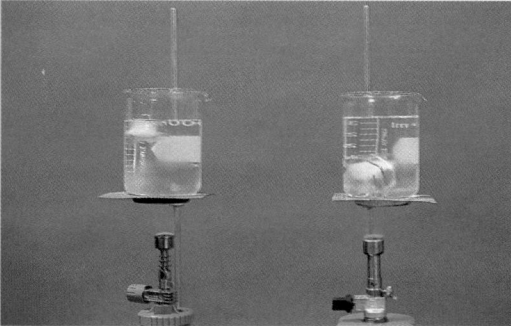

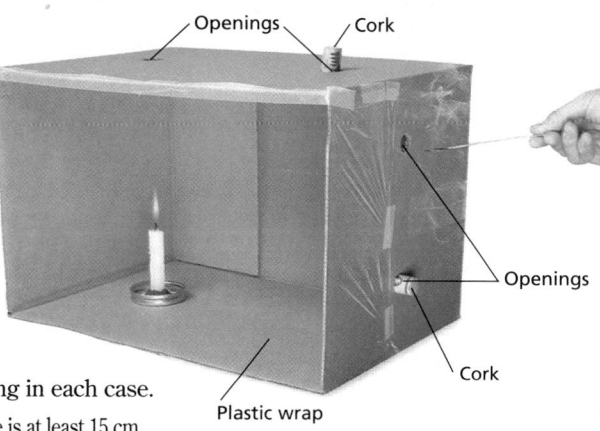

Openings — Cork

Openings

Cork

Plastic wrap

Exploration 3 continued ▶

433

Students should observe that the water in the beaker with the floating ice cube becomes colored throughout.

Sample conclusion: This occurs because the cold water from the melting ice cube sinks and mixes with the water throughout the beaker.

Students should observe that the water in the beaker with the sunken ice cube becomes colored only at the bottom of the beaker.

Sample conclusion: This occurs because the cold water from the melting ice is already at the bottom of the beaker and therefore does not mix with the rest of the water.

ACTIVITY 5

Students should light the wooden splint so that it gives off smoke and can be used as a convection-current indicator. Encourage them to try different combinations of opened and closed holes.

Students should observe that the smoke escapes out of the opening above the candle. If that opening is corked, the smoke escapes out the top opening on the side of the box.

Sample conclusion: The hot air above the candle rises and escapes from the box, allowing cold air to take its place.

ACTIVITY 6

You may wish to have students write their prediction on a slip of paper or in their ScienceLog. Then have them complete the Activity.

Students should observe that the floating ice cube melts first.

Sample conclusion: This happens because the hot water from the bottom of the beaker rises. Ask students to reveal their prediction (and the reasoning behind their prediction) to the class.

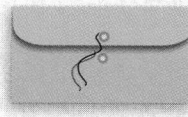 **PORTFOLIO**
After performing Activity 6, encourage students to compare their prediction with their actual results. Suggest that students include an analysis in their Portfolio.

Homework

Once students have completed Activity 6, have them prepare a one-page analysis of their results.

Exploration 3 continued ▶

EXPLORATION 3, continued

ACTIVITY 7

The cans may be secured to the dowel by tying the string to the pull tabs of the cans. Use a binder clip and some string to attach the dowel to the ring clamp.

Students should observe that as the air in one can becomes heated, the equilibrium of the balance is upset, and the heated can rises. After the air has had time to return to its original temperature, the balance returns to its original position. The same effect will take place when the air in the other can is heated.

Sample conclusion: Hot air is lighter than cold air.

ACTIVITY 8

Students should observe that the air is hottest directly above the flame. Students should also observe that matches can be held closest to the bottom of the flame without catching fire.

Sample conclusion: Cold air is pulled in and around the bottom of the flame as the air that is already there becomes heated and rises.

Sample diagram:

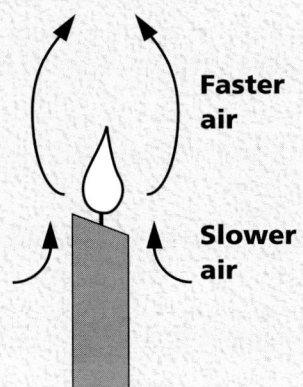

Faster air

Slower air

ACTIVITY 7

You Will Need

- 2 soda cans
- a support stand with a ring clamp
- a wooden dowel
- string
- a binder clip
- a butane burner
- a striker

What to Do

Make a simple balance like the one shown on the right. Make sure that your cans are identical and are empty. Hang each can from the dowel, and slide the center string back and forth on the dowel until the cans are balanced.

With a butane burner, heat the air in one of the cans. What happens? Remove the burner, and do not disturb the balance for a few moments. What happens? Now heat the air in the other can. Explain your results.

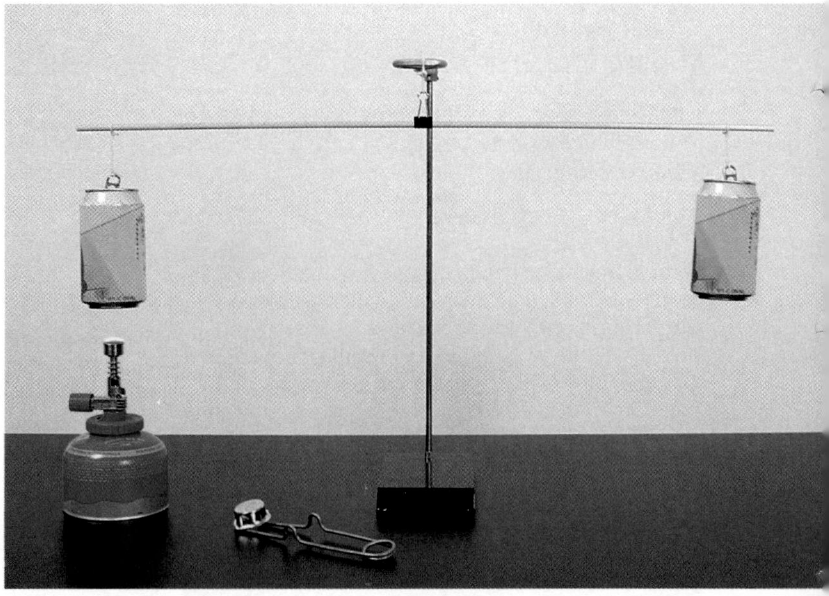

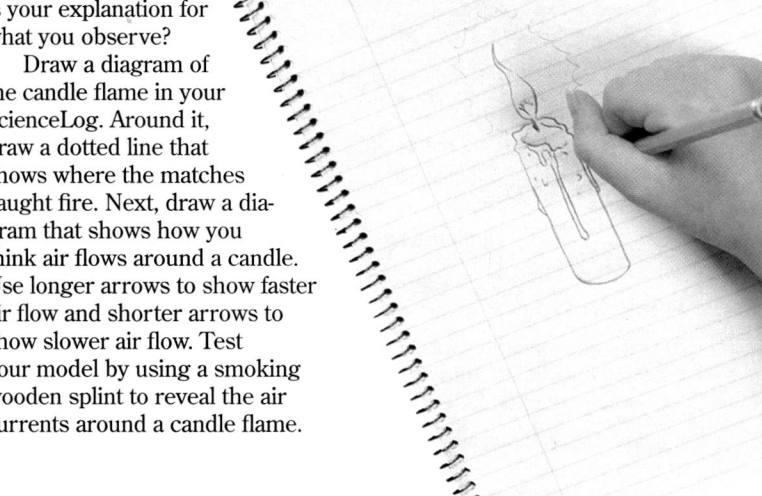

ACTIVITY 8

You Will Need

- a candle
- a metric ruler
- a wooden splint
- modeling clay
- matches
- tongs
- a jar lid

What to Do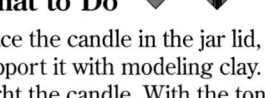

Place the candle in the jar lid, and support it with modeling clay. Light the candle. With the tongs, hold matches at different positions around the flame.

How close to the top of the flame can you put the head of a match before it catches fire? How close to the side of the flame can you put the match before it

catches on fire? What is your explanation for what you observe?

Draw a diagram of the candle flame in your ScienceLog. Around it, draw a dotted line that shows where the matches caught fire. Next, draw a diagram that shows how you think air flows around a candle. Use longer arrows to show faster air flow and shorter arrows to show slower air flow. Test your model by using a smoking wooden splint to reveal the air currents around a candle flame.

434

Did You Know...

In many cases, convection is a more efficient mechanism for transporting heat than is either conduction or radiation. It is the difference in temperature from one point to another that determines whether convection will occur.

Analysis Please!

Hot air rises. Hot liquids rise too. This is why they circulate when heated. You can't always see it happening, however. Here are some problems to help you review what you observed in Exploration 3.

1. Have you ever walked on sand on a sunny day at the beach? If so, then you know what a relief it is to reach the water. The water is much cooler than the sand. This is because the water is capable of absorbing much more heat than the sand without an increase in temperature. Just as the land heats up quickly, it also loses heat quickly when the sun goes down. Again, the temperature of the water doesn't change very much. Copy the diagrams at right into your Science-Log. Predict the air currents in each scene. How might the pattern change from day to night?

2. Carlo the Magnificent explained to his audience that all four bottles contained water, two of them with blue food coloring added. He then pulled the cards from between the bottles—*abracadabra!* How did he arrange this trick? What else could he have done to get the same result?

Before

After

To learn about the relationship between convection currents and the weather, turn to page S168 of the SourceBook.

435

Analysis Please!

You may wish to have students write their responses to the questions in their ScienceLog or involve the class in a discussion of each question.

Answers to
Analysis Please!

1. On a sunny day, the air above the land is warmer than the air above the water. The warmer air above the land rises and cooler air from the water moves in, causing a sea breeze. At night the reverse happens. The land becomes cooler than the water. The warmer air above the water rises and the cooler air from the land moves in to take its place.

2. If the bottle of colored water is cold and the bottle below it is filled with hot water, the colored cold water will sink into the hot water, and the color will spread throughout. If the bottle of colored water is hot and the bottle below it is filled with cold water, the hot water will remain where it is, and the two will not mix.

   Accept all reasonable suggestions for ways that Carlo could have performed the trick. For instance, he could have used colored oil instead of water in the top bottle.

**Answers to Analysis Please!
continued** ▶

3. In the first house, energy rises from the boiler to radiators in each room, where the heat circulates around the room. In the second house, energy rises from the furnace through supply ducts in each room, where the heat circulates and exits through the return duct. In the third house, energy rises from the wood stove and circulates throughout the house through vents.

Sketches will vary, but students should recognize that all three houses use conduction, radiation, and convection.

Answers will vary, but students should demonstrate an understanding of the different types of energy flow.

4. An air conditioner should be installed near the ceiling so that the cold air it generates will sink and spread throughout the room.

5. When a vertical freezer is opened, a large amount of cold air escapes from the bottom while warm air enters at the top. Most of the cold air in a horizontal freezer remains there when the freezer is opened because the air is more dense and does not flow up and out. Therefore, a horizontal freezer is probably the better energy buy.

6. The lampshade rotates as the air heated by the bulb rises and pushes against the vanes in the top of the lampshade.

★ **Transparency 74 is available to accompany the material on this page.**

---

3. Homes can be kept warm in a variety of ways. Several are shown at right. Examine each system, and answer the following questions:

- How does each system work? Describe the flow of energy through each system.
- Make a sketch of each system in your ScienceLog. Label each place that you see conduction, convection, or radiation occurring.
- Which system would you choose if you were building a house? Why? Some things to consider include energy efficiency, amount of space needed, noise level, availability of fuel, and comfort level.

4. Would you install an air conditioner near the floor or near the ceiling? Why?

5. Which is the better energy buy: a vertical freezer or a horizontal one? Why?

6. Study how this magic lamp works. How would you explain to a young child what is happening?

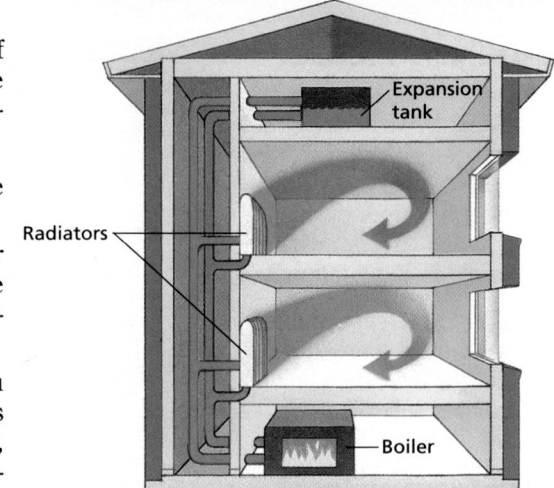

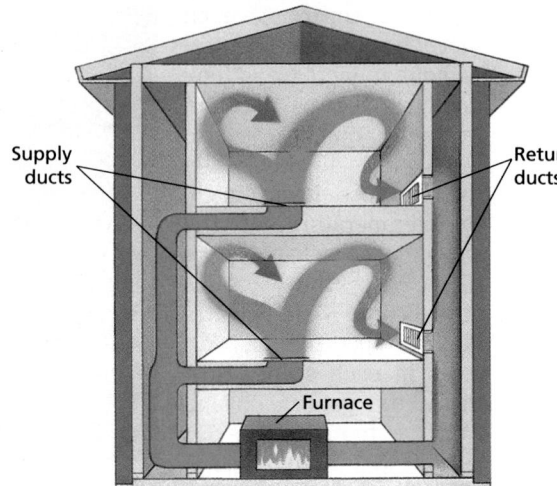

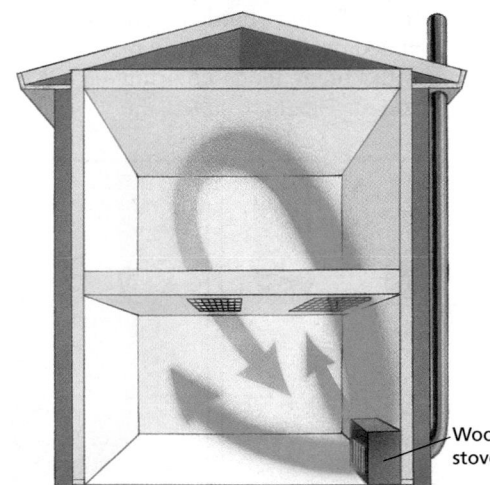

436

FOLLOW-UP

Reteaching
Have students illustrate places where convection currents occur in nature and in their life.

Assessment
Have students draw cartoons to illustrate the difference between conduction and convection. *(Cartoons should demonstrate that conduction is heat transfer by direct contact from one particle to another, and convection is heat transfer by the circulation of heated gas or liquid particles.)*

Extension
Point out to students that when there is a fire in a fireplace, most of the warm air goes up the chimney. Challenge them to improve the design of a fireplace so that more of the heat is used to warm the room.

Closure
Suggest to students that they make a diagram to show how a ceiling fan distributes heat in a room.

Radiation—Heat in a Hurry

Your Chance to Make Headlines!

Newspaper editors know that a good headline should be interesting but also brief and to the point. Pretend that you are a newspaper editor. Read this article carefully. Then think of a good headline for each paragraph. The article is about a very important example of radiation.

Energy From a Star
(adapted from *Whence Energy*
by Dr. Roy L. Bishop)

1. _____?_____

We live on a blue and white planet—a small, round, turquoise gem nestled in the black velvet of space. The energy that keeps our planet alive was either present when the Earth was formed or has been received from space since that time.

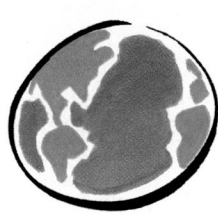

2. _____?_____

Energy reaches the Earth from space in several ways. Meteors streak down through the atmosphere, burning up with great flashes of fire. Cosmic rays bring packets of energy that have traveled through the universe. At night, we can see the stars sending their light energy to us.

3. _____?_____

Most of the energy we receive from space is contributed by the nearest star, the sun. Because it is so close, our sun overwhelms the energy contributions from the other stars, so we sometimes forget that the sun is also a star.

4. _____?_____

On a clear, sunny day, 1 m² of the Earth's surface receives about the same amount of energy per second from the sun as that given off by a 1000 W electric heater. If you hold your hand up to the sun, your hand will receive about 10 J/second of energy (1 J/sec. = 1 W).

5. _____?_____

The most important and immediate use of the sun's energy is the warmth we receive from it. Thanks to the sun, our planet is comfortably warm. Without the sun, the temperatures on the Earth would be almost 300°C lower, approximately the temperature of the space between the stars.

6. _____?_____

It can take anywhere from minutes to millions of years for the energy of sunlight to be transformed into a source of energy that can be used to heat our homes. For instance, sunlight causes moisture to evaporate, resulting in rainfall. This rainfall drains into streams and rivers. The rivers are dammed, and the water is forced through turbines to produce electricity. This electricity is used for many purposes: to run motors, to produce light, and to heat our homes, for example.

437

LESSON 4 ORGANIZER

Time Required
3 class periods

Process Skills
observing, measuring, classifying, analyzing

Theme Connections
Structures, Energy

New Term
Greenhouse effect—the trapping of radiant heat by the Earth's atmosphere

Materials (per student group)
Exploration 4, Question 1: lamp; light bulbs of different wattages;
meter stick; safety goggles; **Question 2:** lamp; 100 W light bulb; metric ruler; alcohol thermometer; square piece of cardboard, about 20 cm × 20 cm; mirror; safety goggles; **Question 3:** lamp; 100 W light bulb; piece of cardboard, about 5 cm × 25 cm; large (thick) magnifying lens; alcohol thermometer; 10 cm of masking tape; safety goggles; **Question 4:** 2 balloons; bottle painted white; bottle painted black; lamp; 100 W light bulb; 5 ice cubes; squares of red, yellow, blue, white, and black

continued ▶

Radiation– Heat in a Hurry

FOCUS

Getting Started

In a dark classroom, have students sit facing away from you. Hold a bright flashlight in front of you so that it points back toward you. One at a time, place paper of various colors in the beam of the flashlight so that the light is reflected out into the classroom. Have students discuss their observations. *(Students should note that white or brightly colored paper reflects the most light and absorbs the least light, while black or darkly colored paper reflects the least light and absorbs the most light.)*

Main Ideas

1. The sun is the source of most of the heat energy available to us.
2. The trapping of radiant heat by the Earth's atmosphere is called the *greenhouse effect.*
3. Light bulbs and other hot objects radiate heat, which can be reflected and focused.

TEACHING STRATEGIES

Your Chance to Make Headlines!

Have students read the entire article to themselves, pausing to write headlines for each paragraph in their ScienceLog. Then have them share and discuss their headlines.

Homework

The Resource Worksheet that is available to accompany the material on this page makes an excellent homework activity (Teaching Resources, page 45).

★ An Activity Worksheet (Teaching Resources, page 47) and Transparency 75 accompany the material on this page.

Answers to Energy From a Star are on the next page. ▶

Possible responses include the following:

1. Energy Around Us
2. Energy Delivery
3. The Nearest Star
4. Energy From Our Sun
5. The Importance of the Sun's Warmth
6. The Sun's Energy Has Many Uses
7. Plants Rely on the Sun Too!
8. Energy Packages
9. Energy in a Hurry!

Primary Source

Description of change: excerpted and adapted from *Whence Energy* by Dr. Roy L. Bishop

Rationale: exerpted to describe how the Earth receives and uses energy; modified to adjust to reading level

Answers to
The Greenhouse Effect, page 439

1. Answers will vary. Some scientists believe it is a serious human-caused problem that is gradually worsening. Others believe that changes in the average global temperature are just natural fluctuations.

2. Major contributors to the greenhouse effect include water vapor, carbon dioxide, and methane.

3. Answers will vary. The greenhouse effect could cause a global increase in temperature. For example, this could result in a change in the sea level or what kinds of crops are produced in a region.

4. Possible solutions include reducing the use of fossil fuels and reducing emissions from automobiles and factories. Because plants reduce the amount of carbon dioxide in the air, planting new trees would also help limit the greenhouse effect.

 A Transparency Worksheet (Teaching Resources, page 48) and Transparency 76 are available to accompany The Greenhouse Effect.

7. _____?_____

Plants use the energy of the sun to grow. In this form it may take months for the energy of the sun to reach us. Plants are the source of food for many animals. We use both plants and animals as sources of food. People need about 100 J of energy per second, or 2000 nutritional Calories per day.

8. _____?_____

When we burn oil, gas, or coal, we are using energy from the sun that was stored long ago. Energy from the sun was captured by plants millions of years ago. These plants were buried and slowly converted into coal, oil, or natural gas. These energy sources will not last forever. There are even fears that most of this energy will be used up within the next century.

9. _____?_____

The sun is 150,000,000 km from the Earth, but energy from the sun reaches the Earth in just 8 minutes! As the sun's energy falls on the Earth, some of the radiant energy is reflected back into space—by clouds or snow, for example. But some of the energy is absorbed by the Earth, and everything on the surface is warmed. Then, when the sun sets, much of this heat is radiated back into space.

438

The Greenhouse Effect

Some of the incoming radiation from the sun is reflected by the upper atmosphere.

Heat from the Earth's surface rises into the atmosphere. Some escapes into space, while the remainder is absorbed by clouds and greenhouse gases.

Earth's atmosphere

Some solar radiation is absorbed by clouds and then is radiated back into space.

The remaining solar radiation is absorbed by the Earth's surface.

⬆ How is this similar to the way in which a greenhouse works?

Have you ever been in a greenhouse on a sunny day? Did you notice how warm it was? Did you wonder why? What happens is that *radiant* energy from the sun passes through the glass of the greenhouse, strikes the ground and the plants, and is absorbed. The plants and ground heat up and then begin to radiate heat energy—but in a slightly different form that cannot easily pass back through the glass, so the heat is trapped!

In many ways the Earth behaves like a giant greenhouse. Only in this case, the glass of the greenhouse is the Earth's atmosphere. Certain atmospheric gases are very good at slowing the flow of radiant energy from the Earth back into space. Carbon dioxide is one of these gases. Carbon dioxide is produced during burning. In the last 200 years we have released huge amounts of carbon dioxide into the atmosphere. The higher level of carbon dioxide in the atmosphere may be increasing the Earth's natural "greenhouse effect." Scientists are concerned that if the current trend continues, the Earth

---

ORGANIZER, *continued*

paper, 10 cm × 10 cm each; alcohol thermometer; variety of colored cards, 5 cm × 25 cm each; 10 cm of masking tape for each colored card; safety goggles (See Advance Preparation on page 387D.); **Question 5:** 3 tin cans of the same size—one painted white, one painted black, and one unpainted; 3 cardboard covers to fit over the cans; 600 mL of hot water; 100 mL graduated cylinder; 3 alcohol thermometers; safety goggles (See Advance Preparation on page 387D.)

Teaching Resources
Resource Worksheet, p. 45
Activity Worksheet, p. 47
Transparency Worksheet, p. 48
Exploration Worksheet, p. 50
Theme Worksheet, p. 54
Transparencies 75 and 76
SourceBook, pp. S166 and S168

could warm up dramatically, causing unforeseen and possibly disastrous results.

Research this issue, and read other articles or books about it. Be sure to consider all sides of the issue. Keep the following questions in mind:

1. How serious do scientists consider an increase in the greenhouse effect to be? Is the problem getting worse?

2. What are the major contributors to the greenhouse effect?

3. What are some problems that the greenhouse effect causes? How might it affect agriculture? the climate? the oceans?

4. What are some possible solutions?

EXPLORATION 4

Reflectors, Absorbers, and Radiators

Radiant heat behaves differently from other forms of heat. In the activities that follow, you will conduct experiments in order to answer a few key questions about this form of heat transfer.

Choose one or two questions to investigate, and then share your results with your classmates.

QUESTION 1

Do light bulbs and other hot objects radiate heat?

You Will Need

- a lamp
- light bulbs of different wattages
- a meter stick

What to Do

Turn on the lamp. Put your hand over the bulb, but do not touch the bulb! Do you feel heat immediately? Turn off the lamp. What do you feel? Can you explain this?

Turn the lamp on again. Put your hand closer to it. What do you feel now? Can you explain this?

Compare what happens when you put your hand to the side of the bulb. Could the heat be traveling by convection? Try this activity with bulbs of various wattages. At what distance can you no longer feel the heat?

QUESTION 2

Can radiant heat be reflected?

You Will Need

- a lamp with a 100 W light bulb
- a square piece of cardboard
- a mirror
- an alcohol thermometer
- a metric ruler

What to Do

Hold your hand about 15 cm from the bulb. Now try to reflect more heat onto your hand by using a mirror. Do you notice a difference? Are you sure? Check with the thermometer.

Next, arrange the lamp, cardboard square, and mirror as shown at right. Can you feel the radiant heat reflected by the mirror?

Exploration 4 continued ▶

439

EXPLORATION 4

SAFETY ALERT Caution students not to put their hand too close to the light bulb. Also caution them to look directly at the light bulb as little as possible.

Answers to Question 1

The heat is felt as soon as the light is turned on. When the light is turned off, the heat is gone almost immediately. Because conduction and convection occur more slowly, the heat must be the result of radiation.

More heat is felt closer to the light because the hand is directly in the path of concentrated radiant energy. The heat felt to the side of the bulb could not be the result of convection because hot air rises. Therefore, the heat is the result of radiation.

As the wattage increases, the same amount of heat can be felt at greater distances. Students should conclude that light bulbs and other hot objects radiate heat.

Answers to Question 2

Students' hands become warmer as a result of the heat being reflected by the mirror. Therefore, the mirror is reflecting the radiant heat emitted from the opposite side of the bulb. The cardboard absorbs the radiant heat that travels directly from the bulb toward the hand. Therefore, any radiant heat felt must have been reflected by the mirror. Students should conclude that radiant heat can be reflected.

Exploration 4 continued ▶

★ An Exploration Worksheet accompanies Exploration 4 (Teaching Resources, page 50).

Theme Connection

Structures

Focus question: How do solar homes use the sun's energy for inexpensive heating? *(Solar homes make maximum use of the light and heat that passes through the windows. There are many windows in a solar home, and the windows are placed to receive as much sunlight as possible.)* The Theme Worksheet on page 54 of the Unit 7 Teaching Resources booklet outlines an excellent activity to accompany this Theme Connection.

Cooperative Learning
EXPLORATION 4

Group size: 3 to 4 students

Group goal: to conduct experiments in order to answer questions about radiation

Positive interdependence: Assign students roles such as primary experimenter, chief recorder, discussion leader, and materials manager and timer.

Individual accountability: Have each student draw a diagram to summarize what he or she has learned about radiation.

QUESTION 3

 Students should be very careful while focusing the light with the magnifying glass. Point out that focused light may burn their skin or eyes.

Answers to
Question 3

The temperature rises as a result of focusing, or concentrating, the light from the light bulb. As a result of their investigations, students should conclude that radiant heat can be focused.

QUESTION 4

For the third activity, students should hold the cards in front of the light for about 1 minute. Make sure that the thermometer starts at room temperature for each card.

Answers to
Question 4

The balloon on the black bottle inflates more than the balloon on the white bottle because the black bottle absorbs more heat, causing the air inside it to expand more.

Another observation could be that the ice cube melts more quickly on the darker paper because dark colors absorb more radiant heat.

The temperature registers higher on dark-colored cards than on light-colored ones because the dark-colored cards absorb more heat than do the light-colored ones.

ENVIRONMENTAL FOCUS

Share the following information with students: Jet exhaust often contains water that is ejected into the upper atmosphere. It has been estimated that 500 supersonic jets could increase the water content of the stratosphere by 50 to 100 percent in a matter of years. Atmospheric water vapor is a contributor to the greenhouse effect.

QUESTION 3

Can radiant heat be focused?

You Will Need

- a lamp with a 100 W light bulb
- a large (thick) magnifying lens
- a thermometer taped to cardboard

What to Do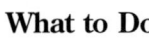

Arrange the lamp and the cardboard as shown below. Move the lens backward and forward until you can see the image of the light bulb on the cardboard. Once you are able to see the image, does the temperature of the cardboard rise? Can you explain your observations?

QUESTION 4

Which colors absorb radiant heat the most? the least?

You Will Need

- a bottle painted white
- a bottle painted black
- 2 balloons
- a lamp with a 100 W light bulb
- a variety of colored cards
- an alcohol thermometer
- tape
- 5 ice cubes
- 1 square each of red, yellow, blue, white, and black paper

What to Do

Put a balloon over the mouth of each bottle. Place a lamp halfway between the bottles (or, if it is a sunny day, put the bottles outside for a while). Observe the balloons. How do you explain what happens?

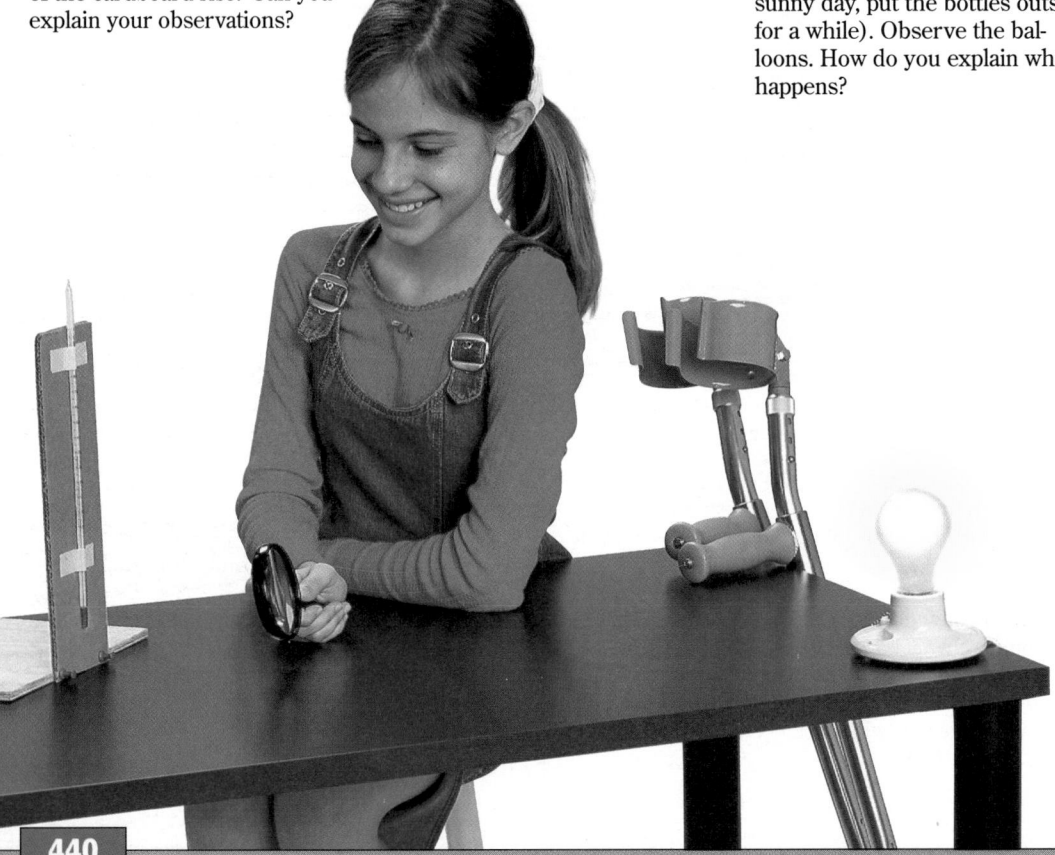

Theme Connection

Energy
Focus question: Why is it important to life on Earth that energy from the sun travels as radiant energy? *(Only radiant energy can travel through empty space. Convection and conduction require the presence of matter to transfer heat energy. Also, the energy from the sun warms the Earth and drives the process of photosynthesis.)*

Meeting Individual Needs

Gifted Learners
Suggest that students make a *solar still*. A solar still is a device that uses heat from the sun to separate impurities, like salt, from water. Point out to students that they may have to do some research to discover how to make their still. Students should set up displays in the classroom similar to the type they would use for a science fair. In their display, encourage students to include some practical uses for their solar still.

Put an ice cube on each square of paper. Place them in sunlight. In your ScienceLog, record the order in which the ice cubes melt.

Next, tape a thermometer to a colored card, and then hold the card up to the lamp. Repeat with the other cards. Remember to keep them the same distance from the lamp. (How long should you wait before reading the thermometer?) In your ScienceLog, record the temperature for each card. How would you explain your results?

QUESTION 5

Which color radiates the most heat?

You Will Need

- 3 tin cans of the same size—one painted white, one painted black, and one unpainted
- 3 cardboard covers to fit over the tin cans
- 3 thermometers
- hot water
- a graduated cylinder

What to Do

Set up the three cans as shown below. Fill each can with the same amount of hot water. You can use hot water from the tap, but be careful not to scald yourself. Water that is 55°C or above will scald you. Record the temperature of each can at 3-minute intervals until the water in all three cans reaches the same temperature. Which cools the quickest? How can you explain what happens?

441

SAFETY ALERT Test the hot water from the tap before class to be sure that it is below 50°C. If it is not, provide students with hot water that is below 50°C for this activity.

Answers to *Question 5*

The water in the black can loses the most heat in the shortest amount of time. The water in the unpainted can loses the least amount of heat. This is because dark colors radiate more heat than do light or metallic colors.

Meeting Individual Needs

Second-Language Learners

Remind students that in this unit they learned about the three ways in which heat flows from one object or place to another object or place—conduction, convection, and radiation. Suggest that they make a bilingual poster to illustrate each of these methods of heat transfer.

Invite volunteers to share their finished poster with the class by reading their information in both English and their native language.

Homework

The Extension activity on this page makes an excellent homework assignment.

FOLLOW-UP

Reteaching

Have students write a poem about how the sun provides the Earth and all living things with energy. Encourage students to use specific examples in their poem.

Assessment

Have students create a mural or bulletin-board display that shows how radiant heat can be reflected, absorbed, and radiated. Suggest that they include drawings of everyday examples of each.

Extension

Suggest that students write a persuasive newspaper article about the greenhouse effect. Encourage students to research the long-term changes associated with the greenhouse effect.

Closure

Set up the following demonstration: Fill a 500 mL Erlenmeyer flask with water. Following the instructions provided in Advance Preparation on page 387D, insert a glass tube into a one-hole rubber stopper.

Place the stopper onto the mouth of the flask so that one end of the glass tube is in the water. Mark the level of the water in the tube. Then turn on a lamp and shine it on the water. Ask students to observe the water in the glass tube. *(It begins to rise.)* Ask them to explain their observations. *(Heat from the lamp passes through the glass flask. Once inside, the heat remains inside. The heat warms the air and water inside the flask, and the increase in temperature causes the water in the tube to rise.)*

Answers to *Challenge Your Thinking*

1. The answers to the paragraph are as follows, listed in the order in which they appear: sun, radiant, magnifying lens, mirror, convection, radiation, conduct, insulators.

2. a. White clothing helps people stay cool in the tropics because white reflects more of the radiant energy from the sun than do darker colors.

b. It is often cooler on clear nights because there are no clouds to stop the Earth from radiating energy into space.

c. Baking a potato in foil allows heat to be conducted from the hot oven to the potato while keeping the potato from radiating heat. This reduces cooking time.

d. A house painted white would reflect the most heat on summer days and would radiate the least heat on cold winter nights.

3. Accept all reasonable responses. Student sketches should indicate the heat source and the direction in which the heat travels.

CROSS-DISCIPLINARY FOCUS

Art

Students might enjoy working together to develop a comic strip to illustrate one of the following topics from the lesson:
- the difference between heat and temperature
- the difference between an insulator and a conductor
- the difference between conduction, convection, and radiation

Suggest to students that they create a central character for their strip. You may wish to display several comic strips from a newspaper so that students can observe different styles. Have students display their finished comic strip.

1. Heat-Transfer Hotshot

Copy the following paragraph into your ScienceLog, and fill in the blanks. If you can do this quickly, then you are a heat-transfer hotshot! Do not write in this book.

> Ninety percent of the Earth's heat comes from the __?__ in the form of __?__ energy. This type of energy can be focused by a __?__. You could also send this type of energy around a corner using a __?__. If you held your hand a short distance above a candle, most of the heat you would feel would be from __?__. If you held your hand to one side of the candle, most of the heat you would feel would be from __?__. Metals at room temperature feel cold to the touch. This is because they __?__ heat away from your skin very quickly. Materials that carry heat very poorly in this way are known as __?__.

2. Problems to Ponder

a. Why is white clothing popular in tropical countries?

b. Why is it sometimes cooler on clear nights than on cloudy nights?

c. Baked potatoes are often wrapped in aluminum foil. Why?

d. On hot summer days, a home should reflect heat. On cold winter nights, a house should radiate as little heat as possible. What color should a house be painted: green, red, black, white, or pink? Why?

3. Hot Dog!

Can you cook a hot dog by conduction? convection? radiation? Sketch how you would cook a hot dog by each method.

★ **You may wish to provide students with the Chapter 23 Review Worksheet that accompanies this Challenge Your Thinking (Teaching Resources, page 56).**

4. Explanations Please!

a. When you make toast with the end of a loaf of bread, the crusty side feels hotter. Is it really hotter? Explain.

b. To accurately measure the temperature of the air, a thermometer must be shaded from the sun. Why?

c. Sand that is added to the snow on a highway not only helps tire traction, but also helps melt the snow. Explain.

5. Professor Duncer's Experiment

Professor Duncer wanted to find out if heat traveled faster through short metal rods or through long metal rods. The picture at right shows his experiment. Can you improve it?

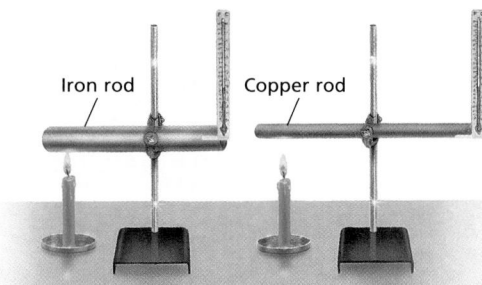

Iron rod   Copper rod

6. Solar Heater

The picture below shows a simple solar collector.

a. Why do you think it is covered with glass?

b. Why do you think the inside is painted black?

c. Why do you think it uses copper tubing?

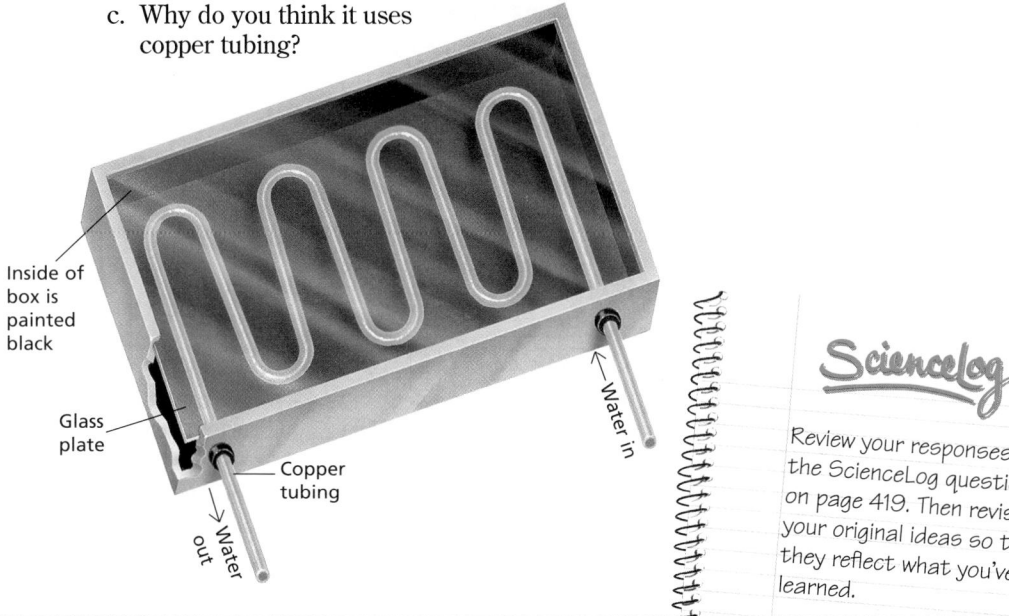

Inside of box is painted black

Glass plate

Copper tubing

Water out

Water in

ScienceLog

Review your responses to the ScienceLog questions on page 419. Then revise your original ideas so that they reflect what you've learned.

4. a. The crusty side feels hotter even though it is the same temperature as the other side of the piece of bread. This is because the crusty side is darker and thus radiates heat more quickly.

b. If a thermometer is not shaded from the sun, it will absorb solar radiation that will heat the liquid inside the thermometer to a temperature higher than that of the air.

c. Sand helps to melt the snow because it converts solar radiation into heat, which it conducts to the snow. Snow by itself would reflect most of the radiant energy.

5. Student answers should reduce the number of variables manipulated to just one—the length of the metal rods.

6. a. The glass admits incoming solar radiation but blocks outgoing radiation.

b. The inside is painted black to absorb as much solar radiation as possible.

c. Because copper conducts heat well and is fairly inexpensive, the copper tubing is a good material for conducting heat to the water efficiently.

ScienceLog

The following are sample revised answers:

1. Heat moves through solids by conduction and through liquids and gases by convection. Radiant energy, such as light energy, can move through empty space and can change into heat energy as it is absorbed by matter. The movement of heat along the fireplace poker is an example of conduction. The movement of heat within the hot drink is an example of convection. The movement of heat to

warm the woman's hands is an example of radiation.

2. • A bathroom rug feels warmer than a tile floor because the tile conducts much more heat away from your feet than does the rug.

• The air heated by the light bulb moves above the bulb in convection currents, while the air to the side does not.

• Radiant energy from the sun passes through transparent materials such as a glass window and changes into heat energy as it is absorbed by objects in the car.

3. It is possible for the water at the top of the test tube to boil even though there is ice at the bottom because of convection currents. Because the ice is not allowed to float to the top of the tube and because the flame is applied to the middle of the tube, little heat reaches the ice by convection. Instead, the water at the top of the tube moves in a circular pattern from the flame to the top of the tube and back to the flame. As the water moves in this circular pattern, it absorbs heat from the flame until it finally boils.

The Big Ideas

The following is a sample unit summary:

When most substances absorb energy, the particles that make up the substance move more quickly, causing the substance to expand and feel warm to the touch. (1) A thermometer works by measuring the expansion of a substance that results from its absorption of heat. The thermometer is calibrated to relate the expansion of the substance to an increase in temperature. (2) Heat is the total amount of energy in a substance. Temperature is an indirect measure of that energy. (3) People get their heat energy from the food they eat. When food is broken down, some of the chemical energy is converted to heat energy during the reaction. The chemical energy contained in foods is measured in calories or joules. (4) The amount of heat energy in a material depends on its mass and its temperature. The greater the amount of a substance, the more heat energy it is able to absorb. (5) Heat travels by conduction, convection, and radiation. (6) Heat travels through solids by conduction, in which the energized particles of a substance are in direct contact with other particles of the same or another substance. Heat travels by convection in fluids (liquids and gases) when the increased energy in one part of the fluid causes that part to rise, creating a current. Radiation is the transfer of energy through empty space. Radiation comes from any heated object including the sun, which is the most important source of energy for life on Earth. (7, 8) The radiant energy from the sun passes through our atmosphere and heats the air, the water, and the ground. Nearly all of the energy that drives the Earth's systems comes from the sun. (9)

Unit 7

TEMPERATURE AND HEAT

SOURCEBOOK

Read the SourceBook to find out how heat is related to the motion of atoms and molecules and how heat is transferred from place to place. The SourceBook also describes the different ways that heat can be measured and how heat from the sun affects the Earth.

Here's what you'll find in the SourceBook:

The Big Ideas

In your ScienceLog, write a summary of this unit, using the following questions as a guide:

1. What are some clues that heat energy has been added to a substance?
2. How does a thermometer work?
3. How do heat and temperature differ? How are they related?
4. Where do people get their heat energy? How is this energy measured?
5. How are mass and heat related?
6. How does heat travel?
7. How does heat transfer differ in solids, fluids, and empty space?
8. What causes convection?
9. How is the Earth affected by the sun's heat?

Checking Your Understanding

1. Each example below presents a heat-related problem. Use your understanding of heat and temperature to write brief explanations of each.
    a. You cannot use a thermometer filled with water to measure temperatures below 0°C or above 100°C.
    b. The metal lid to a jar is stuck. It comes off easily, however, after holding the lid under hot, running water for a few seconds.

 You may wish to provide students with the Unit 7 Review Worksheet that is available to accompany this Making Connections (Teaching Resources, page 63).

Homework

The Unit 7 Activity Worksheet makes an excellent homework assignment (Teaching Resources, page 62). If you choose to use this worksheet in class, Transparency 77 is available to accompany it.

c. Glider pilots experience uplift when flying over plowed fields and houses. They experience downdrafts when flying over water and forests.

d. A paper cup is filled with water and placed over an alcohol lamp. The water in the cup boils, but the paper does not burn.

e. A car is parked in the sunlight with the windows closed. The inside becomes very warm, yet the glass windows feel cool.

f. A potato doesn't burn, but it gives you energy.

g. A motorcycle engine has fins around it.

h. On the way home from the store, Anita keeps ice cream cold by wrapping it in her sweater!

i. In a room heated by a fireplace, most of the air heated by the fire actually leaves the room through the chimney, but people in the room feel warm.

j. Five seconds after being removed from a 250°C oven, a piece of aluminum foil is cool enough to touch with bare hands.

2. For the description below of a thermos bottle, fill in the blanks using the words *conduction, convection,* and *radiation.*

a. The plasctic cup reduces heat transfer by ____?____ .

b. The cap reduces heat transfer by ____?____ .

c. The double-walled glass bottle reduces heat transfer by ____?____ .

d. The vacuum between the glass walls reduces heat transfer by ____?____ .

e. The silvered surface of the inner glass wall reduces heat transfer by ____?____ .

f. The air between the glass bottle and the outside of the thermos reduces heat transfer by ____?____ .

g. The plastic case reduces heat transfer by ____?____ .

3. You are going to bake potatoes in aluminum foil. Aluminum foil has a shiny side and a dull side. Would it make any difference which side is next to the potatoes? Explain.

4. Some people prefer their coffee black. Others prefer it with cream. Which do you think cools faster, black coffee or coffee with cream added? Explain.

5. (concept map) Construct a concept map using the following terms: *conduction, heat, solids, radiation, fluids, convection,* and *space.*

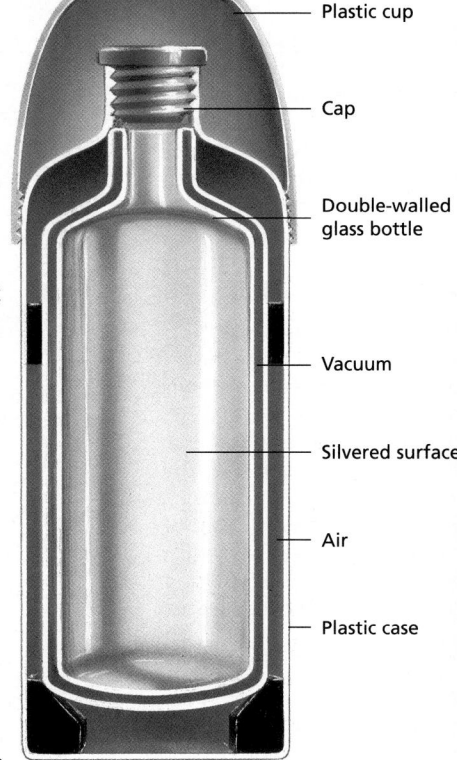

Plastic cup
Cap
Double-walled glass bottle
Vacuum
Silvered surface
Air
Plastic case

445

Answers to
Checking Your Understanding,
continued

d. The water in the cup absorbs the heat energy so quickly that the paper cannot reach a temperature high enough to burn.

e. Radiant energy passes through the glass and is absorbed by the interior of the car.

f. The energy in a potato is stored as chemical (or potential) energy. Chemical reactions within your body release this stored energy, and some of it is converted into heat. Although the potato does not burn, it contains the same amount of energy as would be given off during burning.

g. The fins of a motorcycle engine supply much more surface area, so excess heat can be released more easily.

h. The sweater acts as an insulator to keep the heat out.

i. The people surrounding the fire are warmed by radiant energy.

j. Because the foil is so thin, its surface area is large compared with its volume. Thus, the heat that it absorbs is released quickly to the environment.

2. a. conduction
b. conduction and convection
c. conduction and convection
d. conduction and convection
e. radiation
f. conduction and convection
g. conduction

3. The shiny side reflects heat and should be placed next to the potatoes. The dull side absorbs heat.

4. If all of the substances start out at the same temperature, the black coffee will cool the fastest because it radiates heat energy the quickest.

5. Sample concept map:

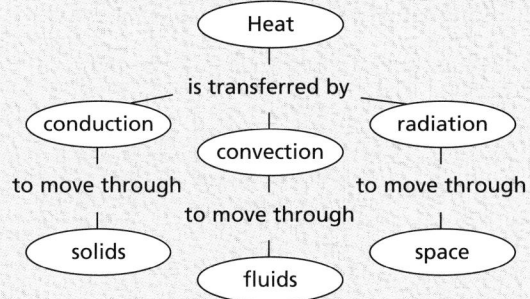

Heat
is transferred by
conduction   radiation
convection
to move through   to move through
to move through
solids   space
fluids

1. a. The water inside the thermometer will freeze at temperatures below 0°C and will boil at temperatures above 100°C.

b. The lid comes off because it expands a greater amount and more quickly than does the glass.

c. Water and forests are capable of absorbing much more heat energy without becoming hot. Plowed fields and houses become hot more quickly, which heats the air above them and causes the air to rise.

Background

Carbon dioxide makes up about half of the heat-retaining gases in the atmosphere. Other gases, however, also contribute to the greenhouse effect. Methane, which is released during the digestive process of livestock and is emitted from landfills and oxygen-depleted marshlands, makes up about 20 percent of the greenhouse gases. Each molecule of methane absorbs infrared radiation 20 times more effectively than does a molecule of carbon dioxide. Another 15 percent of the greenhouse gases are chlorofluorocarbons (CFCs), which are released from refrigerants and aerosol cans.

Although not all scientists agree that global warming is occurring or that it is caused by an increase in greenhouse gases, almost everyone agrees that it would be a good idea to reduce the amount of harmful emissions currently being released into the atmosphere. Some recommendations have been made to reduce the possible risk of global warming. These include instituting international regulation of the use and production of CFCs, planting new trees, preserving existing forests, and restricting excess production of methane gas.

Answers to
Greenhouse Gases in Action

Carbon dioxide is released during the chemical reaction of baking soda and vinegar. Both jars will heat up at about the same rate. However, the jar with the carbon dioxide will take longer to cool down. The carbon dioxide prevents heat from escaping longer than air does, illustrating its role as a greenhouse gas.

Global Warming

What's the hottest summer you can remember? Now imagine that it had been 10 degrees hotter! That's just too hot to handle. Some scientists predict that *global warming* may cause the average temperature of the Earth to increase by 2°C to 10°C by the year 2060.

What It Takes to Warm the Globe

Sunlight enters the Earth's atmosphere and heats the planet. As the heat rises, some of it escapes into space. Gases in the atmosphere trap some of the heat and keep it near the Earth's surface. That keeps the Earth warm. This process is called the *greenhouse effect* because it resembles what happens in a greenhouse on Earth.

The Carbon Connection

One of the most common gases that contribute to the greenhouse effect is carbon dioxide (CO_2). It is released when animals breathe out and when carbon is burned. All plants and animals contain carbon. Those that were buried for millions of years may form fossil fuels such as coal, oil, and natural gas. These fuels release large amounts of CO_2 when burned.

One Hot Hypothesis

Some scientists think that increasing amounts of pollution

▲ Heat is trapped near the Earth's surface in much the same way that heat is trapped in a greenhouse.

are intensifying the greenhouse effect. They predict that this will cause the Earth's temperature to rise and global warming to occur if the amounts of CO_2 and other greenhouse gases are not reduced.

Other scientists argue that it is hard, if not impossible, to predict what will happen. Cloud cover, certain chemical reactions, and other variables may counteract trends in global warming. However, most scientists agree that pollution of the atmosphere should be reduced. An important way to do this is by reducing the amount of CO_2 we add to the atmosphere. This means reducing our use of fossil fuels. It also means increasing the amount of CO_2 absorbed on Earth by preserving forests and planting trees, which remove CO_2 from the air.

Greenhouse Gases in Action

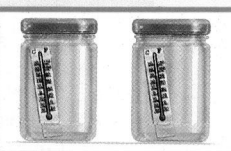

Get two identical glass jars, two outdoor thermometers, some vinegar, and some baking soda. Place an outdoor thermometer and 50 mL of vinegar in each jar. Seal one jar. Place 25 g of baking soda in the second jar and seal it. What greenhouse gas is released by the reaction of vinegar and baking soda? Place the jars in bright sunlight and record the change in temperature over time. How fast do they heat up? How fast does each jar cool down? Explain how a greenhouse gas affects the heating and cooling of the air in the second jar.

Theme Connection

Cycles

Describe the carbon dioxide–oxygen cycle to students. (When animals exhale, carbon dioxide is released into the environment. Plants use carbon dioxide to make food and release oxygen, which animals then breathe.) **Focus question:** Based on this cycle, how would deforestation contribute to global warming? *(With fewer plants, less carbon dioxide may be removed from the atmosphere, adding to the greenhouse effect.)*

Integrating the Sciences

Earth and Life Sciences

Global warming may cause a variety of changes on the Earth. Many scientists predict that global warming may cause the polar icecaps to melt. This could raise the sea level as much as 5 m. Ask: What might happen to coastal cities if the sea level rose 5 m? How might plant and animal life be affected?

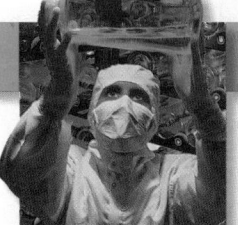

In the Line of Fire

Once a fire dies down, you might see arson investigator Lt. Larry McKee on the scene. He's likely to call on the expertise of someone like Valerie Turner, the director of a forensic arson laboratory. Together, they mix science and detective work to find out what caused the fire.

First on the Scene

"All fires involve chemical reactions," McKee explains. "Fuel and oxygen combine in a chemical reaction called combustion." On the scene, McKee questions witnesses and firefighters about what they saw. He knows, for example, that the color of the smoke can indicate certain chemicals.

McKee explains that fires usually burn "up and out, in a V shape." To find where the V begins, he says, "We work from the area with the least amount of damage to the one with the most damage. This normally leads us to the point of origin." Once the origin has been determined, it's time to call in the dogs!

An Accelerant-Sniffing Canine

"We have what we call an accelerant-sniffing canine. Our canine, named Nikki, has been trained to detect approximately 11 different chemicals." When Nikki arrives on the scene, she sniffs for traces of chemicals, called accelerants, that may have been used to start the fire. When she finds one, she immediately starts to dig at it. At that point, McKee takes a sample from the area and sends it to the lab for analysis.

At the Lab

In Turner's laboratory, the sample is treated so that any accelerants in it are dissolved in a liquid. A small amount of the liquid is then injected into an instrument called a *gas chromatograph.* The instrument heats the liquid, forming a mixture of gases. The gases then are passed through a flame. Turner explains that as each gas passes through the flame, it "causes a fluctuation in an electronic signal, and that's what creates our graphs."

◄ "After the fire is out, I can investigate the fire scene to determine where the fire started and how it started. If it was intentionally set and I'm successful at putting the arson case together, I can get a conviction. That's very satisfying," says Lt. McKee.

▲ Nikki searches for traces of gasoline, kerosene, and other accelerants.

From the graphs, Turner can determine if an accelerant was present in the mixture.

Solving the Case

If the laboratory report indicates that a suspicious accelerant has been found, McKee begins to search for arson suspects. By combining detective work with scientific evidence, fire investigators can successfully catch and convict arsonists.

Fascinating Fire Facts

The temperature of a house fire can reach 980°C! At that temperature, aluminum window frames melt and furniture goes up in flames. Do some research to discover three more facts about fires. Create a display with two or more classmates to illustrate some of your facts.

447

Fascinating Fire Facts

Fire facts might include facts about when most fires take place, what happens to common household objects, and how quickly a house fire can spread. Encourage students to be creative and thorough in making their displays. You may wish to put the displays in prominent positions around the classroom.

Background

In the laboratory, a mass spectrometer is sometimes used to find more detailed information about the chemical make-up of the accelerants. Unlike the gas chromatograph, the mass spectrometer actually breaks down the chemical compounds in the vapor sample. The compounds break apart into characteristic fragments according to their chemical composition. The fragments are then checked against data in a computer to see how closely they match known accelerants. Turner explains, "We can't always identify everything. The sample may contain a lot of chemicals that come from the carpeting or the synthetic products that are present in the household. A lot of the time there are so many chemicals in a sample that we can't make a conclusive statement about what is there."

Extension

Invite a firefighter, arson investigator, or fire chief to visit your classroom. Ask if he or she could bring visual aids such as household items found after a fire or some of the equipment used to detect how and where a fire started. Encourage the class to ask questions about the chemical nature of fires, how fires start, how to prevent them, and how fire investigators use science to determine the cause of fires.

CROSS-DISCIPLINARY FOCUS

Health

Encourage students to investigate the following questions:

1. Why is it a bad idea to throw flour on a kitchen fire? *(The flour is extremely flammable and will ignite.)*

2. Why is it recommended that a person stay low while exiting a burning building or room? *(In a fire, smoke tends to form a layer that thickens from the ceiling down. This upper layer is very dangerous. Not only can the smoke cause severe respiratory problems, but also the air in this layer can reach very high temperatures. Fortunately, below this extremely hot and smoke-filled layer is usually a layer of cooler and relatively smoke-free air.)*

Background

Absolute zero is actually 0 on the Kelvin temperature scale. Temperatures are easily converted from one scale to the other. $K = °C + 273.2$ and $°C = K − 273.2$.

Scientists speculate that as gases approach absolute zero, they condense into a new form of matter. This process is called *Bose-Einstein condensation* and was named in honor of the research and calculations performed by the physicists Satyendra Nath Bose and Albert Einstein. At these very low temperatures, groups of particles move simultaneously in a highly ordered fashion. The gases become superconductive and flow in an unusual manner. Recent experiments at the Fermi lab in Batvia, Illinois, have yielded promising results as scientists struggle to create this new form of matter.

Discussion Questions

1. What is the relationship between temperature and motion? *(The warmer a substance is, the faster its particles move. The colder a substance is, the more slowly its particles move.)*

2. What might happen to matter at absolute zero? *(Students should understand that all movement of matter will effectively stop. You may wish to take this time to explain Bose-Einstein condensation.)*

3. Why is cryogenics useful to biological researchers? *(It can be used to preserve tissues and to test the effects of extremely low temperatures on living tissue.)*

Answers to
Freezing Fun on Your Own

The tap water should freeze first. The salt water should freeze second. The alcohol should not freeze at all because alcohol's freezing point (−117.3°C) is below the temperature of the freezer. Different liquids freeze at different temperatures. (Remind students not to ingest these substances.)

The Deep Freeze

*I*n the dark reaches of outer space, temperatures drop below −270°C. Perhaps the only place colder is a laboratory here on Earth!

The Quest for Zero

All matter is made up of tiny, constantly vibrating particles. Temperature is a measure of the movement of these particles. The colder a substance gets, the less energy its particles have, and the slower the particles move. In theory, at absolute zero (−273°C), all movement of matter should stop. Scientists are working in laboratories to slow matter down so much that the temperature approaches absolute zero.

How Low Can They Go?

Using lasers along with magnets, mirrors, and supercold chemicals, scientists have cooled matter to within a millionth of a degree of absolute zero. In one method, scientists fire lasers at tiny gas

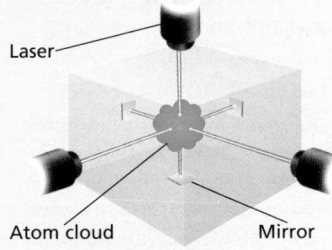

Laser

Atom cloud   Mirror

▲ Lasers are used to hold matter very still. The slower it vibrates, the colder it becomes.

particles inside a special chamber. The lasers hold the particles so still that their temperature approaches −272.999998°C.

To get an idea of what takes place, imagine turning on several garden hoses as high as they go. Then direct the streams of water at a rolling soccer ball so that each stream pushes the ball from a different angle. If the hoses are aimed properly, the ball won't roll in any direction. That's something like what happens to the particles in the scientists' experiment.

Cryogenics—Cold Temperature Technology

Supercold temperatures have led to some supercool technology. Cryosurgery, which is surgery that uses extremely low temperatures, allows doctors to seal off tiny blood vessels during an operation or to freeze diseased cells to destroy them.

Cooling materials to near absolute zero also led to the discovery of superconductors. **Superconductors** are materials that lose all of their electrical resistance when they are cooled to low enough temperatures. Imagine the possibilities for materials that could conduct electricity indefinitely without any energy loss. Unfortunately, it takes a great deal of energy to cool such materials. So right now, applications for superconductors are still just dreams.

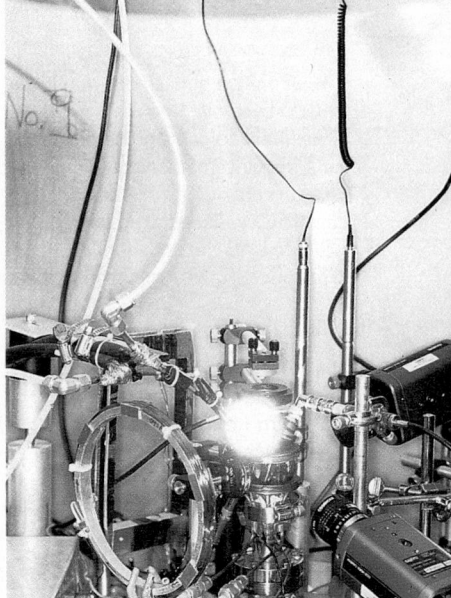

▲ Laser device used to cool matter to nearly absolute zero

Freezing Fun on Your Own

You can try your hand at cryo-investigation. In three separate containers, place 50 mL of tap water, 50 mL of salt water (50 mL of water plus 15 g salt), and 50 mL of rubbing alcohol (isopropanol). Then put all three containers in your freezer at the same time. Check the containers every 5 minutes for 30 minutes. Which liquid froze first? Did all three liquids freeze? Why do you think they froze at different times?

CROSS-DISCIPLINARY FOCUS

Mathematics

Have students create a graph or chart that shows the freezing temperatures of a variety of substances. The chart should include absolute zero as its lowest value so that students can perceive how cold materials are in relation to absolute zero. Students might consider the following substances: water, liquid nitrogen, and liquid hydrogen.

Bronze Casting

*D*o you want to make a lasting impression? Have a sculptor create a bronze cast of your head. Bronze is made by mixing tin with copper. People have used bronze to make utensils, tools, weapons, armor, jewelry, and artwork for over 5000 years.

Ancient History of the Alloy

Ancient metal workers needed a strong material to fashion their knives, swords, and shields. Many used copper and tin separately, but these metals were too weak to be very useful as weapons. Since tin and copper are usually mined together, the first people to create bronze may have combined the two metals accidentally.

Mixing copper and tin together yields a metal that is much stronger than either metal is by itself. When two or more metals are mixed together, the mixture is called an *alloy*. In bronze, tin makes up between 5 and 20 percent of the alloy.

Bronze was first made about 5000 years ago in the Near East. Places like Palestine, Greece, and Egypt began using bronze

▲ These artifacts are from the Bronze Age.

▲ Molten bronze being poured into a plaster cast

extensively at the same time. It took more than a thousand years for the technology to spread through Europe.

One unusually beautiful type of bronze appears to have been developed over 3000 years ago, and it's still made in Japan today. Called *shakudo*, this bronze has a purplish black color and is inlaid with gold, silver, and other metals. Archaeologists have found evidence that links the casting of *shakudo* with that used to produce bronze pieces of Roman, Egyptian, Indian, Anglo-Saxon, and Celtic origin.

The Modern Science of Casting

A bronze casting begins as a wax model. After the wax has been sculpted, it is dipped in plaster and left to harden. The

plaster is then fired in a special oven called a *kiln*. The kiln's temperature reaches 750°C, causing the wax to melt and the plaster to harden into a mold.

At the same time, copper and tin are placed in a furnace and heated to about 1200°C, causing them to melt together. While the plaster is still hot from the kiln, the liquid bronze is poured into the mold. After cooling, the plaster is removed, and a bronze cast is revealed. Polishing adds the final touch to a bronze work of art.

Green Metal?

Many bronze sculptures look greenish. The green color is called patina. Find out what chemical reaction causes patina.

449

Background

Modern bronze sculpture is begun by making a clay model of the sculpture and covering it with plaster of Paris to make a mold. When the plaster has dried, sculptors use one of two methods to complete their sculpture. In the waste-mold method, the clay is scooped out of the plaster mold. The mold is then filled with the molten metal. When the metal cools, the plaster is chipped away to reveal the bronze sculpture. In the piece-mold method, the plaster of Paris mold is assembled in two or more sections over the clay so that the mold may be removed from the clay, filled with the metal, removed from the metal, and then reused.

For extremely large sculptures, wax is used. After the plaster mold is made, wax is poured into it so that the inner surface of the plaster is coated. Then more plaster is added to the mold, sandwiching the wax between the layers of plaster. When the plaster has dried, the mold is placed in a kiln so that the plaster hardens and the wax melts out of the mold. The bronze can then be poured into the cavity between the layers of plaster. When cool, the plaster is chipped away.

CROSS-DISCIPLINARY FOCUS

Art

Contact the art department of a high school, college, or art school in your area, and ask permission for your students to fire some clay sculptures in a kiln. Tell students to be wary of air trapped in their clay. The gas that is trapped in the air pockets will expand in the high temperatures of the kiln and will crack the clay. You may wish to consult an art teacher about safe procedures for firing clay.

Meeting Individual Needs

Second-Language Learners

Have students prepare a display of metal-working terms in both English and their native language. Encourage students to create illustrations of the metals and metal-working processes used in their native culture. Provide a time for students to share their projects with the class.

Answers to
Green Metal?

Patina forms from a chemical reaction involving bronze, oxygen, salt, and sulfur in the air. (The greenish color of the Statue of Liberty is a result of patina on copper.)

Unit 8 OUR CHANGING EARTH

Bibliography for Teachers

Ballard, Robert D. *Exploring Our Living Planet.* Washington, DC: National Geographic Society, 1983.

Glenn, W. H. "Drifting: Continents on the Move." *The Science Teacher,* February 1983.

Kaufman, Jeffrey S., Robert C. Knott, and Lincoln Bergman. *River Cutters.* Berkeley, CA: Lawrence Hall of Science, 1990.

Bibliography for Students

Aylesworth, Thomas G. *Moving Continents: Our Changing Earth.* Hillside, NY: Enslow Publications, 1990.

Bramwell, Martyn. *Glaciers and Ice Caps.* New York, NY: Franklin Watts Inc., 1986.

Brownstone, David M., and Irene M. Franck. *Natural Wonders of America.* New York, NY: Atheneum, 1989.

Unit Overview

In this unit, students are introduced to the long- and short-term changes that shape the surface of the planet. They synthesize their own observations of the world around them into a working model of a changing Earth. In Chapter 24, students are encouraged to organize their prior knowledge about the Earth in terms of the processes that shaped its landforms over geologic time. In Chapter 25, students directly observe physical- and chemical-weathering processes and use their observations to infer patterns of erosion taking place on a larger scale. In Chapter 26, students examine the effects of erosion and deposition by running water, wave action, wind, and glaciers.

Using the Themes

The unifying themes emphasized in this unit are **Structures, Energy,** and **Changes Over Time.** The following information will help you incorporate these themes into your teaching plan. Focus questions that correspond to these themes appear in the margins of this Annotated Teacher's Edition on pages 459, 464, 472, 478, 504, and 511.

The theme of **Structures** is emphasized in the discussion of the various ways the Earth's landforms are structured. The structure of the Earth's interior is also discussed in relation to the theory of plate tectonics.

The theme of **Energy** can be discussed in relation to the forces that change the Earth. Students contemplate the incredible forces necessary to move continents, build mountains, and create volcanoes. Converting heat (thermal) energy to kinetic energy can be discussed in this context. The energy of colliding tectonic plates can also be explored.

Changes Over Time are evident in Chapter 24 in the discussion of the process of plate tectonic and the stages

involved in the formation of the modern continents from the single landmass, Pangaea. In Chapters 24 and 25, students learn how geologists use what they know about the present to construct a history of the Earth's past. In Chapter 26, the roles of running water, wind, and ice in shaping the landscape are explored. **Changes Over Time** can be used as a basis for explaining these processes of erosion and deposition.

Films, Videotapes, Software, and Other Media

Erosion and Weathering: Looking at the Land

Erosion: Leveling the Land
   Two Videotapes
   Britannica
   310 S. Michigan Ave.
   Chicago, IL 60604-9839

Geodynamics Multimedia Database
   CD-ROM
   EME Corporation
   P.O. Box 2805
   Danbury, CT 06810

Glaciers: Ice on the Move
   Videotape
   National Geographic Society
   Educational Services
   Washington, DC 20036

Planet Earth: The Force Within
   Videodisc
   Coronet/MTI
   108 Wilmot Rd.
   Deerfield, IL 60015

Plate Tectonics: An Introduction
   Filmstrip
   National Geographic Society
   Educational Services
   Washington, DC 20036

Using the SourceBook

Unit 8 focuses on the natural forces of weathering, erosion, and deposition. Students learn about the varieties of and differences among these geologic activities, as well as how they work together to shape the Earth's surface. The specific topographic features resulting from these processes are identified and examined.

Unit Organizer

Unit/Chapter	Lesson	Time*	Objectives	Teaching Resources
Unit Opener, p. 450				Science Sleuths: The Missing Beach English/Spanish Audiocassettes Home Connection p. 1
Chapter 24, p. 452	**Lesson 1, Seeing the Sights,** p. 453	3	1. Describe some landforms that exist throughout the world. 2. Explain why this variety exists.	Image and Activity Bank 24-1 Activity Worksheet, p. 3 Exploration Worksheet, p. 4
	Lesson 2, Unlocking a Planet's Past, p. 456	3	1. Appreciate the different types of work that geologists do. 2. Consider the use of observation and inference in the work of geologists. 3. Know the historical development of the theories of continental drift and plate tectonics. 4. Examine the structure of the Earth's interior and its relationship to the theory of plate tectonics.	Image and Activity Bank 24-2 Exploration Worksheet, p. 5 Transparency Worksheet, p. 7 ▼ Exploration Worksheet, p. 9 Transparency Worksheet, p. 11 ▼ Theme Worksheet, p. 13 Transparency 80 Transparency 81
End of Chapter, p. 468				Activity Worksheet, p. 15 Chapter 24 Review Worksheet, p. 16 Chapter 24 Assessment Worksheet, p. 19
Chapter 25, p. 470	**Lesson 1, Evidence of Change,** p. 471	2	1. Identify local geological features that can be used to infer the geological history of the area. 2. Appreciate the methods used by geologists in studying the geological history of an area. 3. Identify changes in the environment. 4. Consider different rates of change. 5. Distinguish between natural and human-caused changes.	Image and Activity Bank 25-1 Discrepant Event Worksheet, p. 22 Exploration Worksheet, p. 23 ▼ Transparency 83 Transparency 84 Exploration Worksheet, p. 25
	Lesson 2, Examining One Type of Change, p. 481	2	1. Identify examples and causes of weathering. 2. Identify factors that determine the rate of weathering. 3. Relate these factors to weathering in the local environment.	Image and Activity Bank 25-2 Exploration Worksheet, p. 26 ▼ Exploration Worksheet, p. 28 Exploration Worksheet, p. 29
	Lesson 3, Gravity—The Great Leveler, p. 486	1	1. Understand changes caused by erosion. 2. Realize that the force of gravity underlies the erosion caused by water, ice, and wind. 3. Distinguish between different types of downslope movements.	Image and Activity Bank 25-3 Exploration Worksheet, p. 33
End of Chapter, p. 491				Chapter 25 Review Worksheet, p. 34 Chapter 25 Assessment Worksheet, p. 36
Chapter 26, p. 493	**Lesson 1, The Power of Water,** p. 494	3	1. Describe the effects of erosion caused by falling rain. 2. Compare soil types on the basis of their porosity and permeability.	Image and Activity Bank 26-1 Transparency Worksheet, p. 38 ▼ Exploration Worksheet, p. 40 Transparency 87
	Lesson 2, Water on the Move, p. 499	1	1. Identify the factors that determine the speed of a river or stream. 2. Relate the energy of moving water to the process of erosion.	Image and Activity Bank 26–2 Exploration Worksheet, p. 42
	Lesson 3, How Rivers Change the Land, p. 502	4	1. Explain how materials are eroded and deposited in a river system. 2. Identify the stages of development in a river.	Image and Activity Bank 26-3 Exploration Worksheet, p. 44 Exploration Worksheet, p. 45
	Lesson 4, Where Land Meets Sea, p. 506	2	1. Identify deltas and estuaries, and explain how they are formed. 2. Understand how ocean waves erode the shoreline and build beaches.	Image and Activity Bank 26-4 Math Practice Worksheet, p. 46
	Lesson 5, Wind—An Invisible River, p. 510	3	1. Identify examples of wind erosion. 2. Compare and contrast wind and water erosion. 3. Relate the shape of sand dunes to the process of dune formation. 4. Explain how human practices can increase wind erosion.	Image and Activity Bank 26-5
	Lesson 6, Glaciers—Rivers of Ice, p. 512	3	1. Describe the origin and locations of glaciers. 2. Explain the formation and movement of glaciers. 3. Provide evidence of past glaciers. 4. Describe formations resulting from glacial action.	Image and Activity Bank 26-6 Transparency Worksheet, p. 48 ▼ Exploration Worksheet, p. 50 Activity Worksheet, p. 51 Exploration Worksheet, p. 52 Transparency 89 Exploration Worksheet, p. 53
End of Chapter, p. 520				Activity Worksheet, p. 54 Chapter 26 Review Worksheet, p. 55 Chapter 26 Assessment Worksheet, p. 58
End of Unit, p. 522				Unit 8 Activity Worksheet, p. 61 Unit 8 Review Worksheet, p. 62 Unit 8 End-of-Unit Assessment, p. 67 Unit 8 Activity Assessment, p. 73 Unit 8 Self-Evaluation of Achievement, p. 76

* Estimated time is given in number of 50-minute class periods. Actual time may vary depending on period length and individual class characteristics.

▼ Transparencies are available to accompany these worksheets. Please refer to the Teaching Transparencies Cross-Reference chart in the Unit 8 Teaching Resources booklet.

Materials Organizer

Chapter	Page	Activity and Materials per Student Group
24	454	**Exploration 1:** poster board; markers; magazines; scissors; glue
	459	**Exploration 2:** Exploration Worksheet on page 5 of the Unit 8 Teaching Resources booklet or tracing paper and pencil; scissors
	462	**Exploration 3:** 2 or 3 L of water; shallow, rectangular pan (30 cm × 40 cm); hot plate; a few drops of dark food coloring; 8 pieces of thin cardboard, about 1 cm × 1 cm; watch or clock with a second hand; safety goggles
25	473	**Exploration 1:** meter stick or metric ruler
	479	**Exploration 2:** meter stick or metric ruler; optional materials: camera with film
	480	**People and Change:** magazines; newspapers
	482	**Exploration 3:** hammer; safety goggles
	484	**Exploration 5, Simulation 1:** rock samples (granite, limestone, sandstone, brick, piece of concrete); magnifying glass; large plastic or metal tray; freezer; 500 mL of water; large beaker or jar; watch or clock with a second hand; **Simulation 2:** whole piece of chalk; 200 mL of 0.1 M hydrochloric acid; two 250 mL beakers; 100 mL graduated cylinder; several paper towels; safety goggles; latex gloves; lab aprons (See Advance Preparation below.)
	489	**Exploration 6:** piece of plywood, about 20 cm × 30 cm; about 3000 cm³ of sand; 2 L of water; shallow, rectangular pan; wood blocks or bricks to prop up the board; pitcher
26	494	**A Numbers Game:** optional materials: meter stick or metric ruler; calculator
	496	**Exploration 1:** three 500 mL beakers; three different soil samples, such as sand, clay, and garden soil; 100 mL of water; 100 mL graduated cylinder; watch or clock with a second hand (See Advance Preparation below.)
	500	**Exploration 2:** watch or clock with a second hand; metric ruler; a few drops of food coloring; wax pencil; 5 L of water; 2 buckets; block of wood; stream table or large pan (about 35 cm × 120 cm) with drain, 2 lengths of rubber tubing (30 cm each), 2 small clamps or clothespins, and 50 cm trough made from a cardboard tube (additional teacher materials: cardboard tube; scissors or knife; see Advance Preparation below.)
	503	**Exploration 3:** about 3000 cm³ of sand; metric ruler; 5 L of water; block of wood; watch or clock with second hand; a few drops of food coloring; stream table or large pan with drain (about 35 cm × 120 cm), 2 lengths of rubber tubing (30 cm each), and 2 small clamps or clothespins
	511	**The Wind Shadow:** large jar; candle; a few matches; jar lid; small ball of modeling clay; safety goggles
	514	**Exploration 6:** 500 mL of water; small tray (about 4 cm deep and 30 cm long); 10 cm of wire; several masses of varying sizes; freezer; 2 supports, such as blocks of wood (additional teacher materials: wire cutters; see Advance Preparation below.)

Advance Preparation

Exploration 5, Simulation 1, page 484: Label each rock sample with a number. **Simulation 2, page 485:** You may also wish to test the chalk in advance to make sure it reacts with the acid.

Exploration 1, page 496: Soil samples may be collected from the schoolyard or brought from home by students. You may wish to obtain some samples from a plant nursery. If possible, sand and clay should be included.

Exploration 2, page 500: You may wish to cut the troughs out of the cardboard tubes using a knife or scissors before students enter the classroom.

Exploration 6, page 514: You may wish to cut the wire into 10 cm strips before class.

Unit Compression

Chapter 24 should be considered core because it introduces the study of geology and the diversity of geologic processes on Earth.

In Chapter 25, inviting a geologist to speak to the class might allow you to cover the material presented in Lesson 1 more quickly. Other ways to save time include omitting The Niagara Story on page 477 and allowing students to choose one of the following activities: Exploration 2, Exploration 4, or Simulation 1 of Exploration 5.

Chapter 26 provides more opportunities for compression. Lessons 1 and 2 should be considered core, but you may wish to omit some of the material that follows based on its relevancy to your local geology. For instance, students along the coast may be particularly interested in Lesson 4, whereas students in inland areas may prefer Lesson 5. Lesson 6, which covers glaciation, may also be omitted, but be aware that students usually find this information quite engaging.

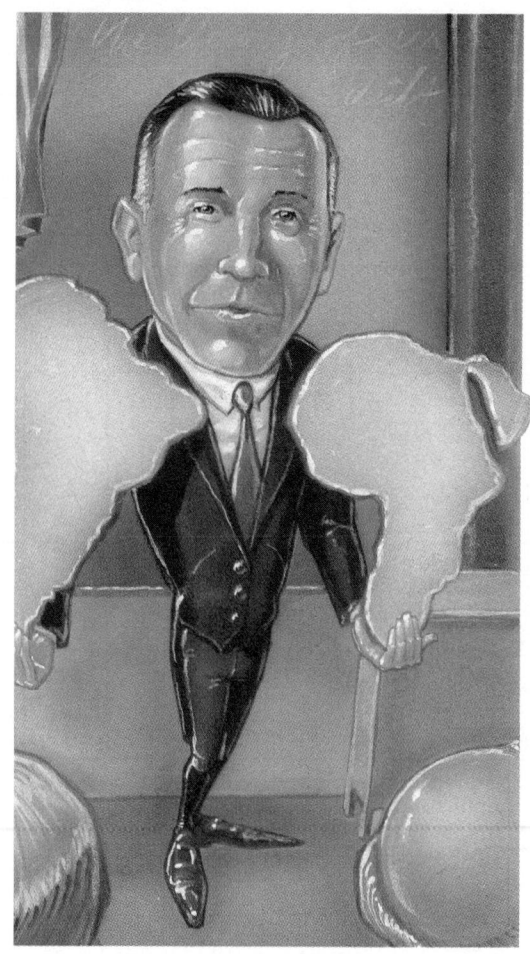

Homework Options

Chapter 24	See Teacher's Edition margin, pp. 454, 459, 460, 461, 462, 464, and 468 Activity Worksheets, pp. 3 and 15 Exploration Worksheets, pp. 5 and 9 Theme Worksheet, p. 13 SourceBook, p. S182
Chapter 25	See Teacher's Edition margin, pp. 470, 473, 475, 476, 479, 482, 487, and 490 Discrepant Event Worksheet, p. 22 Exploration Worksheets, pp. 23, 25, 26, and 28 SourceBook, pp. S185 and S190
Chapter 26	See Teacher's Edition margin, pp. 495, 498, 501, 504, 505, 506, 509, 513, 514, 518, and 520 Exploration Worksheets, pp. 45, 50, 52, and 53 Math Practice Worksheet, p. 46 Activity Worksheets, pp. 51 and 54 SourceBook, pp. S191, S192, S193, S195, S198, and S201
Unit 8	Unit 8 Activity Worksheet, p. 61 SourceBook Activity Worksheet, p. 77

Assessment Planning Guide

Lesson, Chapter, and Unit Assessment	SourceBook Assessment	Ongoing and Activity Assessment	Portfolio and Student-Centered Assessment
Lesson Assessment Follow-Up: see Teacher's Edition margin, pp. 455, 467, 480, 485, 490, 498, 501, 505, 509, 511, and 519 **Chapter Assessment** Chapter 24 Review Worksheet, p. 16 Chapter 24 Assessment Worksheet, p. 19* Chapter 25 Review Worksheet, p. 34 Chapter 25 Assessment Worksheet, p. 36* Chapter 26 Review Worksheet, p. 55 Chapter 26 Assessment Worksheet, p. 58* **Unit Assessment** Unit 8 Review Worksheet, p. 62 End-of-Unit Assessment Worksheet, p. 67*	SourceBook Review Worksheet, p. 79 SourceBook Assessment Worksheet, p. 83* 	Activity Assessment Worksheet, p. 73* **SnackDisc** Ongoing Assessment Checklists ♦ Teacher Evaluation Checklists ♦ Progress Reports ♦	Portfolio: see Teacher's Edition margin, pp. 455, 479, 484, 488, 497, 500, 503, 507, 511, and 519 **SnackDisc** Self-Evaluation Checklists ♦ Peer Evaluation Checklists ♦ Group Evaluation Checklists ♦ Portfolio Evaluation Checklists 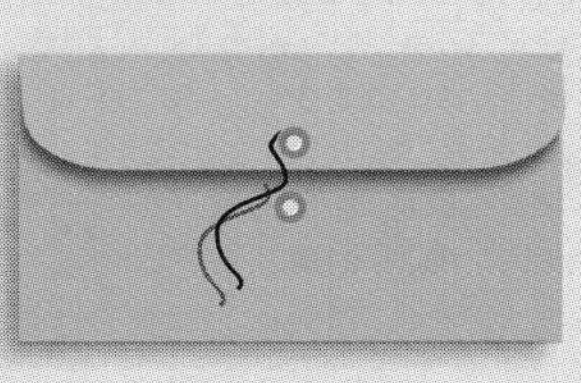 ♦

* Also available on the Test Generator software
♦ Also available in the Assessment Checklists and Rubrics booklet

Science Discovery is a versatile videodisc program that provides a vast array of photos, graphics, motion sequences, and activities for you to introduce into your *SciencePlus* classroom. *Science Discovery* consists of two videodiscs: Science Sleuths and the Image and Activity Bank.

Using the *Science Discovery* Videodiscs

Science Sleuths: The Missing Beach
Side A

The riverside beach next to a summer home has disappeared. The homeowner thinks it might have been stolen. The Science Sleuths must analyze the evidence for themselves to determine the real reason the beach disappeared.

Interviews
1. Setting the scene: Summer homeowner 32761 (play ×2)

2. Upstream neighbor 33494 (play)

3. Farmer on the ridge 34035 (play)

4. Fishing guide 34666 (play)

5. Downstream neighbor 35290 (play)

Documents
6. Fax from hydroengineering firm 35922 (step)

7. Memo to Department of Public Works 35925 (step)

8. Partial minutes from town meeting 35928 (step ×2)

Literature Search
9. Search on the words: BEACH, BOATS, FLOOD, SAND, TECUMSEH RIVER, WAKE 35932

10. Article #1 ("Governor Dedicates Fish Hatchery") 35934

11. Article #2 ("County Dumping Its Sand Bags") 35936

12. Article #3 ("Speed Limit for Party Boats") 35938

Sleuth Information Service
13. Yearly Tecumseh River flow 35940 (step ×6)

14. Historical maps of area 35948 (step ×2)

15. Tecumseh River peak flow 35952

16. Rainfall in the Tecumseh River area 35954

17. Construction-material standards 35956 (step)

Sleuth Lab Tests
18. River-sediment analysis 35959

19. Effect of flow rate on suspended sediment load 35961

20. Suspended sediment from three parts of the river 35965 (step ×2)

21. Sedimentation simulation in Tecumseh River 35969 (step ×2)

Image and Activity Bank
Side A or B

A selection of still images, short videos, and activities is available for you to use as you teach this unit. For a larger selection and detailed instructions, see the Videodisc Resources booklet included with the Teaching Resources materials.

24-1 Seeing the Sights, page 453
Great Sand Dunes, CO 764
An average sand dune is 9 to 15 m high. Shifting seasonal wind patterns pile the sand up higher and higher in this dune field. Here, the highest dunes are over 180 m above the surrounding plain.

Sierra Nevada 1036
The Sierra Nevada is a gigantic igneous rock body called a *batholith*.

Stock; Devils Tower, Wyoming 1000
A *stock* is a small form of igneous intrusion. The surrounding terrain was eroded away, leaving this erosion-resisting rock structure exposed.

◀▮ Step Reverse    Play ▶   Pause ▮▮   Step Forward ▮▶

24-2 Unlocking a Planet's Past, page 456

Continental drift 913–917 (step ×4)
225 million years ago all continents were joined in a landmass called *Pangaea*. (step) By 180 million years ago the continents had begun to drift apart. (step) Arrows show the direction of plate movement. (step) What makes the plates move? Geologists theorize that centers of heat set up currents that push the plates apart in the ocean depths. (step) Continental drift is still occurring today.

25-1 Evidence of Change, page 471

Sediment; compaction 546–548 (step ×2)
When sediment is deposited, the weight of more overlying sediment causes compaction. (step) When the sand, silt, and mud are compacted by the weight above, the layers become thinner. (step) The harder the sediments become. Eventually they turn into rock.

25-2 Examining One Type of Change, page 481

Mesa 594
An erosion remnant in the Southwest is called a *mesa* when it is broader across than it is high. A *butte* is higher than it is broad.

Weathering, mechanical; Yosemite National Park 599
Water runs down a cliff face, freezes in cracks in the rock, and expands, breaking off large pieces of rock from the cliff. This process created this large arch shape.

Weathering, chemical 602
Chemical weathering causes boulders to be rounded, as opposed to the jagged forms caused by mechanical weathering.

25-3 Gravity—The Great Leveler, page 486

Talus slope 614
Talus slopes are made up of rocks and boulders eroded from the mountains. The lake at the bottom of the photo is the result of glacial action.

Soil slump 17717–17969 (play ×2) (Side B only)
This soil slump is caused by the river undercutting its bank, causing a large section of soil to dislodge and slump into the river.

26-1 The Power of Water, page 494

River flow change 10488–10824 (play ×2) (Side B only)
In these time-lapse sequences, the streams have filled up due to the release of water from dams upstream. In nature, streams have low periods during dry seasons but fill during wet seasons.

26-2 Water on the Move, page 499

Waterfall; Yellowstone 678
This is an example of a very young stream.

26-3 How Rivers Change the Land, page 502

Meandering stream with oxbows 664
This slow-moving stream shows signs of many years of meandering.

26-4 Where Land Meets Sea, page 506

Delta 679
Rivers carry large amounts of sediments that are deposited at the river mouth as a delta. This delta is the Mississippi River delta in the Gulf of Mexico.

Coast; cliff 687
This coastline is rapidly eroding due to heavy wave action. As a result, steep cliffs have formed.

26-5 Wind—An Invisible River, page 510

Sandstone arch 608
Sandstone arches are a form of differential erosion in which some parts of rock formations erode faster than others.

Sand dune field; barchan dunes 759
The tiny dark specks between the dunes are shrubs 30 to 60 cm high.

Sand dunes 11974–12973 (Side B only) (play ×3)
Blowing sand moves from the rear of the dune and piles up on the crest. (play) This may look like continuous movement of sand grains; but actually an individual grain skips, lands, and collides with another, causing the second grain to skip. (play) This repeated skipping movement is called *saltation*. (play) Each avalanche down the dune face covers the preceding one, creating layers in the sand. This process causes the dune to move forward.

The dust bowl years 8206–8521 (play ×2) (Side B only)
Drought, overgrazing, and poorly managed cultivation contributed to the formation of dust bowls like this one, which developed in the 1930s in the south-central United States.

26-6 Glaciers—Rivers of Ice, page 512

Glacier, alpine 798
This glacier is probably several hundred meters thick. The ice conceals a deep valley.

Glacier, side-joining 800
If this glacier were to melt away, the valley of the side glacier would be a hanging valley.

Iceberg 803
When glaciers meet the sea, large portions of ice break, or calve, off and become icebergs.

Glacial effects on rocks 816–817 (step)
The action of ice and its load of rock debris wears huge walls of rock into smooth, polished surfaces. (step) Large rock particles in a glacier scraped grooves in this rock as the glacier passed over it.

Glacial boulder; erratic 822
This giant boulder is a great distance from its source. It was left behind by a melting continental glacier. Boulders like these are called *erratics*.

Glacial valley formation 792–797 (step ×5)
These six oil paintings of Yosemite National Park show the process of alpine glacial erosion. (step) The first three pictures give a pre-glacier view. They show a valley being eroded by a stream. (step) Stream erosion carves the valley into canyons. (step) A glacier fills the canyon to its maximum depth. (step) The glacier begins to melt, and the first glacial-carved features are revealed. (step) Yosemite Valley as it appears today, not including the big lake.

Unit 8 OUR CHANGING EARTH

Pose the following question to students: How does the Earth change? Students may describe the variation in weather from one day to the next or changes in a familiar landmark such as a mountain, a waterfall, or a coastline. Other students may describe earthquakes and volcanic eruptions. Write students' suggestions in one column on the chalkboard. Label that column Changes. Ask students: How can you tell that these changes have occurred? Label a second column Evidence, and have students suggest evidence for each of the changes listed. For example, they may know that a hillside has changed because they saw rock slides and debris where there had not been any before.

A good motivating activity is to let students listen to the English/Spanish Audiocassettes as an introduction to the unit. Also, begin the unit by giving Spanish-speaking students a copy of the Spanish Glossary from the Unit 8 Teaching Resources booklet.

Unit 8 OUR CHANGING EARTH

Photographers are forced to retreat as clouds of superheated ash pour from Mount Pinatubo in the Philippines

Connecting to Other Units

This table will help you integrate topics covered in this unit with topics covered in other units.

Unit 1 Science and Technology	Observations and inferences were important in the development of the theory of plate tectonics.
Unit 2 Patterns of Living Things	Animal behavior is often affected by the seasonal changes in the weather.
Unit 3 It's a Small World	The impact of sedimentation, runoff, and freezing on stagnant water can influence germ growth.
Unit 4 Investigating Matter	Many weathering processes, such as glaciation and erosion, involve a change in the state of matter.

In June 1991, 600 years after its last eruption, Mount Pinatubo once again rocked the island of Luzon in the Philippines. Accompanied by nightmarish flashes of colored light, the once dormant volcano exploded in one of the most powerful eruptions of the century. The eruption killed nearly 900 people, destroyed 42,000 homes, and buried 100,000 acres of farmland in ash. Without the advance warning of geologists from the Philippines and the United States, however, many more people would have died.

Despite the violence of volcanic eruptions, volcanic activity can be constructive as well. For example, the Earth's early atmosphere was formed from gases released during volcanic eruptions. The atmosphere, in turn, provided the rains that created the oceans, and it warmed the Earth by capturing the sun's energy. Today, as in the past, volcanic activity helps shape the continents and the sea floor.

What other geological forces affect our planet? How do geologists figure out how the Earth has changed over time? In this unit, you will find that getting to know the changing nature of the Earth takes more than scientific observation and inferences—it takes imagination.

451

Ask students to imagine that they are geologists traveling in the truck featured in the photograph. What would they be thinking? What would they hope to learn by studying volcanic eruptions? What might they learn about the Earth's past? You may wish to have students write a story that answers these questions.

Answers to
In-Text Questions

Other prominent geological processes include erosion by wind and water, sedimentation, scouring by glaciers, and other phenomena associated with tectonic activity, such as earthquakes and geysers.

Geologists can sometimes consult historical records to determine how the Earth has changed over time. But because geological time scales are so immense, a geologist must often resort to observations of current events to infer what happened in the past. For instance, the presence of a fossil in a rock layer could indicate the type of climate that would have been necessary to sustain that kind of organism.

Connecting to Other Units, continued

Unit 5 Chemical Changes	The composition of the Earth's crust is altered by processes such as volcanism at crustal plate boundaries.
Unit 6 Energy and You	Many agents of weathering, such as blowing winds and flowing water, can be useful sources of energy.
Unit 7 Temperature and Heat	The boundaries of crustal plates are regions of intense pressure and heat.

Connecting to Other Chapters

Chapter 24
investigates the geologic diversity of the Earth and introduces the theory of plate tectonics.

Chapter 25
examines evidence of geologic changes and the role that gravity plays in such changes.

Chapter 26
explores how landforms are altered through sedimentation, erosion, and the actions of winds and glaciers.

Prior Knowledge and Misconceptions

Your students' responses to the ScienceLog questions on this page will reveal the kind of information—and misinformation—they bring to this chapter. Use what you find out about your students' knowledge to choose which chapter concepts and activities to emphasize in your teaching.

In addition to having students answer the ScienceLog questions on this page, you may wish to have them perform the following exercise: Separate the class into groups of three or four students, and tell them that they are teams of geologists who must prepare for an expedition to an unexplored region in the Andes. Ask group members to prepare an itinerary, including what they will be looking for, what each group member's task will be, and what the group hopes to find. Remind students that there are no right or wrong answers to this assignment. Collect the papers, but do not grade them. Instead, read them to find out what students know about geology and what geologists do, what misconceptions they may have, and what they find interesting about the subject.

CHAPTER
24
The Changeable Planet

1 ▸ **Has the Earth always looked like it does now? In what ways do you think it has changed?**

2 ▸ **What kinds of things does a geologist do?**

3 ▸ **How do fossils help us to understand the Earth's past?**

4 ▸ **Is it possible for the continents to change their location on the Earth?**

ScienceLog

Think about these questions for a moment, and answer them in your ScienceLog. When you've finished this chapter, you'll have the opportunity to revise your answers based on what you've learned.

452

Seeing the Sights

Where in the World?

"And now it's time for the big drawing—a trip for two to anyplace in the world! Are you ready? Have your tickets handy. Here we go! The winning number is . . ."

Suppose you won. Where in the world would you go? Read the travel posters on pages 454 and 455. Do any of the places they describe appeal to you?

The World Is Full of Different Places

People like to visit new places, if only for a change of scenery. You could have many different reasons for wanting to go to a particular place. Your ideal vacation spot might have a certain type of climate or offer certain types of recreational activities. Or you may be interested in seeing wildlife or experiencing different cultures.

Imagine for a moment that the most important consideration for your trip is to go somewhere with spectacular scenery. Carry out the following steps:

1. Read the travel posters again.

2. In a few sentences, summarize the scenery that each describes.

3. Suggest reasons (or causes) for the shape and form of the different landscapes.

4. Which scenic landscape would you choose for your trip? What are your reasons?

Some of the answers to how landscapes have been formed can be found in the study of **geology**. In this unit, you will make observations about the landscape around you, much like a geologist would. By examining some of the forces that shape the land, you will gain a better understanding of how the world came to be what it is today.

Backpackers on Bright Angel Trail in Grand Canyon National Park, Arizona

453

FOCUS

Getting Started

Explain to students that there are a wide variety of landforms on Earth. Ask students to think of some landforms and to describe them. Students may be interested to know the following facts about Earth: the highest mountain is Mount Everest (8848 m); the lowest point is the Dead Sea (393 m below sea level); the longest river is the Nile River (6690 km); the largest sheet of ice covers 98 percent of Antarctica (with an average thickness of 2200 m); and the largest desert is the Sahara (9,000,000 km^2).

Main Idea

Geology helps explain the reasons for the vast variety of landforms.

TEACHING STRATEGIES

Where in the World?

GUIDED PRACTICE Have students read the travel posters on pages 454 and 455. A world map or globe (preferably showing topography), travel magazines, and brochures may help students choose a location. The purpose of this activity is to have students identify many locations with different geological features.

The World Is Full of Different Places

Ask students to read page 453 silently, and then discuss the reasons why people like to visit different scenic locations. Ask volunteers to read each poster aloud, and then discuss with students the questions in the text. Sample answers for each location can be found on pages 454 and 455.

INDEPENDENT PRACTICE If a slide projector is available, ask students to bring in slides of scenic areas that they have visited, or have students borrow slides from the library to show the class an area that they would like to visit. Encourage students to suggest reasons why the landscapes vary.

LESSON 1 ORGANIZER

Time Required
3 class periods

Process Skills
observing, analyzing, inferring

New Term
Geology—the study of how the Earth formed and how it changes

Materials (per student group)
Exploration 1: poster board; markers; magazines; scissors; glue

Teaching Resources
Activity Worksheet, p. 3
Exploration Worksheet, p. 4

Answers to
The World Is Full of Different Places, page 453

Mountain Memories

2. This travel poster describes an area with high, snowy mountains. There are ice fields, glacial lakes, and deep rocky valleys.

3. This area is at a high elevation. Glaciers cut deep valleys in the land. They left behind many holes that filled with water from the melted snow and ice.

4. Students who choose this location for their trip may enjoy mountain scenery. Related sports include skiing and mountain climbing.

Arizona's Marvel

2. This travel poster describes deep, rock canyons and the ruins of prehistoric Indian cliff dwellings.

3. Rivers flowing in the same location over thousands of years cut deep canyons into the land.

4. Students who choose this location for their trip may enjoy colorful and rugged canyon scenery. Related activities include horseback riding, hiking, and camping.

Answers to The World Is Full of Different Places continued ▶

Homework

The Activity Worksheet that accompanies Seeing the Sights on page 453 makes an excellent homework assignment (Teaching Resources, page 3).

EXPLORATION 1

Answers to
Selling Scenery

Accept all reasonable responses. For each of the four activities, students should emphasize the positive aspects of the scenic landscape they choose, such as recreation or sightseeing opportunities. If students choose to address one or more geologists, their sales pitch should focus on appropriate geological features of the landscape, such as mountains or bodies of water.

MOUNTAIN MEMORIES

ARIZONA'S MARVEL

Banff National Park is an area of startling contrasts. It is an inheritance of unspoiled beauty, ice fields, glacial lakes, and deep valleys cutting between soaring snow-capped mountains. At the same time, it is a place where the visitors can enjoy all the conveniences of modern hotels and motels, excellent restaurants, and wonderful ski facilities.

 BON VOYAGE TRAVEL

Thrill to one of the most stunning scenic attractions in the world—the Grand Canyon. This tour makes it easy for you to visit prehistoric Indian cliff dwellings, the gorgeous red rock formations of Oak Creek Canyon, the fabled city of Prescott, and many other places that are unique to Arizona.

Selling Scenery

How good a salesperson are you? Test your skill. Choose *one* of the activities below.

1. You are a travel agent. What would you say to your clients to persuade them to visit the scenic landscape you chose in step 4 on page 453? Write down the main points you would make.

2. Design a travel brochure to persuade vacationers to visit the area you chose. Highlight the area's scenic attractions.

3. You are a travel agent again, but this time your client is a geologist. How would you change your sales pitch?

4. Now you are a geologist, and you are planning an expedition to a place of your choice. Where would you go? What would be your sales pitch to get others to join you?

454

Cooperative Learning
EXPLORATION 1

Group size: 3 to 4 students
Group goal: to write a sales presentation or design a travel brochure for geologists
Positive interdependence: Assign students roles such as artistic director, copywriter, team manager, and editor.
Individual accountability: Have each student write a self-assessment. Encourage students to review their personal participation in the group and the effectiveness of their contributions.

Remind students of the discussion from the previous page. Stress that their emphasis should be on geological features. Common tourist attractions such as restaurants, galleries, and museums may also be included. In question 3, students should focus strictly on geological attractions. Encourage students to use precise, scientific language.

★ **An Exploration Worksheet is available to accompany Exploration 1 (Teaching Resources, page 4).**

Scandinavian Heart

Scandinavia opens its whole heart to you in this tour that brings you the best of everything Scandinavian: the wonderful cities of Copenhagen, Oslo, and Bergen; Norway's wild mountains and fjords; and Sweden's miles of rolling forest and magnificent lakeside resorts.

BON VOYAGE TRAVEL

Celtic Magic

Feel the magic of Ireland, the Emerald Isle. The meadows of southern Ireland are as green as they say they are. Tour the quiet lakes of Killarney, the peat bogs and ice-scoured landscape of Connemara, and the sandstone mountains of Kerry. This spellbinding tour will take you back in time as you visit medieval castles and Celtic crosses.

BON VOYAGE TRAVEL

Paradise Islands

Experience paradise on Earth. The Hawaiian islands will greet you with tropical breezes, gently swaying palms, and brilliantly colored flowers. Explore the beauty of this volcanic island chain nestled in the deep blue waters of the Pacific. Immerse yourself in Polynesian culture amidst lush green valleys and spectacular waterfalls, or relax on one of Hawaii's black sand beaches.

BON VOYAGE TRAVEL

455

Answers to
The World Is Full of Different Places, continued

Scandinavian Heart
2. This travel poster describes high mountains, many lakes, and large areas of hilly, forested countryside.

3. The ocean waters rose to cover valleys formed by glaciers so that only high mountains remained above water. Rain and snow provide plenty of water for lakes.

4. Students who choose this location for their trip may enjoy ocean, lake, and mountain scenery. Related activities include hiking and boating. Tourist attractions of the major cities nearby may also be mentioned.

Paradise Islands
2. This travel poster describes tropical, volcanic islands that are surrounded by coral reefs. There are many sandy beaches.

3. Volcanoes erupted in the middle of the Pacific Ocean. The lava from these eruptions built up over time to form these islands. The rainy climate helps wear down the rock into sand.

4. Students who choose this location for their trip may enjoy tropical beach scenery. Related activities include boating and snorkeling.

Celtic Magic
2. This travel poster describes flat, grassy areas surrounded by rocky mountains. There are many lakes. The coastline is very rocky, and there are many small bays.

3. The rainy climate in Ireland caused meadows and lakes to form in parts of a rocky, mountainous island.

4. Students who choose this location for their trip may enjoy mountain and pastoral scenery. Related activities include hiking, fishing, and boating.

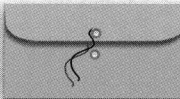

PORTFOLIO
Students may wish to include in their Portfolio a description of the scenic landscape they chose for their trip (page 454, question 4). Encourage students to add reasons for their choice.

FOLLOW-UP

Assessment
Have students write and design a travel poster describing a nearby area. Posters should include specific geologic descriptions of all landforms.

Reteaching
Have students prepare a list of landforms from the travel posters. Answers might include mountains, valleys, lakes, canyons, volcanic islands, ice fields, fjords, cliffs, rock formations, and beaches.

Extension
Have students do research to find out how landforms on the ocean floor compare with those on the Earth's surface. Ask: How are they the same? How are they different? *(Students should conclude that many are similar but that those on land are subject to more weathering.)*

Closure
Have students create a topographical map that highlights prominent geographical features of the local area.

FOCUS

Getting Started

Remind students of the activity in Exploration 4 on page 27 called The Dancing Disk. You may wish to repeat the activity as a demonstration. Ask: How do you think this activity is similar to changes that occur on and beneath the Earth's surface? *(Accept all reasonable responses.)* Hot materials beneath the Earth's surface cause the sections of the surface to move, resulting in the formation of mountains, volcanoes, and earthquakes.

Main Ideas

1. Geologists make inferences about the past based on observations of the present.
2. The theory of continental drift proposes that all of the continents were once joined together and have since gradually drifted apart.
3. The theory of plate tectonics proposes that the Earth's crust is made up of plates that are continually shifting.

TEACHING STRATEGIES

Answer to
In-Text Question

Ⓐ Other sample questions might include the following: How do geologists classify rocks? What is the Earth like far below the surface? Are islands really underwater mountains? How old are glaciers?

Teaching Strategies for Geologists—The Earth's Historians are on the next page. ▶

LESSON 2 · Unlocking a Planet's Past

Geologists— The Earth's Historians

At one time or another, you may have collected rocks. Chances are, something about these rocks caught your eye. Perhaps they were pretty or oddly shaped. Perhaps they contained the fossil remains of ancient plants or animals. Perhaps you wondered how these rocks formed. Geologists also collect rocks, but they go one step further. By looking closely at rocks and the formations they come from, geologists are able to "read" the Earth's history. To the geologist, rocks tell the story of how the Earth came to be what it is today and even of what it might be like millions of years from now. The photographs on these pages give you an idea of some of the different areas of geological study. Start a class list of the questions you and your classmates would like to ask a geologist. Some sample questions are shown on the black board at right. Ⓐ

The types of minerals and fossils in a rock provide the geologist with clues about how the rock was formed.

1. Where do mountains come from?

2. Will the Earth ever erode away completely?

3. What causes earthquakes?

4. Why are fish fossils found on mountains?

5. Will an ice age come again?

6. Why do volcanoes occur only in certain places?

LESSON 2 ORGANIZER

Time Required
3 class periods

Process Skills
observing, inferring, analyzing, communicating

Theme Connection
Structures, Energy

New Terms
Asthenosphere—the region of the mantle directly below the lithosphere, on which the tectonic plates flow
Core—the central portion of the Earth, below the mantle

Crust—the thin, rocky, outer portion of the Earth
Lithosphere—the portion of the Earth consisting of the crust and the upper mantle; it is divided into tectonic plates.
Mantle—the middle layer of the Earth, between the crust and the core
Pangaea—In Wegener's theory, the single continent that drifted apart to become the continents we know today
Plates—large segments of the Earth's crust that are constantly in motion

continued ▶

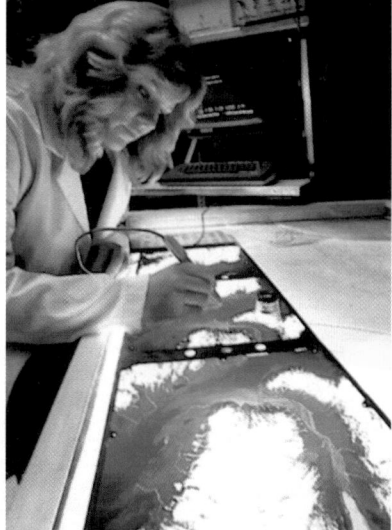

Satellites provide geologists with color-coded images of the Earth's surface.

A volcanologist is a person who studies volcanoes.

Mapping surface features helps geologists record changes over time.

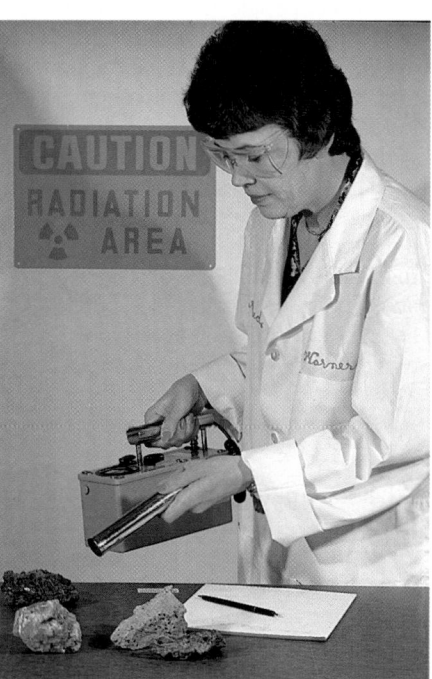

Radioactivity levels can help determine the age of rocks.

457

Geologists—The Earth's Historians, pages 456–457

INDEPENDENT PRACTICE If any students have rock collections, ask them to share these collections with the class. Allow students time to observe the different collections and to ask questions about them.

After students have had time to look at the photographs on pages 456 and 457, divide the class into discussion groups. Use the questions on page 456 and the class list of questions to guide student thinking. Do not worry about correct answers at this point.

Answers to
Geologists—The Earth's Historians, page 456

1. Mountains can be built up by volcanic activity or by collisions between plates of the Earth's crust.

2. Land formations will eventually erode away, but recycling of Earth's materials constantly creates new formations. This means that the Earth's surface may be reshaped by erosion, but the Earth itself will not be destroyed.

3. Movements within the Earth's crust cause parts of the crust to grind together, which creates stress. When the stress becomes too great, the rocks along a crustal boundary fracture. This causes an earthquake.

4. Parts of the Earth's crust that were originally formed on the sea floor can be pushed up when they collide with other parts of the crust.

5. The geologic record shows that ice ages have occurred in cycles. On the basis of that evidence, another ice age will probably occur, although the exact time cannot be predicted.

6. Volcanoes occur where interactions within the Earth's crust release molten magma (from the Earth's core) through the Earth's surface.

ORGANIZER, *continued*

Plate tectonics—theory of how the continents move, based on Wegener's theory of continental drift

Materials (per student group)
Exploration 2: Exploration Worksheet on page 5 of the Unit 8 Teaching Resources booklet or tracing paper and pencil; scissors
Exploration 3: 2 or 3 L of water; shallow, rectangular pan (30 cm × 40 cm); hot plate; a few drops of dark food coloring; 8 pieces of thin cardboard, about 1 cm × 1 cm; watch or clock with a second hand; safety goggles

Teaching Resources
Exploration Worksheets, pp. 5 and 9
Transparency Worksheets, pp. 7 and 11
Theme Worksheet, p. 13
Transparencies 78–81
SourceBook, p. S182

What Happened Here?

How can we find out what event occurred after the event has already taken place? The best we can do is to try to reconstruct the event from the evidence that is left. For the event shown at right, make a list of observations, and then try to explain what you think happened based on these observations. Compare your list with those of your classmates. How closely do you agree? Can you decide which explanation best fits the evidence?

Interpreting the Evidence

Some evidence is easy to interpret. No doubt everyone agreed that a catastrophic event occurred. Some of the evidence in the photo is more difficult to interpret. For example, what caused the incident? How can you decide what actually happened? **A**

In court, much of the evidence comes from witnesses, people who saw or heard the event taking place. Are there witnesses to describe the events shown above? What happens when there are no witnesses? What other kinds of evidence could be gathered? **B**

This is the task that faces geologists. How do they figure out what happened in an area thousands or millions of years ago? There are no witnesses, but there is evidence. What do you think the evidence might be? **C**

Something to Think About

What kind of events are occurring today that might help a geologist interpret the evidence from past events? Add at least three ideas of your own to those pictured.

Over millions of years, rivers carve deep canyons in the landscape.

Volcanoes eject lava and huge clouds of ash.

Constructing Theories

By reading subtle clues and details in the Earth's surface, geologists learn to recognize the evidence that each of these events leaves behind. The more they know about the events happening now and about the records that these events leave behind, the easier it is for them to construct theories about the history of the Earth.

The following Exploration puts you in the shoes of one Earth scientist who used his observations to develop a startling new theory.

EXPLORATION 2

A Continental Jigsaw Puzzle

Imagine that you are a young geologist living in Germany in the early 1900s. Looking at a map of the coasts of Africa and South America, you notice that the coastlines appear to fit together like the pieces of a jigsaw puzzle. You also remember reading about the discovery of fossils of a small lizard called *Mesosaurus* that were found only in eastern South America and southwestern Africa. You then begin to wonder whether all the continents can be made to fit together.

Trace the continents from the map below. Then cut them out, and try to put the pieces together to get the best possible fit. What clues do you have? Are there missing pieces?

Finally, you believe you have found the right fit. You ask yourself, "What does this mean? How can I possibly explain what I think happened in the past?" For days you puzzle over the answer. At last you think you have worked out a possible solution. With great excitement you write a letter to the Royal Geological Society in London and ask them to consider your ideas. What might the letter say? Finish the letter started here, and sign your name as Alfred Wegener.

April 1, 1911

Royal Geological Society
Holywell House, Worship Street
London, EC2A 2EN

My Esteemed Colleagues,

I wish to bring to your attention a matter of great importance. For several months I have been working on an idea…

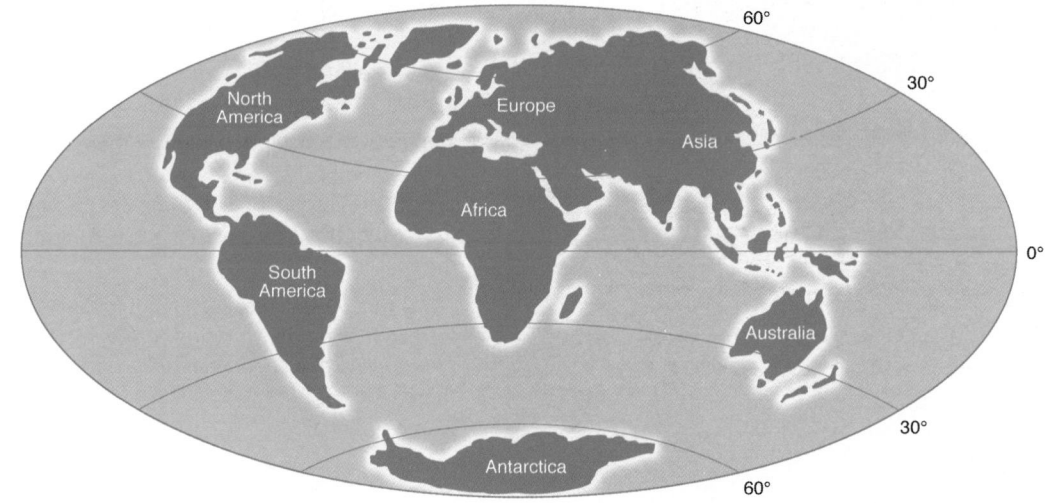

459

Homework

The Exploration Worksheet that accompanies Exploration 2 makes an excellent homework activity (Teaching Resources, page 5).

EXPLORATION 2

Have students study the world map on page 459. Ask them to make continent cutouts using the Exploration Worksheet on page 5 of the Unit 8 Teaching Resources booklet.

As students work with the cutouts, they should find that there are several possible ways of fitting the continents together. Instruct students to study the coastlines for clues. One obvious clue involves the shape of the coastlines of western Africa and eastern South America. There are other obvious clues in the regions near the Mediterranean Sea and Indian Ocean.

When students find gaps in their fit, discuss what may have happened to the "missing" pieces. In many cases, the pieces are actually still there in the form of continental shelves, which are not shown on this map. Other missing pieces may have moved to unexpected locations or may have changed shape as they became attached to other land masses. The currently accepted fit is pictured on page 466 of the Pupil's Edition.

Student letters to the Royal Geological Society should express the idea that the continents as they are known today were once part of a larger landmass. It seems reasonable to conclude that the continents have since moved from their original positions. You may wish to read some of the letters aloud to the class or to display them on a bulletin board.

Theme Connection

Structures

Focus question: How can the structure of the Earth be compared with that of a peach? *(The peach pit could be compared to the Earth's solid inner core. The fleshy part of the fruit could be compared to the mantle, and the skin could be compared to the crust.)*

Putting the Pieces Together

Have student volunteers read the hypothetical dialogue presented on pages 460–461. Then have students form discussion groups and decide whether they agree with Wegener based on the evidence presented. Reassemble the class, and ask one member of each group to share his or her group's decision.

Homework

Have students do library research on *Mesosaurus* and prepare a brief written report on their findings. Reports should include a statement about whether their research supports the findings of Alfred Wegener.

A Transparency Worksheet is available to accompany the material on this page (Teaching Resources, page 7). Transparency 78 is also available.

Putting the Pieces Together

A Debate

The Royal Society, formed in 1662, provided a forum in which scientists and philosophers could gather to examine, discuss, and criticize new discoveries and old theories. Try to imagine the discussion that might have occurred after Alfred Wegener's letter was presented to the group. Perhaps it went something like this:

"What a preposterous idea! How could anyone imagine that all the continents were once joined together? Such an idea is sheer fantasy, a beautiful dream."

"The dream of a great poet, I would say. What a bold and imaginative idea!"

"What does he call this supercontinent? **Pangaea**? I believe that means 'all Earth.' Rather appropriate, wouldn't you say?"

"When was **Pangaea** supposed to have existed?"

"About 300 million years ago, I believe."

"I think that he has been doing too many jigsaw puzzles!"

"But he isn't the first person to suggest the similarities of the coastlines on each side of the Atlantic. The scientist Francis Bacon suggested the idea as early as 1620."

"It is very difficult to imagine what kind of forces could split a continent apart, let alone keep the pieces moving apart."

"Don't be too quick in your judgment. Wegener offers some very convincing evidence."

"Let us examine this evidence!"

"Identical rock formations have been found in Africa and South America, as well as similar fossils. *Mesosaurus* has been found in only two places in the world, Africa and South America."

"That's true, but some think that there was once a land bridge connecting these two continents. Certainly a reptile of that size could never have swum the 2000 miles of ocean separating the continents."

"Only three years ago, coal deposits that formed from ancient tropical forests were found in Antarctica. His theory might help to explain this phenomenon."

"Antarctica could have had a warmer climate many years ago."

"As you know, the fossil plants *Glossopteris*, large seed ferns, are common to southern South America, southern Africa, India, Australia, and Antarctica. If the continents were once joined, these different locations could be explained."

"All that is well and good, but there is a major problem with Wegener's theory. Ocean floors are too rigid to let the continents barge through them. Besides, what kind of force could push continents about the face of the planet?"

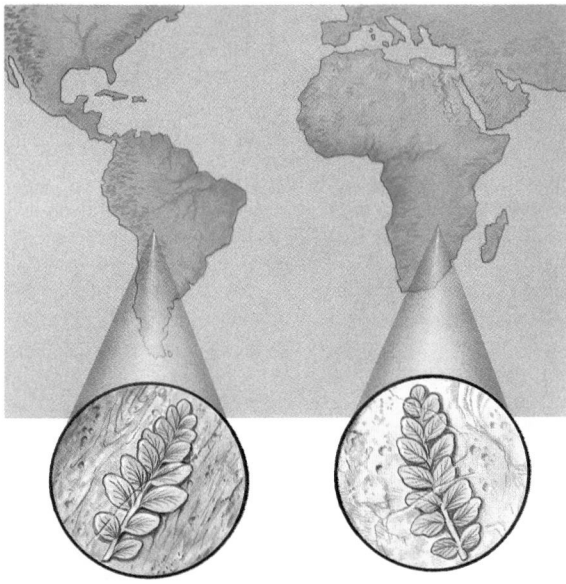

And so the discussion continued as the two sides of the issue were debated.

A Response

How do you think the discussion ended? To help you make a decision, make a list of comments that supported the theory, those that questioned the theory, and those that did neither. When you finish, look over your list and try to decide which comments carry the most weight. What response do you think Wegener received to his letter? Be prepared to defend your viewpoint in a class discussion.

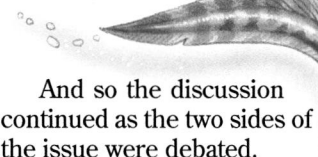

Mesosaurus, a small aquatic lizard, was less than 1 m long.

How the Story Ended

New ideas, particularly controversial ones such as continental drift, are usually met with strong resistance. Why do you think that is? It took 15 years before scientists Ⓐ decided to take a good look at Wegener's theory of continental drift. In 1926, the American Association of Petroleum Geologists organized a meeting to consider Wegener's hypothesis. Sadly for Wegener, the big guns of geology blew it right out of the water. And there the matter stood for another 30 years. During the 1950s and 1960s, new techniques provided new evidence and scientists began to theorize that the Earth's crust is composed of a number of huge **plates**, moving in various directions. By the 1970s, this theory of **plate tectonics** was almost universally accepted by scientists around the world. However, some questions about the theory remain. The question that troubled members of the Royal Society in 1911 is still not fully understood. Do you remember what that question was? The next Exploration may provide you with some clues to the answer.

461

Answers to
Thinking Things Over

1. The food coloring moved in a circular pattern because of convection currents caused by heating the water in the pan.

2. When the water was heated, the cardboard pieces in the center of the pan moved away from each other in a circular pattern. The pieces in the corners of the pan did not move at first unless they were touched by one of the other pieces. Eventually, all of the pieces began to move in circles. Answers will vary. Students may note that in this model, the heat source was closer to some of the pieces than to others. Heat from the Earth's core is more evenly distributed than is the heat in the model.

3. The water represents the molten rock of the Earth's mantle, the cardboard pieces represent tectonic plates, and the hot plate represents the Earth's core.

Homework

The Exploration Worksheet that accompanies Exploration 3 makes an excellent homework activity (Teaching Resources, page 9).

Creating a Model

Although questions still remain, most scientists think that plate movement is caused by convection currents. When a gas or a liquid is heated unevenly, the part that is heated rises. The movement of heated gas or liquid is called a *convection current*. To study processes that they cannot see, scientists often make models. In this Exploration, you will create a model of plate movement by convection current.

You Will Need

- water
- a shallow, rectangular pan (30 cm × 40 cm)
- a hot plate
- dark food coloring
- 8 pieces of thin cardboard, about 1 cm × 1 cm
- a watch or clock with a second hand

What to Do

1. Fill the pan three-quarters full of water. Place the pan on the hot plate.

2. Heat the water over very low heat for 30 seconds. Add a few drops of food coloring to the center of the pan. Write down your observations in your ScienceLog. Turn off the heat.

3. Label the cardboard pieces 1 through 8. Carefully place pieces 1 through 4 in the center of the pan, as close together as possible. Place the rest of the cardboard pieces in the corners of the pan.

4. Turn the hot plate on low. Sketch the pattern of movement for each cardboard piece. Turn off the heat.

Thinking Things Over

1. What happened when you added food coloring to the water? How can you explain your observations?

2. Describe what happened to the cardboard pieces when the water was heated. Do you think this activity is a good model for plate tectonics? Explain.

3. Examine the illustration of the Earth's interior on the next page. What portion of the Earth does each of the following represent: the water, the cardboard pieces, the hot plate?

The Inside Story

The imaginary story that follows has some real-life facts about the structure of the Earth. Follow along and use what you learn to correctly label the illustration of the Earth below. Do not write in this book.

Intrepid adventurer Ricochet Roy climbed aboard the Roto-Earth Craft and prepared for his descent into the previously unexplored depths of the Earth. Up to this point, information about the Earth's interior was known only from seismic data gathered during earthquakes. Roy planned to dig his way to the bottom of the mystery of the Earth's structure.

Day 1
First to the **crust**. By my calculations, this thin, rocky layer makes up only 1 percent of the Earth's total volume. It's sort of like the skin on an apple.

Day 2
Even after digging to a depth equal to the length of 50 football fields, I still haven't made it through the crust. The continental crust is much thicker than the crust under the oceans. Now on to the mantle!

Day 6
As I journey deeper into the Earth, I find that the temperature continues to climb. The upper portion of the **mantle** is still somewhat rocky—together with the crust, this rocky portion is known as the **lithosphere**. The lithosphere is divided into segments known as *tectonic plates*.

Day 17
The area of the mantle beneath the lithosphere makes me think of plastic putty. It's solid rock but is able to flow because the temperature and pressure are greater than at the surface. This area is known as the **asthenosphere**. As the taffy-like asthenosphere flows, it carries the tectonic plates along with it.

Day 25
As I descend into the lower mantle, temperature and pressure continue to increase. The outer **core** is next. Temperatures range from 2200°C to 5000°C. Molten iron and nickel are found here.

Day 30
I am approaching the inner core, which has a temperature of 5000°C. The core is made of iron and nickel, which would normally melt at this temperature. However, the pressure is so great that the core remains solid. The temperature is getting a bit much for me too. Time to return home!

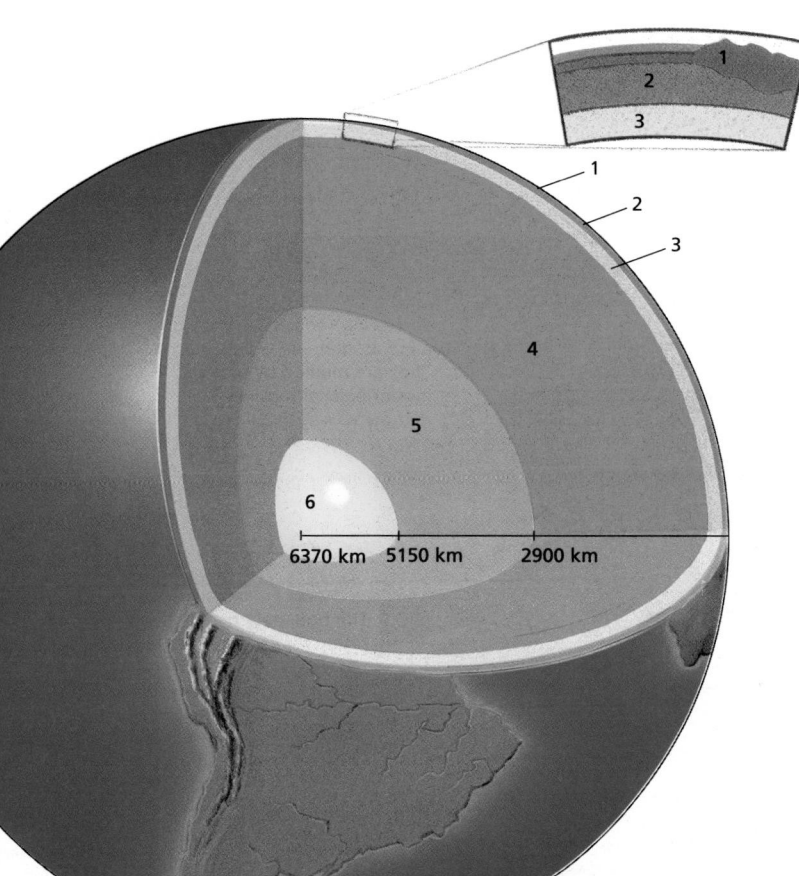

6370 km   5150 km   2900 km

The Inside Story

The journey on this page is intended to aid students in visualizing the overall structure of the Earth. As students read this material, help them to develop a conceptual understanding of the defining characteristics of each layer. The numbers in the illustration follow the order in which each layer appears in the story. You may wish to point out that the time spent in the story traveling through a layer is proportional to the thickness of the layer.

CROSS-DISCIPLINARY FOCUS

Social Studies
Earthquakes affect the economy, architecture, and many other aspects of life in regions near the plate boundaries described in this unit. Scientists estimate that hundreds of small earthquakes occur every day and that larger and more dangerous earthquakes are always a possibility. As a result, cities located in tectonically active areas, such as the Pacific Rim, have special building codes and other laws that are designed to minimize earthquake damage. Have students do research on the architectural and economic impact of earthquakes on an area. Some possibilities to investigate include costs required to make a building earthquake-safe and architectural strategies for earthquake-resistant buildings.

463

Continental Facts

Use these facts to conduct a class discussion. Assign one question to each group of students, and have each group present its analysis and opinions to the class. Students will probably not know the exact answers, so encourage them to make educated guesses based on evidence presented in the text. Involve the class in a discussion of the implications of each fact and the associated questions. Do not worry about presenting more details of these subjects at this point.

Answers to
In-Text Questions

- North America and Europe are on separate tectonic plates; these plates are moving apart.
- Earthquakes occur along plate boundaries because that is where the plates are constantly colliding and slipping over and under one another.
- Rocks on the ocean floor are younger because new rocks are formed from lava flows at the mid-ocean ridges.

ENVIRONMENTAL FOCUS

The Great Rift Valley of Africa marks a spreading center that is found on a continent. Have students identify the rift valley on a map. Ask: What kind of features mark this location? *(Deep valleys and lakes)* What do you think will become of this rift valley in a few million years? *(It will probably continue to widen and eventually become an ocean or inland sea.)*

 A Transparency Worksheet is available to accompany the material on this page (Teaching Resources, page 11). Transparency 79 is also available.

Continental Facts

Imagine that you are a geologist presented with the following facts. How would you respond to the questions posed? Examine the illustration of the Earth's crust, below, to help you find the answers.

- The Atlantic Ocean between North America and Europe is getting wider by 1 cm each year. How is this possible?
- Most of the world's earthquakes occur along plate boundaries. Why?
- Scientists estimate that the oldest rocks on the ocean floor are no more than 180 million years old. However, rocks on land may be up to 4 billion (4000 million) years old. Why is the ocean floor younger?

Obviously, it must take enormous energy to move the Earth's crustal plates. Where does this energy come from? The interior of the Earth consists of hot, molten rock. As the Earth loses heat through the crust, the molten rock beneath the crust circulates in convection currents that carry the crustal plates with them.

As the plates move, they interact with one another at their boundaries. There are three basic types of plate boundaries: where plates are separating, where they are converging (approaching each other), and where they are sliding past each other.

The San Andreas fault results from two plates sliding past one another. Because the plates do not slide smoothly past each other, their movement creates shock waves, or earthquakes. Turn to page S182 of the SourceBook to learn more about plate tectonics.

When two plates collide, the results can be spectacular. For example, if one plate is pulled down, a deep *trench* forms. As this plate reaches the asthenosphere, the rock melts, creating volcanoes. The Andes Mountains were formed in this way.

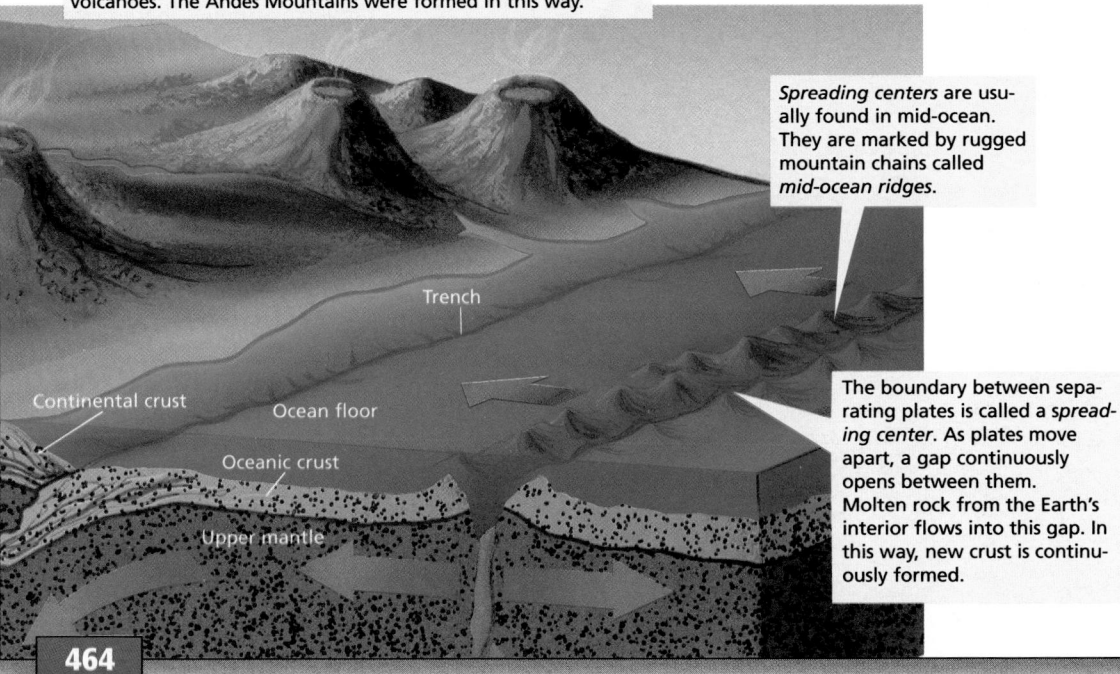

Spreading centers are usually found in mid-ocean. They are marked by rugged mountain chains called *mid-ocean ridges*.

The boundary between separating plates is called a *spreading center*. As plates move apart, a gap continuously opens between them. Molten rock from the Earth's interior flows into this gap. In this way, new crust is continuously formed.

Continental crust
Ocean floor
Oceanic crust
Trench
Upper mantle

464

Did You Know...

The San Andreas fault, which results from the eastern portion of the Pacific plate sliding past the North American plate, runs for 1050 km (650 mi.). Even more amazing is the fact that tunnels drilled for San Francisco's underground transit system pass directly through the fault.

Theme Connection

Energy

Focus question: How is the occurrence of an earthquake along a sliding plate boundary similar to the breaking of a stretched rubber band? *(Energy that is stored in the rocks of the crustal plates is released during an earthquake. Similarly, the energy stored in a stretched rubber band is released when the rubber band breaks.)* The Theme Worksheet on page 13 of the Unit 8 Teaching Resources booklet accompanies this question and makes an excellent homework activity.

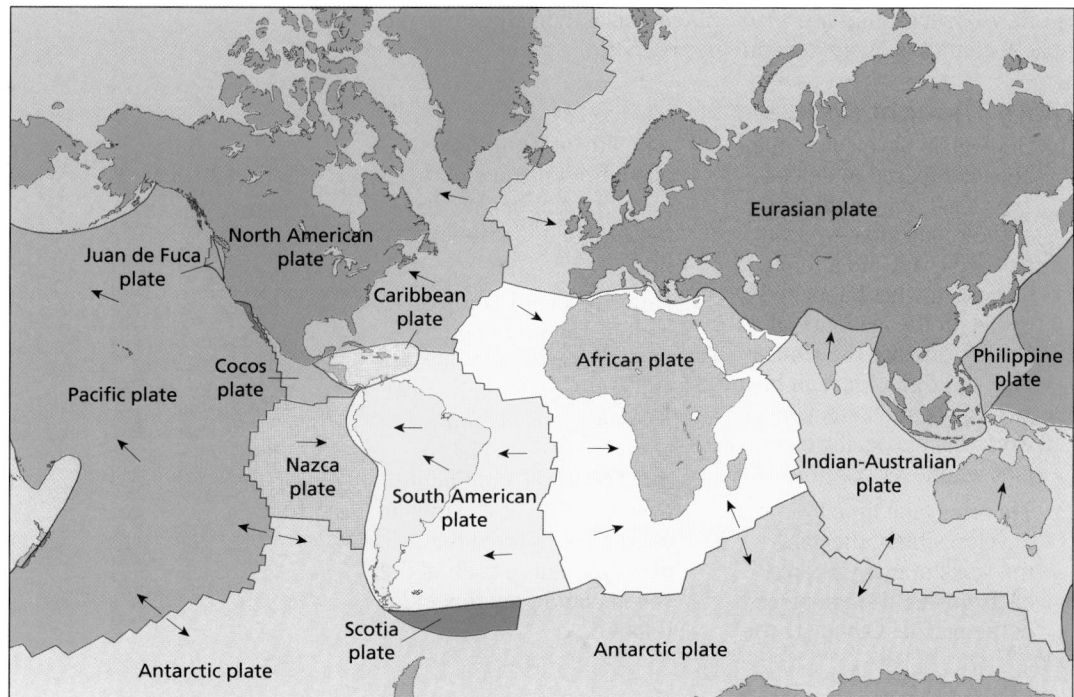

The Seven Major Plates and Their Movement

The theory of how the continents move, which is based on Wegener's theory of continental drift, is called plate tectonics. Use the map above, which shows the seven major plates as well as several smaller plates, to answer the questions that follow.

Questions

1. What are the names of the seven major plates?
2. What do the arrows on the map show?
3. Which major plate has no continent on it?
4. To which continents was North America once joined?
5. How can you explain the coal deposits found in Antarctica?
6. Where are most volcanoes found? How would you explain this phenomenon?

Integrating the Sciences

Earth and Life Sciences

Life around spreading zones on the sea floor, where molten rock flows onto the sea floor from the Earth's interior, is different from life anywhere else on Earth. Ordinarily, food chains of living things depend on energy from the sun. However, the creatures that exist around these deep sea zones (called *hydrothermal vents*) receive energy from chemicals in the water. Have students research how hydrothermal vents occur and what organisms exist around them.

465

Multicultural Extension

Iceland

Tell the class that the Mid-Atlantic Ridge rises above water at only one place, Iceland. Point to Iceland on a map and indicate the Arctic Circle, which includes the northern part of the island. Tell the class that these two features indicate that Iceland is a place of both extreme cold and extreme heat. Ask students to do some research on what life is like in Iceland. Make sure they focus on how geologic processes, such as glaciers, geysers, or volcanic activity, affect the lifestyles of the people who live there.

 Transparency 80 is available to accompany the material on this page.

Answers to
More Thought Required

1. Geologists and detectives are similar in that they use qualitative as well as quantitative data, they use many different kinds of data, and they try to construct theories that include all of the evidence. Geologists and detectives are different in that geologists deal with physical and chemical processes, whereas detectives must also deal with human nature. Detectives may also question eyewitnesses, whereas geologists cannot.

2. A sample time line is as follows:
- 1620—Francis Bacon inferred that continents had once been joined together.
- 1908—Coal deposits were found in Antarctica.
- 1911—Alfred Wegener proposed a single-continent theory; he called the continent Pangaea.
- 1950s—Sea-floor exploration confirmed continental and ocean-floor movement.
- 1960s—Geologists proposed that the Earth's crust is composed of a number of huge plates moving in various directions.
- 1970s—The theory of plate tectonics was almost universally accepted by scientists.
- Today—Geologists continue to refine the theory of plate tectonics as they make new observations about the structure of the Earth and about the processes involved in geological change.

GUIDED PRACTICE   For this exercise, students need the continental cutouts that they made at the beginning of Lesson 2. Have them compare the arrangements of the continents they made in Exploration 2 on page 459 with the arrangement on this page. Discuss and resolve any differences. Ask them if knowing about the underlying process of plate tectonics would have helped them figure out the placements. *(Probably not)* Remind them that Wegener did not know about this process either. Explain that Wegener made an inference based on his observations.

A Further Look at Continental Drift

By the 1970s, the theory of plate tectonics was almost universally accepted by scientists. It has been useful in explaining earthquakes, volcanoes, and mountains. However, questions remain. As new observations are made about the Earth, the theory of plate tectonics is likely to be further refined.

More Thought Required

1. Geologists have sometimes been compared to detectives since they both arrive at the scene after an event has occurred and attempt to explain what happened based on the evidence at hand. Compare the similarities and differences in how geologists and detectives work, using specific examples whenever possible.

2. Theories take time to develop. Often, they are the work of many people, each solving a small piece of the puzzle. Complete the time line at right to compile a brief history of the development of plate tectonics. Do not write in this book.

3. How does your completed continental jigsaw puzzle compare with the one shown below? On the next page, you will see how some geologists picture the drift of the continents over a period of 225 million years. Try to trace the development of each of the seven continents that exist today.

A Developing Theory	
1620	?
1908	?
1911	?
1950s	?
1960s	?
1970s	?
Today	?

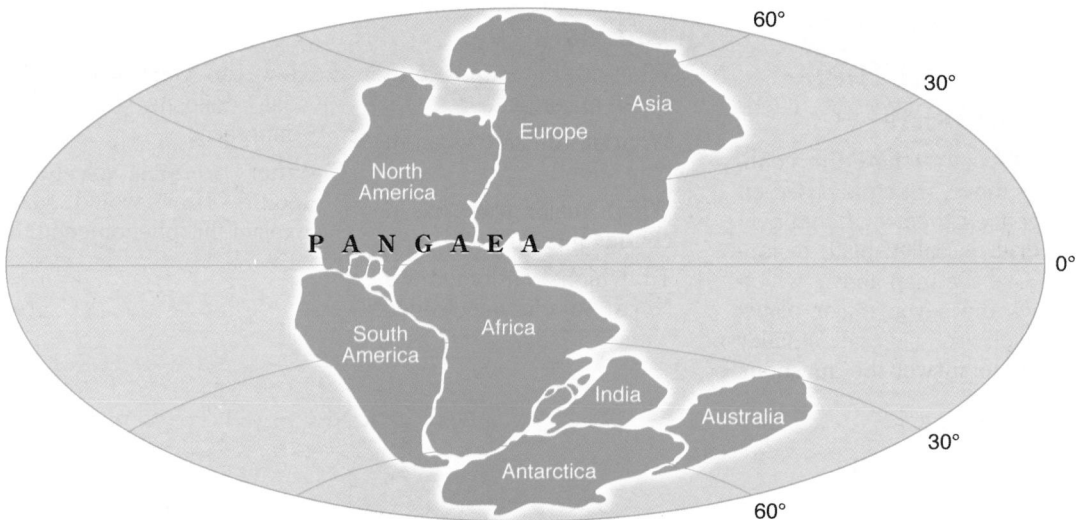

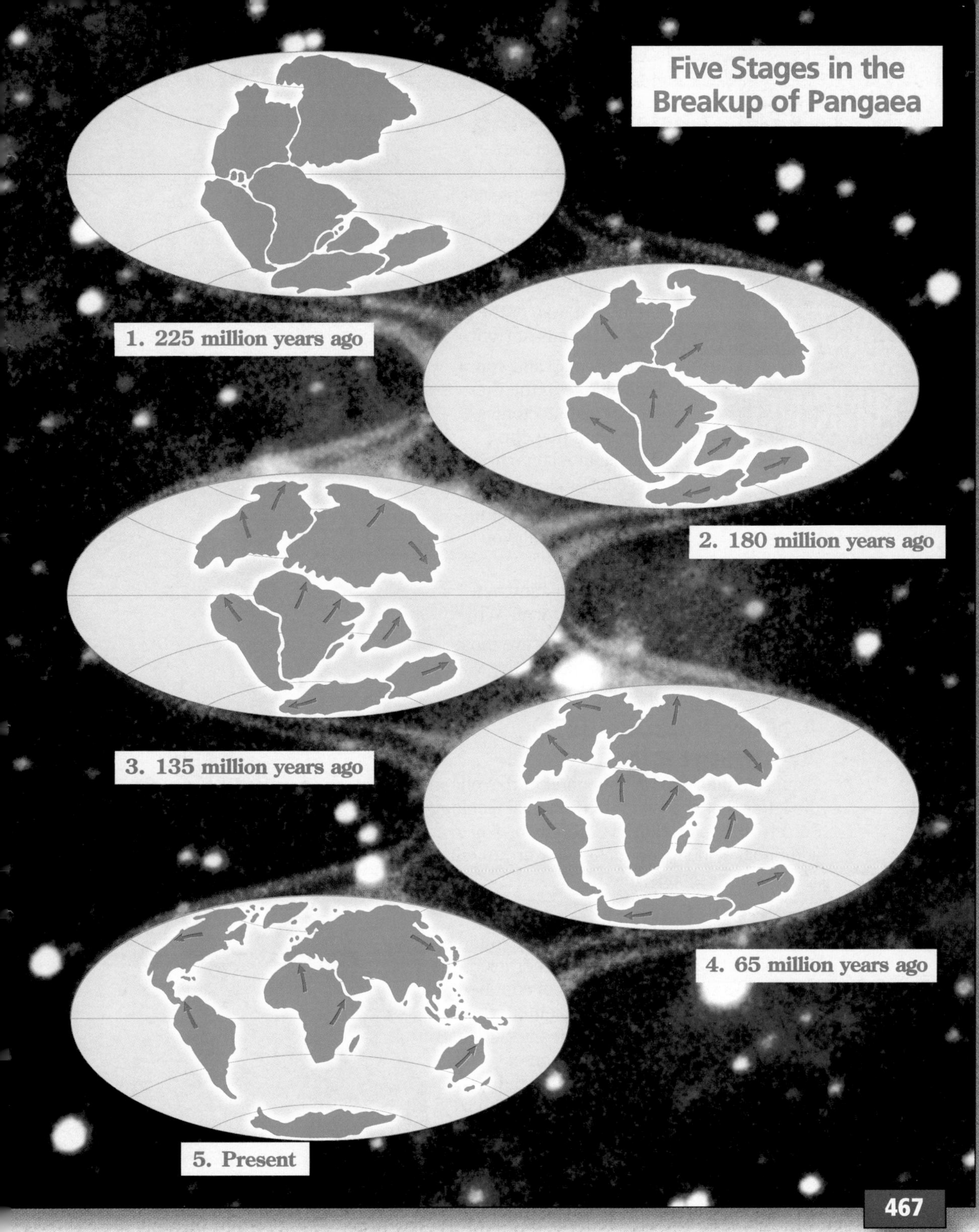

Five Stages in the Breakup of Pangaea

1. 225 million years ago

2. 180 million years ago

3. 135 million years ago

4. 65 million years ago

5. Present

Five Stages in the Breakup of Pangaea

Have students move their continental cutouts apart as indicated in each diagram on page 467 until they have a map of the continents as they exist today.

INDEPENDENT PRACTICE Ask students to draw a picture in their ScienceLog illustrating how future plate movements might change the geography of the world. *(Some predictions might include a wider Atlantic Ocean and a more northerly Africa, Australia, and South America.)*

 Transparency 81 is available to accompany the material on this page.

Meeting Individual Needs

Gifted Learners

Ask: What might have happened to the animals inhabiting Pangaea as it broke up? How might the changing landscape influence the evolution of living things? Have students explain how the breakup might have affected the distribution of animal species. Encourage students to use diagrams or other visual effects when sharing their ideas with the class. *(Student responses should demonstrate an understanding that changes within the environment affect the ability of certain organisms to survive.)*

FOLLOW-UP

Reteaching

Ask students to use a map of the world and their knowledge of plate tectonics to identify 10 cities that they think will be vulnerable to earthquakes in the future. *(Answers may vary, but possible responses include Los Angeles, Tokyo, San Francisco, Seattle, Anchorage, Mexico City, Manila, Naples, Beijing, and Bombay. All cities along the Pacific Rim are likely candidates.)*

Assessment

Have students create models of the Earth's interior with materials of their choice. Students should label and describe each layer.

Extension

Have students research what their local area was like millions of years ago. They can develop a written report and poster describing the climate, living things, and landforms in their area at different points in time.

Closure

There is evidence that several mass extinctions took place during the history of life on Earth. The most recent mass extinction occurred 65 million years ago and brought an end to the age of the dinosaurs. This is thought by some scientists to have been caused by the collision of a massive meteorite with Earth. Explain this to students, and propose the following scenario: You can go back in time and stop the mass extinction of the dinosaurs by diverting the meteor from its collision course with Earth. Ask: What will the world be like in the future if you do this? Have students respond with a short story.

Answers to *Challenge Your Thinking*

1. a. The hot spot beneath the Earth's surface is causing magma to be forced to the surface. This hot spot remains in place while the plate is carried over it, resulting in a series of islands.

b. The islands in the upper left of the diagram are the oldest because they are the farthest away from the hot spot. The movement of the plate has taken the islands that were formed first away from the hot spot.

c. New islands will continue to form as long as the Pacific plate continues to move over the hot spot.

d. The youngest islands are most likely to be volcanically active because they are nearest to the hot spot.

2. In their revised letter, students could include some of the following information in addition to the evidence given in Exploration 2 on page 459:
- Identical rock formations and fossils are found in Africa and South America.
- Coal deposits in Antarctica indicate that this continent once had a warmer climate and may have once been closer to the equator than it is now.
- The fossil plant *Glossopteris* is found in southern South America, southern Africa, India, Australia, and Antarctica.
- The Atlantic Ocean between North America and Europe widens by 1 cm each year. The creation of new crust along the sea floor is marked by underwater mountain chains called mid-ocean ridges.
- The boundaries between tectonic plates are often marked by deep trenches, mountains, and volcanoes.

3. One possible explanation for the location of *Mesosaurus* fossils is that a land bridge connected the continents of Africa and South America at one time.

1. Island Hopping

A *hot spot* is an area of concentrated heat, deep within the Earth's interior. Hot spots melt the rocks of the mantle, which creates a column of magma that rises toward the surface. Examine the map of the Hawaiian Islands at right.

a. Use the map and your knowledge of the theory of plate tectonics to explain how the Hawaiian Islands might have formed.

b. Which islands are the oldest? How can you tell?

c. Do you think any new islands will form? Why?

d. Which islands are most likely to have active volcanoes? Explain.

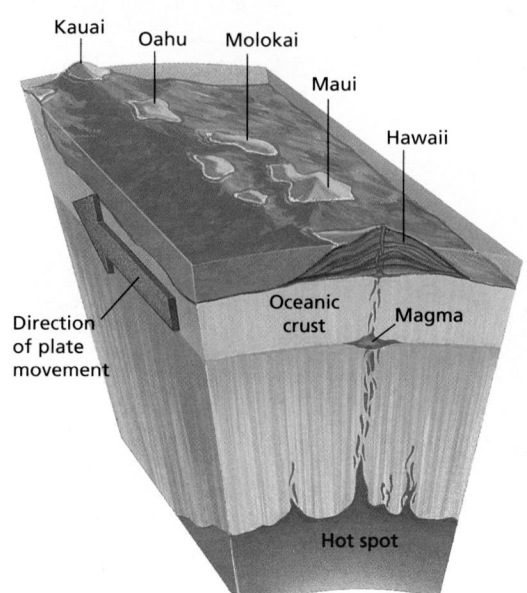

2. A More Persuasive Argument

Using the information contained in this chapter, revise the letter that you wrote to the Royal Geological Society in Exploration 2. Carefully organize any additional evidence to make your argument as persuasive as possible.

3. Lizard Debate

Fossils of the reptile *Mesosaurus* are found only in South America and Africa. Wegener used this evidence to support his theory of continental drift. Imagine that you are a scientist who refuses to accept Wegener's theory. Suggest other explanations for the location of *Mesosaurus* fossils. They certainly didn't have speedboats! Can you come up with a more reasonable explanation?

Homework

The Activity Worksheet on page 15 of the Unit 8 Teaching Resources booklet makes an excellent homework assignment.

You may wish to provide students with the Chapter 24 Review Worksheet that accompanies this Challenge Your Thinking (Teaching Resources, page 16).

4. Photo Mystery!

Carefully examine the photo at right. Make inferences about what caused the scene. How would you check the accuracy of your inferences?

5. Sunken Treasure

Imagine that you are a travel agent. You are offering submarine tours of a spreading center at the bottom of the ocean. Write a brochure or sales pitch designed to convince an average tourist (not a geologist) to take this tour.

Review your responses to the ScienceLog questions on page 452. Then revise your original ideas so that they reflect what you've learned.

469

Answers to
Challenge Your Thinking,
continued

4. Sample answer: The curved layers of rock indicate that the crust in this location has been folded by the collision of continental plates. A geologist could be consulted to determine the accuracy of this inference.

5. Answers will vary but should demonstrate an understanding of spreading centers. Encourage students to be creative in the delivery of their sales pitch.

ScienceLog

The following are sample revised answers:

1. The Earth is constantly changing. Glaciers have created lakes and shaped valleys. Rivers have carved deep canyons in the landscape. Even the shape and location of the continents have changed over time.

2. Geologists do a wide variety of things to learn about the processes that shape the Earth. They study rocks and fossils; they make and use different kinds of maps; they measure earthquakes and document volcanic eruptions; and they constantly refine the theories related to plate tectonics.

3. Fossils help to explain how continents and tectonic plates have moved over time. Fossils can also show that climates in different regions have changed over time.

4. The continents are part of the Earth's crust, which consists of tectonic plates. These plates are moved by convection currents in the molten rock beneath the crust. Geologists theorize that the continents were once joined together in a supercontinent called Pangaea and that the continents are constantly in motion. The movement of continents and tectonic plates is responsible for many geological features, such as mountain ranges, deep trenches in the ocean floor, and volcanoes.

Connecting to Other Chapters

Chapter 24
investigates the geologic diversity of the Earth and introduces the theory of plate tectonics.

Chapter 25
examines evidence of geologic changes and the role that gravity plays in such changes.

Chapter 26
explores how landforms are altered through sedimentation, erosion, and the actions of winds and glaciers.

Prior Knowledge and Misconceptions

Your students' responses to the ScienceLog questions on this page will reveal the kind of information—and misinformation—they bring to this chapter. Use what you find out about your students' knowledge to choose which chapter concepts and activities to emphasize in your teaching.

In addition to having students answer the ScienceLog questions on this page, you may wish to have them complete the following written assignment: Ask students to imagine traveling 10 million years into the future and visiting the site on which the school is built. Ask: Will the school still be standing? What will have happened to it? Will the terrain around the area look the same? What changes will have occurred? Collect student papers, but do not grade them. Instead, read them to find out what students know about the way the Earth changes over time, what misconceptions they have, and what interests them about the subject.

CHAPTER 25
Changes Fast and Slow

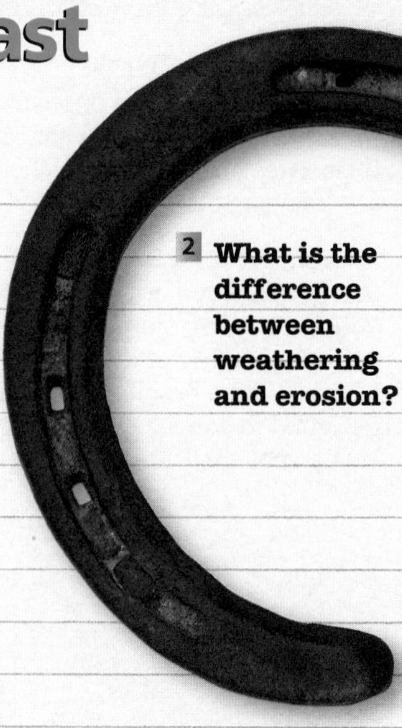

1 What are some of the forces that cause the Earth to change?

2 What is the difference between weathering and erosion?

3 This granite pillar stood in Egypt for 3000 years before it was moved to New York City in 1880. Within 100 years, it had been severely damaged. How?

ScienceLog

Think about these questions for a moment, and answer them in your ScienceLog. When you've finished this chapter, you'll have the opportunity to revise your answers based on what you've learned.

4 What does weathering have to do with the weather?

Homework

The Discrepant Event Worksheet on page 22 of the Unit 8 Teaching Resources booklet makes an excellent homework activity.

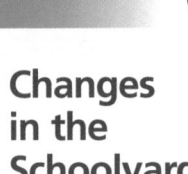

Evidence of Change

Changes in the Schoolyard

A science class became curious about the geological history of their area when several students saw on television that a well-known geologist planned a public appearance in their town. They decided to invite the geologist to visit their school.

This is how they prepared for the visit. They decided to collect clues that the geologist could use to explain the forces that shaped the land where they lived and went to school. They started their investigation in the schoolyard. One student suggested that the location of each clue should be marked on a map. To help the geologist, the map would be mailed to her before she arrived. Then the class discussed the types of clues they should look for.

The discussion went something like this:

"Let's look for fossils. Maybe we can find a dinosaur footprint," Konrad said.

"We found some fossils at camp one summer, but they weren't footprints, just some

Konrad

Kim

Peter

kind of plant," answered Dave.

"There sure are a lot of rocks out there. Do fossils form in rocks?" asked Adriana.

"I've found some shells mixed in with the rocks behind the softball field," Tanu'e said.

"Maybe this schoolyard used to be underwater!" Kim replied with a smile.

Then Paco observed, "There are lots of rocks in front of the school, but the playing

field is flat."

"Maybe it was bulldozed," suggested Peter. "Maybe the rocks were buried underneath. Let's dig down and see. Maybe we'll find some clues under the ground like archaeologists do. Wouldn't it be something to find some old money?"

"Do you suppose that being on top of a big hill is a clue?" asked Heather. "What about the soil; is it a clue?"

471

LESSON 1 ORGANIZER

Time Required
2 class periods

Process Skills
observing, analyzing, inferring

Theme Connections
Changes Over Time, Energy

New Term
Sediments—sand, mud, and other loose material

Materials (per student group)
Exploration 1: meter stick or metric ruler
Exploration 2: meter stick or metric ruler; optional materials: camera with film
People and Change: magazines; newspapers

Teaching Resources
Exploration Worksheets, pp. 23 and 25
Transparencies 82–84

 **LESSON 1** Evidence of Change

FOCUS

Getting Started
Tell students that the Earth has changed in many dramatic ways since its beginning. Challenge each student to write down one piece of evidence that shows that the Earth has changed since its beginning. Collect students' responses. Generate a class discussion as you go through the responses.

Main Ideas
1. The geological history of an area can be determined by observing rocks, soil, landforms, and other features.
2. The Earth is constantly changing.
3. Changes occur at different rates.
4. The actions of people may affect the Earth's surface.

TEACHING STRATEGIES

Changes in the Schoolyard
Call on a volunteer to read aloud the first two paragraphs on page 471. Then choose eight people to role-play the discussion in the text. Invite the class to identify the clues mentioned by the students in the story. (*Possible clues include fossils, shells, flatness or hilliness, underground rocks, and soil.*)

Point out the photographs on pages 472 and 473. Explain what feature each photograph shows and how the feature was formed. Explain that the petrified tree stump was formed when the organic material that made up the stump was replaced by certain minerals. The natural bridge was formed from the roof of a limestone cave by the dissolution of minerals by ground water. The basalt columns were formed when a thick layer of basalt cooled and contracted, forming vertical fractures. The grooves in the rock were carved by the tremendous grinding pressure of a glacier and the scraping of rocks and debris that were dragged along by the flowing ice. The dinosaur footprint was formed when the animal made the footprints in soft mud, and the footprints became buried and fossilized.

Second-Language Learners

Have students build models of the landforms that are prominent in their local area or in their homeland. They could use clay, papier-mâché, wire, construction paper, and many other materials. Ask students to label and define each of the landforms depicted. You may wish to have them display their finished models for the class.

Theme Connection

Changes Over Time

Focus question: How long does it take for a land formation such as the Grand Canyon to form? *(The Colorado River has been eroding the land in northern Arizona for 6 million years. The Grand Canyon, which is the result of the river's work, is 446 km long and between 0.2 and 29 km wide.)*

CROSS-DISCIPLINARY FOCUS

Language Arts

The physical setting or location of a story has a great deal to do with its dramatic impact. Authors must research settings to make their stories realistic. In this chapter, students learn about a wide variety of landforms and physical environments on the Earth's surface. Have the class make a list of settings. Then ask students to think of story ideas that would be appropriate for each setting. Invite students to choose one of these ideas and to develop the idea into a short story. You may wish to have students research the landscape and climate of their settings to add realistic details to their story. Encourage students to combine their story into a booklet with illustrations and a cover.

You may not find geological evidence in your schoolyard as spectacular as the examples shown here, but you may be surprised by what you do find.

A petrified tree stump

This natural bridge is part of the roof of a former limestone cave.

Lava basalt columns

The grooves in this rock were cut by a glacier.

472

Collecting Clues in the Schoolyard

1. Discuss in class how you would help the geologist. Then draw up a preliminary list of clues.

2. Go outside and observe! Add any other clues you think of to your list.

3. Draw a map. Indicate on the map where you found each clue.

4. Is your evidence complete? Do you have enough clues?

Analyze the clues to see what they can tell you about your area in geological terms. If your class is lucky enough to get a geologist to visit your school, ask for help in determining the geological history of your schoolyard.

Clues	Possible history
1.	
2.	
3.	
4.	
etc.	

Other classes might be interested in your results, especially after watching your class working outside. Write a case-study report to describe what you found out about the geology of your schoolyard and the surrounding area. Share your discoveries with other classes.

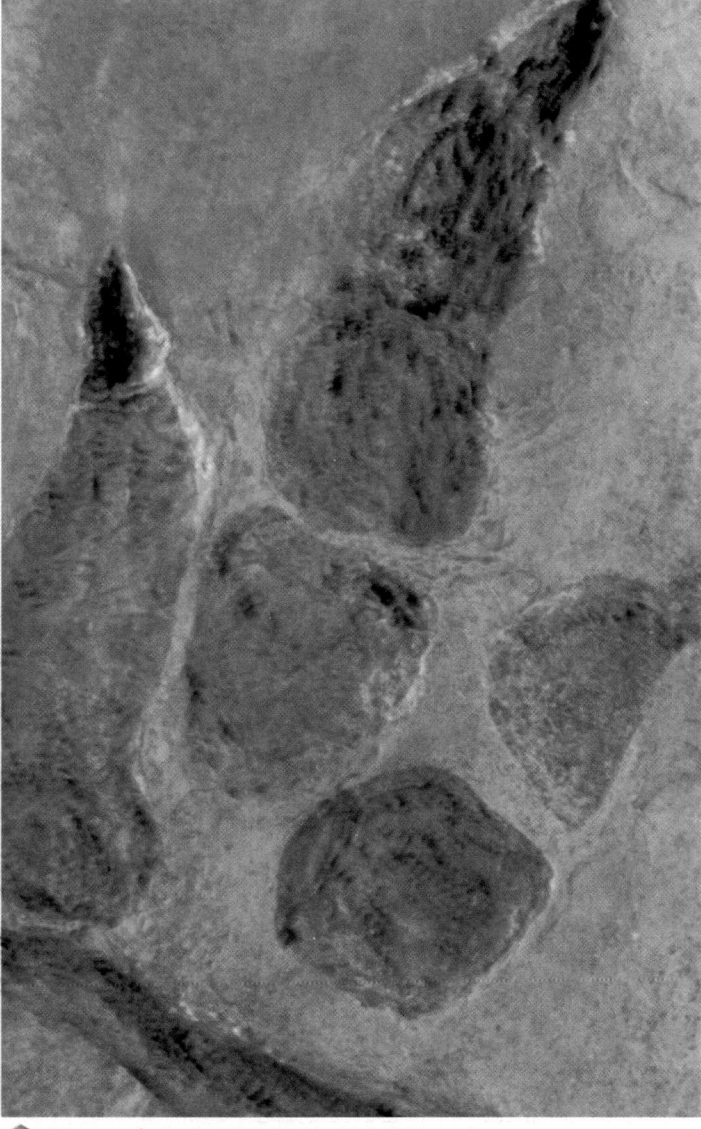

Dinosaur footprint, Connecticut Valley, Massachusetts

Group size: 3 to 4 students

Group goal: to find clues about the geologic history of your schoolyard, document the clues in the form of a map, and make inferences about the significance of the clues

Positive interdependence: Assign students roles such as secretary, map maker, historian, and primary explorer.

Individual accountability: Have each student write a summary of the case-study report. Encourage students to add their comments about the activity.

Before going outside, start a list on the chalkboard of possible clues that students should look for. Ask students to copy the list in their ScienceLog for reference when in the schoolyard. You may also wish to suggest that students collect soil, rock, and vegetation samples. They can study the samples after they have completed their work outside.

GUIDED PRACTICE The following is a list of questions you can use to help students evaluate the evidence that they obtained in Exploration 1:

1. Is the land flat, sloping, or hilly? How can you tell?

2. Is the school at sea level? What proof do you have?

3. Are there a lot of rocks or just a few? How large are they? What do they look like? Can you identify them? Are there rocks below the surface? If so, how far below? Do some stick out partway?

4. Is the soil rocky or level, dense or thin? Can you tell what it is made of?

5. How many living things can you find growing in the schoolyard? Are they large, medium, or small?

6. Is there running water in or near the schoolyard?

7. What human-made features are present in the environment?

Homework

The Exploration Worksheet that accompanies Exploration 1 makes an excellent homework activity (Teaching Resources, page 23). If you choose to use this worksheet in class, Transparency 82 is available to accompany it.

Before beginning the Exploration, arrange for a geologist to visit the class, if possible. Be sure to check that the schoolyard has observable "clues." Use a nearby park if the schoolyard is not suitable. You may also wish to review the geological history of the area.

A visiting geologist could help students collect clues in the schoolyard and then assist with the interpretation of these clues. The geologist could also give a short talk about the nature of his or her work. You may wish for students

to organize their results in the form of a table, as shown on page 473.

If possible, bring in samples of sedimentary rocks, such as coarse-grained sandstone and mud shale. Let students examine the rock samples and hypothesize about how they were formed. Discuss the changes that occurred to turn loose sand or mud into rock. If sedimentary rock samples are not readily available, color photographs from a field guide on rocks and minerals can serve the same purpose.

A Case Study in Change

James Hutton, a Scottish chemist, was born in 1726. He is credited with providing geologists with a framework for thinking about the Earth's history. His method was to learn all that he could about modern deposits of sediments and the processes that formed them. Then he compared these deposits with prehistoric deposits he discovered. In so doing, he made many valuable inferences about the Earth's geological history. Have students silently read the first two paragraphs on page 474. Involve them in a discussion of their perceptions of local geological history. For example, do they think that their community was ever buried under ice or water? Discuss any evidence that students have observed for these changes. Other changes that students may want to discuss include changes due to earthquakes and volcanic eruptions. If these are a part of the geological history of your area, ask students to describe evidence of these changes.

Primary Source

Description of change: excerpted from *A New Discovery of a Vast Country in America,* by Louis Hennepin
Rationale: to focus on Hennepin's description of Niagara Falls

Answer to
In-Text Question

Ⓐ Answers will vary. Possible responses include the following:
* The perspectives are different.
* The shape of the falls in the photograph is considerably different from that in the sketch. There are more curves in the photograph, and they are more exaggerated than those in the sketch.
* The landmass dividing the falls in the drawing is much smaller in the photograph.
* The stream of water on the right-hand side of the drawing is not in the photograph.
* The sketch was made many years ago; the photograph was taken recently.
* Human-made structures, such as highways, railroads, and a bridge are visible in the photograph.

A Case Study in Change

Do you find it hard to believe that the area in which you live was once buried under a thick layer of ice or was once at the bottom of the sea? It is difficult to imagine the landscape being any different from the way it is now. In fact, it wasn't until the late 1700s that scientists began to explore some of the processes of change that must have taken place to produce the landscapes that exist today.

A Scottish geologist and chemist, James Hutton, made careful observations of the Scottish countryside in the late 1700s. He concluded that layers of rock are in fact hardened layers of sand, mud, and other loose material called *sediments.* Hutton is often called the "father of geology" because he developed a method of discovering the Earth's geological history. Hutton believed that Earth processes at work today have also been at work throughout time and that the evidence of this could be seen in the rocks around us.

> **"The present is the key to the past."**
> —James Hutton

Flashback—Niagara Falls, 1678

In 1678 Father Louis Hennepin became the first European explorer to reach Niagara Falls. Read his description of what he saw and heard.

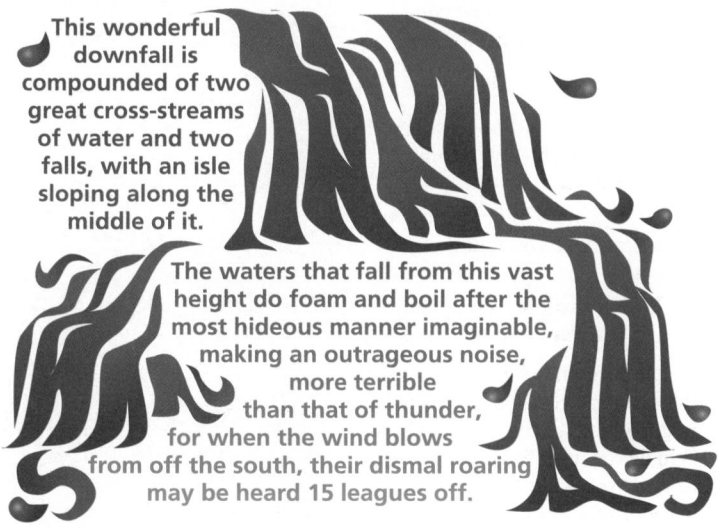

This wonderful downfall is compounded of two great cross-streams of water and two falls, with an isle sloping along the middle of it.

The waters that fall from this vast height do foam and boil after the most hideous manner imaginable, making an outrageous noise, more terrible than that of thunder, for when the wind blows from off the south, their dismal roaring may be heard 15 leagues off.

Compare the sketch below, based on Father Hennepin's description of Niagara Falls, with the contemporary photograph on the next page. Both pictures feature a portion of the falls now known as Horseshoe Falls. Compare the right-hand side of each picture. What differences do you observe between the sketch and the photo? Ⓐ

CROSS-DISCIPLINARY FOCUS

Social Studies
Point out to students that Father Louis Hennepin was not only a missionary but also an important explorer. Encourage interested students to research Father Hennepin's life, including the parts of North America he explored and what important contributions he made. Encourage students to share their findings with the class.

475

Homework

Present students with the following scenario: You have traveled back in time to 1678 and are accompanying Father Hennepin on the first trip by a European to Niagara Falls. Using the sketch on page 474, describe the first sighting of the falls and the reactions of your fellow explorers.

Integrating the Sciences

Earth and Life Sciences

As James Hutton pointed out, "the present is the key to the past." The geological record can tell us what environmental conditions were like at different stages of a region's history, as well as provide us with a fossil record. Have interested students research an extinct species and find out as much as they can about the environment in which these animals are thought to have lived and how the geological record supports these conclusions. Encourage students to share their findings with the class.

Multicultural Extension

Multilingual Naming

Names for surface features come from many different languages. Names for unusual or distinctive landforms that occur in one region sometimes become known by their local name throughout the world. One example is the word *tepui.* Tepui is the name used by natives of modern-day Venezuela for very tall, flat-topped mountains in tropical rain forests. Ask students to research other multilingual names in geology, such as *delta* or the native Hawaiian words *aa, pahoehoe,* and *lava.* Encourage students to find out where these landforms are found and how they affect the culture of the people who live near them.

Did You Know...

Since the 1960s, millions of gallons of the Niagara River have been siphoned off for use in local power plants. Because less water is now traveling over the falls, the rate of erosion is occurring more slowly, and the life span of the falls has been extended.

ENVIRONMENTAL FOCUS

Ask: How can we help preserve Niagara Falls for future generations to enjoy? In 1969 part of the Niagara River was diverted away from the American Falls so that the bedrock could be reinforced with cement. The goal was to lengthen the amount of time that it would take for the river to erode the rock cliffs that form the falls.

Gauging the Change

Call on a volunteer to read the first paragraph, and then direct students' attention to the diagram at the bottom of the page. The water is flowing over the falls and into the Niagara River. Help students visualize the paths of the two waterways. Point out that the measurements made in 1678, 1764, and 1842 were to mark the position of the rock under the falls.

GUIDED PRACTICE   Explain to students that the recession of the falls is a fairly rapid change. Point out that recession, in this case, refers to the wearing away of the rock. Ask the class to name some causes of this change. *(Possible responses include the type of rock and the force of the water.)* Involve the class in a discussion of why the rate of change has slowed since 1842. *(Possible responses include that the volume of water going over the falls has decreased and that the type of rock that is now exposed is more resistant to erosion.)*

Answers to
In-Text Questions

- The water is flowing from south to north, or from the top right of the diagram to the bottom left.
- Horseshoe Falls receded about 340 m since 1678, measured at the curve of the falls.
- American Falls has receded about 25 m since 1842.
- For Horseshoe Falls, the rate of recession between 1678 and the present in meters per year is 340 ÷ 317 = 1.07 m/y. For American Falls, it is 0.107 m/y.
- The erosion may have occurred at different rates because of differences in rock type, volume of water, or speed of the water.
- Over the past 300 years, the water has eroded the rocks to create an elongated shape. The shape has changed from a gentle curve to a deep horseshoe and then to a "bent" horseshoe.
- No, the type of rock and force of the water may vary along the crest, creating differences in recession rates. Students can quantify these differences by measuring the distance between recession lines at different points and calculating the recession rate for those points.
- The shape of the falls will probably continue to curve and elongate.

Gauging the Change

Niagara Falls has changed since Father Hennepin first saw it in 1678. Study the diagram below. Niagara Falls actually consists of two separate falls: American Falls and Horseshoe Falls. Each broken line represents a survey carried out during the year indicated. How have the falls changed over the years?

Answer the following questions in your ScienceLog:

- Which way is the water flowing in the illustration?
- How far did Horseshoe Falls recede (wear away) between 1678 and the present?
- How far has American Falls receded since 1842?
- What is the rate of recession in meters per year for each of the falls?
- Why have Horseshoe Falls and American Falls eroded at different rates?
- How has the shape of the falls changed in the last 300 years?
- Do all parts of the edge change at the same rate? Why or why not?
- What do you infer will happen to the shape of the falls in the future?
- Where do you predict the falls will be in 50 years? in 5000 years?

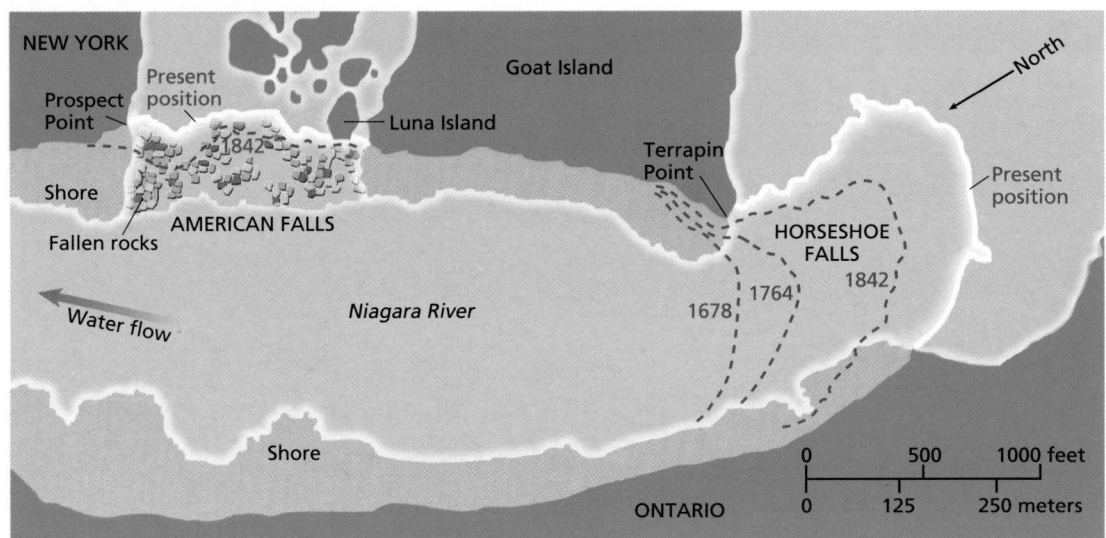

- In 50 years, the falls will probably have receded an additional 50 m and will not appear very different from the way they look now. In 5000 years, they may have receded as much as another 5000 m, and the falls will be much farther upriver.

 Transparency 83 is available to accompany the material on this page.

Homework

Present students with the following scenario: You have traveled back in time to see your neighborhood as it was 300 years ago. Since that time, what kinds of changes have occurred? How do you think this environment will change over the next 300 years?

The Niagara Story

Niagara Falls has not always existed. In fact, in geologic terms Niagara Falls is extremely young—only about 12,000 years old! The falls formed at the end of the last ice age as the huge glaciers that covered the continent retreated northward. Meltwaters caused the newly formed Lake Erie to overflow, forming the Niagara River.

What you have just read is Chapter 1 of The Niagara Story. The illustrations at right and below tell more of the story. Your job is to use these illustrations along with your knowledge of Earth processes to bring the story up to date and to make a prediction about the future. Answer the following questions. They will help you to understand the story.

Points to Ponder

1. How did the Niagara gorge form?
2. Why do the falls not erode away?
3. What will happen to the falls in the future?
4. If the present trend continues, what might eventually happen to Niagara Falls?
5. What might eventually happen to Lake Erie?

Now write two more chapters of The Niagara Story. Chapter 2 should bring the story up to the present. Chapter 3 should include your prediction of future events.

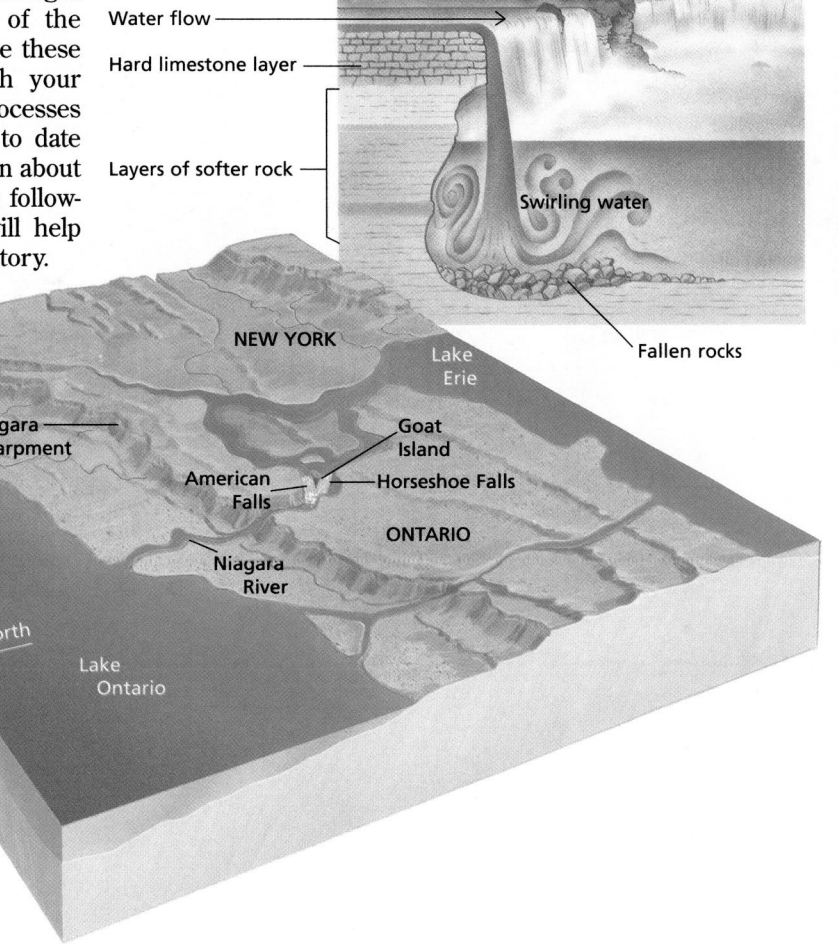

Water flow
Hard limestone layer
Layers of softer rock
Swirling water
Fallen rocks

NEW YORK
Lake Erie
Niagara Escarpment
Goat Island
American Falls
Horseshoe Falls
ONTARIO
Niagara River
North
Lake Ontario

The Niagara Story

Call on a volunteer to read aloud the first two paragraphs to the class. Then direct students' attention to the illustration. Point out the location of Lake Erie, the overflow water from Lake Erie, and the Niagara River. Be sure students visualize this path. Ask them to pinpoint the falls and the gorge and to explain those terms in relation to the diagram. Explain the meaning of the term *escarpment* to students if they cannot infer its meaning from the diagram. (*An escarpment is a steep cliff formed by erosion or by faulting between plates.*)

> ★ **Transparency 84 is available to accompany the material on this page.**

Answers to *Points to Ponder*

1. The flow of water wore away the softer rock beneath the surface limestone. The surface then collapsed. This process occurred over and over again, forming the Niagara gorge.

2. The limestone wears down more slowly than the rock beneath, forming a "lip" for the waterfall.

3. The location of the falls will continue to move upriver.

4. Eventually, the falls will move so far upriver that they will actually cease to exist. Instead there will be a deep gorge reaching from Lake Erie to Lake Ontario.

5. Eventually, Lake Erie might merge with Lake Ontario.

Answer to *In-Text Question*

A Possible chapters for The Niagara Story might be as follows:
Chapter 2: The Niagara River flowed over the hard limestone for many thousands of years. At the edge, under the limestone, there was softer rock. As the water fell over the edge, the softer rock beneath the limestone wore away. With nothing to support it, the edge of hard limestone collapsed. This happened over and over again for many thousands of years. Finally, the Niagara gorge was formed. The edge of the falls continues to move upriver at the rate of approximately 1 m per year.
Chapter 3: As the falls move upriver, they will continue to cut a gorge between New York and Ontario. Eventually, the falls will reach Lake Erie and cease to exist. Niagara gorge will then serve as a channel between Lake Erie and Lake Ontario. Over time, the two lakes may merge as the Niagara Escarpment erodes away.

What Caused the Change?

This exercise gives students an opportunity to review the terms *erosion* and *weathering*. These concepts will be considered in more depth in the next few lessons.

Discuss the agent of change in each photograph. Encourage free discussion and new ideas.

Answers to
What Caused the Change?

Sample answers for the table include the following:

Photo	Change	Cause	Natural	Human
road building	alteration of landscape	person and machine	no	yes
Mayan sculpture	weathering	gases in air, air pollution, and moisture	yes	yes
river in flood	erosion of shore	moving water	yes	no
rock layers	horizontal layering of rock and folding of rock layers	pressure within the Earth	yes	no

Arranging the changes in order from fastest to slowest may lead to some debate. The most likely order is the following: road building, river in flood, weathering of Mayan sculpture, and horizontal layering and folding of rock.

Humans can have the following effects on these changes:

- Humans can speed up or slow down the bulldozing of the land.
- Humans can prevent weathering by applying protective materials to the sculpture, keeping it clean, or preventing the effects of acid rain.
- Humans can slow down shore erosion by building breakwaters and retaining walls. They can reduce the chance of flooding by maintaining plants and trees that would reduce the amount of runoff.
- Humans would have difficulty affecting the horizontal layering of rock.

What Caused the Change?

In each photograph on this page, a change has occurred or is taking place. Your tasks are to (a) identify the change, (b) suggest what has caused the change, and (c) decide whether the change was caused by natural or human forces. Record your conclusions in your ScienceLog in a table with these headings: Change, Cause, Natural, and Human.

Now arrange the changes in order from the fastest to the slowest. Which changes could humans speed up, slow down, or affect in some other way?

478

Theme Connection

Energy

Focus question: How can Earth processes such as flooding be used to provide energy? *(Sample answer: If flood control is necessary, dams can be built. Water flowing through the dams can be used to generate electricity.)*

Multicultural Extension

Civilization in a Floodplain

The annual flooding of the Nile River helped Egypt to become one of the first great civilizations of the world. This flooding deposited fertile soil in the Nile Valley each year. The resulting crops provided the basis for Egyptian civilization.

A Science Project: Recognizing Change

Look for a feature in your back-yard, on the way to school, or around your town that seems to show a change taking place. It might be a crack in a rock, a spot of bare ground, or a gully formed in a hill by running water.

Once you have found a suitable feature, prove that a change is occurring. The ideas below will help you conduct your project scientifically.

Before you begin your investigation, plan ahead. Observe the condition of the feature over a period of time. Compare its condition when you started your investigation with its condition over time. In your ScienceLog,

write down what you see; it's easy to forget details after a while. (Sketches and diagrams often help also.) When you have completed your observations, prepare a report with the following sections:

1. Results of the Investigation (Write down the change you observed.)

2. Investigating the Change (Give details of how you collected your information.)

3. Results Obtained (Include drawings, photos, a data sheet, etc.)

4. Inferences and Summary (What caused the change? Will the change continue to take place? What will eventually result? Can the change be reversed? Can humans have any effect on the change?)

One Student's Investigation

Caroline decided to watch for a change under a downspout at the corner of her house. First she took a photo and made a drawing of the area directly under the downspout. Then she observed the area closely. She noticed that a change occurred after a rainfall. She made measurements of the changed area. Caroline also made notes about some features of the area: the number of rocks; their appearance, size, and position; and any living things nearby. After each rainfall, she made further observations and measurements. Finally, Caroline summarized her findings in a brief report.

This Exploration can be assigned as a long-term science project to be done at home. To be sure that students understand what they are expected to do, spend part of a class period discussing the organization of the project. Have students think about the purpose of their project, the specific research methods they will use, the raw data they plan to collect, and what analysis will need to be done.

To get students started on their outdoor explorations of local landforms, take them on a walk or a field trip to familiarize them with evidence of changes in the landscape. Try to give them a feeling for where they live geologically. Relate what the students see to the processes of erosion, deposition, and plate tectonics.

Have students check with you as each phase is completed. In this way, you can review their progress and suggest improvements.

PORTFOLIO

Suggest that students list their plans and accomplishments for Exploration 2 in a booklet entitled "Progress Log." Students should use the guidelines presented in the text to prepare their report. Encourage students to include a column for comments and difficulties they encountered while doing their research. When students are finished with their log, they can include it in their Portfolio.

People and Change, page 480

INDEPENDENT PRACTICE Divide the class into small groups of four or five students to discuss and classify the 10 changes listed. Students can assemble photos and illustrations of human-caused changes in a poster or folder. Ask them to briefly describe the changes as harmful or beneficial and to give their reasons.

Homework

The Exploration Worksheet that accompanies Exploration 2 makes an excellent homework activity (Teaching Resources, page 25).

Accept all reasonable answers. Encourage students to explain their responses. Possible answers include the following:

1. Irrigating the desert creates a habitat that allows more plants and animals to thrive. However, the water may be coming from sources that already provide habitats for other organisms.

2. Clearing land for housing can cause soil damage and flooding in addition to destroying animal habitats. However, people need housing.

3. Building dams to control floods will prevent the destruction of property and lives. However, dam building alters a river's flow and may destroy fish habitats.

4. Strip mining for coal provides an energy source for many cities and provides many people with employment. However, strip mining destroys the landscape and many animal habitats.

5. Draining swamps to make farmland increases food production and may eliminate many disease-carrying insects that live in swamps. However, draining swamps destroys the ecosystem of many organisms.

6. Clear-cutting the forest provides many useful materials, including paper and lumber. The environmental damage can be lessened (but not eliminated) with reforestation.

7. Plowing land for crops may offer increased food production, but unless wise farming methods are practiced, the soil can become seriously depleted.

8. Using the tides to generate electric power will reduce the dependency on pollution-causing fossil fuels. However, ocean water is replete with marine organisms, and the power plants that would have to be built may destroy the habitat of organisms that live along the shoreline.

9. Constructing breakwaters along the coast will slow the erosion of the shoreline and protect structures built near the water. However, altering the shoreline will alter animal habitats.

10. Damming rivers for hydroelectric power alters a river's flow and may destroy fish habitats, but it decreases the dependency on pollution-causing fossil fuels.

People and Change

Most of the changes examined in this unit are changes brought about by natural causes such as glaciers, earthquakes, flowing water, and gravity. But changes in our environment that are caused by humans also occur all the time. Some would be classified by most people as improvements, but others would not be. Look over the list below. In your ScienceLog, classify each change as either *for the better*, *for the worse*, or *difficult to classify*. Try to see the positive and negative aspects of each situation. Briefly explain your classification of each change.

1. irrigating the desert
2. clearing land to build houses
3. building dams to control floods
4. strip mining for coal
5. draining swamps to make farmland
6. clear-cutting the forest
7. plowing land for crops
8. using the tides to generate electric power
9. constructing breakwaters along the seacoast
10. damming rivers for hydroelectric power

Can you find specific examples of changes like these in newspapers and magazines? Look for articles and photographs that show the effects of human-caused changes.

This bulldozer is clearing land for new home construction.

The Aswan High Dam, Egypt, was built to control floodwaters and to generate electricity.

Strip coal mining in Hazelton, Pennsylvania

480

FOLLOW-UP

Reteaching

Have students write a letter to a friend to explain the methods used by geologists to study the geologic history of an area.

Assessment

Have students analyze a building's design and construction to see how the building can resist changes such as flooding and weathering. *(Students may describe drainage systems, exterior paint or siding, caulking, and weatherstripping.)*

Extension

Have students interview older members of the community about changes in landforms that they may have witnessed, such as those caused by erosion, weathering, and human actions.

Closure

Have students develop skits portraying a series of people who see changes in a local landform over time. For example, the first person could describe the landform as it looks today. The next person could describe the landform as it might appear in 100 years, and so on.

Examining One Type of Change

Detecting Change

A change is taking place in each of the photographs on this page. Can you figure out what it is?

Agents of Change

Were you able to detect that the rocks in each of the photos were being broken or worn down? This process is known as **weathering**. Water, wind, and ice all cause weathering. Human-caused phenomena, such as acid rain, can accelerate the weathering process. Living things such as lichens and mosses can also contribute to weathering. All exposed materials weather, not just rocks. House paint, concrete, glass, brick, and even steel undergo change because of the action of weathering agents. Some materials merely take longer than others to show the effects of weathering.

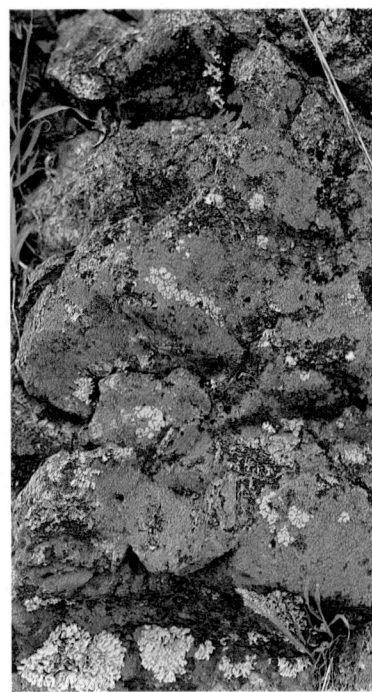

481

Examining One Type of Change

FOCUS

Getting Started

Have each student use a pencil to write his or her name on a piece of paper. Then ask students to erase their names and brush the small eraser particles into a pile. Ask: How might this activity compare with the eroding effects of certain materials on rock? (*Wind blows sand particles against rocks. The grinding action of the sand removes small rock pieces just as the eraser removed the graphite marks from the paper. Over time, more and more of the rock is rubbed away. Soon, only sand and thin soil are left, just as the pile of eraser particles is all that remains of the graphite marks.*)

Main Ideas

1. Weathering is a process by which rocks are broken and worn down by water, ice, air, pollutants, and the activities of plants and animals.
2. The amount and rate of weathering are determined by a variety of factors.

TEACHING STRATEGIES

Answers to
Detecting Change

Sample answers are as follows, running clockwise from the top left:

The desert rock formation shows how differences in rock composition can affect the rates of erosion, causing this strange structure. The lichens on the rocks produce acids that break down rock surfaces. Ice in the freezing stream causes the waterway to expand and erode more of the surrounding land. The gravestone has worn away due to cracks formed by wind, water, and ice.

Teaching Strategies for Agents of Change are on the next page. ▶

LESSON 2 ORGANIZER

Time Required
2 class periods

Process Skills
observing, predicting, analyzing

New Term
Weathering—the process by which exposed surfaces are broken and worn down

Materials (per student group)
Exploration 3: hammer; safety goggles
Exploration 5, Simulation 1: rock samples (granite, limestone, sandstone, brick, piece of concrete); magnifying

glass; large plastic or metal tray; freezer; 500 mL of water; large beaker or jar; watch or clock with a second hand; **Simulation 2:** whole piece of chalk; 200 mL of 0.1 M hydrochloric acid; two 250 mL beakers; 100 mL graduated cylinder; several paper towels; safety goggles; latex gloves; lab aprons (See Advance Preparation on page 449C.)

Teaching Resources
Exploration Worksheets, pp. 26, 28, and 29
Transparency 85
SourceBook, p. S185

Agents of Change, *page 481*

Have students silently read the center column on page 481. You may wish to point out to students that weathering can be classified as physical or chemical. Physical weathering breaks down rocks but does not change their mineral composition. Water is the primary agent of physical weathering. Plants and animals are also involved in physical weathering. Human activities such as irrigation, dam and highway building, mining, and quarrying also break down rocks.

Chemical weathering occurs when chemical reactions occur between the exposed material and substances in the environment. Chemical weathering breaks down rocks and, in the process, changes their minerals into new substances, such as soil. Most chemical weathering is brought about by the action of carbon dioxide, water, and oxygen. Carbon dioxide combines with water to produce a weak acid that reacts chemically with rocks to break them down. Rocks can also be oxidized to form new minerals.

EXPLORATION 3

 Cooperative Learning
EXPLORATION 3

Group size: 3 to 4 students
Group goal: to find evidence that various agents of change are at work in familiar areas
Positive interdependence: Assign students roles such as chief investigator, investigative assistant, reporter, and safety inspector.
Individual accountability: Have each student write a paragraph analyzing how the weathering process in an area they examined could be slowed or prevented. Ask each student to record his or her group's table in his or her ScienceLog before going outside. Instruct each group to choose three different locations to include in the table. Students should record their observations in their ScienceLog and summarize their findings in the Exploration Worksheet on page 26 of the Teaching Resources booklet.

Investigating Rock-Breaking Forces

Do you know how much force you would need to break a rock? Have you ever tried to do so? Some rocks break easily, but tremendous force is usually required. Though you may not realize it, rock-breaking forces are present in your schoolyard and are acting at this very instant. Think of all the places where there is rock, brick, or concrete. Look at some of them on your way home from school. Can you find evidence that forces of change are at work?

Choose three different locations to investigate. Consider the suggestions below as you carry out your study.

What to Do

1. If your school is made of brick, check for rounded corners, cracks, chips, or rough surfaces in the brick. Does the mortar stick out or recede? Does the brick look old? If your school is made of something other than brick, can you find evidence of weathering?

2. Check the foundation of your school. Are there any rough and irregular surfaces, cracks, or broken corners? Compare the foundation on each side of the building. Do you notice any difference in the amount of weathering?

3. Find a rock. Break away a portion to expose a fresh surface. Compare the exposed and fresh surfaces.

Be Careful: Wear safety goggles when chipping or breaking rock.

4. Look for moss or lichen growing on exposed rocks. Gently peel away the lichen and feel the rock surface underneath. Compare this surface with that of a rock without moss or lichen.

5. Examine sidewalks, stone fences, or concrete pillars. Is the concrete cracking, breaking, or crumbling? Are there trees or other plants growing nearby whose root systems may be affecting the concrete in some way?

Keeping a Record

Summarize your findings in your ScienceLog in a table like the one below.

Location	Evidence of weathering	Cause(s)	Comments
sidewalk	cracks, broken corners	frost, tree roots	built in 1985— only a few cracks

---

Homework

The Exploration Worksheets that accompany Explorations 3 and 4 make excellent homework activities (Teaching Resources, pages 26 and 28). Transparency 85 is also available to accompany Exploration 3.

EXPLORATION 4

Monuments to Change

If there is a graveyard in your community that is over 100 years old, you have the makings of an interesting project.

PROBLEM 1

What is the relationship between the amount of weathering and the length of time that a monument has been exposed? Hint: Find stones of different ages made from the same kind of rock. Look for newer ones (0–100 years old), older ones (over 150 years old), and some in between. Compare the amount of weathering in each.

PROBLEM 2

Do different types of stones weather at different rates? Hint: Find several gravestones of approximately the same age but made of different kinds of rock. Compare the amount of weathering in each.

What do you conclude from your study?

How Are They Affected?

If you worked at one of the jobs pictured below—building construction, farming, house painting—how would weathering affect you?

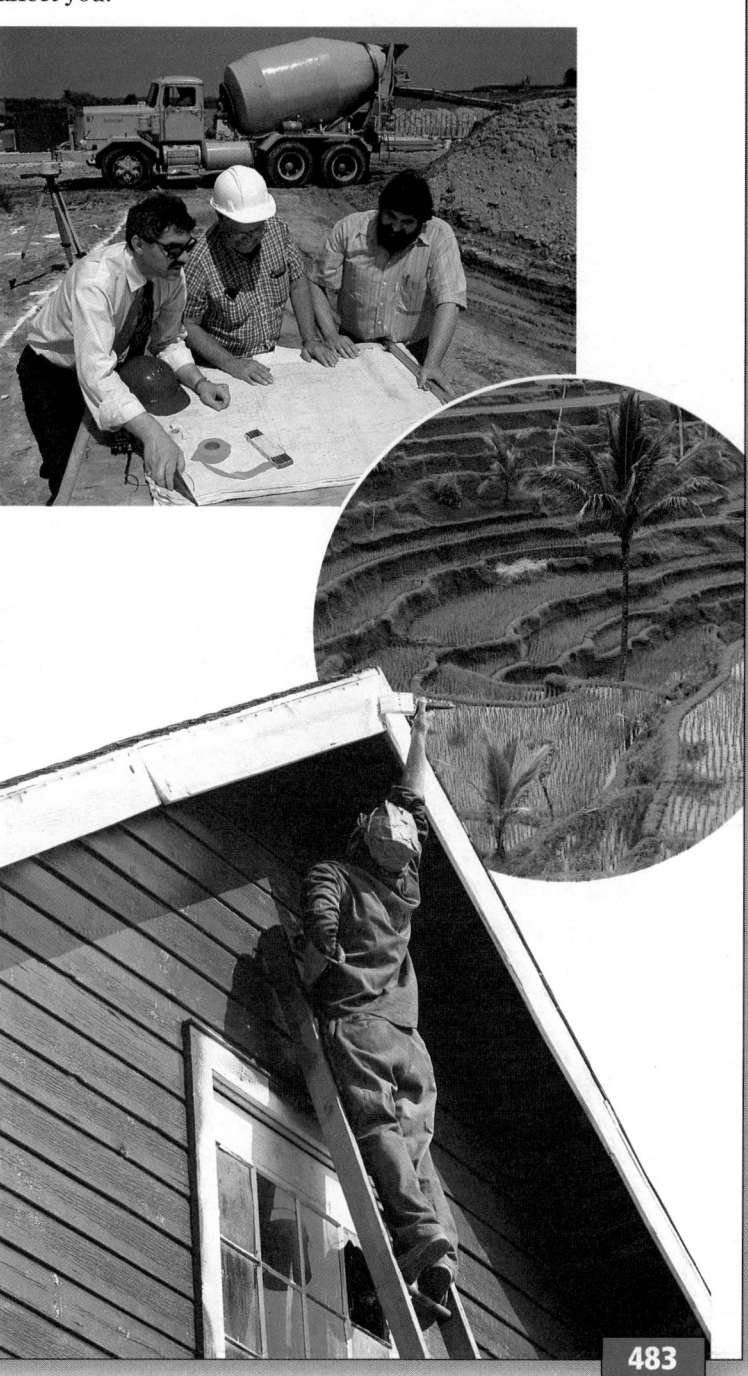

EXPLORATION 4

This activity could be assigned as an individual project or reserved for a class field trip. Remind students that they should treat the land and gravestones with respect.

Answer to
Problem 1

Generally, the longer a gravestone has been exposed, the greater the amount of weathering that has occurred. Remind students to keep a list of the ages of the stones and the amount of weathering as they proceed.

Answers to
Problem 2

Harder rock, such as granite, will probably show less weathering than rock made of marble, even if the ages of the rocks are the same. Again, remind students to take careful notes of their findings. Encourage them to make sketches of their observations as well.

Students may conclude from Exploration 4 that gravestones exposed for longer periods of time will show more weathering. However, hard rocks, such as granite, weather more slowly than soft rocks, such as marble.

Answers to
How Are They Affected?

- The architect and builders view weathering as harmful and try to design houses and use materials that will resist weathering.
- A farmer could consider weathering both helpful and harmful. Weathering is the process that creates the soil used for growing crops, but weathering by wind and water can erode and carry away the soil.
- The house painter works to protect a house against the harmful effects of weathering. He covers the house with a protective coating of paint and seals cracks and crevices where water could enter.

What Is Your Prediction?

Involve students in a discussion of the places in which the amount of weathering would be the least. *(The polar and moon locations would experience the least amount of weathering—the polar regions because of their low humidity and low temperatures and the moon because it has no atmosphere.)*

EXPLORATION 5

Before carrying out each simulation, students are asked to make a prediction about the results of the activity. Have students record their prediction, along with reasons that support it, in their ScienceLog. After the Exploration, have students evaluate whether their prediction was correct.

 Cooperative Learning
EXPLORATION 5

Group size: 3 to 4 students
Group goal: to simulate factors that might affect the speed of weathering
Positive interdependence: Assign students roles such as leader, materials manager, safety monitor, and recorder.
Individual accountability: Have students express what they learned by performing the simulations.

SIMULATION 1

GUIDED PRACTICE After the simulation, have students discuss the role of freezing water in the weathering of rocks. Ask students to explain why a cold, wet climate might have a higher rate of weathering than a hot, dry climate. *(Rocks will weather faster in a humid climate than in a dry climate. Ice formation contributes to weathering. Water is also a catalyst in chemical reactions with the rock.)* Point out to students that water seeps into the cracks in rocks. If the water freezes, its volume expands by about 10 percent. Every time the water thaws and refreezes, it penetrates farther into the rock. More rock surface is then exposed to chemical weathering.

What Is Your Prediction?

Does weathering occur at the same rate all over the world? If you were in the business of selling gargoyles, in which of the following places could you offer the longest guarantee on your product? Make a prediction.

1. polar regions
2. tropics
3. deserts
4. mountains
5. prairies
6. coastal regions
7. on the moon
8. where you live

EXPLORATION 5

Three Simulations

It is very difficult (not to mention very time consuming) to determine the speed of weathering through actual observation. Real rocks take a very long time to break down. Sometimes, however, laboratory activities can *simulate* (imitate) what happens in the environment. Three factors that might affect the speed of weathering are investigated in the simulations that follow.

SIMULATION 1

What effect do freezing and thawing have on rocks? Make a prediction.

You Will Need

- a magnifying glass
- rock samples (granite, limestone, sandstone, brick, and a piece of concrete)
- a tray
- a freezer

What to Do

1. Examine your samples with the magnifying glass. Can you find any cracks where water might enter?
2. Soak each rock sample in water. After several minutes, remove it and place it in a tray in the freezer overnight.

3. The next day, remove the samples from the freezer, and allow the rocks to warm up to room temperature.
4. Examine the samples for any changes. Then repeat the process of soaking, freezing, and thawing. If time allows, repeat the freezing-thawing cycle several times.

Questions

1. Compare what happens to the different samples. Which sample shows the greatest change? the least change?
2. How does this simulation relate to conditions in the environment?

484

Answers to *Questions*

1. Sandstone and concrete should show the greatest change, while granite should show the least.

2. The activity simulates the repeated processes of freezing and thawing caused by yearly weather cycles.

 An Exploration Worksheet is available to accompany Exploration 5 (Teaching Resources, page 29).

SIMULATION 2 *page 485*

 SAFETY ALERT Even when dilute, hydrochloric acid is irritating to the skin and eyes.

 **PORTFOLIO**
Suggest that students include in their Portfolio a chart comparing their predictions with their results for the simulations performed in Exploration 5.

SIMULATION 2

Do small pieces of rock weather faster than large pieces? Make a prediction.

You Will Need

- a whole piece of chalkboard chalk
- dilute hydrochloric acid
- a graduated cylinder
- two 250 mL beakers
- latex gloves

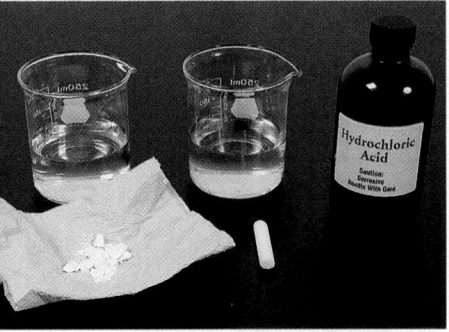

What to Do

1. In addition to your safety goggles, wear latex gloves and an apron during this activity. Break a piece of chalkboard chalk in half. Put aside one half, and break the other half into several small pieces. (You can wrap it in a paper towel and hit it gently with a hammer.)
2. Carefully pour 100 mL of dilute hydrochloric acid into each beaker.
3. Place the large piece of chalk in one beaker and the smaller pieces in the other.
4. Clean up your lab area and wash your hands when you are finished.

Questions

1. In which beaker did the bubbling last the longest?
2. What happened to the chalk in each beaker?
3. What comparison can you make between this investigation and the natural weathering process?

SIMULATION 3

Do some kinds of rocks resist weathering better than others? One class investigated this problem by using marble chips and quartzite. They took 100 g of each rock and placed each in a container about half full of water. After putting on the cover, they shook each container for 5 minutes at a rate determined before the experiment. After draining each container, they found and recorded the mass of the chips and quartzite. They repeated the procedure three more times and found the mass of each sample after each 5-minute period of shaking, for a total of 20 minutes. Their results are recorded in the table below.

Time (min.)	Mass of marble chips remaining (g)	Mass of quartzite chips remaining (g)
0	100.0	100.0
5	98.5	99.7
10	96.9	99.6
15	95.3	99.3
20	92.7	99.2

Questions

1. Which kind of rock lost the greatest amount of mass?
2. What inferences can you make about the resistance of marble to weathering? the resistance of quartzite to weathering?
3. During which time period did the greatest loss of mass from the marble chips occur? Can you explain this?
4. What do you think might happen if you shook each sample for 1 hour?
5. How does this investigation simulate the weathering process?

How Fast Does Weathering Occur in Your Community?

Weathering occurs at different rates, depending on the conditions in the environment. Suggest conditions in the environment that affect the speed of weathering (other than those you have already dealt with, of course). Make a list of all the factors in your area that might affect the speed of weathering.

485

FOLLOW-UP

Reteaching

Display several photos of materials that have been weathered. Ask students to identify the agent of change at work, such as wind, water, ice, or lichens, for each photograph.

Assessment

Tell students they have been asked to design a statue for a park in Montana. Ask: What kind of research would you have to do before you could decide on the type of stone to use? (*Investigate the climate of the area, including the amount of rainfall, humidity, temperature variation, and ice formation.*)

Extension

Place some moist steel wool in a closed plastic container, and have students observe it for several days. Ask: What changes can be seen? Is the "weathered" steel wool as strong as the original? Explain. (*The "weathered" steel is not as strong because corrosion has weakened its fibers.*) Is this an example of physical or chemical weathering? (*Chemical, because the iron combines with oxygen to produce iron oxide, or rust*)

Closure

Have students make a display titled "How Weathering Affects Our Surroundings" using their data from Exploration 5. Students could add visual examples to show how weathering has affected the schoolyard.

Gravity—The Great Leveler

FOCUS

Getting Started

Write the terms *Yellow River, Colorado River,* and *Red River* on the board. Ask students why they think these rivers received these names. *(Students will probably suggest that the names reflect the color of the water.)* Explain to students that the rivers are named for the color of the sediments that they carry. (The word *colorado* is a colloquial Spanish word for "red.")

Main Ideas

1. Erosion is a weathering process in which parts of the Earth's surface are carried away primarily by wind and water.
2. Gravity plays a role in all types of erosion.
3. Down-slope movements include landslides, mudslides, and avalanches.

TEACHING STRATEGIES

Weathering Wonders

After students have read page 486, ask them to look at the photographs carefully. Bryce Canyon, Utah (the large photograph on this page), was formed from many tiny streams, which caused both chemical and physical weathering. The weathered materials were then removed by both wind and water.

Answers to
In-Text Questions

Ⓐ Weathered materials are removed in various ways, primarily by wind and water. Moving water deposits the material downstream or eventually dumps it into a lake or ocean.

Ⓑ The strange rock formations indicate that parts of the rock have been removed. The brown water indicates that the river is carrying much of the sediment with it as it travels downstream.

Weathering Wonders

Where does the debris from weathering go? Imagine how the Earth would look if the debris of billions of years of weathering were piled up where it had formed. Why do you think this doesn't happen? Ⓐ

In these photographs, what indications are there that loose, weathered material has been removed? Ⓑ

LESSON 3 ORGANIZER

Time Required
1 class period

Process Skills
observing, inferring, analyzing

New Term
Erosion—the carrying away of weathered soil, rock, and other materials by gravity, wind, glaciers, or moving water

Materials (per student group)
Exploration 6: piece of plywood, about 20 cm × 30 cm; about 3000 cm³ of sand; 2 L of water; shallow, rectangular pan; wood blocks or bricks to prop up the board; pitcher

Teaching Resources
Exploration Worksheet, p. 33
SourceBook, p. S190

What agents of erosion are at work here? **C**

Erosion and Gravitational Force

The weathering and carrying away of soil, rock, and other materials is known as **erosion**. Erosion is caused by many natural agents. Flowing water, wind, waves, and glaciers continually wear away the land. The most important source of erosion, though, is the flowing water in streams and rivers. Water carries away more rock and soil than all of the other causes of erosion put together. Consider this fact: The Yellow River in China moves over 2000 million tons of sediment per year!

Gravity plays a role in all types of erosion. The Earth exerts a pull on everything on or near the planet's surface. The result of this is that things tend to move downhill until they can move no farther. Boulders and other loose rocks roll or slide downhill. Rivers and streams flow downhill. Glaciers grind their way slowly downhill. Water vapor that is carried to high elevations by air currents eventually falls as rain or snow. Wind redistributes fine particles by picking them up one place and then dropping them somewhere else. None of these erosive processes could happen without gravity.

487

Homework

Have students prepare a report on weathering after visiting an old building, an old statue, or an old stone wall. The report should be highlighted with sketches or photographs and should include a discussion of the weathering agents. Estimates should also be made of how quickly the weathering has occurred.

Answer to
Caption

C The photograph shows a desert dust storm. These storms contribute to wind erosion, especially in deserts where vegetation is sparse.

Erosion and Gravitational Force

GUIDED PRACTICE Call on a volunteer to read aloud the two paragraphs on page 487. Ask another student to explain the term *erosion*. Call on a volunteer to name the causes of erosion. *(Sample answers include gravity, flowing water, wind, waves, and glaciers.)* Invite another student to explain the term *gravity* and how it affects erosion. *(The pull of gravity tends to force things to move downhill.)* Emphasize the role of gravity in the process of erosion, and continue to emphasize it in subsequent lessons.

This is a good opportunity to discuss students' personal observations of erosion in action. Discuss the appearance of rivers when viewed from a distance. The sediments that they carry are often visible and give the river a distinctive color or a muddy appearance.

INDEPENDENT PRACTICE If there are local or regional parks with hiking trails nearby, suggest that students visit these areas. If it is possible, invite a ranger or naturalist to talk to the class about the impact of erosion in park areas.

ENVIRONMENTAL FOCUS

In the 1930s, a region of the central Great Plains, including areas of Texas, Oklahoma, Kansas, Colorado, and New Mexico, became known as the Dust Bowl. This area was wracked by a series of wind storms that led to massive erosion, causing one of the worst national disasters on record. Have students research some of the causes and effects of the Dust Bowl. *(Some factors leading to the Dust Bowl include the planting of wheat that didn't adequately protect the soil against erosion, the overgrazing of pasture land, and a severe drought that began in 1931. One storm in 1934 carried 318 million metric tons of dirt all the way to the East Coast. The dust storms destroyed crops, damaged people's lungs and machinery, and caused a mass migration of people out of the region.)*

GUIDED PRACTICE   After students have read the newspaper account of the Frank slide, help them to visualize the scene during and after the landslide. Ask them to look at the photograph and imagine tons of rocks sliding down the mountain. This will help them to appreciate the effects of this disaster on human life. Remind them that the account of the 1903 disaster would have been communicated by newspapers because television had not yet been invented.

INDEPENDENT PRACTICE   A possible activity to accompany this article could be to have students role-play as on-the-spot reporters. They could report orally or develop written copy describing what they found at the scene.

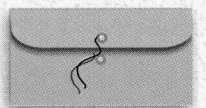

PORTFOLIO

Students may wish to include in their Portfolio their written copy describing what they found at the scene of the landslide disaster. Also encourage students to include their thoughts about whether they would enjoy having a career as a reporter.

Integrating the Sciences

Earth and Physical Sciences

Isaac Newton developed several theories of motion. His first law of motion states that a body in motion tends to stay in motion and that a body at rest tends to stay at rest unless an unbalanced force acts on it. Ask: How might this law apply to landslides and avalanches? What might an unbalanced force be? What forces might resist the flow of an avalanche or a landslide? Ask students to write a short paragraph summarizing their ideas. *(Unbalanced forces such as gravity, wind, and rain may cause an avalanche. An unbalanced force such as friction would resist the flow of an avalanche.)*

Landslide Claims 66

FRANK, ALBERTA, CANADA—
April 29, 1903

At 4:10 A.M. this morning, 35,000 tons of Turtle Mountain broke loose without warning and plummeted into the valley below, cutting a swath of total destruction through the heart of this small rural community and killing 66 people as they slept. Eyewitnesses report that the landslide moved at tremendous speed, leaving no time for escape.

A few lucky survivors were hurled to safety by a blast of air that preceded the landslide as it raced downhill.

Rescue operations are underway. Police, firemen, and members of the community are digging through up to 30 m of soil and rock strewn across the valley floor for a distance of 3.5 km. Officials admit that the chance of finding any survivors is very slim.

It is believed that the slide was triggered by unusually heavy spring rains.

Analyzing the Landslide

1. What types of materials slide down a slope?

2. What might cause them to come loose?

3. Have you ever tried to move a large rock? How easily might a large rock move down a steep hill?

4. Do you think that the landslide could have been prevented?

5. Where else might slides occur?

6. What caused the blast of air that preceded the landslide?

Turn to page S190 of the SourceBook to learn more about erosion.

488

Answers to
Analyzing the Landslide

1. Any loose objects, such as rocks, gravel, soil, and living things, can slide down a slope.

2. These materials can be loosened by water between soil particles, by the alternate freezing and thawing of water, and by earthquake vibrations.

3. Huge rocks could easily move down a steep hill once they are dislodged.

4. The slide itself could not have been prevented. However, if warning signs had been detected early enough, people could have been safely evacuated from the area. Signs of a potential slide include cracks in the ground or in walls and foundations, trees at an angle in the ground, and broken underground pipes.

5. Slides can occur where the incline is steep enough and where soil and rocks are not held firmly in place by vegetation.

6. The force of the rocks and other material breaking loose caused a compression wave to form in the air, like one that forms in an explosion.

A Model Landslide

Each year one of the projects in Mrs. Jefferson's class is to build a model and do an experiment with it. Pam and Sanjay decided to build a model that would help them figure out what causes landslides. Pam and Sanjay thought of the following ways to get the sand to slide:

1. Pour water on the sand.
2. Pile the sand deeper.
3. Make the angle of the board steeper.

You Will Need

- a shallow, rectangular pan
- sand
- water
- a piece of plywood
- wooden blocks

What to Do

Test Pam and Sanjay's methods. What other methods might they have tried? Do all these methods cause the sand to slide down the board? Can you think of any natural occurrences that match your investigations?

489

 An Exploration Worksheet is available to accompany Exploration 6 (Teaching Resources, page 33).

ENVIRONMENTAL FOCUS

Present students with the following problem: Weathering occurs even on your own property. Describe what measures homeowners can take to combat natural weathering processes. *(Student responses may describe benefits of painting, controlling rainwater runoff, creating windbreaks, filling cracks in foundations and walks, planting gardens with winter grasses to prevent soil erosion, or placing stepping stones in paths.)*

Review the concept of a fair test with students. Discuss how to control the variables in this experiment in order to make the comparisons valid. Emphasize that only one variable can be investigated, or changed, at a time and that the others must be controlled, or kept constant. For example, the sand must be piled to the same height each time unless the effect of the sand's depth is being investigated. Similarly, the angle of the board must be kept constant unless board angle is the variable that is being studied.

 Cooperative Learning
EXPLORATION 6

Group size: 3 to 4 students
Group goal: to investigate a different variable of the landslide model and to demonstrate the result to the rest of the class
Positive interdependence: Assign students roles such as facilitator, materials coordinator, reporter, or timekeeper.
Individual accountability: Have each student list the variable that his or her group investigated and give a brief summary of the group's results and conclusions.

Answers to
Exploration 6

Other methods of causing the sand to slide include jarring the board, pushing on the sand, or even using material other than sand (clay, topsoil, or a mixture of sand and gravel). Students will probably answer that all of these methods cause the sand to slide down the board. Natural occurrences that match the investigations include the following:

- Pouring water on sand matches soil saturation after a rain.
- Piling on more sand matches the buildup of sand, snow, or loose rocks.
- Changing the angle of the board matches the natural differences that exist in the degrees of slopes.
- Changing the type of material matches the natural differences among the materials that are present on a slope.
- Jarring the board matches naturally occurring earthquake vibrations.
- Pushing on the sand matches the natural occurrence of a small slide starting a larger one.

Summing Up

This word list should help students to connect all the concepts that they have encountered in their study of down-slope movements.

Answers to
Captions

Ⓐ Avalanches can be started easily. Depending on snowfall and weather conditions, the slightest added weight (such as a clump of snow falling off a tree) or even a sound (such as a gunshot) can start the slide.

Ⓑ A heavy rainfall can cause mud to slide. The water between the soil particles loosens the mud, causing it to flow.

Ⓒ Avalanches are started deliberately in order to prevent an unexpected slide.

FOLLOW-UP

Reteaching

Display several photographs of dramatic changes that have resulted because of gravity. Examples include landslides, mudslides, avalanches, and slumps. Have students choose one of the photographs and write a newspaper article describing what occurred and why.

Assessment

From the information learned in Exploration 6, suggest that students design a device that could prevent landslides.

Extension

Have students research areas where avalanches occur. Ask them to find out how avalanches can be predicted. They can prepare a written report or a poster for classroom display.

Closure

Invite a guest speaker to your class to discuss the impact of soil erosion on farm communities in your state or region. Sources could include agricultural extension offices, local universities, and local farmers.

Other Down-slope Movements

The photographs on this page show examples of down-slope movement.

Summing Up

Write an explanatory paragraph entitled Down-slope Movements. Use as many of the following words as you can in your description: weathering, air, debris, water, angle, rockfall, glacier, erosion, slide, snow, earthquake, slowly, avalanche, mountain, rain, gravity, downhill, boulder, and mud.

⬇ What causes an avalanche? Ⓐ

⬆ What triggered this mudslide? Ⓑ

⬆ Sometimes avalanches are started deliberately. Why do you think this is done? Ⓒ

490

Homework

The Extension activity on this page makes an excellent homework assignment.

CHALLENGE YOUR THINKING

1. Artist's Choice

You are a sculptor, and you have been asked to design a monument that will be placed in downtown New York City. How would you go about picking the type of stone that would be best to use? What would you have to know?

2. Physical or Chemical?

There are two types of weathering: physical and chemical. Limestone dissolving to form caves is an example of chemical weathering. The cracking of rock during a freeze-thaw cycle is an example of physical weathering. Classify the following examples of weathering as *chemical, physical,* or *not enough information to say*.

- A rock splits open after a day in the hot sun.
- A block of sandstone grows soft and crumbly after being exposed to water for a period of time.
- A rock breaks into pieces when it is heated.
- A rock develops a grayish color after being left outside.
- A rock is ground down during a flash flood.
- A rock's surface is crumbly where a lichen grew on it.

491

You may wish to provide students with the Chapter 25 Review Worksheet that is available to accompany this Challenge Your Thinking (Teaching Resources, page 34).

Answers to *Challenge Your Thinking*

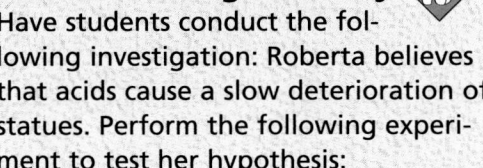

1. A sculptor would need to research the type of climate there, including humidity, temperature changes, and wind, as well as the amount and type of pollution. The sculptor would also need to find out which types of stone resist weathering the best.

2.
- Physical
- Not enough information to say
- Physical
- Chemical
- Physical
- Chemical

Meeting Individual Needs

Learners Having Difficulty
Have students conduct the following investigation: Roberta believes that acids cause a slow deterioration of statues. Perform the following experiment to test her hypothesis:

Add a piece of chalk to a glass one-quarter full of vinegar. The piece of chalk is made of limestone, as are many statues. Which of the following is a correct interpretation of the experiment?

a. The chalk disappears. Since statues do not completely disappear, Roberta's hypothesis must be incorrect.

b. The chalk undergoes a noticeable change. Roberta's hypothesis cannot be confirmed.

c. The chalk changes color just as statues often do. Therefore, Roberta's hypothesis is correct.

d. Rising bubbles and disintegration of the chalk indicate that the chalk has changed into a new substance, confirming Roberta's hypothesis. *(Correct)*

Answers to Challenge Your Thinking continued ►

Answers to
Challenge Your Thinking,
continued

3. Encourage students to write about the sights and sounds of a landslide or avalanche in their story. Student articles should demonstrate an understanding of factors that contribute to landslides, including erosion and gravitational force. A library search might provide an example of a real-life account of a landslide or avalanche.

4. Rip could have noticed whether the swing had been damaged, whether his initials were more difficult to read, whether the stone church showed signs of erosion, how much the tree had grown, and whether the creek had changed shape or size.

5. Encourage students to focus on conveying information effectively. Students may wish to use the Explorations in this chapter to generate ideas for their displays.

The following are sample revised answers:

1. The theory of plate tectonics explains many, but not all, of the processes that shape the Earth. In addition to continental drift, the landscape of the Earth is constantly changing through the processes of weathering and erosion.

2. Weathering refers to the physical and chemical decomposition of rocks. Erosion refers to the movement of the products of weathering.

3. The Egyptian obelisk called Cleopatra's Needle has been damaged by both physical and chemical weathering. Because water expands when it freezes, moisture and cold weather enlarge tiny cracks in rocks and cause the surface to crumble. In addition to this physical weathering, acid rain caused by air pollution reacts with the rock to cause chemical weathering.

4. Weathering results from the forces of weather, such as wind, rain, ice, heat, and humidity.

3. Please Publish
Write a human-interest magazine article entitled "I Survived the Frank Landslide" or "Buried Alive in an Avalanche!" Be sure to include a description of the cause of the disaster.

4. Time Warp
After a 20-year absence, Rip returns to his hometown. He is dumbfounded. "Why, it's just as I left it. It hasn't changed a bit. The old sycamore tree we used to have the swing on is still there by the old stone church. Look—the tree still has my initials scratched in it. It hasn't changed, and neither has the church. The creek is still crystal clear. It hasn't changed . . ." Were things truly just as Rip left them, or should he take a closer look? What would Rip see if he looked more closely?

5. Designer Display
Design a bulletin-board display whose theme is geological changes. Ask yourself the following questions:

- Who is my audience?

- What information do I want to convey?

- How can I best convey the information—through articles? through photographs? through drawings?

- How can I make the display interactive enough to keep my audience interested?

ScienceLog

Review your responses to the ScienceLog questions on page 470. Then revise your original ideas so that they reflect what you've learned.

Water, Wind, and Ice

2 Where does the water go after a rainstorm?

1 Which carries away more rock and soil—water or wind?

3 How many tons of sediment do you think are carried to the ocean each year?

4 How do we know that much of the Earth was once covered by ice?

ScienceLog

Think about these questions for a moment, and answer them in your ScienceLog. When you've finished this chapter, you'll have the opportunity to revise your answers based on what you've learned.

493

Connecting to Other Chapters

> **Chapter 24**
> investigates the geologic diversity of the Earth and introduces the theory of plate tectonics.

> **Chapter 25**
> examines evidence of geologic changes and the role that gravity plays in such changes.

> **Chapter 26**
> explores how landforms are altered through sedimentation, erosion, and the actions of winds and glaciers.

Prior Knowledge and Misconceptions

Your students' responses to the ScienceLog questions on this page will reveal the kind of information—and misinformation—they bring to this chapter. Use what you find out about your students' knowledge to choose which chapter concepts and activities to emphasize in your teaching.

In addition to having students answer the ScienceLog questions on this page, you may wish to have them complete a written assignment. Tell students to think of a river that they have seen and to answer the following questions: How wide is the river? How steep are its banks? What effect has the river had on the terrain of the area? How long do you think this river has been flowing? Why do you think so? Remind students that there are no right or wrong answers to these questions. Collect the papers, but do not grade them. Instead, read them to find out what students know about the formation of rivers, what misconceptions they may have, and what they find interesting about the topic.

LESSON 1

The Power of Water

FOCUS

Getting Started

Write the number *40 quadrillion* on the board (40,000,000,000,000,000). Tell students that this number represents the approximate number of liters of rainwater that washes over the Earth and flows into the seas each year. Ask students what effect this flow of water might have on the land. *(One effect is erosion.)*

Main Ideas

1. Raindrops strike the Earth with great force, breaking off fragments of rock and creating runoff.
2. The volume of water absorbed by the ground is a measure of soil porosity.
3. The rate at which water can pass through a porous material is a measure of the material's permeability.

TEACHING STRATEGIES

Answers to
A Numbers Game

- Students should convert the area of the schoolyard into hectares (1 hectare = 10,000 m²) and multiply this number by 12 million to get the total number of raindrops falling on the schoolyard every second. Finally, they should multiply this number by 3600 to get the total number of raindrops falling on the schoolyard every hour.
- There are 1 ÷ .05 = 20 raindrops per cm³.
- A 10 × 10 × 10 cm container contains 1000 cm³, or 20,000 raindrops.
- The total number of liters equals the total number of raindrops falling on the schoolyard in an hour divided by 20,000.

 A Transparency Worksheet (Teaching Resources, page 38) and Transparency 86 are available to accompany the material on this page.

LESSON 1

The Power of Water

Rain, rain, go away;
Come again some other day.

Have you ever spoken these words as you watched the rain fall? No doubt you wished that the rain would stop, but think of the rhyme in another way. Rain really does go away. Even the heaviest downpour seems to disappear shortly after it strikes the ground. Where does it go? Look closely at the illustration above. To how many different places might a raindrop go? **A**

A Numbers Game

In an average rainfall, over 12 million raindrops fall on every hectare (10,000 m²) of land or surface every second!

- Find out the area of your schoolyard. Then calculate the number of raindrops that would fall on the schoolyard in an hour.

  A raindrop is about 0.05 cm³ in volume.

- How many raindrops per cubic centimeter?
- How many raindrops will fill a 10 × 10 × 10 cm container?
- How many raindrops fell on the total area of your schoolyard in an hour?

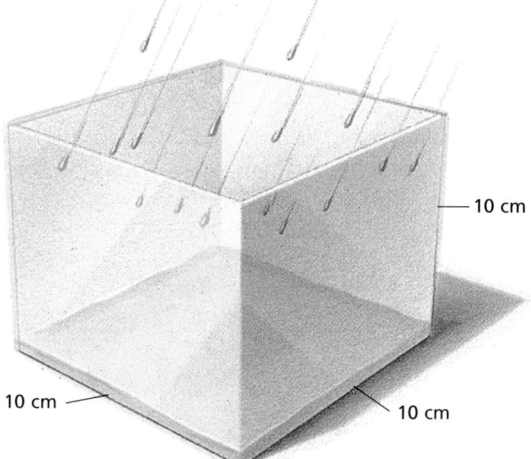

10 cm

10 cm

10 cm

494

LESSON 1 ORGANIZER

Time Required
3 class periods

Process Skills
comparing, analyzing

New Terms
Ground water—water that collects beneath the surface of the Earth
Permeability—how easily water can pass through the open spaces in rock or soil
Porosity—the volume of water that a volume of rock or soil can hold
Runoff—excess water that is prevented from sinking into the ground and thus collects and runs over the surface

Materials (per student group)
A Numbers Game: optional materials: meter stick or metric ruler; calculator
Exploration 1: three 500 mL beakers; three different soil samples, such as sand, clay, and garden soil; 100 mL of water; 100 mL graduated cylinder; watch or clock with a second hand (See Advance Preparation on page 449C.)

Teaching Resources
Transparency Worksheet, page 38
Exploration Worksheet, page 40
Transparency 86–87
SourceBook, p. S191

A raindrop explodes as it strikes the ground and creates a tiny crater in the soil.

Runoff

What effect do all these raindrops have when they strike the Earth? Rain wears away the Earth's surface with astonishing force. Each raindrop is like a miniature hammer pounding the soil with a force of nearly 1 kg/cm². This is enough force to break off tiny fragments from solid rock. Each raindrop that strikes loose soil gouges out a miniature crater. A muddy mixture of water and undissolved particles forms and clogs the surface pores of the ground. The mud prevents the ground from absorbing more water. Gradually, puddles form as more and more rain is prevented from sinking into the ground. Excess water, or **runoff**, collects and flows over the surface. In a hard rain, almost all of the water becomes runoff. The runoff quickly collects into rivulets, which dislodge soil and rock as the rivulets flow. The dislodged material adds to the water's erosive power by scraping the ground like sandpaper. The runoff carries its sediment load as it flows into the nearest stream. The stream, swollen with runoff, tears away at the land as it flows to the river, where it dumps its load. The river itself, swollen with runoff from many streams, flows toward the sea, continuing the erosive process on a much larger scale.

Make a flowchart of runoff. The diagram at right will get you off and running.

A Flowchart of Runoff

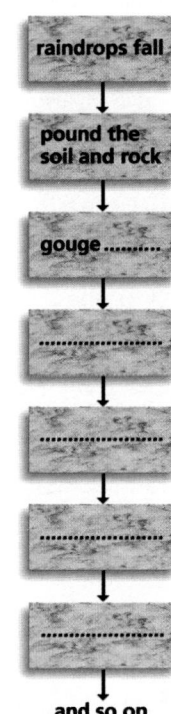

raindrops fall

pound the soil and rock

gouge ....

....

....

....

....

and so on

495

EXPLORATION 1

Cooperative Learning
EXPLORATION 1

Group size: 3 to 4 students

Group goal: to investigate the movement of water through different types of soil

Positive interdependence: Assign students roles such as primary investigator, investigative associate, supply coordinator, and timer.

Individual accountability: Have each student answer the questions asked in the final paragraphs of the Exploration. The questions could be answered either orally in the form of an oral exam or in essays in the students' ScienceLog.

Soil samples may be collected from the schoolyard or brought from home by students. If possible, sand and clay should be included. Three different samples should be adequate. Calibrate the transparent containers in deciliters (100 mL = 1 dL). This will make it easier for students to measure equal amounts of soil. Each student should have a chance to add water to a soil sample and to measure the time it takes to reach the bottom of the container.

Attention should be paid to how the water is added to the soil so that all procedures are similar. Timing should start when the water begins to move through the soil, and it should stop when the water reaches the bottom of the container. The water may reach the bottom by seeping between the soil and the container.

Generally, water will flow faster through soil composed of larger particles. This is because the larger particles do not pack together as tightly as smaller ones, thereby leaving more space through which water can pass.

 An Exploration Worksheet is available to accompany Exploration 1 (Teaching Resources, page 40).

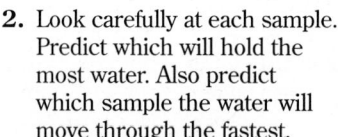

Investigating the Movement of Water in Soil

Do some soils hold more water than others? How fast does water move through soil?

You Will Need

- 3 500 mL beakers
- 3 soil samples
- water
- a graduated cylinder
- a watch with a second hand

What to Do

1. Fill each container to the 400 mL mark with soil taken from several locations (sand, clay, garden soil, etc.). Be careful not to pack down the soil.

2. Look carefully at each sample. Predict which will hold the most water. Also predict which sample the water will move through the fastest.

3. Fill the graduated cylinder with water to the 100 mL mark. Slowly add as much of the 100 mL of water as possible to each soil sample until the soil sample can hold no more. Make note of how much water you used. At the same time, measure how long it takes for the water to reach the bottom of each container.

4. Copy the following table into your ScienceLog, and fill it in as you make your observations:

Type of soil	Time for water to pass through	Volume of water absorbed

5. When you have finished, clean up your area, and wash your hands with soap and water.

Making Sense of Your Results: Porosity and Permeability

Certain soils allow water to pass through them faster than others do. In other words, some soils are more *permeable* than others. When you measure the time it takes for water to pass through a soil, you are measuring the **permeability** of the soil. Which sample was the most permeable? the least permeable?

Because of gravity, rainwater moves downward into tiny spaces in soil and rock. These spaces are called pores. The greater the volume of water that a soil can hold, the more *porous* the soil is. The volume of water that a soil can hold is a measure of the soil's **porosity.** Which sample was the most porous? the least porous? Which type of soil would cause the most runoff during a rainstorm? Explain your answer based on the concepts of porosity and permeability.

Making Sense of Your Results

Reassemble the groups. Ask students to read the material silently. Then involve them in a discussion of the terms *porosity* and *permeability*. Call on volunteers to explain the results of the Exploration in terms of the porosity and permeability of the various soil samples.

Answers to
Making Sense of Your Results

Sand was the most permeable. Clay was the least permeable. Garden soil was the most porous. Clay was the least porous. Clay would cause the most runoff during a rainstorm. Since clay is the least permeable, it does not allow much water to soak into the ground. Since clay is also the least porous, it is unable to absorb much water. The excess rainwater becomes runoff.

Water Below!

Much of the water that reaches the Earth's surface drains into lakes, rivers, and streams. Additional water may evaporate, or it may filter down through the soil until it reaches an *impermeable* layer—a layer through which it cannot pass. This water then collects in the air-filled spaces above the impermeable layer. The upper surface of this water-filled area is known as the *water table*. Water that collects beneath the surface of the Earth is known as **ground water**. For humans, ground water is a vital source of fresh water. Streams and springs are fed by ground water, and so are the wells of towns and farms and the reservoirs of cities. Does your community rely on ground water? What do you think might cause wells to go dry? **Ⓐ**

Soil-water zone

Well

Stream

Water table

Ground-water zone

Rock

497

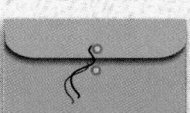

PORTFOLIO
Suggest that students include in their Portfolio their thoughts about the investigative abilities they used during Exploration 1. Encourage students to compare their answers with what they think real-world scientific investigations might be like.

Answers to *In-Text Questions*

Ⓐ Drinking water is supplied in a variety of ways. Some communities get their water supply directly from a nearby river. Others use water from a lake or reservoir. Still other communities tap ground-water supplies by digging wells.

    Water wells may be hundreds of meters deep if the local water table is very far below the surface. To get water from such a deep well requires the use of pumps. A well "goes dry" when the water table drops below the bottom of the well.

GUIDED PRACTICE Have a volunteer read aloud the paragraph on this page. Call on a volunteer to define the term *ground water*. Then direct students' attention to the diagram. Point out the soil-water zone, the water table, and the ground-water zone. The information below will help you explain the different areas.

- **soil-water zone**—Area where water percolates through the soil and filters down into the water table. The soil here is not saturated with water.
- **water table**—The top surface of the ground water in which all of the open spaces are filled with water.
- **ground-water zone**—Water will continue to travel downward in the soil until it reaches an impermeable layer of rock or clay. Water will collect and pool on top of this impermeable layer, filling all of the open spaces with water.

    This might be a good opportunity to discuss problems with our water supply. At the present time, there is great concern over the extent and safety of ground-water supplies in the United States. For many years, garbage and toxic materials have been buried in landfill sites. These sites are natural or human-made depressions in the soil that are sealed at the bottom, filled with waste materials, and then covered over by soil. Unfortunately, the techniques used to seal the bottom of these landfill sites have proven to be inadequate in many cases. This means that toxic materials can contaminate ground-water supplies.

    Another problem is the interconnectedness of ground water. Because many rivers flow under the ground, a leaky landfill could contaminate well water in communities many hundreds of kilometers away.

 Transparency 87 is available to accompany the material on this page.

You Be the Judge

INDEPENDENT PRACTICE Divide the class into three groups. Assign one situation to each group to discuss. Then reassemble the groups and involve the class in a discussion of the situations.

Answer to
Situation 1

The better decision was to put the rows across the hill. Since gravity pulls water downhill, more gullying and soil erosion occurs between rows that go up and down a slope. You could demonstrate this in class by using a stream table.

Answer to
Situation 2

The better decision was to use the straw. Straw absorbs the impact of raindrops, thereby preserving the large pores in the soil. Hence, moisture is retained rather than lost in runoff.

Answer to
Situation 3

The mayor made the wrong decision. Heavy snowfall and spring rain increases the chance of flooding, and the concrete parking lot does not allow the water to soak into the soil. Therefore, the runoff to the already swollen river would increase dramatically.

FOLLOW-UP

Reteaching

Have students make a flowchart like the one on page 495 for each alternative in one of the You Be the Judge situations. Then they should explain how each flowchart illustrates the better decision.

Assessment

Have students compare the qualities of sand, clay, and garden soil. They should describe the characteristics of each type of soil and compare how each interacts with water. The terms *porosity* and *permeability* should be included in their answers.

Extension

Have students research the source of their community's drinking water. Ask

You Be the Judge

Apply what you have discovered to the following three situations. See if you can judge who made the right decision in each situation.

SITUATION 1

Two gardeners were sharing a plot on the side of a hill. When it was time to prepare the rows, they had an argument. One gardener insisted that the rows should go straight up and down the hill. The other gardener argued that the rows should cut across the hill. To settle the argument, they divided the plot and did as they wished. Rainfall was heavy that summer. Who made the right decision? Explain.

SITUATION 2

Another argument arose the next year. After the plants began to grow well, one gardener wanted to put layers of straw between the rows. The other gardener preferred to leave the ground bare. Because they were unable to settle their argument, they divided the plot once again and did as they wished. It was a very dry summer except for a few thunderstorms with heavy rain. Who made the right decision? Explain.

SITUATION 3

A plan to build a shopping mall came before the town council of Mesopotamia, a community built on a plain between two rivers. One of the features of the plan was a huge, concrete parking lot. Several members of the council were concerned that the chances of flooding in the area would be greater if the shopping center and its parking lot were built. Others argued that the town needed the tax money that the project would generate. A vote was taken. The result was a tie. The mayor cast the deciding vote in favor of the application. The summer after the mall was built, there was a heavy thunderstorm that dropped 15 cm of rain in only 2 hours. Did the mayor make the right decision? Explain.

Turn to page S190 of the SourceBook to find out more about runoff and soil erosion.

them to find out what is done to the water to make it safe for drinking before it is piped into their homes. Students could discuss the characteristic taste and smell of local drinking water and relate the characteristics of the local water to its source and processing.

Closure

Have students examine erosion patterns underneath the opening of a drainpipe. Invite students to make a map of the drainage pattern observed.

Homework

The Assessment activity on this page makes an excellent homework assignment.

LESSON 2 · Water on the Move

Rain or snow falling on the land eventually drains away. Streams and rivers catch the water and funnel it downhill. Falling water has energy, so it causes changes as it flows downhill. Rivers cause dramatic changes in their surroundings as they flow downstream. Several streams and rivers are shown on this page. Make as many observations as you can about each stream. Compare them with one another. Rate each stream by its ability to affect its surroundings. Which appears to have the most energy? the least energy? What is the connection between the speed of the stream and its ability to erode its bed and banks? **Ⓐ**

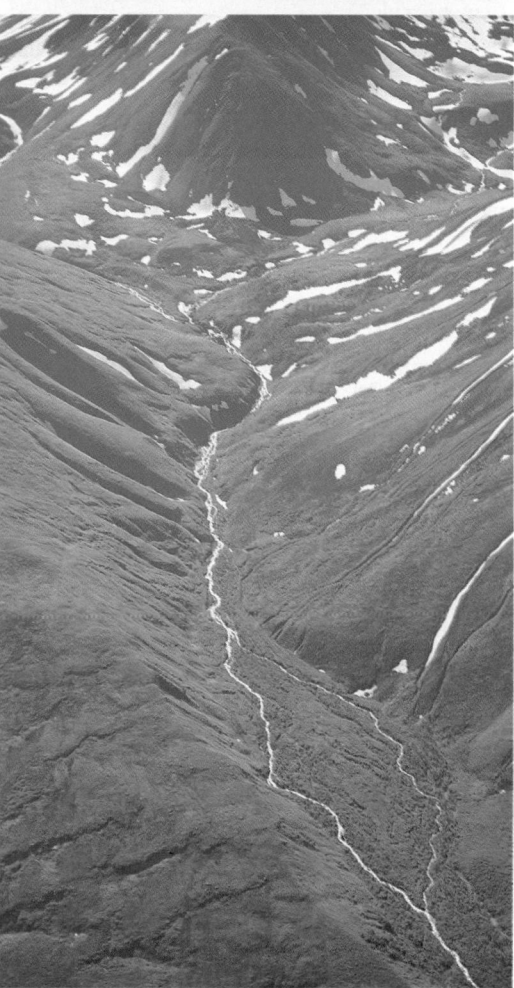

499

LESSON 2 · Water on the Move

FOCUS

Getting Started
Display a map of the United States. Ask one or two volunteers to use tracing paper to trace the Mississippi River and all of the rivers and streams that are connected to it. Ask the class if they can identify the source of the Mississippi (where the river begins) and its mouth (where it ends). Discuss the map by asking questions such as the following: Does the river seem to get larger or smaller as it flows toward its mouth? What contributes to this? What influences the direction in which the river flows?

Main Ideas
1. Water picks up, sorts, and moves dirt and rocks as it flows downhill.
2. The faster water moves in a stream, the more power it has to erode the surface over which it travels.

TEACHING STRATEGIES

Call on a volunteer to read page 499 aloud. Then have students study the stream photographs and list their observations in their ScienceLog. Observations should include a description of each stream's branches and subsequent paths. All streams, whether small or large, affect their surroundings.

Answers to
In-Text Questions

Ⓐ The stream that travels the steepest slope with the greatest volume of water will usually have the most energy. The small stream traveling on ground that is level or nearly level will probably have the least energy. Generally, greater stream speed means greater ability to erode.

LESSON 2 ORGANIZER

Time Required
1 class period

Process Skills
observing, predicting, analyzing

New Terms
none

Materials (per student group)
Exploration 2: watch or clock with a second hand; metric ruler; a few drops of food coloring; wax pencil; 5 L of water; 2 buckets; block of wood;

stream table or large pan (about 35 cm × 120 cm) with drain, 2 lengths of rubber tubing (30 cm each), 2 small clamps or clothespins, and 50 cm trough made from a cardboard tube (additional teacher materials: cardboard tube; scissors or knife; see Advance Preparation on page 449C.)

Teaching Resources
Exploration Worksheet, p. 42
SourceBook, p. S198

EXPLORATION 2

Have students record in their ScienceLog their prediction about the result of each Activity before performing any of the steps. This Exploration may also be done as a class demonstration. If performed by students, it is important to demonstrate the procedure first. Do not allow students to start their siphon by suction. Instead, demonstrate the method of starting a siphon by first immersing it in water. When the siphon is filled with water, hold your finger over the free end and immerse the other end in your water source, which is higher than the pan. Release your finger, and the siphon will start flowing.

Cooperative Learning
EXPLORATION 2

Group size: 3 to 4 students
Group goal: to work together to construct and operate a device to determine what factors cause a stream to flow quickly or slowly and to accurately time the results of the experiment
Positive interdependence: Assign students roles such as timer, materials manager, results tracker, and experiment coordinator.
Individual accountability: Ask each student to answer the following question in their ScienceLog: What determines whether a stream flows quickly or slowly? Ask students to support their reasons with examples from the activities they performed.

Encourage students to record their findings in a chart listing the slope or volume in one column and the travel time in another column. The results of Activities 1 and 2 should confirm that greater slope and greater water volume both produce greater stream speed.

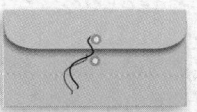

PORTFOLIO
Encourage students to assess their group's ability to cooperate and successfully perform Exploration 2. Students may want to include thoughts about their own contributions to the group's results. Students can then put their assessment and thoughts in their Portfolio.

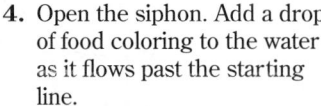

Fast or Slow Streams

What determines whether a stream flows quickly or slowly? Do the following Activities to find out.

You Will Need
- a setup such as the one shown below (Be sure that there is a drain at the lower end of the pan.)
- a metric ruler
- a wax pencil
- food coloring
- a watch with a second hand

ACTIVITY 1

What effect does changing the slope of a stream have on the speed of the water?

What to Do
1. In your ScienceLog, write down a prediction for the question above.
2. Set up the pan of water and the siphon as shown. Adjust the trough so that the upper end is 4 cm high.
3. With a wax pencil, mark a starting line on the side of the trough.

4. Open the siphon. Add a drop of food coloring to the water as it flows past the starting line.
5. Time how long it takes the dye to reach the end of the trough.
6. Change the height of the trough to 8 cm. Then repeat steps 4 and 5. Compare the results.

ACTIVITY 2

What effect does changing the volume of the water in a stream have on the speed of the water?

What to Do
1. In your ScienceLog, write down a prediction for the question above.

2. Set up an experiment to answer this question. Consider the following as you plan what to do:
   - Can you use the same setup as before?
   - How can you increase the volume of water?
   - How will you determine the speed of the water?
   - Will you change the slope, as you did in Activity 1?
   - How will you determine whether the volume affects the speed of the water?

3. Can you think of another setup that might be used to answer this question?

500

Answer to
Activity 2

3. Students may increase the volume of flowing water either by adding another siphon tube or by decreasing the size of the stream channel.

 An Exploration Worksheet is available to accompany Exploration 2 (Teaching Resources, page 42).

CROSS-DISCIPLINARY FOCUS

Mathematics
Nickie's grandmother is 100 years old. If a 1 cm thickness of soil is eroded from the land by water every 300 years, approximately how many millimeters of soil has water eroded in Nickie's grandmother's lifetime?
a. 100 mm
b. 3 mm *(Correct)*
c. 30 mm
d. 300 mm

1. Look at the predictions you made. Does the data that you recorded support your predictions? In your ScienceLog, summarize your data, and present your conclusions.

2. You have identified two variables that affect how fast water flows in a trough. What other variables might affect the flow of a stream in a natural setting?

3. Which stream do you predict can carry the most sediment, a fast-flowing stream or a slow-flowing one? What evidence do you have to support your prediction? Explain why the speed of a stream affects the amount of sediment that it can carry.

Believe It or Not

Did you know that as a result of erosion by flowing water

- 1 km² of land loses between 20 and 1000 tons of rock and soil each year?
- 1 cm of thickness is eroded from the land every 300 years?
- 4 billion tons of sediment are carried to the ocean each year?

How big is your state? From the first item above, figure out the smallest and the greatest amount of material that could be removed each year. Which figure do you think is closer to the actual amount of material lost by your state each year?

Outside the Classroom—A Project

Do you have a stream in your neighborhood? If so, check to see if the water flows quickly or slowly. Which of the variables that you discussed in the previous Exploration exist in the stream? What evidence is there of stream erosion? Write a short report to share with your classmates.

Caution: Do not go in or near a body of water without adult supervision.

501

2. Responses could include the shape of the stream bed, the amount and type of sediment carried by the stream, and the roughness of the stream bed.

3. In general, a fast-flowing stream can carry more sediment than a slow-flowing one because the fast-flowing stream has more energy. Some students may have noticed that sediment accumulates in low-lying areas after a storm. Other evidence includes the formation of deltas at the mouths of rivers.

Believe It or Not

To perform the calculation of the material removed each year from a particular state, have students determine the approximate dimensions of their state by measuring the length and width of the state on a map drawn to a known scale. (Approximate a rectangle for irregularly shaped states.) Students may have to convert from miles to kilometers. One mile equals approximately 1.6 km. The amount lost each year would be a result of the factors listed by the students. Factors could include average rainfall, number of rivers, type of soil, amount and type of underlying rock, and the amount of vegetation.

Outside the Classroom—A Project

Observations of a neighborhood stream will vary, but students should conclude that the speed of the water varies in different parts of the stream.

FOLLOW-UP

Reteaching

Have students draw pictures of a river at different stages of development. Students should emphasize the effect that the developing river has on surrounding land formations.

Assessment

Tell students to imagine that there is a flash flood in the area. The school is located near a stream. Ask: What predictions can you make about the way the water near the school will move? *(On a hill with a large slope, the water would* move very fast; on a flat area, the water would form a stationary body of water.)*

Extension

Have students create a concept map using the following terms: sediment, water, fast-moving, stream, volume, and slope. *(Accept all reasonable concept maps.)*

Closure

Have students write and perform a skit that could demonstrate the erosive force of moving water to younger students.

Homework

The Reteaching activity on this page makes an excellent homework assignment.

LESSON 3
How Rivers Change the Land

FOCUS

Getting Started

Ask students how much water they think an average family uses each day. *(Students may be surprised to learn that the average family in the United States uses 760 L of water per day.)* Our planet is known as the "water planet" because about 70 percent of the Earth's surface is covered by water. However, 97 percent of the water on the Earth is salt water and 2 percent is locked in the frozen icecaps at the poles. This leaves only 1 percent of the Earth's water available for drinking and growing food. Help students to relate this information to the development and characteristics of the world's water systems.

Main Ideas

1. A stream or river slows down and deposits sediments when it reaches a bend, when it widens, or when its slope decreases.
2. The age of a river can be inferred by examining the degree to which the land around it has been worn down and eroded.

TEACHING STRATEGIES

Answers to
In-Text Question

Ⓐ • Lot *A* is the best choice because there is little chance of erosion occurring there.
• Lots *B* and *E* are poor choices because they are both on the outside of a curve, where water tends to move faster, causing erosion.
• Lots *C* and *D* are good choices because they are both on the inside of a curve, but *C* may soon become an island as the water erodes the bank upstream, and *D* will suffer from the wave action of the lake.

LESSON 3
How Rivers Change the Land

Imagine that you have won a contest, and the grand prize is a plot of land of your choice. The choices are shown in the illustration below. Which lot would you choose? Why? Collect evidence to Ⓐ support your choice by using a stream table as described in the next Exploration.

502

LESSON 3 ORGANIZER

Time Required
4 class periods

Process Skills
observing, inferring, analyzing

Theme Connection
Energy

New Terms
Gullies—ditches carved out by moving water
Tributaries—streams that feed into a main stream
Watershed—an area that a river drains

Materials (per student group)
Exploration 3: about 3000 cm³ of sand; metric ruler; 5 L of water; block of wood; watch or clock with second hand; a few drops of food coloring; stream table or large pan with drain (about 35 cm × 120 cm), 2 lengths of rubber tubing (30 cm each), and 2 small clamps or clothespins

Teaching Resources
Exploration Worksheets, pp. 44 and 45
SourceBook, p. S192

A Stream-Table Experiment

You Will Need

- a setup as shown in the photograph below
- sand
- a metric ruler
- food coloring

What to Do

1. Place sand to a depth of 5 cm in the stream table.
2. Wet the sand so that it can be shaped. Form the sand into a gentle slope.
3. Trace an S-shaped path with your finger almost to the bottom of the sand.
4. Turn on the water and adjust the flow so that it moves slowly into the tray.

5. Allow the water to flow down the path for 20 minutes. While the water is flowing, note the following locations:
   a. places where the water slows down or speeds up (food coloring may help here)
   b. places where banks are being eroded
   c. places where sand buildup occurs
   d. changes in the course of the stream

6. Turn off the water, but leave the sand in place.

Pretend that a friend was absent from school during the stream-table experiment. Describe to your friend exactly what you saw during the experiment. Do your observations support your choice of lot?

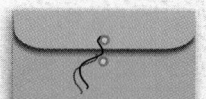

⭐ **An Exploration Worksheet is available to accompany Exploration 3 (Teaching Resources, page 44).**

PORTFOLIO
Students may wish to include comments about Exploration 3 in their Portfolio. Ask: How did this activity help you understand the ways in which water in streams and rivers slows down? Encourage students to include any further questions or concerns.

EXPLORATION 3

Cooperative Learning
EXPLORATION 3

Group size: 3 to 4 students
Group goal: to conduct an experiment and to gather data on stream formation
Positive interdependence: Assign students roles such as primary investigator, supportive assistant, materials manager, and recorder.
Individual accountability: Have each student answer the questions posed on page 502 and the question asked at the end of this Exploration.

Suggest that students keep the volume of water low so that the changes will occur slowly. Have groups set up their stream tables by a sink or use at least two buckets under the table to catch the overflow.

Point out to students that a stream or river slows down when it reaches a bend, when it gets wider, or when its slope decreases. Slower water lacks the energy needed to carry its load, so it drops what it has been transporting. Over time, this process exaggerates the curves of the river. When a curve becomes extremely exaggerated, the river may cut across it, thereby forming an island. This means that a river's shape may change from straight to curved and back again as it ages.

Answer to
In-Text Question

Ⓑ Sample answer: We poured water down an S-shaped channel that we had dug down a sandy slope. The water slowed down around curves and sped up along straight areas. The water picked up sand from the outside of the curves and deposited sand on the inside of the curves. This caused the water channel to get wider and deeper and to curve even more.

If islands do not form within the 20 minutes that students observe the stream, students may conclude that lots C and D on page 502 are also good choices. You may wish to point out why they are not good choices by explaining the effects of other sources of erosion, such as wave action.

Do Rivers Have Age?

Call on a volunteer to read the verse from "The Negro Speaks of Rivers." Students may be interested to know that Langston Hughes wrote this poem the summer after he graduated from high school. Then have students read the paragraphs that follow. The numbers included in the text will be used to answer question 2 at the bottom of page 504. Students may find it simpler to answer question 3 first.

Answers to
Do Rivers Have Age?

1. The stages of river development shown in each figure are as follows:
   A—youth
   B—maturity
   C—old age

2. (1)—A
   (2)—C
   (3)—A
   (4)—B
   (5)—B
   (6)—C

3. Possible responses include the following:

   A This figure shows a youthful river. It has carved, deep, V-shaped valleys with mountains on all sides. It does not bend or curve much. It has not begun to widen and erode its banks.

   B This figure shows a mature river. The distinctive V-shaped valley can still be recognized. There are long tributaries and deep valleys. The river's path is beginning to develop curves. This shows that it is eroding its banks and enlarging its valley.

   C This figure shows an old river. It travels slowly from side to side. It moves along a highly curved path. Erosion has worn away the hillsides.

Primary Source

Description of change: excerpted from "The Negro Speaks of Rivers" by Langston Hughes
Rationale: excerpted to focus on the age of rivers

Do Rivers Have Age?

The Negro Speaks of Rivers
by Langston Hughes

I've known rivers:

I've known rivers ancient as the world and older than the flow of human blood in human veins.
My soul has grown deep like rivers.

In this poem, published in 1921, Langston Hughes suggests that rivers grow old and deep. You might wonder, How can a river be old? How does a river grow?

About 100 years ago, an American geologist named William Davis proposed that rivers follow a distinct pattern of development. Davis compared rivers to living organisms, noting that both change markedly as they age. But the age of a river is not measured in years. The age of a river is measured by how completely the river has eroded its **watershed**, which is the area that it drains.

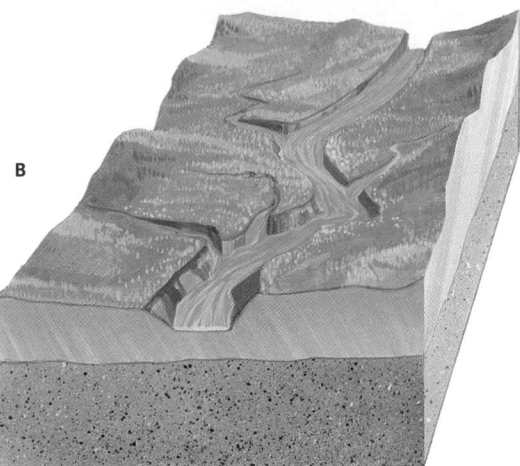

The figures at right illustrate three stages of river development. Trace each one into your ScienceLog. Can you identify the features described in the story that follows?

1. Match each figure with one of the following stages: youth, maturity, or old age.

2. Match the features numbered 1–6 in the text below to one or more of the figures.

3. Write a descriptive sentence about each figure.

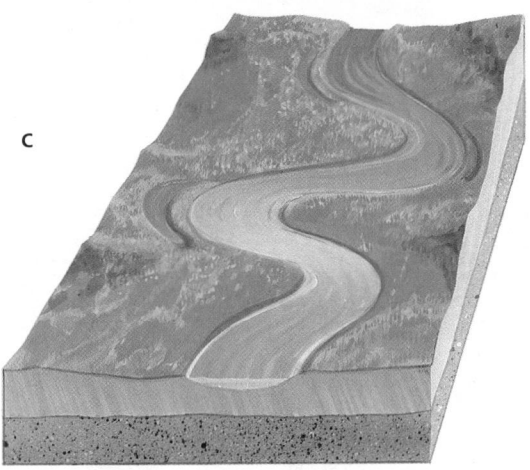

Young rivers (1) have steep banks and narrow valleys and fall in elevation quickly. Old rivers (2) fall very slowly in elevation, have low banks and wide valleys, and tend to meander (travel an S-shaped course).

The life cycle of a river begins when a huge block of land is lifted above sea level. Rainwater running down the fresh slope carves out *gullies*, or ditches (3). These gullies run together to form larger channels, which in turn run together to form still larger channels.

Theme Connection

Energy
Focus question: Is the energy of water in a river converted from one form to another as it flows from its source to its mouth? *(Yes; the higher elevation at which a river begins gives the water potential energy because of its position. As it flows downhill, this potential energy is converted to kinetic energy.)*

Homework

Have students write a poem about a river that they have seen. Encourage students to focus on what qualities the river possesses that make it both unique and interesting.

Over time, rivers carve valleys into the uplifted land. The river continues to develop as its **tributaries** (feeder streams) deepen their channels and cut downward to form steep, V-shaped valleys (4). As the river matures, new tributaries form, and existing tributaries grow.

A young river carries little water because it has few tributaries. It quickly erodes downward. As a river matures, the amount of water that it carries increases, and it begins to erode its valley as well as its bed. Erosion wears away at the valley walls and causes them to become less steep (5). After a long time, erosion wears down the river valley so thoroughly that it is almost the same elevation as the river. At this point, the river can no longer erode downward, and it begins to erode sideways by meandering back and forth across its valley (6). Eventually, the river wears down its watershed into a featureless plain near sea level. The river is now in a state of old age. It flows slowly and carries little sediment.

EXPLORATION 4

Check It Out

1. Is there a river near you? What stage is it in?
2. Many states contain well-known valleys or canyons. Find out about features such as these in your state. Then use a map to find out whether or not rivers flow through them. Try to answer these questions:
   a. Examine the photograph of the Grand Canyon at right. Which came first, the canyon (or valley) or the river?
   b. What is the relationship between a valley, a canyon, and a river?
   c. A river may have flowed through a valley or canyon at one time, even though there isn't a river flowing there now. How can you tell?

Grand Canyon National Park, Arizona

505

EXPLORATION 4

INDEPENDENT PRACTICE This activity is intended to draw students' attention to their local area. Suggest that students talk to available resource people or use reference books at the library to answer these questions. Generally, a river comes first and creates the valley or canyon. If the river disappears, the valley or canyon will remain.

Answers to
Exploration 4

2. **a.** The Colorado River came first. It began forming the Grand Canyon approximately 6 million years ago. Because the Colorado River flows through a hot, dry climate, the effects of weathering occur slowly.
   b. A young river erodes its channel, forming a shallow valley. As the river matures, the valley widens and deepens, forming a canyon.
   c. The valley or canyon may follow a narrow channel where the river once flowed. There may also be accumulated sediments, remains of marine organisms, or markings on the valley or canyon walls indicating the effects of erosion.

Homework

The Exploration Worksheet that accompanies Exploration 4 makes an excellent homework activity (Teaching Resources, page 45).

FOLLOW-UP

Reteaching

Have students review the rivers shown in the illustrations and photographs on page 499. Ask them to classify them as young, mature, or old. Students should support their answers with evidence. *(The photograph at the top of page 499 is of an old river. The bottom two photographs are of young rivers.)*

Assessment

Have students write an essay predicting the future of the land around the local river they studied in Exploration 4.

Extension

Have students do library research to find stories or poems that describe rivers. Students can collect these writings in their ScienceLog. The school librarian should be able to help students with their research.

Closure

Obtain historic as well as current maps of your state. The maps will probably show rivers or streams that are no longer evident above ground or whose appearance has changed. Discuss with students the possible reasons for these changes. Trace the streams that have changed onto one sheet of paper, using different colored markers so that students can examine the patterns of change.

LESSON 4 Where Land Meets Sea

LESSON 4 Where Land Meets Sea

FOCUS

Getting Started

Tell students to imagine that you are a river. Walk through the classroom and pick up an object from the desk of several students. Take the objects to the front of the classroom and dump them on the floor or on a table. Ask students to explain what might have caused you to drop these items. *(You may have just run into a large body of water, such as the Gulf of Mexico.)* As you return the objects to the students, explain that in this lesson they will learn about the ways in which a river or stream might deposit the sediments that it carries.

Main Ideas

1. Coastlines and shorelines are eroded by the action of ocean waves and currents.
2. The mouth of a river can be in the form of a delta or an estuary.
3. Beaches are not permanent features of a landscape. They are deposits of sediments and rock fragments that are moved by the action of water.

TEACHING STRATEGIES

Deltas and Estuaries

The top photograph on page 506 shows the Mississippi Delta. Over many thousands of years, deposited sediment has enlarged the Mississippi Delta and pushed it out into the Gulf of Mexico. The photograph on the bottom shows the Nile Delta in Egypt. The wedge shape is readily apparent from space.

Homework

The Math Practice Worksheet on page 46 of the Unit 8 Teaching Resources booklet makes an excellent homework activity.

Deltas and Estuaries

Eventually, rivers reach the sea. What happens then? The last few lessons contained clues that may help you answer this question. The best clues are in Explorations 2 and 3. Write down your ideas! Now look at the photographs shown on this page. Do the photographs support your ideas? If so, how?

Many rivers deposit a fan-shaped wedge of sediment at their mouth. This deposit is called a **delta**. Why do you think it forms?

The Mississippi River drops 2 million tons of sediment into the Gulf of Mexico each year. What conditions that encourage deltas do you think exist at the mouth of this river? Ⓐ

Predictions Please!

Not all rivers, even those that carry large loads of sediment, deposit deltas. Predict whether or not deltas will form for each of the following sets of conditions:

1. moderate sediment load
   shallow sea floor
   weak tides and wave action

2. large sediment load
   deep sea floor
   strong tides and wave action

3. large sediment load
   moderate tides
   weak wave action

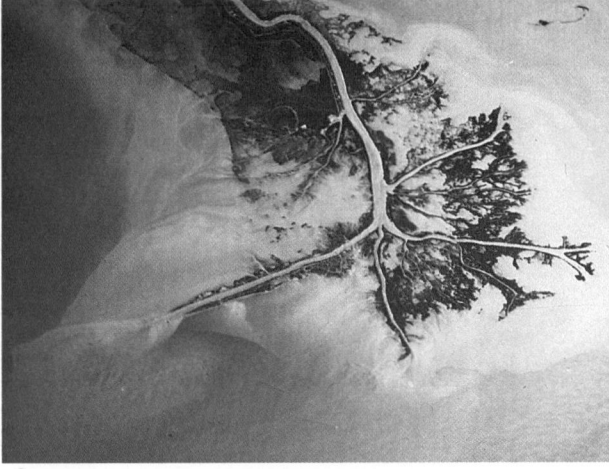

⬆ Mouth of the Mississippi River, United States

⬆ Mouth of the Nile River, Egypt

506

LESSON 4 ORGANIZER

Time Required
2 class periods

Process Skills
observing, comparing, contrasting

New Terms
Delta—fan-shaped deposit of sediment at the mouth of a river
Estuary—a wide river mouth that is submerged in ocean water and extends far inland

Materials (per student group)
none

Teaching Resources
Math Practice Worksheet, p. 46
SourceBook, p. S201

There is another type of meeting place between river and ocean. Look at the photograph of the Chesapeake Bay on the right. Notice that the river has a wide mouth that extends far inland. How would you explain this? Ⓑ

Scientists believe that during the last ice age, much of the Earth's water was trapped in huge glaciers. As a result, the sea level was much lower than it is now. What do you think happened as the climate became warmer? According to one theory, the melting ice caused the sea level to rise, and many valleys near the coast were flooded. These flooded coastal valleys are called **estuaries**. The mouths of most of the rivers along the Atlantic Coast of the United States are estuaries.

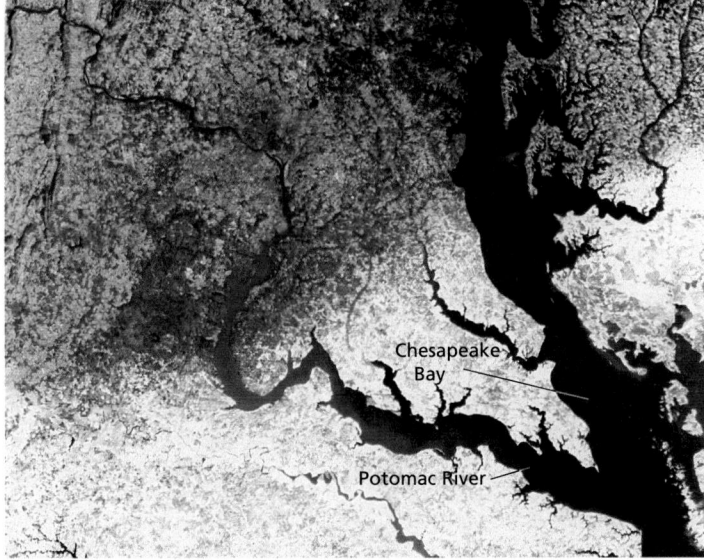

Chesapeake Bay

Potomac River

Estuaries nurture a wide variety of fish, shellfish, and crustaceans. Thus, they are a valuable source of food for humans and wildlife. Many species prized by humans for food require an estuary to reproduce.

Points to Ponder

1. Why do deltas occur only at the mouths of streams and rivers?

2. How do the Nile and Mississippi Deltas differ? What conditions might account for the differences?

3. Do you think that the Mississippi Delta has always been where it is today? (Look carefully at the photo before answering.) Explain.

4. Suppose that a dam was built somewhere upstream on one of the rivers shown on page 506. How would the delta be affected?

5. When Sophia was asked to describe an estuary, she replied that it is a valley that has drowned and that the water in an estuary is not fit to drink. What did she mean?

Egrets feed on fish and small animals in shallow waters and gather by the hundreds in estuaries during breeding season.

507

Answers to
Points to Ponder

1. A delta forms from the buildup of sediment carried by a river. The sediment is deposited at the mouth of a river because the water at the mouth moves slowly and is no longer able to carry the sediment.

2. The Nile Delta is more fan-shaped; the Mississippi Delta is more elongated. Perhaps the sea floor is more shallow and the sand builds up more quickly in the area of the Mississippi Delta.

3. No. It can be inferred from the photograph that the Mississippi Delta has been moving outward into the Gulf of Mexico.

4. If a dam were built upstream, sediment would be trapped there. The delta would not be replenished as rapidly, and it might eventually be washed away.

5. Sophia was referring to the fact that estuaries are river valleys that were flooded by the sea as sea levels rose at the end of the last ice age. The salt water of the estuary is unsafe to drink.

Answers to
In-Text Questions, pages 506–507

Ⓐ The Mississippi River carries a great amount of sediment. It is very long and has many tributaries. The Gulf of Mexico is also relatively shallow near the river's mouth. Finally, the delta is protected from tides and wave action by a series of barrier islands.

Ⓑ Sample answer: The sea level was lower during the last ice age. A river had to cut a new course to the sea through the exposed sea bed. Over time the river eroded its channel to form a deep, wide valley. When the ice age ended, the sea level rose, creating a wide river mouth.

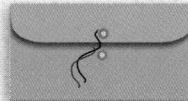

PORTFOLIO
Suggest that students draw pictures of deltas and estuaries to include in their Portfolio.

Answers to
Predictions Please! page 506

The greater the sediment load that a river carries, the more likely a delta will form at the river's mouth. Shallow water also helps in the formation of deltas because less material is required to raise the soil surface above sea level. Strong tides and wave action inhibit the formation of deltas because they erode the deposited soil, stirring it up and spreading it out over the sea floor.

Sample predictions:
1. A delta will form because all the necessary conditions are met.

2. A delta will probably not form because, even though the sediment load is large, the sea floor is deep. Any accumulated material will probably not rise above sea level before it is eroded by the strong waves and tides.

3. A delta probably will form because the sediment load is large and the wave action is weak. However, this also depends on the depth of the water. The tides are moderate, so they will have some effect on the delta.

On the Waterfront

GUIDED PRACTICE Have a volunteer read aloud the top of page 508. Involve students in a discussion of their experiences with waves at the beach. Ask students to focus on their experiences of the force and movement of waves as they break on a beach. Point out that they may have observed sediments and sand suspended in the curl of a breaking wave.

Mention that not all beaches are made of sand. Direct students' attention to the photographs, which show a variety of dramatically different beaches and coastlines.

Have students study the photographs and propose answers to the questions on this page. You can then share the answers that follow with your students.

Answers to
In-Text Questions

Ⓐ Accept all reasonable responses. Students may have experienced similar wave motion in a water park or in the wake of a passing boat. They may have also noticed that waves are usually more powerful in the ocean than in gulfs, bays, or lakes and are strongest during stormy weather. Encourage students to describe their observations of waves as vividly as possible.

Ⓑ Coastlines develop through the interaction of the sea and the land.

One factor involving the land is the movement of tectonic plates that can cause uplift and faulting. Factors involving the sea include changes in water level, erosion, and storms.

Beaches differ because of local differences in wave action and in the types of rocks or sediments present. The rock, sand, or gravel that form beaches can be derived from many kinds of parent rocks or, in some cases, coral.

Beaches are rocky when the adjacent waters are rough. Strong wave action carries away the smaller sand particles, leaving only large rocks and boulders. Beaches are sandy when the adjacent waters are gentle and do not move enough to keep the tiny sand grains suspended. The grains drop onto the coastline, forming a sandy beach.

On the Waterfront

Have you ever played in the ocean waves? Think about the experience. Describe how the force of the waves felt. Did this give you an appreciation for the power of waves? Ⓐ

Day after day and year after year, the shore is constantly eroded by the sea. The power of the ocean waves and currents endlessly shapes and reshapes the shore. The shore is narrow and rocky in some places and wide and sandy in others. Some shores are bordered by high, rugged cliffs; others slope gently to the sea.

How do coastlines develop? What will they look like in the future? Why do beaches differ? Why are some rocky and others sandy? Ⓑ

Think About It

Before you read the next page, think about each of the following questions:

- Which of the beaches pictured on this page and the next probably experiences gentle waves most of the time?
- Which probably experiences many fierce storms?
- Which is probably eroding very quickly?
- What causes a sandy beach to form? Where does the sand come from?

Answers to
Think About It

- The tropical beach, with its fine-grained sand, probably experiences gentle waves most of the time.
- Any beach with little or no sand, such as the cobble beach and rocky coastline, probably experiences fierce storms.
- The sea cliffs, without a beach to absorb the impact of the waves,

probably erode at a very fast rate. Parts of a sea cliff may erode at different rates, resulting in caves and arches (see photograph on page 509). If an arch collapses, sea stacks, as shown in the bottom photograph on this page, may remain.

- A sandy beach forms when sand and sediment (either carried out to sea by a nearby river or formed by wave action) wash up and settle out of the waves along the shoreline.

Location is an important factor in the way a beach develops. Rocky beaches are especially common on coastlines in northern latitudes. Ocean beaches frequently experience large waves. Winter storms may lash the beach with waves more than 5 m high.

Depending on the size of the waves, materials are either removed from or returned to a beach. When large waves attack a beach, they shift sand out to sea and expose the rocks along the shore. Gentle waves gradually shift sand back onto a beach, often to a depth of several meters.

Think of the land that borders bodies of water in your state. Can you locate both sandy and rocky beaches? **C**

Waves can also cause other shoreline features to develop. Can you explain how the formations shown in the accompanying photographs might have formed? **D**

Reteaching

Have students describe the differences and similarities between deltas and estuaries. Suggest that students include drawings if they wish.

Assessment

Provide students with the following scenario: You have been assigned to write an article for a surfing magazine about the beaches shown in the photographs on pages 508–509. What would you say about the waves at each of the places shown?

Extension

Have students choose a major river in the United States. Tell them to research the course of the river and its major features. Suggest that they draw a map showing the river's path and its interesting features, such as waterfalls, heavy rapids, oxbow lakes, a delta, an estuary, or an alluvial fan.

Closure

Many cities are located near the mouths of rivers. Have students write an essay discussing the pros and cons of building a city near a river delta or an estuary.

Homework

The Assessment activity on this page makes an excellent homework assignment.

Integrating the Sciences

Earth, Life, and Physical Sciences

Point out to the class that although the city of New Orleans and much of the Netherlands are below sea level, neither is covered by water. Ask: How do you think this is possible? *(Typically, this is accomplished by a system of dikes and levees.)* Why do you think this land has been reclaimed from the sea? *(Possible responses include the following: to relieve the burden of overpopulation, to create more farmland, or to preserve already existing structures.)* What dangers exist in these places? *(Water erosion is a constant problem. If a barrier breaks, much of the area will be flooded.)*

Answers to
In-Text Questions

C Answers will vary. You may wish to supply the class with some maps on which they can identify different types of beaches. Or have volunteers share any observations that they have made when they have visited beaches in your state.

D The force of ocean waves can cause the development of the shoreline features shown in the photographs. For example, wave action can create very smooth, rounded pebbles or rocks on the beach. Waves can also bore caves in cliffs. Recession of the coastline increases the distance of the sea stacks from the shore. Wave action causes cliffs to recede. Each time water rushes into a crack in the rocks, the air in the crack is compressed. The force of compression enlarges the cracks until parts of the cliff begin to break away.

LESSON 5

Wind— An Invisible River

FOCUS

Getting Started

Share this riddle with students, and ask them to solve it: What glides over the desert, can cut telephone poles in half, is made of a fine cloud of sand and dust, and can turn into a low, thick cloud of heavier particles? *(A sandstorm)*

Main Ideas

1. Wind is a major agent of erosion in dry areas.
2. Human activities can increase wind erosion.

TEACHING STRATEGIES

Answer to
In-Text Question

Ⓐ Sample answer: The wind flows just like an invisible river, sometimes in a steady stream and other times turbulently. Wind currents can carry almost any object downwind, including pieces of the eroded land-scape. When the wind slows down, the load of sand it was carrying set-tles to the ground, similar to a river depositing its sediment where the current slows.

Answers to
Picked Up or Laid Down?

The dunes in the photograph at the top of page 510 show the effect of sand deposition by the wind. These dunes were formed in areas where there was abundant sand and where the wind blew in the same direction for a long period of time. The feature in the bottom photograph on page 510 was created by wind erosion. The material in the rock formation is more resistant to wind erosion than the material that once surrounded it.

LESSON 5

Wind—An Invisible River

What do you think the title above implies? Does wind act like an invisible river? Wind can certainly carry objects from one place to another like a river does. What materials have you seen being blown about by wind? Does the speed of the wind make a difference in the kind of material it can carry? What other factors determine what materials wind can carry? Can it erode the landscape? No matter where you live, winds blow, but the effects of wind are most evident in desert regions and along the seashore. What causes wind to deposit sedi-ment? How is this similar to a river? Ⓐ

A Comparison

In your ScienceLog, copy the following table comparing how rivers and wind affect the landscape. You may complete the table at the end of the lesson when you have more information.

Picked Up or Laid Down?

Look at the photographs on this page, and decide whether materials are being eroded or deposited.

River	Wind
carries large and small material	carries small material

510

LESSON 5 ORGANIZER

Time Required
3 class periods

Process Skills
observing, analyzing, inferring, comparing, contrasting

Theme Connection
Changes Over Time

New Terms
none

Materials (per student group)
The Wind Shadow: large jar; candle; a few matches; jar lid; small ball of modeling clay; safety goggles

Teaching Resources
SourceBook, p. S195

Drifting Sand

One of the most familiar landforms created by the wind is a sand dune. Sand dunes are found in the desert, but they are also common along the seashore. Have you ever played in the sand dunes along a beach? How do you think the dunes formed? The following activity may help you answer this question.

The Wind Shadow

Put a candle in a jar lid, and secure it with modeling clay. Place a large jar on a table with the lit candle immediately behind it. On the side of the jar opposite the candle, put your mouth close to the jar, and blow as hard as possible. Move the candle away from the jar until you can blow it out. Draw a diagram to show what is happening. How does this activity help to explain the formation of sand dunes? Compare your explanation to Nadia's.

Nadia's Explanation

Suppose there was something on the beach, like a rock or a clump of plants. When the wind blows against the rock or the plants, the wind separates and curves around these obstacles. This makes a wind shadow behind the rock or the plants where the wind slows down. When this happens, the wind drops the sand that it is carrying into the wind shadow. Over time, the sand piles up and forms a sand dune.

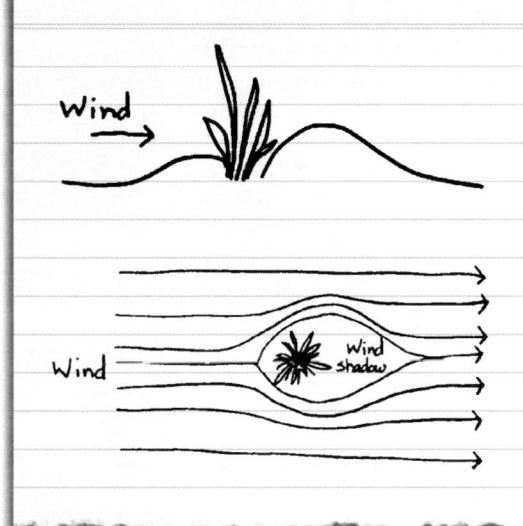

Something to Think About

1. What is a wind shadow? Why does sand collect there?

2. Water is a major agent of erosion. How would you compare it with wind erosion?

3. Do all beaches have sand dunes? Why or why not?

4. In many areas, sand dunes are protected by law. Why is this necessary?

511

Answers to Something to Think About

1. A wind shadow is an area on the leeward side of any obstacle where the wind slows down and deposits its load of sediment.

2. Although both water and wind are major agents of erosion, water is typically a more powerful agent because it is much denser than air and thus has more energy.

3. Not all beaches have sand dunes. In coastal areas with only large pebbles or rocks, the wind is unable to lift and move them, so no dunes are formed.

4. Sand dunes are protected because any loss of vegetation or deformation of the dune could cause the dune to shift, or migrate, into developed areas. In addition, some organisms are indigenous to sand dune environments.

PORTFOLIO
Students may wish to write and include in their Portfolio a short science-fiction story describing what the Earth would be like if it were completely covered with sand.

FOLLOW-UP

Reteaching

Tell students to think about the purpose of sandpaper. Ask: What kinds of tasks do people use sandpaper for? How can a person who is using sandpaper be compared with wind erosion?

Assessment

Have students write a descriptive story about life in a town that is trying to fight migrating sand dunes.

Extension

Have students research the problem of desertification, or have students make and use a wind-direction indicator to determine local wind patterns.

Closure

Encourage students to form small groups and to compose a poem, limerick, or song lyrics about the erosive power of wind.

Answers to
A Comparison, page 510

Sample ScienceLog table:

River	Wind
carries large and small material	carries small material
deposits sediment in deltas	deposits sediment in dunes
the size and amount of material carried depend on the flow rate	the size and amount of material carried depend on the flow rate
obstacles in the river bed decrease the river's flow rate	obstacles on the ground decrease the wind's flow rate

Theme Connection

Changes Over Time

Focus question: How might deforestation lead to an increase in wind erosion and a loss of fertile soil? *(Many forested regions are being harvested or clear-cut for farming. If the trees are not replaced, the loose particles of soil can be carried away by the wind. In regions where rain forests are being cleared for farming or grazing, the soil is not particularly fertile. Once cleared, much of this land is abandoned, leaving it vulnerable to erosion.)*

LESSON
6
Glaciers—Rivers of Ice

FOCUS

Getting Started

Ask students what Glacier National Park in Montana, Banff National Park in Canada, the Mer de Glacé on Mont Blanc in the French Alps, Yakutat Bay in Alaska, and Mount Ranier in Washington all have in common. *(Possible responses include the following: they are all mountainous areas with cold temperatures much of the year; they are all beautiful, scenic areas with many hills and valleys; they all have glaciers.)*

Main Ideas

1. Glaciers are thick sheets of ice that cover cold areas such as polar regions and high mountains.
2. Glaciers are formed when snow accumulates, compacts, and gradually changes to ice.
3. Glaciers flow at different rates. These rates can be measured.
4. As glaciers move, they carry loose, eroded materials and rock fragments scraped from valley walls and floors.
5. Glacial action has created many physical features and landforms.

TEACHING STRATEGIES

Ask a volunteer to read aloud the story of the *Titanic*. At this point, you may wish to discuss the use of robotic submarines, which have been used to locate and film the wreckage of the *Titanic* at its resting place in the north Atlantic Ocean. Pose the question at the end of the first paragraph to ascertain students' prior knowledge about icebergs, glaciers, and their origins. *(Accept all reasonable responses.)*

⭐ **A Transparency Worksheet and Transparency 88 are available to accompany the material on this page (Teaching Resources, page 48).**

On the night of April 14, 1912, while on its maiden voyage from Southampton, England, to New York, a luxury liner called the *Titanic* struck an iceberg. In less than 3 hours, the *Titanic*, a ship that was advertised as unsinkable, plunged to the bottom of the ocean with over 1500 passengers aboard. Where in the world could such a huge chunk of ice have come from?

Icebergs come from thick sheets of ice called **glaciers**. Glaciers cover cold areas like northern Canada, Greenland, and Antarctica, as well as certain high mountains. The iceberg that sank the *Titanic* probably originated in Greenland. Both Greenland and Antarctica are almost completely buried by ice to a depth of over 3000 m in some places! Ice flows slowly outward and downward from the center of these glaciers. Eventually, it spills over the edge of the land into the sea. This is how icebergs are born. The massive glaciers covering Greenland and Antarctica are remnants of the last ice age, which ended about 10,000 years ago. At the peak of the last ice age, about 20,000 years ago, roughly one-third of the Earth's land area was covered by ice many hundreds of meters thick.

Titanic strikes iceberg!

A valley glacier

LESSON 6 ORGANIZER

Time Required 3 class periods

Process Skills inferring, analyzing

New Terms
Glaciers—thick sheets of flowing ice
Meltwater—a stream of water carrying a large load of sediment that flows down from the foot of a glacier
Névé—rounded grains of snow found on the upper end of a glacier
Outwash—a deposit of sediment carried by meltwater, found down the valley from a terminal moraine
Terminal moraine—till that has been deposited, forming the end of a glacier ridge

Till—unsorted sediment left by glaciers

Materials (per student group)
Exploration 6: 500 mL of water; small tray (about 4 cm deep and 30 cm long); 10 cm of wire; several masses of varying sizes; freezer; 2 supports, such as blocks of wood (additional teacher materials: wire cutters; see Advance Preparation on page 449C.)

Teaching Resources
Transparency Worksheet, p. 48
Exploration Worksheets, pp. 50, 52, 53
Activity Worksheet, p. 51
Transparency 88–89
SourceBook, p. S193

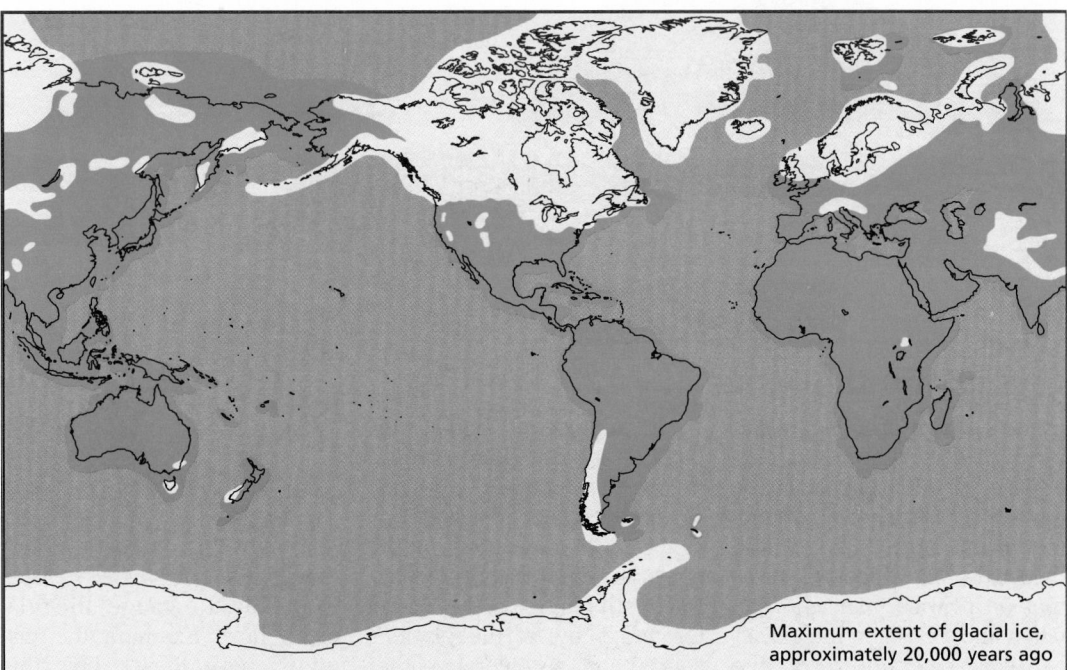

Maximum extent of glacial ice, approximately 20,000 years ago

EXPLORATION 5

How Do Glaciers Form?

Can you turn snow into ice? How would you do it? It happens all the time in glaciers, although it takes quite a while.

To understand the process by which glaciers form, read the sentences below. Beware, though—their order is scrambled! Try to rearrange them into the proper order. Use the accompanying diagram as a guide.

a. The snow particles become rounded pellets due to the added weight of accumulating snow.

b. Evaporation and melting cause the flakes to lose their shape.

c. As the snow builds up, the pressure from the top layers squeezes out the air between the rounded pellets of snow. Icy particles called *névé* (NAY vay) are formed.

d. Star-shaped flakes collect in a fluffy mass with a lot of air caught between them.

e. The névé undergoes further packing and melting until it gradually changes into solid ice.

Did you get the sentences in the right order? If you did, then you can truthfully say that you have got glaciers down cold!

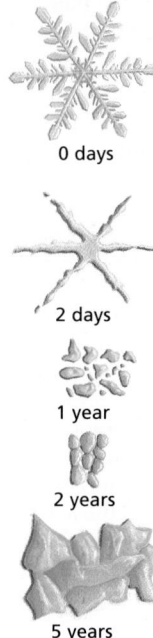

0 days

2 days

1 year

2 years

5 years

513

Use the text and the glaciation map on page 513 to provide background information on continental glaciers. At present, only about 10 percent of the world's total land area is covered by continental glaciers, or icecaps. The continental glacier of Antarctica makes up 85 percent of that area. The continental glacier of Greenland contributes 10 percent, and the rest is located throughout the arctic regions of northern Canada, central Asia, and Iceland.

There are also valley glaciers, such as those shown in the aerial photograph at the bottom of page 512. Have students study this photograph and develop an explanation of how these glaciers formed independently in these mountain valleys. *(Possible student explanation: Valley glaciers form in areas where the snow does not melt from year to year. After several years, the layers of snow are packed together, forming ice. The weight of the ice causes it to spread out. When the ice reaches a slope, gravity causes it to move downward.)*

EXPLORATION 5

GUIDED PRACTICE   If your school is located in an area that receives snow, you can use the characteristics of a snowdrift to help students visualize the process described in this Exploration, or you can obtain a photograph of a snowbank at the side of a freshly cleared road or highway. The photograph should show distinct layering of the snow.

The layering in a snowdrift provides a model for the different stages in the formation of a glacier.

Multicultural Extension

Arctic Cultures

Indigenous peoples of several different cultural groups make their homes in the glacier-covered regions of the Arctic Circle. Have students do research on one or more of these cultural groups and report on the technological adaptations they have made to life among the glaciers and the ways in which they utilize the landscape.

Answers to
Exploration 5

The correct order is as follows:
d, b, a, c, e.

Glaciers on the Move

Involve students in a discussion of both the origin of the glacier that carried George Winkler and its speed of movement. Stress the fact that glacial movement is not regular. The speed can vary from glacier to glacier, within the same glacier, or from year to year.

Ask students to study the diagram at the top of the page and to consider the following questions:

- Which parts of the glacier move the fastest? Why do you think this is so? *(The middle part of the glacier moves the fastest. Friction with the walls and floor of the valley slows down the sides and bottom of the glacier.)*
- Estimate how far the glacier in the diagram moved between 1874 and 1882. On average, how far did it move each year? *(700 m between 1874 and 1882; an average of 87.5 m per year)*
- What is happening at the lower end of the glacier? Why? *(The glacier is receding. It is melting because of an increase in temperature from the heat of the Earth, the lower elevation, and changes in climate.)*

Point out to students that glacial flow has two components: *basal slip* and *internal flow*. Basal slip occurs when the ice at the bottom of a glacier is at or near its melting point as a result of the weight of the snow above and the heat of the Earth below. The thin film of water that forms at the base lubricates the glacier, allowing it to slip over its bed in much the same way that skates allow a skater to glide over the ice. Internal flow is a slow creep that occurs within a glacier as ice crystals slip over one another.

 Transparency 89 is available to accompany the material on this page.

Answer to
In-Text Question

Ⓐ The glacier in the Alps moved a little over 700 m in 8 years, or about 88 m per year. The glacier that carried George Winkler's body moved 1.5 km in 68 years, or about 22 m per year.

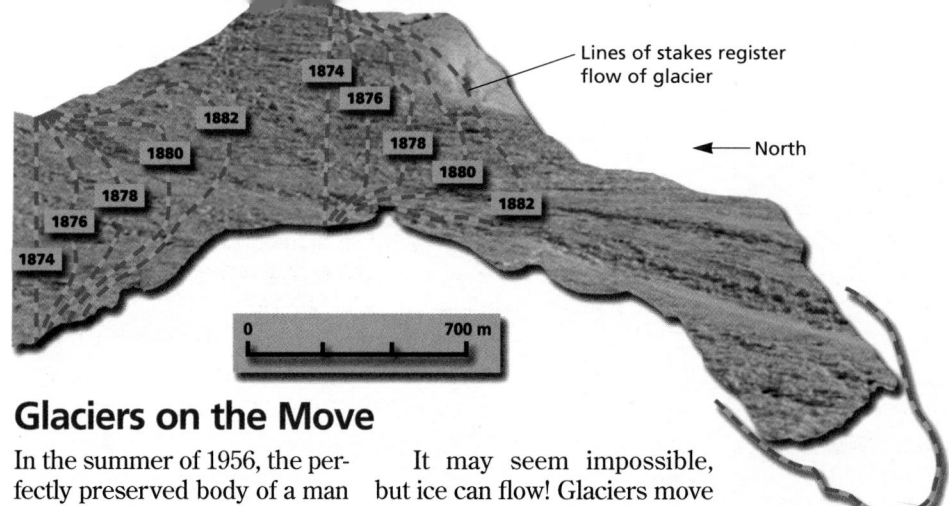

Lines of stakes register flow of glacier

1874 1876 1882 1880 1878 1876 1874 1878 1880 1882

← North

0   700 m

Glaciers on the Move

In the summer of 1956, the perfectly preserved body of a man emerged from a glacier at the foot of the Weisshorn, a mountain in the Swiss Alps. At first, the body was believed to be that of a Swiss man who had fallen into a crack in the glacier in 1946. Later, however, the body was correctly identified as that of a young German mountain climber, George Winkler, who had fallen from the Weisshorn in 1888. In 68 years, the ice had carried Winkler approximately 1.5 km, from the upper part of the glacier to its lower end.

It may seem impossible, but ice can flow! Glaciers move at different speeds, however. Their speed depends on several factors: the steepness of the slope on which the glacier lies, the depth of the ice, the load of ice and snow at the glacier's source, and the general weather conditions in the area. Gravity, however, is what makes the glacier flow. The force of gravity acting on a glacier's huge mass causes the glacier to sag under its own weight and slowly slide downhill.

One way to measure a glacier's speed is to place stakes in a straight line across it at several points. The stakes are later examined to gauge the glacier's flow. Examine the diagram above to see how this works. Two sets of stakes were driven into the ice of the Rhone Valley glacier in the Alps in 1874. Their positions were mapped every 2 years until 1882. Compare the speed of this glacier with the speed of the glacier that carried George Winkler's body for 68 years. Ⓐ

EXPLORATION 6

Can Ice Really Flow?

Can you design an experiment to prove that ice can change its shape or flow without melting? Perhaps you could try this idea. Freeze a layer of ice about 1 cm thick in a tray. Once it has frozen, remove the ice from the tray. In a freezer, support the layer of ice as shown at right. Hang a fairly heavy mass from the middle of the ice with a wire. Observe what happens during the next few days. Explain what you observe. Repeat the experiment with different sized masses. Can you apply your findings to explain how glaciers move?

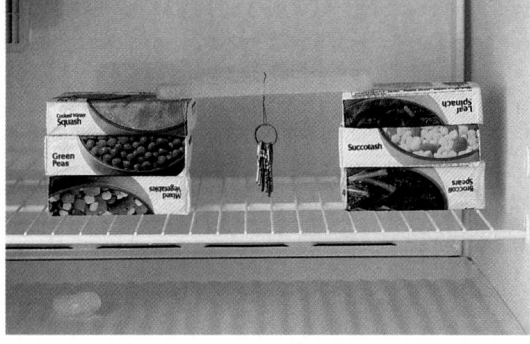

514

EXPLORATION 6

INDEPENDENT PRACTICE   Encourage students to try this activity at home or to design an experiment of their own. Ask them to write a report based on their findings.

Answers to
Exploration 6

Students should observe that the mass will slowly pass through the layer of ice without visibly affecting the shape of the ice. The force that gravity exerts on the mass determines the flow of the ice, just as the force that gravity exerts on the mass of a glacier determines the flow of the glacier.

Homework

The Exploration Worksheet that accompanies Exploration 6 makes an excellent homework assignment (Teaching Resources, page 52).

Developing a Theory of Glaciation

Charpentier and Agassiz

In the summer of 1836, Louis Agassiz, a young Swiss naturalist, visited Bex, a small town in the Alps. He intended to search for fish fossils in limestone beds high in the mountains near Bex. He had been invited to Bex by Johann von Charpentier, the director of the local mine, whom Agassiz had met 20 years earlier. Charpentier wanted to show Agassiz evidence for a theory that he had been developing for some time. Charpentier believed that a type of sediment called *till*, which was found throughout the region on hills, mountains, and lowlands, was left there by glaciers. To Agassiz, the idea that huge sheets of ice had once covered Europe seemed like a fairy tale. However, since his mind was really on fossils, Agassiz didn't argue the point.

Imagine accompanying these men on their first expedition into the mountains. Charpentier starts to present the evidence to support his theory. As they stand beside a large outcrop of rock, Agassiz argues stubbornly against Charpentier's theory.

Agassiz: It's quite obvious that the till formed right where it is by weathering. You can see that these rocks have been shattered.

Charpentier: If weathering is the cause, then why don't the rocks look weathered? These rocks have fresh, unweathered surfaces. Moreover, all of the rocks in this outcrop are different types: there's granite, basalt, limestone, and who knows what else. The rock underneath is sandstone. How can granite be weathered from sandstone?

Continuing up the trail, they come to an immense boulder.

Charpentier: Look at that huge boulder many kilometers from where it should be. How do you think it got here? By wind? by water? It's far too heavy.

Agassiz: Perhaps there was a violent explosion of some sort that bounced the boulders from hill to hill and also churned up the soil to form the till.

Charpentier: Nonsense! That idea is about as good as suggesting that these big rocks were dropped by huge icebergs drifting about in some huge flood. What evidence is there to support such an idea?

 A glacial boulder

515

Charpentier and Agassiz

INDEPENDENT PRACTICE The "icy argument" between Agassiz and Charpentier lends itself to several different teaching approaches. One approach is to perform it as a class dramatization. One student could act as the narrator, while two others play the parts of the two scientists. As an alternative, the class could hold a debate. You could assign a group of students to present each point of view and have the rest of the class listen, discuss, and evaluate the arguments.

GUIDED PRACTICE Take this opportunity to discuss the role of different types of thinking in science. Science is based on collecting evidence, but intuitive thinking and questioning based on the evidence also play an important role.

Students may suggest many possible rejoinders to Charpentier's last argument, on page 516. One possible response might be, "If the ground here was once covered by ice, what happened to it? If it melted away, why didn't the water carry away the till?" There are many other possibilities. Encourage students to examine the evidence and explore all of the possibilities.

The following is background information on Louis Agassiz that you can share with your students. In his later years, Agassiz became so intrigued with glaciation that he abandoned his study of fossil fish and spent 10 years studying alpine glaciation and till. His studies led him to propose that glacial ice once covered Eurasia from the North Pole to the Mediterranean Sea. In 1840 he published a book on his research, entitled *Studies on Glaciers*. The book presented evidence of continental glaciation. The detailed, circumstantial evidence presented in his book began to convince many people who had formerly doubted his theory. In 1846, Agassiz left Europe for the United States, where he studied evidence of North American glaciation.

Ⓐ Have students compare the evidence by making lists in their ScienceLog, as suggested on page 516. Sample lists of evidence:

Agassiz's Evidence
1. The rocks appear shattered.

2. Immense boulders are strewn about.

3. Till is found everywhere, from lowlands to mountains.

Charpentier's Evidence
1. The rocks have fresh, unweathered faces.

2. The underlying rock is sandstone; the outcrop is granite, basalt, limestone, and others.

3. Granite cannot be formed from weathered sandstone.

4. Sea deposits do not bear any resemblance to present-day till.

5. Scratches and grooves can be seen on the bedrock.

6. The bedrock looks polished.

Overall, Charpentier had the stronger argument because he presented more evidence to support his position. Agassiz theorized about what might have happened but did not substantiate his opinions with factual evidence.

Louis Agassiz

Scratches in bedrock

Agassiz: I believe the till is excellent evidence. It could have been deposited by a flood that once covered the entire Earth. That explains why till is found everywhere, from lowlands to mountains.

Till

Charpentier: How can you say that? Haven't you read Hutton? Do you not believe him when he says that the present is the key to the past? Do sea deposits formed today resemble the till around us? Absolutely not! Here's more evidence for glaciation—the scratches and grooves on this bedrock. What forces could have done such a thing? There's only one answer. Rocks that were stuck in the ice gouged out these deep grooves. Have you ever seen bedrock that looks like it has been polished? How do you explain that? Only a glacier could have done that! There's only one explanation: The ground that we are standing on was once covered by ice!

And so the argument continued. How might Agassiz have answered Charpentier? Now play the referee in their debate. Make two columns in your ScienceLog: Agassiz's evidence and Charpentier's evidence. List the evidence presented by both men. Who do you think had the strongest case? Why? Ⓐ

516

CROSS-DISCIPLINARY FOCUS

Social Studies

Provide students with the following list of places: Iceland, Greenland, Antarctica, and Tierra del Fuego. Have them comment on the appropriateness of each name, using what they have learned about glaciation and plate tectonics. They may wish to consult an atlas. *(Both Iceland and Tierra del Fuego lie on spreading centers and thus experience volcanic activity. Yet they are also very cold because they are at extreme latitudes, one near the North Pole and one near the South Pole. Tierra del Fuego lies off the southernmost tip of South America; its name means "Land of Fire."*

Both Greenland and Antarctica are covered with glaciers. Hence, Greenland is not actually very green. The name Antarctica means "opposite of the arctic," but this is in reference to its geographic location at the South Pole, not its climate. In fact, the lowest recorded temperatures in the world have been recorded in Antarctica.)

Glacial Calling Cards

A glacier consists of more than just ice. As a glacier moves, it scrapes off and carries huge amounts of soil and rock. In addition to the loose material that is scraped off, the enormous grinding weight of the glacier wears down any projecting parts of the valley's sides or floor. Boulders sometimes fall from the valley's walls onto the glacier, which then carries the boulders great distances before dropping them. Remember how Charpentier used the large boulder as evidence for his theory?

A glacier is like a giant bulldozer pushing a huge pile of debris before it. When a glacier stops advancing, or terminates, it dumps whatever rock fragments it has been carrying or pushing. This till deposit forms a ridge called a **terminal moraine**. Terminal moraines are mixtures of soil and rock fragments of all sizes, from microscopic particles to boulders.

At the same time, a stream of **meltwater** flows down from the foot of the glacier, carrying a large load of sediment with it. This sediment is deposited in formations called **outwash**. Outwash is found farther down the valley from the terminal moraine.

Now look at the diagram below. Notice where the terminal moraine has formed.

Questions

1. What happens to the projecting bedrock as a glacier moves over it?

2. Where might the glacier pick up material?

3. Why are moraines composed of rocks of different sizes and types?

4. Think back to your stream studies. Why does glacial meltwater carry away only the smaller rock fragments?

5. Why are outwash deposits layered?

Glacial deposits

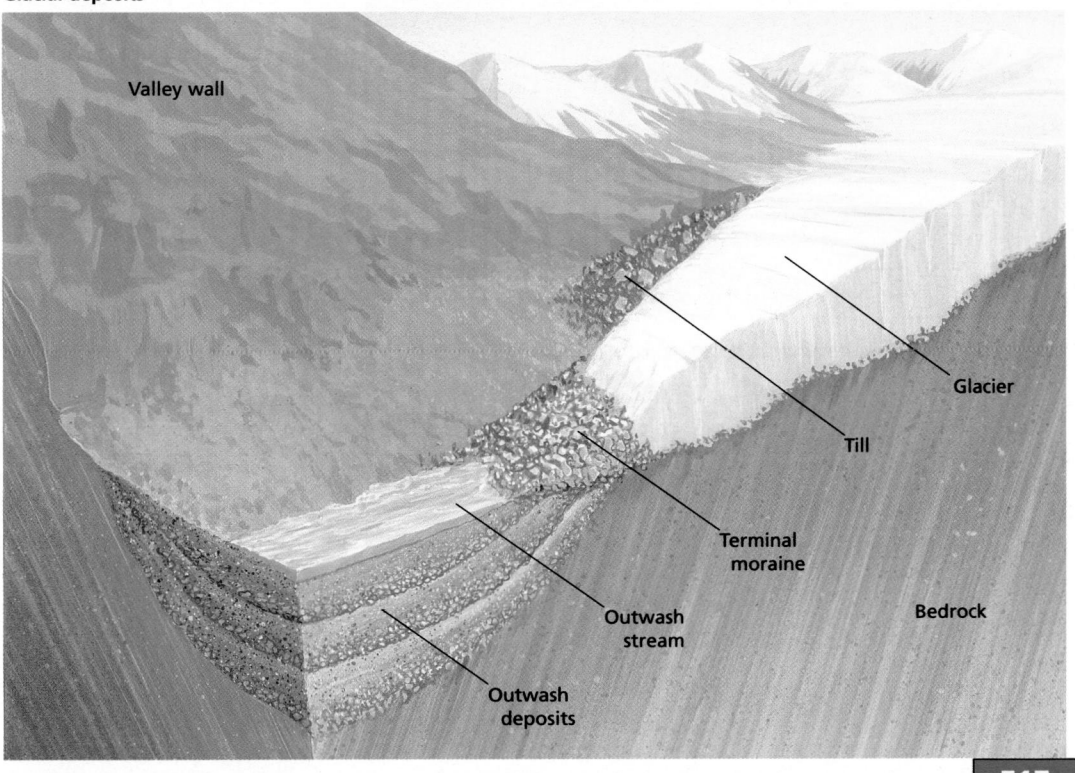

Valley wall

Glacier

Till

Terminal moraine

Bedrock

Outwash stream

Outwash deposits

Glacial Calling Cards

GUIDED PRACTICE   Call on a volunteer to read aloud page 517. Ask students to explain in their own words the meanings of the terms *terminal moraine, meltwater,* and *outwash.* Direct their attention to the diagram of the glacier. Point out the locations of the different glacial features. Be sure that students understand the terms and their locations on the diagram. It may be difficult for students to visualize how glaciers can change the land so dramatically. To demonstrate this, freeze a layer of water in which rocks, pebbles, and gravel have been submerged. Allow the ice to melt until the materials protrude. Then push the ice over some soil in a stream table or across a slab of modeling clay. Finally, allow the ice to melt completely to let students see how glaciers drop their load of assorted rocks and gravel.

To help students understand how rocks can become embedded in solid ice, try the following activity: Place two ice cubes in a shallow pan with a key, coin, or heavy nail on one of them. Put both cubes in the freezer overnight. Have students observe what happens to the piece of metal. If the object is heavy enough, it will become embedded in the ice. A similar process is the embedding of leaves in ice, a phenomenon students may have seen if they live in an area that reaches freezing temperatures in the winter.

Identify any glacial deposits that exist in your area. A field trip to an appropriate site would be an ideal way to build understanding.

Answers to Questions

1. As the glacier continues to move, it wears away at any projecting bedrock.

2. "Hitchhikers" fall from the valley walls, are plucked from the bedrock, and are gouged from the softer parts of rocks and soil.

3. The rocks carried by the glacier come from different sources, such as stream deposits, soil, weathered rock, and bedrock scraped from valley walls. This material is dumped together when the glacier terminates.

4. There is too much material for the meltwater to carry away, and some of the fragments are too heavy.

5. Meltwater flows from the glacier at its point of termination and carries the smaller fragments. The meltwater deposits sorted beds of pebbles, sand, and clay. The amount varies with the rate of flow, causing layers of outwash deposits to form.

Glacial Landforms

This page shows photographs of a variety of glacial landforms. Have students study the photographs and identify those features that they think might have been formed by glacial activity.

The following is a list of glacial landforms and the processes by which they are formed:

- **glaciated valleys**—valleys with a distinct U-shape, gouged out by a glacier. Yosemite Valley is an outstanding example.
- **outwash deposits**—formed of material left behind by the meltwater that comes from the end of the glacier
- **braided streams**—formed in a braided pattern as meltwater from the end of a glacier travels over loose deposits of rock and gravel
- **drumlins**—elongated teardrop-shaped hills formed from glacial sediment. They are formed when material left behind by a previous glacier is "run over" by a later glacier.
- **eskers**—long, winding ridges deposited by glaciers. They are formed by meltwater running in channels beneath the glaciers and depositing sediments.
- **striations**—long, parallel scratches in rocks. Striations are caused by rock fragments embedded in moving ice that are scraped over rocks on the surface.
- **kettle lakes**—round, water-filled depressions in the outwash plain. They form when buried blocks of ice melt.
- **fjords**—glacial valleys that have been filled with water. In many cases, the glaciers are so heavy that they become submerged at the edge of the sea and continue to gouge out an underwater valley.
- **erratic**—a large boulder of a different type of rock than is generally found in the surrounding area. It is carried and left by a glacier.

To make students' understanding of these processes more concrete, use models. For example, the process of forming a kettle lake can be illustrated by burying a large chunk of ice in a stream table, allowing it to melt, and observing the depression that results.

Glacial Landforms

Glaciers leave many changes in their wake. On this page are a few of the many landforms that glaciers can cause. Some are baffling, and some are beautiful. Does the area where you are have any of these glacial landforms?

Glacial valley—valleys with distinct U-shape, gouged out by a glacier

Esker—a long, winding ridge deposited by melt water from glaciers

Fjord—a glacial valley that has been filled with water

518

Homework

The Exploration Worksheet that accompanies Exploration 7 makes an excellent homework activity (Teaching Resources, page 53).

EXPLORATION 7

Glaciers in Your Neighborhood

Look at the map on the next page. Do you live in an area that was once covered by glaciers? If so, then you can probably find evidence of glaciation in your area. Organize a field trip to try to find that evidence.

Depending on your area, you may be able to find several different types of glacial features. Consult your teacher, reference materials, or a geologist to find out what types of glacial features, if any, you are likely to find.

Next, prepare a field guide to use on your trip. It should include descriptions of typical evidence of past glacial activity.

Then, after studying a site, make complete notes and several sketches. Finally, write a field report to sum up your findings.

EXPLORATION 7

If a field trip is not possible, invite a geologist to come to the class to discuss glacial landforms. Slides of glacial formations may be available from a state or national park.

Maximum Extent of Glaciers in the United States

Area covered by glacial ice approximately 20,000 years ago

A Final Note on the Glacial Scientists

Like any good scientist, Agassiz considered Charpentier's new ideas very carefully. Eventually, he came to accept Charpentier's theory. He even went further by becoming an important researcher of the role that glaciers have in shaping the landscape.

What Causes Ice Ages?

Four times within the last 1,300,000 years, glaciers have advanced over large areas of the continents and then mysteriously melted away. Other ice ages have also occurred in the more distant past. Why? What conditions caused ice ages to happen? What might cause another glacial episode?

So far, many theories have been proposed to explain the causes of ice ages, but none has all of the answers. Certainly, a significant drop in the Earth's average yearly temperature (about 6°C) would cause another "deep freeze." Such a drop in temperature might result from a change in the sun's energy output, from an alteration in the rotation of the Earth on its axis, from major changes in the arrangement of the continental plates, or from volcanic dust thrown up into the atmosphere, cutting off the heat from the sun.

If, on the other hand, the greenhouse effect worsens, the Earth's glaciers could begin to melt. What would be the result of this? What could cause the greenhouse effect to worsen?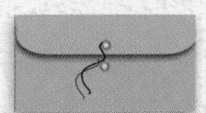

Many scientists point out that the ice age has never really ended and that the Earth is simply in an interglacial period—a kind of global warm spell. As long as the Earth has polar icecaps, they say, it is in an ice age. The ice could come surging back at any time. No one can say for certain what will happen in the future. We can only wait and see.

Have students read the material, and involve them in a discussion of the possible causes of ice ages. Several theories are listed below. Ask students which theory they think makes the most sense and why.

- Astronomical theories focus on the geometry of the Earth's orbital path, the tilt of the Earth, the variations in solar flares, and the occurrence of sunspots.
- Atmospheric theories focus on the Earth's atmosphere and the buildup of dust that blocks the sun's rays.
- Geological theories focus on mountain building and changes in ocean currents that block the relatively warm water from the equator from reaching the poles.

Answers to
In-Text Questions

A Many scientists estimate that the Earth's oceans could rise as much as 5 m if the water now trapped as glacial ice were to melt. Many coastal cities on the Earth would be flooded.

The greenhouse effect could worsen for many reasons. Humans could continue to pollute the air with gases such as carbon monoxide that prevent heat from escaping into space. Volcanoes also contribute to the greenhouse effect by releasing ash into the atmosphere during an eruption. However, some scientists theorize that global warming is a natural fluctuation in the global climate and that the Earth will eventually begin to cool again.

PORTFOLIO

Encourage students to write a short comparison of what they knew about glaciers before studying this lesson and after studying this lesson. Encourage students to include the comparison in their Portfolio.

FOLLOW-UP

Reteaching

As a class project, have students build a model of a glacial landscape based on a topographical map of the area.

Assessment

Present students with the following situation: You have been sent to explore an area where no one has ever been before. What kinds of evidence would you look for to determine if the area had once been covered by a glacier?

Extension

Have students research the Bronze Age man, known as the Iceman, who was found almost completely intact in a glacier in the Alps near the border between Italy and Austria.

Closure

Have students give a presentation to younger students based on their field report from Glaciers in Your Neighborhood.

Answers to *Challenge Your Thinking*

1. The many forks at the headwaters of a river are caused by runoff of excess water. As the waters accumulate to form a river, a deeper and wider valley forms. The waters of the main stream do not fork downstream because of the forces of gravity.

2. Answers will vary. The following are possible responses:

a. During the last ice age, glaciers traveled here from another country, depositing rocks in your backyard.

b. The body was frozen inside a glacier and moved along with the glacier as it flowed.

c. Several causes might explain the disappearance of a beach, but tidal encroachment and erosion from wave action are probably the major causes.

d. The boulder was probably deposited by a glacier. The crack might be the result of weathering from ice wedging into cracks or of blasting during highway construction.

e. Rivers could be salty many miles inland due to river valleys being flooded when the ocean rises above previous levels. This is called the estuary effect.

f. As a glacier moved across the area, the lake could have been gouged out, and the boulder could have been deposited.

g. The water table is the surface of the ground water. If the ground water is reduced, the water table drops below the level of the well, making it go dry.

CHALLENGE YOUR THINKING

1. Take a Hike!

If you followed a river upstream, you would notice that the river forks again and again. However, rivers almost never fork as they flow downstream. Why is this?

2. How Good a Geologist Are You?

To find out, take the following test. A list of observations is given below. Write down each one and give one or more possible explanations for it. Check your answers with your teacher to find out how you did.

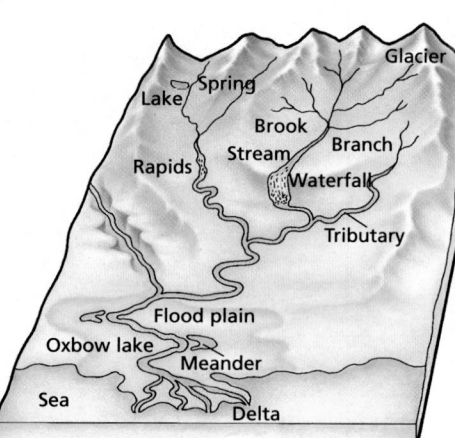

Super Geologist	6 or 7
Geologist	4 or 5
Future Geologist	3 or less

a. Deep scratches are discovered in the surface of highly polished bedrock.

b. A perfectly preserved body is found in ice several kilometers from where it disappeared 75 years before.

c. Over a period of 5 years, a beach disappears.

d. A crack large enough to put your foot into runs the length of a huge boulder lying next to the highway.

e. River water is salty many kilometers inland from the ocean.

f. A huge granite boulder stands alone at the edge of a lake.

g. A well goes dry in mid-August.

520

Homework

The Activity Worksheet on page 54 of the Unit 8 Teaching Resources booklet makes an excellent homework assignment.

⭐ **You may wish to provide students with the Chapter 26 Review Worksheet that accompanies this Challenge Your Thinking (Teaching Resources, page 55).**

3. Threats to Ground Water

Because of your research skills, a journalist has hired you to do background research for a magazine article. Find out why the ground water in parts of the United States may be unsafe to drink. As you do your research, write down information that the journalist could use in the article.

4. Wasting Away?

If streams and rivers are such effective agents of erosion, why have they not reduced the entire United States to sea level?

5. A Heated Debate Over Ice

Sonia and Sunjay were traveling with their parents in the mountains. During their trip they visited a glacier and spent some time walking around on it. An argument broke out when Sonia said, "This glacier is moving, you know." "No way," replied Sunjay. "I can't see or feel it move." Who won the argument? Why?

Review your responses to the ScienceLog questions on page 493. Then revise your original ideas so that they reflect what you've learned.

521

Answers to
Challenge Your Thinking,
continued

3. About 90 percent of the Earth's unfrozen fresh water is stored as ground water. Ground water may become contaminated by salt water, sewage, or pollutants. Pollutants include pesticides, fertilizers, and leaks from hazardous-waste dumps. Encourage students to find information about solutions to the problem of contaminated ground water and about private organizations and government agencies concerned with the problem.

4. While erosion carries away weathered material, new soil is constantly forming, and areas of the Earth's crust are uplifted. New crust is also formed by volcanic activity and by magma filling the gaps between diverging tectonic plates.

5. Sonia was correct in saying that the glacier was moving. By definition, glaciers are moving masses of ice, although the flow of ice is very slow.

ScienceLog

The following are sample revised answers:

1. Water carries away more rock and soil than all of the other agents of erosion combined.

2. Rainwater can become runoff, or it can penetrate the soil to become ground water. Both runoff and ground water contribute to the volume of water carried by rivers and streams.

3. Roughly 8 billion tons of sediment are carried to the oceans each year by rivers. In general, the larger the volume of water is and the faster the water flows, the more sediment a river can carry.

4. Glaciers leave distinct features in the landscape because of their ability to move large amounts of rock and sediment. Glaciers have carried rocks far from their source and have deposited them in piles of till called moraines. Glaciers can carve U-shaped valleys in mountains, create glacial lakes, and polish or scratch bedrock.

The Big Ideas

The following is a sample unit summary:

Geology is the study of the history of the Earth, the forces that have shaped its history, and the forces that will shape its future. (1) Geologists use clues found in the Earth's landforms in the present to deduce what forces and changes occurred in the past. (2) Geologists use their observations of landforms to make inferences about the processes leading to their formation. (3)

For example, the evidence that supports the theory of continental drift comes directly from observations of the Earth's features. This evidence includes the way in which the outlines of the continents seem to fit together like pieces of a puzzle. Also, fossil evidence shows that identical plants and animals lived on widely separated landmasses. Similarities in age and type of rocks, as well as continuity of landforms, indicate that the continents were once joined. (4)

The main features of the Earth's interior are the crust, the mantle, and the core. The lithosphere consists of the crust and part of the upper mantle. The lithosphere is divided into segments known as tectonic plates. The asthenosphere lies beneath the lithosphere. The rock here is plastic-like and is able to flow. The core is often divided into an inner core and an outer core. (5) The floor of the Atlantic Ocean is growing wider because the South American and North American plates are drifting west while the Eurasian and African plates are drifting east. The spreading zone between the plates is called the Mid-Atlantic Ridge and is a region where new crust is formed. (6)

The Earth is changed by many forces. One of the most important forces is that of weathering. The primary agents of weathering are water, wind, ice, and gravity. (7) Water wears down the landscape both physically and chemically. The force of moving water carries away loose debris. This debris then acts like sandpaper, eroding even more of the landscape. When water freezes, the ice that forms within cracks in rocks expands and widens the existing cracks. Portions of the rocks eventually break off. (8) Water and debris flow downhill under the influence of gravity.

Glaciers also move due to the force of gravity. (9) As they move, they carry along with them pieces of loose rock. The glacier polishes the rock surface (bedrock) under it. The debris carried along with the glacier often leaves deep scratches in this polished surface. When the glacier deposits its load of debris, many different types of rock are mixed together, some much too large to have been moved by any force other than that of an immense glacier. The geologic record shows that ice ages have occurred in cycles. (10)

SOURCEBOOK

Turn to the SourceBook for information on the types of rock that make up the Earth and how these different rocks are formed. You can also learn more about erosion and deposition, both above and below the sea.

Here's what you'll find in the SourceBook:

522

Making Connections

The Big Ideas

In your ScienceLog, write a summary of this unit, using the following questions as a guide:

1. What is geology all about?
2. What does the expression "the present is the key to the past" mean?
3. How do geologists use inferences and observations?
4. What is some evidence that supports the theory of continental drift?
5. What are the main features of the Earth's interior?
6. Use your knowledge of the structure of the Earth to explain why the floor of the Atlantic Ocean is growing wider.
7. What are some of the agents of weathering?
8. What role does water play in the weathering process?
9. What role does gravity play?
10. What is some evidence for widespread glaciation in the past?

Checking Your Understanding

1. A famous saying about geology is written in code below. Here is how to crack the code. One letter simply stands for another.

   C H K O O K L P A N E O
   G L O S S O P T E R I S

   Use the example to get started. Can you decipher this famous saying?

   PDA LNAOAJP EO PDA GAU
   PK PDA LWOP—FWIAO DQPPKJ

Answers to Checking Your Understanding

1. The quotation reads as follows: The present is the key to the past. —James Hutton

2. Imagine that all tectonic activity suddenly came to a grinding halt. What would the Earth be like in 10 million years? 100 million years? 1 billion years?

3. Most words have more than one meaning. Read each sentence below and explain the meaning of the underlined term. Then write a sentence to show that you understand the geological or scientific meaning of the underlined term as it was used in this unit.

   Example: Betty and Sergei were in the <u>runoff</u> race.
   Heavy rains can cause great damage due to <u>runoff</u>.

   a. He is in big trouble for having his hand in the <u>till</u>.
   b. How are you <u>weathering</u> the storm?
   c. Why not <u>meander</u> over to my house after supper?
   d. Her election win was a <u>landslide</u> victory.
   e. She received an <u>avalanche</u> of phone calls after she won the trophy.
   f. The baseball missed home <u>plate</u> by a country mile.
   g. Is a <u>water table</u> like a water bed?

   Find other examples of words used in this unit that have both a scientific and a common meaning.

4. How do you think the process of erosion would be affected if
   a. the force of gravity were doubled?
   b. the average global temperature went up by 5°C?
   c. the atmosphere vanished?
   d. the boiling point of water suddenly changed to 10°C?
   e. there were no gravity?
   f. water froze at 25°C?
   g. water were four times heavier (for a given volume)?

   Explain each answer.

5. 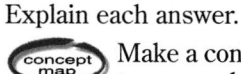 Make a concept map using the following terms or phrases: *waves, ice, gravity, erosion, wind,* and *flowing water.*

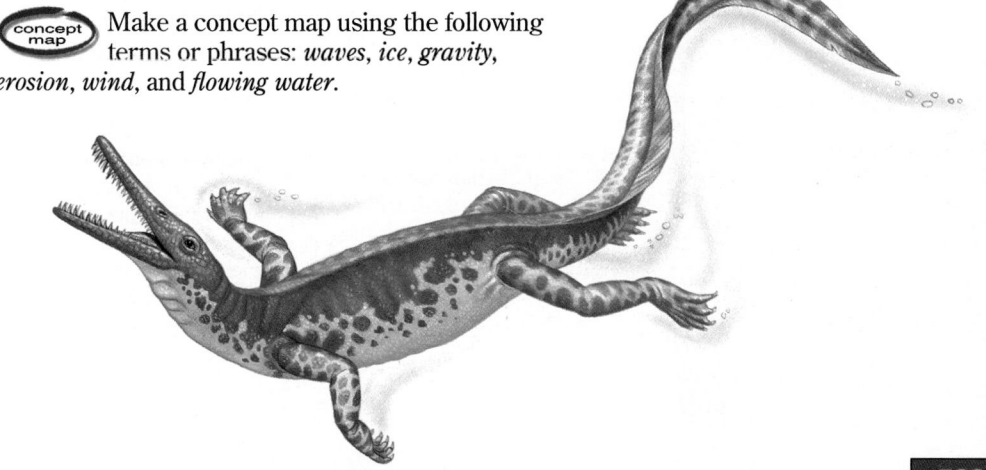

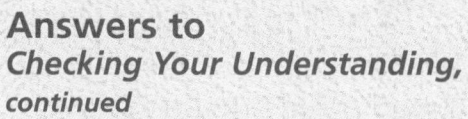

523

Science Snapshot

Background

Geology is the study of the Earth. Two main branches of geology are physical geology (a study of the materials that make up the Earth) and historical geology (a study of the fossil record found in the layers of the Earth). Walter Alvarez is a historical geologist.

One of the things that historical geologists research is the evolution and extinction of species over time. Dinosaurs appeared around 220 million years ago. About 65 million years ago dinosaurs became extinct. Many theories to explain dinosaur extinction have been proposed. Walter Alvarez developed the meteorite theory based on his geological findings.

Discussion Questions

1. What does Walter Alvarez study to support his theory? *(Composition of layers of the Earth's crust, meteorites, and dinosaur fossils)*

2. What made Alvarez think that dinosaurs may have become extinct due to a meteorite? *(He found iridium, which is common in meteorites, in a clay layer that dates back to the time of the extinction of the dinosaurs. There are fewer dinosaur fossils above the iridium layer, so Alvarez speculates that the great extinction happened as the iridium-rich layer appeared.)*

3. What information can geologists gain by studying dinosaur fossils? *(Information about dinosaurs' diet, size, and age)*

Answers to
Dig In!

Other possible theories include a change in the Earth's climate, the explosion of a nearby star that emitted radiation and caused colder temperatures on Earth, evolution of new plant types that could not be digested by plant-eating dinosaurs, and competition with mammals for food. Student opinions will vary.

Science Snapshot

Walter Alvarez (1940–)

Have you ever wondered why the dinosaurs suddenly died out 65 million years ago? If so, then you've pondered one of the great mysteries of science! Walter Alvarez, a geologist, has also spent time thinking about the dinosaurs and has a fascinating theory about what might have happened to them.

An Out-of-This-World Theory

Walter, working with his father, physicist Luis Alvarez, developed the theory that the dinosaurs and many other species died out when a giant meteorite struck the Earth. Walter hypothesized that such an impact would have caused enormous damage, started continent-wide fires, and filled the atmosphere with smoke and dust. The smoke and dust would have blocked out the sun for months. Without sunlight, a cold darkness would have covered the Earth, causing living things to perish on a massive scale.

◄ This crater, called Meteor Crater, is a national landmark in Arizona.

Piecing Together the Extinction Puzzle

Walter began forming this theory in 1980 when he discovered an unusual clay layer in a limestone formation in Italy. This layer was deposited during the time that the dinosaurs became extinct and is rich in *iridium*, an element rare on Earth but common in meteorites.

The clay layer represents a sharp boundary in the rock formation. Below the clay layer, the limestone has many kinds of fossils. These fossils formed before the clay layer was deposited. Above the clay layer, fossils are scarce. Clearly, something happened to kill off many organisms.

Walter believes that a meteorite impact is the best explanation. The clay layer could have formed from the dust and smoke caused by a meteorite impact. To support his theory, Walter needed to locate a meteor crater that was formed around the time the dinosaurs died out. The crater also had to be large enough to cause the disturbances that Walter suggested in his theory.

A good candidate was recently discovered off the coast of the Yucatan peninsula in Mexico. The crater, about 300 km in diameter, could have

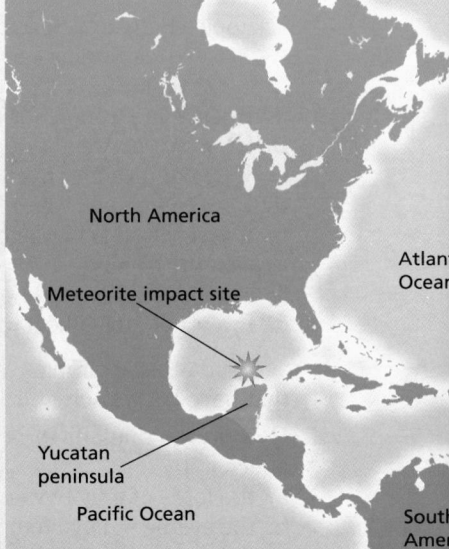

▲ The crater off the coast of the Yucatan peninsula, at 300 km in diameter, may be the imprint of a meteorite that caused the extinction of the dinosaurs.

been formed by a meteorite that landed just before the dinosaurs died out.

Dig In!

Walter Alvarez has developed one theory about dinosaur extinction. Although his ideas are widely supported, some scientists have challenged this theory. Do some research to find out what other explanations have been offered. Which theories are the most convincing? the least convincing?

Meeting Individual Needs

Learners Having Difficulty

Provide students with a variety of materials, such as colored clay and papier-mâché, to make a model of the layers of the Earth that support Alvarez's theory. Be sure they label the layers of limestone. *(The layers should include a layer with no fossils, a clay layer with iridium and soot, and a layer with fossils.)*

SCIENCE and the *Arts*

Discovering Ice-Age Artwork

*I*n 1940, four boys living in France decided to explore a deep, dark cave. As they entered the cave, their oil lamp lit up the cavern walls. Everywhere the light fell, the boys saw painted animals—bulls, horses, and stags. They had just discovered cave paintings that were more than 17,000 years old!

Clues to the Ice Age

Lascaux Cave is one of nearly 200 caves in France and Spain that are filled with prehistoric artwork and artifacts. Scientists study these caves to learn more about life during the last ice age. Leftover bones tell scientists what animals these prehistoric humans ate. Tools, such as picks and scrapers, help scientists picture how they worked. The paintings show many of the animals that lived at that time and give us clues about the kinds of stories that prehistoric humans may have told.

▲ "Reclining Bison" in Cave of Altamira, Spain

Preserving the Past

Unlike other objects from the last ice age, many cave paintings and artifacts have been perfectly preserved. In one cave, scientists even discovered footprints of humans who lived more than 20,000 years ago!

One reason that cave paintings and artifacts are so well preserved is the environment within a cave. Wind, rain, and temperature changes can erode objects that are left in the open. Inside a dry, dark cave, paintings and artifacts are protected from extreme temperature changes and erosion. Because there is so little environmental change within a cave, the artifacts remain intact.

Unfortunately, human visitors can quickly alter a cave's environment. When we enter a cave, our bodies add moisture to the air. Every time we breathe

Cave painting in ▶ Nianx, France

out, we add carbon dioxide to the air. The carbon dioxide and water combine to form a mild acid that harms the paintings on the cavern walls. Light and heat from torches and lanterns cause organisms such as lichens to grow on the paintings. While prehistoric humans could enjoy cave paintings firsthand, modern-day viewers must take every precaution to protect these priceless works of ice-age art.

Be an Art Detective

Scientists are still trying to figure out the clues left by ice-age humans. Take a close look at the pictures on this page. What can you learn about the past from the paintings? Write a story about ice-age life based on your observations.

◀ Cave painting in Lascaux, France

Background

Paleolithic art, also known as Ice-Age art, is some of the earliest known human art. It was produced roughly between 40,000 and 10,000 B.C., during the late Paleolithic period. The best known examples of Paleolithic art are the paintings and engravings that line the walls of limestone caves in France and Spain. These well-preserved artifacts include highly detailed, colored figures of horses, deer, and other animals. The painters used mixes of black, yellow, red, and white pigments that were often combined with minerals or heated to obtain subtle color variations. Black paint came from charcoal and manganese. Red and yellow paint came from red clay, animal blood, and iron compounds. White paint came from clay and lime mud. The pigments were mixed with animal fats or blood to produce a pastelike paint. Instead of brushes, the painters rubbed it on with their fingers or blew it on through hollow bones.

Several dating techniques have been used to pinpoint when the art was created. Some techniques involve carbon dating of artifacts and fossilized pollen found in the caves. A newer technique allows scientists to date the pictures themselves. This method involves an instrument called an accelerator mass spectrometer. The device can separate and count individual carbon isotopes in tiny amounts of the charcoal-based black paints. Analysis of paintings in Altamira, Spain, has indicated that the artwork was painted 14,000 years ago.

Extension

Have students find out about spelunkers—people who explore caves. Have them investigate the type of equipment required, the geographic locations of caves, and the dangers and costs involved in exploring caves.

CROSS-DISCIPLINARY FOCUS

Art

Have students find out about prehistoric art from other areas of the world, such as Asia, Africa, Australia, and South America. Have them compare art materials from different parts of the world. Encourage students to experiment with different materials in order to create their own version of prehistoric or ancient artwork.

Be an Art Detective

Students should observe pictures of ice-age humans hunting or herding flocks of animals. The cave paintings reflect the lives of the people. Stories should include descriptions of human life in prehistoric times.

Background

Most avalanches occur during or just after storms on mountainsides with slopes of more than 30 degrees. A number of factors contribute to avalanches, including the amount and rate of snowfall, the density and shape of the snowflake crystals, the wind speed and direction, the temperature, the depth of snow cover, and the strength of the bonds between snow layers.

There are several different ways in which avalanches start. *Airborne* or *wind avalanches* occur when new snow (loose powder) becomes airborne. It roars down the mountain at speeds of 322 km/h or more. *Slab avalanches* result when packed snow breaks loose. The slabs range in size from a few square meters to several hundred square meters. *Ice avalanches* occur when tips of glaciers break off.

Ninety percent of all avalanches are caused by skiers or hikers who fail to pay attention to warning signs about unsafe conditions. Searching for surviving victims is accomplished by using probes to detect buried people. Dogs are also used in rescue attempts. The Saint Bernard has been specially trained and bred to rescue avalanche victims. According to rescue-attempt statistics in the United States, only 50 percent of avalanche victims survive after being buried for more than 30 minutes.

Extension

Have students investigate the history of the Saint Bernard. Ask them to investigate the following: Did Saint Bernards really wear barrels on their collars? What group of people first trained Saint Bernards for avalanche rescue? *(Saint Bernards were first trained as rescue dogs in the 1600s by a group of monks at the Saint Bernard monastery in the Swiss Alps. The dogs were trained to use their keen sense of smell to locate travelers who were lost or buried in snow banks. In a popular painting, English artist Sir Edwin Landseer portrayed a Saint Bernard with a barrel around its neck. However, the dogs trained by the Saint Bernard monks never carried such barrels.)*

Avalanche Alert!

*I*magine hiking through snowy mountains. Suddenly you hear a booming noise, like the sound of rolling thunder. Across the valley, you see a wall of snow rushing down the slope at 300 km per hour! Petrified, you look up at the mountain peaks directly above your head. How can you tell when an avalanche will strike?

Coming Down the Mountain

An avalanche occurs when a huge amount of snow slides down a mountainside. Some avalanches are relatively small and do little damage. Other avalanches may move as much as 750,000 m³ of snow. An avalanche of that size can bury an entire town!

Sometimes avalanche forecasters or mountain rangers create avalanches on purpose. After a heavy snowstorm, they may set off bombs near mountain highways and resorts so that they can control the timing of the snowslide and prevent future accidents.

Predicting a Super Snowslide

Scientists study the way that snow falls and packs together in different layers in order to figure out when an avalanche will occur. Some forecasters travel to steep mountain areas to dig in the snow and examine the snow layers themselves. It's a dangerous job because an avalanche could happen right where they are standing. Computers help forecasters evaluate the temperature, wind conditions, and depth of the snow, but a good avalanche-prediction program is accurate only about 70 percent of the time.

Artificial Avalanches

An avalanche simulator, such as the one at the University of California at Los Angeles, allows scientists to mimic avalanche conditions in the laboratory. The simulator uses 100,000 tiny plastic beads to represent snow. During an artificial avalanche, these beads drop down a chute. The process helps scientists study how snow layers form and fall on a real mountainside. In the future, that information could help save the lives of people who live under the dangerous threat of sliding snow.

◀ **An avalance in action**

Surviving the Slide

A person buried in a snowslide is not likely to survive. Do the following investigation to find out why. Fill a pail with a liter of sand, and place a small gumdrop on the top of the sand. The gumdrop represents a human being. Use two or three telephone books to prop up a cookie sheet at a steep angle. Spill the contents of the pail down the cookie sheet. What happens to the gumdrop? Why do you think it is so difficult to locate people buried by an avalanche?

CROSS-DISCIPLINARY FOCUS

Mathematics

Have students model avalanches with clay and sand. Ask students to build a clay slope against the side of a box or plastic container. Then have students pour fine, dry sand onto the clay slope and measure the slope of the sand using a protractor. Then have students spray water on the sand until the sand slips. Invite students to create different slopes to determine which angles are the most likely to cause a slide.

Answers to *Surviving the Slide*

Students should observe that the gumdrop becomes buried in the sand. People buried by an avalanche have little oxygen to breathe, and the deeper they are buried, the more difficult it is to rescue them.

Mississippi River Delta in Peril

From Lake Itasca in Minnesota to New Orleans, Louisiana, the Mississippi River journeys 3766 km to empty into the Gulf of Mexico. Each year, it moves 265 million tons of sediment along the way. At the end of the river, the Mississippi River delta forms the largest area of wetlands in North America—but both the delta and the wildlife it supports are disappearing.

Disaster Along the Delta

A river *delta* forms when sediments settle at the mouth of the river. At the Mississippi River delta, the sediments build up and form new land along the Louisiana coastline. The area around the delta is called the wetlands. It has fertile soil, which produces many crops, and a variety of habitats—marsh, fresh water, and salt water—which support many species of plants and animals.

The delta and wetlands are now in serious danger. Large portions of the river bottom have been dredged out to make the river deeper for ship traffic. Underwater channels have been built to control flooding. Sediments that would have been deposited to form new land now pass through the deep channels and are carried far out into the ocean.

Those sediments used to replace the land that was lost every year to erosion. These days, erosion is still happening

▲ The Mississippi River flows from Minnesota, through the Midwest, to the Gulf of Mexico in the southern United States. Up to 80 km² of coastal wetlands are lost each year due to erosion.

as fast as ever, but the river can't keep up anymore. As a result, the Mississippi River delta is disappearing. By 1995, over half of the wetlands were already gone, swept out to sea by the waves along the coast.

Taking Action to Preserve the Delta

If the delta cannot replace the land lost to erosion, the wetlands will soon be completely destroyed. Many plant and animal species will lose their habitats. Highways and railroads will have to be moved, and many people will lose their homes and property.

Since the mid-1980s, local, state, and federal governments, along with Louisiana citizens and businesses, have been monitoring the Mississippi River delta.

Some projects to protect the delta include filling in canals that divert the sediments and even using old Christmas trees as fences to trap the sediments! With the continued efforts of scientists, government leaders, and concerned citizens, the Mississippi River delta stands a good chance of recovering.

Exploring the Delta

A delicate balance between the needs of people and the needs of nature must be maintained for humans and other organisms to successfully share the Mississippi River delta. Find out more about the industries and the organisms that depend on the Mississippi River delta for survival. What will happen to them if we don't take care of the ecosystem?

527

Background

The word *Mississippi* comes from a Native American word that means "big river." The Mississippi River flows for 3766 km from its source in Minnesota to its mouth in the Gulf of Mexico. More than 2897 km of the river can be navigated by ships for commercial, recreational, and transportation uses.

The Mississippi River forms part of the boundaries of 10 different states. As it travels to the gulf, other rivers such as the Ohio and Missouri flow into it and increase the volume of water in the river.

The Mississippi River delta, located at the mouth of the river, covers about 33,700 km². It is responsible for continually re-creating the coastal wetlands that provide habitats for diverse populations of plants and animals. The erosion and sinking of land near the Mississippi River delta has seriously disrupted local ecosystems. Currently, the destruction in this area accounts for 80 percent of the nation's loss of coastal wetlands per year.

Spanish and French explorers used the Mississippi as a route for exploring North America during the 1500s and 1600s. As steamboats came into use in the 1800s, the Mississippi became a critical transportation and trade route. In Mark Twain's *Life on the Mississippi, The Adventures of Huckleberry Finn,* and *Tom Sawyer,* Twain gives detailed accounts of the history, sights, people, and towns along the Mississippi River.

Answers to *Exploring the Delta*

The coastal wetlands provide a home for a variety of plants and animals such as fish, shellfish, and birds. These species and the industries (such as fishing industries) that rely on them will suffer greatly if the wetland ecosystems are not protected. Also, if erosion is severe enough, buildings, highways, phone lines, and pipelines will need to be demolished or relocated.

Meeting Individual Needs

Gifted Learners

Have students find out what major tributaries empty into the Mississippi River and for what states the river forms a boundary. Then instruct them to construct a relief map of the river, its major tributaries, and state boundaries. Encourage students to include as much detail as possible on their maps.

CROSS-DISCIPLINARY FOCUS

Language Arts

Have students read excerpts from the writings of Mark Twain or other authors who have written about the Mississippi River. Then have students write poems or short stories incorporating scientific information about the Mississippi River, the delta, or the wetlands.

IN THIS UNIT

SCIENCE AND TECHNOLOGY

Now that you have been introduced to science and its methods, consider these questions.

1. What is "science"? What does this term suggest to you?
2. How is the study of science helped by separating it into different subject areas?
3. How would you describe the career of a scientist to a younger person interested in becoming a scientist?
4. How do you use some of the skills of science in everyday life?

In this unit, you will find out more about the different areas into which science has been divided and about how practicing scientists use scientific methods while doing research.

S1

SCIENCE IS CURIOSITY

Curiosity may have "killed the cat," but the desire to know is a characteristic also shared by human beings. People have always asked questions about things they have seen and experienced. They ask questions such as: Why did that happen? Where did it come from? How can I use it?

These kinds of questions lead to more questions, which lead to even *more* questions. Asking questions and finding answers is the driving force behind what we call **science**. Think about the following examples of curious events that have caused people to ask questions.

Science

The asking and answering of questions to satisfy curiosity about the natural world

DID YOU KNOW...

that living organisms have been found underneath the ocean floor? Microorganisms have been discovered living 500 m below the floor of the ocean, feeding on organic matter found in sediments that are 4 million years old.

▼ Giant tube worms were discovered in the hot water around ocean vents.

Asking Questions

In 1977, scientists on board the research submarine *Alvin* were exploring the bottom of the Pacific Ocean near the Galápagos Islands. As they were studying some hot-water springs coming from cracks, or *vents*, in the ocean floor, the scientists were surprised to discover a large community of creatures living around the vents.

▲ *Alvin* can carry a pair of scientists to a depth of about 4000 m.

Among them the scientists found blood-red tube worms up to 3 m long. They also saw several other types of *organisms*, or living things, that no one had ever seen before.

But how did these organisms survive? For a long time, scientists thought the sun was the source of energy for all living things on Earth. Yet sunlight does not reach the ocean bottom. The organisms around the hot springs, therefore, must have a different source of energy. For the first time, a whole community of living things that did not get their energy from the sun had been discovered. How can living things exist in the darkness of the deep ocean floor? What do living things need to exist anywhere? Could life exist on other planets?

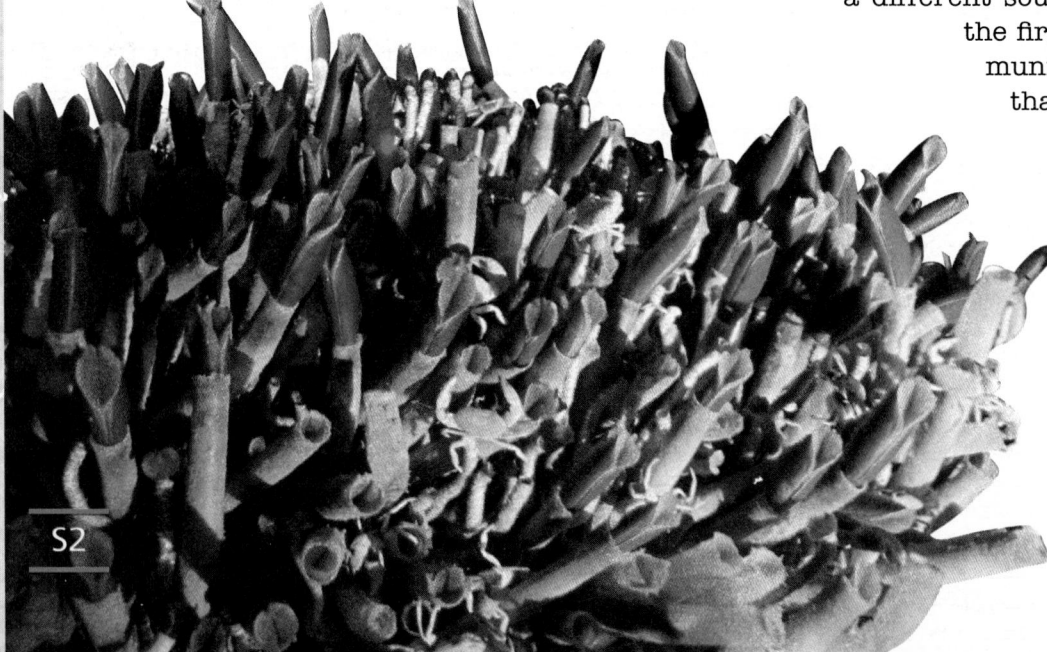

S2

▲ Mount St. Helens before it erupted ...

... and after it erupted. ▲

It was a place for camping, hunting, and fishing. There were green forests, meadows, and lakes. Such was the area around Mount St. Helens in the state of Washington. Then, at 8:32 A.M. on May 18, 1980, the mountain blew its top. About 250 billion kg of rock and dust were blasted out by the eruption. The forests, meadows, lakes, and surrounding areas were turned into a landscape that resembled the surface of the moon. Since that May morning, scientists have been studying Mount St. Helens, searching for answers to many questions.

Our planet is always being changed by powerful forces. Almost every day, we get news of volcanoes, earthquakes, storms, floods, and droughts occurring somewhere on the planet. At some time, these forces will touch the life of every person on Earth. Why do volcanoes erupt? What causes weather and earthquakes? Can earthquakes and volcanic eruptions be predicted?

Finding answers to questions like these could save lives and property.

You are caught outside in a sudden rainstorm. As the rain comes down, your eyes begin to sting. You notice that the rainwater has a sour taste. The falling rain is not pure water. Instead, it seems more like vinegar or lemon juice.

Scientists now call this kind of rain *acid rain*. What causes acid rain? Scientists are trying to find answers to this question, but understanding the cause of acid rain is only the beginning. Many more questions must be answered. How can we protect our soil from acid rain? What does acid rain do to our water supplies? What effect does acid rain have on the health of plants, animals, and people?

▼ The scientist in this photo is testing the acidity of lake water.

S3

Finding Answers

One of the ways to find answers to the types of questions just asked is by doing scientific research. Through observation, scientists have learned that the water around hot-spring vents abounds with tiny organisms that are able to use chemicals in the hot water as a source of energy. Larger organisms then get their energy by feeding on these tiny organisms. Through analysis, the composition of acid rain has been determined. By its composition, the sources of acid rain can be identified.

Many other questions have been answered and much has been learned. Through research, our understanding of the universe increases daily, helping us answer even harder questions. Much more, however, remains to be learned. Many answers need to be found, including answers to questions that have not even been asked yet. The research of scientists can help provide these answers.

Gathering knowledge through observing and asking questions is part of the process of science. But science can also be used to solve practical problems—an endeavor called **technology**. The knowledge gained through scientific discovery, along with the tools and

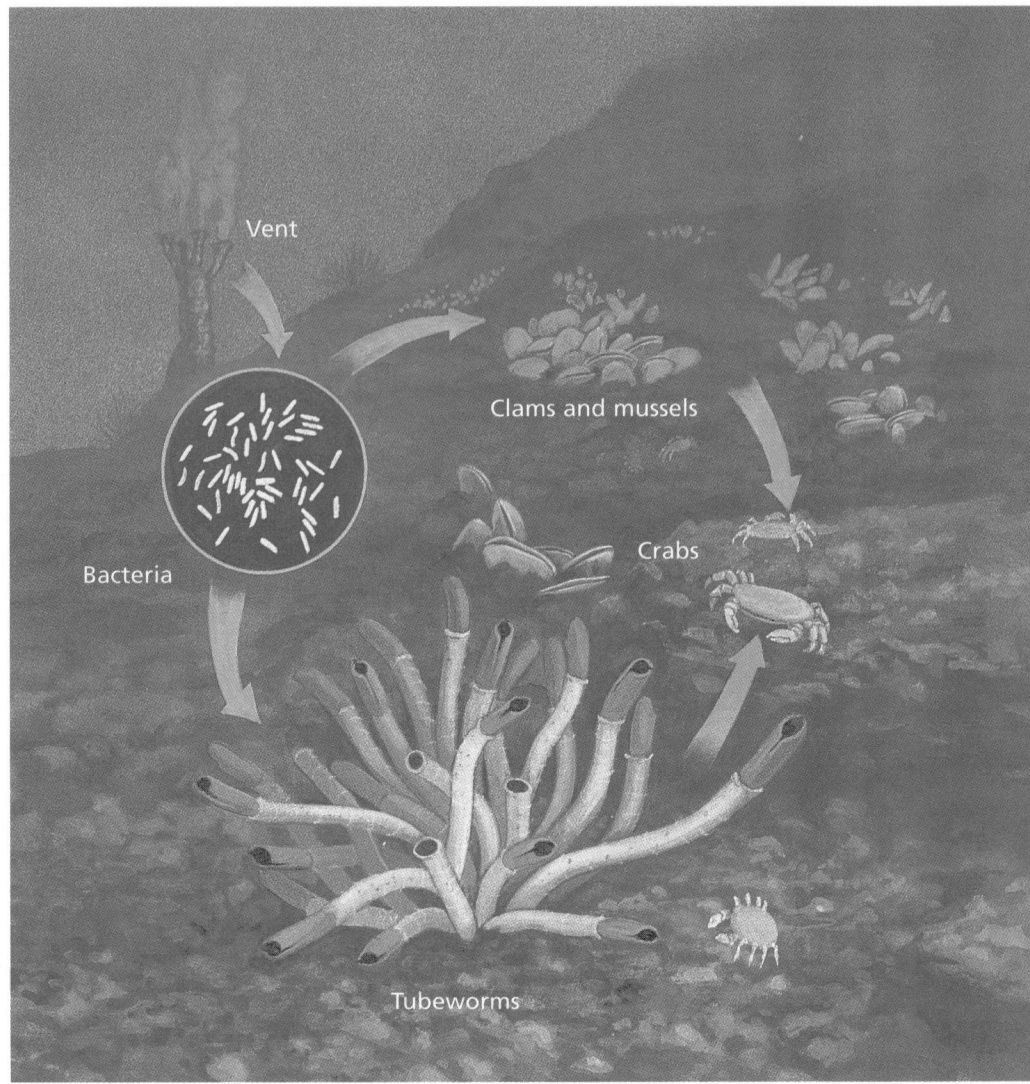

Chemicals provide a source of energy for a whole underwater community.

materials developed through an understanding of science, has led to an increase in technology and an improved quality of life.

Although technology is displayed by a few other species that use tools, science itself is a uniquely human endeavor.

Technology

The application of knowledge, tools, and materials to solve practical problems

S U M M A R Y

Curiosity is the desire to know and learn about things. Because of curiosity, people ask questions. This leads to knowledge and usually to more questions. Science is one way people deal with curiosity and questioning. Technology is the application of the knowledge gathered through science.

S4

WHO IS A SCIENTIST?

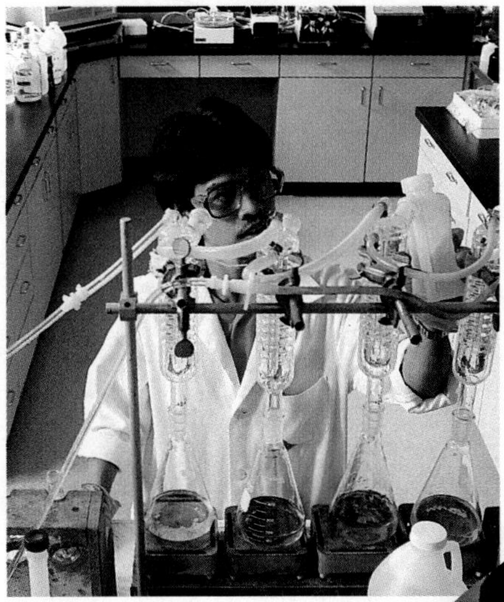

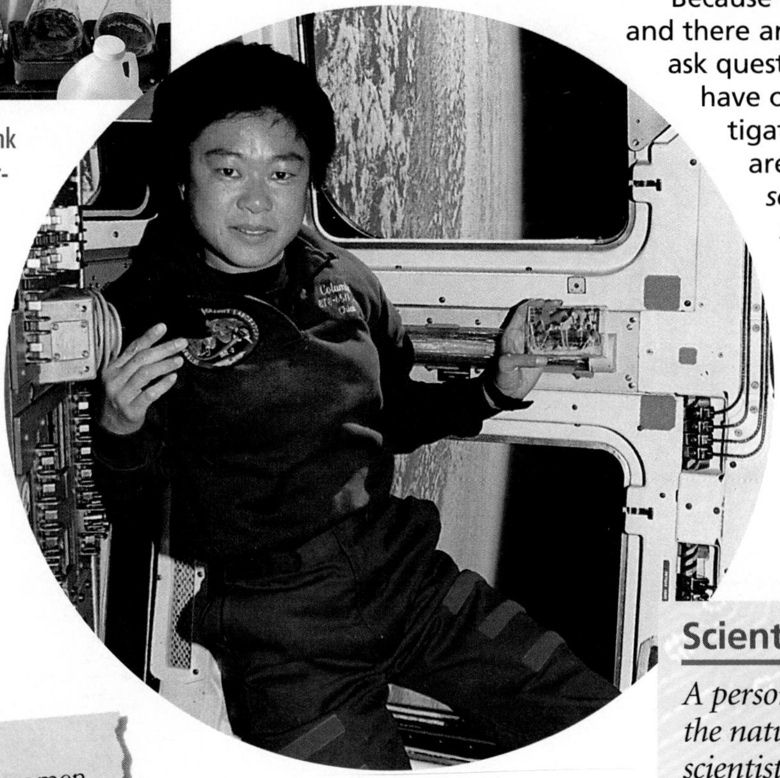

Look at the two people in the photos shown here. Which one do you think is a **scientist**? How can you tell? You might think of a scientist as a person who wears a white coat and works in a laboratory. However, scientists do not always dress the same way, and not all of them work in laboratories. But all scientists *do* have one thing in common—they are all curious about nature. They ask questions and then try to answer their questions.

Are *you* a scientist? Do you ask questions? Are you curious about the world around you? If so, then you could be considered a scientist. Professional scientists, however, have studied the *techniques* of science for many years, and they follow certain *methods* when they ask and answer their questions.

Because the world is so large and there are so many things to ask questions about, scientists have organized their investigations into three major areas: *life science, Earth science,* and *physical science.* The scientists that spend most of their time asking questions in each of these areas are likewise called life scientists, Earth scientists, and physical scientists, respectively.

▲ Many people often think of a scientist as a person who works in a laboratory. But people like Dr. Chiaki Mukai, a NASA astronaut, are also scientists, even though they may work in places that are "out of this world." ▶

Scientist

A person who is curious about the natural world. Professional scientists are trained in certain techniques and methods.

Wanted: Men and women willing to work 9 hours at a time inside a small, metal sphere on the ocean bottom. Must have a background in biological sciences. Duties include measuring water depth and temperature, taking water samples, and photographing and collecting living things from the ocean floor. Must be able to withstand cramped areas, cold temperatures, and long periods of isolation.

What Do Life Scientists Do?

Why would anyone want a job like the one listed to the left? To a scientist, it would be a rewarding opportunity. To a marine biologist, a scientist who studies life in the oceans, it might be the best chance to observe the ocean floor.

Life scientists are people who are involved in *biology*, the study of living things. Most life scientists are called biologists. There are many specialized fields of study in biology. Two of the major areas are *botany* and *zoology*. Botanists are biologists who study plants, while zoologists study animals.

S5

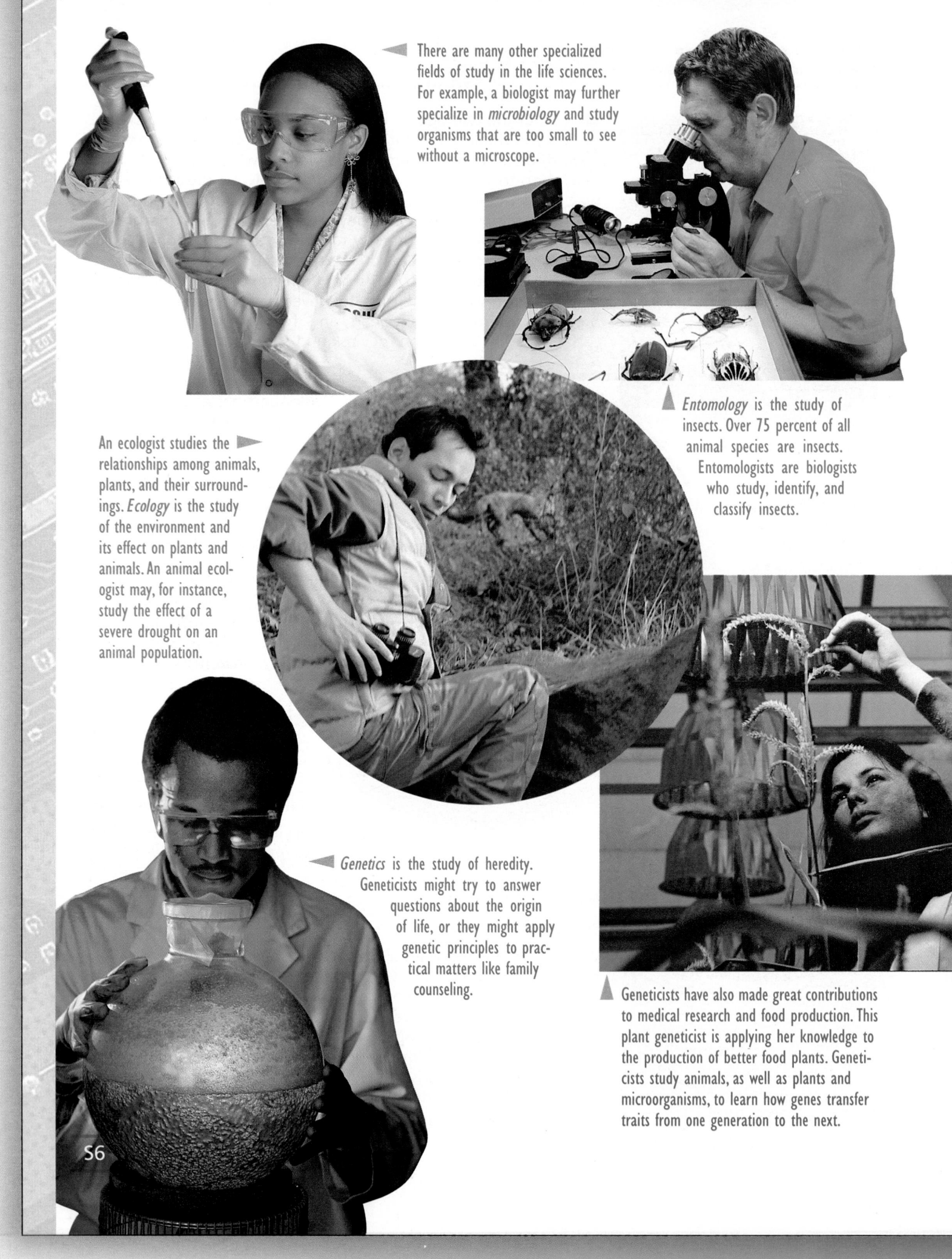

There are many other specialized fields of study in the life sciences. For example, a biologist may further specialize in *microbiology* and study organisms that are too small to see without a microscope.

Entomology is the study of insects. Over 75 percent of all animal species are insects. Entomologists are biologists who study, identify, and classify insects.

An ecologist studies the relationships among animals, plants, and their surroundings. *Ecology* is the study of the environment and its effect on plants and animals. An animal ecologist may, for instance, study the effect of a severe drought on an animal population.

Genetics is the study of heredity. Geneticists might try to answer questions about the origin of life, or they might apply genetic principles to practical matters like family counseling.

Geneticists have also made great contributions to medical research and food production. This plant geneticist is applying her knowledge to the production of better food plants. Geneticists study animals, as well as plants and microorganisms, to learn how genes transfer traits from one generation to the next.

S6

People with careers in health care and medicine are also part of the life sciences. Their work ranges from trying to understand how the human body works or treating it for various diseases to building devices such as artificial hearts and knees that help the body stay healthy.

What Do Earth Scientists Do?

The job described in the ad below is one type of work that an Earth scientist might do. The photo beside it shows an Earth scientist taking samples of *lava*, which is the melted rock that comes from volcanoes. By studying samples of lava and other details of the mountain, Earth scientists can learn much about volcanoes.

Earth science, which is basically the study of our planet, can be divided into four branches. One branch—called *geology*—is the study of the solid surface of the Earth and its interior. Geologists study the matter and the structure of the planet. They do this to find out

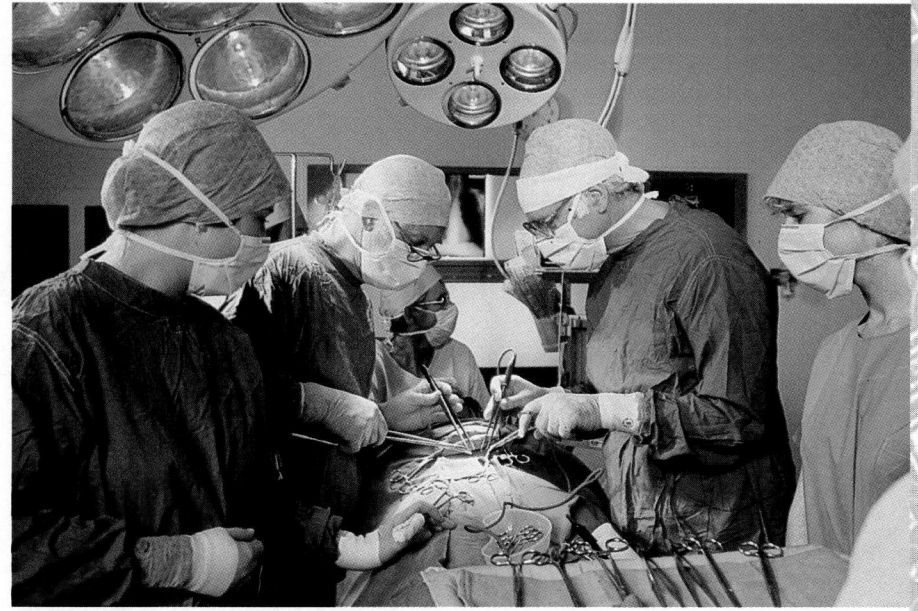

▲ Careers in surgery require a background in biology.

how the Earth was formed and how it changes over time. Some geologists study volcanoes because volcanoes change the surface of the Earth. Volcanoes also give hints as to what lies beneath the surface. Other geologists search for natural resources, such as oil. Oil is important to modern society because it provides energy, plastics, and solvents for our use. In

fact, almost all of the things we use daily come from the Earth.

As you know, the Earth is surrounded by a layer of air called the atmosphere. *Meteorology* is the branch of Earth science that involves the study of this atmosphere. The scientists who work in the field of meteorology are called meteorologists. They also observe weather and study climates.

▲ This volcanologist is wearing special clothing and a mask for protection against the intense heat while using a steel pipe to collect a sample of molten lava.

Wanted:

Men and women for dangerous work on active volcanoes. Must be able to work in extreme heat and be able to lift heavy objects. Backpacking or helicopter-piloting experience a plus. Duties include collecting and studying lava, monitoring subsurface pressures, and forming models for predicting eruptions.

S7

Oil geologists study the Earth's features to find oil deposits.

A geologist uses a laser range finder to measure the movements of parts of Mount St. Helens.

Meteorologists study the weather.

S8

Oceans cover almost three-quarters of the Earth's surface. *Oceanography* is the branch of Earth science that deals with the oceans of the world. Oceanography also includes the study of the Earth's surface below the oceans, the sea floor. The scientists who make these studies are called oceanographers, or marine scientists.

Astronomy, sometimes called space science, is the study of the universe beyond the Earth. Astronomy is included as a branch of Earth science because it shares many questions with other Earth sciences. Astronomers study other planets, the sun and stars, and many other objects in space. Also, from space we can see the Earth itself as never before. Already, pictures of the Earth from orbiting spacecraft have helped locate valuable deposits of minerals and pinpoint sources of pollution.

In many cases, specialists from one or more branches work together on a problem. Wastes from human activity, for example, cause pollution of both air and water. By studying the atmosphere and the oceans, scientists can begin to understand the effects of pollution. They can then try to come up with methods of controlling it.

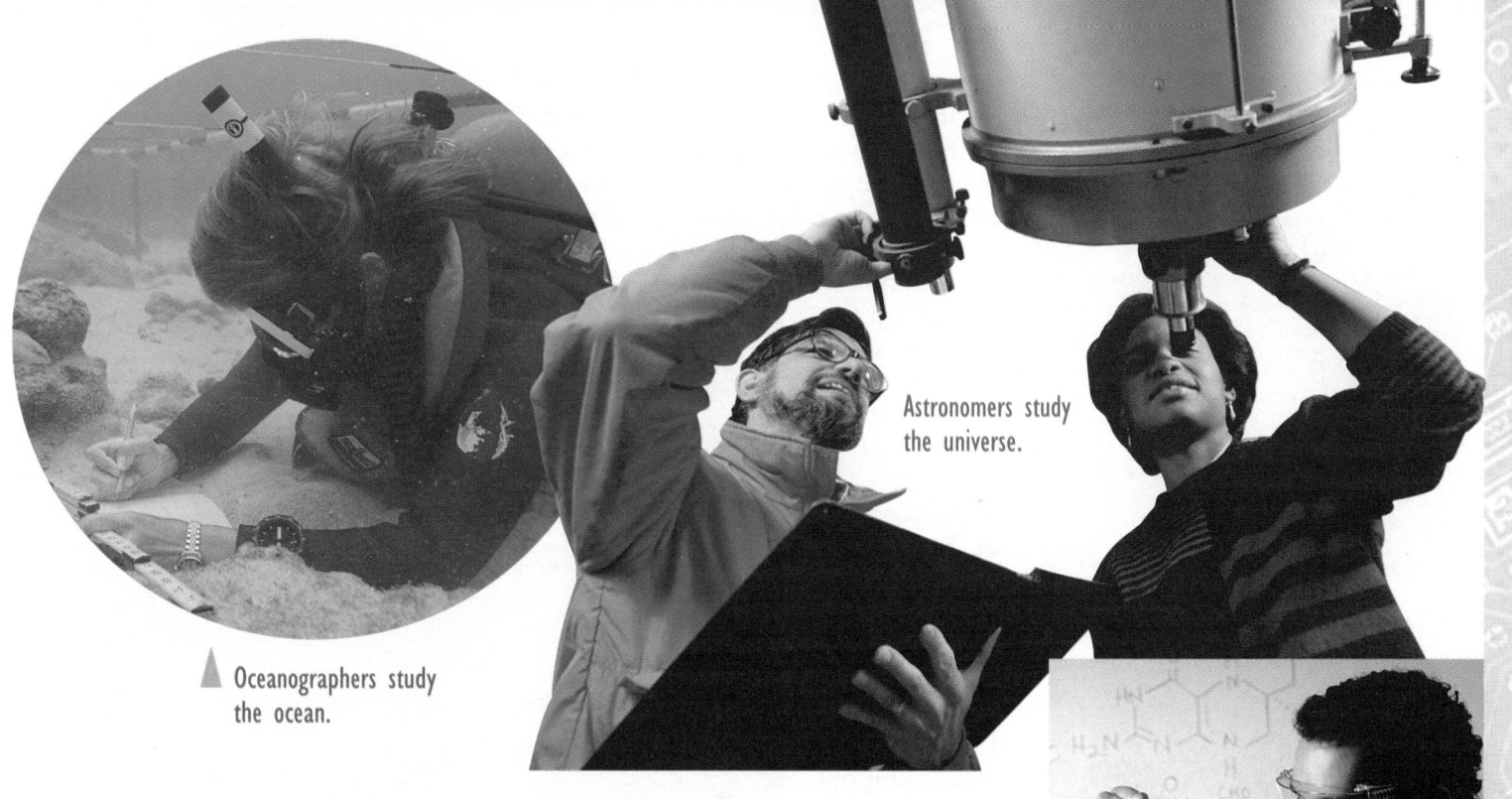

Oceanographers study the ocean.

Astronomers study the universe.

What Do Physical Scientists Do?

Physical science includes the study of both matter and energy. The two main branches of physical science are chemistry and physics. *Chemistry* is the study of all forms of matter and their interactions. *Physics* is the study of energy, its changes, and its relationship to matter.

Chemists and physicists also work in the life sciences and Earth sciences. For example, many biochemists—scientists who study the chemistry of life—work in the field of medicine. Their research may find answers to how different disease-causing organisms affect the body. It may also provide information that can lead to a cure. Other biochemists develop new drugs and test their effectiveness for drug manufacturers.

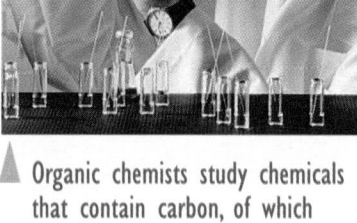

Organic chemists study chemicals that contain carbon, of which there are millions.

Wanted:

Men and women for challenging jobs in the physical sciences. Must be imaginative, detail oriented, and good problem-solvers. Laboratory and field positions available. Duties may include smashing atoms, perfecting musical sounds, or developing new ways to improve the flavor and shelf life of canned foods.

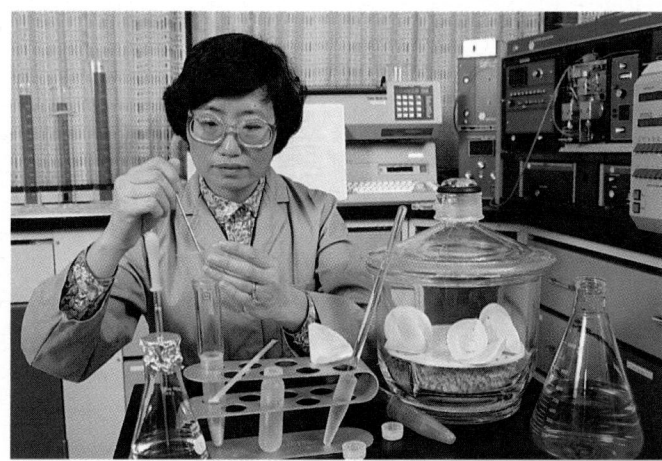

This biochemist works in a medical laboratory.

S9

▲ This chemist is taking air samples that will be analyzed in a laboratory for pollutants.

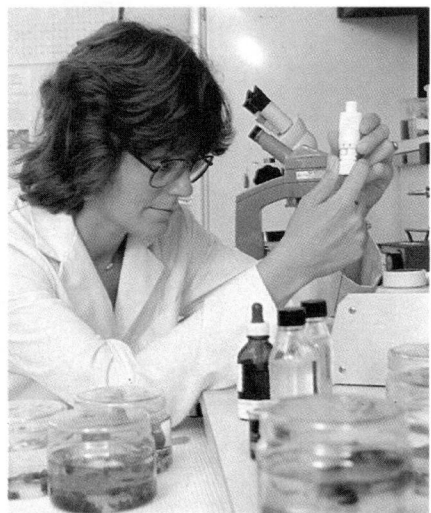

▲ Physical oceanographers apply the principles of physics and chemistry to their study of the oceans. They might study ocean forces (such as waves, currents, and tides) and the physical properties of ocean water (such as density and temperature).

Chemists also work in the chemical-manufacturing industry, making the chemicals that are used in factories that make the products we use. Many work for the oil industry doing research on crude oil and other petroleum products. Other chemists might test air samples for a variety of environmental pollutants.

Physicists study things such as light, mechanical forces, sound, and the structure of and the forces inside atoms, the tiny particles that make up all matter. The physicist in the photo to the right is studying *laser* light—a highly concentrated form of light. Lasers are used in space science and medicine, as well as in the entertainment industry.

▲ Laser light is a form of radiant energy. This concentrated energy can be used to cut metal or perform delicate eye surgery.

Some physicists study sounds—sounds we hear and sounds we cannot hear. Concert halls and theaters are designed by engineers who use the findings of physicists. These engineers work with sound, making sure sound waves combine in pleasing ways. *Sonar* equipment can be used to detect objects underwater. *Ultrasound* is used by doctors to study the inside of the human body. Both sonar and ultrasound use rapidly vibrating (high-frequency) sound waves that are above our range of hearing, but that can be detected by special instruments.

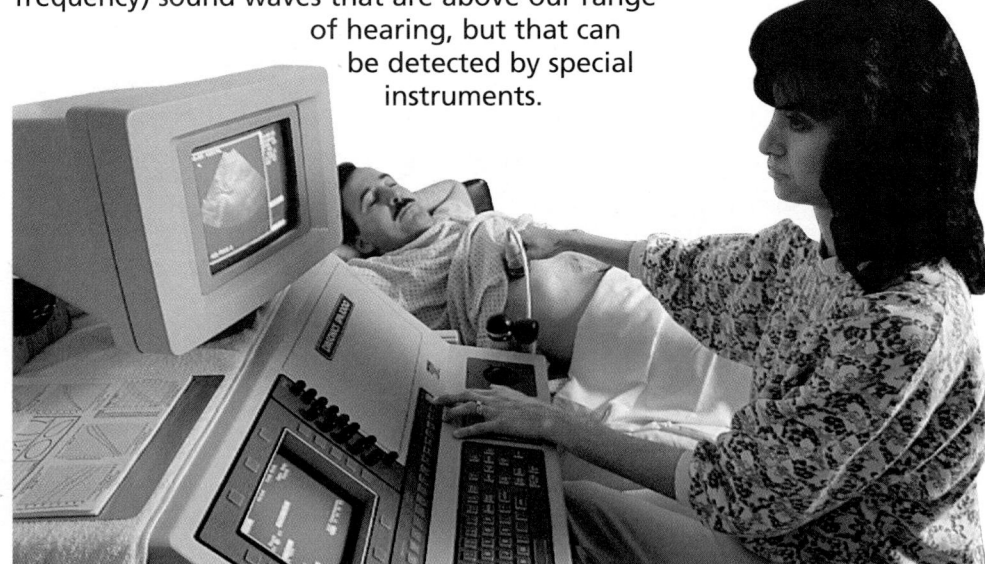

Doctors use ultrasound waves to ▶ form images of the inside of the human body.

S10

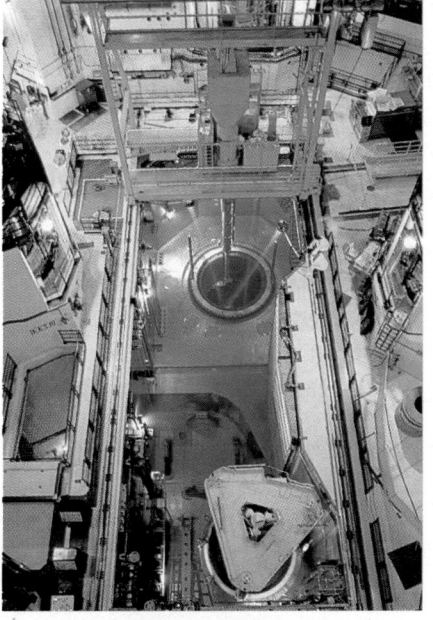

Within a bubble chamber (inset), the collision of atomic particles produces even smaller pieces of matter, as shown by the tracks in this photograph.

Another important branch of physics involves studying the nature of the atom. Physicists who study the atom have identified even smaller particles that are produced when atoms are smashed together. The presence of these particles can be detected by devices such as the bubble chamber.

Knowledge of atoms and how they interact has been applied to the useful production of energy. Fission, the splitting of an atom, is the source of nuclear energy that is currently used in the production of electricity. The tremendous amount of energy that fission releases has been demonstrated by the destructive power of atomic weapons. Fusion, the joining together of two atoms to form a new kind of atom, is thought to be the source of the sun's heat and light. Someday, perhaps, people will be able to control fusion reactions on Earth to provide a safer and cheaper form of nuclear energy.

Inside a nuclear power plant

SUMMARY

People who work as life scientists study living things. They are called biologists and may specialize in a variety of areas. Two major branches of biology are botany and zoology. Earth scientists study the complex planet we live on, as well as its environment in space. The main branches of Earth science are geology, meteorology, oceanography, and astronomy. Physical scientists study chemistry and physics, both of which are composed of many specialized areas. Chemists study matter and its interactions, and physicists study all forms of energy and the nature of atoms.

DID YOU KNOW...

that an atom is not the smallest piece of matter? According to current theory, atoms are made of tiny *quarks* that, in turn, are held together by even smaller particles called *gluons*.

S11

USING SCIENTIFIC METHODS

Scientific method

A way of thinking about nature that involves the use of certain skills to solve problems in an orderly manner

Modern experimental science began about 400 years ago when an Italian scientist named Galileo Galilei (1564–1642) performed experiments to test his ideas about nature. Galileo tried to find basic rules to explain the way things happen in the natural world. Today we call Galileo's way of finding answers a **scientific method**. A scientific method is not a *technique* for always getting the right answer to a problem. Rather, it is a *way of thinking* about and investigating nature to discover more about it.

Galileo used a scientific method to study the motion of falling objects. He believed that scientific understanding of the natural world comes from making observations and doing experiments. He refused to blindly accept ideas that people had believed for thousands of years. One such idea was that heavy objects fall faster than light objects. But Galileo wasn't convinced. He therefore tested various objects, measuring the time it took for each of them to fall. He made careful observations and recorded them. He then carried out experiments to further understand his observations.

Galileo used ramps as a way of studying falling objects. He showed that two balls of different masses will roll down the same ramp in the same amount of time.

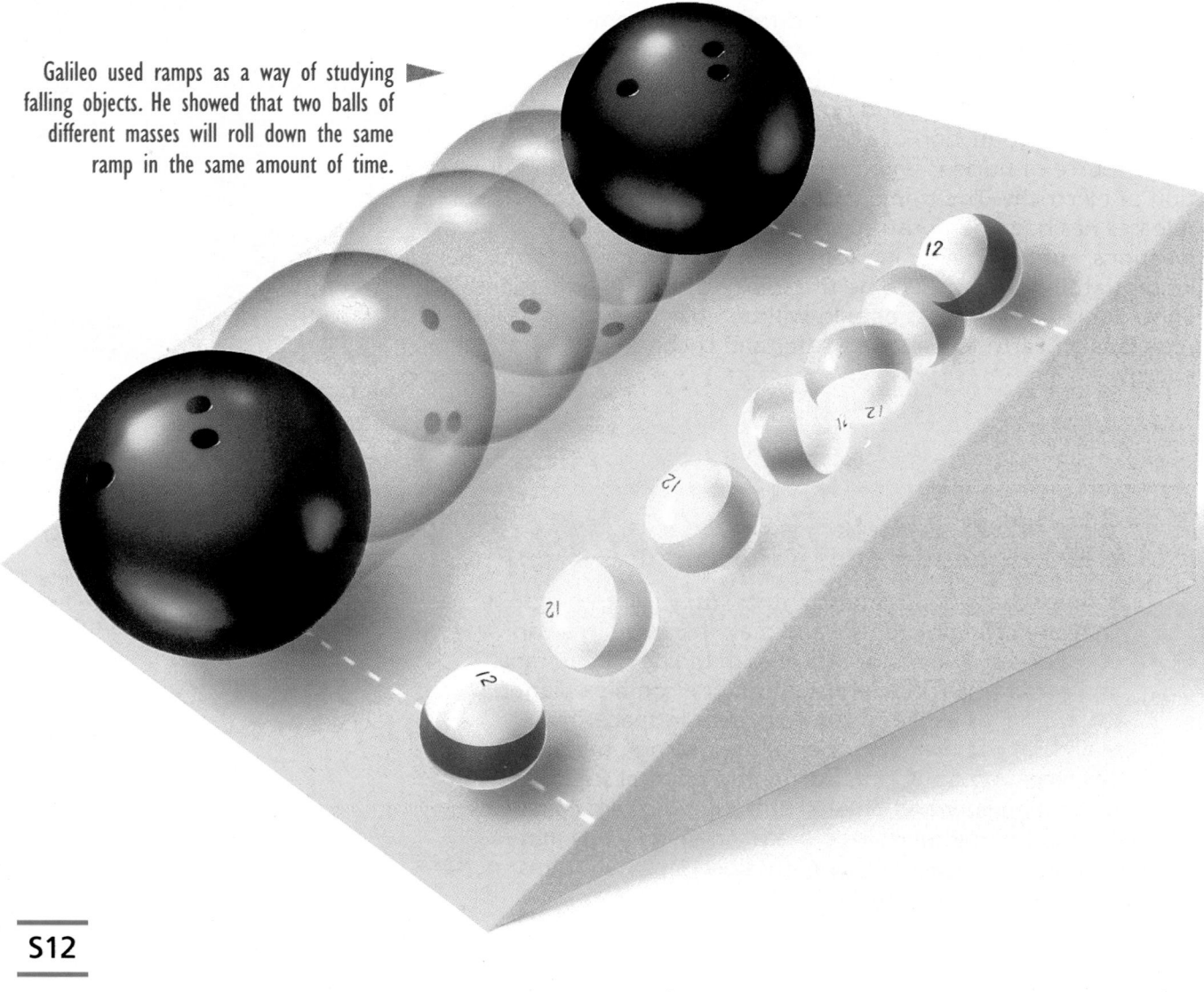

S12

Modern scientists also use scientific methods to study many types of questions. However, not all scientists use all of the same steps, nor do they necessarily use them in the same order. These steps are better thought of as skills that are common to all scientists. These skills can also be used by anyone who wants to solve problems in an orderly, thoughtful way.

Scientists tackle different problems in different ways. Some problems lead to specific solutions. For example, a new type of disease might appear. Its cause must be found and a cure developed. At other times, the purpose of research is to find out more about a general subject, such as the nature of gravity. The following paragraphs detail the steps or skills involved in using a scientific method.

Defining the Problem

The mournful howl of a lone coyote breaks the stillness of the night. At a distant sheep ranch, a rancher hears the howl and reaches for his rifle. "That pest won't kill any of my sheep tonight!" he mutters.

For over a century, the coyote has been the focus of a dispute. Ranchers and farmers claim it kills domestic animals, such as sheep and poultry, for food. Environmental groups say that this does not happen very often. In fact, very little is known about coyotes. If more could be learned about them, perhaps the dispute could be settled.

▲ While other large North American animals are disappearing, coyotes seem to be thriving.

Marc Bekoff and Michael Wells are scientists. They wanted to learn about the behavior of coyotes. As part of this effort, they carried out a 3-year study. One of their goals was to learn why some coyotes lived alone and others formed packs. Bekoff and Wells suspected that some coyotes in the wild switched back and forth from one type of behavior to the other.

First, they stated the problem as a clearly worded question: *What makes some coyotes live alone and others live in packs?* Their next step was to find out what information was already known. Bekoff and Wells went to a library. They found some earlier studies done by scientists on coyotes. They also read about the behavior of animals such as jackals and hyenas because these animals are similar to coyotes. The information from these earlier studies was a starting point from which Bekoff and Wells attempted to find answers to their own questions.

S13

Forming Hypotheses

After collecting basic information, scientists usually suggest an answer to the problem they have defined. This is done by forming a **hypothesis**. A hypothesis is a possible answer to a question. It is sometimes called an "educated guess." A hypothesis is based on the information the scientist has already gathered. It should be the best possible solution to the problem that the scientists can think of. More than one hypothesis may be formed from the same information. Each possible solution must then be tested.

Bekoff and Wells formed a hypothesis that they felt would be a good answer to their question: *The food sources available to coyotes determine whether they live alone or in packs.*

Stating a hypothesis is only one step in using a scientific method. Demonstrating that there is enough evidence to support a hypothesis is another. Scientists *test* their hypotheses by making observations and doing experiments.

Making Observations

When you think of someone using scientific methods, do you think of a scientist in a laboratory mixing chemicals? Science *is* often done in the laboratory, but there are also other ways to use scientific methods. Imagine that you are an astronomer trying to find out how stars form. There is not much that you could do in a laboratory. Your investigation would involve making many **observations** of stars. The coyote study also involved many observations. Bekoff and Wells had to watch the coyotes over a long period of time.

To test their hypothesis, the two scientists went to Grand Teton National Park in Wyoming. From the top of a hill, they observed the behavior of several lone coyotes. They could also see packs of coyotes in the valley below.

Hypothesis

A possible answer to a question based on gathered information

▲ Dr. Marc Bekoff uses a spotting scope to observe coyotes in the wild.

Observation

Any information that we gather by using our senses

Grand Teton National Park

S14

During the summer months, the coyotes hunted and ate field mice, gophers, and ground squirrels. During the winter months, they ate the remains of deer, elk, and moose. These larger animals had died from other causes. Throughout the study, coyotes were never seen attacking any live, large animal. However, this did not *prove* that they never do so.

The information gathered by Bekoff and Wells showed several other things. In the summer, when the coyotes hunted and killed small animals for food, there were fewer coyotes in packs. More coyotes were hunting alone over wider areas.

▲ In summer, a lone coyote hunts for small animals.

◀ In winter, a pack of coyotes feeds on the carcass of a mule deer.

In the winter, when the weather was very cold and snowy, there were more large, dead animals available. More coyotes fed on the same food source, and the coyotes formed packs.

Where the winter supply of large, dead animals was plentiful, the pack size was larger. Where the food was scarce, the packs were smaller. These observations tended to *support* the scientists' hypothesis that food supply determines whether coyotes live alone or in packs.

S15

Performing Experiments

Bekoff and Wells were interested in another aspect of coyote behavior. They wanted to find out how lone coyotes find and capture their prey. The two scientists believed that coyotes use all of their senses while hunting. But were some senses more important than others? How could the importance of sight be compared with the importance of hearing or smell?

The scientists would have liked to watch the animals from their hilltop spot, but the range of a lone coyote may be more than 30 km². They also realized that just watching animals in the wild would not give them enough information to answer their questions. They decided to run an **experiment** under more controllable conditions.

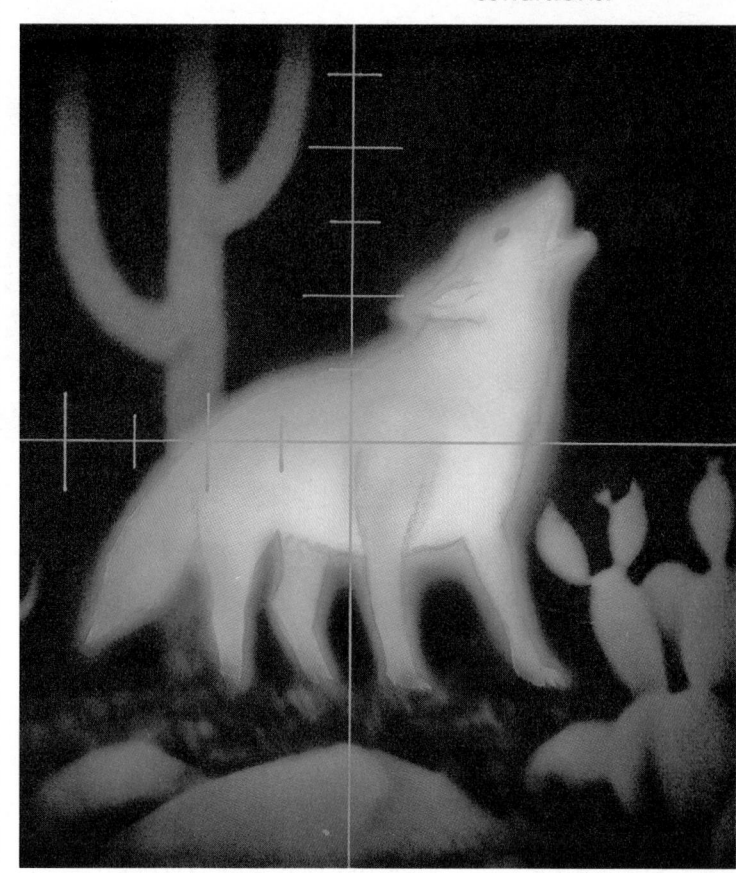

▲ Coyotes were tracked at night to test the importance of vision.

They captured several coyotes and brought them to a university. Each coyote was placed in a large, fenced-in outdoor area. Hidden somewhere in that area was a rabbit. The coyotes were able to use their senses of sight, hearing, and smell to find the rabbit. Bekoff and Wells measured and recorded the amount of time each coyote took to find the rabbit. Since the times they recorded were for the coyotes using all three of these senses, the relative importance of *one* sense could not be identified. This part of the experiment is called the *control*. Control conditions are as close to the natural setting as possible. All other parts of an experiment are compared to the control.

Testing Variables The part of an experiment that is different from the control is called the **variable**. Scientists test only one part of an experiment, or variable, at a time. In this way, they know that the different results are probably caused by that one specific variable. For example, Bekoff and Wells wanted to check how important sight was to coyotes. To do this, they had to eliminate the other two senses coyotes might use.

To do away with the variable of sound, they used a dead rabbit as bait. There would be no noise to help the coyotes locate the rabbit. To eliminate the variable of smell, they flushed the coyotes' nostrils with a nasal spray. The spray temporarily blocked the coyotes' sense of smell. To exclude the variable of sight, they tracked the coyotes' movements on a moonless night, using infrared photography to observe the coyotes in the dark.

One at a time, each of the three variables was tested. Again, the time it took for each coyote to find the rabbit was measured and recorded. Each part of the experiment was run many times. This was done to find an average time. Repeating an experiment several times improves the reliability of its results.

S16

Analyzing Data

Once all of the *data* from an experiment are collected, the data must be organized. This organized information can then be studied. Data in the form of measurements, or **quantitative observations**, can be organized in data tables.

Once data are collected and averaged, results can be analyzed. The results of the coyote experiment showed that sight was the most important sense in locating food. The sense of smell was second in importance, followed by hearing.

Sometimes results are more easily analyzed and understood in the form of graphs. The diagram shown below illustrates the results of the coyote experiment in the form of a *bar graph*. The average time in seconds is given on the left side of the graph.

Coyote Number	Three senses		Sight only		He
	Trials	Avg.	Trials	Avg.	Tr
1.					
2.					
3.					
4.					

Quantitative observations

Observations that involve taking measurements or using numbers

1. Bar **A** shows that it took the coyotes an average of 30 seconds to find the rabbits while using all three senses. This was the control time against which the other parts of the experiment were compared.

2. Bar **B** shows the average time using sight only. The variables of sound and smell had been removed. Using sight only, the time increased to about 55 seconds. This indicated that coyotes used more senses than sight to find their food.

3. When the coyotes were able to use only smell, as shown by Bar **C**, their average time was more than double the control time—about 73 seconds.

4. Bar **D** shows the results when using hearing only. The time increased to 209 seconds. This is almost seven times as long as the control situation.

5. Finally, all three senses were removed. It then took the coyotes an average of 22 *minutes* to find the rabbit. The bar representing 22 minutes would have been too long to fit on the graph. It had to be broken and labeled so that readers would understand that it actually should be much longer.

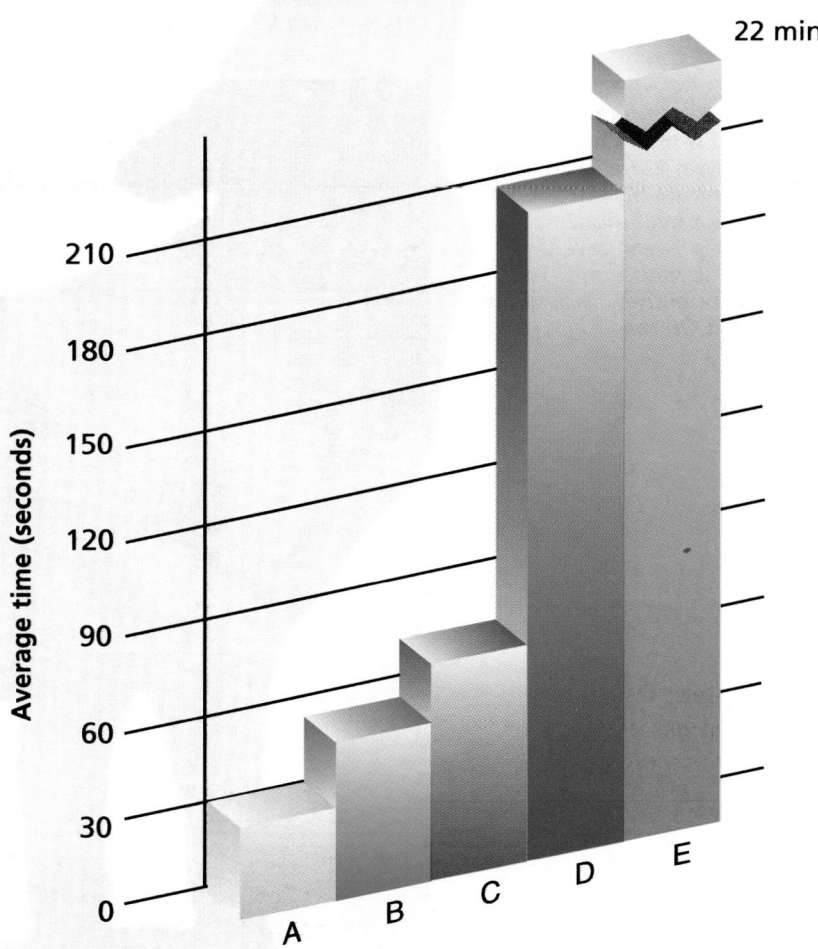

22 min.

S17

You can see from the graph that the use of smell or hearing alone required more time than sight alone. Therefore, Bekoff and Wells concluded that sight is the most important sense for finding prey. Could this be because the wind can distort sounds and disperse smells? Again you can see how some answers can lead to more questions.

Drawing Conclusions

After data are collected and analyzed, conclusions can be made. The results may lead the scientist to believe that the hypothesis was a good one. On the other hand, they may not. If not, the scientist must make a decision. Was there an error in the experimental setup, or was the hypothesis not correct to begin with? In either case, experiments are run many times to check the results. If a hypothesis is not supported by the results, it must be replaced with a new hypothesis.

The hypothesis made by Bekoff and Wells will have to be tested again and again. The experiment must be done with other coyotes, by other researchers, and in other places. Only when these additional experiments verify the original results can the hypothesis be considered reliable.

Communicating Results

From their results, Bekoff and Wells were able to make several general statements. They said that the social behavior of coyotes was related to the food supply. They also found that sight was the most important factor in locating food. They reported their results by writing an article for a scientific journal. Experimental results can also be presented at scientific meetings or in one-on-one discussions with other scientists.

This sharing of results is an important part of science. In this way, other scientists learn what has been done. These scientists can then repeat the experiments to check the reliability of the original results. They may also see ways to apply other scientists' results to their own work and to determine what still needs to be studied.

A scientist shares the results of an experiment with other scientists. ▶

S18

The Changing Nature of Science

As demonstrated by the investigations of Bekoff and Wells, a scientific method is not a "straight line" of steps. It contains loops in which several steps may be repeated over and over. Some steps may not be necessary. For example, one scientist may make observations that lead directly to a conclusion; another may decide that experiments are needed to verify the hypothesis that was made. Furthermore, the steps in a scientific method are not always used in the same order. Experimentation often precedes further observation. Follow the steps in the diagram below, and see how many different ways you could go through the paths of a scientific method.

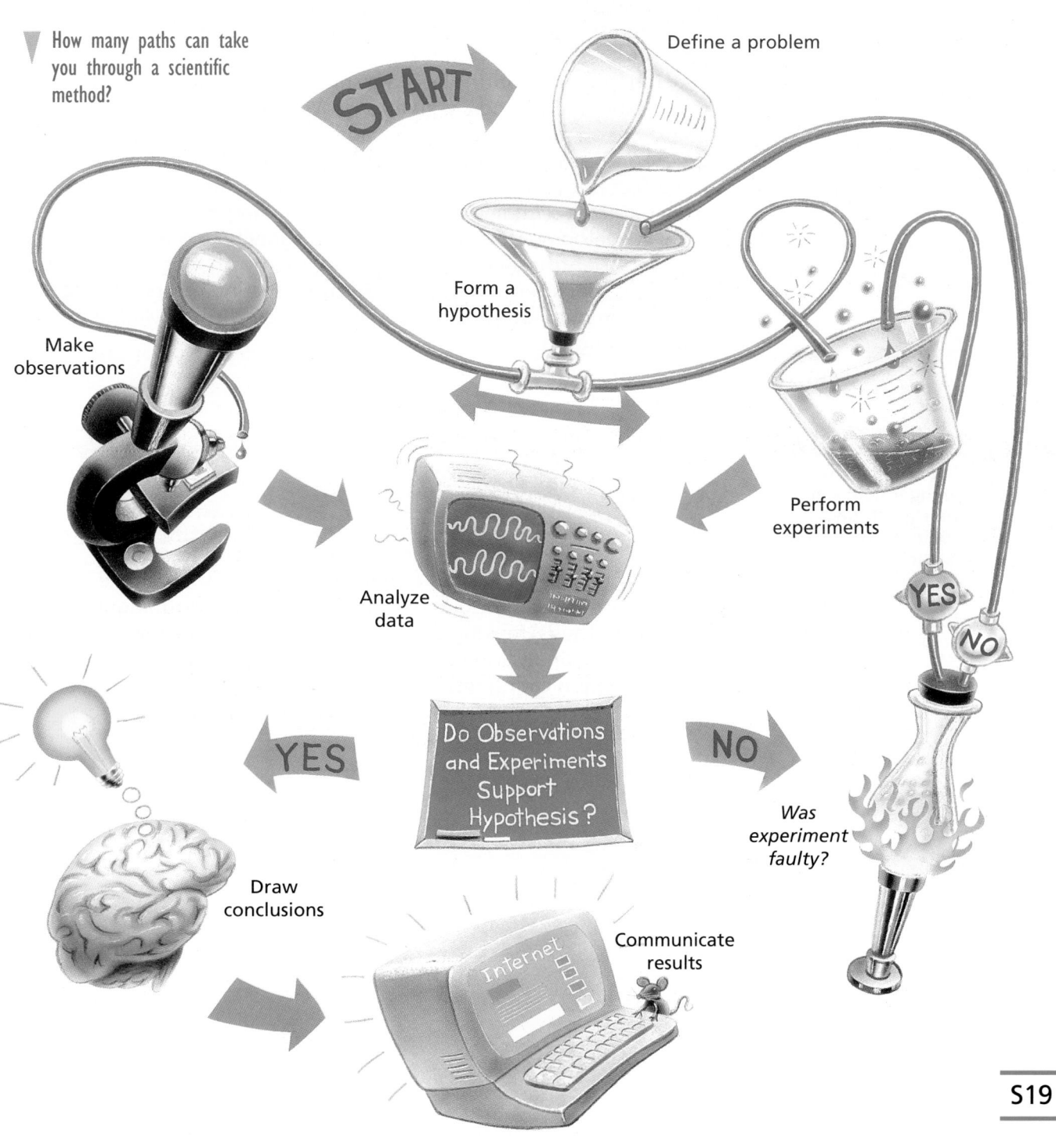

▼ How many paths can take you through a scientific method?

START

Define a problem

Form a hypothesis

Make observations

Analyze data

Perform experiments

YES NO

YES Do Observations and Experiments Support Hypothesis? NO

Draw conclusions

Was experiment faulty?

Communicate results

Internet

S19

Any one hypothesis may also be extended to include related areas of questioning. The hypothesis by Bekoff and Wells might be applied to similar animals like wolves, jackals, and hyenas. If these other animals also show the same behavior, the hypothesis may be accepted by scientists as evidence that supports a **scientific theory**.

A scientific theory is an explanation of why things work the way they do. Theories are used to *explain* the results of different kinds of experiments and also to *predict* the outcome of other experiments. In this way, scientific theories actually guide the process of science. Theories must always be tested. They sometimes turn out to be wrong. For example, the theory that the sun is the only source of energy for life on Earth was shown to be wrong. It had to be changed when communities around hot-water vents in the ocean were discovered.

Scientific theories are often confused with **scientific laws**. A scientific law describes *what* happens during a certain event, not *why* it happens. Scientific laws are summaries of the results of many, many experiments and observations. When results of sets of experiments are always the same, they no longer need to be tested each time. For example, Newton's law of gravity can be used to describe the effect of gravity on any object without testing that specific object. Galileo's studies of the motion of falling objects contributed to the development of Newton's laws of gravity and motion.

Even scientific laws, however, can undergo change. A scientist once said that science moves ahead by correcting errors it made earlier. The body of information we call scientific knowledge is always changing.

Scientific theory

A general statement of why things work based on hypotheses that have been tested many times

Scientific law

A statement of what happens in a certain event based on verified observations and experiments

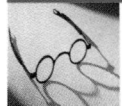
SUMMARY

All scientists answer questions with a sequence of problem-solving skills called a scientific method. Some of these skills include defining the problem, forming a hypothesis, making observations, performing experiments, analyzing data, drawing conclusions, and communicating results. The answers to questions may change as scientists design better experiments and make new observations.

Concept Mapping

The concept map shown here illustrates major ideas in this unit. Complete the map by supplying the missing terms. Then extend your map by answering the additional question below. Write your answers in your ScienceLog. **Do not write in this textbook.**

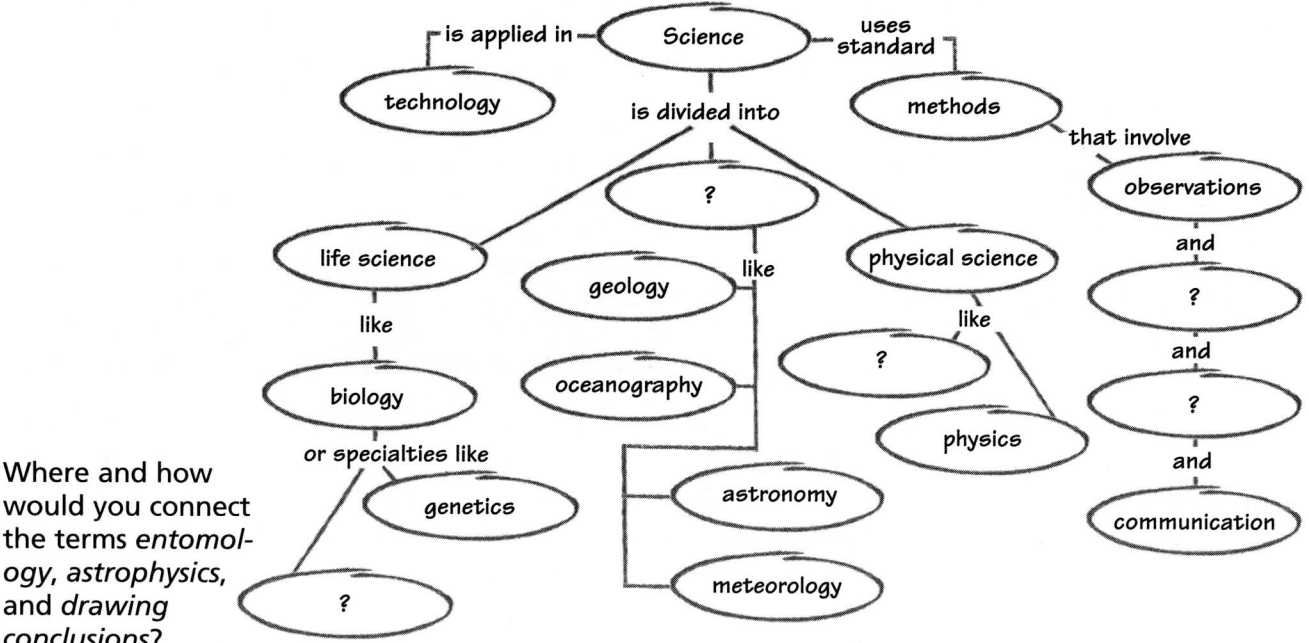

Where and how would you connect the terms *entomology, astrophysics,* and *drawing conclusions*?

Checking Your Understanding

Select the choice that most completely and correctly answers each of the following questions.

1. An Earth scientist would most likely investigate which of the following?
   a. an underwater volcano
   b. an African tree lizard
   c. an unbalanced nuclear force
   d. an insect invasion

2. A scientific method is a
   a. technique for finding answers.
   b. process for doing experiments.
   c. way of thinking about the natural world.
   d. reliable system for recording observations.

3. A *control* for any experiment is necessary because
   a. the scientific method requires one.
   b. it predicts what the results should be.
   c. it establishes a basis for comparison.
   d. scientists want to be in charge.

4. If an experiment supports a hypothesis,
   a. the hypothesis is valid.
   b. the hypothesis may be valid.
   c. the hypothesis is invalid.
   d. the hypothesis is incomplete.

5. When scientists fail to communicate their conclusions,
   a. they get all of the glory for themselves.
   b. their work cannot be considered scientific.
   c. other scientists cannot verify their results.
   d. other scientists must steal their ideas.

S21

Interpreting Graphs

Look at the graph to the right. What general relationship between temperature and rainfall could you infer from the information given? At what point in a scientific method would someone produce a graph such as this? Explain.

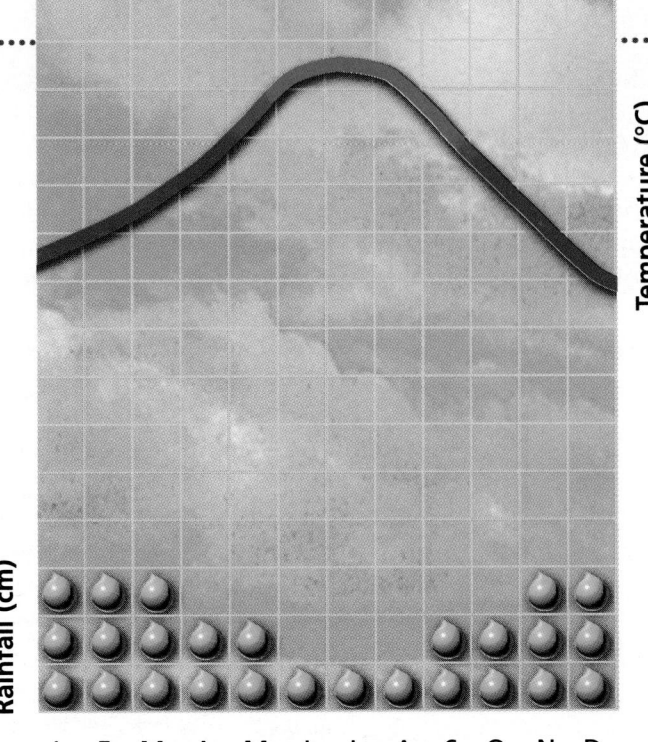

Critical Thinking

Carefully consider the following questions, and write a response in your ScienceLog that indicates your understanding of science.

1. Would the question, How can we use hydrogen to run our automobiles? be a question for science or for technology? Explain.

2. Some people argue that astronomy is *not* an Earth science. Why might they say this?

3. Why is it necessary for scientists in different fields, such as biology and physics, to communicate and work together?

4. Explain why scientists must have at least two samples when they run any experiment.

5. A scientific law might state that the sun will rise tomorrow morning. How might this law be affected if, for some cosmic reason, the Earth stopped rotating?

Portfolio Idea

Imagine that the principal of your school called you to the main office and asked you to explain the meaning of science. What would you say? Think about the science classes you have taken. Would you call what happened in them "science" or not? Make an outline to help you remember the main points you want to bring up. Remember that what you tell the principal may influence how science is taught in your school.

S22

PATTERNS OF LIVING THINGS

Unit 2

IN THIS UNIT

Now that you have learned about some of the characteristics of living things, consider these questions.

1. How are living things classified, and why?
2. Why do living things change over time?
3. What are the main parts of cells, and how do they function?
4. How are cells organized, and how do they function in the human body?

In this unit, you'll take a look at how scientists organize the study of living things and how living things change over time. You will also see how the fundamental units of life combine to form larger living things.

S23

SCIENTIFIC CLASSIFICATION

As a way of making sense of the world around them, people tend to place similar things into groups. For example, in a grocery store, all fresh fruits and vegetables are grouped together on the shelves of one department. These are then separated into smaller groups by the individual type of fruit or vegetable—apples on one shelf, carrots on another—making it easier to find what you want. Putting similar things into groups is called **classification**.

Scientists also group things together. They *classify* many things: rocks and minerals, stars and galaxies, chemical elements and compounds, as well as living organisms. Classification makes it easier for scientists to "keep track" of the natural world. It also helps them understand how different objects are related to each other. This is especially true for living things. Groups of organisms that look similar will often be more closely related than groups of organisms that look different. Take, for example, coyotes, wolves, and lions. Coyotes look more like wolves than lions. It is also true that coyotes are more closely related to wolves than to lions.

Organizing Organisms

The classification of living things is hardly a new idea. Over 2000 years ago, a Greek scientist and philosopher named Aristotle began grouping organisms. He divided living things into two large groups—plants and animals. Then he divided each of these into smaller groups. He classified animals by whether they were land, water, or air animals. His plant categories included herbs, shrubs, and trees.

Other scientists tried different systems of classification. At one time, animals were classified as either useful, harmful, or unnecessary.

Classification

The organization of things into groups according to ways in which they are alike

▼ How did Aristotle's classification scheme work? Does it seem like a useful system?

Living Things

Herbs
Land dwelling
Trees
Shrubs
Water dwelling
Air dwelling

Plants were once grouped by whether they produced fruits, vegetables, fibers, or wood. The system used today was first developed in the mid-1700s by a Swedish botanist named Carolus Linnaeus. Linnaeus classified organisms by comparing similarities of form and structure.

As Aristotle had done, Linnaeus used external similarities for comparison. However, Linnaeus went further. He also considered similarities of the internal organs and the way body systems seemed to perform their functions.

Linnaeus's system had two **kingdoms**: the plant kingdom and the animal kingdom. Kingdoms are the largest groups of organisms in his classification system. They contain the organisms that share only the few fundamental characteristics that distinguish plants from animals.

Linnaeus established other classification categories that he called *class*, *order*, *genus,* and **species**. He placed organisms with the greatest degree of similarity in the same species. Members of the same species are extremely similar in all major respects but may vary somewhat in size, color, and so on. People, for example, are all the same species, even though they may be different in appearance. Since some species are very similar to one another and differ only in a few characteristics, Linnaeus placed these species into a larger group—a genus.

While people come in a variety of shapes, sizes, and colors, they are all members of the same species.

Kingdom

The most general level of the classification system

Species

The most specific level of classification

Each of these organisms belongs to the same kingdom.

S25

Later, two additional categories were added to Linnaeus's classification system, making a total of seven. The classification levels are, in order from largest and most general to smallest and most specific, kingdom, phylum, class, order, family, genus, and species. Each level is less general than the one before it and more general than the one after it. Each classification level contains a collection of related organisms: a species contains similar individuals, a genus contains similar species, a family contains similar genera (plural of genus), and so on.

Modern Classification

An understanding that living organisms change over time and the discovery that traits are transmitted to offspring by means of cellular instructions called *genes* are two developments that have affected the way organisms are classified today. The study of fossil remains has also revealed lifeforms from the past that were different from those of today, including some that have no apparent living relatives. Modern classification systems now include both living and extinct forms of life. Organisms are now classified according to how closely they are related to one another.

These relationships are determined by careful studies of fossils, the biochemicals that make up organisms, their reproductive habits, and the structure of their DNA. The idea that living organisms change over time is a fundamental concept in biology. According to this idea, natural processes cause species to change in response to changing environments. Over long periods of time, some members of a species may change so much that they no longer belong to the same species as their ancestors. Since evidence indicates that the first organisms appeared more than 3 billion years ago, the process of change would have been repeated over and over again. As a result, millions of species now exist. And for every species alive today, it is estimated that a hundred or more once lived but are now extinct.

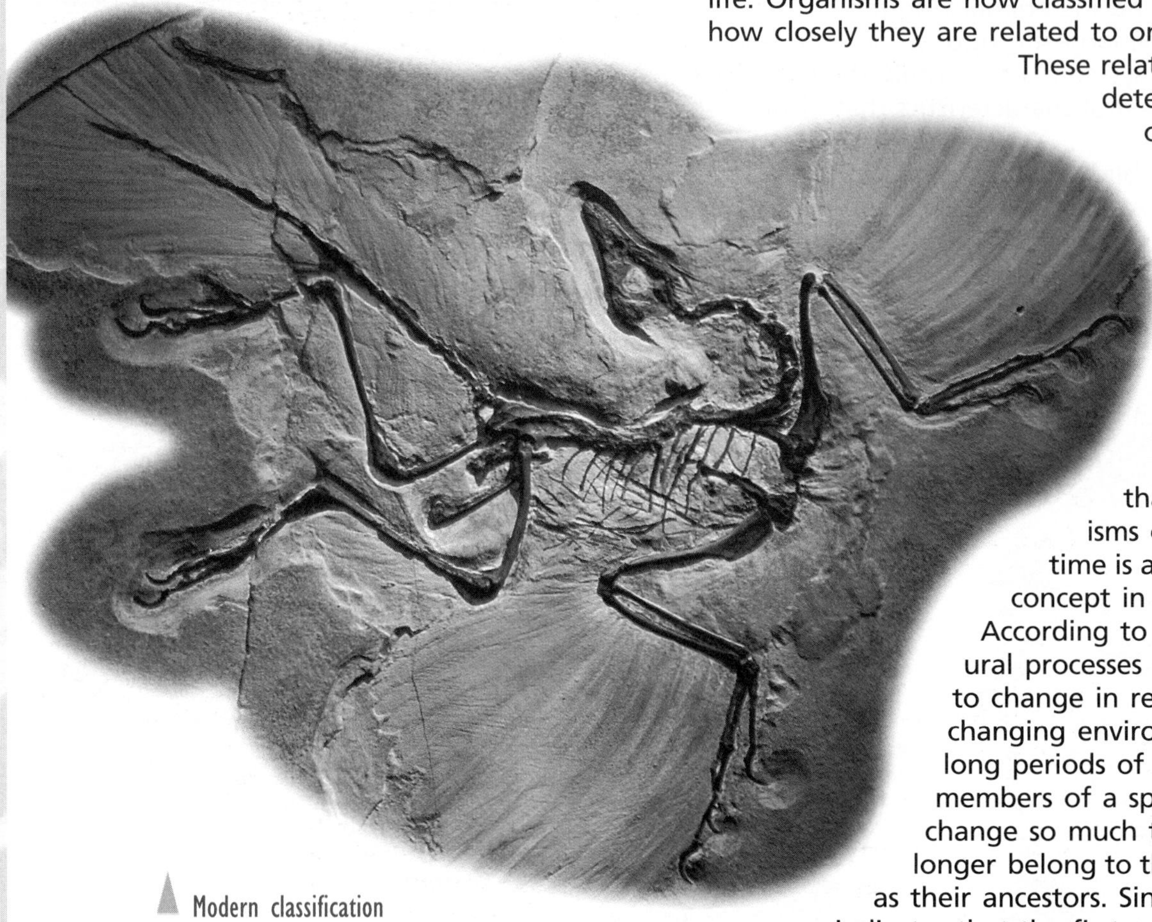

Scientists wondered how life-forms had developed the traits that made them well-adapted to their surroundings. This inquiry led to the development of *theories of evolution* to explain the process by which this change occurred. We will discuss these theories in greater depth in the next section.

▲ Modern classification systems include both living and extinct organisms such as this *Archaeopteryx*—found in a 140-million-year-old German limestone deposit. Look closely. What modern creatures do you think descended from this organism?

S26

Showing the Relationships

As you know, modern scientists look at more than just an organism's structure and how its organ systems function when they attempt to classify it. Taking into account information about the fossils, biochemicals, behavior, and DNA of organisms allows scientists to more accurately classify them and to better understand the evolutionary relationships among groups of organisms.

Scientists often use their knowledge of the evolutionary relationships among organisms to construct what is known as an "evolutionary tree." To understand how an evolutionary tree represents relationships among organisms, think of a real tree. A tree begins as a single shoot that branches again and again as it grows. All of the branches on the tree are "related" because they all originate from the central trunk. Imagine that the entire tree represents a kingdom; a main branch, then, is like a phylum. A major branch off a main branch is like a class. A prominent branch off this major branch is like an order, and so on. Starting at the outermost branches of the tree and working back to the trunk is like going back in time. Looking at an evolutionary tree, like the one shown here, lets you see that all species have a common ancestor somewhere in the past.

Modern scientists are continually adding to

our body of knowledge about organisms. This new information sometimes causes a change in how an organism is classified. For example, new living and fossil organisms are constantly being discovered. Sometimes classification categories must be rearranged to accommodate the new discoveries.

For years, Linnaeus's plant and animal kingdoms were the only ones recognized by scientists. However, new methods of studying organisms have led to the addition of three new kingdoms. In fact, a sixth kingdom is being considered by some biologists at this time.

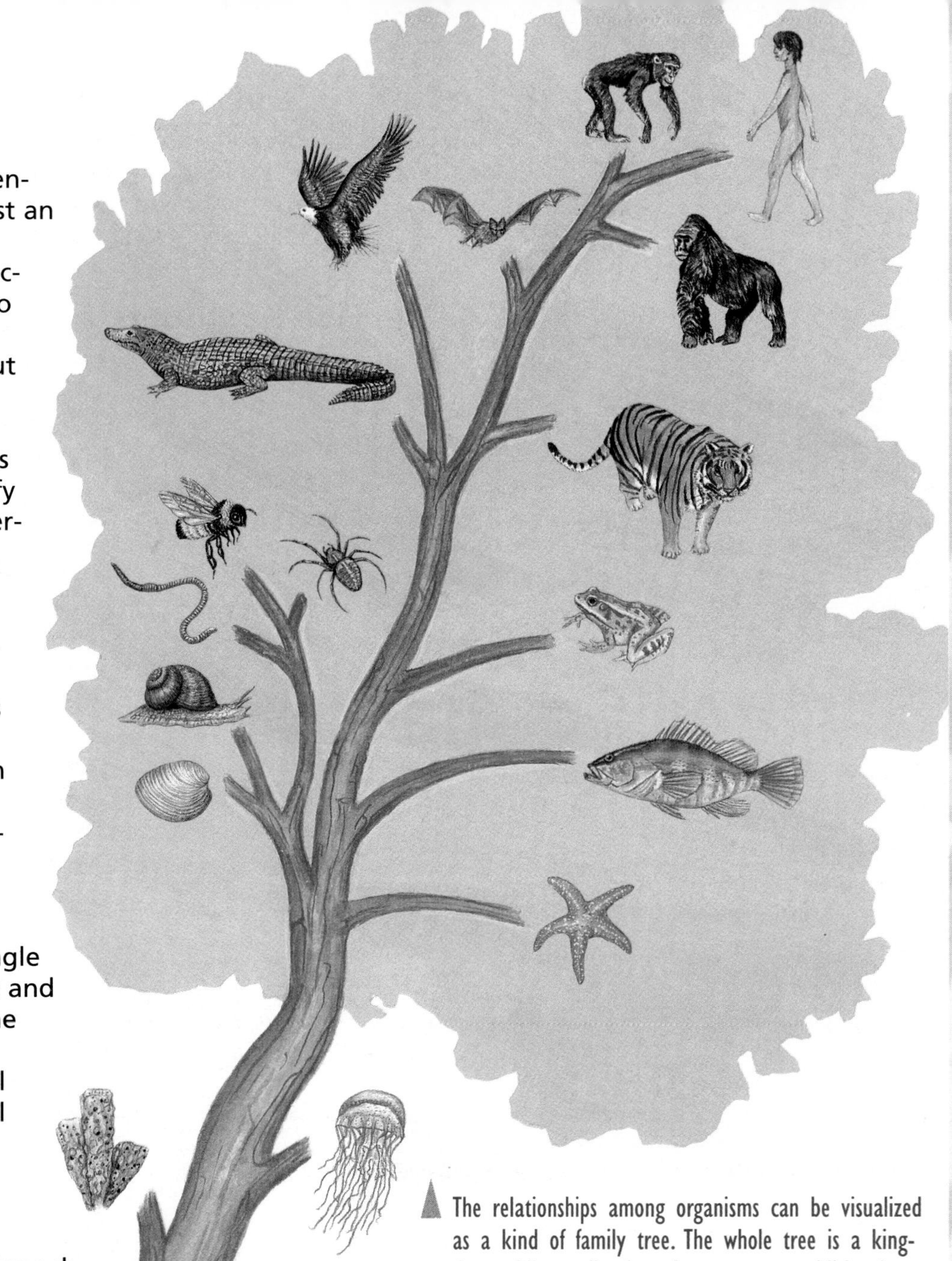

▲ The relationships among organisms can be visualized as a kind of family tree. The whole tree is a kingdom, while smaller branches represent additional classification levels. This tree shows the evolutionary relationships of the major animal groups.

S27

How Many Kingdoms Are There?

Most biologists now recognize five kingdoms. The organisms of each kingdom share certain fundamental characteristics and are distinctly different from the organisms of the other kingdoms.

The Five Kingdoms

Kingdom Plantae

All plants are multicellular. Their cells have nuclei, organelles, and cell walls. Their bodies are organized into organs. Most members make their own food by photosynthesis. Included in this kingdom are ferns, cone-bearing plants, and flowering plants.

Kingdom Fungi

Fungi can be either single-celled or multicellular organisms. Their cells also have nuclei and organelles. The bodies of multicellular forms develop specialized reproductive structures. None of the fungi can make their own food; they must *absorb* food from their surroundings. Included in this kingdom are yeasts, molds, and mushrooms.

Kingdom Animalia

All animals are multicellular organisms. Their cells have nuclei and organelles. All but the simplest forms show the organ system level of organization. All members obtain energy by *ingesting* and digesting food. Included in this kingdom are sponges, worms, insects, fishes, reptiles, birds, and mammals.

Kingdom Protista

Called protists, the members of this kingdom have complex cells with nuclei and organelles. All but a few are single-celled organisms. The multicellular forms show no organization above the tissue level. Some absorb food molecules or ingest small food particles, while others make food by photosynthesis. Included in this kingdom are algae and protozoans.

Kingdom Monera

Members of this kingdom, called *monerans*, are single-celled organisms with very simple cells. This kingdom includes eubacteria (true bacteria), blue-green bacteria, and archaebacteria. Monerans lack the structures found in more recently evolved kingdoms. They either make their own food through photosynthesis or chemosynthesis, or they absorb it from their surroundings. Some can live in environments where there is no oxygen.

Recent discoveries have shown that one group of monerans, the archaebacteria, are quite different at the cellular level, indicating the need for a separate, sixth kingdom. Archaebacteria are interesting because they live in extremely harsh environments that would quickly kill other living things, and they are probably very similar to the first living organisms.

S28

Naming Living Things

Linnaeus made another major contribution to science by establishing a new naming system for living organisms, one that is still used today. He gave each different type of plant and animal a two-part name called its *scientific name* or species name. He used the Latin language because at that time it was understood by educated people throughout Europe.

The scientific name of an organism always has two parts. The first part is the genus name, while the second part is a descriptive term. For example, the sugar maple is called *Acer saccharum*, the red maple is called *Acer rubrum*, and the bigtooth maple is called *Acer grandidentatum*. You have probably figured out that the word *Acer* means "maple" in Latin. Knowing that, you can guess that *saccharum* means "sugar," *rubrum* means "red," and *grandidentatum* means "big tooth." If you look carefully, you may see similarities between these Latin words and English words with similar meanings. These examples are somewhat unusual because the scientific name is not usually a Latin translation of the common name.

In this naming system, each species of living organism has its own two-part name. When scientists want to discuss a particular organism, they use its scientific name because it means the same thing to everyone, regardless of their native language or dialect. As a human, what is your scientific name? What does it mean?

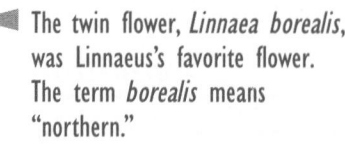
Many small seed-eating birds are commonly called sparrows. The house sparrow shown here, *Passer domesticus*, is more closely related to weaver-finches than to other sparrows. Sometimes called the English sparrow, it is actually found worldwide.

The twin flower, *Linnaea borealis*, was Linnaeus's favorite flower. The term *borealis* means "northern."

Two types of maple tree leaves are shown here—sugar maple (left) and bigtooth maple (right).

SUMMARY

The system scientists use today to classify and name living things was invented by Carolus Linnaeus in the mid-1700s. Once classified only by similarities and differences in form and structure, organisms are now classified by biochemical makeup and DNA content to show their evolutionary relationships. There are seven levels of classification categories for organisms. The most general is the kingdom and the most specific is the species. There are presently five kingdoms in the modern classification system, although some biologists now use six. Each kind of organism, or species, has a two-part scientific name.

LIVING THINGS IN A CHANGING WORLD

Here's a thought experiment for you. Suppose a population of rabbits lives in a certain area. Suddenly, a large river nearby changes course, cutting this population in two. The two groups can no longer have anything to do with each other. Now imagine being able to come back a million years later to observe the descendants of the original rabbits. Do you think that the two groups would still be exactly alike, or would they have changed? Why do you think this is so?

You probably concluded that the two groups of rabbits in the above thought experiment would have changed after a million years of being apart and that they would no longer be exactly alike. If so, you have recognized an important scientific principle. **Evolution**, the idea that things *evolve*, or change through time, is one of the most important concepts of science. It is obvious that most things change with time, and in a world where everything else changes continuously, why would living things be any different? How could anything possibly remain exactly the same over millions and millions of years? Yet how might such change take place?

Evolution

The process by which living organisms change over time

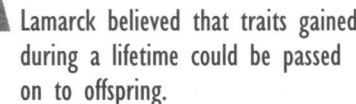
Lamarck believed that traits gained during a lifetime could be passed on to offspring.

Theories of Evolution   According to a theory proposed by the French naturalist Jean Baptiste de Lamarck in the late 1700s, organisms change over time by passing along to their offspring traits that they acquire during their own lifetime. Lamarck believed that organisms develop new traits to help them survive and lose others due to lack of use. These changes are then automatically passed along to their offspring. For example, a giraffe's neck, he said, grows longer from stretching to reach leaves high up on trees. The giraffe's offspring would inherit this trait for a longer neck. In Lamarck's theory, changes occur in organisms when they acquire

S30

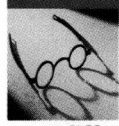 The tallest giraffes can reach vegetation that others cannot reach. This gives them a survival advantage, especially when vegetation becomes scarce. As a consequence, the tall giraffes survive and pass on their traits to offspring.

characteristics through the use or disuse of body structures and then pass on those characteristics to offspring. Because scientists' observations did not support Lamarck's theory, it lost support and was eventually rejected by the scientific community.

Another theory of evolution was developed a few decades later by the English naturalist Charles Darwin. Darwin's theory states that over many generations, the characteristics of organisms change as individuals with certain preexisting traits are "selected" by nature for survival. Darwin spent many years observing the variations in traits among organisms of the same species. He noticed that organisms with certain traits were better suited to surviving in their environment than others who did not have those particular traits. Individuals with the advantage of those traits would survive and pass those traits on to their offspring. For example, the giraffes that already had longer necks, and thus were able to reach higher for food, would be more likely to survive and produce offspring than their shorter-necked relatives. This selective process is repeated generation after generation as offspring with longer necks survive in slightly higher numbers than offspring with shorter necks. In Darwin's theory, such changes are the result of this *natural selection*. Darwin's theory of evolution by natural selection is the currently accepted model of how life evolves on this planet.

Natural selection occurs because of the natural variations that exist between individuals in every population. Every individual of a species is at least slightly different from every other individual. Any trait that confers even a very slight advantage will, after many generations of selection, spread throughout the population.

S31

The Genetic Connection

One early criticism of Darwin's theory—which Darwin himself acknowledged as a serious weakness—was the lack of an explanation of *how* favorable traits were handed down from generation to generation. It seemed logical that unless traits could be transferred, they would become "watered down" with the passage of each generation. The science of genetics provided the missing information needed to explain this apparent problem.

Units of inherited information called **genes** are responsible for an organism's traits. During reproduction, copies of the parents' genes are passed to offspring. Offspring inherit their parents' genes and, therefore, their parents' traits. However, for any of a number of reasons, genes sometimes *mutate*, or undergo changes. The occurrence of mutations helps to explain the appearance of new variations in organisms. Some mutations cause harmful changes to the organism, and natural selection quickly removes organisms with these mutations. Many other muta-tions are harmless and have no effect on the organism as a whole. Some of these, however, may plant the seeds of change by making the offspring slightly different from their parents. Over many generations, tiny changes can accumulate until an organism is quite different from its distant ancestors. At every step along the way, natural selection acts on the traits that exist, amplifying some traits and suppressing others.

Evidence of Evolution The evidence of evolution takes many forms, as shown in the accompanying figures. Perhaps the most convincing evidence comes from examining the **DNA** of different organisms, the complex molecule of which genes are made. Organisms that are closely related have similar DNA. Organisms that are distantly related have less similar DNA. The DNA of humans and chimpanzees, for example, differs by less than 2 percent. Obviously, we are more closely related to a chimp than to a horse, which is no surprise to anyone who has ever seen a live chimpanzee. It is impossible to ignore how much they look and act like humans.

Gene

A segment of a chromosome that carries hereditary information

DNA

Deoxyribonucleic acid; the cellular material that directs the cell and transmits hereditary information between generations

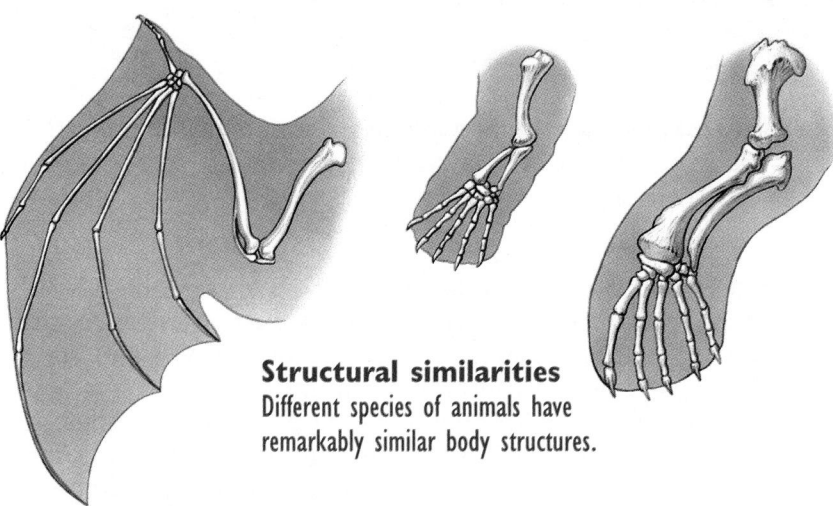

Structural similarities
Different species of animals have remarkably similar body structures.

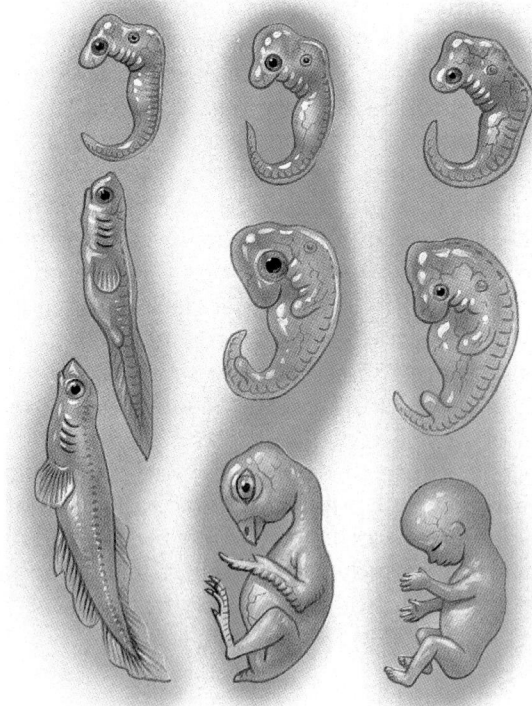

Developmental similarities
Through the early part of their development, many organisms that are not closely related are almost indistinguishable. Which two organisms are more closely related?

There are many popular misconceptions about evolution. One is that humans evolved from apes. In actuality, fossil evidence indicates that humans and apes both evolved from a now-extinct common ancestor that lived about 12 million years ago. Humans followed one evolutionary path, and apes followed another.

Why Species Diverge

What causes one kind of organism to give rise to a new kind of organism? Let's go back to our thought experiment from the beginning of this section. This hypothetical example illustrates a common cause of *speciation*, or the development of new species. As long as a population of related organisms is whole and its members can breed freely, traits that arise through natural selection continue to spread throughout the whole population. But when parts of the population become isolated from each other, they "drift apart" genetically because they are no longer exchanging genes by breeding. Significant change takes time, of course, but over millions or hundreds of millions of years much can happen.

Two forms of the peppered moth are shown. How might natural selection affect these animals?

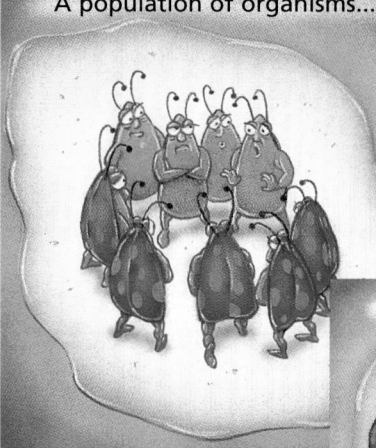

A population of organisms...

is split by some natural event.

When one part of ▶ a population of organisms becomes isolated from the other part, . . .

Over time, the isolated groups evolve apart.

. . . the two groups ▶ "drift apart" genetically over a period of time.

S33

DID YOU KNOW...

that life on Earth had a very close call? The mass extinction of 65 million years ago is thought to have killed 90 percent of all living things.

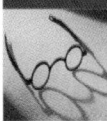

Catastrophic events can also shape the development of different species. Scientists think that about 65 million years ago, the Earth was struck by a large asteroid near what is now the Yucatán peninsula of Mexico. Evidence suggests that the impact and its aftermath caused massive damage to the biosphere, killing off enormous numbers of living things—including the dinosaurs. However, because of certain traits they had, one group of animals—the mammals—survived and thrived in the damaged world. Before the impact, mammals were a minor group, mostly outcompeted by the larger, more powerful dinosaurs. Afterward, mammals spread out into the territory formerly occupied by their now-extinct competitors. Mammals evolved rapidly and within a few million years came to be the dominant large life-form on the planet.

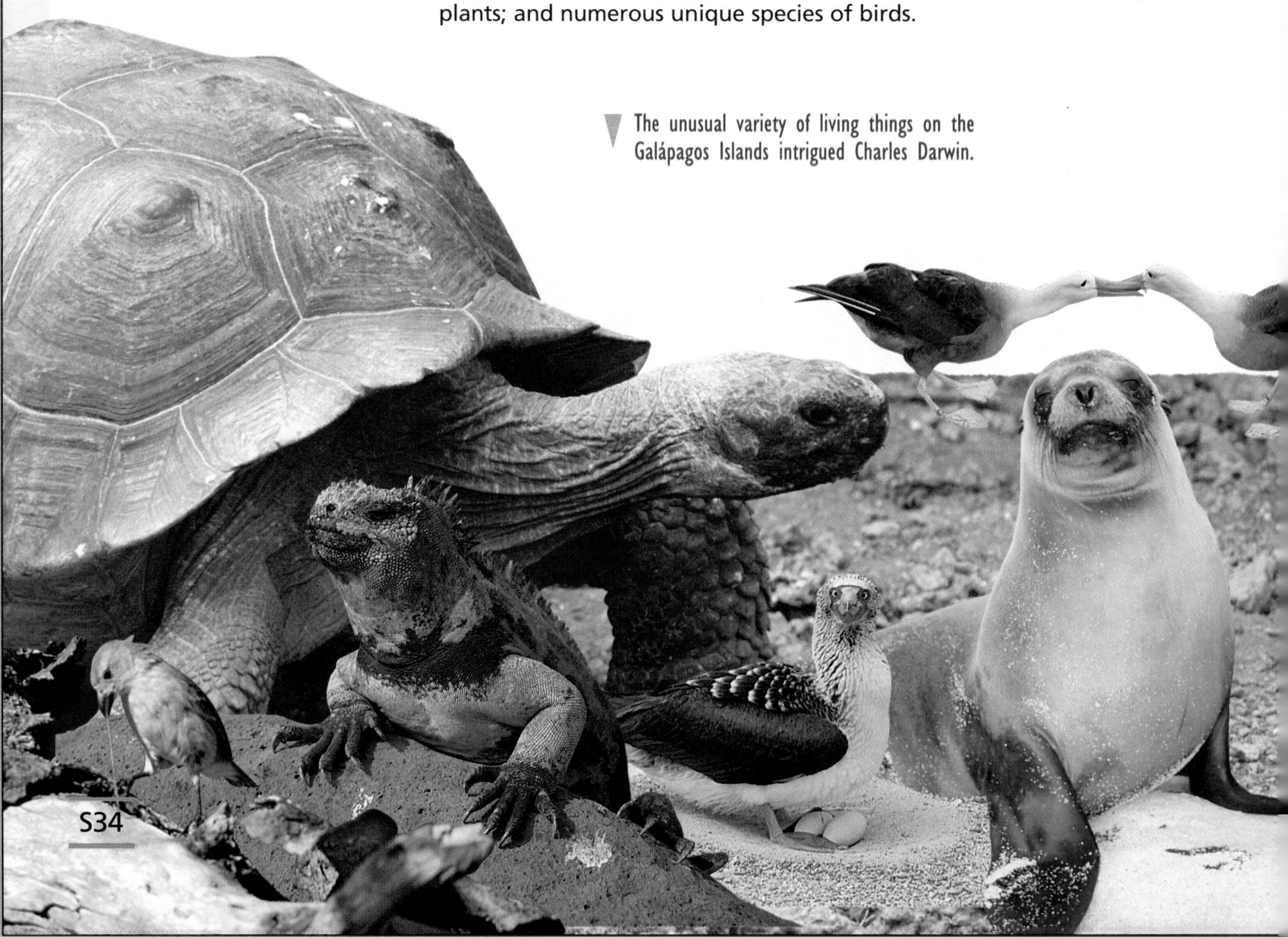

Adaptation

Any characteristic that helps an organism survive and reproduce in its environment

Adaptation and Evolution You have learned how adaptation allows living things to improve their odds of surviving. **Adaptation** results from evolution. Let's look at some examples that show how evolution caused organisms to develop unique adaptations.

The Galápagos Islands, a young volcanic chain about 1000 km off the west coast of South America, are known for their unusual plants and wildlife. Charles Darwin began to formulate his famous theory of evolution after a visit there in the early 1830s. Darwin saw living things that were different from any other on Earth yet clearly similar to species found on the mainland. There were swimming, seaweed-eating iguanas; huge land tortoises; bizarre-looking plants; and numerous unique species of birds.

The unusual variety of living things on the Galápagos Islands intrigued Charles Darwin.

S34

Darwin noted that the islands had a number of different species of finches (a kind of small bird). In spite of the close relatedness among the species, there was little competition for food among them. It turned out that each type of finch favored a certain way of life because of the size and shape of its bill, which determined what foods it could eat. Some finches had large, seed-crunching bills. Others had small, insect-gobbling bills.

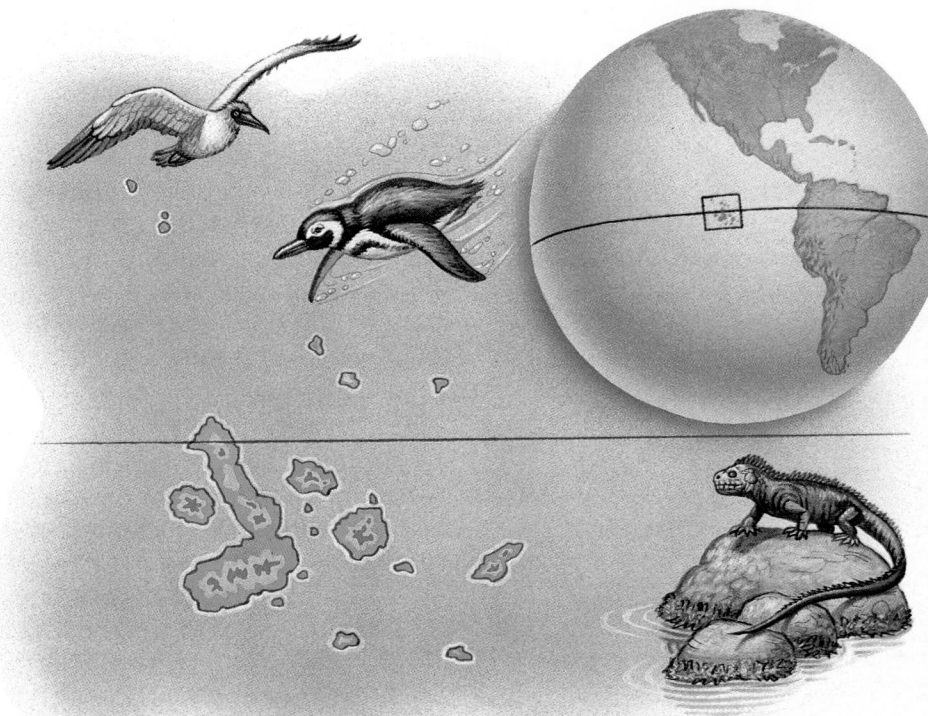

▲ The Galápagos Islands are an isolated chain of volcanic islands in the Pacific Ocean.

Darwin reasoned that the Galápagos Islands had been settled some time in the past by an original flock of finches from the mainland—probably blown there by high winds. Because finches normally do not fly very far over water, once on an island they became effectively isolated from finches on other islands. In their isolation, the finches became adapted through natural selection to the local environment. After many years, the finches changed enough to be considered new species. Over time, some 14 different species developed.

Isolated places like the Galápagos Islands often have unusual life-forms found nowhere else. The isolation of these places keeps them from being easily "colonized" by living things. Eventually, by chance, a few creatures and plants are washed or blown ashore from the mainland. Natural selection shapes the descendants of these original "colonists," causing them to evolve, over millennia, unique adaptations to accommodate the unique local environment.

S U M M A R Y

Evolution by natural selection is the prevailing explanation of how evolution occurs on Earth. The theory of natural selection holds that organisms well suited to the environment survive in greater numbers than those less well suited. The characteristics that allow them to survive are then passed on to offspring. Characteristics are passed from parents to offspring by genes. All living species are descended from earlier species. Related species are descended from a common ancestor species. Natural events can cause existing species to give rise to new species.

CELLS: THE STUFF OF LIFE

Would you believe that there are living things smaller than you can see with your own eyes? Of course you would; with the aid of a microscope you have probably even seen some of them. We take the existence of microscopic organisms for granted, but before the late 1600s people had never imagined that there were living things smaller than the unaided eye could see.

The Cell Theory

Very little was known about the structure of living matter until the development of the light microscope. Then Robert Hooke, an English scientist, made an important discovery in 1665 while using a simple microscope that he designed. He observed tiny, orderly spaces in a thin slice of cork, a type of dead plant material. These spaces reminded him of the small rooms in which monks lived. So he gave the tiny spaces the same name as those small rooms, *cells*. Microscopes were soon used to study all kinds of plant and animal material. Careful examination revealed that all living or once-living materials were composed of similar small units. These units were named **cells**, after the term Robert Hooke first used to describe the spaces in cork.

Studies using microscopes also revealed a world of previously unseen organisms. Some of these tiny creatures were observed to reproduce themselves. This and other observations caused many scientists to doubt the long-held belief that some living organisms came from nonliving substances. By this process, called *spontaneous generation*, maggots were thought to come from rotting meat and mice from old rags and wheat. Although it took another 200 years, scientific experiments and microscopic observations eventually showed that all living things come from other living things.

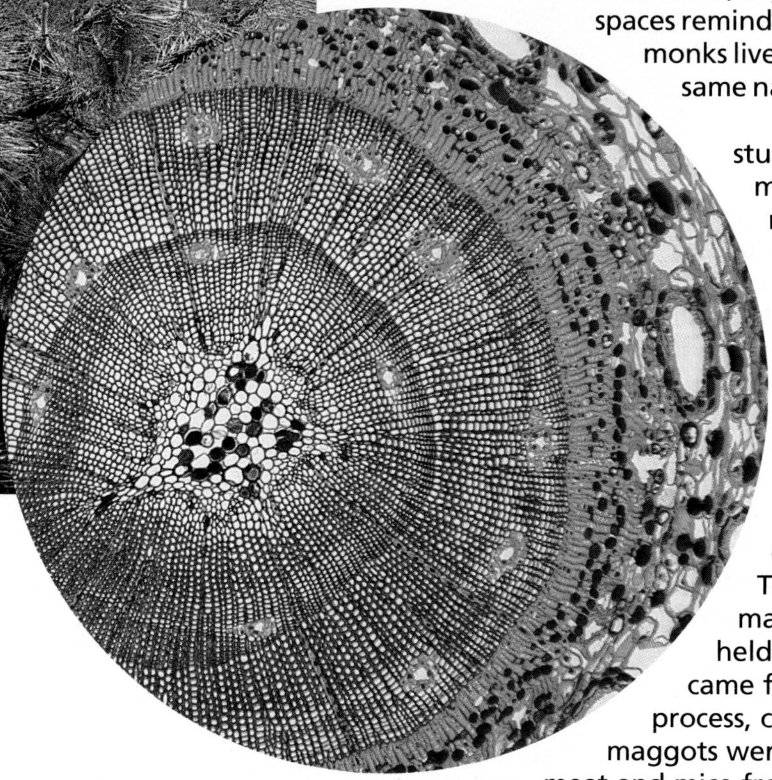

By the late 1830s, a formal theory about the structure and function of all life had been developed. This theory, called the *cell theory*, may be stated as shown to the right.

▲ Robert Hooke examined plant cells such as these.

Cells

The basic units of all living organisms

The Cell Theory

1. All living things are made up of cells.
2. The cell is the basic unit of all living things.
3. Only living cells can produce new living cells.

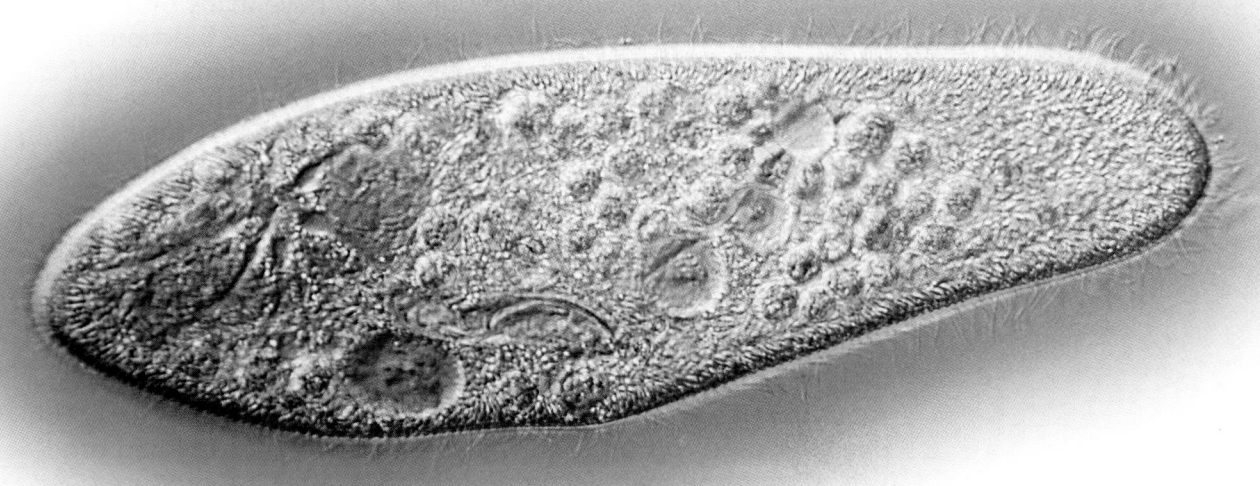

▲ Unlike the empty cells examined by Hooke, this *paramecium* has several different structures. Can you locate its cell membrane, cytoplasm, and nucleus?

What Are Cells Made Of?

Because they were taken from dead plants, Robert Hooke's cork cells were empty. Most of the once-living material had decayed, leaving only the rigid and durable outer parts of the cells. Yet the same early investigators who studied different types of living material with their microscopes found that living cells are not empty. They contain a thick, jelly-like fluid.

Scientists later discovered that this fluid is mostly water, with molecules of many other substances dissolved or suspended in it. The gases oxygen and carbon dioxide are present, along with a variety of minerals and organic (carbon-based) substances. Most of the organic substances are of the following types: carbohydrates (such as sugars and starches), which are sources of energy; fats, which store energy; proteins, which serve as building materials; and DNA, which determines how organisms grow and develop. These substances are produced only by living things.

Cells are the building blocks of living organisms and carry on all of the activities necessary for life. All of the materials necessary to maintain and reproduce a cell are scattered throughout the cell's jelly-like substance, which is surrounded by a covering.

Cells come in many shapes and sizes. Some cells are comparatively simple. Others are quite complex. Almost all complex cells, such as those that make up animals and plants, have the following basic parts:

1. a covering called the **cell membrane**. This membrane completely encases the cell and regulates what materials enter or leave it.
2. a jelly-like substance called **cytoplasm**. Many of the activities of a cell are carried out in the cytoplasm.
3. a structure called the **nucleus**. The nucleus is the "control center" of the cell. All cellular activities, including reproduction, are directed from the nucleus. The nucleus is surrounded by a membrane called the *nuclear membrane*.

Cell membrane

The part of the cell that determines what enters and leaves the cell

Cytoplasm

The jelly-like substance surrounding the nucleus of the cell, where the cell's activities are carried on

Nucleus

The cell's "control center," which directs all of the cell's activities

S37

Plant and animal cells are quite similar and have many of the same structures. These common structures have the same functions in both plant and animal cells. However, plant and animal cells differ in a few important ways.

Animal Cells

While all cells have certain fundamental similarities, let's use animal cells as a basis for comparison. It may help you to understand the function of each part of an animal cell if we compare the cell to the workings of a factory. Fuel and raw materials delivered to a factory enter through gates in its security fence. The factory workers follow a set of directions from the main office as they do their jobs. Fuel is burned to operate generators that provide electrical energy. This energy is used to transform the raw materials into finished products. During the manufacturing process, wastes are produced and must be removed. Some wastes are given off into the air, while others are stored in containers and then removed through the exit gate. Finished products are packed and stored until they can be shipped out of the factory.

The "finished products" of cells are the compounds they use to maintain themselves. During cell division, these products are also used to form the parts of new cells.

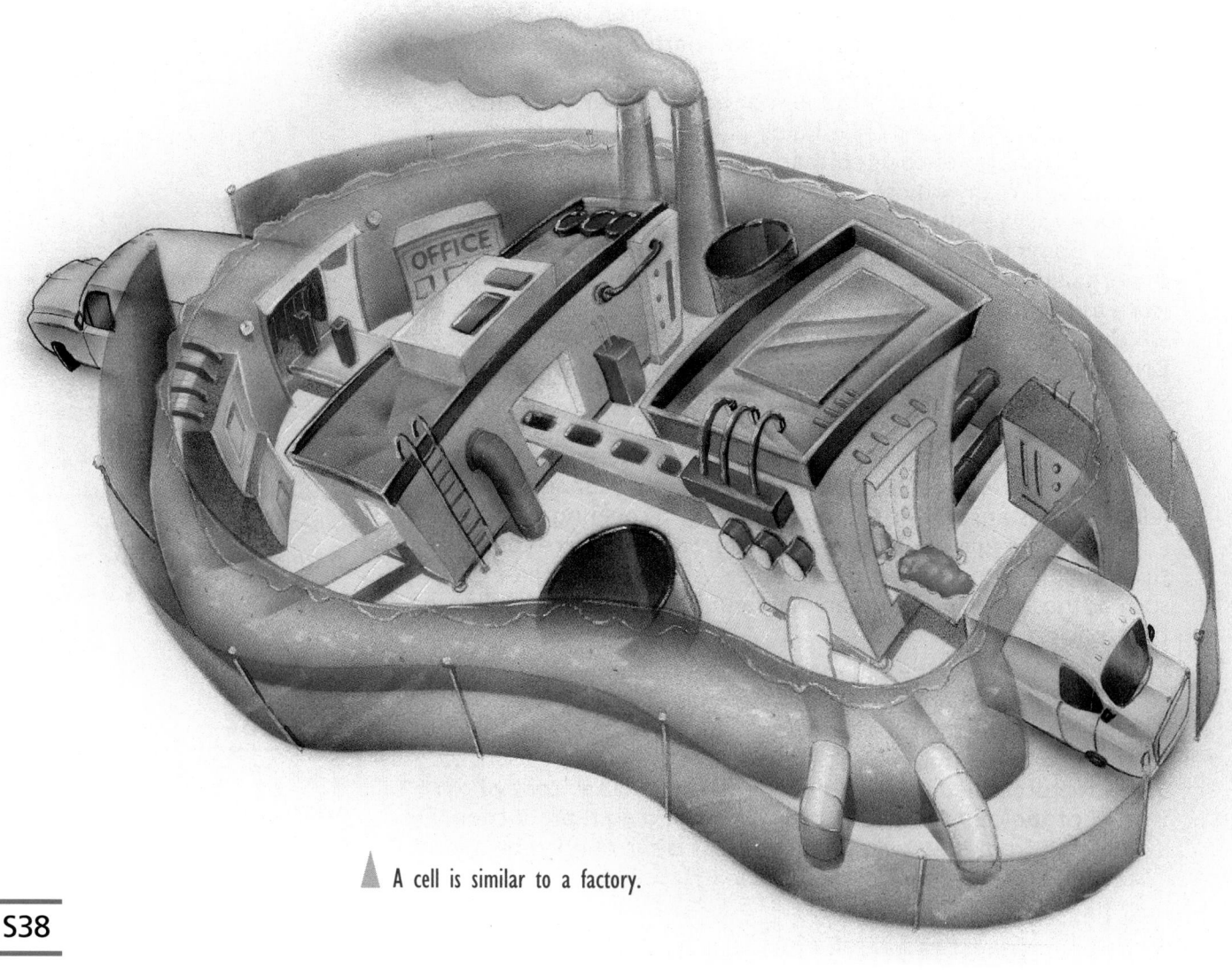

▲ A cell is similar to a factory.

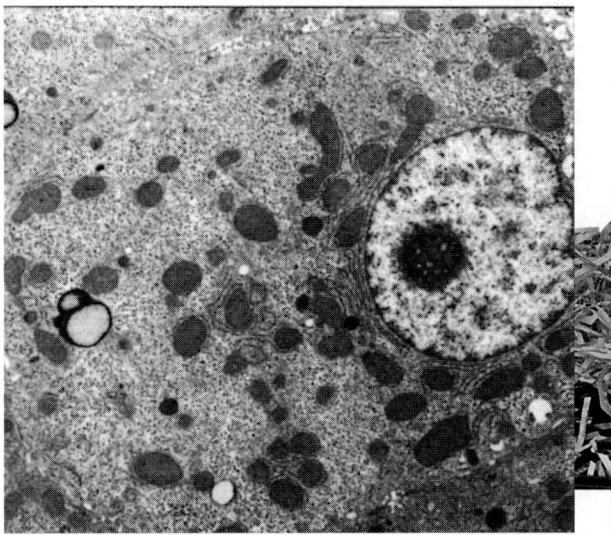

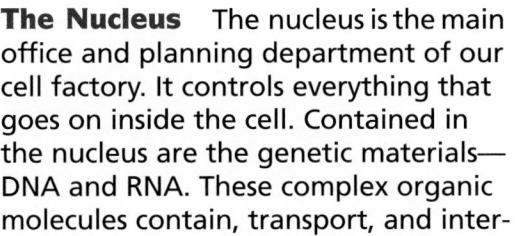

▲ Compare the electron micrograph (magnified 5000 times) with the drawing of a typical animal cell.

The Nucleus The nucleus is the main office and planning department of our cell factory. It controls everything that goes on inside the cell. Contained in the nucleus are the genetic materials— DNA and RNA. These complex organic molecules contain, transport, and interpret the instructions for the reproduction, growth, and development of all living organisms.

Most light microscopes are not powerful enough to show many details inside the nucleus. You may see a dark spot called the *nucleolus* inside the nucleus. The nucleolus is rich in RNA, which functions in the production of proteins.

Throughout the nucleus are structures called **chromosomes**. The chromosomes are made of DNA and contain the instructions for the manufacture of the finished products of the cell. The chromosomes can be compared to the blueprints in the factory's main office. Some blueprints contain the plans for the factory. Another identical factory could be built from them. Blueprints also show how the finished products of the factory are produced and what they will look like.

In a factory, the original blueprints never leave the office. Instead, copies are made and sent from the office to the work area. In a cell, RNA molecules take the instructions found in the DNA and carry them out of the nucleus to be used in making the various products of the cell.

The Cytoplasm The cell's manufacturing processes are carried out in the cytoplasm. In our factory model of a cell, the cytoplasm is the "shop floor." There are many specialized structures located in the cytoplasm. They perform the processes needed to carry out the instructions sent from the nucleus. These tiny structures are called **organelles**, which means "little organs."

Chromosomes

Rod-shaped structures found in the nucleus of the cell that are made of coiled DNA

Organelles

Tiny specialized structures within a cell that perform cell functions

S39

Mitochondria

Structures in the cytoplasm that release energy from food

Ribosomes

Structures in the cytoplasm where proteins are made

Vacuole

A storage site for food, wastes, and other chemicals

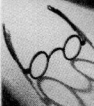

The "power plant" organelles of the cell are called **mitochondria**. Mitochondria provide energy for cell activities through *cellular respiration*, a process that can be compared with burning fuel. Molecules of sugar are among the fuels that are "burned" in the mitochondria. After sugar has been broken down in the mitochondria, usable energy is released to the cell.

Some of this energy provided by the mitochondria is used by tiny, spherical structures called **ribosomes**. Thousands of ribosomes are scattered throughout the cell, attached to a long, winding network that extends throughout the cytoplasm. Ribosomes can be compared to the machines in a factory that manufacture the finished products. Ribosomes are where proteins are made according to instructions carried by RNA. These proteins are some of the most important products that are made from the raw materials brought into the cell.

The organelles that serve as storage areas are called **vacuoles**. Some of these vacuoles store food for future use. Some store chemicals. Others store wastes until they can be removed from the cell.

The living cell is an incredibly complicated, highly dynamic thing. The cytoplasm churns constantly, transporting chemical messages and food to the various parts of the cell and removing waste. New materials are brought in and wastes are excreted as the organelles work nonstop, both manufacturing and consuming the chemicals of life.

DID YOU KNOW...

that scientists think mitochondria and other organelles may have originated when primitive bacteria invaded and became incorporated into larger bacteria? This is because mitochondria contain DNA that is totally separate from the cell's normal set of DNA. Mitochondria also strongly resemble certain primitive bacteria.

Ribosomes
Vacuole
Mitochondrion
Cell membrane
Small molecules, such as water, flow freely through tiny openings in the cell's surface. Larger molecules, such as sugar, must be transported by special structures.
Water
Sugar
Oxygen
S40

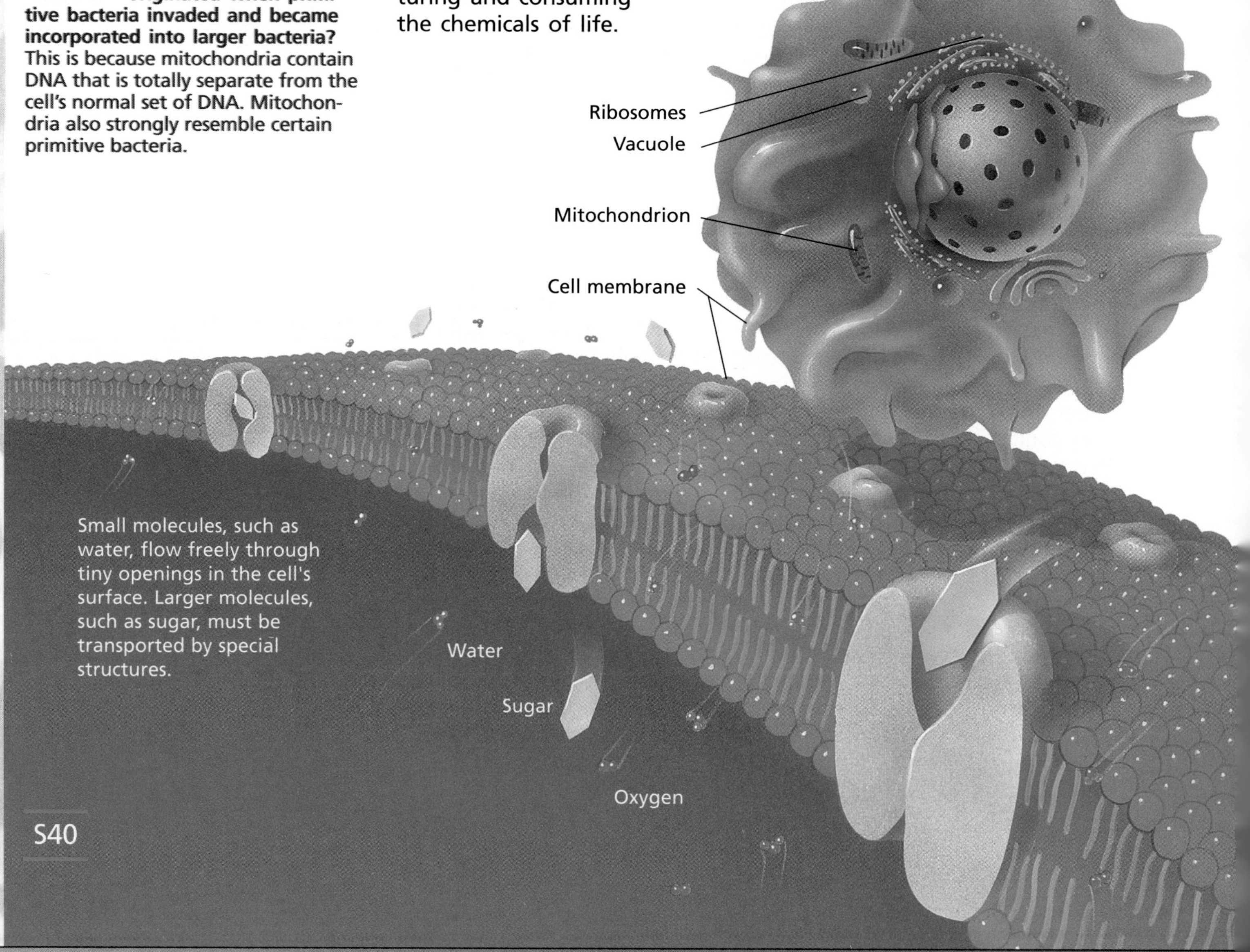

The Cell Membrane   The cell membrane that surrounds the cytoplasm acts like the fence that surrounds a factory. It allows only certain substances to enter. A cell membrane has tiny openings in it that allow small molecules, like water and oxygen, to pass freely through it. Larger molecules, such as sugar molecules, are shuttled into and out of a cell through certain "pathways" in the cell membrane. These pathways can be compared to the gates in a factory fence through which workers and materials enter and leave. Trucks bring in raw materials and take out finished products and wastes through these gates. The cell must expend energy to use this form of transport.

Plant Cells

Plant cells contain almost all of the structures that are found in animal cells. Even so, there are several key differences between plant and animal cells. One of the most obvious differences is the size of the vacuoles. Vacuoles in plant cells are very large. Like the vacuoles in animal cells, the plant-cell vacuoles also act as storage areas. Some of them are filled with a clear fluid that is mostly water but also contains sugar, starch, and protein molecules.

Plant cells also have a thick, firm outer boundary called the **cell wall**. The cell wall supports and protects the plant cell. These thick walls were the outlines of the "cells" that Robert Hooke saw in his thin slice of cork. The cell wall surrounds a plant's cell membrane.

The thick, rigid cell walls remain long after the plant cell dies. In trees, many of these cell walls are arranged in layers. Eventually, these layers form the rigid material we call wood.

Cell wall

The rigid, protective structure that surrounds the plant cell

Compare the photo and drawing of this plant cell with the animal cell on page S39.

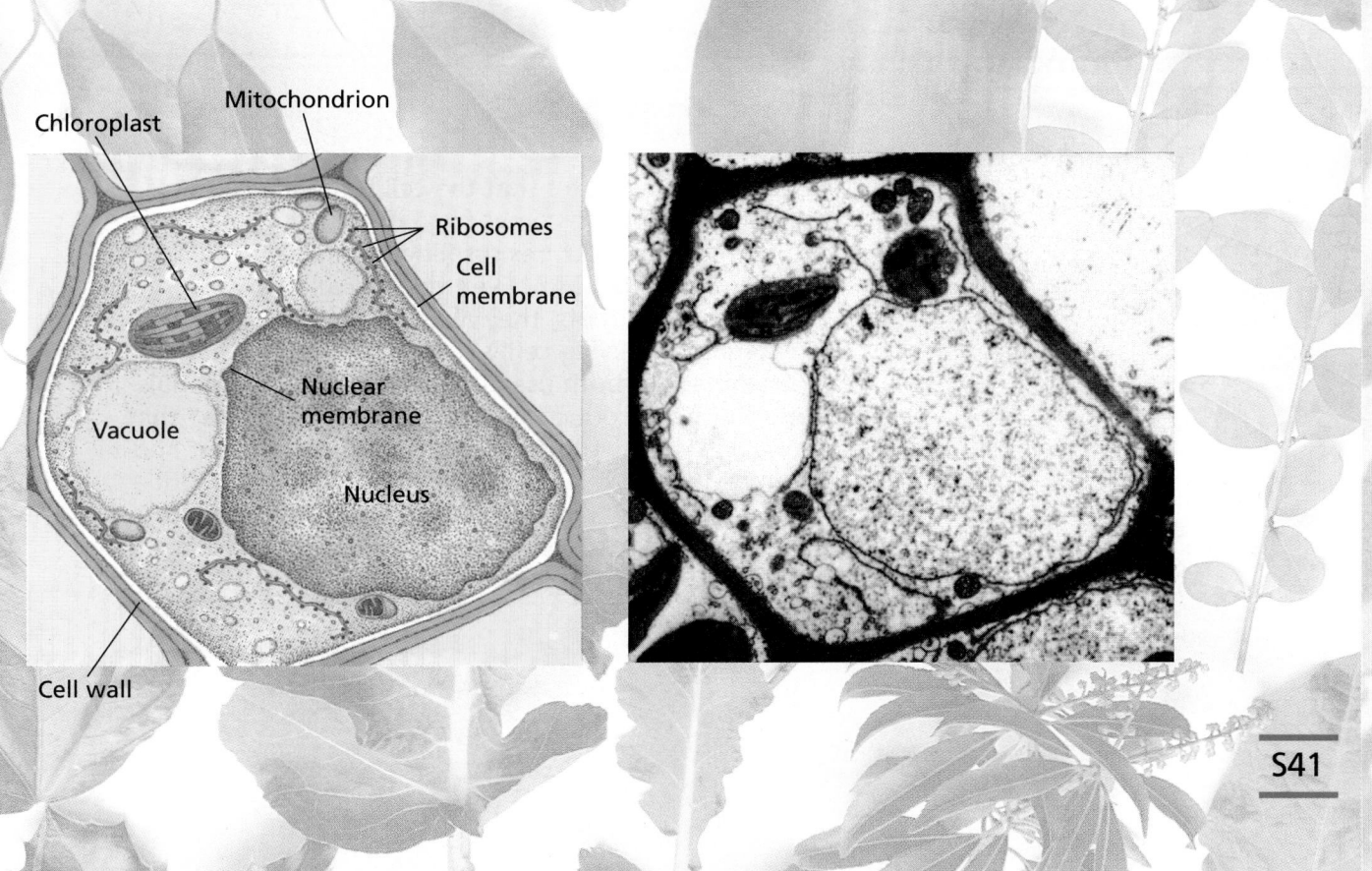

Chloroplast
Mitochondrion
Ribosomes
Cell membrane
Nuclear membrane
Nucleus
Vacuole
Cell wall

The ability to make food is the major functional difference between plant and animal cells. In the cytoplasm of some plant cells, usually those that make up the leaves, there are many small, green organelles called **chloroplasts**.

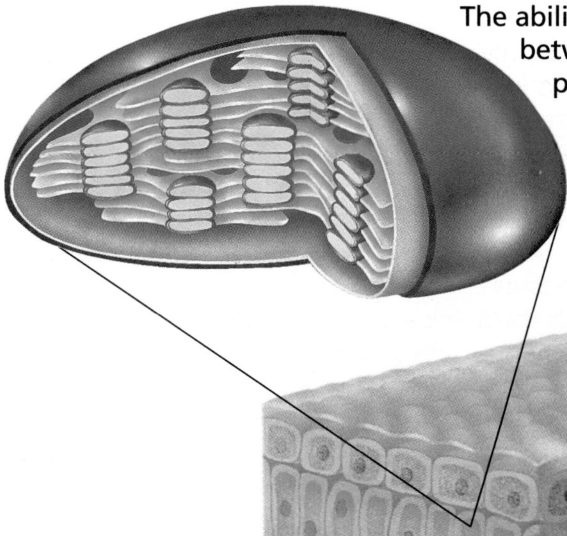

◄ Chloroplasts, found in the leaves of green plants, are the location of photosynthesis.

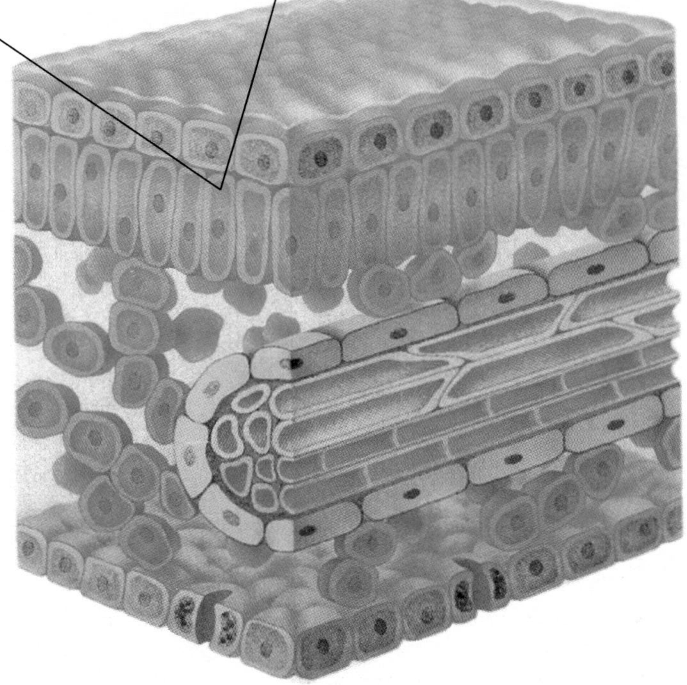

These chloroplasts contain molecules of **chlorophyll**. Chlorophyll enables a plant cell to make its own food by a process called *photosynthesis*. *Photo* means "light" and *synthesis* means "to put together." Photosynthesis uses energy from sunlight to put together water and carbon dioxide to make sugar. Oxygen is also a product of photosynthesis. Animal cells do not have chlorophyll. Therefore, they cannot make their own food.

Chloroplasts

Oval-shaped organelles in plant cells that contain chlorophyll

Chlorophyll

Organic molecules in green plants that absorb sunlight for making food

S U M M A R Y

The cell theory states that the cell is the basic unit of life and that all life comes from other living cells. Cells contain many structures suspended in a jelly-like material called cytoplasm. Even though cells vary in size, shape, and contents, they all have certain basic parts in common. A living cell is similar to a factory. The "machines" of both use energy to turn raw materials into finished products. Both have a "main office" that controls what happens in the "shop" areas. Plant and animal cells have some of the same basic parts. But plant cells have some structures that are not found in animal cells, and they can produce their own food.

S42

LEVELS OF ORGANIZATION

In the previous section we compared the workings of a cell to the workings of a factory. But let's think about the purpose of a factory. What do factories do? Factories make products. Different kinds of factories produce different products. In other words, they specialize. For example, car factories produce cars, shoe factories produce shoes, and still other factories process and package food. Each uses special machinery to produce a certain product. How many different kinds of factories can you think of?

Let's look at some other familiar examples of specialization. Carpenters build houses and offices from wood. Teachers help people learn. Grocers sell food products. Police officers enforce our laws. Doctors treat sicknesses. What is the value of all this specialization? For one thing, it makes a large and complex society like ours possible.

Humans once lived in very small groups in which individuals participated in many different activities necessary for survival—gathering food, finding or making shelter, making clothing, treating wounds, and protecting the group from attacks. But over a period of time, hunters learned that they could kill enough game for many individuals, and other individuals could make more than enough clothing for themselves. Specializing was found to be more efficient because each person did what he or she was best at doing. The hunter could share food with the clothes maker, who would in turn share clothing with the hunter. This specialization allowed groups to become larger, which in turn led to more forms of specialization. Today we live in a highly specialized world in which millions of people are organized into cooperative and functioning groups. Each person plays a role that contributes to the group as a whole.

▼ A complex society is made up of many cooperating specialists. Complex organisms are the same way.

Single-celled organisms are similar to early human groups. Each single cell performs many different functions. But in *multicellular* organisms (organisms made up of many cells), the cells are specialized. Each cell has a special function that benefits the entire organism. In this way, the cells of a multicellular organism depend on one another for survival.

As living things evolved into multicellular organisms, cells began to specialize. The specialized cells of multicellular organisms are commonly grouped together into larger organized structures that carry out life processes more efficiently. Biologists have identified five levels of organization in multicellular organisms.

Cells

The cell itself is the first and simplest level of organization. Most specialized cells have a certain size and shape related to their purpose. For example, humans have two basic types of blood cells and three types of muscle cells. There are also *epithelial* (covering) cells, nerve cells, bone cells, fat cells, and more. Each cell has a special shape that allows it to perform its task.

In plants, there are specialized cells for absorbing water from the soil. Plants also have specialized cells for covering and protecting themselves, transporting water and nutrients, and growing.

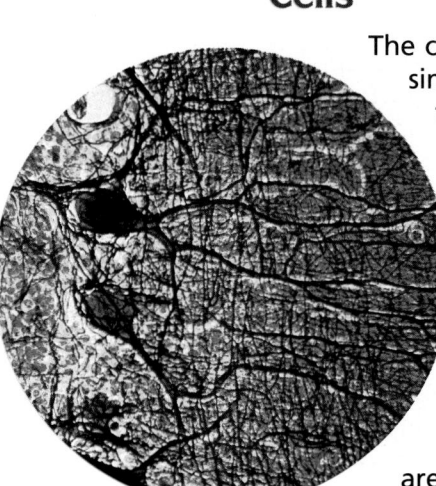

▲ The shapes of cells vary with their function.

Tissues

Specialized cells of the same type are often grouped together to perform their functions. These cell groups are called **tissues**. For example, some groups of cells form tissues that are capable of contracting and therefore causing movement. Other tissues defend the body against infection. Another group of cells forms structures that support the organism. Tissues are the second level of organization.

There are four main types of animal tissue—epithelial, muscle, nervous, and connective. *Epithelial* tissue covers and protects internal and external surfaces of the organism. Muscle tissue allows movement. Nervous tissue helps organisms respond to stimuli in the environment and also signals the muscles to work. One type of connective tissue, blood, helps distribute oxygen and food. Other forms of connective tissue include fat, bone, and cartilage.

Tissue

A group of similar cells with similar functions

S44

Plant cells also form specialized tissues. There is an outer covering of *epidermal* tissue that protects the parts of a plant. Epidermal tissue also controls the movement of oxygen, carbon dioxide, and water vapor into and out of the leaves. Tissues under the epidermal layer perform photosynthesis and store energy. Plants also have tissues that transport water and nutrients and tissues that are responsible for growth.

Organs

Organs represent the third level of organization. An organ is made up of different tissues that work together to perform a special task. The heart is an organ that is made up of muscle, nerve, epithelial, and connective tissues. These tissues work together to pump blood throughout the body. The lungs, which exchange oxygen for carbon dioxide, are organs that contain muscle, epithelial, and connective tissues. Another organ, the stomach, carries out its function of digestion with the aid of strong muscle tissue and epithelial tissue. The brain is composed of nervous, epithelial, and connective tissues. The skin, the largest organ, covers almost the entire body. It also is composed of all four types of tissue. Its major function is protection.

Some organs have more than one function. The male and female reproductive organs—the testes and ovaries—are examples of organs with more than one function. They produce reproductive cells but also produce chemical messengers called *hormones*.

Organ

A group of different tissues working together to perform a specific function

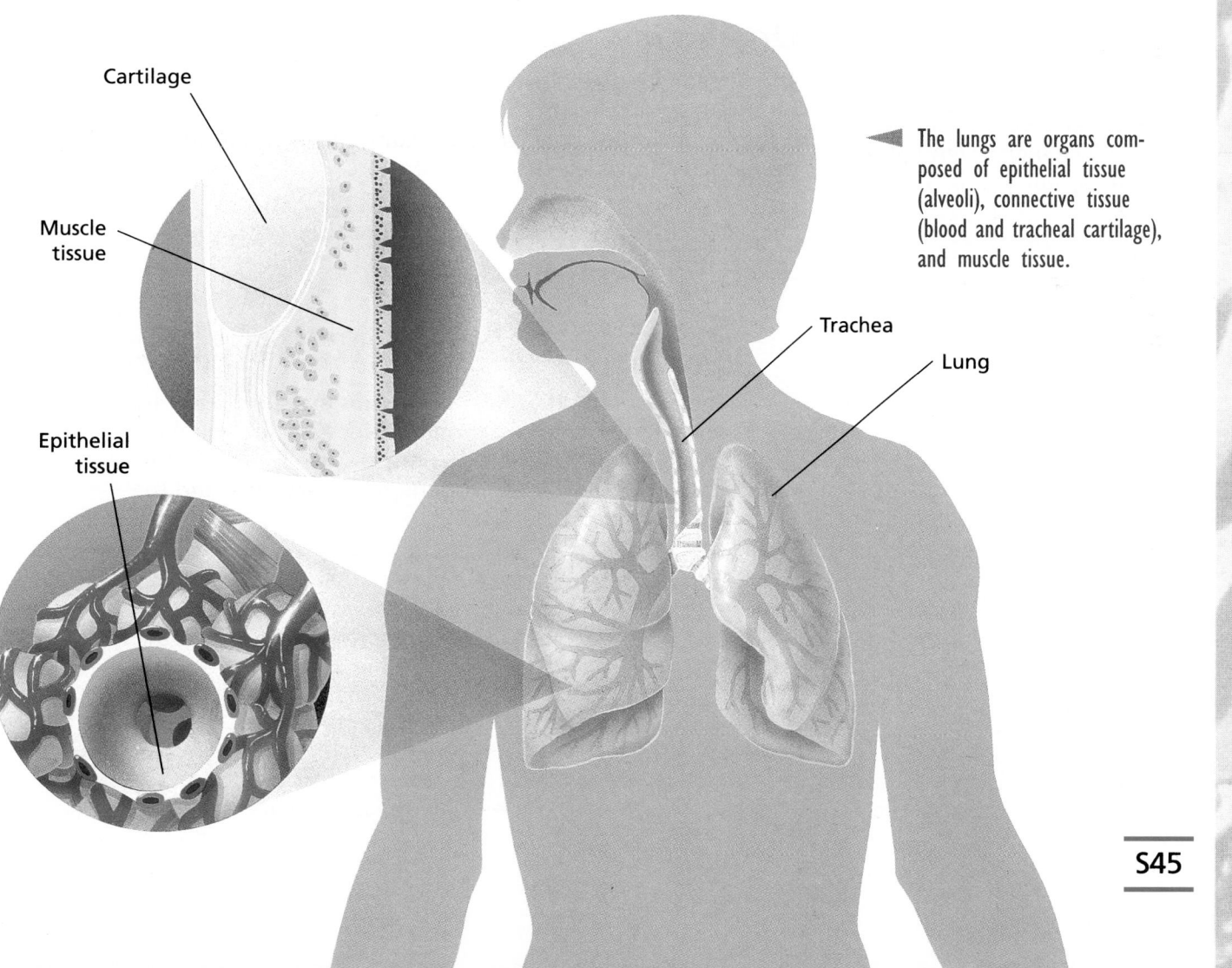

Cartilage

Muscle tissue

Epithelial tissue

◄ The lungs are organs composed of epithelial tissue (alveoli), connective tissue (blood and tracheal cartilage), and muscle tissue.

Trachea

Lung

Plants also have structures similar to animal organs. Leaves are plant organs that make food and exchange gases. The tissues of a leaf include the epidermal tissue that covers the surface, tissues that make food by photosynthesis, and tissues that conduct water and other materials. Roots and stems are plant organs that conduct water, provide support, and store water and food molecules. They contain conducting tissues, covering tissue, and growth tissue. Plants also have reproductive organs. The *pistil* is the female reproductive organ, and the *stamen* is the male reproductive organ in flowering plants.

Organ Systems

A group of organs working together is known as an **organ system**. Organ systems are the fourth level of organization of life. Examples of human organ systems include the circulatory, digestive, nervous, reproductive, and respiratory systems. All complex animals have these same basic systems. However, the organs involved may be different. In the next section, you will learn more about human organ systems.

Plants also have systems that function much like some animal organ systems. The roots, stems, and leaves compose the transport system. Roots and stems function as a support system. Cones and flowers are highly specialized plant structures that contain the reproductive organs and are thus a part of the reproductive system for most plants.

Organism

The fifth and highest level of organization of life is the **organism**. An organism is the sum of all of its cells, tissues, organs, and organ systems. Obviously, only complex multicellular organisms have developed all five levels of organization.

Organ system

A group of related organs performing a major function for an organism

Organism

The highest level of cell organization. An organism is the sum of its cells, tissues, organs, and organ systems.

S U M M A R Y

The cells in multicellular organisms are specialized for certain jobs. Specialization enables each job to be performed more efficiently. There are five levels of organization: cells, tissues, organs, organ systems, and the organism itself.

HUMAN ORGAN SYSTEMS AND GENETICS

The parts of the human body work together so that we can breathe, eat, move, think, protect ourselves from disease, reproduce, and perform other life processes. Understanding the structure and function of your body systems helps you understand how your body carries out its different functions. Each system is extremely complex, so this section includes only some of the important cells, tissues, and organs in each system.

Digestive and Excretory Systems

Your body needs food because food is a source of energy for your cells. However, food must first be changed into a form that cells can use. Changing food into a simpler form that can be used by cells is the job of the **digestive system**.

Entering the System Before your cells can use food, it must be broken down into smaller pieces. This begins when you put food into your mouth, where your teeth tear and grind the food. This is the start of digestion. Follow the path of digestion shown in the figure below as you read along. Your tongue also helps to mash the food while you are chewing. This food is mixed with *saliva*, which moistens it and makes it easier to swallow.

The digestive system begins at the mouth and involves several organs. ▶

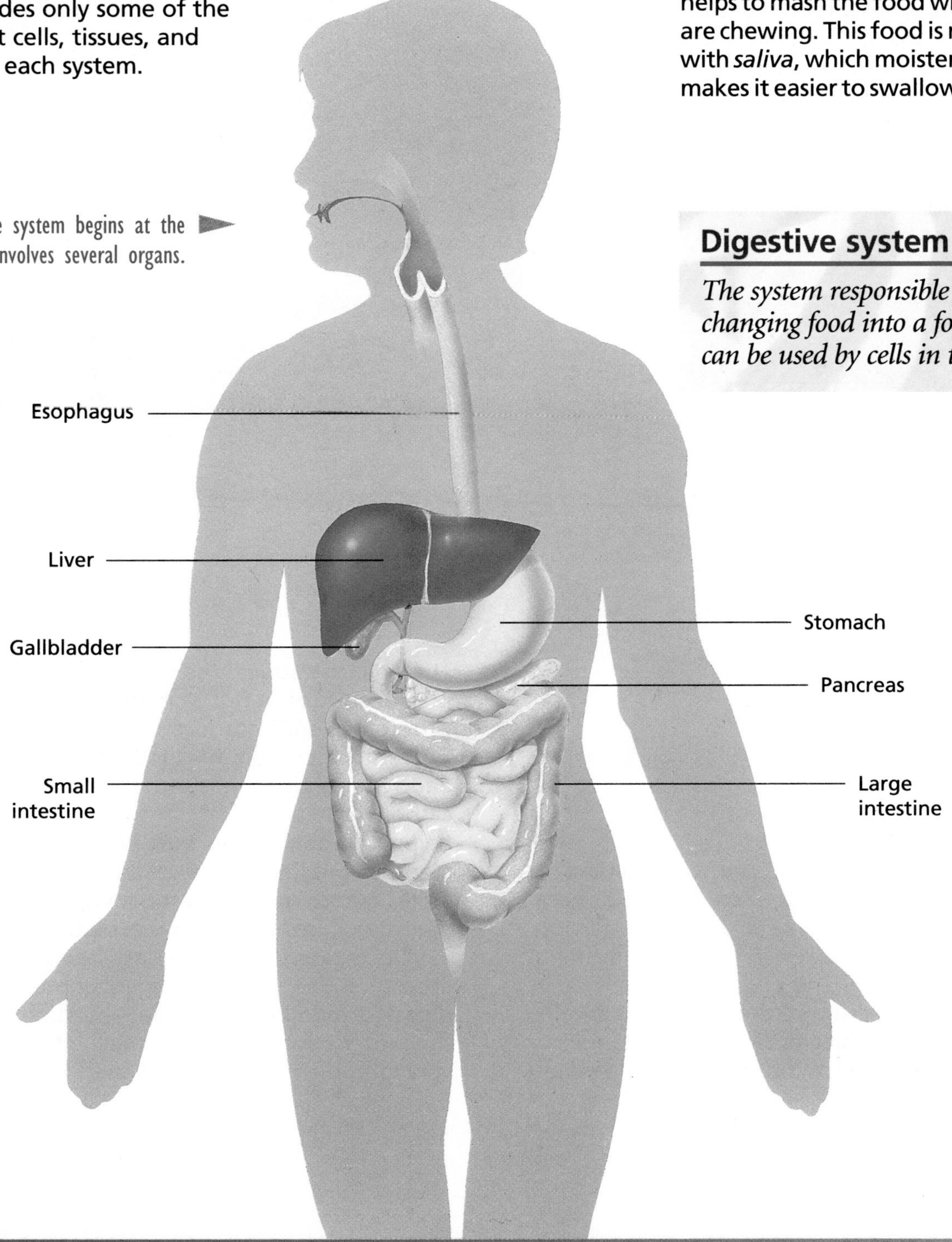

Esophagus

Liver

Gallbladder

Small intestine

Stomach

Pancreas

Large intestine

Digestive system

The system responsible for changing food into a form that can be used by cells in the body

Once swallowed, the food passes down the food tube, called the *esophagus*. Strong muscles along the walls of the esophagus help push the food down to your stomach. The walls of the esophagus and the rest of the digestive system are lined with mucus, a slippery liquid that helps the food move through the digestive system.

The Breakdown of Food From the esophagus, the food enters the stomach. At the point where the esophagus meets the stomach is a specialized ring of muscle called a *sphincter*. When the sphincter tightens, the entrance to the stomach closes—this keeps the contents of the stomach from moving back up the esophagus.

The stomach is a large, muscular sac that stores and digests food. Food stays in the stomach from two to six hours. The movements of the muscles in the stomach help break down the swallowed food even more.

▼ Three sets of muscles cause the stomach to twist and churn, movements that mechanically break up food.

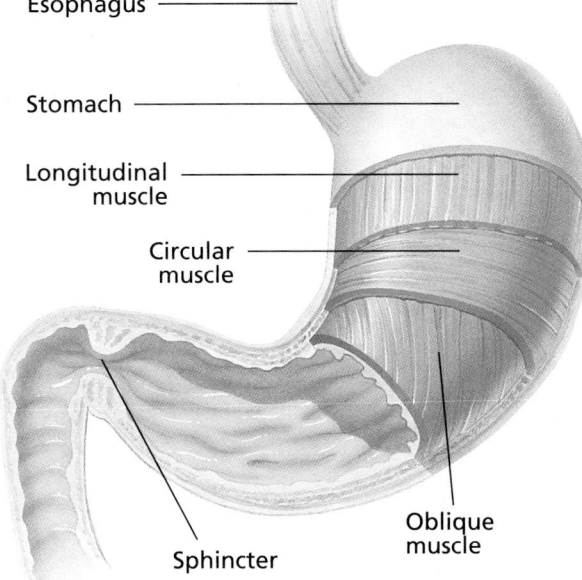

Esophagus

Stomach

Longitudinal muscle

Circular muscle

Sphincter

Oblique muscle

Glands lining the walls of the stomach produce *gastric juices* that contain acid and a protein-digesting chemical. These substances work together to digest the proteins in things like meat, cheese, and eggs.

At the lower end of the stomach is another sphincter. This sphincter slowly releases the contents of the stomach into the *small intestine*.

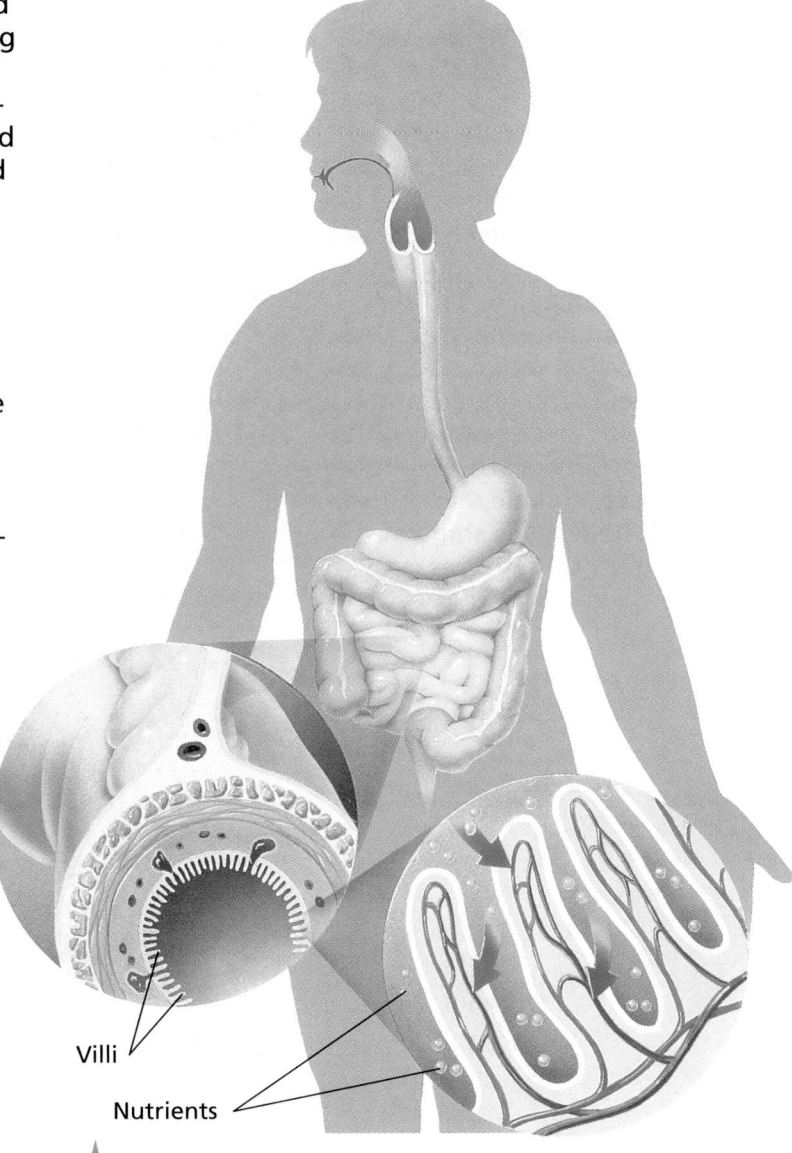

Villi

Nutrients

▲ The villi absorb nutrients and water from the small intestine.

The small intestine is the main digestive organ. It is a coiled tube about 3 cm wide and 7 m long. Different glands produce digestive juices that flow into the small intestine. For example, the *pancreas* produces juices that digest proteins, carbohydrates, and fats. A substance called bile helps to break up fats into tiny droplets. Bile is made in the liver and stored in the *gallbladder*. Other glands in the intestinal walls produce digestive juices that help complete digestion.

Once digestion is complete, the process of absorption begins. The nutrients obtained from digested food leave the small intestine and enter the bloodstream. Since the small intestine is a very long tube that is coiled and folded back onto itself, the area through which nutrients can be absorbed is greatly increased. In addition, *villi* within the small intestine add to its surface area. Villi are tiny, fingerlike structures that cover the inner lining of the small intestine. Nutrients move through the cell membranes of the villi and into the bloodstream, which delivers them to cells throughout the body.

Leaving the System   Undigested bulk materials pass from the small intestine into the large intestine. No nutrient absorp-tion occurs here, but bacteria in the large intestine partially break down the undigested food and produce vitamins B and K. By this point, most of the water has been absorbed by the body, and the substance inside the large intestine has become fairly solid. This material, called feces, is released from the body through the anus.

Excretion of Waste Products
Excess water and cellular waste materials are removed from the body by the **excretory system**. Excretion of cellular wastes occurs in the *kidneys*, which are two bean-shaped organs located behind the stomach. The kidneys and the urinary system are illustrated in the diagram below. Millions of tiny tubes inside the kidneys filter out wastes, sugar, water, and minerals carried by the blood. The kidneys return the sugars, minerals, and some of the water to the blood, where they can be used again. However, waste materials and excess water move out of the kidneys as urine and into a muscular sac called the *urinary bladder*, where the urine is stored until it can be released from the body.

Some wastes leave the body through other organs. Carbon dioxide, a waste product carried by the blood, leaves the body through the lungs. When you perspire, your skin excretes nitrogen wastes, salts, and water in the form of sweat.

Excretory system

The system that eliminates cellular wastes and excess water from the body

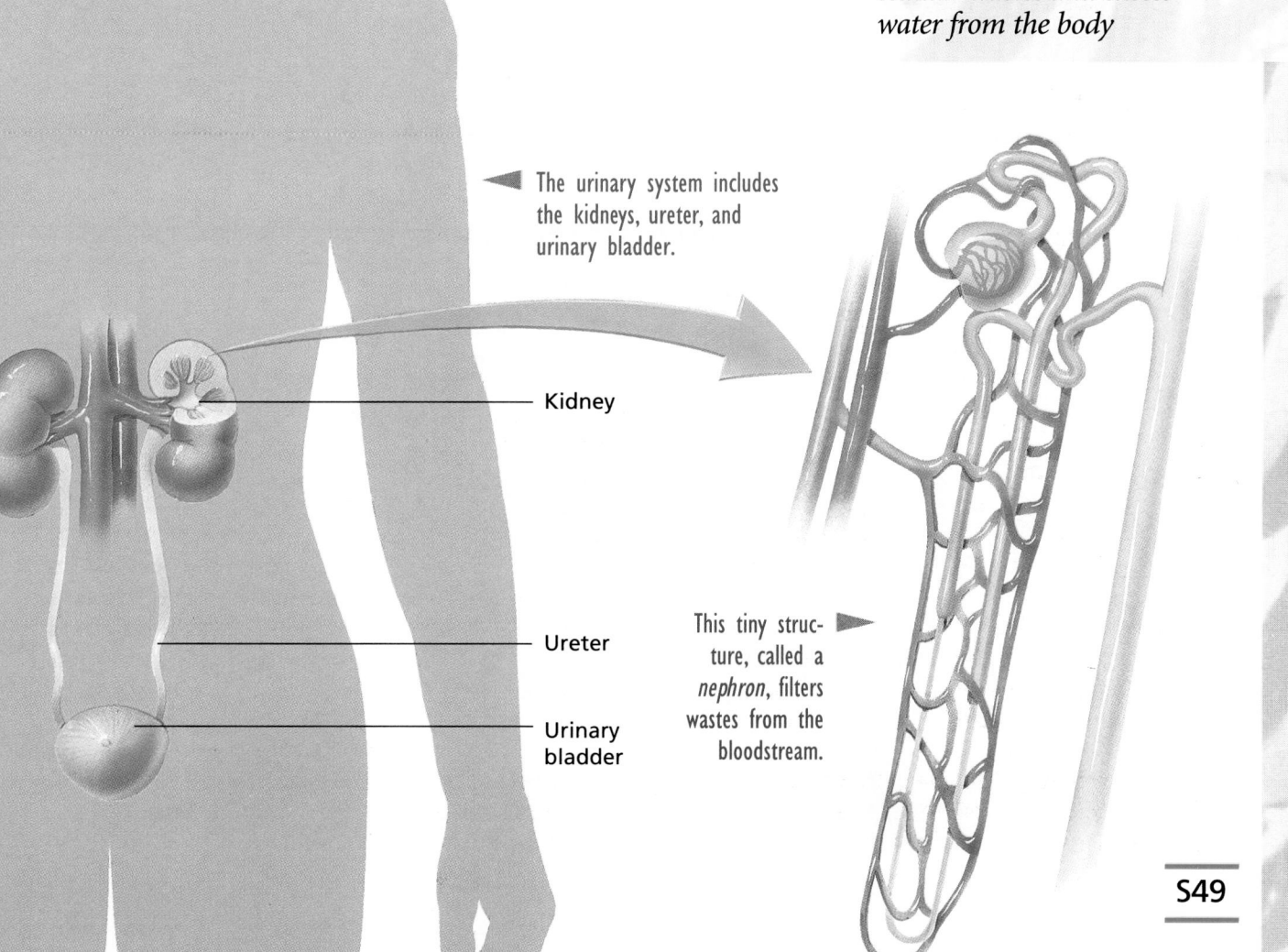

The urinary system includes the kidneys, ureter, and urinary bladder.

Kidney

Ureter

Urinary bladder

This tiny structure, called a *nephron*, filters wastes from the bloodstream.

S49

Circulatory System

The **circulatory system** is responsible for transporting nutrients, oxygen, wastes, and cells that attack infectious organisms. This system consists of the heart, blood vessels, and blood. The heart pumps the blood, and the blood vessels provide a passageway for it. Blood reaches every tissue and organ in the body. *Arteries* carry blood away from the heart. Blood is supplied to the entire body through a single artery—the *aorta*—which exits the heart. The aorta soon divides into smaller branches that divide again and again into smaller and smaller branches as they move away from the heart. Eventually, a blood vessel becomes so small that just one blood cell at a time can pass through. At this point, it is called a *capillary*. Capillary walls are only one cell thick. It is through the thin walls of capillaries that substances are exchanged between the blood and tissues. The blood inside capillaries supplies oxygen and nutrients to tissues and remove wastes.

Circulatory system

The system responsible for transporting nutrients to, and wastes from, the cells of the body

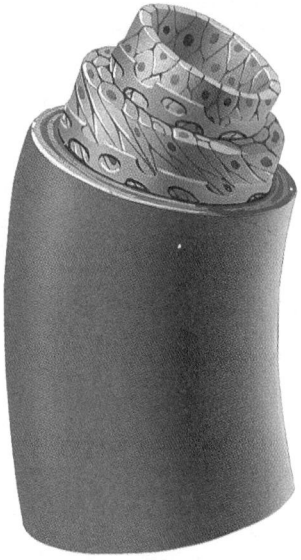

Arteries carry blood away from the heart, have thick walls that are muscular and elastic, and do not have valves.

The circulatory system supplies oxygen and nutrients to all cells of the body.

Capillaries connect small arteries and veins and are the sites of nutrient and gas transfer. Capillaries are microscopic and have walls that are only one cell thick.

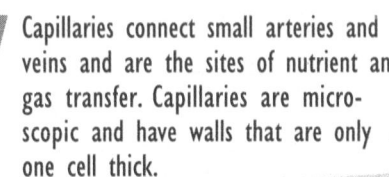

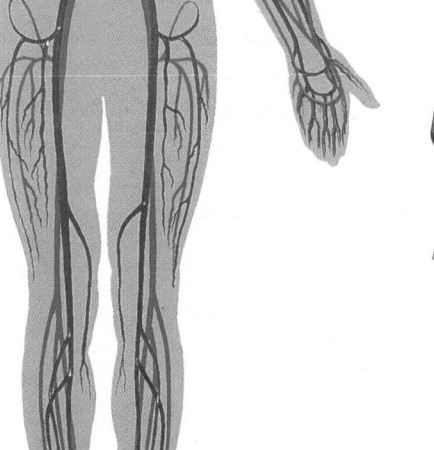

Veins return blood to the heart and have thinner, less muscular walls than arteries. Valves ensure that blood flows in the proper direction.

S50

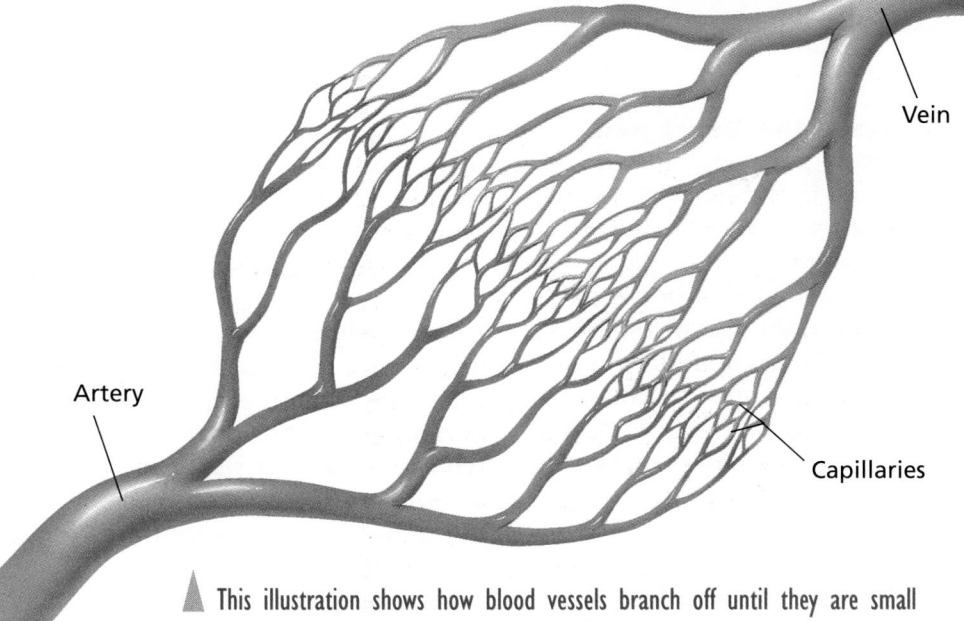

Vein

Artery

Capillaries

This illustration shows how blood vessels branch off until they are small enough to reach every cell in the body.

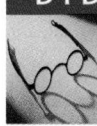
After flowing through the capillaries, the blood returns to the heart through the *veins*. The arrangement of the veins is essentially the reverse of the arteries. Smaller veins join together to form larger and larger veins until the two veins that empty into the heart are formed.

Blood is composed of two basic components: *plasma* and *formed elements*. Plasma constitutes about 55 percent of the blood's volume. It consists of water and dissolved substances such as proteins, nutrients, hormones, minerals, and wastes. Formed elements are the cells that perform various functions. About half the formed elements are red blood cells, which carry oxygen and carbon dioxide. White blood cells, which defend the body against infection, make up about 15 percent. *Platelets*, which aid in clotting, make up the remainder of the formed elements.

The heart is a pear-shaped, hollow muscle about the size of your fist and consists of two upper chambers and two lower chambers. The average adult heart pumps about 5 L of blood per minute. Blood flowing into the heart first flows into the upper chambers, called *atria*

(singular, *atrium*). The contraction of the atria forces the blood into the *ventricles*. The ventricles then contract, forcing the blood from the heart into the large blood vessels at the top of the heart. From there the blood travels to the rest of the body. Valves within the heart open and close like doors to keep blood flowing in only one direction.

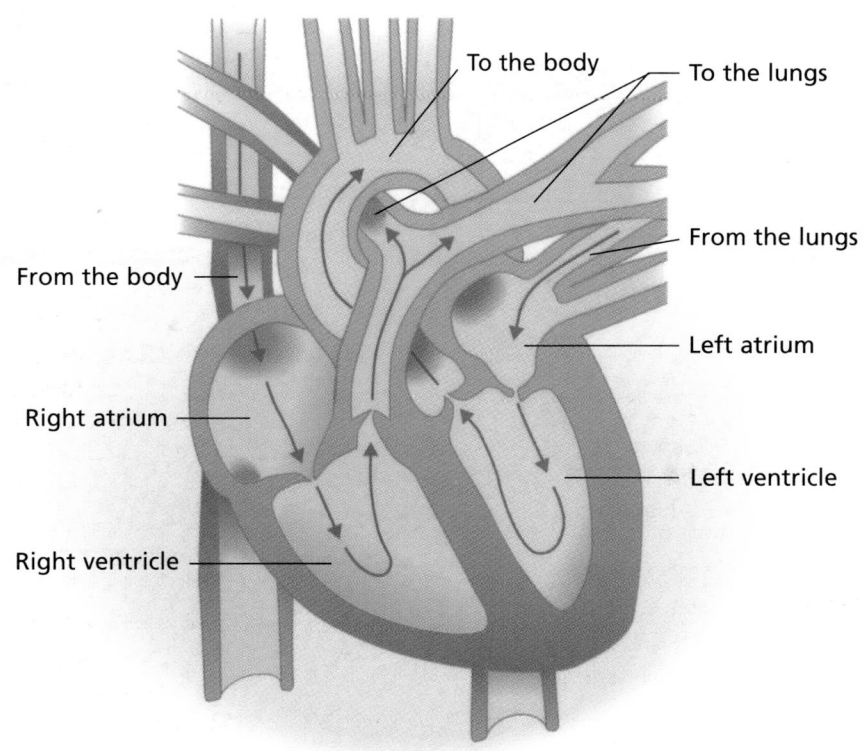

To the body

To the lungs

From the lungs

From the body

Left atrium

Right atrium

Left ventricle

Right ventricle

The human heart has four chambers. Notice the pathway of blood through the heart.

S51

Respiratory System

Cells use oxygen to release the energy stored in food molecules, the chemical products of digestion. In the process, cells produce carbon dioxide as a waste product. The **respiratory system** allows oxygen to be brought to the cells and carbon dioxide to be removed. The respiratory system uses the circulatory system as a way of transporting both oxygen and carbon dioxide.

Respiratory system

The system by which gases enter and leave the body

Between the Heart and Lungs   Blood returning from the body to the heart flows into the right side of the heart. This blood contains little oxygen but is saturated with carbon dioxide. From the right side of the heart, the blood is pumped to the lungs, which enrich the blood with oxygen and remove the carbon dioxide. The oxygenated blood flows back to the heart, entering the left atrium. The left atrium contracts, squeezing the blood into the left ventricle. The left ventricle then contracts, pumping the oxygenated blood from the heart into the aorta, the large artery that leaves the heart. Branches from this artery deliver blood to the entire body.

Respiration   The lungs consist of air passages that branch again and again, eventually ending in millions of tiny, flexible sacs. These air sacs, called *alveoli*, are surrounded by capillaries and have walls that are only one cell thick.

When you inhale, oxygen is supplied to the blood inside the capillaries of your lungs. At the same time, carbon dioxide flows from the capillaries into the alveoli. Breathing out removes that carbon dioxide. This is called *gas exchange*. You breathe in oxygen that gets transported into your blood. After the blood leaves the capillaries of the lungs, it returns to the heart. The oxygen-rich blood is then pumped to the rest of the body.

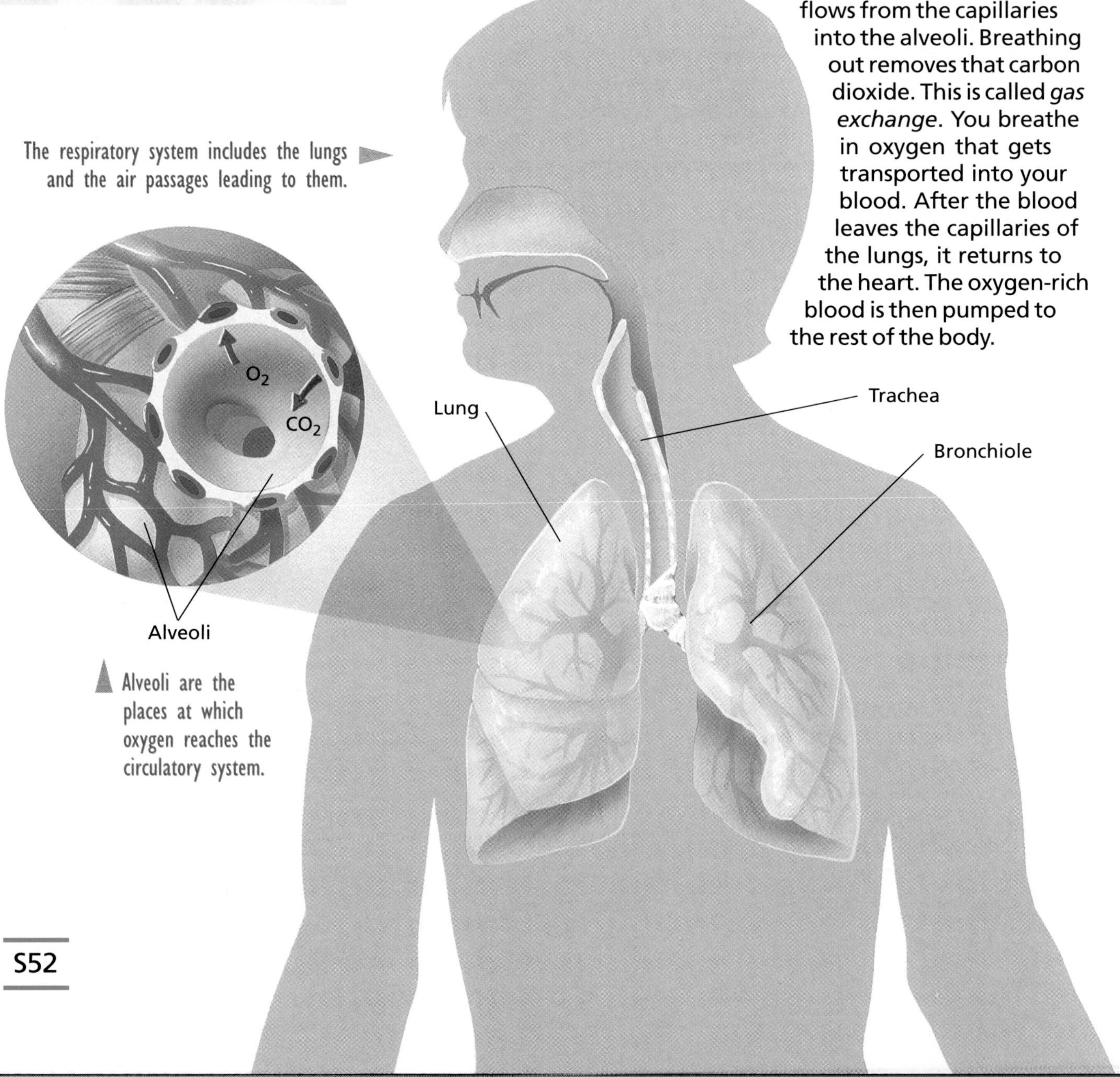

The respiratory system includes the lungs and the air passages leading to them.

O_2

CO_2

Lung

Trachea

Bronchiole

Alveoli

Alveoli are the places at which oxygen reaches the circulatory system.

Integumentary, Skeletal, and Muscular Systems

The integumentary, skeletal, and muscular systems provide your body with protection, support, and movement. The integumentary system includes your skin, the skeletal system includes your bones and joints, and the muscular system includes your muscles.

Integumentary System   Your **integumentary system** contains the largest organ in the human body—your skin. The outer layer of your skin is called the *epidermis*. The cells of the epidermis are broad and flat. Some of these cells are dead and hardened. They form a tough, waterproof shield that most germs cannot penetrate. Beneath the epidermis is the *dermis*. The cells of the dermis are all living. The dermis contains blood vessels, sweat and oil glands, hair follicles, and fat cells. By releasing excess water from the sweat glands, the skin can cool the body. The functions of the skin are summarized below.

Integumentary system

The system responsible for regulating body temperature and protecting it from the outside environment

▼ The skin is a complex organ that performs a variety of functions.

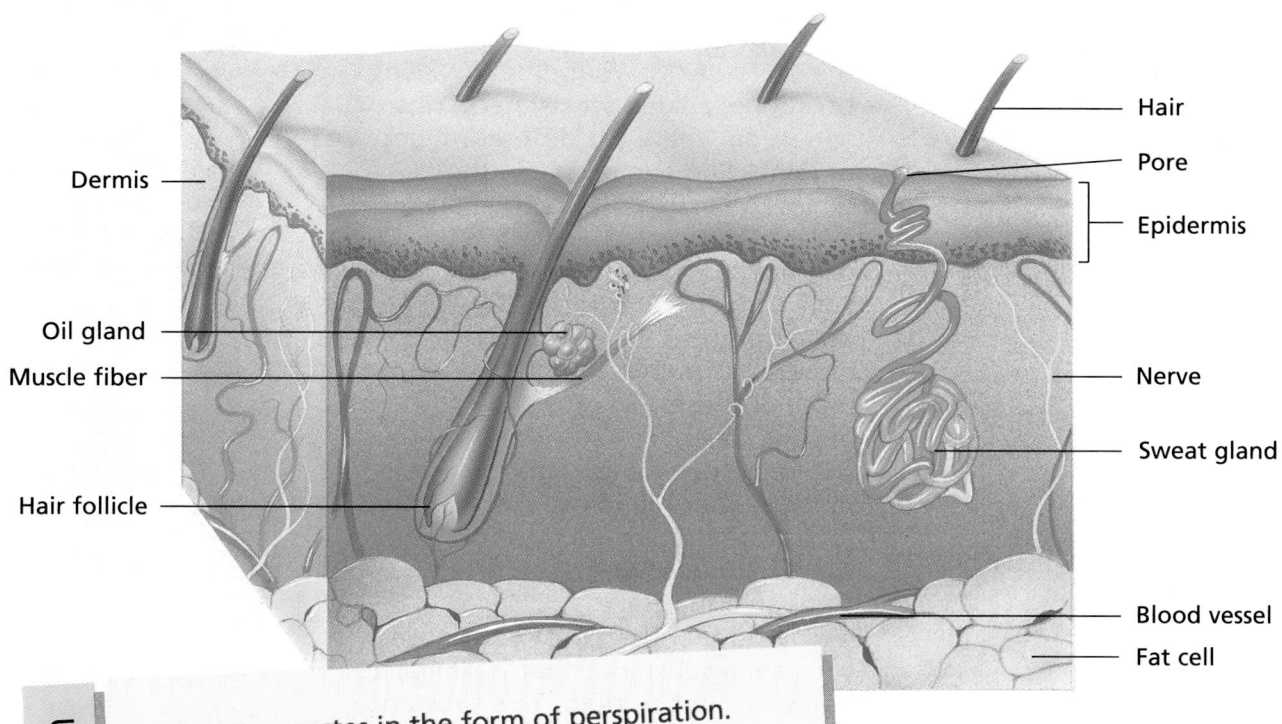

Dermis

Oil gland

Muscle fiber

Hair follicle

Hair

Pore

Epidermis

Nerve

Sweat gland

Blood vessel

Fat cell

Functions of the Skin

1. Excretes wastes in the form of perspiration.
2. Provides protection from germs and injury.
3. Acts as a waterproof covering.
4. Keeps body tissues from drying out.
5. Helps regulate body temperature.
6. Contains sensors for heat, cold, touch, and pain.
7. Stores excess nutrients in the form of fats.
8. Produces vitamin D in the presence of sunlight.

S53

Skeletal system

The system that supports the body with a framework of bones

Skeletal System The framework of your body, or **skeletal system**, consists of your skeleton, which is made up of 206 separate bones. Bones are strong but lightweight structures composed of hard living tissue. They provide protection and support to the body. The outside of the bone is surrounded by a tough membrane that helps repair bone injuries and attaches muscles to the bone. Beneath this membrane is the *bony layer*. The bony layer contains living bone cells that release calcium and phosphorus, which make bones hard. The inside of the bone is filled with soft, fatty *marrow*. Marrow contains many nerves and blood cells. New blood cells are also made in the marrow.

Because bones do not bend, movement can occur only where bones meet. The place where two or more bones meet is called a *joint*. Joints are found throughout the skeleton—in your elbows, knees, backbone, ankles, wrists, neck, ribs, and so on. There are different types of joints. Some joints allow the bones to move more freely than others. For example, the joints in your backbone do not move as much as those in your knees. But your knee joints cannot move as much as your hip joints. And your hip joints cannot move as much as your shoulder joints, which can rotate through a full circle.

If you feel the tip of your nose, you feel a substance that is similar to bone, but not as rigid. This substance is called *cartilage*. Cartilage covers the inside surfaces of many joints and is found in other areas near bony tissue.

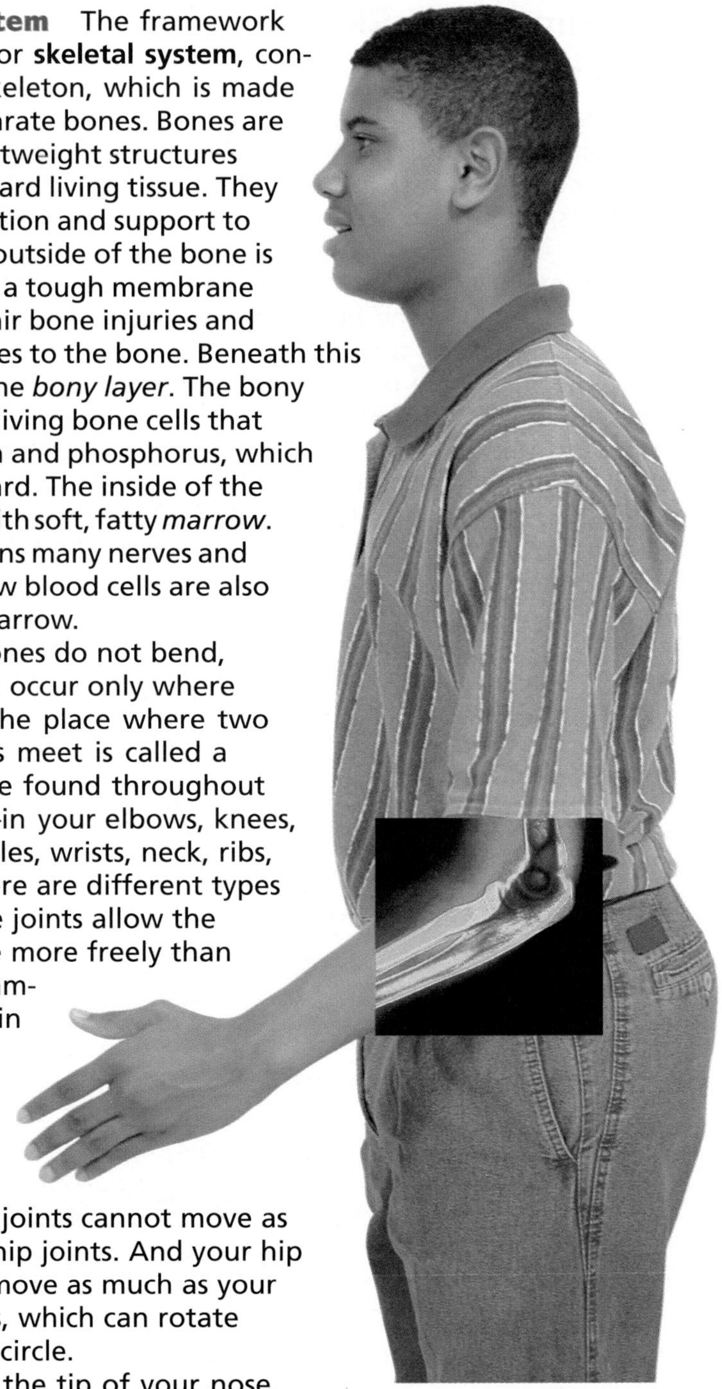

The elbow is one type of joint. What types of movement does it allow?

Muscular system

The system that enables the body to move, the heart to beat, and other organs of the body to function

Muscles in Your Body Your body's **muscular system** consists of three types of muscle: smooth, skeletal, and cardiac. *Smooth muscle* is found in places like the digestive tract and the walls of blood vessels. Smooth muscle tissue is organized into sheets. The movements of smooth muscle are *involuntary*, meaning that you cannot consciously control them.

S54

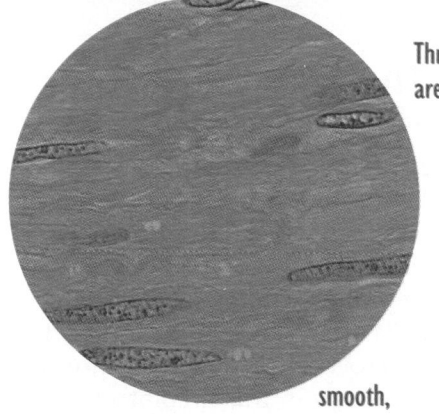

Three types of muscle are shown here:

smooth,

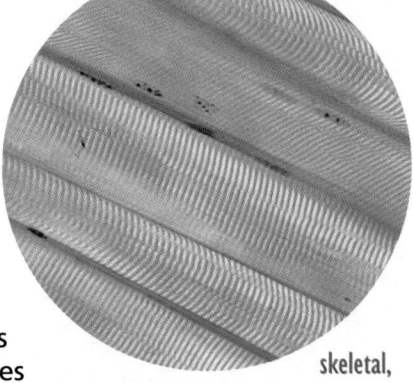

skeletal,

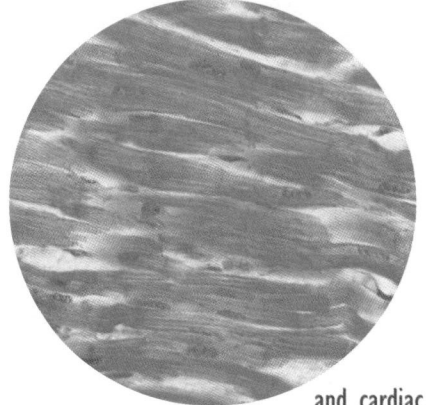

and cardiac.

Skeletal muscle is found in places like your arms and legs. These muscles are joined together in bundles and are generally attached to bones, so that when the muscles contract, the bones move. Skeletal muscle is considered *voluntary* muscle because you can consciously control its activity.

The third type of muscle is *cardiac muscle*, which is found only in the heart. The cells of cardiac muscle contain many mitochondria (the energy-producing organelles) and are tightly woven together. Like smooth muscle, cardiac muscle works involuntarily. Moreover, your cardiac muscles work 24 hours a day. They rest only between heartbeats.

The Nervous System

The nervous system allows an organism to interact with its surroundings. Your **nervous system** is responsible for sensing stimuli from the environment and coordinating the body's response.

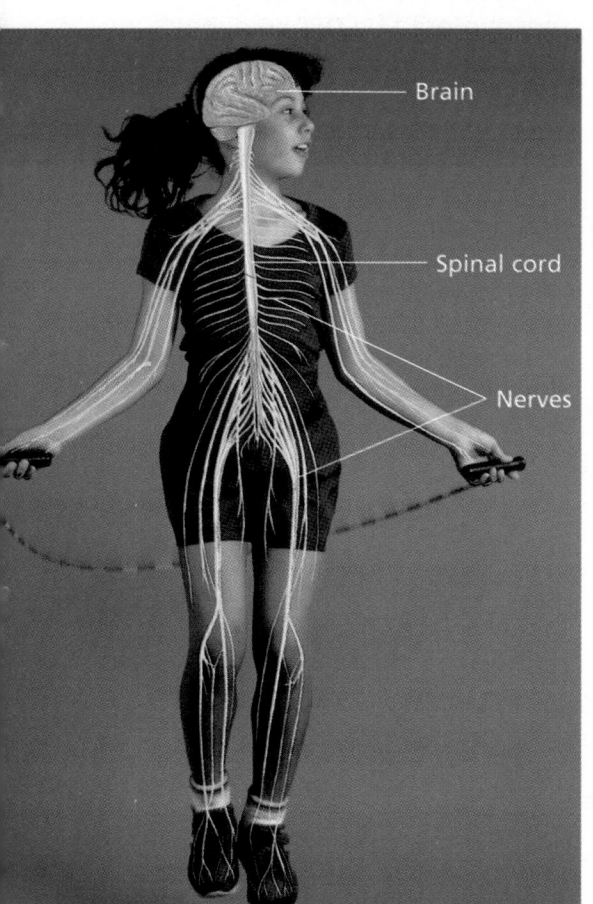

Brain

Spinal cord

Nerves

◀ The human nervous system

Stimulus and Response

When you receive a *stimulus*—see, hear, smell, taste, or feel something—nerves send an electrochemical message to your brain and the rest of your body so that you can respond to the stimulus. Your *response* meets your needs in some way. If your shoes are too tight, you remove or loosen them. You draw back from something hot. When someone calls your name, you turn in the direction of his or her voice. When the sun gets in your eyes, you squint or turn away. The nervous system is made up of the brain, the spinal cord, and nerves. Nerves are made up of cells called *neurons*, which are organized

Nervous system

The system that senses the environment and coordinates the body's response to the environment

much like telephone wires; they connect to every part of your body. The electrochemical impulses from neurons move at a rate of about 100 m per second, about as fast as a race car!

When impulses move from sense organs to the brain or spinal cord, they travel along *sensory neurons*. Impulses that go to the muscles and glands move along *motor neurons*. In the brain and spinal cord, *association neurons* link these neurons and allow sensory neurons to pass impulses to motor neurons.

S55

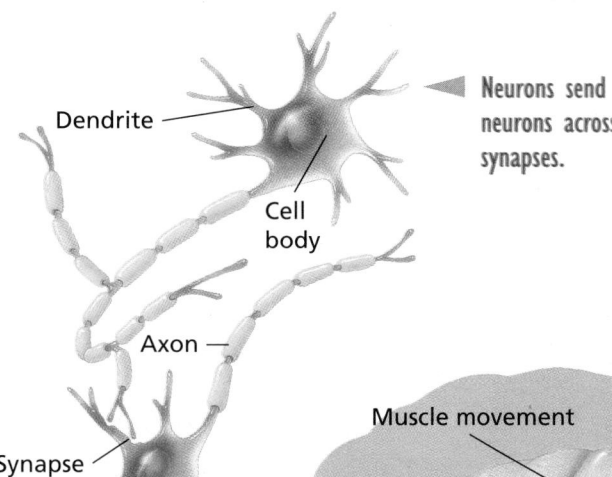

Dendrite

Cell body

Axon —

Synapse

Neurons send signals to other neurons across gaps called synapses.

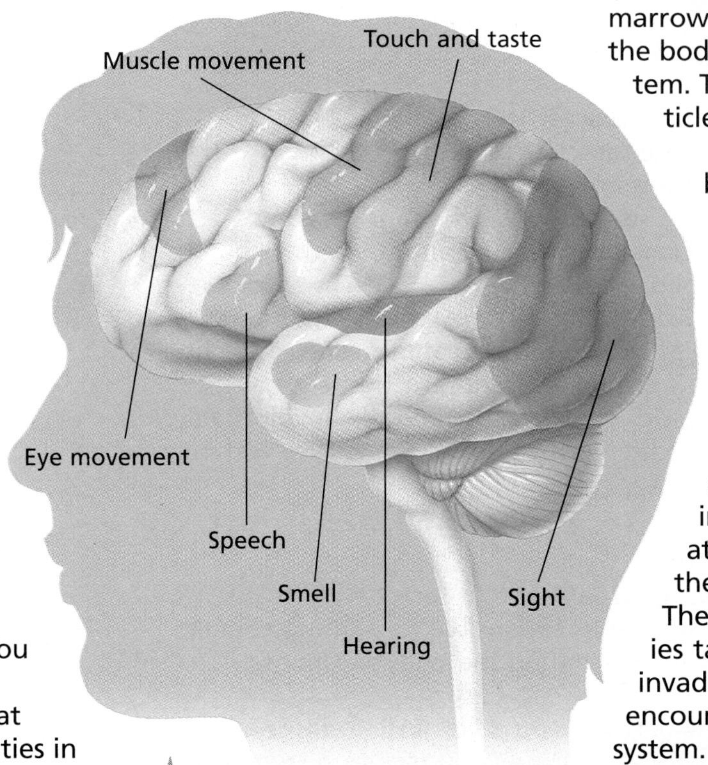

Muscle movement

Touch and taste

Eye movement

Speech

Smell

Hearing

Sight

▲ The human brain contains specific areas for sensing the environment and controlling body functions.

The brain interprets the information from the senses. It also remembers, dreams, imagines, coordinates, and makes decisions. You can see some of the regions of the brain that control particular activities in the accompanying figure. All of this activity requires a lot of energy. In fact, the brain uses 25 percent of the oxygen taken in by the body. It is one of the most active organs and works 24 hours a day to keep your body systems regulated.

Your brain, however, does not have to consciously regulate the body's response to everything it senses. An automatic response, called a *reflex*, allows the body to react involuntarily in certain situations. In a reflex action, the impulse passes from a sensory neuron to a motor neuron without going to the brain first. Instead, the impulse passes through the spinal cord. The brain receives a message after the reflex has occurred. Reflexes include coughing, sneezing, blinking, pulling away from a hot object, and jumping when startled.

The Immune System

The body also has a system to protect itself against disease. White blood cells and chemicals that the white blood cells produce make up the body's **immune system**. White blood cells are made within the bone marrow and travel throughout the body in the circulatory system. They engulf foreign particles and destroy them.

Some kinds of white blood cells help make special chemicals called *antibodies*. Each antibody fights a particular type of microorganism or foreign chemical. The diagram at the top of the next page illustrates an antibody attacking a foreign invader. Antibodies attach to the outside of the invader and destroy it. The production of antibodies takes a few days if the invader has never been encountered by the immune system. Once the disease has been suppressed, a group of white blood cells "remembers" how to make the antibody. If the virus, bacteria, or other foreign substance ever enters the body again, these white blood cells can make the antibodies very quickly. This resistance to disease is called *acquired immunity*. Acquired immunity to some diseases can last a lifetime.

Immune system

The system responsible for protecting the body from invading substances

Bacteria being attacked by a ▶ white blood cell

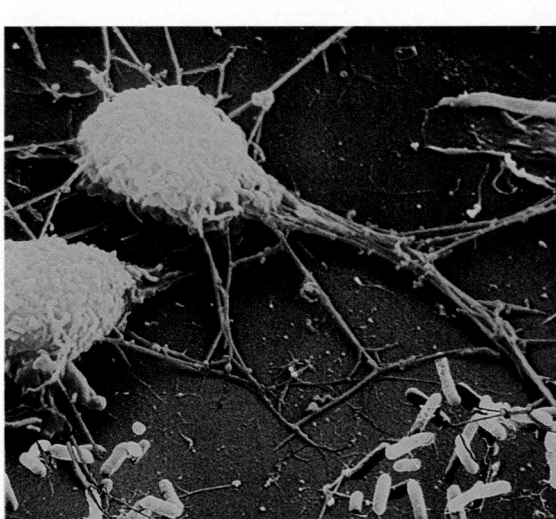

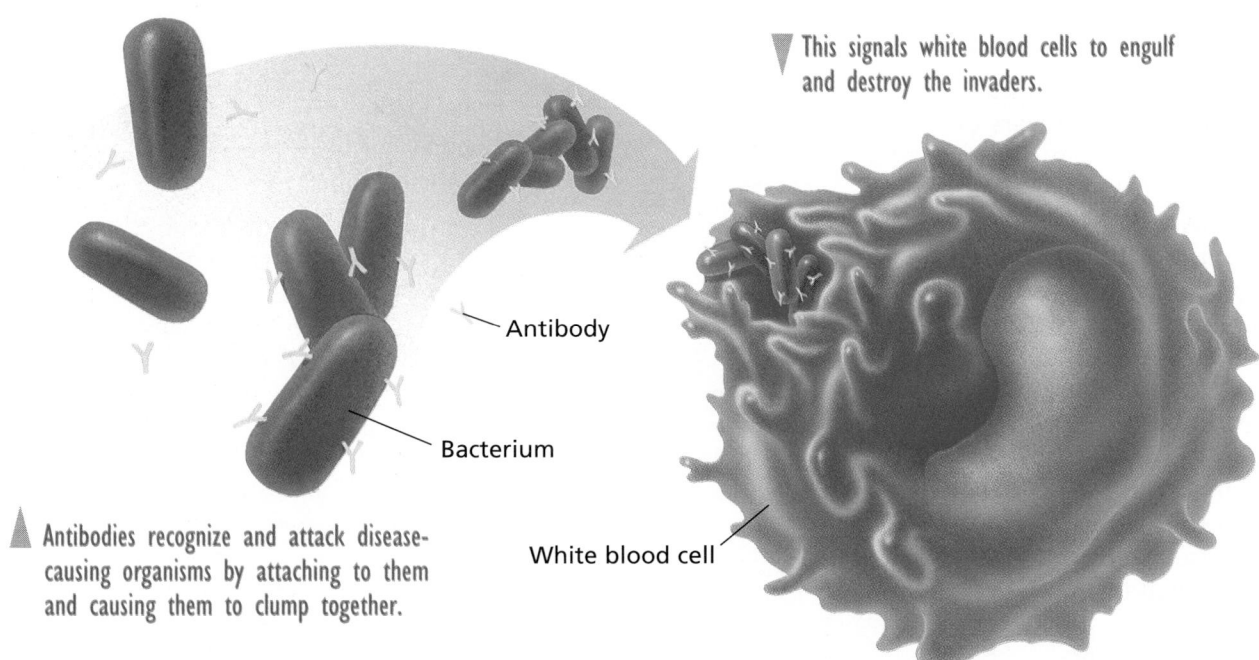

▼ This signals white blood cells to engulf and destroy the invaders.

Antibody

Bacterium

White blood cell

▲ Antibodies recognize and attack disease-causing organisms by attaching to them and causing them to clump together.

AIDS is a disease of the immune system that you have probably heard a great deal about. AIDS stands for *acquired immune deficiency syndrome*. AIDS is caused by a virus called HIV (human immuno-deficiency virus). This virus gets inside certain cells and changes the genetic instructions that the cells follow. Instead of fighting disease and producing antibodies, the infected cells only make more of the virus. People get sick and die from AIDS because their immune systems can no longer fight other diseases.

The spread of AIDS has alarmed many people. However, it does not appear to be possible to get AIDS from being around an infected person in an ordinary, everyday kind of way. The only known way to get HIV is by being exposed to the blood or bodily fluids of a person who has the virus. Most people have acquired the AIDS virus by having sexual intercourse with an infected person, sharing drug-injection equipment with an infected person, or receiving infected blood transfusions.

Although the AIDS epidemic is serious, there is hope. In this country, AIDS occurs predominantly in certain high-risk groups. Furthermore, there are only about three dozen metropolitan areas in the United States with a substantial population of AIDS sufferers. Many experts believe that AIDS can be conquered by concentrating medical resources in those areas. Interestingly, the disease already shows some signs of starting to level off. While there is no cure for AIDS yet, some medicines can help prolong the life of a person with AIDS.

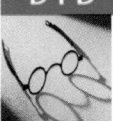
▼ The green dots are AIDS viruses attacking a white blood cell.

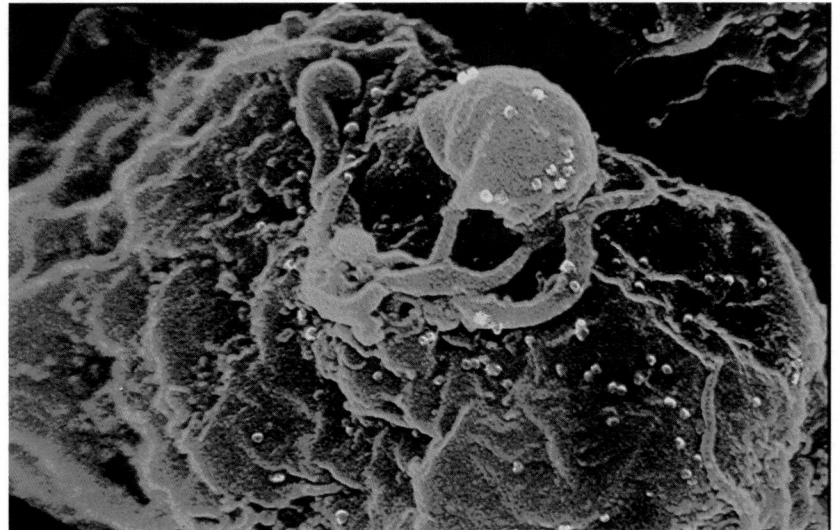

The Endocrine System

It is the bottom of the ninth. Your team has two players on base, you are two runs behind, and there are two players out already. Now it's your turn at bat. As you walk over to the plate your heart begins to pound, your hands start to sweat, and your mouth goes dry. You can feel the nervous energy in the pit of your stomach, and it gives you the extra strength you need to smash that ball over the fence!

This kind of reaction to an exciting or dangerous event is the result of cooperation between two main body systems: the nervous system and the **endocrine system**. The endocrine system consists of a number of glands that manufacture powerful chemicals called *hormones*. Hormones act as chemical messengers. Endocrine glands release the hormones they manufacture directly into the bloodstream. The diagram shown here illustrates where hormones are produced in the body.

During the baseball game described above, a hormone called *adrenaline* was responsible for stimulating the body's responses. Adrenaline is produced in the adrenal glands,

Endocrine system

The system responsible for regulating certain body processes in conjunction with the nervous system

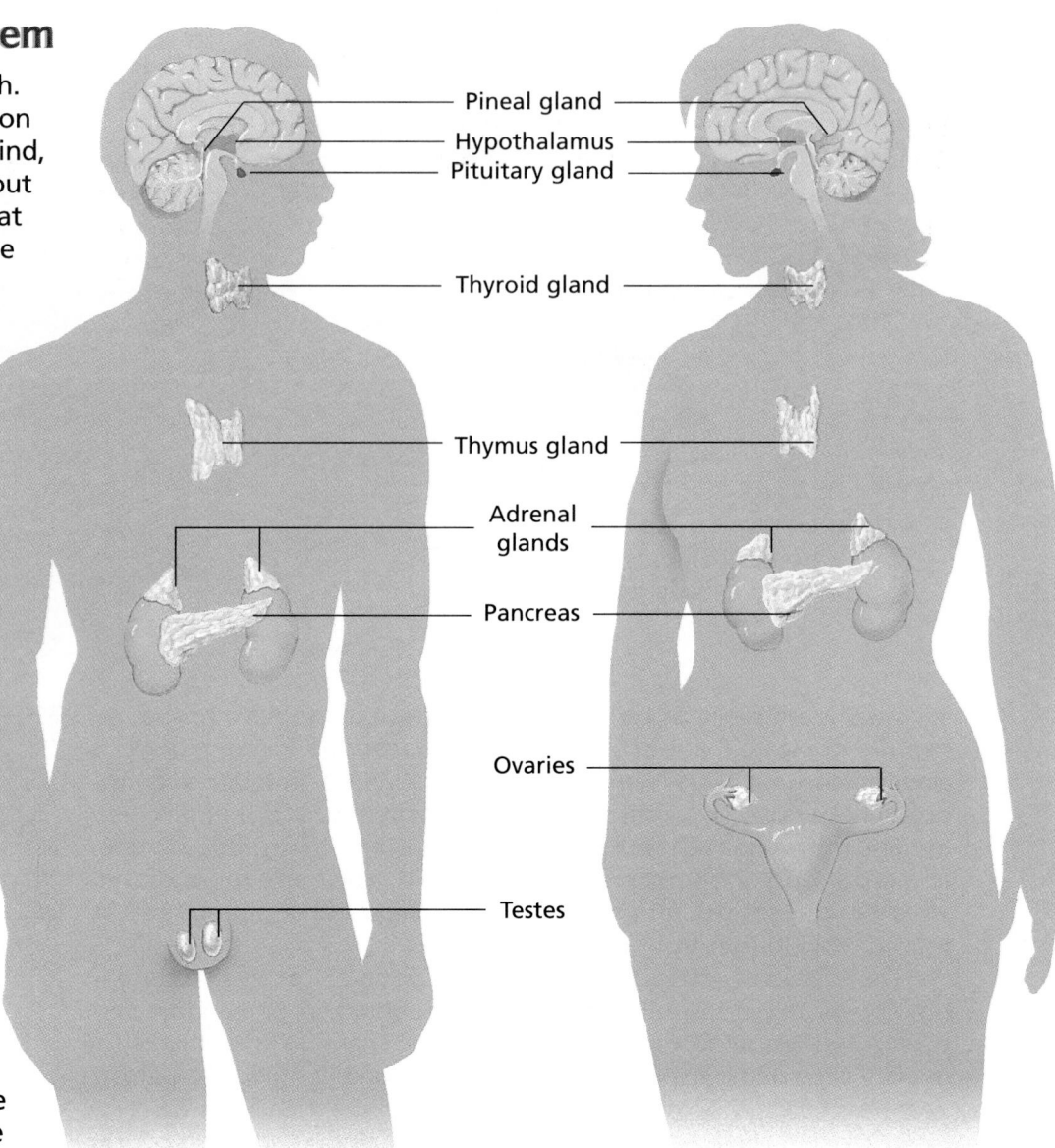

▲ The endocrine system contains glands that monitor and control many body functions.

located just above the kidneys. When you are excited, frightened, or startled, the adrenal glands react by secreting a tiny amount of adrenaline into your bloodstream. In response, your heart beats faster, your lungs work harder, and the amount of sugar in your blood rises. As a result, your body is ready for action—whether it's an exciting turn at bat during a ball game or something much more dangerous. The adrenal response is sometimes called the *fight-or-flight* response. When was the last time you experienced this response? What caused it?

As you can see, hormones are very powerful chemical regulators. Hormones act in three ways: some speed up body processes, some hold back or slow down body processes, and some influence growth and sexual development by producing changes in body structure. For the most part, the body does not store hormones because the glands produce hormones only in small amounts as they are needed.

The Reproductive System

The birth of a new human being is an amazing event. Every baby develops from a fertilized cell into a complex assortment of trillions of specialized cells. The organs of the **reproductive system** are responsible for the conception and birth of human babies. Those same organs also produce the hormones that cause the physical differences in the appearance of men and women. The reproductive system is formed at birth but does not mature fully until *puberty*, during early adolescence.

The Female Body Inside a woman's body are two reproductive organs called *ovaries*. These produce eggs and a female hormone called *estrogen*. Each egg is a single cell that contains half of the genetic information needed for a baby to form. During *ovulation*, a woman releases one of the eggs from her ovaries. The egg moves out of the ovary and into a tube called the *fallopian tube*. After ovulation, the ovary releases a hormone that causes the lining of the uterus to thicken with blood-rich tissue. The egg continues down the

fallopian tube into the *uterus*. The uterus is a pear-shaped organ with thick, muscular walls. It is connected to the outside of a woman's body by the *vagina*, also called the birth canal. If the egg is not fertilized, the uterine lining breaks down and is expelled during the process of *menstruation*.

The Male Body Men have two reproductive organs called *testes*, which produce *sperm* and the male hormone *testosterone*. The testes are located in a saclike structure called the *scrotum*, located on the outside of the man's body. The testes produce a large number of single-celled sperm. Like eggs, each sperm contains half of the information needed for the

baby to develop. A sperm has a head and a tail. The sperm's head carries its cargo of DNA, while the tail lashes vigorously back and forth, causing the sperm to move. A mixture of sperm and nutrient-rich fluids is forcefully ejected from the male's penis (the external reproductive organ) during sexual intercourse.

Reproductive system

The system responsible for passing on hereditary information and producing the next generation

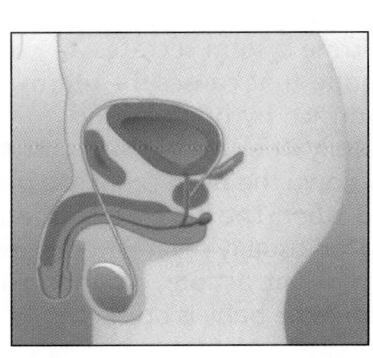

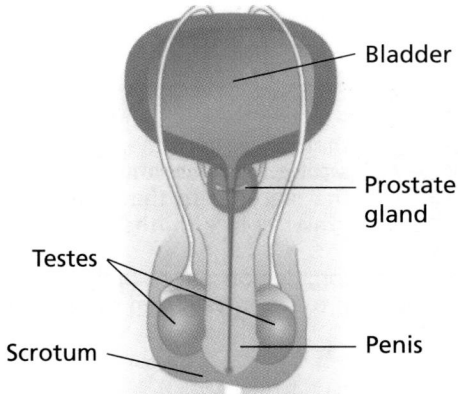

Bladder

Prostate gland

Testes

Scrotum

Penis

▲ Side and front views of the male reproductive system

▼ Side and front views of the female reproductive system

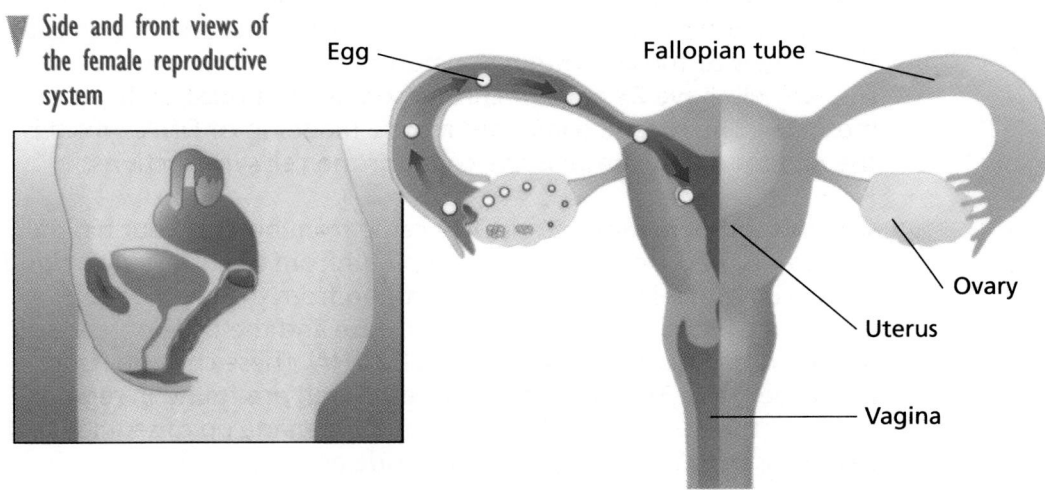

Egg

Fallopian tube

Ovary

Uterus

Vagina

S59

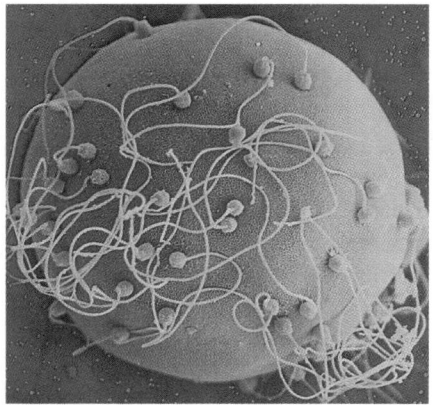

Sperm surround an egg, each one attempting to be the one to fertilize it.

The Reproductive Process   During intercourse, the male releases millions of sperm into the female's vagina. The sperm must then swim up the vagina, through the woman's uterus, and into her fallopian tubes. Although many sperm may reach the egg, only one can *fertilize* it. To fertilize the egg, the sperm must penetrate the egg's membrane. Once the sperm is inside the egg, the genetic materials of the sperm and the egg combine, beginning the development of a new human being.

Once the egg is fertilized, it immediately begins to divide, creating more and more cells. In about seven days, the growing mass of cells, called an *embryo*, reaches the uterus, where it burrows into the thickened uterine wall. There it forms a special organ called a *placenta*. Through the placenta, the woman supplies the embryo with all of its oxygen and nutrients. As the embryo continues to grow, its cells begin to perform specific functions. Some cells begin to beat— these will become heart cells. Other cells form into nerves, muscle, skin, digestive organs, and all of the other kinds of cells found in a human body. After about nine full months of continuous development, the child is ready to be born.

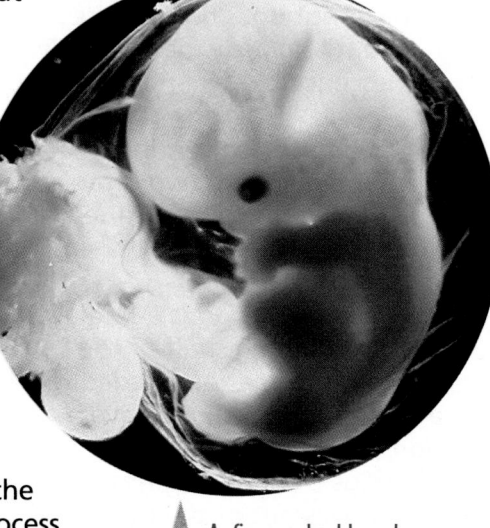

A five-week-old embryo attached to the placenta

When the child is developed enough to be born, the endocrine system secretes a hormone that causes the uterus to contract rhythmically and forcefully. The contractions push the baby down the birth canal in a process called *labor*. The birth process can take a long time. It is usually several hours from the time that the contractions, or labor pains, begin to the time the baby is born.

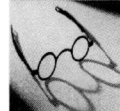

DID YOU KNOW...

that drugs, nicotine, and alcohol can pass through the placenta to the child growing in a mother's uterus?

For this reason, doctors advise pregnant women to stay away from these substances.

Human Genetics

Inside every cell is a set of instructions that guide the reproduction, structure, and function of the cell. The instructions are found inside the nucleus of the cell within structures called chromosomes. Human cells have 23 pairs of chromosomes for a total of 46 chromosomes. The chromosomes are mainly made up of DNA, which is the molecule that contains the code for the cell's instructions.

Genetic Inheritance   Your DNA came from both of your biological parents, half from one parent and half from the other. What kind of information is in your DNA? Everything from the color of your eyes to the type of blood cells you produce and the development of all your organ systems is determined by DNA. These characteristics are called *traits*. The instructions for each trait are found in regions on the chromosome called *genes*. Genes are made up of segments of DNA. This DNA is passed on as cells divide and reproduce.

S60

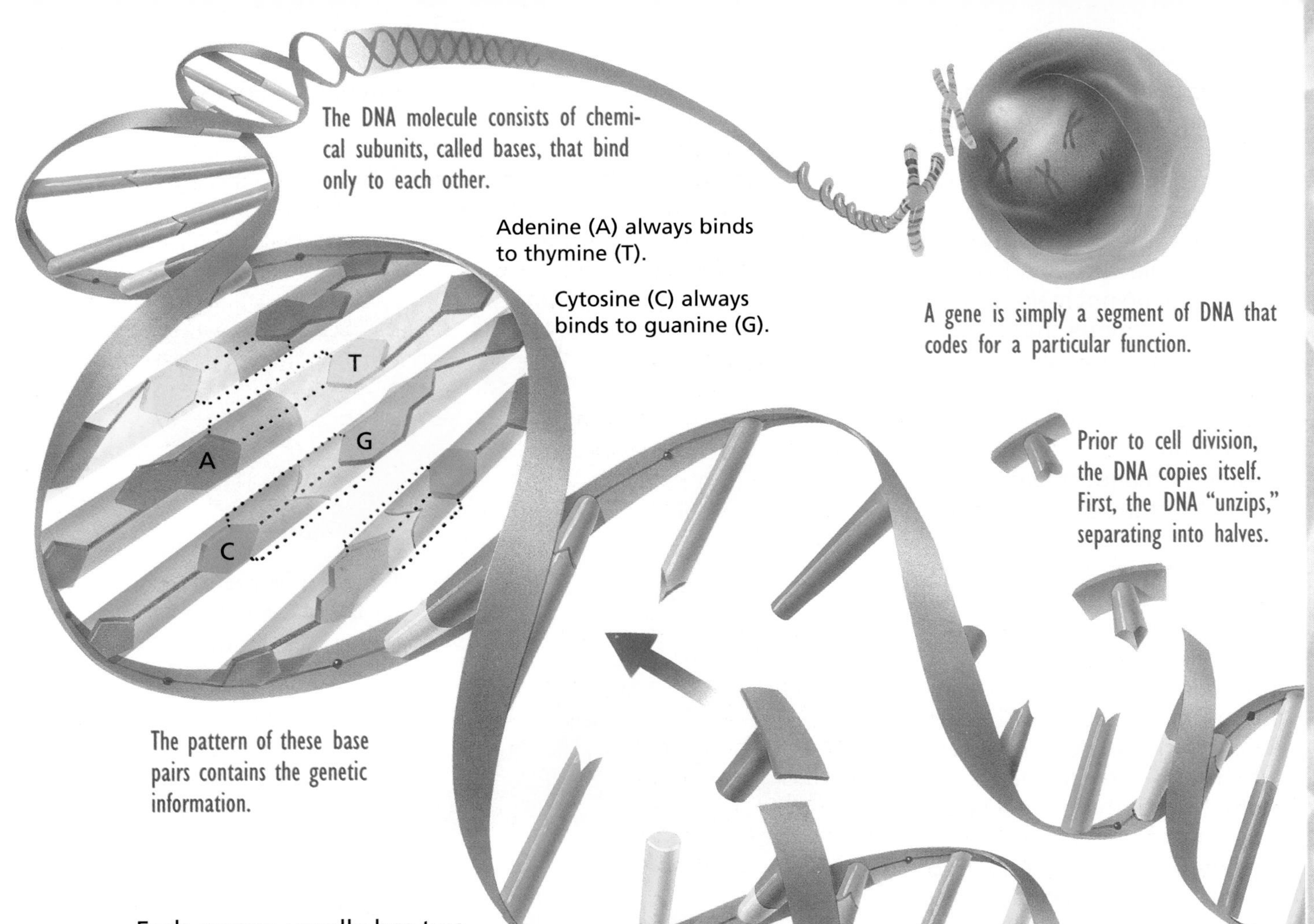

The DNA molecule consists of chemical subunits, called bases, that bind only to each other.

Adenine (A) always binds to thymine (T).

Cytosine (C) always binds to guanine (G).

A gene is simply a segment of DNA that codes for a particular function.

T

G

A

C

The pattern of these base pairs contains the genetic information.

Prior to cell division, the DNA copies itself. First, the DNA "unzips," separating into halves.

Free-floating bases in the cell nucleus are combined with the bases on the exposed DNA halves, . . .

. . . creating two identical copies of the original DNA molecule.

Each person usually has two genes for each trait. One chromosome in a pair has one copy, and the other chromosome has the other copy. When cells combine in fertilization, the new cells contain half of the genes from the mother's chromosomes and half of the genes from the father's chromosomes. Each half is randomly selected. For any trait, the mother's and the father's genes may be identical, or they may be different.

There are usually various forms of any one gene. For some genes, one of their two forms can be *dominant* while the other form is *recessive*. When an individual inherits the dominant form of a gene, the trait expressed by that gene will appear. Take, for example, eye color. There are two basic variations in eye color: brown or blue. Brown is dominant and blue is recessive. If you inherit a gene for blue eyes from one parent and a gene for brown eyes from the other parent, you will have brown eyes because genes for brown eyes are dominant over those for blue eyes. "What about people with green eyes?" you might ask. Eye color, in fact, is not as simple as it first appears. There are actually four different genes that determine eye color. The relative numbers of each gene you inherit determine the shade of color—from light blue to dark brown, with gray, green, and hazel in between.

There are over 100,000 genes in your chromosomes. Genes can combine in a practically unlimited number of ways. There are so many different possible combinations that each person is essentially unique. Except for identical twins, no two people probably have ever had the same set of genes—or ever will.

S61

Studying Genetics A genetic engineer is a scientist who studies and works with genes and DNA. What kinds of things do genetic engineers do? One example is the treatment of diabetes. People with one form of this disease do not make enough insulin, a hormone that regulates how much sugar is in blood. Diabetics require daily injections of insulin from some other source. At one time, insulin for diabetics was collected from the pancreases of pigs. It was an expensive and time-consuming task. Genetic engineers have now solved that problem by putting a piece of human DNA that has the instructions for making insulin into harmless bacteria. The bacteria, which can now manufacture insulin, are allowed to multiply until there are a large number of them. Combined, they produce a large amount of insulin, which is collected, processed, and given to patients by injection. Genetic engineering holds the promise of someday curing diabetes and other inherited diseases by "fixing" the abnormal genes that cause these diseases.

Insulin Processing Through Genetic Engineering

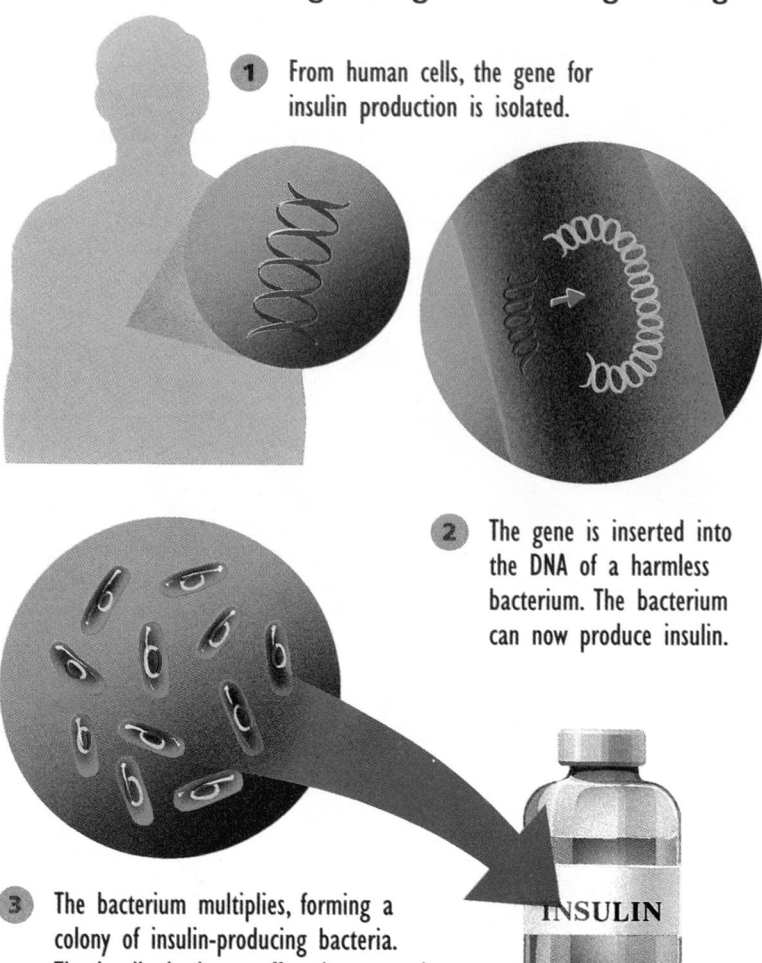

1. From human cells, the gene for insulin production is isolated.

2. The gene is inserted into the DNA of a harmless bacterium. The bacterium can now produce insulin.

3. The bacterium multiplies, forming a colony of insulin-producing bacteria. The insulin is drawn off and processed.

INSULIN

DID YOU KNOW...

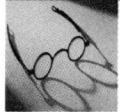

that an ambitious effort to map the entire human genome (set of genes) is now underway?
This project, called the Human Genome Project, is one of the most complex tasks ever tackled by human beings. By identifying the genes responsible for hereditary diseases, scientists hope to be able to develop effective drugs and treatments against these diseases.

S62

SUMMARY

The human body is composed of many organ systems that work together to perform all of the body's functions. The digestive and excretory systems break down food to provide nutrients and eliminate waste products. The circulatory system delivers nutrients and oxygen to all cells of the body and carries away wastes. The respiratory system supplies the body with oxygen and removes carbon dioxide. The integumentary system provides protection for the body, the skeletal system provides support, and the muscular system allows movement. The nervous and endocrine systems serve to sense, control, and regulate body functions. The immune system protects the body from microscopic invaders. The reproductive system transmits genetic instructions to the next generation. Scientists are learning to "engineer" genes to improve the quality of life.

Concept Mapping

The concept map below illustrates major ideas in this unit. Complete the map by supplying the missing terms. Then extend your map by answering the additional question below. Write your answers in your ScienceLog. **Do not write in this textbook.**

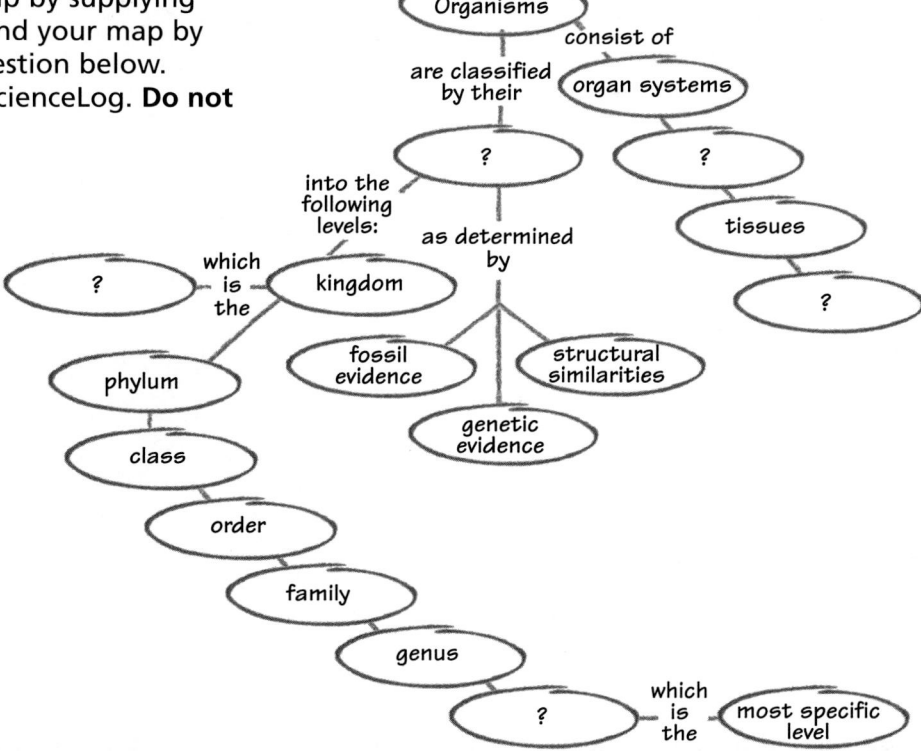

Where and how would you connect the terms *DNA* and *organelles*?

Checking Your Understanding

Select the choice that most completely and correctly answers each of the following questions.

1. Starfish and humans are members of the same
   a. kingdom.
   b. family.
   c. genus.
   d. species.

2. The material that transmits genetic instructions between generations is
   a. mitochondria.
   b. cytoplasm.
   c. DNA.
   d. testosterone.

3. Which of the following is NOT a function of the human skin?
   a. waste removal
   b. temperature regulation
   c. gas exchange
   d. protection

4. The body system enabling you to read and understand this question is the
   a. optic system.
   b. nervous system.
   c. immune system.
   d. educational system.

5. Chromosomes can be compared with
   a. ladders.
   b. security fences.
   c. a shop floor.
   d. blueprints.

6. Which choice correctly lists the levels of organization from simplest to most complex?
   a. organism, organ system, organ, tissue, cell
   b. cell, tissue, organ, organ system, organism
   c. tissue, organ system, organism, cell, organ
   d. organ, organ system, organism, tissue, cell

S63

Interpreting Illustrations

The accompanying illustration compares the forelimbs of a whale and a human. Identify the corresponding parts, and describe how these similarities point to an evolutionary relationship between whales and humans.

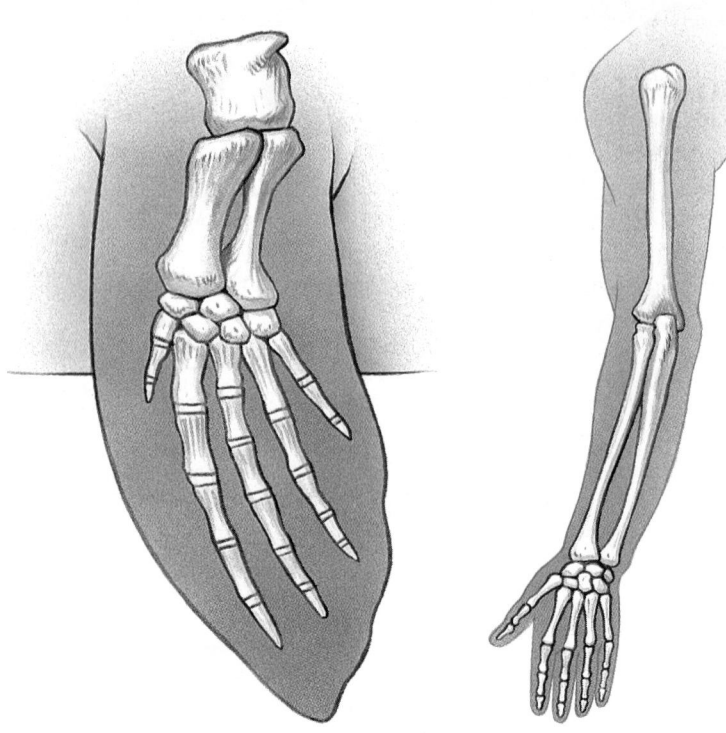

Critical Thinking

Carefully consider the following questions, and write a response in your ScienceLog that indicates your understanding of science.

1. On a late-night television show, a self-proclaimed psychic predicts that four entirely new biological kingdoms will be discovered in the coming year. Is this a realistic prediction? Explain.

2. In a certain population of foxes, 95 percent are reddish in color, and the rest are solid white. Foxes feed by hunting small animals. Suppose that the climate (which is now fairly moderate with little winter snow) changes and becomes cold and snowy for much of the year. How might this affect the ratio mentioned in the first sentence of this question? Explain.

3. Certain microorganisms carry out photosynthesis but are also able to take in food from their surroundings and to move about on their own. Speculate how the discovery of organisms such as these helped to do away with the two-kingdom system of classification.

4. How does the cell wall determine some of the major defining characteristics of plants? Why would a cell wall be a disadvantage in animal cells?

Portfolio Idea

Write a story in which you are an organ system, and describe a typical day in your life. Describe what you do, why you do it, how you work with other organ systems, typical problems that might affect you, and so on. Be creative but factually accurate.

Unit 3

IT'S A SMALL WORLD

Now that you know something about microorganisms and how they can be both harmful and helpful in food preparation, consider these questions.

● **1.** What are infectious diseases, and how are they spread?

2. How are viruses different from other microorganisms?

3. What are the body's levels of defense, and what happens at each level to protect the body from illness?

In this unit, you will take a closer look at diseases, the microorganisms that cause them, and how your body defends itself against them.

S65

MICROORGANISMS AND DISEASE

In Ireland in the 1800s, many people lived on small farms. Potatoes were a major crop because they were easy to grow, were a good food source, and could be stored over the winter. In 1845 and 1846, however, a potato disease called *late blight* destroyed almost the entire potato crop of Ireland in just a few short weeks. Because of this, over 1 million people died of starvation. Many others left their homeland and came to America.

Other tragedies have been caused by diseases that affect food sources or cause illnesses in people. One of the most tragic was the plague of the 1300s called the Black Death, which wiped out one-fourth of Europe's entire population! This deadly disease was spread by rats that infested the villages and towns. But the rats were only a means of transporting the actual cause of the plague—microorganisms.

The disease of a plant forced many people to face the unknown in search of a new life.

Rats can carry fleas whose bite can pass disease-causing microorganisms on to humans. This is how the plague spread throughout medieval Europe.

Germ

Any microorganism that can cause disease

Linking Microorganisms to Disease

The idea that microorganisms can cause disease was developed in the mid-nineteenth century by a French chemist named Louis Pasteur. This idea is often called the *germ theory of disease*. A **germ** is any microorganism that can cause disease. At first this theory, like many others, was not taken seriously. However, the idea gradually gained the support of some famous scientists and physicians of the day.

S66

Joseph Lister, an English surgeon, was one of the first to see the importance of Pasteur's work. Lister used chemicals that killed germs to make his operating rooms safe. He made sure that everything that touched his patients was very clean. His methods prevented infections after surgery. An **infection** is the growth of disease organisms in the body of a host organism. The chemicals that Lister used are called *disinfectants*. Disinfectants kill many of the germs that cause disease.

Robert Koch, a German physician, was the first to demonstrate that microorganisms do cause disease. He studied a disease called anthrax, which affects horses, cows, sheep, and humans. During his investigations, Koch developed a set of rules, or steps, for proving that a specific microorganism is the cause of a specific disease. This set of four rules, called *Koch's postulates*, is still used today. Koch's postulates, a fine example of a scientific method, are illustrated below.

Infection

The growth of disease microorganisms in the body of a host organism

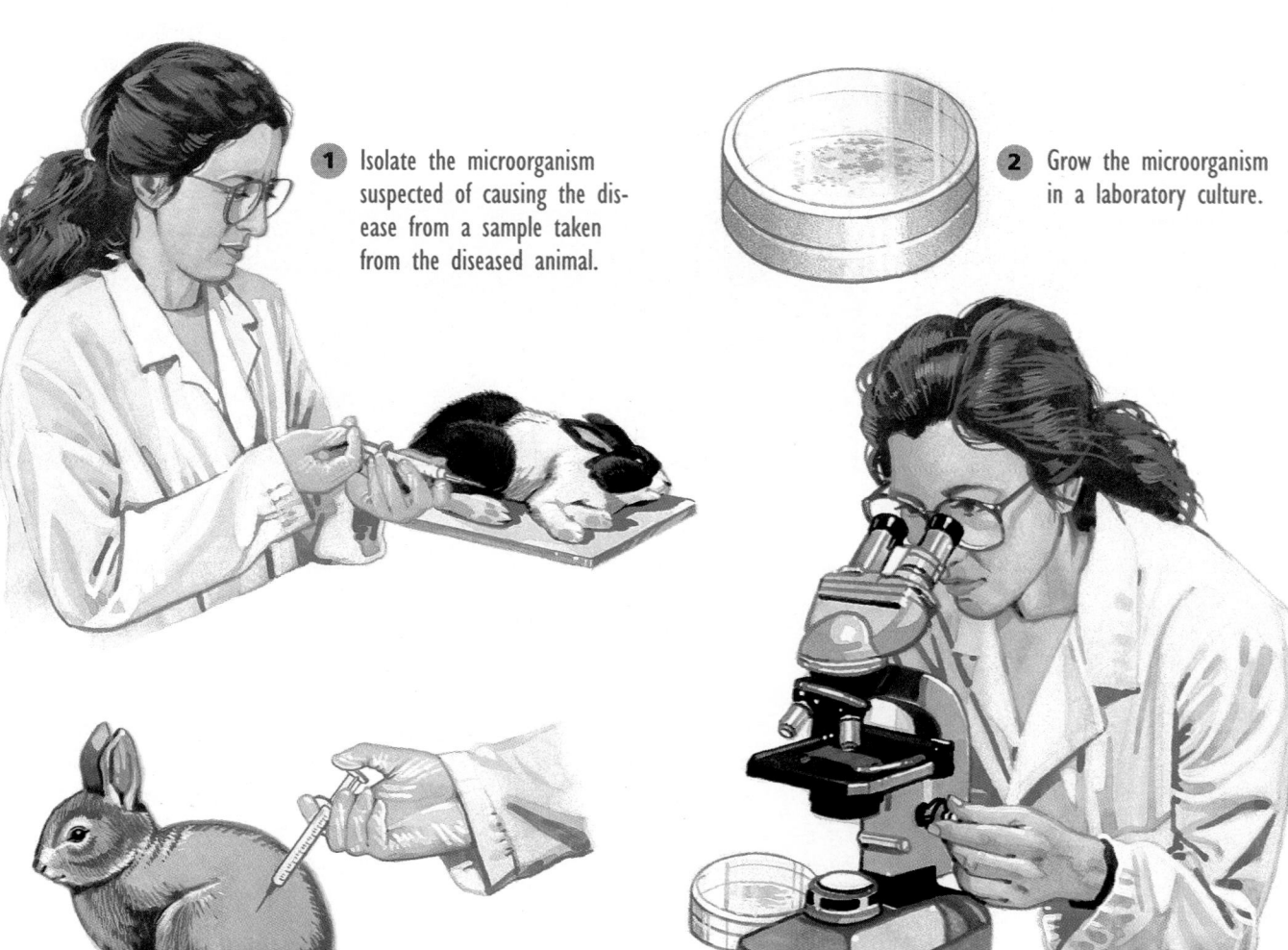

1. Isolate the microorganism suspected of causing the disease from a sample taken from the diseased animal.

2. Grow the microorganism in a laboratory culture.

3. Inject the microorganisms from the laboratory culture into a healthy animal. Examine the animal for the disease.

4. If the animal gets the disease, grow another laboratory culture using a sample from this animal. Check to make sure that the microorganisms that grow in this culture are the same as the microorganisms that were in the original culture.

S67

Infectious Disease

What exactly is disease? A *disease* is a condition in which some part of a living thing is not working properly. Some types of disease, like diabetes, are caused by genetic factors. Other diseases are caused by environmental factors. For example, inhaling large amounts of coal dust can cause coal miners to develop a disease known as black lung. Many diseases are caused by microorganisms. When a disease-causing microorganism lives in a host plant or animal, the microorganism may attack and destroy the host's tissues and make it ill. Some microorganisms make poisons that kill their hosts.

Many diseases caused by microorganisms can be transmitted to other individuals. This type of disease is called an **infectious disease**. Infectious diseases are spread when the microorganisms that cause them are transferred from one host to another. These diseases can be spread by water, air, or direct contact.

▲ Some disease-causing microorganisms live in water and can be spread through the water supply. This happens if wastes from humans and animals are not kept far enough away from the water supply. If people drink that water, they may get one of these diseases. Typhoid is a disease that is spread in this way.

Infectious disease

A disease caused by a microorganism that can spread from one organism to another

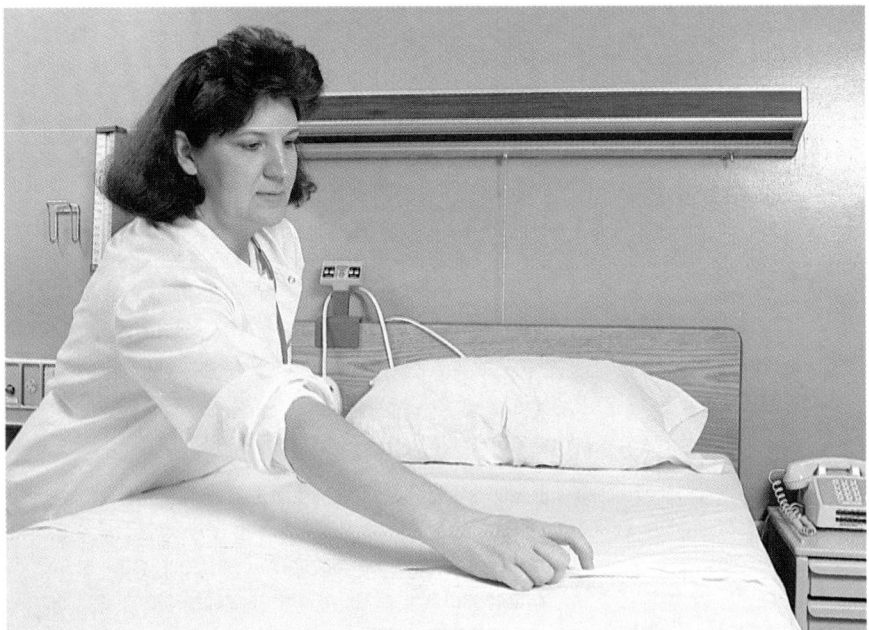

▲ Many diseases are spread by tiny droplets of water in the air. For example, suppose that you have a cold and you sneeze. Millions of the microorganisms that caused your cold are expelled from your mouth on tiny water droplets. Anyone who inhales them could catch your cold. Diphtheria and tuberculosis are also spread in this way.

◄ With certain diseases, touching an infected individual, or even the dishes and bed linens they have used, may cause the spread of the disease. This includes touching or eating food on which disease-causing organisms grow. Colds, the flu, and food poisoning are spread in this way.

Animals can spread diseases to each other and to humans. For example, the common housefly carries and spreads dozens of different kinds of disease microorganisms. Diseases such as malaria and yellow fever are transmitted by the bites of certain mosquitoes. Dogs, cats, wolves, cattle, squirrels, and bats are among the animals that can spread rabies by direct contact.

Even people who do not show symptoms of a disease can still spread a disease. Such people are called *carriers*. Carriers can be just as infectious as people who are ill with the disease. For example, infections such as a sore throat, which is caused by *Staphylococcus* bacteria, can be spread by carriers.

Some animals can spread disease to humans. The housefly, the *Anopheles* mosquito, and the skunk are all capable of transmitting disease.

Causes of Infectious Disease

The organism identified by Koch as causing anthrax was a type of bacteria. However, bacteria are only one kind of microorganism that can cause infectious disease. By using Koch's postulates for isolating disease-causing microorganisms, viruses, protozoans, and fungi have also been identified as agents of disease. Bacteria and viruses are responsible for a long list of human diseases, some of which are shown in the table to the right. Lists just as long could be made of the diseases caused by bacteria and viruses in other animals or in plants.

Bacterial diseases	Viral diseases
bacterial dysentery	chickenpox
bacterial pneumonia	the common cold
boils	fever blisters
bubonic plague	influenza (flu)
gonorrhea	measles
strep throat	mumps
syphilis	polio
tetanus	rabies
tuberculosis	viral pneumonia

S69

Toxin

A poison produced by some organisms that causes harm to other organisms

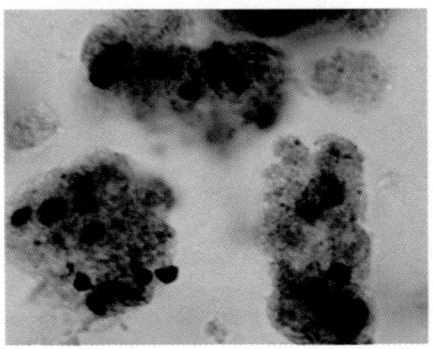

▲ This amoeba is a parasite that causes dysentery in humans.

Bacteria Bacteria can affect the human body in different ways. The bacteria that cause tuberculosis kill the cells and tissues of the host. Other types of bacteria do damage by releasing strong poisons called **toxins** into the host. The bacteria that cause *tetanus* can grow in the host where the skin is badly broken. These bacteria make a toxin that is carried by the bloodstream to the brain. There, it can do enough damage to the brain to cause death.

Viruses Viruses are another major cause of disease. They are also the smallest of the disease-causing microorganisms. Even the largest virus can just barely be seen with a light microscope. This is one of the reasons that viruses were not photographed until the 1930s. Nevertheless, doctors have been treating diseases caused by viruses for over 200 years. These diseases include the common cold, influenza (flu), cold sores, warts, smallpox, and rabies. Diseases such as AIDS and Ebola are caused by viruses that have just recently been discovered.

Protozoans Amebic dysentery, malaria, and sleeping sickness are examples of diseases caused by *protozoans*. Amebic dysentery, a disease characterized by stomach cramps and severe diarrhea, is caused by a species of amoeba. Malaria is spread by the bite of an *Anopheles* mosquito that injects a parasitic protozoan into the bloodstream. Sleeping sickness is caused by a protozoan that is passed from host to host by the bite of the tsetse fly.

Fungi Fungi can also cause diseases, such as athlete's foot, in humans. Other examples of fungal diseases are ringworm and histoplasmosis. Histoplasmosis is a sometimes serious disease that often affects the lungs by causing severe pneumonia. It is caused by a fungus that grows in soil containing bird or bat droppings. Cells of the fungus can enter the body through the digestive or respiratory system.

The moisture in athletic ▶ locker rooms and showers can be ideal for the growth and spread of the fungus responsible for athlete's foot.

Diseases in Plants

Microorganisms also cause diseases in plants. The control of plant diseases is of great economic and social importance. Recall the effect of Ireland's potato disease! Plant diseases that are caused by microorganisms are spread from plant to plant, just like infectious animal diseases are spread between animals.

Bacteria are responsible for many plant diseases. Houseplants and greenhouse-grown crops commonly wilt in moist or wet soil because of a disease called bacterial wilt. A disease of tobacco, which causes the destruction of chlorophyll in a leaf, was found to be caused by a virus. The result of the disease is an irregular pattern (mosaic) on the leaf. Much of the early work on viruses was done to study the cause of this tobacco disease. Many plants are affected by the tobacco mosaic virus (TMV). Amazingly, these viruses cannot be destroyed by burning. In fact, mosaic diseases are often spread among greenhouse crops, such as tomatoes, by people smoking tobacco.

Fungi also cause very serious plant diseases. Wheat rusts, a type of fungi, have caused large losses in wheat crops. A disease caused by another fungus wiped out the chestnut tree in this country in the early part of the twentieth century. More recently, thousands of elm trees have been dying from Dutch elm disease, a fungal disease that is transmitted by beetles.

▲ This tree has a crown gall tumor. The tumor is caused by a bacterium that infects plants.

▲ Dutch elm disease has nearly wiped out this favorite shade tree in North America.

S U M M A R Y

Microorganisms have been shown to be the cause of various diseases. A disease is a condition in which some part of a living thing is not working properly. Infectious diseases, which are caused by bacteria, viruses, protozoans, and fungi, are spread from host to host in several ways. Microorganisms cause infectious diseases in plants as well as in animals and humans.

S71

MORE ABOUT BACTERIA AND VIRUSES

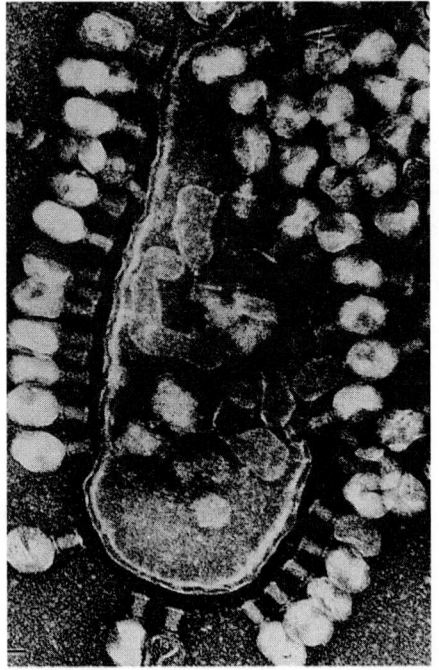

▲ What is this? How big do you think these objects are?

What do you see in the picture to the left? Is it an abstract design printed on a piece of fabric? Is it a group of animals gathered around a watering hole as seen from a helicopter? Is it a squadron of alien spacecraft about to land on Earth in the middle of the night?

Actually, it *is* the beginning of an invasion. It is the invasion of a single bacterium by an army of viruses. Bacteria and viruses are the smallest and most numerous disease-causing microorganisms. Each is unique and interesting in its own way.

Bacteria

Bacteria are among the smallest living things. Some 50,000 bacteria could fit on the head of a pin. Even at a magnification of 1000 times, only their general shape can be seen. Bacteria are everywhere. They have been found in many places where other life-forms cannot exist. Some bacteria have been detected 8 to 10 km up into the atmosphere or deep underground. Others thrive in the hot-water vents on the ocean floor, and some even live inside other living things. Fossil evidence indicates that bacteria have been on Earth for billions of years.

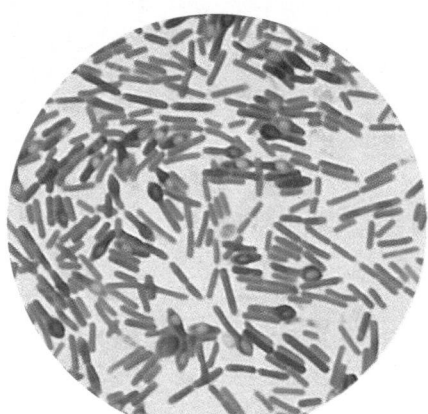

▲ These bacteria, magnified 400 times, cause botulism (food poisoning).

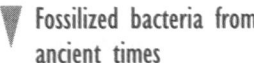

▼ Fossilized bacteria from ancient times

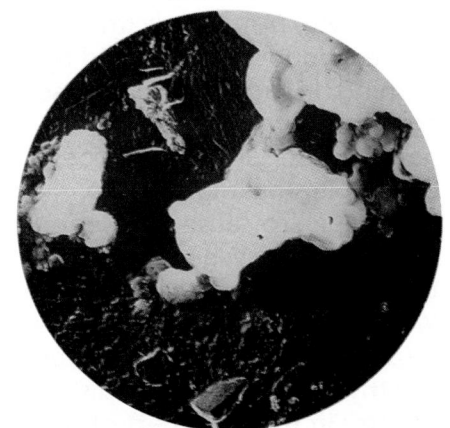

In addition to being adapted to various unusual environments, some bacterial colonies have evolved ways to survive extremely unfavorable conditions. They often do this by forming inactive cells called *spores*. When conditions are better, the spores become active and continue the existence of the colony. Some types of bacteria can survive for over 50 years in this way.

S72

Many bacteria are surrounded by a gelatinous layer called a capsule that protects them. In this illustration, you can see a diagram of a bacterial cell with a capsule and many *flagella*. The flagella are whiplike structures that help the bacterium move in response to changes in its environment. Not all bacteria, however, have all of these structures.

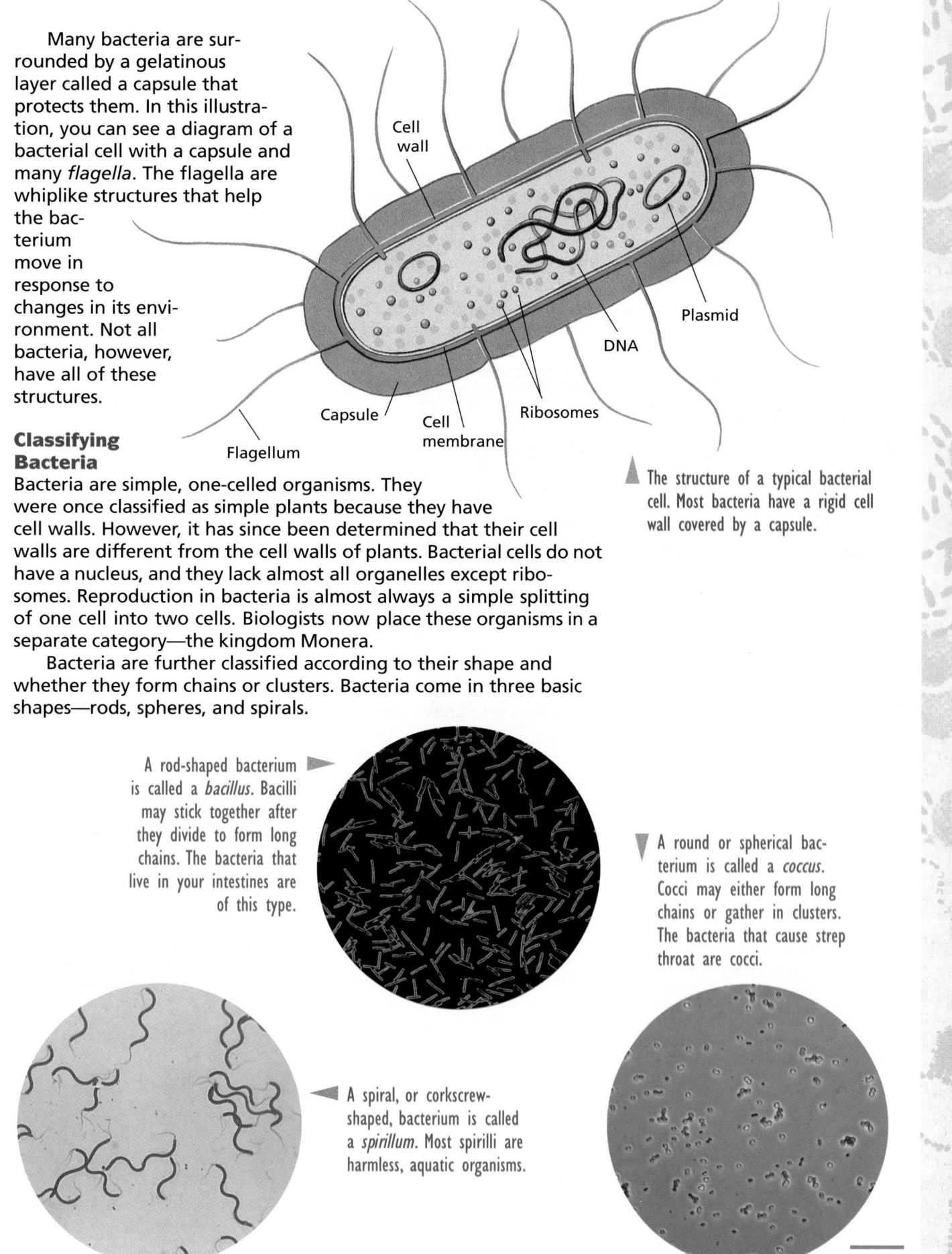

Cell wall

Plasmid

DNA

Capsule

Cell membrane

Ribosomes

Flagellum

▲ The structure of a typical bacterial cell. Most bacteria have a rigid cell wall covered by a capsule.

Classifying Bacteria

Bacteria are simple, one-celled organisms. They were once classified as simple plants because they have cell walls. However, it has since been determined that their cell walls are different from the cell walls of plants. Bacterial cells do not have a nucleus, and they lack almost all organelles except ribosomes. Reproduction in bacteria is almost always a simple splitting of one cell into two cells. Biologists now place these organisms in a separate category—the kingdom Monera.

Bacteria are further classified according to their shape and whether they form chains or clusters. Bacteria come in three basic shapes—rods, spheres, and spirals.

A rod-shaped bacterium is called a *bacillus*. Bacilli may stick together after they divide to form long chains. The bacteria that live in your intestines are of this type. ►

▼ A round or spherical bacterium is called a *coccus*. Cocci may either form long chains or gather in clusters. The bacteria that cause strep throat are cocci.

◄ A spiral, or corkscrew-shaped, bacterium is called a *spirillum*. Most spirilli are harmless, aquatic organisms.

S73

Parasites

Organisms that live in or on other living organisms, usually causing them harm

Saprophytes

Organisms that get food from dead or decaying organisms

How Bacteria Obtain Food   Differences in the ways that bacteria make or obtain food may also be used to classify them. Many bacteria get their food from dead organisms, animal wastes, and rotting vegetation. Others live in or on other living organisims and obtain their food from those organisms. Some of these bacteria are **parasites** that injure their hosts by robbing them of nutrients they would normally receive. But other bacteria live in a host without causing harm. In fact, they may even help. For example, certain plants have bacteria that live inside their roots. These bacteria change nitrogen gas from the air into nitrogen compounds that the plants use for growth. Bacteria that get their food from dead organisms are called **saprophytes**. Many of these bacteria are very useful. Saprophytes break down the tissues of dead organisms, thus releasing their chemicals back into the air and soil.

Some bacteria make their own food. One major group of bacteria, called cyanobacteria, can produce their own food by photosynthesis. The blue-green color of these bacteria comes from the chlorophyll they contain, which enables them to use the sun's energy to make food. Most cyanobacteria live in fresh water. Some contain gas bubbles and can float near the water's surface, where they receive more sunlight. They produce oxygen and also serve as food for other organisms. Other bacteria make their own food without using the energy of the sun. These bacteria get energy by transforming certain compounds that contain nitrogen and sulfur. The bacteria in the hot-water vents on the ocean floor are in this group, as are some of the bacteria found in the soil.

Bacteria that live in this hot spring in Wyoming get energy by transforming sulfur compounds found in the water.

Bacteria and Humans   No matter how clean you are right now, bacteria are living in and on you. They are everywhere! In fact, bacteria live on your skin and in your nostrils, mouth, and large intestine. Bacteria play an important role in the lives of humans. Although some bacteria do cause diseases in humans, most of the bacteria we encounter are harmless.

Some bacteria are useful in a variety of ways. For example, bacteria can be added to milk to produce yogurt, sour cream, cottage cheese, and hard cheeses. Sauerkraut, sour pickles, and vinegar are also made by the action of bacteria.

Cheddar cheese begins as a soupy mixture that includes milk and bacteria.

S74

Bacteria are used to make medicines that help our bodies fight other, harmful bacteria. Bacteria are also useful in the mass production of substances such as human insulin. As you recall, human genes for insulin synthesis can be isolated and spliced into the DNA of bacteria. Because bacteria reproduce rapidly, the new genetic instructions are used to produce insulin in amounts that can be used by doctors in the treatment of patients with diabetes.

Bacteria are in all of the foods we eat. They cause rotting and spoilage as they break down these foods for their own consumption. Because of this, bacteria can cause heavy losses of stored food. Fortunately, refrigeration slows down the growth of bacteria. The heat of cooking also reduces the number of bacteria in food.

Bacteria can contribute to an environmental problem called *eutrophication*. Water that contains chemical fertilizers can permit the rapid growth of algae. When these tiny organisms die, bacteria feed on their remains and soon the population of bacteria explodes in number. Too many bacteria can use up all the oxygen in the water, causing fish and other organisms in the water to die from lack of oxygen.

Viruses

Years ago, smallpox was an easily spread and often fatal disease. It caused severe chills, headaches, fever, nausea, and pains in the back, arms, and legs. Red blisters appeared all over the body, and once they dried up, they left hollow pockmarks. For a long time, scientists could not isolate the microorganisms that caused diseases such as smallpox.

▲ Bacteria abound in lakes affected by pollution.

What Are Viruses? In his work, Pasteur found that the bacteria that caused milk to turn sour could be captured by fine filters. Eventually, all of the disease-causing bacteria that scientists identified could be captured by very fine filters. However, the agents that caused smallpox and other mysterious illnesses, such as rabies, were so small that they were able to pass through these filters. They came to be known as *filterable viruses*, or simply viruses, from a Latin word that means "poison." But not much was known about viruses until the development of the electron microscope in the early 1930s.

▲ This virus, magnified 80,000 times by an electron microscope, causes measles.

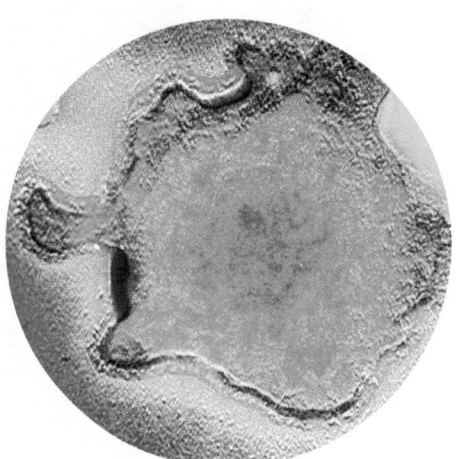

◄ These viruses are the cause of the common cold.

S75

Modern equipment makes it easier to study viruses. Photographs of viruses enlarged over 300,000 times have been made with the help of the electron microscope. These photos show that viruses come in many shapes and sizes. They may be round or shaped like rods, needles, or cubes. They may have many-sided shapes as well. Some even have long, thin, leg-like structures.

Electron microscopy revealed that viruses are not cells. They do not have a nucleus, cytoplasm, or a cell membrane. The simplest viruses seem to be made of just a **nucleic acid** (DNA or RNA) with a coating of protein. Remember from Unit 2 that nucleic acids control the activities in a cell and are the means by which traits are passed on to new cells. Viruses can be classified by the kinds of organisms they affect, as well as by the type of nucleic acid they contain and their shape and size.

Are Viruses Alive? It is difficult to decide if viruses are alive or not. They do not carry on life processes in the way that other living cells do. They do not take in food or give off wastes, and they can reproduce only inside the cells of living organisms. The mumps virus, for example, becomes active only when it comes in contact with the gland cells near a person's jaw. The polio virus multiplies only in one kind of nerve cell in the brain and spinal cord.

Because scientists do not consider them to be alive, viruses are not included in the five-kingdom system of classification. However, they are often included in the study of microorganisms because, like other microorganisms, viruses cannot be seen or identified without the use of a microscope.

Nucleic acid

A chemical that controls activities in a cell and passes on traits to new cells

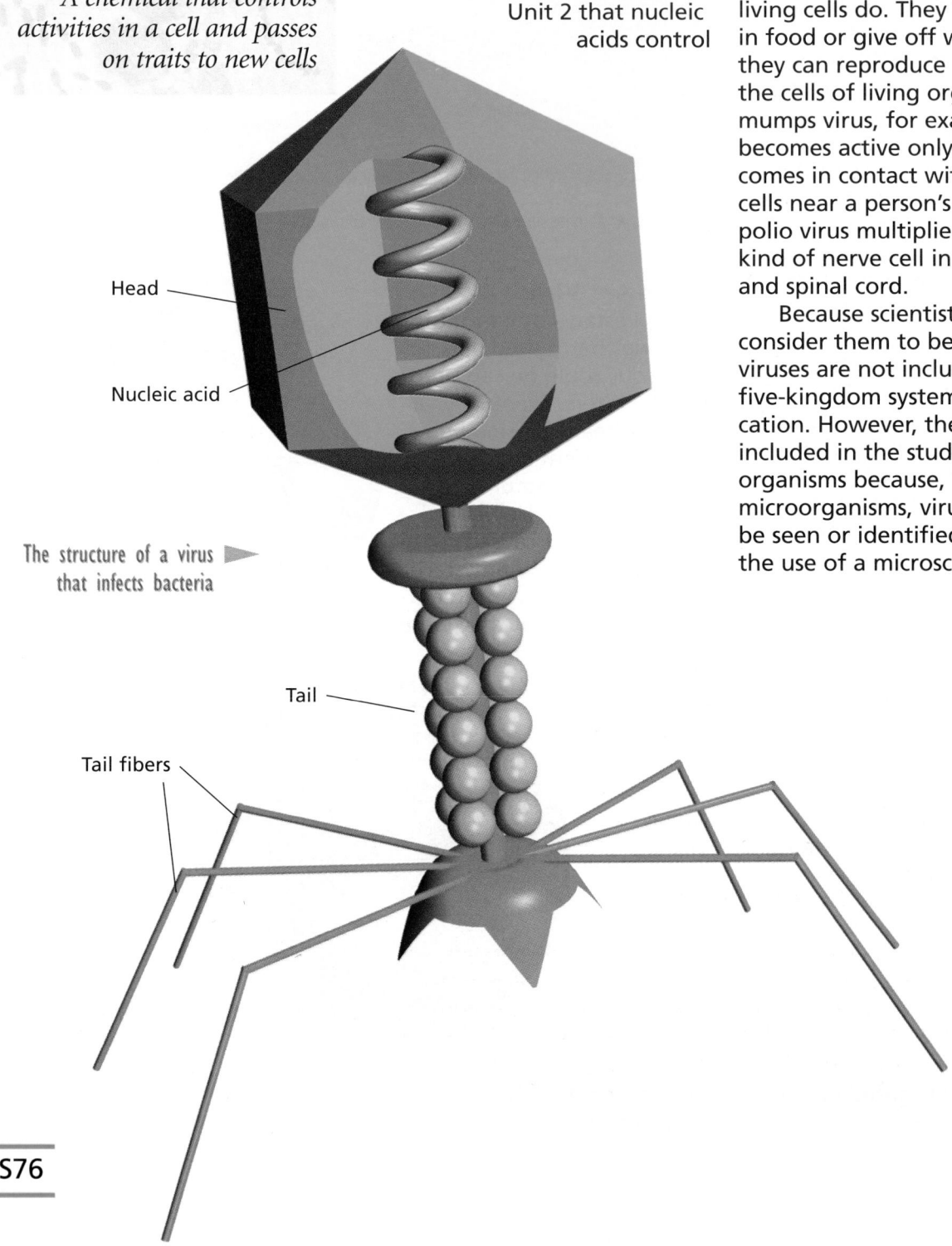

Head

Nucleic acid

The structure of a virus that infects bacteria ▶

Tail

Tail fibers

S76

How Viruses Function

Scientists have learned a lot about viruses by studying a type that attacks bacteria. These viruses have a complex structure that includes "landing legs," which are used when the viruses "touch down" on the surface of bacteria.

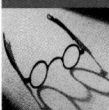
1 When a virus attacks a bacterial cell, the tail fibers of the virus attach to the cell surface.

2 The nucleic acid of the virus is then injected into the cell through the membrane. Once this happens, the nucleic acid may become active, or it may instead enter into an inactive, or *dormant*, state. If the nucleic acid remains inactive, it can be reproduced as the host cell divides.

3 Later, a change in conditions may cause the virus's nucleic acid to become active, causing the host cell to use its own energy and chemicals to make new viral nucleic acids and protein coats.

4 The nucleic acids and protein coats assemble to form hundreds or thousands of new viruses as multiple copies of the original virus are formed within the cell.

5 Soon the cell bursts, and the new viruses are released. These, in turn, attack other bacteria and repeat the process. In this way, additional viruses are formed and spread from cell to cell.

SUMMARY

Bacteria are single-celled organisms that lack nuclei and other organelles. They are classified by their shape and the different ways they obtain food. Bacteria may be harmful or helpful to other organisms. Many viruses are made only of a nucleic acid covered with protein. Viruses do not carry on all of the life processes of living organisms. Therefore, viruses are not usually classified as living organisms. Viruses reproduce only in the cells of living things.

THE BODY'S DEFENSES

The boy in the photo to the left has spent his entire life inside a plastic bubble. He is not allowed to touch anyone, not even his mother or father, because he was born with no natural defense against infectious diseases. As a result, even a simple cold could endanger his life. Fortunately, most of us have a defense system that automatically fights off most of the bacteria and viruses that could harm us. This system consists of several lines of defense.

This boy must be protected from all germs—even those from his family. He has no immunity to disease. ▶

Skin and Mucus

The first line of defense is the skin. Under normal conditions, the skin stops microorganisms from entering the body. However, when the skin is broken, cut, or damaged, germs can enter. That is why it is very important to clean cuts and scrapes. But skin does not cover every surface of the body.

Mucus is a thick, sticky fluid that covers many surfaces inside the body and inside the natural openings of the body. Mucus stops germs from attacking tissue not covered by skin. For example, the inside of the nose is covered by tiny hairs and mucus. These hairs and mucus trap dust and germs from the air you breathe. Sometimes extra mucus is made by the body in response to the presence of foreign substances such as dust, pollen, or germs. If there is a lot of mucus in your nose, you should try blowing it out. Blowing your nose and sneezing helps remove trapped microorganisms. It is important to cover your mouth and nose when you sneeze to prevent the spread of these microorganisms.

If particles in the air are not trapped in your nose, they may still be trapped by another layer of mucus that coats the inside of your windpipe. This mucus, along with the trapped germs, moves upward to the back of the throat, where it is then swallowed. The swallowed mucus travels down the esophagus to the stomach. In the stomach, strong acids kill the microorganisms. These stomach acids also kill most germs brought into the body with food.

Mucus

A thick, sticky fluid covering many surfaces inside the body and in its natural openings

▲ Blowing your nose helps clean out the mucus that traps disease-carrying microorganisms.

S78

White Blood Cells

What happens if you cut your skin and germs enter the cut? Then your second line of defense, the *white blood cells*, becomes active.

White blood cells are one part of your blood. They are made inside some of your bones. Many of them are found in structures called lymph nodes and in the tonsils.

It is believed that damaged tissue, such as a cut, and invading germs both release chemicals. These chemicals attract white blood cells. At the same time, the area around the cut becomes warm and appears red, indicating that the cut has become infected.

White blood cells surround and destroy germs and damaged tissue. This action is similar to the way that an amoeba surrounds its food. The activity of white blood cells stops infection and cleans the area so that proper healing can take place.

The Immune System

In the body, white blood cells have still another way of fighting disease. Some kinds of white blood cells make special chemicals called **antibodies**. Antibodies help in the destruction of microorganisms and other foreign substances. Your body is capable of producing antibodies for just about every kind of germ or foreign substance that exists on Earth.

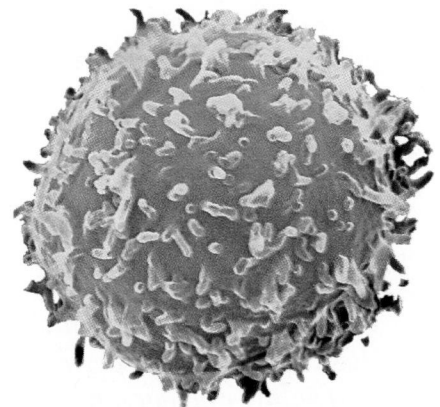

▲ White blood cells, such as that shown here, are the body's second line of defense.

Antibodies

Chemicals that are made by the body and that fight germs or other foreign substances

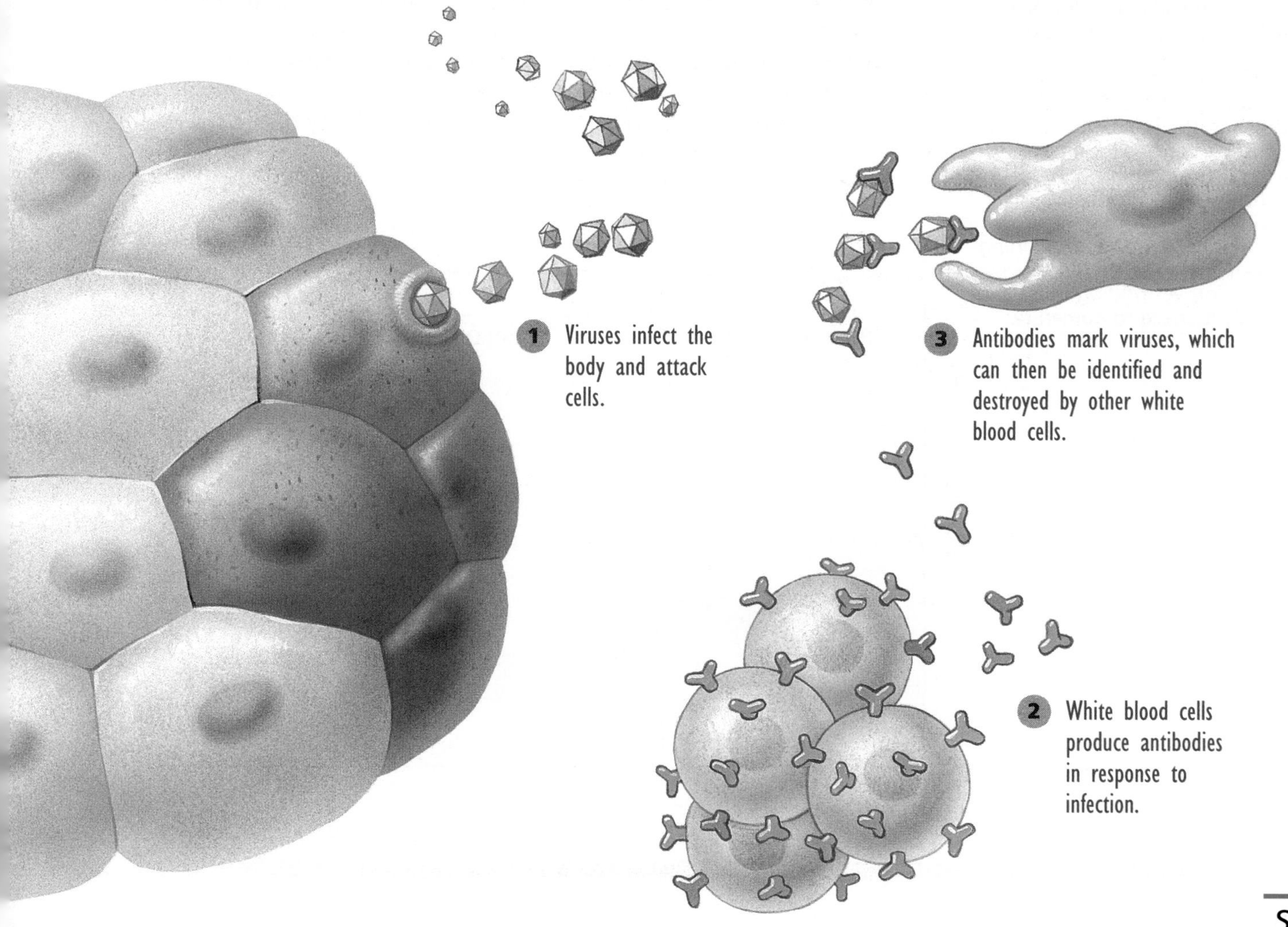

1. Viruses infect the body and attack cells.

3. Antibodies mark viruses, which can then be identified and destroyed by other white blood cells.

2. White blood cells produce antibodies in response to infection.

S79

Immune system

Body system that uses antibodies to seek out and destroy invading microorganisms

Acquired immunity

Resistance to reinfection by a disease after the body has recovered from the original infection

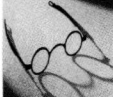
The production of antibodies is a relatively quick process. A few days after an invader has entered the body, a large number of antibodies can usually be found in the blood. This process of antibody production is a function of the **immune system**, the body's third line of defense.

Soon after a disease is successfully stopped, the level of the antibody that fought against it drops. For example, a person who has recovered from chickenpox will have only a small amount of chickenpox antibody left in his or her bloodstream. But a few of the white blood cells that made the chickenpox antibody remain in the bloodstream to fight the chickenpox virus if it returns. These white blood cells "remember" how to make the antibody for chickenpox. If the virus that causes chickenpox enters the body again, these cells will make a lot of new antibodies in a very short time. They will eliminate the virus before it can do any damage and before you become ill. That is why a person usually gets diseases like measles, mumps, whooping cough, scarlet fever, and chickenpox only once. This resistance to a disease is called **acquired immunity**. Acquired immunity to some diseases lasts a lifetime.

Immune System Disorders

Not all diseases that affect the human body are caused by microorganisms. Some diseases, such as various types of heart disease and arthritis, are caused by malfunctions of the immune system. Like other systems in the body, the immune system can malfunction or even break down altogether. Such a breakdown is called an immune disease or disorder. There are three basic types of *immune disorders*: allergies, autoimmune disorders, and immune deficiency disorders.

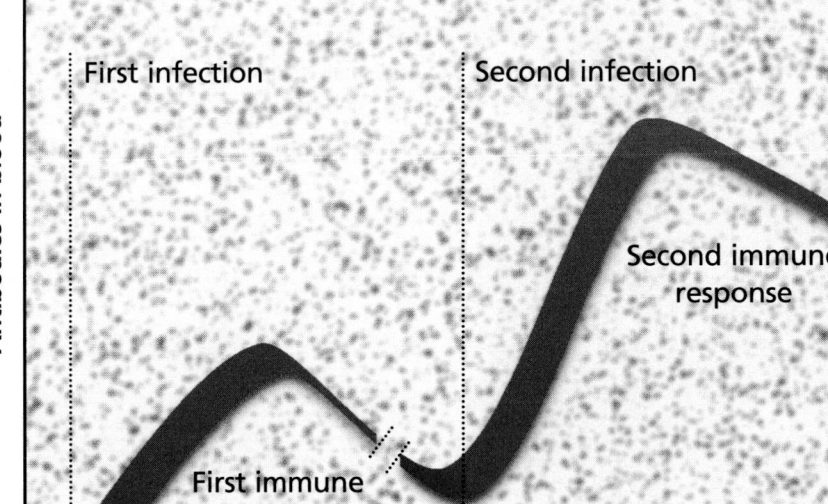

▲ The second time an infection occurs, the immune system responds more quickly and with more antibodies to prevent a second illness.

An **allergy** is a condition in which the immune system reacts to a normally harmless foreign substance, such as pollen or certain foods. In an allergic reaction, the immune system produces antibodies that attack the foreign substance, causing a variety of symptoms: runny nose, sneezing, red and watery eyes, swelling, rashes, and so on. In a few cases, severe allergic reactions can be fatal.

Allergy

A condition in which the immune system's antibodies attack a normally harmless foreign substance

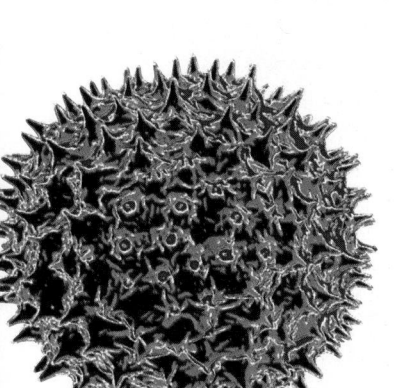
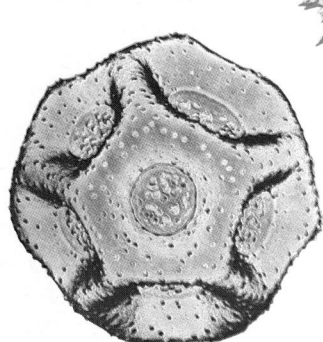
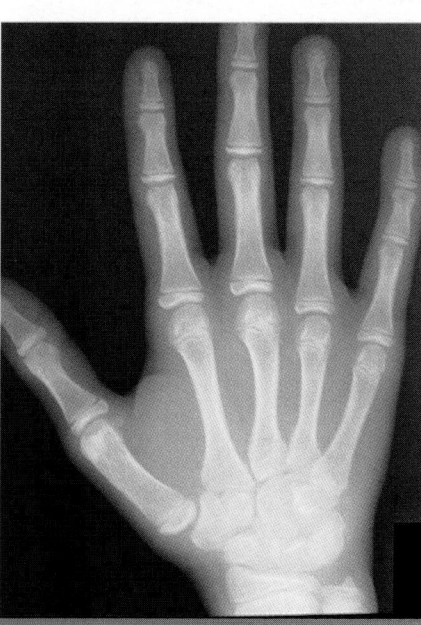

Airborne pollen grains, which come in many shapes and sizes, may trigger an allergic reaction.

An **autoimmune disorder** is a condition in which the body attacks its own tissues with antibodies. There are many different types of autoimmune disorders. Some are not very serious, but others can be life threatening. Some examples of autoimmune disorders are rheumatoid arthritis, multiple sclerosis, and Graves' disease. Some autoimmune disorders are caused by abnormal genes, and others are caused by exposure to certain chemicals. Still others develop after the body has fought off a viral infection.

Autoimmune disorder

A condition in which the immune system's antibodies attack the body's own tissues

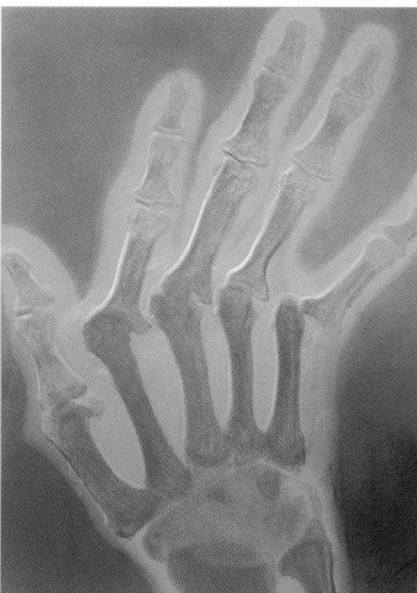

Rheumatoid arthritis is an autoimmune disorder that causes the joints between bones to become swollen and painful. The X ray on the left shows the bones inside a normal hand. On the right is an X ray of a hand severely affected by rheumatoid arthritis. In extreme cases such as this, the disease is crippling.

S81

Immune deficiency disorder

A condition in which the immune system breaks down, leaving the body unprotected from disease

An **immune deficiency disorder** is a condition in which the body is unable to defend itself against invading microorganisms. This condition causes its victims to suffer from repeated infections. Severe cases of immune deficiency may completely disable the immune system. Such cases are almost always fatal. Some immune deficiencies are inherited. Others, such as AIDS, are acquired during a person's lifetime.

The green "dots" in the photograph are human immunodeficiency viruses (HIV), the virus that causes AIDS. They are shown attacking a white blood cell.

SUMMARY

The body has three lines of defense against infectious disease: the skin and mucous membranes, the white blood cells, and the immune system. Immunity to a particular disease often occurs after a person recovers from the illness. Disorders of the immune system include allergies, autoimmune disorders, and immune deficiency disorders. Disorders of the immune system can be inherited or acquired.

Unit CheckUp

Concept Mapping

The concept map below illustrates major ideas in this unit. Complete the map by supplying the words needed to connect the terms. Then extend your map by answering the additional question below. Write your answers in your ScienceLog. **Do not write in this textbook.**

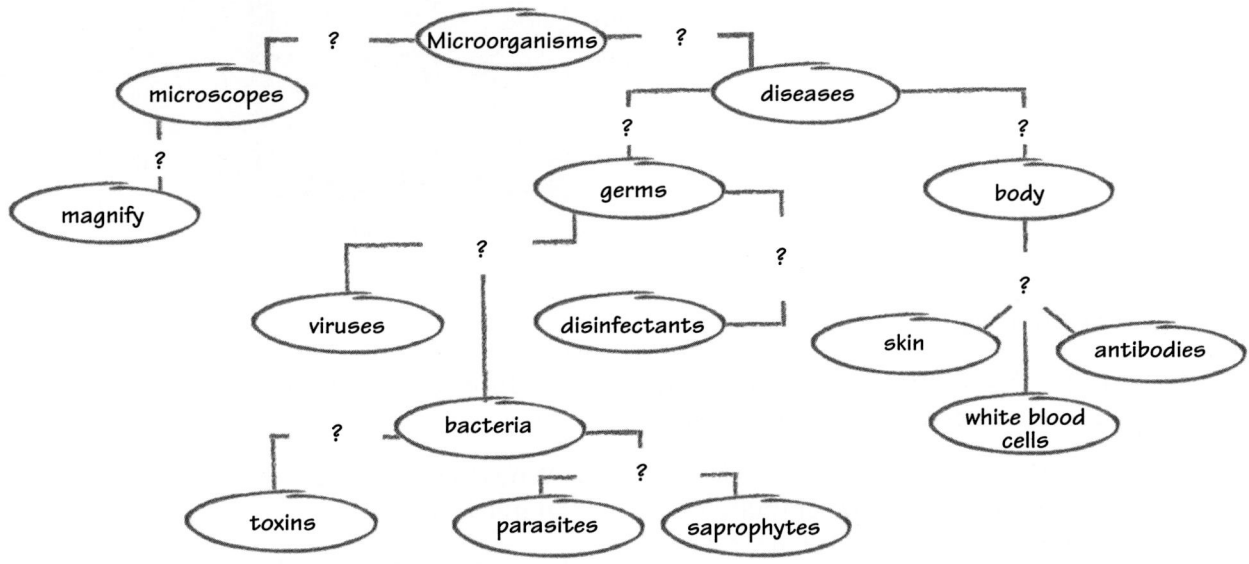

Where and how would you connect the terms *protozoans*, *mucus*, and *nonliving*?

Checking Your Understanding

Select the choice that most completely and correctly answers each of the following questions.

1. The germ theory of disease was first proposed by
   a. Joseph Lister.
   b. Louis Pasteur.
   c. Robert Koch.
   d. Anton van Leeuwenhoek.

2. Koch's postulates represent a scientific method that is designed to
   a. identify disease microorganisms.
   b. prevent infectious diseases.
   c. compare bacteria and viruses.
   d. treat the symptoms of anthrax.

3. Bacilli, cocci, and spirilli are all types of
   a. diseases.
   b. bacteria.
   c. viruses.
   d. protozoans.

4. The third, and last, line of defense that the human body has against disease is
   a. the immune system.
   b. the skin.
   c. white blood cells.
   d. mucus.

5. A condition in which the body attacks its own tissues is called an
   a. allergy.
   b. immune deficiency disorder.
   c. autoimmune disorder.
   d. acquired immunity.

S83

Interpreting Graphs

Look at the graph shown here. It represents the change in the number of viruses located outside an infected bacterial cell over a certain length of time. At the point marked *X*, what stage of virus reproduction is taking place? Explain.

Number of viruses outside the cell

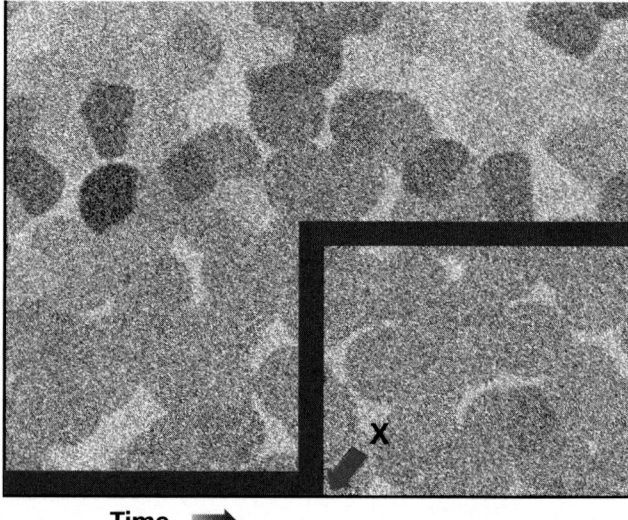

X

Time ➡

Critical Thinking

Carefully consider the following questions, and write a response in your ScienceLog that indicates your understanding of science.

1. The bacteria that cause tetanus thrive in environments that contain low amounts of oxygen. Which type of wound, a deep puncture or a surface cut, would present a better place for the growth of these bacteria? Explain.

2. Suppose you were investigating a micro-organism by using Koch's postulates. Why is it important to use a healthy animal or plant during your tests?

3. Although plants and people are not attacked by the same microorganisms, scientists are still very interested in studying plant diseases. Why?

4. Why do many scientists consider viruses to be nonliving?

5. Consider the effect that AIDS has on the immune system. Why would it be correct to say that the virus that causes AIDS does not *directly* kill people?

P o r t f o l i o   I d e a

Imagine that you are a nineteenth-century American newspaper reporter on assignment in France. You are there to gather information for an article about vaccines and how they will affect the future of humankind. At present, you are busily preparing for an interview with Louis Pasteur, who is the current leading authority on the use of vaccines in preventing disease. Write down the questions you will ask professor Pasteur during your interview. Then answer the questions and use the answers to write your article. (Use your school or public library to learn about vaccines and the research Louis Pasteur did with them.)

S84

Unit 4

IN THIS UNIT

INVESTIGATING MATTER

Now that you have been introduced to the properties of matter, consider these questions.

1. In what ways can matter change?
2. What is density? How is it related to mass and volume?
3. Sugar seems to disappear when you mix it into water. What happens to it?
4. What is matter made of? What is the smallest unit of matter?

In this unit, you will consider the atomic model of matter and how this model of matter fits the observations we make.

S85

MATTER AND MOLECULES

Imagine that you are an astronaut returning to Earth after a trip on a space shuttle. What might you see? From space you have a terrific view. Whole continents are visible. You see only the largest geographic features—the Rocky Mountains, the Great Lakes, Chesapeake Bay. Towns and smaller cities are invisible. You can easily see the curve of the Earth. Once you start your reentry, the ground seems to come up to meet you. You begin to see more detail. Larger structures like malls and factories come into view. Highways become visible, dotted with tiny cars and trucks. Soon you can see houses and lawns, then individual trees, then individual people. As you roll to a stop on the runway at Cape Canaveral you look out the window and see cattail reeds and a great blue heron in the nearby marsh. As you walk to the awaiting van, you see grass, insects, and twigs on the ground around you.

Let's look at matter in a similar way. **Matter** is defined simply as anything that has mass and takes up space. We see the matter around us in much the same way that an astronaut in space sees the surface of the Earth. Only the coarsest details are visible. However, just as the details on Earth's surface become clearer as you get closer to it, a closer look at matter shows that it also is made up of smaller parts.

The closer you get to Earth, the more detail you can see.

Matter

Anything that has mass and takes up space

Particles of Matter

If you "zoom in" on matter, you see that it also has structure not visible from a distance. From a distance of a few centimeters, for example, an ice cube looks white and smooth. You can see its overall structure, but you can't make out much fine detail. Once you zoom in closer, you begin to see much more detail. You see that the surface is not entirely smooth. It has many irregularities. As you move closer still, the surface of the ice cube begins to look as rugged as the surface of the moon. If you continue to zoom in, eventually you begin to see that the ice cube is composed of many tiny, identical particles. Each of these particles is a **molecule**. The word *molecule* comes from the Latin word *molecula,* meaning "little mass." A molecule of water is the smallest particle of water (or ice) that is still water. Molecules are quite small. It would take about 60 million water molecules side by side to stretch across a penny!

What would happen if you could close in on a different substance, such as sugar, in a similar way? Eventually, you would come to the smallest particle that is still sugar. This particle would be called a molecule of sugar.

S86

▲ Just as the returning astronaut sees more and more detail of the Earth's surface, "zooming in" on the matter around us shows us its hidden structure.

Physical Properties of Matter

If you were trying to identify a substance, what clues could you look for? Since you cannot see the particles it is made of, you might first notice its color and texture. You might also note such properties as its hardness, melting point, or boiling point. Properties such as these, which you can observe or measure without changing the composition of the material, are called **physical properties**. You can observe physical properties of matter using only your senses. You rely on physical properties to identify things all the time. Can you think of any other physical properties?

Molecule

The smallest particle of a substance, such as water, that is still identifiable as that substance

Physical property

Any property of matter that can be observed by your senses without altering the composition of the matter

◀ How many different physical properties can you identify in this photograph?

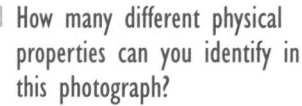

S87

In a liquid, the molecules move about freely, but they are held close together by attractive forces.

In a gas, the molecule's energy completely overcomes the attractive forces. The molecules move about randomly, with large spaces between them.

In a solid, strong attractive forces lock the molecules in place, but they are free to vibrate in all directions.

States of Matter

One easily observed physical property of matter is whether it is solid, liquid, or gas. At ordinary temperatures, all matter exists in one of these three *states*. In each of these states, the molecules of the material behave in different ways. In a *solid,* the molecules are close together in a fixed pattern, like people sitting in the rows of a theater. In a *liquid,* the molecules can change position and move past each other. This is like people milling about the lobby of the theater during intermission. In a *gas,* the molecules are spread far apart, like people scattering as they leave the theater after the performance.

At very high temperatures, substances enter a fourth state of matter, called *plasma*. In a plasma, the molecules have been broken down into electrically charged particles. The plasma state is rare on Earth, but it is, in fact, one of the most common forms of matter in the universe. The sun and other stars are all composed of matter in the plasma state. The glowing gases inside neon signs and fluorescent lights are also examples of plasmas, as are lightning bolts.

Chemical Properties of Matter

Matter, however, is not fully described by its physical properties alone. Different forms of matter can also interact with each other. The ways in which one substance interacts with others are called **chemical properties**. Suppose you dropped an old penny into a jar of vinegar. After a few hours the penny would become shiny. But a penny dropped into water would remain unchanged. Why the difference? The penny in the vinegar becomes shiny because the vinegar *reacts* (interacts) with the surface of the penny. Because of its chemical properties, vinegar reacts with the penny, while water does not.

Changes in Matter

You know from experience that water is not always a liquid. You can freeze liquid water to make solid ice, or you can boil it to change it into a gas. Changes such as these do not alter the fundamental properties of water. They are examples of **physical changes**. The molecules of a substance may be rearranged but are otherwise unaffected by physical changes.

A physical change is any change in matter that does not change the chemical properties of the matter. Changing from one state to another is a physical change. Other examples of a physical change are cutting, grinding, pulverizing, and compressing. Can you think of any others?

Chemical property

Any property of matter that describes how that kind of matter interacts with other kinds of matter

S88

Not all changes in matter, however, are physical changes. Suppose an iron nail lies on damp ground for a few days. It will become rusty. Obviously, the rust comes from the iron somehow, but do rust and iron have the same properties? It turns out that rust and iron are quite different and have many different properties. Iron and rust have different physical properties, such as color, hardness, and melting point.

But this is not what makes them different materials. It is how each acts or reacts with other substances that makes them truly different. For example, certain acids will dissolve iron but leave rust unaffected. The simplest example is how each reacts with water. If you put iron into water, it changes

▲ When water changes from liquid to solid or gas, what type of change takes place?

to rust—it reacts with water. If you put rust into water, it sits there—no reaction. The rusting of iron is an example of a **chemical change**. Chemical change occurs when one substance interacts with another, causing at least one new

substance to form as the chemicals making up the substances are combined in new ways. The new substance has new properties. For example, when vinegar is mixed with baking soda, both the vinegar and the baking soda are changed chemically. One of the new substances produced by this chemical change is the gas that is given off in the bubbles—carbon dioxide.

People sometimes confuse chemical properties and chemical changes. The difference is this: Chemical properties determine how substances interact with each other, while chemical changes are the result of the interaction. Putting it another way: Chemical properties can result in chemical changes.

A few examples of chemical change

S U M M A R Y

Matter is made up of very small parts. The molecules that make up a substance determine its physical and chemical properties. Physical properties are those that can be determined using the senses. Chemical properties determine the way substances interact with each other. A physical change may rearrange a substance in some way, but it leaves the molecules otherwise unaffected. A chemical change occurs when one substance interacts with another to become a new substance.

Physical change

Any change in matter that does not change the chemical properties of the matter

Chemical change

A change in matter in which one substance is changed into another substance

S89

MEASURING MATTER

Jorge is reading an article about gravity in a science magazine. The article mentions that *all* objects with mass exert a gravitational pull and that the greater the mass of the object, the greater the gravitational pull. "Wow," he thinks. "That means I exert a gravitational pull, too. And if I weighed twice as much, my gravitational pull would be twice as strong." Then it occurs to Jorge that he isn't sure what the article means by "mass." Is it the same as weight? If not, what's the difference?

Mass

Mass is a fundamental property of matter. All matter has mass. In fact, matter is anything that has mass and takes up space. The **mass** of an object is a measure of how much "stuff" it contains. The more massive an object, the more matter it contains. In the metric system, the standard unit of mass is the *kilogram* (kg). To measure small masses, the *gram* is generally used.

Mass Versus Weight Although mass and weight are related, they are not the same. While mass is the amount of matter in an object, **weight** is the downward force exerted by an object due to gravity. The mass of an object is constant, but its weight is not. This is because the amount of matter an object contains does not usually change, but gravity can change according to location. For example, the Earth's gravitational pull diminishes as you move farther from the Earth's center. An object would weigh slightly less on top of Mount Everest than at sea level. Here is another example. You may recall that objects weigh about one-sixth as much on the moon as they do on Earth. This is because gravity on the moon is about one-sixth as strong as on Earth. Note that weight is proportional to mass. This is because weight is mass multiplied by the gravitational pull. Weight, being a force, is measured in newtons (N) in the metric system.

Mass

A measure of the amount of matter in an object

Weight

The downward force exerted by an object due to gravity

A teeter-totter is a simple balance that indicates differences in mass.

Your weight depends on the gravitational pull at the spot on which you are standing.

Because gravity is nearly constant over the Earth's surface, the terms *weight* and *mass* are often used interchangeably. But this is scientifically incorrect and sometimes leads to confusion. Here is a way to help you see the difference between mass and weight. In orbit, astronauts on a space shuttle do not feel the effects of gravity. They are basically weightless and can drift about the cabin. Yet are they massless? Of course not! The amount of matter in each astronaut remains unchanged, even though they no longer have weight.

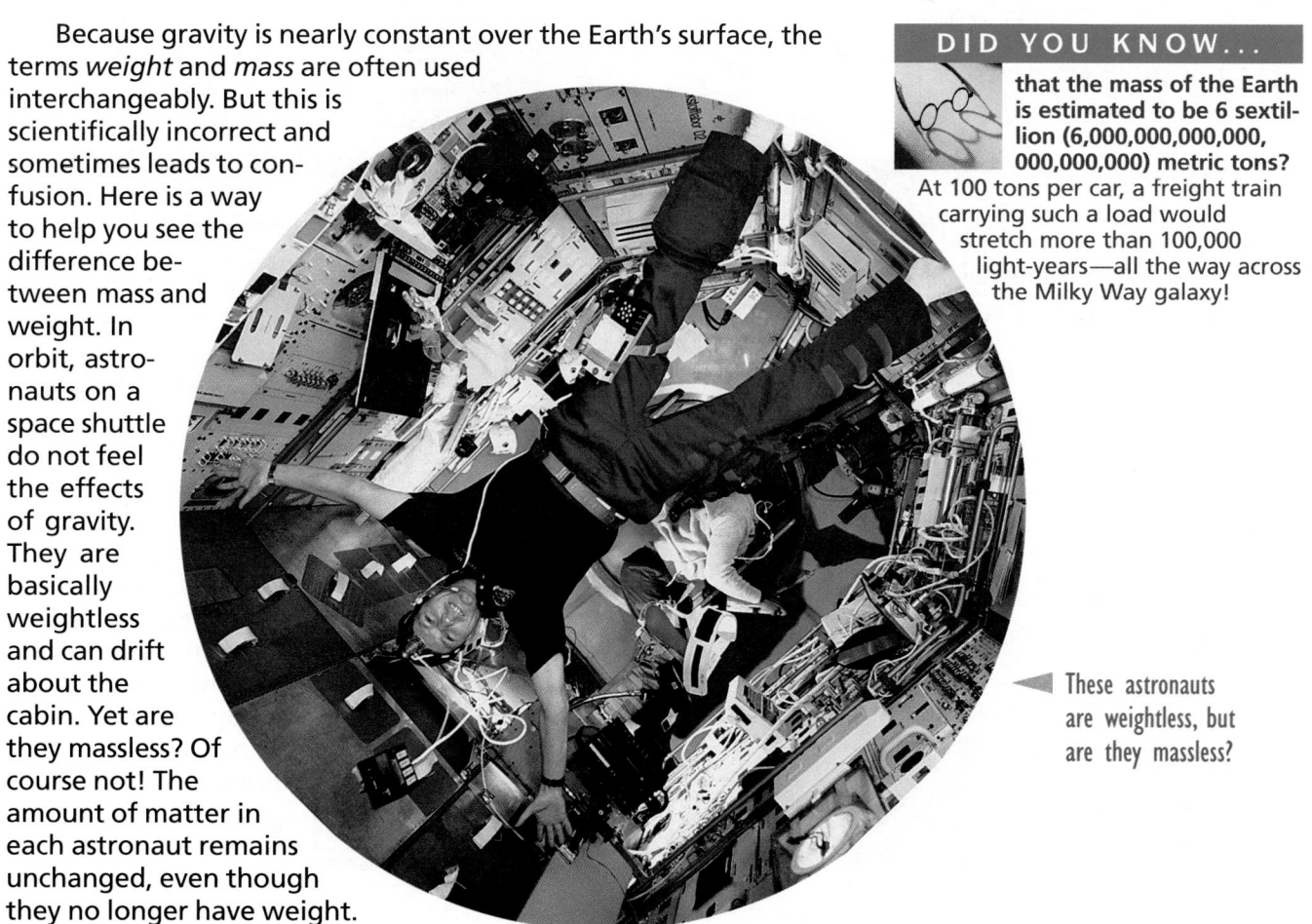

These astronauts are weightless, but are they massless?

S91

Volume

Another fundamental property of matter is that it takes up space. All objects with mass—even the tiniest molecules—take up space. The amount of space taken up by something is called its **volume**. Volume is a physical property. In the United States we commonly measure volume in gallons, quarts, pints, cups, and fluid ounces. In the SI system, volume is measured in cubic meters (m^3), that is, a cube measuring one meter on each side. This is actually quite a large volume (imagine a box with sides as long as a meter-stick), so for most common situations the cubic centimeter (cm^3) is more appropriate.

Volume

The amount of space occupied by an object

Density

The mass of a material per unit of volume

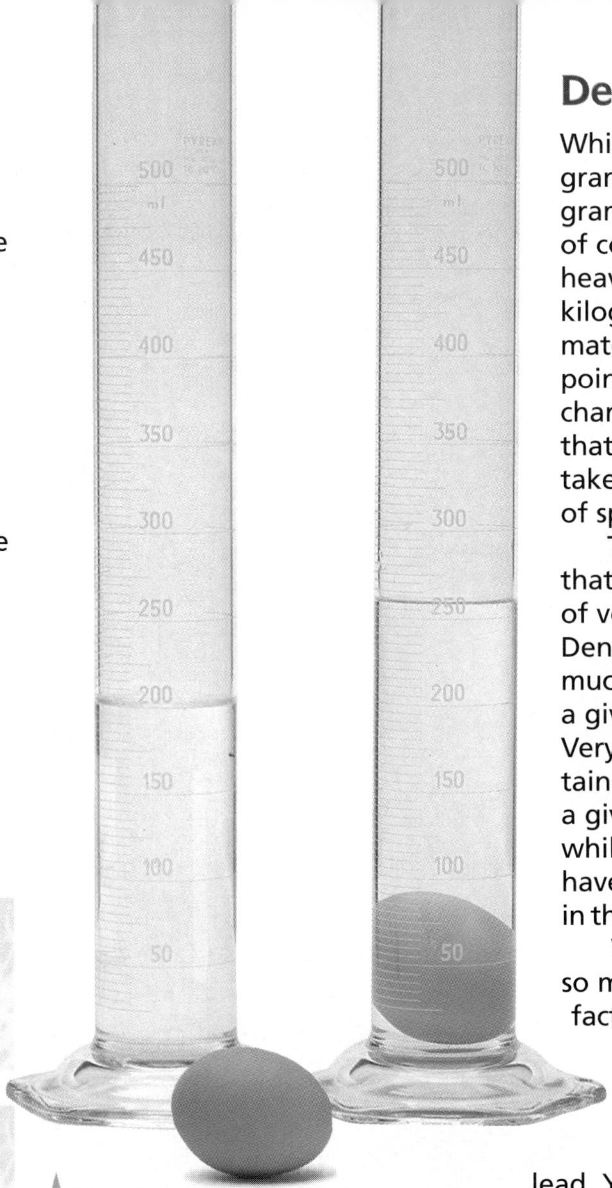

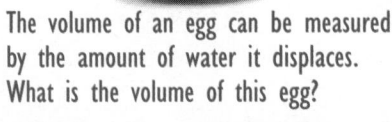

The volume of an egg can be measured by the amount of water it displaces. What is the volume of this egg?

Density

Which is heavier, a kilogram of feathers or a kilogram of lead? The answer, of course, is that neither is heavier—a kilogram is a kilogram, regardless of the material. But this riddle points out a fundamental characteristic of matter—that different materials take up different amounts of space for the same mass.

The amount of mass that a material has per unit of volume is called **density**. Density is a measure of how much matter is packed into a given amount of space. Very dense materials contain a lot of matter in a given space or volume, while very light materials have relatively little matter in the same space or volume.

Why do materials vary so much in density? One factor involves the mass of the individual particles of which the substance is made. Consider the metal lead. You know that lead is very dense. In fact, it is more than 11 times as dense as water. But why? For one thing, atoms of lead have a lot more mass than individual molecules of water. Yet lead is not the densest material around. The metal osmium is the densest material of all. It is almost 23 times denser than water and is twice as dense as lead! A brick-sized chunk of osmium would have a mass of about 25 kg—almost too heavy to lift!

A large cork and a small lead sinker of equal mass have different densities.

The density of a material also depends on its overall structure. Think of a feather. If you look closely at a feather, you see that it is mostly space. Even the central shaft of the feather is hollow to minimize its mass. If you placed a bunch of feathers in a box, their shape would prevent them from packing very tightly. There would still be a lot of space between the feathers. As a consequence, even a large volume of feathers would have a small mass and thus would have a very low density. Other substances are similar. If you examined a slice of balsa wood or cork through a microscope, you would see a network of dead cell walls enclosing a large amount of empty space.

In a similar way, the manner in which the atoms and molecules of a material are arranged affects its density. Some materials have arrangements that result in relatively little space between the atoms or molecules. Others have a lot of space.

Most substances also have different densities in different states. Like most substances, water changes in density when it freezes. Unlike most substances, however, water is actually less dense in the solid state than in the liquid state. The reason is that when water freezes, its molecules lock into a position in which they are farther apart, on average, than when in the liquid state.

▼ In the liquid state, water is more dense than it is in the solid state because its molecules are closer together.

▲ When water freezes, its molecules lock into fixed positions that are farther apart than the molecules are in the liquid state. As a result, ice is less dense than water and thus will float in it.

SUMMARY

The amount of matter a substance contains is its mass. Mass is constant. Weight is the force exerted on an object due to gravity. Mass and weight are proportional because weight is mass multiplied by gravity. The volume of an object is the amount of space it occupies. Density is a measure of the amount of matter per unit of volume. Different materials have different densities. Density depends on the mass of a material's atoms or molecules and their arrangement.

S93

MIXTURES AND SOLUTIONS

The next time you make a glass of instant lemonade, read the list of ingredients. You will see that the lemonade mix is made up of a number of different substances blended together. Some of the individual ingredients you can easily distinguish. Others are mixed in so well that they are practically invisible. It all becomes invisible, though, when you mix it in water. The solids seem to disappear, leaving behind only a little coloring. What happened here?

Mixtures

The lemonade mix in the previous example is a *mixture*, or blend, of different substances. A **mixture** is any form of matter that contains a blend of more than one substance. Mixtures are a very common form of matter.

There are many different kinds of mixtures, but they all have three things in common.

Ingredients: High fructose corn syrup, water, lemon juice concentrate, lemon pulp, sugar, lemon oil.

▲ Many different foods are mixtures. All mixtures contain more than one substance.

Mixtures

1. A mixture is always made of at least two substances.
2. The substances are not changed chemically when they are mixed together.
3. The substances in a mixture can be put together in any proportion.

Because substances in a mixture are chemically unchanged, they can be separated by physical means. For example, in a mixture of salt and pepper, the grains of salt and pepper can be seen and separated. You can separate the dark pepper grains from the light salt grains simply by picking them out one by one. Pepper and salt each have their own properties, which do not change when they are mixed.

Solutions

Once you dissolved the lemonade mix in the water, would you be able to separate its various ingredients from the water very easily? You can no longer see them, but you know that they are still there (at least the sugar and lemon flavor) because you can taste them.

When you dissolve the lemonade mix, its molecules are no longer in the form of a solid. The individual molecules have become completely mixed with the water molecules. But as in other mixtures, the chemical properties of a substance do not change when the substance is dissolved in another.

A mixture that forms when one substance, such as sugar, mixes completely with another substance, such as water, is called a **solution**. Solutions are uniform mixtures of separate substances. In a solution, the substance that is dissolved is broken down into individual molecules. These molecules disperse among the molecules of the substance doing the dissolving.

Mixture

Any form of matter that contains more than one substance and can be separated by physical means

Solution

A mixture formed when one substance mixes completely with another substance

S94

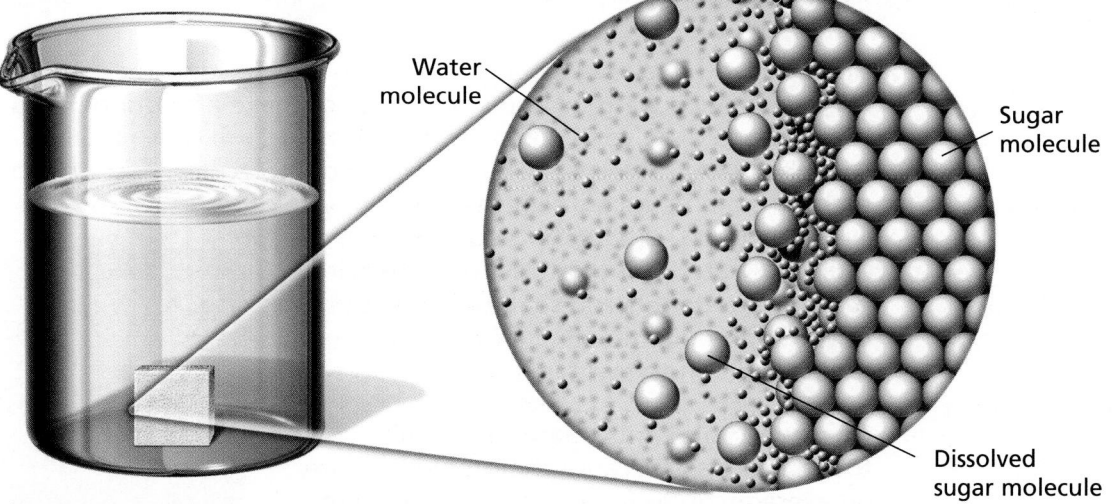

Water molecule

Sugar molecule

Dissolved sugar molecule

▲ When sugar dissolves in water, the sugar molecules mix with the water molecules. Sugar is the solute and water is the solvent.

Something that dissolves to make a solution is called a **solute**. In a solution of sugar and water, the sugar is the solute. The water is called the **solvent**. A solvent is a substance in a solution that does the dissolving.

Solutions can sometimes be separated by heating the mixture to boil away the solvent. In a salt solution, the water will boil away, and the salt will then be left in the form of solid crystals. The solid salt is the same substance it was before it dissolved. The steam from the boiling solution can be captured, cooled, and changed back into liquid water. The water will also be the same as it was before it was mixed with the salt.

Liquid Solutions Using a liquid solvent such as water makes a *liquid solution*. Usually, a solid solute is dissolved in the liquid solvent. For example, if you read the label on a bottle of soda, you will see that it is partly a liquid solution of sugar in water. However, you also know that

Solute

The part of a solution that is dissolved

Solvent

The part of a solution that does the dissolving

soda gives off bubbles of carbon dioxide gas. Therefore, gases must also be able to dissolve in liquids. It is also possible to dissolve one liquid in another. Rubbing alcohol, for example, can dissolve in water, and vice versa.

In most common liquid solutions, the solvent is water. Water is better at dissolving things than any other liquid. For this reason, water is sometimes called the *universal solvent*. It is the solvent in the most abundant liquid solution on Earth—ocean water. The oceans contain about 3.5 percent dissolved material, of which about 85 percent is sodium chloride—sea salt. Gases, such as carbon dioxide, nitrogen, and oxygen, are also solutes of ocean water.

▼ When a salt solution is boiled, the water molecules gain enough energy to leave the liquid state. Solid salt is left behind.

S95

However, you also know that some things will not dissolve in water. For example, cooking oil and water will not make a solution. For a substance to dissolve in water, its particles must mix very closely with the water molecules. The two kinds of molecules must attract each other to be mixed together so closely. Oil will not mix with water because the molecules of water do not attract oil molecules strongly enough. If you shake up a mixture of water and oil, the oil will form small droplets within the water.

Gaseous Solutions Gases easily mix together. This is because their molecules are, on average, quite far apart compared with solids and liquids. A mixture of gases is called a *gaseous solution*. Air is an example of a gaseous solution. It may surprise you to learn that the most abundant gas in air is nitrogen. Because nitrogen is the most abundant gas in air, it is said to be the solvent. The next most abundant gas in air is oxygen. Air also contains small amounts of many other gases.

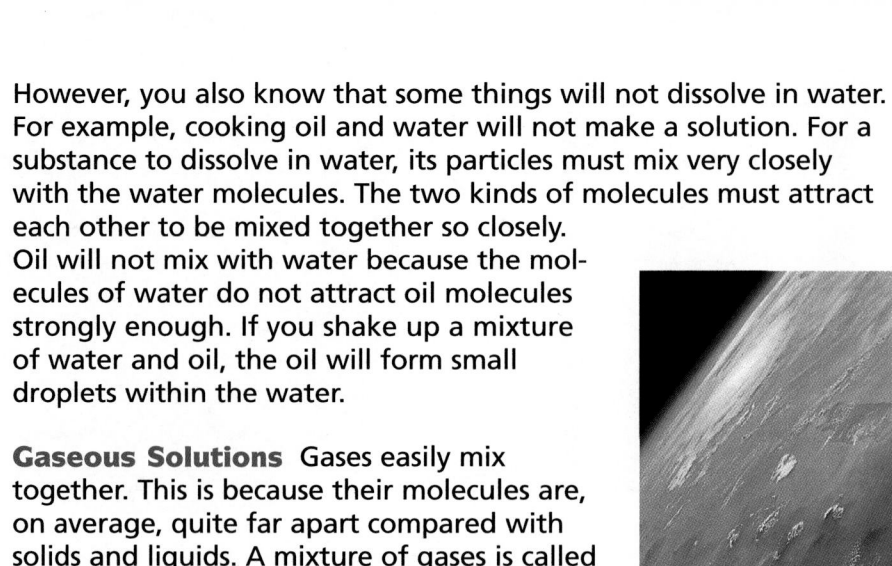

▲ One example of a gaseous solution

Solid Solutions Many solids are really *solid solutions*. For example, steel is a solution of carbon and other added metals in iron. When steel is made, carbon and other metals are dissolved in the molten (liquid) iron. When the mixture cools, the carbon and metals remain dissolved in the solid steel. Solid solutions of metals are called *alloys*. Liquids and gases can also be dissolved in solids.

▼ Some solid solutions you might have seen

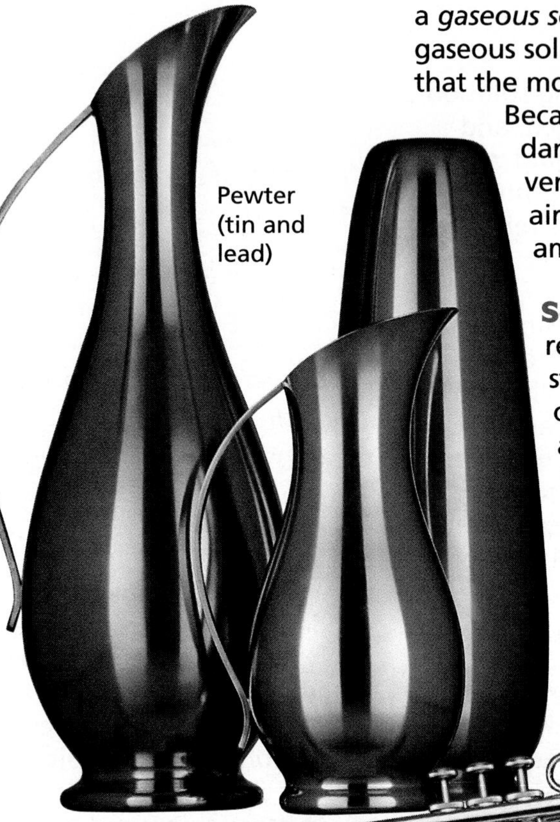

Pewter
(tin and
lead)

Bronze
(copper and tin)

Brass
(copper
and zinc)

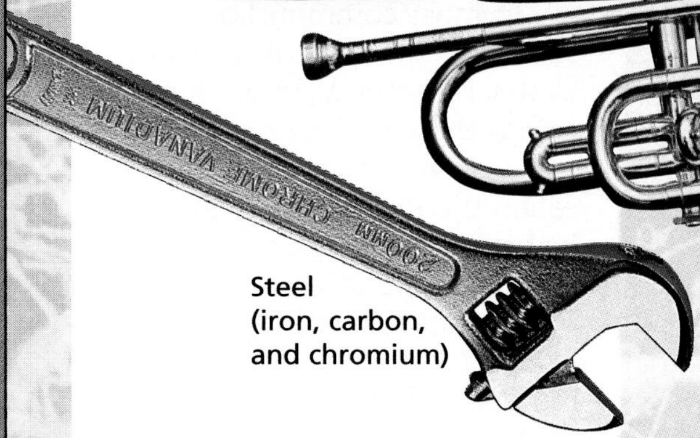

Steel
(iron, carbon,
and chromium)

SUMMARY

A mixture contains more than one kind of substance. A solution is a mixture in which a solute is dissolved in a solvent. In all mixtures, even solutions, the substances of which they are made can be separated by a physical means. Liquids, gases, and solids can all form solutions.

COMPOUNDS AND ELEMENTS

If you were to crush an aspirin tablet, you would end up with smaller particles of aspirin. Crushing cannot separate an aspirin into different substances, nor can any other physical means. In the same way, pure water cannot be broken down (separated into component materials) by any physical means. Substances such as these are called **compounds**.

The properties of a compound are usually different from the properties of the substances that make up the compound. For example, ordinary table salt is a compound formed from an explosive metal (sodium) and a poisonous gas (chlorine).

▲ Aspirin is a compound. Like all compounds, it can be broken down by a chemical change, but not by physical change.

Compound

A substance that can be broken down into simpler parts only by a chemical change

Breaking Down Compounds

You already know that compounds cannot be broken down by any physical means, but is it possible to do so by other methods? In fact, it is. Compounds can be broken down by chemical means. For example, sugar can be chemically broken down by burning it. The sugar, which is made of carbon, oxygen, and hydrogen, turns into carbon dioxide and water vapor. Electricity can also be used to break down some compounds. For example, a process called *electrolysis* uses electricity to break water down into its chemical components. Using equipment such as that shown, an electric current is passed through the water. Bubbles of gas form around the ends of the wires where the current enters and leaves the water. The component gases collect in the test tubes above the electrodes.

Hydrogen collects at one electrode and oxygen collects at the other. Both hydrogen and oxygen have properties completely different from those of water.

◄ When an electric current passes through the solution in the beaker, it breaks down water into two gases: hydrogen and oxygen. The hydrogen and oxygen are collected in the test tubes.

S97

Element

A substance that cannot be broken down by either physical or chemical changes

But what about the carbon, hydrogen, and oxygen in these examples—could these substances be broken down any further? Eventually there comes a point when matter cannot be broken into simpler units by chemical means. Substances such as these are called **elements**.

More than 100 elements have been identified. Of these, 91 are found in nature. The others have all been created artificially. The most common elements in nature are listed in the diagram below. As far as scientists can tell, all matter in the universe is made up of the known natural elements, either alone or in different combinations.

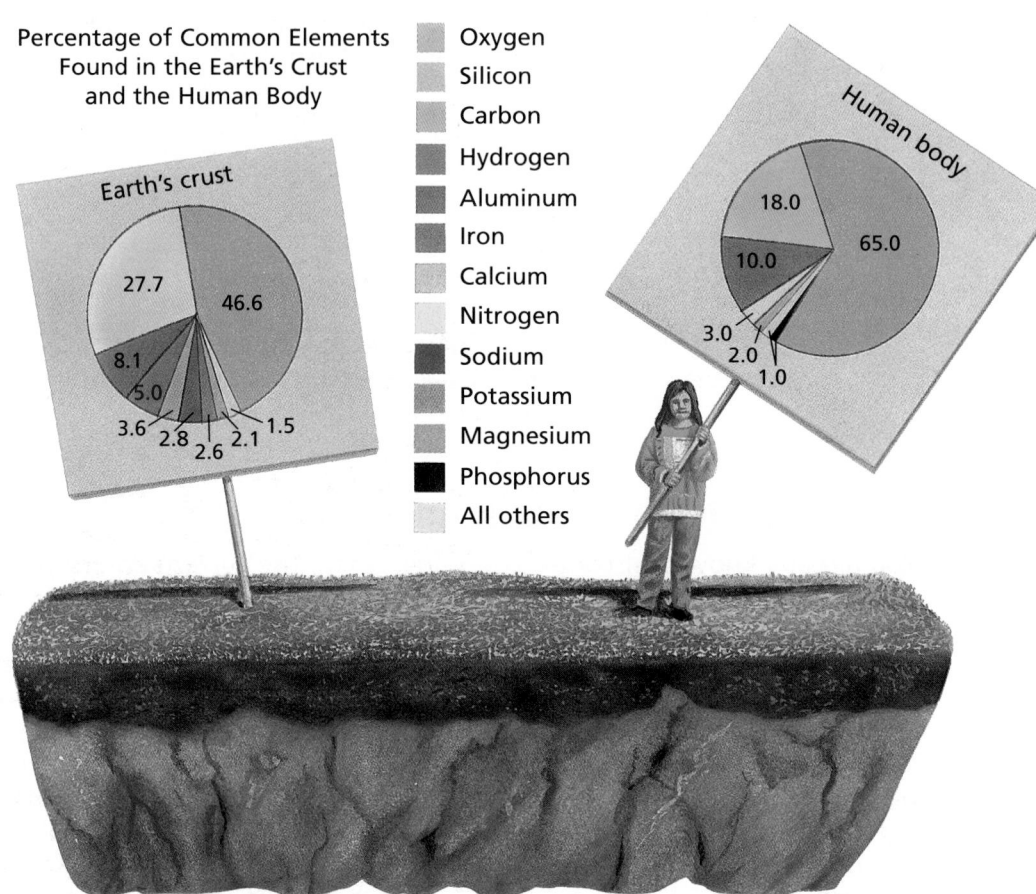

Percentage of Common Elements
Found in the Earth's Crust
and the Human Body

■ Oxygen
■ Silicon
■ Carbon
■ Hydrogen
■ Aluminum
■ Iron
■ Calcium
■ Nitrogen
■ Sodium
■ Potassium
■ Magnesium
■ Phosphorus
■ All others

Earth's crust

27.7   46.6
8.1
5.0
3.6 2.8 2.1 1.5
2.6

Human body

18.0   65.0
10.0
3.0
2.0
1.0

SUMMARY

Compounds are materials that cannot be broken down into simpler substances by physical means. They can, however, be broken down by chemical means. Compounds are formed of elements. Elements are substances that cannot be broken down by either physical or chemical means.

NATURE'S FUNDAMENTAL PARTICLES

About 2300 years ago, two different ideas about the composition of matter were being debated by the scholars of ancient Greece. One Greek thinker, Democritus, believed that everything in the universe was ultimately made up of tiny objects suspended in a void. Democritus described these objects as solid, indivisible balls. He called these tiny balls of matter "atoms."

At the same time, the Greek philosopher Aristotle held to the belief that all matter was made up of four "elements"—earth, air, fire, and water. Wood, for example, was said to be made of the elements fire and earth. When wood burned, *fire* escaped and *earth* remained as ashes. Since Aristotle was a respected philosopher and teacher, his ideas won out over those of Democritus. In fact, Aristotle's ideas prevailed for the next 2000 years.

The ancient Greeks debated the nature of matter. Aristotle believed that everything was composed of the four "elements" earth, air, fire, and water. Democritus thought that everything was composed of tiny indivisible particles. Who won the debate?

S99

sulphur

alumine

potash

oxygen

hydrogen

▲ Dalton and his atoms

Developing a Modern Theory of Matter

During the 1600s and 1700s, scientists began to consider the ideas of Democritus once again. In 1808 an English schoolteacher and chemist named John Dalton presented an atomic theory that combined the findings of several researchers along with some of his own. His theory of matter stated the following:

Matter

- All matter is composed of tiny particles called **atoms**.
- Each element is made up of atoms of the same kind, and the atoms of one element are different from the atoms of all other elements.
- Atoms cannot be divided, created, or destroyed.
- Atoms of elements combine in certain ratios to form compounds.

Atom

The smallest particle of an element; from the Greek word atomos, *meaning "cannot be divided"*

For example, water is made from the atoms of two elements, hydrogen and oxygen. The atoms of hydrogen are different from the atoms of oxygen. The compound water is always made of two parts hydrogen and one part oxygen.

Several of John Dalton's ideas are still used today. But his beliefs that *all* matter is composed of atoms and that atoms could not be divided have since been shown to be not completely accurate in all situations.

Modern Atomic Theory

Look at any of the colored photographs in this book with a strong magnifying glass. You will see that each picture is made up of many small colored dots, as shown in the photos below. From a distance, these dots give the illusion of a solid image. Suppose that you could look at a compound, such as water, in the same way. You would see that the water is also made up of very small identical molecules. If you looked even closer, you would discover that each water molecule is made up of still smaller particles—hydrogen and oxygen.

DID YOU KNOW...

that all atoms are close to the same size?
An atom of gold is actually slightly smaller than an atom of sodium, even though it has almost 10 times the mass!

A picture is made up of small colored dots. Matter is also made up of small particles. ▶

In honor of Democritus, who first developed the idea, we call the smallest particle of an element an atom. The element oxygen, for example, is made up only of oxygen atoms. Hydrogen contains only hydrogen atoms, and carbon contains only carbon atoms. Since there are over 100 elements, there are also over 100 kinds of atoms.

How small is an atom? The smallest things you can see without a microscope are millions or billions of times larger than an atom. For example, think about a soap bubble. The surface of a soap bubble is incredibly thin, much thinner than a human hair. Yet such a bubble is thousands of times thicker than the diameter of an atom.

More Particles By the latter part of the nineteenth century, scientists began to suspect that atoms might not be the smallest form of matter. The results of certain experiments with electricity seemed to suggest the existence of particles smaller than atoms. Then in 1897, J. J. Thomson of England demonstrated that cathode rays were composed of negatively charged particles. Cathode rays are a special form of electricity that flows through low-pressure gases. Television tubes, among other devices, use cathode rays. These particles eventually came to be called **electrons**. Through other experiments, electrons were found to be much smaller than the smallest atom—hydrogen. This confirmed the existence of particles of matter that were not composed of atoms. Electrons became the first of many *subatomic particles* to be discovered.

▲ Graphite and diamond are about as different as night and day. But both are made of carbon atoms.

Electron

The negatively charged subatomic particle, discovered in cathode rays, that is much smaller than a hydrogen atom

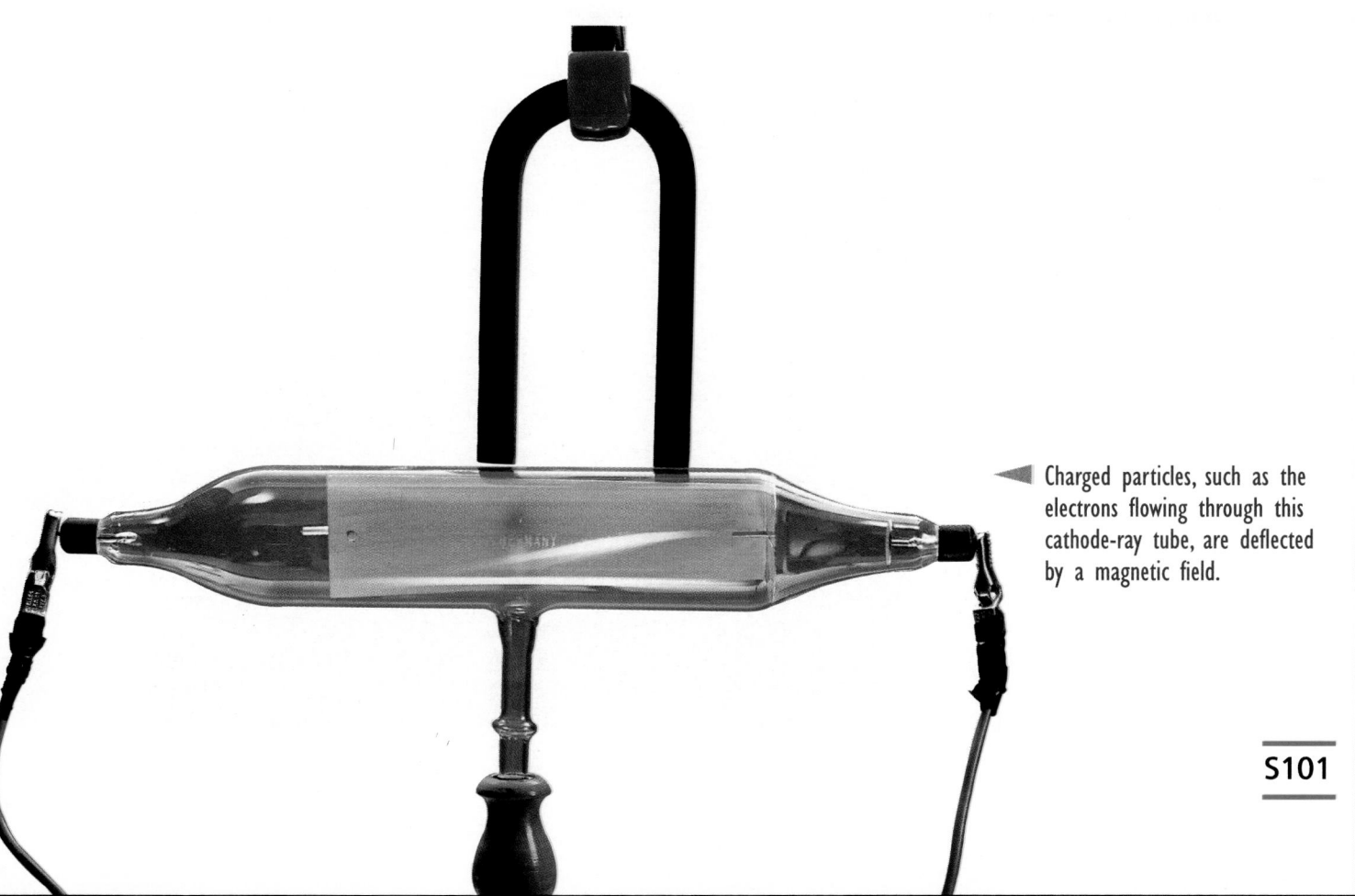

◀ Charged particles, such as the electrons flowing through this cathode-ray tube, are deflected by a magnetic field.

S101

The discovery of two additional subatomic particles soon followed. In 1919 Ernest Rutherford announced the discovery of the **proton**. Rutherford found that the proton has a positive charge equal in force to the negative charge of an electron. But the proton has a mass almost 2000 times as great. The **neutron** was discovered in 1932. A neutron has no electrical charge and has about the same mass as a proton. The existence of such a particle had been predicted by Rutherford in 1920. The findings of Dalton, Rutherford, and other scientists have been used to make a model of the atom. According to this model, atoms consist of a nucleus of protons and neutrons surrounded by a "cloud" of electrons. You will learn more about models of the atom in the next unit.

Particle Accelerators   You may have wondered how scientists are able to determine the structure of objects as tiny as atoms. It is no easy task. To learn about the inner workings of atoms, scientists rely on sophisticated devices called *particle accelerators*. Using particle accelerators, popularly called "atom smashers," physicists explore the strange world of subatomic particles.
Particle accelerators work by accelerating subatomic particles to ultrahigh velocities and then smashing them into each other. By examining the debris from the collisions, scientists have been able to detect even smaller particles—more than 200 so far. These strange new particles have been given whimsical names like *quarks, leptons, bosons,* and *upsilons.*

▲ Ernest Rutherford's experiments suggested the existence of an atomic nucleus consisting of protons.

Proton

The subatomic particle that has a positive charge equal to the negative charge of an electron

Neutron

The subatomic particle with no electric charge and about the same mass as a proton

A particle accelerator is a ▶ ring-shaped apparatus that is usually buried underground, as shown by this photo of Fermilab in Batavia, Illinois.

S102

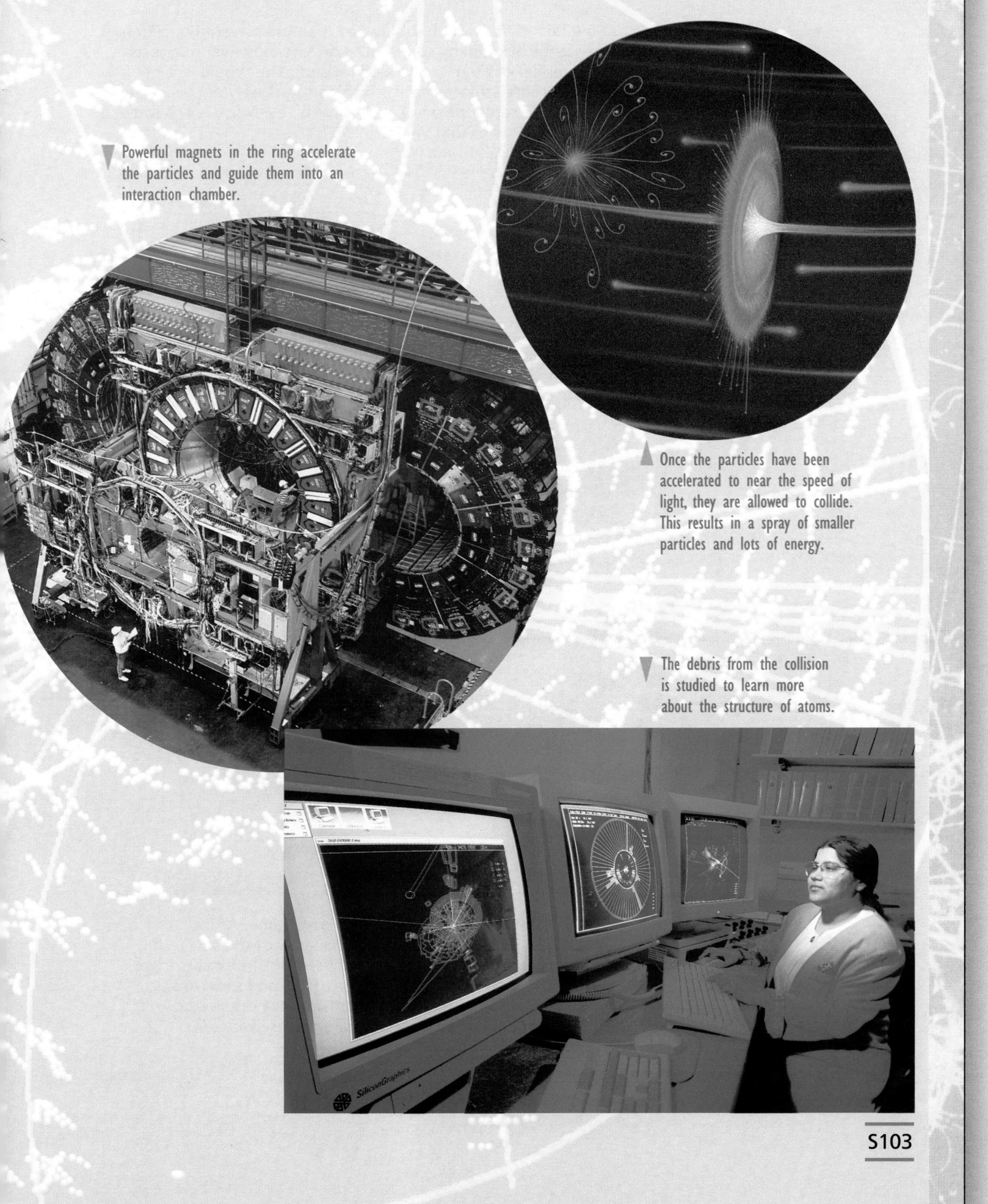

Powerful magnets in the ring accelerate the particles and guide them into an interaction chamber.

Once the particles have been accelerated to near the speed of light, they are allowed to collide. This results in a spray of smaller particles and lots of energy.

The debris from the collision is studied to learn more about the structure of atoms.

S103

How far matter can be subdivided is an unanswered question. The existence of so many subatomic particles, however, suggests that our model of atoms may need revision. This would not be unusual, because science changes constantly as new discoveries are made. Perhaps someone you know—maybe even you—will some-day make discoveries that will refine our view of nature's funda-mental particles.

The future of science may depend on someone like you. ►

SUMMARY

Ideas about atoms have changed over time. Atoms, the basic units of all the elements, are not the smallest form of matter. Atoms are composed of three main subatomic particles: electrons, protons, and neutrons. Using particle accelerators, scientists have discovered that these subatomic particles are themselves com-posed of even smaller particles.

Concept Mapping

Using the terms supplied here, construct a concept map that illustrates major ideas from this unit. Two connectors are given; you must supply the rest. Then extend your concept map by answering the additional question below. Use your ScienceLog. **Do not write in this textbook.**

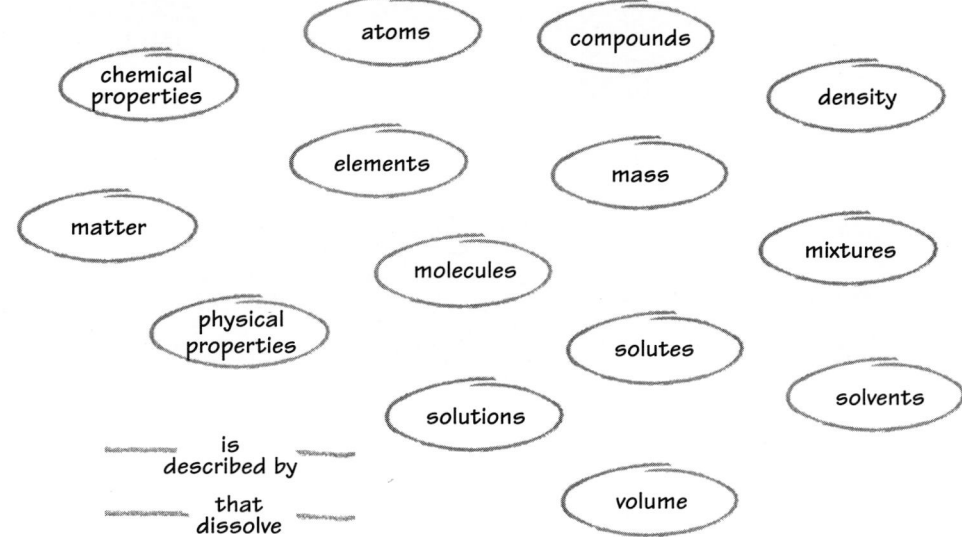

How could you add to your concept map to reflect the finding that atoms are not the smallest particles?

Checking Your Understanding

Select the choice that most completely and correctly answers each of the following questions.

1. Which of the following is NOT an example of a chemical change?
   a. silver tarnishing
   b. a piece of iron corroding
   c. a piece of coal burning
   d. a cookie being crumbled

2. The element phosphorus bursts into flame when exposed to air. This is an example of
   a. a chemical change.
   b. a physical change.
   c. a gaseous solution.
   d. an alloy.

3. The mass of an object
   a. changes with its distance from the Earth's center.
   b. is responsible for its chemical properties.
   c. is a measure of how much matter that object contains.
   d. Both a and c are correct.

4. A molecule is
   a. the smallest particle of matter known.
   b. the smallest unit of matter that can be seen.
   c. the smallest particle of a compound that still has the properties of that compound.
   d. a particle that has no mass or volume.

5. A solution
   a. is a type of mixture.
   b. is always composed of equal parts solvent and solute.
   c. must be separated by chemical means.
   d. All of the above are correct.

Interpreting Illustrations

The illustration below shows the same object measured on the moon with two different devices: a balance and a spring scale. As you can see, the results differ. How would you explain this difference?

Critical Thinking

Carefully consider the following questions, and write a response in your ScienceLog that indicates your understanding of science.

1. People commonly use the term *volume* to describe the capacity of a container. How does this definition of volume differ from the scientific definition?

2. When water undergoes electrolysis, twice as much hydrogen as oxygen is produced (by volume). Use your knowledge of matter to explain this phenomenon.

3. As you know, it is possible to dissolve a solid in a liquid. Does doing so change the solid *into* a liquid? Explain.

4. Why does a mixture of hydrogen and oxygen not have the same properties as a compound composed of these same two elements?

Portfolio Idea

Imagine that a device has been invented that can shrink you by 90 percent at a time. Each time the device is used, you are reduced to 10 percent of your former size. After two uses you would be 1/100 of your original size, after three uses you would be 1/1000 of your original size, and so on. There is no limit to the number of times you can repeat the procedure. You have volunteered to be shrunk to the size of an atom. Write about your experience. How many stages of shrinkage must you undergo? What do you see and experience at each stage? Be creative but scientifically accurate in your descriptions.

S106

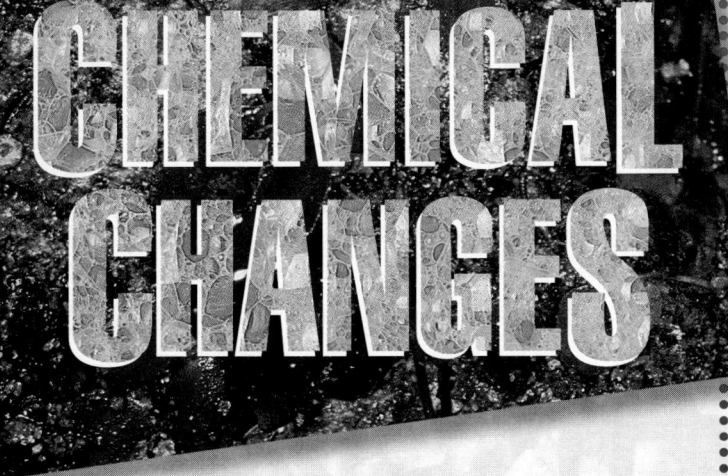

CHEMICAL CHANGES

IN THIS UNIT

Now that you have learned how to identify physical and chemical changes and to carefully observe these changes, consider these questions.

1. How are the basic parts of atoms arranged?
2. How is the periodic table arranged?
3. How is the periodic table used to predict the chemical activity of an element?
4. How do chemical changes occur, and what are some examples of the types of chemical change?

In this unit, you will learn more about atoms, elements, and the way atoms and elements behave.

S107

ATOMIC STRUCTURE

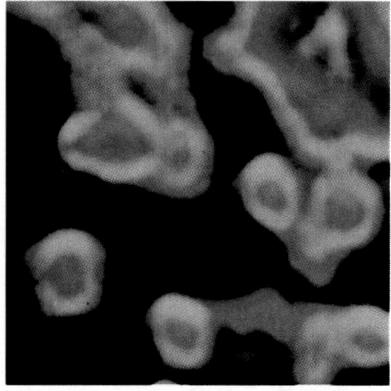

▲ This is an actual image of uranium atoms enlarged 10,000,000 times.

Atoms are very, very small. A typical atom has a diameter from 0.0000002 mm to 0.0000007 mm. As incredible as it may seem, a stack of more than a million atoms would be about as high as the thickness of this page. As you can imagine, an atom is much too small for anyone to see easily, even with a powerful microscope. Therefore, scientists cannot study atoms directly. Instead, they use **scientific models** to explain the nature of atoms. The earliest scientific models of atoms showed them as tiny solid balls, but the discovery of electrons and other subatomic particles showed that these early models were not entirely correct. Models, like theories, often change over time as we learn more about the world around us.

Scientific model

A theoretical representation of something that you cannot observe directly

Developing an Atomic Model

As you learned in Unit 4, the Greek scholar Democritus was the first to suggest the existence of atoms. By 1900, scientists knew that atoms were not the smallest particles of matter. Could these smaller particles possibly be parts of atoms? If so, how might they fit together? For instance, if you had all the parts of a watch, could you put them together to make a watch that works? A similar problem faced the scientists studying atoms in the early 1900s.

▲ The atomic model in 430 B.C. Democritus thought of an atom as a tiny solid ball.

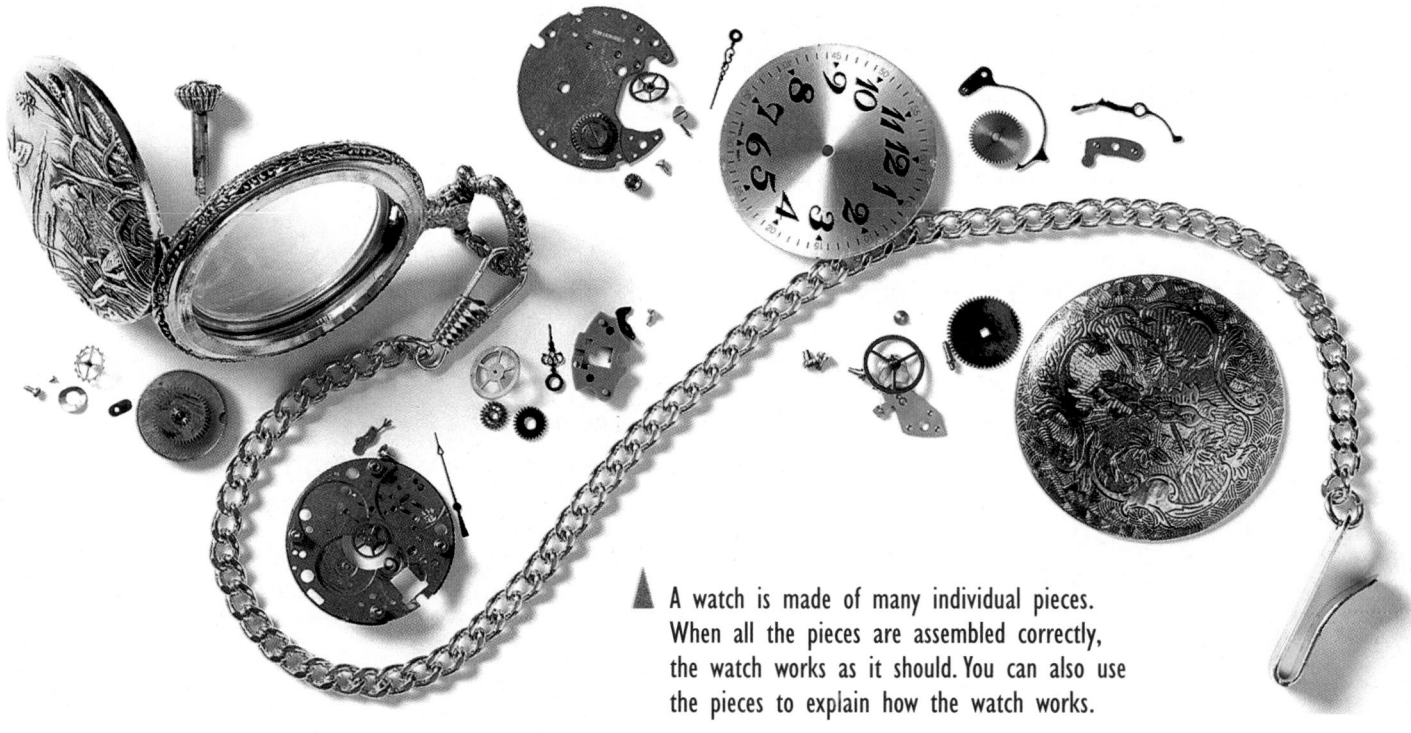

▲ A watch is made of many individual pieces. When all the pieces are assembled correctly, the watch works as it should. You can also use the pieces to explain how the watch works.

Early Models of the Atom In 1897, the physicist J. J. Thomson made an important discovery about the makeup of atoms. Thomson studied cathode rays, which are beams of negatively charged particles. Cathode rays form when electricity flows between two electrodes in a sealed glass tube containing traces of a gas. When Thomson changed either the gas within the tube or the material from which the negative electrode was made, the nature of the cathode-ray particles stayed the same. Thomson concluded that all matter contains the negatively charged particles in cathode rays. Another scientist named these particles *electrons*.

Because most matter has no apparent charge, scientists thought that matter must also contain positive charges. (Remember, a positive charge cancels an equal negative charge.) However, no one knew how electrons, positive charges, and atoms were related. Based on what he knew, J. J. Thomson suggested a model for an atom. He thought that atoms might be balls of positive electricity with electrons scattered within the positive charge.

In 1911, Ernest Rutherford, a scientist from New Zealand working in England, designed a brilliant experiment that led to the first modern scientific model of the atom. The experiment aimed a beam of fast-moving, positively charged particles at a very thin sheet of gold foil. He used gold as the target because it can be made into sheets of foil that are only a few hundred atoms thick. Rutherford wanted to see how the paths of the positively charged particles would change when they hit the gold atoms. But the experiment produced very surprising results. Most of the particles went straight through the gold foil as if nothing were there. A few of the particles bounced off the gold as if it were solid. How could this happen? Electrons, the only particles known to be in atoms at that time, are too light to deflect the heavy, positively charged particles that Rutherford aimed at the gold foil.

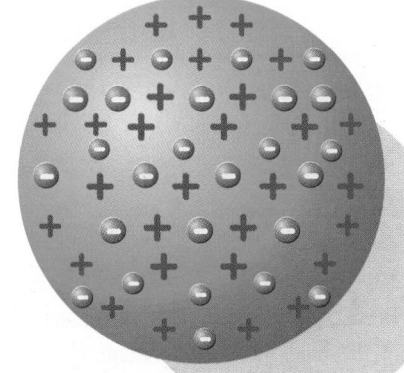

The atomic model in 1898. J. J. Thomson thought that an atom might be a ball of positive electricity with smaller negative charges scattered throughout—much like the berries in a blueberry muffin.

Compare the results of Rutherford's experiment to what happens when you throw pebbles at a chain-link fence. Most of the pebbles go right through the fence because it is mostly empty space. A few pebbles, however, hit the wire and bounce back.

Setup of Rutherford's Gold-Foil Experiment

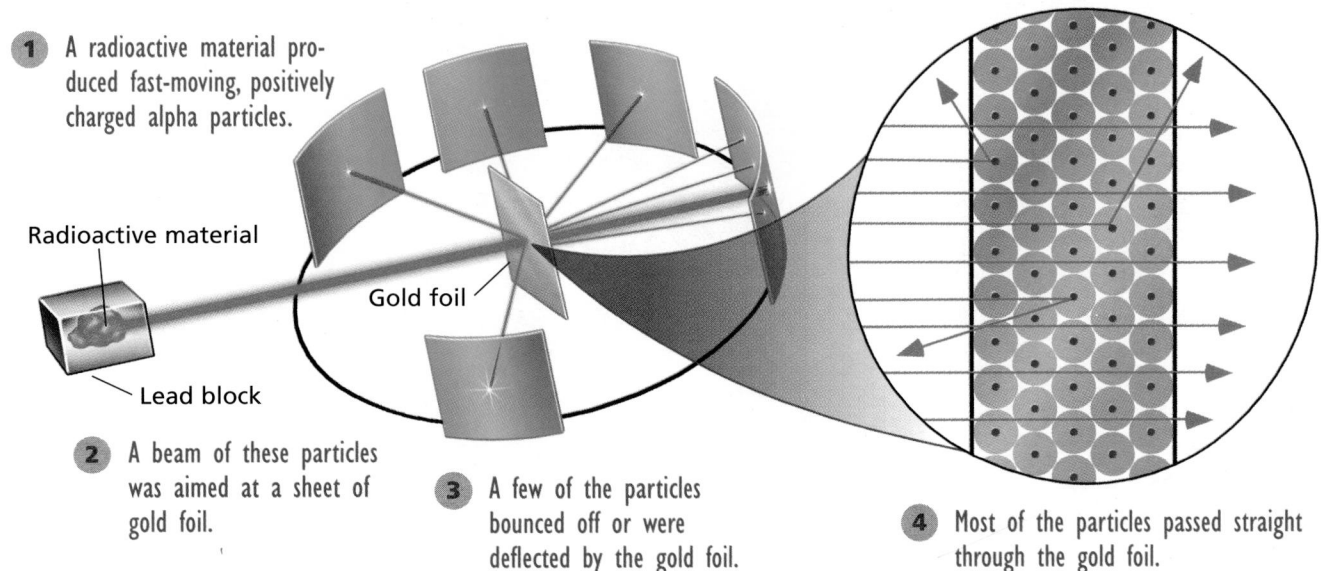

1 A radioactive material produced fast-moving, positively charged alpha particles.

Radioactive material

Gold foil

Lead block

2 A beam of these particles was aimed at a sheet of gold foil.

3 A few of the particles bounced off or were deflected by the gold foil.

4 Most of the particles passed straight through the gold foil.

S109

Nucleus

In an atom, the central core that contains most of the atom's mass

From his experiment, Rutherford inferred two things. First, atoms are mostly empty space. Otherwise, the gold foil would have deflected more particles. Second, the positive charges in an atom could not be evenly distributed. Only a dense concentration of positive charge and mass could deflect the heavy, positively charged particles. Therefore, Rutherford concluded that a tiny core at the center of an atom holds most of the atom's mass. This central core is called the **nucleus**. Electrons were thought to move around the nucleus. Later experiments showed that positively charged protons and neutrons, which have no charge, make up an atom's nucleus.

The Bohr Model of the Atom In 1913, Niels Bohr developed a model to explain how electrons move around an atom's nucleus. Bohr was a Danish scientist who worked in Rutherford's laboratory. Bohr showed each electron traveling in an orbit that is a certain distance from the nucleus. In every atom of a particular element, the electrons always follow the same basic orbits. Bohr drew his model using circles to represent the paths of the electrons. We call this picture of an atom the *Bohr model* of the atom.

Some of Bohr's ideas were confirmed, but his model could not explain all of the properties of electrons and atoms that had been observed. By 1928, other scientists had proposed a different arrangement for electrons. This arrangement, which is based on complex mathematical formulas, better explains the chemical behavior of atoms.

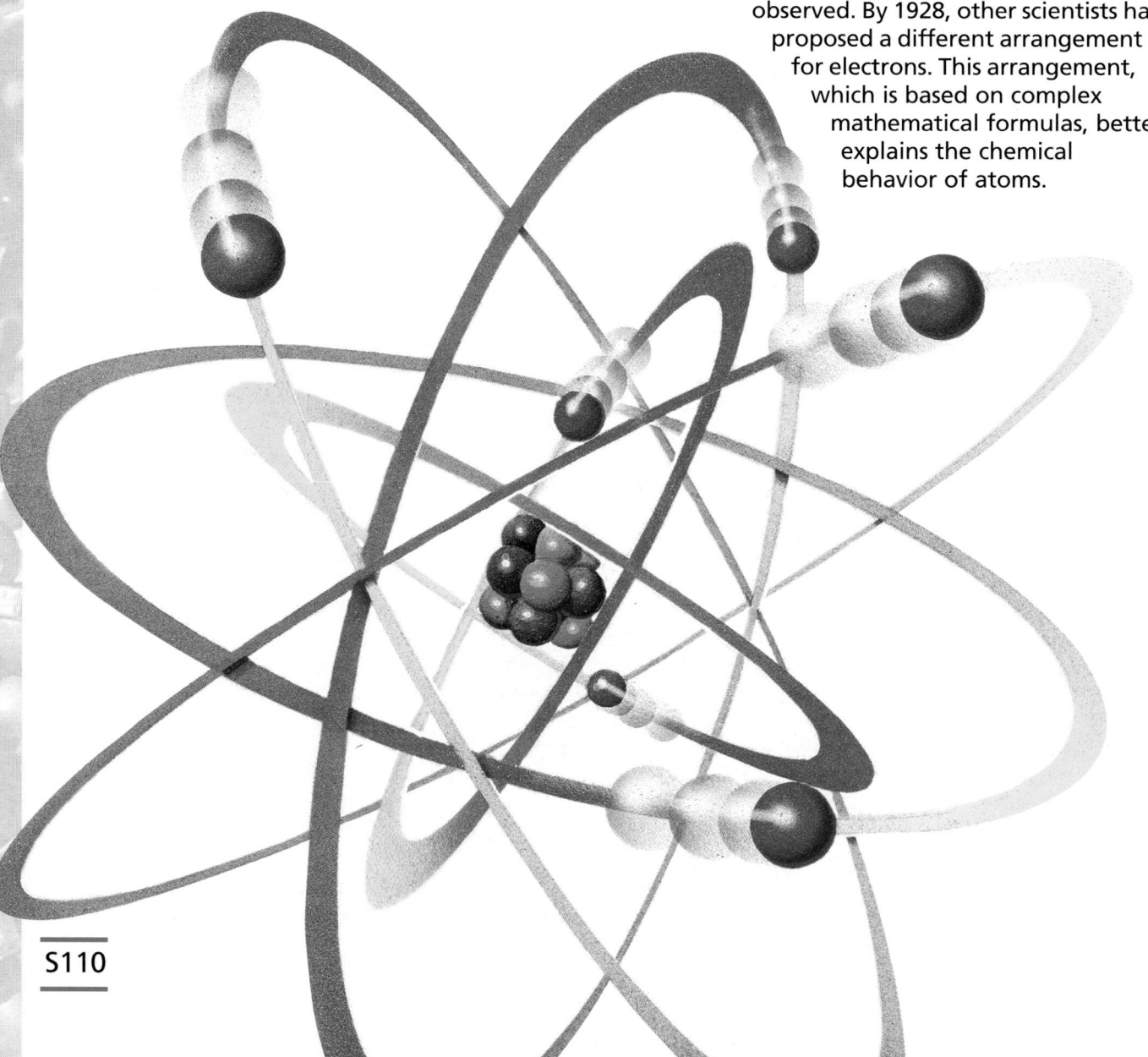

▼ The atomic model in 1913. According to the Bohr model of an atom, electrons move around a nucleus only in certain orbits, much as planets orbit the sun. Electrical attraction between the positive nucleus and the negative electrons keeps the electrons in orbit around an atom's nucleus.

S110

The Atomic Model Today According to the current atomic model, atoms contain three types of particles—protons, neutrons, and electrons. The nucleus holds all of an atom's protons and neutrons. The positively charged protons give the nucleus a positive charge. But instead of picturing the negatively charged electrons as staying in distinct orbits, scientists now picture electrons as being found within a "cloud." The diameter of this **electron cloud** is about 100,000 times greater than the diameter of the nucleus. The speedy electrons move around the nucleus somewhat like a swarm of bees buzzing around a hive. As you might expect, electric forces between the negative electrons and the positive nucleus hold the electrons around an atom's nucleus.

Inside the nucleus, a seemingly impossible situation exists. Protons and neutrons are crowded together in an extremely small space. The positive charge on the protons makes them repel each other. As a result, it seems like the nucleus should fly apart. Why doesn't it? First, the neutrons act like buffers and keep the protons from getting too close together. Second, a great deal of force is needed to hold the nucleus together. This force, called the *strong nuclear force,* holds neutrons and protons together. When the nucleus of an atom splits apart, the energy providing the strong force is released. That energy is called *nuclear energy.*

You may wonder why this book still uses Bohr models to describe individual atoms. It is very hard to represent an atom's electron cloud with a physical shape. Even though the Bohr model is not the way an atom really looks, it gives us a lot of useful information about atoms. For example, we can use the Bohr model to show the numbers of protons, neutrons, and electrons in a particular element's atoms, and we can also use it to show how atoms combine with each other.

Using Atomic Models

An atomic model helps scientists explain why the elements have different properties and why some elements have similar properties. For example, both hydrogen and helium are gases that are lighter than air. But what makes hydrogen explosive and helium unable to burn? Hydrogen combines easily with itself and with other elements to form compounds. Atoms of helium, however, do not combine with atoms of any element, not even other helium atoms. The way that atoms behave appears to depend on the arrangement of their parts. An atomic model also helps scientists predict how elements will combine to form compounds. Before we look at how elements combine, let's see how scientists use atomic models to describe atoms and elements.

Electron cloud

The area surrounding an atom's nucleus where electrons move

The modern atomic model. Today scientists describe atoms as tiny particles with a dense nucleus surrounded by a cloud of electrons. Protons and neutrons in the nucleus make up most of the mass of the atom. Electrons are located about the nucleus within energy levels.

DID YOU KNOW...

that according to modern atomic theory, neutrons and protons each contain three smaller particles called *quarks?* Currently, quarks are considered to be the fundamental particles of matter.

S111

Atomic Number Each element can be identified by the number of protons in its atoms. Hydrogen, for example, has only 1 proton; oxygen has 8 protons. The number of protons in an atom, called the **atomic number** of the atom, establishes the identity of the element and gives its atoms properties unique to that element.

To examine the atomic numbers of the first 20 elements, look at the table to the right. Did you know that atomic numbers tell you more than just the number of protons in the atoms of each of these elements? While defined as the number of protons, the atomic number of an element also tells you how many electrons there are.

Normally, atoms do not have an overall electric charge. The negative charges of the electrons moving around the nucleus exactly cancel the positive charges of the protons in the nucleus. Therefore, the number of electrons around the nucleus must be equal to the number of protons in the nucleus. For example, all hydrogen atoms have one proton in the nucleus. Every hydrogen atom also has one electron moving around its nucleus. Thus, hydrogen atoms are electrically neutral. Oxygen atoms have eight protons and eight electrons. Oxygen atoms are also neutral.

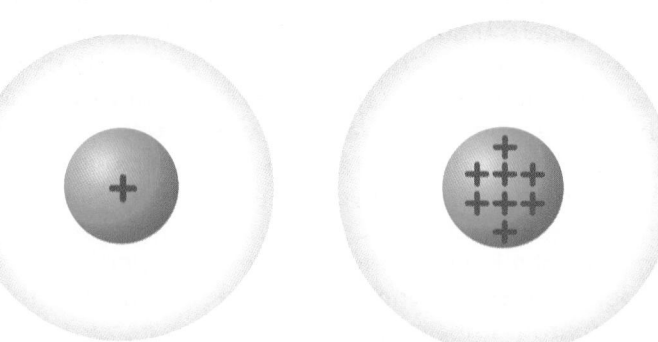

▲ Each element has a different number of protons in its nucleus. As shown here, hydrogen has 1 proton and therefore has an atomic number of 1. Oxygen has 8 protons and an atomic number of 8.

First 20 Elements			
Element	**Atomic number**	**Element**	**Atomic number**
hydrogen	1	sodium	11
helium	2	magnesium	12
lithium	3	aluminum	13
beryllium	4	silicon	14
boron	5	phosphorus	15
carbon	6	sulfur	16
nitrogen	7	chlorine	17
oxygen	8	argon	18
fluorine	9	potassium	19
neon	10	calcium	20

Atomic number

The number of protons in the nucleus of an atom

Electron Arrangement An atom's electrons occupy a region of mostly empty space surrounding the nucleus. As was mentioned, electrons move around the nucleus somewhat like a swarm of bees around a hive. But how are an atom's electrons arranged? Unlike bees, which randomly fly around a hive, electrons tend to be found in certain regions around the nucleus. Each electron has a distinct amount of kinetic energy, and different electrons may have different amounts of this energy.

Electrons with similar amounts of energy occupy an *energy level*. An energy level is a region around the nucleus where electrons with a particular amount of energy are most likely to be found. The one electron in a hydrogen atom is usually located not far from the nucleus in the first energy level. This energy level contains electrons with the lowest amount of energy. The two electrons in helium normally occupy the first energy level as well.

Within an energy level, electrons are found at a certain average distance from the nucleus. An electron can gain or lose energy, but it will change energy levels if it does so.

The diagram below shows the electron arrangement for the atoms of several elements. As you can see, each energy level can hold only a certain number of electrons. Two electrons are the limit for the first energy level. The next energy level holds up to eight electrons. The third energy level holds up to 18 electrons, and the fourth energy level holds up to 32 electrons.

In the table that follows, you can see the number of electrons that occupies each energy level in the atoms of the first 20 elements. Notice that electrons fill the energy levels in a particular order. The first, or lowest, energy level fills first. Next, the second energy level fills, and so on. Notice, however, that before the third energy level fills completely, some electrons enter the fourth energy level.

This illustration shows Bohr models of several atoms, with their electrons in circular orbits. Each circle represents an energy level. Keep in mind that energy levels are really regions of space around the nucleus of an atom where electrons with certain amounts of energy are likely to be found.

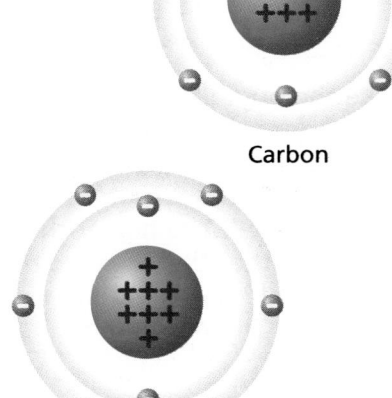

Carbon

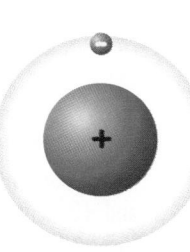

Hydrogen

Helium

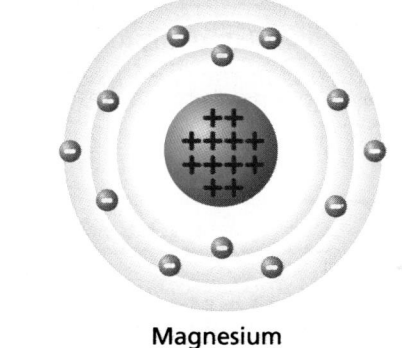

Oxygen

Magnesium

Electron Arrangement of the First 20 Elements		
Atomic number	Element	Number of electrons per energy level
1	hydrogen	1
2	helium	2
3	lithium	2 1
4	beryllium	2 2
5	boron	2 3
6	carbon	2 4
7	nitrogen	2 5
8	oxygen	2 6
9	fluorine	2 7
10	neon	2 8
11	sodium	2 8 1
12	magnesium	2 8 2
13	aluminum	2 8 3
14	silicon	2 8 4
15	phosphorus	2 8 5
16	sulfur	2 8 6
17	chlorine	2 8 7
18	argon	2 8 8
19	potassium	2 8 8 1
20	calcium	2 8 8 2

DID YOU KNOW...

that when the mineral calcite is exposed to the bright light from a camera's flash bulb, it emits light of its own? Atoms within the calcite molecules absorb some of the light, causing electrons to move to a higher energy level. The atoms whose electrons gained energy emit light when those electrons return to their normal energy level.

Mass Number   With the exception of hydrogen, all atoms have both protons and neutrons in the nucleus. Most hydrogen atoms have one proton and no neutrons in the nucleus. The total number of protons and neutrons in an atom's nucleus is called the **mass number** of the atom. You can use the mass number to find the number of neutrons in the nucleus. For example, if you know that the atomic number of oxygen is 8 and its mass number is 16, you could find the number of neutrons in an oxygen nucleus by using the following calculation:

$$
\begin{array}{ll}
\text{mass number} & 16 \ (\text{protons} + \text{neutrons}) \\
\text{atomic number} & \underline{-8} \ (\text{protons}) \\
& 8 \ \ \text{neutrons}
\end{array}
$$

Unlike the number of electrons in an atom, the number of neutrons does not have to equal the number of protons. The figure to the right shows an atom of fluorine. In this case, the fluorine atom has 9 protons and 10 neutrons. This gives a mass number of 19. The higher the atomic number is, the more likely it is that there will be more neutrons than protons. For example, gold has an atomic number of 79 and a mass number of 197. Therefore, it has 79 protons and 112 neutrons in its nucleus.

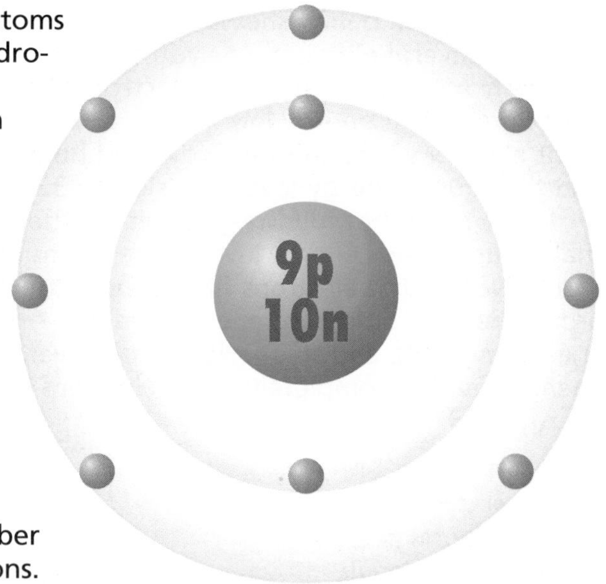

A Bohr model of a fluorine atom shows that the atom has 9 protons and 10 neutrons. Therefore, this fluorine atom has a mass number of 19.

Mass number

Number of protons and neutrons in the nucleus of an atom

Atomic mass

The mass of an atom expressed in atomic mass units

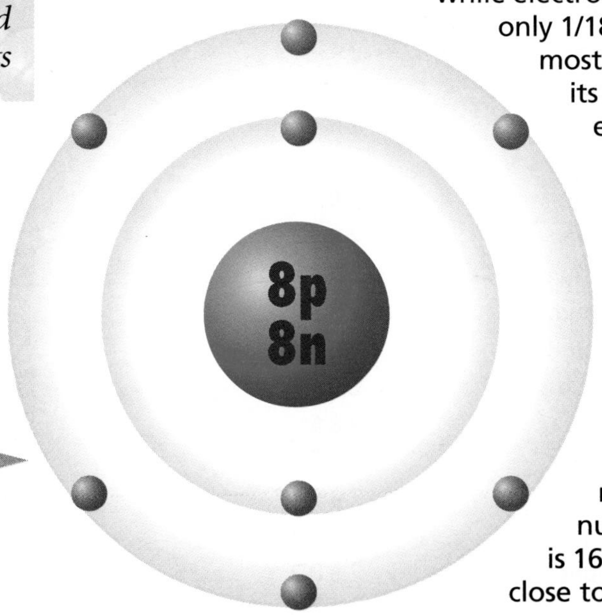

A Bohr model of an oxygen atom ▶ shows that this oxygen atom has eight protons and eight neutrons in its nucleus. The atomic mass of this oxygen atom is about 16 u.

Atomic Mass   The three particles that compose our model atom differ not only in electric charge but also in mass. These particles have very small masses. For instance, if you could count out 600,000,000,000,000,000,000,000 protons and determine their mass, you would find that they have a mass of only about 1 g. The masses of protons, neutrons, and electrons are so small that scientists use a special unit of mass for atoms and their particles. This unit is called the *atomic mass unit* (u).

Protons and neutrons each have an **atomic mass** of about 1 u, while electrons have an atomic mass of only 1/1837 u. You can see why most of an atom's mass is in its nucleus. Altogether, the electrons in any atom have a mass so small that it can usually be ignored. Therefore, the mass of an atom in atomic mass units is about the same as its mass number. For example, an ordinary oxygen atom has 8 protons and 8 neutrons in its nucleus. Thus, the mass number of this oxygen atom is 16, and its atomic mass is close to 16 u.

Isotopes Scientists discovered that different samples of the same element have very small differences in physical properties such as mass and density. This is because not all atoms of the same element have the same number of neutrons. Atoms of the same element that differ in their number of neutrons are called **isotopes**. The isotopes of an element behave the same chemically because they contain the same number of protons and electrons. However, because they have a different number of neutrons, isotopes have different masses and slightly different densities.

A hydrogen atom, for example, may have any one of three arrangements in its nucleus. As you know, most hydrogen atoms have one proton in the nucleus and no neutrons. However, some hydrogen atoms have one proton and one neutron, while a few hydrogen atoms have one proton and two neutrons in the nucleus. Therefore, the element known as hydrogen is made up of three isotopes: hydrogen-1 *(protium),* hydrogen-2 *(deuterium),* and hydrogen-3 *(tritium)*—each one with a single proton.

Almost all elements have isotopes. For example, lithium exists as lithium-6 and lithium-7. In most chemical tables, the atomic mass given for an element is actually a *weighted average* of the atomic masses of all the isotopes of that element found in nature. Analysis shows that 93 percent of all lithium atoms in nature have an atomic mass of 7 u. Only 7 percent have an atomic mass of 6 u. Therefore, the atomic mass of the element lithium is 6.94 u. The atomic mass of the mixture of hydrogen atoms found in nature is 1.008 u.

Isotope

Atoms of the same element that have different atomic masses

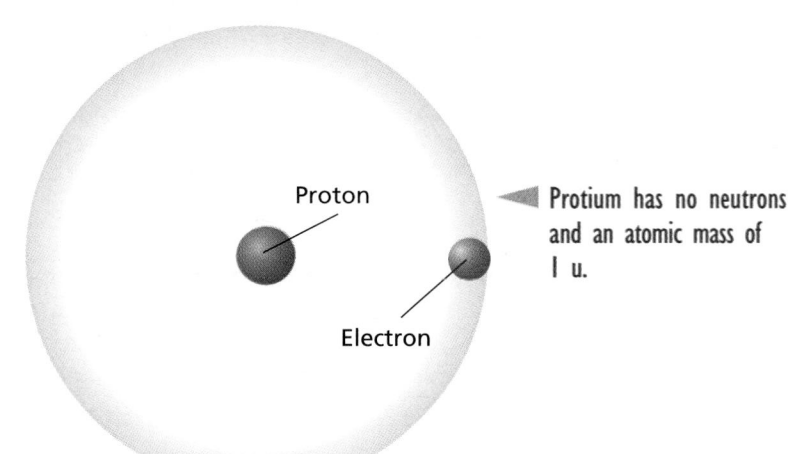

Proton

Electron

◀ Protium has no neutrons and an atomic mass of 1 u.

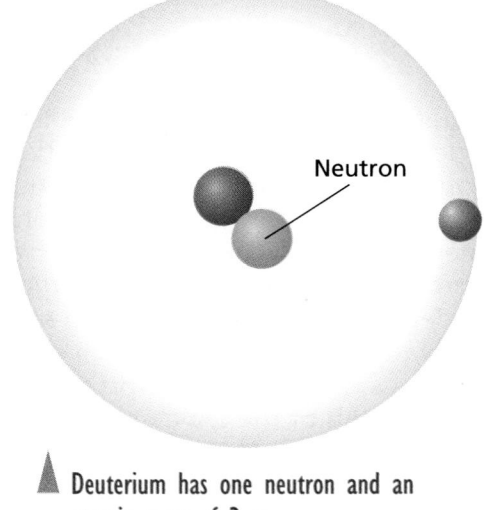

Neutron

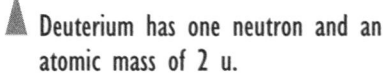
▲ Deuterium has one neutron and an atomic mass of 2 u.

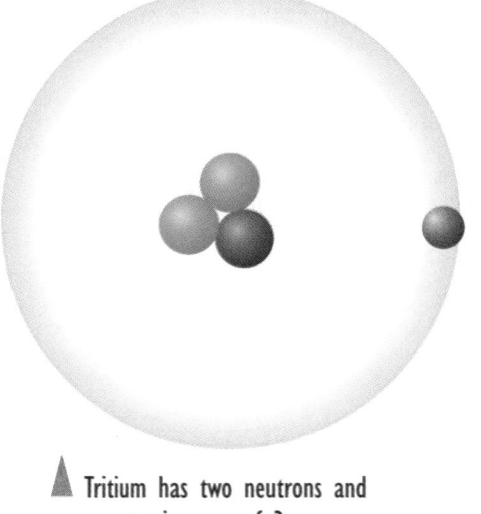
▲ Tritium has two neutrons and an atomic mass of 3 u.

S115

Radioactivity Most isotopes found in nature are stable; they have nuclei that do not change over time. Some isotopes, however, have unstable nuclei that do change over time. What makes some nuclei unstable? A nucleus that contains a certain number of protons needs just the right number of neutrons to be stable. If the nucleus has too many or too few neutrons, it will be unstable and *decay*, or change some of the particles in its nucleus. An atom whose unstable nucleus changes in this manner is said to be *radioactive*.

Radioactivity, the decay of unstable isotopes to more stable isotopes, was discovered by the French scientist Henri Becquerel in 1896. Becquerel observed that substances containing uranium give off rays that pass through paper and affect photographic film. Marie Curie and her husband Pierre became interested in the strange new rays detected by Becquerel. After several years of work, they discovered two previously unknown radioactive elements. They named these elements radium and polonium. Following the pioneering work of Becquerel and the Curies, other elements were found to have radioactive isotopes.

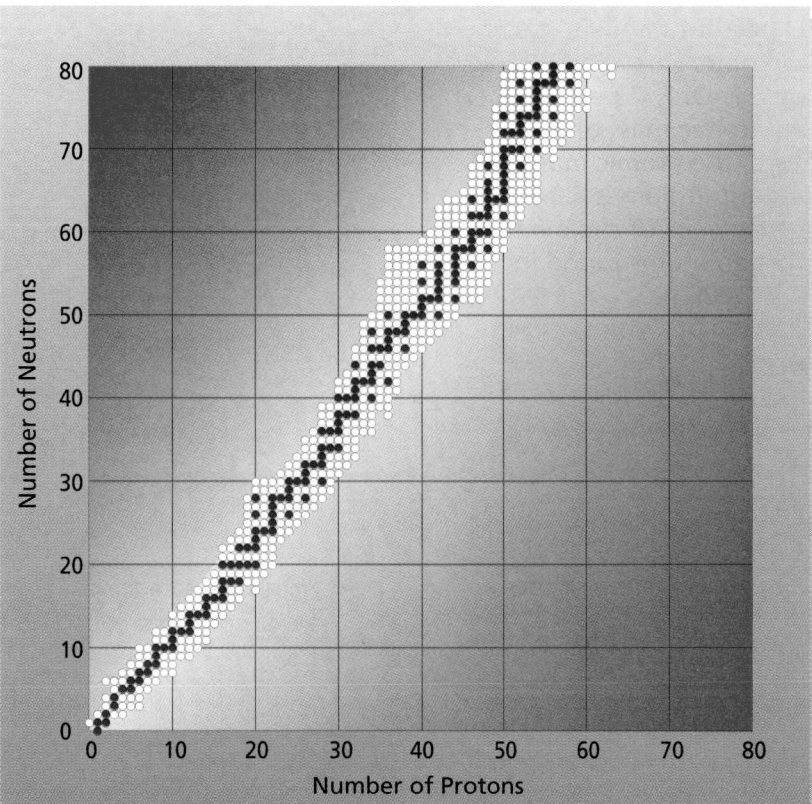

The ratio of neutrons to protons affects nuclear stability. If a nucleus has too many or too few neutrons, it will be unstable. In this graph, solid circles represent stable nuclei, and open circles represent unstable nuclei.

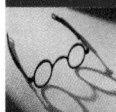
S U M M A R Y

Scientists use models to describe atoms. According to the current model, atoms consist of a nucleus that is made up of protons and neutrons. The nucleus is surrounded by a cloud of electrons that occupy energy levels. These energy levels can contain only a certain number of electrons. Most of the mass of an atom is found in the nucleus. A neutron has almost the same mass as a proton, but an electron has a much smaller mass. The atomic mass is the sum of the protons and neutrons. Different atoms of the same element may have different numbers of neutrons and are called isotopes. Some isotopes are radioactive. The radioactive isotopes of elements contain unstable nuclei.

S116

ELEMENTS AND CHEMICAL ACTIVITY

If you ever watch football games on television, you have probably seen a blimp floating high above the field. A blimp is an airship, which floats because it is filled with a gas that is lighter than air. At one time, airships were filled with hydrogen. But hydrogen is a dangerous gas. It burns rapidly when it combines with oxygen. Some early airships that used hydrogen were destroyed when the hydrogen accidentally ignited. Modern airships are filled with helium gas. Helium is a very stable element and is therefore safe to use in large amounts. Like hydrogen, helium is much lighter than air. Unlike hydrogen, however, helium does not burn because it does not combine with oxygen. In fact, helium is almost totally nonreactive and will not combine with any other element under conditions normally found on Earth.

Many other elements combine with oxygen, although not as explosively as hydrogen. There are other elements that behave like helium and do not combine with other elements. Even though each element has distinct properties, scientists have observed that some elements share similar physical and chemical properties. In this section, you will learn how the structure of atoms affects the properties and chemical behavior of the elements.

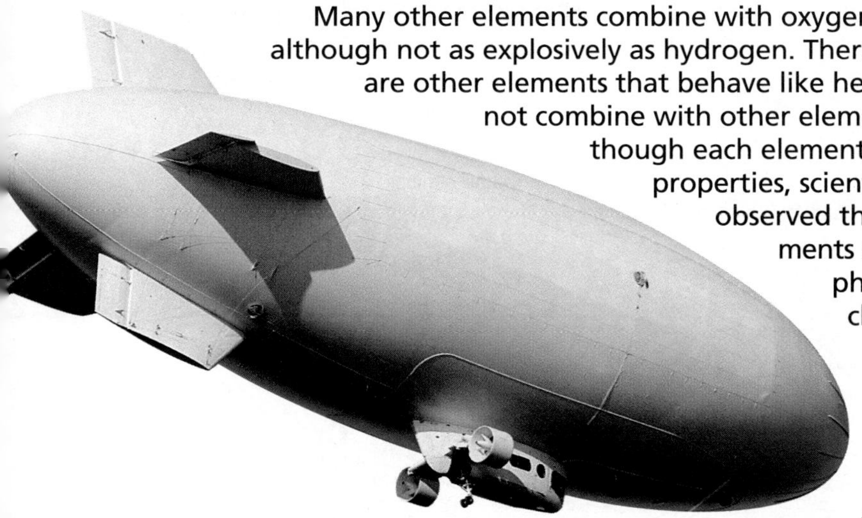

▲ In 1937, an explosion of highly reactive hydrogen gas destroyed the *Hindenburg*, the largest airship ever built.

▲ Helium, which does not burn, is used in all modern airships, such as this blimp.

Classifying the Elements

To date, scientists have discovered a total of 91 naturally occurring elements. Several other elements have been made by artificial means. As scientists were discovering more elements, they began to search for ways to organize these elements. It was found that when the elements were listed by increasing atomic mass, elements with similar properties occurred periodically, or at regular intervals.

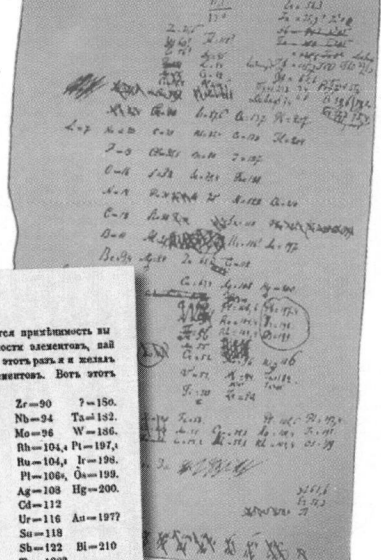

This is a page from Mendeleev's original manuscript.

This is how Mendeleev's periodic table looked when it was published in a journal.

The Periodic Table By the mid-1800s, scientists had discovered and described the properties of many elements using newly developed methods. The atomic masses of most known elements had also been determined. In 1869, a Russian chemist named Dmitri Mendeleev announced that he had developed a chart of the elements that reflected the periodic similarities in their properties.

Mendeleev developed his chart by listing the elements in order of increasing atomic mass. He then divided the list of elements into columns so that elements with similar properties appeared next to each other. To make elements with similar properties fall in the same row, he had to leave some gaps in the table. Mendeleev predicted the discovery of new elements that would fill these blank spaces. A modified version of his chart eventually became known as the **periodic table** of the elements.

Modern periodic tables are arranged differently from the one originally made by Mendeleev. In 1911, scientists discovered that the elements of the periodic table should actually be arranged by increasing atomic number rather than atomic mass. This caused only small changes from Mendeleev's original order. The table itself has also been turned so that Mendeleev's rows are now columns. Compare the modern table with Mendeleev's original.

Periodic table

Chart of the elements arranged according to periodic properties of the elements

Each box in this periodic table contains the name, atomic number, and mass number of an element, plus its letter symbol.

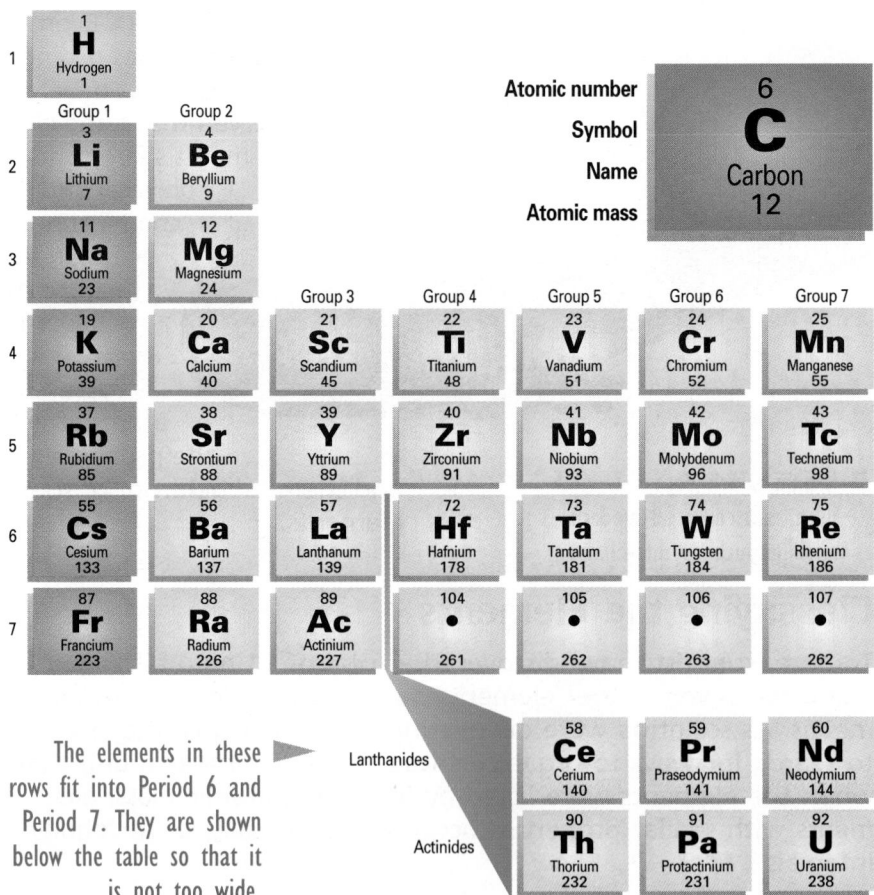

The elements in these rows fit into Period 6 and Period 7. They are shown below the table so that it is not too wide.

Symbols of the Elements Notice that instead of writing out the names of the elements, the modern periodic table uses a symbol for each element. For many years, chemists either wrote out the name of an element or invented a different symbol each time they wanted to represent an element or one of its atoms. Therefore, one element was often referred to in several different ways. Then the Swedish chemist Jöns Jakob Berzelius suggested a standard system of referring to the elements with letters.

The system that Berzelius suggested is still used today. The symbols for the elements consist of one or two letters, taken from the element's name. For example, the symbol for hydrogen is H. The symbol for helium is He. When there are two letters in a symbol, the first letter is capitalized and the second is not. More recently discovered elements tend to be named after famous people or places, like einsteinium and californium. As new elements are discovered, they are assigned new symbols. An international committee of scientists decides the names and symbols of newly discovered elements.

You may notice that some symbols on the periodic table seem to be unrelated to the element's name. For example, Hg is the symbol for mercury, and Au is the symbol for gold. These symbols came from the Latin names for these elements. Latin was the language of science when these substances were identified as elements. Thus, Hg stands for *hydrargyrum,* which is the Latin word for mercury, and Au stands for *aurum,* which is the Latin word for gold.

Elements With Latin Names	
sodium (natrium)	Na
potassium (kalium)	K
copper (cuprum)	Cu
gold (aurum)	Au
silver (argentum)	Ag
iron (ferrum)	Fe
lead (plumbum)	Pb
tin (stannum)	Sn
antimony (stibium)	Sb

▼ Periodic table of the elements. The columns of the periodic table are referred to as *groups,* and the rows are referred to as *periods.*

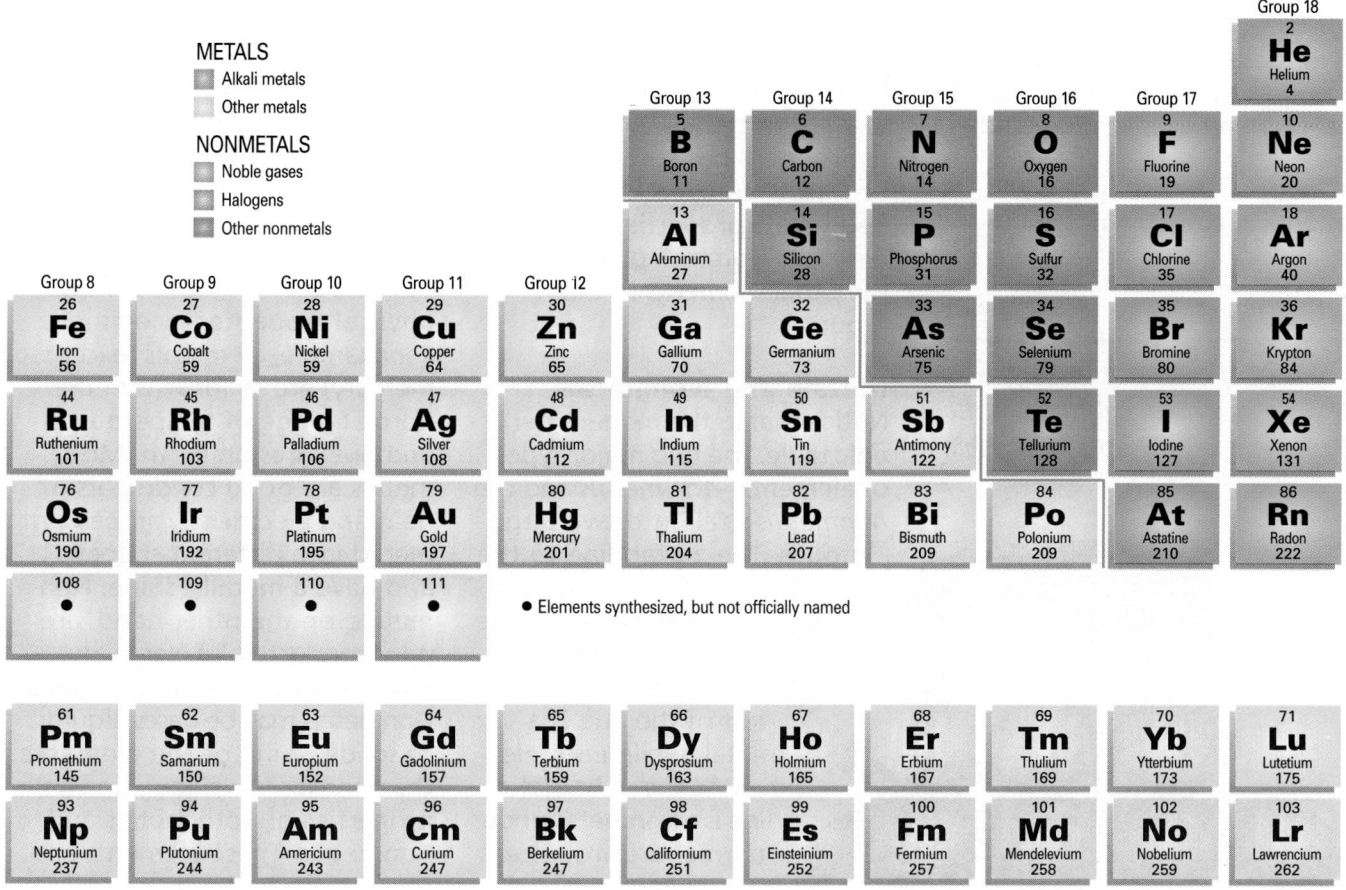

S119

Using the Periodic Table

By inventing the periodic table, Mendeleev developed a valuable tool for scientific research. The periodic table is useful for learning about the chemical elements in the same way that a street map is helpful for finding your way around an unfamiliar city. Because of its arrangement, certain areas of the periodic table contain elements that display similar physical and chemical properties.

Each element has its own position in the periodic table; if you know the chemical properties of two or three neighboring elements, you can predict the properties of the element in question. You can even predict the chemical behavior of elements not yet discovered.

Mendeleev himself predicted the general properties of several elements that had not yet been discovered. When the elements were discovered, they were found to behave almost exactly as Mendeleev had predicted.

Groups By listing the elements in rows by increasing atomic number, scientists were able to make the elements that have similar properties fall into vertical columns. Each vertical column of elements in the periodic table is referred to as a **group**. There are 18 groups in the modern periodic table, numbered from left to right. The elements within a group share more properties with each other than they do with any of the other elements. For example, all elements in Group 1 are metals that combine readily and in a similar manner with other elements. All of the elements in Group 18 are gases that usually do not combine with other elements. Also notice that hydrogen is separated from the rest of the elements in Group 1. Hydrogen is not considered to be a true member of Group 1 because its properties are quite different from the properties of the other elements in that group.

Metals and Nonmetals

Notice that in the modern periodic table, the two major types of elements—the *metals* and the *nonmetals*—fall into two large groups. The zigzag line to the right of the center separates these two large groups. In general, each element to the left of the line is a metal, while each element to the right of the line is a nonmetal under normal circumstances. The elements in each of these sections have several physical properties in common. For example, all metals (except mercury) are solids within the normal range of temperatures and pressures on Earth. Most metals are good conductors of heat and electricity, can be hammered into different shapes, and have a metallic shine. Nonmetals, on the other hand, are poor conductors of electricity and do not have a metallic shine. Nonmetals may be gases, liquids, or solids. Some of the elements that border the zigzag line have properties of both metals and nonmetals. These elements are called *metalloids*.

What characteristics of metals does this copper tea kettle have?

Group

A set of elements arranged in vertical columns in the periodic table

S120

Periods When the periodic table was arranged so that the elements that had similar properties fell into columns, or groups, another pattern emerged. In general, each of the table's horizontal rows contained elements whose atoms have the same number of energy levels occupied by electrons. For example, atoms of hydrogen and helium in the first row have electrons in only one energy level. Atoms of elements in the second row have electrons in two energy levels. Each row of the periodic table is called a **period**. As the numbers to the left of each row indicate, there are seven periods. Therefore, no known element's atoms have more than seven occupied energy levels.

You might be wondering why the rows of the periodic table differ in length. Again, this is how the table had to be arranged to make the elements that share similar properties fall into columns. Still, the length of a row is significant. It indicates the way that electrons fill the energy levels of atoms in that row's elements. In fact, arranging the table in this way helps scientists understand more about how the electrons are arranged in the atoms of the elements.

The two series of elements (the lanthanides and the actinides) located below the rest of the table should actually be included in Periods 6 and 7, between Groups 3 and 4. They are placed below the table to save space. Notice that the last period is not complete. The remaining elements have not yet been identified.

Chemically Active Atoms

Think back to what you learned about the use of hydrogen and helium in airships. Hydrogen burns because it combines with oxygen. Helium does not burn because it does not combine with oxygen. In fact, helium will not take part in any chemical changes. Helium atoms exist separately and are never part of a compound.

The way an atom combines with atoms of other elements is called its **chemical activity**. Hydrogen, which combines readily with other elements, is chemically active. Helium, which does not combine with other elements, is not chemically active. Why do some atoms combine with other atoms, while some do not? Looking at the diagrams on this page might give you a clue.

Electrons and Chemical Activity Recall that hydrogen and helium are the only elements whose atoms have just one occupied energy level. This energy level can hold only two electrons. With two electrons, a helium atom has a completely filled energy level. Could this be why helium atoms do not join with other atoms to form compounds? With only one electron, the energy level of a hydrogen atom is not completely filled. A hydrogen atom needs one more electron to fill the energy level. It gets this electron by combining with many other atoms to form compounds.

Period

A row, or horizontal line, of elements in the periodic table

Hydrogen, with just a single electron in its outer level, is highly reactive.

Notice that when two atoms get close together, the first parts of the atoms to interact are their electron clouds. Also notice that hydrogen and helium have different numbers of electrons in their electron clouds. Perhaps hydrogen and helium atoms behave so differently because they have different numbers of electrons.

Helium, with two electrons in its outer energy level, is stable.

Chemical activity

The tendency of an atom to react with atoms of other elements

S121

As you might suspect, the number and arrangement of an atom's electrons determine an atom's chemical activity. If this is true, then the behavior of atoms with the same number of outer electrons should be alike. The periodic table shows that this assumption is true because the atoms of each element in a group have the same number of electrons in their outer energy level.

Many atoms combine, or form bonds, to obtain a stable number of electrons in their outer energy level. This is why hydrogen atoms, which have an unfilled energy level, are chemically active, but helium atoms, which have a filled energy level, are not. Look back at the table on page S113. Besides helium, which of the elements in this table do you think might be unable to combine with other elements?

To obtain a stable number of electrons in the outer energy

Alkali metals

The group of elements that have atoms with only one electron in their outermost energy level

Halogens

The group of elements that have atoms with seven electrons in their outermost energy level

The Alkali Metals

Atoms that easily lose 1 electron	Atomic number	Stable electron number after losing 1 electron
lithium	3	2
sodium	11	10
potassium	19	18
rubidium	37	36
cesium	55	54
francium	87	86

level, atoms can either gain, lose, or share electrons. For the lighter atoms, two electrons in the outer energy level is a stable number. For most atoms, however, eight is a stable number of electrons for the outer energy level. If an atom has one, two, or three electrons *less* than a stable number, it will tend to add electrons until it has a stable number. If an atom has one, two, or three electrons *more* than a stable number, it will tend to lose electrons until it has a stable number.

For example, consider a lithium atom. The atomic number of lithium is 3. Therefore, a lithium atom has three electrons. As the table on page S113 indicates, there are two electrons in the first energy level and one electron in the second level. Thus, lithium will tend to lose that one electron to reach a stable number of two electrons. The table above lists five other

elements that can also be expected to lose one electron. If you look back at the periodic table on page S118, you will see that these elements are all in Group 1. This group of elements is called the **alkali metals**.

Now consider a fluorine atom. The atomic number of fluorine is 9. Therefore, fluorine has a total of nine electrons. As shown in the top figure on page S114, fluorine has two electrons in its first level and seven electrons in its outer level. Thus fluorine has one electron less than the stable number of eight in its outer energy level. Fluorine adds one electron to get a stable electron arrangement. The table below lists four other elements that also tend to gain one electron to become stable. If you look at the periodic table once more, you will see that these elements are all in Group 17. This group of elements is called the **halogens**.

The Halogens

Atoms that easily gain 1 electron	Atomic number	Stable electron number after gaining 1 electron
fluorine	9	10
chlorine	17	18
bromine	35	36
iodine	53	54
astatine	85	86

Noble Gases Elements that lose or gain electrons readily, like the alkali metals and the halogens, are chemically active. In other words, they take part readily in chemical changes. Atoms with a stable outer energy level, such as helium or neon atoms, do not easily lose or gain electrons. Their electron arrangements are already stable. Because helium atoms do not associate with "common" elements, helium is called a noble gas. There are a total of six noble gases among the known elements. If you look at the periodic table, you will see that all of the noble gases are in Group 18.

Why are the noble gases so stable? Scientists have determined how much energy is needed to remove an electron from the outer level of an atom. These amounts of energy form a pattern. The graph below shows this pattern for the first 20 elements. As you can see, the atoms of the elements helium, neon, and argon need the highest amounts of energy to remove electrons. Therefore, the atoms of these elements have a tighter hold on their electrons than do other atoms.

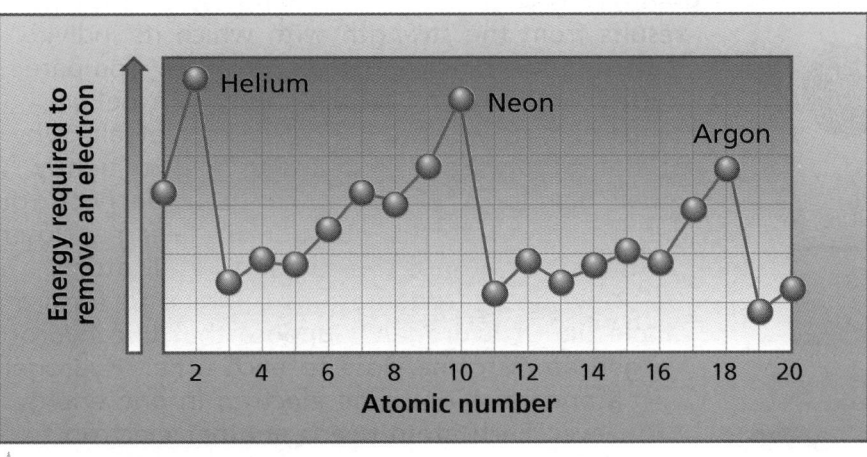

This graph compares the amounts of energy required to remove an electron from atoms of the first 20 elements.

Noble gas

One of the elements in Group 18 of the periodic table, which have a stable number of electrons in the outer energy level

SUMMARY

By listing the known elements in order of atomic mass, scientists discovered that the physical and chemical properties of the elements occur periodically, or in a repeating pattern. Later, scientists found that by arranging the elements in order of increasing atomic number in horizontal rows, the elements that look and act alike could be made to fall into vertical columns. The modern periodic table was developed in this way. Elements whose atoms have the same number of electrons in their outer energy levels have similar characteristics and are in the same group. Elements whose atoms usually have the same number of energy levels occupied by electrons are in the same period. Scientists use the periodic table to predict the chemical activity of the elements. Atoms that combine with other atoms are chemically active. The number and arrangement of an atom's electrons determine its chemical activity. Atoms tend to gain or lose electrons to obtain a stable number of electrons in the outer energy level.

S123

PUTTING ATOMS TOGETHER

Have you ever put together a jigsaw puzzle? At first, the separate pieces seem to have no pattern. But you know that the different pieces will fit together if you can match them properly. In this manner, atoms are somewhat like the pieces of a puzzle. Atoms join together, but only in certain ways.

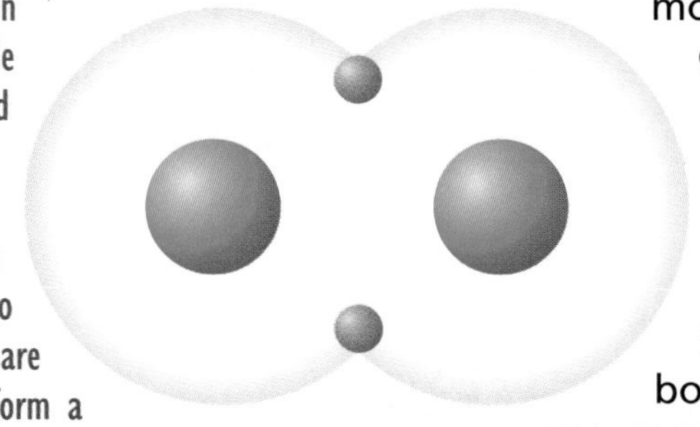

What Holds Atoms Together?

Diamond is the hardest naturally occurring substance known. A diamond crystal is made only of carbon atoms. However, carbon also exists in softer forms, such as the graphite in your pencil. The hardness of a diamond results from the strength with which its individual carbon atoms cling to each other. Compare carbon, in the form of diamond, with helium, whose atoms will not cling together at all. Why do the carbon atoms in diamond hold so tightly to their neighbors, while helium atoms do not? You can find an answer to this question by taking a closer look at hydrogen atoms, the simplest of all atoms.

How do hydrogen atoms combine to form a molecule of hydrogen? Suppose that two hydrogen atoms come close to each other. Hydrogen atoms have only one electron in one energy level. Each atom needs another electron to become stable. Both hydrogen atoms can become stable by *sharing* their electrons. When this happens, a molecule of hydrogen is formed as the shared electrons move about both nuclei. No more than two hydrogen atoms can combine because each atom has only one energy level and can hold only two electrons. In a hydrogen molecule, the energy level of each atom is completely filled by the two shared electrons.

When each of two hydrogen atoms shares its electron with the other, they are joined by a **chemical bond**. A chemical bond is a link between atoms that results from their mutual attraction for electrons. One way in which atoms are held together by a chemical bond is by the attraction of their electrons to the nuclei of both atoms. The figure to the left shows how the electrons in a hydrogen molecule are attracted to both hydrogen nuclei to form a chemical bond.

Now compare the behavior of two hydrogen atoms with that of two helium atoms. The outer energy level of each helium atom is filled with two electrons. Thus a helium atom does not need to gain or lose electrons to become stable. In other words, helium atoms will not form a chemical bond. In nature, helium atoms do not combine with other atoms to form molecules. The same is usually true for the other noble gases. Noble gases are the only elements that exist in nature as free atoms.

▼ When two hydrogen atoms approach each other, their electron clouds begin to interact.

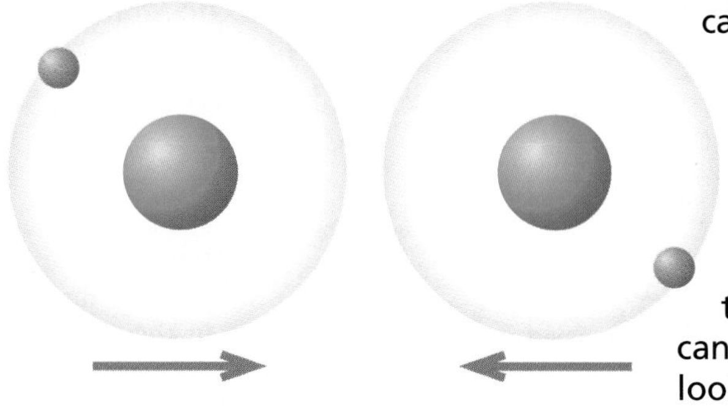

Two hydrogen atoms ▼ are held together in a hydrogen molecule because the shared electrons are attracted to the nuclei of both atoms. The two hydrogen atoms share their electrons to form a hydrogen molecule.

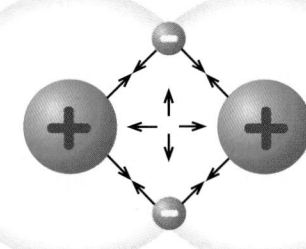

Chemical bond

A link between atoms resulting from the mutual attraction of their nuclei for electrons

S124

Types of Chemical Bonds

Atoms form chemical bonds in two general ways. They either *share* electrons to become stable, or they *gain* or *lose* electrons to become stable. For example, a hydrogen atom has only one electron. Because two electrons are needed to fill hydrogen's only energy level, a single hydrogen atom often shares one electron with another atom, as in the previous example.

Covalent Bonds   Under normal conditions, hydrogen gas exists as molecules rather than as individual atoms. Each atom of this element is missing only one electron to complete its outer energy level. As a result, two hydrogen atoms each share one electron with the other to form a molecule made up of two atoms. When atoms share electrons to fill their outer energy levels, the chemical bond that forms is called a **covalent bond**. The bond between two hydrogen atoms is a covalent bond. Compounds formed by covalent bonding are called *covalent compounds*.

A molecule that consists of two atoms joined by covalent bonds is known as a *diatomic molecule*. All of the elements that normally are found in nature as gases (except the noble gases) exist as diatomic molecules. The halogens, as well as hydrogen, nitrogen, and oxygen, form diatomic molecules.

Some atoms share more than one pair of electrons to form a covalent bond. For example, an oxygen atom has six outer electrons. Each oxygen atom needs two electrons to reach a stable number of eight electrons in its outer energy level. Thus, two oxygen atoms could share two

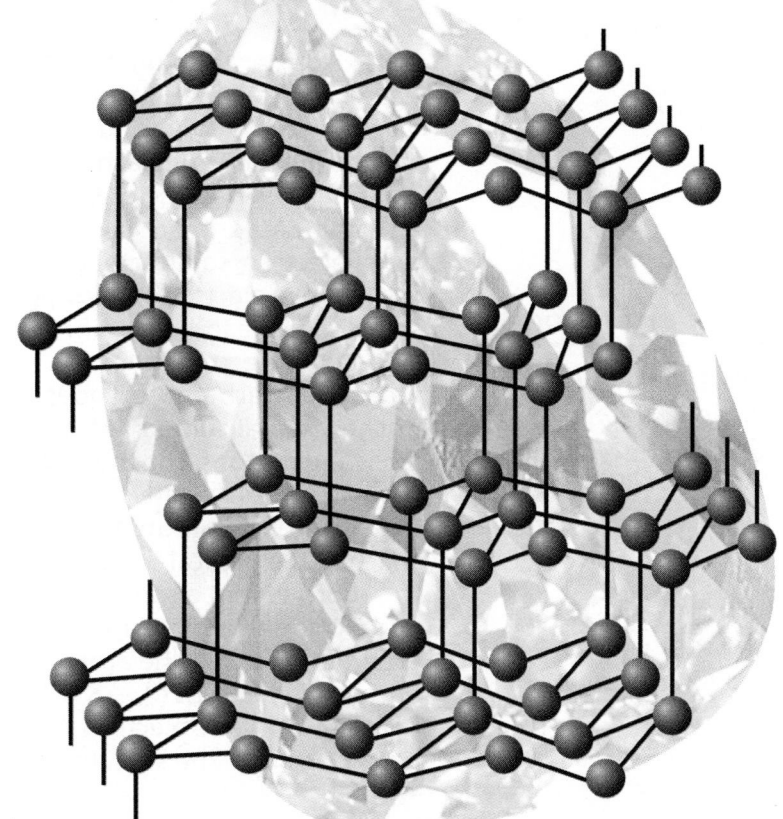

In diamond, each carbon atom is bound to four other carbon atoms.

pairs of electrons to form an oxygen molecule. Carbon atoms have four outer electrons, which they can share to form covalent bonds. However, carbon is not found as diatomic molecules. Elements with more than one electron to share—like phosphorus, sulfur, carbon, and silicon—often form molecules with two or more atoms. Diamonds are so hard because each of the carbon atoms in a diamond shares one electron with each of four other carbon atoms around it.

Covalent bonds also form between atoms of two or more different types. For example, two hydrogen atoms and one oxygen atom can join to make a water molecule. The oxygen atom's outer energy level is filled by sharing an electron from each of the two hydrogen atoms, and each hydrogen atom fills its outer energy level by sharing one electron from the oxygen atom.

DID YOU KNOW...

that the greatest number of different elements known to exist in a single compound is 10?
Though quite a mouthful, the compound's name is dichloro(diiodocadmium)bis[μ-[3-(diphenylphosphino)-1-propanaminato-*N*:*P*]]dinitrosylbis[tris (3,5-dimethyl-1*H*-pyrazolato-N^1) hydroborato(1-)-N^2,N^2,N^2]ditungsten.

Covalent bond

A chemical bond resulting from the sharing of electrons

Ionic bond

A chemical bond resulting from the attraction between positive and negative ions

Ionic Bonds Sodium atoms have 11 electrons—2 in the first level, 8 in the second, and 1 in the third level. Thus, each sodium atom has one outer electron. Rather than sharing its outer electron, sodium tends to just lose one electron to become stable. When a sodium atom gives up an electron, it is no longer neutral because it does not have equal numbers of electrons and protons. The sodium atom becomes a positive *ion*. An ion is an atom or group of atoms that has an electrical charge.

Suppose that a sodium atom and a chlorine atom come together. The chlorine atom, having 17 electrons (2 + 8 + 7), needs to gain one electron to become stable. The sodium atom needs to lose one electron to become stable. Therefore, for both of the atoms to become stable, the sodium atom will lose one outer electron to the chlorine atom. The sodium atom becomes a positive sodium ion, and the chlorine atom becomes a negative ion. The two oppositely charged ions are then electrically attracted to each other. This attraction results in another kind of chemical bond called an **ionic bond**.

Compounds formed with ionic bonds are called *ionic compounds*. Sodium and chlorine form an ionic bond to become sodium chloride, which is common table salt. An ionic compound, like salt, always contains ions formed from atoms of two or more elements. Because one atom must lose electrons and one must gain electrons, an ionic bond cannot form between two atoms of the same element. Why? Recall that the atoms of an element always have the same chemical properties. For two atoms of the same element to form an ionic bond, one would have to lose electrons and the other would have to gain electrons at the same time. In other words, the chemical behavior of two identical atoms would have to be different. Obviously, this cannot happen.

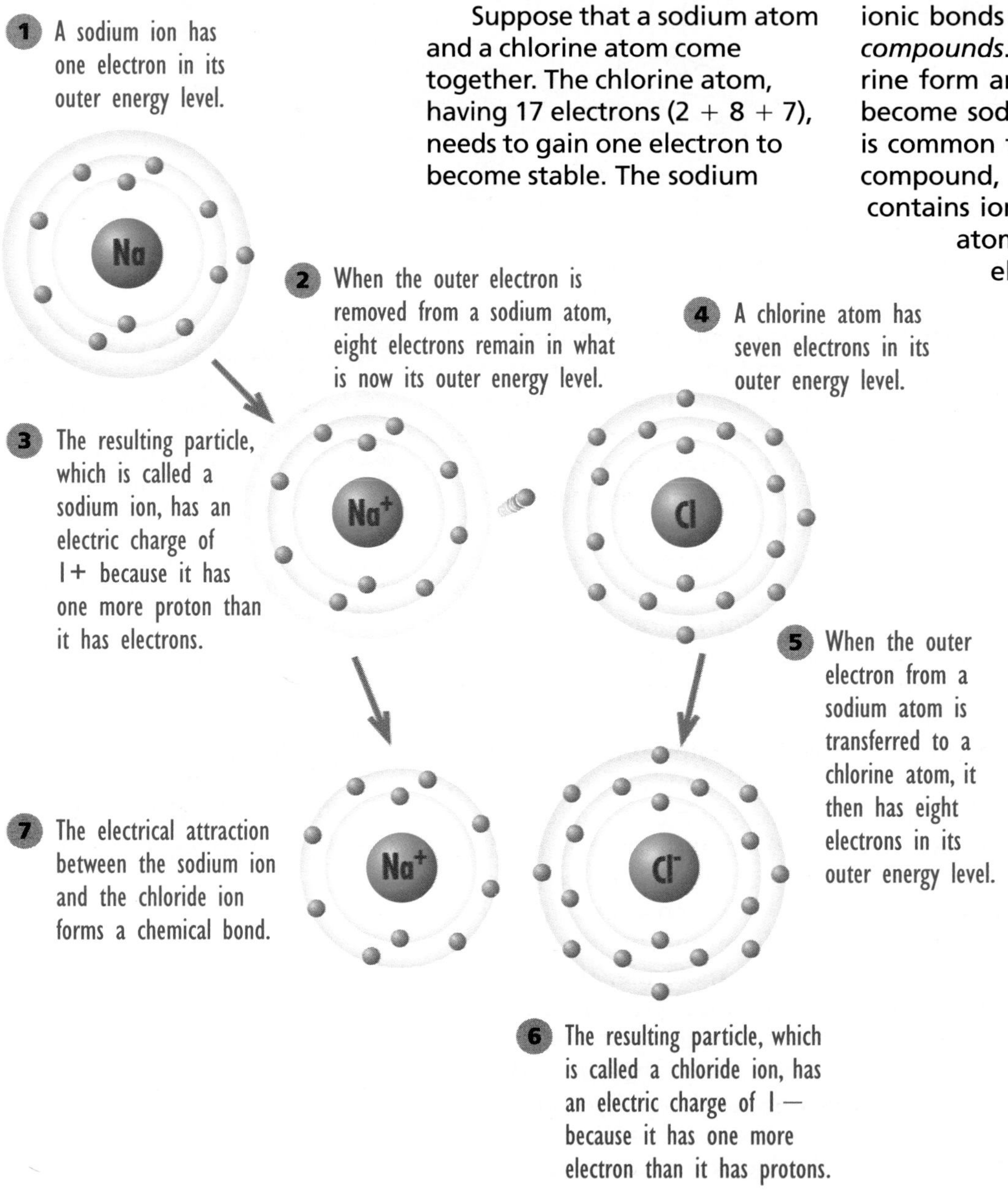

1 A sodium ion has one electron in its outer energy level.

2 When the outer electron is removed from a sodium atom, eight electrons remain in what is now its outer energy level.

3 The resulting particle, which is called a sodium ion, has an electric charge of 1+ because it has one more proton than it has electrons.

4 A chlorine atom has seven electrons in its outer energy level.

5 When the outer electron from a sodium atom is transferred to a chlorine atom, it then has eight electrons in its outer energy level.

7 The electrical attraction between the sodium ion and the chloride ion forms a chemical bond.

6 The resulting particle, which is called a chloride ion, has an electric charge of 1— because it has one more electron than it has protons.

Acids and Bases

The bonds that hold ions together are easily broken when an ionic compound dissolves in water. As the compound dissolves, the positive and negative ions separate from one another and move about freely. Some covalent compounds can separate into ions as well. Acids and bases are compounds that form certain types of ions when they dissolve in water.

Characteristics of Acids In scientific terms, an **acid** is a compound that produces hydrogen ions (H^+) when dissolved in water. Note that a positive hydrogen ion is simply a proton. However, protons are not normally found uncombined in nature. What actually happens is that each H^+ ion combines with a water molecule to form a *hydronium* ion, or H_3O^+. Acids are classified as strong or weak, depending on how many hydronium ions result when they dissolve in water; the stronger the acid, the more hydronium ions that form.

You may think of an acid as a substance that "eats" through metal. Reacting with metals is one characteristic of strong acids. Strong acids can also react with many other substances, including your skin. However, not all acids are that strong. Acids also react with substances such as limestone and marble, releasing carbon dioxide. This reaction causes the deterioration of marble buildings and statues.

One characteristic of acids is that they taste sour. You have certainly tasted many weak acids. In fact, you might have had some of a weak acid for breakfast. Orange juice contains a weak acid. Carbonated drinks also contain a weak acid. Most fruits have some acid content. Oranges and lemons contain citric acid; apples contain malic acid. Green, or unripe, fruits taste sour because they have a greater amount of acid than ripe fruits have. Some of the acid in fruit is converted into sugar as the fruit ripens. Acetic acid, another common weak acid, is found in vinegar.

▲ In nature, the reaction of a naturally occurring weak acid with limestone causes the formation of some caves and sinkholes.

▲ Here you can see hydrochloric acid reacting with limestone. The bubbles are carbon dioxide gas.

Acid

A compound that produces hydrogen ions (H^+) when dissolved in water

▼ Each of these products contains an acid.

Never touch or taste a strong acid.

Base

A compound that produces hydroxide ions (OH⁻) when dissolved in water

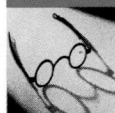
Characteristics of Bases

In scientific terms, a **base** is defined as a compound that produces hydroxide ions (OH^-) when dissolved in water. Like acids, bases can be classified as strong or weak, depending on how many OH^- ions they produce when dissolved in water. The stronger the base, the greater the number of hydroxide ions that are produced. Some examples of such bases are sodium hydroxide and ammonium hydroxide.

One characteristic shared by all bases is that they have a soapy, slippery feel. Many cleaning compounds contain bases. If you have ever gotten soap in your mouth, you are familiar with another characteristic of bases—their bitter taste. Strong bases are just as dangerous as strong acids. They can cause serious burns and must be handled with care. Solutions that contain bases are often designated by the word *alkaline*.

When an acid and a hydroxide base are mixed, they react to form water and a salt. A *salt* is any ionic compound formed when the negative ion from an acid combines with the positive ion from a base. For example, when hydrochloric acid is mixed with sodium hydroxide, the salt sodium chloride is produced.

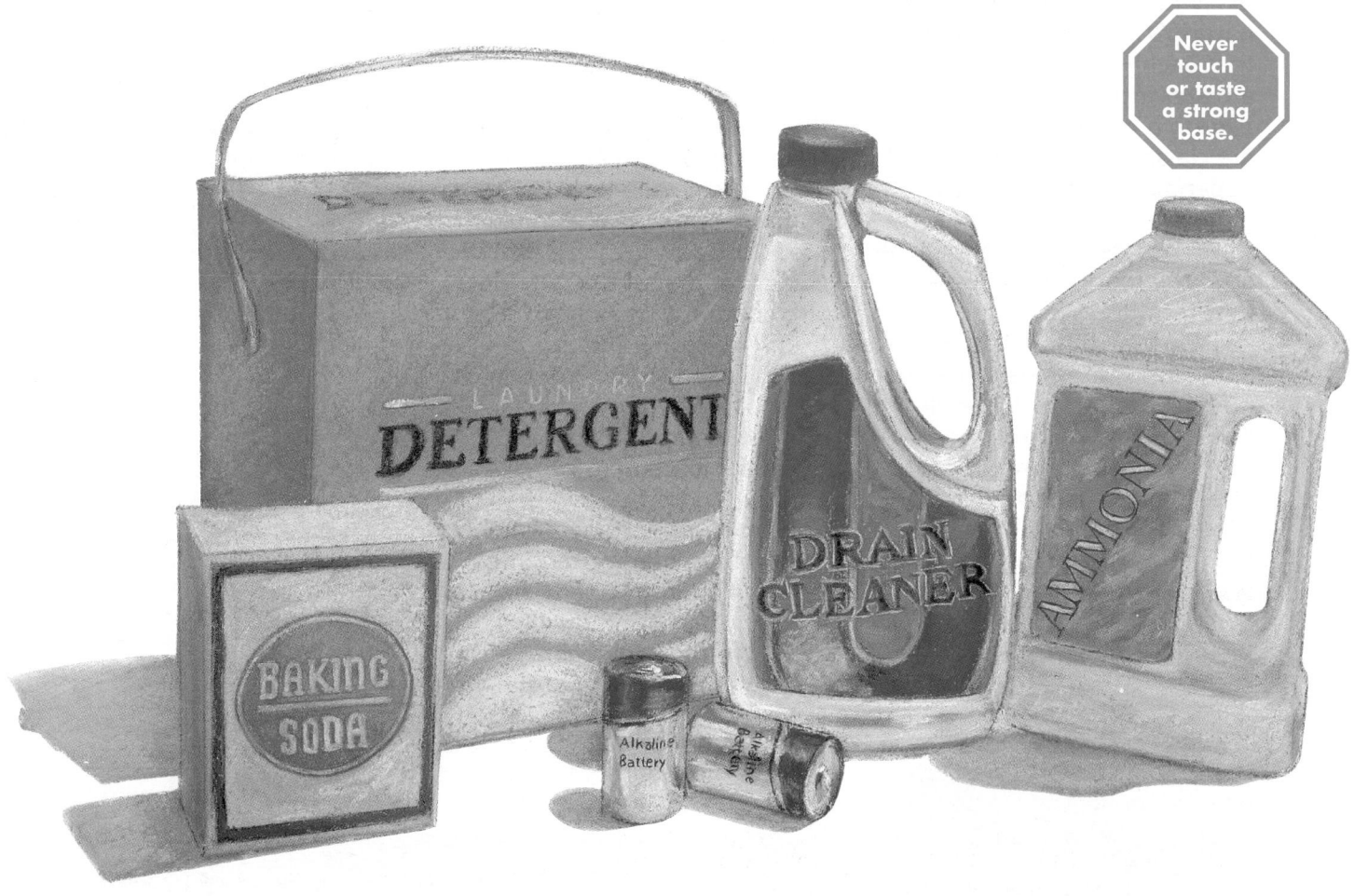

Never touch or taste a strong base.

▲ Each of these products contains a base.

Describing Chemical Compounds

When you mix lemonade, you may make it too sweet by adding too much sugar or too sour by adding too much lemon juice. Like all mixtures, the composition of a solution is not always the same. However, chemical compounds are always made up of the same elements in the same proportions. A pure compound, such as water or sodium chloride, will always be exactly the same no matter how or where it is made.

Chemical Formulas Every element can be represented by its symbol. Thus, symbols for the atoms in a compound can be used to describe the compound. For example, the compound water contains hydrogen atoms and oxygen atoms. The symbols H for hydrogen and O for oxygen can be used to describe water. However, only certain numbers of hydrogen and oxygen atoms combine to make water molecules. Because oxygen atoms need two electrons to fill their outer energy level, the oxygen atom combines with two hydrogen atoms. These numbers must be included in a correct description of water. A description of a chemical compound using symbols and numbers is called a **chemical formula**.

The chemical formula for a molecule of water is H_2O. The number 2 written below the symbol for hydrogen is called a *subscript*. This subscript indicates that there are two atoms of hydrogen in each molecule of water. As you will see, the subscript number 1 is never used as part of a chemical formula. If no subscript number is given after the symbol of an atom, it means

▲ The elements in a particular compound always combine in the same ratio. For example, the compound water consists of molecules that contain two hydrogen atoms covalently bonded to one oxygen atom.

that the molecule contains one atom of that element. Every chemical compound can be described by a chemical formula. For example, remember that two hydrogen atoms can form a diatomic molecule by sharing electrons. The chemical formula for a diatomic molecule of hydrogen is H_2.

Chemical formula

A description of a chemical compound using symbols and numbers

DID YOU KNOW...

that Jöns Jakob Berzelius, who developed the system of letter symbols we use for the elements, was one of the founders of modern chemistry? During a 10-year span, he analyzed more that 2000 chemical compounds. He also determined the atomic masses of 40 elements and discovered 4 new elements.

S129

Describing Chemical Changes

You have already learned that a chemical change is different from a physical change. For example, chopping a large piece of wood into smaller pieces is a physical change, while burning wood in a fireplace is a chemical change. In a chemical change, the atoms in the molecules of one or more substances are rearranged or exchanged to form one or more new substances.

Chemical Reactions How would you describe the burning of charcoal? Burning charcoal, or any other material, produces a chemical change. When a chemical change takes place, the substances that are present before the change are called *reactants*. The reactants are the substances that react or change. The new substances formed as a result of the chemical change are called *products*. Thus, a general description of a chemical change could be written as "reactants yield products." For example, charcoal is made up mostly of carbon atoms. When charcoal burns, the carbon in it combines with oxygen to form carbon dioxide. Carbon and oxygen are the reactants and carbon dioxide is the product. The burning of charcoal can be described as follows:

> carbon plus oxygen yields carbon dioxide

The combination of any element, such as carbon, with oxygen is called *oxidation*. The combining of iron and oxygen to form rust is another example of oxidation. Oxidation is one kind of **chemical reaction**. A chemical reaction is a process in which a chemical change takes place. Burning charcoal is a

chemical reaction because carbon and oxygen are chemically changed into the compound carbon dioxide.

▲ A chemical reaction takes place during the burning of charcoal.

You can usually observe some evidence that a chemical reaction has taken place. For example, many chemical reactions result in an increase or a decrease in the temperature of the reactants. Energy is involved in making and breaking chemical bonds. A change in temperature indicates that energy, in the form of heat, is being released or absorbed as a reaction takes place.

Chemical Equations Another way of writing the reaction of carbon with oxygen is with the chemical formulas for carbon and oxygen:

$$C + O_2 \rightarrow CO_2$$

This description of a chemical reaction is called a **chemical equation**.

Chemical reaction

A process in which a chemical change occurs: reactants yield products

Chemical equation

A description of a chemical reaction using chemical formulas

S130

The chemical equation for the burning of charcoal says that one atom of carbon combines with one molecule of oxygen to form one molecule of carbon dioxide.

In a chemical equation, the chemical formulas for all reactants are written on the left side. Each reactant is separated from the others by a plus sign (+). The chemical formulas for the products are written on the right. Reactants and products are separated by an arrow that means *produces* or *yields*.

Now consider the word equation for the reaction of hydrogen and oxygen to form water:

hydrogen plus oxygen
yields water

Using chemical formulas, the chemical equation for this reaction is written as follows:

$$H_2 + O_2 \rightarrow H_2O$$

However, this equation is not complete. Look at the number of oxygen atoms on each side of the equation. There are two oxygen atoms on the left side of the arrow but only one oxygen atom on the right. One oxygen atom seems to have disappeared. However, scientists have determined that matter cannot be destroyed—only rearranged—during chemical reactions. The same number and type of atoms exist after a chemical reaction as existed before the reaction. The fact that matter cannot be created or destroyed is called the **law of conservation of mass**.

To correct the equation above, you must *balance* it. Because there are two oxygen atoms on the left of the arrow, there must also be two on the right. This problem cannot be solved by changing the formulas

(H_2, O_2, and H_2O). Each of these chemical formulas correctly describes the molecules of hydrogen, oxygen, and water. The only way that there could be two oxygen atoms on the right side is for two water molecules to have formed. You can indicate that the reaction forms two water molecules by placing the number 2 in front of the chemical formula for water, as shown below:

$$H_2 + O_2 \rightarrow 2H_2O$$

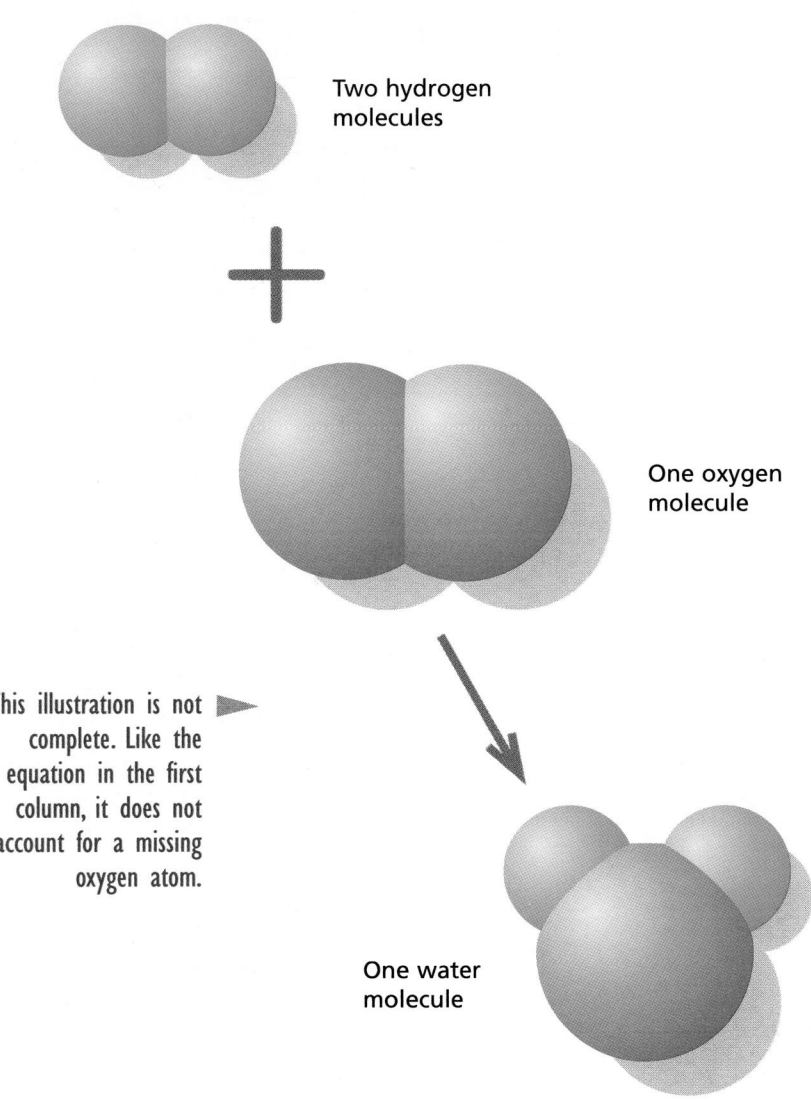

Two hydrogen molecules

One oxygen molecule

This illustration is not complete. Like the equation in the first column, it does not account for a missing oxygen atom.

One water molecule

A number written in front of a formula is called a *coefficient*. A coefficient tells how many chemical compounds are used or made. Now the oxygen atoms are balanced, but the hydrogen atoms are not. There are four hydrogen atoms on the right side but only two hydrogen atoms on the left. This problem can be solved by placing a 2 in front of H_2. With the coefficient 2 before H_2, the equation now reads:

$$2H_2 + O_2 \rightarrow 2H_2O$$

This equation says that two molecules of H_2 react with one molecule of O_2 to yield two molecules of H_2O. Now there are four hydrogen atoms and two oxygen atoms on either side of the equation. The equation is correctly balanced. A balanced equation demonstrates the law of conservation of mass—atoms are neither created nor destroyed in a reaction.

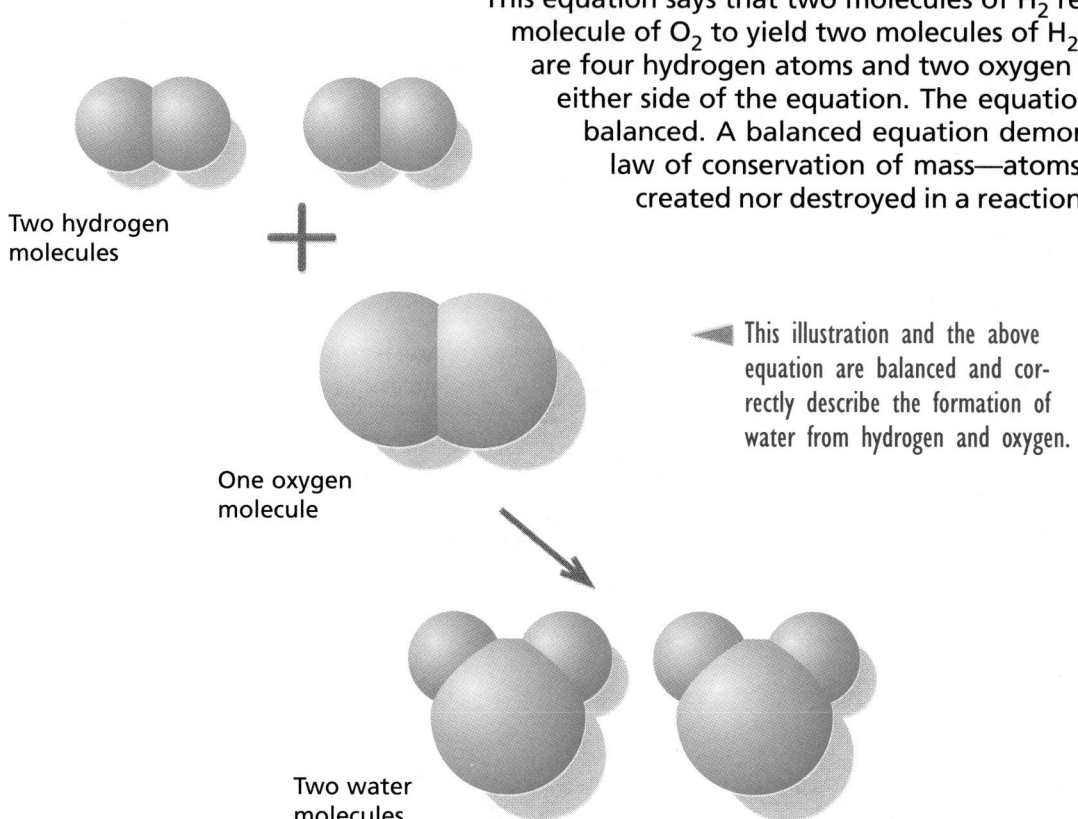

Two hydrogen molecules

One oxygen molecule

Two water molecules

◄ This illustration and the above equation are balanced and correctly describe the formation of water from hydrogen and oxygen.

SUMMARY

Atoms are held together by chemical bonds to form compounds. In a covalent bond, electrons are shared by two or more atoms. In an ionic bond, atoms lose or gain electrons. By sharing, losing, or gaining electrons, atoms achieve a stable number of electrons in the outer energy level. Acids and bases are two types of compounds that form a salt (an ionic compound) when they react. Chemical formulas are used to represent the number and kind of atoms that combine to form a compound. Chemical changes occur when the atoms of substances are rearranged to form new substances. Chemical reactions are events in which chemical changes occur. Chemical equations use chemical formulas to describe chemical reactions.

Concept Mapping

The concept map below illustrates major ideas in this unit. Complete the map by supplying the missing terms. Then extend your map by answering the additional question below. Write your answers in your ScienceLog. **Do not write in this textbook.**

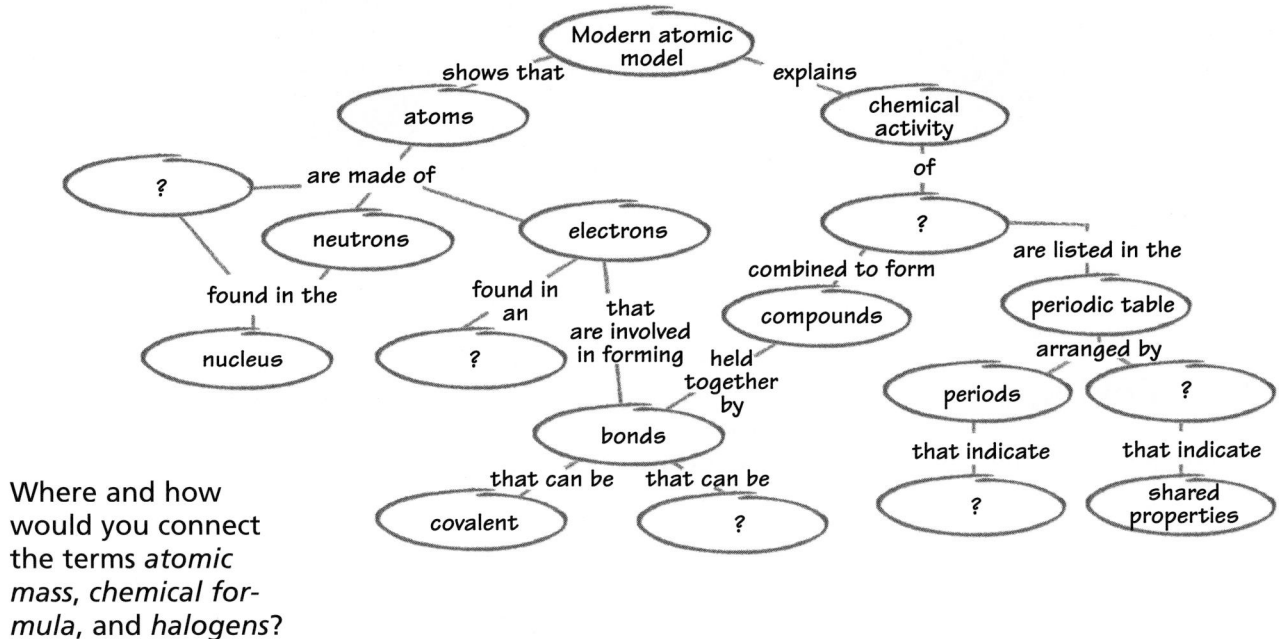

Where and how would you connect the terms *atomic mass*, *chemical formula*, and *halogens*?

Checking Your Understanding

Select the choice that most completely and correctly answers each of the following questions.

1. The modern atomic model states that
   a. an atom is a solid ball of positive electricity with negative charges in it.
   b. an atom's mass is distributed evenly.
   c. most of an atom's nucleus is empty space.
   d. electrons are located in energy levels.

2. An energy level
   a. holds a specific number of electrons.
   b. may hold up to 200 electrons.
   c. is a region within the nucleus.
   d. contains electrons with various amounts of energy.

3. Atoms that are isotopes of the same element
   a. have different numbers of electrons.
   b. have different mass numbers.
   c. have the same atomic mass.
   d. have the same number of neutrons.

4. Which of the following groups of elements is not chemically active?
   a. alkali metals
   b. metalloids
   c. noble gases
   d. halogens

5. If two atoms are held together by sharing three pairs of electrons, they are joined by
   a. covalent bonds.
   b. ionic bonds.
   c. strong nuclear forces.
   d. magnetic attraction.

S133

Interpreting Illustrations

Identify the two atoms represented here by Bohr models as either atoms of the same element or atoms of different elements. Explain.

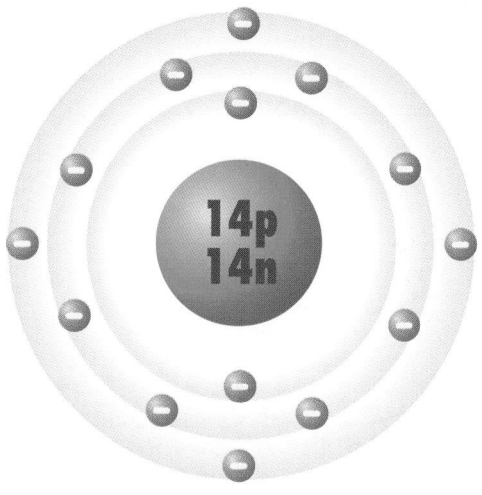
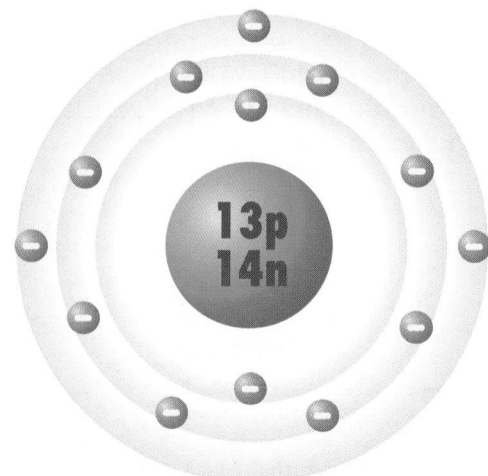

Critical Thinking

Carefully consider the following questions, and write a response in your ScienceLog that indicates your understanding of science.

1. Compare and contrast the masses of protons, neutrons, and electrons.

2. The element potassium has one electron in its outer energy level, whereas the element bromine has seven electrons in its outer energy level. What type of bond do you think will form between a potassium atom and a bromine atom? Why?

3. Why is it acceptable to use the Bohr model of the atom to explain the basic properties of elements even though the model is outdated?

4. How are the elements within a period of the periodic table similar? How are they different?

5. How would you balance the following chemical equation? $Fe + O_2 \rightarrow FeO$

Portfolio Idea

Our view of the atom has changed greatly since the time of Democritus. Write a letter to Democritus describing how scientists now view atomic structure. In your letter, include diagrams that will help Democritus understand your descriptions.

S134

ENERGY and YOU

Now that you have been introduced to several forms and uses of energy, consider these questions.

1. What is energy, and in what forms is it observed?
2. How is the energy of a system conserved?
3. How is energy from the sun passed through a community of organisms?
4. In what ways do humans harness energy from the sun and use it to do work?

In this unit, you will take a closer look at what energy is and how we and other organisms use energy to do work.

S135

WHAT IS ENERGY?

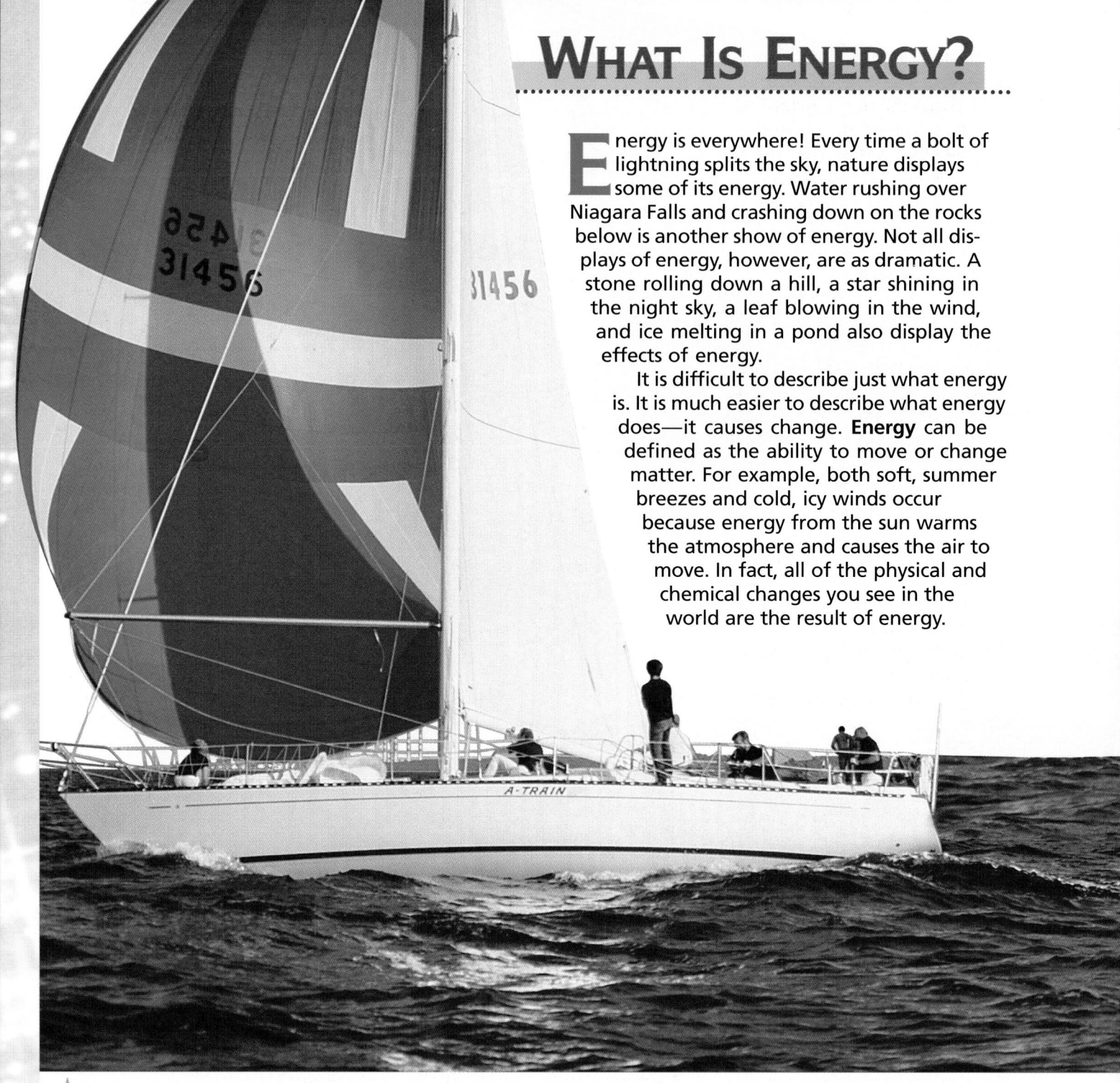

nergy is everywhere! Every time a bolt of lightning splits the sky, nature displays some of its energy. Water rushing over Niagara Falls and crashing down on the rocks below is another show of energy. Not all displays of energy, however, are as dramatic. A stone rolling down a hill, a star shining in the night sky, a leaf blowing in the wind, and ice melting in a pond also display the effects of energy.

It is difficult to describe just what energy is. It is much easier to describe what energy does—it causes change. **Energy** can be defined as the ability to move or change matter. For example, both soft, summer breezes and cold, icy winds occur because energy from the sun warms the atmosphere and causes the air to move. In fact, all of the physical and chemical changes you see in the world are the result of energy.

▲ The energy in the wind blowing against billowing sails is moving this sailboat through the water.

Forms of Energy

Unless you know what to look for, energy can be hard to recognize. We often classify energy as having one of several different forms. Some of these energy forms are known as chemical energy, electrical energy, nuclear energy, radiant energy, and thermal energy. But physicists tell us that each of these familiar types of energy can be described in even more basic terms—as kinetic and potential energy.

Energy

The ability to cause matter to move or change

S136

Kinetic energy is the energy of motion, which can be used directly to do work. **Potential energy** is the energy stored in an object or system because of its position or arrangement. Stored energy is called potential energy because it has the "potential" to make matter move or change. In other words, potential energy can be converted into kinetic energy by physical or chemical changes. All objects can have both kinetic and potential energy. These two basic types of energy are constantly changing from one form to another. Look for examples of kinetic and potential energy in the photographs to the right.

Chemical Energy All matter contains energy in the arrangement of its atoms and molecules. The chemical bonds that hold atoms and molecules together store this energy, which is called **chemical energy**. Since it is stored within bonds, chemical energy is a form of potential energy. Chemical energy can be converted into other forms of energy during a chemical change, which occurs when the atoms of one or more molecules are rearranged and new bonds are formed.

A match, for example, has chemical energy. The head of a match is made of chemicals that release energy when they react. When the match is struck, the chemicals react and produce a flame. As the match burns, the chemical energy in the match head and match stick is conver-

ted into heat and light. Burning releases energy from many substances, such as wood and oil. These substances are called *fuels*.

Food is another type of fuel with chemical energy. Your body's cells release this energy slowly by breaking chemical bonds in food molecules and forming other bonds. This process is similar to burning but is much more controlled. Some of the energy released from your food becomes the heat that keeps your body temperature constant. Much of it, however, is used to enable you to grow, move, and even think.

▲ Energy can be described as either kinetic (due to an object's motion)...

◄ ...or potential (due to an object's position).

◄ Through chemical reactions, the chemical energy stored in a match is converted into heat and light. As the match burns, many chemical bonds are broken, while others are formed.

Kinetic energy

The energy an object has because of its motion

Potential energy

The energy that is stored in an object because of its position

Chemical energy

Energy stored in the arrangement of atoms and molecules

S137

▲ Lightning is an example of electrical energy.

1 Electrons tend to build up in the bottom of storm clouds, giving the bottom of the clouds a negative charge. These charges repel negative charges in nearby objects on the ground.

2 As a result, objects on the ground tend to develop a positive charge.

3 A huge spark results when the cloud's electrons flow through air to a positively charged object. This spark is the lightning that we see.

Electrical energy

Energy of electrical charges as a result of their position or motion

Electrical Energy Have you ever heard the expression "opposites attract"? This is true of electrically charged particles. When a negatively charged particle, such as an electron, is near a positively charged particle, the two particles will move toward one another if they can. On a larger scale, a negative charge develops in a place where electrons collect. A location with too few electrons takes on a positive charge. When two nearby places develop opposite electric charges, an *electric potential* exists between them. When given a path to follow, charged particles will "flow" from one place to the other, resulting in an electric current. Thus, electric potential energy results in the kinetic energy of the moving particles. The energy of these electrically charged particles is called **electrical energy**. A bolt of lightning, like the one seen above, is a vivid example of an electrical energy system.

Power plants produce large quantities of electrical energy with generators. Batteries produce small amounts of electrical energy with chemicals. The batteries that you are familiar with have chemicals that react in ways that create an electric potential between their terminals. When a battery is connected to a radio, for example, the electric potential energy is converted into the kinetic energy of moving particles. Electrons from the negative pole flow to the radio in a wire (conductor) that runs through the radio and then back to the positive pole of the battery. The electric kinetic energy is then converted into sound, which is another type of kinetic energy.

S138

Nuclear Energy The forces that act on the protons and neutrons in the nucleus of an atom are responsible for **nuclear energy**. Nuclear energy is by far the most concentrated form of energy. It is released in large amounts either by splitting atomic nuclei, a process called *fission*, or by forcing nuclei together, a process called *fusion*. The energy that makes the sun shine, for example, comes from the fusion of hydrogen nuclei to make helium. Enormous amounts of heat, light, and other types of radiant energy are produced during this process.

Fission takes place when radioactive isotopes split into isotopes of other elements. Particles called neutrons are given off during this process. The particles released by a fission reaction have a tremendous amount of kinetic energy—enough to split other nuclei and continue the reaction.

Nuclear energy

Energy that is released either by splitting atomic nuclei or by forcing the nuclei of atoms together

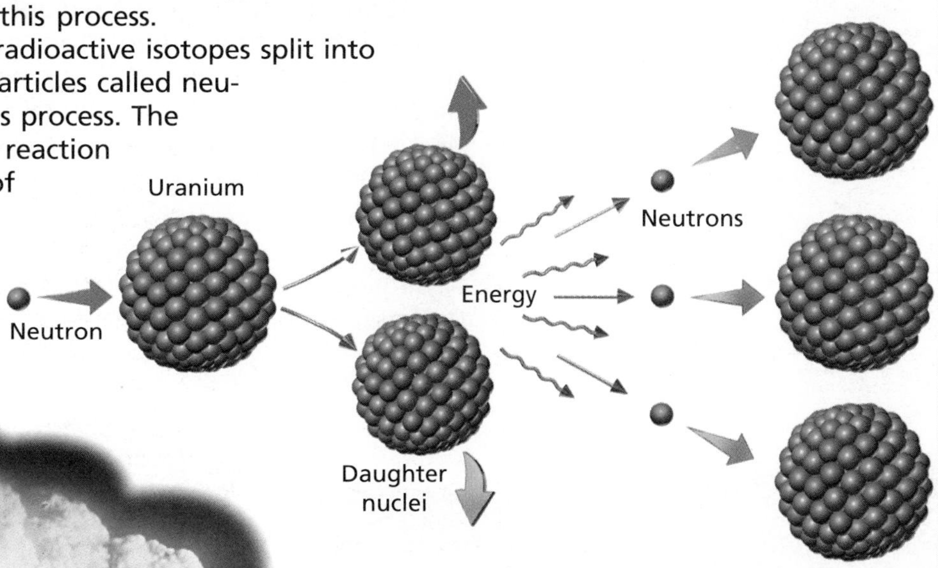

Neutron
Uranium
Neutrons
Energy
Daughter nuclei

▲ Uranium undergoes fission to form two smaller atoms, three neutrons, and a large amount of energy. These neutrons can then cause other uranium atoms to split.

◄ An uncontrolled fission reaction results in a nuclear explosion.

S139

Radiant Energy Sunlight is a familiar example of **radiant energy**. Radiant energy is transmitted in the form of special waves called *electromagnetic waves*. Unlike all other kinds of waves, electromagnetic waves do not need a medium (substance) to travel through. They can even pass through a vacuum such as space. Electromagnetic waves travel extremely fast—about 300,000 km/s, the speed of light. Sunlight, in fact, takes only 8 minutes to travel from the sun to Earth, a distance of 150 million kilometers.

About 90 percent of the radiant energy given off by the sun is either visible light or *infrared rays*. Infrared rays cannot be seen, but they can be felt as heat. Other types of radiant energy that you might recognize are X rays, ultraviolet light, and radio waves. The whole range of radiant energy makes up the *electromagnetic spectrum*, shown below.

Radiant energy

Energy transmitted in the form of electromagnetic waves

1 The electromagnetic spectrum is arranged in order from the waves with the shortest wavelength to those with the longest wavelength.

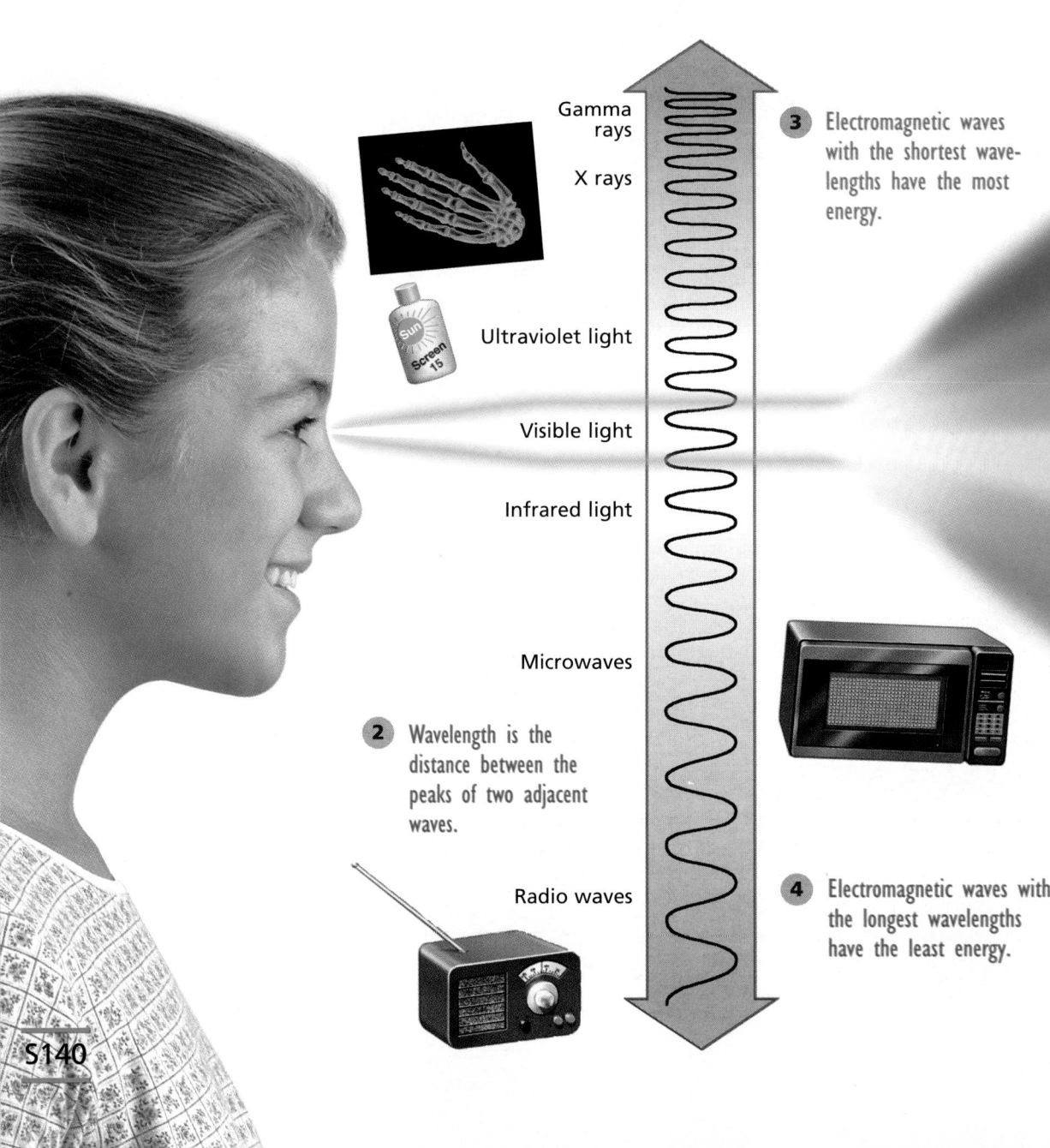

Gamma rays

X rays

3 Electromagnetic waves with the shortest wavelengths have the most energy.

Ultraviolet light

Visible light

Infrared light

Microwaves

2 Wavelength is the distance between the peaks of two adjacent waves.

Radio waves

4 Electromagnetic waves with the longest wavelengths have the least energy.

S140

Thermal Energy   As you know, the atoms and molecules of any substance are constantly in motion, some at faster speeds and some at slower speeds. The total kinetic and potential energy of these moving atoms and molecules is called **thermal energy**. One indication of a substance's thermal energy is how hot or cold it is.

Temperature is a measure of the average amount of kinetic energy the molecules of a substance have. When molecules of a substance gain kinetic energy, the substance becomes hotter. When the molecules lose kinetic energy, the substance becomes cooler.

When two substances have different temperatures, thermal energy, or heat, may flow from one to the other. Heat always flows from warmer objects to cooler objects. Why? The faster-moving molecules of the warmer substance collide with and speed up the slower-moving molecules of the cooler substance, transferring energy to them. Heat and temperature are discussed further in Unit 7.

Thermal energy

Energy due to the motion and arrangement of molecules in a substance

▲ Thermal energy is associated with the motion of the particles that make up matter. The motion of the atoms in this steel causes it to glow red.

S U M M A R Y

Energy is the ability to move or change matter. Energy can be described as either kinetic or potential energy. Kinetic energy is the energy of motion. Potential energy is the energy of position or arrangement. Energy appears in different forms, including chemical energy, electrical energy, nuclear energy, radiant energy, and thermal energy. All forms of energy can be converted into other forms. Chemical energy is the potential energy stored in chemical bonds. Electrical energy is the energy of charged particles. Nuclear energy is released during fission and fusion reactions. Radiant energy takes the form of electromagnetic waves. Thermal energy is the total kinetic and potential energy of the atoms and molecules within a substance.

ENERGY TRANSFORMATIONS

▲ The potential energy of the water at the top of these falls is converted into kinetic energy as the water falls.

Energy transformation

A change in energy from one form to another

Scientists know that all physical and chemical changes involve energy. Energy itself constantly changes from its potential state to its kinetic state or from one form to another. Every change that occurs in the universe involves an energy change. The change of energy from one form to another is called **energy transformation**.

Transformation of Energy From the Sun

Much of the energy on Earth originally came from the sun. In the sun, matter is converted into thermal and radiant energy. Some of the radiant energy that reaches Earth as sunlight is transformed into heat that warms the planet's surface. This warming determines Earth's climate. Directly or indirectly, almost all organisms depend on the sun for their energy. The illustration below shows how organisms use the sun's radiant energy by transforming it into other forms of energy.

Much of the energy we use in our daily lives can also be traced back to the sun. Plants and animals that lived and died millions of years ago have become the coal, oil, and natural gas we now use as fuels. The chemical energy of these fuels was stored in the organisms while they lived and is transformed into heat and light by the process of burning. This energy can then be transformed into electrical energy by a generator, which in turn can be used to run our machines and appliances.

1. Green plants capture energy from sunlight and use it to convert carbon dioxide from the air into food. Chemical bonds in food molecules store energy captured from sunlight.

2. Organisms that cannot make their own food with sunlight depend on the energy stored by plants. By eating plants, for example, an animal is able to maintain its body temperature with the chemical energy in its food.

3. Animals also transform the chemical energy in food into the kinetic energy of their movements.

4. An animal's kinetic energy can be transferred to other objects, such as when a horse carries a rider or pulls a wagon.

S142

Conservation of Energy

An automobile engine converts only about 30 percent of the chemical energy in gasoline into motion. What happens to the other 70 percent? Is it used up in the change? No, it is not. The chemical energy of gasoline is first transformed into light, sound, and heat as it is burned. Only some of the heat is then transformed into the kinetic energy of the moving car; the rest escapes into the air.

During any energy transformation, the total amount of energy involved remains the same. In other words, the energy is *conserved*. All of the energy is still there—it is just in different forms. The *law of conservation of energy* summarizes these observations. This scientific law states the following: *Energy can be changed from one form to another, but it cannot be created or destroyed.*

The law of conservation of energy often seems to defy common sense. For example, the bouncing ball in the illustration above does not continue to bounce to the same height. If energy is conserved, shouldn't the ball's potential energy be the same at the top of each bounce? A bouncing ball does not continue to bounce to the same height because some of its kinetic energy is changed into sound and heat. The energy is not destroyed; it is simply converted into other forms of energy.

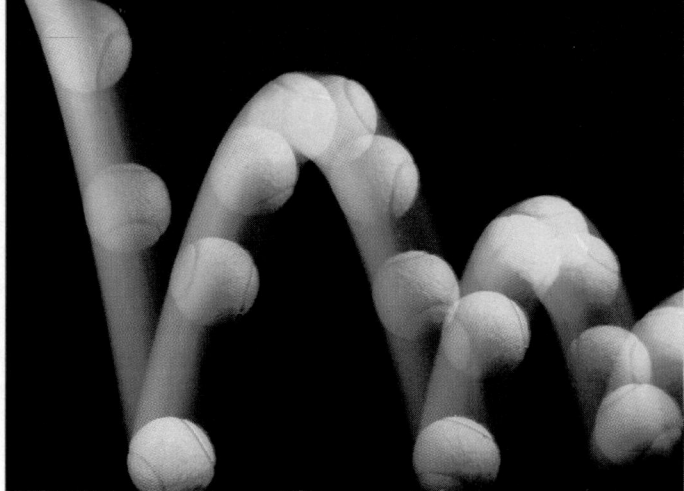

▲ A bouncing ball loses some height with each bounce. The ball makes a noise as it hits the floor. It also gets warmer. Therefore, the ball loses kinetic energy with each bounce and, as a result, does not bounce as high.

Decrease in Useful Energy

Every time energy in a system changes from one form to another, the amount of *useful* energy decreases. Recall that only 30 percent of the energy stored in gasoline becomes the kinetic energy of a moving a car. The other 70 percent is wasted, in a sense, because it is not used to move the car. The heat produced merely warms parts of the car and the air around it.

In any energy transformation, some useful energy is converted into heat that cannot be used to do work. You can observe this heat by touching any of the following: a ball after you bounce it several times, a nail after you hammer it into a piece of wood, or a bicycle pump and tire after you pump air into the tire.

▲ Some of the energy used to pump air into a bicycle tire is converted into heat. In this instance, the heat is not a useful form of energy because it does not do work.

S U M M A R Y

Energy can change from one form to another. During energy transformations, the total amount of energy remains the same because energy cannot be created or destroyed. Some of the energy in a transformation, however, is changed into heat that does no work. Therefore, the total amount of useful energy in a system decreases as energy transformations occur.

S143

ENERGY AND THE ENVIRONMENT

Where do you and all other living things on Earth get energy for life's activities? You know that plants convert energy from the sun into the chemical energy in food. Many animals get energy by eating plants. Still others get energy by eating other animals. When plants and animals die, they become food for organisms such as bacteria and fungi.

From Light to Life

Two complex processes turn the radiant energy in sunlight into the chemical energy needed by living things. Through the process of **photosynthesis,** green plants, algae, and some bacteria capture energy from sunlight and use it to produce the food they need for their activities. During photosynthesis, molecules of water and carbon dioxide are used to make sugars and oxygen. Chemical bonds in the sugars store the energy that was captured from sunlight. That energy can later be used by the plant.

Photosynthesis

The process by which green plants, algae, and some bacteria use light energy, water, and carbon dioxide to make sugars that they use for food

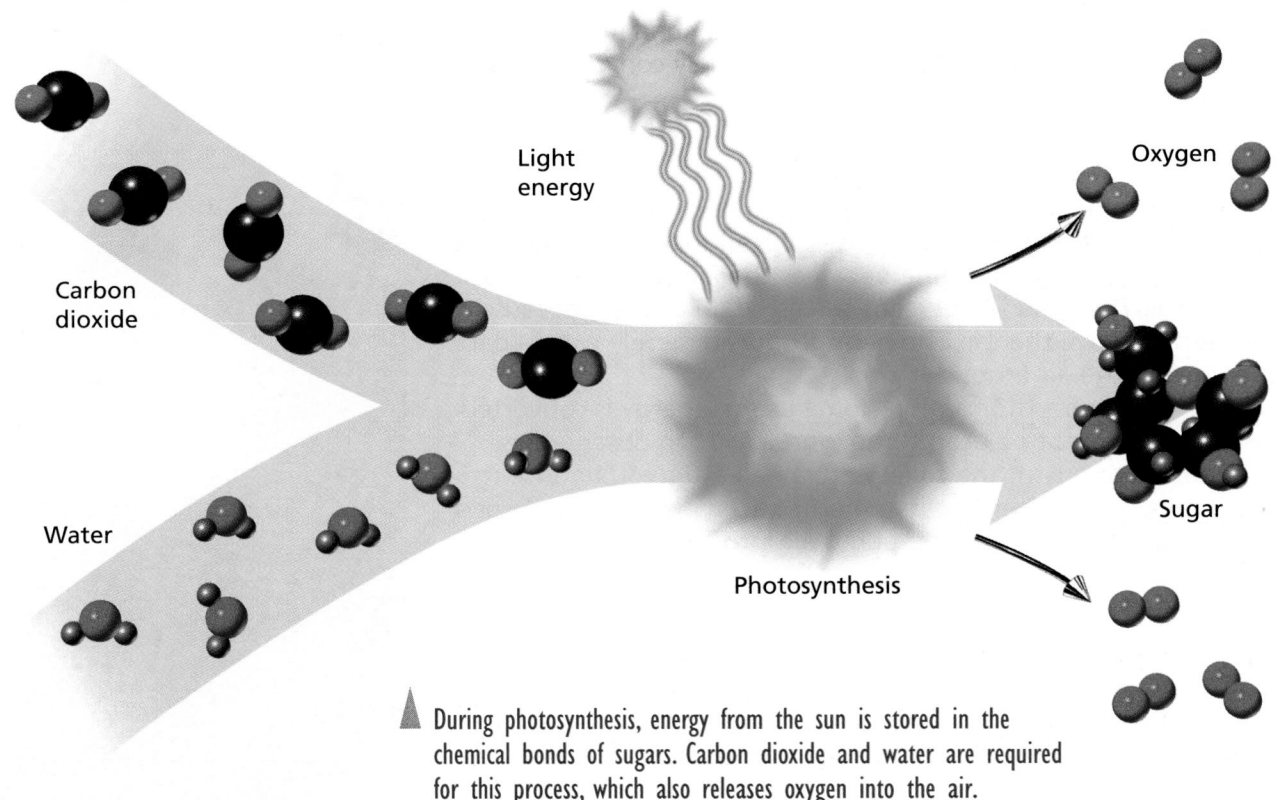

Light energy

Oxygen

Carbon dioxide

Water

Photosynthesis

Sugar

During photosynthesis, energy from the sun is stored in the chemical bonds of sugars. Carbon dioxide and water are required for this process, which also releases oxygen into the air.

Cellular respiration

The process by which the energy stored in sugars is released for use by the cells of organisms

Before plants can use the energy stored in sugars, that energy must be released by a process called **cellular respiration.** During cellular respiration, photosynthesis is essentially reversed. Energy is released as sugars are broken down into water and carbon dioxide. Oxygen is required for maximum energy release. The energy released is then used for the activities of life.

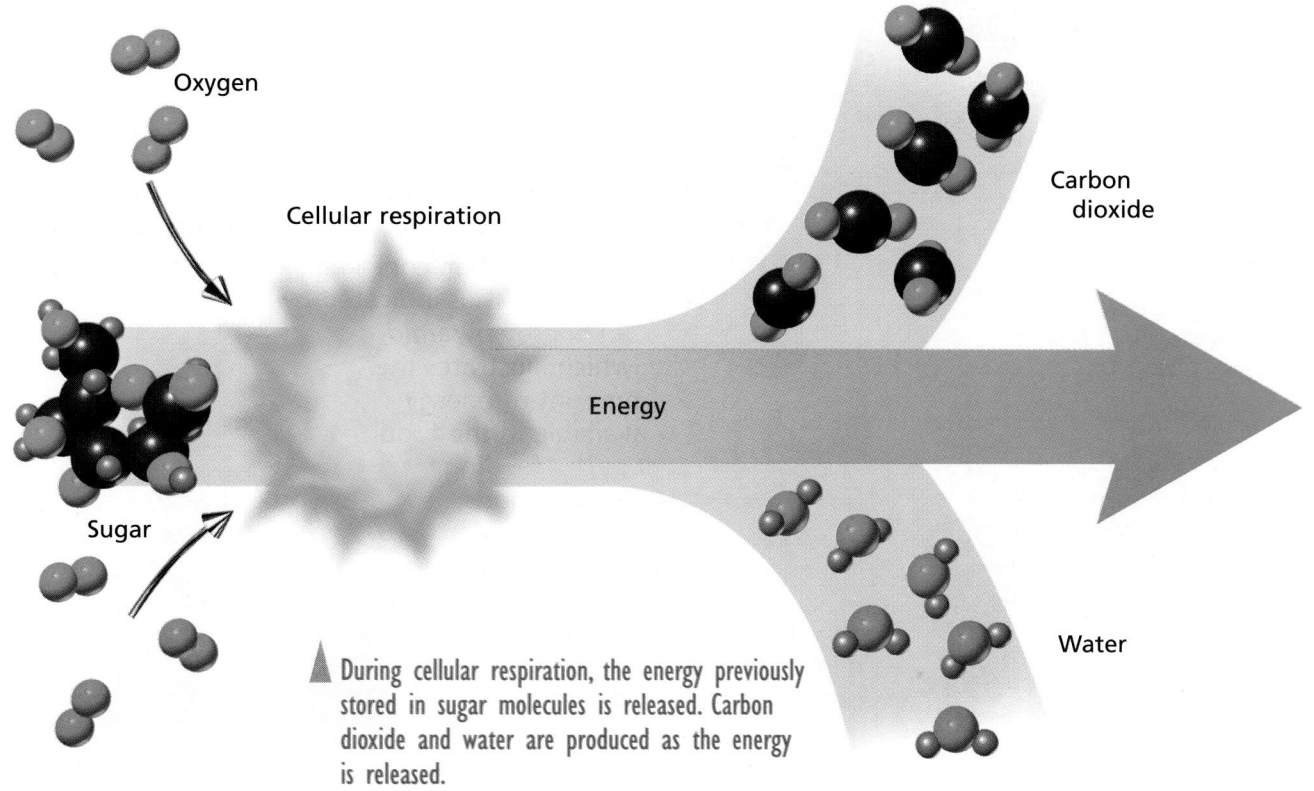

Oxygen

Cellular respiration

Carbon dioxide

Energy

Sugar

Water

▲ During cellular respiration, the energy previously stored in sugar molecules is released. Carbon dioxide and water are produced as the energy is released.

Photosynthesis is also important to animals and other organisms that cannot make their own food. Because most animals eat either plants or animals that have eaten plants, photosynthesis indirectly provides animals with energy. Animals, as well as protists, fungi, and bacteria, also use cellular respiration to release the energy stored in sugars.

Energy Flow in Ecosystems

After being captured by plants during photosynthesis, the sun's energy is transferred through communities of living things. This flow of energy can be seen in patterns of feeding and predation.

Food Chains   Energy originally obtained from the sun is passed through a community of living organisms by means of a *food chain*. The organisms in a food chain that make food using sunlight, carbon dioxide, and water are called *producers*. The organisms that get their food from other organisms are called *consumers*. The figure to the right shows one food chain that could be found in a meadow community. As you can see, energy passes through each organism in the food chain. Each organism uses some of the energy in order to stay alive and stores some of the energy. This stored energy has the potential to become food for the next link in the food chain. However, much of the energy in a food chain is lost to the environment as heat.

▼ In this food chain, energy from the sun reaches the owl after it has passed through the other organisms shown.

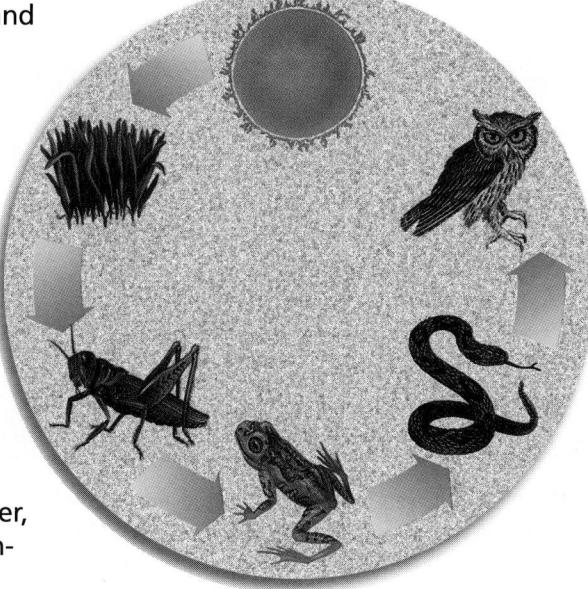

S145

Food Webs and Food Pyramids Most communities have more than one food chain. For example, a complex network of food chains can be identified in a meadow community that includes grasses, insects, birds, mice, snakes, and weasels. This network forms a *food web,* like the one shown to the left.

No matter how complicated a food web gets, each food chain in the web must have more producers than consumers. This pattern is indicated by an *energy pyramid,* which illustrates the amount of energy available in the food at each "feeding level" in a food chain. Producers make up the bottom level of an energy pyramid. The size of the pyramid's base indicates that producers contain the most energy. Animals that eat plants make up the next, smaller level. Animals that eat other animals compose the smallest levels, at the top of the pyramid. If there were more consumers than producers, the pyramid would collapse.

▲ In this food web, energy from the sun may pass through the members of the web in various ways. Connecting lines show possible pathways for energy transfer.

▲ The shape of this food pyramid indicates that many organisms are needed at the base of a food chain to support a few organisms at the top.

▲ This mushroom gets food from the dead matter in this stump. In the process, it returns to the environment nutrients that are needed by plants.

The Final Link After an organism dies, the matter it was made of cycles back into the ecosystem. The materials in dead matter can be used by living things called *decomposers*—organisms that break down waste materials and dead organisms. They return carbon, oxygen, nitrogen, and other chemicals to the ecosystem to be used again. Many bacteria and fungi are decomposers. They use some of the dead and decaying material for their own energy needs and return the rest of the matter to the soil. This decayed organic matter provides nutrients that can be used by plants.

S146

Disrupting the Energy Flow

What happens to the environment when the energy flow in an ecosystem is disrupted? Suppose that a factory was built in a meadow, reducing the amount of grass in the meadow community by 50 percent. All of the organisms in the meadow would be affected.

Deforestation, the loss of trees and other plant life, also disrupts an ecosystem. The demands of our growing human population have led to deforestation. In many areas of the world, forests are cut or burned to provide land for agriculture and urban development. Erosion often follows deforestation. As a result, topsoil and nutrients in the topsoil are washed away by rain. Deforestation and erosion may prevent an ecosystem from supporting healthy communities of living things for decades or even centuries.

Ocean communities are also affected by human activities. For example, overfishing of certain species affects any food web of which that fish species is a part. Pollution can also destroy a coral reef. Since a variety of organisms are needed to support an ecosystem, the loss of key organisms disrupts the movement of energy through a reef ecosystem. This reduces the number of organisms that it can support. If the producers within a coral reef are killed by pollution, the entire reef ecosystem collapses.

Deforestation is responsible for the loss of over 50 million acres of the world's forests every year. This area is about half of the size of California.

A healthy coral reef supports a multitude of organisms. Its health, as well as that of the community it supports, depends on the quality of the water in which it lives.

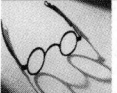

DID YOU KNOW...

that most food chains consist of only four or five types of organisms? So much energy is lost (released as heat) with each transfer of energy that additional feeding levels cannot be supported.

Summary

Producers convert energy from the sun into chemical energy through photosynthesis. All organisms release this energy through cellular respiration and use the energy for life processes. Energy is passed through food chains from producers to consumers. Human activity can disrupt the energy flow in an ecosystem and can affect all of the organisms that make it their home.

S147

ENERGY FOR SOCIETY

Today's society uses great quantities of energy. We use energy to produce light and warmth for our homes; to produce food, clothing, and other products; and to transport people, raw materials, and products from place to place. Most of the energy we use comes from fossil fuels—oil, coal, and natural gas. These fuels are burned to produce heat, which can be used to drive engines and generators. People also obtain energy from other sources, such as wind, moving water, and Earth's internal warmth. As you know, most of the energy we use originally came from the sun. This energy, however, may undergo several transformations before it propels a car down the road or warms a home.

Energy Through the Ages

Humans first relied on their own muscles to do work. Then animals such as oxen, camels, and horses were domesticated and harnessed to do work. The fuel that provided energy for humans and animals to do this work was the food they ate. Wood, of course, was burned for heating and cooking. As other sources of energy were made available, societies grew larger, and more energy was available to do more work.

Moving Water   The energy of moving or falling water was once widely used to turn waterwheels. These wheels first turned mill stones that ground grain into flour. Later, waterwheels were connected to pulleys, belts, or ropes to drive machines. Using a device called a turbine, falling water was eventually used to produce electricity. A *turbine* is a fan-shaped disk or wheel that spins when a substance, such as water, steam, or air, pushes on it. A turbine operates much like a pinwheel in a breeze. Turbines run generators that produce electricity. *Hydroelectric energy* comes from the energy of falling water and is an important source of energy in certain areas of the world. Very large dams, such as the Hoover Dam on the Colorado River and the Grand Coulee Dam on the Columbia River, restrict the flow of the rivers and form large reservoirs of water. When some of this water is allowed to flow through the dam and past turbines, electricity is generated.

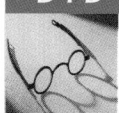

▼ Energy from moving water is captured by this waterwheel.

S148

▲ In a hydroelectric dam, water flows from a reservoir through turbines. Each turbine, in turn, operates a generator that produces electricity.

Wind Energy People also harness wind to do work. Winds are currents of air that form when the sun warms some parts of Earth's surface more than others. Wind transfers some of its kinetic energy to other objects when it blows against them. For example, sails propel sailing ships across the seas by capturing the kinetic energy of the wind.

Windmills also capture the kinetic energy of the wind. When wind strikes the blades of a windmill, the wind's kinetic energy is converted into the spinning motion of the blades. This motion turns shafts and gears that do various kinds of work. Windmills were first used to pump water from the ground and to grind corn and other dried grains. Later, windmills were connected to generators and used to generate electricity. In the early part of this century, thousands of American farms used wind power for pumping water and generating electricity. Some of these early windmills are still at work.

◀ Windmills, such as this one, pump water on many American farms today.

S149

Fossil Fuels When James Watt developed an improved steam engine, a new method of producing energy was introduced. This was a major event in the Industrial Revolution. Heat from a burning fuel was used to boil water and produce steam. The steam was used to run a variety of machines, including locomotives and electrical generators. Early steam engines used wood as a fuel. But wood was soon replaced by coal, oil, and natural gas—fuels that are more plentiful and provide more heat per unit of mass. These fuels are referred to as *fossil fuels* because they are formed from organisms that died millions of years ago.

Coal is a dark brown or black rocklike material that can be burned like wood for energy. Coal is formed from the remains

This piece of coal contains a fossil of a fern that lived millions of years ago.

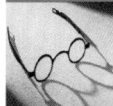
of dead plants that lived in vast swamps millions of years ago. These ancient plants captured and stored radiant energy from the sun. Eventually, the swamps disappeared as they were filled with layers of mud and sand called *sediment*. Over millions of years, the weight of this sediment compressed the dead and decaying plant material into coal.

The use of coal as a fuel made large amounts of energy available to humans. Unfortunately, coal is a very "dirty" fuel. Cities such as Pittsburgh, Pennsylvania, were once literally black with the soot that results from burning coal to provide energy for making steel and producing electricity. Coal still provides about 23 percent of the total amount of energy used in the United States. Although soot is now controlled, the burning of coal is thought to be one of the major causes of acid rain.

Pittsburgh at the turn of the 20th century

S150

By the late 1940s, oil and natural gas had replaced coal as the major fuels for industry. Like coal, oil and natural gas are formed from decaying organisms—in this case, the small plants and micro-organisms that were abundant in ancient, warm, shallow seas. Like the energy in coal, the energy in these fuels originally came from the sun. However, oil and natural gas provide more heat per unit of mass, are easier to handle, burn cleaner, and cost less than coal. At the present time, oil provides about 40 percent and natural gas about 25 percent of all energy used in the United States.

It took millions of years to form the world's deposits of coal, oil, and natural gas. Once they are used up, they are gone. We cannot make any more of these fuels. For this reason, fossil fuels are called *nonrenewable resources.* Experts are not sure just how much coal, oil, and natural gas we have left.

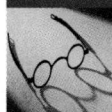

▲ Drilling rigs dotted the landscape during the early years of exploration for oil and natural gas.

Nuclear Energy During the 1960s, nuclear energy was added to the world's energy sources. When atoms are split in a *fission reactor,* great amounts of energy are released as heat. This heat is used to create steam, which is used to operate turbines. The turbines drive generators that produce electricity. Radioactive uranium is the fuel used in most fission reactors. During nuclear fission, nuclei of uranium atoms are split by free neutrons. When each nucleus is split, it releases heat and additional neutrons. These neutrons go on to split other nuclei, causing a *chain reaction.* This chain reaction must be carefully regulated with special equipment and materials.

S151

Many people are concerned about the use of nuclear fission to make electricity. Perhaps the main cause for concern is that nuclear fission produces radioactive wastes that remain dangerous for thousands of years. The safe disposal of these wastes is a major problem. Another concern is the danger of a nuclear accident in a poorly designed reactor, which could release tons of radioactive materials into the environment. The radioactivity in these materials can kill humans and other organisms that are exposed to it. A long-term effect is an increase in cases of cancer. The worst nuclear accident to date occurred in 1986 at Chernobyl in Ukraine, which was then part of the Soviet Union.

This is a computer simulation of the ▶ distribution of radioactivity 10 days after the accident at the Chernobyl nuclear power plant.

The billowing cloud coming from these huge cooling towers is steam, not smoke. The steam comes from water used to cool the hot water flowing through a separate system that is isolated from the outside environment.

The nuclear reactor itself is found inside a containment building that has thick walls of reinforced concrete. The building is designed to prevent the escape of radioactive materials into the environment in the event of a malfunction.

▲ Nuclear reactors, such as this one, use fission reactions to produce electricity. About 8 percent of the electricity currently used in the United States is produced by nuclear reactors.

Nuclear energy can be thought of as a *nonrenewable* energy source because the uranium needed for fission reactors is not limitless. Sooner or later, we may be forced to use other, *renewable resources* for energy production. Renewable resources are available in an almost unlimited supply. Some are also much cleaner than the fuels currently in use. The old sources of wind and moving water are two renewable resources that could replace nonrenewable energy sources. Other sources of energy, including solar energy, geothermal energy, nuclear fusion, and hydrogen fuel, must also be developed as alternative energy sources.

S152

Energy From the Sun

As you know, energy from the sun travels through space as radiant energy, providing our planet with both light and heat. Energy from the sun controls the climate on Earth. The uneven heating of different parts of Earth's surface drives the wind and ocean currents. The sun's warmth also causes the evaporation of water, which then cools and falls as rain or snow. Energy from the sun also makes life on Earth possible. Most producers use carbon dioxide, water, and energy from sunlight to produce the food and oxygen that supports much of the life on Earth. Furthermore, nearly all of the energy we use to run our machines originally came from the sun. For example, fossil fuels such as coal and oil store energy from the sun. Other sources of energy, such as wind, can also be traced back to the sun. The sun's energy can even be used directly to make electricity and run machines.

Solar Energy Energy that comes directly from the sun is called solar energy. People can capture this energy and use it to do work. Solar energy systems are designed to convert the sun's radiant energy into usable forms such as heat or electricity.

Solar heating systems are either *passive* or *active*. **Passive solar heating** systems use radiant energy directly to heat buildings or other structures. Think of a time when you sat in a sunny window. Did you notice how warm it was? This is an example of passive solar heating. The windows of a house let in sunlight that heats the house naturally. **Active solar heating**

These strategically placed windows and skylights collect the sun's energy and heat this home with a passive solar heating system. A solar collector on the roof heats water for household use.

systems use *solar collectors* to convert sunlight into heat. The heat is absorbed by air or water and then distributed throughout the building by a mechanical device, such as a fan or a pump. Excess heat is stored in insulated containers so that it can be used at night or on cloudy days. The house shown above uses both passive and active solar heating.

Solar energy can also be used to make electricity. With devices called *solar cells,* sunlight can be converted directly into electrical energy. However, each cell produces only a small amount of electricity. For this reason, solar cells are particularly useful for appliances in which only small amounts of electricity are needed, such as calculators, radios, and even telephones.

Passive solar heating

The use of simple absorption of radiant energy to warm buildings or other structures

Active solar heating

The use of solar collectors and mechanical devices to gather sunlight, convert it into heat, and distribute that heat

S153

Each solar cell in a solar panel has the ability to generate electricity directly from sunlight.

Some people have described solar energy as the answer to all of our energy needs. While the sun's radiant energy is free, using solar collectors and solar cells for large-scale energy production is not practical at this time. For one thing, solar collectors and solar cells are expensive to produce. Their production also requires large amounts of energy. In addition, to produce enough electricity for a city, large areas of land that receive abundant sunlight must be covered with these devices. It is also important to note that the technology presently used to manufacture solar cells produces toxic wastes that are not easy to dispose of safely.

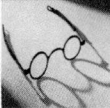

This array of solar collectors is designed to produce enough heat to generate steam.

Many wind turbines that are grouped together are called wind farms. This wind farm is in California.

Wind Turbines Windmills have long been used to generate electricity for individual farms. Today, modern *wind turbines* are being used to generate electricity on a larger scale. A typical wind turbine works by harnessing the kinetic energy of the wind to operate an electric generator. A single wind turbine may be used to generate electricity for a single house. In other applications, wind farms are built to harness enough energy to provide electricity for whole communities.

Like all other sources of energy, wind energy has disadvantages. The major disadvantage is that wind turbines work only when the wind is blowing at a high enough speed. Therefore, they are practical only in areas where the wind blows strongly most of the time. Wind turbines may also be considered unattractive and noisy additions to a neighborhood. Research is currently being done to find more efficient designs for these new windmills.

S154

Energy From Ocean Water

Water flowing down a river is not the only way to use the energy of moving water to produce electricity. Another method uses ocean tides, which involve large masses of moving water. Tides are caused by the gravitational pull of the moon and the sun acting on Earth's surface. Ocean water is the part of Earth's surface most affected by this pull.

In a few locations around the world, special dams are used to trap water within a bay area during high tide. Then, at low tide, the water is allowed to drain through turbines inside the dams. These turbines produce hydroelectric energy.

Research is being conducted on methods to produce hydroelectric energy in still another way—from the movement of ocean waves. Some scientists and engineers believe that the up-and-down motion of waves could be used to compress air or another fluid and force it through turbine-driven generators. Though many technical difficulties still need to be worked out, this energy source may have great potential.

The thermal energy stored in ocean water is yet another possible energy source. The water at the surface of the oceans absorbs large amounts of solar energy. As a result, surface water is much warmer than deeper water. In the future, power plants that extend from the surface into deep water may use temperature differences to generate electricity.

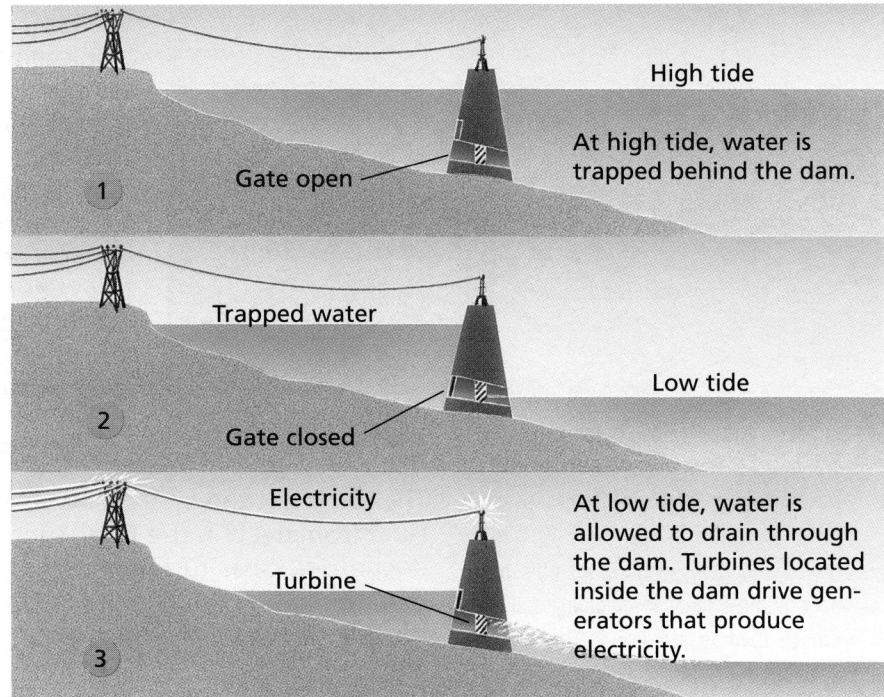

1. At high tide, water is trapped behind the dam. High tide / Gate open

2. Trapped water / Gate closed / Low tide

3. Electricity / Turbine / At low tide, water is allowed to drain through the dam. Turbines located inside the dam drive generators that produce electricity.

The principle of a tidal dam is illustrated here.

DID YOU KNOW...

that energy from ocean tides was first put to use in England in the twelfth century?

A tidal-powered mill that was built in Woodbridge, England, in 1170 has been operating for more than 800 years.

Biomass Fuels

Biomass is a term that refers to plant and animal materials. Wood, for example, is a form of biomass. Because biomass will burn, it can be used as a fuel. In the case of wood, biomass is burned directly to provide heat. In other instances, biomass is converted into another fuel before it is burned. For example, plants such as corn, sugar cane, and sorghum can be used to make alcohol. The alcohol is then mixed with gasoline. This mixture, called *gasohol,* can be burned in automobiles.

Firewood is a readily available form of biomass. However, firewood takes many years to grow and produces pollution when burned.

S155

Methane is another fuel that is a product of biomass. The decomposition of garbage, sewage, or almost any other kind of organic matter produces methane, which can be used as a replacement for natural gas. One approach to producing methane is to fertilize fast-growing water hyacinths with sewage and then collect the methane from the water hyacinths as they decay. This process not only produces methane, but also treats unwanted sewage to reclaim clean water at the same time.

In lakes and streams, water hyacinths are pests that tend to cover the water, making it uninhabitable for most other organisms. When grown in controlled conditions, water hyacinths produce a huge amount of biomass that can be used to produce methane gas.

Other Renewable Sources of Energy

The sun is not the only renewable source of energy available to us. Heat from deep in the Earth's interior can be tapped for energy production. Nuclear fusion is also a promising energy source. Hydrogen, the lightest of the elements, is yet another promising source of fuel for future energy needs.

Geothermal Energy   The Earth's natural, internal warmth is called *geothermal energy.* Deep in the Earth's interior, radioactive nuclei produce heat as they decay. The pressure of thousands of kilometers of rock also produces heat. In some places, geologic forces push large pools of melted rock, or *magma,* close to the Earth's surface. These places are called geothermal hot spots. If the ground above a hot spot allows water to seep into spaces around hot magma, the water becomes very hot. Sometimes this hot water makes its way to the surface, forming geysers or hot springs.

Steam turbines in power plants work just as well with steam from under the Earth's surface as they do with steam from a boiler. Therefore, a plant located above a hot spot can direct the steam through its turbines to generate electricity. In some cases, subsurface rocks above hot spots do not contain water and steam. However, water can be pumped into the rocks above a hot spot to produce the steam necessary for operating turbines. Unfortunately, there are only a few places in the world where geothermal energy production is feasible.

Geysers, such as this one, are produced by geothermal energy.

S156

Nuclear Fusion In addition to the fission of atoms such as uranium, nuclear energy can be obtained by the process of nuclear fusion. Isotopes of hydrogen are used as fuel in a nuclear fusion reaction. Nuclear energy is released when the hydrogen nuclei fuse to make a helium nucleus. A tremendous amount of energy is released when this happens. However, extremely high temperature are necessary to start a fusion reaction. Fusion occurs when temperatures are millions of degrees—like the temperature inside the sun.

Some day, an unlimited supply of energy may come from controlled nuclear fusion, the most intense energy-yielding process presently known. The ability to control the energy of nuclear fusion and to use it to generate electricity has been a goal of scientists for many years. Unfortunately, the amount of energy required to produce, control, and contain a fusion reaction is still much greater than the amount of electrical energy generated. It cannot be used as an efficient energy source at this time. Scientists hope that with advanced technology, controlled nuclear fusion reactions might someday be achieved and be put to practical use.

Hydrogen The element hydrogen could be the ultimate energy source. For example, hydrogen in either its liquid or solid state could be the fuel for nuclear fusion reactions once the technology for controlling these reactions is developed. Hydrogen in its liquid state is presently used as a rocket fuel. In the main engines of the space shuttle, liquid hydrogen (the fuel) and liquid oxygen (an oxidizer) combine explosively in a chemical

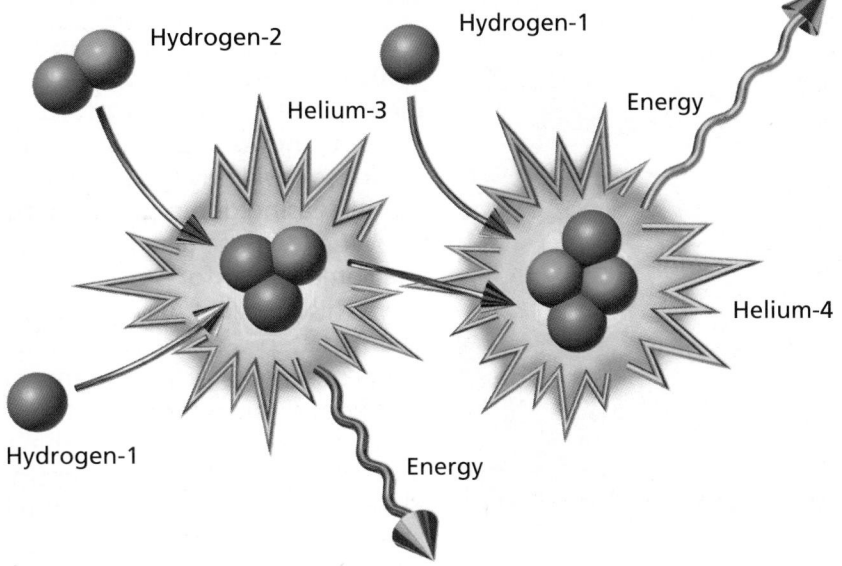

In a fusion reaction, hydrogen isotopes are forced together to form helium. Great amounts of energy are released during this reaction.

reaction that propels the vehicle into space. Some power plants use hydrogen fuel to generate electrical energy. Working models of engines for cars and buses that burn hydrogen fuel have also been developed.

Scientific research has made possible the uses we have for hydrogen today. But there are still many problems involved with its use that must be overcome. Although it is the cleanest-burning fuel available, safely storing highly reactive hydrogen gas is a problem. Another problem with using hydrogen for fuel relates to the water vapor that is formed in the process. If released into the atmosphere on a large scale, this excess water vapor could affect Earth's climate. The water vapor must be captured somehow and, perhaps, recycled. Cheaper methods of producing hydrogen for energy use must also be found. Therefore, the use of hydrogen as an energy source for the future depends on the development of many new technologies.

Liquid hydrogen is the fuel used to lift the space shuttle into orbit.

S157

Using Energy Wisely

The United States alone uses over 17,000,000 barrels of oil, 6,000,000 metric tons of coal, and 200 kg of uranium a day to meet its energy needs. Unfortunately, these resources will eventually run out. They also produce a lot of pollution. As we progress into the future, we will have to make many decisions about how to handle our increasing demand for energy and the pollution it causes.

When we use less energy, we create less pollution. Although the combined efforts of government and industry will be needed to solve this problem, using energy wisely starts with the individual. The first step is *conservation*—using energy efficiently. People can help conserve energy in many ways. Lowering thermostats in the winter and raising them in the summer conserves energy. Energy-efficient fans that circulate the air increase the efficiency of heating and cooling systems. Turning off lights and appliances such as radios and televisions when they are not in use also conserves energy. Using public transportation, such as subways and buses, instead of privately owned vehicles will also lower our overall energy consumption. As an added bonus, conserving energy lowers your energy costs.

Another important way to conserve energy is to **recycle** materials once they are used. Americans throw away a huge amount of garbage every year—an average of 680 kg (1500 lb.) per person. Much of this material, such as paper, iron, steel, glass, aluminum, and many types of plastic, can be recycled and made into new products. By recycling aluminum cans, for example, manufacturers can produce new aluminum products using only 20 percent of the energy needed to mine and obtain the metal from aluminum ore.

The use of oil and gas to meet our energy demands is responsible for much of the pollution of our air, land, and water.

Recycle

To reuse materials by making new products from discarded items

SUMMARY

People first used their own muscles and then used animals to do work. The energy in wind and moving water was then harnessed for society's energy needs. Burning wood, coal, oil, and natural gas provide most of our energy needs now. New sources of energy include solar energy, biomass, ocean water, and modern wind turbines. All of these forms of energy can be traced back to the sun. Other sources of energy include geothermal energy, nuclear fission and fusion, and hydrogen. While reliable, renewable sources of energy are being developed, we should also consider recycling and conserving the nonrenewable energy sources we now use so that they will last longer and pollution will be reduced.

Concept Mapping

The concept map below illustrates major ideas in this unit. Complete the map by supplying the words needed to connect the terms. Then extend your map by answering the additional question below. Write your answers in your ScienceLog. **Do not write in this textbook.**

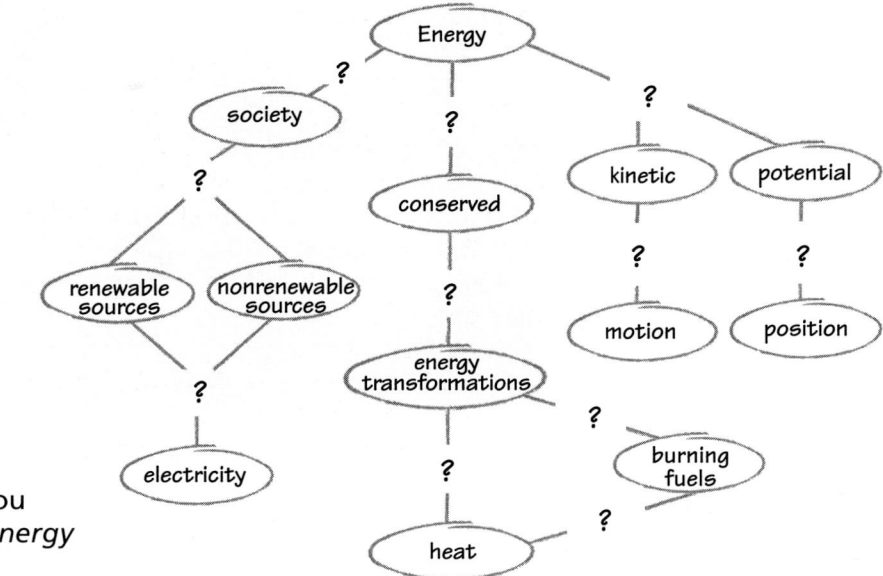

Where and how would you connect the terms *solar energy* and *fossil fuels*?

Checking Your Understanding

Select the choice that most completely and correctly answers each of the following questions.

1. The two most basic forms of energy are potential energy and
   a. nuclear energy.
   b. chemical energy.
   c. kinetic energy.
   d. thermal energy.

2. Which of the following types of energy is NOT limited by the need for a medium in which to travel?
   a. chemical energy
   b. radiant energy
   c. geothermal energy
   d. tidal energy

3. Living things convert radiant energy from the sun into chemical energy by the process of
   a. photosynthesis.
   b. cellular respiration.
   c. deforestation.
   d. recycling.

4. Which of the following energy sources does NOT rely on heat?
   a. tidal energy
   b. solar energy
   c. wind energy
   d. hydroelectric energy

5. The cleanest-burning type of fuel and perhaps the ultimate source of energy for our society's needs is
   a. coal.
   b. natural gas.
   c. uranium.
   d. hydrogen.

S159

Interpreting Photos

Identify three examples of potential energy and three examples of kinetic energy in the photograph shown here.

Critical Thinking

Carefully consider the following questions, and write a response in your ScienceLog that indicates your understanding of science.

1. How does the sun's energy reach Earth?

2. Use the law of conservation of energy to explain what happens when a firecracker explodes.

3. Explain why you would generally find more field mice than foxes in a meadow.

4. Trace the flow of energy from the sun to the kinetic energy of a moving automobile.

5. Compare and contrast renewable sources of energy with nonrenewable sources of energy.

Portfolio Idea

Write a newspaper story about the events of the day on which the world runs out of fossils fuels (about 175 years in the future). In your story, be sure to include information about the history of energy production by humans and about how future generations will meet their energy needs. Also write a headline for your story.

TEMPERATURE AND HEAT

IN THIS UNIT

Now that you have studied the concepts of temperature and heat, consider these questions.

1. What is heat, and how is it related to energy?
2. How does heat move from place to place?
3. What effects does heat have on the Earth?
4. In what ways is heat measured, and how do these methods differ?

In this unit, you will take a closer look at heat as a form of energy, its relationship to temperature, and how they both are measured.

S161

HEAT ENERGY

You know that heat is a form of energy. Heat energy is one of the most common forms of energy. Whether hot or cold, motionless or still, all things contain heat energy. Directly or indirectly, heat energy can be obtained from or transformed into any other form of energy. For example, radiant energy can be changed into heat energy with solar collectors—or by any light-absorbing surface. And a power plant converts heat energy into electrical energy. But what exactly is heat energy?

Heat Energy Is Motion

As you learned in previous units, matter is made up of tiny particles called atoms and molecules. The energy these particles contain causes them to move continuously. The more heat energy a particle has, the faster it moves. For example, a drop of water contains a huge number of individual water molecules, all of which move about in rapid, random motion. When water is heated (absorbs heat energy), its molecules move about faster and bump into each other more often.

Individual water molecules are too small to see, so you cannot see their individual behavior. But you can see the effects of the random motion of large numbers of molecules. For example, if you put a drop of food coloring into a glass of cool water, you will see the coloring spread slowly throughout the water. The constant random motion of the water molecules causes the coloring to spread until it is evenly dispersed in the water. If you try it again with hot water, the coloring will spread much faster. Why do you think this is so?

Even though you cannot see it, the molecules of a steel bar are also in constant random motion, but each in its own little space. When the bar is heated, its molecules gain energy, causing them to vibrate faster and collide more vigorously. As a result, the molecules spread apart farther.

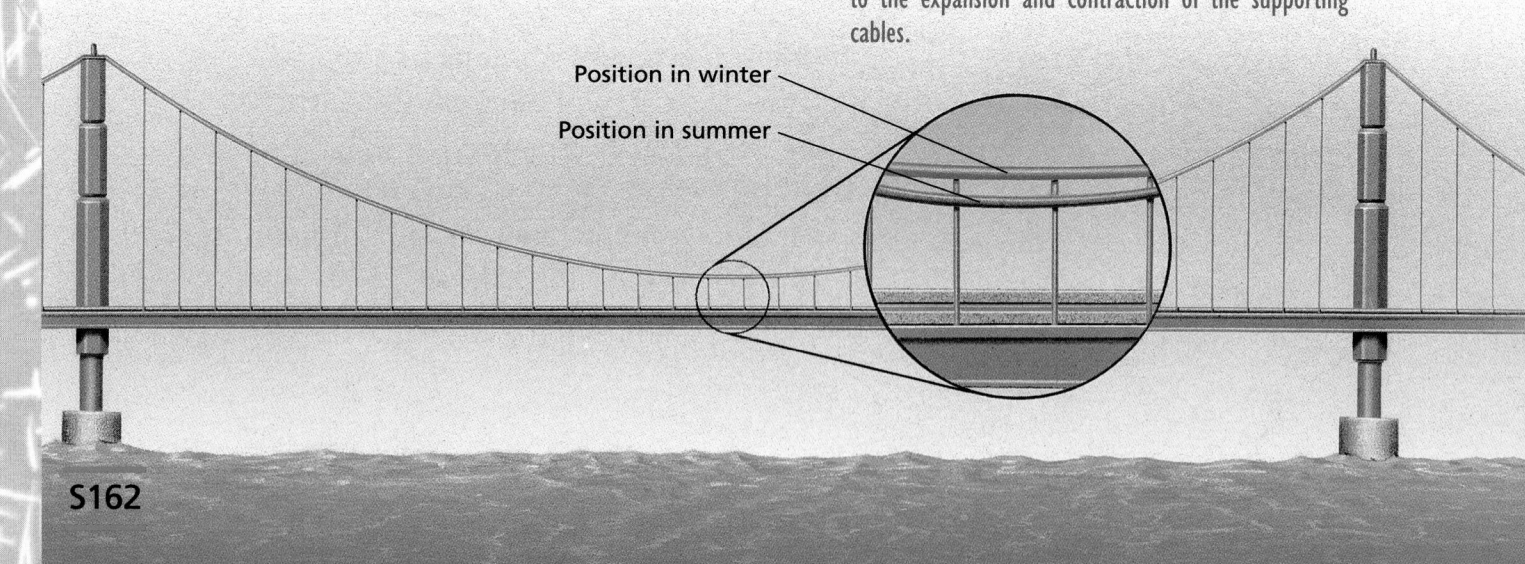

▼ The elevation of this suspension bridge changes noticeably from warm weather to cold weather due to the expansion and contraction of the supporting cables.

Position in winter
Position in summer

S162

The steel bar expands because the increase in speed increases the average distance between its molecules. Cooling the steel decreases the energy of the molecules, reducing their vibration and causing the distance between them to decrease. The bar shrinks. Almost all kinds of matter expand when heated because their particles move faster and the distance between the particles increases.

Heat, Thermal Energy, and Heat Transfer

The term *heat* is often used incorrectly. Sometimes people use the word to mean "hotness." But all things, whether hot or cold, contain heat. **Heat**, or thermal energy, refers to the energy of a substance due to the random movement of its atoms and molecules. This heat energy can then be transferred from one place or object to another. For example, when you sit in the sun, the radiant energy is absorbed and changed into heat energy, and you feel warm. The thermal energy of your skin increases as its temperature rises. If you touch your shoulder, heat is transferred from your shoulder to your hand.

A hot object contains more thermal energy than a cold object of the same size and composition. If they are brought into contact, heat will flow from the hot object to the cold object until the temperature of both is the same. As you will see, this flow of energy can be used to do work.

Putting Heat to Work

Certain machines, called *heat engines*, change thermal energy into mechanical energy that can be used to do work. The first widely used device of this type was the steam engine. A steam engine uses the heat energy produced when coal or wood is burned to convert water into steam. The steam pushes a *piston* back and forth inside a *cylinder*. The conversion of thermal energy into mechanical energy sets the piston and crank into motion. The steam engine's operation is explained below.

Heat

The energy of a material due to the random motion of its particles; thermal energy

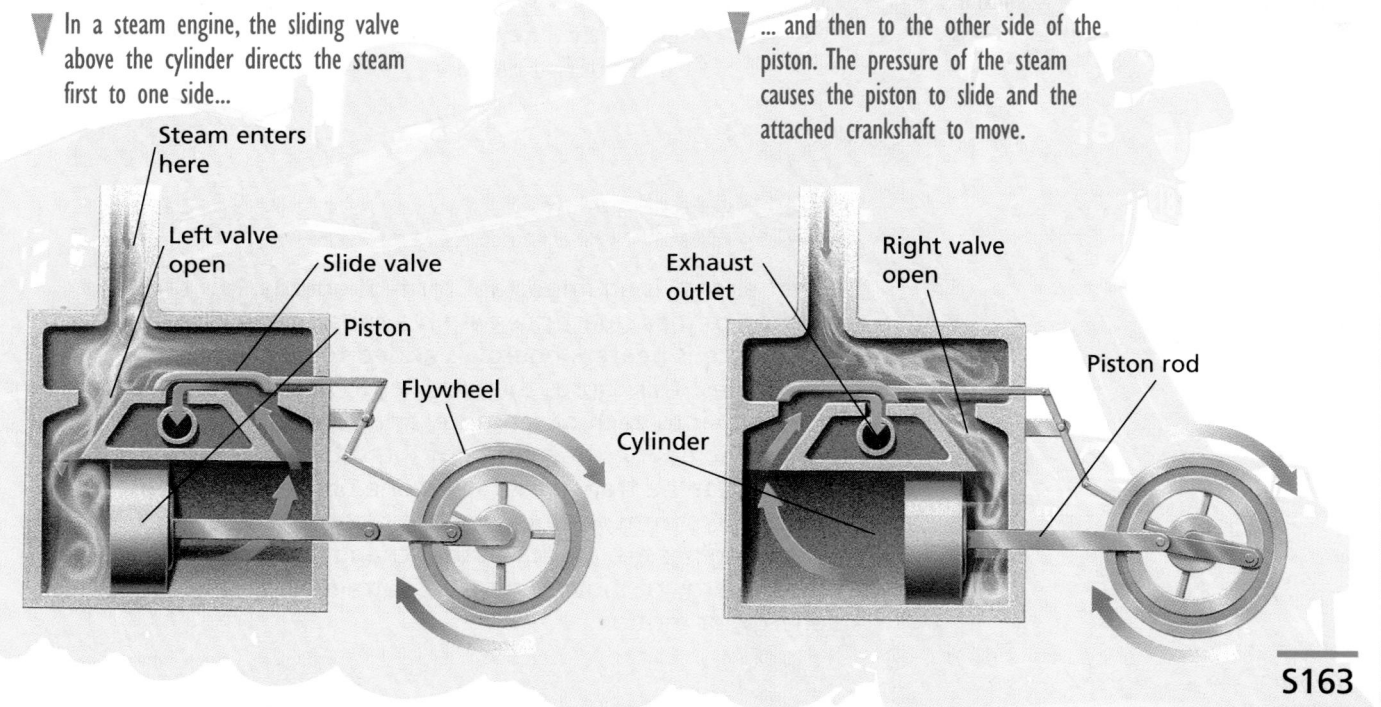

▼ In a steam engine, the sliding valve above the cylinder directs the steam first to one side...

Steam enters here

Left valve open

Slide valve

Piston

Flywheel

▼ ... and then to the other side of the piston. The pressure of the steam causes the piston to slide and the attached crankshaft to move.

Exhaust outlet

Right valve open

Piston rod

Cylinder

S163

The steam engine was an important machine during the Industrial Revolution of the late eighteenth and early nineteenth centuries. The large amount of mechanical energy that the steam engine made available allowed large-scale industrial operations to exist. Today, the most common heat engine is the type found in cars and trucks—the internal combustion engine. As you know, car and truck engines generally use gasoline. This fuel is mixed with air and drawn into the engine, where it is ignited. The burning fuel-air mixture produces hot, high-pressure gases that push down on the pistons and turn the attached crankshaft. Through a series of gears, the spinning motion of the crankshaft turns the wheels of the car.

Another type of heat engine is the jet engine. In a jet engine, fuel is burned to produce a stream of hot gases that shoots at high speed out the back of the aircraft's engine. This jet of hot gases causes a reaction force that pushes the engine (and the plane) forward.

Internal Combustion Engine

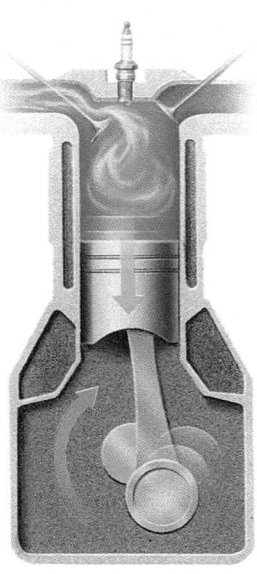 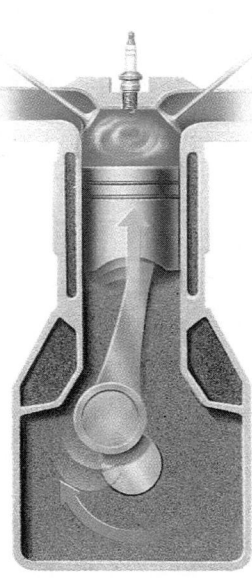  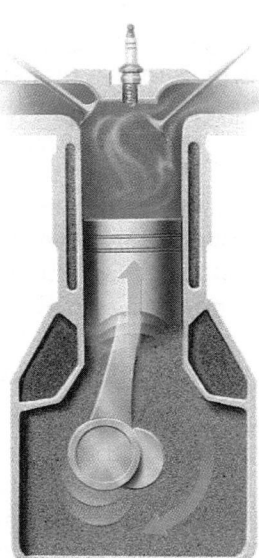

Intake stroke: Momentum carries the piston downward. The downward motion of the piston draws the fuel-air mixture into the combustion chamber through the intake valve.

Compression stroke: The upward motion of the piston compresses the fuel-air mixture.

Power stroke: The spark plug ignites the compressed fuel-air mixture. The hot, high-pressure gases push down on the piston. Part of the heat energy is thus changed into work.

Exhaust stroke: The upward motion of the piston forces the hot exhaust gases through the exhaust valve. Most of the heat is lost.

SUMMARY

Heat energy is an important form of energy. It is the energy of many particles within a substance. When a substance is heated, energy is added to its atoms and molecules. This causes the particles to move faster, to bump into each other more, and to move farther apart, increasing the energy of the substance. Heat energy can be transferred from one object to another. Heat flows from warmer objects to cooler objects. Heat engines are machines that change some of the heat energy of moving particles into mechanical energy, or motion.

S164

TYPES OF HEAT TRANSFER

Harry S. Truman, 33rd president of the United States, once said, "If you can't stand the heat, get out of the kitchen." He didn't mean it literally, of course, but his observation was certainly valid. When you use the stove, it warms up not only the food you are cooking but also the entire kitchen. If all of the burners and the oven are going at once, the kitchen may get uncomfortably hot. What's happening here?

When you cook, heat is transferred from the flame or heating elements to the pan and its contents. Heat is also transferred to the surrounding air. If you were to stand on a ladder, you could feel some of this hot air concentrated near the ceiling. You could also feel the warmth of the flame or heating coil from 2 or 3 ft. away.

How Heat Moves

The kitchen is a good place to see how heat moves from place to place. But exactly how does heat get around? It turns out that heat moves by three different processes: *conduction*, *convection*, and *radiation*.

Conduction   Objects can gain heat by being in contact with a heat source. If you touched the end of the iron bar in the photo to the right, what do you think would happen? The end of the rod is hot, and the heat would instantly be transferred to your finger. The transfer of heat by direct contact is called **conduction**. Conduction is the simplest method of heat transfer. Conduction occurs both within materials and between objects that are in contact. The metal in the bar, like all other substances, is made up of atoms. The flame transfers heat energy to the bar, causing the metal atoms to vibrate vigorously back and forth. The heated atoms of metal then bump their cooler neighbors, causing them to vibrate more vigorously. This continues until all atoms in the bar are moving with equal

▲ The flame transfers heat energy to the bar, causing the metal atoms to vibrate. The energy of this random vibration is transferred from one metal atom to the next by direct contact.

energy. If you touch the bar, you will be burned because heat is transferred from the bar to your skin by conduction.

Most metals conduct heat very well; that's why cooking pans are usually made of metal. However, most nonmetallic materials do not conduct heat as well as metals. Wood, for

example, is a poor conductor of heat. That is why the handles of pans and other metal objects are often made of wood. Materials that are poor conductors of heat are called *insulators*.

Convection   How does heat from a stove reach the water at the top of a pot? The answer is found in the way liquids behave when they are heated. The water at the bottom of the pot becomes hot due to contact with the pot itself—conduction. This increases the random motion of the water molecules and the energy of their collisions. As a result, the space between the water molecules increases and the water becomes less dense. Since the heated water is less dense than the colder water above it, it rises to the top of the pot.

Conduction

Transfer of heat by direct contact

Convection

Transfer of heat by movement of a heated gas or liquid

S165

Transfer of heat by the movement of a heated gas or liquid is called **convection**. Convection occurs when a gas (such as air) or a liquid (such as water) is heated unevenly. Convection is the main method of heat transfer in liquids and gases. It cannot occur in solids, however, because the molecules are not free to move about.

When people think about staying warm on a cold winter night, they may think of a roaring fire in the fireplace. Actually, an open fireplace is not a very good source of heat for a room. Because of convection, most of the heat goes up the chimney rather than into the room. Some heating systems, however, do take advantage of convection. In such systems, air heated by a furnace usually enters each room through a vent near the floor. Due to convection, this heated air rises toward the ceiling while cool air sinks toward the floor. Another vent near the floor collects the cool air to return it to the furnace.

Radiation   If you have ever stood near a large fire, you no doubt noticed that you could feel its heat. The heat might even have been so intense that you could hardly stand it, even though you might have been standing several meters away. How did the energy travel across the gap between you and the flames of the fire?

The answer is **radiation.** Vibrating atoms and molecules in liquids and solids produce electromagnetic waves, which are radiated into space. These waves have longer wavelengths than visible light and are called *infrared waves.* This infrared radiation can be transmitted through space from one object to another. All objects radiate some energy. The higher the temperature, the more intense the radiation. When radiation from a heat source reaches your skin, for example, it is absorbed; it causes increased random motion of atoms and molecules in your skin and makes you feel warm. In this way, heat is transferred from a hot object to a cooler one.

Radiation

Transfer of heat through space by infrared waves

DID YOU KNOW...

that heat from the sun takes several minutes to reach the Earth? Traveling at 300,000 km/s, the sun's radiant energy takes about 8 minutes to travel the 150-million-kilometer distance!

A fireplace heats a room primarily by radiation. Convection currents carry most of the heat of the fire up the chimney and also draw air into the fireplace from the room. Even though they lose so much heat, why are chimneys necessary?

S166

Hindering Heat Transfer

The vacuum bottle, shown in the illustration below, is very effective at keeping its contents insulated. Its design minimizes heat transfer by all three methods.

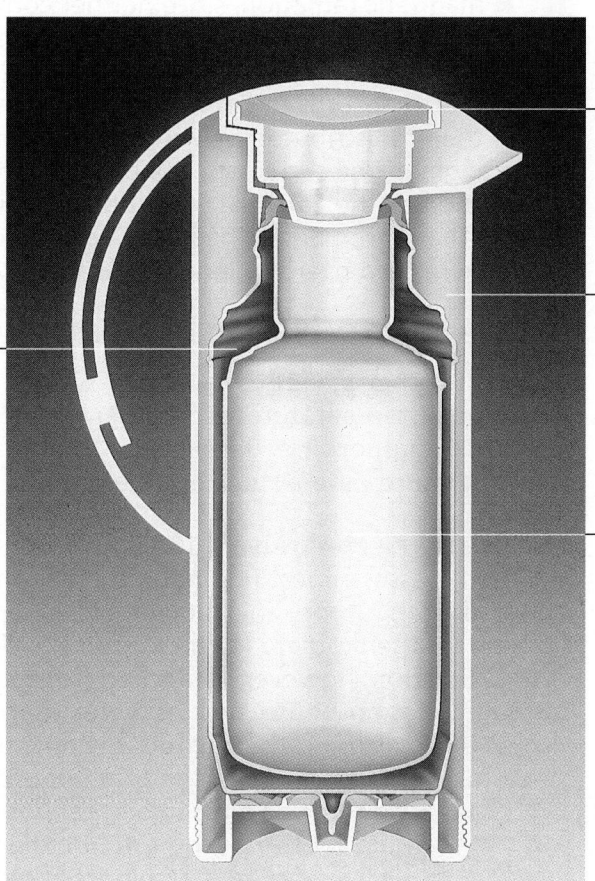

The bottle is covered by a cap with an insulation-filled plug that is also a poor conductor of heat.

An air space between the glass bottle and the outer walls hinders the transfer of heat by conduction. In addition, glass is a poor conductor of heat.

The inner surfaces of the glass are silvered, and so they reflect heat inward, preventing heat transfer by radiation.

The space between the two walls of the double-walled glass bottle is a vacuum.

Since there are very few molecules of any kind present in a vacuum, heat cannot be transferred across the vacuum by either conduction or convection.

▲ A vacuum bottle can keep its contents hot or cold for hours. How do its design features allow it to do this?

S U M M A R Y

Heat energy is transferred in three ways: by conduction, convection, and radiation. Conduction occurs by direct contact within a material or between materials. Solids transfer heat only by conduction. Convection is the transfer of heat by the motion of gases and liquids. Radiation is the transfer of heat energy across space in the form of electromagnetic radiation. Insulators inhibit the flow of heat by any or all of these methods of heat transfer.

HEAT ON EARTH

Heat comes to the Earth from the sun across 150 million kilometers of space. Obviously, the heat is transferred by electromagnetic radiation, since neither convection nor conduction could carry heat across the emptiness of space. You can feel the sun's radiation when you stand in the sunlight.

The Earth absorbs radiation from the sun and is warmed. The heat is then redistributed by winds and ocean currents. Factors such as the angle of the sun's rays, the density of the atmosphere, and the rotation of the Earth all contribute to this movement of heat energy. The movement of heat through the oceans, through the atmosphere, and over the land determines weather and climate.

A Temperate Island in Space

Unlike the other planets of our solar system, the Earth has a moderate range of temperatures. The other planets are either too hot or too cold to support life. Because the Earth has just the right range of temperatures, life flourishes here.

The Global Greenhouse   The light from the sun (mostly visible radiation) passes almost unhindered through the atmosphere. Both land and water absorb this radiant energy and become warm. These areas then radiate infrared waves back into space. A portion of this radiation, however, is blocked (reflected) by some of the gases in the atmosphere, such as water vapor, carbon dioxide, and nitrous oxide, causing the Earth to become even warmer. This warming process is called the *greenhouse effect* because it is similar to the way glass in a greenhouse works. The glass admits light from the sun but keeps most of the infrared waves radiated within from escaping. At present the Earth is warmed about 20°C by the greenhouse effect, an amount necessary to keep our planet habitable. However, the burning of fossil fuels and the clearing of forest has been increasing the concentration of carbon dioxide in the atmosphere. Many scientists predict that the Earth will warm too much as a consequence.

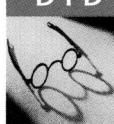

DID YOU KNOW...

that the lowest temperature ever recorded on Earth was −89°C and that the highest temperature recorded was 58°C?

The low temperature (−128°F) was recorded at Vostok Station, Antarctica, and the high temperature (136°F) was recorded in the desert of Libya in North Africa.

▼ Like a greenhouse, the Earth's atmosphere warms up because it lets in light but traps heat.

S168

Convection in the Atmosphere

Because of the Earth's curved shape, its surface, and therefore its atmosphere, are not heated evenly. If the Earth were flat, all areas would receive about the same amount of energy from the sun. But because the Earth's surface is curved, the intensity of sunlight reaching the Earth's surface varies. Polar regions receive much less solar energy than the regions near the equator. Regions around the equator receive direct rays from the sun that provide concentrated energy, while polar regions receive slanting rays that spread the energy over a larger area. Therefore, the air near the equator gets hotter than it does at the poles. Without some way to transfer the heat, the equatorial region would become unbearably hot and the poles would become unbearably cold.

Convection causes air to move between the Earth's cold and warm regions. A gigantic convection pattern redistributes the sun's energy over the surface of the Earth. The illustration above illustrates this convection system. Convection has made it possible for life to exist on much of the land surface of the Earth.

Almost all winds are driven by convection patterns that occur because of uneven heating of the Earth's surface. This uneven heating occurs locally as well as globally. For example, forested regions don't warm up as much as grasslands, and oceans don't warm up as much as forested regions. Where air is warmed, it rises, causing cooler air to move in from the surrounding area. As the air rises, it expands and cools. Eventually it cools enough to sink back to Earth. The circulation pattern formed by the cycle of air

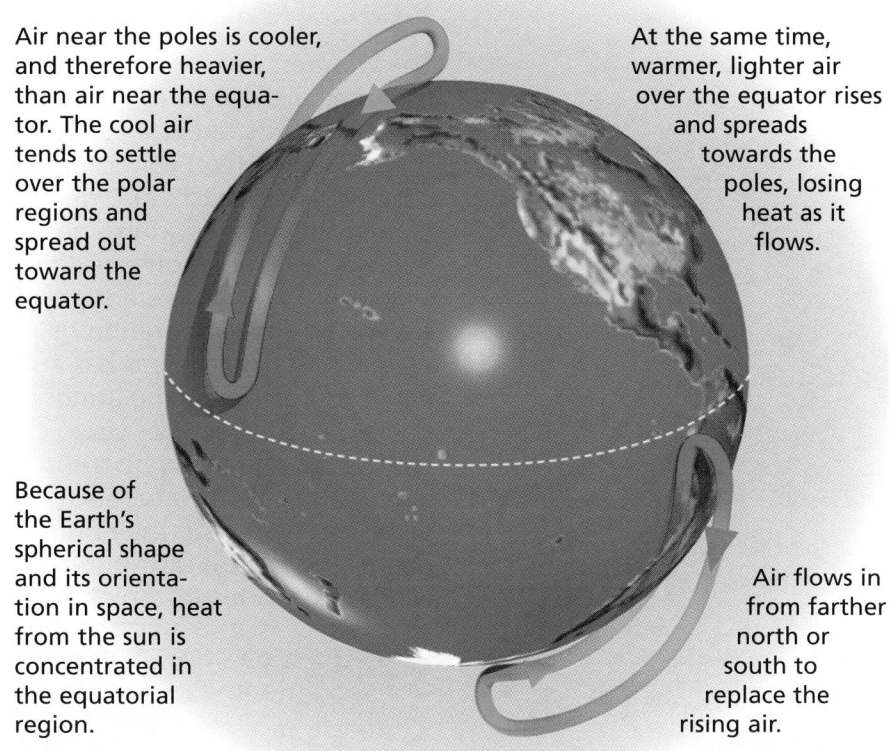

Air near the poles is cooler, and therefore heavier, than air near the equator. The cool air tends to settle over the polar regions and spread out toward the equator.

Because of the Earth's spherical shape and its orientation in space, heat from the sun is concentrated in the equatorial region.

At the same time, warmer, lighter air over the equator rises and spreads towards the poles, losing heat as it flows.

Air flows in from farther north or south to replace the rising air.

heating, rising, cooling and returning to the ground is called a **convection cell.**

Storing Heat   An interesting property of water is that it can absorb large amounts of heat without changing much in temperature. You may have noticed this before. If you heat a large pan of water on the stove, it can take quite a while for the water to boil, even with the burner turned up all the way. And once you turn the burner off, it takes many hours for the water to return to room temperature. If you are a swimmer, you may have noticed that in the spring, several weeks of consistently warm weather must pass before pools, rivers, and lakes are warm enough to swim in.

Water is able to store much more heat energy than land. This property has a huge effect on the Earth's climate, because most of the Earth's surface is covered by water. Large bodies

Convection cell

A complete circle of moving liquid or gas caused by temperature differences

of water tend to moderate the climate of nearby areas. For example, coastal areas typically have narrower temperature ranges, both throughout the day and from season to season, compared with inland areas. In fact, the climate of the Earth as a whole is much more moderate than it would be if there were no oceans. The ocean acts like a storage tank for thermal energy, absorbing and releasing vast amounts of heat. In the winter, when the inflow of solar energy drops, the oceans release heat to the atmosphere, keeping temperatures relatively high. The reverse happens in summer, and the temperature remains moderate.

S169

Heat, Weather, and Climate

A hurricane (above) is an enormous natural heat engine, as shown by the thermal image below.

The warmest ocean waters are found near the equator, where the sun's rays strike the Earth most directly. Here, water temperatures may reach 30°C or higher. High temperatures and intense solar radiation cause the evaporation of large amounts of sea water. The evaporation of water from tropical seas has an important effect on the atmosphere. The air becomes very moist, causing rain to be frequent. This warm water also adds heat energy to the atmosphere. This heat is distributed by winds and storms to cooler regions away from the equator. You've heard of hurricanes, no doubt. Hurricanes (known in other parts of the world as typhoons or cyclones) are giant natural heat engines that feed on the thermal energy contained in tropical seas. These huge storms typically form in the tropics and then veer into the temperate zones, dissipating enormous amounts of tropical heat as they go.

Hurricanes are just one way in which the heat of the sun is redistributed across the Earth. The movement of air and water across the Earth's surface redistributes the Earth's huge stores of heat energy. Previously, you learned that there is a general flow of air between the poles and the equator. The global air circulation pattern, which helps to redistribute the sun's heat, is actually quite a bit more complicated than that general model. Ocean currents on the surface, which are generally driven by prevailing winds, also redistribute heat. Many factors influence global air and ocean circulation, such as the distribution of land and water, major and minor geographic features, the amount of snow and ice covering the surface, the rotation of the Earth, and even the amount of cloud cover.

Heat stored in the ocean modifies the climate of Ireland, whereas the climate of inland Canada at the same latitude remains cold, as these two winter scenes show.

Heat From Underground

As you know, heat from the sun warms the Earth's surface. But the Earth is also warmed from within. The Earth's interior contains much heat. Some of this heat may be left over from the formation of the Earth over 4.5 billion years ago, when it solidified from a molten ball. The Earth most likely formed from countless pieces of cosmic debris drawn together by gravity. As these pieces collided, their gravitational energy was converted into heat. But most of the Earth's internal heat probably results from the decay of radioactive elements in the interior of the Earth. Convection currents in the Earth's interior carry heat from the superhot core (an estimated 5000°C) to the cool crust. These convection currents flow unimaginably slowly. A complete convection cycle may take hundreds of millions of years.

Heat from the Earth's interior flows quite slowly through the crust. This is because 30 or 40 km of solid (and poorly conductive) rock make a very effective insulator. In everyday life we do not see evidence of the heat inside the Earth. But phenomena such as erupting volcanoes, geysers, and bubbling hot springs provide evidence that the inside of the Earth is a very hot place indeed.

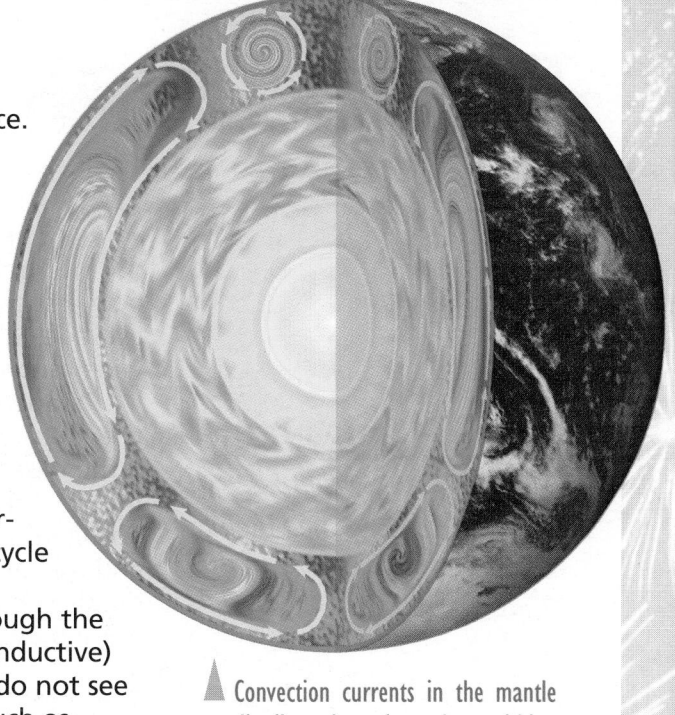

▲ Convection currents in the mantle distribute heat from deep within the Earth.

S U M M A R Y

Radiant energy from the sun is absorbed by air, land, and water, which in turn store and redistribute the heat. The Earth's climate is determined by the amount of energy it absorbs from the sun and how the resulting heat is distributed. The transfer of heat from place to place on the Earth's surface results in weather. The interior of the Earth is quite hot, but little of that heat reaches the Earth's surface because of the crust's insulating properties.

S171

TEMPERATURE AND HEAT

While heat and temperature are related, they are not the same thing. As you have learned, heat is the energy of the random motion of the molecules in an object. The more heat a substance contains, the faster its molecules are moving on average. We say "on average" because in any material there is a wide range of molecular speeds—some faster, some slower.

Temperature is the "hotness" of a substance. It is a measure of the *average* random kinetic energy of the molecules of a substance. If the molecules of an object have a high average kinetic energy—that is, if they are moving very fast—then that object is "hot." An object with low molecular kinetic energy is "cold." Note that cold and warmth are not separate properties. Something that is cold simply contains less heat energy than something that is hot. Our sense organs provide the sensation of hotness and coldness.

Temperature

The "hotness" or "coldness" of a material; determined by the average kinetic energy of the molecules of the material

▲ The molecules of a cold substance have a lower average energy than those of a hot substance.

Though the water in both beakers is the same temperature, the larger beaker contains more heat.

The total amount of heat energy in an object is a function of two things: the average kinetic energy of the molecules of the object and the total number of molecules in the object. Consider the beakers on the hot plates above. The beakers contain exactly the same substance, and the molecules of each have the same average kinetic energy. But which beaker contains more thermal energy? The beaker on the left contains more heat because it is larger and therefore contains more molecules.

Because the larger beaker contains more heat, it could do more work if you had a way to turn its thermal energy into mechanical energy. The liquid inside it could certainly melt more ice than that in the smaller beaker. The heat available to do work is related to the temperature *and* amount of the material present.

S172

Measuring Temperature

How hot is "hot"? How cold is "cold"? If someone were to ask you what the outside temperature was today, could you give an accurate answer? If you stuck your hand outside, could you estimate the temperature of the air? Do you think that everyone in the class would agree with your estimate? If you had to report the temperature to the weather station, could you do so accurately without a measuring instrument?

As you know, a *thermometer* is an instrument used to measure temperature, and it works by undergoing physical changes. These changes result from the increase or decrease in the average random kinetic energy of molecules in the materials that form the thermometer. As you know, most materials expand when heated and contract when cooled. The most commonly used type of thermometer is based on such a change. When a liquid, such as mercury or alcohol, is sealed in an evacuated glass tube, it can be used as a thermometer. These two liquids are commonly used because they expand and contract at a uniform rate and do not boil or freeze within the range of temperatures normally experienced. When a thermometer is heated, its liquid expands more than the glass, so it rises in the tube. Cooling causes the liquid to contract and fall.

Not all thermometers use liquid in a glass tube. A *thermostat*, for example, uses a coiled strip made of two different metals sandwiched together. This strip is called a *bimetallic strip*. The two metals expand and contract at slightly different rates when the temperature changes. A temperature change causes the bimetallic strip to either wind or unwind.

Another kind of thermometer, called a *thermocouple*, measures temperature by its effect on electric charges. Because thermocouples react very quickly, they are used by many medical professionals to take patients' temperatures.

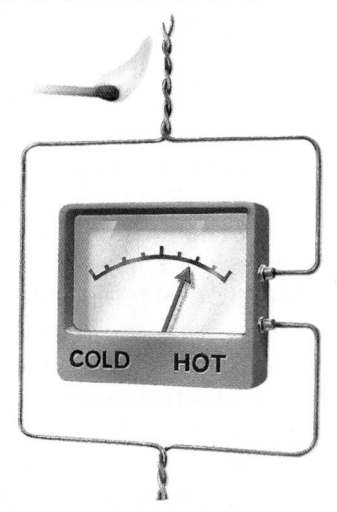

A thermocouple consists of two different types of wire twisted together and connected to a voltmeter. When heat is applied to one junction, voltage is generated. The greater the temperature difference between the two junctions, the greater the voltage.

Thermocouples are suitable for reading high temperatures that would destroy liquid thermometers.

S173

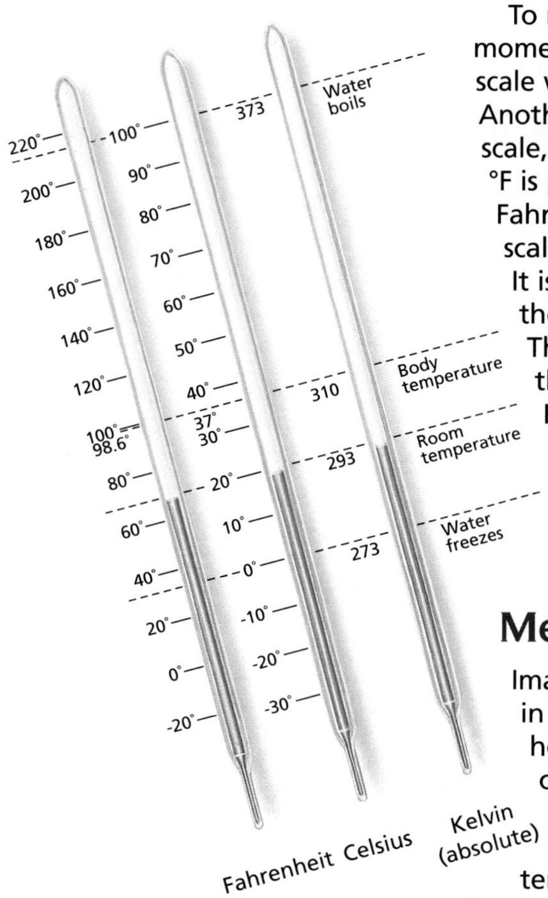

To measure the amount of change in temperature, a thermometer must have a scale. You already know that on the Celsius scale water (at sea level) freezes at 0°C and boils at 100°C. Another scale that is probably familiar to you is the Fahrenheit scale, which is customarily used in the United States. The symbol °F is placed after temperatures measured on this scale. On the Fahrenheit scale, water freezes at 32°F and boils at 212°F. A third scale, called the Kelvin scale, is the official SI temperature scale. It is used by scientists all over the world. See the illustration to the left for a comparison of these three temperature scales. The unit of temperature on the Kelvin, or absolute, scale is the *kelvin* (K). Notice that the degree symbol is not used for Kelvin scale temperatures. On the Kelvin scale, water freezes at 273 K and boils at 373 K. On this scale 0 K, or *absolute zero*, is the temperature at which molecular motion is at a minimum. Absolute zero is the lowest possible temperature. A change of 1 K is equal to a change of 1°C.

Measuring Heat

Imagine that you are making a cup of cocoa and you are in a hurry. How much milk would you put into the pot to heat? Would you fill it or heat just enough milk to fill your cup? You probably realize that it takes more heat, and thus more time, to raise the temperature of a potful of milk than a cupful of milk. When both amounts of milk are the same temperature, the potful of milk has more heat energy than the cupful. But how can you measure a specific amount of heat?

Heat cannot be measured directly by a simple instrument like a thermometer. It must be measured by the effects it produces. For example, the amount of heat given off by a fuel could be measured by comparing the temperature change of a quantity of water that it could heat. If one type of fuel warms a certain amount of water by 1°C and the same amount of another type of fuel warms the same amount of water by 2°C, the second fuel releases twice as much heat per unit of mass.

Scientists may measure heat in units called *calories* (cal). A calorie is the amount of heat needed to raise the temperature of 1 g of water by 1°C. Thus, to raise the temperature of 100 g of water by 1°C, 100 cal are required. Raising the temperature of 100 g of water 20°C requires 2000 cal. It takes about 20,000 cal to heat a cup of water from room temperature to boiling. As you can see, a calorie is a very small quantity of heat. Scientists more commonly use the unit called a *kilocalorie* (kcal) to measure heat. One kilocalorie is equal to 1000 cal. It is the amount of heat needed to raise the temperature of 1 kg of water by 1°C. The kilocalorie is sometimes referred to as the Calorie (with a capital C) or the "big calorie." It is this larger unit that you refer to when "counting Calories," because the energy supplied by food is measured in this unit.

Because heat is now understood to be a form of energy, scientists also measure heat with a unit called the *joule* (J). This is the same unit used to measure other forms of energy as well as to measure work. By definition, 1 cal equals 4.184 J. An instrument known as a *calorimeter* is used to determine how much heat is exchanged between substances.

The Fahrenheit, Celsius, and Kelvin temperature scales are compared in this diagram.

Fahrenheit   Celsius   Kelvin (absolute)

220° 100° 373 Water boils
200° 90°
180° 80°
160° 70°
140° 60°
120° 50°
100° 40° 310 Body temperature
98.6° 37°
100° 30° 293 Room temperature
80° 20°
60° 10° 273 Water freezes
40° 0°
20° -10°
0° -20°
-20° -30°

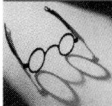
S174

Specific Heat

Heat does not affect the temperature of all materials in the same way. For example, if 100 cal (418.4 J) of heat were added to 100 g of water, the temperature of the water would be raised by 1°C; however, if 100 cal (418.4 J) of heat were transferred to 100 g of iron, the temperature of the iron would be raised by a little over 9°C. To compare the effects of heat on different materials, scientists calculate the *specific heat* of the substance.

Specific heat is the amount of heat needed to raise the temperature of 1 g of a substance by 1°C. The specific heat of water is, by definition, 1.0 cal/g·°C (4.184 J/g·°C). This reads: one calorie per gram per degree Celsius. In other words, specific heat compares the ability of a material to absorb heat with the ability of water to absorb heat. The specific heats of some other materials are given in the table to the right.

Note that water has the highest specific heat on the list.

In fact, very few substances have a higher specific heat than water. This should not surprise you since you already know that water is able to absorb a lot of heat without changing much in temperature. It takes a lot more heat to raise the temperature of a given mass of water than the same mass of almost any other substance. The amount of heat needed to change the temperature of a substance depends on the mass of the substance, its specific heat, and the amount of change in the temperature.

Specific heat

The amount of heat needed to raise the temperature of 1 g of a substance by 1°C

Specific Heat of Some Common Materials		
	(cal/g·°C)	(J/g·°C)
water	1.00	4.19
ice	0.50	2.09
wood	0.42	1.76
aluminum	0.22	0.92
iron	0.11	0.46
copper	0.09	0.39

▲ All of these metal weights have the same mass and were heated to the same temperature (left). However, because they each have a different specific heat, they each melt a different amount of wax (right).

S175

You have probably experienced personally the effects of water's high specific heat. Have you ever felt the difference in temperature between water in an outdoor pool and the concrete walk around it on a hot summer day? Concrete has a much lower specific heat than water. Even though the amount of sunlight per square meter of surface may be the same, the concrete gets too hot to walk on, while the water stays comfortably cool. Furthermore, the sidewalk will cool off quickly once the sun sets, but the water will remain at more or less the same temperature.

The high specific heat of water explains why areas near large bodies of water stay cooler in the summer and warmer in the winter than inland areas. The water has such a high specific heat that its temperature changes comparatively little with the seasons. Thus, the temperature of the overlying air is moderated as well.

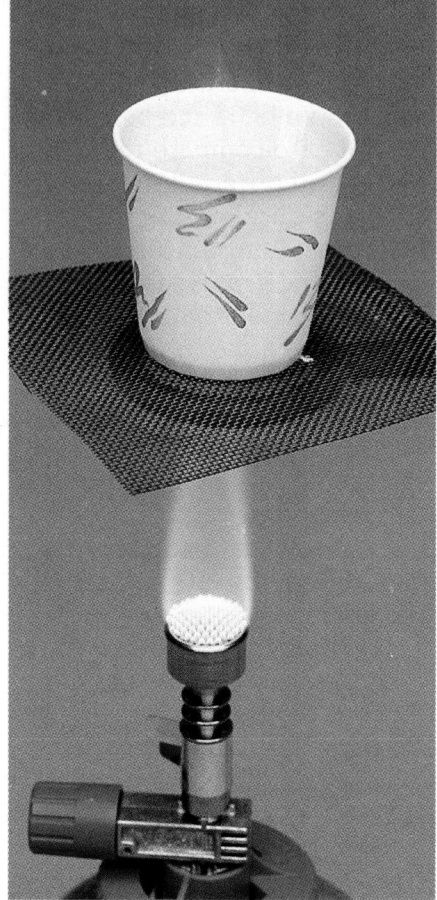

Water is so effective at absorbing heat that this paper cup does not burn, even though the water is starting to boil. ▶

S U M M A R Y

Heat, or thermal energy, is a measure of the total molecular energy of an object. Temperature is a measure of the average random molecular kinetic energy of a substance. Thermometers measure temperature in degrees Celsius, degrees Fahrenheit, or kelvins. The amount of thermal energy in a material depends on its temperature and the amount of material present. Heat is measured in calories or joules. Specific heat is the amount of heat, in calories or joules, required to raise 1 g of a substance by 1°C.

S176

Unit CheckUp

Concept Mapping

Using the terms supplied below, construct a concept map that illustrates major ideas from this unit. Then extend your concept map by answering the additional question below. Use your ScienceLog. **Do not write in this textbook.**

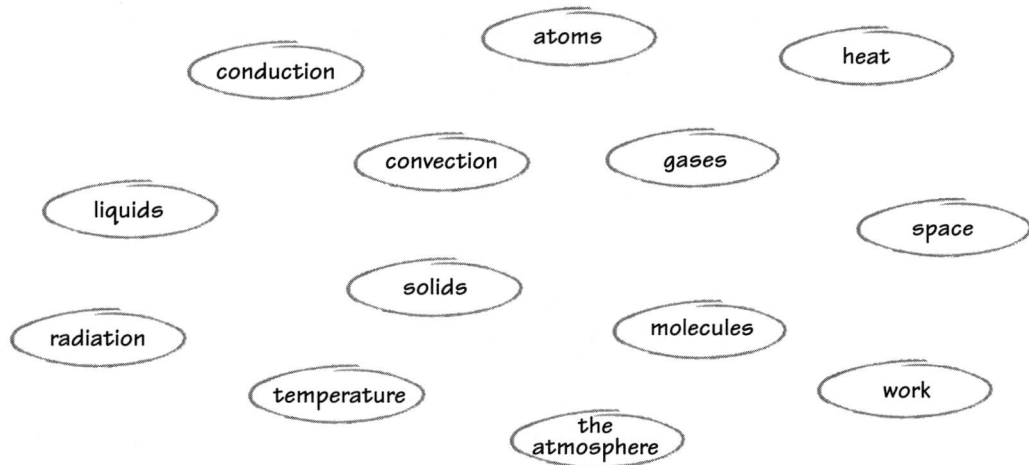

How could you add to your concept map to indicate the difference between a *thermometer* and a *calorimeter*?

Checking Your Understanding

Select the choice that most completely and correctly answers each of the following questions.

1. Which of the following is powered by a heat engine?
   a. an electric train
   b. a windmill
   c. the "stealth" fighter
   d. Both *b* and *c* are correct.

2. You instantly feel heat as a floodlight across the room is switched on. This is an example of
   a. temperature.
   b. conduction.
   c. convection.
   d. radiation.

3. Within a pot of water simmering on a stove, heat is transferred primarily by
   a. conduction.
   b. convection.
   c. radiation.
   d. There is no way to tell.

4. Which statement *best* describes the effect of the oceans on the Earth's climate?
   a. Without oceans, the Earth would be colder.
   b. Without oceans, the Earth would be warmer.
   c. Without oceans, the temperature range would be much greater.
   d. Without oceans, the climate would be less variable.

5. Water has a very high specific heat, meaning that
   a. it absorbs a lot of heat without increasing much in temperature.
   b. only very high temperatures will cause it to increase in temperature.
   c. the amount of heat contained in any given amount of water is comparatively high.
   d. it can be used to do work in specific ways.

S177

Interpreting Illustrations

Examine the diagram of the bimetallic strip and the accompanying information. If the strip is straight at 0°C, what will it look like at a temperature of –20°C? at 25°C? Explain.

Percent expansion/contraction per °C of change
■ Iron — 0.0012   ■ Copper — 0.0017

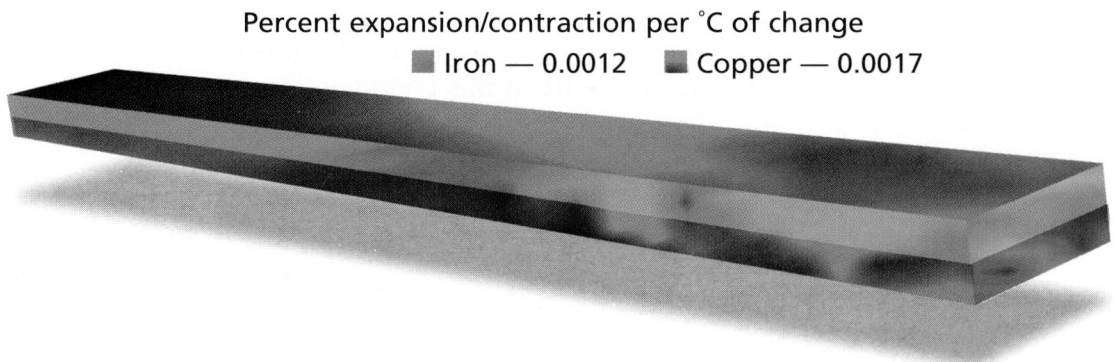

Critical Thinking

Carefully consider the following questions, and write a response in your ScienceLog that indicates your understanding of science.

1. Explain how water's thermal properties make it useful as a coolant.

2. If the atoms of solids are vibrating vigorously, why don't we feel the vibrations?

3. You are warmed by the sun through a sheet of thick glass. Since glass is solid, does this mean the heat you feel is being transferred by conduction? Explain.

4. Since there is a lowest possible temperature—absolute zero—do you think there might also be a highest possible temperature? What might define such a temperature?

5. In the future, the Earth will continue to lose its inner heat. Does this mean that the ground will cool off and the Earth's climate will grow cold? Explain.

Portfolio Idea

Design an experiment to determine whether heat and cold are separate but similar phenomena. In other words, can heat and cold be transmitted in the same way? Describe the experiment in detail, including sketches. Be sure to get your teacher's approval before actually doing the experiment.

S178

OUR CHANGING EARTH

8 Unit

IN THIS UNIT

Now that you have been introduced to the forces that alter the shapes of continents, consider the following questions.

● **1.** Of what types of rock is the Earth formed? How do these different rocks form?

2. Why do landforms wear away? What processes cause this to happen?

3. How does the movement of rock particles relate to Earth's surface features both above and below the sea?

In this unit, you will take a closer look at the processes that shape the Earth's surface.

S179

PROFILE OF A ROCKY PLANET

Unlike the Sun and the giant gas planets of the outer solar system, Earth and its three nearest planetary neighbors, Mercury, Venus, and Mars, are solid bodies with surfaces of rock. For this reason they are known as terrestrial planets. But even though the Earth has a solid, durable surface, it is constantly changing because of forces acting at and below its surface.

Types of Rock

Rock is the stuff of which the Earth is made. What is rock? **Rock** is simply any naturally occurring solid mixture of minerals and other materials. *Minerals*, in turn, are any naturally occurring, inorganic chemical compound found in the Earth. Because there are many minerals in the crust of the Earth, there are also many types of rock. Rocks are broadly classified according to how they are formed.

Igneous Rocks   As you probably know, the interior of the Earth is quite hot. Below a certain depth, the heat is actually intense enough to melt rock. Such melted rock is called *magma*. Rock that forms from cooled magma is called **igneous rock**. Igneous rock can form deep underground or at the surface. Because they are well insulated from the cold surface, igneous rocks that form from magma deep underground cool very slowly. Because they cool slowly, they form relatively large mineral crystals. Such rocks are called *intrusive* igneous rocks because they *intrude* into underground spaces. Not all igneous rocks, however, are intrusive. Earth processes, such as volcanic eruptions, sometimes cause magma to flow onto the Earth's surface. Magma that reaches the Earth's surface is called *lava.* Because it is exposed to cool air or water, lava cools very rapidly and forms rocks with small crystals. These kinds of rocks are called *extrusive* igneous rocks because they are *extruded* (squeezed out) onto the Earth's surface.

Sedimentary Rocks   When rocks are exposed at the Earth's surface, sun, rain, wind, and frost begin to break them apart almost immediately. Pieces of broken rock are then moved to different locations where they collect in low places. Over time, rock fragments can collect to great thicknesses. The weight of the accumulating layers of rock fragments plus the action of ground water causes the rock fragments to become *cemented,* or stuck together, forming solid rock. Rocks that are formed from pieces of other rocks are called **sedimentary rocks**. *Sedimentary* means "made of sediments." Sediments are pieces of rock that have been deposited by water, wind, or ice. One example of a sedimentary rock is sandstone, which consists of sand grains that have become cemented together. Other sedimentary rocks, such as gypsum, are deposited when mineral-rich water evaporates, leaving the minerals behind.

The moon, like the Earth, is composed of rock. Without water or vegetation, its surface of exposed rock is clearly evident.

Igneous rocks

Rocks formed from cooled magma and lava

When lava cools, it forms igneous rock.

S180

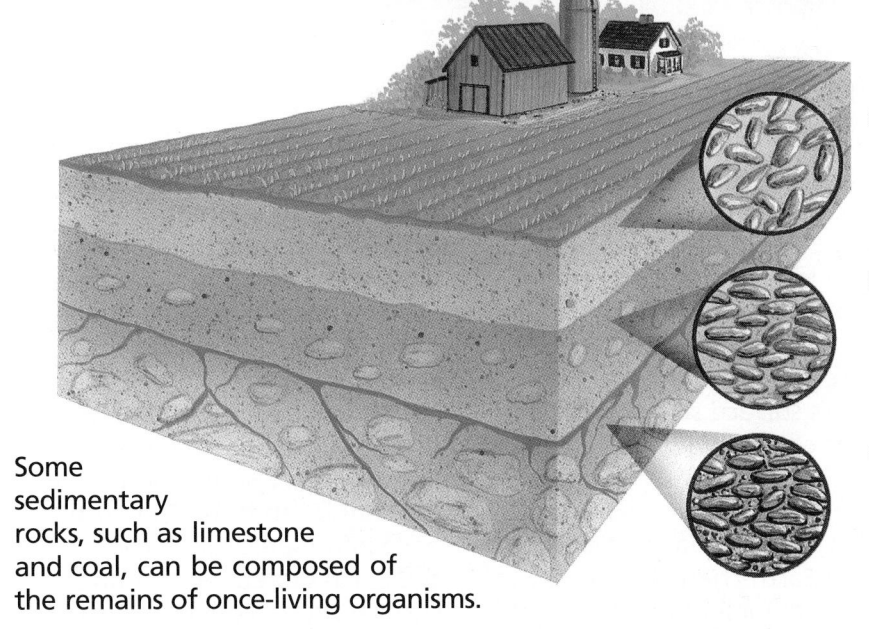

1. Rock fragments collect to form thick layers of sediment.

2. Pressure from the weight of the overlying sediments presses rock grains close together, causing them to stick to each other.

3. Ground water dissolves some of the sediment's minerals. This material is redeposited between the rock grains, forming a cement that binds the rock fragments together.

Some sedimentary rocks, such as limestone and coal, can be composed of the remains of once-living organisms.

Metamorphic Rocks   Suppose a layer of sedimentary rock, for example, is subjected to heat and pressure. Heat and pressure cause the rock to change physically and chemically. What results is a rock that is different in mineral composition and appearance from the original rock. Rocks that form in this way are called **metamorphic rocks**. An example of a metamorphic rock is marble. Marble forms when limestone undergoes *metamorphism*.

The Rock Cycle

Each kind of rock can be changed over time into any other kind of rock. For instance, pieces of igneous rock worn away from a volcano can be deposited elsewhere and may, over time, form sedimentary rock. Earth processes may force this newly formed rock deep underground, where heat and pressure change it into metamorphic rock. Then, if the heat and pressure continue to increase, the metamorphic rock will eventually melt to form magma, which may once again cool to form igneous rock. These processes together are called the **rock cycle**. The diagram of the rock cycle shown below illustrates the various ways in which rocks can be transformed. The rock cycle occurs very slowly. Many millions of years may be required for rocks to go through even one part of the cycle.

Sedimentary rocks

Rocks formed from pieces of other rocks, by evaporation, or the remains of organisms

Metamorphic rocks

Rocks that have undergone changes as a result of intense heat and pressure

Rock cycle

The processes by which one type of rock is transformed into another type of rock

◄ Any type of rock may be changed into other forms by going through the rock cycle.

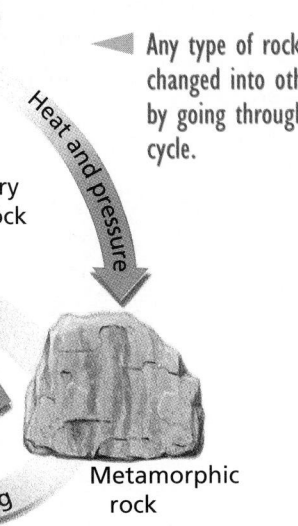

Pressure and cementation

Sediment

Weathering and erosion

Weathering

Weathering/Erosion

Melting

and erosion

Heat and pressure

Heat and pressure

Melting

Sedimentary rock

Cooling

Magma

Melting

Igneous rock

Metamorphic rock

S181

Although all three rock types are found on the Earth's surface, most of the Earth's crust consists of igneous rock. We don't often see these igneous rocks, however, because most of the continents are covered by a relatively thin layer of sedimentary rock. Only in a few places have Earth processes brought the "basement rocks" to the surface where we can see them.

Tectonic plates

Slabs of relatively thin rock that make up the Earth's crust

Building the Earth's Surface

The same Earth processes that drive the rock cycle continuously reshape the Earth's surface. Because of these processes, the Earth looks quite different today than it did a hundred million years ago. What kinds of processes could change the Earth's surface so dramatically? What causes mountains to form? Questions such as these puzzled researchers for many years. The mechanisms behind these Earth-changing processes were not well understood until fairly recently.

For centuries, people had noticed the way that the coasts of South America and Africa have similar outlines, like pieces of a jigsaw puzzle. This led to a revolutionary idea—the two continents must once have been joined together. But how might these two giant landmasses have come to be separated by thousands of kilometers of ocean? Did some unimaginably violent catastrophe take place? Did the continents somehow push their way through the solid rock of the ocean floor?

As you learned earlier, the interior of the Earth is quite hot. Some of it is also *plastic*, meaning that it can flow. The difference in temperature between the Earth's deep interior and its surface causes convection currents to form in this hot, partially plastic material. At the same time, the outer part of the Earth consists of relatively thin slabs of solid rock that ride like rafts on the dense, partially molten rock of the Earth's interior. These slabs of crustal material are called **tectonic plates**.

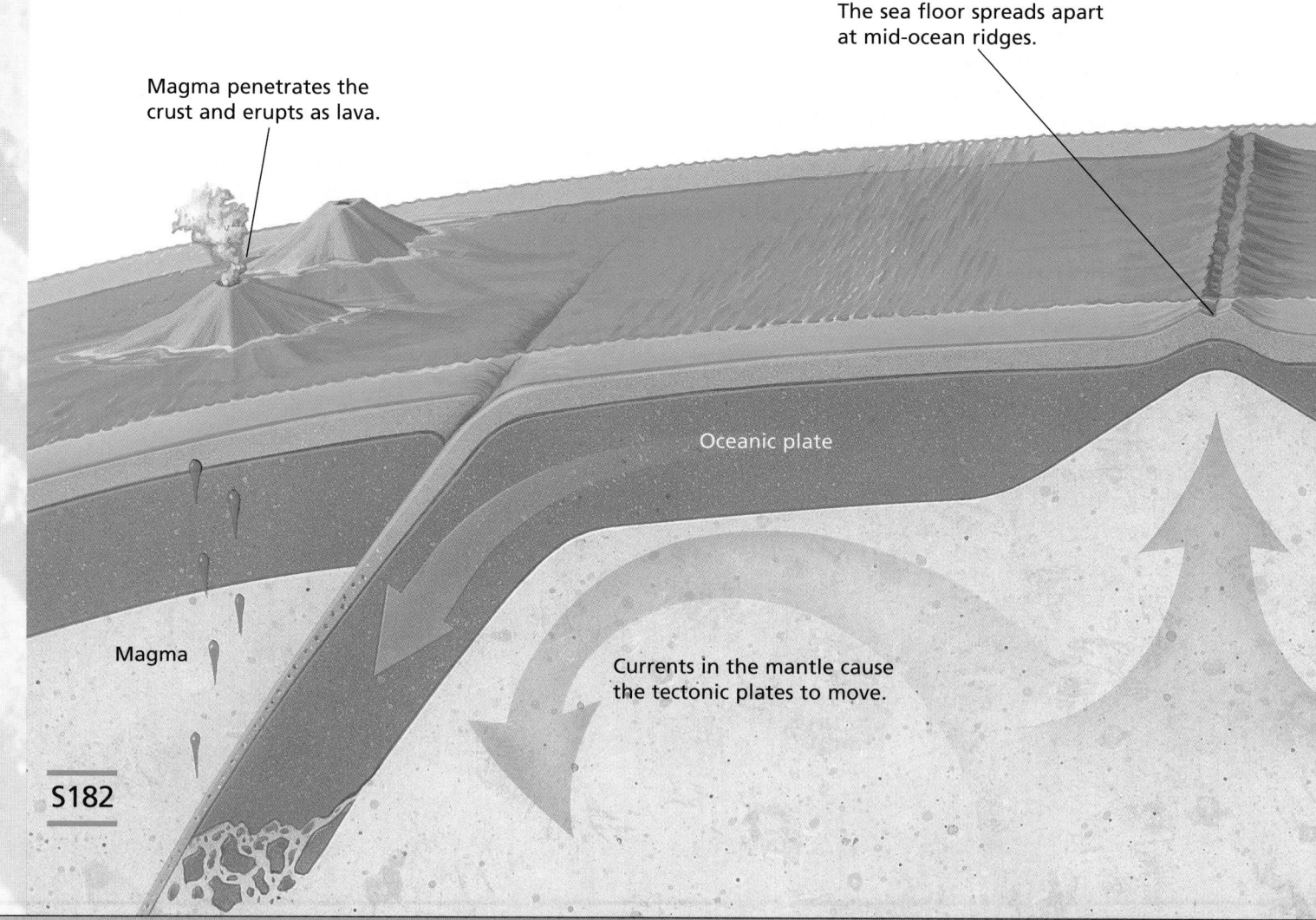

The sea floor spreads apart at mid-ocean ridges.

Magma penetrates the crust and erupts as lava.

Oceanic plate

Magma

Currents in the mantle cause the tectonic plates to move.

Tectonic plates are moved about by the slow churning of the Earth's plastic interior in somewhat the same way that a cork is moved about on the surface of a pot of simmering water. Sometimes plates bump into each other. Sometimes they move away from each other. Sometimes the plates slide past each other. Sometimes a plate is even split in two! The movement and interaction of the tectonic plates continuously reshape the Earth's surface. Of course the process happens very slowly, over millions of years. The plates move only a few centimeters per year on average.

Tectonic forces drive the rock cycle by causing massive changes to the Earth's crust. See if you can spot the changes taking place in the illustration below.

Types of Tectonic Collisions

There are three basic types of tectonic-plate collisions: an oceanic plate and a continental plate can collide, two continental plates can collide, or two oceanic plates can collide. When a plate made of oceanic crust collides with a plate made of continental crust, the oceanic crust (composed of relatively dense basalt) sinks underneath the continental crust (composed of relatively light granite). The sea floor is moved deep into the Earth and eventually melts. The melted sea floor then becomes the "fuel" for volcanoes on land. Similarly, when oceanic plates collide, one sinks beneath the other and a special type of volcanic feature called an *island arc* forms. New Zealand and

Japan are examples of island arcs. As one plate sinks beneath the other, it forms a deep, narrow valley on the ocean floor called a *trench*. Trenches, which can be more than 8 km deep, represent the deepest places on Earth and mark the boundary along which one plate sinks beneath the other.

When two continental plates collide, neither can sink beneath the other, so the force of the collision is spent entirely in buckling the crust. This kind of collision builds huge mountain ranges.

DID YOU KNOW...

that the Himalayas (the highest mountains on Earth) formed from the collision of the Asian and Indian landmasses?
The Himalayas have grown to their current form over the last 40 million years. In fact, they are still growing!

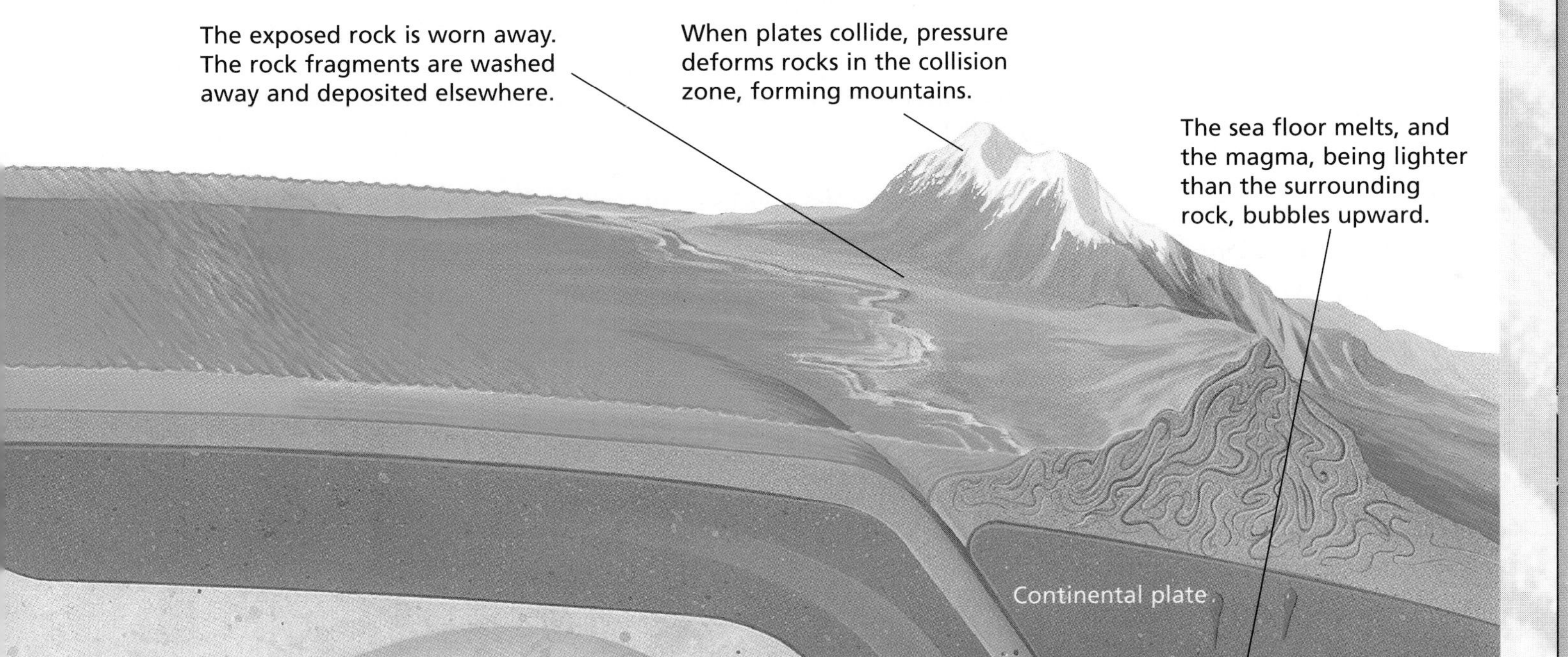

The exposed rock is worn away. The rock fragments are washed away and deposited elsewhere.

When plates collide, pressure deforms rocks in the collision zone, forming mountains.

The sea floor melts, and the magma, being lighter than the surrounding rock, bubbles upward.

Continental plate

The sea floor is moved under the continental plate.

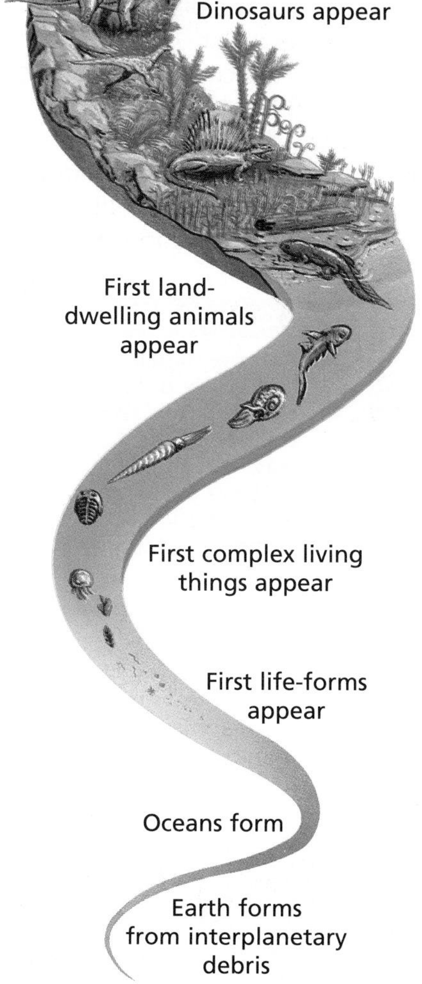

Ice age begins

Dinosaurs
die off

Humans appear

Dinosaurs appear

First land-
dwelling animals
appear

First complex living
things appear

First life-forms
appear

Oceans form

Earth forms
from interplanetary
debris

Geologic Time

Because the Earth's surface changes so little during one lifetime, many people used to think that the Earth had always looked more or less as it does now. But scientific discoveries over the last 200 years have changed that view. Scientists learned how to unlock the secrets of the Earth's past by interpreting the clues left behind in rocks. The evidence in rocks indicates that the Earth is very old and that its surface has changed continuously throughout its history. We don't notice much change because the human life-span is simply too short; Earth processes operate very slowly.

How long is a long time? a year? a hundred years? a thousand? Compared to the age of the Earth, such periods of time are like the blink of an eye. Even a million years—13,000 human lifetimes—is an insignificant span. Consider that what the rocks tell us has happened in just the last 200 million years—about 4 percent of the Earth's history. Continents have split apart and traveled thousands of kilometers. Mountain ranges have risen up and been worn away. The Atlantic Ocean has formed and the Pacific Ocean has gotten smaller. Dinosaurs have come and gone. Mammals have evolved and spread across the Earth.

Geologists use a special kind of "calendar," called the **geologic time scale**, to describe the Earth's history over time. The geologic time scale encompasses 4.5 billion years of Earth's history. The accompanying illustration shows some of the major events of geologic time.

Geologic time scale

The calendar used by scientists to describe events in the Earth's history

S184

S U M M A R Y

The Earth is a planet with a solid surface composed of rock. Three types of rock—igneous, sedimentary, and metamorphic—make up the Earth's surface. Each rock type forms in a different way, but each can be transformed into any other rock type by the processes of the rock cycle. The forces that drive the rock cycle also rearrange the continents, create and destroy ocean basins, and build the Earth's landforms. The Earth is very old and has been reshaped by natural processes continuously throughout geologic time.

WEATHERING

A̲t the same time that Earth's tectonic processes build land-forms, other natural processes tear them down. No matter where you look on the Earth's surface, you see the results of a constant battle between these two types of forces—forces that build up the land and forces that wear down the land.

As you learned in the previous section, movements of the crust determine the basic shapes of the continents and their location on Earth. Such movements produce forces that build mountains by breaking and folding the Earth's crust.

Mountain-building processes are opposed by *weathering* and *erosion*. Weathering is the process by which rocks are broken down, and erosion is the process by which the products of weathering are carried away. These processes work together to break down solid rock and carry it away, thus reshaping the surface of the continents. After they are carried off, the broken rock particles are dropped elsewhere in a process called *deposition*.

Weathering includes all of the processes that break down solid rock. Rocks almost always form deep underground, where they are not exposed to weathering processes. Rocks can be weathered only when they are exposed at the Earth's surface.

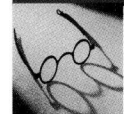

DID YOU KNOW...

the **Appalachian Mountains were once as grand as the Himalayas?** But 200 million years of weathering have reduced them to a fraction of their former size. In some places, the Appalachian Mountains are now little more than hills.

Weathering

All of the processes that break rock into smaller pieces

▼ Over long periods of time, weathering turns youthful mountains like these . . .

▲ . . . into mature mountains like these.

S185

Physical Weathering

Physical weathering

The ways in which rock breaks up into smaller pieces without any chemical change in the materials of the rock

There are a number of ways by which rocks are weathered. One type of weathering is called physical weathering. **Physical weathering** does not cause any change in the composition of the rock. It simply breaks the rock mechanically into smaller pieces. Some of the ways in which physical weathering occurs are discussed below.

Frost Action One type of physical weathering is *frost action*. Frost action takes place when water enters cracks in rock and then expands upon freezing. Frost action is a major form of weathering in temperate and cold climates.

Root Action In vegetated areas, plants can be important agents of physical weathering. The roots of plants work their way into cracks in the rock, enlarging them.

Thermal Expansion On a sunny day, rocks can absorb a great deal of heat. As a result, they get quite hot and expand. At night the heat radiates away and the rock grows cool again and shrinks. This repeated heating and cooling can break the rock apart by creating stress within it.

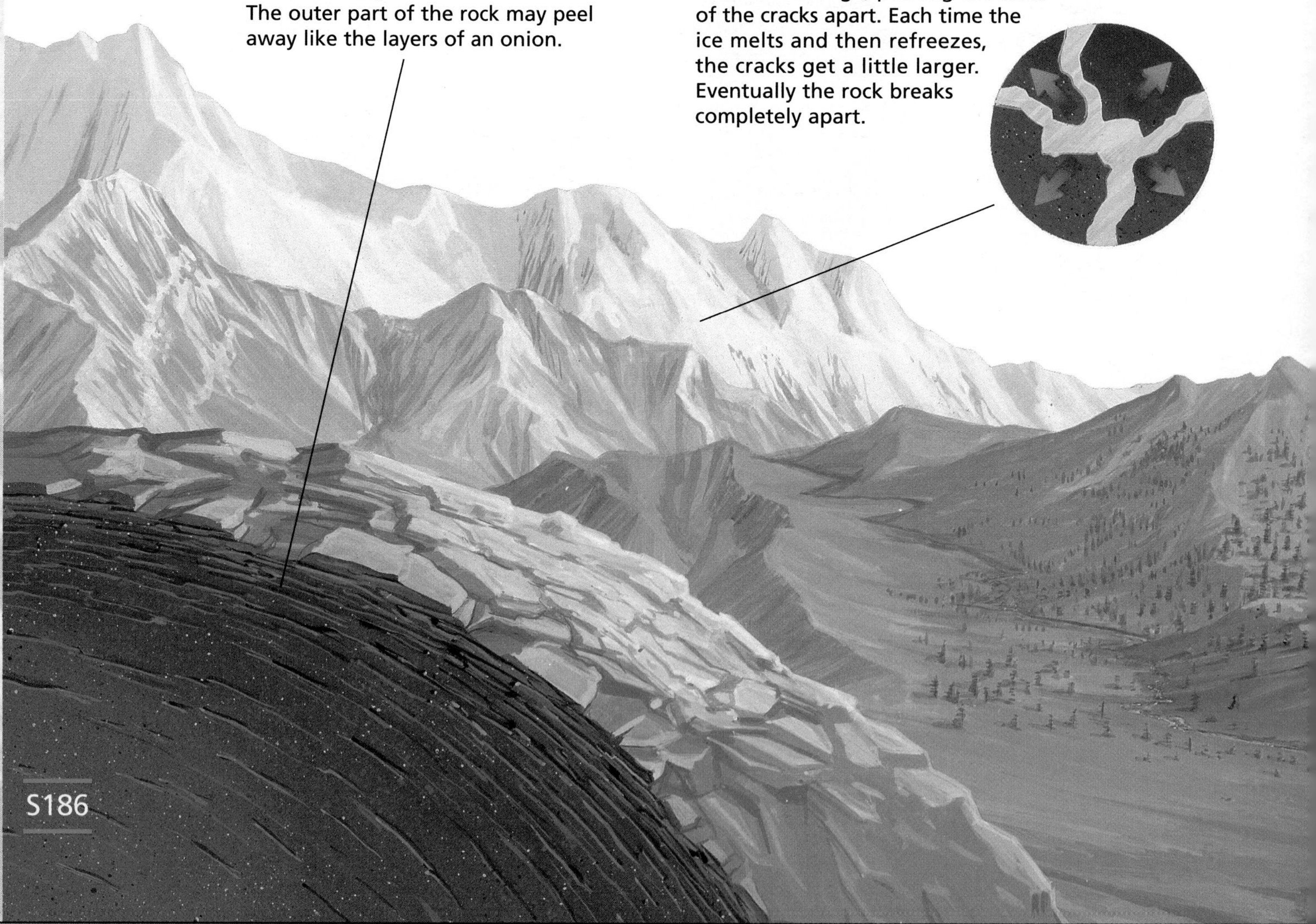

Rebounding When a deeply buried rock is uncovered by erosion, the release of pressure causes the rock to crack parallel to its surface. The outer part of the rock may peel away like the layers of an onion.

Frost Action Water expands when it freezes and, by doing so, produces great pressure. The expanding ice acts like a wedge, pushing the sides of the cracks apart. Each time the ice melts and then refreezes, the cracks get a little larger. Eventually the rock breaks completely apart.

Rebounding   The stage is set for another kind of physical weathering when a mass of rock is buried deep underground. Rock deep beneath the surface is under great pressure from the weight of many kilometers of overlying rock. If the overlying rock wears away, this pressure is reduced, and the exposed rock *rebounds*, or expands, causing it to crack. The outer layers of the rock may peel away like the layers of an onion in a process called *exfoliation*. Exfoliation also results from frost action and the repeated expansion and contraction of rocks as they are heated and cooled. Weathering by exfoliation causes rocks to take on a rounded shape as the layers peel off.

Abrasion   When rocks come into contact with other rocks, such as when they are carried by fast-moving water or blasted by wind-blown sand, they tend to wear away. This process breaks large, jagged rocks into smaller, rounder rocks. The longer that rock fragments undergo this process, called *abrasion*, the smaller and smoother they become.

▲ Half Dome in Yosemite National Park is an example of a large dome of rock produced by exfoliation. (The vertical face was sheared off by glaciers.)

Root Action   Roots from plants of all sizes wedge their way into small cracks in the rock. As the roots grow in length and diameter, they apply constant pressure, enlarging cracks and eventually breaking apart the rock.

Abrasion   Rocks bump into each other when they fall or are carried by streams. The force of the impacts breaks the rock into smaller, less jagged pieces.

Thermal Expansion   Rocks expand when heated and contract when cooled. Rock is made of a mixture of different minerals. Each mineral has its own rate of *thermal expansion.* When heated or cooled, these different minerals in a rock expand and contract by different amounts, stressing the rock. Over time, the repeated stress can cause the rock to crack or even break apart. This cracking aids further weathering by frost or root action.

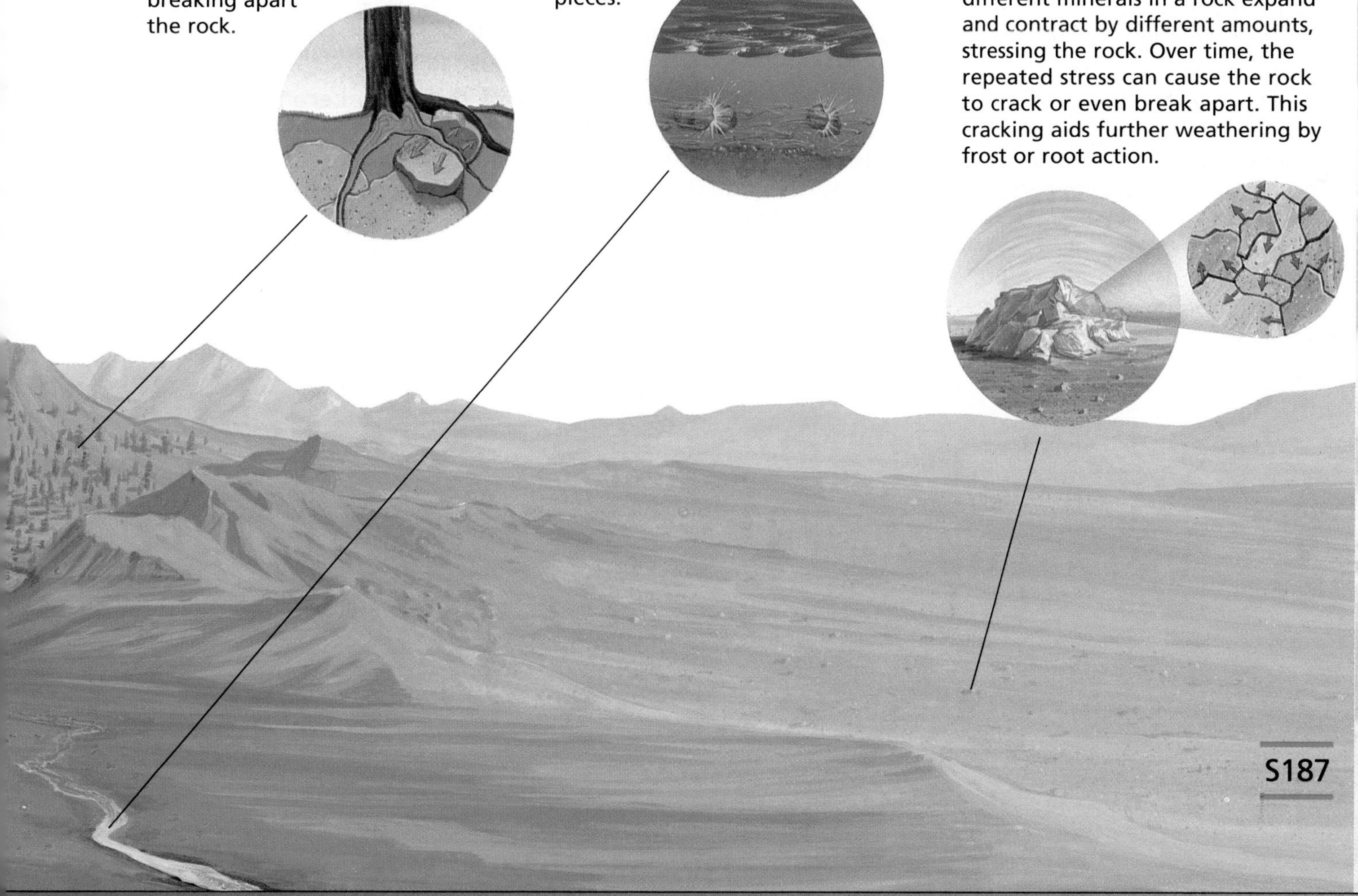

Chemical Weathering

Some weathering processes also cause rocks to change chemically. **Chemical weathering** changes the composition of rock through chemical action. Chemical weathering takes place when minerals in rock are either removed or chemically altered by water, gases in the air, or other substances. Let's look at some of the agents of chemical weathering.

Water There are three ways in which water chemically weathers rocks. Many minerals in rocks can be dissolved in and carried away by water washing over them. Often, the "cement" that holds a sedimentary rock together is dissolved by water. Once this cement is removed, the rock crumbles into the pieces from which it was made.

Some minerals that make up rock are able to absorb water much like a sponge does. This absorbed water increases the size of the mineral grains, straining and weakening the rock. The weakened rock is then more easily broken apart.

In a process called *hydration*, water can combine with certain minerals to form new compounds. For instance, micas and feldspars, which are major mineral components of igneous and metamorphic rocks, are changed

This spectacular landscape in southern China was formed by the action of ground water on limestone. Over time, ground water dissolved gigantic caverns in what was once a continuous layer of limestone. Eventually the cavern ceilings collapsed, leaving former columns as towers.

into clays by hydration. The clays are easily removed by physical weathering.

Oxygen A small amount of oxygen from the air or in water can join with some minerals in rock. These minerals are then chemically changed into new substances through the process of *oxidation*. For example, many minerals contain iron. Oxygen can cause the iron in these compounds to turn to rust. You can often identify an iron-containing

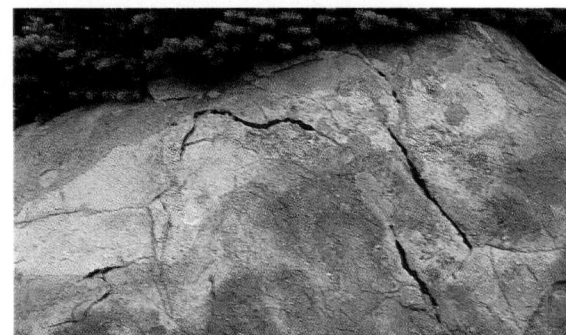

The stains on this rock were caused by the oxidation of its iron-containing minerals.

rock because it becomes streaked with red and yellow coloration due to the chemical weathering of oxidation.

Carbonic Acid Carbon dioxide in the air reacts with water to form a weak acid called *carbonic acid*. This acid can dissolve

Chemical weathering

The ways in which rock breaks down by chemically changing some of the materials in the rock

This underground cavern is a result of chemical weathering.

certain minerals such as calcium carbonate, which makes up limestone. Limestone caves and sinkholes form as a result of this kind of weathering. Sometimes the effect of this kind of weathering forms amazing natural structures.

Living Organisms Chemical weathering can also be caused by acids from living organisms. Lichens, which consist of fungi and algae, grow on rock surfaces and produce an acid that dissolves some of the materials in the rock. Lichens are often called "pioneer plants" because they are one of the first types of living things to grow in a rocky area. By breaking up rock, they make the ground suitable for plant life.

The Formation of Soil

Physical and chemical weathering work in combination to break down rocks. One of the end products of such weathering is called *soil*. Soil contains rock particles and other materials, which form a mixture that supports plant life. The type of parent rock being weathered determines what type of particles a soil contains.

Soil contains pieces of weathered rock in various sizes. The larger pieces of rock are called *gravel*, smaller grains are called *sand*, and the smallest pieces are called *silt* and *clay*. Silt particles can be as fine as flour. Clay particles are even smaller.

Soil ordinarily contains large amounts of an organic material called *humus*. Humus is made up of the decaying remains of dead plants and animals. Humus is a very important part of the soil and is one of the main sources of the nutrients needed for the growth of new plants. Soil without humus is not *fertile*, and therefore cannot support much plant life. Soils differ in the amount of humus they contain. For example, the soil in swamps may be made almost entirely of humus. The soil in deserts, however, may contain little, if any, humus.

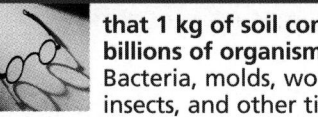
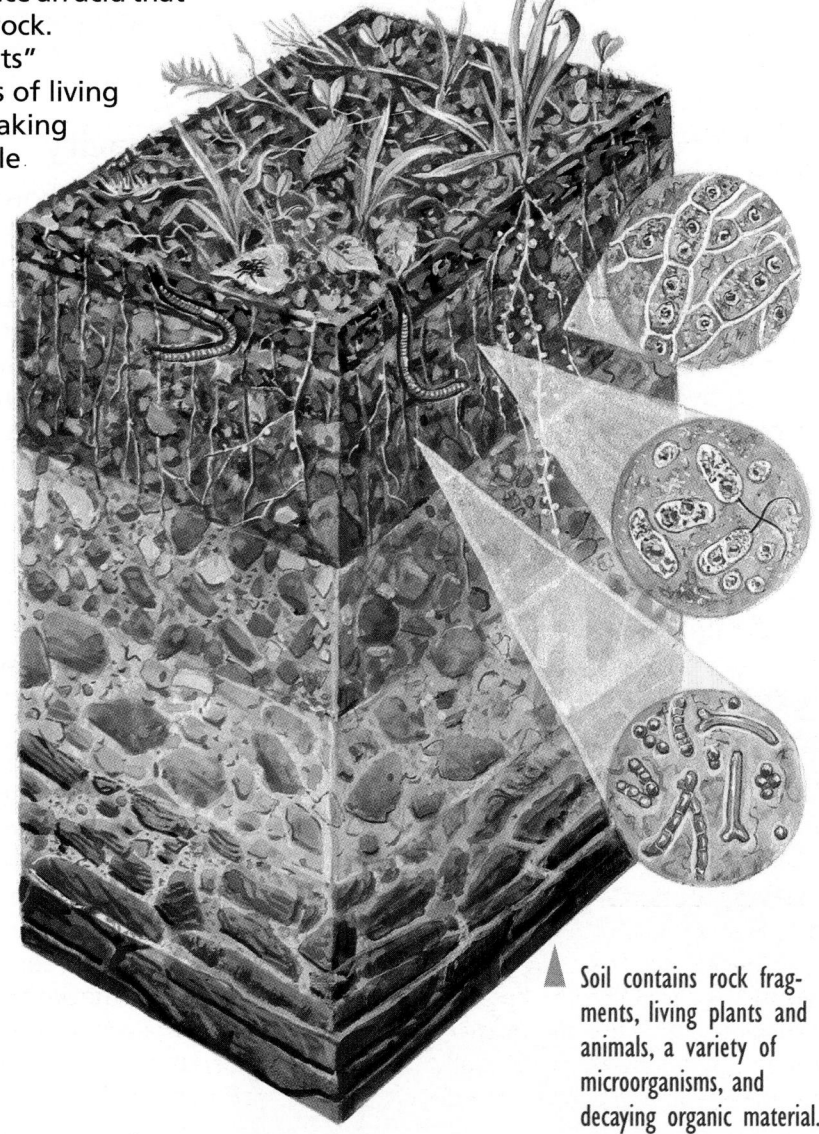

▲ Soil contains rock fragments, living plants and animals, a variety of microorganisms, and decaying organic material.

SUMMARY

The Earth's land surfaces are the result of two opposing forces. Movements of the crust raise the land to form features such as mountains. At the same time, weathering tears down the land. Rocks are broken down into smaller pieces by physical and chemical weathering. Soil is formed from small pieces of weathered rock and the decayed remains of plants and animals.

S189

EROSION

W hat happens to all of the particles of rock that are formed by the processes of weathering? Once rock has been broken down by weathering, the rock particles and soil can be moved by several means: gravity, streams, glaciers, wind, or waves. Everything that happens to cause pieces of rock or soil to be carried away is called **erosion**. Erosion is usually a slow, gradual process that plays a major role in shaping the Earth's landforms.

Erosion

All the processes that carry away soil and weathered rock

Erosion by Gravity

Look at the photo at the bottom of the page. What role did gravity play in making this scene? You probably noticed that there is a pile of rubble at the base of the cliff. Piles of rubble such as this that form at the base of cliffs or steep slopes are called *talus*. Talus forms as bits of rock break loose every now and then and tumble down the slope, pulled by gravity.

Talus normally forms slowly, but gravity can also be an agent of rapid change, such as when a landslide occurs. During a landslide, large amounts of rock and soil break loose from a cliff or slope and rush downhill. A large landslide can move millions of tons of rock in seconds. Landslides are very dangerous because they can occur without warning. Entire towns have been buried by landslides. Earthquakes often cause landslides, as does excess moisture from heavy rains or melting snow.

Masses of weathered rock and soil can also move very slowly, as little as 1 or 2 cm a year. The slow downhill movement of loose rock or soil is called *creep*. It is often hard to tell that creep is taking place because there is usually no obvious change in the way the slope looks. However, careful observation of trees, telephone poles, or fence posts can show evidence of creep. Creep is a very common kind of erosion caused by gravity. It is the main way in which weathered rock moves downhill.

Sometimes part of a slope may slide down more or less in one piece called a *slump*. A slump usually takes place where the slope of the land has been changed due to erosion. For example, the banks of streams often slump into the eroded stream channel. Slumping is also seen along cliffs at the shore where waves have cut into the base of the cliff.

▲ Landslides can occur when heavy rains or earthquakes loosen material on a steep slope.

A pile of talus has ▶ collected at the bottom of these cliffs.

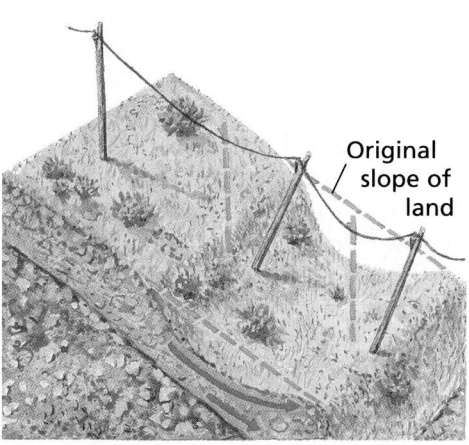

▲ Over time, the uppermost layers of the soil creep downhill.

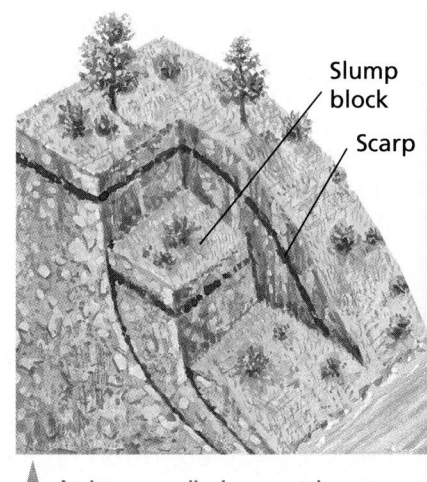

Slump block

Scarp

▲ A slump usually leaves a characteristically curved surface.

Erosion by Streams

Flowing water shapes the Earth's surface by moving soil and rocks of all sizes, even huge boulders. Stream erosion is responsible for many familiar landforms, such as valleys, hills, canyons, mesas, and buttes. Examine the photograph below. If you took a picture at this spot a few million years ago, it would not show the gorge you see now. At one time the San Juan River meandered across a flat landscape. Then tectonic forces pushed the land upward. As the land rose, the river eroded downward, gouging a canyon that got deeper and deeper with the passing of time. After a few million years

Gooseneck Canyon was formed. Another example of stream erosion is the Grand Canyon, carved by the Colorado River. The Grand Canyon, which is about 1.6 km deep and up to 29 km wide, formed at the same time and in the same way as Gooseneck Canyon. In fact, the two canyons are not very far apart. But the greater erosive power of the Colorado River enabled it to dig much deeper into the land. Features such as Gooseneck Canyon and the Grand Canyon show the erosive power of flowing water.

Flowing water erodes the land surface in two basic ways.

One way is when soil and rock materials are dissolved in water and carried away. In most cases, this is a fairly slow process because not much material can dissolve in the water.

Most erosion by streams is done in the second way, when particles of rock and soil are carried away by the water. During times of normal flow, a stream may not carry much rock. But during floods, the power of the water is greatly multiplied. The current tears at loose rock, breaking it off and carrying it away. Huge amounts of small rock particles are picked up and carried, suspended in the current. Larger rocks bounce along the stream bed, pushed by the current. In a large flood, boulders weighing several tons may be rolled along like tumbleweeds. The rocks pound against the stream bed with sledgehammer force, breaking loose other rocks, which are also scooped up by the current and carried along. A single major flood can cause more erosion than centuries of normal flow.

◄ Gooseneck Canyon, on the San Juan River in Utah

S191

From Runoff to Rivers

Erosion by flowing water begins with raindrops that fall on an exposed surface. During heavy rainfall, soil cannot absorb all of the rain as it falls. Therefore, only some of the rain sinks into the soil, while the rest moves over the surface. Water that moves over the surface of the land is called **runoff**. Erosion from runoff takes place most rapidly where the soil is loosely packed or where the slope is steep.

Runoff may come from melting snow and ice as well as from rain. It begins as a shallow layer, or sheet, of water flowing downhill. Pebbles and rocks on the surface quickly break the sheet into tiny streamlets. These streamlets come together to form larger streams, which join together to form still larger streams. In this way, rivers are formed. Streams that flow into a river are called *tributaries*. The area from which a river and its tributaries collect runoff is the river's **watershed**, sometimes called a *drainage basin*.

The flow of a stream or river varies with the amount of runoff. When the amount of runoff is greater than usual, a flood may result. Heavy rain or fast-melting snow can cause floods. Most rivers have a small flood once every year or two; major floods occur less often. Great floods, such as the Mississippi River flood of 1993, occur once or twice every century. Because of their relative rarity, people are often unprepared for great floods. As a result, these events often take lives and cause millions of dollars worth of property damage.

Rate of Erosion   Rivers are often described as being either young, mature, or old. Young rivers flow swiftly down relatively steep slopes and through deep, narrow, V-shaped canyons. They have rocky beds and may have many rapids and waterfalls. Young rivers are very energetic and rapidly erode rock. As the river erodes more and more of the land it flows over, it flattens the slope of its course and is said to age. Old rivers are the end result of the erosive process. Old rivers flow very slowly across broad expanses of nearly level land. These rivers have eroded nearly all that they can erode. Old rivers typically flow across thicknesses of sediments previously deposited by the river. Their channels are wide and flat, and they *meander*, or travel a winding, constantly changing path. Erosion by old rivers is slow to nonexistent. They primarily distribute the materials that are brought into them by their younger tributaries. Mature rivers are intermediate between young and old. They have eroded much of their surroundings but have not yet worn them flat, and they have not yet started to meander.

Runoff

Water flowing over the surface of the land

Watershed

The land area drained by a river and its tributaries

Young rivers typically have rocky, V-shaped valleys.

Old rivers meander across nearly flat floodplains.

S192

Erosion by Glaciers

Most snow that falls on land eventually melts to form runoff, which returns to the sea in rivers. There are places on Earth, however, where it is too cold for all of the snow to melt. As the snow falls year after year, it piles up. The weight of the accumulating snow squeezes the snowflakes together to form ice. If enough snow and ice accumulate, the mass begins to flow downhill due to gravity. Such a mass is called a *glacier*. Glaciers can cause tremendous amounts of erosion.

Types of Glaciers   Some of the greatest rivers on Earth are made of solid ice. Like huge plows, glaciers carve the rock over which they flow. There are two basic types of glaciers: valley glaciers and continental glaciers.

   Valley glaciers are found in mountainous areas all over the world. They are literally rivers of ice flowing through valleys. Like rivers of water, valley glaciers have tributaries. Valley glaciers are fed by snow that falls throughout their watersheds.

▲ Valley glaciers are enormous rivers of ice.

Many small glaciers in high mountain valleys join together to form larger glaciers, which join together to form still larger glaciers.

Much snow falls in the high mountains. Low year-round temperatures ensure that little snow melts. Avalanches carry snow into the valley.

Crevasses form in the brittle ice near the surface of a glacier.

In a manner similar to streams, many small glaciers combine to form a large glacier.

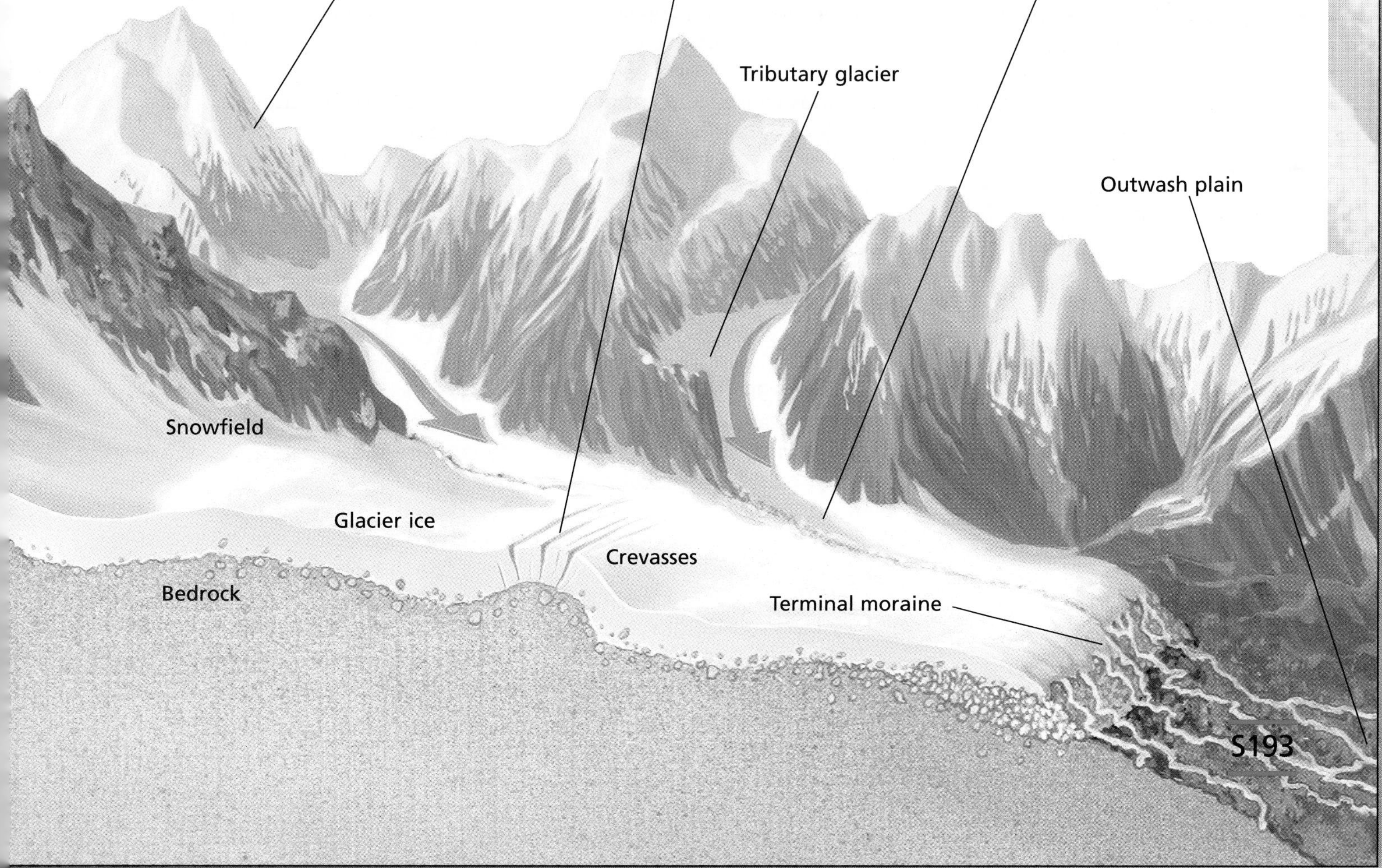

Tributary glacier

Outwash plain

Snowfield

Glacier ice

Crevasses

Bedrock

Terminal moraine

S193

▲ Glaciers carve a number of characteristic landforms.

Glaciers move because ice flows under pressure. The pressure is supplied by the glacier's enormous weight. Pulled by gravity, the glacier creeps downhill. Most glaciers flow quite slowly, less than a meter per day. Glaciers also flow unevenly; the middle flows faster than the sides and the top flows faster than the bottom. Because the ice near the surface is under

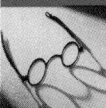
little pressure, it behaves like a brittle solid. As a result, the surface ice often breaks and forms numerous cracks called *crevasses*. When glaciers reach the sea, huge pieces fall off the cracked ice and float away as *icebergs*. Sometimes these icebergs are huge. In March 1995 an iceberg measuring 68 km by 35 km broke off Antarctica.

Continental glaciers are giant sheets of ice that may cover millions of square kilometers. Just 20,000 years ago, continental glaciers that were thousands of meters thick covered almost 30 percent of the Earth's land area, including much of North America. Such a period of extensive glaciation is called an *ice age*. Continental glaciers greatly affect the land they cover. The Great Lakes, for example, were carved from solid rock by continental glaciers.

Today, just two continental glaciers still remain, one in

Greenland and the other in Antarctica. Both Greenland and Antarctica are almost completely covered by ice, which is more than 3000 m thick in many places. The Antarctic ice sheet, in fact, contains 75 percent of the world's fresh water.

The difference between valley glaciers and continental glaciers is one of degree. Continental glaciers form whenever snowfall exceeds melting by a sufficient amount over large areas of the Earth. And if the climate grows colder or snowfall increases, areas that are now occupied by valley glaciers or are free of glaciers could become covered by continental glaciers.

Landforms Carved by Glaciers Glaciers are responsible for the appearance of much of the Earth's surface at high altitudes and latitudes. Glacial action produces very characteristic landforms.

S194

By itself, the tremendous grinding weight of the glacier has great erosive ability, but glaciers also drag along rocks and other debris as they flow. These scrape the underlying rock like the teeth of a file. As glaciers move, the rocks scratch and dig into the land, scraping away solid rock and often leaving deep scars. Given enough time, glaciers can eventually grind the land down to a nearly level surface. For example, much of Canada is bedrock that has been scraped clean and flat by continental glaciation over the last million years.

Glaciers that form on mountains greatly change the shape of those mountains. The changes begin in pockets, called *snowfields*, where snow and ice collect. Frost action breaks rock off mountainsides, and the pieces are carried along when the ice begins to move downhill, carving away more rock. By this process, the mountainside under a snowfield is changed into a rounded, steep-walled basin called a *cirque*. As a glacier grows, it moves down a valley, wearing away its floor and walls. In time, mountain valleys that were originally V-shaped from erosion by streams become U-shaped due to the erosive action of glaciers.

▲ What evidence suggests that these rocks were once beneath a glacier?

Erosion by Wind

Have you ever experienced a strong wind? Sometimes the wind is so strong that it can blow you over unless you lean against its force or hold onto something. It is the wind's energy that you feel. This energy is also able to erode rock and soil.

Like flowing water, the wind can carry pieces of rock. However, the wind moves only small rock particles like sand, silt, and clay. Its effect is greatest in dry, desert climates where many small particles are exposed to the wind. Particles in moist soils cling together, making it harder for the wind to pick them up. In deserts, there is little or no moisture in the soil and very few plants to cover and protect it from the wind. Therefore, the unprotected soil and sand is often blown away. Because of this, most deserts are covered by broken rock—a type of surface called *desert pavement*. Deserts are not, as many people think, always covered by sand dunes. Sand dunes occur only in limited areas in most deserts.

◄ Wind blowing across a desert . . .

▲ . . . removes loose sand and soil . . .

◄ . . . leaving a stony surface called desert pavement.

S195

In dry regions or during droughts, strong winds may cause huge "dust storms." During a large dust storm, 1 km³ of air may carry as much as 900,000 kg of rock particles. A storm covering a large area is able to move many billions of kilograms of rock particles. In the 1930s, decades of poor farming practices followed by several years of severe drought caused the Dust Bowl in the central United States. During the Dust Bowl, wind storms carried away billions of tons of soil and rock particles from Midwestern farms.

How Particles Move Larger particles, like sand, can be moved only short distances through the air. Sand particles are moved across beaches and deserts in a rolling and bouncing motion called *saltation*. By this motion, sand particles are lifted no higher than a couple of meters above the ground. Windblown sand causes weathering by abrasion when the sharp edges of sand grains scrape at and wear away objects with which they come in contact. Since windblown sand particles stay close to the ground, their greatest effect is on rocks at the surface.

Wind pulls individual sand grains into the air, where they are carried for a short distance. By this process, called *saltation*, the individual particles hop along the ground.

This unusual wind-carved formation stands in the desert of Utah

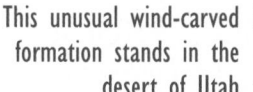

A sea stack off Rialto Beach, Washington

Erosion by Waves

Waves cause significant erosion. When waves reach a shoreline, they often pound against it. During storms, large waves strike the shore with great force. The constant pounding of waves breaks up rocks and washes them away.

Waves break up rocks in two ways. First, waves force water and air into cracks, applying pressure that forces the rock apart. Over time, large blocks of rock are broken up in this way. These rocks may be moved by further wave action and aid in the second method of wave erosion. Broken rocks cause additional erosion when they are smashed against the shore by the waves. The waves slosh the rock fragments around, grinding them against each other and breaking them into smaller and smaller pieces.

The effect that waves have on a shoreline depends on the force of the waves and the type of rock that makes up the shoreline. Interesting structures, such as the sea stacks shown above, form in places where unequal erosion occurs. Sea stacks form when the softer rock is worn away, leaving harder rock standing in columns. Sea cliffs form where waves continuously wear away soft rocks at the base of a slope. This undercuts the slope so that rocks above continuously tumble down. Waves may also hollow out soft rock in a sea cliff to form sea caves.

The White Cliffs of Dover, England, are some of the world's most famous sea cliffs. These cliffs, composed of soft chalk, are eroding at a rate of about 1 m per year.

S U M M A R Y

Gravity drives a number of forms of erosion, some slow (such as creep) and some rapid (such as landslides). Rivers and streams cause erosion by carrying away rock and soil. As rivers wear away more and more of the land that they flow over, they slow down. Glaciers are a major force of erosion in cold regions. Wind can cause erosion in dry regions by picking up small rock fragments. These moving particles may strike against and wear away solid rock. Waves erode shorelines by breaking up the rocks that make up the shore and washing away the fragments.

S197

DEPOSITION

Picture the shore of a river, lake, or ocean. Do you see beaches, sandbars, and dunes? All of these features can be traced back to solid rock that once existed but was broken down by weathering and whose pieces were washed away by erosion. Eventually, waves or currents carried the rock particles to the shore, where they were *deposited*. The **deposition** of eroded materials creates new and constantly changing landforms, helping to shape the continents.

Deposition

All of the processes by which eroded materials are deposited, or dropped

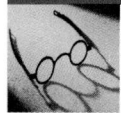
Stream Deposits

Streams and rivers deposit the rock and soil they carry at any place where their flow slows down. If a river encounters an area with less slope (change in elevation) or standing water, it slows down. This reduces the river's ability to transport materials, so some of the rock and soil settles to the river bottom as sediment. The most noticeable stream and river deposits are found at the base of mountains, at the edge of stream channels, and where rivers meet a lake or ocean.

Flooding can be a major cause of deposition in mature and old rivers. The main channels of such rivers are usually flanked by a wide, flat area called a flood plain. Flood plains form from layers of sediment deposited by many repeated floods over many centuries. Older rivers often have wide, flat flood plains that are commonly used for farming because of their rich, deep soil.

Streams and rivers eventually empty into a lake or ocean. When rivers meet a lake or an ocean, they slow quickly and form a wedge-shaped deposit of sediment called a *delta*. Depending on several factors—the amount of sediment carried by the river, the size of the tides, average wave height, currents, water depth, and so on—deltas may have a number of different forms.

The Mississippi River has deposited a classic "bird's foot" delta in the Gulf of Mexico, seen here from space. ▶

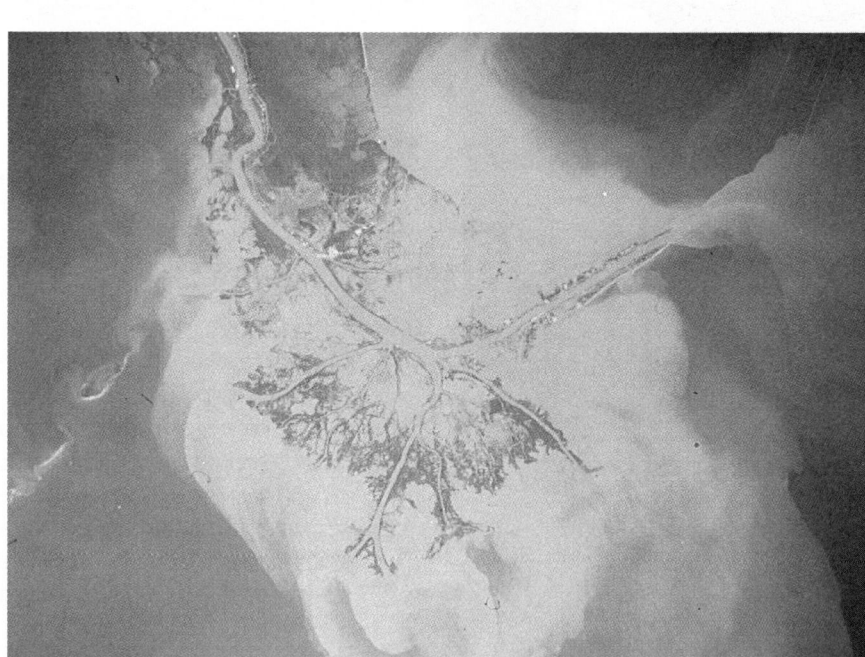

S198

Glacial Deposits

Moving glaciers pick up a large number of rock fragments. When the glacier stops advancing and melts, this material is deposited. Glaciers deposit two kinds of materials. One kind consists of the rocks, sand, and rock dust dropped by the melting ice. This material, called *till,* consists of a chaotic mix of rock fragments of all sizes. Mounds of till, called *moraines,* are deposited as ridges along the sides of a glacier or at its forward edge when it stops advancing.

A second type of material deposited by glaciers is left by streams of meltwater that flow from the glaciers. These materials are called *outwash deposits.* If the meltwater flows onto level ground, the material is deposited in wide, even sheets to form an *outwash plain.* Sometimes blocks of ice break off a melting glacier and become buried in outwash deposits. When these blocks melt, large water-filled holes, called *kettle lakes,* are left. The north-central part of the United States is dotted with tens of thousands of kettle lakes that formed during the ice age that ended about 10,000 years ago.

▲ Glacial meltwater typically deposits a thick blanket of eroded materials to form an outwash plain.

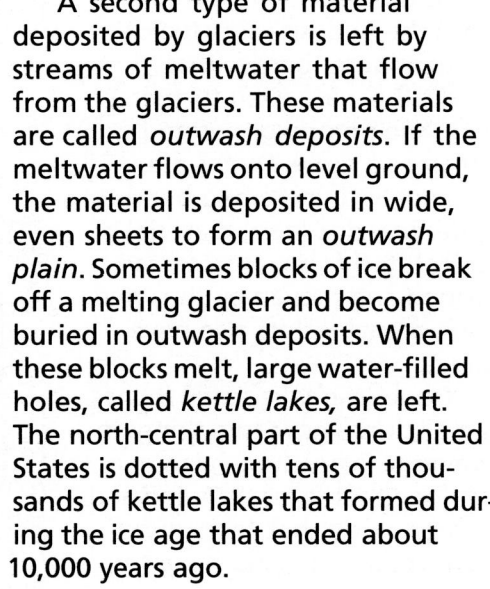

▲ These lakes, in Chippewa County, Wisconsin, were formed by glacial action.

Forms of Deposition

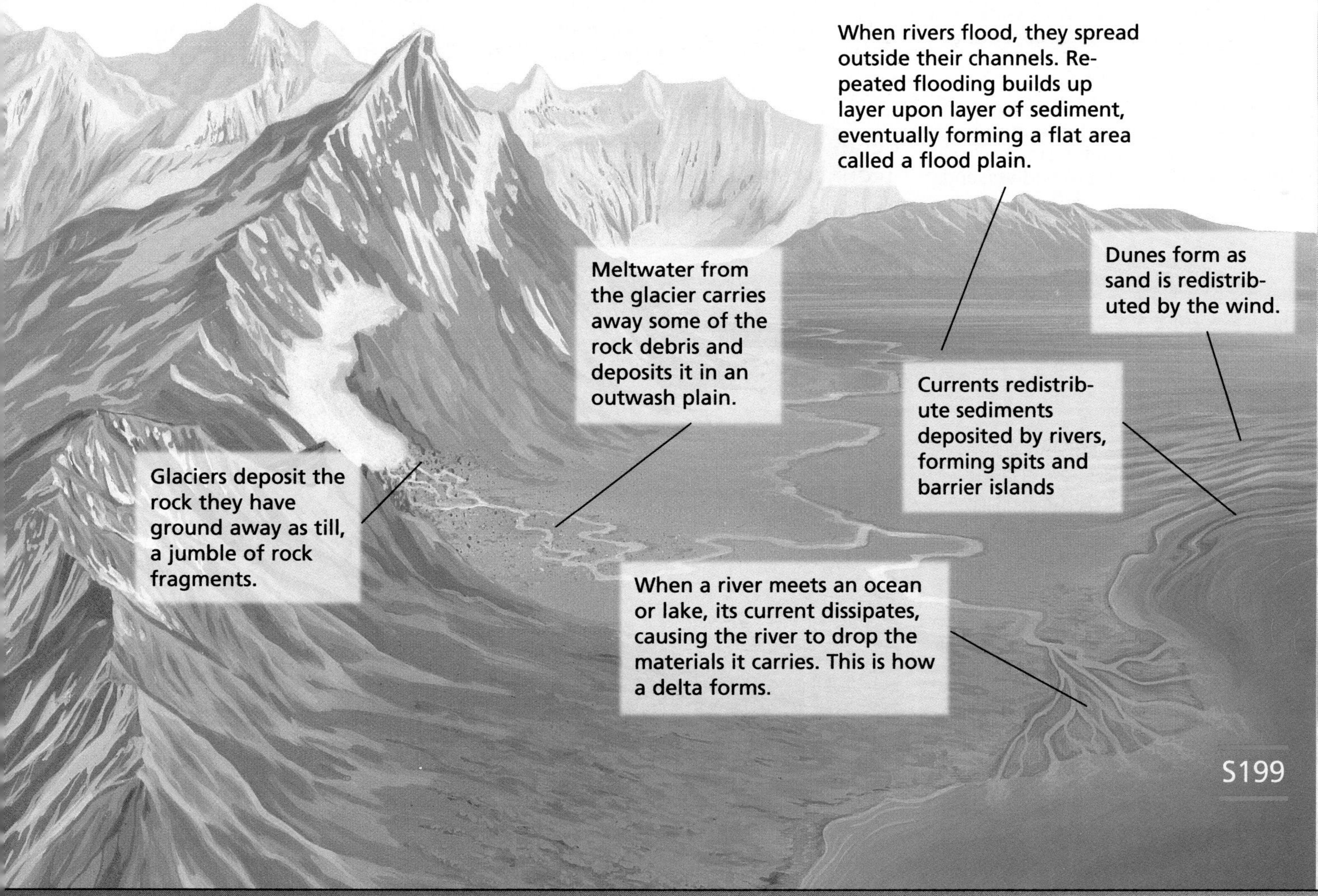

When rivers flood, they spread outside their channels. Repeated flooding builds up layer upon layer of sediment, eventually forming a flat area called a flood plain.

Meltwater from the glacier carries away some of the rock debris and deposits it in an outwash plain.

Dunes form as sand is redistributed by the wind.

Currents redistribute sediments deposited by rivers, forming spits and barrier islands

Glaciers deposit the rock they have ground away as till, a jumble of rock fragments.

When a river meets an ocean or lake, its current dissipates, causing the river to drop the materials it carries. This is how a delta forms.

S199

Wind Deposits

The sand, silt, and dust carried by the wind are deposited when the wind encounters a barrier and slows down. Objects such as bushes or large rocks often form such barriers, called *windbreaks*. A mound of rock particles often builds up around windbreaks. Once the deposit is started, it forms a larger barrier that blocks the wind even further. In time, the mound grows and becomes a *dune*. Dunes are commonly found along beaches. In deserts, dry, loose material is easily lifted from unprotected ground and moved by the wind. Thus, many deserts also have dunes.

Dunes can take many shapes, but all dunes have two things in common. Every dune has a gentle slope that faces the wind and a steep slope that faces away from the wind. Sand moves up

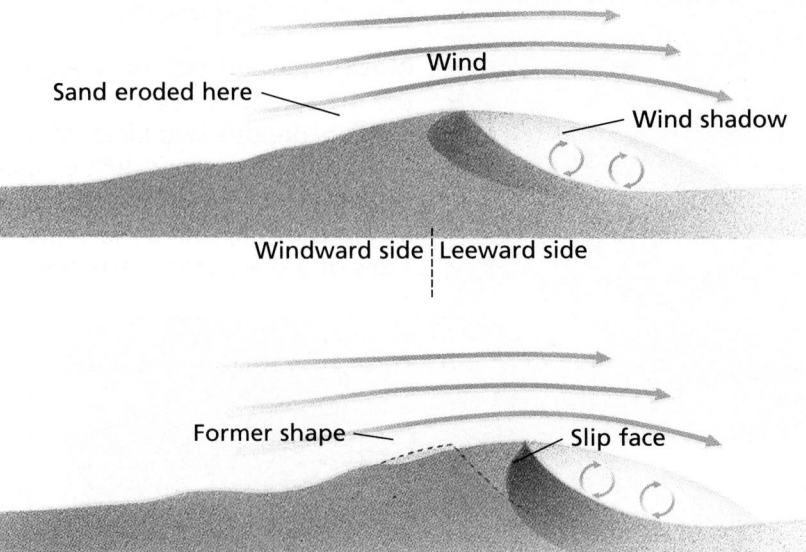

Sand eroded here · Wind · Wind shadow

Windward side ⁝ Leeward side

Former shape · Slip face

The gentle slope of the dune, which faces the wind, is constantly eroded by the wind. Deposition occurs mostly on the leeward side of the dune, which is steep due to constant slumping caused by sand buildup.

to the top of a dune by the rolling and bouncing motion of saltation and then slips down the other side. This causes dunes to *migrate,* or move slowly downwind. Dunes can move over roads, railroad tracks, farmland, and even buildings. Fences, vegetation, and walls are of little use in stopping dunes.

Not all of the particles carried by the wind are deposited in dunes. Smaller and lighter particles are carried farther than sand and soil. When this dust eventually comes to rest, it covers the land like a blanket. Thick layers of wind-blown dust are called *loess*. Deposits of loess are usually tan in color and can be up to 90 m thick, covering hills and valleys. Loess deposits are soft and easily eroded. Much of the Great Plains region of the United States is covered by loess.

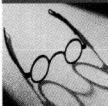

DID YOU KNOW...

the world's largest dune fields are found in the Sahara Desert? While only 20 percent of the desert is covered by sand dunes, the Sahara still contains almost 2 million square kilometers of endless sands.

Loess deposits, such as this one, cover much of the Great Plains. ▶

S200

▲ Waves striking the shallow water near shore dissipate some of their energy, putting the bottom sediments into motion and causing sand bars to form.

Deposition by Ocean Waves and Currents

Wave action can also deposit sediments. Waves, and the currents they produce, deposit sediment to form *beaches*. Beaches are deposits of sand and larger materials, such as pebbles and cobblestones, that are found along shorelines.

Beaches of pebbles and cobblestones occur along rocky coastlines where exposed cliffs are rapidly eroded by high-energy waves. Such beaches are often found in conjunction with sea caves, sea cliffs, and sea stacks. The Pacific Coast of the United States is made up almost entirely of these beaches.

Wide sandy beaches are found along gently sloping coastlines. Such coastlines are typically formed of soft, river-deposited sediments. The Gulf and Atlantic Coasts (south of New England) are made up of these beaches.

Elongated sand ridges, called *sandbars*, are usually found along sandy beaches. Sandbars are built up by wave action. They form in the shallow zone near the beach where the waves dissipate much of their energy. Depending on the slope of the beach and the average wave energy, there may be several sandbars, running generally parallel to the beach at various distances from it. Sandbars shift constantly due to tides and changes in wave action.

Because of the interaction of waves, winds, and tides, coasts that are low and sandy usually develop *barrier islands*. Barrier islands are long, narrow deposits of sand running parallel to the coastline that are similar to sandbars but much larger. Barrier islands may be over 100 km long but only 1–2 km wide. If you look at a map of the United States, you will see that much of the Gulf and Atlantic Coasts of the United States have barrier islands. Barrier islands, like other features of a sandy shoreline, are constantly changed due to the movement of sand by waves, winds, and currents.

▲ Barrier islands, which are found along most low, sandy coasts, are typically long and narrow.

S U M M A R Y

Streams and rivers deposit the rock and soil they erode when their currents slow down. Flood plains and deltas are examples of river-deposited sediments. When glaciers melt, they deposit materials. Deposits of sand and dust collect when wind slows down and drops the material it carries. Many familiar coastal features are formed by the action of waves and currents.

S201

THE UNDERSEA REALM

ou may think that because the ocean floor is covered with water, it is protected from the agents that cause change. This is not the case. Weathering, erosion, and deposition also occur underwater. However, the kinds of changes that occur on the ocean floor are different from those that occur on land. For example, there is no wind in the ocean, so there is no erosion by wind. Nor is there weathering from contact with air. But weathering, erosion, and deposition do take place in this watery environment.

The Undersea Landscape

The ocean floor is a world of its own, and like the land, the ocean floor has distinctive features. Some of these features are similar to those found on land and some are not. But just as on land, on the ocean floor you can find plains, volcanoes, canyons, and towering mountain ranges.

The ocean as it would appear ▶ if all of the water were drained away.

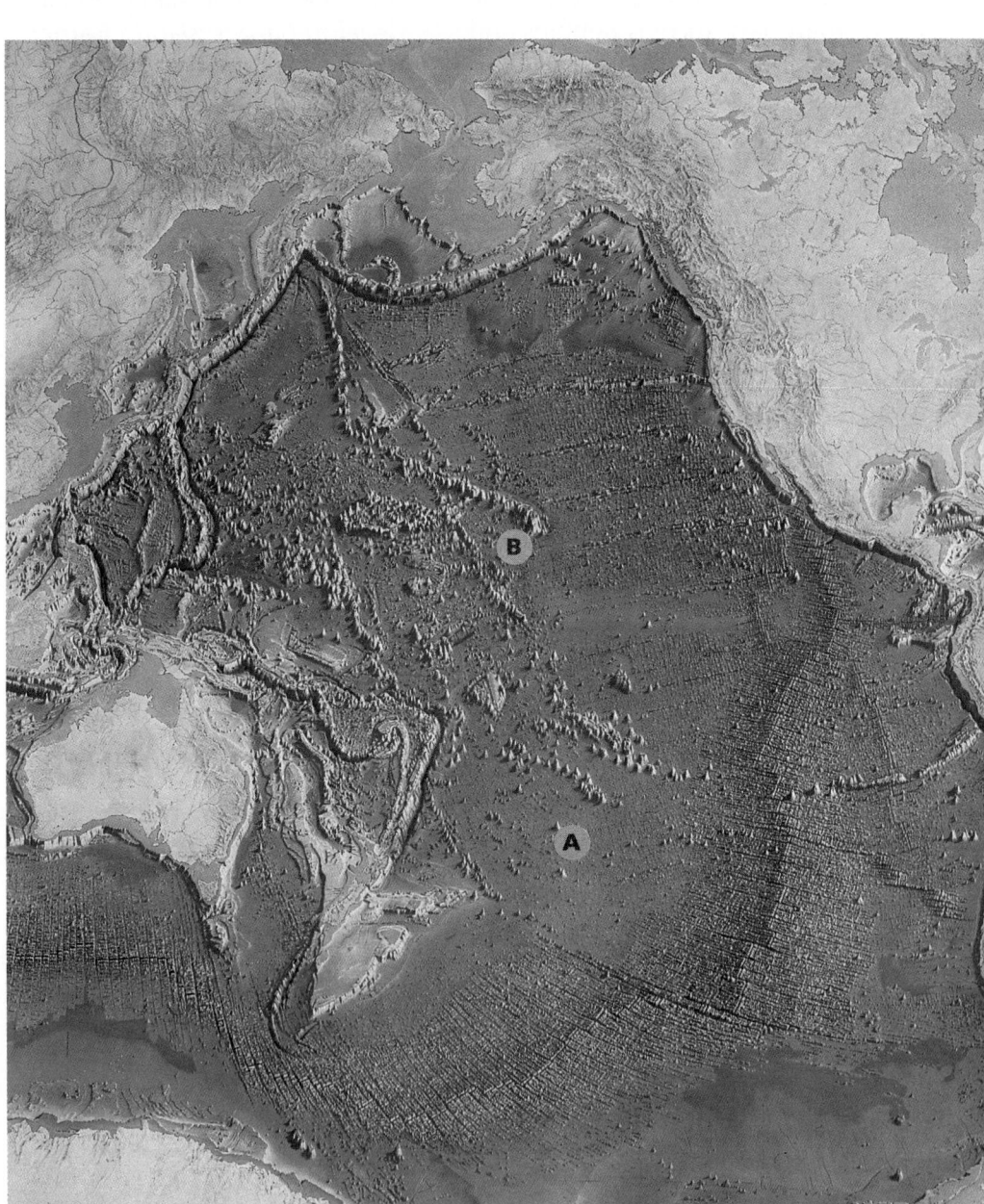

A Abyssal plains consist of thick layers of sediments deposited over millions of years. They are, in fact, the final resting places for sediments from the continents. In areas far from land, sediments accumulate at an incredibly slow rate. Nearly lifeless, the abyssal plains are extremely tranquil places. For thousands of years at a time, nothing may break the absolute stillness.

B Many seamounts are volcanically active. Others, having become inactive, gradually settle beneath the surface. In tropical waters, coral will grow on top of the sinking volcanoes, forming ring-shaped islands known as atolls.

S202

Underwater Mountains The most prominent features of the ocean floor are the *mid-ocean ridges*. Mid-ocean ridges are more-or-less continuous mountain chains circling the Earth much like the seams of a baseball. The ridges, which may stand 3000 m above the surrounding ocean floor, are mostly underwater. However, in a few places, such as the island of Iceland, a mid-ocean ridge actually rises above the ocean's surface.

Shelves and Slopes The continents jut like islands above the deep ocean. Continents do not end at the water's edge however. Rather, they usually extend for some distance underwater before dropping off into the ocean basin. This gently sloping region, called the *continental shelf,* usually ends abruptly. Beyond the edge of the continental shelf is the *continental slope.* The continental slope angles steeply downward in an unbroken descent

to the ocean bottom. At the base of the continental slopes lie the abyssal plains.

The Greatest Plains Also located on the ocean floor are some of the widest open spaces on Earth. Found at depths between 3000 and 6000 m, the *abyssal plains* are vast, extremely flat areas of the ocean floor. In fact, they are the flattest regions on Earth. In some places, the elevation changes less than 3 m over a distance of 1300 km!

Peaking Out Isolated volcanic mountains, called *seamounts,* protrude in spots from the abyssal plains. The Pacific Ocean has many thousands of seamounts. Most do not break the surface, but some do rise above the surface of the ocean to form islands. The big island of Hawaii, in fact, is a seamount that is taller than Mount Everest, measured from base to top.

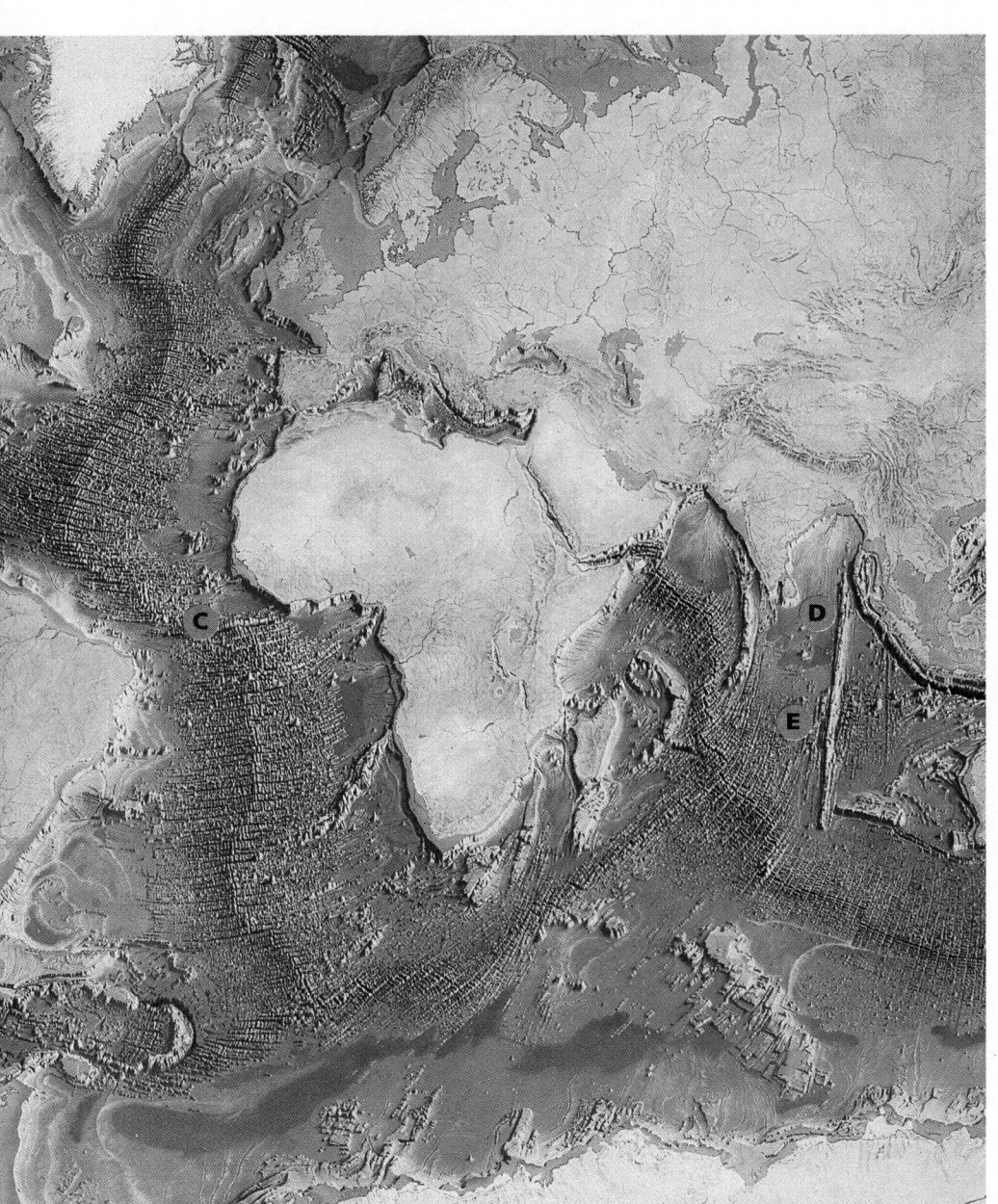

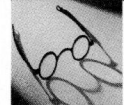

DID YOU KNOW...
that the mid-ocean ridges form the longest mountain chains on Earth? All part of a continuous system, the various mid-ocean ridges extend for more than 65,000 km!

C The mid-ocean ridges wind throughout the world's oceans. They mark the boundary between tectonic plates that are separating. As the crust splits apart, molten rock from the Earth's interior flows into the gap, forming new crust.

D Continental shelves are the underwater portion of the continents. Most of the ocean's living organisms are found here.

E Beyond the continental shelves are the continental slopes. Continental slopes are frequently affected by underwater landslides.

S203

Oceanic Erosion and Deposition

A great deal of sediment is deposited on the continental shelf by rivers and streams. Sediments build up all the way to the edge of the continental shelf and beyond. The outermost edge of the continental shelf is an unstable region. Masses of sediment occasionally slump off and tumble down the continental slope like a slow-motion avalanche. As the mass of sediment falls, it drags the surrounding water along with it. The resulting current of water and sediments often carries weathered materials far out to the abyssal plains, beyond the base of the continental slope. As it gradually slows down, the current drops its sediment load—first the coarse-grained material, and then finer and finer materials.

These offshore currents have tremendous erosive energy. They often move faster than the fastest human could run and may travel for 700 km before finally dissipating in the abyssal plains. If you look at an undersea map, you can see that in many places the continental slope is gouged by deep canyons, some of them as large as the Grand Canyon itself. Interestingly, most appear to be extensions of rivers on land. These undersea canyons are believed to have been formed by the repeated action of these currents.

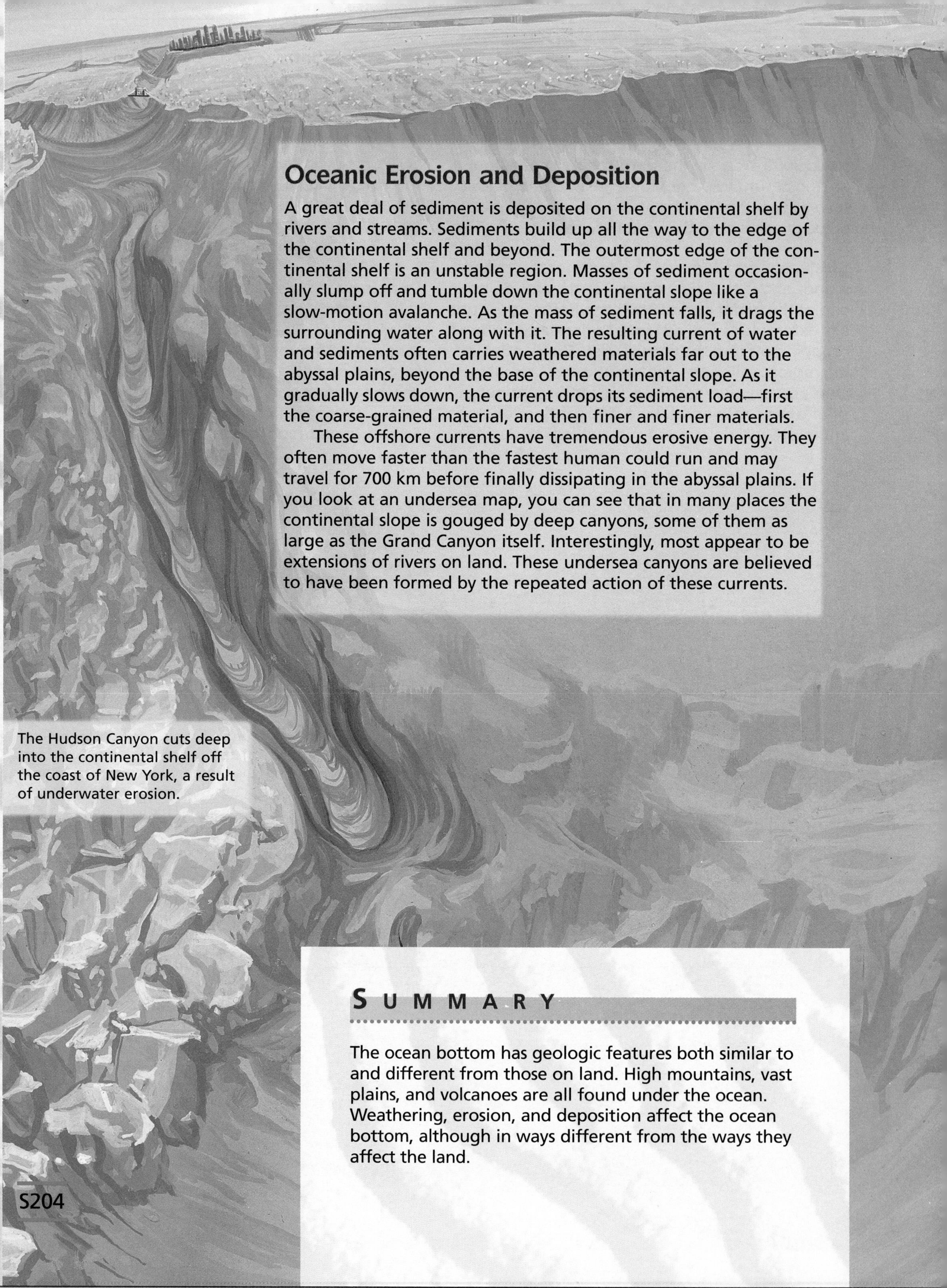

The Hudson Canyon cuts deep into the continental shelf off the coast of New York, a result of underwater erosion.

SUMMARY

The ocean bottom has geologic features both similar to and different from those on land. High mountains, vast plains, and volcanoes are all found under the ocean. Weathering, erosion, and deposition affect the ocean bottom, although in ways different from the ways they affect the land.

S204

Concept Mapping

The concept map below illustrates major ideas in this unit. Complete the map by supplying the missing terms. Then extend your map by answering the additional question below. Write your answers in your ScienceLog.
Do not write in this textbook.

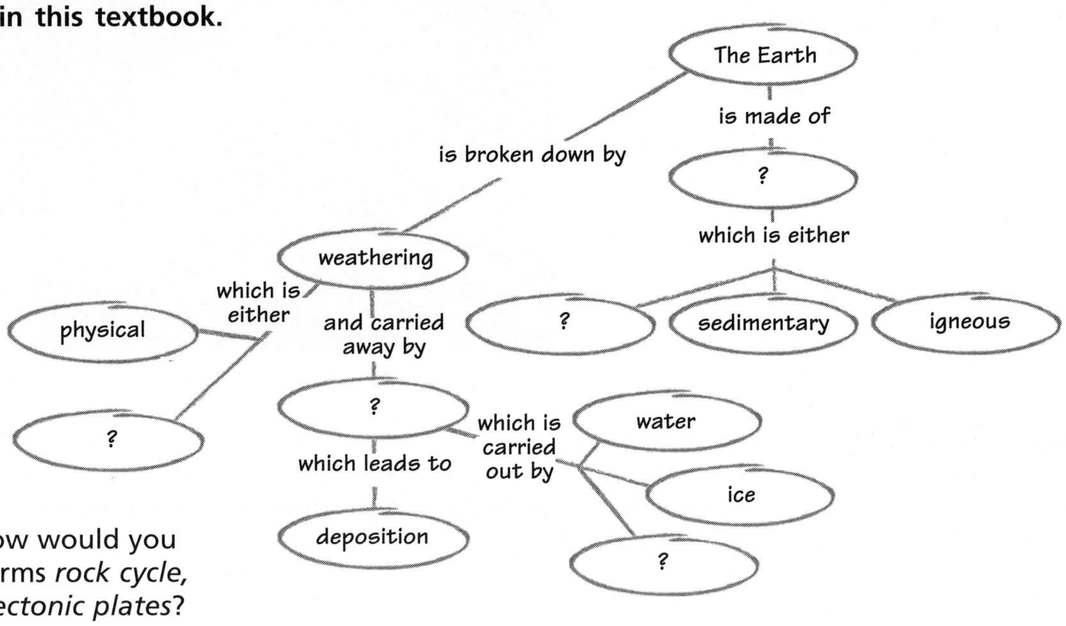

The Earth

is made of

is broken down by

?

which is either

weathering

which is either

physical

and carried away by

?

sedimentary

igneous

?

?

which is carried out by

water

which leads to

ice

deposition

?

Where and how would you include the terms *rock cycle, gravity,* and *tectonic plates*?

Checking Your Understanding

Select the choice that most completely and correctly answers each of the following questions.

1. Which type of rock is formed from molten material?
   a. igneous
   b. sedimentary
   c. metamorphic
   d. magma

2. Which is true of the rock cycle?
   a. Any one rock type can be changed into any other.
   b. Sedimentary rock always changes into metamorphic rock.
   c. Igneous rock always turns into sedimentary rock.
   d. The change from one rock type to another takes several decades.

3. Which would NOT be an example of physical weathering?
   a. A sidewalk cracking in the summer sun
   b. A hole being scoured in a boulder during a sandstorm
   c. A drop of rainwater dissolving a tiny bit of the cement holding sandstone together
   d. A boulder being split apart by the roots of an oak tree

4. The faster a stream flows,
   a. the less likely it is to cause major erosion.
   b. the larger the delta it will deposit.
   c. the more likely it is to have a flood plain.
   d. the greater its erosive ability.

5. The abyssal plains
   a. are often swept by strong currents.
   b. are never found far from land.
   c. are sites of little activity.
   d. are home to many plants and animals.

S205

Interpreting Photos

This photo shows an unusual land feature ▶ that has been formed by the action of erosion on its surface. Using what you have learned in this unit, which type of erosion—gravity, streams, glaciers, wind, or waves—might have been responsible for producing this landmark? Explain.

Critical Thinking

Carefully consider the following questions, and write a response in your ScienceLog that indicates your understanding of science.

1. In what major ways do wind erosion and water erosion differ?

2. Eventually, the Earth may lose so much heat that its interior becomes solid. What effect might this have on the Earth's landforms?

3. Weathering does not occur evenly throughout the world. Which types of climates probably experience the fastest weathering? the slowest? Why?

4. Tests have shown that, as you move away from the center of the mid-ocean ridges, the rocks become steadily older. How does this evidence support the idea that the Earth's crust moves?

5. Lakes are usually short-lived features, geologically speaking. Explain why this is so, using the concepts of erosion and deposition.

Portfolio Idea

As you know, the Earth is estimated to be about 4.5 billion years old. This is a span of time that most people find very difficult to grasp. Design a scenario appropriate for helping an eight-year-old child to understand this span of time, using the major geologic events shown on page S184. For example, you might say "If the Earth's history were compressed into one year, oceans began forming on February . . ." and so on. Don't use this example. Be imaginative, but be accurate.

A

Acid any compound that has a sour taste, turns blue litmus paper red, and reacts with a base to form a salt and water **(293, S127)**

Acquired immunity resistance to reinfection of a disease after the body has recovered from the original infection **(S80)**

Active solar heating the use of solar collectors and mechanical devices to gather sunlight, convert it into heat, and distribute that heat **(S153)**

Adaptation an inherited feature that arises over time and that better enables an organism to survive in a given environment **(106)**

Algae (singular, *alga*) a large group of primitive plants that have chlorophyll but lack true roots, flowers, stems, and leaves. Algae are the chief marine and freshwater plants, ranging in size from single-celled pond scum to giant kelp. **(143)**

Alkali metals the group of elements that have atoms with only one electron in their outermost energy level **(S122)**

Allergy a condition in which the immune system's antibodies attack a normally harmless foreign substance **(S81)**

Alloy a solid mixture of two or more metals, or a mixture of a metal with some other substance **(279)**

Animalcules the name given to the first microorganisms seen by Leeuwenhoek. The word is Latin for "tiny animals." **(135)**

Antibiotic a chemical that is capable of killing or stopping the growth of bacteria **(173, 194)**

Antibody a chemical that is made by the body and that fights germs or other foreign substances **(S79)**

Asthenosphere a layer of the Earth that lies in the upper mantle directly below the lithosphere. Because of high temperatures and pressure, this hard, rocky area is able to flow. **(463)**

Atom the smallest particle of an element; from the Greek word *atomos*, meaning "cannot be divided" **(S100)**

Atomic mass the mass of an atom expressed in atomic mass units **(S114)**

Atomic number the number of protons in the nucleus of an atom **(S112)**

Autoimmune disorder a condition in which the immune system's antibodies attack the body's own tissues **(S81)**

B

Bacteria (singular, *bacterium*) a large class of single-celled microorganisms, considered neither plant nor animal, that belongs to the kingdom Monera. Bacteria cause a number of diseases and contribute to many processes, including decay, fermentation, and soil enrichment. **(144)**

Base a compound that has a bitter taste, turns red litmus paper blue, and reacts with an acid to form a salt and water **(293, S128)**

Bile a brownish yellow or greenish yellow fluid that is secreted by the liver, stored in the gallbladder, and discharged into the small intestine. It helps the body absorb and process fats. **(303)**

Biological property a characteristic that distinguishes living things from nonliving things **(203)**

Biorhythm any of several cycles that relate sleeping, eating, breeding, migratory, or other habits of organisms to changes in sunlight, temperature, tides, or other natural processes **(99)**

Boiling point the temperature at which a substance in its liquid state changes into a gas **(235)**

C

Calorie a unit of measure of heat energy that is used to describe the energy content of food **(415)**

Calorimeter a device that measures the amount of heat energy in a substance **(413)**

Camera obscura a dark enclosure with a small hole in one end, allowing an image to be projected on the opposite end **(313)**

Cancer harmful, uncontrolled growth in animal tissue **(87)**

Catalyst a substance that affects the rate of a reaction without being consumed in the process **(286)**

Cell the smallest unit of living things **(113, S36)**

Cell membrane the outer layer of an animal cell; the part of the cell that controls what enters and leaves the cell **(117, S37)**

Cellular respiration the process by which the energy stored in sugars is released for use by the cells of organisms **(S144)**

Cellulose the primary substance that makes up the cell walls and fibers of plant tissue; important in the manufacture of a variety of products, including paper, fabric, and medicines **(256)**

Cell wall the outer layer of a plant cell **(116, S41)**

Chemical a substance, such as water, salt, or sugar, produced by or used in chemical reactions **(254)**

Chemical activity the tendency of an atom to react with atoms of other elements **(S121)**

Chemical bond a link between atoms resulting from the mutual attraction of their nuclei for electrons **(S124)**

Chemical change a change, such as burning, that results in the formation of a new chemical **(268, S89)**

Chemical energy energy stored in the arrangement of atoms and molecules **(S137)**

Chemical equation a description of a chemical reaction using chemical formulas **(S130)**

Chemical formula a description of a chemical compound using symbols and numbers **(S129)**

Chemical property a characteristic of a substance that describes how the substance behaves in a chemical change **(204, S88)**

Chemical reaction a process in which a chemical change occurs; reactants yield products **(S130)**

Chemical weathering the ways in which rock breaks down by chemically changing some of the materials in the rock **(S188)**

Chlorophyll organic molecules in green plants that absorb sunlight for making food **(S42)**

Chloroplast a small structure found within the cytoplasm of plant and algae cells where sunlight is used to make sugar **(117, S42)**

Chromosome a rod-shaped structure found in the nucleus of a cell that is made of coiled DNA **(S39)**

Circulatory system the system responsible for transporting nutrients to, and wastes from, the cells of the body **(S50)**

Classification the organization of things into groups according to ways in which they are alike **(S24)**

Coldblooded having a body temperature that changes according to the temperature of the surroundings **(104)**

Combustion burning; chemically combining a substance with oxygen so rapidly that heat and light are produced **(299)**

Compost a mixture of decomposing plants and other matter that can be used as a fertilizer and soil conditioner **(153)**

Compound a substance that can be broken down into simpler parts only by a chemical change **(288, S97)**

Condensation the process in which a gas changes to a liquid **(235)**

Conduction the transfer of heat energy through a substance or from one substance to another by direct contact of atoms and molecules **(423, S165)**

Continental drift the theory that the continents were formed when Pangaea, one large landmass, split apart and its pieces began to drift away from each other **(460)**

Controlled experiment an experiment in which all variables but the one being tested are controlled in order to make the experiment fair and the results reliable **(34)**

Convection the transfer of heat in a liquid or gas as groups of molecules move in currents from one region to another **(423, S165)**

Convection cell a complete circle of moving liquid or gas caused by temperature differences **(S169)**

Core the central portion of the Earth that is below the mantle, begins at a depth of about 2900 km,

has temperatures that range from 2200°C to 5000°C, and probably consists of molten iron and nickel. It has a solid inner core and a liquid outer core. **(463)**

Corrosion a type of chemical change, such as that in which iron reacts with oxygen in the process of rusting **(299)**

Covalent bond a chemical bond resulting from the sharing of electrons **(S125)**

Crust the thin, rocky outer layer of the Earth; also known as the Earth's surface **(463)**

Cryptogram a message in a secret code **(12)**

Culture to grow microorganisms by providing the living conditions that will allow them to flourish **(141)**

Current electricity electric charges in motion **(336)**

Cytoplasm the jellylike substance surrounding the nucleus of the cell, where the cell's activities are carried out **(117, S37)**

D

Dead once alive but no longer having any of the signs of life **(63)**

Delta a triangular or fan-shaped area of sediment deposited at the mouth of a river **(506)**

Density the mass of a material per unit of volume **(S92)**

Deposition all of the processes by which eroded materials are deposited, or dropped **(S198)**

Dew point the temperature at which air, or certain other gases, begins to change to a liquid **(235)**

Digestive system the body system responsible for changing food into a form that can be used by cells in the body **(S47)**

Displacement method a method of finding the volume of an object by submerging it in water **(219)**

DNA deoxyribonucleic acid; the cellular material that directs the activities of a cell and transmits hereditary information between generations **(S32)**

Dormancy a sleeplike state that may be triggered by changes in the environment **(104)**

Ductile able to be shaped, bent, or stretched, often by hammering or applying pressure **(279)**

E

Ectotherm an animal whose body temperature closely matches that of its surroundings **(124)**

Ectothermic having a temperature that changes to match the temperature of the surroundings; also called coldblooded **(104)**

Efficiency the relationship between effective work and waste energy produced by a device or system **(326)**

Electrical energy energy of electrical charges as a result of their position or motion **(S138)**

Electrolysis the passing of an electric current through certain substances in order to break

S208

them down, such as when water is broken down into hydrogen and oxygen gases **(284)**

Electron the negatively charged subatomic particle, discovered in cathode rays, that is much smaller than a hydrogen atom **(S101)**

Electron cloud the area surrounding an atom's nucleus where electrons move **(S111)**

Electron microscope a powerful microscope that is able to create magnified images of very small objects, such as bacteria and viruses **(145)**

Element a substance that cannot be broken down during a chemical reaction. It combines with other elements to form compounds. No element can be changed into another element by means of a chemical reaction. **(288, S98)**

Endocrine system the body system responsible for regulating certain body processes in conjunction with the nervous system **(S58)**

Endotherm an animal that maintains a fairly steady body temperature regardless of the temperature of its surroundings **(124)**

Endothermic having a relatively constant temperature regardless of the temperature of the surroundings; also called warmblooded **(101)**

Energy the ability to cause matter to move or to change; the ability to do work **(329, S136)**

Energy transformation a change in energy from one form to another **(S142)**

Enzyme any of a number of large molecules, usually proteins, that are produced by living cells or are made synthetically and that act as catalysts to speed up certain chemical reactions **(303)**

Erosion the carrying away of soil, rock, and other materials on the Earth's surface **(487, S190)**

Estuary a wide, shallow river mouth that usually extends well inland, where fresh water from a river mixes with salt water from an ocean **(507)**

Evaporation the process by which a liquid changes to a gas **(233)**

Evolution the process by which living organisms change over time **(125, S30)**

Excretory system the body system that eliminates cellular wastes and excess water from the body **(S49)**

Experiment an activity designed to test a possible answer to a scientific question **(S16)**

F

Fermentation the breakdown of complex organic molecules, such as when bacteria cause milk to curdle **(161)**

Fertilization occurs when sperm from a male successfully merge with eggs from a female **(87)**

Freezing point the temperature at which a substance in its liquid state changes to a solid **(234)**

Fungi (singular, *fungus*) a kingdom, neither plant nor animal, of organisms that are adapted to absorb food from living or dead matter; includes yeasts, molds, mushrooms, and puffballs **(153)**

G

Gall a harmful growth in the tissue of plants **(87)**

Gallbladder a small sac located under the liver that serves as a storage place for bile until it is needed by the body for digestion **(303)**

Gastric juice a watery, colorless, acidic fluid that is secreted by several glands in the wall of the stomach and consists primarily of hydrochloric acid and pepsin **(305)**

Gene a segment of a chromosome that carries hereditary information **(S32)**

Geologic time scale the calendar used by scientists to describe events in the Earth's history **(S184)**

Geology the scientific study of the Earth's origin, history, and structure and of the changes brought about by natural processes **(453)**

Germ any microorganism that can cause disease **(S66)**

Germinate to sprout; to begin to grow from a seed **(81)**

Glacier a huge mass of ice that slowly flows over a landmass **(512)**

Gravity the force of attraction between objects **(487)**

Greenhouse effect the warming of the Earth's surface and lower atmosphere caused by heat rising from the surface of the Earth and being reflected back toward Earth by gases in the atmosphere **(438)**

Ground water water that collects beneath the surface of the Earth **(497)**

Group a set of elements arranged in vertical columns in the periodic table **(S120)**

Gully a deep ditch carved out of the Earth by running water **(504)**

H

Halogens the group of elements that have atoms with seven electrons in their outermost energy level **(S122)**

Heat the energy of a material due to the random motion of its particles; thermal energy **(406, S163)**

Hot spot an area of concentrated heat deep within the Earth's interior that causes some of the rocks in the Earth's mantle to melt, forming a column of magma that rises toward the surface **(468)**

Hyphae (singular, *hypha*) threadlike or stemlike parts of a fungus **(167)**

Hypothesis a testable explanation for an observation **(35, S14)**

I

Igneous rock rock formed from cooled magma **(S180)**

Immune deficiency disorder a condition in which the immune system breaks down, leaving the body unprotected from disease **(S82)**

S209

Immune system the body system responsible for protecting the body from invading substances **(S56, S80)**

Indicator a substance that is one color in an acid and another color in a base **(293)**

Infection the growth of disease microorganisms in the body of a host organism **(S67)**

Infectious disease a disease caused by a microorganism that can spread from one organism to another **(S68)**

Inference a conclusion that attempts to explain or make sense of observations **(25)**

Infusion a preparation obtained by steeping a substance in a liquid **(140)**

Insulator a material that prevents the flow of heat into or out of something, such as through walls, ceilings, or floors. Insulators can also prevent the transfer of electricity or sound. **(430)**

Insulin a chemical produced by a small gland called the pancreas. It is used to break down sugars during the process of digestion and to treat diabetes. **(8)**

Integumentary system the body system responsible for regulating body temperature and protecting the body from the outside environment **(S53)**

Interface the boundary between two different substances **(230)**

Interpret to explain the meaning of something **(25)**

Ionic bond a chemical bond resulting from the attraction between positive and negative ions **(S126)**

Irritability the ability of an organism to respond to stimuli **(92)**

Isotopes atoms of the same element that have different atomic masses **(S115)**

J

Joule a unit of energy that is approximately equal to the amount of energy required to raise a 100 g mass to a height of 1 m **(413)**

K

Karst landscape a type of terrain characterized by caves, sinkholes, and underground drainage and caused by the erosion of limestone or similar rock **(262)**

Kilowatt-hour the most commonly used unit of energy; equal to the work done by one kilowatt (kW) acting for one hour **(342)**

Kinetic energy energy in motion; the energy that an object has because of its motion **(323, S137)**

Kingdom the largest and most general category that scientists use to classify organisms **(142, S25)**

L

Large intestine an organ of the digestive system through which the food that cannot be digested is passed on, eventually exiting the body as a waste product **(303)**

Lithosphere the solid, rocky, outer layer of the Earth that consists of the crust and the upper mantle and that is divided into tectonic plates **(463)**

Liver a large, reddish brown organ that is located in the upper-right portion of the abdominal cavity in vertebrates. It produces bile, helps form certain proteins, and helps the body process carbohydrates, fats, and proteins. **(303)**

Living presently alive **(63)**

Locomotion the ability of an animal to move on its own from place to place **(66)**

Lung capacity the amount of air the lungs can hold **(220)**

M

Malleable able to be shaped or formed into a number of useful products, often by hammering or applying pressure **(279)**

Mantle the middle layer of the Earth, between the crust and the core **(463)**

Mass a measure of the amount of matter in an object **(222, S90)**

Mass number number of protons and neutrons in the nucleus of an atom **(S114)**

Matter anything that has volume and mass **(202, S86)**

Melting point the temperature at which a substance in its solid state changes to a liquid **(234)**

Meltwater water from melting snow or ice; usually associated with a melting glacier **(517)**

Meniscus the curved surface of a liquid **(215)**

Metamorphic rock rock that has undergone changes as a result of intense heat and pressure **(S181)**

Microbiology the study of microorganisms **(135)**

Micrometer a unit of measure that is equal to one-thousandth of a millimeter, or one-millionth of a meter; also called a *micron* **(136)**

Microorganism a tiny living thing that can be seen only with the aid of a microscope **(133)**

Mid-ocean ridge one of several underwater mountain chains formed in connection with spreading centers **(464)**

Migration movement of animals from one place to another in response to a change of seasons **(100)**

Mitochondria small structures that are found in the cytoplasm of plant and animal cells and that serve as the center of energy production for the cells **(117, S40)**

Mixture any form of matter that contains more than one substance and that can be separated by physical means **(S94)**

Model an illustration, description, small reproduction, or other representation that is used to explain an object, system, or concept **(31)**

Mold a common term for fungi that have a downy or furry appearance and that often occur in the presence of dampness or decay (166)

Molecule the smallest particle of a substance, such as water, that is still identifiable as that substance (S87)

Monera a kingdom of microorganisms, primarily bacteria (144)

Mucus a thick, sticky fluid covering many surfaces inside the body and in its natural openings (S78)

Multicelled composed of more than one cell (139)

Muscular system the body system that enables the body to move, the heart to beat, and other organs of the body to function (S54)

N

Nervous system the body system that senses the environment and coordinates the body's response to the environment (S55)

Neutralization reaction a reaction in which an acid neutralizes a base, forming a salt and water (294)

Neutron the subatomic particle with no electric charge and about the same mass as a proton (S102)

Noble gas one of the elements in Group 18 of the periodic table, which have a stable number of electrons in the outer energy level (S123)

Nonliving never having been alive (63)

Nonrenewable resource a resource, such as coal, oil, or natural gas, that can be used up faster than it can be replenished naturally (370)

Nuclear energy energy that is released either by splitting atomic nuclei or by forcing the nuclei of atoms together (S139)

Nucleic acid a chemical that controls activities in a cell and passes on traits to new cells (S76)

Nucleus (plural, *nuclei*) the control center of a cell's life activities; in an atom, the central core that contains most of the atom's mass (117, S37, S110)

O

Observation any information that we gather by using our senses (19, S14)

Organ a group of different tissues working together to perform a specific function (S45)

Organelle a tiny specialized structure within a cell that performs cell functions (S39)

Organism the highest level of cell organization. An organism is the sum of its cells, tissues, organs, and organ systems. (S46)

Organ system a group of related organs performing a major function for an organism (S46)

Outwash sand and gravel that have been removed from a glacier and deposited beyond the glacier's boundaries by meltwater streams (517)

Ovary a female reproductive gland that forms eggs and, in vertebrates, produces sex hormones (88)

P

Pancreas a long, irregularly shaped gland that is located just below the stomach of vertebrates. It produces a digestive juice that is released into the small intestine and produces insulin and other substances that are released into the bloodstream. (303)

Pangaea according to Alfred Wegener's theory, the single landmass from which all continents stem (460)

Parasite an organism that lives in or on another living organism, usually causing it harm (S74)

Particle model of matter a model that demonstrates how all matter is made of particles (238)

Passive solar heating the use of simple absorption of radiant energy to warm buildings or other structures (S153)

Pasteurization heating a substance to kill unwanted microorganisms while sparing beneficial ones (160)

Penicillin a valuable antibiotic that was discovered by Alexander Fleming. It is obtained primarily from certain kinds of molds and is effective against many disease-causing bacteria. (174)

Pepsin an enzyme that is found in gastric juice and that speeds up the breakdown of protein molecules (305)

Period a row, or horizontal line, of elements in the periodic table (S121)

Periodic table a list of symbols of every known element arranged by groups (columns of elements with similar chemical properties) and periods (rows of elements whose properties resemble each other more closely than those that are far apart) (290, S118)

Permeability a measure of how well water, air, or plant roots penetrate, or pass through, a porous material such as soil (496)

Perpetual motion machine a hypothetical machine that, once started, would run forever without any additional input of energy (330)

Perspire to sweat; to secrete water from small glands in the skin (102)

Petrochemical a chemical that is derived directly or indirectly from petroleum or natural gas (369)

Photosynthesis the process by which green plants, algae, and some bacteria use light energy, water, and carbon dioxide to make sugars that they use for food (S144)

Physical change any change in matter, such as boiling, freezing, or breaking, that does not change the chemical properties of the matter (268, S89)

Physical property any property of matter that can be observed by your senses without altering the composition of the matter (204, S87)

Physical weathering the ways in which rock breaks up into smaller pieces without any chemical change in the materials of the rock (S186)

S211

Plankton tiny, usually microscopic, plants or animals that float in the ocean or in bodies of fresh water; a food source for nearly all aquatic animals **(155)**

Plate tectonics the theory that the Earth is divided into a small number of large, rigid plates that move, causing volcanoes and earthquakes and resulting in changes in the shape and size of ocean basins, mountains, and continents **(461)**

Porosity a measure of the empty space in a material, as determined by the volume of water that can be absorbed by the material **(496)**

Potential energy stored energy **(323, S137)**

Power a measure of the amount of work done in a certain period of time **(342)**

Precipitate a solid formed as the result of a chemical change **(272)**

Prediction telling about something in advance **(33)**

Product a new substance that results from a chemical reaction **(282)**

Property a characteristic that distinguishes one substance from another **(20)**

Protist a type of usually one-celled organism, such as a euglena or a paramecium, that has plantlike or animal-like characteristics or both **(142)**

Protista a kingdom consisting of the protists, which are usually single-celled microorganisms **(142)**

Proton the subatomic particle that has a positive charge equal to the negative charge of an electron **(S102)**

Q

Qualitative observation an observation that does not involve measurements or numbers **(20)**

Quantitative observation an observation that involves measurements and numbers **(20, S17)**

R

R-value a measure of how well a material, such as insulation, can prevent heat flow, with increasing values indicating a greater resistance to heat flow **(359)**

Radiant energy energy transmitted in the form of electromagnetic waves **(421, S140)**

Radiation energy traveling through empty space or through a transparent material without heating the empty space or the transparent material **(423, S166)**

Radioactivity a condition in which an object gives off radiant energy in the form of particles or rays such as alpha or beta particles or gamma rays **(311)**

Reactant a substance that is part of a chemical reaction and that is present when the reaction begins (as opposed to a product, which is created as a result of the reaction) **(282)**

Recycle to reuse materials by making new products from discarded items **(S158)**

Refining a chemical process that leads to the purification of a substance **(255)**

Regeneration the growth of new body parts to replace those lost or damaged **(84)**

Renewable resource a resource that is in great abundance and is continually produced, such as wind or sunlight, and that can be replaced naturally if used wisely **(369)**

Reproduction the process by which organisms produce offspring of their own kind and which is necessary for the continued existence of a species **(87)**

Reproductive system the body system responsible for passing on hereditary information and producing the next generation **(S59)**

Respiratory system the body system by which gases enter and leave the body **(S52)**

Response a reaction to a stimulus **(92)**

Ribosome a structure in the cytoplasm where proteins are made **(S40)**

Rock cycle the process by which one type of rock is transformed into another type of rock **(S181)**

Runoff excess rainwater that is not absorbed by the soil, but instead begins to collect and flow over the surface of the ground **(495, S192)**

S

Saliva a watery solution released by the salivary glands. It serves as an aid to swallowing and begins the process of digestion. **(303)**

Salivary glands three pairs of glands located at the back of the mouth that secrete saliva and help begin the process of digestion **(303)**

Salmonella a genus of bacteria that can cause food poisoning and are often found on eggs and poultry **(171)**

Saprophyte an organism that gets food from dead or decaying organisms **(S74)**

Science knowledge about the natural world that is derived from observations and experiments **(47, S2)**

Scientific law a statement of what happens in a certain event based on verified observations and experiments **(S20)**

Scientific method a way of thinking about nature that involves the use of certain skills to solve problems in an orderly manner **(S12)**

Scientific model a theoretical representation of something that you cannot observe directly **(S108)**

Scientific theory a general statement of why things work based on hypotheses that have been tested many times **(S20)**

Scientist a person who is curious about the natural world and who explores that curiosity by using certain techniques and methods **(S5)**

Sediment fragments of weathered rock, such as sand, mud, and small bits of gravel, that are carried and deposited by wind, water, or ice **(474)**

S212

Sedimentary rock rock formed from pieces of other rocks, by evaporation, or the remains of organisms **(S181)**

Simulate to represent, model, or imitate a situation, system, or process **(484)**

Skeletal system the system that supports the body with a framework of bones **(S54)**

Small intestine an organ of the digestive system that receives food from the stomach and is where water-soluble proteins are absorbed and passed into the bloodstream **(303)**

Solute the part of a solution that is dissolved **(S95)**

Solution a mixture formed when one substance mixes completely with another substance **(S94)**

Solvent the part of a solution that does the dissolving **(S95)**

Species a group of organisms that are able to reproduce together and that resemble each other in appearance, behavior, and internal structure; the most specific level of the classification system **(87, S25)**

Specific heat the amount of heat needed to raise the temperature of 1 g of a substance by 1°C **(S175)**

Spore a type of small reproductive cell that is capable of developing into a new individual. Spores are produced by bacteria, algae, mosses, ferns, and other organisms. **(152)**

Spore case a swelling or sphere that contains spores **(167)**

Spreading center a linear zone usually found in mid-ocean where two adjacent tectonic plates move apart, opening a gap into which molten rock from the Earth's interior continuously flows, thereby creating new crust **(464)**

Static electricity stationary electrical charges that build up on objects **(336)**

Stimulus anything that causes a living thing to react **(92)**

Stomach a large, bag-shaped organ in vertebrates that serves as a storage area for food that passes from the esophagus to the small intestine; one of the principal organs in the digestion process **(303)**

Strip mining the clearing away of topsoil and other portions of the upper layers of the Earth to remove substances such as coal **(375)**

Supernova a very large and very bright exploding star **(7)**

Système International d'Unités (abbreviated SI) the international system of metric units **(211)**

T

Technology the application of knowledge, tools, and materials to solve practical problems **(47, S4)**

Tectonic plate one of several large sections of the Earth's crust that are in constant motion, according to the theory of plate tectonics **(463, S182)**

Temperature the degree of hotness or coldness of a substance that is determined by the average kinetic energy of the molecules of the substance **(391, S172)**

Terminal moraine a ridge of rock formed from the materials left behind by a glacier that has stopped advancing **(517)**

Thermal energy energy due to the motion and arrangement of molecules in a substance **(S141)**

Thermal pollution the dumping of heated waste water into a natural body of water, causing environmental damage due to an increase in water temperature **(127)**

Thermal resistance the R-value of an insulating material; a measure of how well a material can prevent heat flow **(359)**

Thermometer an instrument used to measure temperature. It usually consists of a thin glass tube and a liquid, such as alcohol or mercury, that expands or contracts in the tube as the temperature changes. **(392)**

Thermostat a control device, often found in home heating and air-conditioning systems and refrigerators, that contains a temperature-activated electrical switch **(402)**

Till a type of sediment left behind by glaciers **(515)**

Tissue a group of similar cells with similar functions **(S44)**

Toxin a poison, especially one that is produced by certain organisms and that is harmful to the normal cell functions of other organisms **(167, S70)**

Trench a long, deep valley created when two plates collide and the force of the collision pushes one of the plates downward **(464)**

Tributary a stream or river that flows into a larger river, lake, or other body of water **(504)**

V

Vacuole a small cavity within the cytoplasm of a cell that is bound by a membrane and that serves as a storage area, usually for food, water, or air **(117, S40)**

Variable any factor in an experiment that could affect the results and is therefore tested separately **(34, S16)**

Virus any of various particles that contain genetic material and invade the cells of plants, animals, fungi, and bacteria, often destroying the organism's cells and causing symptoms of disease **(145)**

Volume the amount of space that an object occupies **(214, S92)**

W

Warmblooded having a body temperature that remains relatively constant, regardless of the

S213

temperature of the surroundings **(101)**

Watershed a region drained by a river, river system, or other body of water **(504, S192)**

Weathering a chemical or mechanical process in which rocks exposed to the weather are worn down by water, wind, or ice **(481, S185)**

Weight the downward force exerted by an object due to gravity **(S90)**

Wet mount a thin sample of a cell, placed in water between a microscope slide and a coverslip **(115)**

Word equation a chemical equation that shows the reactants and products of a reaction in words rather than in symbols **(282)**

and safety, 260
signs of change, 270
testing of, 272–274
Chemical sterilization
process of, 182
Chemical weathering
agents of, S188–S189, **S188**
defined, S188
Chemistry
defined, S9
biochemists, work of, S9, **S9**
chemists, work of, S9, **S10**
organic chemists, work of,
S9, **S9**
Chicken
embryo, **88**
ovaries in, 88
reproduction in, 88, **88**
Chlamydomonas, **140**
Chlorella, **143**
Chlorophyll
defined, S42
Chloroplasts
defined, 117, S42
Chromosomes
defined, S39
Circulatory system, 50
defined, S50
Clams and mussels, **83, S4**
Classification
defined, S24
modern methods of, S26
Classifying
of organisms, S24
statements of, 30
Climate
and heat and weather, S170
Coccus, S73, **S73**
Combustion, 447
Compounds
defined, 288, S97
and elements, 288, S97
Compost
defined, 153
making of, 154, **154**
Condensation
defined, 235
Conduction
defined, 423, S165
demonstration of, 424–427,
424–427, S165
Conservation of energy, S143
law of, S142
Continental drift, theory of,
465–466, **465–466**
Continental shelf, S203
Controlled experiment, 34
Convection
defined, 423, S165
demonstration of, 431–434,
S165–S166
Convection cell
defined, S169
Corrosion
defined, 299
example of, 299
Covalent bonds
defined, S125
Coyotes, **S13**
observations of, S13–16,
S13–S16

Crabs, **S4**
Cryogenics, 448
Culture, 141
Curie, Marie, 47, 290, 311
and the discovery of radioactivity, 290, 311
and Henri Becquerel, 311
and Pierre Curie, 311
Curie, Pierre, 311
and the discovery of radioactivity, 311
and Henri Becquerel, 311
and Marie Curie, 311
Cuttlefish, **67**
Cytoplasm
defined, 117, S37
structures in, S40, **S40**

D

Dalton, John
and atomic theory, S100
Darwin, Charles
and evolution, S31
and Galápagos Islands,
S34–S35, **S34–S35**
Data
analysis of, S18
and observations, S17
Davis, William, 504
and geology, 504
DeCoursey, Patricia, 99
and the study of flying squirrels, 99
Deforestation
defined, S147
effects of, S147
Delta
defined, 506, S198
formation of, 506–507, S198
Mississippi River, 506, **506,**
S198
Nile River, **506**
Democritus
and matter, S100
Density
defined, S92
Deposition
defined, S198, **S198**
forms of, S199–S201,
S199–S201
Dew point
defined, 235
Diabetes
treatment of, 8
Digestion
process of, 302–305
Digestive system, 302–305, S47
defined, S47
Dinosaurs
study of, 524
Discovery of penicillin, 174
Disease
and antibiotics, 175
bacterial and viral, S69
carriers, S69
causes of, S68
defined, S68
disinfectants and, S67
infectious, S68
in plants, S71

treatment of, 59
types of, 59, 175, 197
Disinfectants
and disease, S67
Displacement method
defined, 219
DiTullio, Jeff
and the study of bats, 247
DNA
defined, S32
and human genetics,
S60–S61
Dormancy, 104
Dust mites, **150**
Dutch elm disease, S71

E

Earth science
defined, S7
Earthworms
methods of movement, 72
parts of, 72, **72**
response to stimuli, 92–94,
92–94
Ecology
defined, S6
ecologists, work of, S6
Ecosystems
energy flow in, S145
food chains in, S145, **S145**
food pyramids in, S146,
S146
food webs in, S146
Ectothermic (animals)
defined, 104, 124
examples of, **104, 124**
temperature changes in, 104,
124
Electricity
as a form of energy, 336,
S138, **S138**
making, 338–339
measuring electrical energy,
342–343
static, 336
Electron cloud
defined, S111
Electrons
arrangement of, S112–S113
defined, 385, S101
examples of, **S101**
Elements
alkali metals, S122
and chemical activity, S117
classification of, S117–S118
and compounds, 288
defined, 288, S98
examples of, S98
halogens, S122
Jöns Jakob Berzelius and,
S119
noble gas, S123
and periodic table, 290–291,
S118–S119
Elephants
African vs. Asian, 82, **82**
size of, 82, **82**
Embryo
of chicken, **88**
Empedocles
theory on burning, 296

Endangered species
example of, **2–3,** 3
Endocrine system, S58
defined, S58
Endothermic (animals)
defined, 101, 124
examples of, **101, 124**
temperature changes in,
101–102, 124
Energy
chemical, S137, **S137**
conservation, S143, S158
defined, S136
efficiency of, 327
electrical, 336, S138, **S138**
and the environment, S144
in food, 414–416
food energy table, 416
forms of, 323, 412,
S136–S141, S162,
S136–S141
future sources of, 371–373
geothermal, S156, **S156**
history of, 365–367, 413,
S148
kinetic, 323, S137, **S137**
measuring of electrical,
342–343
nuclear, 376–377, S139,
S139, S151–S152, **S152**
and perpetual motion
machines, 330–332,
330–332
potential, 323, S137
pyramids, S146
radiant energy, S140, **S140**
recycling, as a source of,
S158
solar, 373, S153
testing, 319–321
thermal, S141, **S141**
transformation, S142
water power, 373
wind power, 374, S149
Energy transformation
defined, S142
Entomology
defined, S6
entomologists, work of, S6
Environment
and energy, S144
and global warming, 446
and plastics, 249
and power plants, 127
Environmental health
Long, Dr. Irene, 14–15, **14**
research in, 14–15
Enzymes
defined, 303
Erosion
defined, 487, S190
by glaciers, S193
and gravitational force, 487
by gravity, S190
rates of, S192
by streams, S191
by waves, 508–509, S197
by wind, 510–511, S195–S196
Estivation, 104
Estuaries
defined, 507
Euglena, **140**
classification of, 142

S218

Abbreviations used: (t) top, (c) center, (b) bottom, (l) left, (r) right, (bckgd) background, (bdr) border.

PHOTO CREDITS

Photos not otherwise credited below are HRW photos by Sam Dudgeon.

FRONT COVER: (tl), Sinclair Stammen/Science Photo Library/Photo Researchers, Inc.; (br), Antony Mercieca Photo/Photo Researchers, Inc.; (bkgd), Page Overtures

BACK COVER: (tl), Tony Stone; (br), Jeff Smith/FotoSmith/Reptile Solutions of Tucson; (bkgd), Page Overtures

TITLE PAGE: Page i(l-bkgd), Page Overtures; (bl), Jeff Smith/FotoSmith/Reptile Solutions of Tucson

TABLE OF CONTENTS: Page iv(l-bkgd), Letraset/Phototone; iv(bl), (cr-bottom), HRW photo by John Langford; iv(cr-top), NASA; v(l-bkdg), Letraset/Phototone; v(t),(bl), HRW photo by John Langford; v(cr), Rod Plank/Photo Researchers, Inc.; v(br), Grant Heilman/Grant Heilman Photography; vi(l-bkgd), Letraset/Phototone; vi(cr), HRW photo by John Langford; vi(tr), Science Photo Lab/Custom Medical Stock; vii(l-bkgd), Letraset/Phototone; vii(tr), HRW photo by John Langford; vii(bl), Microwork/Phototone; vii(br), HRW photo by Michelle Bridwell; viii(l-bkgd), Letraset/Phototone; viii(t), Superstock; viii(bl),(br), HRW photo by John Langford; viii(cr), Larry Lefever/Grant Heilman Photography, Inc.; ix(l-bkgd), Letraset/Phototone; ix(tl),(cr), (br) HRW photo by Michelle Bridwell; ix(tr), D. Luria/FPG International, Corp.; ix(bl), Keith Gunnar/Bruce Coleman, Inc.; x(l-bkgd), Letraset/Phototone; x(tl), HRW photo by John Langford; x(bl), HRW photo by Michelle Bridwell; x(cr), David Muench; xi(l-bkgd), Letraset/Phototone; xi(tl), John

Mason/International Stock Photo; xi(r), H.R. Bramaz/Peter Arnold, Inc.; xi(bl), S Camazine/S. Billo/Photo Researchers, Inc.

UNIT 1: Page 2(tl-ICON), NASA; 2-3(bkgd), Art Resource; 2-3, Douglas Faulkner/Photo Researchers, Inc.; 4(tl), Pat & Tom Leeson/Photo Researchers, Inc.; 4(cl), Louie Psihoyos/Matrix; 4(bl), Howard Bluestein/Science Source/Photo Researchers, Inc.; 5(cl), James King-Holmes/Science Photo Library/Photo Researchers, Inc.; 5(tr), Breck P. Kent; 5(cr), Merlin Tuttle/Photo Researchers, Inc.; 5(bl), Breck P. Kent; 5(br), Robb Kendrick/Aurora; 6(l), James King-Holmes/Science Photo Library/Photo Researchers, Inc.; 6(r), Breck P. Kent; 7(tr), Vera Lentz/Black Star; 7(bl), David Malin/AP/Wide World Photos; 7(br), Steven Lee/AP/Wide World Photos; 8(t), University of Toronto; 8(b), Will & Deni McIntyre/Photo Researchers, Inc.; 13, HRW photos; 14, NASA; 15(t), Steve Lunetta/PhotoEdit; 15(b), NASA; 18(tr), The Bettmann Archive; 21, HRW photo by John Langford; 25, HRW photo by John Langford; 30,35, HRW photos by John Langford; 38(cl), HRW photo by John Langford; 39(tr), Archive Photos; 39(cl), HRW photo by John Langford; 40,41, HRW photo by John Langford; 44,45,46, HRW photos by John Langford; 49(t), Metro Toronto Reference Library, Picture Collection; 49(b), AT&T Photo Service; 50(tl),50(tr),Metro Toronto Reference Library, Picture Collection; 50(b), Brown Brothers; 50(r), Brown Brothers; 51, N. H. (Dan) Cheatham/DRK Photo; 52, Archive Photos; 53, Nawrocki Stock Photos; 54(l), HRW photo by John Langford; 54(r), Douglas Faulkner/Photo Researchers, Inc.; 55, H. Reinhard/Okapia/Photo Researchers, Inc; 56(tl-icon), Rogge/The Stock Market; 56, morping created by Greg Geisler; 57(tl-icon), Robert Johannsen; 57(bl), Culver Pictures, Inc.; 57(tr), Photri; 58, San Diego Aerospace Museum; 59(bl), Philippe Plailly/Eurelios/ Science Photo

Library/Photo Researchers, Inc.59(tr), Photri; 59(tr), Custom Medical Stock Photo; 59(tl-icon), Howard Sochurek/The Stock Market

UNIT 2: Page 60(tl-ICON), Letraset/Phototone; 60,62(bkgd), Letraset/Phototone; 60-61, Stephen Dalton/Photo Researchers, Inc.; 62(tr), Sinclair Stammers/Science Photo Library/Photo Researchers, Inc.; 62(cr), HRW photo by John Langford; 63(top row, l to r), Photri; A. Blank/Bruce Coleman, Inc.; A. Blank/Bruce Coleman, Inc.; (2nd row, l to r), Photri; Norman Owen Tomalin/Bruce Coleman, Inc.; Photri; Norman Owen Tomalin/Bruce Coleman, Inc.; (3rd row, l to r), Photri; Denise Cupen/Bruce Coleman, Inc; Photri; Norman Owen Tomalin/Bruce Coleman, Inc.; (4th row, l to r), Ken Sherman/Bruce Coleman, Inc.; Norman Owen Tomalin/Bruce Coleman, Inc.; Norman Owen Tomalin/Bruce Coleman, Inc.; 64(tl), HRW Photo; 64(tc), SuperStock; 64(tr), Jim Rosen/TerraPhotographics; 64(cr), E.R. Degginger/Color-Pic; 64(bl,br), HRW photos by John Langford; 65, Christi Carter/Grant Heilman Photography, Inc.; 67(tl), Merlin D. Tuttle/Bat Conservation International; 67(tr), G.I. Bernard/Animals Animals; 67(cr), Anthony Merciea/Photo Researchers, Inc.; 67(bl), Carl Roessler/Animals Animals; 67(br), Adrienne T. Gibson/Animals Animals; 68(tl), David Weintraub/Photo Researchers, Inc.; 68(c,bl), HRW photos by John Langford; 70(t), Runk Schoenberger/Grant Heilman Photography, Inc.; 70(b), SuperStock; 71(t), K.L. Switack/Photo Researchers, Inc.; 71(b), Fran Allan/Animals Animals; 72,73, HRW photo by John Langford; 74(t), Cabisco/Visuals Unlimited; 74(bl), 75(tl,bl), HRW photo by John Langford; 75(tr), Massillon Museum; 76, 77, 78(t), 78-79(b), HRW Photo by John Langford; 80(tr), Donald Specker/Earth Scenes; 80(b), Carson Baldwin, Jr./Earth Scenes; 83(tl), H. Schwind/Okapia/Photo Researchers, Inc.; 83(tr), Andrew J. Martinez/Photo Researchers, Inc.; 83(cl), E.R. Degginger/Color-Pic; 83(cr), Rod Planck/Photo Researchers, Inc.; 84(all), Runk/Shoenberger/Grant Heilman Photography, Inc.; 85(t), Sherman Hines/Masterfile; 85(b),86(l), 86(tr), Runk/ Schoenberger/Grant Heilman Photography, Inc. ; 86(b), Denise Tackett/Tom Stack & Associates; 87(tl), J.H. Robinson/Photo Researchers, Inc.; 87(tr), James Stevenson/Science Photo Library/Photo Researchers, Inc.; 87(b, clockwise from tl), Runk/Rannels/Grant Heilman Photography, Inc.; Rana Temporaria/Animals Animals; Donald Specker/Animals Animals; Runk/Schoenberger/Grant Heilman Photography, Inc.; E. R. Degginger/Color-Pic; Rana Fremporaria/Animals Animals; 88(t-b), Oxford Scientific Films/Animals Animals; Jerome Wexler/Photo Researchers, Inc.; SuperStock; HRW photo by John Langford; 90(t), Lawrence Migdale/Photo Researchers, Inc.; 91(tl), HRW photo by John Langford; 91(tr), J.H. Robinson/Photo Researchers, Inc.; 91(bl), Pat and Tom Leeson/Photo Researchers, Inc.; 91(br), Stephen Dalton/Photo Researchers,Inc.; 93,94, HRW photos by John Langford; 95(tl,tr), Runk/Schoenberger/Grant Heilman Photography, Inc.; 95(bl,br), Bill Ivy; 100, Tom Bean/Tony Stone Images; 101(tl), Arthur C. Smith III/Grant Heilman Photography, Inc.; 101(tr), Stephen J. Krasemann/Photo Researchers, Inc.; 101(bl), Stouffer Prod. Ltd./Animals Animals; 101(br), Colin Milkins/Animals Animals; 102, Grant Heilman/Grant Heilman Photography, Inc.; 103, HRW photos by John Langford; 104, G. Synatzschke/Okapia/Photo Researchers, Inc.; 105(t), Whit Bronaugh; 105(b), Zig Leszczynski/Animals Animals; 108(tl), Runk/Schoenberger/Grant Heilman Photography, Inc.; 108(tr), E.S. Ross; 108(bl), K.G. Preston-Mafham/Animals Animals; 108(br), Stephen Dalton/Photo Researchers, Inc.; 112(tl), J.M. Labat/Jacana/Photo Researchers, Inc.; 112(tr), M.I. Walker/Photo Researchers, Inc; 112(c), HRW Photo by Russell Dian; 112(cr), E. R. Degginger/Color-Pic; 112(bl),114, 115, HRW photos by John Langford; 116(br), M.I. Walker/Photo Researchers, Inc.; 118, E.R. Degginger/Color- Pic; 119(top row, l to r), Runk/Schoenberger/Grant Heilman Photography, Inc.; Phillip A. Harrington/Peter Arnold, Inc., (second row, l to r) Alfred Pasieka/Peter Arnold, Inc.; Bruce Iverson/Bruce Iverson Photomicrography; Phil Degginger/Bruce Coleman, Inc.; (third row, l to r), Ed Reschke/Peter Arnold, Inc.; Ray Simons/Photo Researchers, Inc.; Runk/Schoenberger/Grant Heilman Photography, Inc.; (bottom row, l to r), Ed Reschke/Peter Arnold, Inc.; Grant Heilman/Grant Heilman Photography, Inc.; E.R. Degginger/Color-Pic; 120 (tr) Custom Medical Stock; 121 (tr) HRW photo by Richard Hutchings; 122(l), James L. Amos/Photo Researchers, Inc.; 122(r), Stephen Dalton/Photo Researchers, Inc.; 123(cr),123(c),123(tl),123(t), Runk/Schoenberger/Grant Heilman Photography, Inc.; 123(b), Adrienne T. Gibson/Animals Animals; 124(tl-icon), Ron Kimball; 124(bl), Norbert Wu; 125(tl-icon), Tony Stone Images; 125(bl), AP/WideWorld Photos; 125(inset) Kjell B. Sandved/Photo Researchers, Inc.; 125(inset), E.R. Degginger/Color-Pic; 126(bc), Victoria and Albert Museum, London, Art Resource, NY; 126(tr), Free Library of Philadelphia/Scala/Art Resource, NY; 127(tl-icon), Tony Stone Images; 127(l), Phil Degginger/Tony Stone Images.

UNIT 3: Page 128(tl-ICON),Patricia Barber/Custom Medical Stock; 128(l-bckgd),Letraset/Phototone; 128-129,ManfredKage/PeterArnold,Inc.; 130(tl), Dr.Tony Brain/Science Photo Library/Photo Researchers, Inc.; 130(tr),130(cr), HRW photo by John Langford; 134(br), The Bettmann Archive; 135(tr), Archive Photos; 135(bl), Volker Steger/Peter Arnold Inc.; 135(bc),135(bc), The Bettmann Archive; 136(cr), David Scharf/Peter Arnold Inc; 138(tr),138(tr), Bruce Iverson; 139(cr), 139(br),141(tc),142(tr), HRW photo by John Langford; 142(c), Dr. E.R. Degginger/Color-Pic; 143(cl), Kim Taylor/Bruce Coleman Inc.; 143(cr), Bruce Iverson; 143(bl), Carolina Biological Supply Co.; 143(br), Bruce Iverson; 144(tr), Manfred Kage/Peter Arnold, Inc.; 144(cl), Murti/Science Photo Library/Photo Researchers, Inc.; 144(bl), CNRI/SPL/Photo Researchers, Inc.; 145(tr), David Parker/Photo Researchers, Inc.; 145(c), Scott Camazine/CDC/Photo Researchers, Inc.; 145(cr), NIBSC/Science Photo Library/Photo Researchers; 147(c), Superstock; 148(cl), HRW photo by John Langford; 148(cr), Jeff Foott/Bruce Coleman, Inc.; 150(tr),150(cl),150(br), David Scharf/Peter Arnold; 151(bl), HRW photo by Michelle Bridwell; 152(tr), HRW photo by John Langford; 152(br), Photo Researchers,Inc.; 153, E.R. Degginger/Color-Pic; 155(cr),157(cl), Manfred Kage/Peter Arnold; 157(cr), Gordon R. Williams/Bruce Coleman 157(br), Norbert Wu/Peter Arnold; 157(b), Kevin Aitken/Peter Arnold; 157(bckgd), Jeff Falk/The Stock Shop; 158(tr), SuperStock; 159(tr), Archive Photos; 160(tc), SuperStock; 161(tr), HRW photo by John Langford; 161(br), Andrew Yates; 162(br), HRW photo by Michelle Bridwell; 163(cr),163(cr),163(br), David R. Frazier Photolibrary; 166(t), David Scharf/Peter Arnold; 166(br), HRW photo by John Langford; 167(tr), Bruce Iverson; 167(cr), Photo Researchers, Inc.; 169(bckgd), SuperStock; 170(cl),170(cl),170(cr),170(bl), HRW photos by John Langford; 171(br), Moredun Animal Health Ltd/Science Photo Library/Photo Researchers, Inc.; 172(tr),172(cr), Dr. Tony Brain/Photo Researchers, Inc.; 172(cr), Dr. Tony Brain/Custom Medical Stock; 172(br), Dr. Tony Brain/Photo Researchers, Inc.; 173(b),174(tr), St. Mary"s Hospital Medical School/Science Photo Library/Photo Researchers, Inc.; ,174(c), Noble Proctor/Photo researchers; 174(br), Science Photo Lab/Custom Medical Stock; 175(cl), HRW photo by Michelle Bridwell; 177(br),179(tr), HRW photos by John Langford; 179(br), Custom Medical Stock; 182(t), The Bettman Archives; 183(tr), Dr. Jeremy Burgess/Science Photo Library/Photo Researchers, Inc.; 183(bl), Ed Reschke/Peter Arnold, Inc.; 184(b),186(b),187(t),190(bl), HRW photos by John Langford; 192(l-bkgd), Letraset/Phototone; 192(tr), Manfred Kage/Peter Arnold, Inc.; 193(tr), Biophoto Associates/Science Source/Photo Researchers Inc.; 194(tl-icon), Howard Sochurek/The Stock Market; 194(tr), E.R. Degginger/Bruce Coleman, Inc.; 194(bl), Dr.Kart Lounatmaa/Science Photo Library/Photo Researchers, Inc.; 195(tl-icon) Ron Kimball; 195(c), Photo Researchers, Inc; 196(tl-icon), Rogge/The Stock Market; 196(tr), Manfred Kage/Peter Arnold, Inc.; 196(bl), Dr. Chris Somerville/Science Photo Library/Photo Researchers, Inc.; 197(bl),(tr) Archive Photos; 197(tl), Burt Glinn/Magnum Photos, Inc.

UNIT 4: Page 198-199(inset), HRW photo by John Langford; 198(tl-icon); HRW photo; 198(l-bkgd), Letraset/Phototone; 200(tl), Spencer Swanger/Tom Stack & Associates; 200(tr,bl), HRW photo by Patrick R. Dunn; 203(t), Martin Dohrn/Science Photo Library/Photo Researchers, Inc.; 203(c), Allen B. Smith/Tom Stack & Associates; 203(br), Tom & Pat Leeson/Photo Researchers, Inc.; 204, HRW photo by Patrick R. Dunn; 206, HRW photo by John Langford; 207(tl), Mike Boroff/TexaStock; 207(tc), Smiley/TexaStock; 207(tr), Alan Oddie/PhotoEdit; 207(bl), Tony Freeman/PhotoEdit; 207(br), Tony Freeman/PhotoEdit; 209(l), HRW photo by John Langford; 209(c), Jim Corwin/Photo Researchers, Inc.; 210(tl), Bullaty Lomeo/The Image Bank; 210(br), HRW photo by John Langford; 213,215,216 217, HRW photos by John Langford; 218(tl), William Perry/Focus on Sports; 218(b), HRW Photo by Patrick R. Dunn; 219, HRW photo by John Langford; 222(tr), HRW Photo by Patrick R. Dunn; 222(b), HRW photo by John Langford; 223(tr), Jeanne White/Photo Researchers, Inc.; 226(br), HRW photo by John Langford; 227, Russ Kinne/Comstock; 228(tl,trs), HRW photo by John Langford; 228(br), HRW photo by Patrick R. Dunn; 232(tr), Spencer Swanger/Tom Stack & Associates; 232(c), HRW photo by John Langford; 232(bl), Phil Degginger/H. Armstrong Roberts; 233(tr), Nicholas Foster/The Image Bank; 234 235(t), HRW photos by John Langford; 237(t), David R. Frazier/David R. Frazier Photolibrary; 237(c), HRW photo by John Langford; 238(br), Dwight Ellefsen/SuperStock; 239(tl), Beerman Collection/SuperStock; 239(br), William R. Sallaz/Duomo; 240, Michael Melford/The Image Bank; 241(t), John Lemker/Animals Animals/Earth Scenes; 241(b), Pekka Parviainen/Science Photo Library/Photo Researchers, Inc.; 243, HRW photo by John Langford; 244(tr), HRW photo by John Langford. 246(tl-icon), Tony Stone Images; 246(bl), Sylvia Otte/The Walt Disney Co./Discover Magazine; 246(tr), Fermilab Visual Media Services; 247(tr), R.N. Matheny/The Christian Science Monitor; 248(tl-icon), Tony Stone Images; 248(tr), Kim Heacox/Tony Stone Images; 249(tl-icon), Rogge/The Stock Market; 249(bl), HRW photo by John Langford; 249(tr), Peter Arnold, Inc./GE, Plastics Division.

UNIT 5: 250(tl), HRW photo by Dennis Fagan; 250(bckgd), The Granger Collection, New York; 250-251, Jany Sauvanet/Photo Researchers, Inc.; 252,253, HRW photo by John Langford; 254(br), Joe McDonald/Bruce Coleman, Inc.; 255(tr), Bruce Bander/Photo Researchers, Inc.; 255(c), Joyce Photographics/Photo Researchers, Inc.; 255(cr), Grant Heilman/Grant Heilman Photography, Inc.; 255(bl), HRW photo by Michelle Bridwell; 256(tr), Brown Brothers; 258(clockwise from (tl)), HRW photos by John Langford; 260(c), 260(bc), HRW photo by Michelle Bridwell; 262(b), S.J. Krasemann/Peter Arnold, Inc.; 264(tl), HRW photo by John Langford; 264(cr), Alan Briere/SuperStock; 264(bl), Nik Wheeler; 266(br), 267(bl), 267(br), HRW photo by Michelle Bridwell; 268(cr), HRW photo John Langford; 269(tr), Art Wolfe/Tony Stone Images; 270(br),271(tc), HRW photo by Michelle Bridwell; 272(tl), HRW photo by Michelle Bridwell; 272(cr), Dr. E.R. Degginger; 272(c),273,274, HRW photo by Michelle Bridwell; 276(cr), HRW photo by John Langford; 278(cr), HRW photo by John Langford; 279(tr), Wiley Wales/Adventure Photos; 279(cl), Scott Markewitz/FPG International, Corp.; 279(bl), Phil Degginger/Earth Scenes; 280(b), HRW photo by Michelle Bridwell; 281(both images), HRW photo by John Langford; 284(tr), From the picture by Marcus Stone; 284(br), Gianni Tortoli/Photo Researchers, Inc.; 287(r), Joe Viesti/Viesti Associates, Inc.; 287(l), HRW photo by John Langford; 288(tr), Dennis Hallinan/FPG International, Corp.; 289(tl), Vanessa Vick/Photo Researchers, Inc.; 289(br), Tom Pantages; 290(bl), Brown Brothers; 291(tr), DR. E.R. Degginger; 291(cl), Paul Silverman/Fundamental Photography; 291(cr), Dr. E.R. Degginger; 292(br), Novosti/Science Photo Library/Photo Researchers, Inc; 295(tr), HRW photo by John Langford; 298(tr),298(cr), 298(bc),300(br), HRW photo by Michelle Bridwell; 303(br), microwork/Phototake; 308(l), Archive Photos; 308(r), Jany Sauvant/PhotoResearchers, Inc.; 309(r), Phototake; 310(tl-icon), Rogge/The Stock Market; 310(l), Ghandi Shemdin/Tony

Stone Images; 310(r), H. David Hartman/Archive Photos; 311(tl), Archive Photos; 311(tr), The Everett Collection; 312(tl-icon), Ron Kimball; 313(tr), HRW Photo by Yoav Levy; 313(bl), Joseph Nicephore Niepce/Nicephore Niepce, Gernsheim Collection/HRHRC/Universtiy of Texas at Austin; 313(cl), The Granger Collection

UNIT 6: 314(bkgd), Letraset/Phototone; 314(tl), Jisas/Lockheed/Science Photo Library/Photo Researchers, Inc.; 314-315(inset), (c) Co Rentmeester; 316(tl), Nasa; 316(tr), Fred Bruemmer/Peter Arnold, Inc.; 316(bl), HRW photo by Russell Dian; 317(all), HRW photos by John Langford; 318(tr), Super Stock; 318(cl), Neal Graham/The Stock Shop; 318(cr), Nawrocki Stock Photo; 318(bl), Howard Bluestein/Photo Researchers, Inc.; 319(br),320(bl),321(br), HRW photos by John Langford; 323(tl), Tom Pantages; 323(tc), Nawrocki Stock Photo; 323(cl), Putnum/Peter Arnold, Inc.; 323(tr), The Kobal Collection; 323(br), Jeff and Alexa Henry/Peter Arnold, Inc.; 323(bc), David Lissy/Nawrocki Stock Photo; 323(br), John Elk/Bruce Coleman, Inc.; 325(tr), John Clausen/Mountain Stock; 325(cl),Yoav Levy/Phototake; 325(c),Nick Pawloff/The Image Bank; 325(cr), Frank Rossotto/ The Stock Market; 325(tc), Daniel MacDonald/The Stock Shop; 328(tr), Tom Pantages; 328(clockwise from (tc)), David Weintraub/Photo Researchers, Inc.; William E. Ferguson; 329(tl-top to bottom), Tom Campbell/Nawrocki Stock Photo, Inc.; HRW Photos by Sam Dudgeon; 331(b),335(tl), HRW photos by John Langford; 335(bl),336(br),337(tr),337(bl),HRW photo by John Langford; 343(c), E. R. Degginger/Color-Pic; 344(b),345(c), HRW photos by John Langford; 347(tl), Jeff Greenberg/d/Rainbow; 347(bl), N.Pecnik/Visuals Unlimited; 352(bl), 353(bckgd), 354(t),354(br),HRW photos by John Langford; 359(cl), Daedalus Enterprises, Inc.; 363(tl), Barbara Hofer/Barbara Hofer; 363(cr), C.C. Lockwood/Bruce Coleman Inc.; 363(br), HRW photo by John Langford; 365(tl), Michael Crone; 365(cl), Richard J. Green/Photo Researchers, Inc.; 365(cr-lower),365(cr-upper), The Granger Collection, New York; 365(bl), The Bettmann Archive; 365(br), Stock Montage; 366(tl),366(tr),366(c),366(bl), The Granger Collection, New York; 366(bc),Wide World Photos; 367(tl), AT&T; 367(tr), Everett Collection; 367(cl), Ken Graham/ Nawrocki Stock Photo Inc.; 367(br), Brooklyn Union Gas/Peter Arnold; 369(tc), The Bettmann Archive; 370(tr), Barry L. Runk/Grant Heilman Photography; 372(tr), Joe Rychetnik/Photo Researchers, Inc.; 372(b), Michael Gadomski/Bruce Coleman, Inc.; 373(tr), R. Phillips/The Image Bank; 373(bl), Lowel J. Georgia/Photo Researchers, Inc.; 374(l), Larry Lefever/Grant Heilman Photography, Inc.; 374(br), Alan Pitcairn/Grant Heilman Photography, Inc.; 375(tr), William Felger/Grant Heilman Photography, Inc.; 375(br), Larry Lefever/Grant Heilman Photography, Inc.; 382(tr), Co Rentmeester; 382(cl), Superstock; 384(tr), Kent Wood/Photo Researchers, Inc.; 384(tl-icon), Howard Sochurek/The Stock Market; 385(c) Bob Abraham/The Stock Market; 385(b), Manfred Kage/Peter Arnold, Inc.; 386(tl-icon), Ron Kimball/The Stock Market; 386(tr), Rod Planck/Photo Researchers, Inc.; 386(b), Dr. E.R. Degginger; 387, AIP Emilio Segre' Visual Archives.

UNIT 7: 388(l), HRW photo; 388(r), 389(all), G. Robertson/Auscape International; 390(tl), Charles Thatcher/Tony Stone Worldwide; 390(tr)(cl), HRW photos by John Langford; 390(bl), Mark Burnett/Photo Researchers, Inc.; 391, HRW photo by John Langford; 392(cr), C.S. Nielsen/Bruce Coleman, Inc.; 392(br), Keith Gunnar/ Bruce Coleman, Inc.; 393(cl), Matt Meadows, Peter Arnold, Inc.; 393(tr), D. Luria/FPG International, Corp.; 393(br), Photri Inc.; 394 HRW photo by Michelle Bridwell; 395(tr), HRW photo by Michelle Bridwell; 397, 399, HRW photos by Michelle Bridwell; 401,403, HRW photos by John Langford; 404, HRW photo by Michelle Bridwell; 405(tr), Seattle Times/Liaison International; 405(bl), William Sallaz/Image Bank; 406, HRW photo by Michelle Bridwell; 408, HRW photo by John Langford; 413(tl), Brown Brothers; 418(bl), HRW photo by Michelle Bridwell; 419(tr)(cl), HRW photos by John Langford; 419(bl), HRW by Michelle Bridwell; 420(tr), HRW photo by John Langford; 420(cr), Jonahan T. Wright/Bruce Coleman, Inc.; 420(bl), David M. Grossman/Photo Researchers, Inc.; 422(all),424, HRW photos by Michelle Bridwell; 427(top), HRW photo by John Langford; 427(cr), Canapress Photo Service; 428(cr),429(tr), HRW photos by Michelle Bridwell; 432(br), HRW photos by John Langford; 433(cr), HRW photo by Michelle Bridwell; 435(br), HRW photo by John Langford; 439(all), Michael Mitchell; 440(b), HRW photo by John Langford; 441(cr),(br), Michael Mitchell; 444(l), HRW photo by John Langford; 444(tr), G. Robertson/Auscape International; 446(tl-icon), Tony Stone Images; 446(tr), David Parker/Science Photo Library/Photo Researchers, Inc.; 447(tl-icon), Tony Stone Images; 447(tr,bl), Bob Parker/Austin Fire Investigation; 448(tl-icon), Rogge/The Stock Market; 448, Dan Winters/Discover Magazine; 449(bl), Forester/Wilkinson/ Superstock; 449(inset), Forester/Wilkenson/E.T. Archive, London/Superstock; 449(tr), James H. Karales/Peter Arnold, Inc.

UNIT 8: Page 450(tl-icon), C. Bruce Stoddard/FPG International, Corp.; 450(text), HRW photo by Ken Karp; 450(l-bckd), Letraset/Phototone; 450-451, Alberto Garcia/SABA; 452(cl), Dick Luria/FPG International, Corp.; 452(cl), Tom Bean; 453(tr), HRW photo by John Langford; 453(br), Tom Bean/Tom Bean & Associates; 454(tr), Birgitte Nielson; 454(br), C.E. Nagele/FPG International, Corp.; 455(tl), David Muench; 455(tr), Birgitte Nielson; 455(bc), David Muench; 456, HRW photo by John Langford; 457(tl), Tom Tracy/After Image; 457 (tr), U.S. Geological Service; 457(br), E.R. Degginger/Color-Pic; 458(tr), David L. Langford/UPI/The Bettmann Archives; 458(br), UPI/Bettmann News Photos; 458(bl), Tom Bean/Tom Bean & Associates; 464(tr), Alex S. MacLean/Landslides; 469(tr), E. R. Degginger/Color-Pic; 470(tl), Tom Stack/Tom Stack & Associates; 470(br), Joen Laconetti/Bruce Coleman Inc; 470(bl), The Metropolitan Museum of Art; 471, HRW photo by John Langford; 472(tl), Phil Degginger; 472(tr), Photri; 472(bl), Dr. E.R. Degginger; 472(br), E. R. Degginger/Color-Pic; 473(r), Ward's Natural Science Establishment; 474, Hennepin's First Sight of Niagara Falls, December, 1678, Charles William Jeffereys pen and black ink over pencil, Acc. No. 1972-2C-579X, C-70245, National Archives of Canada; 475(c), Dallas and John Heaton/Stock Boston; 478(br), J. Verheyden/Bruce Coleman, Inc.; 478(tr), Joe Szkodzinski/The Image Bank; 478(cr), Keith Gunnar/Bruce Coleman Inc.; 478 (bckgd), C. Bruce Stoddard/FPG International, Corp.; 479(b), HRW photo by Michelle Bridwell; 480(tr), E. R. Degginger/Color-Pic; 480(cr), E. R. Degginger/Color-Pic; 480(bl), E. R. Degginger/Color-Pic; 481(tr), C.G. Maxwell/Photo Researchers, Inc.; 481(cl), Doris Dewitt/TSW/Click/Chicago; 481(bl), Tom Bean/Tom Bean & Associates; 481(br), E. R. Degginger/Color-Pic; 482, HRW photo by Michelle Bridwell; 483(tr), Phil Degginger; 483(cr), E.R. Degginger; 483(bl), Phil Degginger; 483(br), E.R. Degginger; 485(br), E.R. Degginger; 486(br), Tom Bean; 486(b), David Muench; 487(t), E.R. Degginger; 488(t), AP/WideWorld Photos, Inc.; 488(bl), John Elk/Bruce Coleman, Inc.; 489(bc), HRW photo by John Langford; 490(tr), Doug Menuez/Stock Boston; 490(br), James Balog; 490(bl), Tom Ericson/ Mountain Stock; 491(bl), William H. Mullins/Photo Researchers, Inc.; 493(tr), E.R. Degginger; 493(cl), Tom Bean; 495(t), Official U.S. Navy photograph; 496(cl), 496(bl),498(br), HRW photos by John Langford; 498(bckgd), Alan L. Detrick/Color-Pic; 498(br-bckgd), HRW photo; 499(tr), Fredrick D. Atwood; 499(bl), E.R. Degginger; 499(br), E.R. Degginger; 500(b), HRW photo by John Langford; 501(br), E.R. Degginger; 503(b), HRW photo by John Langford; 505(br), Dr. E.R. Degginger/ Color-Pic; 506(tr), E.R. Degginger; 506(br), NASA; 507(tr), Norbert Wu; 507(br), Landsat/Photri; 508(tr),508(c), Tom Bean; 508(b), E.R. Degginger; 509(tr),509(b), Tom Bean; 509(br),Thomas Wanstall; 510(tr), Phil Degginger; 510(br), Lillian Bolstad/Peter Arnold; 511(l), Norbert Wu; 511(bl), John Elk/Bruce Coleman Inc.; 512(tr), UPI/Bettmann; 512(br), Tom Bean; 514(t), Paolo Koch/Photo Researchers; 514(br), HRW photo by John Langford; 515(b), Tom Bean; 516(br), James P. Blair/National Geographic Image Collection; 516(tl), The Bettmann Archive; 516 (tr), E.R. Degginger; 516(bl), Tom Bean; 518(tl), E.R. Degginger; 518(cl), Tom Bean; 518(bl), Dave Bartruff/FPG International, Corp.; 522(tr) Alberto Garcia/SABA; 522(cl), Charlie Ott/Photo Researchers; 524(tl), Jane Scherr; 524(bl), Tom Till/Tony Stone Images; 525(tc), Tom McHugh/Photo Researchers, Inc.; 525(tr), Archive Photos; 525(bl), D. Mazonowicz/Bruce Coleman, Inc.; 526(tr), W. Bacon/Photo Researchers, Inc; 526(bl), James Balog/Bruce Coleman, Inc.; 527(inset), Superstock; 527(tr), Gay Bumgarner/Tony Stone Images; 527(tl-icon), Tony Stone Images

SOURCEBOOK: Page S1(tl), Douglas Faulkner/Photo Researchers, Inc.; S2(tr)Emory Kristof/National Geographic Society,S2(bl), National Geographic Society; S3(tl), James Sugar/Black Star; S3(tr), Ralph Perry/Black Star; S3(br), Dr. E. R. Degginger/Color-Pic; S5(tl), Pedrick/The Image Works; S5(c), NASA; S6(tl), Robert Reichert/Liaison International; S6(tr),S6(c), HRW photo by Yoav Levy; S6(cr), Phototake/Phototake; S6(bl), Erich Hartmann/Magnum Photos; S7(tr), Photri; S7(bl), U.S.Geological Service; S8(tr), Ralph Perry/Black Star; S8(cl), Stephen J. Krasemann/DRK PHOTO; S8(bl), HRW photo by Russell Dian; S9(tl), Bill Curtsinger; S9(tr), Alan Carey/The Image Works; S9(cr), Will & Deni McIntyre/Photo Researchers, Inc.; S9(br), Michael Heron/Woodfin Camp & Associates; S10(tl), Eric Sander/Gamma-Liaison; S10(tr), John Mason/International Stock Photo; S10(cl), Michal Heron/Woodfin Camp & Associates; S10(br), Joseph Lynch/Medical Images, Inc.; S11(t), Dan McCoy/Rainbow; S11(tl), CERN/ Science Photo Library/Photo Researchers, Inc.; S11(cr), Y. Arthus-Bertrand/Photo Researchers, Inc.; S13(cr), Stephen J. Krasemann/DRK; S14(cl), Michael Wells; S15(tr), Tom Bledsoe/DRK; S15(bl), Tom McHugh/Photo Researchers, Inc.; S18(bc), Tom Tracy/Medichrome/The Stock Shop; S22(tr), HRW photo; S23(tl), Stepen Dalton/Photo Researchers, Inc.; S24(br), HRW; S24(br), Tom McHugh/ Steinhart Aquarium/Photo Researchers, Inc.; S25(tr), D. Klesenski/International Stock Photos; S25(c), Gregory Ochocki/Photo Researchers, Inc.; S25(bl), M. P. L. Fogden/Bruce Coleman, Inc.; S25(bl), Dr. E. R. Degginger/Color-Pic; S25(cl), HRW photo by Russell Dian; S25(bc), HRW; S25(bc), Michael Fogden/Bruce Coleman, Inc.; S25(br), HRW photo by Russell Dian; S25(br), R. Kelly/Photri, Inc.; S25(br), Fred Bavendam/Peter Arnold, Inc.; S26(cl), James L. Amos/Photo Researchers, Inc.; S29(tc), Hällie Flygare/Bruce Coleman, Inc.; S29(tr), Kamal/Photri; S29(c), M. J. Manuel/Photo Researchers, Inc.; S29(cr), Scott T. Smith/Scott T. Smith; S33(tr), Breck P. Kent; S33(cr), Breck P. Kent; S34(cr), Breck P. Kent/Breck P. Kent; S34(bl), Miguel Castro/Photo Researchers; S34(bl), Breck P. Kent/Breck P. Kent; S34(bc), Francois Gohier/Photo Researchers, Inc.; S34(br), Breck P. Kent/Breck P. Kent; S34(br), Breck P. Kent/Breck P. Kent; S36(tl), Horst Schafer/Peter Arnold, Inc.; S36(cl), Dan McCoy/Rainbow; S37(t), M. Abbey/Visuals Unlimited; S39(tr), HRW photo by Richard Haynes; S41(b), HRW photo by Richard Hutchings; S43(b), Art Tilley/FPG International, Corp.; S44(tr), Manfred Kage/Peter Arnold, Inc.; S44(cl), M. I. Walker/Science Source/Photo Researchers, Inc.; S44(cr), Dr. Tony Brain/Science Photo Library/Photo Researchers, Inc.; S54(cr), Martin/Custom Medical Stock Photos; S55(bl), HRW photo; S55(tl), Ed Reschke/Peter Arnold, Inc.; S55(tc), Michael Abbey/Science Source/Photo Researchers, Inc.; S55(tr), Leonard Lessin/Peter Arnold, Inc.; S56(br), Manfred Kage/Peter Arnold; S57(br), Scott Camazine/Photo Researchers, Inc.; S60(tl), David M. Phillips/Photo Researchers, Inc.; S60(cr), Petit Format/Nestle/Photo Researchers, Inc.; S65(tl), Manfred Kage/Peter Arnold, Inc.; S66(tl), Photri, Inc.; S66(cr),Tom McHugh/Photo Researchers, Inc.; S68(tl), R.A. Mittermeier/Bruce Coleman Inc.; S68(c,LtoR),Michael P. Gadomski/Bruce Coleman Inc.; S68(bl), Will & Deni McIntyre/Photo Researchers, Inc.; S69(tl), Dwight Kuhn/Dwight Kuhn; S69(tr), James A. Carmichael Jr./Bruce Coleman Inc.; S69(cl), Larry West/Bruce Coleman Inc.; S70(cl), Arthur M. Siegelman/Visuals Unlimited; S70(br-inset), Manfred Kage/Peter Arnold, Inc.; S71(tc), Coco McCoy/Rainbow; S71(cr), Peter

Ward/Bruce Coleman, Inc.; S72(tl), Lee D. Simon/Science Source/Photo Researchers, Inc.; S72(c), M. Abbey/Photo Researchers, Inc.; S72(br), M. D. Maser/Visuals Unlimited; S73(bl), Runk/Schoenberger/Grant Heilman Photography; S73(bc), Michael Abbey/Science Source/Photo Researchers, Inc.; S73(br), Michael Abbey/Photo Researchers, Inc.; S74(cl), R.N. Mariscal/Bruce Coleman Inc.; S74(c), F. Widdel/Visuals Unlimited; S74(br), Hank Morgan/Rainbow; S75(tr), Ronald L. Sefton/Bruce Coleman Inc.; S75(cr), CNRI/Science Photo Library/Photo Researchers, Inc.; S75(br), Biophoto Associates/Photo Researchers, Inc.; S78(cl), Diane Koos Gentry/Black Star; S78(br), SuperStock; S79(tr), PhotoTone; S81 (top half page-clockwise l to r), (first two) Dr. Jeremy Burgess/Science Photo Library/Photo Researchers, Inc., Ralph C. Eagle/Photo Researchers, Inc., (next two) Dr. Jeremy Burgess/Science Photo Library/Photo Researchers, Inc., S. Camazine/S. Billo/Photo Researchers, Inc.; S81(bl), S. L. Craig, Jr./Bruce Coleman, Inc.; S81(bc), CNRI/Science Photo Library/Photo Researchers, Inc.; S82(cr), NIBSC/Science Photo Library/Photo Researchers, Inc.; S85(tl), HRW photo by John Langford; S86(all), NASA S87(t) (l to r),E.R. Degginger/Color-Pic; S89(tr), David R. Frazier/David R. Frazier Photolibrary; S89(cl), Bob Burch/Bruce Coleman, Inc.; S89(c), HRW photo by Dennis Fagan; S90(b), HRW photo by Russell Dian; S91(bc), Johnson Space Center, NASA; S92(tc,l to r), HRW photo by John Langford; S92(bl), S92(bc, HRW photo by Dennis Fagan; S93(c), Kevin Schafer/Tom Stack & Associates S94(tl),S94(tl,inset), HRW photo by John Langford; S95(bl),S95(bc), Tom Pantages/Tom Pantages;S96(cr), NASA; S96(cl), E.R. Degginger/Color Pic; S96(bl), Phil Degginger/Color-Pic; S96(bc),S96(br), Tom Pantages/Tom Pantages; S97(tl), HRW photo by Sam Dudgeon; S97(bc), Breck P. Kent; S100(tl), New York Public Library; S100(bc), Antoinio Gusmao/Black Star; S101(tr), Paul Silverman/Fundamental Photographs; S101(bl), Richard Megna/Fundamental Photographs; S102(br), Peter Arnold, Inc.; S103(tr),S103(cl),S103(br), Fermilab Visual Media Services; S107(tl), Jany Sauvanet/Photo Researchers, Inc.; S108(tl), Dr. Albert V. Crewe/University of Chicago; S117(tr), Archive Photos; S117(cl), Paul F. Gero/Nawrocki Stock Photos; S118(tl), Mendeleev's original manuscript; S120(tr), S120(bl), HRW photo by John Langford; S125(tr), Wm. Rivelli/Image Bank; S127(tr), Charles D. Winters/Photo Researchers, Inc.; S127(tr-inset), Dave Brown/Nawrocki Stock Photos; S129(tr), John Zoiner; S130(tr), HRW photo by Russell Dian; S135(bkgd) (tl), Co Rentmeester; S136(tl), Four BY Five/SuperStock; S137(tc),SuperStock; S137(tr), Ronald C. Modra/Sports Illustrated/Time Inc.; S138(t), SuperStock; S139(bl), Archive Photos; S140(bl), HRW photo by

John Langford; S141(cr), Woodfin Camp; S142(tl), John Gerlach/Visuals Unlimited; S143(tr), Richard Megna/Fundamental Photographs; S143(c), HRW photo by Sam Dudgeon; S146(bl), M.A Doolottle/Rainbow; S147(tr), SuperStock; S147(cl), Hal Beral/Visuals Unlimited; S148(bl), Dr. E.R. Degginger/Color Pic; S149(tl), John Zoiner; S149(br), Visuals Unlimited; S150(tr), Manfred Kage/Peter Arnold, Inc.; S150(bl), The Bettmann Archive; S151(c), Tom Hollyman/Photo Researchers, Inc.; S152(t), SuperStock; S152(tr), Lawrence Livermore Lab/Science Photo Library/Photo Researchers, Inc.; S153(tr), Linda K. Moore/Rainbow; S154(tl), Douglas Kirkland/Sygma; S154(cr), Lowell Georgia/Science Source/Photo Researchers, Inc.; S154(bl), H.R.Bramaz/Peter Arnold, Inc.; S155(bl), Dan McCoy/Rainbow; S156(tl), John & Karen Hollingsworth; S156(bc), Simon Fraser/Science Photo Library/Photo Researchers, Inc.; S157(br), Frank Rossotto/The Stock Market; S158(tl), SuperStock; S160(tr), Hiroyuki Matsumoto/Black Star; S161(tl), G. Robertson/Auscape International; S161(bkgd), Letraset/Phototone; S162(cl), Joe Sohm/Photo Researchers, Inc.; S163(b), Jerry Driendl/FPG International, Corp.; S165(c), Douglas R. Shane/Photo Researchers, Inc.; S166, HRW photo by John Langford; S167(tc), Thermos Company; S170(tl), Photo Researchers, Inc.NOAA; S170(cl), Photo Researchers, Inc.; S170(bc), Bill Brooks/Bruce Coleman, Inc.; S170(br), Nicholas DeVore III/Bruce Coleman, Inc.; S171(tr), NASA; S171(cl), Jim Steinberg/Photo Researchers, Inc.; S171(cr), Bern Pedit/Breck P. Kent Photos; S171(bl), Breck P. Kent; S173(bl), William E. Ferguson; S175(b),World Book, Inc.; S176(tr), HRW photo by John Langford; S179(tl), Alberto Garcia/SABA; S180(tl), NASA; S180(bl), SuperStock; S185(br), Joseph L. Fontenot/Visuals Unlimited; S185(b), Danny Lehman/Westlight; S187(tr), Breck P. Kent/Breck P. Kent; S188(b), S188(tr), Tom Till/Tom Till Photography; S188(cr), Scott T. Smith/Scott T. SmithS190(cl), Bart Bartholomew/Black Star; S190(br), Breck P. Kent/Breck P. Kent; S191(tr), Jacques Jangoux/Tony Stone World Wide; S191(bl), Tom Till/TomTill Photography; S192(bc), Richard Weiss/Peter Arnold, Inc.; S192(br), Stouffer/Earth Scenes; S193(tr), Danny Lehman/Westlight; S194(tl), Breck P. Kent/Breck P. Kent; S195(tr), Chlaus Lotscher/Alaska Stock; S195(b), Breck P. Kent/Breck P. Kent; S196(br), Charlie Ott/Photo Researchers Inc.; S197(tl), Breck P. Kent/Breck P. Kent; S197(tr), Tom Till/Tom Till Photography; S197(bl), John Sohlden/Visuals Unlimited; S198(br), Visuals Unlimited; S199(tl), Dr. E. R. Degginger/Color-Pic; S199(tc), Tom Bean/DRK Photo; S200(bl), Grant Heilman/Grant Heilman Photography; S201(t), Breck P. Kent/Breck P. Kent; S201(cr), Robert W. Parvin; S202 & 203(b), "World Ocean Map", Bruce C. Heezen and Marie Tharp, 1977,; S206(tr), SuperStock

Abbreviated as follows: (t) top; (b) bottom; (l) left; (r) right; (c) center.

ART CREDITS

All work, unless otherwise noted, contributed by Holt, Rinehart and Winston

TABLE OF CONTENTS: v, Sarah Woodward/Morgan Cain & Associates; vi, Stephen Durke/Washington-Artists' Represents, Inc.; ix, Don Brautigam/Bill Erlacher, Artists Associates; x, Craig Attebery/Jeff Lavaty Artist Agent.

Unit 1: Page 9, James Swanson/Renée Kalish; 10, James Swanson/ Renée Kalish; 11, John Francis; 12, John Francis; 13, Liaison Production Services; 16, Doug Bowles/Diann Roche Represents; 19, Stephen Durke/Washington-Artists' Represents, Inc.; 20(c-l), Reggie Holladay; 20(bl), Stephen Durke/Washington-Artists' Represents, Inc., 22, Doug Bowles/Diann Roche Represents; 25, Stephen Durke/Washington-Artists' Represents, Inc.; 26, Blake Thornton/Rita Marie and Friends; 27, Uhl Studio; 28, Uhl Studio; 29, Rich Bowman/Diann Roche Represents; 31-32, Tim Spransy/Wilson Zumbo Illustration Group, Inc.; 33-34, Wayne Parmenter/Cliff Knecht Artist Representative; 37, Tim Ladwig/Suzanne Craig Represents; 43(cl), Lyne Raff; 43(tr), James Swanson/Renée Kalish; 47, Don Brautigam/Bill Erlacher, Artists Associates; 51, Donna Kae Nelson/Sharon Langley Artist Representatives; 52, Keith Locke/ Suzanne Craig Represents.

Unit 2: Page 65, Lori Anzalone/Jeff Lavaty Artist Agent; 66, Doug Bowles/Diann Roche Represents; 69, Lori Anzalone/Jeff Lavaty Artist Agent; 70, Patrick Gnan/Deborah Wolfe Limited Artists' Representative; 71, Patrick Gnan/Deborah Wolfe Limited Artists' Representative; 72, Walter Stuart/Richard W. Salzman Artist Representative; 76-77, Cary Henrie/James Conrad Artist Representative; 78, Tim Ladwig/Suzanne Craig Represents; 82(t-b), Pedro Julio Gonzalez/Melissa Turk & The Artist Network; 83, Sarah Woodward/Morgan Cain & Associates; 86, Patrick Gnan/Deborah Wolfe Limited Artists' Representative; 92, Reggie Holladay; 96-99, Joyce Kitchell/James Conrad Artist Representative; 102(c), Bob Lange/Diann Roche Represents; 103, Valerie Marsella; 106-107, David Griffin; 109(t-b), Lori Anzalone/Jeff Lavaty Artist Agent; 110, Gary Locke/Suzanne Craig Represents; 111, Guy Wolek; 113, Brad Gaber; 116, Uhl Studio; 117, Morgan Cain & Associates; 124, Ka Botz/Melissa Turk & The Artist Network; 127, Bob Lange/Diann Roche Represents.

Unit 3: Page 131-133, Todd Lockwood/Carol Guenzi Agents; 134, Leo Monahan/ Jae Wagoner, Artists' Representative; 137, Richard Pembroke/American Artists Rep., Inc.; 138, Uhl Studio; 140(blk & wht line art), Cende Courtney-Hill/Morgan Cain & Associates ; 140(color), Liaison Production Services; 142, Gary Yealdhall/ American Artists Rep., Inc.; 144, Blake Thornton/Rita Marie and Friends; 146, Michael Krone; 149, Daniel Vasconcellos; 154, Don Brautigam/Bill Erlacher, Artists Associates; 158, Liz Wheaton/Nadine Hunter; 167, Sarah Woodward/Morgan Cain & Associates; 168, Cende Courtney-Hill/Morgan Cain & Associates; 171, Edd Patton; 175, Cende Courtney-Hill/Morgan Cain & Associates; 176, Dean Kennedy/ Carol Guenzi Agents; 178, Todd Lockwood/Carol Guenzi Agents; 185, Howard Fullmer/Cary & Company; 188, Todd Lockwood/Carol Guenzi Agents; 195, Wendy Smith-Griswold/Melissa Turk & The Artist Network.

Unit 4: Page 201, Keith Locke/Suzanne Craig Represents; 202, Pedro Julio Gonzalez/Melissa Turk & The Artist Network; 205, Aletha Reppel/Suzanne Craig Represents; 208, Leo Monahan/Jae Wagoner, Artists' Representative; 214, Valerie Marsella; 217, Uhl Studio; 220, Barbara Kiwak/Barbara Gordon Associates Ltd.; 221, Don Sullivan; 223, Bob Lange/Diann Roche Represents; 224, Stephen Durke/ Washington-Artists' Represents, Inc.; 225, Uhl Studio; 226, Richard Wehrman; 227, Gary Otteson/Diann Roche Represents; 229, Stephen Durke/Washington-Artists' Represents, Inc.; 230(t-c), Stephen Durke/Washington-Artists' Represents, Inc.; 231, Stephen Durke/Washington-Artists' Represents, Inc.; 233, Tim Spransy/Wilson Zumbo Illustration Group, Inc.; 238, David Merrell/Suzanne Craig Represents; 242, Lyne Raff; 247, James Pfeffer.

Unit 5: Page 257, Holly Cooper; 259, David Merrell/Suzanne Craig Represents; 265, Dennis Jones; 275, Elena Poladian/Creative Freelancers Management Inc.; 282, Don Brautigam/Bill Erlacher, Artists Associates; 283, Aletha Reppel/Suzanne Craig Represents; 285, Lyne Raff; 290-291, Graber Graphics; 295, Holly Cooper; 296-297, Tim Ladwig/Suzanne Craig Represents; 302, Martens & Kiefer/Carol Chislovsky Design Inc.; 306, Uhl Studio; 310, Uhl Studio; 311, Stephen Durke/Washington-Artists' Represents, Inc.; 312, Uhl Studio.

S223

Unit 6: Page 317(bkgd), Reggie Holladay; 319, Gary Otteson/Diann Roche Represents; 320, Uhl Studio; 321, Uhl Studio; 322, Uhl Studio; 324, Gary Locke/Suzanne Craig Represents; 325, Edd Patton; 326, Thomas Hudson/Woody Coleman Presents Inc.; 327-332(b), Norman Bendell/David Goldman Agency; 329(cr), Edd Patton; 330(cl), Uhl Studio; 330(r), Steve Schindler; 333, Richard Wehrman; 337, John Francis; 338, Uhl Studio; 339, Tony Morse/Ivy Glick & Associates; 340, John Francis; 341(t-c), John Francis; 342(tr), Michael Koester/Woody Coleman Presents, Inc.; 342(c), Patrick Gnan/Deborah Wolfe Limited Artists' Representative; 346(t-b), Neal Aspinall/Wilson Zumbo Illustration Group, Inc.; 348-349, Ed Wexler/Carol Chislovsky Design Inc.; 350-351, Gary Locke/Suzanne Craig Represents; 352, John Francis; 355, Doug Bowles/Diann Roche Represents; 356-357, Aletha Reppel/Suzanne Craig Represents; 358(c-b), John Francis; 360, Patrick Gnan/Deborah Wolfe Limited Artists' Representative; 361(t-b), Mary Ross/Rita Gatlin Represents; 362, Michael Krone; 364(t), Rich Bowman/Diann Roche Represents; 365(tl), Michael Krone; 365(tr), Don Brautigam/Bill Erlacher, Artists Associates; 366(cr,br), Patrick Gnan/Deborah Wolfe Limited Artists' Representative; 367(tl), Edd Patton; 371(c), Don Brautigam/Bill Erlacher, Artists Associates; 373, John Francis; 376, Mark Kaufman/Barney Kane & Friends; 377, Mark Kaufman/Barney Kane & Friends; 378, Mark Kaufman/Barney Kane & Friends; 379(tr), Don Brautigam/Bill Erlacher, Artists Associates; 380, Gary Otteson/Diann Roche Represents; 381, Mary Ross/Rita Gatlin Represents; 384, Bob Lange/Diann Roche Represents.

Unit 7: Page 391, Dennis Jones; 392, Brad Gaber; 398, Uhl Studio; 400, Tim Ladwig/Suzanne Craig Represents; 402(r), Valerie Marsella; 402, James Pfeffer; 403, Richard Wehrman; 404, Uhl Studio; 410, Daniel Vasconcellos; 411, Gary Locke/Suzanne Craig Represents; 413(t), Lyne Raff; 413(b), Uhl Studio; 414, Uhl Studio; 417, Keith Locke/Suzanne Craig Represents; 421, Robin Carter/Morgan Cain & Associates; 423, Tim Ladwig/Suzanne Craig Represents; 427, Don Brautigam/Bill Erlacher, Artists Associates; 431-432, Uhl Studio; 435, Scott Johnston; 436, Brad Gaber; 437-438, Mary Ross/Rita Gatlin Represents; 438(tr), Richard Wehrman; 443(cr), Don Brautigam/Bill Erlacher, Artists Associates; 443(b), Brad Gaber; 445, John Francis; 446, Don Brautigam/Bill Erlacher, Artists Associates; 448, James Pfeffer.

Unit 8: Page 459, Liaison Production Services; 460, Keith Locke/Suzanne Craig Represents; 461(t,b), Walter Stuart/Richard W. Salzman Artist Representative; 463, Uhl Studio; 465, GeoSystems Global Corportation; 466, Liaison Production Services; 467, Liaison Production Services; 468(t), Craig Attebery/Jeff Lavaty Artist Agent; 468(b), Dennis Jones; 469, Valerie Marsella; 471, Tim Ladwig/Suzanne Craig Represents; 476, Morgan Cain & Associates;477(cr), Sarah Woodward/Morgan Cain & Associates; 477(b), Craig Attebery/Jeff Lavaty Artist Agent; 484, Daniel Vasconcellos; 488, Richard Wehrman; 491, David Merrell/Suzanne Craig Represents; 494(t), Robin Carter/Morgan Cain & Associates; 494(b), Uhl Studio; 497, Robin Carter/Morgan Cain & Associates; 502, Craig Attebery/Jeff Lavaty Artist Agent; 504, Craig Attebery/Jeff Lavaty Artist Agent; 513(t), GeoSystems Global Corportation; 513(br), Uhl Studio; 517, Craig Attebery/Jeff Lavaty Artist Agent; 519, GeoSystems Global Corportation; 520, Richard Pembroke/American Artists Rep., Inc.; 521, Gary Locke/Suzanne Craig Represents; 523, Lyne Raff; 524, Morgan Cain & Associates.

Sourcebook Illustration Credits

Unit 1: Page S4, Sally J. Bensusen/Visual Science Studio; S12, John Francis; S14, Bob Lange/Diann Roche Represents; S16, Mark Mille/Sharon Langley Artist Representatives S19, Mark Mille/Sharon Langley Artist Representatives.

Unit 2: Page S24(bkgd), Robert Pasternak/Melissa Turk & The Artist Network; S24(plants & animals), Susan Johnston Carlson/Melissa Turk & The Artist Network; S27, Wendy Smith-Griswold/Melissa Turk & The Artist Network; S28(t-b), Wendy Smith-Griswold/Melissa Turk & The Artist Network; S30, Pedro Julio Gonzalez/Melissa Turk & The Artist Network; S31, Pedro Julio Gonzalez/Melissa Turk & The Artist Network; S32, Walter Stuart/Richard W. Salzman Artist Representative; S33(cl-b), Keith Locke/Suzanne Craig Represents; S35, Walter Stuart/Richard W. Salzman Artist Representative; S38, Keith Locke/Suzanne Craig Represents; S39(c), Pfizer, Inc.; S40, Martens & Kiefer/Carol Chislovsky Design Inc.; S41(l), Karen Klomparens/Michigan State University; S42(c), James Dowdalls; S45, Martens & Kiefer/Carol Chislovsky Design Inc.; S47, Martens & Kiefer/Carol Chislovsky Design Inc.; S48(tr-cl), Martens & Kiefer/Carol Chislovsky Design Inc.; S49, Martens & Kiefer/Carol Chislovsky Design Inc.; S50, Martens & Kiefer/Carol Chislovsky Design Inc.; S51(t), Martens & Kiefer/Carol Chislovsky Design Inc.; S52, Martens & Kiefer/Carol Chislovsky Design Inc.; S53, Keith Kasnot; S56(tl-c), Martens & Kiefer/Carol Chislovsky Design Inc.; S57, Martens & Kiefer/Carol Chislovsky Design Inc.; S58, Martens & Kiefer/Carol Chislovsky Design Inc.; S59, Lennart Nilsson/A Child is Born, Dell Publishing Company; S61, Martens & Kiefer/Carol Chislovsky Design Inc.; S62, Martens & Kiefer/Carol Chislovsky Design Inc.; S64, Walter Stuart/Richard W. Salzman Artist Representative.

Unit 3: Page S67, Bob Dorsey; S73, Walter Stuart/Richard W. Salzman Artist Representative; S76, Mitch Grasso/Deborah Wolfe Limited Artists' Representative; S77, Mitch Grasso/Deborah Wolfe Limited Artists' Representative; S79, Walter Stuart/Richard W. Salzman Artist Representative.

Unit 4: Page S88(tl), Uhl Studio; S88(3 molecular insets), Graber Graphics; S91(t), Paul Hess/Garden Studio Illustrators; S93(cr-bl), Graber Graphics; S95(t), Stephen Durke/Washington-Artists' Represents, Inc.; S95(b), Graber Graphics; S97, Graber Graphics; S98, Tim Spransy/Wilson Zumbo Illustration Group, Inc.; S99, Gary Locke/Suzanne Craig Represents; S102, Tim Spransy/Wilson Zumbo Illustration Group, Inc.; S103, Craig Attebery/Jeff Lavaty Artist Agent; S104, Tim Ladwig/Suzanne Craig Represents; S106, Joe McDermott/Koralik Associates.

Unit 5: Page S108, Graber Graphics; S109(tr), Graber Graphics; S109(b), Michael Koester/Woody Coleman Presents, Inc.; S110, Uhl Studio; S111, Uhl Studio; S112, Graber Graphics; S113, Graber Graphics; S114(t-b), Graber Graphics; S115, Graber Graphics; S116, Graber Graphics; S118-S119, Graber Graphics; S121, Graber Graphics; S123, Graber Graphics; S124, Graber Graphics; S125, Graber Graphics; S126, Graber Graphics; S127-S128, James Swanson/Renée Kalish; S129, Graber Graphics; S131, Graber Graphics; S132, Graber Graphics; S134, Graber Graphics.

Unit 6: Page S139, Stephen Durke/ Washington-Artists' Represents, Inc.; S140, Michael Koester/Woody Coleman Presents,Inc.; S142, Tim Ladwig/Suzanne Craig Represents; S144, Stephen Durke/ Washington-Artists' Represents, Inc.; S145(t), Stephen Durke/ Washington-Artists' Represents, Inc.; S145(b) Bob Lange/Diann Roche Represents; S146(t-c), Bob Lange/Diann Roche Represents; S149, John Francis; S155, John Francis; S157, Stephen Durke/ Washington-Artists' Represents, Inc.

Unit 7: Page S162, Uhl Studio; S163, Uhl Studio; S164, Uhl Studio; S165, Graber Graphics; S166, Saul Rosenbaum/Deborah Wolfe Limited Artists' Representative; S168, Richard Wehrman; S171, Stephen Durke/ Washington-Artists' Represents, Inc.; S172, Uhl Studio; S172(2 molecular insets), Graber Graphics; S173, John Francis; S174, Patrick Gnan/Deborah Wolfe Limited Artists' Representative; S178, Bill Geisler.

Unit 8: Page S181(t), Greg Harris/Cornell & McCarthy Artist Representatives; S181(b), Uhl Studio; S182-183, Craig Attebery/Jeff Lavaty Artist Agent; S184, Greg Harris/Cornell & McCarthy Artist Representatives; S186-S187, Craig Attebery/Jeff Lavaty Artist Agent; S189, Norman Adams/Bill Erlacher, Artists Associates; S191, Norman Adams/Bill Erlacher, Artists Associates; S193, Craig Attebery/Jeff Lavaty Artist Agent; S195, Uhl Studio; S196, Uhl Studio; S199, Craig Attebery/Jeff Lavaty Artist Agent; S200, Uhl Studio; S204, Craig Attebery/Jeff Lavaty Artist Agent.

Acknowledgments

For permission to reprint copyrighted material, grateful acknowledgment is made to the following sources:

Roy Bishop: Excerpt (retitled: "Energy from a Star") from *Whence Energy* by Roy Bishop.

Houghton Mifflin Company: From "Riddles in the Dark" from *The Hobbit* by J.R.R. Tolkien. Copyright © 1966 by J.R.R. Tolkien. All rights reserved.

Alfred A. Knopf, Inc.: From "The Negro Speaks of Rivers" from *Selected Poems* by Langston Hughes. Copyright 1926 by Alfred A. Knopf, Inc.; copyright renewed © 1954 by Langston Hughes. "January" from *A Child's Calendar* by john Updike. Copyright © 1965 by John Updike and Nancy Burkert.

S224

Abbreviations used: (t) top, (c) center, (b) bottom, (l) left, (r) right

PHOTO CREDITS

OWNER'S MANUAL: Page T5, T15, T17, T18, T22, T24,T25(b),T28, T29, T36, T41, T44(t), T44(t), T45, T46, T55(r),T56, T58, HRW photo by Sam Dudgeon; T19, T42, T44(b), T49, HRW photo by Daniel Schaefer; T16, T50–T51, HRW photo by Tomas Pantin; T25, T26, T33, T35, T38, T39, T47, T48, HRW photo by John Langford; T55(l), HRW photo by Jack Newkirk; T20, T23, T30(t), HRW photo by Scott Van Osdol; T31, David Young Wolff/Photo Edit; T43, Frontera Fotos/Michelle Bridwell

UNIT 1: Page 1A, HRW photo by Sam Dudgeon; 1C, James King-Holmes/SPL/Photo Researchers; 1C(b), N.H. (Dan) Cheatham/DRK; 1D, HRW photo by John Langford; 56(t), Rogge/The Stock Market; 57(t), Robert Johannsen; 58(t), HRW photo; 59(t), Howard Sochurek/The Stock Market

UNIT 2: Page 59D, HRW photo by Sam Dudgeon; 124(t), Ron Kimball; 125(t) Tony Stone Images; 126(t), HRW photo; 127(t), Tony Stone Images

UNIT 3: Page 127D, HRW photo by Sam Dudgeon; 194(t), Howard Sochurek/The Stock Market; 195(t), Ron Kimball; 196(t), Rogge/The Stock Market; 197(t), Burt Glinn/Magnum Photos, Inc.

UNIT 4: Page 197A, 197B, HRW photo by John Langford; 197D, HRW photo by Sam Dudgeon; 246(t), Tony Stone Images; 247(t), HRW photo; 248(t), Tony Stone Images; 249(t), Rogge/The Stock Market

UNIT 5: Page 249A, Dr. E.R. Degginger/Color-Pic; 249D, HRW photo by John Langford; 310(t), Rogge/The Stock Market; 311(t), Archive Photos; 312(t), Ron Kimball; 313(t), HRW photo

UNIT 6: Page 313D, HRW photo by John Langford; 384(t), Howard Sochurek/The Stock Market; 385(t), HRW photo; 386(t), Ron Kimball; 387(t), AIP Emilio Segre/Visual Archives

UNIT 7: Page 387A, HRW photo by John Langford; 446(t), Tony Stone Images; 447(t), Tony Stone Images; 448(t), Rogge/The Stock Market; 449(t), HRW photo

UNIT 8: Page 449D, HRW photo by Sam Dudgeon; 524(t), Jane Scherr; 525(t), 526(t), HRW photo; 527(t), Tony Stone Images

ART CREDITS

All work, unless otherwise noted, is contributed by Holt, Rinehart and Winston.

Page 1B, John Francis; 1F, Lyne Raff; 59A, Pedro Julio Gonzalez/Melissa Turk & The Artist Network; 59C, Joyce Kitchell/James Conrad Artist Representative; 127C, Blake Thornton/Rita Marie and Friends; 127A, Leo Monahan/Jae Wagoner, Artists' Representative; 197C, Valerie Marsella; 313A, Mary Ross/Rita Gatlin Represents; 313C, Aletha Reppel/Suzanne Craig Represents; 387A, Richard Wehrman; 449A, Dennis Jones; 449C, Keith Locke/Suzanne Craig Represents

Concept Mapping

Sample concept map:

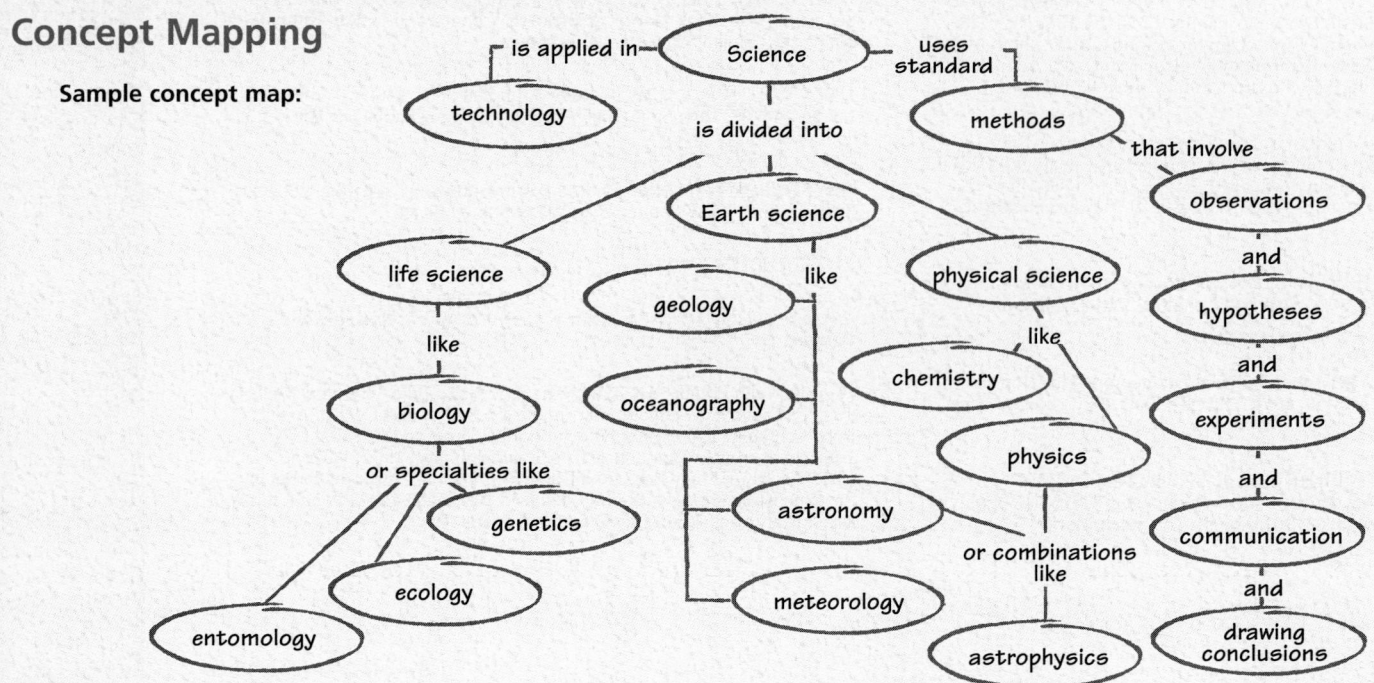

Checking Your Understanding

1. a. an underwater volcano
2. c. way of thinking about the natural world.
3. c. it establishes a basis for comparison.
4. b. the hypothesis may be valid.
5. c. other scientists cannot verify their results.

Interpreting Graphs

As temperature rises, rainfall decreases. As temperature falls, rainfall increases.

This graph might be produced after a scientist made observations of the weather and collected the data. From this graph, the scientist could then draw conclusions about the relationship between temperature and rainfall.

Critical Thinking

1. Technology. The use of scientific knowledge for the development of products useful to human society is called technology.
2. Since astronomy deals with objects other than the Earth itself, it seems strange to call it an Earth science.
3. Many phenomena cannot be studied in relation to a single scientific discipline. When studying the human body, for example, biologists focus on the organ systems, while physicists study the motion of the bones.
4. There must always be a control as well as a variable. One sample must be kept unchanged while the other is adjusted in order to observe the effect of the change.
5. The law would have to be changed. A new description of how the Earth rotates would then need to be determined.

Portfolio Idea

Answers will vary. However, responses should include an understanding that science is something people do and that students can be a part of it.

Concept Mapping

Sample concept map:

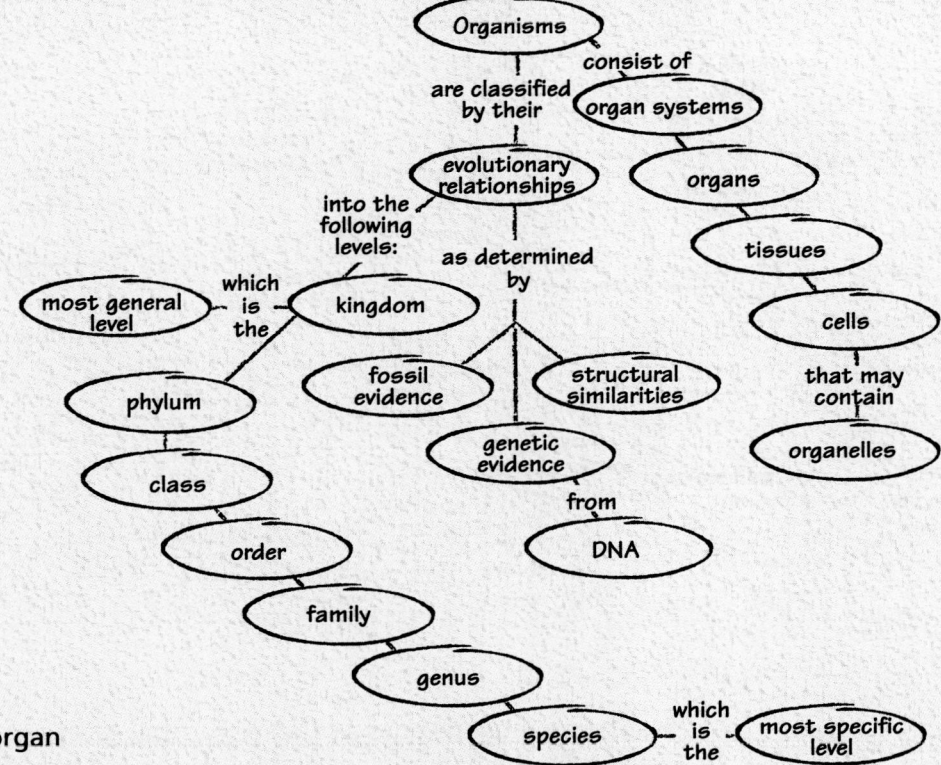

Checking Your Understanding

1. a. kingdom.
2. c. DNA.
3. c. gas exchange
4. b. nervous system.
5. d. blueprints.
6. b. cell, tissue, organ, organ system, organism

Interpreting Illustrations

In this illustration, corresponding parts are labeled with similar colors. Because there are so many corresponding parts, it is unlikely that the forelimbs of these two organisms could have evolved independently.

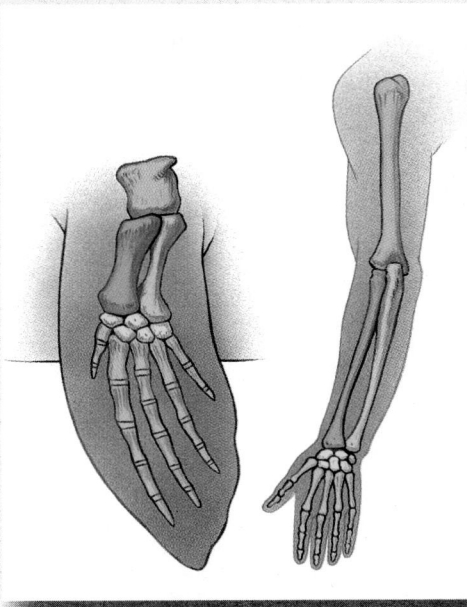

Critical Thinking

1. No; it is not realistic. Kingdom is such a basic biological category that it is unlikely that organisms so fundamentally different from those already known could have been overlooked.
2. The ratio would probably change. As predators, red foxes would be at a severe disadvantage for much of the year because of their coloration, which would stand out against the white snow, thus making it difficult to sneak up on prey. White foxes would not face this problem.
3. Such organisms have traits normally found either in plants or in animals, but not both, thus pointing out the need for additional classification categories.
4. The rigid cell wall provides the plant with structural support. This rigidity would be a disadvantage for animal tissues such as muscles or nerves, which must be very supple.

Portfolio Idea

Answers will vary, but should reflect an understanding of the form and function of the organ system that each student chooses. Accept all reasonable responses.

Concept Mapping

Sample concept map:

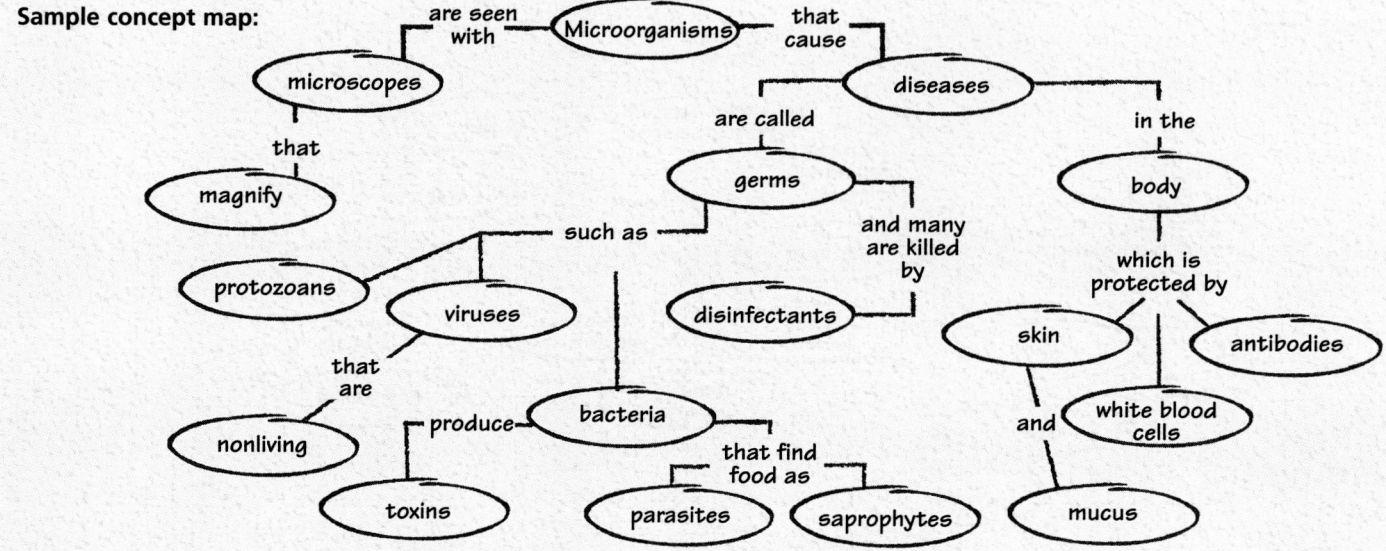

Checking Your Understanding

1. b. Louis Pasteur.
2. a. identify disease microorganisms.
3. b. bacteria.
4. a. the immune system.
5. c. autoimmune disorder.

Interpreting Graphs

At point *X*, the viruses that have replicated inside the cell have broken through the membrane and have been released to infect other cells.

Critical Thinking

1. A deep puncture wound would provide an environment in which tetanus would thrive because the area would have less contact with oxygen. The surface cut would be well exposed to the oxygen in the atmosphere.
2. A healthy animal is needed to show that it is the microorganism in question, and not some other germ, that causes the illness.
3. Microorganisms that attack plants can affect the human food supply, either directly, as in the plants we eat, or indirectly, as in the plants eaten by the animals that we eat. By combating plant disease, we ensure a continued source of nutrition for ourselves.
4. Because viruses are not composed of cells and cannot use energy or reproduce without a living host, they do not fit the accepted definition of a living thing.
5. AIDS reduces the body's ability to fight infections that would ordinarily be overcome by the immune system. People with AIDS therefore die of secondary diseases such as pneumonia.

Portfolio Idea

Questions asked may include the following: (1) What is a vaccine, and when was the first one produced? (2) What does the term *vaccine* mean? (3) What work with vaccines have you (meaning professor Pasteur) done? (4) How will vaccines be used in the future? (5) Will vaccines be developed for every disease that exists?

Answers to these questions should include the following: (1) A vaccine is a dead or weakened germ that gives you immunity to a specific disease. The first vaccine was made in 1798. It was used to prevent smallpox. (2) *Vaccine* is from the Latin word *vacca*, which means "cow." The smallpox vaccine was originally made from the fluid inside the sores of people infected with cowpox, a similar but less dangerous disease than smallpox. (3) Pasteur was the first to vaccinate sheep against anthrax. He was also the first to develop a vaccine for rabies. (4) Student answers will vary. (5) Student answers will vary.

Concept Mapping

Sample concept map:

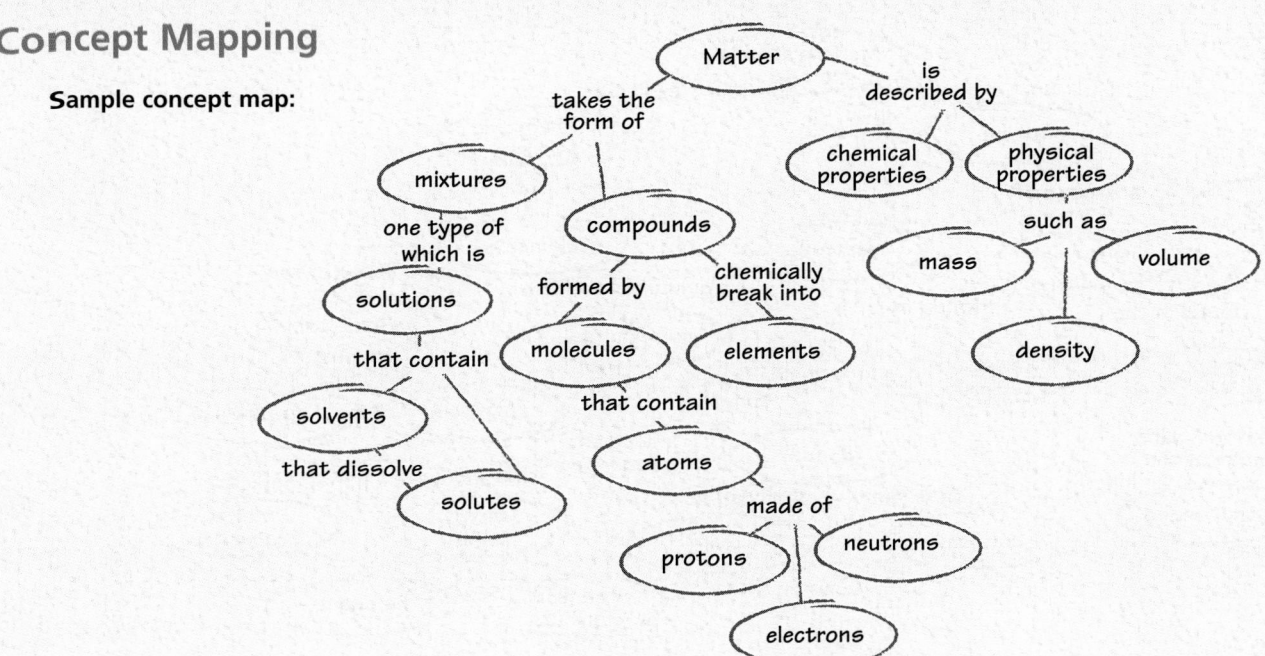

Under "atoms," students could add either *neutrons, protons,* and *electrons* or *subatomic particles.*

Checking Your Understanding

1. d. a cookie being crumbled
2. a. a chemical change.
3. c. is a measure of how much matter that object contains.
4. c. the smallest particle of a compound that still has the properties of that compound
5. a. is a type of mixture.

Interpreting Illustrations

The spring scale measures the weight of the object through the stretching of the spring. Since gravity on the moon is a fraction of the gravity on Earth, the weight, and hence the amount by which the spring is stretched, is also a fraction of that on Earth. On the other hand, the balance measures the mass of the object by comparing its mass with a standard mass. Both masses are equally affected by the lower gravity, so the reading remains unchanged.

Critical Thinking

1. When people commonly refer to the volume of a container, they are referring to how much of a substance the container can hold. A scientist, on the other hand, would be referring to how much space the container occupies.
2. Water molecules consist of two parts hydrogen and one part oxygen. This ratio is preserved when water is broken down.
3. No; the solid is simply broken down into molecules. These tiny particles are able to mix completely with the particles of the liquid so that they seem to disappear.
4. When hydrogen and oxygen are merely mixed, the gases do not interact greatly; each gas preserves its own chemical and physical properties. By contrast, when hydrogen and oxygen react chemically to form a compound, the individual properties of each gas are lost; the compound they form has entirely new properties.

Portfolio Idea

Students should realize that they would have to undergo about nine cycles of shrinkage to reach the approximate size of an atom. Otherwise, student accounts will vary. Student descriptions may note different types of interactions with their environment at different stages of shrinkage. For instance, the forces that result from being struck by air molecules might not be significant at a normal body size; they might seem very strong at much smaller body sizes, however.

Concept Mapping

Sample concept map:

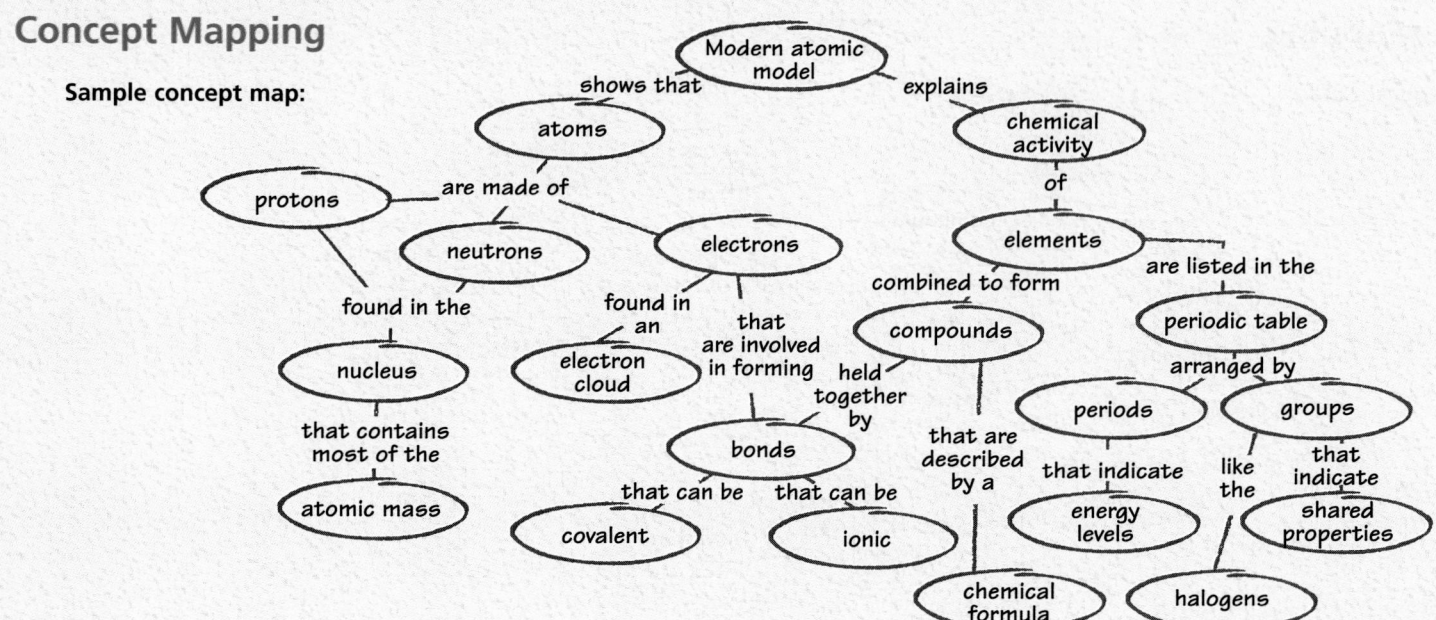

Checking Your Understanding

1. d. electrons are located in energy levels.
2. a. holds a specific number of electrons.
3. b. have different mass numbers.
4. c. noble gases
5. a. covalent bonds.

Interpreting Illustrations

These two atoms are of different elements because they have different numbers of protons.

Critical Thinking

1. The atomic masses of protons and neutrons are about the same; both are 1 u. The atomic mass of an electron is far less—1/1837 u.
2. An ionic bond will form because the potassium atom will probably give up its outer electron to the bromine atom, just as a sodium atom gives up its outer electron to a chlorine atom. Sodium and potassium are in the same group of the periodic table, as are bromine and chlorine, so potassium and bromine should have the same chemical behavior as sodium and chlorine.
3. It can be used to indicate an atom's number of protons and neutrons and their location in the nucleus as well as the total number of electrons and the energy levels in which they are found.
4. The elements within a period of the periodic table have atoms with the same number of energy levels. The elements within the same period differ in their chemical properties, and their atoms differ in the number of electrons in their outer energy levels.
5. $2Fe + O_2 \rightarrow 2FeO$

Portfolio Idea

Student letters may vary, but they should demonstrate an understanding of the atomic model and its development over time. Democritus viewed the atom as a tiny solid ball. The current model views the atom as a dense nucleus of protons and neutrons surrounded by a cloud of electrons in various energy levels. Diagrams similar to those found in the unit are appropriate.

Concept Mapping

Sample concept map:

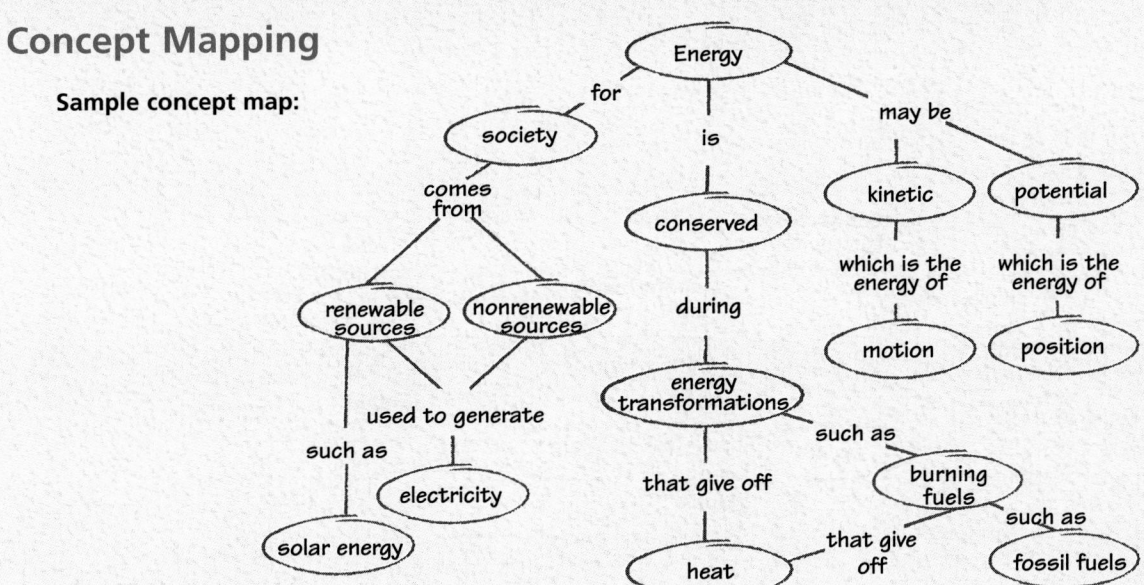

Checking Your Understanding

1. c. kinetic energy.
2. b. radiant energy
3. a. photosynthesis.
4. a. tidal energy
5. d. hydrogen.

Interpreting Photos

Answers will vary. Students should be able to identify stationary objects and all materials as examples of potential (mechanical or chemical) energy. Also, moving objects, electricity, heat, and light are examples of different forms of kinetic energy.

Critical Thinking

1. To reach Earth, the sun's energy travels through space in the form of electromagnetic radiation.
2. When a firecracker explodes, energy stored in the chemicals of the firecracker is not lost but is converted into heat energy, light energy, and the mechanical energy of sound waves and flying debris.
3. Each food chain in an environment must have more producers than consumers. Since foxes are higher up on the food chain than field mice, it takes many field mice to supply the energy requirements of one fox. If there were more foxes than field mice, the food chain would collapse.
4. A long time ago, energy from the sun was taken in and converted into chemical energy by photosynthetic organisms that lived in the ocean. Some of the energy remained in the chemicals that made up the organisms. When these organisms died, their bodies sank to the ocean floor, where they collected and became covered with sediments. Over time, pressure from the layers of sediment converted the remains of the organisms into petroleum. After the petroleum is recovered and refined, some of it is made into gasoline. When gasoline is burned, chemical energy that was originally stored in ancient organisms is converted into the kinetic energy of expanding gases that moves the parts of an automobile and eventually the automobile itself.
5. Renewable sources of energy include solar energy, geothermal heat, biomass, ocean tides, ocean temperature differences, wind, flowing water, hydrogen, and nuclear fusion. Nonrenewable sources include fossil fuels and uranium for nuclear fission. Nonrenewable sources of energy have provided a large amount of energy at a relatively small cost. However, they will eventually run out. Renewable sources have been more difficult and expensive to develop and are not appropriate for every place where energy is needed. However, renewable sources will continue to be available for conversion into the energy that humans need to generate electricity and run their machines.

Portfolio Idea

Students' stories will vary but should address how individuals and governments will deal with having to use other sources of energy to generate electricity and run machines. Background information on how the sources of energy that humans use to do work have changed from human power to nuclear power should also be included.

Concept Mapping

Sample concept map:

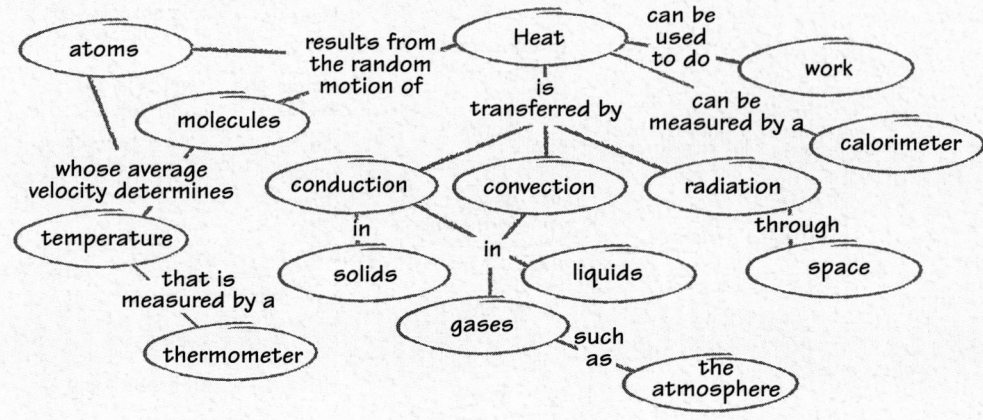

Checking Your Understanding

1. c. the "stealth" fighter
2. d. radiation.
3. b. convection.
4. c. Without oceans, the temperature range would be much greater.
5. a. it absorbs a lot of heat without increasing much in temperature.

Interpreting Illustrations

At –20°C the bimetallic strip will curve upward; at 25°C it will curve downward.

Critical Thinking

1. Water absorbs and releases a great deal of heat without changing much in temperature. Therefore, it can be used to remove the large amounts of heat produced by a car engine, keeping the engine from getting too hot.
2. The molecular motions occur on far too small a scale for our sense organs to detect them as vibrations.

3. No; the glass does not conduct the heat. The molecular structure of the glass is such that radiation simply passes through it.
4. Since heat is a measure of molecular motion, there would be a maximum possible temperature if there were a maximum possible speed for the particles, which, in fact, there is—the speed of light.
5. No; most of the heat we feel is delivered to the Earth from the sun. The Earth's internal heat makes only a small contribution to the climate.

Portfolio Idea

Students' experiments will vary. One possible strategy would be to test whether "cold" could be reflected around a corner in the same way as heat can. Students should use a thermometer to measure the change in temperature that would accompany such a transmission.

Concept Mapping

Sample concept map:

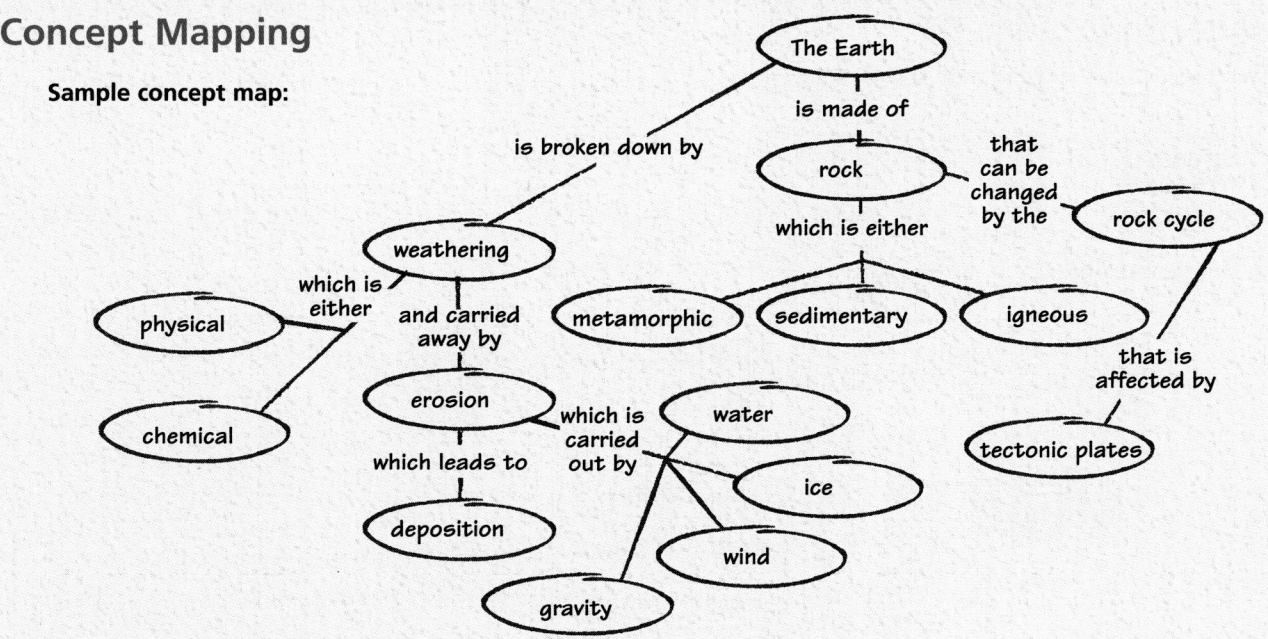

Checking Your Understanding

1. a. igneous
2. a. Any one rock type can be changed into any other.
3. c. A drop of rainwater dissolving a tiny bit of the cement holding sandstone together
4. d. the greater its erosive ability.
5. c. are sites of little activity.

Interpreting Photos

This is a photograph of the Matterhorn, a famous peak that rises 4478 m above sea level in the Pennine Alps between Italy and Switzerland. The smooth, steep sides of the peak were formed by the action of glaciers that scoured its sides, creating a bowl-shaped hollow called a *cirque* just below the peak.

Critical Thinking

1. Water erosion moves both large and small particles. Wind erosion moves only very fine-grained particles.
2. If the Earth becomes solid, then tectonic plates will stop moving and landforms will stop being cre- ated. Eventually the planet's land- forms will wear down to a more or less level plain near sea level.
3. Very warm and moist climates, such as those found in rain forests, probably experience the fastest rates of weathering. Very cold and dry climates, such as those found near the poles, prob- ably experience the slowest rates of weathering. Warm, wet cli- mates experience faster weather- ing because the chemical process- es that break things down work faster in warm conditions than in cold and because water is a major agent of chemical weathering. The reverse is true of cold and dry climates.
4. This evidence suggests that rock is formed at the mid-ocean ridges and then carried away, presum- ably by the motion of the Earth's surface.
5. Lakes are low places on the sur- face. Furthermore, they are us- ually fed by streams or rivers. Because streams and rivers gener- ally carry a certain amount of rock and soil and because sedi- ments tend to settle in low places, the lakes are eventually filled with the products of weathering.

Portfolio Idea

Student examples will vary. Answers should reflect an appropriate analogy of the geologic time scale.

Unit 1 Making Connections, page 55
Checking Your Understanding, question 6

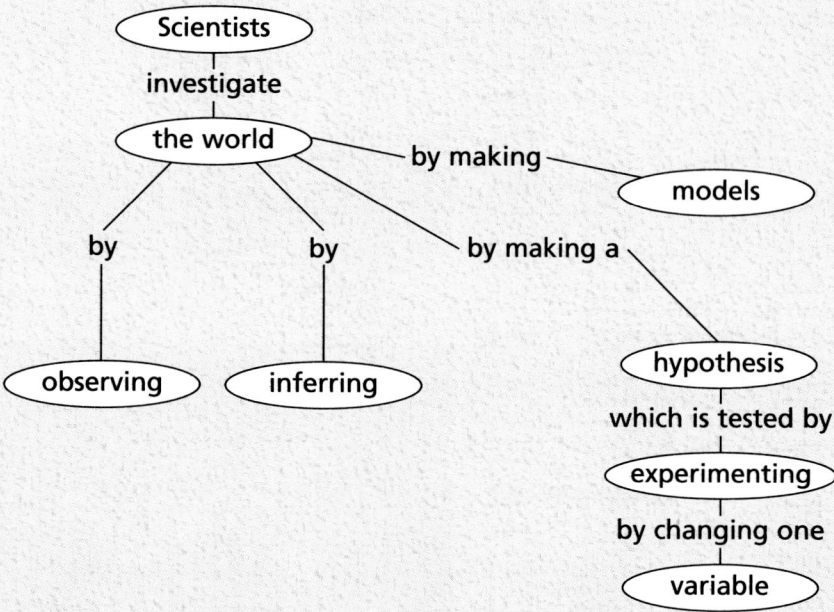

Unit 2, Chapter 4, page 70
Exploration 3, Comparing Structures, question 1

The horse equivalents of human structures are the following: hoof—fingernail; foot bones—finger; bones from hoof to knee—hand; knee—wrist; leg bones above knee—forearm; next joint above knee—elbow; upper leg—upper arm.

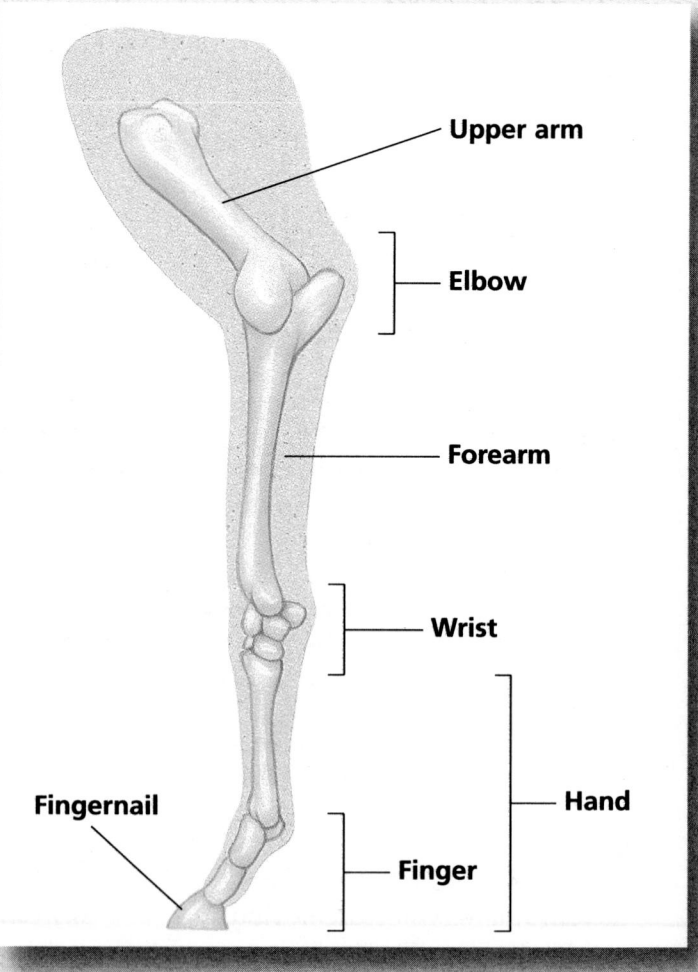

Unit 3, Chapter 8, page 146
Challenge Your Thinking, question 4, part a

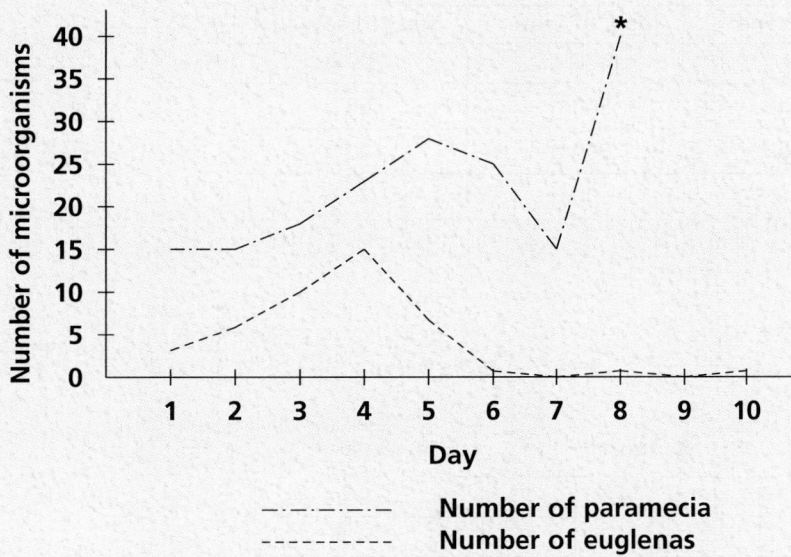

* The number of paramecia from days 9 and 10 cannot be plotted because there were "too many to count."

Unit 4, Chapter 11, page 205
Lesson 2, in-text table

The following are sample answers:

Material	Use	Important property	Type of property
corn	cornflakes	nutritious	chemical
gold	jewelry, electronics	does not corrode	chemical
oil	lubricant	reduces friction	physical
aluminum	building material	lightweight	physical
plastic	building material	can be molded	physical
coal	fuel	burns	chemical
concrete	building material	durable	physical
wool	clothing	traps heat	physical

Unit 4, Chapter 12, page 224
Some Problems to Solve, question 3

Volume	Mass of sand	Mass of water	Mass of sawdust
1000 mL	2800 g	1000 g	700 g
500 mL	1400 g	500 g	350 g
200 mL	560 g	200 g	140 g
100 mL	280 g	100 g	70 g

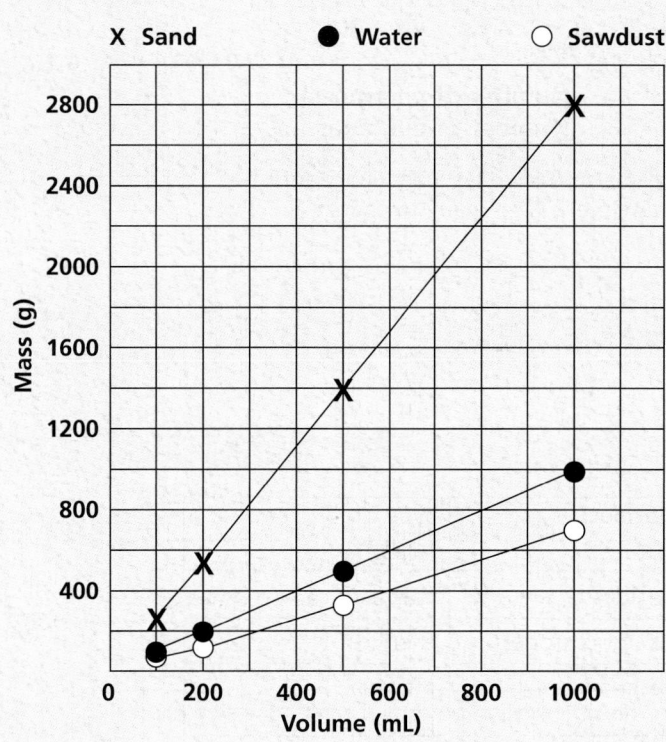

Unit 4, Chapter 12, page 226
Challenge Your Thinking, question 1

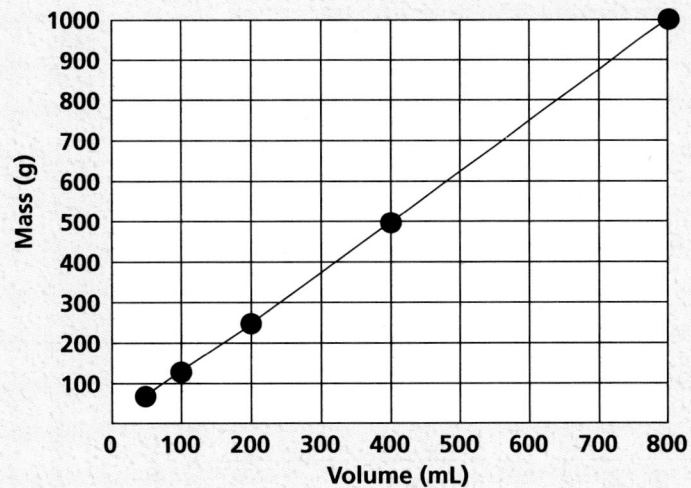

Unit 4, Chapter 13, page 231
What Have You Learned? question 1

Characteristics of States		
	Has a boundary with air and cannot be noticeably compressed	**Has no boundary with air and can be noticeably compressed**
Apparently rigid	solid	--------
Flows	liquid	gas

Unit 4, Chapter 13, page 233
The Language of Changes of State, question 4

Change of state	Terms	Change in temperature
solid to liquid	melting or liquefaction	increase
liquid to solid	freezing or solidification	decrease
liquid to gas	evaporation, vaporization, or boiling	increase
gas to liquid	condensation or liquefaction	decrease
solid to gas	sublimation or vaporization	increase
gas to solid	solidification or crystallization	decrease

Unit 4 Making Connections, page 245
Checking Your Understanding, question 3

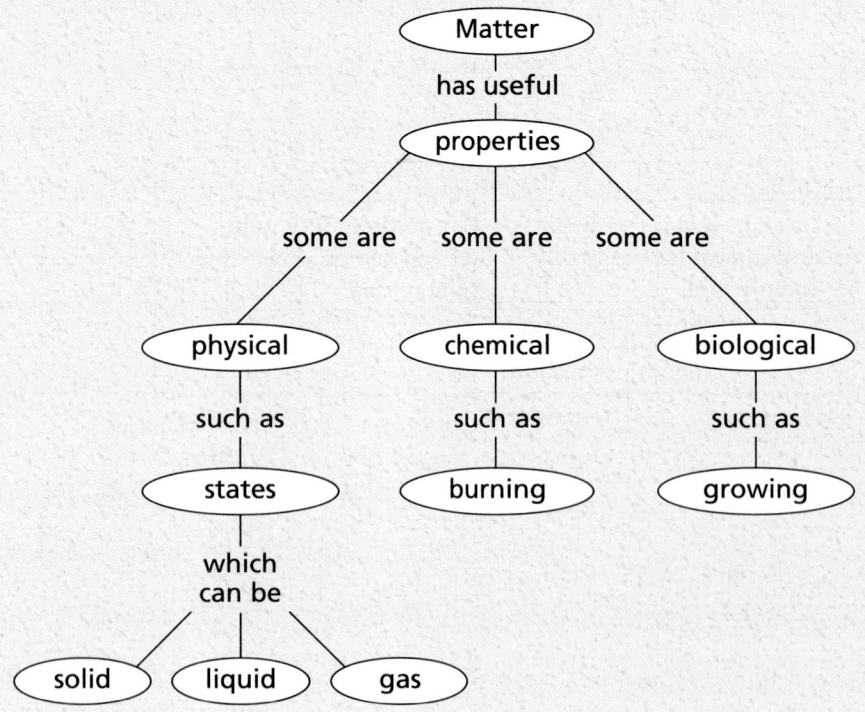

Unit 5, Chapter 14, page 254
Exploration 1, Activity 2

The following is a completed sample table:

Chemical	Result of activity	Scientific idea about this chemical
distilled water	Nothing was left after the water evaporated.	There is nothing in pure water except water.
table salt	The salt dissolved in water.	Salt is the same chemical no matter what its source is.
sugar	The sugar turned black when it was heated.	Sugar burns.
chalkboard chalk	The chalk bubbled when vinegar was added.	Chalk and vinegar produce a gas when combined.
baking soda	The baking soda bubbled when vinegar was added.	Baking soda and vinegar produce a gas when combined.
cellulose	The pieces of paper burned.	Some types of paper burn more quickly than others.

Unit 5, Chapter 15, page 269
Physical and Chemical Words, in-text table

The following are sample answers:

Physical words		Chemical words	
Word	**Example**	**Word**	**Example**
grinding	wheat, rocks	burning	wood, oil
breaking	plates, eggshells	rotting	wood
eroding	soil, sand on a beach, a cliff	rusting	tools, an old car
evaporating	water from a lake, perspiration	reacting	vinegar and baking soda
melting	ice, wax	digesting	food
condensing	water in a terrarium	respiring	living things
drying	clothes	photosynthesizing	plants
freezing	fish, meat, water		
cutting	grass, paper		

Unit 5, Chapter 15, page 270
Exploration 1, in-text table

The following are sample answers:

Change	Physical	Chemical	Reasons or evidence
Limewater turns milky. (Test 5)		✓	1. Color change occurs. 2. A new substance is formed. 3. Change is difficult to reverse.
Blue copper sulfate solution turns paler, and bits of copper metal fall out of solution. (Test 1)		✓	1. Color change occurs. 2. A new substance is formed. 3. Change is difficult to reverse.
Milk turns lumpy in vinegar. (Test 2)		✓	1. A precipitate forms. 2. Change is difficult to reverse.
Eggshell dissolves, and bubbles (of carbon dioxide) are released. (Test 3)		✓	1. Gas is produced. 2. Change is difficult to reverse.
Candle wax melts when heated. (Test 4)	✓		1. Change is easily reversed.
Lemon juice turns brown when heated. (Test 6)		✓	1. Color change occurs. 2. Change is difficult to reverse.
Iodine solution turns purple in the presence of starch. (Test 7)		✓	1. Color change occurs. 2. Change is difficult to reverse.
Bubbles form when baking powder is added to water. (Test 8)		✓	1. Gas is produced. 2. Change is difficult to reverse.

Unit 6, Chapter 18, page 343
Answers to In-Text Questions, question ⓒ

Sample completed table:

Appliance	Power (W)	Power (kW)	Energy used in 1 hour (kWh)
bright light bulb	200 W	0.2 kW	0.2 kWh
clothes dryer	3000 W	3 kW	3 kWh
electric drill	300 W	0.3 kW	0.3 kWh
electric kettle	1500 W	1.5 kW	1.5 kWh
hair dryer	1500 W	1.5 kW	1.5 kWh
iron	1100 W	1.1 kW	1.1 kWh
refrigerator	2000 W	2 kW	2 kWh
stereo	130 W	0.13 kW	0.13 kWh
television set	200 W	0.2 kW	0.2 kWh
toaster	1200 W	1.2 kW	1.2 kWh
vacuum cleaner	900 W	0.9 kW	0.9 kWh
washing machine	80 W	0.08 kW	0.08 kWh

Unit 6 Making Connections, page 383
Checking Your Understanding, question 3

The following are sample answers:

Machine	Efficiency	Desired form of energy	Forms of waste energy
toaster	90%	heat	light, excess heat
steamboat	10%	mechanical	excess heat, sound
hair dryer	85%	heat	sound
car engine	30%	mechanical, electrical	heat, sound

Unit 8 Making Connections, page 523
Checking Your Understanding, question 5

Sample concept map:

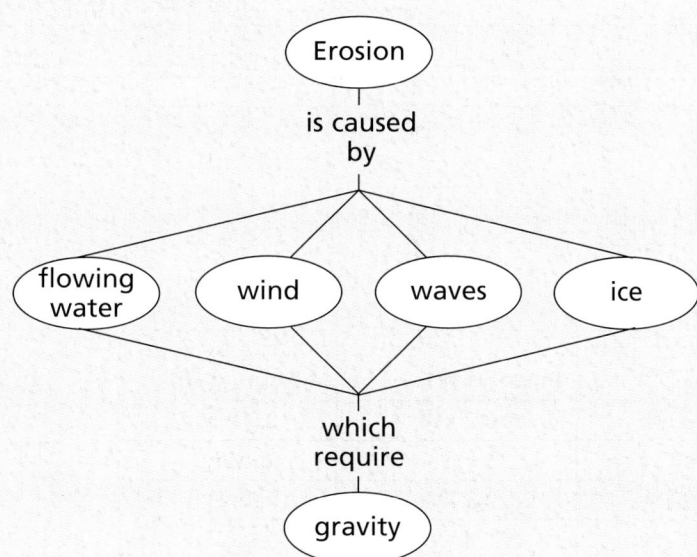